Antibiotics

Volume III

Mechanism of Action of Antimicrobial and Antitumor Agents

Edited by

John W. Corcoran and Fred E. Hahn

Assisted by J. F. Snell and K. L. Arora

Springer-Verlag Berlin · Heidelberg · New York 1975

Professor JOHN W. CORCORAN, Ph.D., Northwestern University, Department of Biochemistry, The Medical and Dental Schools, 303 E. Chicago Avenue, Chicago, Illinois 60611/USA

Professor FRED E. HAHN, Ph.D., Department of the Army, Walter Reed Army Institute of Research, Walter Reed Army Medical Center, Washington, D.C. 20012/USA

ISBN-13: 978-3-642-46306-8   e-ISBN-13: 978-3-642-46304-4
DOI: 10.1007/ 978-3-642-46304-4

Typesetting, printing and bookbinding: Stürtz AG, Würzburg

Preface

This volume is the third in the series devoted to Antibiotics initiated by Springer Verlag in 1967. The first two volumes were devoted to the Mode of Action of Antibiotics and Biogenesis, respectively and were received graciously. During the intervening years these two works have been used often by research workers and students alike and have been quoted extensively. Although a number of other excellent treatises on antibiotics have appeared, the Springer series has set a standard for thoroughness and quality that meets the need of the scientific community.

It is against this background that the present Editors set about the preparation of a third volume in the Series on Antibiotics. Since the appearance of Volume I, also dealing with Mechanism of Action, tremendous strides have been made in the depth and breadth of our knowledge of molecular biology, microbial chemistry and molecular pharmacology and of their direct application to studies on the mode of action of drugs. The field of molecular biology itself was in its relative infancy during the preceding decade and the unique role played by many antibiotics in the development of our understanding of nucleic acid synthesis and function and its relationship to protein synthesis and cell physiology has led rapidly to a very precise understanding of how many of these same antibiotics inhibit susceptible cells. Thus, a few years since the preparation of Volume I we feel that a new volume on antibiotic action may be offered which is almost completely new in its content and which may prove useful to student, teacher and research worker alike.

A definition of classification of antimicrobial agents in terms of their origin is now largely academic. Accordingly, the selection of topics was based on the state of our knowledge of mechanisms of action. With few exceptions authors could be found to write on the major antimicrobial agents meeting this criterion and all elected are primary authorities on their subject. The Editors' efforts were aimed at assembling a volume whose various chapters have been prepared within one year of submission to the publisher and with a very few exceptions this goal has been achieved. When the first version of a chapter was written much earlier than desired, the authors have been encouraged to make any necessary revisions and most have done this. The result offered now is a treatise containing forty-six separate chapters dealing with both natural and synthetic antimicrobial agents (here again the distinction is somewhat academic, for at least half a dozen substances are of both natural and synthetic origin). The chapters are well supported by citations of the original source literature (over 3500) and subject distribution reflects today's trend toward increased research on anti-tumor agents. Some thirty-three percent of the chapters deal with this subject. In addition, a broad spectrum of information is presented in this volume on the mode of action of certain well-known antimalarial, antitubercular and antiparasitic compounds.

Of the forty-six topics presented, twenty-eight are new to the Springer series on Antibiotics. Others are dealt with again because there has been a significant advance in our understanding of their mode of action since the initial treatment.

The editors have attempted only a general organization of the subjects included in the present Volume. Then guidelines have been suggested by the limited number of categories of the modes of action known; e.g. 1. inhibitors of nucleic acid bio-synthesis either at the level of the template or by some other mechanisms, 2. inhibitors of protein synthesis at the ribosomal or translational level, 3. interference with the formation or integrity of the cell wall or membrane, 4. inhibition of specific enzyme reactions in the intermediary metabolism of the cell. Because of the paucity of subject material, the last two topics are arranged as one unit together with some miscellaneous subjects where significant knowledge has yet to lead to an unambiguous decision regarding the mode of action.

The Editors have included, indeed encouraged, consideration of antimicrobial agents which are not or may never be useful clinical chemotherapeutic agents. This is done without apology since such agents illustrate general principles of the interaction of antimicrobial drugs with either the host or infectious cell and indeed much of our present knowledge of molecular biology and antibiotic action has come from the study of such substances.

In summary, the Editors wish to thank their throughout-the-world colleagues for innumerable instances of help and advice during the conception, planning and preparation of Volume III in the series *Antibiotics*. They apologize for the obvious omissions of desirable new chapters (e.g. quinine, nitrofurans, para-aminosalicylic acid etc.) and for the lack of updated treatments of certain more familiar substances (e.g. actinomycin, chromomycin, hadacidin, rubiflavin, streptonigrin etc.). When primary authorities could not be enlisted we have decided in favor of deferral to a future new volume on the Mode of Action of Antibiotics. Like the editors of Volumes I and II we have attempted to maintain the highest standards for authorship and editorial review. It is for you, the intended recipient of this our most sincere effort to evaluate the result.

JOHN W. CORCORAN
FRED E. HAHN

Chicago/Washington, Summer 1974

Acknowledgements

Preparation of this third volume in the series on antibiotics published by Springer Verlag was begun under the editorship of Professor J. F. SNELL. With his being unable to continue in this effort, the responsibility was assumed by the present editors. They wish to express their deep appreciation to Dr. KONRAD F. SPRINGER for his continued interest, encouragement and understanding during the long period of planning and the recent period of manuscript pieparation. Above all, the editors wish to thank their colleague Dr. KASTURI LAL ARORA for his extraordinary support in the detailed preparation of this volume. Dr. ARORA carefully surveyed the abstract and original literature and advised them during the planning phase for this work and again during the period of manuscript preparation and review he helped immeasurably in the maintenance of contact with both authors and publisher. His painstaking and scholarly review of each manuscript prior to submission to one or the other of the editors made their task much simpler and permitted the review process to focus on the scientific aspects of each chapter. The support of Dr. ARORA has made this volume a reality.

Contents

Section I. Interference with Nucleic Acid Biosynthesis

Section II. Interference with Protein Biosynthesis

Section III. Interference with Cell Wall/Membrane Biosynthesis, Specific Enzyme Systems and Those in Which Mode of Action Not Known with Certainty

Contributors

NITYA ANAND, Division of Medicinal Chemistry, Central Drug Research Institute, Lucknow/India

DAVID APIRION, Department of Microbiology, Washington University Medical School, St. Louis, Missouri 63110/USA

F. ARCAMONE, Istituto Ricerche Di Base Farmitalia, Milano/Italy

PINAKILAL BHATTACHARYYA, Department of Biology, Washington University, St. Louis, Missouri 63130/USA

JAMES W. BODLEY, Department of Biochemistry, University of Minnesota, Minneapolis, Minnesota 55455/USA

JAMES J. BURCHALL, The Wellcome Research Laboratories, Burroughs Wellcome Co., 3030 Cornwallis Road, Research Triangle Park, North Carolina 27709/USA

JENNIE CIAK, Department of the Army, Walter Reed Army Institute of Research, Walter Reed Army Medical Center, Washington, D.C. 20012/USA

THOMAS M. COOK, Department of Microbiology, University of Maryland, College Park, Maryland 20742/USA

A. DI MARCO, Division of Experimental Oncology B, Istituto Nazionale Per Lo Studio E La Cura Dei Tumori, Via Venezian, 1, 20133 Milano/Italy

DENNIS DOHNER, Department of Microbiology, Washington University Medical School, St. Louis, Missouri 63110/USA

G. F. GAUSE, Institute of New Antibiotics, Bolshaia Pirogovskaia, 11, Moscow/USSR

IRVING H. GOLDBERG, Department of Pharmacology, Harvard Medical School, 25 Shattuck Street, Boston, Massachusetts 02115/USA

WILLIAM A. GOSS, Sterling-Winthrop Research Institute, Division of Sterling Drug Inc., Rensselaer, New York 12144/USA

A. P. GROLLMAN, Department of Pharmacology, Albert Einstein College of Medicine of Yeshiva University, 1300 Morris Park Avenue, Bronx, New York 10461/USA

FRED E. HAHN, Department of Molecular Biology, Walter Reed Army Institute of Research, Walter Reed Army Medical Center, Washington, D.C. 20012/USA

ERICH HIRSCHBERG, College of Medicine and Dentistry of New Jersey, New Jersey Medical School, 100 Bergen Street, Newark, New Jersey 07103/USA

SUSAN B. HORWITZ, Albert Einstein College of Medicine of Yeshiva University, 1300 Morris Park Avenue, Bronx, New York 10461/USA

FLOYD M. HUBER, Antibiotic Fermentation Technology, Eli Lilly and Company, Indianapolis, Indiana 46206/USA

Z. JARKOVSKY, Department of Pharmacology, Albert Einstein College of Medicine of Yeshiva University, 1300 Morris Park Avenue, Bronx, New York 10461/USA

D. C. JORDAN, Department of Microbiology, College of Biological Science, University of Guelph, Guelph, Ontario/Canada

KEN KATAGIRI, Shionogi Research Laboratory, Shionogi and Co., Ltd., Fukushima-ku, Osaka 553/Japan

F. KNÜSEL, Biological Research Laboratories, Pharmacology Division, Ciba-Geigy, Ltd., 4002 Basel/Switzerland

KURT W. KOHN, National Cancer Institute, Department of Health, Education, and Welfare, Public Health Service, National Institutes of Health, Bldg. 37, Rm. 5B27, Bethesda, Maryland 20014/USA

C. R. KRISHNA MURTI, Division of Biochemistry, Central Drug Research Institute, Lucknow/India

Z. KURYLO-BOROWSKA, Department of Biochemistry, The Rockefeller University, New York, 10021/USA

GERALD MEDOFF, Departments of Microbiology and Medicine, Infectious Disease Division, Washington University School of Medicine, S. Louis, Missouri 63110/USA

B. A. NEWTON, Medical Research Council, Biochemical Parasitology Unit, Molteno Institute, University of Cambridge, Cambridge/Great Britain

NANCY L. OLEINICK, Departments of Radiology and Biochemistry, Case Western Reserve University, School of Medicine, 2220 Circle Drive, Cleveland, Ohio 44106/USA

JOHN G. OLENICK, Department of Molecular Biology, Department of the Army, Walter Reed Army Institute of Research, Walter Reed Army Medical Center, Washington, D.C, 20012/USA

W. PACHE, Pharmaceutical Division, Chemical Research, Sandoz Ltd. 4002 Basel/Switzerland

SIDNEY PESTKA, Roche Institute of Molecular Biology, Nutley, New Jersey 07110/USA

KARL PORALLA, Lehrstuhl für Mikrobiologie II der Universität Tübingen, Berghof, 7400 Tübingen-Lustenau/Fed. Rep. Germany

P. E. REYNOLDS, Sub-department of Chemical Microbiology, Department of Biochemistry, University of Cambridge/Great Britain

KOSABURO SATO, Shionogi Research Laboratory, Shionogi and Co., Ltd., Fukushima-ku, Osaka 553/Japan

DAVID SCHLESSINGER, Departments of Microbiology and Medicine, Infectious Disease Division, Washington University School of Medicine, St. Louis, Missouri 63110/USA

HAROLD T. SHIGEURA, Merck Institute for Therapeutic Research, Rahway, New Jersey 07065/USA

SIMON SILVER, Department of Biology, Washington University, St. Louis, Missouri 63130/USA

M. STAEHELIN, Ciba-Geigy AG, Bau 125/717 Klybeck, 4000 Basel/Switzerland

NOBUO TANAKA, Institute of Applied Microbiology, The University of Tokyo, Bunyko-ku, Tokyo/Japan

HAMAO UMEZAWA, Microbial Chemistry Research Foundation, Institute of Microbial Chemistry, 14-23 Kamiosaki 3-Chrome, Shinagawa-ku, Tokyo/Japan

DAVID VAZQUEZ, C.S.I.C. Centro De Investigaciones Biologicas, Instituto De Biologia Celular, Velazquez, 144, Madrid-6/Spain

MICHAEL WARING, University of Cambridge, Department of Pharmacology, Medical School, Hills Road, Cambridge/Great Britain

WALTER WEHRLI, Ciba-Geigy AG, Bau 125/717 Klybeck, 4000 Basel/Switzerland

LOUIS W. WENDT, Department of Biology, Montana State University, Bozeman, Montana 59715/USA

ALAN DAVID WOLFE, Department of Molecular Biology, Walter Reed Army Institute of Research, Washington, D.C. 20012/USA

TADASHI YOSHIDA, Shionogi Research Laboratory, Shionogi and Co., Ltd., Fukushima-ku, Osaka 553/Japan

MUNEHIKO YUKIOKA, Department of Biochemistry, Osaka City University Medical School, 1-4-54, Asahi-machi, Abeno-ku, Osaka 545/Japan

W. ZIMMERMANN, Biological Research Laboratories, Pharmacology Division, Ciba-Geigy, Ltd., 4002 Basel/Switzerland

F. ZUNINO, Division of Experimental Oncology B, Istituto Nazionale Per Lo Studio E La Cura Dei Tumori, Via Venezian, 1, 20133 Milano/Italy

I. Interference with Nucleic Acid Biosynthesis

Anthramycin

Kurt W. Kohn

Origin

Anthramycin was derived from a thermophilic actinomycete that was isolated from subtropical soils in the 1950's by M.D. TENDLER. TENDLER searched for the production of antitumor agents by these highly aerobic organisms on the hypothesis that such organisms might produce inhibitors of anaerobic metabolism (Chem. Eng. News, Oct. 31,p. 42 1966). The most active of the several antitumor principles found in fermentation broths of these organisms was called "refuin" (from the Hebrew "refuah," meaning a medicine). Refuin, obtained from a fermentation beer by precipitation with ammonium sulfate and by fractional crystallization, was active against sarcoma 180, adenocarcinoma 755, and Ehrlich ascites tumor upon intraperitoneal or subcutaneous injection in mice. The material also inhibited some, but not all, Gram-positive bacteria (TENDLER and KORMAN, 1963). LEIMGRUBER et al. (1965a) isolated a pure crystalline antibiotic from a fermentation broth of *Streptomyces refuineus* var. *thermotolerans* (NRRL 3143) and assigned the name Anthramycin. The isolation was accomplished on the basis of assay for its anti-bacterial activity.

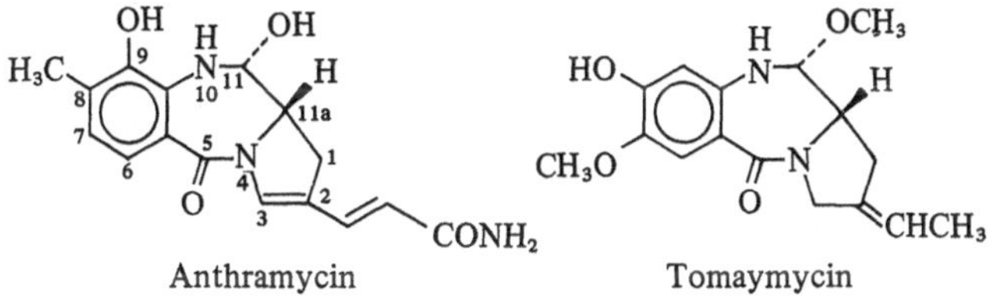

Fig. 1.

Chemistry

The structure of anthramycin (Fig. 1) has been confirmed by synthesis (LEIMGRUBER et al., 1965b; LEIMGRUBER et al., 1968). The absolute configuration at C_{11a} is the same as in L-hydroxyproline. The structure determination was rendered difficult by the instability of anthramycin. Therefore, instead of using anthramycin itself, the initial structure determination was performed on the more stable 11-methyl ether derivative which was formed upon recrystallization of anthramycin from hot methanol.

Anthramycin is readily interconverted with 11-epi-anthramycin, anthramycin-11-methyl ether (and its epimer), and 10, 11-anhydroanthramycin. In aqueous solution, anthramycin is in equilibrium with its epimer and with anhydroanthramycin (LEIMGRUBER *et al.*, 1965a; LEIMGRUBER, personal communication). Anthramycin-11-methyl ether spontaneously hydrolyses in neutral aqueous solution at room temperature; the hydrolysis is catalysed by hydrogen ions and by phosphate (KOHN *et al.*, 1974).

The absorption spectrum of anthramycin in aqueous solution has a maximum at 333 nm, where the molar extinction coefficient is 35,000. The spectrum of anthramycin-11-methyl ether is similar, but there is a slight shoulder on the long-wavelength side of the band; the maximum difference appears at 375 nm where the extinction is increased about 20% relative to anthramycin.

Reaction With DNA

The anthramycin molecule does not have any of the structural features that have heretofore been identified as responsible for tight binding to DNA. The molecule nevertheless reacts specifically with DNA to form a nearly irreversible complex. The interaction of anthramycin with DNA differs from that of most other DNA-binding antibiotics in that the reaction is slow, requiring a period of minutes to hours, and in that the complex does not readily dissociate (KOHN and SPEARS, 1970). Considering the modest size of the anthramycin molecule, these features suggest that the binding is covalent. The molecular mechanism of this binding however is still conjectural.

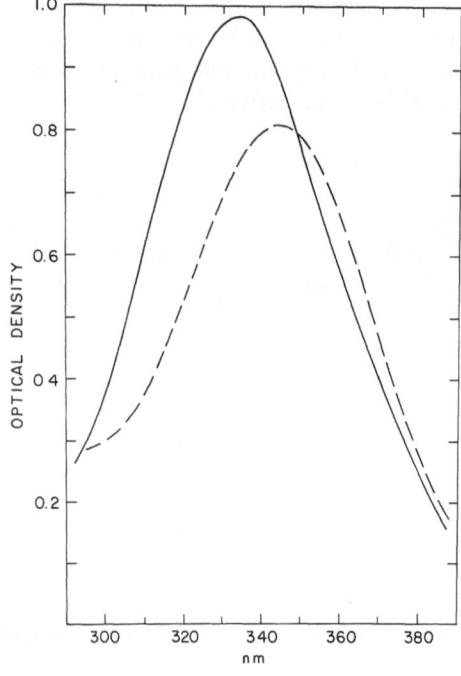

Fig. 2. Absorption spectra of free and DNA-bound anthramycin. (—) 2.8×10^{-5} M anthramycin methyl ether in 0.01 M sodium phosphate, 0.5 mM EDTA (pH 7.0) after complete hydrolysis. (----) same after complete reactions with 1.5 mM calf thymus DNA. (DNA present also in reference cell.) (After KOHN and SPEARS, 1970)

Reaction with DNA causes hypochromic and bathochromic changes in the ultraviolet absorption spectrum of anthramycin (Fig. 2). The absorption peak is shifted from 333 to 343 nm (STEFANOVIC, 1968; KOHN *et al.*, 1968; BATES *et al.*, 1969). The anthramycin-binding capacity of DNA is approximately 1 anthramycin per 10 nucleotide units. Although denatured DNA reacts much more slowly than does native DNA, there is little or no reduction in the stoichiometry (KOHN and SPEARS, 1970).

The kinetics of the reaction with DNA have been studied both spectrophotometrically and by solvent extraction, and the two methods have given consistent results. An example of the time course of the reaction is shown in Fig. 3. The kinetics of the reaction exhibit dependences on the concentrations of both anthramycin and DNA. In addition, the reaction rate is proportional to hydrogen ion concentration. In the reaction with native DNA, the rate decreases with increasing salt concentration. In the case of denatured DNA, the reaction is much slower than with native DNA and, as the salt concentration is raised from very low levels, the rate first increases and then decreases. This indicates that the reaction depends upon the presence of the proper conformation in the DNA (KOHN and SPEARS, 1970; GLAUBIGER and KOHN, 1974).

Both anthramycin and its 11-methyl ether appear to be able to react with DNA directly; however the reaction of the methyl ether is slower (Fig. 3). Accordingly, when the methyl ether is partially hydrolysed by preincubation in aqueous

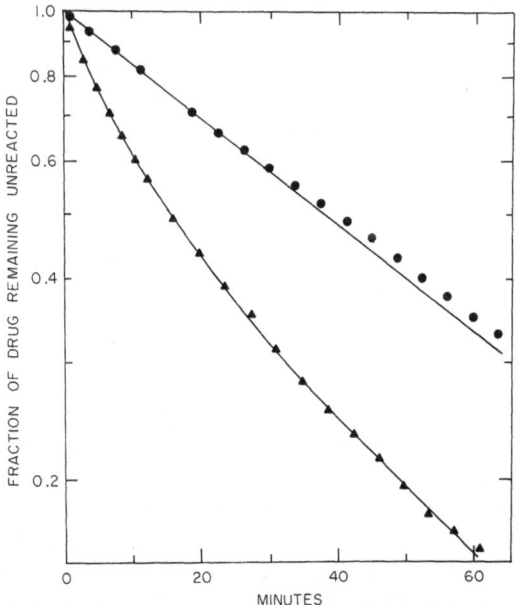

Fig. 3. Time-course of the reaction of anthramycin (lower curve) or anthramycin methyl ether (upper curve) with DNA. Anthramycin methyl ether was freshly dissolved and immediately mixed with DNA at zero time; its hydrolysis rate under these conditions but in the absence of DNA corresponded to a half-time of 27 min. The lower curve was obtained by pre-incubating the aqueous solution of anthramycin methyl ether until hydrolysis was complete and then mixing with DNA. Solvent: 0.08 M triethanolamine buffer, 0.02 M NaCl, 0.4 mM EDTA, pH 7.34 (KOHN *et al.*, 1974)

solution, the initial rate of reaction with DNA increases linearly with the extent of hydrolysis (KOHN et al., 1974).

The reactivity of the anthramycin C_{11} position and the observations that both the hydrolysis of the 11-methyl ether and the reaction with DNA are acid-catalysed suggest that C_{11} may be a site of binding to DNA.

Anthramycin reacts specifically with DNA containing guanine. No reaction has been detected with RNA, mononucleotides, or with poly-nucleotides not containing guanine (KOHN et al., 1968; STEFANOVIC, 1968; KOHN and SPEARS, 1970; GLAUBIGER and KOHN, 1974).

The anthramycin-DNA complex has an unusually high stability, and survives conditions that dissociate the DNA complexes of most other antibiotics. The anthramycin-DNA complex does dissociate, however, when the pH is reduced below 3.

The chemical stability of anthramycin is increased by binding to DNA. Conversely, anthramycin binding increases the stability of the DNA helix, as indicated by an increase in T_m. Strand separation is not inhibited, however, showing that there is no inter-strand crosslinking. After strand separation of a DNA-anthramycin complex by alkali, the anthramycin remains bound to the separated strands (KOHN and SPEARS, 1970).

Anthramycin binding decreases the buoyant density of DNA in CsCl, and increases the intrinsic viscosity of high molecular weight DNA. Increased viscosity is one of the effects produced by intercalation between base-pairs in DNA. Unlike intercalating agents such as acridines and actinomycin (MÜLLER and CROTHERS, 1968), however, the effect on viscosity is greatly diminished by sonic fragmentation of DNA to a low molecular weight. Also, unlike with intercalating agents, the sedimentation constant of DNA is not reduced. The possibility of intercalation has further been excluded by electric dichroism measurements which indicate that the plane of the anthramycin chromophore deviates from the plane of the DNA bases by at least 40° (GLAUBIGER and KOHN, 1974). The anthramycin-DNA complex therefore appears to differ from previously characterized modes of binding of small molecules to DNA. The viscosity and sedimentation data indicate that anthramycin causes stiffening without lengthening of the DNA helix.

Whereas free anthramycin has a pKa at about 8.7, the anthramycin-DNA complex has no detectable acid dissociation. Hence the anthramycin proton that dissociates with pK 8.7, presumably from the phenolic group at position 9, is either lost or has its ionization suppressed in the complex. When anthramycin binds to DNA there is neither a net gain nor a net loss of a proton (KOHN et al., 1974). Therefore, if a proton is lost from position 9, there must be a compensatory uptake of a proton elsewhere. Consistent with the involvement of position 9 in the binding, the circular dichroism spectrum of bound anthramycin has a monophasic pattern similar to that of dissociated anthramycin at high pH, and unlike the biphasic pattern of free anthramycin at pH 7 (GLAUBIGER and KOHN, 1974).

Anthramycin-reacted DNA is inactivated as a template for RNA and DNA polymerase reactions and as a substrate for nuclease action (BATES et al., 1969; HORWITZ and GROLLMAN, 1968).

Effects on Macromolecular Syntheses

In most of the biological studies, anthramycin-11-methyl ether (AME) has been used rather than anthramycin. No differences have been reported between the biological actions of the two substances, however, and the name "anthramycin" will be used even when the material employed was the methyl ether[1].

Anthramycin inhibits RNA and DNA synthesis without affecting protein synthesis, as gauged by the rates of incorporation of precursors. In L1210 cell cultures, RNA synthesis appears to be inhibited to a slightly greater degree than is DNA synthesis (Fig. 4), the difference being much less than in the case of actinomycin (KOHN et al., 1968). In Ehrlich ascites cells in vitro, low concentrations of anthramycin inhibited RNA synthesis more than DNA synthesis; for example, 1 µM anthramycin inhibited uridine incorporation into RNA 54% and into DNA 29% (BATES et al., 1969). Similar results have been obtained in HeLa cells (HORWITZ and GROLLMAN, 1968). In Ehrlich ascites tumor in mice, RNA and DNA synthesis were inhibited equally; the inhibition was nearly complete after an intraperitoneal dose of 0.5 mg/kg anthramycin while protein synthesis was unaffected even at 1.5 mg/kg (BATES et al., 1969).

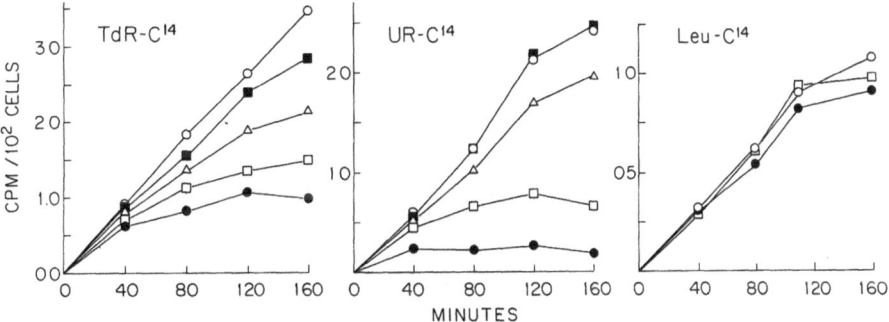

Fig. 4. Effect of anthramycin on incorporation of precursors into DNA, RNA, and protein by cultured L1210 cells. Anthramycin methyl ether and radioactive precursors were added at zero time. Anthramycin concentration; o——o, zero; ■——■ 0.03 µM; △——△ 0.09 µM; □——□ 0.28 µM; ●——● 0.86 µM. (After KOHN et al., 1968)

The inhibition of RNA synthesis in L1210 cell cultures has a gradual onset and increases progressively with time (Fig. 5a). This differs from the case of actinomycin which has a prompt inhibitory effect.

Fig. 5b relates the effect on RNA synthesis to the killing effect on the cells.

Whereas many inhibitors of RNA synthesis selectively inhibit the synthesis of ribosomal precursor RNA in nucleoli, anthramycin inhibits nucleolar RNA synthesis to the same degree as nucleoplasmic RNA synthesis. The chain lengths of both types of RNA synthesized in the presence of anthramycin, however, are reduced (Fig. 6). This may be due to premature termination of RNA chain growth or to the selective inhibition of synthesis of the longer chains. Anthramycin

[1]Anthramycin is most commonly supplied as anthramycin-11-methyl ether hydrate, molecular weight 347; this molecular weight will be implied when the dosage is stated on a weight basis.

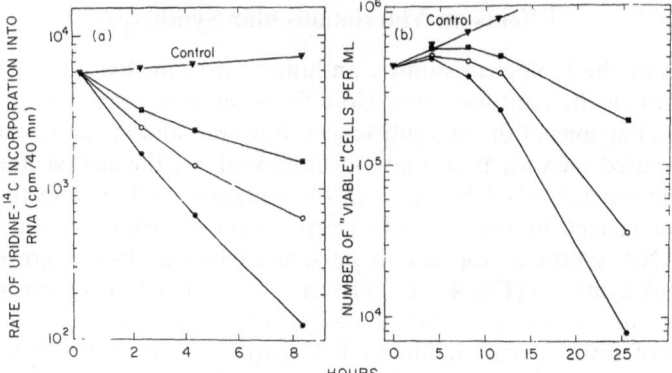

Fig. 5. Effect of anthramycin: a On rate of incorporation of uridine-C^{14} into RNA, and b on survival of morphologically intact cells as determined by phase microscopy and trypan blue staining. L1210 cell culture. Anthramycin concentrations: ▼ zero, ■ 0.14 μM, ○ 0.21 μM, ● 0.3 μM. (After KANN and KOHN, 1972a)

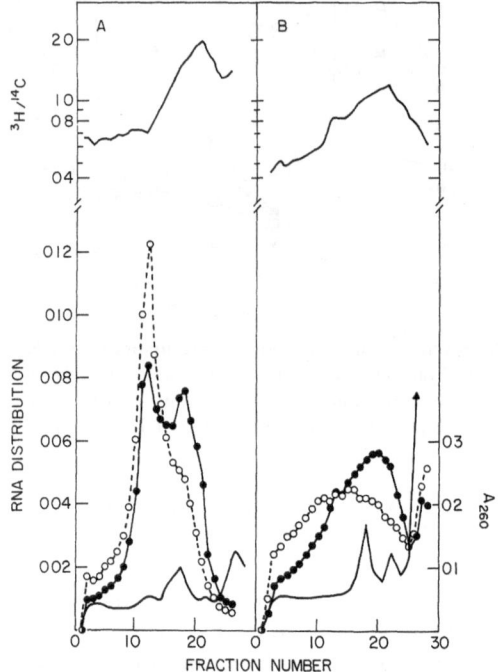

Fig. 6. Effects of anthramycin on sedimentation of newly synthesized RNA in L1210 cells. Sedimentation is from right to left. A Nucleolar fraction. B Nucleoplasmic fraction. Cells were exposed to 2 μM anthramycin for 90 min, then to uridine-6-^{3}H (●——●) for 10 min; they were then mixed with control cells which had been labeled for 10 min. with uridine-2-^{14}C (○----○). The RNA sedimentation distributions are normalized to unit area under each curve. Overall inhibitions of uridine incorporation were: nucleolar RNA, 78%; nucleoplasmic RNA, 82%. (From KANN and KOHN, 1972a)

does not inhibit the processing of nucleolar RNA (SNYDER *et al.*, 1971). In these respects, the effects of anthramycin resembles those of nitrogen mustard, and may reflect the nearly irreversible nature of its binding to DNA (KANN and KOHN, 1972b).

Biological Effects

Anthramycin has been found to be active against a variety of experimental neoplasms. The drug was tolerated upon daily intraperitoneal, subcutaneous, or oral administration at dosages of 0.06, 0.25 and 1.0 mg/kg, respectively in adult albino mice. Treatment at these dosages for 8 days produced anti-tumor responses against Ehrlich solid or ascites tumor and sarcoma 180, but not against leukemia L1210 (GRUNBERG *et al.*, 1966). A low level of activity against L1210 was demonstrated, however, at a dosage of 0.2 mg/kg for 10 days, which produced a 50% increase in median survival time and an LD_{10} level of toxicity (ADAMSON *et al.*, 1968). Greater increases in median survival time were obtained with other mouse leukemia lines: L5178Y, P388, and K1964. The variant line, P388/38280, which is resistant to terephthalanilides and to actinomycin, was noted to be resistant also to anthramycin. Anti-tumor responses were also seen against the rat carcinosarcoma, Walker 256, heterotransplants of the human lines H. Ep. No. 3 and H. Ad. No. 1 (GRUNBERG *et al.*, 1966), and against a plasma cell tumor (ADAMSON *et al.*, 1968).

In normal mice, anthramycin caused cellular necrosis in duodenal crypts and pancreatic acini. Changes were prominent in cell nucleoli. Nucleolar segregation into granular and fibrillar components occurred in several cell types, but not in hepatocytes (HARRIS *et al.*, 1968).

Antibacterial effects are produced *in vitro* against *E. coli* and *Sarcina lutea* at anthramycin concentrations of 10 μg/ml (HORWITZ and GROLLMAN, 1968). No chemotherapeutic effects were obtained, however, against a variety of experimental infections (GRUNBERG *et al.*, 1966). Among experimental protozoal and helminth infections, therapeutic activity was reported against *Trichomonas vaginalis*, *Endomeba histolytica* and pinworms (GRUNBERG *et al.*, 1966). There was no activity against fungi or viruses.

Anthramycin causes a permanent loss of chloroplasts in *Euglena* when applied at slightly sublethal concentrations (EBRINGER, 1971).

Anthramycin has a chemosterilizant action on male houseflies (HORWITZ *et al.*, 1971).

Structure-Activity Relations

The cytotoxic and chemotherapeutic activities of anthramycin derivatives have in all reported cases been correlated with the ability of the derivatives to react with DNA. Therefore it is reasonable to assume that these activities are, in fact, due to the reaction with DNA. The published data are summarized in Table 1. It is seen that activity is retained after etherification or epimerization at position 11 or after elimination of H_2O between positions 10 and 11. This conforms with the convertibility of these species to anthramycin in aqueous

Table 1. *Structure-activity relations of anthramycin derivatives*

Derivative	Activity	Systems and reference
Anthramycin	+	a, b
11-epimer	+	b
11-methyl ether (both epimers)	+	a
11-ethyl ether	+	a
11-benzyl ether	+	a
10,11-anhydro-	+	a, b
Terminal — CONH$_2$ replaced by – CN	+	b
9-methyl ether	–	a, b
11-keto-	–	b
10-acetyl-	–	b
9-desoxy-2,3,1′,2′-tetrahydro-	–	a
2,3-dihydro- and tail at position 2 replaced by —OH or —OAc	–	b

[a] Spectral change with DNA, antibacterial activity, anti-tumor activity (STEFANOVIC, 1968).
[b] Inhibition of RNA polymerase reaction and chemosterilizant action in houseflies (HORWITZ et al., 1971).

solution (see Chemistry). On the other hand, a more permanent change at position 11 by oxidation to the ketone as well as substitutions at positions 9 and 10 resulted in biological inactivity. Conversion of the terminal —CONH$_2$ to —CN retained activity, but major changes in the side-chain region of the molecule may inactivate.

Structure analysis has recently disclosed an antibiotic relative of anthramycin: tomaymycin (Fig. 1), an antibiotic produced by *Streptomyces achromogenes* var. *Tomaymycetis* (KARIYONE et al., 1971). The major structural differences from anthramycin are the location of the phenolic group at position 8 instead of 9, the presence of a methoxy group at position 7, and a major change in the side-chain region of the molecule. The absolute configuration at position 11 a is the same as in anthramycin. As in the case of anthramycin, the 11-keto derivative is inactive. The antibiotic has been stated to inhibit bacteriophages and neoplasms.

References

ADAMSON, R. H., L. G. HART, V. T. DEVITA, and V. T. OLIVERIO: Antitumor activity and some pharmacologic properties of anthramycin methyl-ether. Cancer Res. **28**, 343 (1968).

BATES, H. M., W. KUENZIG, and W. B. WATSON: Studies on the mechanism of action of anthramycin methyl ether, a new antitumor antibiotic. Cancer Res. **29**, 2195 (1969).

EBRINGER, L.: Action of inhibitors of nucleic acid synthesis on Euglena. Experientia **27**, 586 (1971).

GLAUBIGER, D., and K. W. KOHN: The reaction of anthramycin with DNA: Properties of the complex. Biochim. Biophys. Acta, in press (1974).

GRUNBERG, E., H. N. PRINCE, E. TITSWORTH, G. BESKID, and M. D. TENDLER: Chemotherapeutic properties of anthramycin. Chemotherapia **11**, 249 (1966).

HARRIS, C., H. GRADY, and D. SVOBODA: Segregation of the nucleolus produced by anthramycin. Cancer Res. **28**, 81 (1968).

HORWITZ, S. B., S. C. CHANG, A. P. GROLLMAN, and A. B. BORKOVEC: Chemosterilizant action of anthramycin: A proposed mechanism. Science **174**, 159 (1971).

HORWITZ, S. B., and A. P. GROLLMAN: Interaction of small molecules with nucleic acids. I. Mode of action of anthramycin. Antimicrobial Agents Chemoth. **21** (1968).

KANN, H. E., JR., and K. W. KOHN: Effects of anthramycin and actinomycin on RNA synthesis patterns in L1210 cells. J. Cellular Comp. Physiol. **79**, 331 (1972a).

KANN, H. E., JR., and K. W. KOHN: Effects of DNA-reactive drugs on RNA synthesis patterns in L1210 cells. Molec. Pharmacol., in press (1972b).

KARIYONE, K., YAZAWA, and M. KOHSAKA: Structure of tomaymycin and oxotomaymycin. Chem. & Pharm. Bull. (Tokyo) **19**, 2289 (1971).

KOHN, K. W., V. H. BONO, JR., and H. E. KANN: Anthramycin, a new type of DNA-inhibiting antibiotic: Reaction with DNA and effects on nucleic acid synthesis in mouse leukemia cells. Biochim. Biophys. Acta **155**, 121 (1968).

KOHN, K. W., D. GLAUBIGER, and C.L. SPEARS: The reaction of anthramycin with DNA: Studies of kinetics and mechanism: Biochim. Biophys. Acta, in press (1974).

KOHN, K. W., and C. L. SPEARS: Reaction of anthramycin with DNA. J. Mol. Biol. **51**, 551 (1970).

LEIMGRUBER, W., A. D. BATCHO, and R. C. CZAJKOWSKI: Total synthesis of anthramycin. J. Am. Chem. Soc. **90**, 5641 (1968).

LEIMGRUBER, W., A. D. BATCHO, and F. SCHENKER: Structure of anthramycin. J. Am. Chem. Soc. **87**, 5793 (1965b).

LEIMGRUBER, W., V. STEFANOVIC, F. SCHENKER, A. KARR, and J. BERGER: Isolation and characterization of anthramycin, a new antitumor antibiotic. J. Am. Chem. Soc. **87**, 5791 (1965a).

MÜLLER, W., and D. M. CROTHERS: Studies of the binding of actinomycin and related compounds to DNA. J. Mol. Biol. **35**, 251 (1968).

SNYDER, A. L., H. E. KANN, JR., and K. W. KOHN: Inhibition of the processing of ribosomal precursor RNA by intercalating agents. J. Mol. Biol. **58**, 555 (1971).

STEFANOVIC, V.: Spectrophotometric studies of the interaction of anthramycin with DNA. Biochem. Pharmacol. **17**, 315 (1968).

TENDLER, M. D., and S. KORMAN: 'Refuin': A non-cytotoxic carcinostatic compound proliferated by a thermophilic actinomycete. Nature **199**, 501 (1963).

3'-Amino-3'-Deoxyadenosine

Harold T. Shigeura

AMMANN and SAFFERMAN (1958) observed that the culture filtrate of *Helminthosporium* sp. inhibited mitosis of meristematic cells in the onion root. They also discovered that intraperitoneal administration of the same filtrate increased the survival time of mice bearing Gardner lymphosarcoma. Whether or not the same substance was responsible for the activities in the two test systems was however not clear. PUGH, *et al.* (1962) then prepared some crude concentrate of the antitumor factor from the filtrate of *Helminthosporium* sp. No. 215 (Culture Collection of the Institute of Microbiology, Rutgers University) and examined its growth-inhibitory property against several lines of ascites tumor cells in mice and against a number of microorganisms. Their results showed that the concentrate when administered intraperitoneally was carcinostatic against adenocarcinoma S_3A, Gardner lymphosarcoma 6C3HED, Ehrlich Lettré (hyperdiploid) carcinoma and sarcoma 180. It was, however, not active against any of the four ascitic tumors when given by the intravenous, subcutaneous or oral routes. The material showed no antimicrobial activity against several species of gram-positive and gram-negative bacteria, fungi and yeasts.

The antitumor activity against Ehrlich Lettré hyperdiploid carcinoma was subsequently used by GERBER and LECHEVALIER (1962) as an assay in their successful effort to isolate the active substance from the filtrate. Their studies showed that the bulk of the antitumor factor in the filtrate was not extractable with butanol but retained by strongly basic or strongly acidic ion exchange resin columns. The substance was partially purified by use of sulfonated polystyrene ion exchange column followed by paper chromatography with dilute ammonia. The material at this stage of purification was found to absorb radiation at 260 mμ; the amphoteric properties and ultraviolet light absorption spectrum suggested the presence of a purine nucleus. A final purification step on cellulose column eluted with dilute ammonia yielded a crystalline product with a molecular formula of $C_{10}H_{14}N_6O_3$. The active material was subsequently found to be identical with 3'-amino-3'deoxyadenosine, a nucleoside prepared earlier in another laboratory by chemical synthesis (BAKER *et al.*, 1955b). A comparison of physical and chemical properties of the synthetic and the naturally obtained substances confirmed its identity. The structural resemblance of this new nucleoside antibiotic to the aminonucleoside of puromycin, a potent nephrotoxic compound, and to 3'-deoxyadenosine (cordycepin), a nucleoside antibiotic isolated from the culture filtrate of *Cordyceps militaris* (CUNNINGHAM *et al.*, 1951) can be seen in Fig. 1.

Soon after the announcement of the discovery of 3'-amino-3'-deoxyadenosine by GERBER and LECHEVALIER (1962), the isolation and characterization of this antibiotic from *Cordyceps militaris* was reported from another laboratory (GUAR-

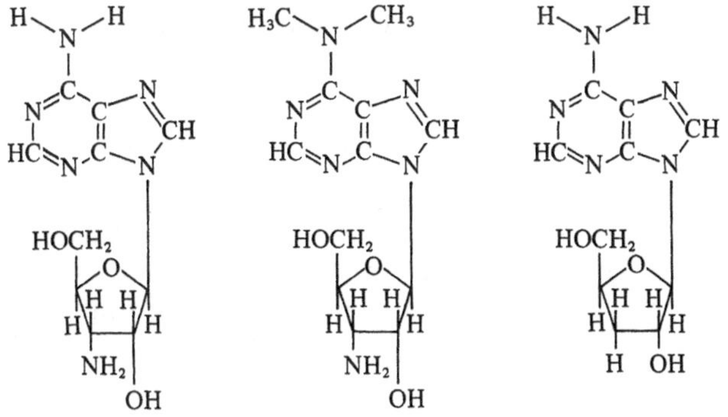

3′-Amino-3′-deoxyadenosine | Puromycin Aminonucleoside | Cordycepin

Fig. 1. Comparative structures of purine 3′-deoxyribonucleosides

INO and KREDICH, 1963). In the latter study, the 70% ethanol extract of the mycelial pad was dried and the water soluble materials were separated by means of Dowex-50 (NH_4^+) column chromatography. The unknown substance in the second peak was concentrated to a small volume. The solid material thus formed upon recrystallization in H_2O, melted at 275°–278° as did an authentic sample of 3′-amino-3′-deoxyadenosine. Comparative studies of the nucleosides and the ribose moieties obtained upon acid hydrolysis by paper chromatography indicated that the nucleoside isolated from the extracts of *C. militaris* was identical with 3′-amino-3′-deoxyadenosine. The antibiotic has since been isolated also from *Aspergillus nidulans* (SUHALDOLNIK, 1970).

3′-Amino-3′-deoxyadenosine (3′-NH_2-3′-dA) did not inhibit the growth of the following bacteria and fungi: *Staphylococcus aureus* strains Le Compte and Valentine, *Escherichia coli* strains AVZ and Monod, *Pseudomonas aeruginosa* Ingant, *Enterococcus* Fanjou, *Sarcina lutea* ATCC 10702, *Bacillus subtilis* strains CHP and ATCC 6633, *Aspergillus* niger 4823, *Penicillium* sp. 4847 and *Trichophyton mentagrophytes* 4805. The material, however, was very active against two species of yeasts, *Cryptococcus neoformans* 4806 and *Candida albicans* (GERBER and LECHEVALIER, 1962).

PUGH and GERBER (1963) followed with an extensive study of the antitumor property of 3′-NH_2-3′-dA. The acute toxicity of the substance in albino male mice, expressed as the LD_{50} were as follows: intraperitoneal and intravenous routes, 28 mg/kg; oral route, 301 mg/kg; and subcutaneous route, 33 mg/kg. The deoxynucleoside inhibited the growth of ascitic adenocarcinoma S_3A in mice following intraperitoneal injection administered 24 hours after tumor implantation. There was also a significant increase in the survival time of treated mice bearing Ehrlich ascites carcinoma as compared to saline-treated tumor control animals. The compound, however, was only weakly active against ascitic Gardner lymphosarcoma and sarcoma 180 tumors. The material as reported previously by PUGH *et al.* (1962) was again found to be inactive when administered by the intravenous, oral or subcutaneous routes. It is of interest to note the additional finding that 3′-NH_2-3′-dA exhibited more antitumor activity against ascitic adenocarcinoma S_3A than did the structurally related puromycin aminonucleoside.

Biosynthesis of 3′-Amino-3′-Deoxyadenosine

Using *Helminthosporium* sp. 215, Chassy and Suhaldolnik (1969) demonstrated that uniformly ^{14}C-labeled adenosine was incorporated into 3′-amino-3′-deoxyadenosine without cleavage of the N-riboside bond; whereas, uniformly ^{14}C-labeled 3′-deoxyadenosine was not converted to the aminonucleoside.

Metabolism

In conjunction with the studies on 3′-deoxyadenosine (cordycepin) in progress at that time in this laboratory, a sample of 3′-NH_2-3′-dA was requested and obtained through the courtesy of Dr. Nancy N. Gerber of Rutgers University. Our interest in this nucleoside was centered not only on its previously demonstrated antitumor property but also on its obvious structural resemblance to 3′-deoxyadenosine (3′-dA). The latter antibiotic had been shown earlier to adversely affect the growth of bacteria (Cunningham *et al.*, 1951) and of Ehrlich ascites tumor cells (Jagger *et al.*, 1961); evidences obtained later indicating that suppression of growth was due to the inhibition of nucleic acid synthesis by 3′-dA (Klenow, 1961; Rottman and Guarino, 1964; Guarino, 1967). A detailed study on the mechanism of action of 3′-dA has been reported from this laboratory (Shigeura and Boxer, 1964a; Shigeura and Gordon, 1964b). For these reasons, particularly because of the demonstration by Pugh and Gerber (1963) of the beneficial effects of 3′-NH_2-3′-dA in mice bearing Ehrlich ascites carcinoma cells and the lack of activity against certain microbial and tumor cells, the metabolism and function of the antibiotic was investigated in this laboratory (Shigeura *et al.*, 1966a).

Suspensions of Ehrlich ascites cells in suitable buffered media containing glucose and radioactive hypoxanthine or orotic acid were incubated at 37° with or without 3′-NH_2-3′-dA. After 30 minutes, the cold TCA-insoluble RNA was hydrolyzed in dilute KOH and the ribonucleotides were separated by Dowex-1 formate chromatography and measured for radioactivity. The results showed that in the presence of 0.9 mM 3′-NH_2-3′-dA, the incorporation of labeled hypoxanthine into AMP-2′, 3′-PO_4 and GMP-2′,3′-PO_4 were inhibited by 83% and 58%, respectively. Similarly, the uptake of orotic acid into UMP-2′,3′-PO_4 was suppressed by 54%. These initial results suggested that the inhibitory action of the antibiotic on Ehrlich ascites cells growing in the mouse observed by Pugh and Gerber (1963) may primarily be due to the inhibition of RNA synthesis. The observations were also reminiscent of the effects of 3′-deoxyadenosine on RNA synthesis reported from this laboratory in 1964 (Shigeura and Gordon). On an equimolar basis, the two antibiotics appeared to be equally active against Ehrlich ascites cells. Hence, a comparative study was indicated early in our work.

Whether or not 3′-NH_2-3′-dA could be phosphorylated and thereby mimic the action of adenosine ribonucleotides in inhibiting the synthesis of 5-phosphoribosylamine in whole ascites cells was next investigated. The results demonstrated that the antibiotic at 1.0 mM inhibited the reaction by 74%. 3′-Deoxyadenosine and 2-fluoroadenosine, nucleosides previously shown to be phosphorylated to the 5′-triphosphate level (KLENOW, 1963; SHIGEURA et al., 1965), inhibited the formation of 5-phosphoribosylamine by 92% and 94%, respectively. These observations suggested that the inhibitory effect of 3′-NH_2-3′-dA on 5-phosphoribosylamine synthesis could be due to the formation and consequently to the action of the 5′-triphosphate.

Biosynthesis of 3′-NH_2-3′-dATP would imply that the nucleoside itself was at least not extensively degraded by whole Ehrlich ascites cells; for example, by deamination followed by glycosidic cleavage to yield hypoxanthine. This possibility was examined and the results showed that 3′NH_2-3′-dA was not deaminated by intact Ehrlich ascites cells. Similarly, 3′-deoxyadenosine and adenosine were also unaffected. Inosine, on the other hand, was readily cleaved to yield hypoxanthine as one of the products. It should be noted, however, that 3′-NH_2-3′-dA has been demonstrated by other workers to be deaminated with ease by purified enzyme. In this regard, FREDERICKSEN (1965) showed that the nucleoside was deaminated by calf intestine adenosine deaminase, the K_m of the substrate was 1.67×10^{-4} M. CORY and SUHALDOLNIK (1965) observed with the same enzyme that the initial velocity of deamination of 3′-NH_2-3′-dA was about 1/2 that of adenosine. Similar results were obtained by BLOCK et al. (1967).

The information obtained thus far indicated that 3′-NH_2-3′-dA was not degraded by intact Ehrlich ascites cells but probably metabolized to the phosphorylated derivatives. A study of the phosphorylation of adenosine compared to those of a number of adenosine analogues, both natural and synthetic, has been done by LINDBERG et al. (1967). The partially purified adenosine kinase from rat liver and Ehrlich ascites tumor cells used were free of adenosine deaminase and almost free of adenosine triphosphatase activity. With liver enzyme, the K_m and velocity of conversion of 3′-NH_2-3′-dA to the 5′-monophosphate were 6.1×10^{-4} M and 0.31, respectively. The corresponding values for adenosine were 1.6×10^{-6} M and 1.00. The antibiotic was indeed phosphorylated by adenosine kinase but not as well as the natural substrate.

Mode of Action

In our subsequent study on the metabolism of 3′-NH_2-3′-dA by whole Ehrlich ascites cells (SHIGEURA et al., 1966a), the formation of phosphorylated derivatives was investigated by a method described by KLENOW (1963). The tumor cells were incubated at 37° in a buffered medium containing glucose and the nucleoside under study. After 3 hours, the perchloric acid-soluble fraction of the cells was chromatographed on Dowex-1 formate column by elution with a gradient solution of ammonium formate. Three major ultraviolet light absorbing zones were obtained and all three fractions showed absorption spectra similar to that of 3′-NH_2-3′-dA. The molecular extinction coefficient of the nucleoside was determined to be 14,300. Using this ε-value for the nucleosides, the three fractions were found to contain 1.13, 1.97 and 3.05 µmoles of phosphate/µmole of nucleoside. After treatment with 1N HCl for 10 minutes at 100°, the µmole of acid-labile phosphate/μmole of nucleoside for the fractions were found to be 0, 1.03 and 1.97, respectively; each fraction consumed 0.96 μmole of periodate when tested by the spectrophotometric method. The fractions were therefore characterized as 3'-amino-3'-deoxyadenosine-5'-mono-, 5' di- and 5'-triphosphate, respectively. The fractions corresponding to the di- and tri-phosphates migrated considerably slower than ADP or ATP on paper electrophoresis at pH 3.5. The presence of an additional amino group on 3'-NH₂-3'-dA would account for the difference in mobility. Under these conditions, ATP and 3'-dATP showed the same electrophoretic mobility. In this particular experiment, the micromoles of 5'-mono-, 5'-di- and 5'-triphosphate formed were 1.25, 1.93 and 3.18, respectively.

The most obvious experiment—an examination of the possible effect of 3'-NH₂-3'-dATP on RNA polymerase—was then done. Using DNA-dependent RNA polymerase from *Micrococcus lysodeikticus* (prepared according to NAKA-MOTO *et al.*, 1964), it was soon apparent that the triphosphate markedly suppressed the incorporation of each of the four nucleotide triphosphates into RNA. When the ratio of concentration of ATP to 3'-NH₂-3'-dATP was 2.5, the percent inhibition was about 80. 3'NH₂-3'-dADP, as expected, was ineffective. The inhibitory activity of this antibiotic thus resembled that of the 5'-triphosphate of 3'-deoxyadenosine (cordycepin) which was previously found not only to inhibit DNA-dependent RNA polymerase but also managed to be incorporated into the nascent RNA polymer (SHIGEURA and BOXER, 1964; SHIGEURA and GORDON, 1965). It was demonstrated at that time that the incorporation of 3'-dATP in lieu of ATP into a growing RNA chain prevented further elongation of that particular molecule. The reason given for the termination of RNA synthesis was the lack of hydroxyl group on carbon 3 of the deoxyribose moiety at 3'-dA. The absence of a hydroxyl group on carbon 3 of 3'-NH₂-3'-dA could therefore also be responsible for its inhibitory effect on RNA synthesis. This interesting possibility led to a series of comparative studies of the two nucleoside analogues as described below.

The effects of the two triphosphates on DNA-dependent Poly Adenylate formation by *Micrococcus lysodeikticus* RNA polymerase were first examined. At equimolar concentrations of 3'-dATP and 3'-NH₂-3'-dATP, the reaction was inhibited by 90 and 91%, respectively (SHIGEURA *et al.*, 1966a). A close similarity in the action of the two inhibitors was also evident in an experiment in which DNA-dependent RNA synthesis was studied. The test system contained 100 mμmoles of each of the four natural ribonucleotide triphosphate, 60 μg of calf thymus DNA, 50 μg protein of *Micrococcus lysodeikticus* RNA polymerase and other essential co-factors. The amount of UTP-¹⁴C incorporation in the presence of varying amounts of inhibitors was measured. The mμmoles of UTP-¹⁴C incorporated in the presence of 0, 10, 20, 40 and 80 mμmoles of 3'-NH₂-3'-dATP were 7.1, 2.9, 2.0, 1.4 and 1.0, respectively. The corresponding values with 3'-dATP were 7.1, 2.9, 1.9, 1.3 and 1.0, respectively.

When a study of the kinetics of incorporation of UTP-¹⁴C in the presence of equimolar amounts of the two analogue triphosphates was done, the results showed that there was essentially no difference in the reaction rates during the first 20 minutes.

The information obtained supported the view that 3'-NH₂-3'-dATP, like 3'-dATP, was incorporated into nascent RNA chains and thus prevented further chain elongation. Another test of this concept was an experiment in which 3'NH$_2$-3'-dATP or 3'-dATP was added to a RNA polymerase system lacking only ATP. In such a mixture, addition of increasing amounts of either analogue triphosphate should not, according to the view expressed above, stimulate the incorporation of any of the three remaining natural ribonucleotide triphosphates. In a complete system containing 50 mμmoles of each of the four ribonucleotide triphosphates, 5.6 mμmoles of GTP-^{14}C were incorporated. In the absence of ATP, the amount of GTP-^{14}C incorporated was 0.11 mμmoles. In the absence of ATP, but in the presence of 20 and 50 mμmoles of 3'-NH$_2$-3'-dATP, the amounts of GTP-^{14}C incorporated were 0.06 and 0.06, respectively. The corresponding values with 3'-dATP were 0.06 and 0.11. Consistent with the hypothesis proposed, the analogue triphosphates indeed did not permit RNA synthesis. Moreover, the inhibitory activities of the two analogue triphosphates were essentially identical in all of the systems examined. Thus, on the basis of the striking resemblance in the activities of 3'-NH$_2$-3'-dATP to those of 3'-dATP in the several RNA polymerase systems examined, it was concluded that the mechanism of action of these two analogues of ATP were essentially similar.

More recently (TRUMAN and KLENOW, 1968), another study of the metabolism of 3'-amino-3'-deoxyadenosine in Ehrlich ascites tumor cells was reported. The investigators also observed that the nucleoside antibiotic inhibited nucleic acid synthesis in ascites cells. The incorporation of adenine-^{14}C or uridine-^{14}C into DNA and RNA was inhibited by 70–75% in the presence of 0.3 μmole of the analogue per ml and about 95% in the presence of 2.2 μmole per ml. In a manner similar to their experiments with 3'-deoxyadenosine (KLENOW, 1963a), the phosphorylated derivatives were isolated. 3'-NH$_2$-3'-dATP was then shown to inhibit the incorporation of labelled ATP into RNA by DNA-dependent RNA polymerase from Ehrlich ascites tumor cells. The degree of inhibition was approximately the same as that found with an equivalent concentration of 3'-dATP, a finding in agreement with those reported earlier with RNA polymerase from *Micrococcus lysodeikticus* (SHIGEURA *et al.*, 1966a). The authors concluded, as we had done earlier, that the incorporation of 3'-NH$_2$-3'-dATP into nascent RNA chains inhibited further chain elongation.

While the mechanism of action of 3'-NH$_2$-3'-dA has been clearly defined, an interesting differential effect of the antibiotic on the biosynthesis of various RNA molecules in Ehrlich ascites cells was reported by TRUMAN and FREDERICKSEN (1969). The formation of 28S and 18S cytoplasmic RNA were markedly suppressed, while that of 50–60S nuclear RNA was only slightly inhibited. It was suggested that the reason may be due to the presence of different species of RNA polymerase in various cellular fractions or to compartmentation of nucleoside triphosphates around different regions of DNA.

Puromycin Aminonucleoside
and Other Structurally Related Nucleosides

Puromycin aminonucleoside (PAN), originally synthesized by BAKER *et al.*, (1955a). has been known for sometime to produce a nephrotic syndrome when injected into rats (FRENCK *et al.*, 1955). Although a clear explanation of this phenomenon is still lacking, an extensive study on the metabolism of PAN in mammalian cells has recently been done by Farnham and Dubin (1965, 1967). With the intent of investigating the effects of the nucleoside on nucleic acid synthesis in mouse fibroblasts (L-cells), they observed that PAN preferentially inhibited ribosomal RNA synthesis. Messenger RNA, DNA and protein synthesis were virtually unaffected. Furthermore, the inhibitory action of PAN was not prevented by adenosine. A notable finding that may explain the inhibitory action of PAN was the observation that L-cells metabolized significant amounts of this material to compounds tentatively identified as phosphorylated 3′-amino-3′-deoxyadenosine derivatives. With Ehrlich ascites cells, (Shigeura et al, 1966a), the inhibitory effect of PAN was considerably less than that of 3′-NH$_2$-3′-dA and moreover, there was no evidence of direct phosphorylation of PAN, or of the formation of 3′-amino-3′-deoxyadenosine derivatives. The inability of Ehrlich ascites cells to demethylate PAN may explain the differences in metabolism of PAN by Ehrlich ascites cells and by mouse fibroblasts.

Since the inactivity of 6-aminonucleoside analogues in some of the cellular systems has been ascribed to the ease with which these nucleosides are deaminated, attempts have been made in these laboratories (Walton, et al., 1965) to chemically synthesize related compounds that will withstand degradation but still retain the potent inhibitory activity of the parent molecule. To this end, several new analogues of deoxyadenosine have been prepared, and their structure-activity relationship and structural specificity for phosphorylation have been examined (Shigeura et al., 1966b; Shigeura and Sampson, 1967b).

A study of the metabolism of 3′-C-methyladenosine was particularly interesting because this compound is the first derivative of adenosine with a branched-chain pentose. This nucleoside was neither deaminated nor cleaved by whole Ehrlich ascites cells. When tritiated 3′-C-methyl-adenosine was incubated with ascites cells at 37° for 3 hours and substances soluble in cold 10% perchloric acid were fractionated by ion-exchange chromatography, a radioactive ultraviolet light absorbing material was eluted from the region corresponding to nucleotide triphosphate. This substance showed an absorption spectrum similar to that of adenosine or 3′-C-methyladenosine (λ_{max}258, λ_{min}227, 250 mμ/260 mμ=0.79, 280 mμ/260 mμ = 0.16) at pH 7 and contained 3.2 μmoles of phosphorus per μmole of adenosine. On paper electrophoresis at pH 3.5, the mobility of the substance was found to be comparable to that of ATP. These results, in addition to the observation that this analogue was not deaminated or cleaved, indicated that the radioactive triphosphate obtained was 3′-C-methyl-adenosine-5′-triphosphate.

In a similar system, the 6-methylamino analogues (6-methylaminopurine ribonucleoside, 6-methylaminopurine-2′-deoxyribonucleoside and 6-methylaminopurine-3′deoxyribonucleoside) were metabolized only to the 5′-monophosphate level and showed minimal effect on RNA synthesis. The lack of further phosphorylation was substantiated by the demonstration that all three of the 6-methylaminopurine nucleotides did not serve as substrate for crystalline rabbit muscle myokinase. However with KB cells, 6-methylaminopurine 3′deoxyribonucleoside was demethylated to 3′-deoxyadenosine. The latter compound was subsequently converted to 1. 3′-dATP and incorporated into RNA and 2. deaminated to 3'-deoxyinosine and incorporated into RNA *via* hyproxanthine. The inhibitory property of 6-methylaminopurine-3'-deoxyribonucleoside against KB cells thus appeared to be due primarily to the ability of these cells to convert the relatively inactive nucleoside to a potent inhibitor (3'-dATP). It may be noted here that a parallel type of metabolism of PAN by L-cells could also explain in part the observations of FARNHAM and DUBIN (1967). As opposed to the pathway of activation, KB cells were also able to detoxify part of 3'-dA by deamination and subsequent glycolytic cleavage to a normal substance, hypoxanthine. Metabolic pathways of this nature may account for the resistance by some tumor cells and microorganisms to 3'-NH_2-3'-dA observed by PUGH and GERBER (1963). Nucleosides with dimethyl or ethyl groups at the 6-N position were not metabolized by Ehrlich ascites cells and hence were essentially inactive against nucleic acid synthesis.

Some of the conclusions drawn from these studies with Ehrlich ascites cells were that the most active nucleosides possessed a free amino group at the 6-N position, a structure apparently required for conversion to the 5'-triphosphates. Nucleosides alkylated at the 6-N position, although comparatively resistant to degradation, were either not phosphorylated at all or phosphorylated only to the 5'-monophosphate level. KB cells an the other hand were, able to either activate a comparatively inert nucleoside by demethylation or inactivate a potent inhibitor by deamination (SHIGEURA and SAMPSON, 1967a; SHIGEURA and SAMPSON, 1967b).

References

AMMANN, C.A. and R.S. SAFFERMAN: The onion test as a possible screening method for antitumor agents. Antibiot. & Chemotherapy **8**, 1 (1958).

BAKER, B.R., J.P. JOSEPH, and J.H. WILLIAMS: Puromycin. Synthetic studies, VII. Partial synthesis of amino acid analogues. J. Am. Chem. Soc. **77**, 1 (1955).

BAKER, B.R., R.E. SCHAUB, and H.M. KISSMAN: Puromycin. Synthetic Studies. XV. 3'-amino-3'-deoxyadenosine. J. Am. Chem. Soc. **77**, 5911 (1955).

BLOCK, A., M.J. ROBINS, and J.R. MCCARTHY: The role of the 5'-hydroxyl group of adenosine in determining substrate specificity for adenosine deaminase. J. Med. Chem. **10**, 908 (1967).

CHASSY, B.M., and R.J. SUHALDOLNIK: Nucleoside Antibiotics. V. The biosynthesis and interconversion of 3'-amino-3'-deoxyadenosine and 3'-acetamido-3'-deoxyadenosine by *Helminthosporium* sp. 215. Biochim. Biophys. Acta **182**, 316 (1969).

CORY, J.G., and R.J. SUHALDOLNIK: Structural requirement of nucleosides for binding by adenosine deaminase. Biochemistry **4**, 1729 (1965).

CUNNINGHAM, K.G., S.A. HUTCHINSON, W. MANSON, and F.S. SPRING: Cordycepin, a metabolic product from cultures of *Cordyceps militaris* Linn). Part I. Isolation and characterization. J. Chem. Soc. **1951**, 2299.

FARNHAM, A.E., and D.T. DUBIN: Effect of puromycin aminonucleoside on RNA synthesis in L-cells. J. Mol. Biol. **14**, 55 (1965).

FARNHAM, A.E., and D.T. DUBIN: Studies on the mechanism of action of puromycin aminonucleoside in L-cells. Biochim. Biophys. Acta **138**, 35 (1967).

FREDERICKSEN, S.: Specificity of adenosine deaminase toward adenosine and 2'-deoxyadenosine analogues. Arch. Biochem. Biophys. **113**, 383 (1965).

FRENK, S., I. ANTONOWICS, J.M. CRAIG, and J. METCOFF: Experimental nephrotic syndrome induced in rats by aminonucleoside. Renal lesions and body electrolyte composition. Proc. Soc. Expt. Biol. Med. **89**, 424 (1955).

GERBER, N.N., and H.A. LECHEVALIER: 3'-Amino-3'-deoxyadenosine, an antitumor agent from *Helminthosporium* sp. J. Org. Chem. **27**, 1731 (1962).

Guarino, A., and N. M. Kredich: Isolation and identification of 3'-amino-3'-deoxyadenosine from *Cordyceps militaris*. Biochim. Biophys. Acta **68**, 317 (1963).

Guarino, A. J.: Antibiotics, Vol. I, Ed. by D. Gottlieb and P. D. Shaw. Berlin-Heidelberg-New York: Springer 1967.

Jagger, D. V., N. M. Kredich, and A. J. Guarino: Inhibition of Ehrlich ascites tumor growth by cordycepin. Cancer Res. **21**, 216 (1961).

Klenow, H.: Effect of cordycepin on the incorporation of ^{32}P- orthophosphate into the nucleic acids of ascites tumor cells *in vitro*. Biochem. Biophys. Res. Commun. **5**, 156 (1961).

Klenow, H.: Formation of the mono-, di- and triphosphate of cordycepin in Ehrlich ascites tumor cells *In vitro*. Biochim. Biophys. Acta **76**, 347 (1963a).

Klenow, H.: Inhibition of cordycepin and 2-deoxyglucose in the incorporation of ^{32}P-orthophosphate into the nucleic acids of Ehrlich ascites tumor cells *In vitro*. Biochim. Biophys. Acta **76**, 354 (1963b).

Lindberg, B., H. Klenow, and K. Hansen: Some properties of partially purified mammalian adenosine kinase. J. Biol. Chem. **242**, 350 (1967).

Nakamoto, T., C. F. Fox, and S. B. Weiss: Enzymatic Synthesis of Ribonucleic Acid. I. Preparation of ribonucleic acid polymerase from extracts of *Micrococcus lysodeikticus*. J. Biol. Chem. **239**, 107 (1964).

Pugh, L. H., and N. N. Gerber: The effect of 3'-amino-3'-deoxyadenosine against ascitic tumors of mice. Cancer Res. **23**, 640 (1963).

Pugh, L. H., H. A. Lechevalier, and M. Solotorovsky: Antitumor activity of a substance produced by a strain of *Helminthosporium*. Antibiot. & Chemotherapy **12**, 310 (1962).

Rottman, F., and A. J. Guarino: Studies on the inhibition of *Bacillus subtilis* growth by cordycepin. Biochim. Biophys. Acta **80**, 632 (1964a).

Rottman, F., and A. J. Guarino: The inhibition of purine biosynthesis *de nova* in *Bacillus subtilis* by cordycepin. Biochim. Biophys. Acta **80**, 640 (1964b).

Shigeura, H. T., and G. E. Boxer: Incorporation of 3'-deoxyadenosine-5'-triphosphate into RNA by RNA polymerase from *Micrococcus lysodeikticus*. Biochem. Biophys. Res. Commun. **17**, 758 (1964a).

Shigeura, H. T., G. E. Boxer, M. L. Meloni, and S. D. Sampson: Structure-activity relationship of some purine 3'-deoxyribonucleosides. Biochemistry **5**, 994 (1966a).

Shigeura, H. T., G. E. Boxer, S. D. Sampson, and M. L. Meloni: Metabolism of 2-fluoroadenosine by Ehrlich ascites cells. Arch. Biochem. Biophys. **111**, 713 (1965).

Shigeura, H. T., and C. N. Gordon: The effects of 3'-deoxyadenosine on the synthesis of ribonucleic acid. J. Biol. Chem. **240**, 806 (1964b).

Shigeura, H. T., and S. D. Sampson: Utilization of 6-methylamino-9-(3'-deoxy-β-D-ribofuranosyl) purine by KB cells. Biochim. Biophys. Acta **138**, 26 (1967a).

Shigeura, H. T., and S. D. Sampson: Structural basis of phosphorylation of adenosine congeners. Nature **215**, 419 (1967b).

Shigeura, H. T., S. D. Sampson, and M. L. Meloni: Limited phosphorylation of some 6-methylaminopurine nucleosides. Arch. Biochem. Biophys. **115**, 462 (1966b).

Suhaldolnik, R. J.: Nucleoside Antibiotics. New York: John Wiley & Sons, 1970.

Truman, J. T., and S. Fredericksen: Effect of 3'-deoxyadenosine and 3'-amino-3'-deoxyadenosine on the labelling of RNA sub-species in Ehrlich ascites tumor cells. Biochim. Biophys. Acta **182**, 36 (1969).

Truman, J. T., and H. Klenow: Effect of 3'-amino-3'-deoxyadenosine on nucleic acid synthesis in Ehrlich ascites tumor cells. Mol. Pharmacol. **4**, 77 (1968).

Walton, E., F. W. Holly, G. E. Boxer, R. F. Nutt, and S. R. Jenkins: 3'-Deoxynucleosides. II. Purine 3'-deoxynucleosides. J. Med. Chem. **8**, 659 (1965).

Bleomycin

Hamao Umezawa

Origin, Chemistry, and Artificial Bleomycins

Bleomycin was discovered by UMEZAWA *et al.* (1966a) in a study of antibiotics resembling phleomycin which earlier had been found by MAEDA *et al.* (1956). Bleomycin is produced by *Streptomyces verticillus,* extracted by successive application of cation exchange resin process, carbon and alumina chromatography, and obtained in the copper-chelated form. Both copper-chelated and free bleomycins show the same activity in inhibiting growth of bacterial and animal cells. The antibiotic thus obtained is a mixture which is separated by CM-Sephadex C-25 column chromatography with a gradient of ammonium formate from 0.05 M to 1.0 M into bleomycins $A_1, A_2, A_3, A_4, A_5, A_6, B_1, B_2, B_3, B_4$ and B_5 (UMEZAWA *et al.*, 1966b). The bleomycin-producing streptomyces also produces bleomycins demethyl-A_2, A_2'-a, A_2'-b, B_1' and B_6. Among these, the structures of copper-free bleomycins A_1, demethyl-A_2, A_2, A_2'-a, A_2'-b, A_5, A_6, B_2 and B_4 have been studied. These bleomycins are different from one another in the terminal amine moiety as shown in Table 1 (UMEZAWA, 1971). TAKITA *et al.* (1972b), proposed the structures as shown in Fig. 1.

Table 1. *Amine moieties of various bleomycins*

The amine of A_1:

$$-NH-CH_2-CH_2-CH_2-\overset{\overset{\displaystyle O}{\|}}{S}-CH_3$$

The amine of demethyl A_2:
$$-NH-CH_2-CH_2-CH_2-S-CH_3$$

The amine of A_2:

$$-NH-CH_2-CH_2-CH_2-\overset{+}{\underset{\underset{\displaystyle CH_3}{|}}{S}}-CH_3$$

The amine of A_2'-a:
$$-NH-CH_2-CH_2-CH_2-CH_2-NH_2$$

The amine of A_2'-b:
$$-NH-CH_2-CH_2-CH_2-NH_2$$

The amine of B_2:
$$-NH-CH_2-CH_2-CH_2-CH_2-NH-\overset{\overset{\displaystyle NH}{\|}}{C}-NH_2$$

The amine of A_5:
$$-NH-CH_2-CH_2-CH_2-NH-CH_2-CH_2-CH_2-CH_2-NH_2$$

The amine of B_4:
$$-NH-CH_2-CH_2-CH_2-CH_2-NH-\overset{\overset{\displaystyle NH}{\|}}{C}-NH-CH_2-CH_2-CH_2-CH_2-NH-\overset{\overset{\displaystyle NH}{\|}}{C}-NH_2$$

The amine of A_6:
$$-NH-CH_2-CH_2-CH_2-NH-CH_2-CH_2-CH_2-CH_2-NH-CH_2-CH_2-CH_2-NH_2$$

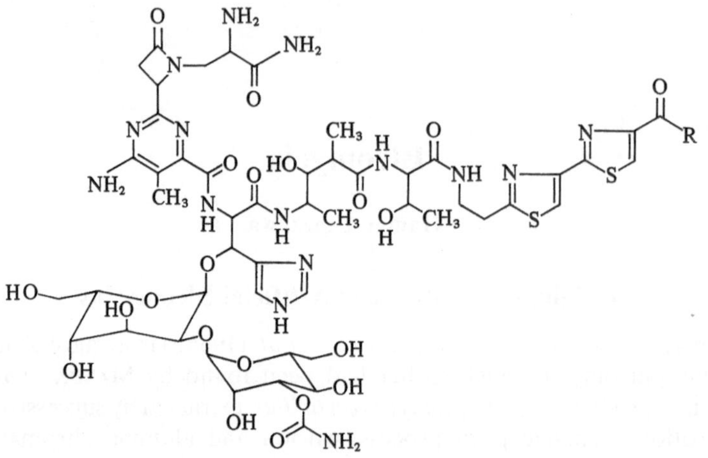

Fig. 1. Structure of bleomycin (R: terminal amine). Bleomycinic acid: R=OH

Hydrolysis of bleomycin A_2 gives β-amino-β-(4-amino-6-carboxy-5-methyl-pyrimidin-2-yl)propionic acid (Muraoka *et al.*, 1970; Yoshioka *et al.*, 1972), β-amino-L-alanine (Takita *et al.*, 1968), L-*erythro*-β-hydroxyhistidine (Takita *et al.*, 1971; Koyama *et al.*, 1973), 4-amino-3-hydroxy-2-methyl-*n*-valeric acid (Takita *et al.*, 1968), L-threonine (Takita *et al.*, 1968), 2′-(2-aminoethyl)-2,4′-bithiazole-4-carboxylic acid (Koyama *et al.*, 1968; Zee-Cheng and Cheng, 1970) and (3-aminopropyl)-dimethylsulfonium (Umezawa, 1971). Refluxing of bleo-mycins in methanol with a strong acidic resin (Amberlyst 15) gives methyl α- and β-L-gulopyranoside and methyl 3-O-carbamoyl-α-D-mannopyranoside (Takita *et al.*, 1969). Gas chromatography of trimethylsilyl derivatives of the methanolysis product indicates the presence of the same sugar moiety in all bleomycins. Omoto *et al.* (1972) isolated the whole sugar moiety, 2-O-(3-O-carbamoyl-α-D-mannosyl)-L-gulose, by hydrolysis of bleomycin A_2 in 0.3 N sulfuric acid at 81–83° for 6 hours. This sugar moiety is cleaved from the hydroxyl group of β-hydroxyhisti-dine moiety in bleomycin by β-elimination which occurs in 0.1 N sodium hydrox-ide for 7 days. Refluxing of a methylated bleomycin in 1.0 N hydrochloric acid for 18 hours gives β-aminoalanine betaine amide (Takita *et al.*, 1972b).

Partial hydrolysis of bleomycin gives the tripeptide S and the tripeptide A. Hydrolysis of the tripeptide S yields threonine, 2′-(2-aminoethyl)-2,4′-bithiazole-4-carboxylic acid and (3-aminopropyl)dimethylsulfonium. Hydrolysis of the tri-peptide A gives β-amino-β-(4-amino-6-carboxy-5-methylpyrimidin-2-yl)propionic acid, β-aminoalanine, β-hydroxyhistidine and 4-amino-3-hydroxy-2-methyl-*n*-valeric acid. N-bromosuccinimide oxidation of bleomycin gives the tetrapeptide S which gives 4-amino-3-hydroxy-2-methyl-*n*-valeric acid, threonine, 2′-(2-amino-ethyl)-2,4′-bithiazole-4-carboxylic acid and (3-aminopropyl)dimethylsulfonium on hydrolysis.

Structures of phleomycin D_1 and E as shown in Fig. 2 were proposed by Takita *et al.* (1972b). Various phleomycins are separated from one another by CM-Sephadex C-25 column chromatography with a gradient of ammonium for-mate (Ikekawa *et al.*, 1964). Among the various phleomycins, D_2, F, G′ are

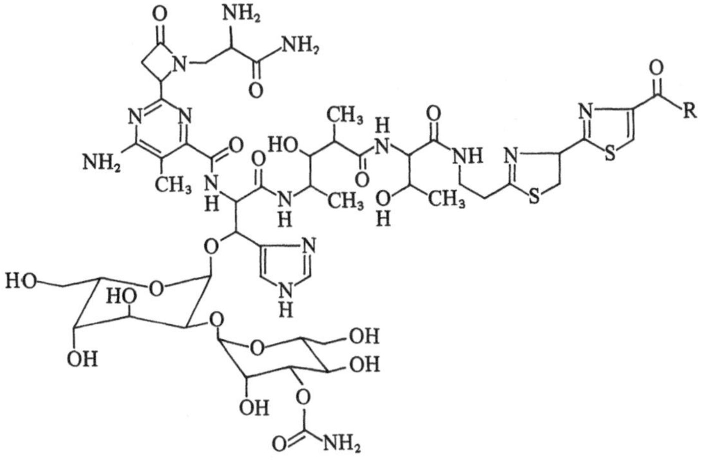

Fig. 2. Structure of phleomycin D_1

$$R = NH-(CH_2)_4-NH-C\overset{\displaystyle NH}{\underset{\displaystyle NH_2}{\diagup}}$$

Phleomycinic acid: $R = OH$

identical with bleomycins B_2, B_4 and B_6 respectively. Phleomycins D_1, E and G show a lower adsorbance at 292 nm than bleomycins.

The total hydrolysis of phleomycin D_1 gives the same hydrolysis products as obtained from bleomycin B_2 except for 2'-(2-aminoethyl)-2,4'-bithiazole-4-carboxylic acid, and instead, it gives β-alanine and 2-acetylthiazole-4-carboxylic acid. Oxidation of phleomycin D_1 and E with manganese dioxide gives bleomycins B_2 and B_4 respectively. Phleomycins D_1 and E thus, can be called dihydrobleomycins B_2 and B_4. Phleomycin G is dihydrobleomycin B_6.

Zorbamycin and zorbonomycin (ARGOUDELIS et al., 1971) and YA-56 antibiotic (ITO et al., 1971; ITO et al., 1972) are structurally related to phleomycin and bleomycin. The hydrolysis products of YA-56 gives 4-amino-3,6-dihydroxy-2-methylcaproic acid, β-hydroxy-L-valine and 6-deoxy-L-gulose, instead of 4-amino-3-hydroxy-2-methyl-n-valeric acid, L-threonine and L-gulose as obtained from phleomycin D_1.

As described above, various bleomycins are different in the terminal amine moiety. Addition of these terminal amines except for the amines of A_1, demethyl-A_2, and A_6 causes the production of the bleomycin containing the amine added and suppresses the production of the other bleomycins. Addition of spermine, the amine of A_6, induces the production of A_5 containing spermidine. The artificial bleomycins containing the amines shown in Table 2 are obtained by addition of the amines to a fermentation medium (UMEZAWA, 1971).

Bleomycinic acid (Fig. 1) is the main part of bleomycin molecule common to all bleomycins. Acylagmatine amidohydrolase of a Fusarium hydrolyzes bleomycin B_2 to bleomycinic acid and agmatine (UMEZAWA et al., 1973). Bleomycinic

Table 2. *Amines incorporated into the terminal amine moiety of bleomycins*

$H_2N-CH_2-CH_3$

$H_2N-CH_2-CH_2-CH_2-NH_2$

$H_2N-CH_2-\underset{\underset{NH_2}{|}}{CH}-CH_3$

$(Bl^a = R-NH-CH_2-\underset{\underset{NH_2}{|}}{CH}-CH_3)$

$H_2N-(CH_2)_3-NH-CH_3$

$(Bl^a = R-NH-(CH_2)_3-NH-CH_3)$

$H_2N-(CH_2)_3-N\overset{\diagup CH_3}{\diagdown CH_3}$

$H_2N-(CH_2)_3-N(CH_3)_3$

$H_2N-(CH_2)_3-\overset{+}{N}H-(CH_2)_3-N(CH_3)_2$

$H_2N-(CH_2)_3-\underset{\underset{CH_3}{|}}{N}-(CH_2)_3-NH_2$

$H_2N-(CH_2)_3-NH-CH(CH_3)-(CH_2)_2-NH_2$

$(Bl^a = R-NH-(CH_3)_3-NH-CH(CH_3)-(CH_2)_2-NH_2)$

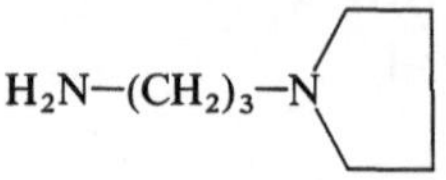

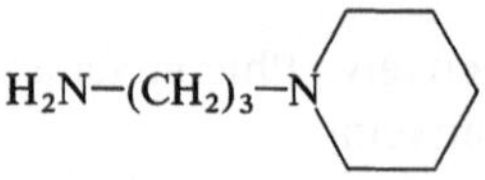

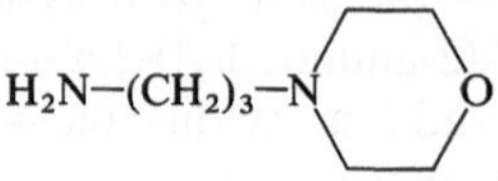

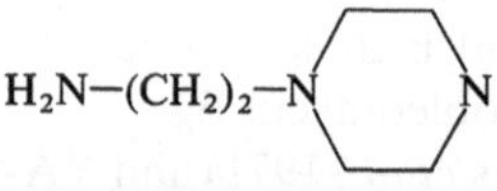

$H_2N-(CH_2)_3-N$⟨pyrrolidine⟩

$H_2N-(CH_2)_3-N$⟨piperidine⟩

$H_2N-(CH_2)_3-N$⟨morpholine O⟩

$H_2N-(CH_2)_2-N$⟨piperazine N⟩

$H_2N-(CH_2)_3-NH-(CH_2)_3-OH$

$H_2N-(CH_2)_3-NH-(CH_2)_3-OCH_3$

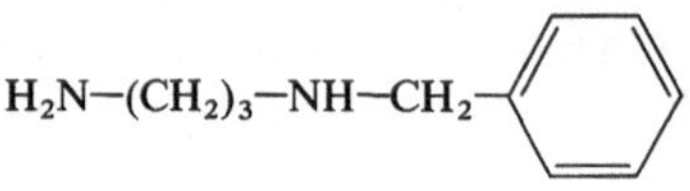

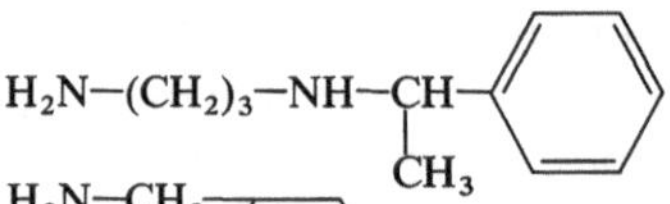

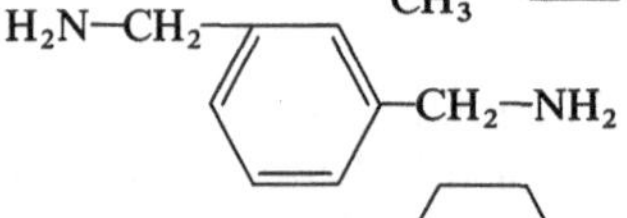

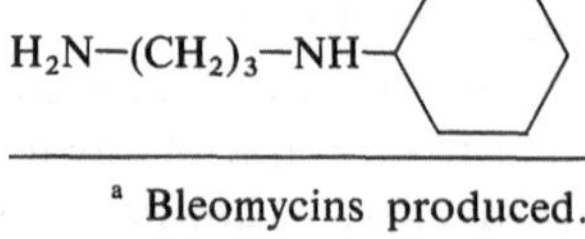

$H_2N-(CH_2)_3-NH-CH_2-$⟨phenyl⟩

$H_2N-(CH_2)_3-NH-\underset{\underset{CH_3}{|}}{CH}-$⟨phenyl⟩

H_2N-CH_2-⟨phenyl⟩$-CH_2-NH_2$

$H_2N-(CH_2)_3-NH-$⟨cyclohexyl⟩

---

^{a} Bleomycins produced.

acid is also obtained by reaction of cyanogen bromide on bleomycin demethyl-A_2 (TAKITA et al., 1973). By this chemical transformation, can be obtained those which are not obtained by fermentation.

Growth Inhibition and Toxicity

The growth inhibitory properties of individual copper-chelated bleomycins have been described by UMEZAWA et al. (1968) and are shown in Table 3. Copper-free bleomycins show the same activity as the copper-chelated ones. A mixture of bleomycins containing 50–60% A_2, 25–32% B_2, 1–6% demethyl-A_2, and A_1, A_2'-a, A_2'-b, and A_5 as minor components have been studied for its activity and clinical effects. Bleomycin inhibited HeLa cells, L-cells, Yoshida rat sarcoma cells and the 50% inhibition concentration was around 5-10 µg/ml. It inhibited ascites and solid forms of Ehrlich carcinoma, and the minimum effective dose was about 0.39–0.78 mg/kg/day for 10 days (ISHIZUKA et al., 1966). TAKEUCHI and YAMAMOTO (1968) observed the inhibition of Rous sarcoma virus-induced mouse ascites sarcoma, when F_1 hybrid mice were employed as host animals. It showed no effect on virus-producing Friend ascites tumor and mouse ascites plasmacytoma. Generally, rat ascites hepatoma was resistant to bleomycin, but SATO et al. (1969) found a strain of rat ascites hepatoma (AH66) which was sensitive to bleomycin but resistant to alkylating agents.

LD_{50} values of a bleomycin mixture for male mice were 210 mg/kg (intravenous), 312 mg/kg (intraperitoneal) and 200 mg/kg (subcutaneous) and for rats were 168 mg by intraperitoneal and the subcutaneous routes respectively. LD_{50} of bleomycin A_2 and B_2 to mice by the intravenous injection were 228 mg/kg and 290 mg/kg respectively. When given daily for 30 days intraperitoneally to rats, 24.3 mg/kg was lethal, all animals died in 11–15 days after the injections, but with a daily dose of 8.1 mg/kg most of rats survived. All were alive by a daily 2.7 mg/kg dose. The chronic toxicity appeared as loss of hair and nail destruction (ICHIKAWA et al., 1967).

Table 3. Antibacterial spectra of bleomycins A_1, A_2, A_3, A_4, A_5, A_6, B_1, B_2, B_3, B_4, B_5

Bleo-mycin	M.I.C. µg/ml								
	Staph. 209 P	Kl. pneu- moniae	Sar. lutea	B. subtilis PCl 219	E. coli	S. typhi 63	Sh. flexneri EW 2	Myc. 607	Myc. phlei
A_1	5.0	0.6	1.2	0.6	2.5	0.6	0.6	0.3	0.04
A_2	3.1	0.1	0.2	0.1	0.8	0.4	0.4	0.4	0.025
A_3	0.1	0.2	0.2	0.012	0.1	0.05	0.1	0.2	0.012
A_4	0.4	0.4	0.4	0.05	0.4	0.4	0.4	0.4	0.05
A_5	0.2	0.025	0.2	0.025	0.025	0.025	0.12	0.1	0.012
A_6	0.2	0.025	0.2	0.003	0.25	0.012	0.012	0.1	0.006
B_1	12.5	1.5	50.0	0.8	0.4	0.1	0.2	0.4	0.025
B_2	0.8	1.5	0.8	0.05	0.2	0.2	0.2	0.1	0.025
B_3	0.2	1.5	0.2	0.025	0.4	0.8	0.4	0.2	0.012
B_4	0.1	3.1	0.1	0.05	0.8	0.1	0.2	0.05	0.006
B_5	0.2	1.5	0.2	0.0125	0.4	0.2	0.8	0.2	0.0125

The intravenous injection of 6 mg/kg of a bleomycin mixture twice a week to dogs showed a therapeutic effect against lymphosarcoma on vaginae (Ishizuka et al., 1967). It was injected for a total of 19 times. Toxic signs appeared as the hardening of foot pad and loss of hair. Toxicity to dogs, especially to the lung was reported by Thompson et al., 1972. Phleomycin caused irreversible renal toxicity (Ishizuka et al., 1966). However, the bleomycin mixture showed a reversible hepatotoxicity (Ishizuka et al., 1967). All phleomycins are positive in Sakaguchi reaction, and therefore, the strong renal toxicity may be due to their terminal amine moieties.

Bleomycin is used for treatment of squamous cell carcinoma and Hodgkin's disease.

Mechanism of Effect of Bleomycin on Squamous Cell Carcinoma and of the Toxicity to Skin and Lung

Ichikawa (1967) discovered the clinical effect of bleomycin on squamous cell carcinoma in human, and Ichikawa et al. (1970) confirmed that bleomycin inhibits squamous cell carcinoma in mouse skin induced by 20-methylcholanth-rene, but it does not inhibit sarcoma in mouse skin induced by the same agent. Umezawa et al. (1972) studied the mechanism of the effect on squamous cell carcinoma.

Distribution of ^{3}H-bleomycin in various organs of the mice was examined one hour after the subcutaneous injection by assaying radioactivity and antibacterial activity. In general, the organ extracts except for those of skin and lung did not show the antibacterial activity, indicating the rapid inactivation of the antibiotic. This inactivating activity was lower in the skin and lung than in the other organs. The distribution of ^{3}H-bleomycin in mouse squamous cell carcinoma and mouse sarcoma induced by 20-methylcholanthrene has been studied in three experiments and 15.9–22.7 µg/g of bleomycin was detected by radioactivity in the carcinoma and a lower concentration such as 4.5–14.4 µg/g in the sarcoma. A more marked difference in the content of bleomycin with respect to antibacterial activity i. e. 11.0–17.2 µg/g in the carcinoma and 0–2.8 µg/g in the sarcoma was also seen. These results indicate that the sarcoma is more active in reducing or inactivating bleomycin activity than is the carcinoma, and the antibiotic is taken up by the carcinoma in higher concentrations than by the sarcoma.

All organs contained an enzyme which inactivated bleomycins. By this enzyme bleomycin B_2 was most rapidly inactivated, but bleomycin A_2 and A_5 were also inactivated. This enzyme liberated one mole of ammonia from carboxamide of bleomycins (Fig. 1) (Umezawa, 1971).

Mouse squamous cell carcinoma and mouse skin sarcoma induced by 20-methylcholanthrene were homogenized, and the protein containing the enzyme was extracted by 105,000 G centrifugation, protamine treatment of the supernatant, precipitation with 35–60% saturation by ammonium sulfate and dialysis. The bleomycin-inactivating activity of the protein thus prepared was determined (Umezawa et al., 1972) and activity of the enzyme obtained from the sarcoma was significantly higher than that obtained from the carcinoma.

Distribution of ^{3}H-bleomycin in organs of the mice of different ages, (3 weeks, 5 weeks and 28 weeks after birth) has been examined, injecting 49.71–54.28 mg/kg of ^{3}H-bleomycin subcutaneously. The organs of the youngest group of mice showed the greatest activity in inactivating bleomycin, and none of the active form was detected except in the tissue of the injection site. The inactivating activity of lung and skin was the weakest in the oldest mice, and 50–64% and 46–49% of bleomycin remained in the active form in lung and skin respectively. These results are in agreement with the clinical observation that toxicity of bleomycin appears in the skin and lung and more frequently in older patients than the younger ones.

Most of antitumor compounds show an immunosuppressive effect, but bleomycin did not (YAMAKI et al., 1969; OHNO et al., 1971) and this lack of immunosuppressive effect may be due to the rapid inactivation of the antibiotic in spleen and lymphoid tissues.

Information on distribution and inactivation in various organs may be useful to predict the organ in which the toxicity may appear and the type of a human tumor which may respond to this compound. Bleomycin containing

$$-NH-(CH_2)_3-NH-(CH_2)_4-NH-C(=NH)-CH_2-C_6H_4-Cl(p)$$

in the amine moiety was reported to show high concentrations not only in the skin and lung but also in other organs (UMEZAWA, 1972). The concentrations (μg/g) determined by the antibacterial activity 1 hour after subcutaneous injection of 100 mg/kg to young mice were as follows: its Cu^{++} chelating form: liver 15, spleen 15, stomach 18, lung 42, skin 103; its Cu^{++}-free form: liver 4.2, spleen 1.0, stomach 18, lung 15.3, skin 18.9. This bleomycin caused damage to lymphatic tissue, and prolonged the survival period (130%) of mice with L-1210 leukemia against which other bleomycins were ineffective.

Mode of Action

Bleomycin inhibited growth of bacterial and various cultured mammalian cells, and copper-free and copper-chelated bleomycin showed the same activity (UMEZAWA et al., 1966; ISHIZUKA et al., 1967; UMEZAWA et al., 1968; SUZUKI et al., 1968; PITTILO et al., 1971). Copper-free bleomycin is employed clinically and the modes of action have been studied mostly with a copper-free bleomycin mixture and copper-free bleomycin A_2. Bleomycins caused the induction of a bacteriophage in a lysogenic strain (AOKI and SAKAI, 1967; HAIDLE et al., 1972c).

A low concentration of bleomycin which did not show a significant inhibition of incorporation of ^{3}H-thymidine into DNA inhibited cell division of HeLa cells (KUNIMOTO et al., 1967). It elongated bacterial cells and enlarged HeLa cells (SUZUKI et al., 1968). Polynuclear giant cells appeared in HeLa and L cell cultures which were exposed to bleomycin. During treatment of VX2 epidermoid carcinoma by bleomycin, the population of the stemline cells decreased, population of cells with twice the DNA content increased, and a few cells containing 4 times the DNA content appeared (NAGATSU et al., 1971). The mitotic index decreased immediately after the start of the bleomycin treatment. The effect on HeLa S3 cells was dependent on the phase of its growth cycle (TERASIMA

and Umezawa, 1970). During treatment of Ehrlich ascites carcinoma by bleomycin, the cell cycle was blocked and cells accumulated in G_2 (Nagatsu et al.,
1972). A portion of the cells thus accumulated were severely injured. Another
portion of the blocked G_2 population transferred to the S phase of the next
higher cell cycle without undergoing cytokinesis. The result of analysis of Chinese
hamster ovary cells exposed to bleomycin indicated that the cells in mitosis
were most sensitive and the order of sensitivity of cells in the other phases
was G_2, early S, late S and G_1 (Barranco and Humphrey, 1971).

Dose response curve and the death rate curve of cells at various hours of
the exposure are interesting. Both the curves were upward-concave (Barranco
et al., 1971). Terasima (1972) also found an interesting phenomenon that, if
L cells which had been exposed to bleomycin are washed and incubated in a
bleomycin-free medium, the survival response returns to the original level. This
indicates that bleomycin causes damage and at the same time induces resistance
which disappears rapidly during incubation in a bleomycin-free medium.

Biochemically, bleomycin inhibits incorporation of ^{3}H-thymidine into DNA
of intact cells. Another biochemical occurrence in cells exposed to bleomycin
is the fragmentation of DNA. DNA fragmentation in L cells was shown at as
low a concentration as 0.1 µg/ml. The fragmentation caused by a relatively high
concentration such as 10 µg/ml was rapidly repaired, when the cells were cultivated
in a bleomycin-free medium (Terasima et al., 1970). The effect of bleomycin
on murine epidermoid carcinoma was enhanced by a simultaneous X-irradiation
(Jorgensen, 1972).

In vitro, bleomycin does not inhibit the DNA polymerase reaction. On the
contrary, the antibiotic stimulated DNA polymerase at the beginning of the
reaction (Yamaki et al., 1971). Bleomycin causes a single strand scission of DNA
of E. coli B in vitro in the presence of 2-mercaptoethanol (Nagai et al., 1969a;
Suzuki et al., 1969) or in the presence of hydrogen peroxide (Nagai et al., 1969b).
In order to demonstrate the fragmentation clearly, it was necessary to dialyze
the reaction mixture before alkaline sucrose density gradient centrifugation and
generally 8–40 µg/ml of bleomycin A_2 was required. The lowest active concentration of bleomycin A_2 observed in this reaction was 1.6 µg/ml (Suzuki
et al., 1970). This effect of bleomycin was inhibited by Cu^{++}, Zn^{++} and EDTA.
Strand scission of poly (dG) poly (dC) and poly d(AT) poly d(TA) was also
caused by 40 µg/ml of bleomycin A_2 in the reaction mixture containing 2-mercap-
toethanol (1 mM) (Nagai et al., 1969a). The infectivity of the single-stranded
circular DNA (about 10^{-3} µg/ml) of ϕX174 phage to E. coli spheroplast was
lost after exposure to 1.0 µg/ml of bleomycin in the absence of a sulfhydryl
compound (Shirakawa et al., 1971). The effect of bleomycin causing a single
strand scission of the phage DNA was also shown by a decrease in the sedimentation velocity. This action was enhanced by a sulfhydryl compound (1 mM)
and hydrogen peroxide and counteracted by Cu^{++}. The fragmentation of B. subtilis DNA by bleomycin in the presence of 2-mercaptoethanol has also been
reported (Haidle, 1971).

[3]H-bleomycin A_2 (100 µg/ml) was shown to bind with ^{14}C-thymine-labeled
DNA (E. coli) (250 µg/ml) after incubation for 2 hours. The molar ratio of bleomycin to nucleotides in the complex was raised by 2-mercaptoethanol. The binding

was reduced by Cu^{++} (0.1 mM), Zn^{++} (0.1 mM) and EDTA (1 mM) (SUZUKI *et al.*, 1970).

The reaction of bleomycin B$_2$ with SV40 virus DNA was studied. The virus was prepared by successive application of cesium chloride density gradient centrifugation and sucrose density gradient centrifugation. The reaction caused a nick in DNA at the almost equal rate at 0° and 37°C (UMEZAWA *et al.*, 1973). The rate of the reaction was higher at 10° and 20° than at 0° or at 37°. The reaction was stopped by adding dilute sodium hydroxide solution. In this case, 0.6 μg/ml of bleomycin showed 50% nicking of DNA and the nicking was shown even at 0.2 μg/ml. Addition of 2-mercaptoethanol enhanced the reaction. A copper-free phleomycin also caused nicks in DNA of SV40 virus. However, copper-chelated bleomycin and copper-chelated phleomycin did not react with SV40 DNA.

MÜLLER *et al.* (1972a) confirmed that the reaction of bleomycin (100 μg/ml) with poly d(AT) · poly d(TA) (0.01 μmole) or poly (dA) · poly (dT) (40 μg/ml) in the presence of dithiothreitol (5–50 mM) releases thymine. The thymine was detected by high pressure liquid chromatography. Thymine liberation was more marked in the case of poly d(AT) · poly d(TA).

Thymine was also liberated from native DNA of *M. lysodeikticus*, *E. coli*, herring, *Cl. perfringens*, and denatured DNA of herring, but not from poly (dI) · poly (dT) and poly (dT). The amounts of thymine liberated per mg of these DNAs were proportional to the sum of dA and dT in DNA. Radical scavengers including EDTA did not influence the reaction. FAD substituted for dithiothreitol in this reaction. A possibility of activation of bleomycin by dithiothreitol was suggested by MÜLLER *et al.* (1972a). These authors reported that L-arginase inactivated bleomycin in the presence of dithiothreitol. A sulfhydryl compound has been known to inactivate bleomycin, and UMEZAWA *et al.* have confirmed that this inactivation is dependent on the presence of a trace amount of ferric ion or of ferrous ion and air. MÜLLER *et al.* (1972a) also reported the liberation of a ribose aldehyde group during the action of bleomycin on DNA. HAIDLE *et al.* (1972b) observed liberation of all bases by the action of bleomycin (12–50 mg/ml) on base-labeled DNA (about 100 μg/ml) of *B. subtilis* 168 in the presence of 25–33 mM of 2-mercaptoethanol. Nucleosides or mononucleotides were not liberated. In this reaction, 80% of DNA became soluble in trichloroacetic acid, but only a small amount (5–10%) of the bases was released. Release of uracil by the reaction of bleomycin (10 mg/ml) with uridine-labeled bacteriophage PBS-1 DNA (98 μg/ml) was observed in the presence of 20 mM 2-mercaptoethanol (HAIDLE *et al.*, 1972b). Bleomycin shows no effect on RNA.

Bleomycin decreased Tm of DNA (*E. coli*, Salmon sperm) in reaction mixtures containing 2-mercaptoethanol (1 mM) or hydrogen peroxide (0.01 mM) (NAGAI *et al.*, 1969c). A decrease of Tm of poly d(G).poly d(C) was observed, but no Tm decrease of poly d(AT) · poly d(TA) was seen. Copper-free bleomycin and copper-free phleomycin caused a single strand scission of DNA, but copper-chelated bleomycin and phleomycin did not. Copper-chelated phleomycin increased Tm but not copper-chelated bleomycin. As described above, phleomycin is different from bleomycin in its dihydrothiazole moiety. Therefore, this moiety in the copper-chelated phleomycin may be involved in either binding with DNA or in changing the conformation of the phleomycin molecule. The failure of copper-chelated bleomycin to cause DNA strand scission indicates that the reactive group may be masked by copper chelation. The shift of the infrared band from 1710 cm^{-1} to 1720 cm^{-1} by the copper chelation and absence of pKá 7.3 in copper-chelated bleomycin suggest that the carbamoyl group and α-amino group of β-aminoalanine moiety in bleomycin may be involved in copper chelation. Bleomycin B$_2$ which was inactivated by a liver enzyme as described in the previous section caused the strand scission of SV40 virus DNA, but a 20 times higher concentration was required than of active bleomycin B$_2$ (UMEZAWA et al., 1973). This suggests that the carboxamide favors the reaction which causes strand scission of DNA.

When bleomycin is preincubated with template DNA (T$_4$ phage), the DNA-dependent RNA polymerase reaction of E. coli is inhibited (TANAKA, 1970). This inhibition is thought to be secondary due to the fragmentation of the template.

Bleomycin A$_2$ (2 µg/ml) stimulated pancreatic DNase activity (YAMAKI et al., 1971). This stimulation may be explained by fragmenting action of bleomycin and suggests that DNA fragmentation in vivo caused by bleomycin, might be enhanced by DNase.

Bleomycin inhibited DNA-ligase prepared from rat ascites hepatoma 130 cells at a low concentration such as 0.1 µg/ml which did not cause strand scission (MIYAKI et al., 1971). The action of DNA ligase prepared from T$_4$ phage-infected E. coli B on E. coli, DNA nicked by pancreatic DNase, was also inhibited by bleomycin A$_2$ (1.6 µg/ml) (YAMAKI et al., 1971).

UV sensitive (AB 1885), and UV and X-ray sensitive (AB 2463) mutants were as sensitive to bleomycin as the isogenic strain (AB 1157) devoid of radiation sensitivity (ENDO, 1970). Bleomycin-sensitivity of E. coli was suggested to be controlled by two mechanisms, membrane permeability and repair of DNA. A bleomycin-sensitive strain, BM 2A (blm$^-$) was obtained from AB 1157, and from this mutant was obtained a UV-sensitive mutant BM 2A-9 (blm$^-$ rec$^-$). The mutant BM 2A−9 was far more sensitive to bleomycin than BM 2A (YAMAGAMI et al., 1974). In the gene map, the blm$^-$ marker is located in the region of membrane clusters. A recombination-deficient strain of B. subtilis behaved differently in the presence of bleomycin than did the respective wild type organisms. The wild type cells were able to recover from lower levels of bleomycin while the rec$^-$ strain was not (SAUNDERS and SCHULTZ, 1972).

Papers reporting the effect on membrane have appeared recently. Mixtures of copper-chelated phleomycins acted on bacterial membranes and released membrane-associated DNA (REITER et al., 1972). Exposure of rat hepatoma cells to bleomycin A$_2$ caused liberation of DNA from a DNA-membrane complex (ONO et al., 1972). It is not certain, whether this action is due to the reaction of bleomycin with DNA or not. However, membrane effects are an interesting observation when searching for the primary action of bleomycin on intact cells.

In a previous section, the mechanism of the selective effect of bleomycin on squamous cell carcinoma was described. The mechanism of the selective effect on Hodgkin's disease is thought to be different and may be related to the mechanism of action of bleomycin. MÜLLER et al. (1972) reported that bleomycin (1 µg/ml) inhibits DNA-dependent DNA polymerase obtained from Rauscher murine leukemia in the reaction mixture containing dithiothreitol but a 100 times higher concentration of this antibiotic was required to inhibit RNA-dependent DNA polymerase from the same source; bleomycin does not inhibit DNA-dependent DNA polymerase activity from *E. coli* and mouse liver.

As described above, copper-free bleomycin causes single strand scissions of DNA, liberating thymine or other bases depending on antibiotic concentrations; the drug lowers Tm of DNA, inhibits the DNA-ligase reaction, inhibits DNA-dependent DNA polymerase of RNA virus and releases DNA from membrane-DNA complex. However, the chemical reaction mechanism of bleomycin on DNA, the relationships among each of the effects described above, the primary action against intact cells, and the mode of action of copper-chelated bleomycin remain topics for future studies.

References

AOKI, H., and H. SAKAI: Antiphage property as a screening test for antitumor agents. J. Antibiotics (Tokyo) **20 A**, 87 (1967).

ARGOUDELIS, A. D., M. E. BERGEY, and T. R. PYKE: Zorbamycin and related antibiotics. I. Production, isolation and characterization. J. Antibiotics (Tokyo) **24**, 543 (1971).

BARRANCO, S. C., and R. M. HUMPHREY: The effects of bleomycin on survival and cell progression in Chinese hamster cells in vitro. Cancer Res. **31**, 1218 (1971).

ENDO, H.: Qualitative difference between bleomycin and radiation effects on cell viability. J. Antibiotics (Tokyo) **23**, 508 (1970).

HAIDLE, C. W.: Fragmentation of deoxynucleic acid by bleomycin. Mol. Pharmacol. **7**, 645 (1971).

HAIDLE, C. W., M. T. KUO, and K. K. WEISS: Nucleic acid-specificity of bleomycin. Biochem. Pharmacol. **21**, 3308 (1972a).

HAIDLE, C. W., K. K. WEISS, and M. T. KUO: Release of free bases from deoxyribonucleic acid after reaction with bleomycin. Mol. Pharmacol. **8**, 531 (1972b).

HAIDLE, C. W., K. K. WEISS, and M. L. MACE, JR.: Induction of bacteriophage by bleomycin. Biochem. Biophys. Res. Commun. **48**, 1179 (1972c).

ICHIKAWA, T.: Clinical study of a new antitumor antibiotic, bleomycin. Proceedings of the 5th International Congress (Vienna) of Chemotherapy, A IV-4/35 (1967).

ICHIKAWA, T., A. MATSUDA, K. MIYAMOTO, M. TSUBOSAKI, T. KAIHARA, K. SAKAMOTO, and H. UMEZAWA: Biological studies on bleomycin A. J. Antibiotics (Tokyo) **20A**, 149 (1967).

ICHIKAWA, T., H. UMEZAWA, S. OHASHI, T. TAKEUCHI, M. ISHIZUKA, and S. HORI: Animal experiments confirming the specific effect of bleomycin against squamous cell carcinoma. Proceedings of the 6th International Congress (Tokyo) of Chemotherapy, vol. 2, p. 315 (1970).

IKEKAWA, T., F. IWAMI, H. HIRANAKA, and H. UMEZAWA: Separation of phleomycin components and their properties. J. Antibiotics (Tokyo) **17A**, 194 (1964).

ISHIZUKA, M., H. TAKAYAMA, T. TAKEUCHI, and H. UMEZAWA: Studies on antitumor activity, antimicrobial activity and toxicity of phleomycin. J. Antibiotics (Tokyo) **19A**, 260 (1966).

ISHIZUKA, M., H. TAKAYAMA, T. TAKEUCHI, and H. UMEZAWA: Activity and toxicity of bleomycin. J. Antibiotics (Tokyo) **20A**, 15 (1967).

ITO, Y., Y. OHASHI, Y. EGAWA, T. YAMAGUCHI, T. FURUMAI, K. ENOMOTO, and T. OKUDA: Antibiotic YA 56, a new family of phleomycin-bleomycin group antibiotics. J. Antibiotics (Tokyo) **24**, 727 (1971).

ITO, Y., Y. OHASHI, S. KAWABE, H. ABE, and T. OKUDA: β-Hydroxy-L-valine, a constitutional amino acid of antibiotics YA 56 X and Y. J. Antibiotics (Tokyo) **25**, 360 (1972).

JORGENSEN, S. J.: Time-dose relationships in combined bleomycin treatment and radiotherapy. Europ. J. Cancer **8**, 531 (1972).

KOYAMA, G., H. NAKAMURA, Y. MURAOKA, T. TAKITA, K. MAEDA, H. UMEZAWA, and Y. IITAKA: The chemistry of bleomycin. II. The molecular and crystal structure of a sulfur-containing chromophoric amino acid. Tetrahedron Letters No. 44, 4635 (1968).

KOYAMA, G., H. NAKAMURA, Y. MURAOKA, T. TAKITA, K. MAEDA, H. UMEZAWA, and Y. IITAKA: The chemistry of bleomycin. X The stereochemistry and crystal structure of β-hydroxyhistidin, an amine component of bleomycin. J. Antibiotics (Tokyo) **26**, 109 (1973).

Kunimoto, T., M. Hori, and H. Umezawa: Modes of action of phleomycin, bleomycin and formycin on HeLa S3 cells in synchronized culture. J. Antibiotics (Tokyo) **20A**, 277 (1967).

Maeda, K., H. Kosaka, K. Yagishita, and H. Umezawa: A new antibiotic, phleomycin. J. Antibiotics (Tokyo) **9A**, 82 (1956).

Miyaki, M., T. Ono, and H. Umezawa: Inhibition of ligase reaction by bleomycin. J. Antibiotics (Tokyo) **24**, 587 (1971).

Müller, W. E. G., Z. Yamazaki, H. Breter, and R. K. Zahn: Action of bleomycin on DNA and RNA. Europ. J. Biochem. **31**, 518 (1972a).

Müller, W. E. G., Z. Yamazaki, and R. K. Zahn: Bleomycin, a selective inhibitor of DNA-dependent DNA polymerase from oncogenic RNA viruses. Biochem. Biophys. Res. Commun. **46**, 1667 (1972b).

Muraoka, Y., T. Takita, K. Maeda, and H. Umezawa: Chemistry of bleomycin. IV. The structure of amine component II of bleomycin A$_2$. J. Antibiotics (Tokyo) **23**, 252 (1970).

Muraoka, Y., T. Takita, K. Maeda, and H. Umezawa: Chemistry of bleomycin. VI. Selective cleavage of bleomycin A$_2$ by N-bromosuccinimide. J. Antibiotics (Tokyo) **25**, 185 (1972).

Nagai, K., H. Suzuki, N. Tanaka, and H. Umezawa: Decrease of melting temperature and single strand scission of DNA by bleomycin in the presence of 2-mercaptoethanol. J. Antibiotics (Tokyo) **22**, 569 (1969a).

Nagai, K., H. Suzuki, N. Tanaka, and H. Umezawa: Decrease of melting temperature and single strand scission of DNA by bleomycin in the presence of hydrogen peroxide. J. Antibiotics (Tokyo) **22**, 624 (1969b).

Nagai, K., H. Yamaki, H. Suzuki, N. Tanaka, and H. Umezawa: The combined effects of bleomycin and sulfhydryl compounds on the thermal denaturation of DNA. Biochim. Biophys. Acta **179**, 165 (1969c).

Nagatsu, M., T. Okagaki, R. M. Richart and A. Lambert: Effects of bleomycin on nuclear DNA in transplantable VX-2 carcinoma of rabbit. Cancer Res. **31**, 992 (1971).

Nagatsu, M., R. M. Richart, and A. Lambert: Effects of bleomycin on the cell cycle of Ehrlich ascites carcinoma. Cancer Res. **32**, 1966 (1972).

Ohno, R., H. Nishiwaki, K. Kawashima, T. Uetani, M. Hirano, M. Miura, and K. Yamada: Lack of immunosuppressive effect of bleomycin on the primary response of mice to sheep red blood cells. Gann **62**, 267 (1971).

Omoto, S., T. Takita, K. Maeda, H. Umezawa, and S. Umezawa: The chemistry of bleomycin. VIII. The structure of the sugar moiety of bleomycin A$_2$. J. Antibiotics (Tokyo) **25**, 752 (1972).

Ono, T., M. Miyaki, and J. Kuroda: Studies on bleomycin-DNA. Report at the Annual Meeting of the Japan Cancer Society, at Nagoya, November, 1972.

Pittillo, R. F., C. Woolley, and L. S. Rice: Bleomycin, an antitumor antibiotic: Improved microbiological assay and tissue distribution studies in normal mice. Appl. Microbiol. **22**, 564 (1971).

Reiter, H., M. Milewsky and P. Kelley: Mode of action of phleomycin on *B. subtilis*. J. Bacteriol. **111**, 586 (1972).

Sato, H., and H. Ichimura: Effect of bleomycin on rat ascites hepatoma. Igaku No Ayumi (Progress in Medicine) **69**, 669 (1969) [in Japanese].

Saunders, P. P., and G. A. Schultz: Mechanism of action of bleomycin-I. Bacterial growth studies. Biochem. Pharmocol. **21**, 1657 (1972).

Shirakawa, I., M. Azegami, S. Ishii, and H. Umezawa: Reaction of bleomycin with DNA. Strand scission of DNA in the absence of sulfhydryl or peroxide compounds. J. Antibiotics (Tokyo) **24**, 761 (1971).

Suzuki, H., K. Nagai, E. Akutsu, H. Yamaki, N. Tanaka, and H. Umezawa: On the mechanism of action of bleomycin. Strand scission of DNA caused by bleomycin and its binding to DNA in vitro. J. Antibiotics (Tokyo) **23**, 473 (1970).

Suzuki, H., K. Nagai, H. Yamaki, N. Tanaka, and H. Umezawa: Mechanism of action of bleomycin. Studies with the growing culture of bacterial and tumor cells. J. Antibiotics (Tokyo) **21**, 379 (1968).

Suzuki, H., K. Nagai, H. Yamaki, N. Tanaka, and H. Umezawa: On the mechanism of action of bleomycin: Scission of DNA strands in vitro and in vivo. J. Antibiotics (Tokyo) **22**, 446 (1969).

Takeuchi, M., and T. Yamamoto: Effects of bleomycin on mouse transplantable tumors. J. Antibiotics (Tokyo) **21**, 631 (1968).

TAKITA, T., A. FUJII, T. FUKUOKA, and H. UMEZAWA: Chemical cleavage of bleomycin to bleomycinic acid and synthesis of new bleomycins. J. Antibiotics (Tokyo) **26**, 252 (1973).

TAKITA, T., K. MAEDA, H. UMEZAWA, S. OMOTO, and S. UMEZAWA: Chemistry of bleomycin. III. The sugar moieties of bleomycin A$_2$. J. Antibiotics (Tokyo) **22**, 237 (1969).

TAKITA, T., Y. MURAOKA, A. FUJII, H. ITOH, K. MAEDA, and H. UMEZAWA: The structure of the sulfur-containing chromophore of phleomycin, and chemical transformation of phleomycin to bleomycin. J. Antibiotics (Tokyo) **25**, 197 (1972a).

TAKITA, T., Y. MURAOKA, K. MAEDA, and H. UMEZAWA: Chemical studies on bleomycin. I. The acid hydrolysis products of bleomycin A$_2$. J. Antibiotics (Tokyo) **21**, 79 (1968).

TAKITA, T., Y. MURAOKA, T. YOSHIOKA, A. FUJII, K. MAEDA, and H. UMEZAWA: The chemistry of bleomycin. IX. The structures of bleomycin and phleomycin. J. Antibiotics (Tokyo) **25**, 755 (1972b).

TAKITA, T., T. YOSHIOKA, Y. MURAOKA, K. MAEDA, and H. UMEZAWA: Chemistry of bleomycin. V. Revised structure of an amine component of bleomycin A$_2$. J. Antibiotics (Tokyo) **24**, 795 (1971).

TANAKA, N.: Inhibition of transcription by pluramycin and bleomycin. J. Antibiotics **23**, 523 (1970).

TANAKA, N., H. YAMAGUCHI, and H. UMEZAWA: Mechanism of action of phleomycin. I. Selective inhibition of the DNA synthesis in *E. coli* and in HeLa cells. J. Antibiotics (Tokyo) **16A**, 86 (1963).

TERASIMA, T., Y. TAKABE, T. KATSUMATA, M. WATANABE, and H. UMEZAWA: Effect of bleomycin on mammalian cell survival. J. Natl. Cancer. Inst. **49**, 1093 (1972).

TERASIMA, T., and H. UMEZAWA: Lethal effect of bleomycin on cultured mammalian cells. J. Antibiotics (Tokyo) **23**, 300 (1970).

TERASIMA, T., M. YASUKAWA, and H. UMEZAWA: Breaks and rejoining of DNA in cultured mammalian cells treated with bleomycin. Gann **61**, 513 (1970).

THOMPSON, G.R., J.R. BAKER, R.W. FLEISCHMAN, H. ROSENKRANTZ, U.H. SCHAEPPI, D.A. COONEY, and R.D. DAVIS: Preclinical toxicologic evaluation of bleomycin (NSC 125 066), a new antitumor antibiotic. Toxicol. Appl. Pharmacol. **22**, 544 (1972).

UMEZAWA, H.: Natural and artificial bleomycins. Chemistry and antitumor activity. Pure Appl. Chem. **28**, 665 (1971).

UMEZAWA, H.: Studies of bleomycin: biosynthesis and function. Euchem Conference on Antibiotics, August 28 to September 2, Aarhus, Denmark, 1972.

UMEZAWA, H., H. ASAKURA, and M. HORI: A single strand scission of super coiled SV40 DNA by bleomycin. J. Antibiotics (Tokyo) **26**, 521 (1973).

UMEZAWA, H., M. ISHIZUKA, K. KIMURA, J. IWANAGA, and T. TAKEUCHI: Biological studies on individual bleomycins. J. Antibiotics (Tokyo) **21**, 592 (1968).

UMEZAWA, H., K. MAEDA, T. TAKEUCHI, and Y. OKAMI: New antibiotics, bleomycin A and B. J. Antibiotics (Tokyo) **19A**, 200 (1966a).

UMEZAWA, H., Y. SUHARA, T. TAKITA, and K. MAEDA: Purification of bleomycins. J. Antibiotics (Tokyo) **19A**, 210 (1966b).

UMEZAWA, H., Y. TAKAHASHI, A. FUJII, T. SAINO, T. SHIRAI, and T. TAKITA: Preparation of bleomycinic acid: hydrolysis of bleomycin B$_2$ by acylagmatine amidohydrolase of a Fusarium sp. J. Antibiotics (Tokyo) **26**, 117 (1973).

UMEZAWA, H., T. TAKEUCHI, S. HORI, T. SAWA, M. ISHIZUKA, T. ICHIKAWA, and T. KOMAI: Studies on the mechanism of antitumor effect of bleomycin on squamous cell carcinoma. J. Antibiotics (Tokyo) **25**, 409 (1972).

YAMAGAMI, H., M. ISHIZAWA, and H. ENDO: Phenotypic and genetic characteristics of bleomycin-sensitive strain of *Escherichia coli*. Gann (Tokyo) **65**, 61 (1974).

YAMAKI, H., H. SUZUKII, K. NAGAI, N. TANAKA, and H. UMEZAWA: Effect of bleomycin A$_2$ on deoxyribonuclease, DNA polymerase and ligase reactions. J. Antibiotics (Tokyo) **24**, 178 (1971).

YAMAKI, H., N. TANAKA, and H. UMEZAWA: Effects of several tumor-inhibitory antibiotics on immunological responses. J. Antibiotics (Tokyo) **22**, 315 (1969).

YOSHIOKA, T., Y. MURAOKA, T. TAKITA, K. MAEDA, and H. UMEZAWA: Chemistry of bleomycin. VII. Synthesis of β-amino-β-(4-amino-6-carboxy-5-methylpyrimidin-2yl)-propionic acid, an amine component of bleomycin. J. Antibiotics (Tokyo) **25**, 625 (1972).

ZEE-CHENG, K.Y., and C. CHENG: Synthesis of 2′-(2-aminoethyl)-2′,4′-bithiazole-4-carboxylic acid, a component of the antitumor antibiotic, bleomycin. J. Heterocyclic Chem. **7**, 1439 (1970).

Berenil: A Trypanocide with Selective Activity against Extranuclear DNA

B. A. Newton

Berenil (Diminazene aceturate, originally named Babesin) is an aromatic dia-midin (4,4'-(diazoamino) benzamidine: Mol.wt. 281.32), it was originally synthesised by JENSCH (1954, 1955) and is marketed as the diaceturate (acetyl glycinate) derivative (Fig. 1) by Farbwerke Hoechst, AG. Biological trials showed that berenil has trypanocidal and babesicidal activity (BAUER, 1955 a and b; FUSSGANGER, 1955), and from 1955 onwards it has been used increasingly for the treatment of bovine trypanosomiasis in Africa: it has also been used in clinical trials for treatment of sleeping sickness in man. Initially it was believed that berenil was rapidly excreted from animals, thus it was valued as a curative rather than a prophylactic drug, however subsequent work, discussed below, suggests that berenil may have some short-lived prophylactic activity. Additional virtues of berenil are that its use does not readily give rise to drug resistance and that it is frequently active against strains of trypanosomes which have developed resistance against other commonly used phenanthridine and aminoquinaldine trypanocides such as ethidium bromide and antrycide. Recently, berenil has attracted attention as a selective inhibitor of kinetoplast DNA synthesis in trypanosomes and it seems that it may become a valuable probe for the further investigation of the mechanism of replication of covalently closed circular DNA molecules.

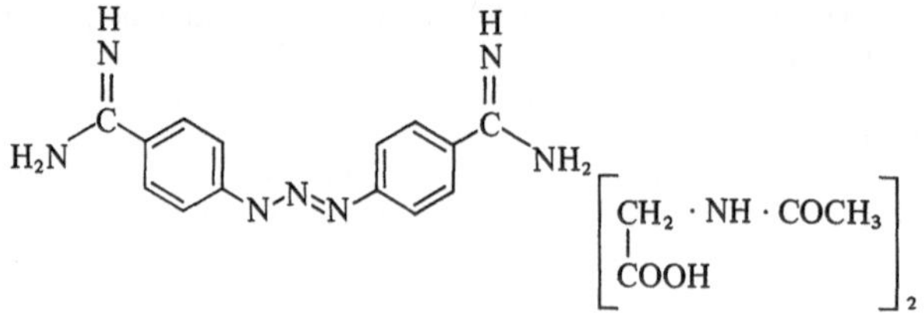

Fig. 1. Berenil diaceturate

Development and Chemical Properties

Berenil was developed by a systematic dissection of an earlier, rather toxic trypanocide known as surfen C (Fig. 2). This investigation, described in detail by JENSCH (1958), led to the examination of compounds in which the 4 aminoquinaldine nucleus was linked to guanidino or amidino-substituted phenyl moieties. It was first found that the phenyl ring could be omitted without loss of activity

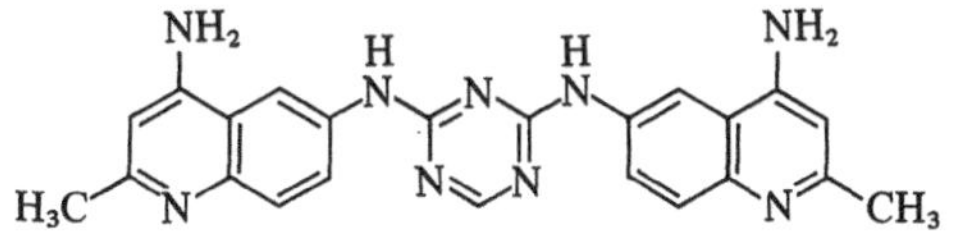

Fig. 2. Surfen C

and subsequently that if the 4-aminoquinaldine nucleus was replaced by a second guanidino- or amidino-phenyl group there was an improved tissue tolerance without loss of activity. These findings coupled with knowledge of factors affecting the trypanocidal activity of aromatic diamidines led logically to the synthesis of a number of symmetrical molecules containing amidino- or guanidino-phenyl moieties separated by a variety of inter-ring "bridges"; maximal trypanocidal activity was found when two amidino-phenyl moieties were joined by a triazine "bridge".

Berenil crystallizes from methanol as brick-red crystals which decompose at 164–165 °C. The diaceturate, which is yellow, dissolves readily in water to give a clear amber-colored solution and a 7% (w/v) solution may readily be obtained at room temperature. Neutral aqueous solutions are stable for only a few days at room temperature, the drug undergoing a rearrangement which results in the formation of an aminoazo derivative from the original triazine structure (Dr. H. LOEWE—personal communication); this rearrangement is accelerated at lower pH values and under more acid conditions the molecule is split to give 4-amino-benzamidine and 4-amino-phenyl-diazonium chloride (Fig. 3). Breakdown to the amino-azo derivative under neutral or weakly acid conditions is accompanied by a shift in the absorption maximum of the solution from 370 nm to 300 nm (Fig. 4) and a loss of trypanocidal activity. Since 1959 commercial preparations of berenil have contained antipyrine (phenyldimethyl pyrazo-

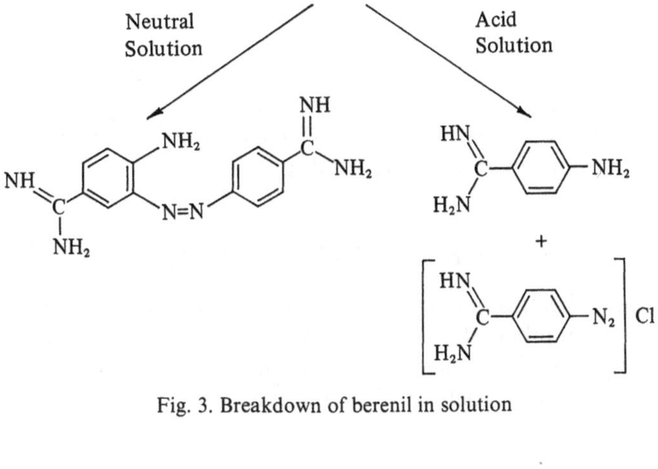

Fig. 3. Breakdown of berenil in solution

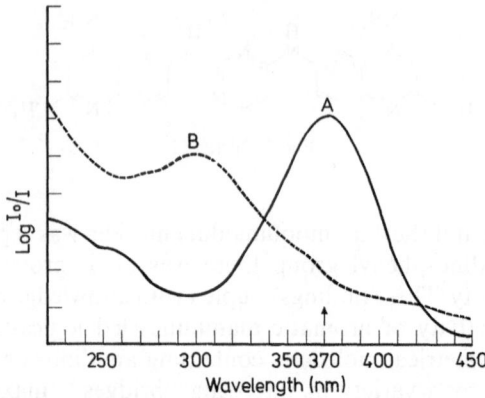

Fig. 4. Absorption spectrum of berenil (curve A) and of the amino-azo derivative (curve B) which
is formed by rearrangement in neutral aqueous solutions

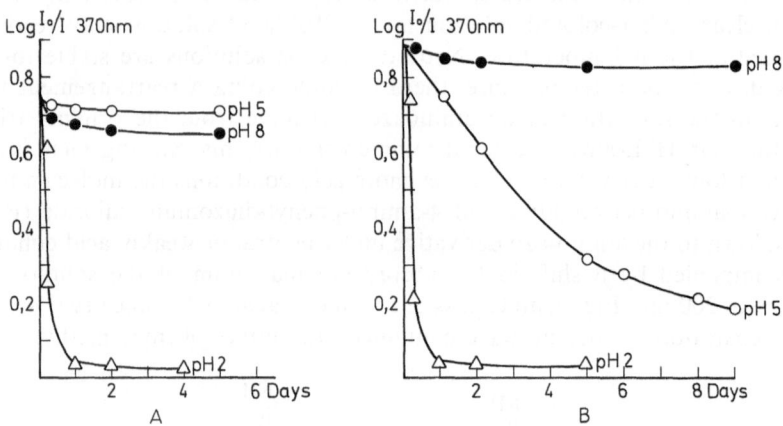

Fig. 5. Effect of pH on the breakdown of berenil diaceturate in the presence (A) or absence (B)
of antipyrine

lone) as a stabilizing agent mixed in approximately equal quantities by weight
with berenil diaceturate: aqueous solutions (ca pH 5) of these preparations remain
stable for 10–15 days at room temperature. Fig. 5 shows the relative stabilites
of berenil diaceturate in the presence or absence of antipyrine at three pH values
as judged by the absoprtion of the solution at 370 nm (NEWTON, unpublished
observations).

Growth Inhibitory Activity

In addition to being a potent trypanocide and babesicide, berenil also inhibits
the growth of a number of species of bacteria and fungi. As shown in Table
1 the sensitivity of different species of pro- and eukaryotic organisms to berenil
varies widely: the factors determining the sensitivity of a particular species are

Table 1. *Sensitivity of various organisms to berenil*

	Gram reaction	Concn for complete inhibition of growth (µg berenil dia-ceturate/ml)
a) Bacteria[a]		
Brucella	−	10
Vibrio Fetus	−	10
Streptococcus agalactiae	+	20
Streptococcus lanceolatus	+	20
Streptococcus haemolyticus	+	70
Listeria monocytogenes	+	100
Staphylococcus aureus	+	250
Escherichia coli	−	250
Corynebacterium diphtheriae	+	250
Haemophilus pertussis	−	1 000
Clostridium welchii	+	1 000
Proteus vulgaris	−	1 000
Pseudomonas aeruginosa	−	1 000
b) Fungi[a]		
Sporotrichon schenki		10
Blastomyces dermatitidis		30
Cryptococcus neoformans		100
Epidermophyton		100
Candida albicans		500
Trichophyton		1 000
c) Protozoa—grown in culture[b]		
Trypanosoma mega		10

	Host animal	Curative dose µg/kg	Site of infection
d) Protozoa *in vivo*[a]			
Babesia canis	dog	1.0	intra muscular
Trypanosoma congolense	dog	1–2.5	intra muscular
Trypanosoma congolense	mouse	2.5	sub-cutaneous
Trypanosoma gambiense	mouse	5.0	sub-cutaneous
Trypanosoma brucei	mouse	25.0	sub-cutaneous
Trypanosoma lewisi	rat	40.0	sub-cutaneous
Trypanosoma rhodesiense	mouse	50.0	sub-cutaneous

[a] FUSSGANGER and BAUER (1958).
[b] NEWTON (1972).

not yet understood. In the case of bacteria, there is no obvious correlation between drug sensitivity and the Gram reaction of organisms. More detailed investigations of the effect of berenil on *Streptococcus agalactiae* (NEWTON, unpublished observations) have shown that cells in the logarithmic phase of growth are much less sensitive to the drug than stationary phase cells (Fig. 6). It seems likely that this difference results from a change in cell permeability to berenil at different

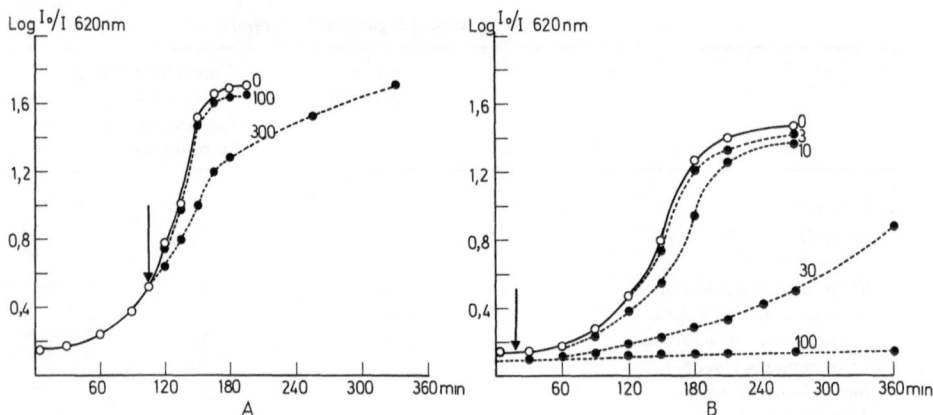

Fig. 6. Effect of berenil on the growth of *Streptococcus agalactiae*. Drug added to culture in (A) the logarithmic phase of growth (B) lag phase. Figures indicate drug concentration (µg/ml)

stages of the growth cycle since the characteristic blue fluorescence of berenil can be readily detected in cells by U.V. microscopy within a short time of adding the drug to cultures which are in the stationary or lag phase of a growth cycle but not when added to cells in the logarithmic phase of growth. No information about the uptake of berenil by different species of trypanosomes is available, and the finding that, in small animals, the drug is more effective against *Trypanosoma congolense* than against *Trypanosoma brucei*, *Trypanosoma rhodesiense* or *Trypanosoma lewisi* (Table 1) remains unexplained. It is interesting that these species have been reported to show equal sensitivity to berenil *in vitro* (VAN HOEVE and GRAINGE 1966).

Use as a Trypanocide

Berenil has been used on an increasing scale as a curative drug for the treatment of bovine trypanosomiasis in Africa since the first clinical trials were reported by BAUER in 1955. During the period 1957–1961 for example, consumption of the drug in Keyna alone rose from 2,000 to 190,000 doses per annum (FAIRCLOUGH, 1963) and in Northern Nigeria the drug largely replaced other trypanocides. The main reasons for the success of berenil have been summarised by WILLIAMSON (1970): it has a higher therapeutic index in cattle than other curative or prophylactic drugs in current use; it is generally active against infections resistant to other drugs and in more than ten years of use over a large area of Africa it has not given rise to any serious degree of drug resistance. The main disadvantage of berenil is its lack of prophylactic activity; this characteristic and the drug's inability to give rise to resistant strains have generally been assumed to be due to the rapidity with which it is excreted from treated animals. FAIRCLOUGH (1963) concluded that "a dose of 7 mg berenil/kg remained in an adult bovine animal long enough to protect it from reinfection for only 24 hours", however, biological assays of berenil levels in bovine blood have indicated that small amounts of the drug can be detected for up to three weeks after an intramuscular injection of 7 mg/kg and that repeated berenil treatment of previously uninfected animals can protect them for some weeks when they are introduced into a tsetse fly area of high risk (VAN HOEVE *et al.*, 1965). These different conclusions may result from the fact that cattle in these field trials could have received highly variable inocula of trypanosomes. More recent work (GITATHA and MANDLIN 1968) indicates that cattle receiving 7 mg/kg are only protected against a standard inoculum (10^6 organisms in 1 ml of mouse blood) of *T. congolense* for 96 hours.

Although berenil was introduced for veterinary use, limited clinical trials carried out in Africa in recent years suggest that the drug may have some value when administered orally or intramuscularly for the treatment of *Trypanosoma gambiense* and *Trypanosoma rhodesiense* sleeping sickness in man (HUTCHISON and WATSON, 1962; DE RAADT *et al.*, 1966; ONYANGO *et al.*, 1969). It has been used alone for the treatment of early cases and in conjunction with melarsoprol for the treatment of cases showing nervous system involvement. However, it must be emphasised that exhaustive toxicity tests on berenil have not yet been completed and, at the present time, the prerequisites for a WHO-sponsored preclinical trial of the drug in humans have not been met (WHO, 1969).

Toxicity

Available data indicate that marked differences exist in the tolerance of berenil by different animals. Rats tolerate the oral administration of 4000 ppm of berenil in their solid food (estimated to be equivalent to 400µg/kg/day) for nine months without obvious abnormalities. The maximum tolerated dose for mice is *ca* 100 mg/kg when administered subcutaneously; in larger animals the maximum tolerated intramuscular dose is considerably less, *ca* 40 mg/kg in rabbits and 8–10 mg/kg in sheep and cattle. Subcutaneous doses as high as 21 mg/kg have been given to cattle without sign of toxicity (FAIRCLOUGH, 1963). In European trials, dogs given 1–3 intramuscular injections of 7–15 mg/kg body weight or prolonged oral doses of 20–60 mg/kg/day showed severe clinical signs of damage to the central nervous system and generally died within four days; at *post mortem* the brain was edematous and had punctate haemorrhages. However, similar trials on mongrel dogs in Africa (OGADA, 1969) showed that the majority tolerated upto 15 mg berenil/kg given over a period of a week. Berenil is believed not to pass the blood brain barrier when given orally or parenterally and thus it is not effective in the treatment of late cases in human trypanosomiasis. Experiments with *Cercopithecus neglectus* monkeys (OGADA and NGULI, 1968) indicate that intrathecal administration of 1 mg/kg can be tolerated by these animals and observations for four months after treatment revealed no abnormalities. Paralysis of the limbs was observed following doses of 1.2–1.8 mg/kg and doses of 2 mg/kg and above were rapidly lethal.

The Action of Berenil on Trypanosomes

1. Morphological Changes

BAUER (1958) reported that berenil is rapidly and irreversibly bound by trypanosomes and suggested that a detailed examination of the morphological changes
induced by the drug in these organisms would be rewarding. However, it was
more than a decade before this suggestion was followed up.

In 1964 KILLICK-KENDRICK observed that berenil treatment of a *Trypanosoma
evansi* infection in a horse gave rise to a population of organisms which apparently
lacked kinetoplasts. (The kinetoplast of trypanosomes, and other flagellates belonging to the order *Kinetoplastida,* is a Feulgen-positive structure located near
the base of the flagellum; it is now regarded as a specialised, DNA-containing,
region of the single branched-tubular mitochondrion which characterizes these
organisms.) (See SIMPSON [1972] for a recent review.) Discussing this effect of berenil, KILLICK-KENDRICK favored the view that the drug only killed trypanosomes
which contained a kinetoplast and thus selected for the naturally occurring dyskinetoplastic organisms (i.e. those organisms which appear to lack a kinetoplast
in preparations stained with FEULGEN or GIEMSA) which were known to be present
in small numbers in the original population of *T. evansi.* However, subsequent
works suggests that this is not so and there is now convincing evidence that
berenil, under certain conditions, gives rise to dyskinetoplastic trypanosomes
by selectively inhibiting the replication of kinetoplast DNA.

Ultra-violet microscopy of *Trypanosoma brucei* isolated from rats before and
one hour after administration of a curative dose of berenil (25 mg/kg i.p.) revealed
a brilliant blue fluorescence in the kinetoplast and nucleus of the drug-treated
organisms (NEWTON and LE PAGE, 1967). Further experiments with *T. brucei*
grown in experimental animals or in culture and *Trypanosoma mega* grown in
culture, established that exposure of organisms to low concentrations of berenil
results in fluorescence in the kinetoplast only, while in the presence of higher
drug concentrations fluorescence occurs first in the kinetoplast and subsequently
in the nucleus. Electron microscopy of *T. rhodesiense* after exposure to curative
concentrations of berenil revealed a "fragmentation of the kinetoplast core"
but no apparent changes in the nucleus or nucleolus (MACADAM and WILLIAMSON,
1969). The rapid localisation of berenil in DNA-containing organelles led to
an investigation of the possibility that the drug might be a selective inhibitor
of DNA replication and that it might do so by direct interaction with DNA.

2. Evidence for the Formation of a Berenil/DNA Complex

Addition of DNA to a solution of berenil causes a shift in the absorption
spectrum of the drug to longer wavelengths, λ max being shifted from 370 nm
to 380 nm (NEWTON, 1967). This metachromatic shift can be taken as evidence
for the formation of a drug/DNA complex under these particular conditions
and use has been made of this effect to measure drug/DNA binding ratios by
the method of PEACOCKE and SKERRETT (1956). Calf thymus DNA and DNA
isolated from a number of bacteria and protozoa were found to bind one molecule
of berenil for every 4–5 nucleotides; heat denaturation of DNA before the addition

of drug doubled the number of drug-binding sites. Addition of RNA, synthetic homopolymers, and poly-AU copolymer to berenil solutions all produced a spectral shift and it was found that the amount of drug bound by these substances was the same as that bound by heat-denatured DNA. No evidence was obtained for the formation of complexes between berenil and mononucleotides.

Reversible interactions between macromolecules and substances of low molecular weight can be detected by the gel filtration method of HUMMEL and DREYER (1962). This technique was adapted by NEWTON (1972) to study DNA/berenil interactions and the results, in close agreement with spectrophotometric studies, indicates a binding of one drug molecule for every four or five nucleotides. Further evidence for the formation of berenil/DNA complexes has come from studies of the buoyant density of DNA by ultracentrifugation in cesium chloride gradients in the presence or absence of drug. Berenil decreases the density of DNA dramatically (NEWTON, 1967 and 1972) and in experiments using DNA's of differing base composition it was found that the decrease in density was dependent upon both the DNA/drug ratio and the base composition of the DNA, the density shift in the presence of a given concentration of berenil increasing with the A+T content of the DNA, suggesting that the drug may bind preferentially to this particular base pair. Similar conclusions were reached by FESTY et al., (1970) from a study of the effect of berenil on the sedimentation of DNA.

These results provide convincing evidence that berenil can form complexes with DNA in vitro. Evidence that the drug may behave similarly in vivo has come from a study of DNA extracted from berenil-treated trypanosomes (NEWTON, 1967); such DNA was found to have a lower buoyant density than DNA from control organisms. This type of experiment is always open to the criticism that the drug may have become associated with DNA during the extraction procedure but ultrastructural studies discussed below support the conclusion that berenil does bind to DNA in intact cells.

3. Some Characteristics of the Berenil/DNA Complex

The structure of the complex formed between berenil and purified DNA has not yet been elucidated. The drug does not produce a measurable increase in the viscosity of DNA solutions (NEWTON, 1967) nor does it cause an uncoiling of the double helix of $\phi \times 174$ R.F. DNA (WARING, 1970); thus it is unlikely that it intercalates between adjacent base pairs in DNA molecules in the manner postulated by LERMAN (1961) for planar heterocyclic structures such as acridines and phenanthridines (see WARING, WOLFE, this volume).

The presence of terminal amino groups in the berenil molecule suggests the possibility of hydrogen bonding between these groups and the charged oxygen atoms in DNA phosphate groups, but the apparent stability of the berenil/DNA complex in solutions of high ionic strength indicates that other types of bonding must be involved. The combination of berenil with double-stranded DNA results in a stabilization of the helix as judged by the temperature required to denature the molecule (NEWTON 1967) but renaturation experiments failed to produce any evidence that the drug can cross link complementary DNA strands in a manner similar to the antibiotic mitomycin (IYER and SZYBALSKI, 1963).

4. Evidence for the Selective Inhibition
of Kinetoplast DNA Synthesis

In vitro experiments with *Trypanosoma mega* have shown that the addition of berenil (3–10 µg/ml) to cultures growing in a peptone-haematin-glucose medium results in the loss of kinetoplast DNA, as judged by acridine orange staining, from about 40 per cent of the population after one generation (Table 2). This finding indicates that the drug is either inhibiting kinetoplast DNA replication or is modifying this DNA in some way so that, in about half the population, it is no longer stained by acridine orange. When DNA, released from control and berenil-treated cells by a detergent-lysis technique

Table 2. *Effect of berenil concentration on growth of Trypanosoma mega and appearance of dyskineto-plastic organisms (for conditions see text)*

Berenil (ng/ml)	Organism × 10^{-4}/ml			% of total population lacking kinetoplast
	To	22 hr	44 hr	
0	157	270	520	4
3		250	330	37
10		180	260	43
30		150	290	32
100		160	160	9

(NEWTON, 1972), was separated into nuclear and kinetoplast components by centrifugation in cesium chloride gradients it was found that the amount of kinetoplast-DNA extracted from drug-treated cells was about half that obtained from control cells (Table 3). While this finding could be due to a change in the properties of DNA in drug treated cells which prevents a proportion of it from being extracted by the procedures used, it is also consistent with a selective inhibition of kinetoplast-DNA replication. Attempts to confirm the latter possibility by autoradiography of cells labelled with tritiated thymidine in the presence or absence of berenil failed due to the juxtaposition of the kinetoplast and nucleus in *T. mega*. However, further evidence for selective inhibition of kinetoplast DNA synthesis has been obtained by studying the effect of berenil on the incorporation of 5-bromodeoxyuridine (BUDR) into the DNA of *T. mega* (NEWTON, 1967).

Table 3. *Effect of berenil (100 µg/ml) on growth and DNA content of Trypanosoma mega*

Time (hrs)	Control			+Berenil (100 µg/ml)		
	cell count × 10^{-6}	kineto-plast DNA % of total	nuclear DNA % of total	cell count × 10^{-6}	kineto-plast DNA % of total	nuclear DNA % of total
0	4.1	24.0	76	4.2	23.6	76.4
24	9.4	24.8	75.2	7.1	16.2	83.8
48	19.5	24.4	75.6	9.2	11.8	88.2

In control organisms this analogue is incorporated into both nuclear and kineto-plast DNA and the incorporation can be detected as an increase in the buoyant density of the DNA. After growth for one generation time in the presence of the analogue both DNA components were found to be "half heavy" indicating that both replicate semi-conservatively. When flagellates were pretreated with berenil for a time sufficient to produce fluorescence in the kinetoplast but not in the nucleus and were then transferred to a drug-free medium containing BUDR and allowed to grow for 24 hours it was found that the buoyant density of nuclear DNA only was increased. While the selective inhibition of BUDR incorpo-ration into kinetoplast-DNA does not necessarily indicate that berenil has inhibi-ted the synthesis of this DNA completely, this finding, taken together with the other observations described above, strongly suggests that this is so.

Evidence of berenil-induced changes in the molecular organisation of kineto-plast-DNA of *Trypanosoma cruzi* has been obtained by Brack *et al.* (1972 b). These workers examined the ultrastructure of the kinetoplast of this organism after four days growth in a disphasic medium consisting of blood agar and tissue culture medium 199 to which berenil was added at a concentration of 2 µg/ml. They found that the double layered pattern of kinetoplast DNA which is characteristic of *T. cruzi* became re-arranged to form elongated fibrillar bodies: similar changes were induced by another aromatic diamidine, hydroxystilbami-dine. When purified kinetoplast DNA from berenil-treated *T. curzi* was examined in the electron microscope, long "lamp-brush-like" structures, often linked to complex associations of circular molecules, were observed and many of the small circular DNA molecules appeared as double branched structures. These latter forms, which the authors assume to be replicating molecules according to the model of Cairns (1963), were seen only rarely in kinetoplast DNA preparations from control organisms: they estimate that berenil treatment increased the propor-tion of these replicating forms by a factor of 10^3 (Brack *et al.*, 1972 a). This effect of berenil has made possible the observation of a large number of double branched circular molecules and the analysis of some of their morphological and physical characteristics. More than 800 double branched circles were measu-red; their mean contour length was found to be equal to that of nonreplicating circles and measurement of the replicating segments showed that the length distri-bution of these segments was neither random nor Gaussian but fell into groups which correspond to multiples of 15% (i.e. about 0.075 µm) of the total contour length (0.50µm). These results suggest that berenil does not block the replication of kinetoplast DNA at initiation but at different specific points which are regularly distributed round the circular DNA molecules. If this interpretation proves cor-rect, berenil may become a valuable research tool for the further investgation of kinetoplast DNA replication and perhaps also for studies of the information content of this DNA.

Relationships between Chemical Structure and Biological Activity

The instability of berenil in neutral and acidic aqueous solutions has already been referred to. The amino-azo derivative (Fig. 3) formed when the drug undergoes molecular rearrangement in neutral solution is without trypanocidal activity and has no detectable effect on the buoyant density of DNA (Newton, 1967). This finding suggests that the spacing of the amidino groups of berenil may be critical in the formation of a DNA/drug complex. In an attempt to investigate this possibility Newton (1972) compared the trypanocidal activity of a number of compounds closely related to berenil with their ability to form complexes with DNA. The structure of two of the most interesting of these are compared with berenil in Fig. 7. These compounds differed markedly in their ability to

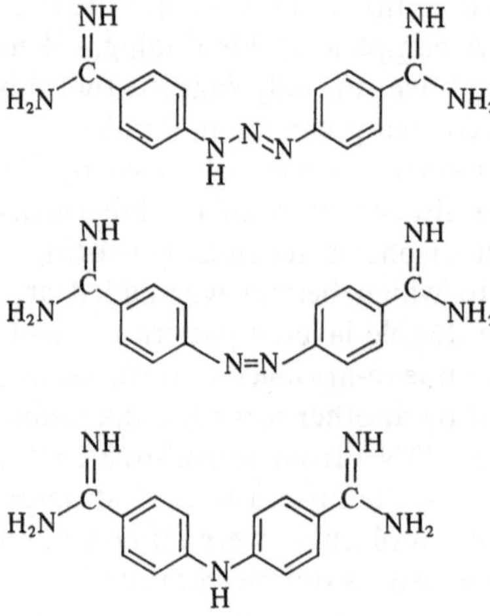

Fig. 7

inhibit the *in vitro* growth of *Trypanosoma mega*, the minimum growth inhibitory concetrations (i.e. the lowest concentration in which no growth oçcurred after 72 hrs. incubation at 25° C in a peptone, yeast extract, haemin, glucose medium with an initial inoculum of 2×10^6 organisms/ml) were 10^{-5} M, 2×10^{-4} M, and 4×10^{-6} M for berenil, compound 1 and compound 2 respectively. Compounds 1 and 2 resembled berenil in forming complexes with purified DNA *in vitro* but ultra-violet microscopy of drug-treated organisms showed that, in contrast to berenil, treatment of *T. mega* with compound 1 for periods of upto five hours produced no fluorescence in the kinetoplast or nucleus of flagellates whereas exposure to compound 2 resulted in fluorescence in both organelles within minutes. Ultracentrifugation of DNA extracted from flagellates grown in the presence of compound 1 or compound 2 for 24 hrs. showed that compound 1 caused no detectable increase in the buoyant density of either nuclear or kinetoplast DNA whereas compound 2 decreased the density of both types of DNA by about the same amount. Thus it appears that removal of one nitrogen atom from the triazine bridge of berenil reduces the growth inhibitory activity and the ability of the drug to bind to DNA *in vivo;* removal of a second nitrogen atom restores growth inhibitory activity and ability to combine with intracellular DNA, but in contrast to berenil, compound 2 does not appear to act selectively on kinetoplast DNA. A study of molecular models of these compounds has shown that removal of one nitrogen atom from berenil reduces the minimum distance between the terminal amidino carbon atoms from *ca* 13A to 5A whereas removal of a second nitrogen atom results in the molecule opening out again to give a distance of 11A between these carbon atoms. Further study of these and related compounds should lead to a better understanding of the molecular basis of the interaction between berenil and DNA.

Resistance

As already stated, one of the virtues of berenil as a trypanocide is that resistance develops relatively infrequently. However, berenil resistant strains of *Trypanosoma vivax* and *Trypanosoma congolense* have been detected in Africa and it is clear that they can result both from direct exposure to berenil and from the use of other drugs such as antrycide and phenanthridines (WHITESIDE, 1962; MACLENNON and JONES-DAVIES 1967; JONES-DAVIES, 1967). GRAY and ROBERTS (1971 a and b) have shown that resistant strains can be transmitted by tsetse flies to both cattle and to trypanosomiasis-resistant vertebrate hosts such as duikers (*Sylvicapra grimmia*) and gazelle (*Gazelle rufifrons*) and cyclical transmission has been continued for periods of upto 29 months without loss of resistance. The implications of these results are clear and it is only a matter of time before berenil resistance becomes a major problem in the field.

Attempts to produce berenil resistance experimentally in cattle or in laboratory animals have met with little succes. (FUSSGANGER *et al.,* 1962). Direct exposure of trypanosomes to berenil has been reported to give rise to small increases (4–5 times) in resistance (WHITESIDE, 1963). Berenil resistance has been observed in *Trypanosoma congolense* made resistant to antrycide (HAWKING, 1963) and FULTON and GRANT (1955) reported that an old laboratory strain of *Trypanosoma rhodesiense* made resistant to stilbamidine was completely resistant to berenil at the maximum dose tolerated by mice, they estimated that this corresponded to an increase in resistance of more than 80 times.

No detailed biochemical studies of berenil resistant trypanosomes have yet been made. Preliminary investigations (NEWTON, unpublished observations) have failed to detect differences in the interaction of berenil with DNA extracted from a sensitive and a resistant strain of *T. vivax* isolated in Nigeria and there is reason to think that the development of resistance may be associated with changes in the cell surface and/or permeability barriers of trypanosomes which influence the uptake of the drug.

Acknowledgement. The author wishes to thank Professor Dr. W.H. WAGNER and Dr. H. LOEWE (Hoechst Farbwerke, AG) for supplies of berenil and related compounds.

References

BAUER, F.: Ergebnisse der klinischen Prüfung von Berenil. Veterinär-Medizinische Nachr. No. 3, 152 (1955a).
BAUER, F.: Trypanosomen- und Babesien-Erkrankungen in Afrika und ihre Behandlung mit dem neuen Präparat „Berenil". Z. Tropenmed. Parasitol. **6,** 129 (1955b).

BAUER, F.: Über den Wirkungsmechanismus des Berenil (4,4′-Diamidinodiazoaminobenzol) bei *Trypanosoma congolense*. Zentr. Bakteriol. Parasitenk., Abt. I, Orig. **172**, 605 (1958).

BAUER, F.: The development of drug-resistance to berenil in *Trypanosoma congolense*. Vet. Rec. **74**, 265 (1962).

BRACK, C. H., E. DELAIN, and G. RIOU: Replicating, covalently closed, circular DNA from kinetoplasts of *Trypanosoma cruzi*. Proc. Natl. Acad. Sci. U.S. **69**, 1642 (1972a).

BRACK, C. H., E. DELAIN, G. RIOU, and B. FESTY: Molecular organization of the kinetoplast DNA of *Trypanosoma cruzi* treated with berenil, a DNA interacting drug. J. Ultrastruct. Res. **39**, 568 (1972b).

CAIRNS, J.: The bacterial chromosome and its manner of replication as seen by autoradiography, J. Mol. Biol. **6**, 208 (1963).

FAIRCLOUGH, R.: Observations on the use of Berenil against trypanosomiasis in cattle in Kenya. Vet. Rec. **75**, 1107 (1963).

FESTY, B., A. M. LALLEMANT, G. RIOU, C. BRACK et E. DELAIN: Mecanisme d'action des diamidines trypanocides. Importance de la composition en bases dan l'association berenil-polynucleotides. Compt. Rend. **271**, 684 (1970).

FULTON, J. D., and GRANT, P. T.: The preparation of a strain of *Trypanosoma rhodesiense* resistant to stilbamidine and some observations on its nature. Exp. Parasitol. **4**, 377 (1955).

FUSSGANGER, R.: Berenil in der Veterinärmedizin. Veterinär-Medizinische Nachr. No. 3, 146 (1955).

FUSSGANGER, R., u. F. BAUER: Berenil, ein neues Chemotherapeuticum in der Veterinärmedizin. Medizin und Chemie **6**, 504 (1958).

FUSSGANGER, R., and F. BAUER: Investigations on berenil resistance of trypanosomes. Vet. Rec. **72**, 1118 (1960).

GITATHA, S. K., and I. MAUDLIN: Some investigations on the effect of berenil in cattle trypanosomiasis. East African Trypanosomiasis Organisation Report p. 88 (1968).

GRAY, A. R., and C. J. ROBERTS: The cyclical transmission of strains of *Trypanosoma congolense* and *T. vivax* resistant to normal therapeutic doses of trypanocidal drugs. Parasitology **63**, 67 (1971a).

GRAY, A. R., and C. J. ROBERTS: The stability of resistance to diminazene aceturate and quinapyramine sulphate in a strain of *Trypanosoma vivax* during cyclical transmission through antelope. Parasitology **63**, 163 (1971b)

HAWKING, F.: Drug resistance of *Trypanosoma congolense* and other trypanosomes to quinapyramine, phenanthridines, berenil and other compounds in mice. Ann. Trop. Med. Parasitol. **57**, 262 (1963).

HOEVE, K. VAN, M. P. CUNNINGHAM, and E. B. GRAINGE: Some observations on the treatment of cattle with berenil. International Scientific Committee for Trypanosomiasis Research, p. 27 (1965).

HOEVE, K. VAN, and E. B. GRAINGE: Berenil sensitivity *in vitro* of trypansomes of the *T. brucei* sub group and *T. congolense* group. East African Trypansomiasis Organisation Report, p. 63 (1966).

HUMMEL, J. P., and W. J. DREYER: Measurement of protein—binding phenomena by gel filtration. Biochim. Biophys. Acta **63**, 530 (1962).

HUTCHINSON, M. P., and H. J. C. WATSON: Berenil in the treatment of *Trypanosoma gambiense* infection in man. Trans. Roy. Soc. Trop. Med. Hyg. **56**, 227 (1962).

IYER, V. N., and W. SZYBALSKI: A molecular mechanism of mitomycin action: linking of complementary DNA strands. Proc. Natl. Acad. Sci. U.S. **50**, 355 (1963).

JENSCH, H.: Basic diazoaminobenzene compounds. U.S. Patent No. 2673197 (1954).

JENSCH, H.: 4,4′-Diamidino-diazoaminobenzol, ein neues Mittel gegen Trypanosomen- und Babesien-Infektionen. Arzneimittel-Forsch. **5**, 634 (1955).

JENSCH, H.: Über neue Typen von Guanyl-Verbindungen. Medizin und Chemie **6**, 134 (1958).

JONES-DAVIES, W. J.: A berenil-resistant strain of *Trypanosoma vivax* in cattle. Vet. Rec. **81**, 567 (1967).

KILLICK-KENDRICK, R.: The apparent loss of the kinetoplast of *Trypanosoma evansi* after treatment of an experimentally infected horse with berenil. Ann. Trop. Med. Parasitol. **58**, 841 (1964).

LERMAN, L. S.: Structural consideration in the interaction of DNA and acridines. J. Mol. Biol. **3**, 18 (1961).

MACADAM, R. R., and J. WILLIAMSON: Lesions in the fine structure of *Trypanosoma rhodesiense* specifically associated with drug treatment. Trans. Roy. Soc. Trop. Med. Hyg. **63**, 421 (1969).

MACLENNAN, K.J.R., and W.J. JONES-DAVIES: The occurrence of a berenil-resistant Trypanosoma-congolense strain in Northern Nigeria. Vet. Rec. **80**, 389 (1967).

NEWTON, B.A.: Interaction of berenil with deoxyribonucleic acid and some characteristics of the berenil-deoxyribonucleic acid complex. Biochem. J. **105**, 50P (1967).

NEWTON, B.A.: Recent studies on the mechanism of action of berenil (Diminazene) and related compounds. In: Comparative biochemistry of parasites, ed. H. VAN DEN BOSSCHE, p. 127. New York: Academic Press 1972.

NEWTON, B.A., and R.W.F. LE PAGE: Preferential inhibition of extranuclear deoxyribonucleic acid synthesis by the trypanocide berenil. Biochem. J. **105**, 50P (1967).

OGADA, T.: Toxicity of intramuscular berenil in dogs. East African Trypanosomiasis Organisation Report, p. 121 (1969).

OGADA, T., and K.N. NGULI: The maximum dose of berenil considered safe when given intrathecially to monkeys. East African Trypanosomiasis Organisation Report, p. 85 (1968).

ONYANGO, R.J., N.M. BAILEY., R.W. OKACH., E.K. MWANGI, and T. OGANDA: The use of berenil for the treatment of early cases of human trypanosomiasis. East African Trypansomiasis Research Organisation Report, p. 120 (1969).

PEACOCKE, A.R., and J.N.H. SKERRETT: The interaction of amino acridines with nucleic acids. Trans. Faraday Soc. **52**, 261 (1956).

RAADT, P. DE, K. VAN HOEVE, N.M. BAILEY, and E.N. KENYANJUI: Observations on the use of berenil in the treatment of human trypanosomiasis. East African Trypanosomiasis Organisation Report p. 60 (1965).

SIMPSON, L.: The kinetoplast of hemoflagellates. Intern. Rev. Cytol. **32**, 139 (1972).

WARING, M.J.: Variation of the supercoils in closed circular DNA by binding of antibiotics and drugs: evidence for molecular models involving intercalation. J. Mol. Biol. **54**, 247 (1970).

WHITESIDE, E.F.: Interactions between drugs, trypanosomes and cattle in the field. In: Drugs, parasites and hosts: a symposium on relations between chemotherapeutic drugs, infecting organisms and hosts, eds. L.G. GOODWIN and R.H. NIMMO-SMITH, p. 116. London: Churchill 1962.

WHITESIDE, E.F.: A strain of *Trypanosoma congolense* directly resistant to berenil. J. Comp. Pathol. Therap. **73**, 167 (1963).

WHO: African trypanosomiasis. World Health Organ. Tech. Rept. Ser. No. 434 (1969).

WILLIAMSON, J.: Review of chemotherapeutic and chemoprophylactic agents. In: The African trypanosomiases, p. 125, ed. H.W. MULLIGAN. London: George Allen & Unwin, 1970.

Camptothecin

Susan B. Horwitz*

Camptothecin (I)[1], a cytotoxic alkaloid, was originally isolated from the bark and stem wood of *Camptotheca acuminata* (family NYSSACEAE), a tree indigenous to China (WALL *et al.*,1966). The structure of camptothecin has been established (WALL *et al.*, 1966; MC PHAIL and SIM, 1968), and the compound was shown to have antitumor properties in experimental animals (WALL *et al.*, 1966; VENDITTI and ABBOTT, 1967; DE WYS *et al.*, 1968).

Interest in camptothecin as an antineoplastic agent stimulated investigations of its mode of action in mammalian cells and their viruses. The present chapter reviews these studies, with particular emphasis on its effects on nucleic acid synthesis.

Chemical Synthesis and Pharmacology

The prospect of using camptothecin for treatment of human malignancies, together with the limited supply of the natural product, encouraged chemists to develop a practical method for synthesis of the alkaloid. These efforts resulted in a number of publications concerned with the chemical synthesis of camptothecin and its analogues (WENKERT *et al.*, 1967; KEPLER *et al.*, 1969; WANI *et al.*, 1970; SHAMMA and NOVAK, 1969; KAMETANI *et al.*, 1970; WINTERFELDT and RADUNZ, 1971; LIAO *et al.*, 1971; BIESLER, 1971; DANISHEFSKY *et al.*, 1971; WARNEKE and WINTERFELDT, 1972; DANISHEFSKY *et al.*, 1973a; DANISHEFSKY *et al.*, 1973b; KENDE *et al.*, 1970). Several laboratories have independently completed a total sythesis of camptothecin (STORCK and SCHULTZ, 1971; VOLKMANN *et al.*, 1971; WANI *et al.*, 1972; BOCH *et al.*, 1972).

A sensitive method has been developed for the measurement of camptothecin following extraction of the drug from urine and plasma (HART, 1969). The assay is based on the intense fluorescence of camptothecin at 434 mμ in ethanol after excitation at 370 mμ; 0.005 µg/ml can be detected by this procedure. Using this method, the plasma disappearance and urinary recovery of camptothecin has been measured in human patients (GOTTLIEB *et al.*, 1970). The mean urine recovery of the unmetabolized drug was 17.5 % during the first 48 hours after intravenous administration. The plasma half-life in these patients was unusually long, which could be due to the considerable plasma binding of camptothecin demonstrated *in vitro*. No circulating metabolites of camptothecin were observed by this fluoro-

*Career Development Awardee (K4-GM-11, 147) of the United States Public Health Service. Research that originated in the author's laboratory was supported by the US Public Health Service (Grant CA-10666).
[1] Roman numerals refer to structural formulas in Fig. 1.

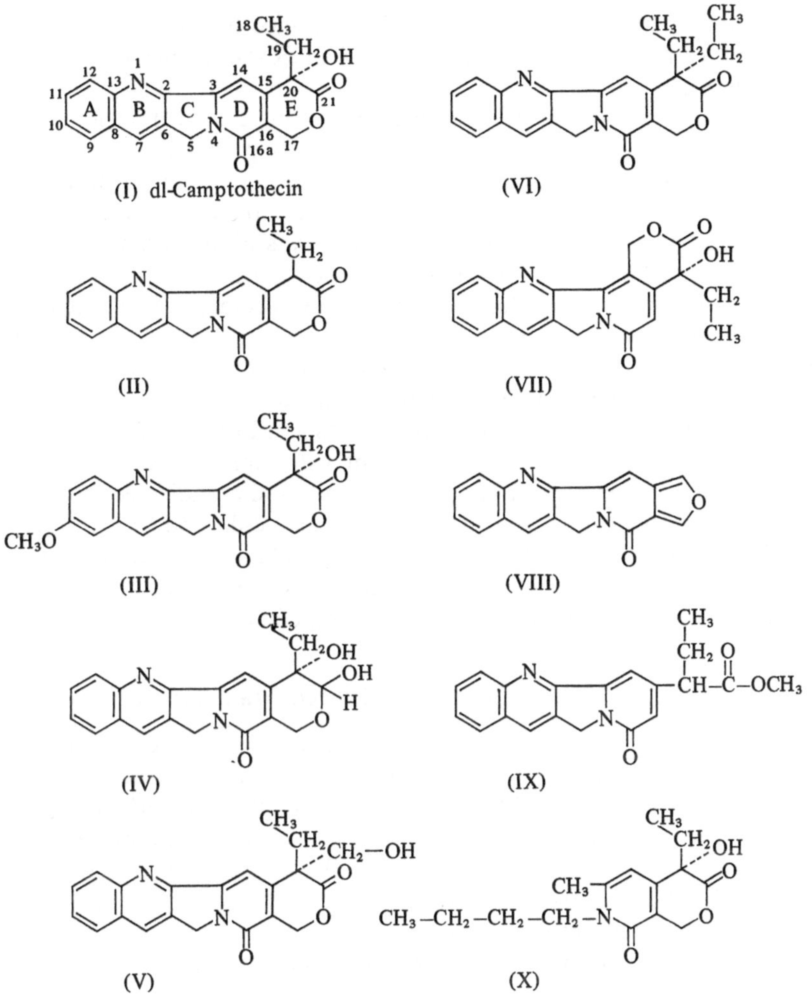

Fig. 1. Structural formulae of camptothecin and analogues

metric method. Camptothecin has a half-life of 35 minutes in mice having L1210 leukemia following injection of a single dose of camptothecin (25 mg/kg) (KESSEL, 1971). In these studies, the loss of camptothecin from both ascitic fluid and ascitic cells was followed.

Antitumor Properties

Camptothecin increases the survival time of rodents having various experimental leukemias. The alkaloid originally showed strong antitumor activity against mouse leukemia L1210 and Walker 256 rat carcinosarcoma (WALL et al., 1966; VENDITTI and ABBOTT, 1967; DE WYS et al., 1968). Camptothecin is also active against a number of murine lymphocytic leukemias and leukemias resistant to other antitumor drugs (GALLO *et al.*, 1971). In mice having leukemias L5178Y and K1964, camptothecin treatment resulted in 70–90% survivors six months after death of all controls.

The preliminary clinical evaluation of camptothecin (GOTTLIEB *et al.*, 1970) was encouraging. Positive responses were noted, primarily in patients with advanced gastrointestinal carcinoma, although duration of the response was brief. This initial study evaluated the effects of single intravenous injections of the drug given at intervals of 2—4 weeks. Toxicity was observed and noted in this group of patients but was not emphasized in the report.

Subsequent clinical trials have demonstrated dose-dependent hematopoietic depression, diarrhea, alopecia, and cystitis; and little, if any, clinical benefit in patients with advanced gastrointestinal adenocarcinoma (MOERTEL *et al.*, 1972). It has also been concluded that camptothecin has no value in the treatment of advanced disseminated melanoma (GOTTLIEB and LUCE, 1972).

Tolerance to daily and weekly treatment with camptothecin was evaluated in patients with cancer because data from animal studies indicated that the effect of camptothecin against L1210 leukemia was schedule-dependent (MUGGIA *et al.*, 1972). This study concluded that there was no clinical benefit to any of the patients on the daily schedule and transient objective responses were noted in only two of the fifteen patients on the weekly schedule.

Effects of Camptothecin on Cultured Mammalian Cells

Camptothecin is a potent inhibitor of nucleic acid synthesis in HeLa cells (HORWITZ *et al.*, 1970; 1971) and in L1210 cells (KESSEL, 1971a). In HeLa cells, the synthesis of DNA and RNA is inhibited by 50% by 5 µM camptothecin while the rate of protein synthesis is essentially unaffected by 100µM concentrations of the drug (Fig. 2). In one study (BOSMANN, 1970), significant inhibition of nucleic acid synthesis was observed in HeLa cells but only when high concentrations of camptothecin (1 mg/ml) were used. Camptothecin has essentially no effect on the uridine and thymidine pools in HeLa cells nor does it significantly

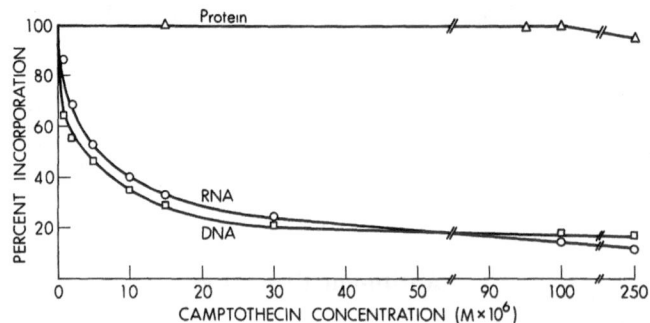

Fig. 2. Effect of various concentrations of camptothecin on synthesis of protein, RNA, and DNA in HeLa cells. Cells were incubated for 60 m. at 37 C, and the incorporation of radioactive leucine, uridine, or thymidine into TCA acid-insoluble material was used to calculate the rates of synthesis of protein (△—△), RNA ○—○), and DNA (□—□) from HORWITZ, *et al.*, 1971. Reproduced by permission of Academic Press, Inc.)

affect enzymatic activity of DNA and RNA polymerases prepared from HeLa cells or of RNA polymerase prepared from *E. coli* (HORWITZ *et al.*, 1971).

An unusual characteristic of camptothecin is the reversibility of the inhibition of nucleic acid synthesis in HeLa cells (HORWITZ *et al.*, 1971) and in L1210 cells (KESSEL, 1971a; 1971b). If HeLa cells are treated with camptothecin for as long as an hour and then washed and resuspended in drug-free medium, uridine incorporation into RNA resumes at a normal rate within minutes after resuspension. Inhibition of DNA synthesis is partially reversed by similar procedures, the extent of reversibility being dependent on the length of time the cells are exposed to the alkaloid.

Camptothecin inhibits the synthesis of ribosomal RNA in HeLa cells to a much grater extent than 4–5S RNA (HORWITZ *et al.*, 1971). Analysis of the RNA of camptothecin-treated cells on sucrose gradients showed little 28S and 18S ribosomal RNA present, while there was only partial inhibition of the RNA in the 4–5S region. Studies on the differential effects of camptothecin on the synthesis and processing of the various classes of nuclear RNA in HeLa cells have demonstrated that the synthesis of precursor ribosomal RNA in the nucleolus is more sensitive to camptothecin than that of heterogeneously sedimenting nuclear RNA. The maturation of 45S RNA to 32S RNA occurs normally in the nucleolus, but the appearance of 28S RNA in the cytoplasm is inhibited by camptothecin (WU *et al.*, 1971). RNA synthesized in HeLa cells in the presence of camptothecin consists of shortened RNA molecules of both heterogeneous nuclear RNA and nucleolar RNA (ABELSON and PENMAN, 1972). Although camptothecin interrupted formation of high molecular weight nuclear RNA in HeLa cells, mitochondrial RNA synthesis was not inhibited (PERLMAN *et al.*, 1973). In leukemia L1210 cells treated with camptothecin, RNA chains approached normal lengths if sufficient time was allowed (KANN and KOHN, 1972). Acrylamide gel electrophoresis demonstrated that camptothecin partially inhibits the synthesis of 4S RNA but does not affect the synthesis of 5S RNA in HeLa cells (ABELSON and PENMAN, 1972). Different results have been obtained by other investigators using L1210 cells. Synthesis of ribosomal precursor and heterogeneous nuclear RNA were both extensively inhibited in L1210 cells, but the maturation of 45S RNA to cytoplasmic 18S and 28S RNA was unaffected by camptothecin. The synthesis of both 4S and 5S RNA was unaffected by camptothecin in L1210 cells (KESSEL, 1971b).

Camptothecin induces single strand breaks in cellular DNA when HeLa cells are incubated with the drug (HORWITZ and HORWITZ, 1971) (Fig. 3). Total DNA, measured as acid insoluble material, remains unchanged in camptothecin-treated HeLa cells, but analysis on alkaline sucrose gradients demonstrates a conversion of cellular DNA to a lower molecular weight form. This degradation of cellular DNA can occur within 10 minutes after exposure of HeLa cells to camptothecin, even at 4° (HORWITZ *et al.*, 1971). This process is reversible after removal of the drug. DNA in L1210 cells is also degraded after treatment of these cells with camptothecin (SPATARO and KESSEL, 1972).

Camptothecin is a much stronger inhibitor of DNA synthesis than RNA synthesis in human lymphocytes stimulated by phytohemagglutinin (GALLO *et al.*, 1971). In these cells, the drug caused a G_2 "lesion" which prevented subsequent

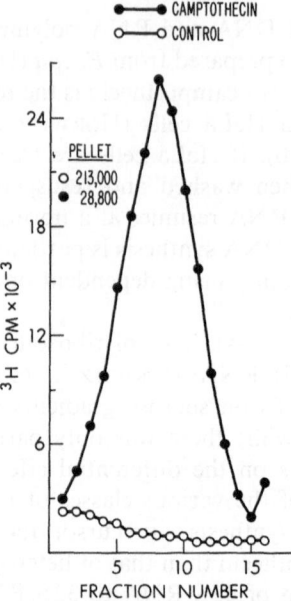

Fig. 3. Sedimentation of DNA from HeLa cells (o—o) and camptothecin-treated HeLa cells (●—●) in alkaline sucrose gradients. The "pellet" represents the acid insoluble ^{3}H-thymidine in the 2.5 cm remaining at the bottom of the gradient. Sedimentation is from right to left (from HORWITZ and HORWITZ, 1971. Reproduced by permission of Academic Press, Inc.)

mitosis. It was concluded that cell proliferation is needed for the antitumor effect of camptothecin. L1210 cells are most sensitive to the inhibitory effects of camptothecin in the S phase of the cell cycle (KESSEL et al., 1972), as are Don cells, a Chinese hamster fibroblast line (LI et al., 1972), and HeLa cells (HORWITZ and HORWITZ, 1973). Camptothecin was tested for its effects on mammalian cell cycle traverse in synchronized cultures of Chinese hamster cells (TOBEY, 1972). The drug allowed initiation of DNA synthesis, but it prevented cells from progressing to mitosis.

Camptothecin has no effect on nucleic acid synthesis in isolated rat liver mitochondria, rat brain mitochondria or E. coli (BOSMANN, 1970).

Effects of Camptothecin on Mammalian Viruses

Camptothecin is an effective inhibitor of the replication of adenovirus type 2 (HORWITZ and BRAYTON, 1972) and vaccinia virus (HORWITZ et al., 1972). The drug prevents the morphogenesis of adenovirus by inhibiting viral DNA synthesis. In the presence of camptothecin, intracellular viral DNA is cleaved to smaller acid-insoluble pieces which are rapidly repaired after removal of the drug.

The synthesis of vaccinia DNA and messenger RNA in HeLa cells was inhibited by camptothecin. The activity of RNA polymerase, as tested in isolated vaccinia cores, was not affected by the drug. This is similar to the observations made in uninfected HeLa cells where nucleic acid synthesis is inhibited by camptothecin but isolated DNA and RNA polymerases are unaffected (HORWITZ et al., 1971).

Camptothecin inhibits the replication of two DNA viruses, adenovirus and vaccinia virus, that replicate in the nucleus and cytoplasm of HeLa cells, respectively. The drug has no effect on the replication of poliovirus, an RNA virus that replicates in HeLa cells (HORWITZ et al., 1972). Therefore, it would seem that camptothecin is active only when DNA serves as a template for nucleic acid synthesis.

Structure-Activity Relationships

Derivatives and analogues of camptothecin (Fig. 1) were tested to determine their inhibitory effects on RNA synthesis and their ability to convert DNA to a lower molecular weight species in HeLa cells (Table 1). Compounds I, II, III and IV were prepared by WALL (1969); V, VII, IX and X by DANISHEFSKY et al. (1973a, b); VI by BOCH et al. (1972); and VIII by KENDE et al. (1970). Most of these compounds, including camptothecin, inhibit RNA synthesis in HeLa cells by 50% or more at concentrations between 1 and 10 μM. The lactol analogue (IV) was less active but still inhibited the rate of RNA synthesis by 50% at a concentration of 30 μM. The least active member of this group (X) contains only the D and E rings and failed to inhibit significantly RNA synthesis at a concentration of 70μM.

Conversion of cellular DNA to a lower molecular weight species seems to be more sensitive to structural alterations in the E ring of camptothecin than inhibition of RNA synthesis. Of the compounds listed in Table 1, only camptothecin (I), desoxycamptothecin (II) and 10-methoxycamptothecin (III) significantly altered the molecular weight of HeLa cell DNA. The structural requirements

Table 1. *Effects of camptothecin analogues on RNA synthesis in HeLa cells and on sedimentation of HeLa cell DNA. The rate of RNA synthesis was determined by measuring the uptake of labeled uridine into cold trichloracetic acid-insoluble material. To determine the effect of the drug on the sedimentation of DNA, camptothecin analogues, at a final concentration of 20 μM, were incubated for 10 min. at 37° with HeLa cells containing tritium-labeled DNA. The size of cellular DNA was analyzed by alkaline sucrose density gradient centrifugation*

Inhibitor added	50% Inhibition of RNA synthesis (μM)	"High molecular weight" DNA (%)
None	—	93
Camptothecin (I)	1	14
Desoxycamptothecin (II)	2	20
10-Methoxycamptothecin (III)	5	5
Camptothecin Lactol (IV)	30	77
Homocamptothecin (V)	1	62
Diethylcamptothecin (VI)	6	88
Isocamptothecin (VII)	8	84
Furan Derivative (VIII)	4	81
Camptothecin Monoester (IX)	5	89
Camptothecin Analog-1 (X)	[a]	93

[a] 40% inhibition at 70 μM.

at position 20 are stringent. Replacement of the hydroxyl group at that position by a hydroxyl methyl group (homocamptothecin, V) or an ethyl group (VI) decreases biological activity. Isocamptothecin (VII) is less active than camptothecin, although the reactivities of the E ring are similar.

Conversion of the ketone group at position 21 to a hydroxyl group (IV) results in loss of activity, as does the introduction of a new E ring (VIII) or the conversion of the E ring to a monoester (IX). Compound X lacks the A, B, and C rings of camptothecin. This analogue is a poor inhibitor of RNA synthesis and has no effect on cellular DNA. It appears from these structure-activity studies that the fragmentation of DNA is particularly sensitive to alterations in the E ring; however, the D and E rings alone are not sufficient for biological activity.

Of the compounds that have been tested as inhibitors of tumor growth in experimental animals (I, II, III, IV, VIII, X), a correlation exists between suppression of tumor growth and the ability to fragment DNA. Except for compounds IV and X, all of the analogues listed in Table 1 inhibit RNA synthesis in HeLa cells and, barring metabolism to inactive species, may be expected to be cytotoxic in animals. However, the ability to inhibit RNA synthesis does not enable one to predict antitumor activity in experimental animals. Camptothecin and III are active antitumor agents (HARTWELL and ABBOTT, 1969), whereas IV (HARTWELL and ABBOTT, 1969), VIII (KENDE, personal communication), and X (DANISHEFSKY et al., 1973a) are inactive.

A discrepancy exists in the literature concerning the antitumor properties of desoxycamptothecin. The desoxy analogue is stated to be totally inactive (WALL, 1969), although another report (HARTWELL and ABBOTT, 1969) has referred to it as an active agent. We find that desoxycamptothecin is active in HeLa cells and might be expected to act as an antitumor agent; however, intracellular metabolism to camptothecin has not been excluded in our experiments.

Mechanism of Action

The rapid inhibition of nucleic acid synthesis and the fragmentation of cellular DNA may underlie the cytotoxic effects of camptothecin. Fragmentation of DNA may create small pieces of template, inadequate for synthesis of DNA. Use of this altered template for RNA synthesis could be responsible for the abnormal RNA synthesized in the nucleus.

The mechanism by which the postulated degradation of DNA occurs is not clear. The drug may directly induce cleavage of DNA; alternatively, it may bind to DNA, thus rendering it susceptible to the action of endonucleases. Camptothecin produces a combination of structural changes in ME-180 tissue culture cells that are unlike those caused by any other drug thus far tested (RECHER et al., 1972). Although the precise mechanism of action of camptothecin is not totally understood, its properties differ from other established inhibitors of nucleic acid synthesis.

Summary

The plant alkaloid, camptothecin, demonstrates potent antitumor properties and inhibits the synthesis of DNA and certain species of RNA in mammalian cells. Only transfer RNA and short nucleoplasmic RNA are synthesized in the presence of camptothecin. Inhibition of RNA synthesis is completely reversed by removing the drug from the medium, while DNA synthesis is only partially reversed under the same conditions. DNA, isolated from cells inhibited by camptothecin and analyzed on alkaline sucrose gradients, has a lower sedimentation constant than DNA isolated from untreated cells. Upon removal of the drug from the medium, cellular DNA sediments normally under alkaline conditions. Camptothecin inhibits replication of adenovirus and vaccinia virus. The properties of camptothecin make it a promising tool for the study of macromolecular synthesis in animal cells and their DNA viruses.

Acknowledgement. I am grateful to Dr. ARTHUR P. GROLLMAN for many helpful discussions during the preparation of this manuscript.

References

ABELSON, H.T., and S.PENMAN: Selective interruption of high molecular weight RNA synthesis in HeLa cells by camptothecin. Nature New Biol. **237**, 144 (1972).

BIESLER, J.A.: Potential antitumor agents. I. Analogs of camptothecin. J. med. Chem. **14**, 1116 (1971).

BOCH, M., T.KORTH, J.M.NELKE, D.PIKE, H.RADUNZ, u. E.WINTERFELDT: Die biogenetisch orientierte Totalsynthese von DL-Camptothecin und 7-Chlor-camptothecin. Chem. Ber. **105**, 2126 (1972).

BOSMANN, H.B.: Camptothecin inhibits macromolecular synthesis in mammalian cells but not in isolated mitochondria or *E. coli*. Biochem. Biophys. Res. Commun. **41**, 1412 (1970).

DANISHEFSKY, S., S.J. ETHEREDGE, R.VOLKMANN, J.EGGLER, and J.QUICK: Nucleophilic additions to allenes. A new synthesis of α-pyridones. J.Am.Chem.Soc. **93**, 5575 (1971).

DANISHEFSKY, S., J.QUICK, and S.B.HORWITZ: Synthesis and biological activity in the camptothecin series. Tetrahedron Lett. 2525 (1973a).

DANISHEFSKY, S., R.VOLKMANN, and S.B.HORWITZ: Isocamptothecin. Tetrahedron Lett. 2521 (1973b).

DE WYS, W.D., S.R.HUMPHREYS, and A.GOLDIN: Studies on therapeutic effectiveness of drugs with tumor weight and survival time indices of Walker 256 carcinoma. Cancer Chemotherapy Rept. **52**, 229 (1968).

GALLO, R.C., J.WHANG-PENG, and R.H.ADAMSON: Studies on antitumor activity, mechanism of action, and cell cycle effects of camptothecin. J.Natl.CancerInst. **46**, 789 (1971).

GOTTLIEB, J.A., A.M.GUARINO, J.B.CALL, V.T.OLIVERIO, and J.B.BLOCK: Preliminary pharmacologic and clinical evaluation of camptothecin sodium (NSC-100880). Cancer Chemotherapy Rept. **54**, 461 (1970).

GOTTLIEB, J.A., and J.K.LUCE: Treatment of malignant melanoma with camptothecin (NSC 100880). Cancer Chemotherapy Rept. **56**, 103 (1972).

HART, L.G., J.B.CALL, and V.T.OLIVERIO: A fluorometric method for determination of camptothecin in plasma and urine. Cancer Chemotherapy Rept. **53**, 211 (1969).

HARTWELL, J.L., and B.J.ABBOTT: Antineoplastic principles in plants: recent developments in the field. In: Advan. Pharmacol. **7**, 137 (1969).

HORWITZ, M.S., and C.BRAYTON: Camptothecin: Mechanism of inhibition of adenovirus formation. Virology **48**, 690 (1972).

HORWITZ, M.S., and S.B.HORWITZ: Intracellular degradation of HeLa and adenovirus type 2 DNA induced by camptothecin. Biochem. Biophys. Res. Commun. **45**, 723 (1971).

HORWITZ, S.B., C.CHANG, and A.P.GROLLMAN: Mechanism of action of camptothecin. Pharmacologist **12**, 1283 (1970).

HORWITZ, S. B., C. CHANG, and A. P. GROLLMAN: Studies on camptothecin: I. effects on nucleic acid and protein synthesis. Mol. Pharmacol. **7**, 632 (1971).

HORWITZ, S. B., C. CHANG, and A. P. GROLLMAN: Studies on camptothecin: II. antiviral action. Antimicr. Ag. Chemother. **2**, 395 (1972).

HORWITZ, S. B., and M. S. HORWITZ: Camptothecin: breakage and repair of DNA during the cell cycle. Cancer Res. **33**, 2834 (1973).

KAMETANI, T., H. NEMOTO, H. TAKEDA, and S. TAKANO. Studies on the synthesis of heterocyclic compounds — CCCXXI. Synthetic approach to camptothecin. Tetrahedron **26**, 5753 (1970).

KANN, JR., H. E., and K. W. KOHN: Effects of deoxyribonucleic acid — reactive drugs on ribonucleic acid synthesis in leukemia L1210 cells. Mol. Pharmacol. **8**, 551 (1972).

KENDE, A. S., I. J. BENTLEY, R. W. DRAPER, J. K. JENKINS, M. JOYEUX, and I. KUBO: Tetrahedron Lett. 1307 (1973).

KEPLER, J. A., MC. WANI, J. N. MCNAULL, M. E. WALL, and S. G. LEVINE: Plant antitumor agents. IV. An approach toward the synthesis of camptothecin. J. Org. Chem. **34**, 3853 (1969).

KESSEL, D.: Some determinants of camptothecin responsiveness in leukemia L1210 cells. Cancer Res. **31**, 1883 (1971a).

KESSEL, D.: Effects of camptothecin on RNA synthesis in leukemia L1210 cells. Biochim. Biophys. Acta **246**, 225 (1971b).

KESSEL, D., H. B. BOSMANN, and K. LOHR: Camptothecin effects on DNA synthesis in murine leukemia cells. Biochim. Biophys. Acta **269**, 210 (1972).

LI, L. H., T. J. FRASER, E. J. OLIN, and B. K. BHUYAN: Action of camptothecin on mammalian cells in culture. Cancer Res. **32**, 2643 (1972).

LIAO, T. K., W. H. NYBERG, and C. C. CHENG: Total synthesis of camptothecin. I. Synthesis of ethyl 8-(α-chlorobutyryloxymethyl)-7,9-dioxo-7,8,9,11-tetrahydroindolizino [1,2-a] quinoline-8-carboxylate and related tetracyclic compounds (1). J. Heterocyclic Chem. **8**, 373 (1971).

MCPHAIL, A. T., and G. A. SIM: The structure of camptothecin: x-ray analysis of camptothecin iodoacetate. J. Chem. Soc. **1968**, 923.

MOERTEL, C. G., A. J. SCHUTT, R. J. REITMEIER, and R. G. HAHN: Phase II study of camptothecin (NSC-100880) in the treatment of advanced gastrointestinal cancer. Cancer Chemotherapy Rept. **56**, 95 (1972).

MUGGIA, F. M., P. J. CREAVEN, H. H. HANSEN, M. H. COHEN, and O. S. SELAWRY: Phase I clinical trial of weekly and daily treatment with camptothecin (NSC-100880): correlation with preclinical studies. Cancer Chemotherapy Rept. **56**, 515 (1972).

PERLMAN, S., H. T. ABELSON, and S. PENMAN: Mitochondrial protein synthesis: RNA with the properties of eukaryotic messenger RNA. Proc. Natl. Acad. Sci. U.S. **70**, 350 (1973).

RECHER, L., H. CHAN, L. BRIGGS, and N. PARRY: Ultrastructural changes inducible with the plant alkaloid camptothecin. Cancer Res. **32**, 2495 (1972).

SHAMMA, M., and L. NOVAK: Synthetic approaches to camptothecin. Tetrahedron **25**, 2275 (1969).

SHAMMA, M., and L. NOVAK: The preparation of some tricyclic analogs of camptothecin. Collection Czech. Chem. Commun. **35**, 3280 (1970).

SPATARO, A., and D. KESSEL: Studies on camptothecin-induced degradation and apparent reaggregation of DNA from L1210 cells. Biochem. Biophys. Res. Commun. **48**, 643 (1972).

STORK, G., and A. G. SCHULTZ: The total synthesis of dl-camptothecin. J. Am. Chem. Soc. **93**, 4074 (1971).

TOBEY, R. A.,: Effects of cytosine arabinoside, daunomycin, mithramycin, azacytidine, adriamycin, and camptothecin on mammalian cell cycle traverse. Cancer Res. **32**, 3720 (1972).

VENDITTI, J. M., and B. J. ABBOTT: Studies on oncolytic agents from natural sources. Correlations of activity against animal tumors and clinical effectiveness. Lloydia **30**, 332 (1967).

VOLKMANN, R., S. DANISHEFSKY, J. EGGLER, and D. M. SOLOMON: A total synthesis of dl-camptothecin. J. Am. Chem. Soc. **93**, 5576 (1971).

WALL, M. E.: Alkaloids with anti-tumor activity. In: International symposium on the Biochemistry and Physiology of Alkaloids, Halle. Berlin: Academic Press 1969.

WALL, M. E., M. C. WANI, C. E. COOK, K. H. PALMER, A. T. MCPHAIL, and G. A. SIM: Plant antitumor agents. I. The isolation and structure of camptothecin, a novel alkaloidal leukemia and tumor inhibitor from *Camptotheca acuminata*. J. Am. Chem. Soc. **88**, 3888 (1966).

WANI, M. C., H. F. CAMPBELL, G. A. BRINE, J. A. KEPLER, M. E. WALL, and S. G. LEVINE: Plant antitumor agents. IX. The total synthesis of dl-camptothecin. J. Am. Chem. Soc. **94**, 3631 (1972).

WANI, M.C., J.A. KEPLER, J.B. THOMPSON, M.E. WALL, and S.G. LEVINE: Plant antitumor agents: alkaloids: Synthesis of a pentacyclic camptothecin precursor. Chem. Commun. 404 (1970).

WARNEKE, J., u. E. WINTERFELDT: die autoxydative Indol-chinolon-Umwandlung eines Camptothecin-Modells. Chem. Ber. **105**, 2120 (1972).

WENKERT, E., K.G. DAVE, R.G. LEWIS, and P.W. SPRAGUE: General methods of synthesis of indole alkaloids. VI. Synthesis of dl-corynantheidine and a camptothecin model. J. Am. Chem. Soc. **89**, 6741 (1967).

WINTERFELDT, E., and H. RADUNZ: A convenient route to the camptothecin chromophore. Chem. Commun. 374 (1971).

WU, R.S., A. KUMAR, and J.R. WARNER: Ribosome formation is blocked by camptothecin, a reversible inhibitor of RNA synthesis. Proc. Natl. Acad. Sci. U.S. **68**, 3009 (1971).

Chloroquine (Resochin)

Fred E. Hahn

I. Introduction

Chloroquine (Resochin) was synthesized by ANDERSAG in 1934 (rev. by COAT-NEY, 1963) and became the prototype of the 7-halogenated 4-aminoquinoline antimalarials which, as a group, were patented by ANDERSAG et al. (1939). The chemical structure of chloroquine is shown in Fig. 1. It is a substituent of quinoline gleaned from the molecule of quinine, with a dicationic aliphatic chain which had been found earlier (SCHULEMANN, 1932) grossly to enhance the antimalarial potency of methylene blue; this potency had been predicted by PAUL EHRLICH and subsequently demonstrated clinically (GUTTMANN and EHRLICH, 1891). Chloroquine is structurally related to quinacrine (Atebrin) which had been previously developed as a synthetic antimalarial (MIETZSCH and MAUSS, 1930) and from which it differs only by the deletion of a methoxy-phenyl group. Chloroquine might also be considered a position isomer of pamaquine (Plasmochin), an 8-aminoquinoline (SCHULEMANN, 1932), although the electronegative substituent, —OCH$_3$, in pamaquine occupies position 6 (rather than 7) of the quinoline ring system.

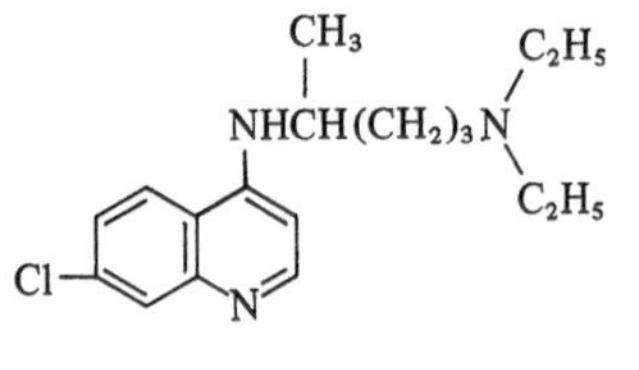

Fig. 1. Structure of chloroquine

Chloroquine was rediscovered in the United States during the course of an antimalarial drug development program in World War II and emerged from a set of 80 active compounds (out of some 16,000 tested) as the most important candidate which received its clinical trial in man in 1944 and was introduced into practical medicine in 1947. The convoluted history of the discoveries of chloroquine (1934–1944) has been interestingly traced by COATNEY (1963). To this, only one detail must be added: the chloroquine patent (ANDERSAG et al., 1939) was received at the U.S. Patent Office on December 14, 1939, and was, hence, available for study at the inception of the U.S. malaria drug program of World War II.

For more than one decade, chloroquine remained the most important drug for the radical cure of malaria until, first in Latin America (1961) and subsequently in Southeast Asia, cases of chloroquine-resistant falciparum malaria made their appearance (rev. by POWELL and TIGERTT, 1968). At the time of this writing, chloroquine-resistant falciparum malaria occurs in tropical Latin America and Southeast Asia but not in Africa.

The emergence of chloroquine-resistant falciparum malaria stimulated a considerable effort in research on the mode and mechanism of action of the drug as well as on the nature of the resistance phenomenon. While the antimicrobial action of chloroquine is one thoroughly investigated topic in molecular pharmacology, the subject has not remained entirely without controversy which has been nourished through the study of the drug's effects by inappropriate methods. Impediments to the elucidation of chloroquine's mode of antiplasmodial action have been the lack of systematic information on the biochemistry of plasmodia and of the molecular determinants of the complicated life cycle of these organisms as well as considerable technical difficulties in routinely producing sufficient quantities of free and metabolically functional plasmodia for biochemical study.

Although the principal medical use of chloroquine is in the chemotherapy of malaria for which the drug was originally developed, and of hepatic amebiasis (LANE, 1951; SODEMAN et al., 1951), it is also used as an anti-inflammatory pharmacological substance in the treatment of rheumatoid arthritis and lupus erythematosus as well as in the treatment of photoallergic reactions such as porphyria cutanea tarda (VOGLER et al., 1970). Beyond mentioning that 1. the anti-inflammatory action of chloroquine has been ascribed to effects of the drug on lysosomes (FILKINS, 1969), 2. the action on lupus erythematosus to the ability of chloroquine to complex with DNA (STOLLAR and LEVINE, 1963) [i.e. to act as a haptene inhibitor in the lupus anti-DNA antibody vs. DNA reaction, and 3. the anti-porphyria action to a drug-mediated release of large quantities of uroporphyrine following transient hapatocellular injury [caused by chloroquine], the consideration of pharmacological effects of chloroquine falls outside the scope of this article. For this the reader is referred to a review of SAMS (1967).

The toxicity of chloroquine was originally overrated after one human trial in four paretics by F. SIOLI (rev. by COATNEY, 1963). In fact, when the drug is given orally to man for the treatment of acute attacks of vivax or falciparum malaria in a staggered regimen administering not more than a total of 3 g over a three day period, chloroquine usually causes only transient headache, gastrointestinal disturbances and pruritus. Prolonged administration of 250 to 750 mg daily for many month or even years in the treatment of other diseases, can result (one case in 1000 to 2000 patients) in retinopathy, characterized by loss of central vision, granular pigmentation and retinal arterial constriction (GILES and HENDERSON, 1965). Chloroquine has been implicated (HART and NAUNTON, 1964) in the development of fetal abnormalities, and D.W. WILLIAMS, in the laboratory of this author (unpublished), has shown a significant reduction in male fertility of rats by the drug.

II. Chloroquine as a Growth Inhibitor

Chloroquine acts on the asexual blood forms of malarial parasites and produces radical cure of malaria infections. From an early postinfection tissue phase,
a form of plasmodia, called trophozoite, enters the blood stream and invades
erythrocytes. After some growth, these organisms (now called schizonts) produce
a multitude of new nuclei and eventually split up into a corresponding multitude
of new asexual forms which, upon rupture of the red cells, enter other erythrocytes
and repeat the process of schizogeny. Finally, certain parasites (merozoites)
become sexually differentiated as male and female gametocytes. Chloroquine
inhibits intraerythrocytic schizogeny of all types of human malaria and also
kills the gametocytes of *Plasmodium vivax* and *Plasmodium malariae* but *not*
those of *Plasmodium falciparum* (COVELL et al., 1955).

At 1.5×10^{-3} M or 6×10^{-5} M, chloroquine inhibits *in vitro* the growth of
Entamoeba histolytica in egg or liver media (H. W. BROWN through CONAN, 1949).
The drug has little effect on the growth of several bacteria (KRADOLFER and
NEIPP, 1958; ROBINSON et al., 1959; HOUBA and ADAM, 1964) but was found
to be bactericidal for *Bacillus megaterium* at 10^{-3} M (CIAK and HAHN, 1966).
Hydroxy-chloroquine exhibits marked effects on the growth of *Nocardia asteroides*
and *Streptomyces pelletieri,* lesser effects on *Achorion schoenleini, Coccidioides
immitis* and *Paracoccidioides brasiliensis* but was non-active against 16 other
fungi (ANSEL and THIBAUT, 1964). The growth of various chick embryo cells
in culture was inhibited by 1.3×10^{-5} M chloroquine (LIPP, 1964), and GABOUREL
(1963) found similar inhibitions of four mammalian cell lines *in vitro*, by hydroxy-
chloroquine.

Among genetic effects of chloroquine are the inhibition of the metamorphosis
of tadpoles in the presence of 2×10^{-5} M in the water (HOPKINSON and JACKSON,
1964), low frequency elimination of the kanamycin-resistance determinant from
an R-factor harbored by *Escherichia coli* (HAHN and CIAK, 1971) and similar
elimination of determinants of resistance to kanamycin, chloramphenicol and
ampicillin from the same R-factor after its transfer into *Salmonella typhimurium*
(HAHN and CIAK, unpublished results). Furthermore, chloroquine inhibits type
transformation in *Bacillus subtilis* (STOLLAR and LEVINE, 1963) by complexing
with free transforming DNA.

III. Morphological Observations on Chloroquine-Exposed Plasmodia

In the hope to gain insight into the mode of action of chloroquine, the
morphology of drug-exposed malarial parasites has been studied under the light
microscope as well as in electron micrographs, and one group of workers has
given this field of endeavor the somewhat ambitious designation as "morphologic
pharmacology in malaria." Before discussing various observations and their interpretations by the observers, it must be pointed out that the resolutions of the
light microscope and of electron microscopy (~ 50 Å for sections of an average
1000 Å thickness) are not commensurate to the molecular dimensions of chloroquine (MW 320 daltons), and that observations of alterations in plasmodial
anatomy are more likely to reveal sequelae on a more complicated level of biologi-

cal organization than primary biochemical drug effects at the molecular level. In the field of mode-of-action studies, there exists only one instance, viz., disorganization of bacterial cell walls following the action of penicillin (DUGUID, 1946), in which morphological observations have correctly forecast the eventual explanation of the action of a chemotherapeutic drug at the molecular level (PARK and STROMINGER, 1957). If one considers the conjunction of inadequate resolution and of the sampling problem in electron microscopy, doubts must be expressed that the conceptual and technical approaches of cellular pathology are appropriate to solving problems of modes of action of chemotherapeutic drugs such as chloroquine.

Since early light-microscopic observations (THURSTON, 1953), observers have agreed that the first and most conspicuous change in plasmodial morphology following exposure to chloroquine [and to quinacrine (BOCK, 1939)], is an aggregation of hemozoine, i.e. of the "malaria pigment" which erythrocytic forms of plasmodia accumulate as the undigestable end-product of the proteolysis of hemoglobin. This hemozoine aggregation can be observed as early as 30 minutes after institution of chloroquine therapy in mice infected with P. berghei and is complete after 4 hours (MACOMBER and SPRINZ, 1967). Electron micrographs show that the hemozoine aggregates appear as "pigment bars" (LADDA, 1966) which are "somewhat crystalloid rather than amorphous as in the untreated parasites" [P. gallinaceum] (AIKAWA and BEAUDOIN, 1969). Several workers have speculated that hemozoine aggregation in chloroquine-exposed plasmodia was related to the mode of action of the drug or to plasmodial resistance to chloroquine. After discovering that chloroquine forms an adduct with hematin, thought to be a constituent of "malaria pigment", SCHUELER and CANTRELL (1964) as well as COHEN et al. (1964) speculated that the accumulation of excess amounts of hemozoine might enable plasmodia to resist the action of chloroquine through sequestration of the drug. Conversely, MACOMBER and SPRINZ (1967) have conjectured that the formation of hemozoine by the parasites may "serve to concentrate the drug within the parasite" and would account for the selective toxicity of chloroquine for the blood forms of plasmodia. PETERS (1970) finally postulated that "the reactivity of hemoglobin breakdown products with chloroquine must play a part in the mode of action of this compound" and that this reactivity in conjunction with effects of chloroquine on plasmodial lysosomes should provide "the answer to the question of how chloroquine really works." Any role of hemozoine in the action of chloroquine is, however, rendered problematic by recent findings (LADDA and SPRINZ, 1969) that in the NYU-2 strain of P. berghei the amount of pigment formed does not correlate with chloroquine sensitivity nor with the accumulation of ^{14}C-chloroquine by these plasmodia.

Another morphological observation on chloroquine-exposed plasmodia which appears to be reproducible is an enlargement of plasmodial vesicles, sometimes called "food vacuoles", but probably representing equivalents of the lysosomes of mammalian cells (WARHURST and HOCKLEY, 1967). LADDA (1966) mentions markedly enlarged vesicles with membranous alterations (P. berghei); others (MACOMBER and SPRINZ, 1967) talk about "progressive change in these vesicles" reaching a peak 4 hours after drug administration, and AIKAWA and BEAUDOIN (1969) report enlarged and compartmentalized "food vacuoles" within one hour of exposure of *P. gallinaceum* to chloroquine and speculate that the drug is initially concentrated in these vesicles before "it leaks out" to reach its site of action.

HOMEWOOD *et al.* (1972) also draw on the putative analogy between mammalian lysosomes and plasmodial "food vacuoles" in offering a speculation on chloroquine's antimalarial action according to which the drug enters these vesicles, raises their internal pH, thereby inhibits the proteolysis of globin and, in this manner, prevents the growth of plasmodia; chloroquine-caused increases in intravesicular pH are also being held responsible for the observed aggregation of the "malaria pigment". This is an egregious example of a combination of morphological observation and physiological speculation employed to project a proposition of chloroquine's mode of action which hardly can be considered a scientific working hypothesis because it fails to account for existing knowledge of the biochemical effects, the molecular pharmacology and the structure-activity relationships of the drug.

Changes in the nuclear morphology of chloroquine-exposed plasmodia have also been observed (LADDA and ARNOLD, 1965). Under the influence of the drug, "the granularfilamentous pattern of the nucleoplasm becomes more compact and densely granulated. The ribonucleoprotein particles become aggregated into a compact 'nucleolar-like' structure (LADDA, 1966)." AIKAWA and BEAUDOIN (1969) observed in *P. gallinaceum* that, upon exposure to chloroquine, nucleus and nucleolus became more electron-dense and that the nuclear membrane became obscure. Eventually, "the nucleus became so dark that it was difficult to distinguish it from the darkened cytoplasm (AIKAWA and BEAUDOIN, 1969)." The three articles discuss morphological observations in the light of the well-known binding of chloroquine to DNA (discussed below).

Ribosomes first increase in electron density and then disintegrate, and plasmodial mitochondria swell under the influence of chloroquine and form myelin figures (AIKAWA and BEAUDOIN, 1969).

When presented with the reviewed inventory of morphological changes in chloroquine-exposed plasmodia, it is difficult, if not downright impossible, to assemble these observations as to causes and effects into a plausible proposal of the mode of action of the drug. Interpretations which have been made by authors of these various contributions, draw heavily on biochemical and physiological data such as the reactions of chloroquine with ferrihemic acid, with DNA, or with mammalian lysosomes. In conclusion, one must concede that a "morphologic pharmacology in malaria" does not exist and that the investigation of the mode and mechanism of action of the drug has only marginally benefited from the morphological approach, derived from cellular pathology.

IV. Effects of Chloroquine on Macromolecular Biosyntheses

In Vivo and *In Vitro*

This section reviews effects of chloroquine on DNA, RNA and protein biosyntheses in plasmodia, mammalian cells in culture and bacteria as well as in certain cell-free reaction systems. Such an integrative review is indicated not only because of the close similarity of the actions of the drug on cells of different taxonomic origin but also because the field has developed in an integrated manner.

SCHELLENBERG and COATNEY (1961) investigated the effect of chloroquine on the incorporation of ^{32}P-orthophosphate into DNA and RNA of chick blood infected with *P. gallinaceum* and into the nucleic acids of *P. berghei* in rat erythrocytes *in vitro*. They obtained log-dosage vs. per cent inhibition (LDR) curves which showed that the incorporations of phosphate into DNA and RNA were inhibited to similar extents [parenthetically, the log dosage-response correlations for the actions of quinine and of quinacrine were similar to those for chloroquine]. Failure to incorporate phosphate into nucleic acids could, of course, have signalled a variety of underlying metabolic inhibitions, while the typical LDR curves indicate that the observed inhibitions are due to reversible occupancy of sites of action according to the law of mass action.

The discovery that chloroquine inhibits the DNA-dependent DNA and RNA polymerase reactions in competition with template DNA *in vitro* (COHEN and YIELDING, 1965a) virtually eliminated from consideration all conceivable effects of the drug on intermediary metabolic reactions which feed nucleic acid biosyntheses and, in conjunction with biophysical data on the binding of chloroquine to DNA (reviewed below), suggested that the drug acts as a DNA template poison on the replication of DNA and the transcription of RNA (O'BRIEN et al., 1966a) by stabilizing the double helix to strand separation.

Indeed, when the two calf-thymus DNA-primed polymerase reactions, catalyzed by DNA polymerase I and RNA polymerase, were tested under standardized identical assay conditions (O'BRIEN et al., 1966b), a typical LDR curve was obtained for the inhibition of the DNA polymerase reaction by chloroquine with an ED_{50} of 2×10^{-4} M, while the RNA polymerase reaction was susceptible to inhibition by chloroquine at approximately 10 times higher concentrations. R. D. ESTENSEN found, in the laboratory of this author, the same inhibitory effects of chloroquine when DNA extracted from *P. berghei* was used as a primer and template (unpublished).

A mode of action study of hydroxychloroquine in L-fibroblasts showed that at 2×10^{-5} M the drug inhibited DNA biosynthesis entirely and RNA protein biosynthesis to the same but lesser extent with an insignificant increase in cell count (GABOUREL, 1963). When data from these biochemical analyses were supplemented by measurements of radioisotope incorporations into DNA, RNA and protein, the extents of inhibition by chloroquine were similar for all three processes and averaged 94 per cent over an experimental period of 50 hours.

CIAK and HAHN (1966) used their chloroquine-sensitive strain of *B. megaterium* for a mode of action study and found that at 10^{-3} M the drug was rapidly bactericidal, causing an exponential decline in the number of viable organisms by > 3 decadic logarithms per one doubling time of a drug-free control culture. By chemical assay, DNA and RNA biosynthesis were completely inhibited. Furthermore, the drug caused a rapid disintegration of the ribosomal RNA and, consequently, of the ribosomal particles themselves which accounts for the observation that protein synthesis failed in chloroquine-treated *B. megaterium*. Two points require comment: 1. The concentration of chloroquine (10^{-3} M) which produced effects in *B. megaterium* would have seemed excessive until it was found that sensitive plasmodium/erythrocyte systems accumulate the drug several hundred times from media of *in vitro* experiments (*P. knowlesi:* Polet and Barr, 1969; *P. berghei:* Fitch, 1969; *P. falciparum:* Fitch, 1970) or from the blood stream of mice infected with *P. berghei* (Macomber *et al.,* 1966). This brings the effective concentration of chloroquine in experiments with intraerythrocytic malarial parasites to the antibacterial range. 2. While the dissimilation of riboso-mal RNA and the disassembly of the ribosomal particles seemed at the time of its discovery (Ciak and Hahn, 1966) to be a specific effect in *B. megaterium*, Warhurst and Williamson (1970) have demonstrated a similar dissimilation of ribosomal RNA in *P. knowlesi* isolated from the blood of monkeys which had been treated with chloroquine for as short a period as *one hour*.

Using intraerythrocytic *P. knowlesi in vitro* during its 20 hour period of syn-chronous schizogeny, Polet and Barr (1968a) showed that chloroquine (at 10^{-7} M *in the medium*) inhibited DNA synthesis strongly, RNA biosynthesis almost to the same extent and protein synthesis appreciably less. The inhibition of DNA biosynthesis was most marked for the first 10 hours during which untreated trophozoites reach a phase just prior to nuclear segmentation. The authors discussed their observation of partially continuing protein biosynthesis in the light of the disassembly of bacterial ribosomes by chloroquine (Ciak and Hahn, 1966) without conclusion. The issue has become important in view of the results of Warhurst and Williamson (1970). In the latter work, monkeys, infected with *P. knowlesi*, received 18 mg/kg of chloroquine base for 1 hour which produced dissimilation of ribosomal RNA in the parasites. Perhaps the very low concentration of the drug in the experiments of Polet and Barr (10^{-7} M) did not suffice to accumulate enough chloroquine in the erythrocyte-plasmodial system to affect the plasmodial ribosomes.

Chloroquine inhibits nucleic acid biosynthesis also in erythrocytic *P. berghei in vitro*; this was shown as effects of the drug on the incorporation of adenosine-8-^{3}H into the nucleic acids of the organism after the authors (Van Dyke *et al.,* 1969) had demonstrated that the label was incorporated into both RNA and DNA. Van Dyke *et al.* (1969) noted that LDR curves for the inhibitions of nucleic acid biosynthesis in *P. berghei* by chloroquine and other drugs which bind to DNA by intercalation were "strikingly similar" to the curves for chloro-quine, quinacrine and quinine from the isolated polymerase studies of O'Brien *et al.* (1966b). The successful extension of these studies to suspensions of *P. berghei* which had been freed from erythrocytes constituted a major methodologi-cal progress in biochemical studies on malarial parasites (Lantz and Van Dyke, 1971). The incorporations of adenosine monophosphate-^{3}H into DNA and RNA of such free *P. berghei* were inhibited by chloroquine to identical extents. The LDR curve for inhibition of DNA biosynthesis by the drug was very similar to that for the inhibition of the calf-thymus-*E. coli* DNA-polymerase I reaction (O'Brien *et al.,* 1966b) with an identical ED_{50} of 2×10^{-4} M

The apparent discrepancy between the concentrations of chloroquine inhibit-ing nucleic acid biosyntheses in *P. knowlesi* (Polet and Barr, 1968a) and in *P. berghei* (Van Dyke *et al.,* 1969) is possibly explained by differences in the developmental biology of the two plasmodia and in the design of the experiments by the two groups of workers. *P. knowlesi*, going through one complete and

synchronous schizogenic phase, showed strong inhibition of nucleic acid biosynthesis during exposure to chloroquine for the first 10 hours (POLET and BARR,
1968a), three of which were required to complete the concentration of the drug
in the red cell/*P. knowlesi* system (POLET and BARR, 1969). VAN DYKE and his
associates (1969) limited the exposure of *P. berghei* to chloroquine to 30 minutes;
higher concentrations of the drug may have been required to facilitate its entry
into the plasmodia and to inhibit less chloroquine-sensitive phases of the asyn-
chronously developing parasites.

Most recently, the influence of chloroquine on incorporations of tritiated
adenosine, uridine and methionine into *P. berghei* was observed radioautographically, employing an unusual experimental design (THEAKSTON *et al.*, 1972). Mice,
infected with *P. berghei*, were divided into two populations; one received 50 mg/kg
of chloroquine phosphate for one hour and the other remained untreated. Blood
was collected from both groups and incubated *in vitro* with the labelled compounds
for one hour. After that, the blood cells were collected, washed by centrifugation,
suspended in saline and injected intravenously into infected and, in some cases,
uninfected animals. After 1, 4 and 24 hours, blood samples from these animals
were processed for eventual radioautography in an electron microscope. No
chloroquine was added to blood samples during the one hour *in vitro* pulse
with radioactive compounds.

The principal result of this work (THEAKSTON *et al.*, 1972) is a large and
specific accumulation of radioactivity from ^{3}H-adenosine in the nuclei of *P.
berghei* which was inhibited to 95–97 per cent by chloroquine. The data of
MACOMBER *et al.* (1966) permit one to estimate that the amount of chloroquine
in packed parasitized mouse erythrocytes 1 hour after injecting 50 mg/kg i.p.
will have been of the order of 50 µg/ml of sedimented red cells, i.e. $\sim 5 \times 10^{-4}$ M
in intraerythrocytic water. Since for the *in vitro* isotope exposure, infected and
chloroquine-containing mouse blood was diluted 4:1, THEAKSTON *et al.* (1972)
maintained, in fact, a drug concentration during the 1 hour radioactive metabolite
pulse which was close to a chemotherapeutic concentration.

BÜNGENER and NIELSEN (1967) have shown by radioautography that ^{3}H-C-5-
uridine was taken up very weakly if at all by rat erythrocytes parasitized by
P. berghei; POLET and BARR (1968b) did not dicover any significant uptake
(radiochemically) of ^{14}C-uridine by intraerythrocytic *P. knowlesi,* and VAN DYKE
et al. (1970) found no incorporation of tritiated uridine into the nucleic acids
of *P. berghei*. The uptake of radioactivity from ^{3}H-uridine into chloroquine-
treated *P. berghei* and its preferred localization in the "limiting membrane"
and the endoplasmic reticulum (THEAKSTON *et al.*, 1970), hence, is an observation
which is difficult to evaluate and does not appear to involve the utilization
of uridine for RNA biosynthesis. There is a body of opinion (BÜNGENER and
NIELSEN, 1967; VAN DYKE *et al.*, 1970) that plasmodia synthesize their own supply
of pyrimidines. One enzyme in pyrimidine synthesis, dihydroorotic acid dehydrogenase, has been shown to be elaborated by *P. berghei* (KROOTH *et al.*, 1969),
and POLET and BARR (1968a and b) have shown that orotic acid, a pyrimidine precursor, is utilized for DNA and RNA biosynthesis by *P. knowlesi* and
that both macromolecular biosyntheses from orotate are inhibited by chloroquine.

In summary, in all cellular entities which have been found to be subject to growth inhibition and cytocidal action by chloroquine, the prominent biochemical effect of the drug is an inhibition of DNA biosynthesis with (sometimes) lesser actions on RNA and protein biosyntheses. Dissimilation of ribosomes, first discovered in bacteria (CIAK and HAHN, 1966), also occurs in plasmodia (WARHURST and WILLIAMSON, 1970). In view of the key importance of nuclear DNA biosynthesis during schizogeny in which from 8 to 16 new nuclei appear exponentially with time (POLET and BARR, 1968 b), the chemotherapeutic effect of chloroquine as a blood schizontocide is explained by the drug's blockade of DNA biosynthesis. The two other practical blood schizontocides, quinacrine and quinine, like chloroquine, inhibit the DNA-primed nucleic acid polymerase reactions (O'BRIEN et al., 1966 b), and quinacrine is bactericidal because it blocks DNA biosynthesis (CIAK and HAHN, 1967) preferentially.

V. Molecular Pharmacology of Chloroquine: The DNA-Chloroquine Complex

Chloroquine belongs to a heterogenous group of substances, antibiotics, alkaloids, synthetic drugs and dyes which form complexes with DNA and exert antimicrobial effects by interfering with DNA's template functions (reviewed by HAHN, 1971). Those who seek an explanation of chloroquine's selective toxicity for erythrocytic forms of malarial parasites in some structural or conformational peculiarities of plasmodial DNA per se might argue (HAWKING, 1969) that the study of complexes of the drug with DNAs from other biological sources "is probably not a good guide to its reactions with the analogous constituent of a parasite cell." However (CIAK and HAHN, 1966), "the molecular architecture of double stranded DNA is evidently universal, and susceptibility or resistance to chloroquine cannot be explained on the basis of structural or compositional differences between the DNAs of susceptible or resistant cells" but may be "based upon the capacity of susceptible cells to permit passage and accumulation of critical concentrations of the drug." From data of WHITFELD (1952, 1953), WALSH and SHERMAN (1968) have derived for the DNA of P. berghei a base composition of 40 per cent $G+C$ and 60 per cent $A+T$ which resembles the base composition of calf thymus DNA (CHARGAFF, 1955) that has been used in many studies with chloroquine; the nucleic acid polymerase reactions in vitro using DNA from P. berghei as a template are as susceptible to inhibition by chloroquine as the same reactions using calf thymus DNA (R.D. ESTENSEN, unpublished). Unless one postulates subtle differences in polymorphic domains between plasmodial and calf thymus DNAs, there exist no biochemical and biophysical reasons to insist that biophysical studies on DNA-chloroquine interactions require the use of plasmodial DNA in order to have validity for considerations of the molecular pharmacology of the drug.

The most direct and simple methods of demonstrating the existence of a DNA-ligand complex and of determining the ligand binding parameters are analytical ultracentrifugation and equilibrium dialysis. Chloroquine cosedimented with DNA (ALLISON et al., 1965) and, in equilibrium dialysis, bound to DNA with a stoichiometry of one drug molecule per approximately 5 bases and a calculated association constant of 4.5×10^3 M^{-1} (STOLLAR and LEVINE, 1963). By a similar method which avoids use of dialysis membranes, the DNA-chloroquine complex was ultracentrifuged into a preconstructed sucrose layer and the concentration of drug remaining in the supernatant fluid was determined spectrophotometrically. These experiments (BLODGETT and YIELDING, 1968) used an excess of DNA and yielded an apparent association constant of 1.4×10^4 M^{-1}. Chloroquine has been observed qualitatively to displace methyl green from its stable complex with DNA (KURNICK and RADCLIFFE, 1962). KREY and HAHN (1971) have developed methyl green displacement analysis into a quantitative method and found that the drug displaces maximally 66.3 per cent of methyl green from calf thymus DNA with a first order rate constant which places the drug between proflavine and acridine orange.

PARKER and Irvin (1952) demonstrated first that DNA alters the absorption spectrum of chloroquine, concluded that the drug interacted with DNA and suggested that such interaction "may be important in the antimalarial activity of chloroquine." As this work was carried out before the double helical nature of DNA was recognized, the authors did not offer suggestions concerning the structure of the DNA-chloroquine complex.

Spectrophotometric titration was used to determine the stoichiometry and apparent association constants of the drug's binding to four DNAs of different base compositions (COHEN and YIELDING, 1965b). The curved adsorption isotherms obtained were interpreted to reflect two different classes of binding sites and processes for which stoichiometries of one drug molecule per two or one base pair(s) and apparent association constants of 10^4 and $10^3 \times M^{-1}$ were estimated. Base composition had no significant influence on the magnitudes of these parameters and denatured calf thymus DNA bound, by the weaker process, 50 per cent more chloroquine than native DNA. Since the fluorescence of chloroquine at 390 nm (excited by light of 330 nm) is quenched by DNA, a fluorometric titration of 3.3×10^{-5} M chloroquine with DNA, up to a concentration $\sim 10^{-3}$ M component nucleotides showed that DNA decreased the fluorescence of the drug in direct proportion to the amounts added (STOLLAR and LEVINE, 1963).

Genuine single-stranded DNA, i.e. DNA which has been heat-denatured in the presence of 3.3×10^{-1} formaldehyde, has no influence on the absorption spectrum of chloroquine beyond slight nonspecific changes such as are also produced by an acidic polysaccharide, heparin (ALLISON et al., 1965), or by DNA itself at 98° C. Single-stranded DNA also fails to induce a Cotton effect in chloroquine's optical rotatory dispersion spectrum (HAHN and KREY, 1969).

Changes induced in the absorption spectrum of chloroquine by duplex polydeoxyribonucleotides are dependent upon the presence of guanine. Poly dIdC and poly dAdT do not alter the spectrum of the drug while the changes produced by poly dGdC resemble those produced by an equimolar concentration of native DNA (O'BRIEN et al., 1966). While these observations might suggest guanine-specific binding of chloroquine to DNA, the drug stabilizes poly dAT to thermal denaturation as effectively as it stabilizes DNA itself (ALLISON et al., 1965), producing in 5×10^{-3} M Tris-HCl ΔT_ms of $+15°$C. This shift of the thermal denaturation profile ("melting curve") of DNA to higher temperatures is of the same magnitude as that caused by spermine (TABOR, 1962) and can be attributed to the dicationic side chain of the drug molecule, interacting electrostatically with the phosphates of DNA, and bridging the minor groove of the double helix (LIQUORI *et al., 1967*).

The results of hydrodynamic experiments with the DNA-chloroquine complex have furnished evidence that the drug is intercalated between base pairs of double-helical DNA (O'BRIEN *et al.*, 1966a). The viscosity of chloroquine-complexed DNA is greater than that of DNA alone, and the drug decreases the sedimentation coefficient of DNA. A viscosity *increase* of DNA is produced upon intercalation because this increases the length and rigidity of the double helix, but the sedimentation coefficient *decreases* because the *mass per unit length* of DNA is decreased when a molecule is inserted whose molecular weight is less than the aggregate molecular weight of one base pair (LERMAN, 1961). Maximal intercalation into DNA occurs with a statistical frequency of one intercalant per two base pairs and with increase in the length of DNA by 50 per cent (CAIRNS, 1962). The stoichiometry of chloroquine's strong binding of one drug per four and a half bases (STOLLAR and LEVINE, 1963; COHEN and YIELDING, 1965b) is in good agreement with the statistical estimate of maximally one intercalant per 4.2 bases (CAIRNS, 1962). Measurements of the linear dichroism of the flow oriented DNA-chloroquine complex (O'BRIEN *et al.*, 1966) have shown that the signs and magnitudes of the dichroic effects of the DNA bases and of chloroquine in the complex are of equal signs and magnitudes, hence, demonstrating (LERMAN, 1963) that the quinoline ring system of chloroquine and the purine and pyrimidine rings are coplanar in the DNA-drug complex. Finally, WARING (1970) has applied a test for intercalation to the interaction of chloroquine with circular supercoiled DNA; the drug produced the typical hydrodynamic indications of conformational changes in this type of DNA which result from intercalation.

A physical consequence of chloroquine's intercalation into DNA is that energy, supplied to chloroquine-complexed DNA as radiation of 254 nm, is transferred to chloroquine. This was measured at room temperature as an inhibition of photochemical pyrimidine dimer formation extending over a critical distance of 8 base pairs, or at 77° K by sensitized phosphorescence of chloroquine. The triplet-triplet energy transfer on which phosphorescence is based, requires close physical proximity between the energy donor (bases) and the energy acceptor (chloroquine) which is provided by intercalation (SUTHERLAND and SUTHERLAND, 1969).

HAHN *et al.* (1966) have proposed a structure of the DNA-chloroquine complex (Fig. 2) in which the 7-chloroquinoline ring is intercalated between base pairs and the cationic aliphatic side chain protrudes beyond the contour of the double helix and bridges the minor groove by electrostatic attraction to phosphates of both strands. An alternate model has assumed stacking of chloroquine in the major groove of DNA, not involving intercalation or bridging (YIELDING *et al.*, 1971). That model is not supported by experimental evidence and can not be considered a working hypothesis since it fails to account for the various lines of hydrodynamic evidence of intercalation (O'BRIEN *et al.*, 1966a; WARING, 1970).

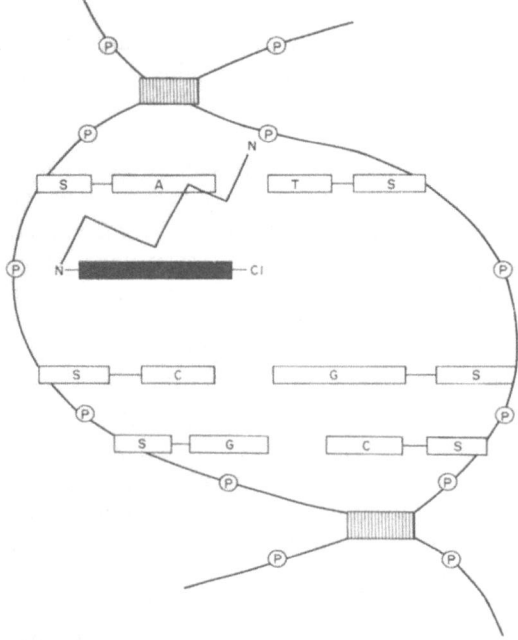

Fig. 2. Structure of the DNA-chloroquine complex (proposed by HAHN *et al.*, 1966, drawing by BASS *et al.*, 1971)

VI. Structure-Activity Relationships for Chloroquine

Structure-activity relationships in the chloroquine series reflect the relationship which exists between the drug and its bioreceptor, DNA. When "activity" is defined as the ability of chloroquine and its congeners to eliminate an intracellular parasite, secondary pharmacological events such as distribution, transport or metabolic dissimilation of drugs might occlude valid structure-activity rules by causing false negative results.

O'BRIEN and HAHN (1966) based an analysis of structure-antimalarial activity relationships in the chloroquine series of compounds on the following premises. 1. Chloroquine forms the complex with DNA whose stucture is shown in Fig. 2. 2. This interferes with DNA's template function in DNA replication, especially during schizogeny of plasmodia (reviewed in Section IV). 3. Structure-antimalarial activity rules in the chloroquine series should, therefore, reflect rules for compounds to form the intercalative DNA complex, Fig. 2.

ALBERT *et al.* (1949) had found that the antimicrobial potencies of N-heterocyclic bases depend upon 1. A flat aromatic ring system of a minimal planar area of 28 Å^2 (such as that of quinoline) and 2. The presence of basic substituents which, at physiological pH, are protonated to minimally 50 per cent. It is now generally recognized (HAHN, 1971) that these conditions are requirements for intercalation binding to DNA.

The most important structural features in the chloroquine series which determine both binding to DNA and antimalarial activity are an electronegative substituent on position 7 (or 6) of the quinoline ring and a non-primary diamino
side chain attached to position 4 of the ring in which the two positive charges
are separated by approximately 7.5 Å (O'BRIEN and HAHN, 1966).

These qualitative results have been subjected to extensive regression analysis,
using a free energy related structure-activity model, with the intent of testing
the hypothesis that binding of substances of the chloroquine series to DNA
(as in Fig. 2) is, in fact, the molecular mechanism underlying their antimalarial
activity (BASS et al., 1971). The results of this analysis are in complete agreement
with the structure-antimalarial activity rules of O'BRIEN and HAHN (1966) and
their rationalization by the structure of the DNA-chloroquine complex (Fig. 2);
they suggest as additional refinements 1. That the size of the alkyl substituents
of the terminal chain nitrogen modifies the ability of this N to engage in electrostatic binding to DNA phosphates and 2. That different sizes of different electronegative subtituents on position 7 of the quinoline ring will sterically impose ("chock")
different orientations of this ring relative to the double helix. This, in turn,
might modify the manner in which the side chain is able to bind to the sugar-
phosphate backbones of DNA.

The importance of the conclusions of O'BRIEN and HAHN (1966) and of
BASS et al. (1971) lies in the fact that they combine biophysical knowledge of
the binding of chloroquine to DNA, biochemical knowledge of the inhibition
of DNA biosynthesis in schizogeny, and chemotherapeutic test data on antimalarial activity from separate experimentations into one coherent logical structure
whose consistency furnishes intrinsic proof of the correctness of the underlying
hypothesis.

An interesting problem in structure-activity relationships which still awaits
solution concerns the difference in action between chloroquine and other 4-
aminoquinolines on one hand and 8-aminoquinolines on the other. Primaquine,
for example (see J. G. OLENICK, this volume), does not intercalate into DNA
(MORRIS et al. 1970), is a specific inhibitor of *protein* biosynthesis (OLENICK
and HAHN, 1972) with little influence on nucleic acid biosynthesis in plasmodia
(SCHELLENBERG and COATNEY, 1961), is not resisted by chloroquine-resistant *P.
berghei* and does not compete with chloroquine for association with *P. berghei*
(FITCH, 1972). From these data it is clear that the antimalarial 4-aminoquinolines
and 8-aminoquinolines despite their seeming structural similarities have unrelated
receptor sites and, hence, unrelated modes of action and, for this reason, address
themselves to different phases of the plasmodial life cycle: primaquine is not
schizontocidal but acts on gametocytes and on exoerythrocytic forms of plasmodia.

VII. Chloroquine Resistance and Association
of the Drug with Plasmodia

A distinction must be made between the spontaneous emergence of resistance
to chloroquine in *P. falciparum* and chloroquine resistance in *P. berghei* produced
experimentally by serially propagating the organism for many months in mice

which receive gradually increasing doses of the drug. Resistance of *P. falciparum* to chloroquine is a stable genetic property and persisted, for example, through 21 blood passages in gibbons (CADIGAN *et al.* as cited by POWELL and TIGERTT, 1968). Experimentally induced resistance of *P. berghei* to chloroquine is unstable; it is rapidly lost in serial blood passages in mice not receiving chloroquine, and is immediately lost when resistant *P. berghei* are transferred from mouse to hamster (TRAGER *et al.*, 1967).

Chloroquine resistance in *P. falciparum* is difficult to explain by the mutation-selection hypothesis and the spread of the resistance phenomenon resembles an epidemiological pattern which is reminiscent of the spread of bacterial R-factors which confer drug resistance upon their recipient cells. "Should a phenomenon akin to episomal drug resistance be operative in malaria, this would go a long way to explain the apparent rapidity with which multiple drug resistance appears to be spreading in *P. falciparum* in parts of South America and Southeast Asia (PETERS, 1970)." Systematic methods for the genetic analysis of plasmodia do not exist. However, a transfer has been shown for pyrimethamine resistance from *P. vinckei* to *P. berghei* during concurrent infection in mice; the isolation and characterization of dehydrofolate reductases from *P. vinckei*, from *P. berghei* and from the latter organism after resistance transfer had occurred, leaves no doubt that the three enzymes are different from each other and suggests that a genetic transfer may have altered the structural gene in *P. berghei* which is coded for the enzyme (FERONE *et al.*, 1970). The question of whether intraspecies transfer of chlorquine resistance (by whatever genetic mechanism) occurs in *P. falciparum* must be asked.

A phenotypic explanation of chloroquine resistance of *P. berghei* has been given by MACOMBER *et al.* 1966). These authors, in referring to the formation of DNA-chloroquine complexes and to the ensuing inhibition of DNA biosynthesis, hypothesized that "susceptibility or resistance to chloroquine cannot be explained on the basis of structural or compositional differences between the DNAs of susceptible or resistant cells but may depend upon differences in capabilities of such cells to permit passage or accumulation of critical concentrations of the drug." This was successfully tested in the erythrocytes of mice, parasitized with sensitive or resistant *P. berghei* and treated chemotherapeutically with ^{14}C-chloroquine. The arithmetic dosage-chloroquine uptake curves from this study are shown in Fig. 3. The insignificant uptake of chloroquine into normal erythrocytes and uptake of the drug into liver and spleen were proportional to the concentrations of ^{14}C-chloroquine injected into mice i.p., as one would expect from diffusion. The uptake curves for erythrocytes parasitized by chloroquine-sensitive (CSS) or resistant (CRS) *P. berghei* can be transformed into conventional LDR curves which suggests that they express binding of the drug to bioreceptors. At the maximal chemotherapeutic dose of 40 mg/kg chloroquine, the concentration attained in erythrocytes parasitized with resistant *P. berghei* was no larger than the concentration in red cells containing sensitive *P. berghei* upon administration of only 4 mg/kg. Under the experimental conditions employed to obtain the data in Fig. 3 and at the highest dosage of chloroquine (40 mg/kg), the concentration of the drug in mouse plasma was approximately 3×10^{-6} M. The ability of the erythrocyte/CSS *P. berghei* system to accumulate chloroquine

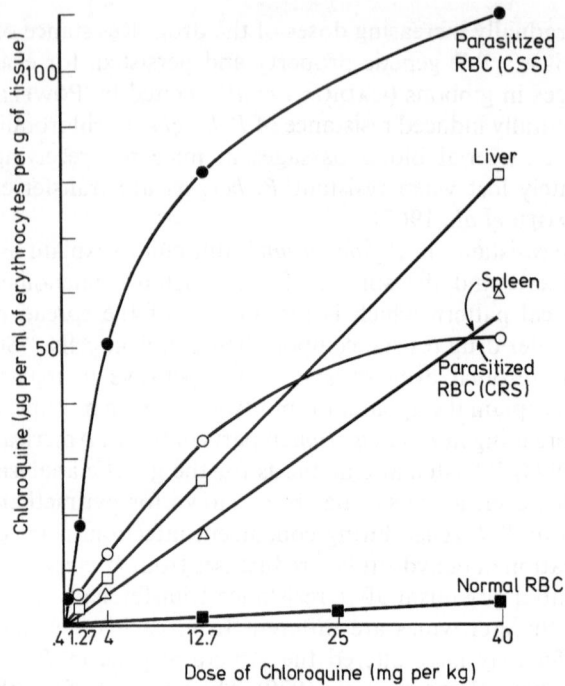

Fig. 3. Correlation between the dosage of chloroquine (i.p. in mice and the uptake of the drug); CSS chloroquine-sensitive and CRS chloroquine-resistant *Plasmodium berghei* (after MACOMBER *et al.*, 1966)

300 times (calculated for intraerythrocytic water) appears to be a prerequisite for the antiplasmodial action of the drug; the impaired ability of the erythrocyte/ CRS *P. berghei* system to accumulate the drug explains the resistance of these plasmodia to chloroquine. The plasma concentrations of chloroquine in man under malaria chemotherapy range from 3×10^{-7} M to 10^{-6} M (COVELL *et al.*, 1955) and it can be assumed that also in human malarias the drug must be concentrated by red cell-plasmodial system in order to attain active concentrations.

That intercalative chemicals become localized in the nuclei of intraerythrocytic plasmodia has been demonstrated for *P. gallinaceum* (DUTTA, 1969) by fluorescence staining *in vitro* with acridine orange. The green flourescence of the plasmodial (and avian erythrocytic) nuclei did not appear when the microscopic preparations were pre-treated with deoxyribonuclease for hydrolysis of DNA. Injection of acridine orange into chicks, infected with *P. gallinaceum*, resulted in preferential fluorescent staining of the plasmodial nucleoli.

Radioautographic evidence in favor of a speculation that chloroquine becomes associated with the "food vacuoles" of *P. berghei* has recently been sought by AIKAWA (1972) who injected mice parasitized by chloroquine-sensitive *P. berghei* i.p. with tritiated chloroquine, collected the erythrocytes and processed them for electronmicroscopic radioautography. Unfortunately, no parallel studies with chloroquine-resistant *P. berghei* were carried out. Convoluted strand-like structures were seen in the projection of, or tangentially to, the plasmodial vesicles which are reminiscent of similar appearances in the cytoplasm and the limiting membrane of chloroquine-treated *P. berghei* which had been labelled with tritiated uridine (THEAKSTON *et al.,* 1972). The relationship between these two sets of observations is difficult to evaluate. On the basis of AIKAWA's methodology and based on an amount of 10^{-13} g of DNA per trophozoite nucleus, it can be estimated that the specific radioactivity of ^{3}H-chloroquine would need to be two to three orders of magnitude higher than AIKAWA's preparation (75.7 mCi per 10^{-3} mole) in order to detect radioautographically a significant binding of chloroquine to DNA in nuclear sections of 500 Å thickness even if granulation from uranyl staining were experimentally excluded.

Other published experiments on plasmodial binding of chloroquine have been carried out with intraerythrocytic parasites *in vitro*. FITCH (1969) suspended mouse erythrocytes, either uninfected or parasitized with chloroquine-sensitive or -resistant *P. berghei* in different concentrations of ^{14}C-chloroquine, separated, after two hours, the red cells from the medium and determined the amounts of chloroquine in both fractions. The technique might be compared to a fairly rapid equilibrium dialysis. Red cells, parasitized by sensitive plasmodia, concentrated chloroquine from the medium (containing 10^{-8} M drug) 600 times, those parasitized with resistant plasmodia 100 times and normal mouse erythrocytes 14 times with estimated apparent association constants of the magnitudes 10^8, 10^5 and $10^3 \times M^{-1}$. Marked or impaired accumulations of chloroquine again accounted for the difference between chloroquine-senstivity and resistance in *P. berghei*; so did the apparent failure of the erythrocyte/CRS *P. berghei* system to adsorb the drug with the high apparent association constant of $10^8 \times M^{-1}$.

The same experimental method has been applied to the study of chloroquine binding to owl monkey erythrocyte/*P. falciparum* systems when these plasmodia were either sensitive or resistant to the drug (FITCH, 1970). The results are comparable to those for mouse erythrocytes/*P. berghei*. The significance of these observations lies in the fact that the genetically stable chloroquine resistance of *P. falciparum* appears to be phenotypically expressed in a similar manner as the unstable resistance of *P. berghei*.

One important question raised by the work of FITCH (1969, 1970) is, of course, that of the nature of plasmodial chloroquine receptor sites which bind the drug with apparent association constants of 10^7 or $10^8 \times M^{-1}$; FITCH (1969) speculated that this "receptor site is possibly the site of action" of the drug. Fractionation of *P. berghei* and a biochemical inventory of ^{14}C- chloroquine in the fractions have not yielded conclusive results but have implicated a sediment of membranes (some perhaps of erythrocytic origin) and DNA as containing binding sites for the drug (KRAMER and MATUSIK, 1971). Efforts at eliminating DNA binding sites for the drug by treatment of the sediments with deoxyribonuclease I were inconclusive because (as the authors suspected) the hydrolysis of DNA by such enzymes is inhibited by bound chloroquine (KURNICK and RADCLIFFE, 1962; HOLBROOK *et al.,* 1971).

Circumstantial evidence, implicating plasmodial DNA as containing the high affinity binding site for chloroquine, has been furnished subsequently by FITCH (1972) who tested the ability of 12 antimalarial drugs to compete with chloroquine for high-affinity association sites in intraerythrocytic chloroquine-sensitive *P. berghei*. Three 4-aminoquinolines and one γ-aminopyridine (whose built-on coplanar p-chlorophenyl residue makes it a chloroquinoline equivalent) as well as quinacrine displaced chloroquine competitively with apparent association constants which, like that of chloroquine itself, were of the order of $10^7 \times M^{-1}$; quinine and three related heterocyclic synthetic aminomethanols displaced the drug with association constants of the order of $10^6 \times M^{-1}$ (the exception being a phenanthrene aminomethanol which was bound more strongly than chloroquine itself). Pyrimethamine ($10^5 \times M^{-1}$), primaquine and 4,4'-diaminophenylsulfone (no displacements) were essentially non-active. FITCH stated that "competition of a drug with chloroquine for binding to the high-affinity receptor of *P. berghei* is evidence that both drugs are interacting reversibly with the same binding sites.... It is apparent that the binding site on the receptor will admit molecules possessing a 30 to 40 Å^2 planar ring system (ALBERT et al., 1949) with a protonated group either in the ring or nearby."

It is difficult to escape the conclusion that the results of FITCH (1972) and his explicit citation of the structural rules for intercalation (see Section V) as explaining competition of chloroquine binding, point to his answer to the question of the nature of the plasmodial chloroquine receptor sites as being on plasmodial DNA. The low apparent association constant ($4.5 \times 10^3 \, M^{-1}$) measured with calf thymus DNA by equilibrium dialysis (STOLLAR and LEVINE, 1963) against free chloroquine concentrations of 10^{-4} to $10^{-3} \, M$ and the similarly low constants (10^4 and $10^3 \times M^{-1}$) derived from spectrophotometric titrations of chloroquine ($4.8 \times 10^{-5} \, M$) with several DNAs (COHEN and YIELDING, 1965 b) do not argue against such a possibility on biophysical grounds. The equilibrium dialysis experiments will have saturated all potential chloroquine binding sites on DNA, and the extrapolation to the ordinate of adsorption isotherms from spectrophotometric titration for very low values of bound chloroquine molecules is beset with "enormous errors" (COHEN and YIELDING, 1965 b) which renders the calculation of apparent association constants for strong binding processes highly uncertain.

Assuming that DNA of chloroquine-sensitive plasmodia possesses high affinity binding sites for the drug while the DNAs of resistant *P. berghei* or *P. falciparum* do not is, however, difficult to rationalize in terms of biochemical genetics if one were to consider compositional or architectural features of such DNAs as determinants of such differences. A reasonable hypotheses would be to assume differences in occupancy of plasmodial DNAs by histones or regulatory proteins which leave the strong chloroquine binding sites free in sensitive organisms but occupy these sites in resistant plasmodia. One conceivable alternative would be the existence of a high-affinity chloroquine binding on a membrane-bound permease system which is different in sensitive and in chloroquine-resistant plasmodia.

VIII. Conclusions

Chloroquine acts as an antimalarial drug because of its effect as a blood schizontocide. Its inability to act also as a tissue schizontocide, such as chloroguanide or pyrimethamine, may be more directly related to the distribution of these drugs in the mammalian organism than to putative biochemical differences in plasmodia between nucleic acid biosyntheses in tissue schizogeny and intra-erythrocytic schizogeny; pyrimethamine, for example, does not accumulate in the plasma. The mode of action of chloroquine as a blood schizontocide is a preferential inhibition of plasmodial DNA biosynthesis. Differences between chloroquine sensitivity and chloroquine resistance of plasmodia are results of strong or impaired abilities of erythrocyte/plasmodial systems to accumulate biochemically active concentrations of the drug. Chloroquine forms molecular complexes with DNAs and inhibits DNA-dependent nucleic acid polymerase reactions *in vivo* in competition with added DNA. The structure of the DNA-chloroquine complex, consistent with biophysical data, the parallelism between those structural rules which govern intercalative DNA binding and those which govern antimalarial potency, and, finally, the quantitative test of this body of data by regression analysis, leave little room for doubt that intercalation binding of chloroquine to plasmodial DNA is the molecular mechanism underlying inhibition of schizontal DNA synthesis and, hence, the chemotherapeutic action of chloroquine. While chemical nature of high-affinity chloroquine binding sites in drug-sensitive plasmodia has not yet been directly determined, competition experiments for the occupancy of these binding sites have shown that chemical structures which satisfy prerequisites for intercalation binding to double-helical DNA do compete with chloroquine for plasmodial binding sites while antimalarial drugs which do not intercalate do not so compete.

References

AIKAWA, M.: High-resolution autoradiography of malarial parasites treated with ³H-chloroquine. Am. J. Path. **67**, 277 (1972).

AIKAWA, M., and R.L. BEAUDOIN: Effects of chloroquine on the morphology of the erythrocytic stages of *Plasmodium gallinaceum*. Am. J. Trop. Med. Hyg. **18**, 166 (1969).

ALBERT, A., S.D. RUBBO, and M.I. BURVILL: The influence of chemical constitution an antibacterial activity. IV. A survey of heterocyclic bases with special reference to benzquinolines, phenanthridines, benzacridines, quinolines and pyridines. Brit. J. Exptl. Path. **30**, 159 (1949).

ALLISON, J.L., R.L. O'BRIEN, and F.E. HAHN: DNA: Reaction with chloroquine. Science **149**, 1111 (1965).

ANDERSAG, H., S. BREITNER u. H. JUNG: Verfahren zur Darstellung von in 4-Stellung basisch substituierte Aminogruppen enthaltenden Chinolinverbindungen. Germ. Pat. 683–692 (1939).

ANSEL, M., et M. THIBAUT: Recherches sur l'action antifungique in vitro du sulfate d'hydroxy-chloroquine. Compt. Rend. Soc. Biol. **158**, 1050 (1964).

BASS, G.E., D.R. HUDSON, J.E. PARKER, and W.P. PURCELL: Mechanism of antimalarial activity of chloroquine analogs from quantitative structure-activity studies. Free energy related model. J. Med. Chem. **14**, 275 (1971).

BLODGETT, L.W., and K.L. YIELDING: Comparison of chloroquine binding to DNA and polyadenylic and polyguanylic acids. Biochim. Biophys. Acta **169**, 451 (1968).

BOCK, E.: Über die morphologische Veränderungen menschlicher Malariaparasiten durch Atebrineinwirkung. Arch. Schiffs- u. Tropenhyg. **43**, 209 (1939).

BÜNGENER, W., u. G. NIELSEN: Nukleinsäurestoffwechsel bei experimenteller Malaria. 1. Untersuchungen über den Einbau von Thymidin, Uridin und Adenosin in Malariaparasiten (*Plasmodium berghei* und *Plasmodium vinckei*). Z. Tropenmed. Parasitol. **18**, 456 (1967).

CAIRNS, J.: The application of radioautography to the study of DNA viruses. Cold Spring Harb. Symp. Quant. Biol. **27**, 311 (1962).

CHARGAFF, E.: Isolation and composition of the deoxypentose nucleic acids and of the corresponding nucleoproteins. The nuclei acids, I.E. CHARGAFF and J.N. DAVIDSON, eds. New York: Adademic Press 1955.

Ciak, J., and F.E. Hahn: Chloroquine: Mode of action. Science 151, 347 (1966).

Ciak, J., and F.E. Hahn: Quinacrine (Atebrin): Mode of action. Science 156, 655 (1967).

Coatney, G.R.: Pitfalls in a discovery: The chronicle of chloroquine. Am. J. Trop. Med. Hyg. 12, 121 (1963).

Cohen, S.N., K.O. Phifer, and K.L. Yielding: Complex formation between chloroquine and ferrihaemic acid in vitro, and its effect on the antimalarial action of chloroquine. Nature 202, 805 (1964).

Cohen, S.N., and K.L. Yielding: Inhibition of DNA and RNA polymerase reactions by chloroquine. Proc Natl. Acad. Sci. U.S. 54, 521 (1965a).

Cohen, S.N., and K.L. Yielding: Spectrophotometric studies of the interaction of chloroquine with deoxyribonucleic acid. J. Biol. Chem. 240, 3123 (1965b).

Conan, N.J.: The tratment of hepatic amebiasis with chloroquine. Am J. Med. 6, 309 (1949).

Covell, G., G.R. Coatney, J.W. Field, and J. Singh: Chemotherapy of malaria. World Health Organ. Monogr. No. 27, Geneva, 1955.

Duguid, J.P.: The sensitivity of bacteria to the action of penicillin. Edinburgh Med. J. 53, 401 (1946).

Dutta, G.P.: Acridine orange staining and flourescence of nucleic acids in plasmodia and associated host erythrocytes. Stain Technol. 44, 223 (1969).

Ferone, R., O'Shea, and M. Yoeli: Altered dihydrofolate reductase associated with drug-resistance transfer between rodent plasmodia. Science 167, 1263 (1970).

Filkins, J.P.: Comparison of in vivo and in vitro effects of chloroquine on hepatic lysosomes. Biochem. Pharmacol. 18, 2655 (1969).

Fitch, C.D.: Chloroquine resistance im malaria: A deficiency of chloroquine binding. Proc. Natl. Acad. Sci. U.S. 64, 1181 (1969).

Fitch, C.D.: Plasmodium falciparum in owl monkeys: Drug resistance and cholorquine binding capacity. Science 169, 289 (1970).

Fitch, C.D.: Chloroquine resistance in malaria: Drug binding and cross resistance patterns. Proc. Helminth. Soc. Wash. 39, 265 (1972).

Gabourel, J.D.: Effects of hydroxychloroquine on the growth of mammalian cells in vitro. J. Pharmacol. Exptl. Therap. 141, 122 (1963).

Giles, C.L., and F.W. Henderson: The ocular toxicity of chloroquine therapy. Am. J. Med. Sci. 249, 230 (1965).

Guttmann, P., u. P. Ehrlich: Über die Wirkung des Methylenblau bei Malaria. Berlin. Klin. Wochschr. 28, 953 (1891).

Hahn, F.E.: Complexes of biologically active substances with nucleic acids — yesterday, today, tomorrow. Progr. Mol. Subcell. Biol. 2, 1 (1971).

Hahn, F.E., and J. Ciak: Elimination of bacterial episomes by DNA-complexing compounds. Ann. N.Y. Acad. Sci. 182, 295 (1971).

Hahn, F.E., and A.K. Krey: Deoxyribonucleic acid-induced anomalous optical rotatory dispersion of antimalarial drugs and dyes. Antimicrobial Agents Chemotherapy 1968, 15 (1969).

Hahn, F.E., R.L. O'Brien, J. Ciak, J.L. Allison, and J.G. Olenick: Studies on modes of action of chloroquine, quinacrine and quinine and on chloroquine resistance. Military Med. 131, 1071 (1966).

Hart, C.W., and R.F. Naunton: The ototoxicity of chloroquine phosphate. Arch. Otolaryngol. 80, 407 (1964).

Hawking, F.: Comments on host-parasite-drug interactions. Military Med. 134, 1007 (1969).

Holbrook, D.J., L.P. Whichard, C.R. Morris, and L.A. White: Interaction of antimalarial amino-quinolines (Primaquine, pentaquine and chloroquine) with nucleic acids, and effects on various enzyme reactions in vitro. Progr. Mol. Subcell. Biol. 2, 113 (1971).

Homewood, C.A., D.C. Warhurst, W. Peters, and V.C. Baggaley: Lysosomes, pH and the anti-malarial action of chloroquine. Nature 235, 50 (1972).

Hopkinson, L., and F.L. Jackson: Effects of chloroquine on growth metabolism. Nature 202, 27 (1964).

Houba, V., a M. Adam: Příspěvek k mechanismu působení resochinu. Časopis Lékařů Českých 103, 540 (1964).

Kradolfer, F., and L. Neipp: Experimental studies on amebicidal, antibacterial and antiparasitic phenanthroline compounds. Antibiot. & Chemotherapy 8, 297 (1958).

KRAMER, P.A., and J.E. MATUSIK: Location of chloroquine binding sites in *Plasmodium berghei*. Biochem. Pharmacol. **20**, 1619 (1971).

KREY, A.K., and F.E. HAHN: Methyl green-DNA complex: Displacement of dye by DNA-binding substances. Proc. First Europ. Biophys. Congr. **1**, 223 (1971).

KROOTH, R.S., K.D. WUU, and R. MA: Dihydroorotic acid dehydrogenase: Introduction into erythrocyte by the malaria parasite. Science **164**, 1973 (1969).

KURNICK, N.B., and I.E. RADCLIFFE: Reaction between DNA and quinacrine and other antimalarials. J. Lab. Clin. Med. **60**, 669 (1962).

LADDA, R.: Morphologic observation on the effect of antimalarial agents on the erythrocytic forms of *Plasmodium berghei in vitro*. Military Med. **131**, 993 (1966).

LADDA, R., et J. ARNOLD: Inclusion intronucléaire dans le tophozoite de la forme érythrocytique du *Plasmodium berghei* chex le Rat par absorption de chloroquine. Compt. Rend. **260**, 6991 (1965).

LADDA, R., and H. SPRINZ: Chloroquine sensitivity and pigment formation in rodent malaria. Proc. Soc. Exptl. Biol. Med. **130**, 524 (1969).

LANE, R.: The treatment of hepatic amoebiasis with chloroquine. J. Trop. Med. Hyg. **54**, 198 (1951).

LANTZ, C.H., and K. VAN DYKE: Studies concerning the mechanism of action of antimalarial drugs. II. Inhibition of the incorporation of adenosine-5'-monophosphate-^{3}H into nucleic acids of erythrocyte-free malarial parasites. Biochem. Pharmacol. **20**, 1157 (1971).

LERMAN, L.S.: Structural considerations in the interaction of DNA and acridines. J. Mol. Biol. **3**, 18 (1961).

LERMAN, L.S.: The structure of the DNA-acridine complex. Proc. Natl. Acad. Sci. U.S. **49**, 94 (1963).

LIPP, R.: Untersuchungen über den Hemmeffekt des Chloroquin (Resochin) auf die Zellproliferation *in vitro*. Arch. Klin. Exptl. Dermatol. **218**, 228 (1964).

LIQUORI, A.M., L. COSTANTINO, V. CRESCENZI, V. ELIA, E. GIGLIO, R. PULITI, M. DE SANTIS SAVINO, and V. VITAGLIANO: Complexes between DNA and polyamines: A molecular model. J. Mol. Biol. **24**, 113 (1967).

MACOMBER, P.B., R.L. O'BRIEN, and F.E. HAHN: Chloroquine: Physiological basis of drug resistance in *Plasmodium berghei*. Science **152**, 1374 (1966).

MACOMBER, P.B., and H. SPRINZ: Morphological effects of chloroquine on *Plasmodium berghei* in mice. Nature **214**, 937 (1967).

MIETZSCH, F., and H. MAUSS: Acridine derivatives. Germ. Pat. 553, 072 (1930).

MORRIS, C.R., L.V. ANDREW, L.P. WHICHARD, and D.J. HOLBROOK, JR.: The binding of antimalarial aminoquinolines to nucleic acids and polynucleotides. Mol. Pharmacol. **6**, 240 (1970).

O'BRIEN, R.L., J.L. ALLISON, and F.E. HAHN: Evidence for intercalation of chloroquine into DNA. Biochim. Biophys. Acta **129**, 622 (1966a).

O'BRIEN, R.L., and F.E. HAHN: Chloroquine structural requirements for binding to deoxyribonucleic acid and antimalarial activity. Antimicrobial Agents Chemotherapy **1965**, 315 (1966).

O'BRIEN, R.L., J.G. OLENICK, and F.E. HAHN: Reactions of quinine, chloroquine and quinacrine with DNA and their effects on the DNA and RNA polymerase reactions. Proc. Natl. Acad. Sci. U.S. **55**, 1511 (1966b).

OLENICK, J.G., and F.E. HAHN: Mode of action of primaquine: Preferential inhibition of protein biosynthesis in *Bacillus megaterium*. Antimicrobial Agents Chemotherapy **1**, 259 (1972).

PARK, J.T., and J.L. STROMINGER: Mode of action of penicillin. Science **125**, 99 (1957).

PARKER, F.S., and J.L. IRVIN: The interaction of chloroquine with nucleic acids and nucleoproteins. J. Biol. Chem. **199**, 897 (1952).

PETERS, W.: Chemotherapy and drug resistance in malaria. London and New York: Academic Press 1970.

POLET, H., and C.F. BARR: Chloroquine and dihydroquinine. *In vitro* studies of their antimalarial effect upon *Plasmodium knowlesi*. J. Pharmacol. Exptl. Therap. **164**, 380 (1968a).

POLET, H., and C.F. BARR: DNA, RNA and protein synthesis in erythrocytic forms of *Plasmodium knowlesi*. Am. J. Trop. Med. Hyg. **17**, 672 (1968b).

POLET, H., and C.F. BARR: Uptake of chloroquine-3-H^3 by *Plasmodium knowlesi in vitro*. J. Pharmacol. Exptl. Therap. **168**, 187 (1969).

POWELL, R.D., and W.D. TIGERTT: Drug resistance of parasites causing human malaria. Ann. Rev. Med. **19**, 81 (1968).

Robinson, L. B., T. M. Brown, and R. Wichelhausen: The effect of erythromycin and antimalarial compounds on pleuropneumonia-like organisms. Antibiot. & Chemotherapy **9**, 111 (1959).

Sams, W. M.: Chloroquine: Mechanism of action. Proc. Staff Meetings Mayo Clinic **42**, 300 (1967).

Schellenberg, K. A., and G. R. Coatney: The influence of antimalarial drugs on nucleic acid synthesis in *Plasmodium gallinaceum* and *Plasmodium berghei*. Biochem. Pharmacol. **6**, 143 (1961).

Schueler, F. W., and W. F. Cantrell: Antagonism of the antimalarial action of chloroquine by ferriheamate and an hypothesis for the mechanism of chloroquine resistance. J. Pharmacol. Exptl. Therap. **143**, 278 (1964).

Schulemann, W.: Synthetic anti-malarial preparations. Proc. Roy. Soc. Med. **25**, 897 (1932).

Sodeman, W. A., A. A. Doerner, E. M. Gordon, and C. M. Gillikin: Chloroquine in hepatic amebiasis. Ann. Internal. Med. **35**, 331 (1951).

Stollar, D., and L. Levine: Antibodies to denatured deoxyribonucleic acid in lupus erythematosus serum. V. Mechanism of DNA-anti-DNA inhibition by chloroquine. Arch. Biochem. Biophys. **101**, 335 (1963).

Sutherland, J. C., and B. M. Sutherland: Energy transfer in the DNA-chloroquine complex. Biochim. Biophys. Acta **190**, 545 (1969).

Tabor. H.: The protective effect of spermine and other polyamines against heat denaturation of deoxyribonucleic acid. Biochemistry **1**, 496 (1962).

Theakston, R. D. G., S. N. Ali and G. A. Moore: Electron microscope autoradiographic studies on the effects of chloroquine on the uptake of tritiated nucleosides and methionine by *Plasmodium berghei*. Ann. Trop. Med. Parasitol. **66**, 295 (1972).

Thurston, J. P.: The morphology of *Plasmodium berghei* before and after treatment with drugs. Trans. Roy. Soc. Trop. Med. Hyg. **47**, 248 (1953).

Trager, W., R. Klatt, and S. Smith: Loss of chloroquine resistance an transfer of *Plasmodium berghei* from mouse to hamster. J. Parasitol. **53**, 1111 (1967).

Van Dyke, K., C. Szustkiewicz, C. H. Lantz, and L. H. Saxe: Studies concerning the mechanism of action of antimalarial drugs—Inhibition of the incorporation of adenosine-8-^{3}H into nucleic acids of *Plasmodium berghei*. Biochem. Pharmacol. **18**, 1417 (1969).

Van Dyke, K., G. C. Tremblay, C. H. Lantz, and C. Szustkiewicz: The source of purines and pyrimidines in *Plasmodium berghei*. Am. J. Trop. Med. Hyg. **19**, 202 (1970).

Vogler, W. R., J. T. Galambos, and S. Olansky: Biochemical effects of chloroquine therapy in porphyria cutanea tarda. Am. J. Med. **49**, 319 (1970).

Walsh, C. J., and I. W. Sherman: Isolation, characterization and sythesis of DNA from a malaria parasite. J. Protozool. **15**, 503 (1968).

Warhurst, D. C., and D. J. Hockley: Mode of action of chloroquine on *Plasmodium berghei* and *P. cynomolgi*. Nature **214**, 935 (1967).

Warhurst, D. C., and J. Williamson: Ribonucleic acid from *Plasmodium knowlesi* before and after chloroquine treatment. Chem.-Biol. Interact. **2**, 89 (1970).

Waring, M.: Variation of the supercoils in closed circular DNA by binding of antibiotics and drugs: Evidence for molecular models involving intercalation. J. Mol. Biol. **54**, 247 (1970).

Whitfeld, P. R.: Nucleic acids in erythrocytic stages of a malaria parasite. Nature **169**, 751 (1952).

Whitfeld, P. R.: The nucleic acids of the malaria parasite, *Plasmodium berghei*. Australian J. Biol. Sci. **6**, 234 (1953).

Yielding, K. L., L. W. Blodgett, H. Sternglanz, and D. Gaudin: Chloroquine binding to nucleic acids: Characteristics, biological consequences and a proposed binding model for the interaction. Progr. Mol. Subcell. Biol. **2**, 69 (1971).

Distamycin A and Netropsin

Fred E. Hahn

I. Distamycin A

Distamycin A is a fermentation product of *Streptomyces distallicus* (ARCAMONE *et al.*, 1958). Its chemical structure (Fig. 1) has been determined by organic chemical methods and confirmed by total synthesis (ARCAMONE *et al.*, 1964; PENCO *et al.*, 1967). The compound has a distinctive ultraviolet absorption spectrum (ARCAMONE *et al.*, 1964) and is optically non-active (DiMARCO *et al.*, 1962; ARCAMONE *et al.*, 1964). The absence of intrinsic Cotton effects indicates that the N-methyl-pyrrole chromophores of the antibiotic molecule in solution are not asymmetrically perturbed but more precise information on the configuration of distamycin A in physiological solution is lacking. Experimental studies of the molecular pharmacology of distamycin A are hampered by the instability of the antibiotic.

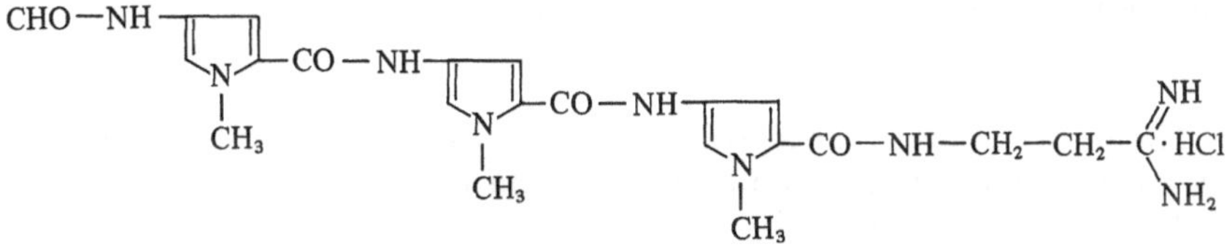

Fig. 1. Structure of distamycin A

Distamycin A as a Growth Inhibitor

Distamycin A, like its congeners in fermentation mixtures, exhibits antifungal activity (ARCAMONE *et al.*, 1964). The antibiotic has significant effects on ascites tumors such as Ehrlich and S 180 in mice; at a daily dose of 50 mg/kg, it increased mean survival times by more than 70 per cent. It also caused marked decrease in the growth of solid tumors, Ehrlich's adenocarcinoma, sarcoma 180, Walker carcinoma and Oberling-Guérin-Guérin myeloma (DiMARCO *et al.*, 1962).

Distamycin has no effect on the growth of *Escherichia coli* (DiMARCO *et al.*, 1963a; SANFILIPPO *et al.*, 1966; HOLLDORF *et al.*, 1970; SANFILIPPO, 1971) or *Salmonellae* (HOLLDORF *et al.*, 1970; SANFILIPPO, 1971) when these bacteria are inoculated into liquid media whose nutrients can be utilized by constitutively synthesized bacterial enzymes. Growth on carbon sources for whose biochemical utilization the test bacteria must synthesize *ad hoc* induced enzymes is decreased or prevented by the antibiotic (SANFILIPPO *et al.*, 1966; HOLLDORF *et al.*, 1970). The literature does not report data from systematic comparative screening of distamycin A as an antimicrobial agent.

Table 1. *Antiviral activity of distamycin A*

Virus	Host	Inhibitory concentration	Reference
T1	E. coli K12	10 µg/ml (2×10^{-5} M)	1
T1, T2	E. coli K12	10 µg/ml (2×10^{-5} M)	2
Vaccinia WR	Mouse embryonic cells	$ED_{50} = 6$ µg/ml (1.25×10^{-5} M)	3, 5
Herpes simplex HF	KB cells	$ED_{50} = 6$ µg/ml (1.25×10^{-5} M)	3
Vaccinia	Mouse embryonic cells	$ED_{50} = 6$ µg/ml (1.25×10^{-5} M)	5
Vaccinia	Chick embryonic cells	$ED_{50} = 6$ µg/ml (1.25×10^{-5} M)	
Vaccinia	HeLa cells	$ED_{50} = 10$ µg/ml (2×10^{-5} M)	5, 6
Shope fibroma	Rabbit lung fibroblasts	5 µg/ml	8
Herpes, canine	Rabbit lung fibroblasts	15 µg/ml	8
Herpes, equine	Rabbit lung fibroblasts	15 µg/ml	8
Herpes, feline	Rabbit lung fibroblasts	15 µg/ml	8
Murine Sarcoma	Mouse embryonic cells	$ED_{50} = 6$ µg/ml (1.25×10^{-5} M)	9
Vaccinia	Rabbit skin	ointment, repeated	4
Vaccinia	Rabbit eye	ointment, repeated	4
Herpes	Rabbit eye	ointment, repeated	6
Keratoconjunctivitis	Rabbit eye	ointment, repeated	6
Mouse hepatitis	Mice	suspension, repeated	7
Shope fibroma	Rabbits	5 mg/kg	8

References: 1. DiMARCO *et al.* (1963a). 2. DiMARCO *et al.* (1963b). 3. WERNER *et al.* (1964). 4. CASAZZA and GHIONE (1964). 5. VERINI and GHIONE (1964). 6. CASAZZA *et al.* (1966). 7. FOURNEL *et al.* (1966). 8. DE RATULD and WERNER (1970). 9. CASAZZA *et al.* (1972, in press).

The most important antibiotic effects of distamycin A are directed against DNA-containing viruses. Representative data from the literature are listed in Table 1.

In man, distamycin A ointments have been used successfully for the topical treatment of chickenpox, herpes zoster and eruptions resulting from smallpox vaccination (BASSETTI, 1968, 1969, 1971; SCARZELLA, 1970; TOSCANO and PASCHETTA, 1971).

The toxicity of distamycin A for animals depends on the route of administration. In mice, the LD_{50} of intravenous distamycin A is 75 mg/kg while the LD_{50} of the antibiotic, given intraperitoneally, is 500 mg/kg. Mice tolerate daily doses of 50–75 mg/kg subcutaneously while rats are more sensitive (DiMARCO *et al.*, 1962).

The morphological effects of distamycin A in cultured mammalian cells include decreases in mitotic index in chick heart and rat fibroblasts, but large increases in mitotic index with a high percentage of metaphase cells in several tumor cells; such cells showed chromosome clusters, aberrant chromosomes and chromosome bridges (DiMARCO *et al.*, 1964).

Effects of Distamycin A on Bacterial Virus Infections

The antiviral action of distamycin A has been studied in *Escherichia coli* K12 infected with DNA-containing viruses of the T-series (DiMARCO *et al.*, 1963a,

1963b). At concentrations from 1 to 100 µg/ml, i.e. up to 2×10^{-4} M, distamycin A has no influence on the growth of the host bacteria but prevents their infection with phages. The antibiotic does not inactivate T1 or T2 upon contact but interferes with the physical adsorption of the viruses to the bacterial surface; it also inhibits the adsorption of the viruses to isolated bacterial cell walls. Since the inhibition of attachment to bacteria shows only slight temperature dependence, it can be inferred that it is the initial adsorption rather than the following irreversible binding which is a target process of distamycin A's action; the effect of the antibiotic on phage adsorption is not antagonized by monovalent or divalent inorganic ions. Distamycin does not cause a lasting change in the bacterial surface which would render it incapable of binding phages: bacteria which have been grown in the presence of the antibiotic support normal infection and proliferation of phages T1, T2 or T4 after they have been collected and resuspended in fresh antibiotic-free medium.

When distamycin A was added to T1-infected bacteria at a time when the viral DNA had penetrated into the cells ($+15$ min), the intracellular proliferation of the phage progeny was not inhibited by the antibiotic: artificial bacterial lysis with a KCN-glycine mixture liberated the normal phage progeny.

Distamycin A causes, however, inhibition of natural lysis so that no "burst" occurs. Artificial lysis with KCN-glycine after 60 min revealed a marked increase in the virus yield per infected cell byond the normal "burst size". When the antibiotic is removed from phage infected *E. coli*, spontaneous lysis occurs immediately. The lysis inhibition by the antibiotic has been ascribed either to an interference with the virus-induced synthesis of the lytic system ("una interferenza specifica a livello dei sistemi litici scatenati dall'infezione fagica") or, alternately, to a modification of the bacterial surface (DiMARCO *et al.*, 1963b). It should be noted that one general inhibitor of protein biosynthesis, chloramphenicol, when added to T1-infected *E. coli* at $+10$ min decreases the phage yield by >90 per cent but does not inhibit or delay the occurrence of the "burst", i.e. of lysis and liberation of progeny phage (BOZEMAN *et al.*, 1954). This suggests that 10 min after infection, the induction of all essential protein syntheses has occurred and is difficult to reconcile with the idea that lysis inhibition by distamycin A, added at $+15$ min, resulted from an inhibition of *induction of protein synthesis*. The mechanisms of lysis and of lysis inhibition are incompletely understood, but there is agreement that the induced synthesis of a special lysozyme is one essential but not a sufficient precondition for lysis to occur. DiMARCO *et al.* (1963b) found that distamycin was without effect on the activity of egg-white lysozyme on bacterial cell walls.

Effect of Distamycin A on Induction of Enzyme Synthesis

A fairly general inhibition of the induction of enzyme syntheses by distamycin A has been demonstrated in non-infected *E. coli* (SANFILIPPO *et al.*, 1966; HOLLDORF *et al.*, 1970). Growth of non-adapted bacteria on carbon sources such as arabinose, xylose, rhamnose, trehalose, maltose, lactose, serine, deoxyribose-5-phosphate or deoxycytidine was strongly or completely inhibited by distamycin A; addition of the antibiotic to *adapted* cultures did not inhibit their growth on lactose, arabinose, xylose, rhamnose, trehalose and maltose. *E. coli* which was adapted to lactose was subject to growth inhibition by distamycin A when supplied with arabinose or trehalose as the sole sources of carbon. When *E. coli* had been subjected to 18 hours of bacteriostasis owing to the inability of the bacteria to metabolize arabinose in the presence of distamycin A, the addition of glucose supported cultural growth.

The induced synthesis of β-galactosidase and its inhibition by the antibiotic was demonstrated by measuring the ability of the bacteria to produce o-nitrophenol from o-nitrophenyl-β-galactoside. The induction of the synthesis of β-galactosidase is more strongly inhibited by distamycin A when lactose is used as an inducer by comparison to induction with isopropyl-β-thiogalactoside (HOLL-DORF *et al.*, 1970); this latter substance is a "gratuitous inducer" which the bacteria do not metabolize.

This is not the only instance in which the inhibition of the induction of enzyme synthesis by distamycin is modified by the metabolic utility of the inducing substrate. The induction of enzymes, necessary to utilize serine or deoxycytidine as sources of *carbon*, is more strongly inhibited by the antibiotic than the induction of enzymes which mobilize the identical compounds as sources of *nitrogen*. HOLL-DORF *et al.* (1970) have speculated that such differences result from regulatory disturbances caused by distamycin A in the formation, and, hence, availability, of cyclic AMP. The authors reported without documentation that in their hands distamycin A inhibited the accumulation of cyclic AMP in glucose starved *E. coli* as well as stimulations of induced enzyme syntheses by cyclic AMP.

Effects of Distamycin A
on Cell-Free Nucleic Acid Polymerizations

The inhibition of induction of enzyme synthesis by the antibiotic might alternately be related to observations that distamycin A acts as a DNA template poison and inhibits the cell-free synthesis of RNA by RNA-polymerase. This has been shown for the calf thymus DNA-dependent *E. coli* RNA-polymerase reaction (CHANDRA *et al.*, 1970). The inhibition of this reaction is dependent upon the dosage of distamycin A and is more pronounced with native DNA as the template than with denatured DNA which has been heated to 95° C and then rapidly cooled. Preincubation of native template DNA with the antibiotic for up to eight minutes increases the inhibition in subsequently performed RNA-polymerase assays. By an ingenious double-labelling technique it has been shown (PU-SCHENDORF *et al.*, 1971) that the antibiotic inhibits preferentially the *initiation* of RNA synthesis rather than the subsequent elongation phase. Indeed, when distamycin A is added to a reaction system which is actively transcribing RNA from DNA, inhibition occurs only after a delay of four minutes which suggests that the antibiotic does not inhibit the synthesis itself but the reinitiation of another round of RNA transcription. By density gradient centrifugation it has been shown that the antibiotic interferes with the attachment of RNA polymerase to SV40 DNA. The initiation complex of DNA, polymerase, and a ribonucleoside triphosphate is not dissociated by distamycin A (PUSCHENDORF *et al.*, 1971). The authors attribute the action of the antibiotic to an occupancy of A-T rich regions of template DNA which are known to exist at the initiation sites of RNA transcription (biophysical findings which suggest a preference of distamycin A for A-T rich DNA segments are discussed below).

The concentrations of the antibiotic at which the RNA-polymerase reaction *in vitro* is inhibited, range from 5×10^{-6} to 5×10^{-5} M. At similar concentrations, distamycin also inhibits the DNA polymerase I reaction (PUSCHENDORF and GRUNICKE, 1969; ZIMMER *et al.*, 1971b) catalyzed by an enzyme preparation from Ehrlich ascites tumor cells and using calf thymus DNA as a primer and template. Preexposure of DNA to distamycin A produces inhibitions similar to those which are measured when the antibiotic is supplied to the complete reaction mixture. Increasing concentrations of DNA reverse, increasingly, the antibiotic's (7.7×10^{-6} M) inhibition of the DNA polymerase reaction. Fig. 2 represents a double-reciprocal diagram which the author of this review has constructed from data of PUSCHENDORF and GRUNICKE (1969). For the non-inhibited as well as for the distamycin A-inhibited reactions, 1/reaction product vs. 1/DNA-concentration yield straight lines which intersect on the ordinate where 1/DNA concentration $=0$. This means that the extrapolation of DNA concentration to infinity completely reverses the inhibition of the polymerase reaction by distamycin A and constitutes enzymological proof that in this reaction the antibiotic acts as a DNA template poison. Up to a concentration of 40 μg/ml of template DNA, the amount of product polymer was directly proportional to the template DNA concentration and the two lines (no distamycin A or 7.7×10^{-6} M distamycin) had *different* slopes with an *intercept* on the origin of the coordinate system (PUSCHENDORF and GRUNICKE, 1969). According to ACKERMANN and POTTER (1949), this signals a reversible type of inhibition.

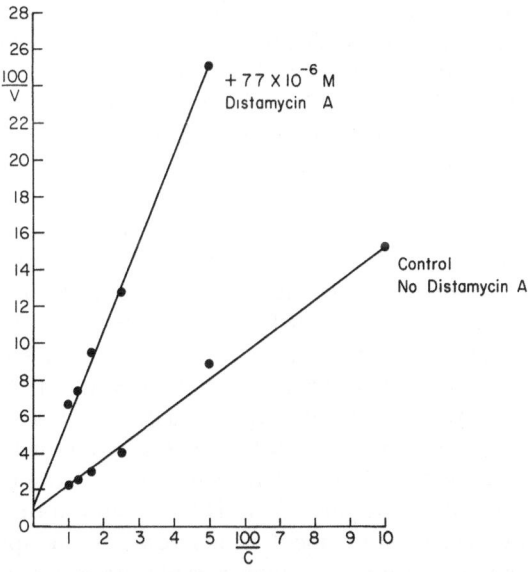

Fig. 2. Competitive inhibition of the DNA polymerase reaction by distamycin A. c: DNA concentration; V: velocity of reaction considered proportional to quantity of reaction product. After data of PUSCHENDORF and GRUNICKE (1969)

DNA polymerase I is thought to catalyze "repair synthesis" of DNA but not DNA replication *in vivo*. The ability of distamycin A to inhibit the conjugational transfer of R-factors in *E. coli* and *Salmonella* (SANFILIPPO, 1971) is probably the result of an effect of the antibiotic on the bacterial surface, perhaps similar to that which prevents bacterial virus adsorption rather than on episomal DNA biosynthesis. At the current state of biochemical knowledge, inhibition of the DNA polymerase I reaction by distamycin A *in vitro* furnishes one indicator that the antibiotic binds to DNA but the importance of this inhibition for the physiology of distamycin A-exposed cells or microorganisms is not apparent.

Of greater biological significance may be the ability of the antibiotic to inhibit the DNA polymerase reaction catalyzed by an enzyme of the RNA-containing Rous sarcoma virus (KOTLER and BECKER, 1971) or of murine sarcoma virus (CHANDRA *et al.,* 1972 b). At 4×10^{-5} M, distamycin A inhibited the intrinsic murine virus DNA polymerase reaction by 30 per cent; the same reaction, slightly stimulated by addition of poly rA-$(dT)_8$, was inhibited to the same extent. RNA-containing tumor viruses possess an RNA-dependent DNA-polymerase known as "reverse transcriptase". The operational definition of this enzyme is that of a polymerase which synthesizes DNA on a single-stranded RNA template (SCHLOM *et al.,* 1971). Since, in the work with distamycin A no external single-stranded ribopolynucleotide template was supplied and the formation of a virus RNA-product DNA hybrid was not demonstrated, the conclusion that distamycin A inhibited the reverse transcriptase reaction is tentative.

The DNA-Distamycin A Interaction

The interaction of distamycin A with nucleic acids, especially with DNAs, has been investigated in detail. The formation of a calf thymus DNA-distamycin A complex is accompanied by a bathochromic ("red") shift of the absorption band of the antibiotic from 303 to 321 nm (KREY and HAHN, 1970); others have recorded a shift to close to 340 nm (ZIMMER *et al.,* 1971 b). The intensity of distamycin A's absoption band is not significantly altered when the DNA-complex is formed. This, as well as the instability of the antibiotic in solution, renders it nonfeasible to carry out spectrophotometric titrations of distamycin A with DNA and to derive numerical values for the stoichiometry of the binding as well as for the apparent association constant(s). That the antibiotic binds very strongly to DNA is borne out by the facts that 10^{-1} M NaCl, 10^{-2} M Mg acetate or 6 M urea do not reverse the DNA-induced bathochromic shift in distamycin A's absorption spectrum (KREY and HAHN, 1970); likewise, dialysis of the DNA-distamycin A complex against buffer or 1% sodium lauryl sulfate or digestion of the complexed DNA with deoxyribonuclease I were without effect (HAHN and KREY, 1971 a, b). Only subsequent digestion with snake venom phosphodiesterase liberated part of the DNA-bound antibiotic (KREY *et al.,* in press). A non-destructive treatment which restores the spectrum of unbound distamycin A is the addition of sodium perchlorate to a concentration of 7.2 M (ZIMMER *et al.,* 1971a). At similar high inorganic ion concentrations, the positive Cotton effect of calf thymus DNA at 290 nm and the negative effect at 260 nm show drastic decreases in molecular amplitudes; this has been attributed to a change in the number of base pairs per turn of the double helix (TUNIS and HEARST, 1968). Whether such a change in DNA conformation results in a physical dissociation of distamycin A from the double helix or, in contrast, the antibiotic remains bound while the structure of the DNA-distamycin A complex is altered, cannot be decided on the basis of available information. Distamycin A can be extracted from DNA with aqueous phenol (KREY et al., 1973).

Like other drugs which bind to DNA, distamycin A displaces methyl green from its complex with DNA (KREY and HAHN, 1970, 1971; ZUNINO and DiMARCO, 1972). At similar molar concentrations of methyl green, bound to DNA, and of free distamycin A, the displacement reaction is of second order with time with a rate constant which places the antibiotic between daunomycin and chloroquine close to miracil D, and an end point of 93 per cent methyl green displaced from calf thymus DNA (KREY and HAHN, 1971). Distamycin A, hence, belongs to a sub-set of methyl green-displacing drugs which all are bound to DNA strongly and for which differences in the reaction constants have only minor influence on the end point of the displacement reactions (KREY and HAHN, in preparation).

An absorption difference spectrum of complexed DNA (distamycin A-DNA minus distamycin A alone) (CHANDRA et al., 1970) shows a slight decrease in DNA's absorption peak and a minor broadening by comparison to the absorption spectrum of DNA alone. Since the antibiotic absorbs light in the same area as DNA with an absorption maximum at 237 nm and a minimum at 266 nm, difference spectroscopy (CHANDRA et al., 1970) can not distinguish between 1. an effect of DNA on the spectrum of the antibiotic, 2. an effect of the antibiotic on the spectrum of DNA, and 3. a combination of the two effects.

Analogous considerations apply to attempts at obtaining distamycin A-DNA minus DNA optical rotatory dispersion (ORD) difference spectra (KREY and HAHN, 1970). The DNA-induced Cotton effects in the ORD spectrum of the antibiotic, that by itself is optically non-active, have been measured by several workers (HAHN and KREY, 1970; ZIMMER and LUCK, 1970). The DNA-induced effects have molecular amplitudes several times as large as the intrinsic Cotton effects of calf thymus DNA. The transition in DNA-bound distamycin A's spectrum at > 320 nm is rendered optically active by calf thymus DNA. This gives rise to the first positive Cotton effect with the peak at 350 nm and the trough at 315 nm.

It is reasonable to assume that the second transition of distamycin A's spectrum at 237 nm will, likewise, be rendered optically active by DNA. This may account for the large trough of the DNA-antibiotic complex at 250 nm and for the crossover at > 230 nm. The small positive Cotton effect at 260 nm may represent one visible contribution of DNA's intrinsic ORD spectrum to that of the complex. An ORD difference spectrum (DNA-distamycin A minus DNA alone) might give an erroneous impression of the DNA-induced Cotton effects in the ORD spectrum of the drug (KREY and HAHN, 1970) because one can not experimentally determine if and to what extent the antibiotic modifies the intrinsic ORD spectrum of DNA. Examples of such modifications are slight decreases in molecular amplitudes produced by steroidal diamines (MAHLER et al., 1968), or large increases

caused by polylysine (COHEN and KIDSON, 1968). Because of the unknown contributions of the potentially altered ORD spectrum of DNA to the spectrum of
the DNA-distamycin A complex, the spectrum of the complex is difficult to
interpret. Some workers (ZIMMER and LUCK, 1970; ZIMMER et al., 1971 a) have
emphasized possible changes in DNA's intrinsic ORD spectrum and have offered
extensive speculations concerning underlying conformational changes. It is perhaps more conservative to restrict the interpretation of the ORD spectrum of
the complex to the above considerations of DNA-induced Cotton effects and
to inquire into structural reasons why the chromophores of distamycin A become
asymmetrically perturbed.

Flow dichroism experiments were, therefore, carried out with the DNA-distamycin A complex in an effort to gain insight into the orientation of the antibiotic's
chromophores relative to the base pairs of duplex DNA (KREY and HAHN, 1970).
The absorption of plane-polarized light of 259 nm (for the DNA bases) and
of 320 nm (for distamycin's chromophores) was measured in flow-oriented and
in stationary solutions of the DNA-antibiotic complex at settings of the polarizer
which were systematically varied through a 180° semi-arc. Flow dichroism is
expressed as the absorbance of the flowing solution minus that of the stationary
solution and this difference divided by the latter. The flow dichroism of the
purines + pyrimidines and of distamycin A's chromophores in the complex have
identical magnitudes but opposite signs (KREY and HAHN, 1970). If the direction
of the transition moment of the 321 nm absorption band of DNA-complexed
distamycin A were in the planes of the N-methylpyrroles, the result of these
experiments would mean that these planes are arranged approximately perpendicularly to the planes of the base pairs of DNA in the complex. Since this direction
is not known, the experiments merely indicate that the N-methylpyrroles are
placed in an orderly array relative to the direction of the flow orientation of
the DNA-antibiotic complex. It has been suggested (KREY and HAHN, 1970)
that this highly ordered structure of the complex gives rise to the positive DNA-
induced Cotton effect which is possibly conformational in origin and may result
from an alignment of the N-methylpyrroles with the helix. This alignment is
probably not the result of intercalation binding: upon being complexed with
distamycin A at graded concentrations, the viscosity of calf thymus DNA decreases until a ratio of one antibiotic molecule per 7 base pairs, i.e. one helical turn,
is attained beyond which it again increases to approach the viscosity of the
DNA preparation alone (ZIMMER et al., 1971 a). Intercalation binding increases
the length of linear DNA and renders it more rigid; this causes an increase
in viscosity (LERMAN, 1961). One specific test for intercalation binding of distamycin A has been carried out in the laboratory of the author of this review (KREY
et al., 1973) by measuring the viscosity changes induced by the antibiotic in
circular supercoiled DNA; the results have eliminated the possibility of intercalation of distamycin A's ring systems between levels of base pairs of DNA.

Distamycin A causes large shifts in the thermal denaturation profile ("melting
curve") of calf thymus DNA to higher temperatures (CHANDRA et al., 1970).
The difference (ΔT_m) between the midpoints of the melting curves of DNA alone
and that of an antibiotic-DNA mixture of a molar ratio of distamycin A/
mononucleotide of 1.0 was +20° at a monovalent cation concentration of

2×10^{-2} M. Increasing concentrations of the antibiotic produced systematic increases in the magnitude of the thermal hyperchromic shift and in the "cooperativity" of the denaturation process. While all these effects are typical for most DNA-drug complexes, their magnitudes are unusually large for distamycin A. It has been observed (KREY, unpublished data) that heating distamycin A alone through a temperature interval from 25–93° causes an appreciable decrease in UV-absorbance which may be attributed to the instability of the antibiotic in solution; since distamycin A has a molar extinction coefficient of $\varepsilon \sim 2 \times 10^4$ at 260 nm, one must consider uncorrected melting curves (CHANDRA et al., 1970) to be only approximations which do not take into account that liberation of the antibiotic from DNA and its thermal decomposition will contribute to the measured absorbances at 260 nm.

The magnitudes of the distamycin A-caused T_ms and of the molecular amplitudes of DNA-induced Cotton effects in the ORD spectrum of the antibiotic are functions of the base composition of different DNAs (ZIMMER et al., 1971a): both parameters increase systematically with increases in the A-T content of DNAs tested. In fact, ΔT_m was inversely proportional to the G-C content of four DNAs containing from 30 to 70 per cent G-C.

KREY et al. (1973) have studied the interaction of distamycin A and the synthetic duplex polymers, poly d(A-T), poly dA-dT, poly dI-dC and poly dG-dC. The bathochromic shift in the antibiotic's absorption spectrum, produced by all polymers not containing G, was similar to that produced by calf-thymus DNA (HAHN and KREY, 1970); poly dG-dC caused only a minor red shift but decreased the intensity of distamycin A's absorption peak at 311 nm. The molecular amplitudes of polymer-induced Cotton effects were large for all tested polymers, not containing G, but a significant small effect was caused by poly dG-dC. In melting experiments all polymers, not containing G, exhibited large and similar ΔT_ms in the presence of distamycin A, while the antibiotic produced a smaller displacement of the thermal denaturation profile of poly dG-dC. These results[1] suggest that the presence of G in DNA or in deoxyribopolynucleotides precludes a strong interaction with distamycin A. One structurally unrelated antibiotic, nogalamycin, which binds to DNA, also causes ΔT_ms which are a function of the A-T content and raises T_m of poly d(A-T) strongly, *that of poly dA-dT much less,* and that of poly dG-dC not at all (BHUYAN and SMITH, 1965).

A preference of antibiotics (distamycin A, netropsin, nogalamycin) for interacting with A-T-rich segments of DNA may either represent an affinity for A or T, a direct exclusion by G, or an affinity for those regions of DNA which, by virtue of their A-T abundance, exist in a peculiar secondary structure. That intramolecular differences in secondary structure of double-helical DNA in solution exist has been shown by X-ray scattering studies (BRAM, 1971a). When synthetic duplex polydeoxyribonucleotides were examined by the same technique (BRAM, 1971b), it was discovered that poly dG-dC has a unique structure, although no model of it has as yet been proposed.

---

[1] It should be noted that distamycin A inhibited the template activity of poly (dA-dT) approximately 5 times stronger than that of poly (dI-dC) in tumor virus DNA polymerase assays (CHANDRA et al., 1972b).

Distamycin A also binds to "denatured" DNA, i. e. to DNA which has been heated to effect strand separation and then rapidly cooled. Such DNA exists as random coils of predominantly single-stranded pieces which are held together by incidental hydrogen bonds. Distamycin A, when added to denatured DNA, exhibits, in "melting" experiments, a cooperative transition with a hyperchromic shift of +39 per cent at 260 nm similar to that which is typical in melting experiments with uncomplexed duplex DNA (CHANDRA et al., 1970). KREY and HAHN (in preparation) have observed that DNA-distamycin A mixtures which have been heated in conventional melting experiments and left to cool show the above cooperative transition upon re-heating. This effect is reversed upon second cooling and is reproduced upon second heating so that the hyperchromic heating curve is superimposed upon the hypochromic cooling curve without hysteresis. Evidently the formation and thermal dissociation of an ordered dista- mycin A-denatured DNA complex is a spontaneous and reversible process.

Denatured DNA induces Cotton effects in the ORD spectrum of distamycin A which show approximately half the molecular amplitudes of those induced by native DNA (ZIMMER and LUCK, 1970). This is also the case for single-stranded DNA which is prepared by heating in the presence of formaldehyde and cannot reassociate with the formation of hydrogen bonds (KREY and HAHN, 1970). The template activity of denatured DNA in the DNA-polymerase reaction (PUSCHEN- DORF and GRUNICKE, 1969) as well as in the RNA-polymerase reaction (CHANDRA et al., 1970) is somewhat less impaired by distamycin A than the same reactions using native DNA as a template. In biochemical terms, it can be inferred that the mechanism of inhibition of nucleic acid polymerase reactions by the antibiotic is probably not one of inhibiting progressive or local strand separations as the result of a stabilization of the duplex structure. In terms of molecular pharmaco- logy, any hypothesis which attempts to explain the binding of distamycin A to DNA and any conceivable structural models of DNA-antibiotic complexes will have to take into account that distamycin A forms a highly ordered complex also with denatured or with genuine single-stranded DNA.

Distamycin A and RNA

Distamycin A has not been demonstrated to interact with RNA. The absorp- tion spectrum of the antibiotic is not changed by rRNA (ZIMMER et al., 1971 a), tRNA does not induce Cotton effects in the ORD spectrum of distamycin A (ZIMMER and LUCK, 1970) and the antibiotic has no influence on the melting profile of yeast tRNA (ZIMMER et al., 1970). Distamycin A does inhibit, however, the polymerization of amino acids in cell-free E. coli ribosome-polynucleotide systems (ZIMMER et al., 1970).The poly U-dependent polymerization of phenyl- alanine and the poly C-dependent polymerization of proline required $> 10^{-4}$ M distamycin to produce 50 per cent inhibition and complete blockade only occurred at 10^{-3} M antibiotic. In contrast, the poly A-dependent polymerization of lysin was more sensitive to the antibiotic with an ED_{50} of the order of 5×10^{-5} M. Reversal studies showed that the inhibition of polyphenylalanine formation by 5×10^{-4} M distamycin A was unaffected by increasing concentrations of tRNA or of ribosomes but was systematically reversed by increasing the concentration of poly U from 100 µg/ml to 800 µg/ml. It is unfortunate that these reversal experiments have been carried out in a system of low distamycin A sensitivity instead of in the poly A system and that no detailed studies on the interaction of poly rA and the antibiotic have been reported[2]. While, at acid pHs, poly rA forms a well-defined double-helical structure, the polymer organizes itself at neutral pH into base-stacked regions and exhibits marked optical rotatory properties and 50 per cent hypochromism relative to riboadenosyl monophosphate (LENG and FELSENFELD, 1966). At pH 7.8 in the polymerization experiments, distamycin A may have interacted with secondary structure regions of poly rA and prevented the polymer from acting as a messenger. An interaction of the formyl group of distamycin A with adenine has recently been postulated (CHANDRA et al., 1972c).

Structure-activity relationships of compounds of the distamycin A series and the general biological significance of effects of the antibiotics are discussed at the end of this review.

II. Netropsin

Netropsin is a fermentation product of *Streptomyces netropsis* (FINLAY et al., 1951). Its structure (Fig. 3) has been determined by organic chemical methods (WALLER et al., 1957). The antibiotic has an absorption spectrum similar to that of distamycin A and is optically non-active (FINLAY et al., 1951). Like distamycin A, netropsin is unstable in solution.

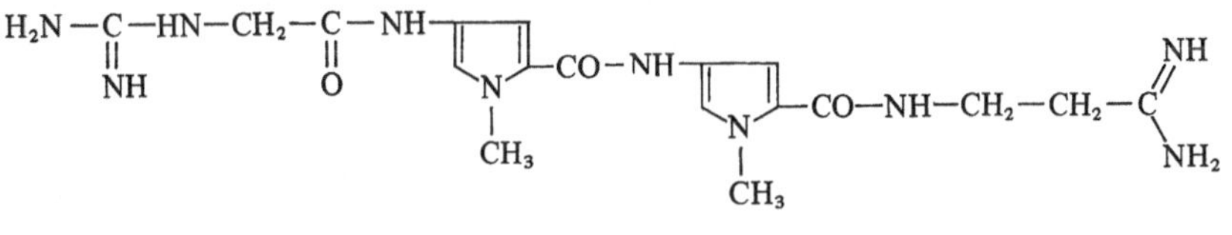

$$H_2N-\underset{\underset{NH}{\|}}{C}-HN-CH_2-\underset{\underset{O}{\|}}{C}-NH-\underset{\underset{CH_3}{|}}{N}-CO-NH-\underset{\underset{CH_3}{|}}{N}-CO-NH-CH_2-CH_2-\underset{\underset{NH_2}{}}{\overset{NH}{C}}$$

Fig. 3. Structure of netropsin

Netropsin inhibits the growth *in vitro* of several bacteria and of *C. albicans* (Table 2). Chemotherapy experiments in mice infected with *S. typhosa* or *Ps. aeruginosa* produced no survivors and no increase in survival times (FINLAY et al., 1951). *In vitro* growth inhibition of *E. coli* by netropsin is partially reversed by phenylalanine, lysine, or by a mixture of 8 amino acids, whereas a mixture of 18 amino acids causes complete reversal (ZYGMUNT, 1961). *S. aureus* which had become resistant to a netropsin-like antibiotic 12782, exhibited cross-resistance to netropsin (MURAVEISKAYA et al., 1966). Resistance to netropsin itself has been produced in *E. coli* by carrying cultures through multiple transfers

---

[2] CHANDRA *et al.* (1971) have stated that in equilibrium dialysis experiments radioactive distamycin A was found to bind to polyribonucleotides.

Table 2. *Inhibition of microbial growth on nutrient agar plates by netropsin* (FINLAY et al., 1951)

Organism	Netropsin HCl in μg/ml
Staphylococcus albus	3
Staphylococcus aureus	5
Bacillus subtilis	5
Aerobacter aerogenes	5
Shigella paradysenteriae	7
Escherichia coli	7
Salmonella pullorum	8
Salmonella paratyphi B	10
Salmonella typhosa	10
Klebsiella pneumoniae	10
Salmonella paratyphi A	20
Bacillus mycoides	20
Proteus vulgaris	40
Candida albicans	90
Pseudomonas aeruginosa	1 000

in liquid media containing increasing concentrations of the antibiotic (SCHABEL, 1958). The antibiotic has a dramatic therapeutic effect on experimental vaccinia virus infections in mice but is without action against infections with RNA-containing viruses such as influenza, Western equine encephalomyelitis or poliomyelitis (SCHABEL et al., 1953). Netropsin is also effective against Rauscher virus-induced lymphoid leukemia in mice (CHIRIGOS et al., 1963).

The toxicity of netropsin administered by different routes to mice has been determined (Table 3). The antibiotic also has toxicity to clothes moth larvae and black carpet beetle (NAKAMURA et al., 1964).

Table 3. *Mouse toxicity of netropsin HCl* [a]

Route of administration	LD_0, mg/kg	LD_{50}, mg/kg
Intravenous injection	10	17
Subcutaneous injection	30	70
Oral	300	> 300

[a] Data from FINLAY et al., 1951.

The mode of action of netropsin has been studied in *E. coli* (HAUPT and THRUM, 1971). Bacterial growth (turbidimetrically) was slightly inhibited at 10^{-5} M antibiotic and progressively at 5×10^{-5} M and 10^{-4} M. Turbidimetry of bacterial density measures predominantly bacterial proteins. At 5×10^{-5} M, netropsin inhibited progressively the incorporation of ^{14}C-phenylalanine into the proteins of *E.coli*; at the same concentration the incorporation of ^{14}C-uracil was blocked after 20 min and the incorporation of ^{14}C-thymine was most strongly inhibited. The antibiotic evidently acts as an inhibitor of RNA and DNA biosynthesis.

Netropsin inhibits the polymerization of amino acids in *E. coli* ribosome-polynucleotide cell-free systems in a manner similar to that of distamycin but is, molecule for molecule, less active. The ED_{50} for the poly U-stimulated formation of polyphenylalanine was 5×10^{-4} M, for the poly C-stimulated polymerization of proline of the same order and for the poly A-stimulated formation of polylysine $>5 \times 10^{-5}$ M (ZIMMER *et al.*, 1970a). Like for distamycin A, the inhibition of phenylalanine polymerization by netropsin was subject to reversal by poly U; at 800 µg/ml of poly U, the antibiotic at 10^{-3} M was without effect. No studies on the interaction of netropsin and synthetic polyribonucleotides have been published. The antibiotic has no influence on the "melting" of tRNA (ZIMMER *et al.*, 1970) and produces only slight changes in the ORD spectra of tRNA and rRNA without any indication of the transition in netropsin's spectrum at 296 nm becoming optically active (ZIMMER *et al.*, 1971a). It is not impossible that reversals of the inhibitions of polyphenylalanine formation by distamycin A or netropsin at exceedingly high concentrations of poly U may be results of aggregations with the antibiotic at polymer concentrations close to 1 mg/ml, i.e. of a physical removal of the inhibitors from the reaction mixtures.

Netropsin inhibits the DNA polymerase I (Ehrlich ascites cells) reaction with calf thymus DNA as a template in a manner very similar to that of distamycin A with only minor differences in inhibitory activity depending on whether native or denatured DNA is used (ZIMMER *et al.*, 1971b). The RNA polymerase (*E.coli*) reaction with calf thymus DNA as the template is somewhat less sensitive to inhibition by netropsin than is the DNA polymerase I reaction but, again, it makes little difference whether native or denatured DNA is used (ZIMMER *et al.*, 1971b).

The inhibition of the DNA-dependent nucleic acid polymerase reactions can be attributed to the strong binding of the antibiotic to DNA. More direct evidence for the template toxicity of netropsin is, however, lacking since neither reversal experiments with graded DNA concentrations, which lend themselves to data handling such as in Fig. 2, nor preincubation experiments of DNA with the antibiotic, followed by polymerase assays, have been reported. The assumption that netropsin acts as a DNA template poison rests on the analogy between the structures of netropsin and distamycin A and on analogies between the interactions of DNA with either antibiotic. Comparative DNA-binding studies for netropsin and distamycin A have been reported by Zimmer and his associates (ZIMMER *et al.*, 1970a; ZIMMER *et al.*, 1970b; ZIMMER and LUCK, 1970; ZIMMER *et al.*, 1971a; ZIMMER *et al.*, 1971b).

The absorption maximum of netropsin at 296 nm shifts to 325 nm when the antibiotic interacts with calf thymus DNA (ZIMMER *et al.*, 1970b; ZIMMER *et al.*, 1971b). DNA renders this transition optically active. The first (positive) DNA-induced Cotton effect in netropsin's ORD spectrum with a peak at 342 nm, a trough at 300 nm and a cross-over point at 315 nm, shows much smaller molecular amplitudes than the first Cotton effect which DNA induces in the ORD spectrum of distamycin A (ZIMMER and LUCK, 1970). Netropsin causes shifts in the thermal denaturation profile of DNA to higher temperatures, increases the magnitude of the hyperchromic endpoint and renders the melting more strongly cooperative (ZIMMER *et al.*, 1970a; 1971b). These observations are entirely comparable to those on the melting of the DNA-distamycin A complex (Chandra et al., 1970) and so is the finding that the ΔT_ms caused by netropsin are a function of the A-T content of different DNAs tested (Zimmer et al., 1971a). The progressive liberation of netropsin from its DNA complex by heating has been followed by recording the decrease in absorbance at 325 nm with temperature. This function shows a sharp transition to a greater temperature dependence at the T_m of the DNA-netropsin complex (Zimmer et al., 1971b).

Netropsin also interacts with denatured DNA, renders its melting behavior cooperative (Zimmer et al., 1971a) and shows induced Cotton effects which, molecule for molecule, have slightly lesser molecular amplitudes than the effects which denatured DNA induces in the ORD spectrum of distamycin A (Zimmer and Luck, 1970). In fact, all optical manifestation of the interactions of netropsin and distamycin A reveal far-reaching analogies, and the discussion of these phenomena, given above for distamycin A, also holds true *cum grano salis* for DNA-netropsin complexes.

A fundamental difference exists, however, in the one hydrodynamic property, viscosity, that has been measured for the two DNA-antibiotic complexes. While distamycin A produces a systematic *decrease* in calf thymus DNA's viscosity up to a ratio of one antibiotic molecule per 7 base pairs, netropsin *increases* the viscosity of DNA with the attainment of a plateau at approximately the same drug/nucleotide ratio at which distamycin A produces the maximal decrease in viscosity (Zimmer et al., 1971a). While such an increase might signal the formation of an intercalation complex, such an inference is not warranted in the absence of additional supporting biophysical data such as sedimentation constants of DNA with and without netropsin, flow-dichroism results, titrations (Blake and Peacocke, 1968) which yield information on different binding processes, their stoichiometries and their apparent association constants and foremost the hydrodynamic effects of netropsin on circular supercoiled DNA which offers the most conclusive test of intercalation (Waring, 1970).

III. Structure-Activity Relationships in the Distamycin A-Netropsin Series

The common features in the structure of distamycin A (Fig. 1) and netropsin (Fig. 3) are the presence of (3 or 2) peptidically linked N-methylpyrrole rings and of the propionamidine side-chain. One can assume that these structural components are responsible for analogies in the binding of the two antibiotics to DNA and in their antineoplastic and anti DNA-virus effects reviewed above. However, no comparative studies on the biological effects of the two antibiotics, side-by-side, have been reported. Whether the antibacterial action of netropsin (Finlay et al., 1951) which can, in E. coli, be reversed by amino acids (Zygmunt, 1961) as contrasted to the failure of distamycin A to inhibit the growth of *E.coli* in conventional media, is a function of the absence of one N-methylpyrrole or the presence of the guanidinoacetamidino group in netropsin or of the presence of the formyl group in distamycin A, is difficult to evaluate.

A homologous series of distamycin A analogues with from 2 to 5 peptidically linked N-methylpyrrole rings (distamycins/2–5) has been studied with respect to the ability to inhibit the (*E. coli*) RNA polymerase reaction with calf thymus DNA as the template. The inhibition of the reaction increased with increasing numbers of N-methylpyrrole rings in the antibiotic molecule. The stabilizing action on DNA against thermal denaturation also increases with the number of ring systems (ZIMMER *et al.*, 1970b). Using the method of KREY and HAHN (1970) of measuring the displacement of methyl green from its complex with DNA by distamycin, ZUNINO and DI MARCO (1972) have measured this displacement for distamycin A, distamycin/4 and distamycin/5. The second order reaction rate obtained for distamycin/5 was significantly greater than that for distamycin A. Optical parameters of DNA binding also increase with increasing numbers of ring systems (ZIMMER *et al.*, 1972).

The progressively stronger interaction with DNA and the increased inhibition of the DNA-dependent RNA polymerase reaction explain, perhaps, the ability of distamycins/2–5, progressively, to inhibit the multiplication of vaccinia virus in HeLa cells with increasing numbers of N-methylpyrrole rings in the molecules. A similar dependence on these numbers was observed for the inhibition of the growth of murine sarcoma virus (Moloney) in mouse fibroblast cultures (CASAZZA *et al.*, 1972, in press). It has also been demonstrated (CHANDRA *et al.*, 1972b) 1972b) that the DNA polymerase reaction catalyzed by murine sarcoma virus (Moloney) is increasingly inhibited by distamycins/2, 3 and 5 in the presence of various templates.

The role of the side chains of distamycin A in the activities of the antibiotic has also been studied (CHANDRA *et al.*, 1972a; CHANDRA *et al.*, 1972c). Substitutions in the formyl or propionamidine side chains lead to decreases in the inhibition of the RNA polymerase reaction. CHANDRA *et al.* (1972a) stated that the influence of these substituted distamycins on the melting behavior of DNA was "in good agreement" with the biological and biochemical test data. Specifically, the substitution of the formyl group by a cyclopentylpropionyl residue causes a total loss of the antiviral activity of distamycin A while a 53 per cent inhibition of the RNA polymerase reaction remains.

The cyclopentylpropionyl derivative has also been tested in comparison to distamycin A itself as to the ability of both compounds to inhibit amino acid polymerizations in cell-free ribosome-polynucleotide systems. As shown previously (ZIMMER *et al.*, 1970), the poly A-stimulated formation of polylysine was more sensitive to inhibition by distamycin A then were the poly U-stimulated polymerization of phenylalanine and the poly C-stimulated formation of polyproline. The last two reactions were inhibited to approximately the same extents by distamycin A and by its cyclopentylpropionyl derivative (CHANDRA *et al.*, 1972c) but the derivative had only 37 per cent of distamycin A's inhibitory activity in the poly A system. From this result as well as from the preference of distamycin A for A-T pairs, reviewed above, it has been conjectured that the antibiotic binds with its formyl residue to adenine (CHANDRA *et al.*, 1972c).

The only derivative of netropsin which has come under study is a degradation product in which the guanidinoacetamidino group is replaced by -NH_2; this substance did not change T_m of calf thymus DNA and rendered the DNA melting curve marginally more cooperative (ZIMMER *et al.*, 1970a); it also showed slight DNA-induced Cotton effects (ZIMMER *et al.*, 1970b). Quaternizing the amino group did not change the influence of the derivative on DNA's melting behavior (ZIMMER et al., 1972).

Of theoretical interest in the consideration of structure-activity relationships in the distamycin-netropsin series is the antibiotic noformicin (Fig. 4), a fermentation product of *Nocardia formica*. It resembles netropsin with the notable differences that it possesses only one pyrrolidine ring system and that the propionamidine side chain is missing. Noformicin inhibits the growth of RNA-containing plant and mammalian viruses (SIDWELL et al., 1968) but also strongly that of vaccinia virus (SETO and SAITO, 1966). The mode of action of the antibiotic is unknown. If the similarities of noformicin to netropsin as concerns chemical structure and antiviral activity are more than fortuitous, this might suggest that netropsin (and perhaps distamycin A) are derivatives of a simpler prototype structure.

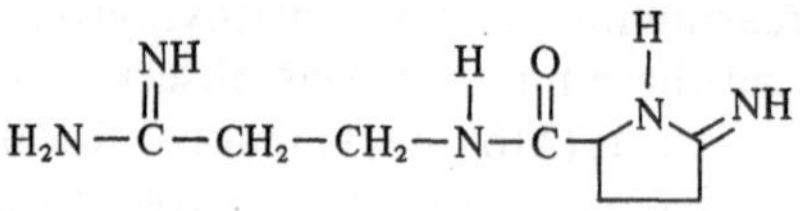

Fig. 4. Structure of noformicin

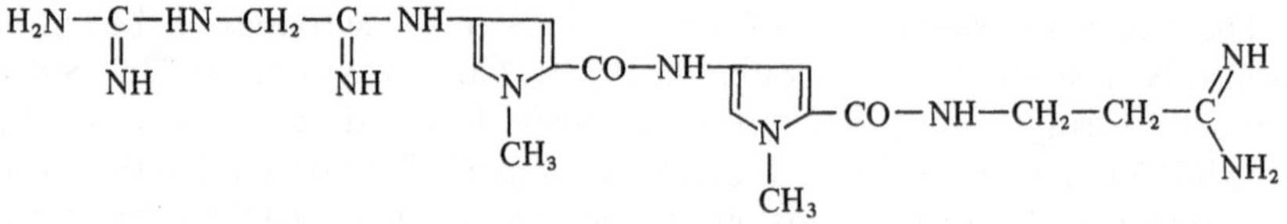

Fig. 5 Structure of congocidine

Another antibiotic, structurally related to netropsin, is congocidine (Fig. 5). It inhibits irreversibly the replication of Shope fibroma and vaccinia viruses in mammalian cell cultures in contrast to distamycin A whose inhibitory effect was reversed when the antibiotic was removed from the infected cells within 24 hours after the initiation of treatment (BECKER et al., 1972). It is reasonable to assume (BECKER et al., 1972) that congocidine, like netropsin and distamycin A, binds to AT-rich segments of DNA. Vaccinia virus DNA has a A+T content of 64 per cent.

The results of structure-activity studies in the distamycin A series encourage one to make the following summary. 1. For distamycins/2–5 the antiviral activities as well as the DNA-template toxicities are a function of the number of N-methyl-pyrrole rings in the molecule. 2. The presence of the formyl group is an absolute requirement for antiviral activity as well as for the *specific* action of distamycin A on the ribosome-poly A system which polymerizes lysine. 3. Replacement of the propionamidine side-chain with triple-protonated residues of different dimensions reduces antiviral and anti-RNA-polymerase activities to one-half or less of that of distamycin A. 4. In most test systems, distamycin/5 is the most active compound.

IV. Discussion

The finding that netropsin in a potent antibacterial agent *in vitro* (FINLAY *et al.*, 1951) offered the opportunity of carrying out systematic mode-of-action studies with this antibiotic. While no data have been published on the response of bacteria to the dosage of netropsin, i. e. no response vs. log dosage correlation, and no distinction between bacteriostatic and bactericidal action has been made, a preliminary inventory of macromolecular biosyntheses and their inhibitions by netropsin has been reported (HAUPT and THRUM, 1971) with the result that netropsin can be considered an inhibitor of nucleic acid biosyntheses, evidently through its function as a DNA template poison.

It is difficult to understand how such an action is antagonized by an increasingly complete complement of amino acids (ZYGMUNT, 1961). The nutrient agar plates on which FINLAY *et al.* (1951) tested for antibacterial potency of netropsin will undoubtedly have contained a complete amino acid complement; while the antibiotic inhibited the growth of *E. coli* on these plates at 7 µg/ml, ZYGMUNT (1961) reported that in liquid mineral medium, growth inhibition of *E. coli* by 5 µg/ml of netropsin was reversed by the addition of beef and yeast extracts, protein hydrolyzates and amino acid mixtures of defined compositions. Evidently, additional and more detailed studies of the antibacterial action of netropsin will need to be carried out in order to elucidate the mechanism of action of this antibiotic in bacteria.

Since distamycin A is an antiviral antibiotic, the clue to its mode of action must lie in its interference with constitutive or virus-induced biochemical reactions in the respective host cells or with the few reactions which are catalyzed by certain viral enzymes.

For the *E.coli* T-phage systems, the antiviral effects of distamycin A have been traced to both an inhibition of the physical attachment of the virus particles to the bacterial surface and a lysis-inhibition (DIMARCO *et al.*, 1963 b). It is not clear if the second effect is due to an inhibition of the *formation* of the lytic system or of the activity of the *existing* system. Infected cells which are centrifuged from a distamycin A-containing medium and resuspended in antibiotic-free broth, lyse immediately which suggests that they possess a complete lytic system which was formed, *but not active* in the presence of distamycin A.

Although the bacterial virus studies (DIMARCO *et al.*, 1963 b) appear to have led directly into investigations of the effects of the antibiotic on inductions of enzyme syntheses in *E. coli* (SANFILIPPO *et al.*, 1966), it is not clear that the results of this investigation have a bearing upon the anti-T-phage effects of distamycin A. One might, perhaps, consider the antibiotic a vicarious repressor of inducible messenger RNA synthesis by virtue of its occupance of A-T-rich segments of DNA which are abundant in initiation sites of RNA biosynthesis (LETALAER and JEANTEUR, 1971) and also may play a role in the binding of specific repressor proteins to their respective operator genes (LIN and RIGGS, 1970). In fact, PUSCHENDORF *et al.* (1971) did show that it was the *initiation* rather than the *elongation* phase in cell-free RNA biosynthesis which was inhibited by distamycin A and that the susceptible process is the initial attachment of the RNA polymerase enzyme to the initiation site of a virus DNA. However, the idea of distamycin A's being a vicarious repressor is difficult to bring into consonance with the findings of HOLLDORF et al. (1970) that the extent of inhibition of induction of enzyme syntheses is strongly modified by the nutritional utility of the inducing substrates; in fact, the induction of β-galactosidase in E. coli by a gratuitous inducer, which is not a substrate, was not inhibited by distamycin A.

The effects of the antibiotic on the replication of mammalian DNA viruses have not been biochemically studied in infected cell cultures; it has been observed (CASAZZA et al., 1972, in press) that distamycin A, in order to be effective, must be added to herpes virus-infected HeLa cells not later than 3 hours after the infection which has a latent period of 8 hours and an exponential phase of virus production from 8 to 15 hours post infection. The authors speculate that distamycin A "blocks essential steps prior to DNA replication."

The effects of the antibiotic on RNA-containing tumor viruses are perhaps explained by observations that it inhibits a virus-associated DNA polymerase. Although such inhibitions have been demonstrated in experiments to which no external templates had been added (as well as in experiments with added templates) and in which the intrinsic DNA-synthesizing capacity, for example, of Friends leukemia virions, could be greatly reduced by treatment with ribonuclease (CHANDRA et al., 1972b), conclusive proof of an effect of distamycin A in RNA-dependent DNA polymerase ("reverse transcriptase") is required, and the failure of the antibiotic demonstrably to interact with RNAs, reviewed above, renders the question of an RNA-template toxicity of distamycin A in the RNA-dependent DNA-polymerase reaction rather problematic.

Assuming tacitly or explicitly, as numerous authors do, that the biological and biochemical effects of distamycin A are consequences of the formation of DNA-antibiotic complexes, the intensive investigation of the interaction of distamycin A and nucleic acids or polynucleotides, as reviewed above, has left two important questions unanswered: 1. Is the antibiotic bound to DNA by one or more than one process? and 2. What is the structure of the DNA-distamycin A complex?

Kinetics of binding of the antibiotic to DNA are not known, except for one observation that maximal inhibition of the RNA-polymerase reaction required preincubation of the template DNA with distamycin A for 8 minutes at 25° (CHANDRA et al., 1970). The displacement of methyl green from DNA by distamycin is of second order with time (KREY and HAHN, 1970; ZUNINO and DIMARCO, 1972). Such observation could not be expected if different binding processes with different kinetics would jointly contribute to the displacement of methyl green. A desorption of distamycin A bound to DNA-cellulose in affinity chromatography has been reported (ZUNINO and DIMARCO, 1972) to occur in two phases, one produced by elution with a NaCl gradient from 0 to 2 M and the second one in a subsequent elution with an urea gradient from 0 to 7 M (in 2 M NaCl). Since similar salt and urea concentrations do not restore the absorption spectrum of free distamycin A when the DNA-distamycin A complex is exposed to these chemicals in solution (HAHN and KREY, 1970), it can be assumed that the drastic handling of DNA during the preparation of the DNA-cellulose adsorbent and the resulting formation of the cellulose-DNA adduct (LITMAN, 1968) change the binding properties of DNA for distamycin A in such a manner that elution of the drug from the adsorbent is possible. The inference from these elution studies of a bimodal attachment of distamycin A to DNA should be regarded with caution until confirmation by independent experimental methods has been obtained.

The structures of the DNA complexes with distamycins and with netropsin have neither been studied by X-ray diffraction methods nor by nuclear magnetic resonance spectroscopy. Without the results of such analyses, any molecular models of these complexes will remain conjectural.

The nearly complete analogy between the biophysical manifestations of the binding of distamycin A and of netropsin to DNA implicates the N-methylpyrrole rings and the propionamidine side chain as equivalent participants in the structures of DNA-complexes. For netropsin which possesses the guanidineacetamidino residue as an additional and important DNA-binding group, ZIMMER et al., (1971a) have suggested a model of the DNA-antibiotic complex in which netropsin is electrostatically attracted to DNA phosphates with both electropositive side chains and the N-methylpyrrole rings "become oriented (perhaps in the same plane) to the bases by contacting them." This idea is further expanded through a consideration of the increases which netropsin produces in the relative viscosity of DNA by commenting that "an increase in $[\eta]$ generally may be interpreted in terms of conformational changes resulting in an increase in length of the molecule and/or in its rigidity." The authors appear to imply an intercalation model of the DNA-netropsin complex. By the same reasoning, this review has ruled out the formation of an intercalation complex of DNA with distamycin A because it *decreases* DNA's relative viscosity.

Clearly, the opposite behavior of the two antibiotics with respect to viscosity changes induced in DNA, constitutes the principal problem in designing models for the DNA-antibiotic complexes which must account for the parallelism in other biophysical properties of these complexes, as far as they have been measured side by side.

The author acknowledges, gratefully, the courtesy of Drs. CHANDRA, ZIMMER, and DiMARCO in making some of their results available prior to publication and of Farmitalia, Milano, in providing extensive documentation on distamycin A.

References

ACKERMANN, W.W., and V.R. POTTER: Enzyme inhibition in relation to chemotherapy. Proc. Soc. Exptl. Biol. Med. **72**, 1 (1949).

ARCAMONE, F., F. BIZIOLI, G. CANEVAZZI, and A.GREIN: Distamycin and distacin. German Pat. 1,027,667, 1958, through Chem. Abstr. **55**, 2012f (1961).

ARCAMONE, F., S. PENCO, P. OREZZI, V. NICOLELLA, and A. PIRELLI: Structure and synthesis of distamycin A. Nature **203**, 1064 (1964).

BASSETTI, D.: La distamicina: un nuovo antibiotico ad azione antivirale. Prima sperimentazione clinica nella varicella. Giorn. Mal. Infett. Paras. **20**, 827 (1968).

BASSETTI, D.: La distamicina A nella terapia di gravissime forme di herpes zoster. Giorn. Mal. Infett. Paras. **21**, 849 (1969).

BASSETTI, D.: Our multiannual experience on the use of a new antiviral antibiotic "distamycin A". Abstr. VIIth Int. Congr. Chemother. **1**, 5/7 (1971).

BECKER, Y., Y. ASHER, and Z. ZAKAY-RONES: Congocidine and distamycin A, antipoxvirus antibiotics. Antimicrobial Agents Chemotherapy 1, 483 (1972).

BHUYAN, B.K., and C.G. SMITH: Differential interaction of nogalamycin with DNA of varying base compositions. Proc. Natl. Acad. Sci. U.S. 54, 566 (1965).

BLAKE, A., and A.R. PEACOCKE: The interaction of aminoacridines with nucleic acids. Biopolymers 6, 1225 (1968).

BOZEMAN, F.M., C.L. WISSEMAN, H.E. HOPPS, and J.X. DANAUSKAS: Action of chloramphenicol on T1 bacteriophage. I. Inhibition of intracellular multiplication. J. Bacteriol. 67, 530 (1954).

BRAM, S.: Secondary structure of DNA depends on base composition. Nature New Biol. 232, 174 (1971a).

BRAM, S.: Polynucleotide polymorphism in solution. Nature New Biol. 233, 161 (1971b).

CASAZZA, A.M., A. FIORETTI, M. GHIONE, M. SOLDATI, and A.M. PIRELLI: Distamycin A, a new antiviral antibiotic. Antimicrobial Agents Chemotherapy 1965, 593 (1966).

CASAZZA, A.M., and M. GHIONE: Therapeutic action of distamycin A on vaccinia virus infections in vivo. Chemotherapia 9, 80 (1964).

CHANDRA, P., A. GÖTZ, A. WACKER, M.A. VERINI, A.M. CASAZZA, A. FIORETTI, F. ARCAMONE, and M. GHIONE: Some structural requirements for the antibiotic action of distamycins. FEBS Letters 16, 249 (1971).

CHANDRA, P., A. GÖTZ, A. WACKER, M.A. VERINI, A.M. CASAZZA, A. FIORETTI, F. ARCAMONE, and M. GHIONE: Some structural requirements for the antibiotic action of distamycins. II. Structural modification of the side chains in distamycin A molecule. FEBS Letters 19, 327 (1972a).

CHANDRA, P., A. GÖTZ, A. WACKER, F. ZUNINO, A. DIMARCO, M.A. VERINI, A.M. CASAZZA, A. FIORETTI, F. ARCAMONE, and M. GHIONE: Some structural requirements for the antibiotic action of distamycins. III. Possible interaction of formyl group of distamycin side chain with adenine. Hoppe-Seylers Z. Physiol. Chem. 353, 393 (1972c).

CHANDRA, P., CH. ZIMMER, and H. THRUM: Effect of distamycin A on the structure and template activity of DNA in RNA-polymerase system. FEBS Letters 7, 90 (1970).

CHANDRA, P., F. ZUNINO, A. GÖTZ, A. WACKER, D. GERICKE, A. DIMARCO, A.M. CASAZZA, and F. GIULIANI: Template specific inhibition of DNA polymerase from RNA tumor viruses by distamycin A and its structural analogues. FEBS Letters 21, 154 (1972b).

CHIRIGOS, M.A., F.J. RAUSCHER, I.A. KAMEL, G.R. FANNING, and A. GOLDIN: Studies with the murine leukemogenic Rauscher virus. II. Chemotherapy of virus-induced lymphoid leukemia. Cancer Res. 23, 1646 (1963).

COHEN, P., and C. KIDSON: Conformational analysis of DNA-poly-L-lysine complexes by optical rotatory dispersion. J. Mol. Biol. 35, 241 (1968).

DIMARCO, A., M. GHIONE, A. MIGLIACCI, E. MORVILLO e A. SANFILIPPO: Studi sul meccanismo dell'azione antifagica dell'antibiotico distamicina. Giorn. Microbiol. 11, 87 (1963b).

DIMARCO, A., M. GHIONE, A. SANFILIPPO, and E. MORVILLO: Selective inhibition of the multiplication of phage T1 in E. coli K12. Experientia 19, 134 (1963a).

DIMARCO, A., M. GOETANI, P. OREZZI, T. SCOTTI, and F. ARCAMONE: Experimental studies on distamycin A – A new antibiotic with cytotoxic activity. Cancer Chemotherapy Rept. No. 18, 15 (1962).

DIMARCO, A., M. SOLDATI, and A. FIORETTI: Antimitotic activity of antibiotic distamycin A. Acta, Unio Intern. Contra Cancrum 20, 423 (1964).

FINLAY, A.C., F.A. HOCHSTEIN, B.A. SOBIN, and F.X. MURPHY: Netropsin, a new antibiotic produced by a streptomyces. J. Am. Chem. Soc. 73, 341 (1951).

FOURNEL, J., P. GANTER, F. KOENIG, Y. DE RATULD, and G.H. WERNER: Antiviral activity of distamycin A. Antimicrobial Agents Chemotherapy 1965, 599 (1966).

HAHN, F.E., and A.K. KREY: Complex of DNA with the antibiotic, distamycin A. Federation Proc. 30, 1095 Abs. (1971a).

HAHN, F.E., and A.K. KREY: Complex of DNA with the antibiotic distamycin A. Abstr. VII th Int. Congr. Chemother. 1, 11/11 (1971b).

HAUPT, I., u. H. THRUM: Wirkung von Netropsin auf das Wachstum, die Nucleinsäure- und Proteinbiosynthese von Escherichia coli. Z. Allgem. Mikrobiol. 11, 457 (1971).

HOLLDORF, A.W., B. FRIEBE u. M. STOBER: Zur Wirkung von Distamycin A auf die Induction von Enzymen in E. coli. Zentr. Bakt. Parasitenk. 212, 265 (1970).

KOTLER, M., and Y. BECKER: Rifampicin and distamycin A as inhibitors of Rous sarcoma reverse transcriptase. Nature New Biol. **234**, 212 (1971).

KREY, A. K., R. G. ALLISON, and F. E. HAHN: Interactions of the antibiotic, distamycin A, with native DNA and with synthetic duplex polydeoxyribonucleotides. FEBS Letters, **29**, 58 (1973).

KREY, A. K., and F. E. HAHN: Studies on the complex of distamycin A with calf thymus DNA. FEBS Letters **10**, 175 (1970).

KREY, A. K., and F. E. HAHN: Methyl green-DNA complex: Displacement of dye by DNA-binding substances. Proc. First Europ. Biophys. Congr. **1**, 223 (1971).

LENG, M., and G. FELSENFELD: A study of polyadenylic acid at neutral pH. J. Mol. Biol. **15**, 455 (1966).

LERMAN, L. S.: Structural considerations in the interaction of DNA and acridines. J. Mol. Biol. **3**, 18 (1961).

LE TALAER, J. Y., and PH. JEANTEUR: Preferential binding of *E. coli* RNA-polymerase to A-T-rich sequences of bacteriophage lambda DNA. FEBS Letters **12**, 253 (1971).

LIN, S-Y, and A. D. RIGGS: *Lac* repressor binding to DNA not containing the *lac* operator and to synthetic poly dAT. Nature **228**, 1184 (1970).

LITMAN, R. M.: A deoxyribonucleic acid polymerase from *Micrococcus luteus (Micrococcus lysodeikticus)* isolated on deoxyribonucleic acid-cellulose. J. Biol. Chem. **243**, 6222 (1968).

MAHLER, H. R., G. GREEN, R. GOUTAREL, and Q. KHUONG-HUU: Nucleic acid-small molecule interactions. VII. Further characterization of deoxyribonucleic acid-diamino steroid complexes. Biochemistry **7**, 1568 (1968).

MURAVEISKAYA, V. A., N. S. PEVZNER, S. P. SHAPOVALOVA, S. T. FILIPPOS'YAN, R. S. UHKOLINA, V. K. KOVALENKOVA, N. A. NACHAEVA, M. A. SVESHNIKOVA, M. F. LAVROVA, G. A. BABENKO: Antibiotik 12782 iz gruppy netropsina. Antibiotiki **11**, 234 (1966).

NAKAMURA, S., H. YONEHARA, and H. UMEZAWA: On the structure of netropsin. J. Antibiotics (Tokyo) **17**, 220 (1964).

PENCO, S., S. REDAELLI and F. ARCAMONE: Distamicina A. Nota II. Sintesi totale. Gazz. Chim Ital. **97**, 1110 (1967).

PUSCHENDORF, B., and H. GRUNICKE: Effect of distamycin A on the template activity of DNA in a DNA polymerase system. FEBS Letters **4**, 355 (1969).

PUSCHENDORF, B., E. PETERSEN, H. WOLF, H. WERCHAU, and H. GRUNICKE: Studies on the effect of distamycin A on the DNA dependent RNA polymerase system. Biochim. Biophys. Res. Commun. **43**, 617 (1971).

RATULD, Y. DE, and G. H. WERNER: *In vitro* and *in vivo* studies on the inhibitory activity of distamycin A on DNA viruses. Antimicrob. & Anticancer Chemotherapy (Tokyo) **2**, 14 (1970).

SANFILIPPO, A.: Activity of distamycin A on the transfer of drug resistance in Gram-negative clinical isolates. Ann. N. Y. Acad. Sci. **182**, 322 (1971).

SANFILIPPO, A., E. MORVILLO, and M. GHIONE: Activity of distamycin A on the induction of adaptive enzymes in *Escherichia coli*. J. Gen. Microbiol. **43**, 369 (1966).

SCARZELLA, M.: La distamicina A nel trattamento delle localizzazioni cutanee del virus erpetico e delle complicanze della vaccinazione jenneriana. Minerva Pediat. **22**, 1444 (1970).

SCHABEL, F. M.: Utility of drug-resistant organisms in cancer chemotherapy studies. Ann. N. Y. Acad. Sci. **76**, 442 (1958).

SCHABEL, F. M., W. R. LASTER, R. W. BROCKMAN, and H. E. SKIPPER: Observations on antiviral activity of netropsin. Proc. Soc. Exptl. Biol. Med. **83**, 1 (1953).

SCHLOM, J., S. SPIEGELMAN, and D. MOORE: RNA-dependent DNA polymerase in virus-like particles isolated from human milk. Nature **231**, 97 (1971).

SETO, Y., and K. SAITO: Antiviral effect of noformicin. Nippon Kagaku Ryohogakukai Zasshi **14**, 457 (1966), through Chem. Abstr. **66**, 45184s, 1967.

SIDWELL, R. W., G. J. DIXON, S. M. SELLERS, and F. M. SCHABEL: *In vivo* antiviral properties of biologically active compounds. II. Studies with influenza and vaccinia viruses. Appl. Microbiol. **16**, 370 (1968).

THRUM, H., I. HAUPT, G. BRADLER, CH. ZIMMER, and K. E. REINERT: Netropsin and distamycin A: Relations between antibiotic properties, action on nucleic acid and protein biosynthesis *in vivo* and interaction with nucleic acids *in vitro*. Abstr. VIIth Int. Congr. Chemotherap. **1**, A-11-9 (1971).

Toscano, F., and G. Paschetta: A double-blind study on the therapeutic activity of distamycin A on the skin manifestations of some viruses. Abst. VIIth Int. Congr. Chemotherap. **1**, 5/8 (1971).

Tunis, M-J. B., and J. E. Hearst: Optical rotatory dispersion of DNA in concentrated salt solutions. Biopolymers **6**, 1218 (1968).

Verini, M. A., and M. Ghione: Activity of distamycin A on vaccinia infection of cell cultures. Chemotherapia **9**, 144 (1964).

Waller, C. W., C. F. Wolf, W. J. Stein, and B. L. Hutchings: The structure of antibiotic T-1384. J. Am. Chem. Soc. **79**, 1265 (1957).

Waring, M.: Variation of the supercoils in closed circular DNA by binding of antibiotics and drugs: Evidence for molecular models involving intercalation. J. Mol. Biol. **54**, 247 (1970).

Werner, G. H., P. Ganter, and Y. de Ratuld: Studies on the antiviral activity of distamycin A. Chemotherapia **9**, 65 (1964).

Zimmer, Ch., I. Haupt u. H. Thrum: Einfluß von Netropsin und Distamycin A auf die zellfreie Proteinsynthese und die Struktur der Nukleinsäuren. Int. Symp. Wirkungsmech. Fungizid., Antibiot. und Cytostat., S. 61. Berlin: Akademie Verlag. 1970a.

Zimmer, Ch., and G. Luck: Optical rotatory dispersion properties of nucleic acid complexes with the oligopeptide antibiotics distamycin A and netropsin. FEBS Letters **10**, 339 (1970).

Zimmer, Ch., G. Luck, and H. Thrum: Changes in the DNA secondary structure by interaction with oligopeptide antibiotics: Thermal melting, ORD and CD of DNA complexes with netropsin and distamycin A derivatives. Studia biophysica **24/25**, 311 (1970b).

Zimmer, Ch., G. Luck, H. Thrum, and Ch. Pitra: Binding of analogues of the antibiotics distamycin A and netropsin to native DNA. Eur. J. Biochem. **26**, 81 (1972).

Zimmer, Ch., B. Puschendorf, H. Grunicke, P. Chandra, and H. Venner: Influence of netropsin and distamycin A on the secondary structure and template activity of DNA. Eur. J. Biochem. **21**, 269 (1971b).

Zimmer, Ch., K. E. Reinert, G. Luck, U. Wähnert, G. Löber, and H. Thrum: Interaction of the oligopeptide antibiotics netropsin and distamycin A with nucleic acids. J. Mol. Biol. **58**, 329 (1971a).

Zunino, F., and A. DiMarco: Studies on the interaction of distamycin A and its derivatives with DNA. Biochem. Pharmacol. **21**, 867 (1972).

Zygmunt, W. A.: Reversal of netropsin inhibition of growth in *E. coli*. Biochem. Biophys. Res. Commun. **6**, 324 (1961).

Daunomycin (Daunorubicin) and Adriamycin and Structural Analogues: Biological Activity and Mechanism of Action

A. Di Marco, F. Arcamone, and F. Zunino

Introduction

Daunomycin (NSC-82151) is an antibiotic isolated from cultures of *Streptomyces peucetius* (DI MARCO *et al.*, 1963; DI MARCO *et al.*, 1964b). While weakly active against some microorganisms, the drug inhibits the multiplication of bacterial and animal viruses and exhibits a high cytotoxic activity against normal and neoplastic cells (DI MARCO *et al.*, 1964a).

The active product was isolated as crystalline hydrochloride and was shown (ARCAMONE *et al.*, 1964a; ARCAMONE *et al.*, 1964b) to be a glycoside antibiotic of the anthracycline group, such as rhodomycin, cinerubin, pyrromycin, rutilantine.

A product with the same chemical composition was independently isolated from *Streptomyces coeruleorubidus* (DUBOST *et al.*, 1963) and named rubidomycin.

Daunomycin is identical with a constituent (c) of the rubomycin complex (GAUSE, 1966; BRACHNIKOVA *et al.*, 1966). Other antibiotics closely related to daunomycin have been subsequently discovered, namely 14-hydroxy-daunomycin (Adriamycin) (ARCAMONE *et al.*, 1969b), 13-dihydro-daunomycin and daunosaminyl-daunomycin (ARCAMONE *et al.*, 1969c; LUNEL and PREUD'HOMME, 1969). Adriamycin (NSC-123127) is of particular importance owing to its activity against a variety of tumors (DI MARCO *et al.*, 1969).

Chemical Structures

Daunomycin can be hydrolyzed with dilute acids to give a red crystalline substance (daunomycinone), for which a partial structure was presented (ARCAMONE *et al.*, 1964b), and a water soluble basic compound, daunosamine. This compound was shown to be a new aminosugar, i.e. 2,3,6-trideoxy-3-amino-L-lyxohexose (ARCAMONE *et al.*, 1964a). Daunosamine structure and stereochemistry were later confirmed by the stereospecific synthesis of derivatives of the D-enantiomer (RICHARDSON, 1967; BAER *et al.*, 1969), and of daunosamine itself (MARSH *et al.*, 1967). The structure and absolute configuration of daunomycin are now known in detail (ARCAMONE *et al.*, 1968a; ARCAMONE *et al.*, 1968b), and are represented in I. (Fig. 1). Ring A has been found to prefer the half-chair conformation, on the basis of the values of the coupling constants $J_{H-7, H-8}$ and $J_{H-7, H-8'}$, indicating a pseudo-equatorial orientation of H-7, in the nuclear magnetic resonance spectrum of N-acetyl-daunomycin, as well as of other daunomycinone derivatives (ARCAMONE et al., 1970) (Fig. 1).

Structure and stereochemistry of daunomycin have been confirmed by X-ray analysis of N-bromoacetyl-daunomycin. These studies have shown the six-membered daunosamine ring in the chair conformation oriented perpendicularly to the planar chromophore (ANGIULI et al., 1971).

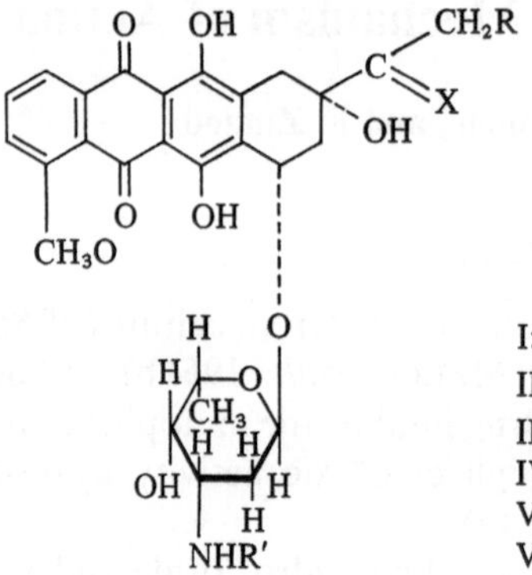

	R	R'	X
I:	H	H	O
II:	OH	H	O
III:	H	H	NOH
IV:	H	H	NNHCONH₂
V:	H	H	NNHCSNH₂
VI:	H	COCH₂NHC(NH)NH₂	O

Fig. 1. Chemical structure of daunomycin (I) and of adriamycin (II) and derivatives

Other Derivatives of Daunomycin

Some semisynthetic derivatives of daunomycin were subjected to biological investigations. These compounds (Fig. 1) include derivatives of the side chain C-13 carbonyl group such as daunomycin oxime, semicarbazone and thiosemicarbazone (III-V), as well as N-acetylated compounds, such as N-acetyl-daunomycin (ARCAMONE et al., 1968a; ARCAMONE et al., 1970) and N-guanidinoacetyl-daunomycin (VI).

Two new glycosides of daunomycinone, namely 7α-(α-D-glucopyranosyl)-daunomycinone and 7α-(2-amino-2-deoxy-α-D-glucopyranosyl)-daunomycinone have been prepared by PENCO (1968).

Adriamycin (II) is a metabolite of S. peucetius var. caesius (ARCAMONE et al., 1969b). This compound gave, on acid hydrolysis, the aglycone, adriamycinone, and the amino-sugar, daunosamine. The structure of adriamycinone (14-hydroxy-daunomycinone) was established, inter alia, on the basis of 1. the identity of adriamycinone and bisanhydroadriamycinone chromophores with those of daunomycinone and of bisanhydro-daunomycinone, 2. the n.m.r. spectrum of pentaacetyl-adriamycinone, and 3. the comparison of the mass spectrum of adriamycinone and daunomycinone. The position of the glycosidic linkage was deduced from the formation of 7-deoxy-adriamycinone on hydrogenolysis of II. The attribution of the absolute stereochemistry to C-7, C-9 and C-1' was established from the comparison of molecular rotations and of circular dichroism curves of compounds I and II as well as from biogenetic arguments (ARCAMONE et al., 1969a). II has also been obtained by chemical synthesis from I (ARCAMONE et al., 1969c; ARCAMONE et al., 1969e).

Growth Inhibitory Properties

1. Microorganisms

Daunomycin has a moderate inhibitory effect on the growth of some gram-positive and gram-negative bacteria, and fungi (for a detailed review of the literature before 1967 consult DI MARCO, 1967).

Different strains of the same organism such as *E. coli* K12 and *E. coli* B have different susceptibilities.

Bacteria, resistant to daunomycin have been obtained by cultivation in presence of the antibiotic (PITTILLO and HUNT, 1968). Various strains of *Streptococcus faecalis* adapted to grow in the presence of high concentration of daunomycin but not permanently resistant to this antibiotic, acquired permanent cross-resistance to 6 MP and 1-β-D-arabinofuranosylcytosine. On the other hand, a strain of *S. faecalis* resistant to 6 MP was more sensitive to daunomycin than the original strain. Several purines, notably xanthine, reversed the growth inhibition of *S. faecalis* by daunomycin.

2. Viruses

An interference of daunomycin with the multiplication of bacteriophages was reported by SANFILIPPO and MAZZOLENI (1964) and by PARISI and SOLLER (1964).

These observations were confirmed by CALENDI *et al.* (1966) who did not observe any inhibiting effect of daunomycin on the multiplication of two single-stranded DNA bacteriophages (EC9 and S13) at a concentration of daunomycin that gives a 99.9% inhibition of the plaque formation caused by T6 double-stranded DNA phage.

The interference is not related to a direct inactivation of phage particles, or to an inhibition of their absorption to the bacterial cell (SANFILIPPO *et al.*, 1964; PARISI *et al.*, 1964).

The inhibiting effect of daunomycin is also exerted on other double-stranded DNA viruses. It was shown that daunomycin protects cells from Herpes simplex virus (HSV) when given before or a few hours after infection (DI MARCO *et al.*, 1968) during the intracellular phase of virus replication but that it is not effective when added 4 hours after the infection, when mature virions appear. Autoradiographic studies on the interference of daunomycin on the rate of DNA and RNA synthesis in non-infected cells show that HSV replication is inhibited at doses active on nucleolar RNA but not on extranucleolar RNA and on DNA synthesis.

The inhibiting effect of daunomycin on the replication of DNA viruses has been confirmed for herpes simplex and vaccinia viruses by COHEN, HARLEY, and REES (1969).

Further studies by COHEN *et al.* (1972) suggest that, in vaccinia virus infected cells, daunomycin does not prevent uncoating or affect earlier stages of the infection and seems to be more effective in preventing DNA synthesis from the parental than from the progeny genome. In fact, when 5 μg/ml were added at the time of infection, daunomycin completely prevented vaccinia DNA synthesis, but synthesis of DNA polymerase and replicative proteins were unaffected and failed to "switch-off" in the normal way. When daunomycin was removed 5 hours after the onset of infection, DNA synthesis recovered but at a reduced rate. Similarly, DNA synthesis was reduced but not prevented when the drug was added 5 hours after the infection.

Daunomycin also exerts a virucidal effect on DNA viruses (herpes virus, vaccinia virus); this effect increases after exposure of the daunomycin-virus mixtures to light (Verini et al., 1968). As regards activity on animal RNA-viruses, Cohen et al. (1969) noted absence of antiviral activity on influenza virus and a reduced virus production in HeLa cells infected with poliovirus. Brahic and Tamalet (1969) have observed an increase of Sindbis virus replication, following daunomycin treatment. However, daunomycin exerts a virucidal effect on one RNA virus (Newcastle Disease Virus) providing that the daunomycin-NDV mixtures were exposed to light (Verini et al., 1968). Similarly to other inhibitors of DNA or of DNA-dependent RNA synthesis, daunomycin, when added to the incubation medium at different times before and after infection, inhibits the "foci" formation induced by Moloney Sarcoma Virus (MSV) on mouse embryo cells (Casazza et al., 1972) and, in parallel, the replication of MSV (Casazza and Soranzo, unpublished observations).

Activity on Normal and Neoplastic Mammalian Cells in Vitro

By evaluation of the number of mitotic figures (mitotic index) (MI), or of the proliferative activity of cells it was shown that daunomycin has a strong inhibiting effect on the *in vitro* growth of a number of cell strains. From Table 1 it appears that both normal and neoplastic cells are affected. Adriamycin has a comparable effect on many of these cell strains (Table 2).

Cell damage induced by daunomycin is mainly nuclear, such as, in resting cells, a finely granular appearance of the chromatin and marked alterations in the shape and size of the nucleoli, while cytoplasmic changes, such as vacuolization, are moderate and appear only after prolonged treatment with high doses.

The damage to the nucleolar structure appears soon after daunomycin addition to the medium and can be followed by electron microscopy: HeLa cells, after 2 hours of incubation, in the presence of 1 µg/ml daunomycin show disorganization and fragmentation of the nucleolar filamentous structure with detachment of the RNA granules (Fig. 2A, 2B) (Dorigotti, 1964). After 24 hours of treatment the nucleolus appears as a very compact structure with appearance of vacuoles (Fig. 3A, 3B). The injuries to the nucleolar structure should be considered a consequence of the inhibition caused by daunomycin on ribosomal RNA synthesis.

During mitosis, chromosomal damage, such as fragmentations, and mitotic aberrations (mainly anaphasic bridges) have been described (Di Marco et al., 1964) (Fig. 4). If a sufficient amount of daunomycin is added to rat fibroblast cultures, it is possible to observe, by phase microscopy, an immediate block of the mitotic process and an anomalous scattering of the chromosomes within the cell.

Table 1. *Inhibitory effect of daunomycin on different cell strains*

Cell strain	Species	Methods of tissue culture	50% inhibiting dose µg/ml			References
			M.I.	C.P.	DNA synthesis	
HeLa cells	human	tissue fragments	0.1			1
HeLa cells	human	monolayer	0.1	0.5	0.9	2
HeLa cells S-3	human	monolayer		0.01	1	3
Fibroblasts	rat	tissue fragments	0.02			1
Helius strain	human	tissue fragments	0.1			1
KB	human	monolayer	0.1			1
Walker ca.	rat	tissue fragments	0.1			1
Bone marrow	mouse	tissue fragments	0.1			1
Blood lymphocytes	human	stationary culture	0.03			4
C3H/f (lymphoid tissue)	mouse	monolayer	0.6	0.95		5
JLS-V6	mouse	monolayer	0.001	1		5
JLS-V8	mouse	monolayer	0.004	0.1		5
PF	human	monolayer	0.004	0.9		5
MS (MSV-M-sarcoma)	mouse	monolayer	0.001	0.87		5
3T3 BALB/c	mouse	monolayer	0.06	0.1		5
3T12 BALB/c	mouse	monolayer	0.01	0.1		5
SVT12 BALB/c	mouse	monolayer	0.02	0.1		5

M.I. = Mitotic index; C.P. = Cell proliferation.

References: 1. DiMarco et al. (1964c). 2. DiMarco et al. (1971b). 3. Kim et al. (1968). 4. Whang-Peng et al. (1969). 5. Necco and Dasdia; to be published.

Table 2. *Inhibitory effect of adriamycin on different cell strains*

Cell strain	Species	Methods of tissue culture	50% inhibiting dose µg/ml			References
			M.I.	C.P.	DNA synthesis	
HeLa cells S-3	human	monolayer		0.01	1	1
HeLa cells	human	monolayer	0.05	0.9	2	2
C3H/f (lymphoid tissue)	mouse	monolayer	0.04	2		3
JLS-V6	mouse	monolayer	0.045	1		3
JLS-V8	mouse	monolayer	1	0.1		3
PF	human	monolayer	0.035	1		3
MS (MSV-M-sarcoma)	mouse	monolayer	0.001	2		3
3T3 BALB/c	mouse	monolayer	0.1	0.1		3
3T12 BALB/c	mouse	monolayer	0.05	0.4		3
SVT2 BALB/c	mouse	monolayer	0.02	0.4		3

M.I. = Mitotic index; C.P. = Cell proliferation.

References: 1. Kim et al. (1972). 2. DiMarco et al. (1971b). 3. Necco and Dasdia, to be published.

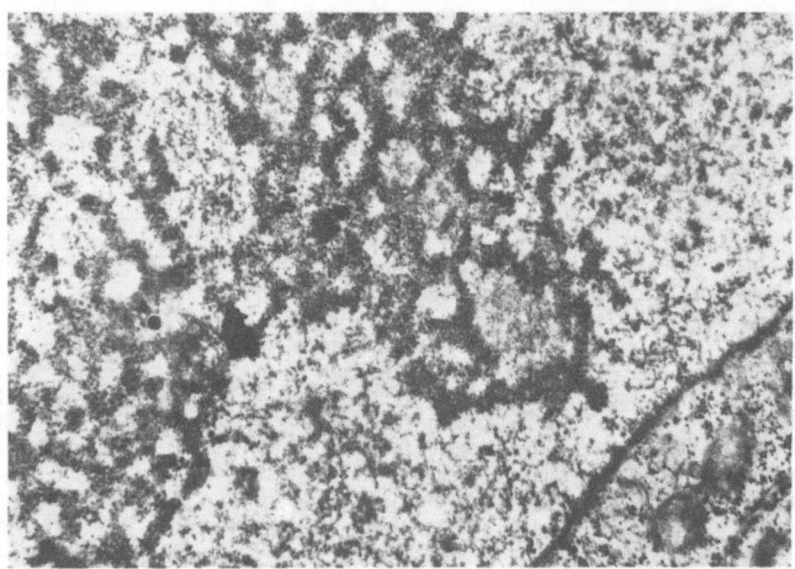

Fig. 2A. Normal HeLa cell showing the filamentous appearance of the nucleolus (from Dorigotti, 1964)

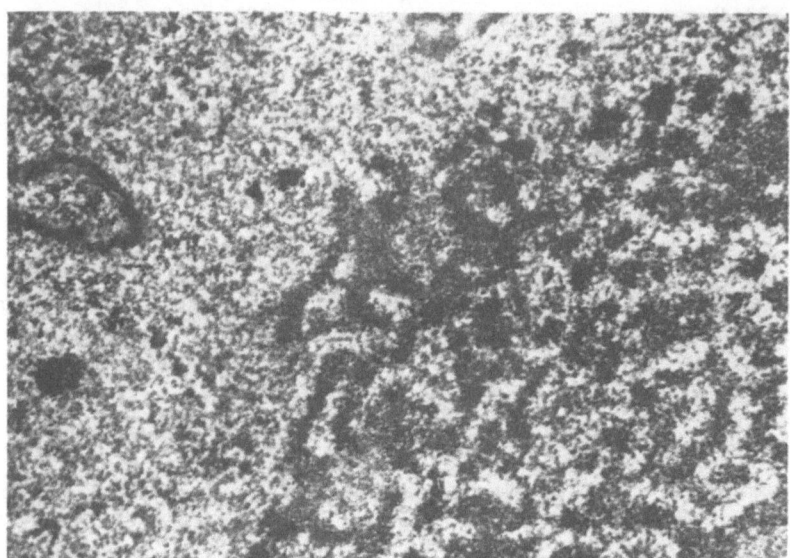

Fig. 2B. HeLa cell after 2 hrs exposure to 1 µg daunomycin/ml. Nucleolonema is fragmented and shows a progressive loss of RNA granules (from Dorigotti, 1964)

▶

Fig. 3A. HeLa cell after 18 hrs exposure to 1 µg daunomycin/ml. The nucleolus is changed into a dense and compact body showing large areas in which RNA granules have completely disappeared (from Dorigtti, 1964)

Fig. 3B. HeLa cell after 48 hrs exposure to 1 µg daunomycin/ml (from Dorigotti, 1964)

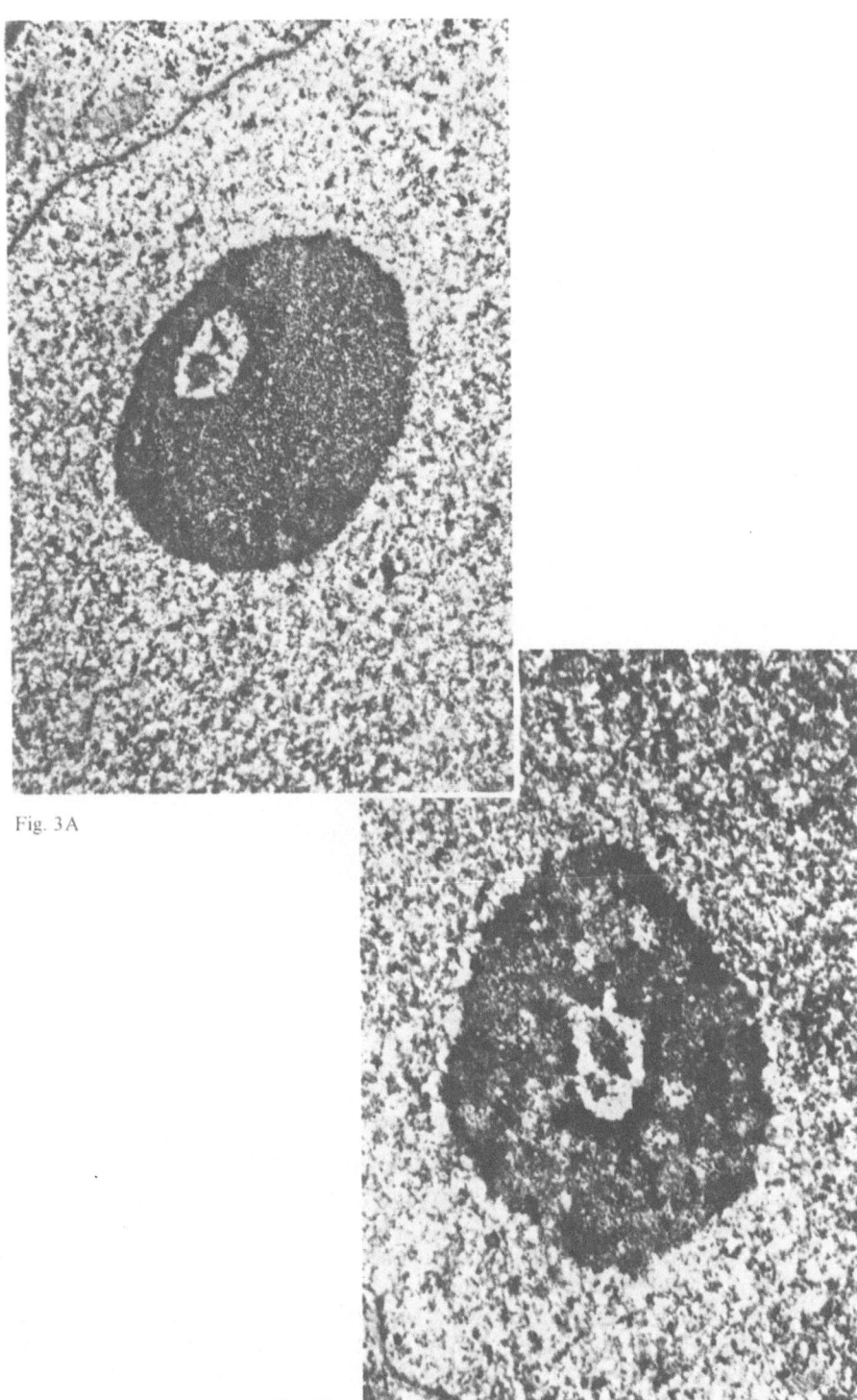

Fig. 3 A

Fig. 3 B

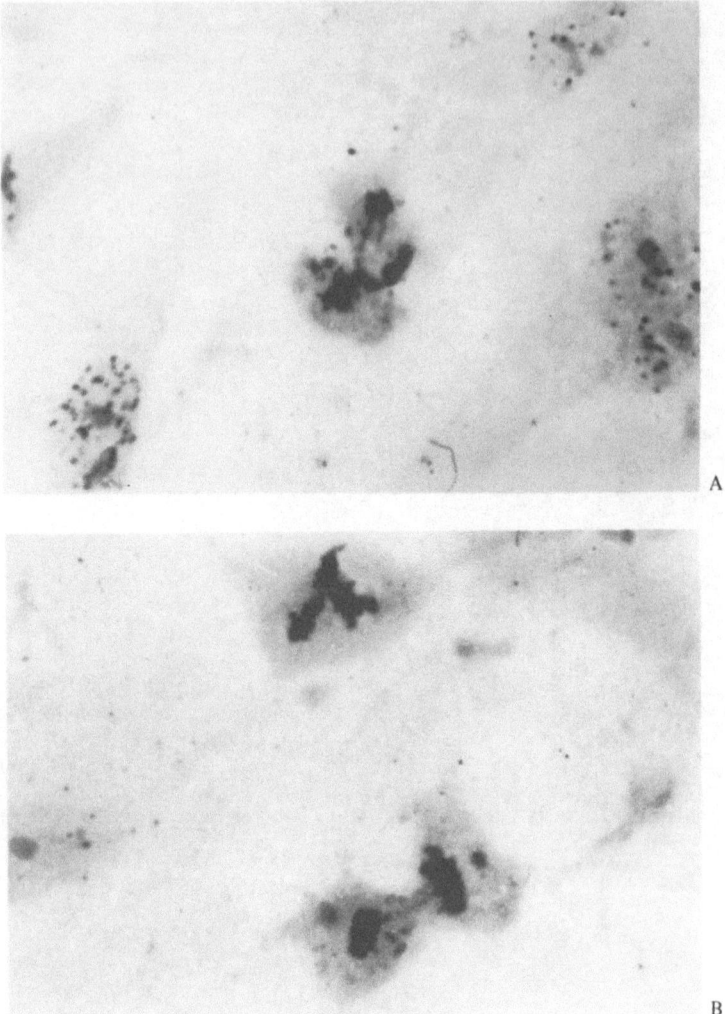

Fig. 4A and B. Mitotic anomalies in rat fibroblasts treated with 0.1 µg/ml. of daunomycin (from Di Marco, 1970)

Activity on Experimental Tumors

Daunomycin has a strong inhibitory effect on ascites tumor (Di Marco, 1970); solid tumors show a lesser susceptibility to daunomycin when compared to the ascitic forms.

Daunomycin induces a significant increase of survival time in mice bearing L1210 leukemia (Venditti *et al.*, 1966).

Adriamycin induces an increase in survival time greater than daunomycin in animals bearing Ehrlich ascites tumor (Di Marco *et al.*, 1969) or L1210 leukemia (Sandberg *et al.*, 1970) and exerts a high inhibitory effect on the growth of many other experimental tumors (Di Marco *et al.*, 1969; Di Marco, 1971).

The effectiveness of adriamycin in sarcoma 180 is about twice that of daunomycin, the therapeutic index being 1.21 in the case of adriamycin and 0.67 in the case of daunomycin. An important antitumor effect is observed with different treatment schedules also in transplantable and spontaneous mammary carcinoma in C3H mice (DI MARCO et al., 1972a).

In MSV-induced tumors in 16-day old mice, high doses of daunomycin (CASAZZA et al., 1971) cause a transient inhibition of tumor growth, followed by strong recurrence of tumors, which progresses to death, while in untreated controls, complete and permanent tumor regression is observed. Tumor recurrence is rare after daunomycin treatment in 20-day old mice and is absent after daunomycin or adriamycin treatment in mice 28-day old at the time of infection. Tumor recurrence can be attributed to the action of daunomycin on the induction and multiplication of immunocompetent cells, able to react with the virus or virus-induced antigens in tumor cells. When compared to daunomycin, adriamycin was more effective in inhibiting tumor growth, while tumor recurrence occured later and to a lesser extent. When administered before the infection, daunomycin completely inhibited tumor regression, while adriamycin did not affect this process; under these experimental conditions, the difference between the two antibiotics is particularly evident.

Toxicology and Pharmacology

A number of studies on the toxicity of daunomycin (DI MARCO et al., 1964a; DI MARCO, 1967) and adriamycin (DI MARCO et al., 1969; BERTAZZOLI et al., 1970; BERTAZZOLI et al., 1972) has been published. As it can be expected for drugs which interfere with nucleic acids synthesis, the tissues most susceptible to the two antibiotics are rapidly proliferating such as intestinal mucosa, lymphoid organs and bone marrow. Few signs of cardiac toxicity have been seen in experimental animals; some EKG alterations have been observed in the hamster (HERMAN et al., 1969; HERMAN et al., 1970) and in the monkey (BURKA et al., 1970) possibly mediated by a central mechanism involving the sympathetic nervous system (BURKA et al., 1970; HERMAN et al., 1970).

No impairment of oxidative phosphorylation by heart mitochondria has been observed. A lesion characteristic to daunomycin, is the development of a nephrotic syndrome in rats given one high single dose (STERNBERG and PHILIPS, 1967; STERNBERG, 1970).

As in the case of most antitumor agents, daunomycin and adriamycin also have been shown to be immuno suppressive (CASAZZA et al., 1971; NATALE and MOCARELLI, 1969; ISETTA et al., 1971).

When administered subcutaneously to infant mice and rats, both daunomycin and adriamycin induce fibrosarcomas at site of inoculation (CASAZZA, unpublished observations). When administered intravenously to young rats, daunomycin induces mammary and other carcinomas while rats given adriamycin develop prevalently benignant neoplasms (BERTAZZOLI, personal communication). Both drugs have proved to be slightly or not teratogenic for chick, mouse and rabbit embryos (BERTAZZOLI, personal communication).

The distribution and excretion of daunomycin and adriamycin have been studied using tritiated drugs, or taking advantage of the fluorescence of these antibiotics (Rusconi et al., 1968; Bachur et al., 1970; Di Fronzo et al., 1971; Di Fronzo and Gambetta, 1971; Picone and Traine, 1970; Dusonchet et al., 1971; Yesair et al., 1971).

Both drugs are rapidly cleared from the blood stream and fixed by different tissues and organs. In rodents, high drug levels are maintained over a long period of time especially in bone marrow, spleen and lymphoid tissues. The two antibiotics do not appear to cross the blood-brain barrier.

Both drugs are slowly excreted into urine and the bile. In rats, the excretion of adriamycin equivalents was about one third that of daunomycin equivalents (Yesair et al., 1971). In mice, 50% excretion of the drug occurs in about 24 hours in the case of daunomycin, and in about 32 hours in the case of adriamycin; moreover, drug accumulation in tissues is more pronounced after adriamycin than after daunomycin subacute treatment (Di Fronzo et al., 1971). The calcula-ted "concentration × time" (C × t) in the case adriamycin equivalents has been demonstrated in all tissues to be several times greater than that found for dauno-mycin or its metabolite (Yesair et al., 1971). Preliminary work (Di Marco et al., 1967b; Di Marco and Rusconi, 1967) provided evidence of extensive metabolism of daunomycin by liver homogenates. The main metabolic product showed chro-matographic and spectral properties very similar to the substance obtained from daunomycin by the action of sodium dithionite; therefore, it was assumed that the metabolic transformation consisted of a reductive modification of the chromo-phore, associated with the breakage of the glycosidic linkage. Besides the described transformation product, a small amount of another substance with a different Rf was observed in thin layer chromatography. More recent studies (Bachur and Gee, 1971; Bachur, 1971; Yesair et al., 1971) have cleared up the major pathway of daunorubicin metabolism in tissue homogenates.

Daunorubicinol is the only daunomycin metabolite present in the urine and bile of mice, rats and man (Bachur, 1971). The presence of the hydroxyl group at position 14 of the chromophore protects against the metabolic transformation; as a matter of fact, no metabolic product, like daunorubicinol, was observed by incubation of adriamycin with tissue homogenates or in excreted materials (Yesair et al., 1971; Di Fronzo et al., 1971). These findings are consistent with the higher toxicity and activity of adriamycin in vivo. Daunorubicinol has a significant cytotoxic activity in vitro and was also shown to be active in vivo against P 338 lymphocytic leukemia (Yesair et al., 1971). It could, therefore, contribute to the antitumor and pharmacological activity of daunomycin.

Clinical Toxicity

The most frequent signs of toxicity caused in patients by both antibiotics (Bernard et al., 1969) are stomatitis, alopecia, bone marrow depression and gastrointestinal disturbaces. No significant kidney damage has been reported. A peculiar toxic effect of these drugs is a cardiopulmonary syndrome (Bonadonna and Monfardini, 1969) consisting of tachycardia, with or without arrhythmia, hypotension, gallop rhythm, tachypnea, and congestive heart failure, not relieved by digitalis. The mechanism of this effect has not as yet been elucidated; no constant EKG alteration has been reported. Damage to the intrinsic cardiac neurons has been recently observed (SMITH, 1969).

Cardiac toxicity is more frequent when high doses (more than 25–30 mg/kg) are given, and in elderly patients: when an intermittent, rather than a daily dose schedule is used, cardiac toxicity is rarely observed. Both drugs are vesicant if extravasated; intravenous injection through the tube of a running infusion is indicated.

Therapeutic Indications

The main therapeutic indications of daunomycin and adriamycin are leukemias, especially acute lymphocytic leukemia in children. Owing to the possible occurrence of cardiac toxicity, the drugs are used, alone or in combination with other antitumor agents, mainly for remission induction and reinforcement dosing during remissions rather than for maintenance therapy. Both drugs are also active on lymphomatous neoplasms (Hodgkin's disease, lymphosarcoma, reticular cell sarcoma).

Daunomycin has proved to have little or no effect on solid tumors, whereas adriamycin (BONADONNA et al., 1970) appears to be temporarily effective in a range of childhood tumors such as Ewing's sarcoma, neublastoma, Wilm's tumor, osteogenic sarcoma and some anaplastic sarcomas. In adults, evidence of some chemotherapeutic effectiveness has been obtained in breast cancer, seminoma, gestional choriocarcinoma, rhabdomiosarcoma and other soft tissue sarcomas, transitional cell carcinoma of the bladder, and lung cancer.

Mode of Action

The fluorescence of daunomycin allows to show the penetration of this substance into the cell and its fixation in the nuclear structure (Fig. 5). The radiochemical data confirm that a large part of daunomycin which enters the cell is taken up by the nucleus; however, some labelling is also found in mitochondrial and microsomal fractions (RUSCONI, unpublished observations). After extraction, according to Crampton, it was shown (CALENDI et al., 1965) that nearly all the antibiotic is found in the nucleoprotein fraction extracted with 2 M NaCl.

Experiments with ^{3}H-daunomycin confirmed that after a short period of in vitro incubation, a large part of daunomycin is bound to the nucleoprotein fraction.

The uptake of daunomycin by HeLa cells is dependent upon the concentration of the drug in the incubation medium (RUSCONI and DI MARCO, 1969) (Fig. 6). In particular, in synchronized cultures of rat fibroblasts, it was observed (SILVESTRINI et al., 1970) that the uptake of ^{3}H-daunomycin occurs through the regenerative cycle, and reaches its maximum during DNA replication as revealed by ^{3}H-thymidine uptake curve (Fig. 7).

In cultures of mammalian cells, daunomycin, like actinomycin, strongly inhibits adenine-8-^{14}C incorporation into RNA (RUSCONI and CALENDI, 1966). An attempt to calculate the quantitative relationship between the amount of dauno-

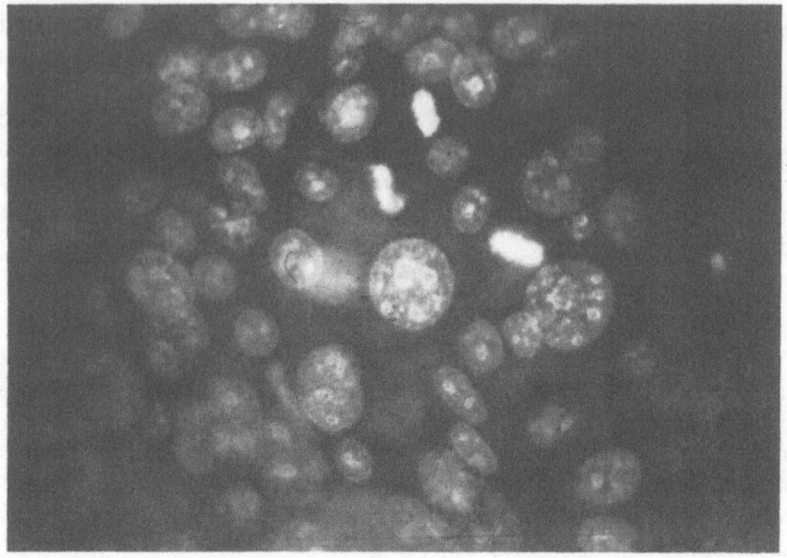

Fig. 5. Daunomycin fluorescence in HeLa cells treated for 5 min with 5 µg/ml. (from Di Marco, 1970)

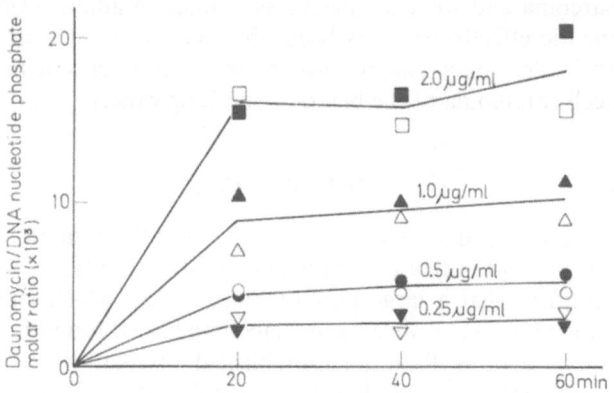

Fig. 6. Time-course of daunomycin-³H uptake by HeLa cells in the presence of different concentration of tritiated daunomycin. The uptake is given as moles of daunomycin bound per mole of DNA nucleotide phosphate of the cells. Closed symbols refer to the samples in which the number of cells (1.2×10^5) was twice as great as in samples indicated by open symbols (from Rusconi *et al.*, 1969)

mycin binding and the inhibition of uridine incorporation (Rusconi *et al.*, 1969) led to the conclusion that a 50 percent inhibition was attained when one molecule of daunomycin was bound per 62 to 100 molecules of DNA nucleotide phosphate (Fig. 8). The interference with the incorporation of labeled precursors into RNA should be related to the inhibition exerted by daunomycin on the activity of DNA-dependent RNA polymerase (Hartmann *et al.*, 1964; Ward *et al.*, 1965); this will be discussed in the next section.

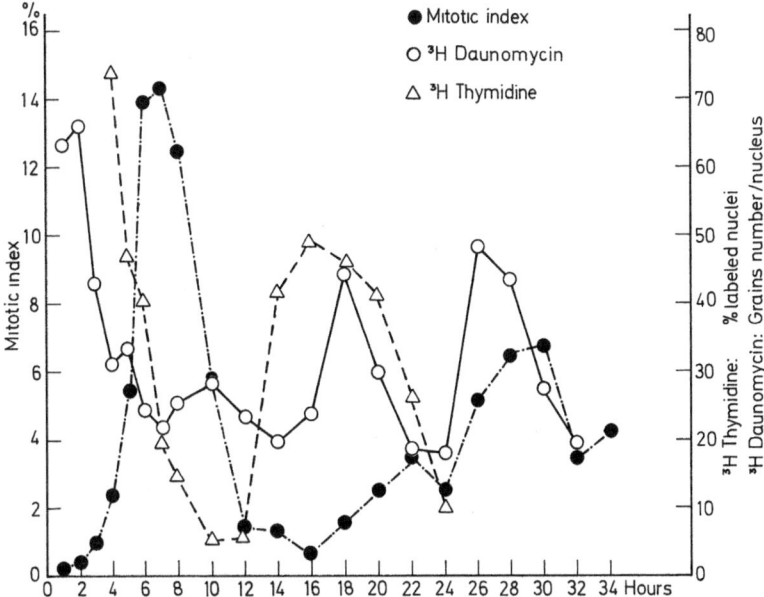

Fig. 7. Mitotic index, ³H-daunomycin uptake and ³H-thymidine incorporation in rat fibroblasts partially synchronized (from DI MARCO, 1970)

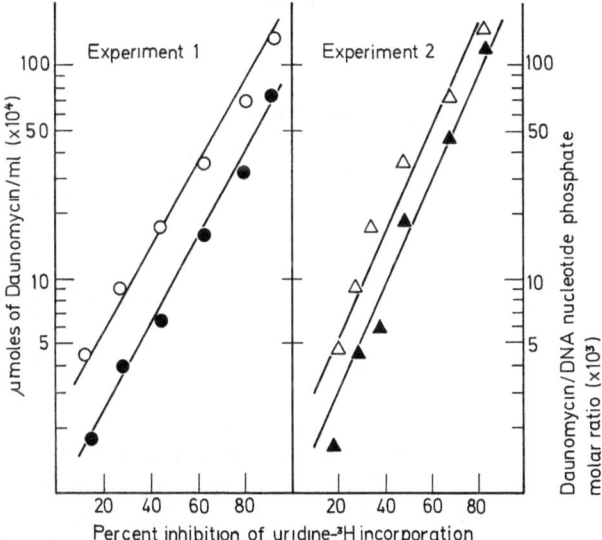

Fig. 8. Dose-response curves of HeLa cells incubated in the presence of ³H-uridine and increasing concentrations of daunomycin (open symbols, ordinate, left scale). Values of daunomycin are in μ moles/ml. Under the same experimental condition, HeLa cells were also incubated without ³H-uridine but in the presence of labelled daunomycin. The amount of the antibiotic taken up by cells is reported in ordinate, (right scale, closed symbols). Values are in moles of ³H-daunomycin per mole of DNA nucleotide phosphate of the cells. Data in ordinate are plotted against the percent inhibition of ³H-uridine incorporation (abscissa). Cells incubated with ³H-daunomycin were collected without any washing by centrifugation (exper. 1) or by filtration through glass fiber discs (exper. 2) (from RUSCONI et al., 1969)

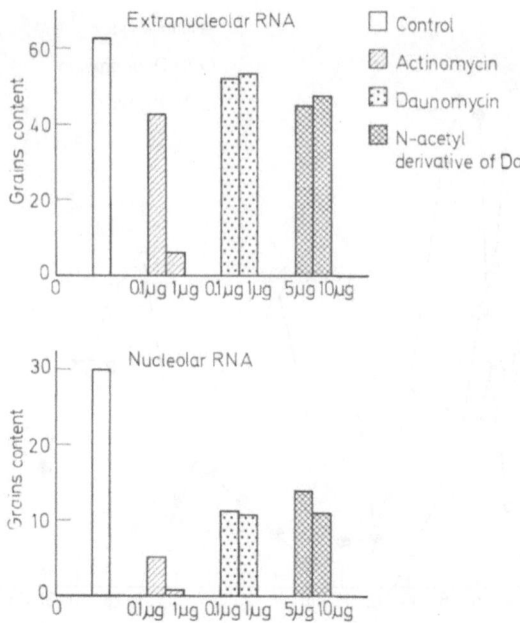

Fig. 9. Uridine-³H incorporation into extranucleolar RNA and nucleolar RNA in control cells and in cells treated with different antibiotics after a 45-min treatment and a 30-min contact with the precursor (from Di Marco et al., 1965)

Autoradiographic studies carried out in *in vitro* cultures of HeLa cells (Di Marco *et al.*, 1965a; Di Marco *et al.*, 1965b) showed a higher susceptibility to daunomycin of uridine-³H incorporation into nucleolar RNA than into extranucleolar RNA (Fig. 9), similarly to what previously observed by Perry (1963) with actinomycin D.

These observations account for the results of Rusconi *et al.*, (1966), showing a high degree of inhibition of synthesis of ribosomal RNA. An RNA fraction, characterized by a base ratio $G+C/A+U$ of about 1, tentatively identified as messenger RNA, appeared to be less sensitive. These results were confirmed by a recent report (Crook *et al.*, 1972). At concentrations of antibiotic that do not completely suppress RNA synthesis, daunomycin inhibited ribosomal RNA synthesis but only partly that of "heterogenous RNA" which was altered in physical characteristics. The preferential inhibition of ribosomal RNA synthesis is similar to that produced by small doses of actinomycin D or mitomycin C (Penman *et al.*, 1968) and of nogalamycin (Ellen and Rhode, 1970). In these cases, the greater sensitivity of the synthesis of ribosomal RNA appears to be a property of the specific transcription process for this species of RNA and is not linked to the specific mechanism of action of the antibiotics. The action of actinomycin or nogalamycin may reflect gene or operon size differences, rather than DNA base specificity.

Interference with DNA synthesis was observed by following the incorporation of thymidine-³H into nucleic acids of cell suspension of Yoshida ascites hepatoma

(RUSCONI and CALENDI, 1964) and in cultured HeLa cells (RUSCONI *et al.*, 1969). The inhibiting effect of daunomycin, cinerubin and toyomycin on the incorporation of nucleic acid precursors into DNA is correlated with the interference by these substances with the activity of DNA-dependent DNA polymerase (HARTMANN *et al.*, 1964).

Conflicting results were reported about the relative activity of the drug on the incorporation of labeled precursors into RNA and DNA *in vivo* or in cultured cells (RUSCONI *et al.*, 1964; RUSCONI *et al.*, 1966; SILVESTRINI *et al.*, 1963; SILVESTRINI and GAETANI, 1963; THEOLOGIDES *et al.*, 1968).

In experiments in cultured HeLa cells (RUSCONI and DI MARCO, 1969) it was demonstrated that experimental conditions, such as the time of addition of daunomycin to the medium or the cell population density, can affect the relative sensitivity of RNA and DNA synthesis to daunomycin. When, however, the antibiotic and the precursors are added to the culture medium at the same time, there are no significant differences in the relative inhibition of precursors uptake in DNA or RNA (Fig. 10).

In a recent report (SILVESTRINI *et al.*, 1970), the influence of daunomycin on nucleic acids synthesis, during different phases of the cell cycle in cultures of rat fibroblasts, synchronized by thymidine excess, was studied. DNA or RNA synthesis was determined by autoradiographic methods. The effect of daunomycin is more evident on DNA synthesis which occurs in the late S phase, tentatively identified with heterochromatin duplication (SIMARD, 1967) and on RNA synthesis which occurs in the middle of G_1 and in the G_2 phase just before the mitotic peak. Clearly, the inhibition of these important metabolic events can explain the arrest of cellular reproduction; however, an immediate antimitotic activity which seems to be unrelated to the inhibition of DNA and RNA synthesis is also present.

KIM *et al.* (1968) observed that the lethal effect of daunomycin on a synchronously dividing population of HeLa cells is higher in S phase than in G_1 or G_2. By analogy of the action of daunomycin and actinomycin D on HeLa cells,

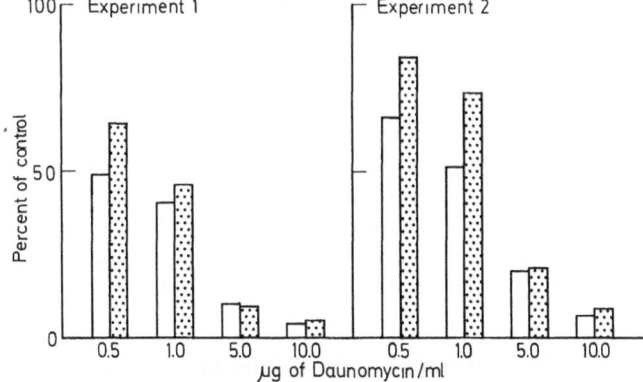

Fig. 10. Action of increasing concentrations of daunomycin on the incorporation of uridine-^{3}H ("clear bars") and thymidine-^{3}H ("shaded bars"). In experiment 2, in the same incubation volume a number of cells (1.1×10^5) twice higher than in experiment 1 was used. The drug and the precursors were added together at the beginning of the incubation time (60 min) (from RUSCONI *et al.*, 1969)

this suggests that "accessibility of DNA in chromosomal structures and repair processes may vary during the division cycle and cell death may be the outcome of at least two reactions which would be differently affected by different DNA-binding drugs". As a strong binding to DNA of daunomycin occurs, in ordinary conditions, only with double-stranded DNA, one might assume that only during the process of replication the drug could find access to the DNA. In this case, a few molecules of drug could irreversibly damage all chromosomal structure.

Resistance to Daunomycin

BIEDLER and RIEHM (1970) observed in chinese hamster cells cross resistance between actinomycin D and daunomycin; this seems noteworthy in view of the many similarities in the mechanism of action of the two antibiotics. However, cell lines resistant to actinomycin D are resistant also to other drugs, such as mithramycin, vinblastine, vincristine and pirromycin. The degree of resistance to actinomycin D was inversely related to the degree of nuclear labeling by actinomycin D–^{3}H and to inhibition of uridine–^{3}H incorporation by the antibiotic. According to these authors, their results support the hypothesis that the development of resistance to actinomycin D in hamster cells is due to qualitative difference in cell membranes, resulting in decreased permeability to daunomycin and other compounds.

Cross-resistance between daunomycin and Vinca alkaloids was observed *in vivo* in lines of Ehrlich ascites tumor in which a permanent resistance to daunomycin developed as a result of long term treatment with subinhibiting doses (DANO, 1971). The subline of leukemia P815 selected for resistance to vinblastine was shown also to be resistant to daunomycin (KESSEL et al., 1968). These authors demonstrated also a good correlation between daunomycin sensitivity *in vivo* of a transplantable murine leukemia (P388) and cell ability to retain the drug. Neither cross-resistance nor collateral sensitivity to methotrexate or 6 MP was observed by DANO (1971) in Ehrlich ascites, resistant to daunomycin; a collateral sensitivity to Ara-C and BCNU was explained on the basis of the slower growth of the resistant by comparison to the sensitive line. The cross-resistance between daunomycin, actinomycin and Vinca alkaloids may be understood as a result of a limited permeability of cells to these drugs. Other unknown factors may contribute, however, to the altered sensitivity of cells to the drug. In fact, in a resistant subline of Ehrlich ascites tumor, the difference in uptake of labeled daunomycin is not large enough to explain the difference in inhibition of nucleic acid synthesis (DANO, personal communication).

Complete cross-resistance to adriamycin was observed in daunomycin resistant sublines of L1210 leukemia (HOSHINO et al., 1971); this appears to confirm that the modes of action of the two antibiotics are very similar.

Mechanism of Action: Binding to DNA

The main biochemical effect of daunomycin and related compounds is concerned with nucleic acid synthesis. The binding of daunomycin to DNA is thought to be responsible for its interference with template DNA function.

The data which must be accounted for in any formulation of action of dauno-mycin at a fundamental level, are derived from several different approaches and are presented below.

Daunomycin forms strong complex with DNA which is associated with 1. Hypochromic and bathochromic visible changes in the antibiotic, 2. Reduction of antibiotic fluorescence (CALENDI *et al.*, 1965), 3. Increase in the melting temperature of DNA (KERSTEN and KERSTEN, 1965; ZUNINO *et al.*, 1972), 4. Decrease in the buoyant density of DNA (KERSTEN *et al.*, 1966), 5. Increase in the intrinsic viscosity of DNA (ZUNINO, 1971), 6. Decrease of the sedimentation coefficient of DNA (CALENDI *et al.*, 1965; KERSTEN *et al.*, 1966), 7. Protection of DNA from degradation by nucleolytic enzymes (ZELEZNICK, and SWEENEY, 1967; KERSTEN and KERSTEN, 1968). The stability of the daunomycin complex with DNA was shown by exhaustive dialysis experiments (WHITE and WHITE, 1969). The anthracycline cannot be removed from DNA by CsCl in high concentrations (KERSTEN *et al.*, 1966) (under these conditions, the acridines are displaced from DNA).

However, daunomycin is removed from DNA by Mg^{2+} (CALENDI *et al.*, 1965) and it can be extracted from an aqueous solution of the complex by organic solvents such as phenol or n-butanol (DI MARCO *et al.*, 1972); at high ionic strengths, the binding of anthracyclines to DNA is reduced (CALENDI *et al.*, 1965; BHUYAN and SMITH, 1965). Complex formation appears to involve electrostatic as well as other stronger interactions.

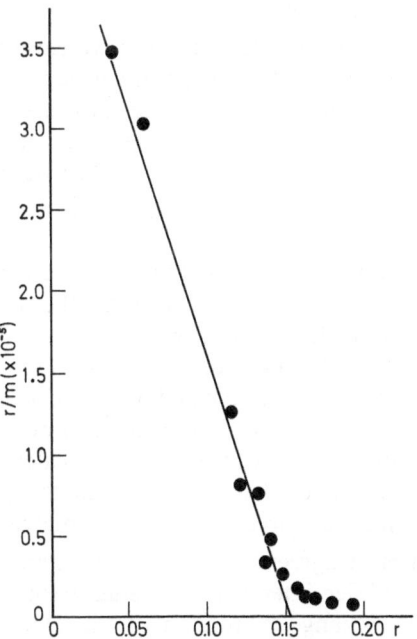

Fig. 11. Scatchard plot of the binding of daunomycin to calf thymus DNA. *r* moles of bound daunomycin/total moles DNA nucleotides; *m* moles free daunomycin. Data were obtained from equilibrium dialysis studies using ^{3}H-daunomycin (from ZUNINO *et al.*, 1972)

The shape of the isotherm (Fig. 11) for the binding of daunomycin to DNA (Zunino et al., 1972) and the chromatographic behaviour of daunomycin on DNA-cellulose column (Zunino, 1971) suggest the existence of more than one class of binding sites for the antibiotic on DNA. At least two modes of interaction between antibiotic molecules and DNA are distinguished; the "strongly" bound antibiotic molecules are thought to be intercalated between base pairs of the double helix; the "weakly" bound antibiotic molecules are conjectured to be attached to DNA by means of electrostatic interaction, involving the DNA phosphate groups and the daunomycin amino group.

The association constant (Zunino et al., 1972) for the stronger binding process ($K_{ap} = 3 \times 10^6 \text{ M}^{-1}$) is of the same order of magnitude as those found for acridine dyes (Blake and Peacocke, 1968) and actinomycin D (Müller and Crothers, 1968). Similar values were derived from polarographic data (Berg and Eckardt, 1970).

The equilibrium binding data derived from spectrophotometric analysis and equilibrium dialysis (Zunino et al., 1972) are consistent with the proposed intercalation model.

Additional evidence in support of intercalation comes from viscosity measurements (Zunino, 1971) and X-ray-diffraction studies (Pigram et al., 1972).

Detailed molecular model-binding studies (Pigram et al., 1972) (Fig. 12) have shown that the intercalation model for the daunomycin complex with DNA is sterically possible. In this model, the amino-sugar of the antibiotic is in the large groove of the DNA and the hydrophobic faces of the base-pairs and the drug overlap extensively.

The amino-sugars lie at the side of the groove close to the sugar-phosphate chain, enabling the positively charged amino group to interact strongly with the second DNA phosphate apart from the intercalation site. It is proposed that the hydroxyl attached to the satured ring of the daunomycin chromophore interacts with the first phosphate by means of a hydrogen bond.

Consistent with the intercalation model are experiments in which daunomycin was found to cause the uncoiling of the supercoiled structure of closed circular DNA in the same qualitative fashion as other drugs believed to intercalate (Waring, 1970). In this study, the values of v_c (the number of drug molecules bound per nucleotide at the equivalence point) were used to calculate ϕ, the apparent unwinding angle per bound drug molecule, based upon $\phi = 12°$ for ethidium bromide. For daunomycin, the calculated value of ϕ is about 5°, yet 12° is supposed to be a minimum for an intercalative process. Also for proflavine, hycathone and nogalamycin, the values of ϕ were calculated to be significantly less than 12°. The discrepancy was attributed to the persistence of a certain proportion of the bound drug molecules in a non intercalated state.

This observation is in agreement with results reported by Saucier et al., (1971). Assuming a modified intercalation model (Paoletti and Le Pecq, 1971), it was shown that the change in torsion of the DNA helix is variable with the intercalated drug. The apparent change of the torsion of the DNA helix (4°) induced by the binding of each daunomycin molecule is about three times smaller than the one observed for ethidium bromide (12°).

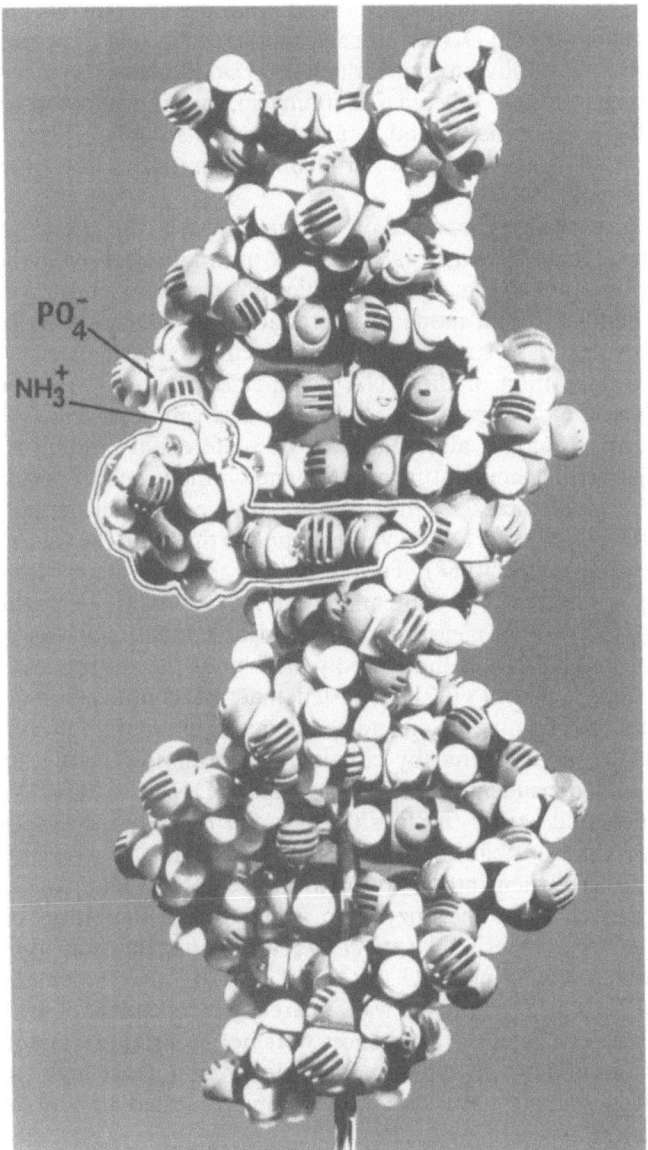

PO$_4^-$

NH$_3^+$

Fig. 12. Model of daunomycin-DNA complex (from PIGRAM *et al.*, 1972)

Like proflavine, nogalamycin and daunomycin are able to completely displace methyl green from DNA (ZELEZNICK *et al.*, 1967; DI MARCO *et al.*, 1971 b). The binding sites on DNA for daunomycin and actinomycin are different, as judged by the difference in the sensitivities of their complexes with DNA to Hg^{2+} (RUSCONI, 1966).

Competition between daunomycin and sibiromycin (DUDNIK and NETYSKA 1972) for binding sites on DNA is not so pronounced as that between sibiromycin and actinomycin D or olivomycin A which show similar base specificty for guanine in their binding to DNA. Whereas, actinomycin, mithramycin and chromomycin are thought to preferentially bind to native double-stranded DNA of high G-C content, it is still uncertain whether daunomycin shows any base preference in binding to native DNA.

Disparate results were reported on base specificity in the interaction of dauno-mycin and nogalamycin with DNA. These two antibiotics belong to the anthracyc-line group and possess many similarities in structure (tetrahydrotetracene-quinones). Similarly, their mode of interaction with DNA shows common fea-tures. Concerning base specificity, nogalamycin has been extensively studied. Nogalamycin inhibition of the RNA polymerase reaction has been correlated with the presence of alternating adenine and thymine in the DNA template (BHUYAN et al., 1965). It also manifests its preferential binding to dAT pairs by raising the melting temperature of DNA and synthetic helical polydeoxynucleo-tides by ΔTm which increase with increasing dAT content in the polymer. How-ever, poly dG·dC-directed DNA synthesis, unlike that with the RNA polymerase, was inhibited (HONIKEL, 1968). To add further to these inconsistencies, studies with DNAs of different GC contents revealed that binding of nogalamycin and daunomycin to DNA, as measured by its buoyant density, increases with higher GC contents in the DNA, although poly dAT binds the antibiotics.

In addition, KERSTEN (1971) reported that the conformational changes of the chromophore of daunomycin and nogalamycin induced upon complex forma-tion with DNA, as measured by circular dichroism spectra of the antibiotics, are only sligthly dependent on the GC content of DNA.

From these and earlier findings (KERSTEN et al., 1966), KERSTEN concluded that anthracycline antibiotics do not show preference or specificty for bases or base sequences when interacting with native DNA. This conclusion differs from that derived from the enzymatic study on the inhibition of the DNA-dependent RNA synthesis (WARD et al., 1965); nogalamycin did not impair the template function of poly dG · dC or of poly dI · dC. In contrast, inhibition of RNA synthesis by daunomycin was apparently independent of the base compo-sition of DNA. The apparent differences were not so striking when more concen-trated solutions of daunomycin were used. However, at these high concentrations of daunomycin, the most sensitive of the templates tested all contained adenine-thymine base pairs.

Recently, template specific inhibition of DNA polymerase from RNA tumor virus by daunomycin and its derivatives has been reported (CHANDRA et al., 1972). The inhibition exerted by these antibiotics upon the enzyme reaction is selectively dependent on the type of the primer-template used in the assay; it is much more pronounced when poly (dA-dT) or poly rA · oligo dT are used as primer-template, as compared to poly (dI·dC). In some cases, a stimulation of dGMP–^{3}H incorporation was observed. These observations lead to conclusions similar to those obtained with nogalamycin (BHUYAN et al., 1965) suggesting a preferential interaction of daunomycin with A-T base pairs of DNA. It is not possible from the information at hand to decide whether the presence of adenine-thymidine base pairs in a DNA is a necessary pre-requisite to the binding of daunomycin. The role of base composition and of conformations of DNA as determinants in binding reaction needs to be further elucidated.

The availability of some natural and synthetic daunomycin derivatives has led to some information on the structural requirements of the antibiotic for biological activity and complex formation with DNA. The tested derivatives are listed in Table 3. In a preliminary report (DI MARCO et al., 1971b), the relationship between the interaction of these compounds with DNA and their biological activity was studied.

Changes in different portions of the daunomycin molecule may affect its inhibitory properties to a greater or lesser extent. The intact amino sugar residue is essential for biological activity and for the ability to form stable complexes with DNA.

Substitution or modification in the acetyl side chain at position 9 of the chromophore causes only minor changes in daunomycin activity.

Complex formation with DNA appears to account for the biological properties of the antibiotics, as suggested by a preliminary study on physical interaction with DNA (DI MARCO et al., 1971b).

Recently (ZUNINO et al., 1972) the affinities for DNA of daunomycin and its derivatives have been determined by equilibrium measurements. The results of binding equilibrium confirm the picture provided by the preliminary report. N-acetylation of the sugar residue of daunomycin markedly reduces the affinity for DNA.

The importance of the amino sugar residue for the binding reaction was emphasized by experiments in which substitution of daunosamine for N-guanidine-acetyl-daunosamine or for D-glucosamine considerably reduced the apparent binding constants. The lower affinity of the compound containing D-glucosamine, instead of daunosamine, suggests the importance of the structure and of the stereochemistry of the amino sugar moiety in the binding reaction. It has been shown (ARCAMONE et al., 1964) that daunosamine, the amino sugar moiety of daunomycin, is an L-sugar, characterized by L-lyxo conformation.

Also the inhibitory properties of daunomycin derivatives in the DNA-dependent RNA polymerase reaction correlate well with their ability to bind to DNA (CHANDRA, personal communication).

The intercalation model may be utilized to explain the mechanism of the inhibitory activity of daunomycin on bacterial and animal viruses. In fact, from the previously discussed data, it seems that inhibition of the function or of replication of the viral genome needs the formation of a strong bond of the antibiotic with the nucleic acid and, from physico-chemical studies it is known that this occurs only with double helical DNA (ZUNINO et al., 1972).

The lack of activity on single-stranded DNA (CALENDI et al., 1966) or RNA (COHEN et al., 1969) viruses appears to be a simple consequence of inability to form this stable bond.

From data on the inhibition of vaccinia virus multiplication (COHEN et al., 1972), it was concluded that daunomycin was more effective in preventing DNA synthesis from the parental than from the progeny genomes. Since daunomycin did not affect the synthesis by DNA polymerase directed by the parental genome,

Table 3. *Biological activity of daunomycin and its derivatives*[a]

Compound	R[b]	X[b]	Amino sugar	50% Inhibiting doses		DNA synthesis (M × 10⁶)	cell proliferation (M × 10⁶)
				mitotic index			
				2 hr (M × 10⁶)	8 hr (M × 10⁶)		
Daunomycin · HCl	H	O	daunosamine	0.44	0.18	1.6	0.89
Adriamycin · HCl	OH	O	daunosamine	0.43	0.09	3.4	1.9
4'-daunosaminyl-daunomycin · 2 HCl	H	O	4'-Daunosaminyl-daunosamine(?)	0.96	0.27	2.7	5.5
Daunomycin-thiosemicarbazone	H	NNHCSNH₂	daunosamine	1.60	1.20	>8.3	>8.3
Daunomycin · oxime · HCl	H	NOH	daunosamine	4.30	4.70	>8.6	>8.6
Daunomycin semicarbazone	H	NNHCONH₂	daunosamine	5.10	2.60	>8.5	>8.5
13-dihydro-daunomycin · HCl	H	HOH	daunosamine	3.50	1.20	8.8	7.9
N-guanidine-acetyl-daunomycin · HCl	H	O	N-guanidine-acetyl-daunosamine	>7.60	>7.60	>7.6	>7.6
2-amino-2-deoxy-glucosyl-daunomycinone · HCl	H	O	2-amino-2-α-D-deoxy-glucose	>8.80	8.80	8.8	>8.8
n-acetyl-daunomycin · HCl	H	O	n-acetyl-daunosamine	>8.30	>8.30	>8.3	>8.3

[a] From DiMarco *et al.* (1971).　[b] See the chemical structure in Fig. 1.

COHEN suggested that daunomycin prevents the initiation of vaccinia DNA synthesis from the parental genome by a direct or indirect effect on an initiation site but not by intercalation. As the observations cited show that daunomycin affects the replication of the parental genome but not of the progeny genomes, it is our feeling that this different effect should not be ascribed to a different kind of binding of daunomycin to DNA but to a different state of DNA in replication.

Assuming that DNA polymerase starts its action on 3'-hydroxyl-end groups of double-stranded segments of DNA, partially opened by the activity of replication-nuclease, and moves along the single-stranded segment, the presence of the intercalating drug and its tendency to tie together the two strands of DNA could interrupt this process.

Alternatively, it is possible that the binding of the antibiotic to DNA, and the alteration in the steriochemical conformation of DNA, as described for proflavine by WARING (1968), could cause a steric hindrance to the formation of the hypothetical DNA: DNA polymerase complex.

The same mechanism should be operative for both DNA polymerase and DNA-dependent RNA polymerase as daunomycin(like proflavine and ethidium bromide) inhibits the two processes nearly equally, in contrast to actinomycin, which is more active on DNA-dependent RNA polymerase.

The inhibiting effect of daunomycin on the multiplication of oncorna viruses may be explained by the ability of daunomycin to inhibit the viral DNA polymerase (MÜLLER et al., 1971; HIRSCHMAN, 1971). Daunomycin has been shown to be a potent inhibitor of the early phase of viral RNA-dependent DNA synthesis (apparently forming RNA: DNA duplexes) (APPLE and HASKELL, 1971).

Inhibition exerted by different structural analogues of daunomycin against virus associated enzymatic activities correlates well with their ability to bind to DNA (CHANDRA et al., 1972).

As it was previously mentioned, exposure to polychromatic light of mixtures of daunomycin and some viruses such as vaccinia virus (VR), HF strains of Herpes simplex virus, increases noticeably the virucidal activity of daunomycin (VERINI et al., 1968). The increase is dependent on the drug concentration and intensity of light; the virucidal activity is a linear function of the exposure to light. Others have observed photoinactivation of dye-sensitized DNA and RNA viruses (SPIKES, 1968). Daunomycin had no activity in darkness on a representative RNA virus such as NDV. The exposure to light induced the appearance of an antiviral activity also on this virus.

Like reducible organic dyes, aromatic hydrocarbons, porphyrins and numerous drugs (SPIKES, 1968), daunomycin, a strongly colored molecule, might sensitize photodynamic effects. For the dye-sensitized inactivation of viruses, a great deal of evidence has accumulated which suggests that photooxidation of nucleic acids is involved. Studies of the physical characteristics of DNA irradiated (visible light) in the presence of daunomycin indicate that irradiation leads to marked damage of nucleic acid (ZUNINO, unpublished observations). These observations have been interpreted as reflecting the photooxidation of nucleic acid components. In a preliminary study of U.V.-irradiation of nucleic acids-daunomycin complex, DI MARCO et al., (1972b) observed that during irradiation a stable combination of daunomycin with native and denatured DNA was formed. Various methods were used to test the stability of the DNA-daunomycin irradiated complex: dialysis, gel filtration, silver ion, solvent extraction and enzymatic digestion. The firmness of binding raises the possibility of the formation of a covalent bond between daunomycin and nucleic acids, as consequence of irradiation.

References

ANGIULI, R., E. FORESTI, L. RIVA DI SANSEVERINO, N. W. ISAACS, O. KENNARD, W. D. S. MOTHERWELL, D. L. WAMPLER, and F. ARCAMONE: Structure of daunomycin; X-ray analysis of daunomycin solvate. Nature New Biol. **234**, 78 (1971).

APPLE, M. H., and C. M. HASKELL: Potent inhibition of sarcoma virus RNA-directed RNA:DNA duplex synthesis and arrest of ascites murine leukemia and sarcoma in vivo by anthracyclines. Physiol. Chem. Physics **3**, 307 (1971).

ARCAMONE, F., W. BARBIERI, G. FRANCESCHI e S. PENCO: Trasformazione della daunomicina in adriamicina. Chim. Ind. (Milan) **51**, 834 (1969a).

ARCAMONE, F., G. CASSINELLI, G. FANTINI, A. GREIN, P. OREZZI, C. POL, and C. SPALLA: Adriamycin, 14-hydroxydaunomycin, a new antitumor antibiotic from *S. peucetius* var. *caesius*. Biotechnol. Bioeng. **11**, 1101 (1969b).

ARCAMONE, F., G. CASSINELLI, G. FRANCESCHI, R. MONDELLI, P. OREZZI e S. PENCO: Struttura e stereochimica della daunomicina. Gazz. Chim. Ital. **100**, 949 (1970).

ARCAMONE, F., G. CASSINELLI, G. FRANCESCHI, P. OREZZI, and R. MONDELLI: The absolute stereochemistry of daunomycin. Tetrahedron Letters **1968a**, 3353.

ARCAMONE, F., G. CASSINELLI, P. OREZZI, G. FRANCESCHI, and R. MONDELLI: Daunomycin. II. The structure and steoreo-chemistry of daunosamine. J. Am. Chem. Soc. **86**, 5335 (1964a).

ARCAMONE, F., G. FRANCESCHI, P. OREZZI, C. CASSINELLI, W. BARBIERI, and R. MONDELLI: Daunomycin. I. The structure of daunomycinone. J. Am. Chem. Soc. **86**, 5334 (1964b).

ARCAMONE, F., G. FRANCESCHI, P. OREZZI, S. PENCO, and R. MONDELLI: The structure of daunomycin. Tetrahedron Letters **1968b**, 3349.

ARCAMONE, F., G. FRANCESCHI et S. PENCO: Substances antibiotiques et procédé pour leur préparation. Belg. Patent 732,968 (1969c).

ARCAMONE, F., G. FRANCESCHI et S. PENCO: Procédé de préparation de l'adriamycine et de l'adriamycinone. Belg. Patent 731,398 (1969d).

ARCAMONE, F., G. FRANCESCHI, S. PENCO, and A. SELVA: Adriamycin (14-hydroxydaunomycin), a novel antitumor antibiotic. Tetrahedron Letters **13**, 1003 (1969e).

BACHUR, N. R.: Daunorubicinol, a major metabolite of daunorubicin: isolation from human urine and enzymatic reactions. J. Pharmacol. Expl. Therap. **177**, 573 (1971).

BACHUR, N. R., and M. GEE: Daunorubicinol metabolism by rat tissue preparations. J. Pharmacol. Exptl. Therap. **177**, 657 (1971).

BACHUR, N. R., A. L. MOORE, J. B. BERNSTEIN, and A. LIU: Tissue distribution and disposition of daunomycin (NSC-82151) in mice: Fluorometric and isotopic methods. Cancer Chemotherapy Rept. **54**, 89 (1970).

BAER, H. H., K. CAPEK, and M. C. COOK: Synthesis of 3-acetamide 2,3,6-trideoxy-D-lyxo-hexose (N-acetyl-D-daunosamine) and its D-*arabino* isomer. Can. J. Chem. **47**, 89 (1969).

BERG, H., u. K. ECKARDT: Zur Interaktion von anthracyclin und anthracyclinonen mit DNS. Z. Naturforsch. **25b**, 4 (1970).

BERNARD, J., R. PAUL, M. BOIRON, C. JACQUILLAT, u. R. MARAL: Rubidomycin. Berlin-Heidelberg-New York: Springer 1969.

BERTAZZOLI, C., T. CHIELI, G. FERNI, G. RICEVUTI, and E. SOLCIA: Chronic toxicity of adriamycin. A new antineoplastic antibiotic. Toxicol. Appl. Pharmacol. **21**, 287 (1972).

BERTAZZOLI, C., T. CHIELI, M. GRANDI, and G. RICEVUTI: Adriamycin toxicity data. Experientia **26**, 389 (1970).

BHUYAN, B. K., and C. G. SMITH: Differential interaction of nogalamycin with DNA of varying base composition. Proc. Natl. Acad. Sci. U.S. **54**, 566 (1965).

BIEDLER, J.L., and H. RIEHM: Cellular resistance to actinomycin D in chinese hamster cells in vitro: Cross-resistance, radio-autographic, and cytogenetic studies. Cancer Res. 30, 1174 (1970).

BLAKE, A., and A.R. PEACOCKE: The interaction of amino acridines with nucleic acids. Biopolymers 6, 1225 (1968).

BONADONNA, G., and S. MONFARDINI: Cardiac toxicity of daunomycin. Lancet 1969-I, 837.

BONADONNA, G., S. MONFARDINI, M. DE LENA, F. FOSSATI-BELLANI, and G. BERETTA: Phase I and preliminary phase II evaluation of adriamycin (NSC-123, 127). Cancer Res. 30, 2572 (1970).

BRACHNIKOVA, M.G., N.V. KOSTANTINOVA, F.A. POMASKOVA, and B.M. ZACHAROV: Physico-chemical properties of antitumor antibiotic rubidomycin, produced by Act. coeruleorubidus. Antibiotiki 11, 763 (1966).

BRAHIC, M., and TAMALET, J.: Effects of actinomycin D and rubidomycin on sindbis virus replication. Compt. Rend. Soc. Biol. 162, 1557 (1969).

BURKA, B., E.H. HERMAN, and H. VICK: Role of the sympathetic nervous system in daunomycin-induced arrhythmia in the monkey. Brit. J. Pharmacol. 39, 501 (1970).

CALENDI, E., R. DETTORI, and M.G. NERI: Filamentous sex-specific bacteriophages of E. Coli K 12. IV-Studies on physico-chemical characteristics of bacteriophage Ec9. Giorn. Microbiol. 14, 227 (1966).

CALENDI, E., A. DI MARCO, M. REGGIANI, B.M. SCARPINATO, and L. VALENTINI: On physico-chemical interactions between daunomycin and nucleic acids. Biochim. Biophys. Acta 103, 25 (1965).

CASAZZA, A.M., A. DI MARCO, and G. DI CUONZO: Interference of daunomycin and adriamycin on the growth and regression of murine sarcoma virus (Moloney)-induced tumors in mice. Cancer Res. 31, 1971 (1971).

CASAZZA, A.M., C. GAMBARUCCI, and R. SILVESTRINI: Effects of daunomycin and adriamycin on the murine sarcoma virus (Moloney) in vitro. Europ. J. Clin. Biol. Res. 17, 622 (1972).

CHANDRA, P., F. ZUNINO, A. GÖTZ, D. GERICKE, R. THORBECK, and A. DI MARCO: Specific inhibition of DNA-polymerase from RNA tumor viruses by some new daunomycin derivatives. FEBS Letters 21, 264 (1972).

COHEN, A., L.E. CROOK, and K.R. REES: Effect of daunomycin Vaccinia nucleic acid synthesis. Int. Virol. 2, 2nd Int. Congr. Virol., Budapest 1971, pp. 282. Basel: Karger 1971.

COHEN, A., E.M. HARLEY, and K.R. REES: Antiviral effect of daunomycin. Nature 222, 36 (1969).

CROOK, L.E., K.R. REES, and A. COHEN: Effect of daunomycin on HeLa cell nucleic acid synthesis. Biochem. Pharmacol. 21, 281 (1972).

DANO, K.: Development of resistance to daunomycin (NSC-82151) in Ehrlich ascites tumor. Cancer Chemotherapy Rept. 55, 133 (1971).

DI FRONZO, G., and R.A. GAMBETTA: "In vivo" studies on the distrubtion of ³H-daunomycin in tumors and in different tissues of the mouse. Europ. J. Clin. Biol. Res. 16, 50 (1971a).

DI FRONZO, G., R.A. GAMBETTA, and L. LENAZ: Distribution and metabolism of adriamycin in mice. Comparison with daunomycin. Europ. J. Clin. Biol. Res. 16, 572 (1971b).

DI MARCO, A.: Daunomycin and related antibiotics. Antibiotics, vol. I., p. 190 (eds. Gottlieb and Shaw). Berlin-Heidelberg-New York: Springer 1967a.

DI MARCO, A.: La daunoblastina, Milano: Minerva Medica 1970.

DI MARCO, A.: Adriamycin. The therapeutic activity on experimental tumors. International Symposium on Adriamycin, Milano, 1971a.

DI MARCO, A., G. BORETTI e A. RUSCONI: Trasformazione metabolica della daunomicina da parte di estratti di tessuti. Il Farmaco 7, 535 (1967b).

DI MARCO, A., M. GAETANI, L. DORIGOTTI, M. SOLDATI, and O. BELLINI: Daunomycin: A new antibiotic with antitumor activity. Cancer Chemotherapy Rept. 38, 31 (1964a).

DI MARCO, A., M. GAETANI, P. OREZZI, B. SCARPINATO, R. SILVESTRINI, M. SOLDATI, T. DASDIA, and L. VALENTINI: Daunomycin, a new antibiotic of the rhodomycin group. Nature 201, 706 (1964b).

DI MARCO, A., M. GAETANI, P. OREZZI, and M. SOLDATI: Antitumor activity of a new antibiotic: Daunomycin, IIIrd Int. Congr. of Chemistry, Stuttgart, 1963.

DI MARCO, A., M. GAETANI, and B. SCARPINATO: Adriamycin (NSC-123, 127): A new antibiotic with antitumor activity. Cancer Chemotherapy Rept. 53, 33 (1969).

DI MARCO, A., L. LENAZ, A.M. CASAZZA, and B. SCARPINATO: Activity of adriamycin (NSC-123, 127) and daunomycin (NSC-82151) on mouse mammary carcinoma. Cancer Chemotherapy Rept. 54, 153 (1972a).

Di Marco, A., and A. Rusconi: Action mechanism and metabolic transformation of daunomycin. Gann Monographies 2, 23 (1967c).

Di Marco, A., R. Silvestrini, T. Dasdia, and S. Di Marco: Nucleolar and chromosomal origin of nuclear RNA and its passage to cytoplasm. Riv. Ital. Istochimica 11, 211 (1965a).

Di Marco, A., R. Silvestrini, S. Di Marco, and T. Dasdia: Inhibition effect of the new cytotoxic antibiotic daunomycin on nucleic acids and mitotic activity of HeLa cells. J. Cell Biol. 27, 545 (1965b).

Di Marco, A., M. Soldati, A. Fioretti, and T. Dasdia: Activity of daunomycin, a new antitumor antibiotic, on normal and neoplastic cells grown in vitro. Cancer Chemotherapy Rept. 38, 39 (1964c).

Di Marco, A., M. Terni, R. Silvestrini, B. Scarpinato, E. Biagioli, and A. Antonelli: Effect of daunomycin on Herpes Virus hominis in human cell. Giorn. Microbiol. 16, 25 (1968).

Di Marco, A., F. Zunino, P. Orezzi, and R. A. Gambetta: Interaction of daunomycin with nucleic acids. Effect of photo-irradiation of the complex. Experientia 28, 327 (1972b).

Di Marco, A., F. Zunino, R. Silvestrini, C. Gambarucci, and R. A. Gambetta: Interaction of some daunomycin derivatives with DNA and their biological activity. Biochem. Pharmacol. 20, 1323 (1971b).

Dorigotti, L.: Studio al microscopio elettrico delle modificazioni indotte dalla daunomicina sulle cellule HeLa. Tumori 50, 117 (1964).

Dubost, M., P. Ganter, R. Maral, L. Ninet, S. Pinnet, J. Preud'Homme et G. H. Werner: Un nouvel antibiotique a propriétés antitumorales. Compt. Rend. 257, 1813 (1963).

Dudnik, Yu. V., and E. M. Netyksa: Interaction of sibiromycin with DNA. Exclusion of actinomycin D and olivomycin A from complexes with DNA. Antibiotiki 17, 44 (1972).

Dusonchet, L., N. Gebbia, and F. Gerbiasi: Spectrophotofluorometric characterization of adriamycin, a new antitumor drug. Pharmacol. Res. Commun. 3, 55 (1971).

Ellen, K. A. O., and S. L. Rhode: Selective inhibition of ribosomal RNA synthesis in HeLa cells by nogalamycin, a dA.dT binding antibiotic. Biochim. Biophys. Acta 209, 415 (1970).

Gause, G. F.: Aspects of antibiotics research. Chem. & Ind. (London) 36, 1506 (1966).

Hartmann, G., H. Goller, K. Koschel, W. Kersten u. H. Kersten: Hemmung der DNA-abhängigen RNA- und DNA-Synthese durch Antibiotica. Biochem. Z. 341, 126 (1964).

Herman, E. H., P. Schein, and R. M. Farmar: Comparative cardiac toxicity of daunomycin in three rodent species. Proc. Soc. Exptl. Biol. Med. 130, 1098 (1969).

Herman, E. H., P. Schein, and R. M. Farmar: Influence of pharmacologic or physicologic pretreatment on acute daunomycin cardiac toxicity in the hamster. Toxicol. Appl. Pharmacol. 16, 335 (1970).

Honikel, K.: Wirkung von Antibiotica auf die DNA-gesteurte DNA-Polymerase-Reaktion. PhD. Thesis, Univ. Würzburg, 1968.

Hoshino, A., T. Kato, H. Amo, and K. Ota: Antitumor effects of adriamycin on Yoshida rat sarcoma and L 1210 mouse leukemia. Cross resistance and combination chemotherapy. International Symposium on Adriamycin, Milano, 1971.

Isetta, A. M., C. Intini, and L. Soldati: On the immunodepressive action of adriamycin. Experientia 27, 202 (1971).

Kersten, W.: Inhibition of RNA synthesis by quinone antibiotics. In: Progress in molecular and subcellular biology. vol. 2, 48. Berlin-Heidelberg-New York: Springer 1971.

Kersten, W., u. H. Kersten: Die Bindung von Daunomycin, Cinerubin und Chromomycin A$_3$ und Nucleinsäuren. Biochem. Z. 341, 174 (1965).

Kersten, W., and H. Kersten: Interaction of antibiotics with nucleic acids. In: Molecular association in biology, p. 289. New York: Academic Press 1968.

Kersten, W., H. Kersten, and W. Szybalski: Physico-chemical properties of complexes between DNA and antibiotics which affect RNA synthesis. Biochemistry 5, 236 (1966).

Kessel, D., V. Botterill, and L. Wodinsky: Uptake and retention of daunomycin by mouse leukemic cells as factors in drug response. Cancer Res. 28, 938 (1968).

Kim, K. H., A. S. Gelbard, B. Djordjevic, S. H. Kim, and A. G. Perez: Action of daunomycin on the nucleic acid metabolism and viability of HeLa cells. Cancer Res. 28, 2437 (1968).

Kim, S. H., and J. H. Kim: Lethal effect of adriamycin on the division of HeLa cells. Cancer Res. 32, 323 (1972).

Lunel, J., u. J. Preud'Homme: Deutsches Patentamt Offenlegungsschrift 1911249, Okt. 2, (1969).

MARSH, J.P., Jr., L.W. MOSHER, E.M. ACTION, and L. GOODMAN: The synthesis of daunosamine. Chem. Commun. 973 (1967).

MÜLLER, W., and D.M. CROTHERS: Studies of the binding of actinomycin and related compounds to DNA. J. Mol. Biol. **35**, 251 (1968).

NATALE, N., e P. MOCARELLI: Depressione immunitaria da daunomicina. Tumori **55**, 409 (1969).

PAOLETTI, J., and J.B. LE PECQ: The change of the torsion of the DNA helix caused by intercalation. I. A discussion of the two different possibilities, winding and unwinding. Biochimie **53**, 969 (1971).

PARISI, B., and A. SOLLER: Studies on the antiphage activity of daunomycin. Giorn. Microbiol. **12**, 183 (1964).

PENCO, S.: New glycosides of daunomycinone. Chim. Ind. (Milan) **50**, 908 (1968).

PENMAN, S., C. VESCO, and M. PENMAN: Localization and kinetics of formation of nuclear heterodisperse RNA, cytoplasmic heterodisperse RNA and polyribosome-associated messenger RNA in HeLa cells. J. Mol. Biol. **34**, 49 (1968).

PERRY, R.R.: Selective effects of actinomycin D in the intracellular distribution of RNA synthesis in tissue culture cells. Exptl. Cell Res. **29**, 400 (1963).

PICONE, M.A., and A. TRAINE: Pharmacokinetic characteristics of daunomycin, an antibiotic with antitumor action. Arzneimittel-Forsch. **20**, 88 (1970).

PIGRAM, W.J., W. FULLER, and L.D. HAMILTON: Stereochemistry of intercalation: Interaction of daunomycin with DNA. Nature New Biol. **235**, 17 (1972).

PITTILLO, R.F., and D.E. HUNT: Biologic activity of daunomycin (NSC-82151) in microbial systems. Cancer Chemotherapy Rept. **52**, 707 (1968).

RICHARDSON, A.C.: The synthesis of N-benzoyl-D-daunosamine. Carbohydrate Res. **4**, 422 (1967).

RUSCONI, A.: Different binding sites in DNA for actinomycin and daunomycin. Biochim. Biophys. Acta **123**, 627 (1966).

RUSCONI, A., e E. CALENDI: Azione della daunomicina sulla sintesi nucleica in cellule di epatoma. Tumori **50**, 261 (1964).

RUSCONI, A., and E. CALENDI: Action of daunomycin on nucleic acid metabolism. Biochim. Biophys. Acta **119**, 413 (1966).

RUSCONI, A., G. DI FRONZO, and A. DI MARCO: Distribution of tritiated daunomycin (NSC-82151) in normal rats. Cancer Chemotherapy Rept. **52**, 331 (1968).

RUSCONI, A., and A. DI MARCO: Inhibition of nucleic acid synthesis by daunomycin and its relationship to the uptake of the drug in HeLa cells. Cancer Res. **29**, 1509 (1969).

SANDBERG, J.S., F.L. HOWSDEN, A. DI MARCO, and A. GOLDIN: Comparison of the antileukemic effect in mice of adriamycin (NSC-123, 127) with daunomycin (NSC-82151). Cancer Chemotherapy Rept. **54**, 1 (1970).

SANFILIPPO, A., e E. MAZZOLENI: Attività antifagica dell'antibiotico daunomicina. Giorn. Microbiol. **12**, 83 (1964).

SAUCIER, J.M., B. FESTY, and J.B. LE PECQ: The change of the torsion of the DNA helix caused by intercalation. II. Measurement of the relative change of torsion induced by various intercalating drugs. Biochimie **53**, 973 (1971).

SILVESTRINI, R., A. DI MARCO, and T. DASDIA: Interference of daunomycin with metabolic events of the cell cycle in synchronized cultures of rat fibroblasts. Cancer Res. **30**, 966 (1970).

SILVESTRINI, R., A. DI MARCO, S. DI MARCO, and T. DASDIA: The action of daunomycin on the metabolism of nucleic acids of normal and neoplastic cells growing in vitro. Tumori **49**, 399 (1963a).

SILVESTRINI, R., and M. GAETANI: The action of daunomycin on the metabolism of nucleic acids of Ehrlich ascites tumor cells. Tumori **49**, 389 (1963b).

SIMARD, R.: The binding of actinomycin D-^{3}H to heterochromatin as studied by quantitative high resolution radioautography. J. Cell Biol. **35**, 716 (1967).

SMITH, B.: Damage to intrinsic cardiac nervous by rubidomycin (daunorubicine). Brit. Heart J. **31**, 607 (1969).

SPIKES, J.D.: Photophysiology, vol. III, p. 33–64. New York: Academic Press 1968.

STERNBERG, S.S.: Cross-striated fibrils and other ultrastructural alterations in glomeruli of rats with daunomycin nephrosis. Lab. Invest. **23**, 39 (1970).

STERNBERG, S.S., and S.F. PHILIPS: Biphasic intoxication and nephrotic syndrome in rats given daunomycin. Proc. Am. Ass. Cancer Res. **8**, 64 (1967).

THEOLOGIDES, A., J.M. YARBRO, and B.J. KENNEDY: Daunomycin inhibition of DNA and RNA synthesis. Cancer **21**, 16 (1968).

VENDITTI, J. M., J. ABBOT, A. DI MARCO, and A. GOLDIN: Effectiveness of daunomycin (NSC-82151) against experimental tumors. Cancer Chemotherapy Rept. **50,** 659 (1966).

VERINI, M. A., A. M. CASAZZA, A. FIORETTI, F. RODENGHI, and M. GHIONE: Photodynamic action of daunomycin. II. Effect on animal viruses. Giorn. Microbiol. **16,** 55 (1968).

WARD, D., E. REICH, and I. H. GOLDBERG: Base specificity in the interaction of polynucleotides with antibiotics drugs. Science **149,** 1259 (1965).

WARING, M. J.: Drugs which affect the structure and function of DNA. Nature **219,** 1320 (1968).

WARING, M. J.: Variation of the supercoilds in closed circular DNA by binding of antibiotics and drugs: Evidence for molecular models involving intercalation. J. Mol. Biol. **54,** 247 (1970).

WHANG-PENG, J., B. G. LEVENTHAL, J. W. ADAMSON, and S. PERRY: The effect of daunomycin on human cells in vivo and in vitro. Cancer **23,** 113 (1969).

WHITE, H. L., and J. R. WHITE: Hedamycin and rubiflavin complexes with deoxyribonucleic acid and other polynucleotides. Biochemistry **8,** 1030 (1969).

YESAIR, D. W., M. A. ASBELL, R. BRUNI, F. J. BULLOCK, and E. SCHWARTZBACH: Pharmacokinetics and metabolism of adriamycin and daunomycin. Internation Symposium on Adriamycin, Milano, 1971.

ZELEZNICK, L. D., and C. M. WEENEY: Inhibition of deoxyribonuclease action by nogalamycin and U-12241 by their interaction with DNA. Arch. Biochem. Biophys. **120,** 292 (1967).

ZUNINO, F.: Studies on the mode of interaction of daunomycin with DNA. FEBS Letters **18,** 249 (1971).

ZUNINO, F., R. A. GAMBETTA, A. DI MARCO, and A. ZACCARA: Interaction of daunomycin and its derivatives with DNA. Biochim. Biophys. Acta, **277,** 489 (1972).

Edeines*

Z. Kurylo-Borowska

Edeines A and B are produced by *Bacillus brevis* Vm4 (KURYLO-BOROWSKA, 1959a and b; KURYLO-BOROWSKA and TATUM, 1966). In addition to these two antibiotics the strain 587 of the same organism produces small quantities of edeines C and D as well (CHAMARA and BOROWSKI, 1966). Edeines are strongly basic and colorless compounds soluble in water and insoluble in most organic solvents.

The Chemical Structure

Edeines A and B are linear oligopeptides of a molecular weight 730 and 772 respectively, but in the process of isolation and purification, some dimerisation can occur. Edeines A and B are composed of five amino acid residues and an organic base: spermidine (edeine A) or guanylspermidine (edeine B). The amino acid fragments are: glycine, isoserine, α,β-diaminopropionic acid, 2,6 - diamino-7-hydroxyazelaic acid and isotyrosine (β-tyrosine) (RONCARI et al., 1966; HETTINGER et al., 1968; HETTINGER and CRAIG, 1968), (Fig. 1).

Of the four optically active amino acids: isoserine, α,β-diaminopropionic acid and 2,6-diamino-7-hydroxyazelaic acid are of L-configuration. The configuration of β-tyrosine is unknown. Edeine D is an analog of edeine A in which the β-tyrosine residue is replaced by β-phenyl-β-alanine (WOJCIECHOWSKA et al., 1972).

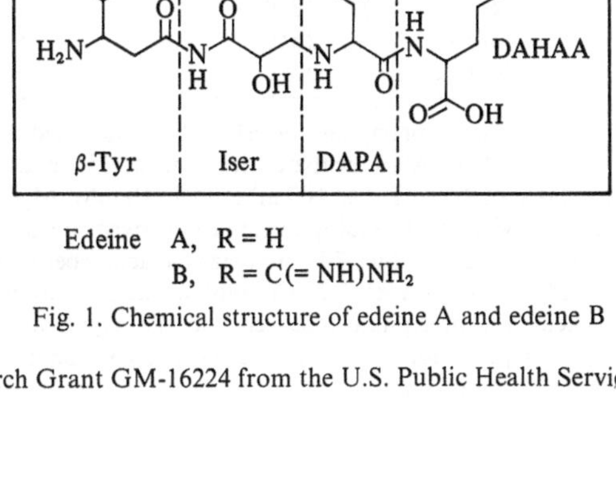

Edeine A, R = H
           B, R = C(= NH)NH₂

Fig. 1. Chemical structure of edeine A and edeine B

*Supported by Research Grant GM-16224 from the U.S. Public Health Service.

Structure of edeine C is unknown. Maximum of absorption of edeine A and B in the pH 6.0 to 8.0 is at 272 nm and the molar absorbancy 3600.

Edeines A and B are mixtures of edeine A_1, A_2, B_1 and B_2. Edeines A_2 and B_2 are products of intramolecular isomerization of edeines A_1 and B_1. The isoserine residue of edeines A_2 and B_2 is linked to the β-amino group of α,β-diaminopropionic acid instead of to the α-amino group, as it is in biologically active edeines A_1 and B_1 (HETTINGER and CRAIG, 1970).

Biosynthesis

Edeines are synthesized by a spore forming *Bacillus brevis* Vm4. The compounds are produced and released into the medium by the cells in the postlogarithmic phase of growth. Once the enzyme proteins necessary for the synthesis of edeines are formed, the synthesis is not affected by inhibitors of protein synthesis (KURYLO-BOROWSKA and TATUM, 1966).

The biosynthesis of edeines requires a combination of enzyme proteins for activation and then polymerization of activated amino acids into a peptide chain, as well as for addition of the non-amino acids moieties.

A cell-free extract of *Bacillus brevis* Vm4, on 25–55% saturation of ammonium sulfate, precipitates the enzymes which are active in the formation of edeines A and B. The activity of the isolated enzymes changes with age of the cells, and data presented in Table 1 would show that synthesis of edeine B precedes that of edeine A. The most active enzymes for synthesis of both edeines are obtained from 12–16 hour old cultures.

Cell-free synthesis of edeines occurs in a system consisting of a multi-enzyme-complex which rules out any major role of ribosomes or RNA in syntheses. However, as in the biosynthesis of proteins the reaction requires ATP and Mg^{2+}

Table 1. *Synthesis of edeine as a function of bacterial growth time*[a]

Time of growth, hrs.	Incorporation of (^{3}H) glycine and (^{14}C) tyrosine into			
	Edeine A		Edeine B	
	glycine (cpm)	tyrosine (cpm)	glycine (cpm)	tyrosine (cpm)
6	none	none	none	none
8	none	none	1900	1400
12	2450	2300	4500	4050
16	2850	2650	5200	4850
20	none	none	3300	2800

[a] 25–55% saturation $(NH_4)_2SO_4$ precipitate of cell extract was used as enzyme source in the reaction mixture. 2–4 mg of proteins were incubated for 1 hr. at 35° in the reaction mixture consisting of 125 mM Tris·HCl buffer pH 7.8, 100 mM KCl, 10 mM $MgCl_2$, 5 mM β-mercaptoethanol, 2.5 mM ATP, 2.5 mM phosphoenolpyruvate, 25 µg pyruvate kinase, 2 µmoles of each: isoserine, diaminopropionic acid, 2,6-amino, 7-hydroxyazelaic acid, spermidine, (^{3}H) glycine, (^{14}C) tyrosine, uniformly labelled (sp. act. of each 0.2 µCi/µmole). The reaction was stopped by addition of 200 µl glacial acetic acid, the precipitate removed and the supernatant chromatographed on paper (Whatman 3MM) in isopropanol-ammonia-water (4:1:1 v/v). Spots of edeines were cut out and counted in a liquid scintillation counter.

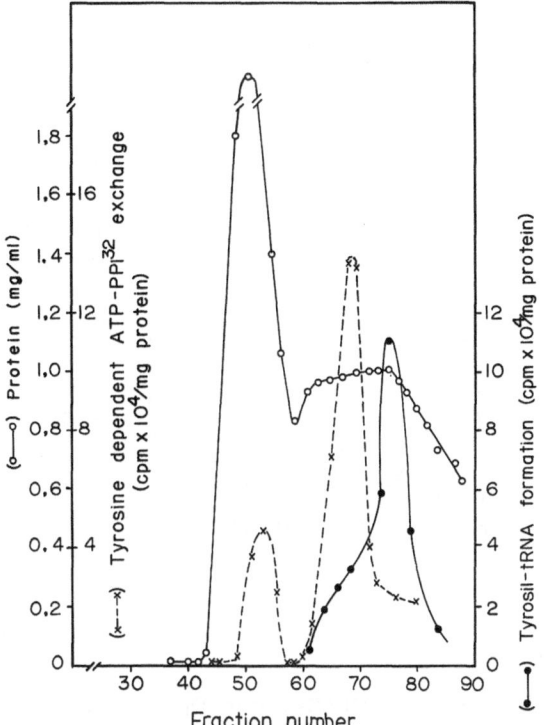

Fig. 2. Purification of enzymes (25–55% ammonium sulfate precipitate) on Sephadex G-200. The column (2.5 × 70 cm) was eluted with 0.1 M Tris · HCl buffer pH 7.6. Fractions of 3 ml were collected

for the activation of the amino acid constituents of edeines. By column chromatography on Sephadex G-200 two main enzyme fractions: Fraction I (Mol.wt. approximately 280,000) and Fraction II (Mol.wt approximately 100,000) can be obtained from the enzymes precipiated with 25–55% saturation of ammonium sulfate (Fig. 2).

These two fractions activate the amino acid constituents of edeine A and B, but both are required for the formation of the whole molecule of edeine A or B. Fractions I and II can be purified by DEAE cellulose column chromatography with a concentration gradient of KCl (Figs. 3 and 4). The procedure also separates further these enzymes, which activate amino acid constituents of edeines, from other activating enzymes (Table 2). From Fraction I, 0.22–0.27 M KCl elutes the enzymes which activate α-tyrosine and glycine, whereas 0.27–0.35 M KCl elutes the β-tyrosine and diaminopropionic acid activating enzymes.

From Fraction II, 0.20–0.25 M KCl elutes the enzymes which activate α-tyrosine, glycine, 2,6-diamino-7-hydroxyazelaic acid and some isoserine. The overlapping enzymes for activation of α-tyrosine and glycine can be removed from this preparation by a repeated fractionation on a DEAE cellulose column with 0.25 M KCl (Fig. 5). Activation of amino acids was measured by exchange of ATP-^{32}PPi as described by CALENDAR and BERG (1966).

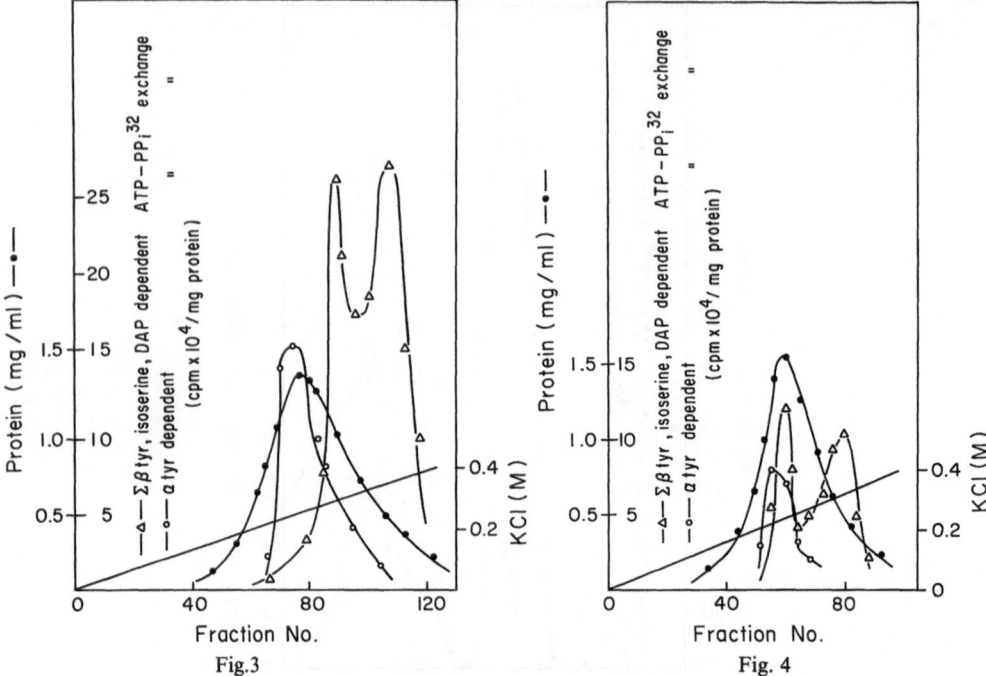

Fig. 3. Purification of Fraction I on DEAE-cellulose. Dialyzed enzymes (Fractions 50–60 from Sephadex G-200) were applied to a column (2.5 × 30 cm) previously equilibrated with 0.1 MTris · HCl buffer pH 7.6 in 1 mM β-mercaptoethanol. A linear gradient of 0.1–0.4 M KCl in the above buffer was used for elution. Fractions of 3.5 ml were collected

Fig. 4. Purification of Fraction II on DEAE-cellulose (Fractions 60–75 from Sephadex G-200). The experimental conditions were as described for Fig. 3

Studying the mechanism of biosynthesis of gramicidins and tyrocidins as a model for antibiotic peptide synthesis, LIPMANN (1971) has shown that the activated constituent amino acids yield thioester-linked fragments which are precursors in polymerization.

The presence of 4′-phosphopantetheine bound to the heavy enzyme fraction of tyrocidin-synthesizing complex led him to the hypothesis that this compound plays a role in the synthesis of polypeptide antibiotics, similar to that in the biosynthesis of fatty-acids, where it performs a switch or translocation of the constituent fragments by alternative condensation and transthiolation.

Since the activation and polymerization of amino acids occurs on the same protein, the whole system for the biosynthesis of antibiotic peptides seems to be very compact. In LIPMANN's opinion this mechanism of synthesis represents a link in the process of evolution between the condensation of two-carbon fragments to β-keto acids in fatty acid synthesis and of amino acids to polypeptides in ribosomal protein synthesis. The experimental data concerning the biosynthesis of edeines support the view that for the formation of peptide antibiotics the

Table 2. *Amino acid dependent ATP-*^{32}*PP$_i$ exchange*

Fractions: I and II from a Sephadex G-200 column were fractionated on a DEAE cellulose column in a linear gradient of 0.1–0.4 M KCl in 0.1 M Tris·HCl buffer pH 7.6

Enzyme	ATP^{32} m μmole/mg Protein/30 min formed in the presence of 1 μmole of					
	α-tyr	β-tyr	isoser	DAP	DAHAA	Gly
Fraction I						
0.22–27 M KCl	121.5	27.9	13.5	35.1	7.5	144.5
0.27–35 M KCl	0	145.0	0	174.4	0	0
Fraction II						
0.20–0.25 M KCl	143.1	21.6	101.2	0	82.0	171.5
0.25–0.35 M KCl	0	0	624.5	229.5	0	110.7

DAP—diaminopropionic acid.

DAHAA—2,6-diamino,7-hydroxyazelaic acid.

Amino acid dependent exchange of ATP with ^{32}PPi was measured in 1 ml of reaction mixture consisting of: 100 mM Tris·HCl buffer pH 8.0, 5 mM MgCl$_2$, 2 mM ATP, 2 mM sodium pyrophosphate (sp. act. 10^5 cpm/μmole), 10 mM β-mercaptoethanol, 10 mM potassium fluoride, 0.1 mg bovine serum albumin, 2 mM amino acids and 0.2 mg enzyme. The reaction was carried out for 15 min at 37°, then stopped by addition of 0.7 ml cold 0.4 M Na$_4$P$_2$O$_7$ in 15% perchloric acid. The ATP was adsorbed to charcoal by addition of 0.1 ml of a 15% suspension of acid washed Norit A. The charcoal suspension was filtered through a GF/A glass filter and the filter was washed 5 times with 5 ml portions of cold distilled water. The filters were dried and counted in a liquid scintillation counter.

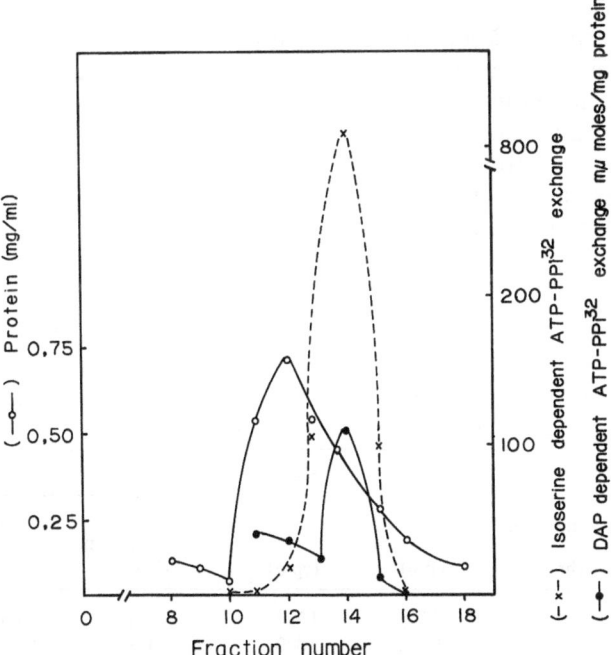

Fig. 5. Fractionation of amino acid activating enzymes on DEAE-cellulose. Enzymes eluted from Fraction II with 0.20–0.25 M KCl (Fig. 4) were pooled and precipitated with 60% ammonium sulfate. The precipitate was dialyzed against 0.1 M Tris·HCl buffer pH 7.6 and applied to a column (as in Fig. 3). The column was eluted with 0.25 M KCl

carboxyl-activated amino acids, bound by a thiol linkage to the enzymes, are transferred to the acceptor part of the multienzyme complex where the polymerization occurs (Kurylo-Borowska and Sedkowska, in press). It would be premature, however, to infer that the formation of whole antibiotics such as edeine A and B follows this pathway, since the mechanism of addition of spermidine, or glycyl-spermidine to the peptide fragment is not yet known.

Inhibition of Growth

Edeines A and B have a broad-spectrum of activity. They inhibit growth of Gram-positive and Gram-negative bacteria (Kurylo-Borowska, 1959 a), mycoplasmas (Borysiewicz, 1966), fungi and yeasts as well as of some mammalian neoplastic cells in tissue culture (Kurylo-Borowska, 1962).

The bacteriostatic concentrations of edeines are 4–8 µg/ml for *E. coli, Shigella typhimurium, Serratia marcescens, Sarcina lutea, Mycobacterium phlei, Bacillus subtilis* and *Streptococus hemolyticus*. Higher concentrations are bacteriocidal.

Toxicity for Mammalian Hosts

Edeine B exhibits a relatively high toxicity in animals. Upon a single intravenous injection, the LD50 in mice is approximately 200 mg/kg. 3–6 minutes after intravenous injection of ^{3}H-edeine B, blood level of the antibiotic (blood drawn from the eye vein) was at a maximum. No measurable amounts of edeine could be detected in brain, lung and spleen 3 min to 24 hours after injection. Table 3 presents data on the amounts of ^{3}H-edeine B found in kidneys, liver and urine after intraperitoneal injection and shows that approximately 50% of the compound is excreted in the urine within 8 hours. The chromatographic behavior as well as the antibiotic activity of the excreted compound have shown that edeine is released in an unchanged form (Kurlyo-Borowska, unpublished).

Table 3. *Distribution of* ^{3}H *edeine B in mouse organs at various times after intraperitoneal injection of 120 mg/kg (specific activity: 10000 cpm/mg)*

Time after injection	Liver	Kidneys	Urine	
			Time of collection	counts
(hr)	(cpm)	(cpm)	(hr)	(cpm)
4	75	1740	0–4	1800
8	495	3800	0–8	5000
24	90	680	20–22	1000

Male white mice (18 gm) were injected with ^{3}H edeine B. Radioactivity was measured on aliquots of organs and expressed per animal.

Mechanism of Action

Since no specific differences in the spectrum of activity of edeines A and B were detected, most of the studies described below were carried out only with edeine A. *In vivo,* edeine reversibly inhibits DNA synthesis of Gram-positive and Gram-negative bacteria (KURYLO-BOROWSKA, 1964). Whereas RNA synthesis is not affected by edeine, inhibition of protein synthesis occurs at concentrations exceeding about 10-fold those necessary for the inhibition of DNA synthesis (Fig. 6) (KURYLO-BOROWSKA and SZER, 1972).

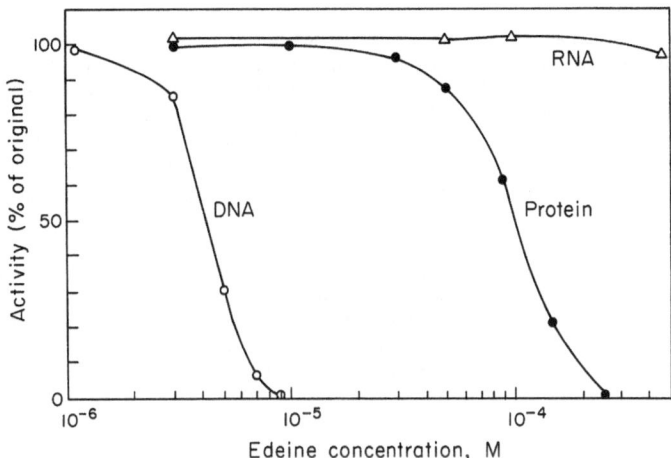

Fig. 6. Effects of edeine A on DNA, RNA and protein synthesis in plasmolysed cells of *E. coli* —o— ^{3}H TMP incorporation. —●— ^{14}C lysine incorporation. —△— ^{3}H GTP incorporation. Edeine A_1: 187 units/μmole. Synthesis of DNA was measured by incorporation of ^{3}H TMP as described for Table 4. RNA synthesis measured by incorporation of ^{3}H GTP. The reaction mixture (0.05 ml) contained 5×10^8 plasmolysed cells, 10 mM Tris·HCl buffer pH 7.8, 10 mM KCl, 5 mM MgSO$_4$, 0.2 mM dithiothreitol, 50 μM ATP, CTP, UTP and ^{3}H GTP (spec. act. 100 Ci/mole). The reaction was carried out for 15 min at 37° and then stopped by the addition of 2 ml of cold 5% trichloroacetic acid. After 2 hr in the cold, the mixture was filtered on Millipore filters, washed 3 times and the radioactivity measured. Protein synthesis was measured by incorporation of ^{14}C lysine. The reaction mixture contained in 0.1 ml: 10^9 plasmolysed cells, 20 mM HCl buffer pH 7.8, 50 mM KCl, 50 mM NH$_4$Cl, 10 mM MgSO$_4$, 0.2 mM dithiothreitol, 50 μg creatine phosphate, 1 μg creatine phosphokinase, 0.1 mM ATP, CTP, GTP and UTP, 25 μM ^{14}C lysine (spec. act. 50 Ci/mole) and 25 μM each of the other 19 amino acids. After 15 min incubation at 37° the reaction was stopped by the addition of 2 ml of cold 10% trichloroacetic acid. After 1 hr, in the cold the precipitate was collected. The precipitate was suspended in 2 ml 5% trichloroacetic acid, heated for 15 min at 90° and cooled in ice. The mixture was filtered on Millipore filters, washed and counted

Effect on DNA Synthesis

Since DNA synthesis promoted by DNA polymerase I is not inhibited by edeine, its effect on DNA polymerase II and III was studied (KURYLO-BOROWSKA and SZER, 1972) in mutants of *Escherichia coli* (P3478 and D110) (DE LUCIA and CAIRNS, 1969) which replicate normally but lack (or contain not more than

Table 4. *Effect of edeine A on various DNA synthesizing systems of E. coli P3478 and D110*

Edeine; Biological activity units/µmole	Concentrations causing 50% inhibition of (^{3}H) TMP incorporation, µM		
	DNA-membrane complex (P3478)	Soluble polymerases (P3478)	Plasmolysed cells (D110)
187.0	3.6	4.8	3.8
70.0	12.0	15.0	—
36.0	46.0	—	—
4.4	450.0	240.0	300.0
1.4	790.0	620.0	—

Assays of biological activity of edeine were carried out on agar plates inoculated with *Bacillus subtilis*. Inhibition zones obtained by a standard cylinder plate diffusion method (0.1 ml of solution) were measured in duplicate. An inhibition zone 23 mm in diameter corresponds to 1 unit of activity. Concentrations of edeine were estimated spectrophotometrically: $E_{272} = 3600$. The incorporation of (^{3}H) TMP was measured in 0.15 ml of the reaction mixture consisting of 20 mM Tris · HCl buffer pH 7.2, 50 mM KCl, 5 mM $MgSO_4$, 0.2 mM dithiothreitol, 32 µM dATP, dCTP, dGTP, (^{3}H) TTP (sp. act. 70 Ci/mole). 20–50 µl DNA-membrane complex (corresponding to 2×10^9 to 5×10^9 cells), or 50 µl plasmolysed cells (1×10^9 cells) were assayed in the presence of 0.5 and 1.3 mM ATP respectively. The soluble polymerases (0.1 mg protein) were assayed in the presence of 15 µg activated salmon sperm DNA (activated according to method of APOSHIAN and KORNBERG, 1962). Samples were incubated at 37° for 15 min and the reaction was stopped by addition of 0.3 ml of cold 10% TCA in 1% $Na_4P_2O_7$. Samples were kept in crushed ice for 60 min and afterwards the precipitate was collected on Whatman GF/A glass filters and washed 3 times with 5% TCA in 1% $Na_4P_2O_7$. Filters were dried and radioactivity measured in a liquid scintillation counter.

2%) of DNA polymerase I. The incorporation of ^{3}H TMP into DNA in the presence of edeine was tested in 3 different synthesizing systems prepared from the above bacteria (Table 4):

1. DNA-membrane complex (STRATLING and KNIPPERS, 1971).
2. Crude and partially purified extracts of DNA polymerases II and III (KORNBERG and GEFTER, 1970).
3. Plasmolysed cells (WICKNER and HURWITZ, 1972).

The results presented in Table 5 show that incorporation of ^{3}H TMP into DNA carried out by the DNA-membrane complex is linear for nearly 15 minutes. The addition of 5 µM edeine at different times of the reaction, starting from 0 to 8 minutes immediately interrupts the incorporation of ^{3}H TMP. On the other hand DNA polymerase I can utilize the endogenous DNA of the DNA-membrane complex for the repair of DNA, under conditions in which the replicative activity of this complex is completely abolished (Table 5).

Edeine appears to be a specific inhibitor of bacterial DNA polymerases, since the concentrations which inhibit the synthesis of host DNA have only slight effects on the intracellular growth and burst size of bacteriophage T4B. At these concentrations (0.04–0.08 unit/ml) an effect on the maturation, expressed by a prolonged time of bacteriophage growth is observed (TABACZYNSKI and JABLONSKA, 1970). Edeine has no effect on absorption of bacteriophages even at high concentrations (1–5 units/ml). These concentrations strongly inhibit replication of T-even phages but this can be partially reversed by the removal of

Table 5. *Effect of edeine on* ^3H *TMP incorporation into DNA by DNA-membrane complex of E. coli D110*

Time (min)	^{3}H TMP incorporated		
	Control (cpm)	Edeine 5 µM (cpm)	DNA polym I +edeine (5 µM) (cpm)
0	100	100	100
2	800	100	600
5	1200	100	1200
12	3200	100	2800
		added 5 min after onset of the reaction	
0	100	—	—
5	800	—	—
10	2800	800	2600

The conditions of the reaction were as described in Table 4. The incorporation of ^{3}H TMP into DNA was measured in the presence of 0.5 mM ATP. DNA polymerase I and edeine (5 µM) were added separately and together at time 0 and 5 min after the onset of the reaction as indicated.

the drug. In contrast to T-even phages, the replication of f2 phages is inhibited irreversibly. The same inhibition (50%) is obtained when 1.5 u/ml of edeine is added to a culture of *Escherichia coli* K13 simultaneously with the bacteriophage or 15 minutes after its absorption. This rules out the assumption that edeine affects the synthesis of a viral coat protein (KURYLO-BOROWSKA, unpublished).

Effect on Protein Synthesis

Although the inhibition of protein synthesis *in vivo* has not appeared to be as pronounced as the inhibition of DNA synthesis, it was demonstrated that edeine is a strong inhibitor of protein synthesis *in vitro*. It inhibits the synthesis of protein, directed by synthetic and viral messengers (HIEROWSKI and KURYLO-BOROWSKA, 1965) due to the interference with the messenger-directed binding of aminoacyl-tRNA to the ribosomes (KURYLO-BOROWSKA and HIEROWSKI, 1965). Edeine inhibits both the enzymatic and non-enzymatic binding of aminoacyl-tRNA to ribosomes. The inhibitory activity of the antibiotic depends on the concentration of magnesium ions. At low Mg^{2+} concentration (3–5 mM) edeine inhibits the enzymatic as well as non enzymatic binding of aminoacyl to the A and P sites of ribosomes. At 10–15 mM Mg^{2+} concentration, the A site becomes inaccessible for edeine, whereas the sensitivity of the P site remains unaltered. Edeine inhibits binding of phenylalanyl-tRNA to 30S subunits more strongly than binding to 70S ribosomes, even at 15–20 mM Mg^{2+}. This suggests that a partial or open A site on the 30S subunit is more accessible to edeine than is the full A site on the 70S ribosome. Studies of the AUG (or phage RNA) and initiation factor-dependent ribosomal binding of fMet-tRNA (enzymatic binding) have shown that this binding can be almost completely inhibited by 5 µM edeine, even at higher concentrations of Mg^{2+}.

The high affinity of edeine for ribosomal binding sites is manifested in the ability of the antibiotic to remove an aminoacyl-tRNA prebound to the AUG codon. Edeine is capable of removing an aminoacyl-tRNA prebound to the trinucleotide codon, but not to a hexa- or polynucleotidic messenger. The residual aminoacyl-tRNA bound to 70S ribosomes in the presence of edeine does not react with puromycin, which shows that its location is in the aminoacyl (A) site (Szer and Kurylo-Borowska. 1970).

Studies of the interaction of edeine with ribosomal subunits have suggested that there is an initiator (I) site on the 30S subunit, which upon formation of the 70S ribosomal unit, overlaps with, and is reinforced by, a peptidyl, or donor site on the 50S subunit. Edeine blocks the initiator (I) site and specifically inhibits formation of the 30S initiation complex. It also blocks the P site of the 50S ribosomal subunit, but this block does not interfere with the synthesis and release of f-Met-puromycin. The 50S component of the P site is blocked by edeine in free ribosomes only. Once the 70S initiation complex is formed, the A-site is created and accepts incoming aminoacyl-tRNA's from the ternary complex of aminoacyl-tRNA-Tu-GTP, this site is insensitive to edeine. On the other hand, in the absence of initiator, t-RNA, transfer of the phenylalanyl-tRNA from the ternary complex to 70S ribosomes is inhibited by edeine (Weissbach et al., 1971). In the case of complete inhibition of aminoacyl-tRNA binding to ribosomes the ratio of edeine: ribosome is approximately 3–4 edeine molecules per 70S particle and 2–3 molecules per 30S particle.

Structure-Activity Relationship

Edeines A and B are natural products with chemical structures different from other peptides (Fig. 1). They are mixtures of biologically active compounds edeine A_1 and B_1, and the inactive isomers edeine A_2 and B_2. The biological activity of edeines A_1 and B_1 depends on their structural integrity and the presence of the free β-amino group of the α, β-diaminopropionic acid with its α-amino group linked in the peptide bond. The isomerization of the isoseryl-α-diaminopropionyl linkage to isoseryl-β-diaminopropionyl, which occurs during fermentation as well as during prolonged storage of the compound, causes inactivation. The substitution of the β-amino group of the diaminopropionyl moiety by a guanido group also leads to the formation of inactive edeines (Hettinger, et al., 1968). Edeine A_1 and B_1 are resistant to attack by most proteolytic enzymes. It was found however that carboxypeptidase B could cleave the peptide bond between α,β-diaminopropionic acid and 2,6-diamino-7-hydroxyazelaic acid of edeine A_2 and B_2. The two peptide fragments obtained are biologically inactive (Hettinger and Craig, 1970). Edeines A_1 and B_1 have approximately equal antibiotic activities and substitution of the terminal amino group of spermidine by a guanido residue seems to produce only a nominal change in the structural and functional characteristics of the antibiotics. It is worth mentioning that in cells of B. brevis Vm4, edeine B_1 (a compound having a lower isoelectric point than edeine A_1) is synthesized before edeine A_1 (Kurylo-Borowska and Tatum, 1970). For studies of the mechanism of inhibition of DNA (Kurylo-Borowska and Szer, 1972; Szer and Kurylo-Borowska, 1972) and protein synthesis by edeines it was necessary to distinguish between effects related directly to the structural features of the antibiotics and any possible effects related to their polycationic nature.

However since many of the chemical properties of edeines are influenced by their polyamine components, it is difficult to explain the functional role of spermidine or guanylspermidine in the specific biological activity of these compounds. It has been shown that edeine A_1, a preparation of high biological activity, inhibits DNA synthesis in concentrations as low as 10^{-6} M. The data presented in Table 4 show that the inhibitory concentrations of edeine increase sharply as the biological activity of preparations decline. The inhibitory concentrations of mixtures of active preparations of edeine and its inactive form, as well as the effect of polyamines on the *in vitro* synthesis of DNA were estimated. It was shown that, to inhibit DNA synthesis by spermine or spermidine as well as by the inactive isomers of edeine, relatively high concentrations of these compounds are required (KURYLO-BOROWSKA and SZER, 1972). Thus, a polycationic, presumably unspecific effect of edeine A_2 and B_2 can be observed at concentrations exceeding 500-fold the effective concentration of edeines A_1 and B_1. This conclusion is in agreement with that previously drawn, which was based on studies of the effect of edeines and polyamines on the binding of aminoacyl-tRNA to ribosomal binding sites (SZER and Kurylo-Borowska, 1972). The authors have shown that the binding of phenylalanyl-tRNA to 70S ribosomes was inhibited by edeine A_1 and B_1 at a molar ratio of $3:1$ and $4:1$ respectively, whereas the inactive isomers had no effect even at a ratio of $60:1$.

Edeine activity is due primarily to the fact that the isoserine moiety is linked to the α-β-diaminopropionic acid, leaving the β-amino group free. The basic nature of the molecule attributable to spermidine appears not to be as important in edeine activity but may play an important role in ribosomal binding.

Acknowledgements. The author wishes to express her thanks to Dr. E. L. TATUM for stimulating discussions and for criticism of the manuscript, to Miss M. PAZDZIOR for assistance in part of the experimental work.

References

APOSHIAN, H. V., and A. KORNBERG: Enzymatic synthesis of deoxyribonucleic acid. IX. The polymerase formed after T2 bacteriophage infection of *Escherichia coli*: a new enzyme. J. Biol. Chem. **237**, 519 (1962).

BOROWSKI, E., H. CHMARA, and E. JARECZEK-MORAWSKA: The antibiotic edeine VI. Paper and thin layer chromatography of components of the edeine complex. Biochim. Biophys. Acta **130**, 560 (1966).

BORYSIEWICZ, J.: Effect of various inhibitors of protein and deoxyribonucleic acid synthesis on the growth of Mycoplasmas. Appl. Microbiol. **14**, 1049 (1966).

CALENDAR, R., and P. BERG: Purification and physical characterization of tyrosyl ribonucleic acid synthetases from *Escherichia coli* and *Bacillus subtilis*. Biochemistry **5**, 1681 (1966).

CHMARA, H., and E. BOROWSKI: Isolation of induced high activity mutant of *Bacillus brevis* Vm4. Acta microbiol. Polon. **15**, 223 (1966).

DELUCIA, P., and J. CAIRNS: Isolation of an *E. coli* strain with a mutation affecting DNA polymerase. Nature **224**, 1164 (1969).

HETTINGER, T. P., and L. C. CRAIG: Edeine II. The composition of the antibiotic peptide edeine A. Biochemistry **7**, 4147 (1968).

HETTINGER, T.P., and L.C. CRAIG: Edeine. IV. Structures of the antibiotic peptides edeine A_1 and B_1. Biochemistry 9, 1224 (1970).

HETTINGER, T.P., Z. KURYLO-BOROWSKA, and L.C. CRAIG: Edeine. III. The composition of the antibiotic peptide edine B. Biochemistry 7, 4153 (1968).

HIEROWSKI, M., and Z. KURYLO-BOROWSKA: On the mode of action of edeine. I. Effect of edeine on the synthesis of polyphenylalanine in a cell-free system. Biochim. Biophys. Acta 95, 578 (1965).

KORNBERG, T., and M.L. GEFTER: DNA synthesis in cell-free extracts of a DNA-polymerase-defective mutant. Biochim. Biophys. Res. Commun. 40, 1348 (1970).

KURYLO-BOROWSKA, Z.: Antibiotical properties of the strain Bacillus brevis Vm4. Bull. State Inst. Marine and Trop. Med., Gdansk, Poland 10, 83 (1959a).

KURYLO-BOROWSKA, Z.: Isolation and properties of pure edeine an antibiotic of the strain Bacillus brevis Vm4. Bull. State Inst. Marine and Trop. Med., Gdansk, Poland 10. 151 (1959b).

KURYLO-BOROWSKA, Z.: On the mode of action of edeine. Biochim. Biophys. Acta 61, 897 (1962).

KURYLO-BOROWSKA, Z.: On the mode of action of edeine. Effect of edeine the bacterial DNA. Biochim. Biophys. Acta 87, 305 (1964).

KURYLO-BOROWSKA, Z., and M. HIEROWSKI: On the mode of action of edeine. II. Studies of the binding of edeine to Escherichia coli ribosomes. Biochim. Biophys. Acta 95, 590 (1965).

KURYLO-BOROWSKA, Z., and J. SEDKOWSKA: Biosynthesis of edeine: Fractionation and characterization of enzymes responsible for biosynthesis of edeine A and B. Biochim. Biophys. Acta (in press).

KURYLO-BOROWSKA, Z., and W. SZER: Inhibition of bacterial DNA synthesis by edeine. Effect on Escherichia coli mutants lacking DNA polymerase I. Biochim. Biophys. Acta 287, 236 (1972).

KURYLO-BOROWSKA, Z., and E.L. TATUM: Biosynthesis of edeine by Bacillus brevis Vm4 in vivo and in vitro. Biochim. Biophys. Acta 114, 206 (1966).

KURYLO-BOROWSKA, Z., and E.L. TATUM: Cell-free biosynthesis of edeine. Purification of the amino acid activating enzymes. In: Progress in antimicrobial and anticancer chemotherapy, p. 1 123, vol. II (1970). Tokyo, Japan: University of Tokyo Press 1970.

LIPMANN, F.: Attempts to map a process evolution of peptide biosynthesis. Synthesis of peptide antibiotics from thiol-linked amino acids parallels fatty acid synthesis. Science 173, 875 (1971).

RONCARI, G., Z. KURYLO-BOROWSKA, and L.C. CRAIG: On the chemical nature of the antibiotic edeine. Biochemistry 5, 2153 (1966).

STRATLING, W., and R. KNIPPERS: Properties of the DNA synthesizing activity in DNA-membrane complexes from bacterial cell extracts. Eur. J. Biochem. 20, 330 (1971).

SZER, W., and Z. KURYLO-BOROWSKA: Effect of edeine on aminoacyl-tRNA binding to ribosome and its relationship to ribosomal binding sites. Biochim. Biophys. Acta 224, 477 (1970).

SZER, W., and Z. KURYLO-BOROWSKA: Interactions of edeine with bacterial ribosomal subunits. Selective inhibition of aminoacyl-tRNA binding sites. Biochim. Biophys Acta 259, 357 (1972).

TABACZYNSKI, M.M., and E. JABLOSKA: The effect of edeine on intracellular growth and mutation of bacteriophage T4B. Acta Microbiol. Polon., Ser. A, 2, 169 (1970).

WEISSBACH, H., Z. KURYLO-BOROWSKA, and W. SZER: Effect of edeine on aminoacyl-tRNA binding to ribosomes. Arch. Biochem. Biophys. 146, 356 (1971).

WICKNER, R.B., and J. HURWITZ: DNA replication in Escherichia coli made permeable by treatment with high sucrose. Biochim. Biophys. Res. Commun. 47, 202 (1972).

WOJCIECHOWSKA, H., J. CIARKOWSKI, H. CHMARA, and E. BOROWSKI: The antibiotic edeine. IX. The isolation and composition of edeine D. Experientia 28/12, 1423 (1972).

Ethidium and Propidium

Michael Waring

Ethidium and propidium (Fig. 1) are phenanthridinium compounds which have attracted a great deal of interest in recent years because of their usefulness as tools for the experimental study of nucleic acids. Of the two ethidium is much the best known since it is readily available and inexpensive, being a well-established drug in widespread use for the chemotherapy of trypanosomiasis. Propidium is not employed in chemotherapeutic practice; it has only recently become of interest for its value in a procedure for the isolation of closed circular duplex DNA (see below). Consequently the bulk of this review will be concerned with ethidium, but reference will be made to the limited information about propidium where possible.

Ethidium is generally available as the bromide; propidium as the (di-)iodide. So far as is known the nature of the salt form does not affect their metabolic and biochemical properties. Both drugs were synthesised some twenty years ago in the course of an intensive programme to develop phenanthridines as effective trypanocidal agents (WATKINS, 1952; WATKINS and WOOLFE, 1952). Some of the history surrounding the development of phenanthridines has been summarised by HAWKING (1963), NEWTON (1963, 1964), and ALBERT (1968). Ethidium (sometimes referred to in the earlier literature as "homidium") originated as a striking improvement on the existing drug dimidium (the 5-methyl homologue of ethidium, Fig. 1) endowed with 10 to 50-fold improved activity against four strains of trypanosomes, at no expense in terms of toxicity to mice (WATKINS and WOOLFE,

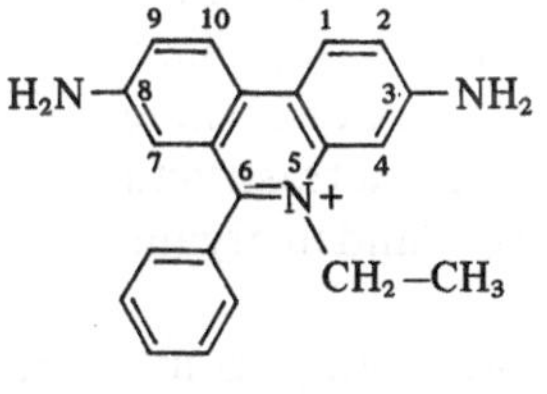

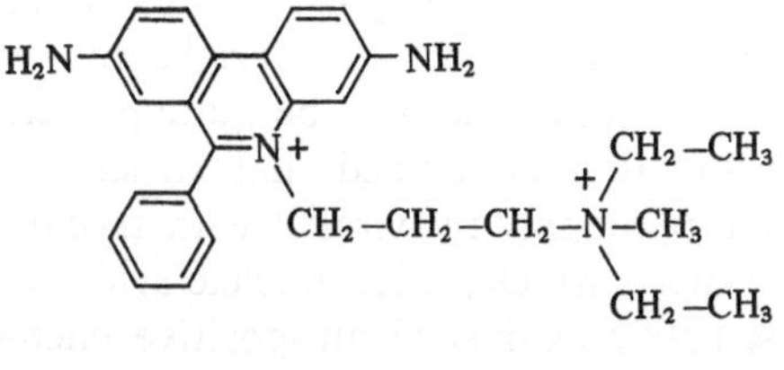

Fig. 1. Structural formulae of ethidium and propidium. The numbering of the phenanthridine ring system is in accord with the International Union of Chemistry

1952) and with improved tolerance in cattle (WOOLFE, 1956a). Among the very many phenanthridines synthesised and tested for trypanocidal (WOOLFE, 1956a, b) or antibacterial (SEAMAN and WOODBINE, 1954) activity at that time only ethidium proved outstanding. Subsequent efforts led to the synthesis of phenanthridines bearing substituents typical of other known trypanocides. Two of these, prothidium and isometamidium, have also found chemotherapeutic application; their structures and trypanocidal activities, in relation to those of other antitrypanosomal agents, are described in the reviews of HAWKING (1963) and NEWTON (1963, 1964).

Since the mechanism of action of ethidium has been covered in numerous reviews (HAWKING, 1963; NEWTON, 1963, 1964; WARING, 1966a, 1968a, 1972) only a brief outline of its biochemical effects need be given here so that attention may be concentrated on recent studies of its interaction with nucleic acids.

Growth-Inhibitory Properties

Ethidium is active against a number of species of trypanosomes *in vivo* and *in vitro*. This property forms the basis of its principal medical use, which is in the veterinary field in tropical countries, for the treatment and prophylaxis of trypanosomiasis in cattle. The onset of its trypanocidal effect is characteristically slow, both *in vivo* and *in vitro*, as if a period of growth in the presence of the drug were mandatory for the manifestation of its lethal effect (HAWKING, 1963; NEWTON, 1964). NEWTON (1957) investigated this slow onset of growth-inhibition using the trypanosomid flagellate *Strigomonas (Crithidia) oncopelti* cultured *in vitro;* he found an initial rapid binding of ^{14}C-ethidium by the cells which did not seem to correlate with growth inhibition, followed by a slower uptake which paralleled a progressive decrease in the growth rate of the organisms. Possible explanations for these effects were considered by HAWKING (1963) and NEWTON (1964), and it seems likely that the more recently discovered selective action of ethidium on the kinetoplast of trypanosomes (described below) may be relevant in this context. Morphological changes in trypanosomes treated with phenanthridines include the development of basophilic refractile granules in the cytoplasm (NEWTON, 1964) and the appearance of dyskinetoplastic forms (DELAIN and RIOU, 1969). Such alterations in cell ultrastructure are not peculiar to phenanthridine-treated cells; they are also found in organisms exposed to acridines and other trypanocidal drugs.

As might be expected from its known ability to interfere with nucleic acid synthesis, ethidium is also inhibitory to the growth of bacteria, viruses and mammalian cells. In an extensive survey of the antibacterial activity of phenanthridines SEAMAN and WOODBINE (1954) found many to be quite powerfully active against *Streptococcus pyogenes,* and they concluded that structure—activity relations for antibacterial activity run roughly in parallel with trypanocidal potency. In common with many other drugs which inhibit nucleic acid synthesis, phenanthridines are generally most active against Gram-positive bacteria (DICKINSON *et al.,* 1953); TOMCHICK and MANDEL, 1964). The latter authors noted pronounced morphological alterations in *Bacillus cereus* exposed to ethidium; upon Gram-staining the cells appeared more granular and were elongated in shape. Some marked antiviral effects of phenanthridines, including inhibition of bacteriophage replication, were reported quite early by DICKINSON *et al.* (1953). More recently ethidium was shown to inhibit the growth of herpes virus in cell cultures (VILA-GINES and ATANASIU, 1967) and to be active against RNA viruses (SPRECHER-GOLDBERGER, 1965); in both these systems inactivation of virus particles *via* a photosensitizing action of ethidium was observed. The development of trachoma agent, a procaryotic obligate intracellular parasite, is also sensitive to quite low concentrations of ethidium (BECKER and ASHER, 1972).

The growth-inhibitory activity of ethidium towards mammalian cells has been studied by many workers, mainly in the context of experiments where the drug was used as a tool to suppress various aspects of mitochondrial function (see below). Although ethidium is not considered a practical antitumour agent it can certainly inhibit the growth of tumour cells, and KANDASWAMY and HENDER-SON (1962a) reported a markedly enhanced carcinostatic effect of ethidium against Ehrlich ascites tumour in mice when administered in conjunction with azaserine. ARLETT (1970) found that the survival of gamma-irradiated Chinese hamster cells was reduced by exposure to ethidium during post-irradiation incubation, suggesting interference with enzymic radiation repair mechanisms.

No doubt cell permeability plays a major role in determining the response of different cell types to ethidium, especially in respect of its selective toxicity towards trypanosomes. The cationic charge on the molecule must present something of an obstacle to its passage across cell membranes and, being a quaternary compound, this problem cannot be surmounted by dissociation of an H^+ ion to yield an uncharged form with much higher lipid solubility as commonly occurs with other drugs such as acridines (ALBERT, 1968). However, an uncharged form of quaternary phenanthridines can arise by attack of a hydroxyl ion on the double bond to the ring nitrogen yielding a pseudo-base form as indicated for ethidium in Fig. 2. In aqueous solution at neutral pH the equilibrium lies far to the left, but the small proportion of pseudo-base present could be a major determinant of the rate of penetration of the drug into cells. In support of this contention WATKINS (1952) pointed out that changing the quaternising group from methyl to ethyl causes a 2–3 fold increase in the concentration of pseudo-base at pH 6–9, which could account for the strikingly improved therapeutic efficacy of ethidium compared to dimidium, its predecessor. The existence of the pseudo-base form of phenanthridines is a feature which deserves to be borne in mind when considering sophisticated experiments on physico-chemical interactions between these drugs and their receptors.

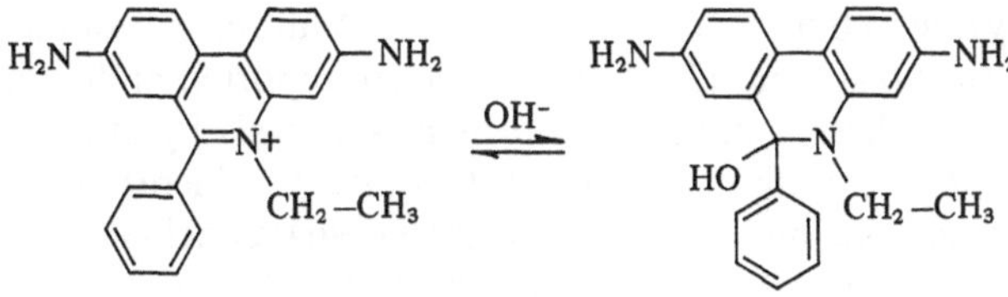

Fig. 2. Formation of a pseudo-base by ethidium

Mammalian Toxicity

An excellent account of the pharmacology, distribution and toxicity of ethidium has been assembled by HAWKING (1963). The drug is poorly absorbed by mouth, so that the LD_{50} to mice is about 150 mg/kg by intraperitoneal injection but about 1500 mg/kg by mouth. Its penetration into the cerebrospinal fluid is rather poor. Ethidium is excreted fairly rapidly in the urine in unchanged form (HAWKING, 1963; KANDASWAMY and HENDERSON, 1963). It does not cause the delayed toxicity in cattle, attributed to photosensitization, which was a problem with earlier phenanthridines. Acute toxicity of ethidium is due to cardiovascular or respiratory failure; chronic toxicity is associated with a number of symptoms including liver necrosis.

Propidium is more toxic than ethidium; LD_{50} doses for mice by subcutaneous injection of the two drugs are 16 and 110 mg/kg respectively (WOOLFE, 1956a, b). Propidium is practically devoid of trypanocidal activity (WOOLFE, 1956b).

Interference with Nucleic Acid Synthesis in vivo

The basic biochemistry of ethidium action *in vivo* has been investigated in a variety of systems. NEWTON (1957) found that addition of the drug to logarithmically-growing *Strigomonas oncopelti* did not immediately inhibit growth but allowed at least a doubling in the number of organisms to occur before multiplication finally ceased; during this period the DNA content of the cells fell to half its normal value whilst the RNA content remained approximately constant. With washed suspensions of the flagellate a preferential effect on DNA synthesis was confirmed: it was inhibited promptly while RNA and protein synthesis continued for a while before inhibition was evident. In yeast, KERRIDGE (1958) found that the minimal growth-inhibitory concentration of ethidium caused powerful inhibition of both DNA and RNA synthesis while protein synthesis was much less affected. TOMCHICK and MANDEL (1964) studied the action of ethidium on *Bacillus cereus* and *Escherichia coli*; in the *Bacillus* the syntheses of RNA and DNA were both affected much more than any other cellular functions, but in *E. coli* a preferential action on DNA synthesis was apparent. In the latter bacterium there was some evidence of uncoupling of oxidative phosphorylation. GRINSTED (1969) employed ethidium as a tool to inhibit RNA synthesis and reveal decay of pulse-labelled RNA in a thermophilic *Bacillus*. He found that it completely inhibited ^{3}H-uridine incorporation within seconds of addition to the culture and, at relatively low drug concentrations, had little effect on the fate of the pulse-labelled RNA. At higher concentrations it interfered with the normal decay of unstable RNA, attributable to secondary interference with protein synthesis, most probably due to binding of ethidium to RNA and/or ribosomes. In mammalian cells KANDASWAMY and HENDERSON (1962b) reported preferential inhibition of nucleic acid synthesis compared to protein synthesis by ethidium; they were impressed by apparent differential effects of ethidium on incorporation of different nucleic acid precursors, but RAZIN and MAGER (1964) concluded that any anomalous effects of ethidium on purine nucleotide metabolism were probably a reflection of accumulation of the drug in mitochondria with consequent impairment of respiration and uncoupling of oxidative phosphorylation. More recently SNYDER et al. (1971) found that ethidium, in common with other DNA-binding drugs (but not all), appears to exert a direct effect on the nucleolar processing of ribosomal precursor 45S RNA in mouse lymphoma cells.

Thus far the effects of ethidium in vivo would appear to be wholly unremarkable—the picture which emerges is of a slightly messy and not very selective inhibitor of nucleic acid synthesis, with a few hints of side-effects on mitochondria. This picture has changed radically within recent years due to the growing realisation that ethidium is endowed with remarkably selective properties where extranuclear DNA, or cell organelles containing extranuclear DNA, are concerned. Ethidium provokes loss ("curing") of bacterial plasmids such as antibiotic resistance factors and F (sex) factors of E. coli, selectively interfering with their replication (BOUANCHAUD et al., 1968; NISHIMURA et al., 1971), it preferentially attacks the kinetoplast of trypanosomes (VAN ASSEL and STEINERT, 1968; DELAIN and RIOU, 1969), it causes massive induction of a mitochondrial DNA polymerase in Tetrahymena (WESTERGAARD et al., 1970), and it efficiently converts yeast cells to respiration-deficient ϱ^- (petite) mutants (SLONIMSKY et al. 1968). Although similar phenomena have been reported with aminoacridines, in most instances ethidium has proved to be more selective in that the effects are larger and/or manifested at ethidium concentrations substantially lower than those required to produce more generalised interference with cell growth and metabolism of the sort quoted above.

The induction of the ϱ^- mutation in yeasts has attracted particular attention. It is an irreversible non-Mendelian cytoplasmic mutation which leads to eventual disappearance of several respiratory enzymes such as cytochromes a/a_3 and b and enzymes of the tricarboxylic acid cycle (DE DEKEN, 1966; ROODYN and WILKIE, 1968). The drug-induced mutagenesis occurs in the absence of cell division, and the mother cells are quantitatively and rapidly converted to the ϱ^- genotype (SLONIMSKY et al., 1968; PERLMAN and MAHLER, 1971a), though a premutational stage exists during which the cells can be rescued from ethidium-induced mutagenesis by a critical heat treatment (PERLMAN and MAHLER, 1971b). The earliest detectable mutational event is a rapid cessation of synthesis of cytochrome oxidase and cyt a/a_3, coincident with a blockage of synthesis and transcription of the mitochondrial DNA which is rapidly degraded (PERLMAN and MAHLER, 1971a; GOLDRING et al., 1970; FUKUHARA and KUJAWA, 1970). The primary mutagenised cells are very heterogeneous; some lack mitochondrial DNA altogether whereas others retain detectable amounts (NAGLEY and LINNANE, 1972). During subsequent growth of the mutagenised cells in the absence of drug, novel species of mitochondrial DNA appear which are typically less dense than the lost parental mitochondrial DNA in buoyant CsCl gradients, reflecting a marked difference in base-composition (PERLMAN and MAHLER, 1971a). The presence of nalidixic acid during exposure of the cells to ethidium prevents the induction of the ϱ^- genotype, but as yet the mechanism of this protective effect is unknown (HOLLENBERG and BORST, 1971; VIDOVÁ and KOVÁČ, 1972; WHITTAKER et al., 1972). With yeasts which are not susceptible to the ϱ^- mutation (petite negative), treatment with ethidium leads to a reversible phenotypic mimicking of the ϱ^- respiratory-deficient state (KELLERMAN et al., 1969; LUHA et al., 1971).

These observations point to a remarkably selective attack by ethidium on mitochondrial structure and function in yeast. Its effect on the kinetoplast of trypanosomes falls into the same category, for this organelle is involved in differentiation to produce mitochondria during the crithidial stage of the life-cycle. Growth of trypanosomes in the presence of ethidium leads to a selective inhibition of kinetoplast DNA synthesis, with the eventual production of dyskinetoplastic cells which lack the organelle altogether (VAN ASSEL and STEINERT, 1968; DELAIN and RIOU, 1969). In *Acetabularia* (HEILPORN and LIMBOSCH, 1971) and cells of higher organisms the same selectivity for mitochondria is apparent: synthesis of cytochrome oxidase (cyt a/a_3) and mitochondrial DNA is inhibited, together with other mitochondrion-associated enzymes, and in the electron microscope mitochondrial profiles appear bloated, with cristae reduced or even obliterated (NASS, 1970; RADSAK et al., 1971; MILNER, 1972). Ethidium, in fact, is now widely used as a selective tool in the study of nucleic acid synthesis by eucaryotic cells since it inhibits the synthesis of mitochondrial RNA and DNA at concentrations which have little or no effect on synthesis of the corresponding nuclear and cytoplasmic species (ZYLBER et al. 1969; ZYLBER and PENMAN, 1969; NASS, 1970, 1972; RADSAK et al., 1971; KOCK and VON PFEIL, 1971). Pre-existing mitochondrial DNA appears altered by exposure to ethidium, but it does not suffer massive degradation as occurs in the yeast system (KOCK and VON PFEIL, 1971; SMITH et al., 1971; NASS, 1972).

What may be the basis for this selectivity? One feature which all the above examples share in common is that they involve selective interference with extra-chromosomal DNA. Moreover, in each case the DNA is known to occur partly or wholly in the form of closed circular duplex molecules. If the existence of these DNAs in circular form is the significant common denominator it may be postulated that the selectivity of action of ethidium derives from preferential attack directly on circular DNA molecules or on some common feature of the way circular DNA is organised inside living cells (WARING, 1972). Sufficient is understood about the interaction of ethidium with circular DNA (see below) to make this an attractive possibility. It is consistent with two recent findings: 1. the presence of ethidium at a concentration which does not inhibit growth induces the appearance of abnormal oligomeric circular DNA molecules in the kinetoplast of *Trypanosoma cruzi* (RIOU and DELAIN, 1969), and 2. treatment of cultured human or mouse cells with ethidium causes, within a few hours, a 3–5 fold increase in the superhelix density of pre-existing circular mitochondrial DNA molecules, as would be expected if the drug had become intercalated into this DNA *in vivo* (SMITH et al., 1971). It is tempting to speculate further that the molecular basis for the selectivity lies in preferential binding of ethidium to circular DNA due to the enhanced affinity for intercalation of ethidium displayed by DNA containing negative superhelical turns *in vitro* (see below). This hypothesis would demand that the circular molecules be actually super coiled *in vivo*, for which there is as yet no direct evidence, but the suggestion merits consideration. Alternative suggestions, e.g. that interaction of ethidium with the mitochondrial membrane (AZZI and SANTATO, 1971) might facilitate selective uptake of the drug by such organelles, do not exclude the notion that preferential binding to the DNA because of its circularity is important. Both effects, as well as others, might ultimately prove relevant.

Binding to Nucleic Acids and Inhibition of Enzymes

The formation of a complex between a phenanthridine (dimidium) and DNA was noted as long ago as 1950 by BROWNLEE *et al.* (1950). A little later the idea that dimidium might act "by combining with the nucleic acids of the organism so as to disrupt both its metabolism and reproduction" was put forward by SEAMAN and WOODBINE (1953). The first explicit description of an ethidium-DNA complex was made by ELLIOTT (1963), who also showed that ethidium is a powerful inhibitor of DNA polymerase I from *E. coli*. It was then found to inhibit the DNA-dependent RNA polymerase of *E. coli* (WARING, 1964, 1965a), its efficacy in inhibiting the DNA and RNA polymerases being much the same when referred to the molar ratio of drug to DNA nucleotides (WARING, 1966a). In the RNA polymerase system presumptive evidence that the inhibition was due to binding of ethidium to the DNA template was obtained by showing that the inhibition could be partially reversed by increasing the amount of DNA present while keeping the concentrations of enzyme, substrates etc. constant. This conclusion was put on a firmer basis by estimating the level of binding of ethidium to the DNA template under conditions where the concentrations of ethidium and DNA were varied either independently or together (WARING, 1964, 1965a), resulting in the plot reproduced in Fig. 3a. It was also shown that incorporation of all four nucleotides into the RNA product was inhibited to the same extent (Fig. 3b). Thus in the presence of ethidium the enzyme still synthesises RNA

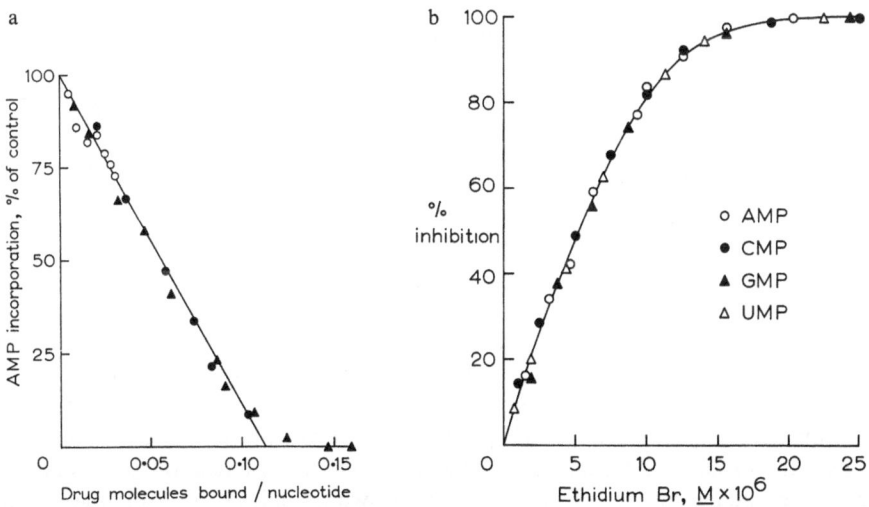

Fig. 3. Inhibition of RNA polymerase by ethidium bromide. The purified *E. coli* enzyme was assayed with bacteriophage T2 DNA as template. In a the reaction was followed by incorporation of AMP from ^{14}C-ATP into trichloroacetic acid-insoluble material. Triangles show results from experiments in which the drug concentration was varied. Circles are from two different sets of experiments where the drug concentration was held constant and the concentration of T2 DNA was varied. Binding of ethidium to the template DNA was estimated using binding constants determined spectrophotometrically by WARING (1965b). In b the incorporation of each of the four nucleotides into RNA was measured; AMP (o), CMP (●), GMP (▲) and UMP (△). (From WARING, 1965a)

whose base-composition is strictly determined by that of the DNA template, but the rate of synthesis is decreased in direct proportion to the amount of ethidium bound to the template. Fig. 3a shows complete inhibition occurring at a binding ratio of 0.11 (equivalent to one ethidium molecule bound per 4.5 DNA base-pairs) but this value may not have absolute significance, as previously discussed (WARING, 1965a). The most likely interpretation of the linear relationship between inhibition and template binding is that ethidium molecules complexed at random intervals along the DNA molecule act as transient blocks to the progression of the polymerase enzyme along its template.

Ethidium has also been shown to inhibit numerous other enzymes concerned with nucleic acid metabolism, including DNAase (ERON and MCAUSLAN, 1966), the RNA-templated RNA replicases of bacteriophages Qβ and R17 (SAFFHILL et al., 1970; IGARASHI and BISSONNETTE, 1971) and, most recently, the DNA polymerases (reverse transcriptases) of oncogenic viruses (see, for example, FRIDLENDER and WEISSBACH, 1971; HIRSCHMAN, 1971). With the latter enzymes some interesting variations in the inhibitory activity of ethidium towards different synthetic polynucleotide templates have been reported (FRIDLENDER and WEISSBACH, 1971). In all these systems the inhibition by ethidium may be attributed to binding of the drug to the polynucleotide template or substrate. An entertaining experiment on evolution *in vitro* was conducted by SAFFHILL et al. (1970) with the Qβ RNA replicase system, in which Qβ RNA variants resistant to ethidium bromide were selected for by repeated replication of the RNA in the presence of mildly inhibitory concentrations of the drug.

Properties of Ethidium-Nucleic Acid Complexes

The formation of a complex between ethidium and DNA is readily observed with the naked eye because of the large metachromatic shift in the visible absorption spectrum of the drug which occurs on binding (Fig. 4). As increasing amounts of DNA are added, the original maximum at 480 nm of the drug absorption spectrum (yellow-orange to the eye) is progressively red-shifted to about 520 nm when all the drug is bound (bright pink to the eye), and a well-defined isosbestic point occurs at 510 nm. This spectral shift is more dramatic than is seen with other DNA-binding drugs such as acridines or actinomycin, and provides a simple means of assessing complex formation by quantitative spectrophotometry (WARING, 1965b) which has been widely employed in subsequent studies of ethidium-nucleic acid complexes. The reversibility of ethidium-DNA interaction was demonstrated by WARING (1965b) using an isotope-dilution method, but complexes are not easily dissociated by simple dialysis. For practical purposes dissociation is best achieved by passing the complex down a small cation-exchange column (WARING, 1965b) or by extraction with organic solvents (BAUER and VINOGRAD, 1971).

The metachromatic effect associated with ethidium-DNA binding is accompanied by a strikingly enhanced fluorescence quantum yield, which was employed by LE PECQ and PAOLETTI (1967) to characterise the complex. Their results are in substantial agreement with those of WARING (1965b), and the combined findings of both studies are summarised below.

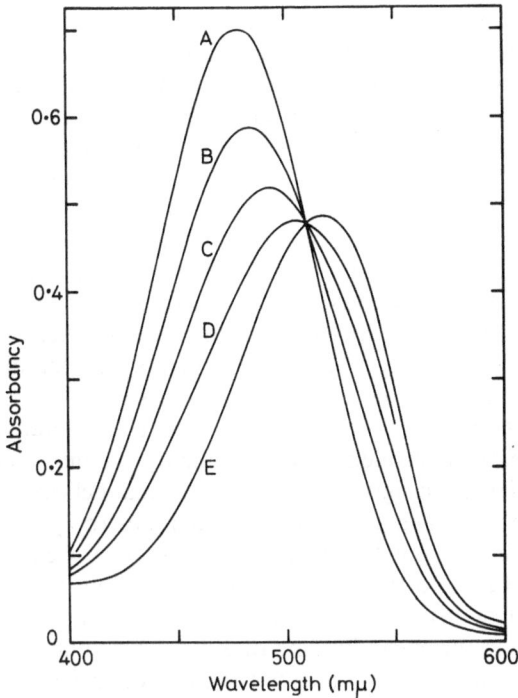

Fig. 4. Shift of the absorption spectrum of ethidium bromide in the presence of DNA. *A* spectrum
of the drug alone in buffer; *B–E* with increasing concentrations of T2 DNA added.
(From WARING, 1965 b)

1. With native double-helical DNA there is a strong (primary) binding process
which is saturated at a binding ratio v (drug molecules bound per nucleotide)
of about 0.2, corresponding to one ethidium molecule bound per 2.5 nucleotide-
pairs in the DNA.

2. Further binding of ethidium can occur by a weaker (secondary) process,
characterised by a similar spectral shift but lower fluorescence quantum yield,
leading eventually to the precipitation of an electrostatically neutral 1 : 1 complex
at $v = 1.0$.

3. Both primary and secondary binding are sensitive to increases in ionic
strength, the latter much more so than the former, particularly if divalent cations
are added. At high salt concentrations the secondary binding is practically totally
suppressed, but binding by the primary process still occurs with no change in
the number of available sites—only the affinity constant is lowered. Binding
is essentially independent of pH within the range where the double-helical structure is stable.

4. The number of sites for strong (primary) binding does not vary with the
base-composition of the DNA, at least over the range covered by naturally-
occurring DNAs (35–72% guanine plus cytosine content).

5. Binding of ethidium raises the thermal denaturation temperature of DNA
and lowers its buoyant density in CsCl density gradients (WARING, 1965 b, 1966 a;
LE PECQ and PAOLETTI, 1967).

In the course of the above work it was observed that ethidium also forms complexes with RNA and heat-denatured or single-stranded DNA (WARING, 1965b; LE PECQ and PAOLETTI, 1967). The specificity of interaction between ethidium and nucleic acids was further investigated in various model systems; it was found that spectral changes similar to those which occur on binding to DNA were produced by nucleotides (especially purine nucleotides), apurinic and apyrimidinic DNA, and the synthetic polyribonucleotides poly rA, poly rI and poly rU but not poly rC (WARD et al., 1965; WARING, 1966b). Even polyvinyl sulphate induced a spectral shift (LE PECQ and PAOLETTI, 1967), though ribitol teichoic acid did not (WARING, 1966b). However, in all these model tests the interaction with the drug appeared to be much weaker than the binding to DNA. In the cases of poly rA, poly rI and poly rU binding curves were determined which did not show the characteristic strong (primary) interaction seen with DNA; instead a relatively weak, apparently cooperative, binding took place which is probably equivalent to the secondary binding process with DNA (WARING, 1966b). The most significant finding was that helical complexes formed between pairs of complementary polyribonucleotides (poly rA.poly rU; poly rI.poly rC; and poly rA.poly rI) bound ethidium very well, yielding binding curves quite comparable to those seen with DNA. From this result it may be concluded that helical structure stabilised by hydrogen-bonded base-pairing is mandatory for strong (primary) binding of ethidium to nucleic acids (WARING, 1966b). In the light of these experiments the interaction between ethidium and RNA or heat-denatured DNA is interpreted as binding of the drug to the "hairpin" helical regions formed by intra-strand interactions in these polynucleotides (WARING, 1965b, 1966b; LE PECQ and PAOLETTI, 1967). The weaker or secondary binding processes seen with ethidium are generally agreed to represent stacking phenomena, mediated largely by electrostatic forces, of the type described by STONE and BRADLEY (1961) for acridine orange.

Ethidium has proved a valuable probe for investigating the secondary (and tertiary) structures of nucleic acids, notably circular DNA (see below), 5S rRNA (GRAY and SAUNDERS, 1971), and tRNA (see, for example, BITTMAN, 1969; CANTOR et al., 1971; TRITTON and MOHR, 1971). BITTMAN (1969) found the ethidium-tRNA complex to be similar to the DNA complex in several respects, including the frequency of sites for strong (primary) binding which would correspond to approximately 15 bound drug molecules per tRNA molecule 80 nucleotides long; these sites are presumably located in the various helical stems of the tRNA molecule. He also studied the kinetics of ethidium-tRNA interaction and found a fast binding reaction, assumed to represent electrostatic binding near the negatively charged phosphates, followed by several slow unimolecular steps of which one would probably correspond to intercalation. CANTOR et al. (1971) and TRITTON and MOHR (1971) worked with low drug concentrations where only one ethidium molecule was bound per tRNA molecule, presumably intercalated into a single site in one of the helical stems. Ethidium has also been used to probe the structure of nucleoproteins. It has generally been found that the number of sites for ethidium binding in native or reconstituted nucleohistone is substantially lower than the number in free DNA, and that further binding sites become exposed as successive proteins are stripped off; the association constant for the drug binding reaction may or may not be affected (OLINS, 1969; ANGERER and MOUDRIANAKIS, 1972; LURQUIN and SELIGY, 1972). WOLFE *et al.* (1972) found that ethidium and propidium were significantly more effective than other nucleic acid-binding drugs in labilising bacterial ribosomes towards thermal degradation.

The enhanced fluorescence of ethidium when bound to nucleic acids has led to a number of applications. Although the physical basis for the enhanced quantum yield is uncertain [LE PECQ and PAOLETTI (1967) consider that it arises from immersion of the drug in a hydrophobic environment; BURN (1969) attributes it to a change in planarity of the chromophore], the effect has nevertheless proved useful in developing sensitive methods for the determination of nucleic acids (LE PECQ and PAOLETTI, 1966; VAN DYKE and SZUSTKIEWICZ, 1968) and for assay of their associated enzymes (LE PECQ, 1971).

The Intercalation Model

The idea that drug molecules possessing planar polycyclic aromatic ring systems might bind to DNA by intercalation, i.e. insertion between the stacked base-pairs of the double helix, was first developed by LERMAN (1961). His intercalation model was proposed to account for the binding of aminoacridines to DNA. Its main experimental support came from observations on X-ray diffraction patterns of proflavine-DNA complexes, together with the findings that binding of proflavine increases the viscosity and lowers the sedimentation coefficient of DNA.

Similarities between the binding of aminoacridines and ethidium to DNA led FULLER and WARING (1964) to undertake an X-ray diffraction and molecular model-building study on the ethidium-DNA complex, culminating in the development of an intercalation model as illustrated in Fig. 5. This model shares many of the features of LERMAN's model: the tricyclic phenanthridine ring system lies perpendicular to the helix axis in van der Waals contact with the base-pairs above and below; it is effectively shielded from contact with the surrounding medium and occupies a 3.4Å space measured along the helix axis. The phenyl and ethyl groups of the ethidium molecule, which are approximately perpendicular to the plane of the phenanthridine ring (HOSPITAL and BUSETTA, 1969), lie in one of the grooves of the helix. The geometry of the base pairs and their positioning with respect to the helix axis are unchanged apart from the 3.4 Å displacement along that axis. The model differs from LERMAN's in suggesting the formation of hydrogen bonds from the primary amino groups of the drug chromophore to phosphate oxygens in the complementary DNA strands, and in respect of the helix unwinding angle which is 12° (FULLER and WARING, 1964). This latter difference is important, as will be seen.

The intercalation model for the strong (primary) binding of ethidium to nucleic acids is supported by the following evidence.

1. X-ray diffraction patterns from oriented fibres of the ethidium-DNA complex reveal loss of well-defined layer lines, retention of the 3.4 Å reflection corresponding to normal stacking of the base-pairs, and a decreased average molecular diameter (FULLER and WARING, 1964). The latter point argues against external attachment of the drug to the helix.

Fig. 5. An intercalation model for the ethidium-DNA complex. On the left is a space-filling (CPK) model of an 18 base-pair segment of a normal B-form DNA helix. At right the same model is shown containing a single ethidium molecule intercalated between two G-C base-pairs half way up. The phenyl and ethyl groups of the drug molecule can be seen projecting out into the narrow groove of the helix; in the precise model of FULLER and WARING (1964) the siting of these groups in the wide groove was preferred, but either mode of orientation appears to be stereochemically feasible. The model shows the extension and local uncoiling of the DNA which occurs due to binding of the ethidium; both photographs are at the same magnification and the lower 9 base-pairs are viewed from the same point

2. Ethidium increases the viscosity and lowers the sedimentation coefficient of DNA (cf. the results for nicked DNA circles in Fig. 6) (WARING, 1964; LE PECQ and PAOLETTI, 1967). These effects are attributable to lengthening and stiffening of the helix.

3. Electron microscopy reveals a 27% increase in the measured length of bacteriophage lambda DNA on binding a saturating level of ethidium (FREIFELDER, 1971). The contour length of ^{3}H-labelled bacteriophage T2 DNA is also increased on binding ethidium, as measured by autoradiography (L. ARONOW and M.J. WARING, unpublished).

4. The primary amino groups of dimidium are shielded from electrophilic attack by HNO_2 when the drug is bound to DNA (LERMAN, 1964).

5. Flow dichroism of the ethidium-DNA complex indicates that the plane of the phenanthridine ring system is parallel to the base-pairs, and perpendicular to the long axis of the DNA molecule (LE PECQ and PAOLETTI, 1967).

6. The variation in magnitude of circular dichroism spectra of ethidium-DNA complexes with the amount of drug bound suggests a common binding position with proflavine and is consistent with intercalation models involving H-bond formation by the chromophore amino groups (DALGLEISH et al., 1971).

7. Binding of ethidium shows a strong preference for two-stranded polynucleotide structures, as described above. The three-stranded poly rA.2 poly rU structure is specifically destabilised by the presence of ethidium in favour of the formation of double-helical poly rA.poly rU. The latter helix is strongly stabilised by binding of ethidium (M.J. WARING, unpublished observations).

8. The kinetics of binding of ethidium to tRNA reveal the occurrence of several steps, some characterised by time constants of the order expected for an intercalation reaction (BITTMAN, 1969; TRITTON and MOHR, 1971).

9. Upfield shifts of uridine proton resonances and selective line-broadening of ethidium proton resonances were observed by KREISHMAN et al. (1971) in a PMR study of the interaction between ethidium and UpU or poly U, indicative of the formation of intercalated complexes. This conclusion is not inconsistent with the failure to detect strong binding to poly U in optical studies (WARING, 1966b) because PMR studies require the use of reactants at concentrations 2–3 orders of magnitude higher than those employed in optical work.

10. Binding of ethidium removes and reverses the supercoiling of closed circular duplex DNA (CRAWFORD and WARING, 1967a; BAUER and VINOGRAD, 1968; see also below), in exact agreement with predictions based on the local unwinding of the helix required by the intercalation model of FULLER and WARING (1964).

While this body of evidence constitutes impressive support for an intercalation model there remain some outstanding problems. In the first place there is no obvious reason why strong (primary, and presumably intercalative) binding of ethidium to DNA should be limited to not more than one drug molecule per 2–2.5 base-pairs. The popular explanation is to postulate a neighbour-exclusion model whereby binding to sites adjacent to one already occupied is forbidden; mathematically this idea fits very nicely with the binding data (for references see WARING, 1972). However, there is as yet no direct experimental proof that neighbour-exclusion is correct. FULLER and WARING (1964) were unable to show steric hindrance preventing binding of a second ethidium molecule immediately adjacent to one already bound, though GILBERT and CLAVERIE (1968) produced an interesting theoretical model to explain the limited binding in terms of electrostatic interactions in the intercalated complex.

A more pressing problem concerns the exact alteration in the winding of the DNA helix caused by intercalation of ethidium. In the LERMAN (1961) and FULLER and WARING (1964) models constraints were applied to limit the permissible distortion of the helix in an acceptable scheme. As a result, LERMAN'S original (1961) model required a local uncoiling of 45° per intercalated ligand, subsequently revised to 36° (LERMAN, 1964). In the FULLER and WARING model of 1964 the *minimum* unwinding required to accommodate ethidium was found to be 12°, and this value was proposed in the interests of maintaining maximal separation between charged phosphate groups. PAOLETTI and LE PECQ (1971a) attempted to estimate the angle by measuring resonance energy transfer between ethidium molecules bound to DNA under saturating conditions; they reached the surprising conclusion that binding of each ethidium molecule to DNA *over-winds* the helix by 13° ± 4°. Several assumptions were required to arrive at this estimate, including the idea of neighbour exclusion, and on this basis they built an intercalation model for ethidium in which an over-winding of approximately 12° occurred for each drug molecule bound (PAOLETTI and LE PECQ, 1971a). An additional feature of this model was that it required a rotation in the DNA sugar-phosphate backbone which could provide an explanation for neighbour exclusion (PAOLETTI and LE PECQ, 1971b). However, since several important assumptions were involved in estimating the proposed over-winding angle, and others have presented fluorescence anisotropy data which differ from those employed by PAOLETTI and LE PECQ (1971a) (GENEST and WAHL, 1972), it is difficult to assess the validity of the over-winding model at the present time. Its most important consequence, if the model should ultimately prove correct, would be that the currently accepted sense of supercoiling in naturally-occurring closed circular duplex DNAs would have to be reversed (see below). Since the great weight of opinion presently favours the idea that these DNAs are under-wound, with negative superhelical turns, the classical unwinding intercalation models for binding of ethidium (and all other intercalating agents) to DNA remain the most likely. Definitive evidence may have to await X-ray diffraction analysis of a crystalline intercalation complex. The recent solution of the crystal structure of a 1:2 actinomycin:deoxyguanosine complex (JAIN and SOBELL, 1972) represents an important step in this direction; it leads to an unwinding intercalation model for actinomycin-DNA binding (SOBELL and JAIN, 1972)[1].

Interaction with Circular DNA

The discovery of closed circular duplex DNA, pioneered by Vinograd and his colleagues (reviewed in BAUER and VINOGRAD, 1971), opened a new area

[1] A reappraisal of the fluorescence depolarization results of PAOLETTI and LE PECQ (1971a) by PIGRAM, FULLER and DAVIES [J. Mol. Biol. **80**, 361 (1973)] refutes the suggestion of PAOLETTI and LE PECQ that intercalation of ethidium *over*-winds the DNA helix by approximately 13°. It is concluded that the helix is unwound by approximately 12° as proposed by FULLER and WARING (1964).

of investigation in which ethidium played an indispensable part, and indeed continues to do so. When examined *in vitro,* naturally-occurring circular DNA molecules have a supercoiled tertiary structure which is believed to originate from a small deficiency of turns in the fundamental Watson-Crick helical structure. The topological properties of these circles are such that any environmental influence causing an alteration in the number of base-pairs per turn of the helix is necessarily accompanied by a corresponding change in the extent of supercoiling. Since the supercoiled state of circular DNA is readily evidenced in such macromolecular properties as its sedimentation coefficient, changes in these properties provide a sensitive means of monitoring alterations in the winding of the helix caused, among other things, by the binding of drugs.

The first experimental evidence that binding of an intercalating drug to closed circular DNA affects its supercoiling was reported by CRAWFORD and WARING (1967) in a study of the interaction between ethidium and polyoma virus DNA. It was found that intercalation of ethidium first removed and then reversed the supercoiling of polyoma DNA, precisely in accord with expectations for the binding of a helix-unwinding ligand to underwound (negatively superhelical) circular DNA. BAUER and VINOGRAD (1968) then reported exactly similar findings with ethidium and SV-40 virus DNA. In Fig. 6 the results of such an experiment with bacteriophage ϕX174 replicative form DNA are shown. As the level of drug binding rises, the initially high sedimentation coefficient of the closed circular molecules falls, reflecting progressive loss of their negative (right-handed) supercoils. At a critical level of binding v_c (just below 0.04 drug molecules bound per nucleotide in Fig. 6) the S_{20} reaches a minimum (the equivalence point) where the supercoils have been completely removed and the closed circles sediment as relaxed, untwisted circular molecules. Further drug binding leads to a rapid rise in S_{20} as the circles develop reversed (left-handed) supercoils. Under the same conditions the S_{20} of nicked circles shows only a gradual fall because any unwinding associated with drug binding is transmitted to the nick(s) where free rotation of one DNA strand about the other can occur; consequently these

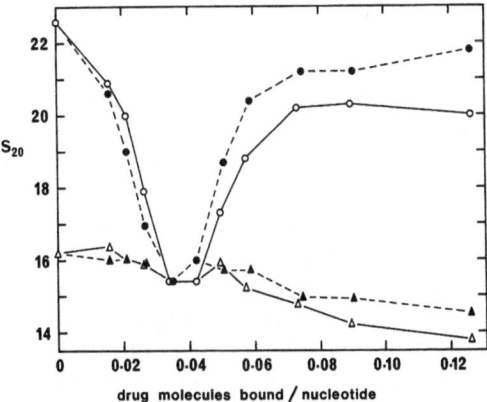

Fig. 6. Removal and reversal of the supercoils of closed circular ϕX174 replicative form DNA by ethidium ($\bullet$) and propidium ($\circ$). The DNA preparation contained 39% nicked circular molecules, whose sedimentation is shown by triangular symbols. (From WARING, 1970)

molecules are not constrained to adopt supercoils and their S_{20} merely reveals the steady decrease characteristic of intercalation which occurs with ordinary linear DNA. This same decrease is also present in the curve for the closed circles, but is overshadowed by the much larger changes in S_{20} due to variation of the supercoiling (Upholt et al., 1971).

At the equivalence point, the accumulated drug-induced unwinding just balances the original number of supercoiling turns in the closed circles. This is, in effect, the end-point of a titration and the two quantities are related by the equation (Bauer and Vinograd, 1968)

$$\tau_0 = -\frac{N\phi v_c}{2\pi} \tag{1}$$

where τ_0 is the number of superhelical turns in the DNA, N is the number of nucleotides, ϕ is the unwinding angle per bound drug molecule in radians ($\pi/15$ radians for a 12° unwinding), and v_c is the binding ratio at equivalence as defined above. Thus, provided that the intercalation unwinding angle is known or can be assumed the number of superhelical turns in closed circular DNA can be estimated from a determination of v_c. By this means Crawford and Waring (1967) calculated that polyoma DNA contains approximately 12 negative superhelical turns in dilute salt solution at neutral pH, assuming the unwinding angle per bound ethidium molecule to be 12° as proposed by Fuller and Waring (1964). This method of estimating the supercoiling of closed circular DNAs, relying upon the 12° unwinding angle for intercalation of ethidium, has now become standard practice in the study of circular DNA and because of its simplicity is the method of choice for estimating τ_0 and changes in superhelix density (reviewed by Waring, 1969; Bauer and Vinograd, 1971). It has led to a number of significant findings, notably that τ_0 is approximately constant (within a factor of 1.5–2) per unit length of DNA for natural circular DNAs of widely differing molecular weight (see, for example, Wang, 1969a), and that the average rotation angle of the DNA helix varies significantly with changes in temperature and salt concentration (Wang, 1969b; Upholt et al., 1971).

Eq. (1) also provides a critical test for the validity of the proposed 12° unwinding angle for ethidium if the supercoiling of circular DNA can be determined by independent means. The most important comparison available to date is the estimate of τ_0 for SV-40 DNA equal to -14.7 ± 1.7 superhelical turns determined by the ethidium intercalation method (Bauer and Vinograd, 1968, revised in Bauer and Vinograd, 1970a) compared with -15 ± 1 superhelical turns for the closely related polyoma DNA determined by an alkaline titration procedure (Vinograd et al., 1968), both in buoyant CsCl at neutral pH. These estimates are in striking agreement, and such other data as are available also support the conclusion that a 12° unwinding angle for intercalation of ethidium is not seriously in error (Waring, 1969; Gray et al., 1971).

A further application of the phenomenon illustrated in Fig. 6 is to compare the unwinding angles associated with intercalation of different drugs. It is easy to see from Eq. (1) that for two different drugs characterised by unwinding angles ϕ_1 and ϕ_2 the observed equivalence points v_{c_1} and v_{c_2} determined with the same DNA under identical conditions must be related by the equation

$$v_{c_1} \phi_1 = v_{c_2} \phi_2.$$

Fig. 6 shows that the equivalence points for ethidium and propidium are not significantly different; thus the helix unwinding angle for intercalation of propidium must be the same as that for ethidium and presumably 12°. This sort of approach has been extended to a large number of DNA-binding antibiotics and drugs as a potential criterion for intercalation, and unwinding angles determined (WARING, 1970, 1971, 1972). In particular, it has been shown that the effect of actinomycin binding on the winding of the DNA helix is qualitatively and quantitatively the same as that of ethidium (WARING, 1968b, 1970; WANG, 1971). Thus to the extent that the X-ray diffraction-based unwinding intercalation model for actinomycin binding is correct (SOBELL and JAIN, 1972) the over-winding intercalation model of PAOLETTI and LE PECQ (1971a, b) can be eliminated.

The adoption of a superhelical structure by closed circular DNA involves changes in free energy, which have been estimated by BAUER and VINOGRAD (1970b) from an analysis of ethidium-binding data. When ethidium binds to the native, negatively superhelical molecule it relieves a certain amount of the superhelical stress and consequently a portion of the superhelix free energy is released and contributes to the total drug-binding free energy change; the net result is an enhanced affinity for ethidium binding compared to an equivalent non-circular DNA molecule (BAUER and VINOGRAD, 1970b). DAVIDSON (1972) has calculated that, for a typical drug-free natural closed circular DNA, the equilibrium constant for binding of a 12°-unwinding ligand such as ethidium would be enhanced by a factor of 2.6. The possibility that this effect might account, at least in part, for the selective attack of ethidium on circular DNA-containing structures *in vivo* has already been mentioned; it is further discussed by BAUER and VINOGRAD (1971).

At the other end of the scale, the introduction of reversed supercoils by binding of an intercalating drug beyond the equivalence ratio involves free energy changes of the opposite sense, and consequently the affinity for binding to closed circles falls below that of non-circular DNA. Eventually the free energy terms for drug binding and reversed supercoil formation balance, with the result that natural closed circular DNAs bind less ethidium at saturation than do nicked circles or linear molecules (BAUER and VINOGRAD, 1968, 1970a, b). Since binding of ethidium to DNA lowers its buoyant density in CsCl gradients, this means that in density gradients containing a saturating concentration of ethidium bromide closed circles band at a denser position than other DNAs. The buoyant density separation is, in fact, linearly related to the superhelix density of the closed circular molecules (GRAY et al., 1971). This phenomenon forms the basis of the extremely valuable CsCl-ethidium bromide density gradient method of RADLOFF et al. (1967) for isolating closed circular DNA from natural sources (Fig. 7). Through the use of such gradients the existence of catenated circular DNA molecules was first detected (HUDSON and VINOGRAD, 1967).

It is in this connection that propidium iodide is of interest. HUDSON et al. (1969) found that it produced buoyant density separations in CsCl gradients 1.8 times larger than those produced by ethidium, thus giving improved resolution

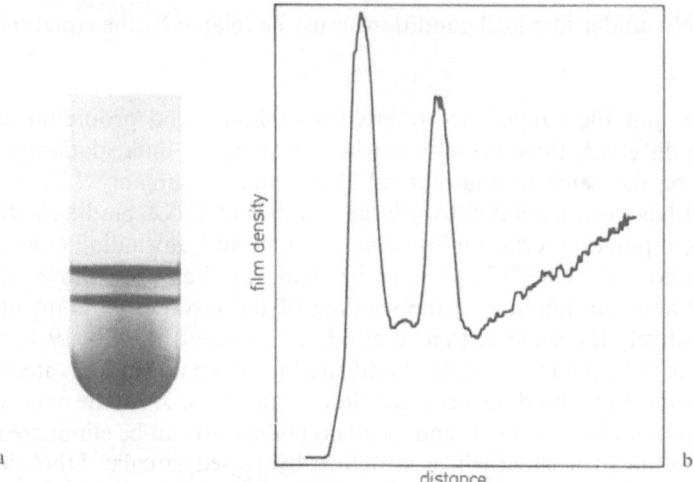

Fig. 7. Mitochondrial DNA from human leukaemic leucocytes in a CsCl-ethidium bromide density gradient. a Fluorescence photograph of the centrifuge tube illuminated with near-ultraviolet light. b A microdensitometer tracing showing band positions. The left hand side of the tracing corresponds to the upper part of the tube. The upper band contains nicked circular mitochondrial DNA and some linear nuclear DNA. The lower band contains closed circular molecules of which 9% were dimers. A third band, midway between the other two, is visible containing largely catenated dimers in which one monomer molecule is nicked. (From CLAYTON and VINOGRAD, 1967)

of closed circles from nicked circles and linear DNA molecules in preparative work. The reason for the increased buoyant separation has not yet been established (BAUER and VINOGRAD, 1971), but it evidently does not derive from an increased unwinding angle (Fig. 6; WARING, 1970).

Structure-Activity Relations for the Intercalation Reaction

The interaction between phenanthridinium drugs related to ethidium and circular DNA has recently been investigated with a view to determining the influence of substituents on the intercalation reaction (WAKELIN and WARING, 1974). A summary of part of the results is given in Table 1. The main conclusions which can be drawn from these findings are as follows.

Substituents at Position 5 (R_4). The nature of the quaternising group has little or no effect on the helix unwinding angle (compare ethidium and dimidium, or phenidium and M&B 3016; also ethidium and propidium in Fig. 6). All of these compounds bind strongly to DNA, and no significant differences in the binding parameters were detected.

Substituents at Position 6 (R_3). Removal of the phenyl group leads to a reduced unwinding angle (M&B 2421) which reflects drastic alterations in the behaviour of the drug. Precipitation of the DNA occurred at any ionic strength below 0.2, presumably due to formation of a neutral 1:1 complex. Lacking a phenyl substituent the whole molecule is practically planar and can readily form stacked aggregates; this probably explains the precipitation effect and could

Table 1. *Unwinding of closed circular duplex DNA by phenanthridinium drugs*

Compound	R_1	R_2	R_3	R_4	Ionic strength	Equivalence binding ratio (v_c)	Unwinding angle (ϕ)
Ethidium	NH_2	NH_2	C_6H_5	C_2H_5	0.02	0.050 ±0.007	(12°)
					0.036	0.051 ±0.006	(12°)
					0.10	0.055 ±0.005	(12°)
					0.50	0.048 ±0.008	(12°)
Dimidium	NH_2	NH_2	C_6H_5	CH_3	0.02	0.052 ±0.010	11.5°
M&B 2421	NH_2	NH_2	H	CH_3	0.50	0.069 ±0.006	8.3°
M&B 3492	NH_2	NH_2	C_6H_4—COO^- (p)	CH_3	0.02	0.051 ±0.006	11.8°
M&B 4594	$NHCOCH_3$	NH_2	C_6H_5	CH_3	0.02	0.074 ±0.007	8.1°
Phenidium	NH_2	H	C_6H_4—NH_2 (p)	CH_3	0.036	0.070 ±0.006	8.6°
M&B 3016	NH_2	H	C_6H_4—NH_2 (p)	C_2H_5	0.036	0.057 ±0.008	10.6°
M&B 3427	H	H	C_6H_5	CH_3	0.02	0.078 ±0.013	7.7°
					0.10	≧0.145	≦4.6°
M&B 1765	Br	Br	C_6H_5	CH_3	0.02	0.050 ±0.009	12.0°
					0.10	≧0.092	≦7.2°

Drugs were generously provided by Drs R. SLACK and S. S. BERG of May and Baker Ltd. and Dr G. WOOLFE of Boots Pure Drug Co. Ltd. Bacteriophage PM2 DNA was used throughout, with buffers containing 0.02 M HEPES, 0.1 mM EDTA and NaCl to the indicated ionic strength (pH 7.0), except for ionic strength 0.036 which was 0.05 M tris-HCl buffer (pH 7.9). Variation in supercoiling was determined by ultracentrifugation, and drug binding by spectrophotometry, as described by WARING (1970). Binding was determined by equilibrium dialysis in certain cases. (From WAKELIN and WARING, 1974.)

account for the lowered ϕ in terms of persistence of externally bound non-intercalated drug at higher salt concentrations (cf. WARING, 1970).

Introduction of a *para*-carboxylate group on the phenyl ring yields an amphoteric neutral molecule (M&B 3492); the unwinding angle is not affected but the strength of binding is substantially reduced, very likely due to a charge effect.

Substituents at positions 8 and 3 (R_1 and R_2). Blocking the 8-amino with an acetyl group (M&B 4594) lowers the unwinding angle without much change in the binding energy; however, the isosbestic spectral behaviour of ethidium and dimidium is lost. Removal of the 3-amino group appears to reduce the unwinding angle somewhat though the binding remains quite strong (phenidium and M&B 3016); it is assumed that the presence of a *para*-amino group on the phenyl ring (uncharged at pH 7) is relatively inconsequential.

Complete removal of both primary amino groups (M&B 3427) results in a definitely lower unwinding angle and a 10–20 fold reduction in the binding constant. Thus the presence of the two amino groups contributes some 1.4–1.7 Kcal/mole of free energy to the stability of the complex. It is tempting to attribute this stabilization to participation of the amino groups in hydrogen bonding as suggested by FULLER and WARING (1964), but other interpretations are also possible (WAKELIN and WARING, 1974). More suprisingly, replacement of both amino groups by bromine atoms yields a derivative (M&B 1765) which still intercalates satisfactorily with the same unwinding angle as ethidium and dimidium; again the binding constant is reduced 20–30 fold. The availability of this heavy-atom derivative, which seems to form a complex with the same geometrical characteristics as that formed by the diamino compounds, promises to be of great value in future studies of intercalation complexes. At 0.10 ionic strength the binding of the des-amino and dibromo compounds was too weak to obtain a usable equivalence point.

To sum up the structure-activity correlations, it appears that quite drastic modifications to the ethidium structure can be made without abolishing its intercalative potential. In particular, it is now clear that the primary amino groups of the chromophore are by no means mandatory for the intercalation reaction. These conclusions are in broad agreement with earlier observations that antibiotics and drugs of quite diverse structure are capable of binding to DNA in this fashion (WARING, 1970, 1972).

References

ALBERT, A.: Selective toxicity, 2nd ed. London: Methuen 1968.

ANGERER, L. M., and E. N. MOUDRIANAKIS: Interaction of ethidium bromide with whole and selectively deproteinized deoxynucleoproteins from calf thymus. J. Mol. Biol. **63**, 505 (1972).

ARLETT, C. F.: The influence of post-irradiation conditions on the survival of Chinese hamster cells after gamma irradiation. Intern. J. Radiation Biol. **17**, 515 (1970).

AZZI, A., and M. SANTATO: Interaction of ethidium with the mitochondrial membrane: cooperative binding and energy-linked changes. Biochem. Biophys. Res. Commun. **44**, 211 (1971).

BAUER, W., and J. VINOGRAD: The interaction of closed circular DNA with intercalative dyes. I. The superhelix density of SV-40 DNA in the presence and absence of dye. J. Mol. Biol. **33**, 141 (1968).

BAUER, W., and J. VINOGRAD: The interaction of closed circular DNA with intercalative dyes. III. Dependence of the buoyant density upon superhelix density and base-composition. J. Mol. Biol. **54**, 281 (1970a).

BAUER, W., and J. VINOGRAD: The interaction of closed circular DNA with intercalative dyes. II. The free energy of superhelix formation in SV–40 DNA. J. Mol. Biol. **47**, 419 (1970b).

BAUER, W., and J. VINOGRAD: The use of intercalative dyes in the study of closed circular DNA. Progr. Mol. Subcell. Biol. **2**, 181 (1971).

BECKER, Y., and Y. ASHER: Obligate parasitism of trachoma agent: lack of trachoma development in ethidium bromide-treated cells. Antimicrobial Agents Chemotherapy **1**, 171 (1972).

BITTMAN, R.: Studies of the binding of ethidium bromide to tRNA: absorption, fluorescence, ultracentrifugation and kinetic investigations. J. Mol. Biol. **46**, 251 (1969).

BOUANCHAUD, D.H., M.R. SCAVIZZI, and Y.A. CHABBERT: Elimination by ethidium bromide of antibiotic resistance in enterobacteria and staphylococci. J. Gen. Microbiol. **54**, 333 (1968).

BROWNLEE, G., M.D. GOSS, L.G. GOODWIN, M. WOODBINE, and L.P. WALLS: The chemotherapeutic action of phenanthridine compounds. I. *Trypanosoma congolense* and *Trypanosoma rhodesiense*. Brit. J. Pharmacol. **5**, 261 (1950).

BURNS, V.W.F.: Fluorescence decay time characteristics of the complex between ethidium bromide and nucleic acids. Arch. Biochem. Biophys. **133**, 420 (1969).

CANTOR, C.R., K. BEARDSLEY, J. NELSON, T. TAO, and K.W. CHIN: Studies on tRNA structure using covalently and noncovalently bound fluorescent dyes. Progr. Mol. Subcell. Biol. **2**, 297 (1971).

CLAYTON, D.A., and J. VINOGRAD: Circular dimer and catenate forms of mitochondrial DNA in human leukaemic leucocytes. Nature **216**, 652 (1967).

CRAWFORD, L.V., and M.J. WARING: Supercoiling of polyoma virus DNA measured by its interaction with ethidium bromide. J. Mol. Biol. **25**, 23 (1967).

DALGLEISH, D.G., A.R. PEACOCKE, G. FEY, and C. HARVEY: The circular dichroism in the ultraviolet of aminoacridines and ethidium bromide bound to DNA. Biopolymers **10**, 1853 (1971).

DAVIDSON, N.: Effect of DNA length on the free energy of binding of an unwinding ligand to a supercoiled DNA. J. Mol. Biol. **66**, 307 (1972).

DEKEN, R.H. DE: The Crabtree effect and its relation to the petite mutation. J. Gen. Microbiol. **44**, 157 (1966).

DELAIN, E., et G. RIOU: Ultrastructure des altérations du DNA du kinétoplaste de *Trypanosoma cruzi* traité par le bromure d'éthidium. Compt. Rend. **268D**, 1327 (1969).

DICKINSON, L., B.H. CHANTRILL, G.W. INKLEY, and M.J. THOMPSON: The antiviral action of phenanthridinium compounds. Brit. J. Pharmacol. **8**, 139 (1953).

ELLIOTT, W.H.: The effects of antimicrobial agents on DNA polymerase. Biochem. J. **86**, 562 (1963).

ERON, L.J., and B.R. MCAUSLAN: Inhibition of deoxyribonuclease action by actinomycin D and ethidium bromide. Biochim. Biophys. Acta **114**, 633 (1966).

FREIFELDER, D.: Electron microscopic study of the ethidium bromide-DNA complex. J. Mol. Biol. **60**, 401 (1971).

FRIDLENDER, B., and A. WEISSBACH: DNA polymerases of tumor virus: specific effect of ethidium bromide on the use of different synthetic templates. Proc. Natl. Acad. Sci. U.S. **68**, 3116 (1971).

FUKUHARA, H., and C. KUJAWA: Selective inhibition of the *in vivo* transcription of mitochondrial DNA by ethidium bromide and by acriflavine. Biochem. Biophys. Res. Commun. **41**, 1002 (1970).

FULLER, W., and M.J. WARING: A molecular model for the interaction of ethidium bromide with DNA. Ber. Bunsenges. Physik. Chem. **68**, 805 (1964).

GENEST, D., et P. WAHL: Etude des transferts d'énergie dans le complexe DNA- bromure d'éthidium au moyen du déclin de l'anisotropie de fluorescence. Biochim. Biophys. Acta **259**, 175 (1972).

GILBERT, M., and P. CLAVERIE: A theoretical study of the electrostatic interactions in the intercalation model of the DNA-dye complex. J. Theoret. Biol. **18**, 330 (1968).

GOLDRING, E.S., L.I. GROSSMAN, D. KRUPNICK, D.R. CRYER, and J. MARMUR: The petite mutation in yeast: loss of mitochondrial DNA during induction of petites with ethidium bromide. J. Mol. Biol. **52**, 323 (1970).

GRAY, P.N., and G.F. SAUNDERS: Binding of ethidium bromide to 5S ribosomal RNA. Biochim. Biophys. Acta **254**, 60 (1971).

GRAY, H.B. JR., W.B. UPHOLT, and J. VINOGRAD: A buoyant method for the determination of the superhelix density of closed circular DNA. J. Mol. Biol. **62**, 1 (1971).

Grinsted, J.: Antimicrobial drugs and RNA. Biochim. Biophys. Acta 179, 268 (1969).

Hawking, F.: Chemotherapy of trypanosomiasis. In: Experimental chemotherapy (eds. R. J. Schnitzer and F. Hawking), vol. 1, p. 129. New York: Academic Press, 1963.

Heilporn, V., et S. Limbosch: Les effets du bromure d'éthidium sur *Acetabularia mediterranea*. Biochim. Biophys. Acta 240, 94 (1971).

Hirschman, S. Z.: Inhibitors of DNA polymerases of murine leukaemia viruses: activity of ethidium bromide. Science 173, 441 (1971).

Hollenberg, C. P., and P. Borst: Conditions that prevent ϱ^- induction by ethidium bromide. Biochem. Biophys. Res. Commun. 45, 1250 (1971).

Hospital, M., et B. Busetta: Structure cristalline et moléculaire du bromhydrate d'éthidium. Compt. Rend. 268C, 1232 (1969).

Hudson, B., W. B. Upholt, J. Devinny, and J. Vinograd: The use of an ethidium analogue in the dye-buoyant density procedure for the isolation of closed circular DNA: the variation of the superhelix density of mitochondrial DNA. Proc. Natl. Acad. Sci. U.S. 62, 813 (1969).

Hudson, B., and J. Vinograd: Catenated circular DNA molecules in HeLa cell mitochondria. Nature 216, 647 (1967).

Igarashi, S. J., and R. P. Bissonnette: Effects of salts and antibiotics on the R17 RNA replicase reaction. J. Biochem. (Tokyo) 70, 835 (1971).

Jain, S. C., and H. M. Sobell: Stereochemistry of actinomycin binding to DNA. I. Refinement and further structural details of the actinomycin-deoxyguanosine crystalline complex. J. Mol. Biol. 68, 1 (1972).

Kandaswamy, T. S., and J. F. Henderson: Inhibition of ascites tumour growth by the trypanocide, ethidium bromide, in combination with azaserine. Nature 195, 85 (1962a).

Kandaswamy, T. S., and J. F. Henderson: Intracellular differentiation of purine ribonucleotides derived from endogenous and exogenous sources. Biochim. Biophys. Acta 61, 86 (1962b).

Kandaswamy, T. S., and J. F. Henderson: The metabolism of ethidium bromide in normal and neoplastic tissues. Cancer Res. 23, 250 (1963).

Kellerman, G. M., D. R. Biggs, and A. W. Linnane: Biogenesis of mitochondria. XI. A comparison of the effects of growth-limiting oxygen tension, intercalating agents, and antibiotics on the obligate aerobe *Candida parapsilosis*. J. Cell Biol. 42, 378 (1969).

Kerridge, D.: The effect of actidione and other antifungal agents on nucleic acid and protein synthesis in *Saccharomyces carlsbergensis*. J. Gen. Microbiol. 19, 497 (1958).

Kock, J., and H. von Pfeil: Interference of ethidium bromide with the formation of supercoiled mitochondrial DNA. FEBS Letters 18, 172 (1971).

Kreishman, G. P., S. I. Chan, and W. Bauer: Proton magnetic resonance study of the interaction of ethidium bromide with several uracil residues, uridylyl (3'-5') uridine and polyuridylic acid. J. Mol. Biol. 61, 45 (1971).

LePecq, J. B.: Use of ethidium bromide for separation and determination of nucleic acids of various conformational forms and measurement of their associated enzymes. Methods Biochem. Analysis 20, 41 (1971).

LePecq, J. B., and C. Paoletti: A new fluorometric method for RNA and erythrocyte determination. Anal. Biochem. 17, 100 (1966).

LePecq, J. B., and C. Paoletti: A fluorescent complex between ethidium bromide and nucleic acids. Physical-chemical characterization. J. Mol. Biol. 27, 87 (1967).

Lerman, L. S.: Structural considerations in the interaction of DNA and acridines. J. Mol. Biol. 3, 18 (1961).

Lerman, L. S.: Acridine mutagens and DNA structure. J. Cellular Comp. Physiol. 64, Suppl. 1, 1 (1964).

Luha, A. A., L. E. Sarcoe, and P. A. Whittaker: Biosynthesis of yeast mitochondria. Drug effects on the petite negative yeast *Kluyveromyces lactis*. Biochem. Biophys. Res. Commun. 44, 396 (1971).

Lurquin, P. F., and V. L. Seligy: Binding of ethidium bromide to avian erythrocyte chromatin. Biochem. Biophys. Res. Commun. 46, 1399 (1972).

Milner, A. J.: Corticotrophin-induced differentiation of mitochondria in rat adrenal cortical cells grown in primary tissue culture: effects of ethidium bromide. J. Endocrinol. 52, 541 (1972).

Nagley, P., and A. W. Linnane: Biogenesis of mitochondria. XXI. Studies on the nature of the mitochondrial genome in yeast: the degenerative effects of ethidium bromide on mitochondrial genetic information in a respiratory competent strain. J. Mol. Biol. 66, 181 (1972).

NASS, M. M. K.: Abnormal DNA patterns in animal mitochondria: ethidium bromide-induced break-down of closed circular DNA and conditions leading to oligomer accumulation. Proc. Natl. Acad. Sci. U.S. **67**, 1926 (1970).

NASS, M. M. K.: Differential effects of ethidium bromide on mitochondrial and nuclear DNA synthesis *in vivo* in cultured mammalian cells. Exptl. Cell. Res. **72**, 211 (1972).

NEWTON, B. A.: The mode of action of phenanthridines: the effect of ethidium bromide on cell division and nucleic acid synthesis. J. Gen. Microbiol. **17**, 718 (1957).

NEWTON, B. A.: Trypanocidal agents. In: Metabolic inhibitors (eds. R. M. Hochster and J. H. Quastel), vol. 2, p. 285. New York: Academic Press. 1963.

NEWTON, B. A.: Mechanisms of action of phenanthridine and aminoquinaldine trypanocides. Advan. Chemotherapy **1**, 35 (1964).

NISHIMURA, Y., L. CARO, C. M. BERG, and Y. HIROTA: Chromosome replication in *Escherichia coli*. IV. Control of chromosome replication and cell division by an integrated episome. J. Mol. Biol. **55**, 441 (1971).

OLINS, D. E.: Interaction of lysine-rich histones and DNA. J. Mol. Biol. **43**, 439 (1969).

PAOLETTI, J., and J. B. LEPECQ: Resonance energy transfer between ethidium bromide molecules bound to nucleic acids: does intercalation wind or unwind the DNA helix? J. Mol. Biol. **59**, 43 (1971a).

PAOLETTI, J., and J. B. LEPECQ: The change of the torsion of the DNA helix caused by intercalation. I. A discussion of the two different possibilities, winding or unwinding. Biochimie **53**, 969 (1971b).

PERLMAN, P. S., and H. R. MAHLER: Molecular consequences of ethidium bromide mutagenesis. Nature New Biol. **231**, 12 (1971a).

PERLMAN, P. S., and H. R. MAHLER: A premutational state induced in yeast by ethidium bromide. Biochem. Biophys. Res. Commun. **44**, 261 (1971b).

RADLOFF, R., W. BAUER, and J. VINOGRAD: A dye-buoyant density method for the detection and isolation of closed circular duplex DNA: the closed circular DNA in HeLa cells. Proc. Natl. Acad. Sci. U.S. **57**, 1514 (1967).

RADSAK, K., K. KATO, N. SATO, and H. KOPROWSKI: Effect of ethidium bromide on mitochondrial DNA and cytochrome synthesis in HeLa cells. Exptl. Cell. Res. **66**, 410 (1971).

RAZIN, A., and J. MAGER: Studies on the mechanism of the inhibitory effect of ethidium bromide on purine nucleotide metabolism. Israel J. Chem. **2**, 5 (1964).

RIOU, G., and E. DELAIN: Abnormal circular DNA molecules induced by ethidium bromide in the kinetoplast of *Trypanosoma cruzi*. Proc. Natl. Acad. Sci. U.S. **64**, 618 (1969).

ROODYN, D. B., and D. WILKIE: The biogenesis of mitochondria. London: Methuen 1968.

SAFFHILL, R., H. SCHNEIDER-BERNLOEHR, L. E. ORGEL, and S. SPIEGELMAN: *In vitro* selection of bacteriophage Qβ RNA variants resistant to ethidium bromide. J. Mol. Biol. **51**, 531 (1970).

SEAMAN, A., and M. WOODBINE: Some aspects of the antibacterial effects of phenanthridine compounds. Atti del VI Congresso Internationale di Microbiologia, **1**, 636 (1953). Staderini, Rome.

SEAMAN, A., and M. WOODBINE: The antibacterial activity of phenanthridine compounds. Brit. J. Pharmacol. **9**, 265 (1954).

SLONIMSKI, P. P., G. PERRODIN, and J. H. CROFT: Ethidium bromide-induced mutation of yeast mito-chondria: complete transformation of cells into respiratory deficient non-chromosomal 'petites'. Biochem. Biophys. Res. Commun. **30**, 232 (1968).

SMITH, C. A., J. M. JORDAN, and J. VINOGRAD: *In vivo* effects of intercalating drugs on the superhelix density of mitochondrial DNA isolated from human and mouse cells in culture. J. Mol. Biol. **59**, 255 (1971).

SNYDER, A. L., H. E. KANN, Jr., and K. W. KOHN: Inhibition of the processing of ribosomal precursor RNA by intercalating agents. J. Mol. Biol. **58**, 555 (1971).

SOBELL, H. M., and S. C. JAIN: Stereochemistry of actinomycin binding to DNA. II. Detailed molecular model of actinomycin-DNA complex and its implications. J. Mol. Biol. **68**, 21 (1972).

SPRECHER-GOLDBERGER, S.: Photosensitization of RNA viruses by ethidium chloride. Acta Virologica (English ed.) **9**, 385 (1965).

STONE, A. L., and D. F. BRADLEY: Aggregation of acridine orange bound to polyanions: the stacking tendency of DNAs. J. Am. Chem. Soc. **83**, 3627 (1961).

TOMCHICK, R., and H. G. MANDEL: Biochemical effects of ethidium bromide in microorganisms. J. Gen. Microbiol. **36**, 225 (1964).

TRITTON, T. R., and S. C. MOHR: Relaxation kinetics of the binding of ethidium bromide to unfractionated yeast tRNA at low dye/phosphate ratio. Biochem. Biophys. Res. Commun. **45**, 1240 (1971).

UPHOLT, W. B., H. B. GRAY, JR., and J. VINOGRAD: Sedimentation velocity behaviour of closed circular SV-40 DNA as a function of superhelix density, ionic strength, counterion and temperature. J. Mol. Biol. **61**, 21 (1971).

VAN ASSEL, S., et M. STEINERT: Inhibition sélective de la réplication du DNA kinétoplastique des trypanosomides par le bromure d'éthidium. Arch. Intern. Physiol. Biochim. **76**, 388 (1968).

VAN DYKÉ, K., and C. SZUSTKIEWICZ: Automated systems for the fluorometric determination of nucleic acids by the ethidium bromide technique. Anal. Biochem. **23**, 109 (1968).

VIDOVÁ, M., and L. KOVÁČ: Nalidixic acid prevents the induction of yeast cytoplasmic respiration-deficient mutants by intercalating drugs. FEBS Letters **22**, 347 (1972).

VILAGINES, R., and ATANASIU, P.: Action of ethidium bromide on growth of herpes virus in cell cultures. Nature **215**, 87 (1967).

VINOGRAD, J., J. LEBOWITZ, and R. WATSON: Early and late helix-coil transitions in closed circular DNA. The number of superhelical turns in polyoma DNA. J. Mol. Biol. **33**, 173 (1968).

WAKELIN, L. P. G., and M. J. WARING: Mol. Pharmacol. **10**, (1974) in press.

WANG, J. C.: Degree of superhelicity of covalently closed cyclic DNAs from *E. coli*. J. Mol. Biol. **43**, 263 (1969a).

WANG, J. C.: Variation of the average rotation angle of the DNA helix and the superhelical turns of covalently closed cyclic lambda DNA. J. Mol. Biol. **43**, 25 (1969b).

WANG, J. C.: Unwinding of DNA by actinomycin D binding. Biochim Biophys. Acta **232**, 246 (1971).

WARD, D., E. REICH, and I. H. GOLDBERG: Base specificity in the interaction of polynucleotides with antibiotic drugs. Science **149**, 1259 (1965).

WARING, M. J.: Complex formation with DNA and inhibition of *Escherichia coli* RNA polymerase by ethidium bromide. Biochim. Biophys. Acta **87**, 358 (1964).

WARING, M. J.: The effects of antimicrobial agents on RNA polymerase. Mol. Pharmacol. **1**, 1 (1965a).

WARING, M. J.: Complex formation between ethidium bromide and nucleic acids. J. Mol. Biol. **13**, 269 (1965b).

WARING, M. J.: Cross-linking and intercalation in nucleic acids. Symp. Soc. Gen. Microbiol. **16**, 235 (1966a).

WARING, M. J.: Structural requirements for the binding of ethidium to nucleic acids. Biochim. Biophys. Acta **114**, 234 (1966b).

WARING, M. J.: Drugs which affect the structure and function of DNA. Nature **219**, 1320 (1968a).

WARING, M. J.: Uncoiling of bacteriophage φ X 174 replicative form DNA by ethidium, daunomycin and actinomycin. Biochem. J. **109**, 28P (1968b).

WARING, M. J.: Nucleic acids. Ann. Rept. Chem. Soc. for 1968, **65B**, 551 (1969).

WARING, M. J.: Variation of the supercoils in closed circular DNA by binding of antibiotics and drugs: evidence for molecular models involving intercalation. J. Mol. Biol. **54**, 247 (1970).

WARING, M. J.: Binding of drugs to supercoiled circular DNA: evidence for and against intercalation. Progr. Mol. Subcell. Biol. **2**, 216 (1971).

WARING, M. J.: Inhibitors of nucleic acid synthesis. In: The molecular basis of antibiotic action, ed. by E. F. Gale, E. Cundliffe, P. E. Reynolds, M. H. Richmond and M. J. Waring, p. 173. London: Wiley, 1972.

WATKINS, T. I.: Trypanocides of the phenanthridine series. I. The effect of changing the quaternary grouping in dimidium bromide. J. Chem. Soc. **1952**, 3059.

WATKINS, T. I., and G. WOOLFE: Effect of changing the quaternising group on the trypanocidal activity of dimidium bromide. Nature **169**, 506 (1952).

WESTERGAARD, O., K. A. MARCKER, and J. KEIDING: Induction of a mitochondrial DNA polymerase in *Tetrahymena*. Nature **227**, 708 (1970).

WHITTAKER, P. A., R. C. HAMMOND, and A. A. LUHA: Mechanism of mitochondrial mutation in yeast. Nature New Biol. **238**, 266 (1972).

WOLFE, A. D., R. G. ALLISON, and F. E. HAHN: Labilizing action of intercalating drugs and dyes on bacterial ribosomes. Biochemistry **11**, 1569 (1972).

WOOLFE, G.: Trypanocidal action of phenanthridine compounds: effect of changing the quaternary groups of known trypanocides. Brit. J. Pharmacol. **11**, 330 (1956a).

WOOLFE, G.: Trypanocidal action of phenanthridine compounds: further 2:7 diamino phenanthridinium compounds. Brit. J. Pharmacol. **11**, 334 (1956b).

ZYLBER, E., and S. PENMAN: Mitochondrial-associated 4S RNA synthesis inhibition by ethidium bromide. J. Mol. Biol. **46**, 201 (1969).

ZYLBER, E., C. VESCO, and S. PENMAN: Selective inhibition of the synthesis of mitochondria-associated RNA by ethidium bromide. J. Mol. Biol. **44**, 195 (1969).

Kanchanomycin *

Irving H. Goldberg

Kanchanomycin is a yellow, very sparingly soluble in water antibiotic of unknown structure produced by a *Streptomyces* species; it is bactericidal and tumoricidal for neoplastic cells in culture at extremely low concentrations (LIU *et al.*, 1963; BATEMAN *et al.*, 1965). In HeLa cell culture assays, it has a cytotoxic end point of 0.002 µg/ml and a cellular lethal end point of 0.05 µg/ml. In this system, kanchanomycin is more cytotoxic than streptonigrin, mitomycin C or actinomycin D. On the other hand, kanchanomycin has been found to have only slight or no antitumor activity against various transplanted tumors in rodents. In man, severe thromophlebitis at the site of intravenous injection has limited extensive clinical trials.

Kanchanomycin is of particular interest because it appears to inhibit *in vitro* RNA and DNA synthesis in two distinctive ways. The antibiotic complexes with polynucleotides in the presence of stoichiometric amounts of divalent cation in a two-step time-dependent reaction (Figs. 1, 2 and 3) (FRIEDMAN *et al.*, 1969a;

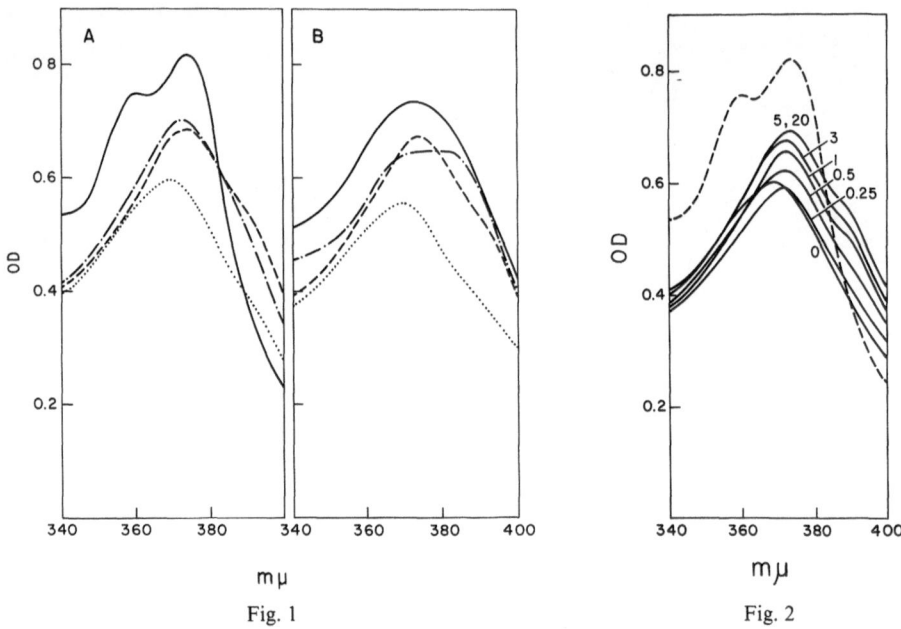

Fig. 1

Fig. 2

* This work was supported by U.S. Public Health Service Research Grant GM 12573 from the National Institutes of Health.

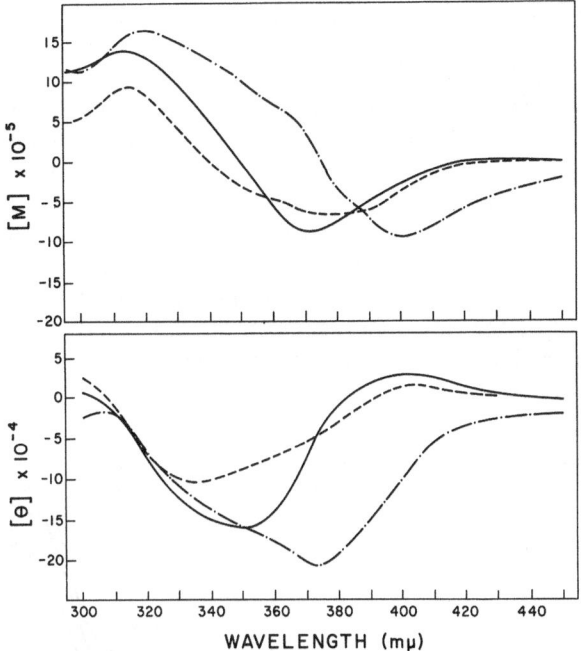

Fig. 3. Optical rotatory dispersion and circular dichroism spectra of the interaction of kanchanomycin with native calf thymus DNA. (---) 33 μM kanchanomycin and 80 μM Mg^{2+}; (——) 33 μM kanchanomycin, 440 μM native calf thymus DNA, and 80 μM Mg^{2+} without incubation; (—·—) same after 20-hrs incubation. (FRIEDMAN, *et al.*, 1969b)

FRIEDMAN *et al.*, 1969b). An initial complex (I) forms immediately and changes with time to a second complex (II) with different spectral and chemical properties. ORD, circular dichroism (CD), and visible spectral data support the view that kanchanomycin first combines stoichiometrically with Mg^{2+} and this complex then interacts with DNA or other polynucleotides (Fig. 4). Such experiments have shown that in complex I formation with polyadenylic acid there are as many binding sites for the antibiotic per polynucleotide as there are bases. The binding of kanchanomycin is very strong; in fact, precipitated kanchanomycin in aqueous media can be resolubilized by DNA in the presence of Mg^{2+}. Further-

◄

Fig. 1. Effect of DNA and Mg^{2+} on the spectrum of kanchanomycin. All samples contained 33 μM kanchanomycin, 0.01 M Tris (pH 7.5), and either 10% dimethylformamide (A) or 1.5% dimethylformamide (B). The following additional components were present: (——) no additional component or 0.44 mM calf thymus DNA; (—·—) 80 μM $MgCl_2$; (·····) 0.44 mM calf thymus DNA and 80 μM $MgCl_2$ with the spectrum taken immediately after mixing; (---) 0.44 mM calf thymus DNA and 80 μM $MgCl_2$ after incubation for 20 hrs at 37°. (FRIEDMAN *et al.*, 1969a)

Fig. 2. Time-dependent changes in the spectrum of kanchanomycin in the presence of DNA and Mg^{2+}. (---) Spectrum of a 1.0 ml sample containing 33 μM kanchanomycin, 0.44 mM calf thymus DNA, 0.01 M Tris (pH 7.5), and 10% dimethylformamide; (——) to the sample was added 80 mμmoles of $MgCl_2$ (in 5 μl) and spectra were recorded after 0-, 0.25-, 0.50-, 1-, 3-, 5- and 20-hrs incubation at 37°. (FRIEDMAN *et al.*, 1969a)

more, yellow antibiotic-bearing DNA strands, precipitated by ethanol, do not lose their color upon attempted extraction with organic solvents in which free kanchanomycin itself is readily soluble.

pH titration studies suggest that Mg^{2+} interacts with the first dissociable group of kanchanomycin (between pH 8.6 and 9.0) and that complex II formation involves a second dissociable group (between pH 9.0 and 10.0) (Fig. 5). Of all the polynucleotides studied, only polyadenylic acid forms a stable complex I which does not get converted to complex II. The formation of complex II appears directly to involve the nucleotide bases since bromination of polyuridylic acid, which destroys the aromaticity of the pyrimidine ring, permits formation of complex I but not its conversion into complex II. Kanchanomycin (in the second complex) increases the sedimentation rate and the viscosity, but not the melting temperature of DNA.

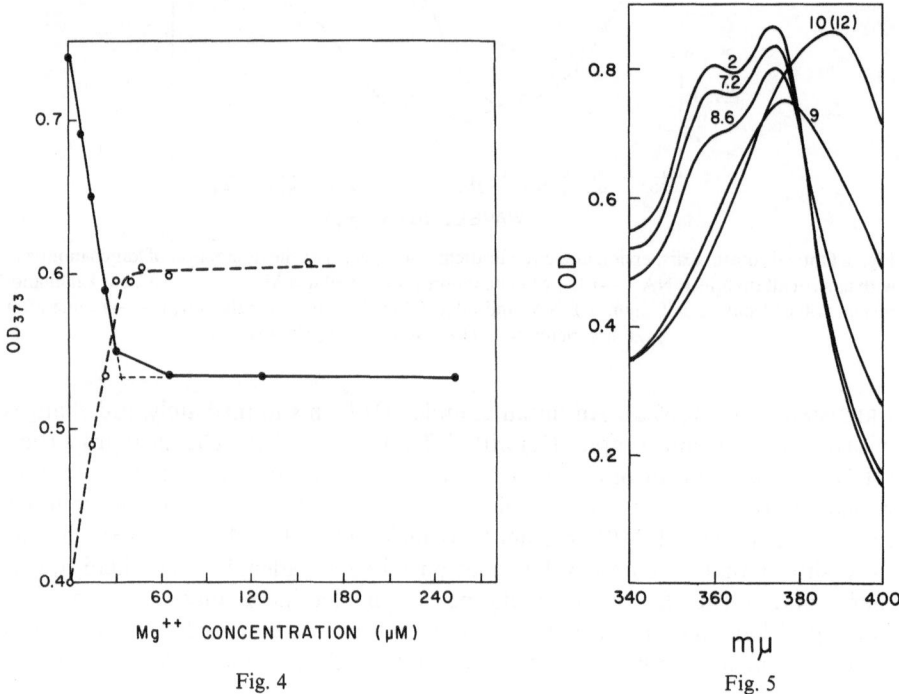

Fig. 4   Fig. 5

Fig. 4. Determination of the Mg^{2+} concentration necessary for the complete interaction of kanchanomycin with polynucleotides. (●-●-) Each sample contained 30 µM kanchanomycin, 0.44 mM poly A, 0.01 M Tris (pH 7.5), 10% dimethylformamide, and the concentration of $MgCl_2$ indicated. The OD_{373} was read immediately after mixing. (o-o) Each sample contained 30 µM kanchanomycin, 0.44 mM calf thymus DNA, 0.01 M Tris (pH 7.5), 10% dimethylformamide, and the concentration of $MgCl_2$ indicated on the graph. The OD_{373} was read after samples had incubated 20 hrs at 37°. (FRIEDMAN et al., 1969a)

Fig. 5. Effect of pH on the spectrum of kanchanomycin. The spectrum of 33 µM kanchanomycin was recorded in each of the following solutions: 0.01 M HCl (pH 2), 0.01 M Tris (pH 7.2 and pH 8.6), 0.01 M sodium carbonate (pH 9.0 and 10.0), and 0.01 M NaOH (pH 12.0). All solutions contained 10% dimethylformamide. (FRIEDMAN et al., 1969a)

Complex I formation is probably electrostatic, involving the negatively charged sugar-phosphate backbone of the polynucleotides. This is suggested by the absence of a specific base or secondary structural requirement of the polynucleotide for interaction and by the rapid dissociation or complete inhibition of formation of complex I by high salt concentration (or EDTA). Complex II, which forms slowly over a period of 5 hr and depends on the base composition of the polynucleotide, is also reversible but much more slowly, not by high salt concentration but by EDTA (Fig. 6). It appears that a change in the structure of the complex takes place which protects the Mg^{2+} from the chelating agent.

In the mechanism of action of kanchanomycin (JOEL et al., 1970) are features common to the two main groups of antibiotics which inhibit nucleic acid polymerization by either inactivating the template DNA (e.g. actinomycin) or the enzyme (e.g. rifamycin). Kanchanomycin strongly inhibits both the RNA and DNA

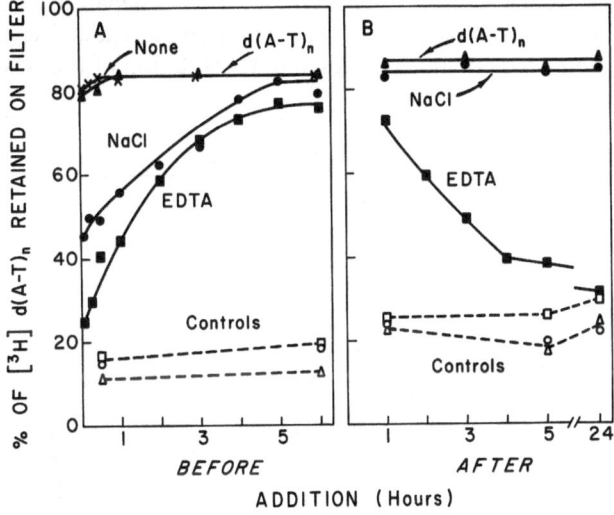

Fig. 6. The dissociability of the d(A-T)$_n^1$-kanchanomycin complex with added d(A-T)$_n$, NaCl, or EDTA. (A) A solution containing 80 µM [^{3}H]d(A-T)$_n$, 1.6 µM kanchanomycin, 0.16 mM MgCl$_2$, 0.01 M Tris (pH 7.5), and 8% dimethylformamide was incubated at 25°. A control solution contained no kanchanomycin. At the times indicated 20 µl samples were withdrawn and incubated 30 min at 25° with 20 µl of the solutions indicated below and then filtered. (x-x-) 0.16 mM MgCl$_2$, 0.01 M Tris (pH 7.5) and 8% dimethylformamide (except 5, 15, 30 and 60 min which were filtered immediately). (▲-▲-) 400 µM d(A-T)$_n$ (unlabeled), 0.8 mM MgCl$_2$, 0.01 M Tris (pH 7.5) and 8% dimethylformamide. (●-●-) 4 M NaCl, 0.16 mM MgCl$_2$, 0.01 M Tris (pH 7.5) and 8% dimethylfomamide. (■-■-) 16 mM EDTA, 0.01 M Tris (pH 7.5) and 8% dimethylformamide. The open symbols connected by dotted lines indicate the corresponding control values. (B) A solution containing 80 µM [^{3}H]d(A-T)$_n$, 1.6 µM kanchanomycin, 0.16 mM MgCl$_2$, 0.01 M Tris (pH 7.5) and 8% dimethylformamide was incubated 24 hrs at 25°. A control solution contained no kanchanomycin. At 24 hrs samples were withdrawn, mixed with an equal volume of the solutions indicated above, and incubated at 25°. At the times indicated 30 µl aliquots were withdrawn and filtered through Millipore filters which retain polynucleotide bound to antibiotic. The same symbols and solutions apply to B as used in A. At the end of the initial 24 hrs incubation, 83% of the [^{3}H]d(A-T)$_n$ was retainable on Millipore filters. (FRIEDMAN et al., 1969a)

---

^{1} Abbreviation used is: d(A-T)$_n$, the strictly alternating deoxyadenylate-deoxythymidylate copolymer.

polymerase reactions, the former being somewhat more sensitive than the latter. Native and heat-denatured DNA-directed RNA synthesis are affected at any time of addition of the antibiotic. The inhibition of DNA synthesis by kanchano-mycin can be overcome by increasing the concentration of DNA template (Fig. 7) but is unaffected by increasing the concentration of DNA polymerase (Fig. 8). In sharp contrast, the inhibition of RNA synthesis by the antibiotic is not over-come by the addition of more DNA template (Fig. 7) which has been incubated with a constant amount of kanchanomycin, although it is reversed by addition of template which has not been previously exposed to the antibiotic. The inhibition of RNA synthesis by kanchanomycin, however, can be overcome by increasing the concentration of the RNA polymerase (Fig. 8). Thus, a double reciprocal plot of the kinetics of inhibition shows that kanchanomycin acts as a competitive inhibitor of *DNA* in DNA synthesis, but as a competitive inhibitor of the RNA *polymerase* in RNA synthesis (Fig. 9).

Kanchanomycin resembles luteoskyrin, the potent hepatotoxin extracted from the fungus mat (*Penicillium islandicum*) associated with *yellow* rice, in some aspects of its interaction with polynucleotides and its inhibition of the RNA polymerase

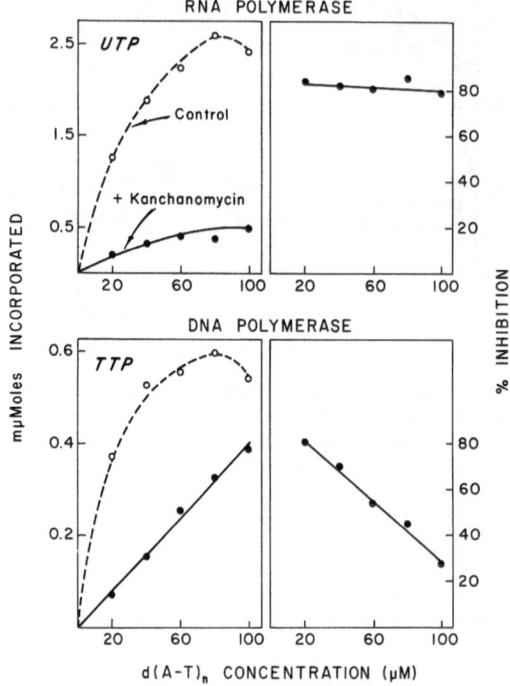

Fig. 7. Effect of increasing d(A-T)$_n$ concentration on the inhibition of RNA and DNA synthesis by kanchanomycin. The pre-incubation mixtures contained 0 or 6 μM kanchanomycin, 0.48 mM MgCl$_2$ and d(A-T)$_n$ at a concentration 4 times the final concentration given on the graph. The pre-incubation period was 2 hrs. Each assay for RNA synthesis contained 2 units of RNA polymerase. Each assay for DNA synthesis contained 0.42 unit of DNA polymerase and was incubated for 10 min. (JOEL *et al.*, 1970)

(URAGUCHI et al., 1961; SHIBATA et al., 1968; UENO et al., 1967a; OHBA and FROMAGEOT, 1967; OHBA and FROMAGEOT, 1968; SENTENAC et al., 1967; UENO et al., 1967b). Each requires divalent cation for complex formation with nucleic acids and forms two types of complexes, although the requirements in the polynucleotide for complex formation differ in essential details. The inhibition of RNA synthesis by the two agents appears not to be due solely to the binding of the inhibitor to the DNA, but must in addition involve inactivation of the polymerase in the complex (SENTENAC et al., 1967; JOEL et al., 1970). It is possible that the enzyme alters the DNA structure locally so as to lead to selective binding of the antibiotic. On the other hand, it is also possible that the enzyme is attracted to the sites (not necessarily the same as those for the initiation of RNA synthesis) on the DNA where the inhibitor is located. When more enzyme is added, inhibition owing to the antibiotic is overcome because the RNA polymerase is able to seek out sites free of the inhibitor. DNA not previously exposed to kanchanomycin, however, can overcome the inhibition because it acts as an effective competitor of the DNA bearing kanchanomycin for the RNA polymerase. This may result from a change in the physical state of the complexed DNA which makes it less accessible to the enzyme. By contrast, the DNA polymerase is not attracted preferentially to sites on the DNA populated by kanchanomycin. Instead, it

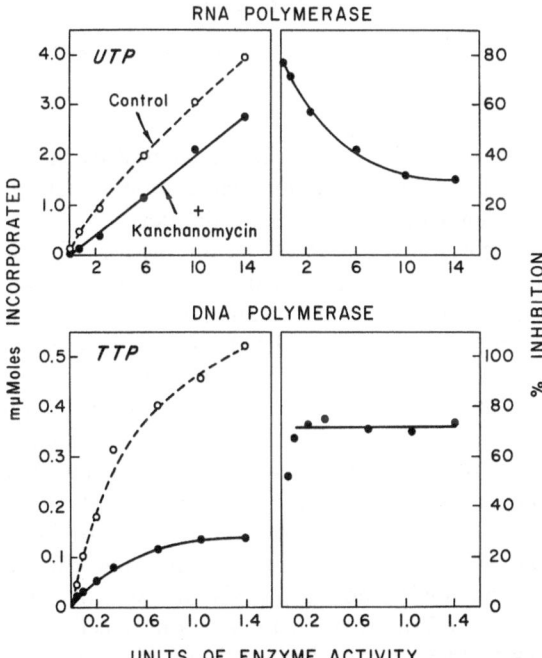

Fig. 8. Effect of increasing enzyme concentration on the inhibition of RNA and DNA synthesis by kanchanomycin with $d(A-T)_n$ as template. The pre-incubation mixtures contained 80 µM $d(A-T)_n$, 0 or 3.2 µM kanchanomycin, and 0.16 mM $MgCl_2$. The pre-incubation period was 2 hrs. The units of RNA or DNA polymerase present in the subsequent assays are indicated on the graph. Assays receiving less than the maximum volume of enzyme-containing solution received appropriate volumes of the medium (minus enzyme) in which the enzyme was stored or diluted. (JOEL et al., 1970)

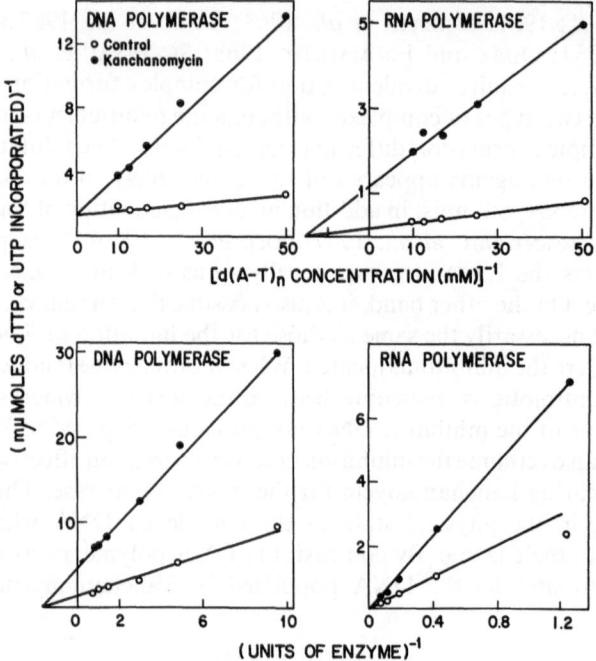

Fig. 9. Lineweaver-Burk plot of data from Figs. 7 and 8. (JOEL et al., 1970)

is the template function of the DNA which is altered by the antibiotic. In these two different types of actions, kanchanomycin combines features of the template-inactivating and the enzyme-inactivating antibiotics.

The molecular and structural nature of the interaction of kanchanomycin with DNA which leads to its interference with the function of both the RNA and DNA polymerase is not yet known. Kanchanomycin does not inhibit DNA replication by preventing separation of the DNA strands, since it does not increase the temperature of thermal transition of DNA (FRIEDMAN et al., 1969a). Whatever the nature of the interaction may be, i.e. intercalation, binding to the sugar-phosphate backbone, or attachment to the grooves of the DNA, which interferes with the function of the DNA polymerase, the direct inactivation of the RNA polymerase by kanchanomycin bound to template suggests that functional groups of the antibiotic may reside in that groove of DNA where this enzyme functions.

It is not certain that kanchanomycin exerts its toxic effect on cells primarily by a direct action on nucleic acid synthesis. In fact, we have found that this antibiotic rapidly shuts off many synthetic reactions in cells simultaneously (MIT-SUGI and GOLDBERG, 1966, unpublished results). This may result from the inhibition by kanchanomycin of oxidative phosphorylation such as has been found with mammalian mitochondria (FRIEDMAN and GOLDBERG, 1967, unpublished results).

The structure of Kanchanomycin has been published recently. [A. I. GUREVICH et al., The structure of albofungin. Tetrahedron Letters. No. **18**, 1751 (1972);

K. FUKUSHEMA *et al.*, Identity of antibiotic P-42-1 elaborated by *Actinomyces trimemacerara* with kanchanomycin and albofungin. J. Antibiotics **26**, 65 (1973)].

References

BATEMAN, J. R., A. A. MARSH, and J. L. STEINFELD: Kanchanomycin (NSC-62773): A phase I study. Cancer Chemotherapy Rept. No. **44**, 25 (1965).

FRIEDMAN, P. A., P. B. JOEL, and I. H. GOLDBERG: Interaction of kanchanomycin with nucleic acids. I. Physical properties of the complex. Biochemistry **8**, 1535 (1969a)

FRIEDMAN, P. A., T.-K. LI, and I. H. GOLDBERG: Interaction of kanchanomycin with nucleic acids. II. Optical rotatory dispersion and circular dichroism. Biochemistry **8**, 1545 (1969b).

JOEL, P. B., P. A. FRIEDMAN, and I. H. GOLDBERG: Interaction of kanchanomycin with nucleic acids. III. Contrasts in the mechanisms of inhibition of ribonucleic acid and deoxyribonucleic acid polymerase reactions. Biochemistry **9**, 4421 (1970).

LIU, W.-C., W. P. CULLEN, and K. V. RAO: BA-180265: a new cytotoxic antibiotic. Antimicrob. Agents Chemotherapy **1962**, 767.

OHBA, Y., et P. FROMAGEOT: Interactions entre pigments et acides nucleiques. 4. Complexes specifiques et non specifiques entre la luteoskyrine, les ions magnesium et les acides nucleiques. Eur. J. Biochem. **1**, 147 (1967).

OHBA, Y., et P. FROMAGEOT: Interactions entre pigments et acides nucleiques. 5. Stoechiometrie du complexe I forme entre la luteoskyrine, les ions magnesium et les acides nucleiques. Eur. J. Biochem. **6**, 98 (1968).

SENTENAC, A., A. RUET et P. FROMAGEOT: Interaction entre pigments et acides nucleiques. III. Inhibition du fonctionnement *in vitro* de la RNA polymerase de *E. coli* par la luteoskyrine. Bull. Soc. Chim. Biol. **49**, 247 (1967).

SHIBATA, S., Y. OGIHARA, N. KOBAYASHI, S. SEO, and I. KITAGAWA: The revised structures of luteo-skyrin, rubroskyrin and rugulosin. Tetrahedron Letters **27**, 3179 (1968).

UENO, Y., A. PLATEL et P. FROMAGEOT: Interaction entre pigments et acides nucleiques. II Interaction *in vitro* entre la luteoskyrine et le DNA de thymus de veau. Biochim. Biophys. Acta **134**, 27 (1967a).

UENO, Y., I. UENO, K. ITO, and T. TATSUNO: Impairments of RNA synthesis in Ehrlich ascites tumour by luteoskyrin, a hepatotoxic pigment of *Penicillium islandicum* sopp. Experientia **23**, 1001 (1967b).

URAGUCHI, K., T. TATSUNO, F. SAKAI, M. TSUKIOKA, Y. SAKAI, O. YONEMITSU, H. ITO, M. MIYAKE, M. SAITO, M. ENOMOTO, T. SHIKATA, and T. ISHIKO: Isolation of two toxic agents, luteoskyrin and chlorine-containing peptide, from the metabolites of *Penicillium islandicum* sopp, with some properties thereof. Japan J. Exptl. Med. **31**, 19 (1961).

Nalidixic Acid—Mode of Action

William A. Goss and Thomas M. Cook

Introduction

Nalidixic acid (Fig. 1) [1-ethyl-1,4-dihydro-7-methyl-4-oxo-1,8-naphthyri-dine-3-carboxylic acid] is a potent antibacterial agent selected from a large group of 1,8-naphthyridines synthesized by LESHER et al. (1962). This synthetic program was initiated by the discovery of the antibacterial activity of 7-chloro-1,4-dihydro-1-ethyl-4-oxoquinoline-3-carboxylic acid which was obtained from the mother liquors during purification of chloroquine.

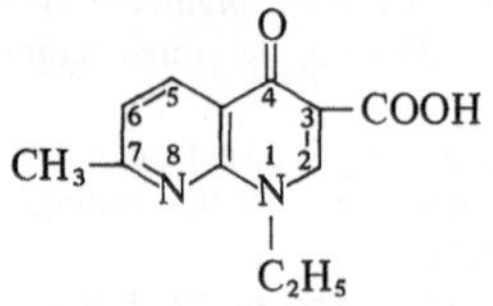

Nalidixic Acid
1-Ethyl-1, 4-dihydro-7-methyl-4-oxo
1,8-naphthyridine-3-carboxylic acid

Fig. 1

The relative *in vitro* and *in vivo* activities of a series of substituted naphthyri-dines indicated that the 1-ethyl, 7-methyl derivative (nalidixic acid) was the most active (Table 1). Substitution of the carboxyl group at the 3-position with $COOC_2H_5$, $COOCH_2N(CH_3)_2$, $CONH_2$, $CONHCH_2COOH$, or CN resulted in decreased *in vitro* activity indicating that the carboxyl function is essential for optimal antibacterial action. The chemotherapeutic activity in mice of most esters was similar to the corresponding acid, most probably due to hydrolysis by host esterases (LESHER, 1963; LESHER, personal communication).

General Properties

Nalidixic acid (NAL) is a heat-stable, weak, organic acid. It is relatively insoluble in water but highly soluble in aqueous bases (i.e. NaOH) and polar organic solvents. This white, crystalline solid has a molecular weight of 232.3 and an empirical formula of $C_{12}H_{12}N_2O_3$. Complexing cations (i.e. Ca^{++}, Mg^{++}) form water-insoluble compounds.

NAL is particularly active against gram-negative bacteria *in vitro* and *in vivo* (DEITZ et al., 1964; FROELICH and DEITZ, 1963). Representative minimal inhibitory concentrations are given in Table 2. Nalidixic acid has been particularly

Table 1. *Effect of 1- and 7-position substitution of 1,8-naphthyridines on antibacterial activity*

$$R_2 \underset{7\ 8}{\overset{6\ 5}{\boxed{}}} \underset{N}{\overset{}{}} \underset{\underset{R_1}{|}}{N} \overset{4\ 3}{\underset{1\ 2}{\boxed{}}} \overset{O}{} -COOH$$

R_1 (where $R_2=CH_3$)	Relative activity		R_2 (where $R_1=C_2H_5$)	Relative activity	
	in vitro[a]	in vivo[b]		in vitro[a]	in vivo[b]
H	0	0	H	0	++
CH_3	0	++	CH_3 (NAL)	++++	++++
C_2H_5 (NAL)	++++	++++	C_2H_5	+++	+++
C_3H_7	+++	+++	C_3H_7	+++	0
i-C_3H_7	+	+	i-C_3H_7	+++	0
$CH_2CH=CH_2$	+++	++	$CH_2C_6H_5$	+	0
C_4H_9	0	0	CH_2CN	+++	++
CH_2CF_3	++++	++	COOH	0	0
CH_2COOH	0	0	CH_2OH (Hydroxy-NAL)	++++	++++
CH_2CH_2OH	+++	++			
$CH_2CH_2N(C_2H_5)$	0	0			
CH_2OCH_3	+	++			

[a] Minimal Inhibitory Concentration, MIC, µg/ml against either *Salmonella typhi* or *Escherichia coli* where $0=\geqq 100$; $+=\geqq 50$; $++=\geqq 10$; $+++=\geqq 5$; and $++++=<5$ µg/ml.
[b] Oral activity (ED_{50}) vs. lethal *Klebsiella pneumoniae* infection in mice where $0=>300$; $+=300-200$; $++=200-100$; $+++=100-50$; $++++=<50$ mg/kg/day.
(From LESHER, 1962; LESHER, personal communication.)

Table 2. *In vitro profile of nalidixic acid against representative bacteria*

Challenge organism (No. strains)	Range of minimal inhibitory concentrations (µg/ml)
Escherichia coli (8)	3.0–7.5
Aerobacter aerogenes (3)	1.6–2.5
Klebsiella pneumoniae (1)	2.0
Proteus mirabilis (2)	2.5–7.5
Proteus vulgaris (2)	7.5–10.0
Salmonella enteriditis (2)	2.5–10.0
Shigella spp. (3)	2.5–5.0
Pseudomonas aeruginosa (4)	80.0–500.0
Staphylococcus aureus (2)	25–50
Streptococcus pyogenes (1)	100.0
Streptococcus faecalis (1)	500.0
Diplococcus pneumoniae (1)	250.0
Bacillus subtilis (2)	5.0

From Goss (1969).

useful in the clinical management of urinary tract infections due to *Proteus mirabilis, P. morganii, P. vulgaris, P. rettgeri, Escherichia coli, Aerobacter* sp., or *Klebsiella* sp. (STAMEY, 1969, 1970; HARRISON and COX, 1970).

NAL is very well tolerated in man and laboratory animals. The LD_{50} in mice of NAL is 3,300 mg/kg by the oral route and 176 mg/kg when given intravenously. The drug is also well tolerated in monkeys at oral doses as high as 450 to 900 mg/kg. In all of the safety tests conducted including chronic administration to rats and monkeys for one year, no toxicological events were detected which one would relate to inhibition of DNA synthesis in mammalian systems (H. P. DROBECK, personal communication).

DEITZ et al. (1964) found that the course of *in vitro* resistance development to NAL by several gram-negative organisms was stepwise. BUCHBINDER et al. (1963) and BARLOW (1963) reported that some strains of susceptible organisms develop resistance rapidly, whereas other strains develop resistance in a "stepwise" manner. In an experimental *E. coli* pyelonephritis infection in mice, O'CONNOR et al. (1970) were unable to isolate NAL-resistant organisms from the kidneys or urine of animals treated over extended periods with suboptimal or optimal levels of the drug. BUCHBINDER et al. (1963); BARLOW (1963), and RONALD et al. (1966) have reported that bacterial resistance to NAL occurs in man during therapy.

NAL-resistant strains of pathogenic and nonpathogenic *E. coli* were found to be hemolytic on blood-agar medium. WALTON and SMITH (1969) postulated that the NAL locus and hemolysin locus are closely linked. HELLING and KUKORA (1971) found that mutants of *E. coli* K-12, selected for their resistance to NAL, were deficient in isocitrate dehydrogenase. HELLING and ADAMS (1970) reported that *E. coli* K-12 mutants resistant to NAL also required either adenine or mixture of amino acids. The significance of these findings and their relationship to the mode of action of, or resistance to, NAL are not clear at the present time.

NAL-resistance is chromosomal in nature and has been mapped in *E. coli* K-12 at 42.5 ± 0.5 and 51 ± 1 min in *nalA*^{r} mutants (growth at 40 µg/ml or higher) and *nalB*^{r} mutants (growth at 4 µg/ml but no growth at 10 µg/ml), respectively (HANE and WOOD, 1969). YOSHIKAWA (1971) also concluded that *nal*^{r} was chromosomal and not associated with an R-factor. Recent studies by BOURGUIGNON and STERNGLANZ (personal communication) have shown that *nal*B is a transport mutant whereas *nal*A is not. MITSUHASHI (1967, 1971) and DATTA (1971) were unable to detect NAL resistance associated with R-factors in gram-negative bacteria from clinical sources.

It is interesting to note that man, animals and certain microorganisms biotransform NAL to the corresponding 7-hydroxy-methyl naphthyridine metabolite (Fig. 2) (MCCHESNEY et al., 1964; HAMILTON et al., 1969; ČÁPEK et al., 1969). The *in vitro* antibacterial profile of this metabolite is indistinguishable from that of nalidixic acid itself (Table 3) (GOSS, 1969). Hydroxy-NAL is also the major active component found in the urine of man (see PORTMANN et al., 1966 for a detailed description of the pharmacokinetic studies in man).

Table 3. *Comparison of the minimal inhibitory concentrations (MIC) of nalidixic acid (NAL) and hydroxymethyl nalidixic acid (HO-NAL)*

Test organisms	MIC, µg/ml	
	NAL	HO-NAL
Escherichia coli No. 198 (ATCC 11229)	3.0	3.0
Escherichia coli Vogel	3.0	3.0
Escherichia coli ATCC 15764	6.0	3.0
Escherichia coli VA 12716	5.0	4.0
Salmonella typhosa ATCC 6539	3.0	4.0
Salmonella typhimurium No. 88	3.0	3.0
Salmonella typhimurium ME-222	3.0	4.0
Salmonella typhimurium MA-918	4.0	5.0
Shigella flexneri B-1070	3.0	3.0
Proteus mirabilis MGH-1	6.0	6.0
Proteus rettgeri 3141	12.0	12.0
Klebsiella pneumoniae ATCC 10872	2.0	3.0
Klebsiella pneumoniae 39645	8.0	8.0
Pasteurella boviseptica No. 1 (Harvard)	1.0	2.0
Aerobacter aerogenes MGH-1	3.0	6.0
Serratia marcescens SW-2	8.0	8.0
Serratia marcescens SW-5	2.0	4.0
Pseudomonas aeruginosa 211	300.0	300.0

From Goss (1969).

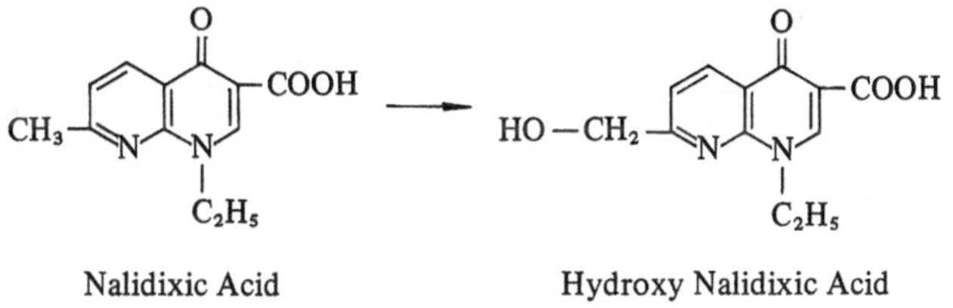

Nalidixic Acid   Hydroxy Nalidixic Acid

Fig. 2

Mechanism of Action on Bacteria

Goss *et al.* (1964) reported that treatment of exponentially growing cultures of *E. coli* with low levels of nalidixic acid markedly reduced the growth rates, whereas higher levels caused drastic reductions in the number of viable cells (Fig. 3). Despite the rapid loss of viability, there were no corresponding decreases in the turbidities of treated cultures; there was actually an increase in turbidity at bactericidal levels. Microscopic examination of exposed cultures revealed extremely elongated, serpentine forms (Fig. 4). The synchrony between the initiation

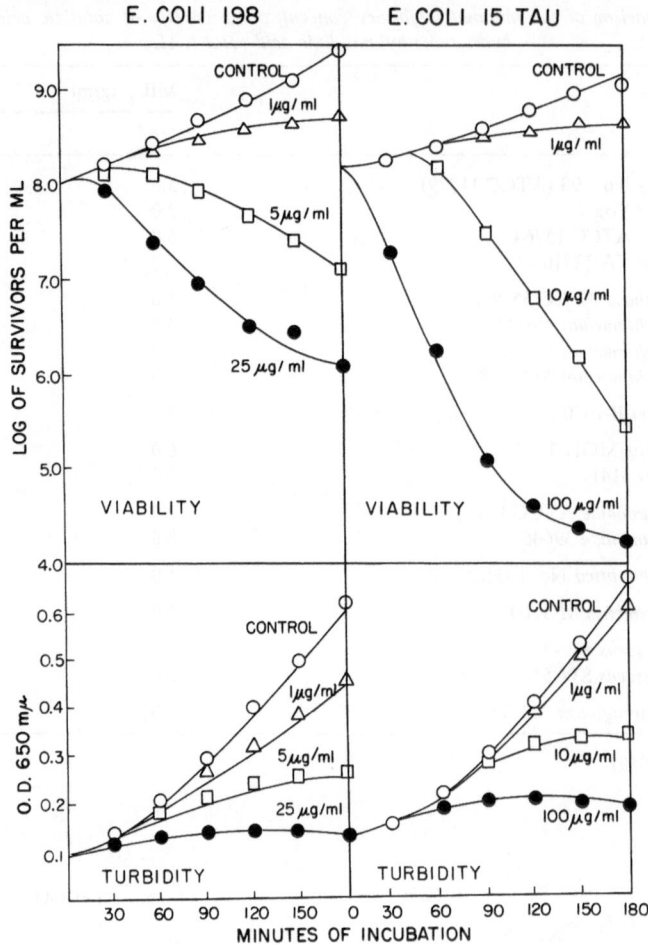

Fig. 3. Growth inhibition of *Escherichia coli* by nalidixic acid: comparison of viability and turbidity. (From Goss *et al.*, 1964)

of growth in the control cultures and of death in the treated cultures was clearly not coincidental. Chemical analyses of cellular constituents provided direct evidence for deficient DNA synthesis, which in the presence of competent cytoplasmic (RNA and protein) synthesis, resulted in unbalanced bacterial metabolism and death of the cell. These observations were similar to those seen when certain strains of *E. coli* undergo "thymineless death" (Goss *et al.*, 1965; BAUERNFEIND and GRÜMMER, 1965).

These preliminary observations on the selective inhibition of DNA synthesis by nalidixic acid in *E. coli* were confirmed subsequently employing radioisotopic techniques (Goss *et al.*, 1965). Using the polyauxotroph *E. coli* 15 TAU, it was found that nalidixic acid prevented the incorporation of thymine into DNA but had only minimal effects on L-arginine and uracil incorporation into protein

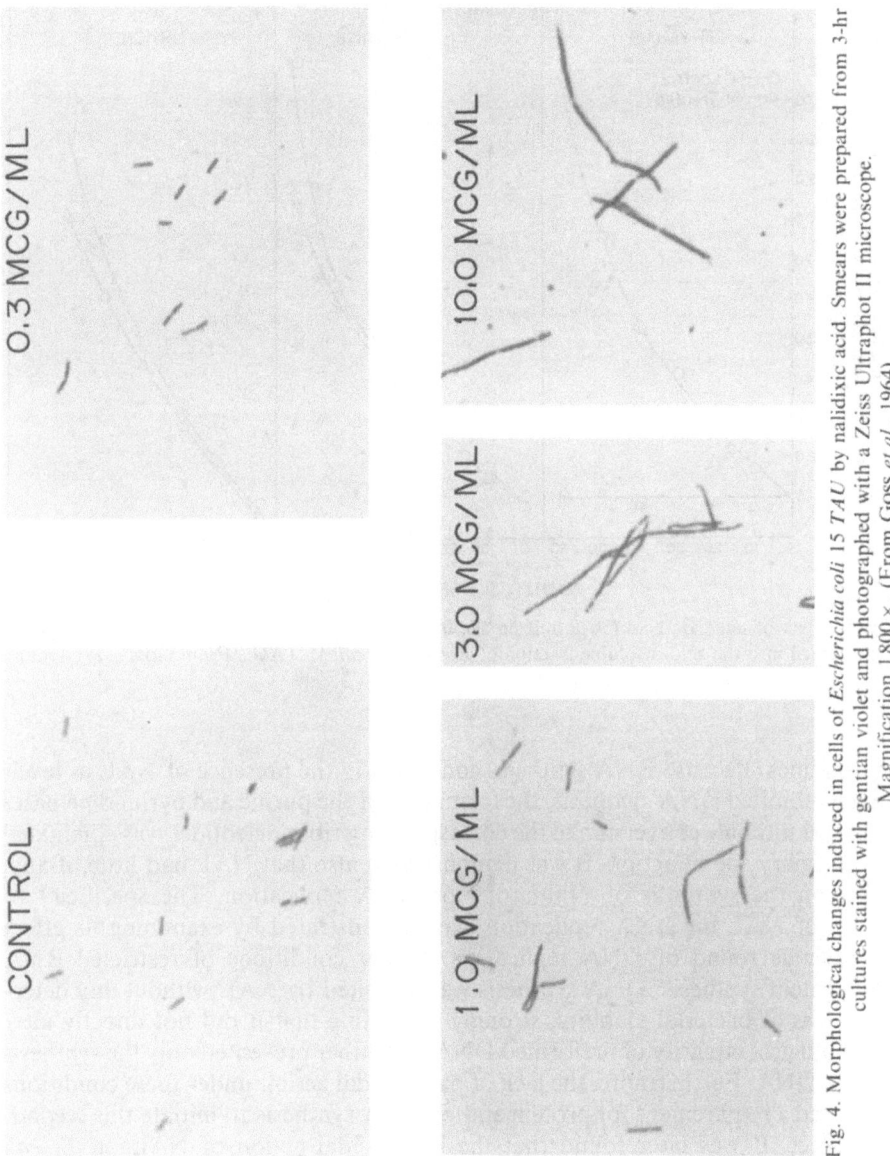

Fig. 4. Morphological changes induced in cells of *Escherichia coli* 15 *TAU* by nalidixic acid. Smears were prepared from 3-hr cultures stained with gentian violet and photographed with a Zeiss Ultraphot II microscope. Magnification 1 800 ×. (From Goss *et al.*, 1964)

and RNA, respectively (Fig. 5). DNA synthesis could be restored by simply washing NAL-treated cultures, suggesting that the drug was not firmly bound to the target site(s). DNA synthesis was also inhibited when *E. coli* 15 TAU was exposed to hydroxy-NAL (Goss, 1969).

Goss *et al.* (1965) examined the effect of nalidixic acid on the incorporation of adenine and uracil into RNA by a wild-type strain of *E. coli*. In the presence of NAL, both bases were incorporated into RNA but not into DNA (Fig. 6). This excluded any selective effect of the drug on incorporation of purines or

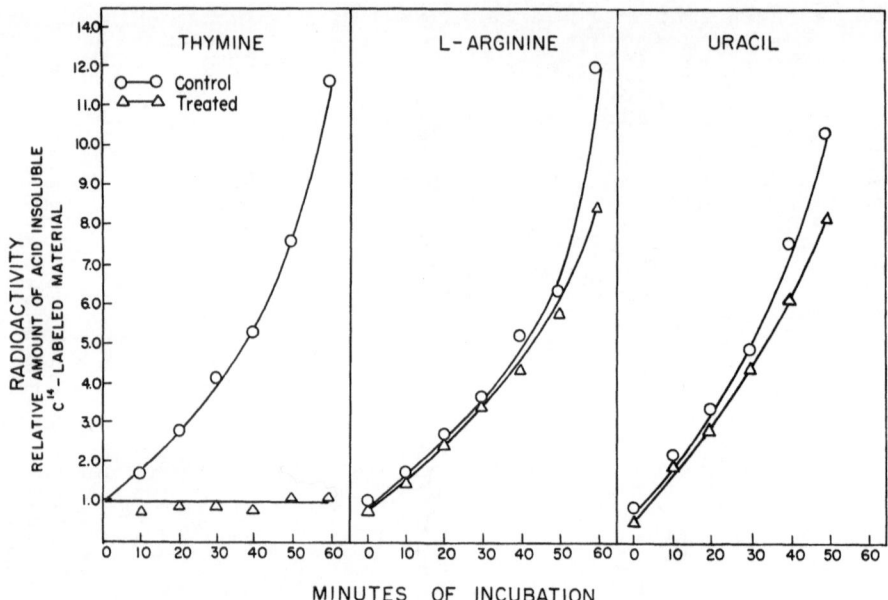

Fig. 5. Effect of nalidixic acid (50 µg/ml) on the incorporation of ^{14}C-labeled thymine, L-arginine, and uracil into the acid-insoluble fraction of *Escherichia coli* 15 TAU. (From Goss *et al.*, 1965)

pyrimidines. Because RNA synthesis continued in the presence of NAL at levels which inhibited DNA synthesis, the formation of the purine and pyrimidine bases and their ultimate conversion to the corresponding ribonucleotides was eliminated as a primary site of action. It was demonstrated also that NAL had little, if any, effect on the synthesis of "initiator" of DNA replication. The specificity of action of NAL on DNA replication was demonstrated by examining its effect on a single round of DNA replication. Under conditions of restricted RNA and protein synthesis, DNA synthesis was inhibited by NAL without any detectable loss of bacterial viability, strongly suggesting that it did not directly alter the biological integrity of preformed DNA but rather prevented only the synthesis of new DNA. Furthermore, the lack of bactericidal action under these conditions indicated a requirement for protein and/or RNA synthesis to initiate this secondary effect. It was later found that the bactericidal action of NAL on *E. coli* can be prevented or controlled by the regulation of protein and RNA synthesis. This was achieved by the use of bacteriostatic inhibitors, selected nutritional conditions as well as temperature-regulated growth rates (Deitz *et al.,* 1966; Bauernfeind and Grümmer, 1965; Cummings and Kusy, 1969, 1970).

Cook *et al.* (1966a) demonstrated the similarities between the susceptibility of the gram-positive bacterium, *Bacillus subtilis,* and *E. coli* to NAL with respect to selective inhibition of DNA synthesis, "unbalanced" metabolism and morphological abnormalities.

Boyle and Jones (1970) reported that uridine, in addition to protecting thymidine against degradation, reversed partially (70 to 80%) the inhibition of DNA

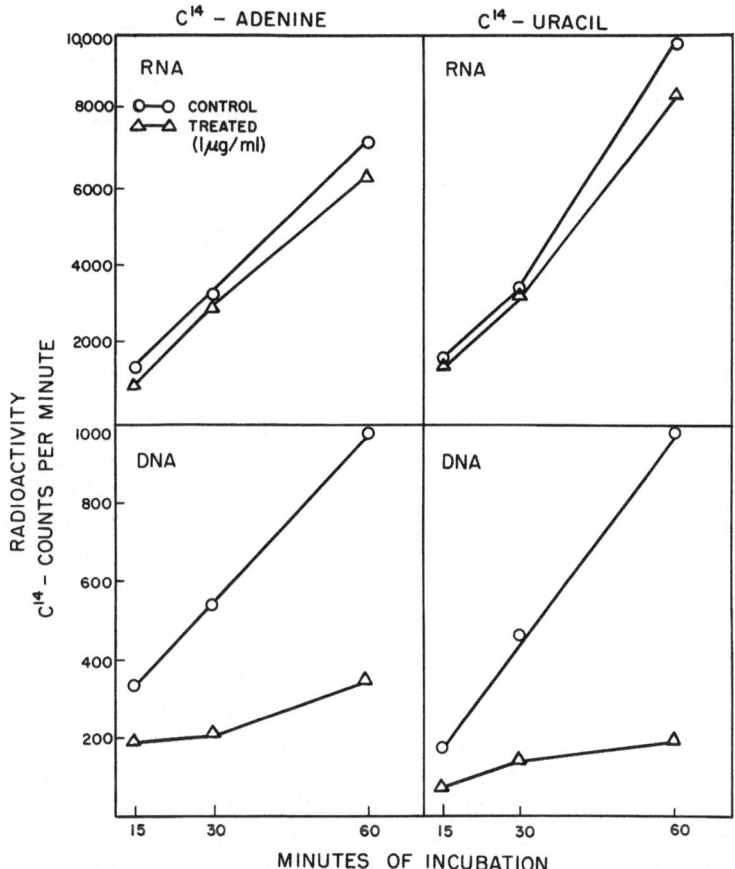

Fig. 6. Effect of nalidixic acid (1.0 μg/ml) on the uptake and incorporation of ^{14}C-labeled adenine and ^{14}C-labeled uracil into RNA and DNA by *Escherichia coli* 198. (From Goss *et al.*, 1965)

synthesis by NAL in thymineless auxotrophs of *E. coli*. No reversal of NAL action by uridine was observed in any of the prototrophic strains examined. The significance of these findings relative to the mode of action of NAL is unknown.

The loss of bacterial viability, formation of serpentine forms, specific inhibition of DNA synthesis by NAL, reversibility of this action, etc., have been observed in several other laboratories (TAKETO and WATANABE, 1967; BARBOUR, 1967; COHEN, 1968; CLARK, 1968; BOUCK and ADELBERG, 1970; WINSHELL and ROSEN-KRANZ, 1970; BEHKI and LESLEY, 1972; PEDRINI *et al.*, 1972).

In addition, the effect of NAL has been examined in a variety of mutant strains deficient in recombination and DNA repair systems. Results of these studies show that sensitivity to the bactericidal action of NAL is influenced by defects in these repair mechanisms. KANTOR and DEERING (1968) found that inhibition of DNA synthesis by NAL was equivalent in *E. coli* B and B/r, but that the B strain was considerably more susceptible to its lethal effects. A u.v.-sensitive strain, Bsl, unable to repair thymine dimers induced by irradiation, also was more resistant to NAL than strain B. Attempts to reverse the lethal effects of NAL on strain B with visible light (photoreactivation) were unsuccessful. Furthermore, pantoyl lactone or liquid holding treatments aided division and colony formation in NAL-treated cultures of B, but not B/r. Further supporting data showing increased resistance to the bactericidal action of NAL associated with *uvr* and *exr*A mutations was supplied by GREEN et al. (1969, 1970). However, *rec*A strains (deficient in genetic recombination and DNA repair enzymes, especially after irradiation) do not appear to be significantly more resistant to NAL. The reason for this difference is not known. The production of thymine dimers obviously cannot be a major mode of action of this drug, and these findings clearly separate the primary inhibition of DNA synthesis from the secondary bactericidal effect.

HOLLOM and PRITCHARD (1965) reported that NAL specifically and reversibly blocked DNA synthesis in strains of *E. coli* K-12. Formation of recombinants during mating was completely inhibited by NAL. A major contribution to this effect was found to be a reversible inhibition of chromosome transfer. BARBOUR (1967), FISHER and FISHER (1968), BOUCK and ADELBERG (1970), and HANE (1971) have confirmed and extended these observations in an attempt to elucidate the mechanisms of conjugation (for a review of this area, see CURTISS, 1969).

NAL effectively inhibited the multiplication of T2r phage in *E. coli* (TAKETO and WATANABE, 1967). This inhibition was reversed by transfer of the infected cells to drug-free medium. Prior treatment of the host bacterium did not prevent growth of the virus. The spectrum of antiphage activity included not only conventional double-stranded DNA phages but also single-stranded DNA phages such as ϕX174 and ϕR; RNA phages were not inhibited. WATANABE and AUGUST (1967) found NAL extremely useful for the isolation of RNA phages in the presence of large numbers of DNA phages. NAL was also found to inhibit viral as well as host DNA synthesis in *Pseudomonas* BAL-31 infected with the lipid-containing bacteriophage PM 2 (DATTA et al., 1971); these investigators concluded that NAL may act at a site or process which is common to both host and phage DNA replication (BRAUNSTEIN et al., 1971).

Consistent with the results of COOK et al. (1966a), GAGE and FUJITA (1969) reported that NAL inhibited host cell DNA synthesis in *B. subtilis* but had little effect on synthesis of viral DNA in cells infected with the bacteriophage SPO1. SPO1 is unusual in that its DNA contains hydroxy-methyluracil in place of thymine. However, in a more extensive study, BAIRD et al. (1972) investigated a series of coliphages as well as *B. subtilis* bacteriophages and concluded that no correlation existed between inhibition of bacteriophage growth and unusual bases in the bacteriophage DNA.

Little is known about effects of NAL on DNA synthesis other than the semi-conservative mode of replication, but apparently NAL will inhibit "extensive repair" synthesis in ultraviolet-irradiated *E. coli* 15 T⁻ (EBERLE and MASKER, 1971). When u.v.-treated cells were exposed to bromouracil and ^{14}C-thymine, isotope was incorporated initially into a "light" DNA fraction, and later appeared at densities intermediate between "light" and "hybrid". With added NAL (20 µg/ml) there is a small initial incorporation into light DNA, but no shift to intermediate densities. In the rather different case of *Micrococcus radiodurans,* repair of X-ray induced breaks in DNA, shown by restoration of normal centrifugation patterns in alkaline sucrose gradients, was not prevented by NAL except at levels high enough to also reduce protein synthesis by 99% (DRIEDGER and GRAYS-TON, 1971). It may be of significance that NAL exerted only a bacteriostatic, and not bactericidal, effect on *M. radiodurans* at the levels employed (20–200 µg/ml).

Action of NAL in Subcellular Systems

Having established that the primary effect of NAL in the intact, living bacterium is an immediate, selective and reversible cessation of DNA synthesis, one next is led to ask exactly how this inhibition is accomplished at the molecular level. It must be acknowledged that the precise molecular mechanism has not been delineated at this time. Recent findings (PEDRINI *et al.,* 1972; SCHALLER *et al.,* 1972) suggest that the answer lies in a still-unidentified step in bacterial chromosome replication as it occurs in the physiological complex of enzymes and DNA attached to the cell membrane. Insensitivity to NAL previously reported for certain *in vitro* DNA synthesizing systems and "permeabilized" cell systems (BOYLE *et al.,* 1969; OKAZAKI *et al.,* 1970; WINSHELL and ROSENKRANZ, 1970) now can be appreciated as merely an indication of their non-physiological character.

Any one of the steps in bacterial chromosome replication conceivably could be a candidate for the primary NAL target, but a number of these appear definitely to have been ruled out. These include 1. binding to the DNA template, 2. production of precursors, and 3. activity of various known enzymes specifically involved in DNA synthesis.

1. NAL does not show detectable binding to DNA *in vitro.* The ultraviolet absorption spectrum of NAL is not altered when it is mixed with calf-thymus DNA; and when a mixture of ^{14}C-labeled NAL and DNA are centrifuged through a sucrose or CsCl density gradient, there is no increment of radioactivity associated with the DNA bands (J. BOYLE and W. GOSS, unpublished). G. BOURGUIGNON and R. STERNGLANZ (personal communication) also have found no change in the absorption spectrum of NAL in the presence of DNA. They also observed that addition of NAL caused no alteration of the Tm of native DNA in thermal denaturation tests. In equilibrium dialysis experiments they found negligible association of ^{3}H-NAL with DNA, indicating that binding constant would have to be less than 10. BOYLE *et al.* (1969) eliminated the possibility that NAL might act *in vivo* as a bifunctional alkylating agent like mitomycin C, cross-linking the complementary strands of DNA. No cross-linkage was seen in DNA extracted from NAL-treated cells, whereas mitomycin cross-linking was readily demonstrable.

2. Biosynthesis of purine and pyrimidine precursors appears to be unaffected by NAL. None of the metabolic pathways leading to formation of ribonucleotides can be involved, inasmuch as RNA synthesis continues unabated in NAL-treated bacteria. A direct test of the effect of NAL on the reduction of ribonucleotides to the corresponding deoxy-ribonucleotides does not appear to have been made,

but this seems unlikely as the NAL target in view of other results to be described. Also, WINSHELL and ROSENKRANZ (1970) found no antagonism by added exogenous deoxyribosides on the NAL inhibition of DNA synthesis under conditions supposedly permitting their entry into the cell. NAL does not prevent phosphorylation of deoxyribonucleoside mono- and diphosphates to the triphosphates in cell-free extracts, nor does it affect deoxyribosyl transferase activity (BOYLE et al., 1969). The activity of thymidine kinase likewise appears to be unaffected (WINSHELL and ROSENKRANZ, 1970).

3. In cell-free extracts, or with partially purified preparations, NAL does not inhibit the known enzymes directly concerned with DNA polymerization. BOYLE et al. (1969), showed that Kornberg polymerase (DNA polymerase I) partially purified from E. coli was not inhibited by NAL. Subsequently, PEDRINI et al. (1972) have shown that the following enzymes are not NAL-sensitive: DNA polymerase I, exonuclease I, II and III, and endonuclease I from E. coli; polynucleotide ligase and DNA methyl-transferase from E. coli infected with T_4; and a DNA polymerase activity from Bacillus subtilis.

Isolation of E. coli mutants (pol Al⁻)which form little or no DNA polymerase I has led to recognition of other DNA synthesizing activities (DNA polymerase II and III) which are more likely to be responsible for bacterial chromosome replication in vivo (GOULIAN, 1971; CAMPBELL et al., 1972). Isolated "membrane fractions" prepared from muramidase lysates of pol Al mutants carry on an in vitro DNA synthesis with certain similarities to in vivo replication, including intermediate formation of Okazaki fragments. The activity, presumably involving DNA polymerase II and ligase, is dependent on the presence of all 4 deoxyribonucleoside triphosphates and ATP, but does not require added DNA primer. This synthesis is not inhibited by NAL (50 µg/ml) [OKAZAKI et al., 1970; PEDRINI et al., 1972]. From this, one can conclude that DNA polymerase II activity per se is not sensitive to NAL, and apparently this has been confirmed using partially purified enzyme preparations (R. KNIPPERS, personal communication cited by SCHALLER et al., 1972).

One possible explanation for a lack of NAL activity in these in vitro DNA synthesizing systems would be if the actual inhibitor were not NAL itself, but some metabolically altered form of the molecule. An examination of this point using ³H-Nal failed to show any evidence of a NAL metabolite in E. coli cells (J. BOYLE and W. GOSS, unpublished; also G. BOURGUIGNON and R. STERNGLANZ, personal communication). Such negative results, while casting reasonable doubt on the metabolite hypothesis, do not provide conclusive proof. Conceivably the binding of a single molecule of inhibitor to the replication fork might stop DNA synthesis, so that the amount of "active form" needed would be undetectable. It seems more profitable to assume for the moment that NAL per se is in fact the proximal inhibitor, and that NAL-insensitive DNA-synthesis results from derangement of the in vivo mechanism during laboratory manipulations.

Recently, two NAL-sensitive in vitro DNA synthesizing systems have been reported, and these should provide a resolution of the problem of the NAL target. PEDRINI et al. (1972) found that a brief toluene treatment (2 minutes) of cells of a pol Al mutant of E. coli yielded a preparation catalyzing an ATP-dependent incorporation of deoxyribonucleotides into acid-insoluble polymer.

This synthesis could be inhibited by NAL, but was approximately 10 times less sensitive than *in vivo* replication; the reaction was inhibited by approximately 80% when NAL (50 µg/ml) was added. Prolonging the toluene treatment progressively lowered both the overall DNA synthesizing activity and its sensitivity to NAL. The residual activity remaining after a 20-min toluene treatment was completely unaffected by the drug, the NAL site presumably having been inactivated by the solvent. These authors conclude that sensitivity to NAL is an important criterion in judging the extent to which an *in vitro* DNA synthesizing system has been altered from the original physiological state.

On this basis the concentrated lysate system described by SCHALLER *et al.* (1972) closely resembles the *in vivo* condition and holds promise for further detailed study of DNA replication and the action of NAL at the molecular level. Minimal alteration and disorganization of the replication system was achieved by osmotic lysis of muramidase-treated cells of a *pol* A1 mutant (also defective in endonuclease I) spread on a cellophane membrane. DNA synthesis by this system was semi-conservative, as shown by bromouracil density-transfer data; was dependent on ATP, all 4 deoxyribonucleoside triphosphates and KCl; and involved formation and joining of Okazaki pieces. The *in vitro* system was sensitive to NAL and other appropriate inhibitors; with 100 µg NAL/ml, the rate of DNA synthesis was reduced by >80%. Synthetic activity of the lysates was slower than the *in vivo* rate, but reflected the number of growing points (forks) present in the DNA at the time of lysis. As previously shown (BOYLE *et al.*, 1967) NAL-treated cells have enhanced rates of *in vivo* synthesis after removal of the drug, presumably because of premature initiation which yields extra forks. The lysates prepared from such NAL-treated cells also exhibited increased rates of *in vitro* synthesis as expected. It is interesting that lysates prepared from NAL-resistant mutants (presumably *nal*A) show DNA synthesis which is not inhibited even by high levels of NAL. This suggests the possibility that the product of the *nal*A gene could be an unrecognized participant in normal replication whose function is blocked by NAL.

One can only speculate on the real function of the NAL target, regardless of whether or not it is actually the *nal*A gene product. It has been suggested (H. REITER, personal communication) that NAL action might be concerned with DNA-membrane attachment. This is an attractive possibility, since J. BOYLE and W. GOSS (unpublished) have been able to demonstrate *in vitro* binding of ^{14}C-NAL to the DNA-membrane complex prepared from *E. coli* by Brij 58 treatment (GODSON and SINSHEIMER, 1967). Disaggregation of the complex with deoxycholate abolished its NAL-binding capacity.

Evidence for NAL effects on the binding of DNA to the membrane is still preliminary and contradictory. G. BOURGUIGNON and R. STERNGLANZ (personal communication) could find no indication that NAL interferes with DNA-membrane attachment. R. FENWICK and R. CURTISS (personal communication) suggest, on the basis of their work with conjugal transfer of an R-factor (R64-11) from a ts$_{DNA}$ *E. coli* donor to DNA-less minicells, that NAL may block replication by causing a (reversible) binding of DNA to the membrane. When NAL (10 µg/ml) is added to the mating cells (at 42 C to stop chromosome replication in the donor), plasmid replication and transfer is greatly decreased, as might have been expected, but also the amount of plasmid DNA bound to membranes of the donor cells is dramatically increased.

These studies of Pedrini *et al.* (1972) and Schaller *et al.* (1972), while not precisely defining the action of NAL at the molecular level, have clearly indicated the relevant area of search. This unique drug, functioning as an inhibitor of bacterial chromosome replication only under truly physiological conditions, now affords a critical test of the validity of *in vitro* systems which should permit new insight into the replication process.

Secondary Consequences of NAL Action

Several secondary effects flow from the primary event of blockage of chromosome replication in NAL-inhibited bacteria. These include prevention of cell division (i.e. septation), degradation of DNA and alteration of membrane components while NAL is still present, and synchronous premature reinitiation of chromosome replication after NAL is removed. NAL treatment induces prophage in lysogenic bacteria and is possibly mutagenic under some conditions.

Boyle *et al.* (1967) found that exposure to NAL in a nutritionally complete medium permitted a greater than normal synthesis of DNA when cells of *E. coli* 15 TAU subsequently were transferred to a nutritionally deficient medium which restricted RNA and protein synthesis. Transfer of control cells from complete medium (+T, +AU) to the arginine- and uracil-deficient (+T, −AU) medium resulted in a 40–50% increase in DNA content, indicating completion of only those chromosomes being replicated at the time of transfer. Prior treatment with NAL resulted in much greater increases, the extent of which was dependent on the duration of the NAL treatment (Fig. 7). These results, which are similar to the effects of thymine starvation (Pritchard and Lark, 1964), indicate that NAL treatment alters the normal DNA replication pattern and that DNA synthesis after removal of NAL proceeds from additional sites on the chromosome.

The "excessive" DNA synthesis in the absence of protein synthesis has been observed by others (Pritchard *et al.*, 1969; Kogoma and Lark,1970). Pritchard *et al.* (1969) obtained data indicating that the new replication sites operative after NAL treatment were at the origin rather than elsewhere on the chromosome, and that the "extra" DNA synthesis represented premature initiation from the origin. This was confirmed by Burger (1971) using density transfer techniques and by Ward *et al.* (1970) using an independent and ingenious method based on preferential mutagenesis by N-methyl-N′-nitro-N-nitrosoguanidine (NTG) of those genes being replicated at the time of NTG treatment. Application of this experimental system to synchronous cultures (Ward and Glaser, 1970) showed that the ability to re-initiate after treatment and removal of NAL was a function of cell "age" rather than duration of NAL treatment *per se*, this capacity being attained at the cell age when a new replication cycle would have been initiated had DNA synthesis not been inhibited. Further, it appeared that DNA replication does not continue at the old growing forks when such premature initiation occurs.

Another indication of the synchronizing effect of NAL treatments has been obtained by J. Boyle and W. Goss (unpublished). As shown in Fig. 8., when *E. coli* 15 TAU was treated 40 min with NAL in complete medium (+T, +AU,

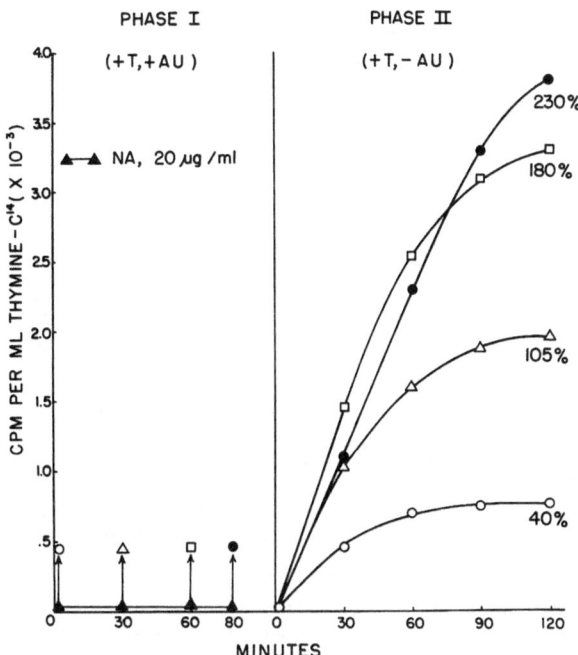

Fig. 7. Dependence of the rate and extent of excessive DNA synthesis in *Escherichia coli* 15 TAU on the duration of exposure to nalidixic acid (NA). An exponentially growing culture of *E. coli* 15 TAU, previously labeled with ^{14}C-thymine (0.01 μc/ml) for several generations was exposed to nalidixic acid for 80 min (Phase I). At 0 (o), 30 (△), 60 (□), and 80 (●) min, 20-ml portions were removed, filtered, washed, and transferred to drug-free (+T, −AU) medium containing ^{14}C-thymine at the same specific activity (Phase II). Samples were removed to ice-cold, 10% trichloroacetic acid for determination of radioactivity in the acid-insolubles. The arrows indicate the time of transfer to Phase II. (From BOYLE *et al.*, 1967)

Phase I) and transferred to drug-free deficient (+T, −AU, Phase II) medium, the expected "extra" synthesis of DNA occurred (200% increase compared with 45% in control cells). When arginine and uracil then were added back (Phase III), DNA synthesis lagged in cells previously treated with NAL. After a delay of almost 60 minutes, two successive, largely synchronous cell divisions took place in the NAL cells, but not in the controls.

Cell division generally is found to be coordinated with completion of chromosome replication, and treatments, selectively interrupting DNA synthesis, also halt septation. This leads to the previously mentioned formation of elongated cells in NAL-treated cultures (Goss *et al.*, 1964). DONACHIE (1969) has found that, while DNA synthesis resumed immediately after removal of NAL or re-addition of thymine to thymine-starved cells, there was a delay in cell division. This delay increased with the duration of DNA inhibition, but did not exceed one generation time. The first division was partially synchronized, but thereafter division was asynchronous. These data were interpreted as resulting from a requirement for completion of chromosome replication before septum formation,

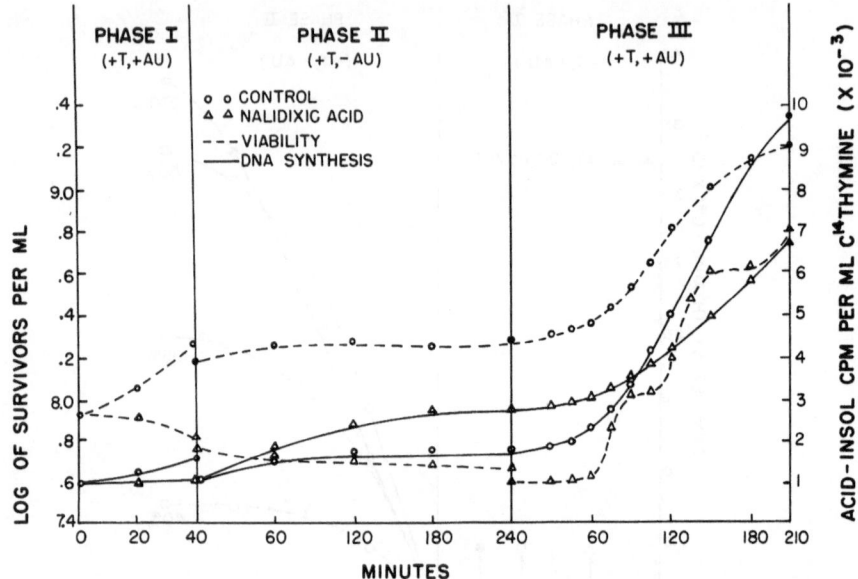

Fig. 8. Relationship between cell division and DNA synthesis during Phase III after treatment of
E. coli 15 TAU with 20 µg/ml nalidixic acid (NAL) (J. V. Boyle and W. A. Goss, unpublished)

coupled with premature initiation of new rounds of synthesis from the replication origin when the inhibition is released.

It has been possible to interfere with this normal linkage of division and DNA replication. Filaments or "snakes" are produced when DNA synthesis is stopped by shifting of temperature-sensitive mutants to the non-permissive temperature. In certain of these mutants, septation is not completely prevented, and this leads to production of "cells" lacking DNA (Hirota *et al.*, 1968; Inouye, 1969). Inouye (1969) makes a good case for the occurrence of division without completion of the chromosome. DNA synthesis ceased immediately upon shifting to 41 C, but cell numbers increased 2.5 fold. Ability to divide at 41 C was not limited to cells in any particular stage of the division cycle. Thymine starvation prevented division at 30 C, but the cells would divide at 41 C even if starved for thymine. In contrast, no division occurred when DNA synthesis was halted with NAL at either temperature. Thus, in this mutant, halting of DNA synthesis *per se* is insufficient to prevent septation at 41 C, and this led Inouye (1969) to suggest that NAL acts not only to inhibit DNA synthesis, but also in some (unspecified) manner to block septum formation.

This again focuses attention on the cell membrane, and there is some evidence pointing to alterations of the membrane in NAL-treated cells. Normark and Westling (1971) found that *envA* mutants of *E. coli* take up more gentian violet (and several other inhibitors) and are more susceptible to muramidase lysis without EDTA than the wild type. In *envA* mutants inhibition of DNA synthesis by NAL caused a marked decrease in uptake of gentian violet and an increased resistance to muramidase. INOUYE and PARDEE (1970) have found changes in at least 2 membrane proteins after inhibition of DNA synthesis by NAL. One protein (peak Y) was significantly diminished in the NAL-treated culture and another (peak X) was increased over the control.

Changes in DNA-associated proteins have been observed after inhibition of DNA synthesis by NAL. MASKER and EBERLE (1971) used DNA-cellulose chromatography to examine alterations in proteins which bind to DNA in *E. coli* 15T⁻. They found one or two of the DNA-binding proteins of the soluble fraction (but not the highspeed pellet) were greatly diminished after a 90-minute (but not 20-minute) treatment with NAL. The function(s) of these DNA-binding proteins is not known but possibly they might be concerned with regulation of DNA synthesis or with stabilization of the chromosome against nuclease attack.

COOK *et al.* (1966b) found that during incubation of *E. coli* 15 TAU with NAL in nutritionally complete medium, a significant portion of the DNA was converted to acid-soluble fragments. The extent of this solubilization was correlated with the concentration of NAL and time of exposure, and was antagonized by conditions restricting RNA and protein synthesis (nutritional deficiency; chloramphenicol; 2,4-dinitro-phenol). DNA degradation also was observed during NAL treatment of *B. subtilis* ATCC 6051 (COOK *et al.*, 1966a). RAMAREDDY and REITER (1969) have shown that in *B. subtilis* 168, this degradation begins at or near the replicating fork and continues sequentially along both strands (i.e. "daughter" and template) from the most recently synthesized to the "older", previously replicated portions of the chromosome. Comparison of the kinetics of DNA degradation in cells with uniformly labeled DNA and in cells receiving a 3-minute pulse of ³H-thymine immediately before NAL showed a relatively greater fraction of radioactivity to be solubilized in the pulse-labeled cells. In contrast, a similar study by BEHKI and LESLEY (1972) showed no such preferential degradation of newly replicated DNA in *E. coli* 15 T⁻. In this culture, DNA degradation in the presence of NAL apparently was not initiated at the replication fork and did not proceed sequentially from newly replicated to older regions, but was random.

It is clear that DNA degradation is influenced by the particular genotype of the organism employed. GRIGG (1970) has shown that stability of DNA during NAL treatment differs in *E. coli* strain B and certain of its dark-repair mutants. In general, these ultraviolet-sensitive mutants show greater DNA degradation than strain B when treated with NAL. This excessive DNA degradation, particularly with strain Bsl, is not easily reconciled with the supposedly greater resistance of such mutants to the killing effects of NAL (DONCH *et al.*, 1968; GREEN *et al.*, 1965).

One factor complicating all of this is the uncertain contribution of known or unknown prophages or other episomes possibly induced by the treatments employed. NAL treatment has been shown to induce prophage to enter the lytic cycle in lysogenic bacteria (TAKETO and WATANABE, 1967; GEISSLER, 1967; GERMAN *et al.*, 1969). COWLISHAW and GINOZA (1970) showed that induction of λ in *E. coli* K-12 was related to NAL concentration and duration of exposure, and occurred only when the cultures were actively growing at the time of NAL treatment. Induction was reduced or prevented by use of non-growing stationary phase cells, amino acid starvation or addition of chloramphenicol. At high levels, NAL interferes with λ replication and reduces burst size, thereby reducing the apparent inducing effect.

Cook et al. (1966c) found that with growing cultures of a streptomycin-dependent mutant of E. coli (ATCC 11143, Sd-4), addition of NAL resulted in a time- and concentration-related increase in streptomycin-independent revertants. Both the absolute numbers per unit volume and the relative frequency among survivors increased, and this was considered indicative of a mutagenic effect of the NAL treatment. As this effect was observed only under growth conditions, one might argue that, in the absence of demonstrated NAL-sensitivity of the revertants, this could be merely a selective effect of the drug permitting outgrowth of spontaneous revertants. Until these preliminary observations are confirmed and extended to other systems, no final conclusions regarding possible mutagenicity of NAL can be drawn.

Action of NAL on Eukaryotic Microorganisms

With certain eukaryotic microorganisms, NAL has been shown to produce effects which differ somewhat from those occurring in bacteria. There is a greater effect on "cytoplasmic" organelles, such as mitochondria and chloroplasts, indicating that DNA replication in such organelles may differ from that of the nucleus, and resemble that of prokaryotes in susceptibility to inhibitors, if not also in other ways.

In Saccharomyces cerevisiae, NAL treatments give a partial and sometimes transient inhibition of growth and DNA synthesis (Wehr et al., 1970) but produce much stronger effects on mitochondria. Mahler et al. (1968) found that NAL prevented derepression of respiratory functions, presumably by mitochondrial inhibition of DNA synthesis. Luha et al. (1971) found that lysates of NAL-treated cultures of Kluyveromyces (Saccharomyces) lactis were lacking in mitochondrial DNA. Another indication of an effect on mitochondrial DNA synthesis is the observation that NAL blocked induction of petites by ethidium bromide (Whittaker et al., 1972; Hollenberg and Borst, 1971; Wallis et al., 1972).

Conklin and Chou (1972) reported that in the protozoan, Tetrahymena pyriformis, 8.5×10^{-4} M NAL (200 µg/ml) produced a 52% inhibition of ^{14}C-thymidine incorporation into acid-insoluble material, but no data was presented relating to differential effects on mitochondria or loss of viability, nor was it determined whether the inhibition of DNA synthesis by NAL was selective or accompanied by blockage of RNA and protein synthesis.

NAL produces a selective inhibition of chloroplast replication in Euglena gracilis var. bacillaris strain Z (Lyman, 1967), and can be used to obtain permanently "bleached" strains (Schiff et al., 1971). No loss of viability or change of growth rate occurred when NAL (50 µg/ml) was added to cultures growing with or without illumination, but large numbers of white colonies were observed upon plating of the cells grown with light plus NAL. Proplastid replication in the dark apparently was not affected by NAL as readily as chloroplast replication. Cells grown in the dark with NAL for 60 hours still gave about 85% green colonies, whereas light-grown cultures plus NAL formed 95% white colonies by 24 hours. Also, NAL did not halt the development of chloroplasts from proplastids which takes place when dark-grown cells are illuminated under non-growth conditions. The bleaching action of NAL was confirmed by EBRINGER (1970, 1972) and shown to be correlated with progressively abnormal appearance of plastids and decrease in chlorophyll content of cells, culminating in the disappearance of plastids. By way of contrast, two prokaryotic blue-green algae, *Anacystis nidulans* and *Anabaena doliolum* have been shown (MADAN and KUMAR, 1971) to grow, after a lag period, in the presence of very high concentrations of NAL (5 mg/ml) with no killing or bleaching effects.

Effect of NAL on Mammalian Tissue Cultures

ROSENKRANZ and LAMBEK (1965) reported that the DNA isolated from human diploid cells grown in tissue culture in the presence 50 μg/ml of NAL was structurally modified. They suggested, based on analytical CsCl density gradient centrifugation, that the altered DNA was cross-linked and possessed properties resembling those of nucleic acids isolated from cells treated with alkylating agents.

In contrast to these findings, BERENBAUM *et al.* (1965) found that NAL at 1 000 μg/ml did not significantly affect *in vitro* the uptake of thymidine and uracil by mouse thymocytes. STENCHEVER *et al.* (1970) studied the effect of NAL on human leukocytes and concluded that there was neither evidence of chromosome damage nor morphological abnormalities at concentrations as high as 100 μg/ml.

J. BOYLE, F. PANCIC and W. GOSS (unpublished) studied the effect of nalidixic acid on mammalian cells cultured *in vitro*. Treatment of Hela cells, monkey heart cells, human amnion cells and primary monkey kidney cells with NAL resulted in a concentration-dependent, partial inhibition of DNA, RNA, and protein synthesis. In contrast to the effects in *E. coli*, no selective inhibition of DNA synthesis was observed in any of the cell lines studied. The partial inhibition of DNA, RNA, and protein synthesis was accompanied by a partial inhibition of cell division. Among those cell lines studied, primary monkey kidney cells were most sensitive to the drug. After 48 hours exposure to 100 μg/ml of NAL, DNA, RNA, and protein synthesis were inhibited 67%, 70%, and 57%, respectively, whereas cell division was inhibited 57%. After 24 hours exposure, this inhibition was rapidly reversed when the medium was removed and replaced with drug-free medium. Analytical CsCl density gradient centrifugation demonstrated that the DNA obtained from monkey heart cells after treatment with NAL was not cross-linked.

It is clear from the foregoing discussion that these results are incompatible with those reported by ROSENKRANZ and LAMBEK (1965). Although NAL inhibits DNA replication by bacteria, bacteriophages, and certain TC cell lines (non-selectively), it did not alkylate DNA in any of these systems. If NAL caused direct chromosomal changes, one would certainly expect to see characteristic toxicities in man and animals, which is not the case.

Conclusions

The one fact of overriding importance emerging from the many studies described is that NAL selectively blocks DNA replication in susceptible bacteria in a manner as yet undefined. This primary event may result in a number of sequellae, the nature of which depends on the particular genotype of the bacterium and environmental conditions imposed. Caution must be urged against indiscriminate use of NAL as a specific inhibitor of DNA synthesis in poorly defined systems. Proper interpretation of results demands that the investigator take into account these secondary processes, which may include unbalanced growth, DNA degradation, prophage induction, and alterations of control mechanisms for DNA synthesis.

Recent studies on the interaction of NAL and the DNA-membrane complex suggest that discovery of the elusive NAL target may be imminent. The identification of this target undoubtedly will reveal a new component involved in bacterial chromosome replication.

Acknowledgments. The authors gratefully acknowledge the assistance of the Library Staff and, particularly, Mrs. A. Weis for conducting the literature search, Dr. J. Boyle for reviewing the manuscript and making valuable suggestions, and the many investigators who contributed unpublished data, all of which helped the authors to present an as up-to-date review as possible. The authors also thank Miss Mary Frazier for her secretarial assistance.

References

Baird, J.P., G.J. Bourguignon, and R. Sternglanz: Effect of nalidixic acid on the growth of deoxyribonucleic acid bacteriophages. J. Virol. **9**, 17 (1972).

Barbour, S.D.: Effect of nalidixic acid on conjugational transfer and expression of episomal *Lac* genes in *Escherichia coli* K12. J. Mol. Biol. **28**, 373 (1967).

Barlow, A.M.: Nalidixic acid in infections of urinary tract. Brit. Med. J. No. 5368, 1308 (1963)

Bauernfeind, A., and G. Grümmer: Biochemical effects of nalidixic acid on *Escherichia coli*. Chemotherapia **10**, 95 (1965/66).

Behki, R.M., and S.M. Lesley: Deoxyribonucleic acid degradation and the lethal effect by myxin in *Escherichia coli*. J. Bacteriol. **108**, 250 (1972).

Berenbaum, M.C., C.A. Brown, and W.A. Cope: Failure of nalidixic acid to inhibit nucleic acid synthesis in mouse thymocytes and to suppress immune responses. Brit. Empire Commonwealth Congress, Annual Report **2**, 277 (1965).

Bouck, N., and E. Adelberg: Mechanism of action of nalidixic acid on conjugating bacteria. J. Bacteriol. **102**, 688 (1970).

Boyle, J.V., T.M. Cook, and W.A. Goss: Mechanism of action of nalidixic acid on *Escherichia coli*. VI. Cell-free studies. J. Bacteriol. **97**, 230 (1969).

Boyle, J.V., W.A. Goss, and T.M. Cook: Induction of excessive deoxyribonucleic acid synthesis in *Escherichia coli* by nalidixic acid. J. Bacteriol. **94**, 1664 (1967).

Boyle, J.V. and M.E. Jones: Effects of ribonucleosides on thymidine incorporation: selective reversal of the inhibition of deoxyribonucleic acid synthesis in thymineless auxotrophs of *Escherichia coli*. J. Bacteriol. **104**, 264 (1970).

Braunstein, S.N., A. Datta, and R.M. Franklin: Structure and synthesis of a lipid-containing bacteriophage. VIII. Effect of nalidixic acid on some membrane-associated activities of control and infected cells. Virology **46**, 161 (1971).

Buchbinder, M., J.C. Webb, L. Anderson, and W.R. McCabe: Laboratory studies and clinical pharmacology of nalidixic acid (Win 18,320). Antimicrobial Agents Chemotherapy **1962**, 308 (1963).

Burger, R.M.: Toluene-treated *Escherichia coli* replicate only that DNA which was about to be replicated *in vivo*. Proc. Natl. Acad. Sci. U.S. **68**, 2124 (1971).

CAMPBELL, J.L., L. SOLL, and C.C. RICHARDSON: Isolation and partial characterization of a mutant of *Escherichia coli* deficient in DNA polymerase II. Proc. Natl. Acad. Sci. U.S. **69**, 2090 (1972).

ČAPEK, A., A. ŠIMEK, E.SVÁTEK, and M. BUDĚŠÍNSKÝ: Antibacterially active substances. IV. Microbial hydroxylation of nalidixic acid. Folia Microbiol. (Prague) **14**, 557 (1969).

CLARK, D.J.: The regulation of DNA replication and cell division in *E. coli* B/r. Replication of DNA in Micro-organisms. Cold Spring Harbor Symp. Quant. Biol. **33**, 823–838 (1968).

COHEN, A., W.D. FISHER, R. CURTISS III, and H.I. ADLER: The properties of DNA transferred to minicells during conjugation. Replication of DNA in Micro-organisms. Cold Spring Harbor Symp. Quant. Biol. **33**, 635–641 (1968).

CONKLIN, K.A., and S.C. CHOU: The effects of antimalarial drugs on uptake and incorporation of macromolecular precursors by *Tetrahymena pyriformis*. J. Pharmacol. Exptl. Therap. **180**, 158 (1972).

COOK, T.M., K.G. BROWN, J.V. BOYLE, and W.A. GOSS: Bactericidal action of nalidixic acid on *Bacillus subtilis*. J. Bacteriol. **92**, 1510 (1966a).

COOK, T.M., W.H. DEITZ, and W.A. GOSS: Mechanism of action of nalidixic acid on *Escherichia coli*. IV. Effects on the stability of cellular constituents. J. Bacteriol. **91**, 774 (1966b).

COOK, T.M., W.A. GOSS, and W.H. DEITZ: Mechanism of action of nalidixic acid on *Escherichia coli*. V. Possible mutagenic effect. J. Bacteriol. **91**, 780 (1966c).

COWLISHAW, J., and W. GINOZA: Induction of λ prophage by nalidixic acid. Virology **41**, 244 (1970).

CUMMINGS, D.J., and A.R. KUSY: Thymineless death in *Escherichia coli*: inactivation and recovery. J. Bacteriol. **99**, 558 (1969).

CUMMINGS, D.J., and A.R. KUSY: Thymineless death in *Escherichia coli*: deoxyribonucleic acid replication and the immune state. J. Bacteriol. **102**, 106 (1970).

CURTISS, III, R.: Bacterial conjugation. Ann. Rev. Microbiol. **23**, 69 (1969).

DATTA, N.: R factors in *Escherichia coli*. Ann. N.Y. Acad. Sci. **182**, 59 (1971).

DATTA, A., S. BRAUNSTEIN, and R.M. FRANKLIN: Structure and synthesis of a lipid-containing bacteriophage. VI. The spectrum of cytoplasmic and membrane-associated proteins in *Pseudomonas* BAL 31 during replication of bacteriophage PM2. Virology **43**, 696 (1971).

DEITZ, W.H., J.H. BAILEY, and E.J. FROELICH: In vitro antibacterial properties of nalidixic acid, a new drug active against gram-negative organisms. Antimicrobial Agents Chemotherapy **1963**, 583 (1964).

DEITZ, W.H., T.M. COOK, and W.A. GOSS: Mechanism of action of nalidixic acid on *Escherichia coli*. III. Conditions required for lethality. J. Bacteriol. **91**, 768 (1966).

DONACHIE, W.D.: Control of cell division in *Escherichia coli*: Experiments with thymine starvation. J. Bacteriol. **100**, 260 (1969).

DONCH, J., M.H.L. GREEN, and J. GREENBERG: Interaction of the *exr* and *lon* genes in *Escherichia coli*. J. Bacteriol. **96**, 1704 (1968).

DRIEDGER, A.A., and M.J. GRAYSTON: The effects of nalidixic acid on X-ray-induced DNA degradation and repair in *Micrococcus radiodurans*. Can. J. Microbiol. **17**, 501 (1971).

EBERLE, H., and W. MASKER: Effect of nalidixic acid on semiconservative replication and repair synthesis after ultraviolet irradiation in *Escherichia coli*. J. Bacteriol. **105**, 908 (1971).

EBRINGER, L.: The action of nalidixic acid on Euglena plastids. J. Gen. Microbiol. **61**, 141 (1970).

EBRINGER, L.: Are plastids derived from prokaryotic micro-organisms? Action of antibiotics on chloroplasts of *Euglena gracilis*. J. Gen. Microbiol. **71**, 35 (1972).

FISHER, K.W., and M.B. FISHER: Nalidixic acid inhibition of DNA transfer in *Escherichia coli* K12. Replication of DNA in Micro-organisms. Cold Spring Harbor Symp. Quant. Biol. **33**, 629–633 (1968).

FROELICH, E.J., and W.H. DEITZ: Antibacterial properties of nalidixic acid, a new drug effective against gram-negative microorganisms. IIIrd International Congress of Chemotherapy, International Society of Chemotherapy Proceedings, presented 1963.

GAGE, L.P., and D.J. FUJITA: Effect of nalidixic acid on deoxyribonucleic acid synthesis in bacteriophage SP01-infected *Bacillus subtilis*. J. Bacteriol. **98**, 96 (1969).

GEISSLER, E.: Untersuchungen über den Mechanismus der Induktion lysogener Bakterien. XI. Der Einfluß von DNS-Synthesehemmern auf lysogene Bakterien. Biol. Zentr. **86**, Suppl., 55 (1967).

GERMAN, A., J. PANOUSE-PERRIN, and A.C. ARDOUIN: A study of the mutagenic action of nalidixic acid on the twort phage of staphylococcus. Compt. Rend. **268**, 1821 (1969).

Godson, G. N., and R. L. Sinsheimer: Lysis of *Escherichia coli* with a neutral detergent. Biochim. Biophys. Acta **149**, 476 (1967).

Goss, W. A.: Nalidixic acid: Biological profile. Urologia Panamericana **1**, 103 (1969).

Goss, W. A., W. H. Deitz, and T. M. Cook: Mechanism of action of nalidixic acid on *Escherichia coli*. J. Bacteriol. **88**, 1112 (1964).

Goss, W. A., W. H. Deitz, and T. M. Cook: Mechanism of action of nalidixic acid on *Escherichia coli*. II. Inhibition of deoxyribonucleic acid synthesis. J. Bacteriol. **89**, 1068 (1965).

Goulian, M.: Biosynthesis of DNA. In: Annual review of biochemistry (ed. E. E. Snell, P. D. Boyer, A. Meister, and R. L. Sinsheimer), p. 855–898. Palo Alto: Annual Reviews Inc. 1971.

Green, M. H. L., J. Donch, Y. S. Chung, and J. Greenberg: Effect of inhibition of DNA synthesis on u. v. sensitive Bs strains of *Escherichia coli*. Genet. Res. **14**, 111 (1969).

Green, M. H. L., J. Donch, and J. Greenberg: Effect of inhibitors of DNA synthesis on uv-sensitive derivatives of *Escherichia coli* Strain K-12. Mutation Res. **9**, 149 (1970).

Grigg, G. W.: Stability of DNA in dark-repair mutants of *E. coli* B treated with nalidixic acid. J. Gen. Microbiol. **61**, 21 (1970).

Hamilton, P. B., D. Rosi, G. P. Peruzzotti, and E. D. Nielson: Microbiological metabolism of naphthyridines. Appl. Microbiol. **17**, 237 (1969).

Hane, M. W.: Some effects of nalidixic acid on conjugation in *E. coli* K-12. J. Bacteriol. **105**, 46 (1971).

Hane, M. W., and T. H. Wood: *Escherichia coli* K-12 mutants resistant to nalidixic acid: genetic mapping and dominance studies. J. Bacteriol. **99**, 238 (1969).

Harrison, L. H., and C. E. Cox: Bacteriologic and pharmacodynamic aspects of nalidixic acid. J. Urol. **104**, 908 (1970).

Helling, R. B., and B. S. Adams: Nalidixic acid-resistant auxotrophs of *E. coli*. J. Bacteriol. **104**, 1027 (1970).

Helling, R. B., and J. S. Kukora: Nalidixic acid-resistant mutants of *E. coli* deficient in isocitrate dehydrogenase. J. Bacteriol. **105**, 1224 (1971).

Hirota, Y., F. Jacob, A. Ryter, G. Buttin, and T. Nakai: On the process of cellular division in *Escherichia coli*. I. Asymmetrical cell division and production of deoxyribonucleic acid-less bacteria. J. Mol. Biol. **35**, 175 (1968).

Hollenberg, C. P., and P. Borst: Conditions that prevent $\bar{p}$ induction by ethidium bromide. Biochem. Biophys. Res. Commun. **45**, 1250 (1971).

Hollom, S., and R. H. Pritchard: Effect of inhibition of DNA synthesis on mating in *Escherichia coli* K12. Genet. Res. **6**, 479 (1965).

Inouye, M.: Unlinking of cell division from deoxyribonucleic acid replication in a temperature-sensitive deoxyribonucleic acid synthesis mutant of *Escherichia coli*. J. Bacteriol. **99**, 842 (1969).

Inouye, M., and A. B. Pardee: Changes of membrane proteins and their relation to deoxyribonucleic acid synthesis and cell division of *Escherichia coli*. J. Biol. Chem. **245**, 5813 (1970).

Kantor, G. J., and R. A. Deering: Effect of nalidixic acid and hydroxyurea on division ability of *Escherichia coli fil*$^+$ and *lon*$^-$ strains. J. Bacteriol. **95**, 520 (1968).

Kogoma, T., and K. G. Lark: DNA replication in *Escherichia coli*: Replication in absence of protein synthesis after replication inhibition. J. Mol. Biol. **52**, 143 (1970).

Lesher, G. Y.: Nalidixic acid, 1-ethyl-1,4-dihydro-7-methyl-4-oxo-1,8-naphthyridine-3-carboxylic acid. A new antibacterial agent. IIIrd International Congress of Chemotherapy. International Society of Chemotherapy Proceedings. Presented 1963.

Lesher, G. Y., E. J. Froelich, M. D. Gruett, J. H. Bailey, and R. P. Brundage: 1,8-naphthyridine derivatives. A new class of chemotherapeutic agents. J. Med. Pharm. Chem. **5**, 1063 (1962).

Luha, A. A., L. E. Sarcoe, and P. A. Whittaker: Biosynthesis of yeast mitochondria. Drug effects on the petite negative yeast *Kluyveromyces lactis*. Biochem. Biophys. Res. Commun. **44**, 396 (1971).

Lyman, H.: Specific inhibition of chloroplast replication in *Euglena gracilis* by nalidixic acid. J. Cell Biol. **35**, 726 (1967).

Madan, V., and H. D. Kumar: Action of nalidixic acid and hydroxyurea on two blue-green algae. Zeitschr. Allg. Mikrobiol. **11**, 495 (1971).

Mahler, H. R., P. Perlman, C. Henson, and C. Weber: Selective effects of chloramphenicol, cycloheximide and nalidixic acid on the biosynthesis of respiratory enzymes in yeast. Biochem. Biophys. Res. Commun. **31**, 474 (1968).

MASKER, W.E., and H. EBERLE: DNA replication and proteins that bind to DNA. Proc. Natl. Acad. Sci. U.S. **68**, 2549 (1971).

McCHESNEY, E.W., E.J. FROELICH, G.Y. LESHER, A.V.R. CRAIN, and D. ROSI: Absorption, excretion, and metabolism of a new antibacterial agent, nalidixic acid. Toxicol. Appl. Pharmacol. **6**, 292 (1964).

MITSUHASHI, S.: Epidemiology and genetics of R factors. The Problems of Drug-Resistant Pathogenic Bacteria, Ann. N.Y. Acad. Sci. **182**, 141 (1971).

MITSUHASHI, S., H. HASHIMOTO, R. EGAWA, T. TAKAKA, and Y. NAGAI: Drug resistance of enteric bacteria. IX. Distribution of R-factors in gram-negative bacteria from clinical sources. J. Bacteriol. **93**, 1242 (1967).

NORMARK, S., and B. WESTLING: Nature of the penetration barrrier in *Escherichia coli* K-12: Effect of macromolecular inhibition on penetrability in strains containing the *env*A gene. J. Bacteriol. **108**, 45 (1971).

O'CONNOR, J.R., R.A. DOBSON, and W.A. GOSS: Profile and treatment with nalidixic acid of induced *Escherichia coli* pyelonephritis in mice. Xth Internat. Cong. for Microbiol., Mexico D.F., VIII, p. 101 (1970).

OKAZAKI, R., K. SUGIMOTO, T. OKAZAKI, Y. IMAE, and A. SUGINO: DNA chain growth: *in vivo* and *in vitro* synthesis in a DNA polymerase negative mutant of *E. coli*. Nature **228**, 223 (1970).

PEDRINI, A.M., D. GEROLDI, A. SICCARDI, and A. FALASCHI: Studies on the mode of action of nalidixic acid. Eur. J. Biochem. **25**, 359 (1972).

PORTMANN, G.A., E.W. McCHESNEY, H. STANDER, and W.E. MOORE: Pharmacokinetic model for nalidixic acid in man. II. Parameters for absorption, metabolism, and elimination. J. Pharm. Sci **55**, 72 (1966).

PRITCHARD, R.H., P.T. BARTH, and J. COLLINS: Control of DNA synthesis in bacteria. In: Microbial growth. Nineteenth Symposium of the Society for General Microbiology, p. 263.   : Cambridge University Press 1969.

PRITCHARD, R.H., and K.G. LARK: Induction of replication by thymine starvation at the chromosome origin. J. Mol. Biol. **9**, 288 (1964).

RAMAREDDY, G., and H. REITER: Specific loss of newly replicated deoxyribonucleic acid in nalidixic acid-treated *Bacillus subtilis* 168. J. Bacteriol. **100**, 724 (1969).

RONALD, A.R., M. TURCK, and R.G. PETERSDORF: A critical evaluation of nalidixic acid in urinary-tract infections. New Engl. J. Med. **275**, 1081 (1966).

ROSENKRANZ, H.S., and C. LAMBEK: *In vivo* effect of nalidixic acid (NegGram) on the DNA of human diploid cells in tissue culture. Proc. Soc. Exptl. Biol. Med. **120**, 549 (1965).

SCHALLER, H., B. OTTO, V. NÜSSLEIN, J. HUF, R. HERRMANN, and F. BONHOEFFER: Deoxyribonucleic acid replication *in vitro*. J. Mol. Biol. **63**, 183 (1972).

SCHIFF, J.A., H. LYMAN, and G.K. RUSSELL: Chapter 12: Isolation of mutants from *Euglena gracilis*. In: Methods in enzymology, eds. COLOWICK, S.P. and N.O. KAPLAN, vol. XXIII, Photosynthesis, Part A. ed. SAN PIETRO, A., p. 155. New York: Academic Press 1971.

STAMEY, T.A.: Observations on the clinical use of nalidixic acid. Postgrad. Med. **47**, Suppl., 21 (1971).

STAMEY, T.A., N.J. NEMOY, and M. HIGGINS: The clinical use of nalidixic acid. A review and some observations. Invest. Urol. **6**, 582 (1969).

STENCHEVER, M.A., W. POWELL, and J.A. JARVIS: Effect of nalidixic acid on human chromosome integrity. Am. J. Obstet. Gynecol. **107**, 329 (1970).

TAKETO, A., and H. WATANABE: Effect of nalidixic acid on the growth of bacterial viruses. J. Biochem. **61**, 520 (1967).

WALLIS, O.C., P. OTTOLENGHI, and P.A. WHITTAKER: Induction of petite mutants in yeast by starvation in glycerol. Biochem. J. **127**, 46P (1972).

WALTON, J.R., and D.H. SMITH: Hemolysin production in *E.coli* associated with nalidixic acid resistance. Antimicrobial Agents Chemotherapy *1968*, 54 (1969).

WARD, C.B., and D.A. GLASER: Control of initiation of DNA synthesis in *E. coli*. Proc. Natl. Acad. Sci. U.S. **67**, 255 (1970).

WARD, C.B., M.W. HANE, and D.A. GLASER: Synchronous reinitiation of chromosome replication in *E. coli* B-r after nalidixic acid treatment. Proc. Natl. Acad. Sci. U.S. **66**, 365 (1970).

WATANABE, M., and J.T. AUGUST: Methods for selecting RNA bacteriophage. In: Methods in virology (eds. K. MARAMOROSCH and H. KOPROWSKI), vol. III p. 337. New York and London: Academic Press 1967.

WEHR, C. T., R. D. KUDRNA, and L. W. PARKS: Effect of putative deoxyribonucleic acid inhibitors on macromolecular synthesis in *Saccharomyces cerevisiae*. J. Bacteriol. **102**, 636 (1970).

WHITTAKER, P. A., R. C. HAMMOND, and A. A. LUHA: Mechanism of mitochondrial mutation in yeast. Nature New Biol. **238**, 266 (1972).

WINSHELL, E. B., and H. S. ROSENKRANZ: Nalidixic acid and the metabolism of *Escherichia coli*. J. Bacteriol. **104**, 1168 (1970).

YOSHIKAWA, M.: Drug sensitivity and mutability to drug resistance associated with the presence of an R factor. Genet. Res. **17**, 1 (1971).

Olivomycin, Chromomycin, and Mithramycin

G. F. Gause

The complete stereochemical structure of olivomycin was published by BAK-HAEVA *et al.* in 1967, and in 1968 BERLIN *et al.*, elucidated structural relations between olivomycin, mithramycin and chromomycin, which are shown in Table 1.

Table 1. *Structural relations between olivomycin, mithramycin and chromomycin*

Antibiotic	Aglycone	Sugars				
		Olivomycose	D-Mycarose	Olivomose	Olivose	Oliose
Olivomycin	Olivin	+	−	+	+	+
Chromomycin	Chromomycinone	+	−	+	+	+
Mithramycin	Chromomycinone	−	+	−	+	+

Olivomycin and chromomycin possess the same sugars, but different (homologous) aglycones. Whereas the aglycones of chromomycin and mithramycin are identical, the latter antibiotic contains three sugars instead of four and the olivomycose present in other antibiotics is replaced in that by its diastereomer, d-mycarose (Fig. 1).

Partial cross resistance observed in Staphylococci and in tumor cells to olivomycin, mithramycin and chromomycin appears to indicate a similar mechanism of action of the three antibiotics (CHORIN and SHAPOVALOVA, 1966). *Staphylococcus aureus* growing in media containing increasing concentrations of olivomycin had their resistance enhanced by 125 times after 10 transfers and similarly cells of Sarcoma 37 became 267 times more resistant to olivomycin after 16 passages on mice treated with increasing concentrations of the antibiotic. Such resistant strains showed partial cross resistance to chromomycin and mithramycin, as shown in Table 2.

Table 2. *Partial cross resistance in staphylococci and in tumor cells to olivomycin, mithramycin and chromomycin*

	Increase of resistance (times)	
	Staphylococci	Sarcoma 37 cells
Olivomycin	125	267
Chromomycin	63	−
Mithramycin	17	25

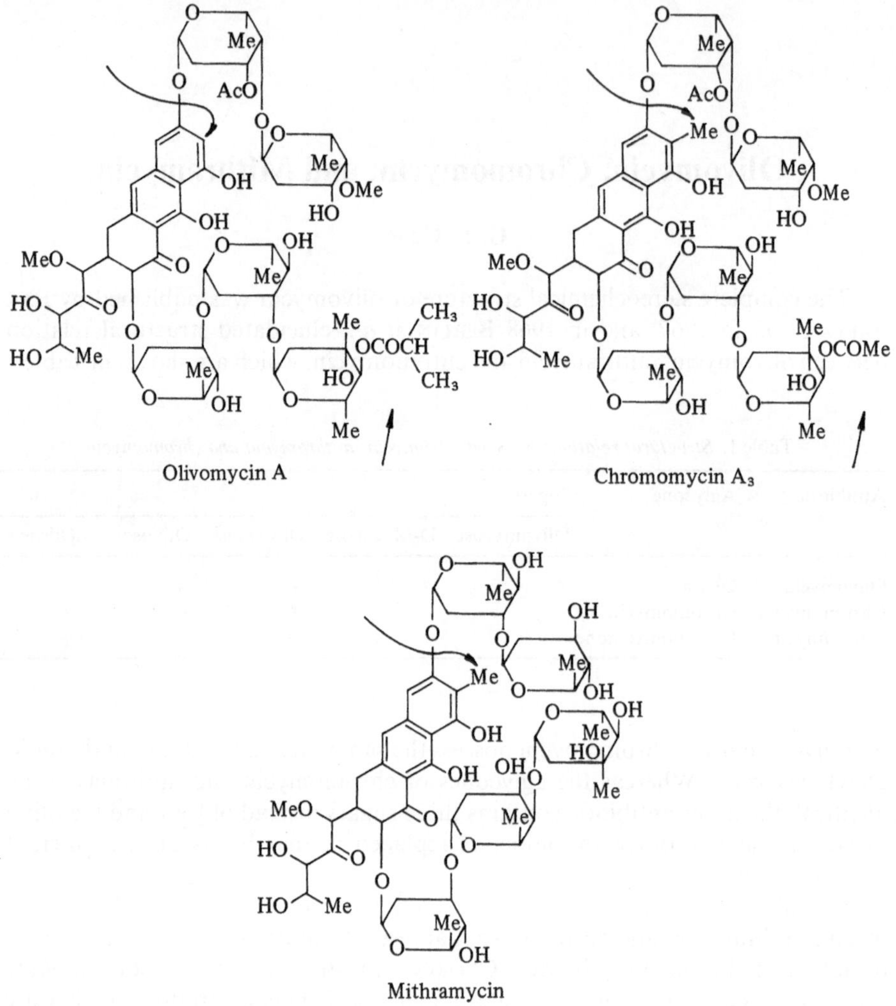

Fig. 1. Structures of olivomycin A, chromomycin A₃ and mithramycin. (From SEDOV et al., 1969)

Of particular interest is the study of the effect of mithramycin on mouse glioma (KENNEDY et al., 1968) where, as in other cases, the drug appears to exert its cytotoxic action by inhibition of RNA synthesis. Both in the mouse glioma and in mouse liver, RNA synthesis was studied at different times following exposure to mithramycin, and a marked difference in the recovery rates of the two tissues from the inhibitory effect of mithramycin on RNA synthesis was observed. Whereas the liver rapidly recovered its capacity for RNA synthesis, recovery by the tumor cells was delayed. This selective vulnerability of the nucleic acid synthesis in tumor cells, as evidenced by a relatively slow recovery rate of the latter from the deleterious effect of inhibitors of this group could be of considerable interest toward current use of these antibiotics in cancer chemotherapy.

Olivomycin is clinically effective against disseminated testicular neoplasms (ASTRAKHAN and GARIN, 1970) and in the treatment of tonsillar tumors (SMIRNOVA, 1969). Mithramycin has given favorable results in the therapy of disseminated germinal testicular cancer (KENNEDY, 1970).

Previously it had been observed that all three antibiotics preferentially inhibit RNA synthesis and that this effect was due to a complex formation between the antibiotic and DNA. CHOCHLOVA (1969) studied in greater detail the selective effect of olivomycin upon RNA biosynthesis in the liver, kidney, spleen and sarcoma of rats *in vivo*.

BEHR *et al.* (1969) investigated interaction of chromomycin with DNA and the complex formation between the antibiotic and DNA was shown to depend on the guanine content and on base pairing of the polydeoxynucleotides. A limit of association is reached when one chromomycin molecule is bound per four nucleotide base pairs. It was observed that the antibiotic had a much less inhibitory effect on the RNA polymerase reaction when heat-denatured DNA was used as the template. This suggested a substantially reduced numer of binding sites in the denatured DNA if complex formation is really responsible for the inhibitory effect. In fact denatured DNA proved to be much less effective in producing shifts in the absorption spectrum of the antibiotic and spectrophotometric titration indicated that the heat-denatured calf thymus DNA contained only 0.07 apparent binding sites per base pair compared to 0.19 of the native DNA (BEHR *et al.*, 1969). Similar findings have been made in the equilibrium dialysis experiments. Heat-denatured DNA still contains considerable portions of base paired regions. Consequently these measurements do not allow one to decide whether base pairing is an absolute requirement for complex formation. To resolve this question, BEHR *et al.* (1969) studied the binding of chromomycin to poly (dG) . poly (dC) and its single stranded componets, poly dC and poly dG. As was expected from the base specificity of the complex formation, poly dC was completely inactive in producing changes in the absorption spectrum. On the other hand poly dG fulfills the base requirement for binding. However, no spectral shift occurred on addition of this polynucleotide to a solution of chromomycin. In contrast to its single stranded components, poly (dG) . poly (dC) binds up to one molecule of chromomycin per four base pairs. These results suggest base pairing in DNA as a necessary prerequisite for complex formation with chromomycin and are also consistent with the hypothesis, that during binding the chromophor of the antibiotic intercalates between base pairs of the DNA helix as has been proposed for the binding of actinomycin to DNA. However, WARING (1971) noted that interaction of chromomycin and mithramycin with DNA differs significantly from the pattern typically seen with intercalating drugs, and the very bulky sugar substituents on their chromophors make an intercalative mode of binding seem unlikely.

BEHR *et al.* (1969) also observed that the rate of complex formation is independent of the size of the sugar side chains of the antibiotic and of the nature and base composition of the DNA. In contrast the rate of complex dissociation increases with a decrease in the size of the sugar side chains. The inhibition by chromomycin and its derivatives, of DNA and RNA polymerase reactions, exhibits similar features and probably the dissociation rate of the complex determines the inhibitory activity. Chromomycin, by complex formation, protects the guanine rich segments of DNA very efficiently against attack by nucleases. After enzymatic hydrolysis, the antibiotic containing base paired oligonucleotide fragments remain in the reaction mixture and can be isolated by electrophoresis. Chromomycin is not covalently bound to these fragments and can be removed by a thorough extraction with ether.

Actinomycin is known to bind to the guanine residues of DNA and this inhibits synthesis of RNA. This is supported by the observation that whereas actinomycin inhibits DNA viruses, the RNA viruses are not affected by it. If mithramycin inhibits RNA synthesis in a manner analogous to actinomycin, one should expect inhibition of only the DNA viruses but not RNA viruses. In actual fact, SMITH et al. (1966) observed that mithramycin prevented the multiplication of the DNA viruses, pseudorabies, and murine cytomegalovirus, but not the RNA viruses, encephalomyocarditis and polio, in tissue culture. The inhibition was fully reversed by removing mithramycin early in the virus growth cycle. The data indicated that the antibiotic reversibly inhibited synthesis of the viral DNA-directed RNA. It is of considerable interest that chromomycin A_3 also does not inhibit RNA-dependent RNA polymerase in the phage Q_β (HARUNA et al., 1970). Chromomycin does not affect the activity of reverse transcriptase of an oncogenic RNA virus (MULLER et al., 1971).

On the basis of similarity of action of olivomycin and related products to actinomycin, it would seem reasonable to suggest that the former binds to the DNA in a manner analogous to actinomycin. Some further similarity in the effects of olivomycin and actinomycin on initiation and growth of RNA chains catalyzed by RNA polymerase has been observed (GAUSE et al., 1968). Olivomycin is a powerful inhibitor of DNA transcription in the reaction catalyzed by RNA polymerase. Recently, the RNA polymerase reactions was shown to proceed in several relatively independent stages, and it was interesting to determine which of these steps, chain initiation or chain elongation, would be inhibited by olivomycin. This can be distinguished by studies on the sedimentation behavior of RNA polymerase reaction products; inhibition of initiation would not cause a difference in the sedimentation coefficient of RNA synthesized in the presence or absence of olivomycin whereas inhibition of chain growth would result in a decrease

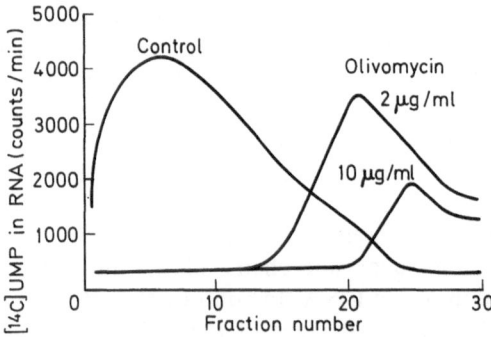

Fig. 2. Effect of olivomycin on the sedimentation profile of RNA synthesized in the RNA polymerase reaction *in vitro*. (From GAUSE et al., 1968)

of the sedimentation coefficient of the synthesized RNA. Fig. 2 shows the effect of olivomycin on the sedimentation profile of RNA synthesized in the RNA polymerase reaction *in vitro* and it is obvious that the sedimentation coefficient decreases in the presence of the antibiotic.

Another approach was to compare the effect exerted by olivomycin on the incorporation of terminally-labeled ATP or GTP, reflecting chain initiation, and on the incorporation of non-terminally labeled nucleoside triphosphate, reflecting total RNA synthesis. It was found that olivomycin is a preferential inhibitor of RNA chain growth as compared with chain initiation. This inhibition of RNA molecules elongation would be caused by the antibiotic molecules bound to the DNA with the progression of polymerase along the template. In this respect the action of olivomycin seems to be similar to that of actinomycin D, which is also an inhibitor of RNA chain elongation, and different from the action of rifamycin and streptovaricin, which inhibit the initiation of RNA molecules on DNA template.

Although common binding sites probably exist in the DNA molecule for olivomycin, chromomycin and mithramycin, there nevertheless may be certain sequences to which each member of this group could be bound specifically. It should be mentioned in this connection that mithramycin preferentially inhibits the incorporation of ^{14}C-labeled uracil into fractions of RNA of low molecular weigth, presumably 5 S RNA (KERSTEN *et al.*, 1967). However, preferential inhibition of this kind was not observed with chromomycin. It is clear that such aspects of possible subtle differences in the mode of action of olivomycin, chromomycin and mithramycin require further study.

References

ASTRAKHAN, V.I., and A.M. GARIN: Olivomycin in therapy of patient with malignant tumors of testicle. Antibiotiki **15**, 837 (1970).

BAKHAEVA, G.P., Y.A. BERLIN, O.A. CHUPRUNOVA, M.N. KOLOSOV, G.Y. PECK, L.A. PIOTROVICH, M.M. SHEMYAKIN, and I.V. VASINA: The stereochemistry of olivomycins. Chem. Commun. **1**, 10 (1967).

BEHR, W., K. HONIKEL, and G. HARTMANN: Interaction of the RNA polymerase inhibitor chromomycin with DNA. Eur. J. Biochem. **9**, 82 (1969).

BERLIN, Y.A., O.A. KISELEVA, M.N. KOLOSOV, M.M. SHEMYAKIN, V.S. SOIFER, I.V. VASINA, and I.V. YARTSEVA: Aureolic acid group of antitumor antibiotics. Nature **218**, 193 (1968).

CHOCHLOVA, D.S.: The effect of olivomycin upon RNA biosynthesis in rat tissues *in vivo*. Vopr. Med. Khim. **15**, 280 (1969).

CHORIN, V.A., and S.P. SHAPOVALOVA: Cross-resistance in a group of antitumor antibiotics belonging to olivomycin-mithramycin type. Antibiotiki **11**, 239 (1966).

GAUSE, G.G., N.P. LOSHKAREVA, and I.B. ZBARSKY: Effect of olivomycin and echinomycin on initiation and growth of RNA chains catalyzed by RNA polymerase. Biochim. Biophys. Acta **166**, 752 (1968).

HARUNA, I., I. WATANABE, Y. YAMADA, and K. NAGAOKA: Specific inhibitors for RNA replicase. Proc. 6th Intern. Congr. Chemoth. (Tokyo) **2**, 796 (1970).

KENNEDY, B.J.: Mithramycin therapy in advanced testicular neoplasms. Cancer **26**, 755 (1970).

KENNEDY, B.J., J.W. YARBRO, V. KICKERTZ, and M. SANDBERG-WOLLHEIM: Effect of mithramycin on mouse glioma. Cancer Res. **28**, 91 (1968).

KERSTEN, W., H. KERSTEN, F. STEINER, and B. EMMERICH: The effect of chromomycin and mithramycin on the synthesis of DNA and ribonucleic acids. Z. Physiol. Chem. **348**, 1415 (1967).

Muller, W., R. Zahn, and H. Seidel: Inhibitors acting on nucleic acid synthesis in an oncogenic RNA virus. Nature New Biol. **232**, 143 (1971).

Sedov, K. A., I. V. Sorokina, Y. A. Berlin, and M. N. Kolosov: Olivomycin and related antibiotics. XXIII. Antibiotiki **14**, 721 (1969).

Smirnova, I. N.: Results of olivomycin treatment of tonsillar tumors. Antibiotiki **14**, 271 (1969).

Smith, R. D., D. Henson, J. Gehrke, and J. R. Barton: Reversible inhibition of DNA virus replication with mithramycin. Proc. Soc. Exptl. Biol. Med. **121**, 209 (1966).

Waring, M.: Binding of drugs to supercoiled circular DNA: evidence for an against intercalation. Progr. Mol. Subcell. Biol. **2**, 216 (1971).

Quinacrine and Other Acridines

Alan David Wolfe

Introduction

Quinacrine, also known as atebrin, atabrine, mepacrine and acrichin, is a 9-alkylaminoacridine (Fig. 1), and was developed as one of a group of potential antimalarial drugs (MIETZSCH and MAUSS, 1930), apparently patterned after successful antimalarials possessing alkyl side chains as, for example, pamaquine (GOODMAN and GILMAN, 1960; The Acridines, A. ALBERT, 1966). Quinacrine was introduced clinically in April, 1932 (MAUSS and MIETZSCH, 1936), but despite rapid professional acceptance, its structure remained masked by inadequate patent disclosure until its degradation and synthesis by Russian scientists (CHELINTSEV et al., 1934). Previously, acridines, particularly proflavine (BROWNING and GILMOUR, 1913) and its 10-methyl acridinium derivative, in conjunction known as acriflavine, had been used as antibacterial agents, but only with the advent of quinacrine and additional 9-alkylaminoacridine derivatives, were acridines developed into highly effective antimalarial drugs. During the Second World War, the Allies, separated from their sources of quinine, turned to the synthetic compound, quinacrine, for both prophylaxis and overt therapy of malaria. As a result of the War, of the prevalence of malaria, and of the many cellular and molecular effects of quinacrine, a vast literature has developed, and for detailed knowledge, the reader should consult the very excellent and comprehensive monographs, The Acridines by A. ALBERT (2nd edition, St Martin's Press, New York, 1966), Malaria Parasites and Other Haemosporidia, by P. C. C. GARNHAM (Blackwell Scientific Publications, Oxford, 1966), and Chemotherapy and Drug Resistance in Malaria, by W. PETERS (Academic Press, London and New York, 1970). These monographs have materially facilitated the writing of this review.

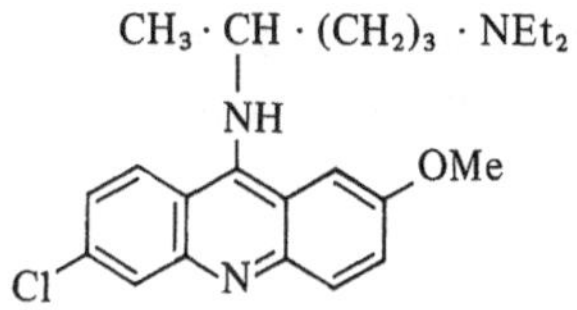

$$CH_3 \cdot CH \cdot (CH_2)_3 \cdot NEt_2$$

Fig. 1

Quinacrine, d,1,-9-(4-Diethylamino-1-methylbutylamino)-7-methoxy-3-chloroacridene, is a planar heterocyclic compound possessing a molecular weight of 400 daltons. The drug is available as the sparingly water-soluble base, or the highly water-soluble doubly hydrated dihydrochloride or methanesulfonate salt. The molecule has a high dipole moment and possesses a pK at 7.7 attributed to protonization of the resonating aminoacridine system, and a second pK at 10.28 corresponding to protonization of the diethylamino nitrogen atom (CHRIS-TOPHERS, 1937; A. ALBERT, 1966). Quinacrine is yellow and in aqueous solution has absorption maxima at 280 nm, 424 nm and 444 nm (BISSEL et al., 1945; DRUM-MOND et al., 1965). Solutions of quinacrine fluoresce, with excitation maxima at 285 nm and 420 nm, and an emission maximum at 500 nm at pH 11 (UDENFRIED et al., 1957), but the literature contains many absorption and fluorescence spectra of the drug under different conditions of pH and in many solvents. The fluorescence of quinacrine is strongly influenced by the anionic environment (AZZI et al., 1971). Preparations of quinacrine are racemic mixtures, but the d and l isomers may be separated (CHELINTSEV and OSETROVA, 1940) and some differential biological activity of them has been noted (ALPATOV, 1946). Crystals of quinacrine contain stacked acridine rings (COURSEILLE et al., 1971) and the side chain has approximately three times the length of the acridine ring (DRUMMOND et al., 1965). Hückel molecular orbital calculations showed quinacrine to be a better electron donor and acceptor than quinoline antimalarials, while a high π electron charge density occurred on the ring nitrogen (SINGER and PURCELL, 1967). The nitrogen mustard derivative of quinacrine has been synthesized (JONES et al., 1957) and binds covalently to DNA (CASPERSSON et al., 1968).

Medical Uses

In addition to the use of quinacrine in malaria, a use which has markedly diminished with the acceptance of its 4-aminoquinoline analog, chloroquine, (CQ), as the drug of choice, quinacrine is currently employed to eliminate protozoan flagellates and tapeworms of the *Taenia* and *Hymenolopis* species (CULBERTSON, 1940). Quinacrine has considerable and growing utility in antibiotic therapy, since it reduces the rate of mutation to antibiotic resistance (SEVAG and ASHTON, 1964). The drug also reduces the rate of serous effusions in cancer (ULTMANN et al., 1963) and relieves inflammation related to rheumatoid arthritis and lupus erythmatosus (PAGE, 1951; A. ALBERT, 1966, on previous Russian contributions). Quinacrine inhibits the primary immune response (ZELEZNICK et. al 1969), and analysis of specific regions of human chromosomes has been materially advanced through study of quinacrine binding and fluorescence (CASPERSSON et al., 1968). Episodes of petit mal epilepsy have been treated with quinacrine (SIBLEY et al., 1962).

Experimental Observations

Quinacrine suppresses the growth, particularly when given prophylactically, of eastern equine encephalitis and other viruses, and also of organisms of the psittacosis and lymphogranuloma groups, although nitroacridines are more potent against the latter organisms (EATON et al., 1974; BURNEY and GOLUB, 1948; HURST et al., 1952a; HURST et al., 1952b; GREENHALGH, et al., 1956). Growth of polio virus (IYKS, 1963) and fungi (ANSEL and THIBAULT, 1965) were inhibited by quinacrine, while tumor growth was suppressed (HARTWELL *et al.*, 1946), and tumor tissues were found to have a high affinity for the drug (BRIL-MAYER *et al.*, 1955).

Quinacrine concentrates in lyosomes (ALLISON and YOUNG, 1964) and binds to nucleoli (FEDORKO and HIRSCH, 1969), such that in conjunction with laser radiation it has been used as a probe for nucleolar function (BERNS *et al.*, 1970). Acridine orange also concentrates in lysosomes (ALLISON and Young, 1964) and has been used for many years as a vital stain. Additionally, the binding of acridine orange and proflavine to acidic polysaccharides, partially comprising the outer envelope of many cells, has been studied in detail (STONE and BRADLEY, 1966). In a most interesting experiment, potentially providing further understanding of factors controlling cellular uptake of acridines, a permanent increase in the binding of acriflavine occurred in three mammalian cell lines after parental exposure and incorporation of calf thymus DNA or SV 40 virus (ROTH *et al.*, 1970).

Toxicity

While intravenous administration of quinacrine results in a transitory high concentration in the lungs, the drug preferentially concentrates in liver, spleen and kidneys. Quinacrine is eliminated slowly from the body over a period of weeks (TROPP and WEISE, 1933). The toxic effects of the drug, though relatively mild, were recognized soon after its clinical introduction. These effects may be divided into three broad categories: 1. effects on the skin and tissue (THONARD-NEUMANN, 1932), including discoloration, pruritis, dermatitis, sweating, fever and muscle pain; 2. gastrointestinal upset (THONARD-NEUMANN, 1932; HECHT, 1933), including nausea, vomiting, cramps and diarrhea; and 3. effects on the brain and central nervous system (HECHT, 1933), inculding vertigo, headache, insomnia, ocular disturbances (HENKIND and ROTHFIELD, 1963), epileptic convulsions and toxic psychoses (NEWELL and LIDZ, 1946). Instances of aplastic anemia (PALMER and SAWITSKY, 1953) and eosinophilia (RUSSELL, 1945) have been reported. A yellow discoloration and a green fluorescence of the skin has been attributed to accumulation of quinacrine in the epithelial and keratinous tissues (GERKE, 1948; MILLER *et al.*, 1950).

Quinacrine and Malaria

The malarial parasite has two phylogenetically different hosts, 1. The mosquito, which injects the malarial sporozoite into 2. The vertebrate, in which the sporozoite initially differentiates within parenchymal cells of the liver. Eventually a tissue schizont is produced and this, in turn, segments to yield merozoites, i.e., forms which enter the blood stream to infect erythrocytes, or to reinfect tissue, there potentially to lie dormant. Upon erythrocyte infection, the merozoite speeds the composite cellular metabolism approximately seventy times (SILVER-MAN *et al.*, 1944) while assuming a ring form and developing an amoeboid cytoplasm. This form, the trophozoite, may feed upon the erythrocyte by an engulfment process termed phagotrophy (RUDZINSKA and TRAGER, 1959; PETERS *et al.* 1965), utilizes hemoglobin as a source of all amino acids, except for isoleucine

(POLET and BARR, 1968a). The hematin residue from the digestion of hemoglobin *via* the intermediate ferrihaemic acid, accumulates in vesicles which pinch off from the "food vacuoles" and combines with peptides to form the characteristic plasmodial pigment mixture, haemozoin (MOULDER and EVANS, 1946; DEEGAN and MAEGRITH, 1956a and b; RUDZINSKA and TRAGER, 1959; SCHUELER and CANTRELL, 1964; COHEN *et al* 1964; SHERMAN *et al* 1965). Soon, however, plasmodial growth slows, vacuoles disappear and the trophozoite nucleus begins dividing by mitosis to produce the schizont, which at maturity contains many nuclei. The cytoplasm then segments to produce merozoites or, alternately differentiates to form gametocytes which renew the plasmodial cycle through the mosquito vector. All erythrocytic forms of the plasmodium except the mature gametocytes are sensitive to quinacrine; thus the ring, the trophozoite and the developing and mature schizonts are suppressed, destroyed and eliminated from erythrocytes in the presence of quinacrine (MUHLENS and FISCHER, 1932).

In 1934, James observed and described the effects of quinacrine upon erythrocytes in benign tertian and quartan malaria. He observed aggregation and eventual elimination from the parasite of pigment, a loss of normal shape and a vacuolization of the cytoplasm. The nucleus distended and the chromatin became progressively less dense and eventually assumed a diffuse, fibrous character until staining could reveal only "light dots". Quinacrine-induced changes (HUHNE, 1942) included swelling, vacuolization and destruction of the cytoplasm, detachment of nuclei from the ring wall, aggregation of pigment and swelling and fragmentation of chromatin.

A time course study by means of fluorescence microscopy (BOCK and OSTERLIN, 1939) demonstrated that erythrocytic plasmodial forms concentrated quinacrine which then exerted a direct effect, quinacrine appeared in parasites in ten minutes. Thirty minutes after quinacrine administration, clumping of pigment had commenced (BOCK and OSTERLIN, 1939; MACKERRAS and ERCOLE, 1949). Six hours after administration, rings had been reduced in size and chromatin was pressed along the red cell periphery. Schizonts were abnormal with large, diffuse nuclei, and clumped pigment had been extruded from the cell and from the erythrocytes (MACKERRAS and ERCOLE, 1949). By sixteen hours, ring nuclei were detached from the remaining cytoplasm of the now wispy ring, while a thousand-fold reduction in sensitive erythrocytic forms had occurred during a twenty-eight hour exposure to quinacrine (MACKERRAS and ERCOLE, 1949).

Of particular interest are detailed electron micrographic studies of the effect of quinacrine on L cell mouse fibroblasts (FEDORKO and HIRSCH, 1969). Cytoplasmic changes, including induction of multiple "food vacuoles," were observed to occur upon exposure of aliquots of these cells to quinacrine for two hours. More important, exposure to quinacrine resulted in profound changes in the nucleolus. Within thirty minutes incubation, nucleolar fragmentation had commenced with segregation of constituents of the nucleolus, including creation of zones of chromatin, protein and unidentified fibrillar and granular components. Many dense structures and clusters of smaller bodies were scattered throughout the nucleus. Chromatin distended in the nucleolar membrane. Later changes included complete nucleolar disorganization and fragmentation. Upon withdrawal of quinacrine, recovery occurred slowly but abnormalities persisted for more than 24 hours. The basis for the effects of quinacrine in L cell mouse fibroblasts was considered to be the binding of the drug to DNA, with consequent inhibition of RNA synthesis.

Biochemical changes, induced in plasmodia by quinacrine, include suppression of respiration (FULTON and CHRISTOPHERS, 1938) and inhibition of nucleic acid synthesis (SCHELLENBERG and COATNEY, 1961). The discovery that quinacrine irreversibly competed with the prosthetic group of cytochrome reductase (HAAS, 1944) and, more generally, with flavin containing yellow enzymes (WRIGHT and SABINE, 1944; HAAS, 1944) gave assurance that the antirespiratory effects of quinacrine possessed a distinct biochemical basis, but an inability to obtain a direct correlation (COGGESHALL and MAIER, 1941) between *in vivo* activity of quinacrine, and *in vitro* inhibition of respiration by the drug led eventually to the demonstration (ALBERT and MARSHALL, 1948) that acridine inhibition of respiration was more closely related to toxicity to the host than to antimalarial action. Such findings discouraged further consideration of inhibition of plasmodial respiration as the mode of action of quinacrine.

In contrast, although JAMES (1934) had described quinacrine-induced destruction of plasmodial chromatin, it was not until well after the discovery (CLARKE, 1952) that quinine inhibited incorporation of P^{32} into the nucleic acids of plasmodia that similar effects of quinacrine were sought (SCHELLENBERG and COATNEY, 1961). Concentrations of 10^{-5} M quinacrine were found to reduce by approximately 80% the incorporation of P^{32} into the DNA and RNA of erythrocytes parasitized with *P. gallinaceum*. Similar results were obtained in rat erythrocytes parasitized with *P. berghei*. Since quinacrine was as effective an inhibitor of DNA and RNA synthesis *in vitro* as was chloroquine or quinine, the inhibitions *in vivo* suggested analogies in the modes of action of all three drugs. The findings of SCHELLENBERG and COATNEY (1961) were confirmed and extended through studies on the uptake and incorporation of tritiated adenine, a nucleic acid precursor, into parasitized erythrocytes, and erythrocyte-free *P. berghei* (VAN DYKE et al., 1969; VAN DYKE and SZUSTKIEWICZ, 1969; VAN DYKE et al., 1970a; LANTZ and VAN DYKE, 1971). Quinacrine not only reduced the incorporation of such precursors into plasmodial DNA but also partially suppressed the uptake of these precursors into the erythrocyte, a potentially important effect since plasmodia do not synthesize purines but only pyrimidines (VAN DYKE et al., 1970b). Thus, two effects were observed: 1. Partial inhibiton of transport of purine nucleic acid building blocks, and 2. Partial inhibition of DNA synthesis *per se*. The relation between these effects requires further investigation. Quinacrine has also been observed to inhibit *in vitro* aminoacylation of plasmodial transfer RNA (ILAN and ILAN, 1969).

Resistance

Resistance to quinacrine appeared during the Second World War when a severe epidemic of *P. falciparum* occurred in New Guinea despite continuous prophylaxis and chemotherapy with quinacrine (FAIRLEY, 1945). Additional outbreaks have occured since in South America and Vietnam (MODELL, 1968). A resistant strain, the M (mepacrine) strain of *P. berghei*, was developed experimentally (PETERS, 1965a; PETERS, 1966) under continuous quinacrine pressure. These plasmodia lacked normal pigment grains but contained pigment granules demonstrable by electron microscopy and, in addition, contained multiple, small "food vacuoles." Upon withdrawal of quinacrine, the M strain reverted to sensitivity and to a morphology typical of normal plasmodia. The M strain exhibited cross-resistance to chloroquine and quinine; conversely, cross-resistance to quinacrine occurred (PETERS, 1965b; JACOBS, 1965) in chloroquine and quinine-resistant strains. Plasmodia resistant to quinacrine and chloroquine possess a similar appearance (PETERS, 1965a, b; PETERS et al., 1965) with many small cytoplasmic vacuoles and pigment granules; these strains as well as a multiple drug-resistant strain (PETERS, 1965c) grow more slowly than wild type plasmodia. In parallel, erythrocytes parasitized with chloroquine-resistant P. berghei concentrate less chloroquine than do erythrocytes parasitized with sensitive P. berghei (MACOMBER et al., 1966), suggesting that resistance or sensitivity is a function of the amount of drug concentrated by the erythrocyte plasmodium system. Erythrocytes parasitized with chloroquine-sensitive P. knowlesi and exposed to ^{3}H-chloroquine (POLET and BARR, 1969) exhibited a release of tritium which was related to the external concentrations of quinacrine, chloroquine or dihydro-quinine, suggesting to the authors that these drugs were in a dynamic equilibrium with the concentrated chloroquine. Were this conclusion valid then the assumption could be made that quinacrine was actively concentrated by the identical plasmodial-erythrocytic system and, indeed, that resistance to quinacrine might well be a function of a lesser uptake into the plasmodium.

Structure-Activity Relationships: Malaria

For acridines structure-activity relationships depend to an unusual degree upon the target organism: highly potent antibacterial acridines including proflavine, 9-aminoacridine, acridine yellow and the nitroacridines possess little antimalarial activity, while quinacrine, an outstanding antimalarial drug, exerts only moderate inhibition of bacterial growth. Information on the antimalarial potency of acridines has been obtained from three principal sources: 1. Survey of Antimalarial Drugs, 1941–1945, by F. Y. WISELOGLE, 2. The Acridines, by A. ALBERT, and 3. reports of O. MAGHIDSON, his colleagues (1936a and b) and other Russian investigators. Evalution of antimalarial drugs has normally been made in quinine equivalents, the weight of quinine necessary to produce a similar effect. Quinacrine has a quinine equivalent of three and is, therefore, three times as potent as quinine on a weight/weight basis. In recent years, avian test screens have been supplemented with, or replaced by, mammalian plasmodial screening systems.

The prototype structure for acridine antimalarial activity is a 9-aminoalkyl derivative of quinacrine with the presence in the drug of the 3-Cl and 7-OCH$_3$ substituents. As reported by WISELOGLE, this structure was 50% as active as quinacrine, while previous German research (The Acridines, by A. ALBERT, 1966), comparing the lowest effective doses of respective compounds, showed this compound to possess approximately 6% of the activity of quinacrine. The dealkylated compound, 3-Cl, 7-OCH$_3$, 9-NH$_2$ acridine possesses little potency and placement of the alkylamino side chain in other ring positions yields non-active compounds. The presence of a cationic 9-aminoalkyl side chain of from 2 to 8 carbon atoms is, therefore, an indispensable requirement for antimalarial action, and only a minor decrease in activity occurred when the chain was lengthened beyond 5 carbons. The side chain may include phenyl, substituted phenyl, quinoline and saturated ring systems. Absence of an alkyl group on the 1-carbon of the side chain reduced activity, possibly through facilitation of hydrolysis, while little loss in potency occurred upon substitution with larger alkyl groups. The terminal secondary amine is essential for high activity, although ring nitrogen atoms suffice, but the absence of cationic substituents, or their substitution by other reactive groups, diminished antimalarial potency, as did side chain heteroatoms.

Ring substitutions which increase activity as compared to the fundamental alkylamino acridine, include substitution of either the Cl atom or a cyano group in position 3, or of the Cl atom in position 1. Placement of the halogens F or I, or of a methyl group, at the 3-position also increases potency but the same substituents in positions 2 and 4 diminish antimalarial action. Conversely, a 2-nitro group increased potency, while a 3-nitro group reduced activity. Alkoxy groups, particularly methoxy or ethoxy, at position 7, do not affect potency but reduce toxicity to the host. Multiple variation in ring substituents has not yielded unusually potent drugs.

Many attempts have been made to rationalize the role of these substituents and the side chain, both with respect to their nature and location, but no hypothesis has gained unqualified acceptance. It appears, however, that ortho-and para-directing substituents at ring position 3 generally produce highly active antimalarials despite the potency of the cyano group at this position. ALBERT has suggested that an electronegative group or atom capable of inducing a high electron density on the 9 carbon atom of the ring may produce maximally effective antimalarials, the potential function of such high electron density being to increase ring ionization.

Tetrahymena

Tetrahymena pyriformis is another protozoan in which the mode of action of quinacrine was investigated. Quinacrine proved the most lethal to *Tetrahymena* of a group of antimalarial drugs including primaquine, quinine and chloroquine (CLANCY, 1968). Synchronized cell division in *Tetrahymena* was reduced to a minimum by 15 µg per ml quinacrine (CHOU, *et al.,* 1968), while sythesis of DNA, which had continued at a reduced rate for thirty minutes, ceased completely (CHOU and RAMANATHAN, 1968). RNA and protein synthesis continued at reduced rates for the entire eighty-minute experimental period, and uptake of acetate continued linearly but was reduced by eighty percent. Despite this intense reduction in the cellular uptake of acetate, quinacrine did not suppress respiration of *Tetrahymena* (CONKLIN *et al.,* 1971). These results are most consistent with the hypothesis that quinacrine inhibits cell replication in *Tetrahymena pyriformis* through inhibition of DNA synthesis (CHOU and RAMANATHAN, 1968).

Bacteria

Acridines were found to have antibacterial properties in 1913 (BROWNING and GILMORE) and by 1945 more than 130 acridines had been tested in this respect (ALBERT et al., 1945). BROWNING, ALBERT, and their respective colleagues determined dose responses and structure-activity relationships, and analyzed the response of more than twenty bacterial species to selected acridines. During the 1940's, the basis was established for decades of research with respect to the growth-inhibitory properties of acridines, their target sites and the structure of these sites. Gram-positive bacteria appeared, generally, more sensitive to acridines than gram-negative bacteria, although it was suggested (ALBERT et al., 1945) that the data reflected the greater sensitivity of more fastidious organisms rather than their tinctorial characteristics. Perhaps the most important clue to the mode of action of acridines was found with the observation that RNA could reverse acridine-induced bacteriostasis (McILWAIN, 1941). Subsequently, quinacrine, among a number of acridines, was shown to have antibacterial activity (LAWRENCE, 1943; GOETSCHIUS and LAWRENCE, 1944; SILVERMAN and EVANS, JR., 1943) which could be reversed by cationic polyamines, including spermine and spermidine (SILVERMAN and EVANS, JR., 1943, 1944; MILLER and PETERS, 1945) or inorganic divalent cations (SILVERMAN, 1948), and by RNA (FITZGERALD and LEE, 1946). The antibacterial potency of quinacrine was weak, however, compared to the more potent acridines, 3-6-diamino 4-5-dimethylacridine and 9-amino-3-nitroacridine (ALBERT et al., 1945; WOLFE et al., 1971 a). Acridines were also discovered to inhibit the growth of E. coli bacteriophages (FITZGERALD and BABBITT, 1946) but the potency of quinacrine was low compared to that of proflavine. The sensitivity of phage proved greater than the sensitivity of the bacterium to the acridine, so that E. coli could grow in concentrations of acridine which were inhibitory to the phage (FITZGERALD and LEE, 1946; FITZGERALD and BABBITT, 1946; FOSTER, 1948). Differential acridine sensitivity appeared to exist even among different bacteriophage species: the T-even phages exhibited greater sensitivity to proflavine than the T-odd phages (FOSTER, 1948). Both RNA and DNA reversed proflavine inhibition of bacteriophage formation and it was inferred that proflavine interfered with a late step in the assembly of the phage, a step possibly involving DNA (FOSTER, 1948). Proflavine was then shown to induce elongated forms of Aerobacter aerogenes (B. lactis aerogenes) (CALDWELL and HINSHELWOOD, 1950). Concurrently, however, antibiotics came into common use and interest in acridines as antibacterial agents waned, while termination of the World War II halted intensive antimalarial research with the result that investigation into the mode of antimicrobial action of acridines, and quinacrine in particular, was reduced, and scientists turned to use of these drugs as blocking agents and molecular probes.

Binding and Uptake of Acridines by Bacteria

Acridines bind to E. coli in the order: acriflavine > proflavine > 9-aminoacridine(9-AA) > quinacrine (SILVER, 1967; SILVER et al., 1968), although the quantity of any single acridine bound appears to be a strain characteristic and is probably even a species characteristic. Thus, *E. coli* B bound twice as much acriflavine at low concentration as did *E. coli* K-12 (SILVER *et al.*, 1968). Binding has been considered reversible and most likely external as well as irreversible (internal) (NAKAMURA, 1966; SILVER *et al.*, 1968). External receptors in *Micrococcus lysodeikticus* were thought to be wall and membrane polyanions, including mucopolysaccharides (BEERS, JR., 1964) and probably were involved in distinguishing rough and smooth forms of strains of motile *Salmonella* (BERNSTEIN and LEDERBERG, 1955). Internal receptors are generally considered to be DNA and RNA, at least in prokaryotes. Strains of *E. coli* possess a gene, denoted Mb (methylene blue) (SUGINO, 1966) or AF (acriflavine) (NAKAMURA, 1965). AF^s bacteria bound more acriflavine reversibly than did AF^r bacteria (Nakamura, 1966) suggesting that cell response was initially a function of the external receptors which were expressed unusually slowly in mating studies. Resistance was dominant to sensitivity in diploid heterozygotes (SUGINO, 1966; NAKAMURA, 1965).

Acridines are thought to enter cells passively in contrast to entry by active transport, since low temperatures, arsenate and cyanide stimulated rather than suppressed reversible acriflavine binding (NAKAMURA, 1966; SILVER *et al.*, 1968). Additionally, colicin-tolerant (tol) VIII mutants possessed an increased capacity to bind acriflavine, compared to tol$^-$ cells; this increased binding has been attributed to a possession of altered receptor sites in a hypersensitive membrane (NAGEL DE ZWEIG and LURIA, 1967). Lethal agents and treatments which destroy cellular permeability barriers, such as toluene, phenethyl alcohol or heating to 100 C, increased irreversible binding of acriflavine, as did a benign treatment, alteration of the pH from 7.0 to 8.5. A considerable increase in the irreversible binding of acriflavine and a small increase in the irreversible binding of quinacrine, occurred in *E. coli* upon bacteriophage infection, although uptake was a function of the type of phage employed; the T-even bacteriophages stimulated greater uptake of acridine and, therefore, appeared more sensitive than the T-odd phages (FOSTER, 1948; SILVER, 1965; HESSLER, 1965; SILVER, 1967; SILVER *et al.*, 1968). *In toto*, these observations point to genetically determined membrane receptors or carriers, possessing, in part, as a function of their pK's, a greater or lesser affinity for basic dyes.

Biochemical Studies

Quinacrine preferentially suppresses DNA synthesis in *E. coli* (CIAK and HAHN, 1967) and RNA synthesis in *B. cereus* (SELIGMAN and MANDEL, 1971), suggesting that 1. the antibacterial mode of action of quinacrine is selective suppression of nucleic acid synthesis, and 2. the mechanism of action of quinacrine involves drug binding to DNA. In *E. coli,* a concentration of 2×10^{-4}M quinacrine inhibited cell division and induced formation of filaments (CIAK and HAHN, 1967) characteristic of unbalanced growth and preferential inhibition of the synthesis of DNA (BARNER and COHEN, 1954; GOSS *et al.,* 1964). Colorimetric analysis showed that the rate of DNA replication was reduced by 26%, while RNA synthesis continued undiminished. Synthesis of protein, determined by measurement of the incorporation of ^{14}C-phenylalanine, was suppressed by 40%, an unexpectedly strong effect. A concentration of 8×10^{-4} M quinacrine caused exponential loss of viability, did not lead to formation of filaments, and totally inhibited DNA replication, while synthesis of RNA and protein continued at strongly diminished rates. It is concluded that the mode of action of quinacrine in *E. coli* was "impairment of DNA synthesis" which occurred through reaction of quinacrine with double-stranded DNA. Support for this conclusion has been found in the effects of Nitroakridin 3582 (WOLFE *et al.*, 1971a) on *E. coli*. This compound inhibits DNA synthesis preferentially and also induces formation of filaments at a bacteriostatic concentration. Probit analysis of dose-responses not only revealed preferential inhibition of DNA synthesis by NA 3582, but demonstrated a correlation between such inhibition and NA 3582-induced reduction in bacterial viability. The lesser inhibitions of RNA and protein synthesis showed a mutual correlation. These results indicate that NA 3582 suppresses cell replication throught inhibition of DNA synthesis and inhibits protein synthesis through suppression of messenger RNA transcription, the latter similar to an action of proflavine in *E. coli* (WOESE *et al.*, 1963).

In *B. cereus*, on the contrary, exposure of cultures to a concentration of 4×10^{-4} M quinacrine (SELIGMAN and MANDEL, 1971) permitted continuation of the synthesis of DNA and protein at rates similar to the rate of increase in bacterial density, with a concurrent increase in m-RNA formation, but produced a selective and specific inhibition of the synthesis of ribosomes; a partial inhibition of the synthesis of r-RNA was also observed. Colorimetric analysis of the increase in RNA coupled with studies on the incorporation of radioactive purines indicated that quinacrine reduced utilization of exogenous purines for RNA synthesis, although a large increase in intracellular ATP occurred. Quinacrine additionally induced a chain-like association of *B. cereus*, a delayed loss in viability accompanied by a reduction in turbidity, [similar to the effects of NA 3582 and other acridines (WOLFE *et al.*, 1971a; WEINBERG *et al.*, 1958)] and a reduction in the Gram-positive character. Total inhibition of both DNA and RNA synthesis and destruction of ribosomes have been observed (CIAK and HAHN, 1966) upon exposure of *B. megaterium* to the quinacrine homolog chloroquine, while dissimilation of RNA occurred in both Gram-negative and Gram-positive bacteria exposed to proflavine (SOFFER and GROS, 1964; GRINSTED, 1969).

Extrachromosomal DNA and Quinacrine

The suppression of the replication of bacteriophage and, additionally, of plasmids and of the transfer of the latter, are important consequences of the binding of acridines to nucleic acids. This importance became manifest upon the discovery that acridines, including quinacrine, inhibited the replication of bacteriophages of *E. Coli* (FITZGERALD and BABBITT, 1946; FITZGERALD and LEE 1946); RNA added to the culture medium relieved such inhibition. Additional research in the 1940's revealed the T-even bacteriophages of *E. coli* to be more susceptible to acridines than the T-odd phages, while proflavine was inferred to inhibit a late step in the assemblage of bacteriophage (FOSTER, 1948). Bacteriophages resistant to acridines were also discovered (FITZGERALD and LEE, 1946; FOSTER, 1948).

During the 1950's the effect of proflavine upon the replication of phage was studied comprehensively and it was found that phage assemblage was inhibited by low concentrations of proflavine without significant suppression of macromolecular synthesis. The ability to separate phage heads from the tail antigens, reacting with neutralizing antibody, and the existence in bacteria and in bacterial lysates of doughnut shaped forms similar to phage heads and heads "leaking" DNA, suggested that the DNA and protein were not properly aligned (DEMARS et al., 1953; DEMARS, 1955; KELLENBERGER and SECHAUD, 1957; KELLENBERGER and KELLENBERGER, 1957; KELLENBERGER et al., 1958; KELLENBERGER et al., 1959). Staphylococcal bacteriophages were also sensitive to acridines (HOTCHINS, 1951); the polyamines spermidine, agmatine and putrescine, as well as RNA, reversed proflavine inhibition of this bacteriophage assemblage (KAY, 1959).

Acridine suppression of bacteriophage replication has been found to be a function of selected acridines (FITZGERALD and BABBITT, 1946; HESSLER, 1963; SILVER, 1967) and of the genetic characteristics of the phage (FITZGERALD and LEE, 1946; SILVER, 1967). Thus, total inhibition of phage T2 formation required approximately 90 µg/ml quinacrine, $2^1/_2$ µg/ml proflavine and 0.3 µg/ml acriflavine (HESSLER, 1963), while a concentration of approximately 1 µg/ml 9-aminoacridine was required totally to inhibit bacteriophage T4 (PIECHOWSKI and SUSMAN, 1967). Phage-induced bacterial uptake of these acridines has been observed to be greatest with acriflavine and least with quinacrine (SILVER, 1967), suggesting that intracellular acridine molecules are similarly inhibitory.

Three processes were considered (SUSMAN et al., 1965; PIECHOWSKI and SUSMAN, 1967) to have been inhibited in phage T4 infection by 9-aminoacridine: 1. a terminal step in the assembly of mature phage, 2. an "acridine sensitive maturation clock" and 3. lysis of the bacterium. Most significantly, however, low concentrations of 9-aminoacridine prevented the conversion to mature phage of an intermediate replicating complex possessing a sedimentation coefficient of 200 s (FRANKEL, 1966) while, more recently, the compound was found (ALTMAN and LERMAN, 1970) to prevent detachment of the parental DNA from the newly synthesized DNA comprising the replicative complex so that the final phage packaging was inhibited and lower molecular weight DNA accumulated. Possession of a resistance marker by the phage relieved 9-aminoacridine inhibition of the detachment of the respective DNAs in the replicative complex and permitted production of mature phage. However, additional synthesis of DNA was necessary for production of phage after drug withdrawal, suggesting a need for removal of distorted or altered DNA.

Mutants resistant to quinacrine and 9-aminoacridine are denoted q (PRATT et al., 1961; HESSLER, 1963), and mutants resistant to proflavine and acriflavine are denoted pr in phage T2 and ac in phage T4 (HESSLER, 1963; EDGAR and EPSTEIN, 1961). Resistance to 9-aminoacridine apparently involves mutation rendering a replicative intermediate insensitive to this drug (ALTMAN and LERMAN, 1970), while pr and ac mutants are considered to reduce bacterial permeability to these compounds (HESSLER, 1965; SILVER, 1965). Since wild-type bacteriophages stimulate bacterial uptake of these compounds (SILVER, 1967), pr and ac mutants must be considered to possess a reduced capacity to stimulate uptake. Pr and q genes are not linked (HESSLER, 1965) although some cross-resistance was

observed. Pr and q revertants produced by nitrous acid or proflavine mutagenesis
are nonspecific and different suppressor mutants, which are thought to increase
phage sensitivity by alteration of "early enzymes" so as to synthesize DNA
of greater acridine-sensitivity (HESSLER et al, 1967).

Symmetrical diaminoacridines photodynamically inactivate bacteriophage
(YAMAMOTO, 1958; FRASER and MAHLER, 1961), although aliphatic diamines,
most effectively cadaverine, which contains two nitrogen atoms separated by
five -CH_2- groups, suppress such inactivation. Tobacco mosaic (CHESSIN, 1960)
and polio viruses (CROWTHER and MELNICK, 1961) are photodynamically inactivated by acridine orange while proflavine (SCHAFFER, 1962; GENDON, 1963) inactivated polio virus photodynamically and also induced mutations. A protective
effect of proflavine on DNA, exposed to ultraviolet radiation, has also been
observed (BEUKERS, 1965; SETLOW and CARRIER, 1967; SETLOW and SETLOW,
1967), while recently quinacrine was found to be the most potent of a group
of acridines which protect infectious DNA of phage d φ 4 exposed to UV radiation
(ROINISHVILI, 1970). UV radiation causes the formation cyclobutyl pyrimidine
dimers in DNA and energy transfer from the DNA singlet state to acridines
is thought to be involved in a diminished frequency of dimer formation (SUTHER-
LAND and SUTHERLAND, 1969).

In E. coli a facultative association of the sex factor, F, occurs with respect
to the bacterial chromosome but in its autonomous form F'-lac is sensitive to
acridines during bacterial replication (HIROTA and IJIMA, 1957; HIROTA, 1960).
Thus, acridine orange, proflavine and quinacrine suppressed transfer of, and
eliminated F'-lac (HAHN and CIAK, 1971) and quinacrine was found to have
considerably reduced the quantity of this satellite DNA. Antibiotic resistance
factors, R, are considered to be composed of DNA in the form of a resistance
transfer factor, RTF (WATANABE and FUKUSAWA, 1961; NAKAYA et al., 1960),
linked to determinants of the segregant characteristics r. Elimination of the activity
of both RTF and all antibiotic markers, as well as of the correlative satellite
DNA band, was observed to occur upon treatment of bacteria with acridines
(MITSUHASHI et al., 1961; ROWND et al., 1966). Treatment with quinacrine (LEVY
and WATANABE, 1966) also suppressed transfer of, and eliminated R, and induced
segregation of r determinants in E. coli RS-2 (HAHN and CIAK, 1971). One strain
of bacteria harboring an R factor, however, has been found to be more sensitive
to quinacrine than R⁻ bacteria (YOSHIKAWA and SEVAG, 1967).

Acridines are frameshift mutagens in bacteriophages (ORGEL and BRENNER,
1961; CRICK et al., 1961) and acridine mustards including the quinacrine mustard,
ICR-170, are strong frameshift mutagens in bacteria (AMES and WHITFIELD, JR.,
1966; MALLING, 1967), while quinacrine itself caused only weak frameshift
reversion in Salmonellae (AMES and WHITFIELD, 1960). Acridine mutagenesis in
bacteriophage T 2 differed from mutagenesis induced by agents which alter bases
chemically (ORGEL and BRENNER, 1961; BRENNER et al., 1961) in that the original
phenotype could be restored upon an opposite deletion or addition of an intracistronic nucleotide, or upon a sequence of three such events which would restore
the original reading frame (CRICK et al., 1961). Frameshift mutagenesis was
considered to occur upon temporary stabilization of mispaired DNA regions
which resulted from acridine intercalation (STREISINGER et al., 1966), in contrast

to an earlier suggestion (LERMAN, 1964) that a DNA exchange and recombination process was involved.

Conversely, quinacrine is a potent antimutagen in both *E. coli* and *Staphylococci*, and markedly reduces the emergence of resistance to certain antibiotics, including streptomycin, sulfathiazole, tetracycline, erythromycin and chloramphenicol (SEVAG, 1964; SEVAG and ASHTON, 1964; reviewed by DE COURCY, 1971); in the presence of quinacrine, a reduction also occurred in the frequency of mutations caused by 2-aminopurine and mutator genes (JOHNSON and BACH, 1966; ZAMENHOF, 1969; reviewed by BACH and JOHNSON, 1971). Antimutagenesis is not limited to bacteria. Quinacrine also reduces the frequency of resistance to 8-azaguanine in Detroit-98 human stern marrow cells, and to cytarabine in L1210 murine leukemic cells (JOHNSON and BACH, 1969; BACH, 1969). Intercalation is not a necessary condition for antimutagenesis, however, since two nonintercalating DNA binders, spermine, albeit at considerably greater concentrations than quinacrine (SEVAG and DRABBLE, 1962), and diazepines (non-planar heterocycles) also reduce the mutation frequency (BACH and JOHNSON, 1971). Potential mechanisms for mutagenesis are: 1. creation of hydrophobic regions in DNA (BACH and JOHNSON, 1971), 2. creation of a DNA-membrane-ligand complex similar to heterocyclic stabilization of lysosome membranes (BACH and JOHNSON, 1971). and 3. reduction in the energy available for alteration of bases in DNA (after SUTHERLAND and SUTHERLAND, 1969).

Acridines and phenanthridines (reviewed by MAHLER *et al.*, 1971) also induce formation of respiratory deficient "petite" colonies in yeast, *viz.*, the ϱ^+-ϱ^- "mutation" (EPHRUSSI *et al.*, 1949; SLONIMSKI *et al.*, 1968) and a dyskinetoplastic state in *Trypanosomatidae* (STEINERT and VAN ASSEL, 1967; STEINERT, 1969). Acriflavine has been the acridine of choice in studies with yeast but quinacrine exerted only a weak effect in trypanosomes which was attributed to poor uptake of the drug by the kinetoplast (HAWKING, 1938). In yeast, the "petite" colonies have been considered to be of mutational origin (EPHRUSSI *et al.*, 1949), while the action of acridines and ethidium bromide upon the specialized mitochondrion of trypanosomes, the kinetoplast, has more recently been attributed to inhibition of DNA replication (GUTTMAN and EISENMAN, 1965; STEINERT and VAN ASSEL, 1967; STEINERT, 1969). In both types of organisms, cell growth is required for acridine-induced mitochondrial alteration and, also, of ethidium induction of the dyskinetoplastic state, although incubation of yeast cells with ethidium in nutrient-free medium is sufficient to induce "petites" (SLONIMSKI *et al.*, 1968).

Structural and biochemical changes involved in "petite" colony formation consist at least in part of, 1. a loss of integrity of the inner mitochondrial membrane (absence of cristae) (YOTSUYANAGI, 1962; MAHLER and PERLMAN, 1972a), 2. an almost complete disappearance of cytochrome oxidase and of cytochromes A-A$_3$ and b (SLONIMSKI and EPHRUSSI, 1949; SLONIMSKI and HIRSCH, 1949; LINNANE and STILL, 1956; SHERMAN and SLONIMSKI, 1964; AVERS *et al.*, 1965; PERLMAN and MAHLER, 1970; MAHLER and PERLMAN, 1971), and 3. a reduction in the buoyant density of mitochondrial DNA resulting from an increase in the AT/GC ratio (MEHROTRA and MAHLER, 1968; MOUNOLOU *et al.*, 1966). Additionally, ethidium bromide causes, as a function of time and concentration, a reduction in the molecular weight and an eventual destruction of the mitochondrial DNA (MEHROTRA and MAHLER, 1968; GOLDRING et al., 1970; GOLDRING et al., 1971; MAHLER et al., 1971)

A unified hypothesis for plasmid elimination or destruction of extrachromosomal DNA is lacking, but most likely the mechanism involves the tertiary structure of these DNA molecules and their attachment to cellular membranes. Thus, F'-lac (FREIFELDER, 1968), R (NISIOKA et al., 1969), colicin ogens (BAZARAL and HELINSKI, 1968) and yeast mitochondrial DNA (HOLLENBERG et al., 1970) have been found to be covalently closed circles and it has been suggested (HAHN and CIAK, 1971) that elimination of F'-lac and R by intercalants involves creation of "unnatural left-handed supercoils" which can not be replicated. An alternate hypothesis to explain the greater sensitivity of cytoplasmic replicons than that of chromosomal DNA to intercalants speculated on the existence of different enzymes for replication of different DNA molecules (JACOB et al., 1963); while a third explanation is suggested by the discovery that intercalants, including quinacrine, ethidium bromide, actinomycin and ellipticine, stimulate a yeast mitochondrial enzymatic activity resulting in hydrolysis of DNA, while noninter-calating (WARING, 1970) DNA binders such as berenil, hydroxystilbamadine and spermidine are ineffective (PAOLETTI et al., 1972). Berenil also induces formation of "petite" colonies (PERLMAN and MAHLER, 1973 a and b), but characteristics of induction by berenil, acriflavine and ethidium bromide are different and even reversible, the latter by the former two.

Structure-Activity Relationships: Bacteria

The antibacterial potency of acridines is primarily dependent upon the extent to which the compound is present as a nuclear cation at physiological pH and temperature (BROWNING et al., 1919; RUBBO et al., 1942; ALBERT et al., 1945; The Acridines, by A. ALBERT, 1966). Positively charged cationic acridines are antibacterial (ALBERT et al., 1945; The Acridines, by A. ALBERT, 1966), while differences in ionization due principally to the electron donor capacity and resonance capability of substituent groups result in acridines of different potency. To be effective at pH 7.3 and 37°, an aminoacridine must be ionized to 60% or more; zwitterionic and anionic acridines show little activity. Antibacterial potency also depend upon a planar area of 38 $Å^2$; nonplanarity of even one ring, as in the tetrahydroacridine derivative tacrine, markedly reduces potency (BROWNING et al., 1921; BROWNING et al., 1922; ALBERT et al., 1945; The A cridines, by A. ALBERT, 1966). Smaller, highly ionized N-heterocycles, for example, pyridines and quinolines, do not generally suppress bacterial growth. The requirements for planarity and charge are explicable in terms of the intercalation model in which the acridine is stabilized between the planes of the DNA bases by Van der Waal's forces and charge attraction between cationic acridines and the phosphoric acid groups of the helix backbone.

Amino and nitro groups are the most potency enhancing substituents, but side chains, such as that in quinacrine, reduced potency (GOETSCHIUS and LAW-RENCE, 1944; ALBERT et al., 1945) which may be ascribed to hindrance of interaction between the acridine ring and its target sites or to the lesser stoichiometry of weak interactions apparently possessed by such molecules; both quinacrine and NA 3582 engage in external binding only to the extent of one drug molecule for every three DNA phosphates (KREY et al., 1973; WOLFE et al., 1971 b).

Interaction with Nucleic Acids

Considerable evidence had already accumulated suggesting the importance of the interaction between acridines and DNA when MICHAELIS (1947) described optical properties of basic dyes upon binding to nucleic acids. He inferred a monomeric binding of basic dyes, exemplified by pyronine and toluidine blue, to nucleic acids as indicated by nucleic acid-induced absorption decreases and shifts in the spectra of dyes from shorter to longer wavelengths, i.e. by bathochromic shifts characteristic of a decrease in dye-dye interaction upon conversion of dimers and larger aggregates to monomers. Such shifts were suggested to occur when the dyes were bound to the phosphoric acid groups of the nucleic acids in such manner as to be positioned between the bases.

Optical studies with the octyl homolog of quinacrine (IRVIN et al., 1949; PARKER and IRVIN, 1952) and other acridines, including proflavine (MORTHLAND et al., 1954) revealed such spectral changes, the magnitudes of which were directly related to the extent of polymerization of DNA or RNA, and inversely related to the ionic strength of the medium (IRVIN and IRVIN, 1954). Proflavine apparently bound more strongly to DNA than to RNA (MORTHLAND et al., 1954) while the stronger binding to the nucleic acids of the acridine than of the quinoline moiety suggested that Van der Waals forces as well as Coulomb attractions were involved (PARKER and IRVIN, 1952; IRVIN and IRVIN, 1954). Thereupon, a detailed study (PEACOCKE and SKERRETT, 1956) by spectrophotometric titrations and equilibrium dialysis of the interaction between proflavine and DNA suggested that proflavine became bound to DNA by two modes, one strong and one weak, with the stronger mode possessing an apparent association constant of 1.3×10^6 M^{-1} occurring with a stoichiometry of one dye molecule for every 4.5 phosphate groups, and the weaker process possessing an apparent association constant of 4.9×10^4 M^{-1} involving binding of one proflavine molecule for each phosphate. Both processes were attenuated although to different extents by increases in ionic strength. The suppressive effect of Mg^{2+} was greater than could be accounted for on the basis of ionic strength alone. The strong process was considered to involve formation of a structure consisting of single molecules of proflavine and DNA while the weaker process was thought to produce a structure of aggregates of proflavine bound to DNA. Soon thereafter the aggregative binding was investigated with acridine orange and considered to represent external stacking of the dye bound to phosphates with considerable dye-dye interactions (BRADLEY and WOLF, 1959).

The culmination of the studies of acridine binding to DNA was the advancement of the intercalation hypothesis (LERMAN, 1961) which is based upon changes in the hydrodynamic properties and x-ray diffraction patterns of DNA when complexed with acridine, proflavine or acridine orange. DNA, so complexed, exhibits a large increase in viscosity and a decrease in its rate of sedimentation in the presence of a singly charged dye. These results and low-angle x-ray scattering studies of DNA-proflavine complexes in solution (LUZZATI et al., 1961) were consistent with a binding model in which charged, single acridine molecules are inserted between nucleotide pairs in DNA with a consequent unwinding and extension (CAIRNS, 1962) of the helix to form a long, relatively stiff rod whose mass per unit length is considerably reduced. Such reduction causes a decrease in the sedimentation coefficient while the lengthening of the helix causes an increase in DNA's viscosity. The amino groups of such DNA complexed acridines, including proflavine, underwent interaction with nitrous acid at a rate 1/20th that of free amino-acridines, suggesting that the Van der Waals contours of these molecules were buried within the DNA (LERMAN, 1964 a and b). Additionally, X-ray fiber diffraction patterns suggested that the 3.4 Å spacing between planar structures of molecules of the B-configuration was maintained while the bases remained perpendicular to the long axis of the helix. A disordering of the backbone of the helix was inferred and the data were most consistent with a model in which acridines were intercalated by extension of the helical backbone.

An extensive X-ray diffraction and optical study (NEVILLE and DAVIES, 1966) of the complexes of DNA with proflavine and acridine orange provided further evidence for intercalation of acridines, but simultaneously indicated that a portion of the bound ligands were not intercalated, consistent with the suggestion of external attachment of acridines to DNA (PEACOCKE and SKERRETT, 1956; BRADLEY and WOLF, 1959). A comparison of the thermal denaturation profiles (GERSCH and JORDAN, 1965) of DNA-acridine complexes with those of free DNA showed an increase in the Tm indicative of helix stabilization by acridines, while free energy calculations suggested the validity of the intercalation model. These calculations were based upon consideration of the forces thought to stabilize an intercalated acridine-DNA complex, including charge, permanent dipole and induced dipole interactions. Additionally, the mechanism by which acridines intercalate was deduced (LI and CROTHERS, 1969a) through temperature-jump relaxation experiments to involve an initial external binding of acridines to helical phosphates. Thus, proflavine was considered to complex with DNA by: 1. intercalation, 2. monomeric and external binding to phosphoric acid groups of the helix, and 3. external binding to phosphates with interaction between dye molecules to form stacked aggregates. External stacking of proflavine to bacteriophage T2 DNA is considerably less (LI and CROTHERS, 1969b) by comparison to the stacking of proflavine with other DNAs; this has been attributed to glucosylation of the T2 DNA. This finding is of interest since the external binding of both quinacrine and NA 3582 (KREY et al., 1973; WOLFE et al., 1971b) to DNA show a stoichiometry of approximately 0.33, leading to the inference that projecting substituents of either the DNA or the acridines may hinder stacking.

The intercalation complex is considered by LERMAN to consist of the planar acridine ring buried between DNA base pairs but a number of revisions of this complex model have been suggested. The property of denatured DNA to bind by a similar process 30% to 50% more proflavine and other acridines led to the proposal that the intercalated acridine ring was positioned between successive bases in a single DNA strand, while the charged ring nitrogen interacted with the phosphate of the helical backbone (DRUMMOND et al., 1965; PRITCHARD et al., 1966; ICHIMURA et al., 1969). Such a structure could also explain the apparently partial intercalation of the hydroaromatic acridine derivative,

9-amino-1,2,3,4 tetrahydroacridine. LERMAN had also postulated that intercalation was accompanied by an *un*winding of the helix, but more recently a comprehensive. analysis and discussion (PAOLETTI and LE PECQ, 1971) has resulted in the
suggestion that intercalation is accompanied by winding of the helix.

Discoveries (DULBECCO and VOGT, 1963; WEIL and VINOGRAD, 1963) that
many biological species and organelles possess covalently closed, circular DNA
led to the recognition that intercalants are able to cause more profound changes
in DNA structure than had been assumed. Thus, intercalant hydrodynamic
titrations of covalently closed, supercoiled DNA show (CRAWFORD and WARING,
1967; BAUER and VINOGRAD, 1968; WARING, 1970) transition or equivalence
points at which the DNA possesses the open circular form.

Finally, the stoichiometry of apparently maximal intercalation, one drug molecule for every four bases or two base pairs has led to a "neighbor exclusion"
hypothesis (CROTHERS, 1968; CROTHERS, 1971).

The complex of quinacrine with DNA has been intensively studied and has
been shown to possess two modes of binding, one strong and intercalative, and
the second weaker and external (KURNICK and RADCLIFFE, 1962; LERMAN, 1963;
DRUMMOND et al., 1965; KREY et al., 1973). The stronger process occurs with
a stoichiometry of one molecule of quinacrine for four nucleotides, and a
$K_a = 1.2 \times 10^6$ M^{-1}, while the weaker process occurs with a stoichiometry of
one molecule of quinacrine for three nucleotides and possesses a
$K_a = 4.6 \times 10^4$ M^{-1} (KURNICK and RADCLIFFE, 1962; KREY et al., 1973). These
conclusions were based upon decreases in the rates of sedimentation and increases
in the intrinsic viscosity of complexes of quinacrine with DNA (KURNICK and
RADCLIFFE, 1962) as well as on results of a variety of experiments including
equilibrium dialysis, inhibition of enzymatic DNA hydrolysis, spectrophotometric
and fluorometric titrations, thermal denaturation and quinacrine-induced release
of methyl green from its complex with DNA (KURNICK and RADCLIFFE, 1962;
LE PECQ et al., 1962; LERMAN, 1963; DRUMMOND et al., 1965; WOLFE et al.,
1971 b; KREY et al., 1973). Proof not only of the intercalation of quinacrine
into DNA but of acridines more generally was adduced by flow dichroism and
fluorescence polarization studies which showed that the acridine ring of quinacrine
was aligned to within 30° of the perpendicular of the long axis of the helix
(LERMAN, 1963). The property of quinacrine, chloroquine and spermine to increase
the T_m of DNA to the same extent has led to the assumption (O'BRIEN and
HAHN, 1965; HAHN et al., 1966) that the cationic side chain of quinacrine, in
an orientation similar to that of spermine (MAHLER and MEHROTRA, 1963), spans
the minor groove of the DNA helix to bind to phosphoric acid groups of both
strands. More recently, induction of different Cotton effects in the ORD spectrum
of quinacrine by single-and double-stranded DNAs suggested that quinacrine
might be used as a probe for helicity in nucleic acids (HAHN and KREY, 1968;
KREY et al., 1973).

Fluorescence has been a useful tool for study of acridine-DNA interactions
(LÖBERT and ACHTERT, 1969) and particularly in analysis of human chromosomes
(CASPERSSON et al., 1968). Thus, while acridine orange fluoresces with equal
intensity regardless of the base composition of the nucleic acid to which it is
bound (WEISBLUM and DE HASETH, 1972), quinacrine and proflavine exhibit con-

siderably enhanced fluorescence upon combination with double-stranded polymers rich in A-T or A-U base pairs, in contrast to fluorescence quenching which occurrs in the presence of polymers high in G-C base pairs (TUBBS et al., 1964; THOMES et al., 1969; WEISBLUM and DE HASETH, 1972; KREY et al., 1973). Quinacrine-mustard which covalently binds to DNA does not require double-stranded polymers but merely single-stranded polymers containing A, T or U to exhibit a similar pattern of fluorescence (MICHELSON et al., 1972).

Binding to RNA

The findings that RNA could reverse the growth inhibitory effect of acridines, and alter their spectra (MCILWAIN, 1941; FITZGERALD and LEE, 1945; IRVIN et al., 1949), indicated that RNAs could interact with acridines, and the discovery of the enzyme polynucleotide phosphorylase (GRUNBERG-MANAGO and OCHOA, 1955) led to the study of the interaction between acridine orange and a single RNA-like polymer. Thus, acridine orange was found to form two complexes with poly A, the second of which specifically involves interaction with the adenine ring (STEINER and BEERS, 1958; BEERS et al., 1958; STEINER and BEERS, 1959). Acridine orange also inhibits polynucleotide phosphorylase. The duplex poly A-poly U induces changes in the spectrum of acridine orange, consistent with dye-dye interaction and external stacking (BRADLEY and WOLF, 1959).

The influence of acridines on the in vitro synthesis of protein and on the RNA constituents of these systems has received considerable attention during the past decade. Quinacrine was found to inhibit amino acid incorporation into chloroplast ribosomes (SISSAKIAN et al., 1965) while ribosomes from yeast were observed to alter the spectrum of acridine orange (MORGAN and RHOADS, 1965). Additionally, proflavine suppressed the aminoacylation of transfer RNA (tRNA) with isoleucine (WERENNE et al., 1966). Formation of a complex between acridine orange or proflavine, and 18 s or 28 s RNA extracted from Krebs II cells was deduced from spectral and thermal studies (SEMMEL and DAUNE, 1967). An initial strong binding, considered intercalation, was followed by an apparent structural deformation which permitted further dye binding, while an additional complex was thought to involve external binding of the dyes to phosphates with additional Van der Waals interaction between the dye molecules.

Proflavine inhibited synthesis of polyphenylalanine in a system of rat liver origin (WEINSTEIN and FINKELSTEIN, 1967). Such inhibition was reduced by the addition to the reaction system of graded concentrations of tRNA, ribosomes or poly U. In an accompanying study (FINKELSTEIN and WEINSTEIN, 1967), tRNA was shown to reduce the fluorescence of proflavine to an extent almost equal to that caused by DNA; tRNA also produced bathochromics shift in the spectrum of proflavine. Additionally, the Tm of tRNA increased in the presence of proflavine; all these changes suggested that proflavine intercalated into the base-paired regions of tRNA, a conclusion independently reached by other investigators (GROSJEAN et al., 1968). Poly A, in particular, both reduced absorption and induced a hypsochromic shift in the spectrum of proflavine, while in a buffer of low ionic strength, poly U and poly C also partially reduced the intensity of the visible absorption spectrum of proflavine, indicating that these polynucleotides could bind proflavine and that this acridine could engage in external stacking to poly A (FINKELSTEIN and WEINSTEIN, 1967). Thus, proflavine binds to both single and double-stranded RNA, by intercalation and external ionic binding.

Acridines and other intercalants, particularly ethidium bromide (BOLLEN *et al.,* 1970), have recently been used as molecular probes for the structure and mechanism of function of ribosomes of both prokaryotic and eukaryotic origin. Lack of understanding of the structure and function of ribosomal RNA (rRNA), specifically, has suggested use of selective RNA probes. Thus, shifts in the absorption spectrum of acridine orange were used (FURANO *et al.,* 1966) to estimate the extent of helicity in *E. coli* ribosomes, although these authors concluded that RNA *in situ* possessed little helical structure, contrary to current estimates (ATTARDI and AMALDI, 1970). Acridine orange and intercalants in general tend to reduce ribosomal stability as a function of the concentration of monovalent and divalent cations (COTTER *et al.,* 1967; PERMOGOROV and SLADKOVA, 1968; HULTIN and SJOQVIST, 1969; HULTIN, 1969; HULTIN, 1970; HULTIN, 1972; WOLFE *et al.,* 1972).

The 60 s ribosomal subunit of 80 s rat liver and other mammalian ribosomes, in the presence of the acridines quinacrine, proflavine and acridine orange, undergoes a time- and temperature-dependent reversible unfolding such that the normally unreactive protein p10 becomes accessible to chymotrypsin, thermolysin and procion blue (HULTIN and SJOQVIST, 1969; HULTIN, 1969; HULTIN, 1970; HULTIN, 1972). The property of quinacrine to uncover p10 at 25 C was concentration-dependent and exhibited an inflection at approximately the drug concentration necessary to saturate all hypothetical intercalation sites in the RNA of the ribosomes (HULTIN, 1970).

A relation between the property of compounds to intercalate into supercoiled DNA and the rates at which the compounds induce thermal degradation of *E. coli* ribosomes, has been observed (WOLFE *et al.,* 1972). Intercalants reduced the thermal stability of *E. coli* ribosomes in an order parallel to the fraction of bound drug estimated to be intercalated into supercoiled DNA at its equivalence point (WARING, 1970). The phenanthridines propidium iodine and ethidium bromide were the most effective intercalant destabilizers with proflavine and quinacrine most effective among the acridines; a linear relationship has been observed between intercalant-induced disassemblage of ribosomes and the planar area of such intercalants (WOLFE, 1971). A further indication that quinacrine alters ribosomal properties is that quinacrine inhibits polyphenylalanine synthesis (WOLFE and HAHN, 1972). Acridine orange was found to associate 30 s and 50 s *E. coli* ribosomal subunits in the absence of magnesium (TAL *et al.,* 1973).

Inhibition of Enzyme Reactions

Quinacrine, proflavine and other acridines inhibit numerous enzyme reactions, most particularly those which require a nucleic acid template, or the participation of a flavin coenzyme. In addition, quinacrine inhibits cholinesterase while proflavine inhibits the proteolytic enzyme chymotrypsin. The type of synthesis of DNA carried out by the enzyme, DNA polymerase, and synthesis of RNA, catalyzed by RNA polymerase, are inhibited by quinacrine (O'BRIEN *et al.,* 1966), by proflavine (HURWITZ et al., 1962) and by NA 3582 (WOLFE et al., 1971 b). The apparent complexity of DNA replicative biosynthesis suggests that acridines may inhibit a number of enzyme reactions dependent upon template integrity and accessibility; thus, quinacrine inhibited synthesis of DNA carried out by toluenized E. coli pol A (WOLFE, 1970).

The drug generally inhibits enzymes which utilize a flavin coenzyme, for example, d-amino acid oxidase and "cytochrome reductase" (WRIGHT and SABINE, 1944; HAAS, 1944) and such inhibition most probably accounts for the property of quinacrine to uncouple flavin-dependent oxidative phosphorylation systems. Quinacrine forms a complex with flavin mononucleotide (HEMKER and HULSMAN, 1960) but neither the formation of this complex nor of one with adenine nucleotides, nor the inhibition of cholinesterase activity account for the growth inhibitory properties of quinacrine (ALBERT and MARSHALL, 1948; IRVIN and IRVIN, 1954; WRIGHT and SABINE, 1948).

Among other acridines, proflavine and 1-, 2-, or 3-aminoacridine, but not quinacrine, inhibit competitively the proteolytic enzyme chymotrypsin (WALLACE et al., 1963; BERNHARD et al., 1966), while 9-AA activated the enzyme (WALLACE et al., 1966). Structure-activity relationships were reminiscent of those for the antibacterial action of acridines (ALBERT et al., 1945; The Acridines, by A. ALBERT, 1966).

Conclusion

Three hypotheses of the mode of action of quinacrine have merited serious consideration since the drug was first synthesized: 1. inhibition of respiration, 2. inhibition of membrane function, and 3. inhibition of nucleic acid synthesis. The overwhelming mass of evidence favors the idea that quinacrine and other acridines suppress microbial growth through interference with the function of nucleic acid templates.

References

ALBERT, A.: The acridines. St Martin's Press, 2nd edition. New York: 1966.

ALBERT, A., and P. B. MARSHALL: Absence of correlation between respiratory inhibition of plasmodia by acridines and antimalarial action. Nature 161, 1008 (1948).

ALBERT, A., S. D. RUBBO, J. J. GOLDACRE, M. E. DAVEY, and J. D. STONE: The influence of chemical constitution on antibacterial activity. Part II. A general survey of the acridine series. Brit. J. Exptl. Path. 26, 160 (1945).

ALLISON, A. C., and M. R. YOUNG: Uptake of dyes and drugs by living cells in culture. Life Sci. 3 (12), 1407 (1964).

ALPATOV, V. V.: Effect of specific action of optical isomers of mepacrine on dextral and sinistral strains of Bacillus mycoides. Nature 158, 838 (1946).

ALTMAN, S., and L. S. LERMAN: Effects of 9-aminoacridine on bacteriophage T4 deoxyribonucleic acid synthesis. J. Mol. Biol. 50, 263 (1970).

AMES, B. N., and H. J. WHITFIELD, JR.: Frameshift mutagenesis in salmonella. Cold Spring Harbor Symp. Quant. Biol. 31, 221 (1966).

ANSEL, M., and M. THIBAUT: Antifungal activity of mepacrine hydrochloride in vitro. Therapie 20 (4), 1027 (1965).

ATTARDI, G., and F. AMALDI: Structure and synthesis of ribosomal RNA. Ann. Rev. Biochem. 39, 183 (1970).

AVERS, C. J., C. R. PFEFFER, and M. W. RANCOURT: Acriflavine induction of different kinds of "petite" mitochondrial populations in *Saccharomyces cerevisiae*. J. Bacteriol. **90**, 481 (1965).

AZZI, A., A. FABBRO, M. SANTATO, and P. L. GHERARDINI: Energy transduction in mitochondrial fragments: Interaction of the membrane with acridine dyes. Eur. J. Biochem. **21**, 404 (1971).

BACH, M. K.: Reduction in the frequency of mutation to resistance to cytarabine in L1210 murine leukemic cells by treatment with quinacrine hydrochloride. Cancer Res. **29**, 1881 (1969).

BACH, M. K., and H. G. JOHNSON: In: Progress in molecular and subcellular Biology, Vol. 2, F. E. HAHN, ed. Berlin-Heidelberg-New York: Springer 1971.

BARNER, H. D., and S. S. COHEN: The induction of thymine synthesis by T2 infection of a thymine requiring mutant of *Escherichia coli*. J. Bacteriol. **68**, 80 (1954).

BAUER, W., and J. VINOGRAD: The interaction of closed circular DNA with intercalative dyes. I. The superhelix density of SV40 DNA in the presence and absence of dye. J. Mol. Biol. **33**, 141 (1968).

BAZARAL, M., and D. R. HELINSKI: Circular forms of colocinogenic factors E1, E2 and E3 from *Escherichia coli*. J. Mol. Biol. **36**, 185 (1968).

BEERS, R., JR.: Acridine orange binding by *Micrococcus lysodeikticus*. J. Bact. **88**, 1249 (1964).

BEERS, R. F., D. D. HENDLEY, and R. F. STEINER: Inhibition and activation of polynucleotide phosphorylase through the formation of complexes between acridine orange and polynucleotides. Nature **182**, 242 (1958).

BERNHARD, S. A., B. F. LEE, and Z. H. TASHJIAN: On the interaction of the active site of α chymotrypsin with chromophores: Proflavine binding and enzyme conformation during catalysis. J. Mol. Biol. **18**, 405 (1966).

BERNS, M. W., S. EL-KADI, R. S. OLSON, and D. E. ROUNDS: Laser photosensitization and metabolic inhibition of tissue culture cells treated with quinacrine hydrochloride. Life Sci. (II) **9**, 1061 (1970).

BERNSTEIN, A., and J. LEDERBERG: Agglutination of motile salmonellas by acridines. J. Bacteriol. **69**, 142 (1955).

BEUKERS, R.: The effect of proflavine on U.V.-induced dimerization of thymine in DNA. Photochemistry and Photobiology **4**, 935 (1965).

BISSEL, H., H. W. MOELLER, and L. D. SEIF: The spectrophotometric determination of quinacrine hydrochloride (atabrine). J. Am. Pharm. Assoc. **34**, 291 (1945).

BOCK, E., et M. OSTERLIN: Über einige fluoreszenzmikroskopische Beobachtungen. Zentr. Bakteriol. Parasitenk. Abt. I Orig. **143**, 306 (1938-9).

BOLLEN, A., A. HERZOG, A. FAVRE, J. THIBAULT, and F. GROS: Fluorescence studies on the 30 S-ribosome assembly process. FEBS Letters **11**, 49 (1970).

BRADLEY, D. F., and M. K. WOLF: Aggregation of dyes bound to polyanions. Proc. Natl. Acad. Sci. U. S. **45**, 947 (1959).

BRENNER, S., F. R. S. L. BARNETT, F. H. C. CRICK, and A. ORGEL: The theory of mutagenesis. J. Mol. Biol. **3**, 121 (1961).

BRILMAYER, C., A. KOHLER, A. MACK, and K. STORDEUR: The affinity of certain dyes for normal and neoplastic tissue, as shown by quantitative extractions. Z. Krebsforsch. **60**, 334 (1955).

BROWNING, C. H., J. B. COHEN, R. GAUNT, and R. GULBRANSEN: Relationships between antiseptic action and chemical constituents with special reference to compounds of the pyridine, quinoline, acridine and phenazine series. Proc. Roy. Soc. (London) **93 B**, 329 (1922).

BROWNING, C. H., J. B. COHEN, and R. GULBRANSEN: Further observations on the relation between chemical constituents and antiseptic action. J. Pathol. Bacteriol. **24**, 127 (1921).

BROWNING, C. H., and W. GILMOUR: Bactericidal action and chemical constitution with special reference to basic benzene derivatives. J. Pathol. Bacteriol. **18**, 144 (1913).

BROWNING, C. H., R. GULBRANSEN, and E. L. KENNAWAY: Hydrogen-ion concentration and antiseptic potency, with special reference to the action of acridine compounds. J. Pathol. Bacteriol. **23**, 106 (1919).

BURNEY, T. E., and O. J. GOLUB: The effect of certain enzyme inhibitors on the activity and growth of psittacosis virus. J. Immunol. **60**, 213 (1948).

CAIRNS, J.: The application of autoradiography to the study of DNA viruses. Cold Spring Harbor Symp. Quant. Biol. **27**, 311 (1962).

CALDWELL, P. C., and C. HINSHELWOOD. The nucleic acid content of *Bact. lactis aerogenes*. J. Chem. Soc. **1950**, 1415.

CASPERSSON, T., S. FARBER, G.E. FOLEY, J. KUDYNOWSKI, E.J. MODEST, E. SIMONSSON, V. WAGH, and L. ZECH: Chemical differentiation along metaphase chromosomes. Exptl. Cell Res. **49**, 219 (1968).

CHAMBRON, J., M. DAUNE et C. SADRON: Etude thermodynamique de l'interaction de la proflavine avec l'acide desoxyribonucleique. I. Etude par equilibre de dialyse. Biochim. Biophys. Acta **123**, 306 (1966).

CHELINTSEV, G.V., I.L. KNUNYANTZ, and Z.V. BENEVOLENSKAYA: The structure and synthesis of new antimalarial substances. The structure of "atebrin". Compt. Rend. Acad. Sci. U.S.S.R. (N.S.) I, **63** (1934) through Chem. Abstr. **28**, 2126 (1934).

CHELINTSEV, G.V., and E.D. OSETROVA: Separation of atebrin into optical isomers. J. Gen. Chem. U.S.S.R. **10**, 1978 (1940) through Chem. Abstr. **35**, 4029 (1941).

CHESSIN, M.: Photodynamic inactivation of infectious nucleic acid. Science **132**, 1840 (1960).

CHOU, S.C., and S. RAMANATHAN: Quinacrine: Site of inhibition of synchronized cell division in tetrahymena. Life Sci. **7**, 1053 (1968).

CHOU, S.C., S. RAMANATHAN, and W.C. CUTTING: Quinacrine: Inhibition of synchronized cell division in Tetrahymena. Pharmacology **1**, 60 (1968).

CHRISTOPHERS, S.R.: Dissociation constants and solubilities of bases of antimalarial compounds. I. Quinine. II. Atebrin. Ann. Trop. Med. Parasitol. **31**, 43 (1937).

CIAK, J., and F.E. HAHN: Chloroquine: Mode of action. Science **151**, 347 (1966).

CIAK, J., and F.E. HAHN: Quinacrine (Atebrin): Mode of action. Science **156**, 655 (1967).

CLANCY, C.F.: The lethal effect of certain antimalarial drugs on *Tetrahymena pyriformis*. Am. J. Trop. Med. Hyg. **17**, 359 (1968).

CLARKE, D.H.: Use of phosphorus^{32} in studies on *Plasmodium gallinaceum*. II. Studies on conditions affecting growth in intact cells and in lysates. J. Exptl. Med. **96**, 451 (1952).

COGGESHALL, L.T., and J. MAIER: Determination of the activity of various drugs against the malaria parasite. J. Infect. Diseases **69**, 108 (1941).

COHEN, S.N., K.O. PHIFER, and K.L. YIELDING: Complex formation between chloroquine and ferrihaemic acid *in vitro* and its effect on the antimalarial action of chloroquine. Nature **202**, 805 (1964).

CONKLIN, K.A., S.C. CHOU, and P. HEU: Quinine: Effect of *Tetrahymena pyriformis*. 3. Energetics of isolated mitochondria in the presence of quinine and other antimalarial drugs. Biochem. Pharmacol. **20**, 1877 (1971).

COTTER, R.I., P. MCPHIE, and W.B. GRATZER: Internal organization of the ribosome. Nature **216**, 864 (1967).

COURSEILLE, C., B. BUSETTA, and M. HOSPITAL: Crystalline structure of quinacrine. Compt. Rend. Ser. C **272**, 34 (1971).

CRAWFORD, L.V., and M.J. WARING: Supercoiling of polyoma virus DNA by its interaction with ethidium bromide. J. Mol. Biol. **25**, 23 (1967).

CRICK, F.H.C., L. BARNETT, S. BRENNER, and R.J. WATTS-TOBIN: General nature of the genetic code. Nature **192**, 1227 (1961).

CROTHERS, D.M.: Calculation of binding isotherms for heterogeneous polymers. Biopolymers **6**, 575 (1968).

CROTHERS, D.M.: In: Progress in molecular and subcellular biology, vol. 2, F.E. HAHN, ed. Berlin-Heidelberg-New York: Springer 1971.

CROWTHER, D., and J.L. MELNICK: The incorporation of neutral red and acridine orange into developing poliovirus particles making them photosensitive. Virology **14**, 11 (1961).

CULBERTSON, J.T.: Elimination of the tapeworm *Hymenolopis fraterna* from mice by administration of atebrin. J. Pharmacol. **70**, 309 (1940).

DE COURCY, S.J., JR.: In: Progress in molecular and subcellular biology, vol. 2, F.E. HAHN, ed. Berlin-Heidelberg-New York: Springer 1971.

DEEGAN, T., and B.G. MAEGRITH: Studies on the nature of malarial pigment (Haemozoin). I. The pigment of the simian species *Plasmodium knowlesi* and *P. cynomolgi*. Ann. Trop. Med. Parasitol. **50**, 194 (1956a).

DEEGAN, T., and B.G. MAEGRITH: Studies on the nature of malarial pigment (Haemozoin). II. The pigment of the human species *Plasmodium falciparum* and *P. malariae*. Ann. Trop. Med. Parasitol. **50**, 212 (1956b).

DE MARS, R.I.: The production of phage-related materials when bacteriophage development is interrupted by proflavine. Virology **1**, 83 (1955).

DE MARS, R.I., S.E. LURIA, H. FISHER, and C. LEVINTHAL: The production of incomplete bacterio-phage particles by the action of proflavine and the properties of the incomplete particles. Ann. Inst. Pasteur **84**, 113 (1953).

DRUMMOND, D.S., V.F.W. SIMPSON-GILDEMEISTER, and A.R. PEACOCKE: Interaction of aminoacri-dines with deoxyribonucleic acid: Effects of ionic strength, denaturation and structure. Biopo-lymers **3**, 135 (1965).

DULBECCO, R., and M. VOGT: Evidence for a ring structure of polyoma virus DNA. Proc. Natl. Acad. Sci. U.S. **50**, 236 (1963).

EATON, M.D., A. VAN ALLEN, and A. WIENER: Action of acridines on agents of the psittacosis-lymphogranuloma group. Proc. Soc. Exptl. Med. Biol. **66**, 141 (1947).

EDGAR, R.S., and R.H. EPSTEIN: Inactivation by ultraviolet light of an acriflavine-sensitive gene function in phage T4D. Science **134**, 327 (1961).

EPHRUSSI, B., H. HOTTINGER et A. CHIMENES: Action de l'acriflavine sur les levures. I. La mutation "petite colonie". Ann. Inst. Pasteur. **76**, 351 (1949).

FAIRLEY, N.H.: Chemotherapeutic suppression and prophylaxis in malaria. Trans. Roy. Soc. Trop. Med. Hyg. **38**, 311 (1945).

FEDORKO, M.E., and J.G. HIRSCH: Nucleolar fragmentation in L cells exposed to quinacrine *in vitro*. Cancer Res. **29**, 918 (1969).

FINKELSTEIN, R., and I.B. WEINSTEIN: Proflavine binding to transfer ribonucleic acid, synthetic ribonucleic acids and deoxyribonucleic acid. J. Biol. Chem. **242**, 3762 (1967).

FITZGERALD, R.J., and D. BABBITT: Studies on bacterial viruses. I. The effect of certain compounds on the lysis of *Escherichia coli* by bacteriophage. J. Immunol. **52**, 121 (1946).

FITZGERALD, R.J., and M.E. LEE: Studies on bacterial viruses. II. Observations on the mode of action of acridines in inhibiting lysis of virus-infected bacteria. J. Immunol. **52**, 127 (1946).

FOSTER, R.A.C.: An analysis of the action of proflavine on bacteriophage growth. J. Bacteriol. **56**, 795 (1948).

FRANKEL, F.R.: Studies on the nature of replicating DNA in T4-infected *Escherichia coli*. J. Mol. Biol. **18**, 127 (1966).

FRASER, D., and H.R. MAHLER: Studies in partially resolved bacteriophage-host systems. VII. Dia-mines, dyes, empty phage heads and protoplast-infecting agent. Biochim. Biophys. Acta **53**, 199 (1961).

FREIFELDER, D.: Studies on *Escherichia coli* sex factors. III. Covalently closed F' Lac DNA molecules. J. Mol. Biol. **34**, 31 (1968).

FULTON, J.D., and S.R. CHRISTOPHERS: The inhibitive effect of drugs upon oxygen uptake by trypano-somes *(Trypanosoma rhodesiense)* and malaria parasites *(Plasmodium knowlesi)*. Ann. Trop. Med. Parasitol. **32**, 77 (1938).

FURANO, A.V., D.F. BRADLEY, and L.G. CHILDERS: The conformation of the ribonucleic acid in ribosomes. Biochemistry **5**, 3044 (1966).

GARNHAM, P.C.C.: Malaria parasites and other haemosporidia. Oxford: Blackwell Scientific Publica-tions 1966.

GENDON, Y.Z.: Induction of mutations in poliomyelitis virus by direct action of proflavine on virus RNA. Vopr. Virusol. **8**, 542 (1963).

GERKE, P.Y.: Analysis of quinacrine deposits in human skin. Vestn. Venerol. Dermatol. **5**, 21 (1948) through Chem. Abstr. **43**, 2701 F (1949).

GERSCH, N.F., and D.O. JORDAN: Interaction of DNA with aminoacridines. J. Mol. Biol. **13**, 138 (1965).

GOETCHIUS, G.R., and C.A. LAWRENCE: The antibacterial effects of various acridine compounds. J. Lab. Clin. Med. **29**, 134 (1944).

GOLDRING, E.S., L.I. GROSSMAN, D. KRUPNICK, D.R. CRYER, and J. MARMUR: The petite mutation in yeast. Loss of mitochondrial deoxyribonucleic acid during induction of petites with ethidium bromide. J. Mol. Biol. **52**, 323 (1970).

GOLDRING, E.S., L.I. GROSSMAN, and J. MARMUR: Petite mutation in yeast. II. Isolation of mutants containing mitochondrial deoxyribonucleic acid of reduced size. J. Bacteriol. **107**, 377 (1971).

GOODMAN, L.S., and A. GILMAN: The pharmacological basis of therapeutics, 2nd edition. New York: The Macmillan Company 1960.

GOSS, W.A., W.H. DEITZ, and T.M. COOK: Mechanism of action of nalidixic acid on *E. coli* I. J. Bacteriol. **88**, 1112 (1964).

GREENHALGH, N., R. HULL, and E.W. HURST: The antiviral activity of acridines in eastern equine encephalomyelitis, Rift Valley fever and psittacosis in mice and lymphogranuloma venereum in chick embryos. Brit. J. Pharmacol. **11**, 220 (1956).

GRINSTED, J.: Antimicrobial drugs and RNA. Biochim. Biophys. Acta **179**, 268 (1969).

GROSJEAN, H., J. WERENNE, and H. CHANTRENNE: The binding of proflavine to transfer ribonucleic acid: Dependence on secondary structure. Biochim. Biophys. Acta **166**, 616 (1968).

GRUNBERG-MANAGO, M., and S. OCHOA: Enzymatic exchange of P^{32} orthophosphate with ADP and IDP. Federation Proc. **14**, 221 (1955).

GUTTMAN, H.N., and R.N. EISENMAN: Acriflavin-induced loss of kinetoplast deoxyribonucleic acid in Crithidia fasciculata (Culex pipiens strain). Nature **207**, 1341 (1965).

HAAS, E.: The effect of atabrine and quinine on isolated respiratory enzymes. J. Biol. Chem. **155**, 321 (1944).

HAHN, F.E., and J. CIAK: The problems of drug-resistant pathogenic bacteria. Elimination of bacterial episomes by DNA-complexing compounds. Ann. N.Y. Acad. Sci. **182**, 295 (1971).

HAHN, F.E., and A. KREY: Deoxyribonucleic acid-induced anomalous optical rotatory dispersion of antimalarial drugs and dyes. Antimicrobial Agents Chemotherapy **1968**, 15 (1969).

HAHN, F.E., R.L. O'BRIEN, J. CIAK, J.L. ALLISON, and J.G. OLENICK: Studies on modes of action of chloroquine, quinacrine and quinine and on chloroquine resistance. Military Med. **131**, 1071 (1966).

HARTWELL, J.L., M.J. SHEAR, J.M. JOHNSON, and S.R.L. KORNBERG: Selection and synthesis of organic compounds (for action against tumors in mice). Cancer Res. **6**, Proc. 489 (1946).

HAWKING, F.: Trypanocidal action of atebrin in relation to absorption by trypanosomes. Ann. Trop. Med. Parasitol. **32**, 383 (1938).

HECHT, G.: Pharmacological studies on atebrin. Arch. Exptl. Pathol. Pharmakol. **170**, 328 (1933).

HEMKER, H.C., and W.C. HÜLSMANN: Inhibition of enzymes by atebrin. Biochim. Biophys. Acta **44**, 175 (1960).

HENKIND, P., and N. ROTHFIELD: Ocular abnormalities in patients treated with synthetic antimalarial drugs. New Engl. J. Med. **269**, 433 (1963).

HESSLER, A.Y.: Acridine-resistant mutants of T2H bacteriophage. Genetics **48**, 1107 (1963).

HESSLER, A.Y.: Acridine resistance in bacteriophage T2H as a function of dye penetration measured by mutagenesis and photoinactivation. Genetics **52**, 711 (1965).

HESSLER, A.Y., M.B. BAYLOR, and J.P. BAIRD: Acridine sensitivity of bacteriophage T2H in *Escherichia coli*. J. Virol. **1**, 543 (1967).

HIROTA, Y.: The effect of acridine dyes on mating type factors in *Escherichia coli*. Proc. Natl. Acad. Sci. U.S. **46**, 57 (1960).

HIROTA, Y., and T. IIJIMA: Acriflavine as an effective agent for eliminating F factor in *E. coli* K12. Nature **180**, 655 (1957).

HOLLENBERG, C.P., P. BORST, and E.J. VAN BRUGGEN: Mitochondrial DNA. V. A 25 μ closed circular duplex DNA molecule in wild-type yeast mitochondria. Structure and genetic complexity. Biochim. Biophys. Acta **209**, 1 (1970).

HOTCHINS, J.E.: Influence of acridines on the interaction of *Staphylococcus aureus* and *Staphylococcus* K phage. J. Gen. Microbiol. **5**, 609 (1951).

HUHNE, W.: The morphological changes in *Plasmodium falciparum* under the influence of atebrin. Deut. Trop. Z. **46**, 385 (1942).

HULTIN, T.: The use of procion blue as a molecular probe in the study of ribosome structure. Eur. J. Biochem. **9**, 579 (1969).

HULTIN, T.: Effects of aminoacridines and related compounds on the conformation of rat-liver ribosomes. Chem. Biol. Interact. **2**, 61 (1970).

HULTIN, T.: Evidence for disulfide interaction *in situ* between two adjacent proteins in mammalian 60-S ribosomal subunits. Biochim. Biophys. Acta **269**, 118 (1972).

HULTIN, T., and A. SJOQVIST: A site of reversible conformational alteration in rat liver ribosomes. Biochim. Biophys. Acta **182**, 147 (1969).

HURST, E.W., P. MELVIN, and J.M. PETERS: The prevention of encephalitis due to the virus of eastern equine encephalomyelitis and louping-ill. Experiments with trypan red, mepacrine and many other substances. Brit. J. Pharmacol. **7**, 455 (1952a).

Hurst, E. W., J. M. Peters, and P. Melvin: The action of mepacrine and trypan red in a number of virus diseases. Brit. J. Pharmacol. **7**, 473 (1952b).

Hurwitz, J., J. J. Furth, M. Malamy, and M. Alexander: The role of deoxyribonucleic acid in ribonucleic acid synthesis. III. The inhibition of the enzymatic synthesis of ribonucleic acid and deoxyribonucleic by actinomycin D and proflavine. Proc. Natl. Acad. Sci. U. S. **48**, 1222 (1962).

Ichimura, S., M. Zama, H. Fujita, and T. Ito: The nature of strong binding between acridine orange and deoxyribonucleic acid as revealed by equilibrium dialysis and thermal renaturation. Biochim. Biophys. Acta **190**, 116 (1969).

Ilan, J., and J. Ilan: Aminoacyl transfer ribonucleic acid synthetases from cell-free extract of *Plasmodium berghei*. Science **164**, 560 (1969).

Irvin, J. L., and E. M. Irvin: The interaction of a 9-aminoacridine derivative with nucleic acids and nucleoproteins. J. Biol. Chem. **206**, 39 (1954).

Irvin, J. L., and E. M. Irvin: The interaction of quinacrine with adenine nucleotides. J. Biol. Chem. **210**, 45 (1954).

Irvin, J. L., E. M. Irvin, and F. S. Parker: The interaction of antimalarials with nucleic acids. Science **110**, 426 (1949).

Iyks, S. R.: Formation of noninfectious particles of poliomyelitis virus in a tissue culture treated with certain acridine compounds. Tr. Mosk. Nauchn.-Issled. Inst. Vakts. Sycorotok **18**, 179 (1963) through Chem. Abstr. **63**, 603ZC (1965).

Jacob, F., S. Brenner, and F. Cuzin: On the regulation of DNA replication in bacteria. Cold Spring Harbor Symp. Quant. Biol. **28**, 329 (1963).

Jacobs, R. L.: Selections of strains of *Plasmodium berghei* resistant to quinine, chloroquine and pyrimethamine. J. Parasitol. **51**, 481 (1965).

James, S. P.: The direct effect of atebrin on the parasites of benign tertian and quartan malaria. Trans. Roy. Soc. Trop. Med. Hyg. **28**, 3 (1934–35).

Johnson, H. G., and M. K. Bach: The antimutagenic action of polyamines: Suppression of the mutagenic action of an *E. coli* mutator gene and of 2-aminopurine. Proc. Natl. Acad. Sci. U.S. **55**, 1453 (1966).

Johnson, H. G., and M. K. Bach: Apparent antimutagenic activity of quinacrine hydrochloride in Detroit-98 human sternal marrow cells grown in culture. Cancer Res. **29**, 1367 (1969).

Jones, R., Jr., C. Price, and A. K. Sen: Nitrogen mustards related to chloroquine, pamaquine and quinacrine. J. Org. Chem. **22**, 783 (1957).

Kay, D.: The inhibition of bacteriophage multiplication by proflavine and its reversal by certain polyamines. Biochem. J. **73**, 149 (1959).

Kellenberger, E., and G. Kellenberger: Electron microscopical studies of phage multiplication. III. Observation of single cell bursts. Virology **3**, 275 (1957).

Kellenberger, E., A. Ryter, and J. Sechaud, Electron microscope study of DNA-containing plasms. II. Vegetative and mature phage DNA as compared with normal bacterial nucleoids in different physiological states. J. Biophys. Biochem. Cytol. **4**, 671 (1968).

Kellenberger, E., and J. Sechaud: Electron microscopical studies of phage multiplication. II. Production of phage-related structures during multiplication of phage T2 and T4. Virology **3**, 256 (1957).

Kellenberger, E., J. Sechaud, and A. Ryter: Electron microscopical studies of phage multiplication. IV. The establishment of the DNA pool of vegetative phage and the maturation of phage particles. Virology **8**, 478 (1959).

Krey, A., R. G. Allison, and F. E. Hahn: In preparation (1973).

Kurnick N. B., and I. E. Radcliffe: Reaction between DNA and quinacrine and other antimalarials. J. Lab. Clin. Med. **60**, 669 (1962).

Kushner, D. J., and S. R. Khan: Proflavine uptake and release in sensitive and resistant *Escherichia coli*. J. Bacteriol. **96**, 1103 (1968).

Lantz, C. H., and K. Van Dyke: Studies concerning the mechanism of action of antimalarial drugs. II. Inhibition of the incorporation of adenosine-5-monophosphate-H[3] into nucleic acids of erythrocyte-free malarial parasites. Biochem. Pharmacol. **20**, 1157 (1971).

Lawrence, C. A.: Effects of quinine, atebrin and substituted acridine compounds on gram-negative bacteria *in vitro*. Proc. Soc. Exptl. Biol. Med. **52**, 90 (1943).

Le Pecq, J. B., J. Y. Le Talaer, B. Festy, and R. Truhaut: Inhibition of deoxyribonuclease (DNAase) by complexing deoxyribonucleic acid with dyes. Compt. Rend. **254**, 3918 (1962).

LERMAN, L. S.: Structural considerations in the interaction of DNA acridines. J. Mol. Biol. **3**, 18 (1961).

LERMAN, L. S.: The structure of the DNA-acridine complex. Proc. Natl. Acad. Sci. U.S. **49**, 94 (1963).

LERMAN, L. S.: Amino group reactivity in DNA-aminoacridine complexes. J. Mol. Biol. **10**, 367 (1964a).

LERMAN, L. S.: Acridine mutagens and DNA structure. J. Cellular Comp. Physiol. **64**, Suppl. 1 (1964b).

LEVY, S. B., and T. WATANABE: Mepacrine and transfer of R factors. Lancet 1966-II, 1138.

LI, H. J., and D. M. CROTHERS: Relaxation studies of the proflavine-DNA complex: The kinetics of an intercalation reaction. J. Mol. Biol. **39**, 461 (1969a).

LI, H. J., and D. M. CROTHERS: Studies of the optical properties of the proflavine-DNA complex. Biopolymers **8**, 217 (1969b).

LINNANE, A. W., and J. L. STILL: A reexamination of some problems of "petite" yeast. Australian J. Sci. **19**, 165 (1956).

LOBERT, G., and G. ACHTERT: On the complex formation of acridine dyes with DNA. VII. Dependence of the binding on the dye structure. Biopolymers **8**, 595 (1969).

LUZZATI, J., F. MASSON, and L. S. LERMAN: Interaction of DNA and proflavine: A small-angle x-ray scattering study. J. Mol. Biol. **3**, 634 (1961).

MACKERRAS, M. J., and Q. N. ERCOLE: Some observations on the action of quinine, atebrin and plasmoquine on plasmodium vivax. Trans. Roy. Soc. Trop. Med. Hyg. **42**, 443 (1949).

MACOMBER, P. B., R. L. O'BRIEN, and F. E. HAHN: Chloroquine: Physiological basis of drug resistance in *Plasmodium berghei*. Science **152**, 1374 (1966).

MAGIDSON, O. YU., and A. M. GRIGOROVSKII: Acridine compounds and their antimalarial action. I. Chem. Ber. **69B**, 396 (1936a).

MAGIDSON, O. YU., and A. I. TRAVIN: Acridine compounds and their antimalarial action. II. Compounds with cyano and methylmercapto groups. Chem. Ber. **69B**, 537 (1936b).

MAHLER, H. R., and B. D. MEHROTRA: The interaction of nucleic acids with diamines. Biochim. Biophys. Acta **68**, 211 (1963).

MAHLER, H. R., B. D. MEHROTRA, and P. S. PERLMAN: In: Progress in Molecular and Subcellular Biology; Vol. 2, F. E. HAHN, ed. Berlin-Heidelberg-New York: Springer 1971.

MAHLER, H. R., and P. S. PERLMAN: Mitochondrio-genesis analyzed by blocks on mitochondrial translation and transcription. Biochemistry **10**, 2979 (1971).

MAHLER, H. R., and P. S. PERLMAN: Effects of mutagenic treatment by ethidium bromide on cellular and mitochondrial phenotype. Arch. Biochem. Biophys. **148**, 115 (1972a).

MALLING, H. V.: The mutagenicity of the acridine mustard (ICR-170) and the structurally related compounds in Neurospora. Mutation Res. **4**, 265 (1967).

MAUSS, I. H., and F. MIETZSCH: Note on the work of O. Yu. Magidson and A. M. Grigorovskii: Acridine compounds and their antimalarial action. Chem. Ber. **69B**, 641 (1936).

MCILWAIN, H.: A nutritional investigation of the antibacterial action of acriflavine. Biochem. J. **35**, 1311 (1941).

MEHROTRA, B. D., and H. R. MAHLER: Characterization of some unusual DNA's from the mitochondria from certain "petite" strains of Saccharomyces cerevisiae. Arch. Biochem. **128**, 685 (1968).

MICHAELIS, L.: The nature of the interaction of nucleic acids and nuclei with basic dyestuffs. Cold Spring Harbor Symp. Quant. Biol. **12**, 131 (1947).

MICHELSON, A. M., C. MONNY, and A. KOVOOR: Action of quinacrine mustard on polynucleotides. Biochimie **54**, 1129 (1972).

MIETZSCH, F., and H. MAUSS (inventors): I. G. Farbenind. A.-G. German Patent 553,072 (1930).

MILLER, A. K., and L. PETERS: The antagonism by spermine and spermidine of the antibacterial action of quinacrine and other drugs. Arch. Biochem. Biophys. **6**, 281 (1945).

MILLER, O. B., F. HERRMANN, and J. RUBIN: The effects of mepacrine hydrochloride (atebrin) upon the human skin. J. Invest. Dermatol. **15**, 445 (1950).

MITSUHASHI, S., K. HARADA, and M. KAMEDA: Elimination of transmissible drug-resistance by treatment with acriflavine. Nature **189**, 947 (1961).

MODELL, W.: Malaria and victory in Vietnam. The first battle against drug-resistant malignant malaria is described. Science **162**, 1346 (1968).

MORGAN, R. A., and D. J. RHOADS: Binding of acridine orange to yeast ribosomes. Biochim. Biophys. Acta **102**, 311 (1965).

MORTHLAND, F. W., P. P. H. DeBRUYN, and N. H. SMITH: Spectrophotometric studies on the interaction of nucleic acids with amino acridines and other basic dyes. Exptl. Cell Res. **7**, 201 (1954).

MOULDER, J. W., and E. A. EVANS: The biochemistry of the malaria parasite. VI. Studies on the nitrogen metabolism of the malaria parasite. J. Biol. Chem. **164**, 145 (1946).

MOUNOLOW, J. C., H. JAKOB, and P. P. SLONIMSKI: Mitochondrial DNA from yeast "petite" mutants: Specific changes of buoyant density corresponding to different cytoplasmic mutations. BBRC **24**, 218 (1966).

MÜHLENS, P., and O. FISHER: The treatment of malaria with atebrin. Arch. Schiffs-Tropen Hyg. **36**, 196 (1932).

NAGEL DE ZWAIG, R., and S. E. LURIA: Genetics and physiology of colicin-tolerant mutants of *Escherichia coli*. J. Bacteriol. **94**, 1112 (1967).

NAKAMURA, H.: Gene-controlled resistance to acriflavine and other basic dyes in *Escherichia coli*. J. Bacteriol. **90**, 8 (1965).

NAKAMURA, H.: Acriflavine binding capacity of *Escherichia coli* in relation to acriflavine sensitivity and metabolic activity. J. Bacteriol. **92**, 1447 (1966).

NAKAYA, R., A. NAKAMURA, and Y. MURATA: Resistance transfer agents in *Shigella*. Biochem. Biophys. Res. Commun. **3**, 654 (1960).

NEVILLE, D. M., JR., and D. R. DAVIES: The interaction of acridine dyes with DNA: An X-ray diffraction and optical investigation. J. Mol. Biol. **17**, 57 (1966).

NEWELL, H. W., and T. LIDZ: The toxicity of atebrin to the central nervous system. Am. J. Psychiat. **102**, 805 (1946).

NISIOKA, T., M. MITANI, and R. CLOWES: Composite circular forms of R factor deoxyribonucleic acid molecules. J. Bacteriol. **97**, 376 (1969).

O'BRIEN, R. L., and F. E. HAHN: Chloroquine structural requirements for binding to deoxyribonucleic acid and antimalarial activity. Antimicrobial Agents Chemotherapy **1965**, 315.

O'BRIEN, R. L., J. G. OLENICK, and F. E. HAHN: Reactions of quinine, chloroquine and quinacrine with DNA and their effects on the DNA and RNA polymerase reactions. Proc. Soc. Acad. Sci. U. S. **55**, 1511 (1966).

ORGEL, A., and S. BRENNER: Mutagenesis of bacteriophage T4 by acridines. J. Mol. Biol. **3**, 762 (1961).

PAGE, F.: Treatment of lupus erythmatosus with mepacrine. Lancet **261**, 755 (1951).

PALMER, L. G., and A. SAWITSKY: Fatal aplastic anemia following quinacrine therapy in chronic discoid lupus erythmatosus. J. Am. Med. Assoc. **153**, 1172 (1953).

PAOLETTI, C., H. COUDER, and M. GUERINEAU: A yeast mitochondrial deoxyribonuclease stimulated by ethidium bromide. Biochem. Biophys. Res. Commun. **48**, 950 (1972).

PAOLETTI, J., and J. LE PECQ: Resonance energy transfer between ethidium bromide molecules bound to nucleic acids. J. Mol. Biol. **59**, 43 (1971).

PARKER, F. S., and J. L. IRVIN: The interaction of chloroquine with nucleic acids and nucleoproteins. J. Biol. Chem. **199**, 897 (1952).

PEACOCKE, A. R., and J. N. H. SKERRETT: The interaction of aminoacridines with nucleic acids. Trans. Faraday Soc. **52**, 261 (1956).

PERLMAN, P. S., and H. R. MAHLER: (2) Formation of yeast mitochondria. III. Biochemical properties of mitochondria isolated from a cytoplasmic petit mutant of yeast. J. Bioenerget. **1**, 113 (1970).

PERLMAN, P. S., and H. R. MAHLER: Induction of respiratory deficient mutants in saccharomyces cerevisiae by Berenil. I. Berenil, a novel nonintercalating mutagen. Mol. Gen. Genet. **121**, 294 (1973a).

PERLMAN, P. S., and H. R. MAHLER: Induction of respiratory deficient mutants in saccharomyces cerevisiae by Berenil. II. Characteristics of the process. Mol. Gen. Genet. **121**, 306 (1973b).

PERMOGOROV, V. I., and I. A. SLADKOVA: Optical assay of free phospate groups in ribosomes of *Escherichia coli*. Mol. Biol. **2**, 276 (1968).

PETERS, W.: Mepacrine- and primaquine-resistant strains of *Plasmodium berghei* Vincke and Lips, 1948. Nature **208**, 293 (1965a).

PETERS, W.: Drug resistance in *Plasmodium berghei* Vincke and Lips, 1948. I. Chloroquine resistance. Exptl. Parasitol. **17**, 80 (1965b).

PETERS, W.: Drug resistance in *Plasmodium berghei* Vincke and Lips, 1948. III. Multiple drug resistance. Exptl. Parasitol. **17**, 97 (1965c).

PETERS, W.: Drug responses of mepacrine- and primaquine-resistant strains of *Plasmodium berghei*. Ann. Trop. Med. Parasitol. **60**, 25 (1966).

Peters, W.: Chemotherapy and Drug Resistance in Malaria. London and New York: Academic Press 1970.

Peters, W., K.A. Fletcher, and W. Staubli: Phagotrophy and pigment formation in a chloroplast resistant strain of *Plasmodium berghei*. Trans. Roy. Soc. Trop. Med. Hyg. **59**, 2 (1965).

Piechowski, M.M., and M. Susman: Acridine-resistance in phage T4D. Genetics **56**, 133 (1967).

Polet, H., and C.F. Barr: DNA, RNA and protein synthesis in erythrocytic forms of *Plasmodium knowlesi*. Am. J. Trop. Med. Hyg. **17**, 672 (1968a).

Polet, H., and C.F. Barr: Uptake of chloroquine-3-H^3 by *Plasmodium knowlesi in vitro*. J. Pharmacol. Exptl. Therap. **168**, 187 (1969).

Pratt, D., G.S. Stent, and P.D. Harriman: Stabilization to ^{32}P decay and onset of DNA replication of T4 bacteriophage. J. Mol. Biol. **3**, 409 (1961).

Pritchard, N.J., A. Blake, and A.R. Peacocke: Modified intercalation model for the interaction of amino acridines and DNA. Nature **212**, 1360 (1966).

Roinishvili, E.S.: Inactivation of infectious DNA of bacteriophage dφ4 with ultraviolet radiation and phage protection with acridine dyes. Zh. Mikrobiol. Epidemiol. i. Immunobiol. **47**, 133 (1970).

Roth, D., M. Manjon, and M. London: Lasting changes in acriflavine binding induced in mammalian cells by exogenous DNA. Exptl. Cell Res. **53**, 101 (1968).

Rownd, R., R. Nakaya, and A. Nakamura: Molecular nature of the drug-resistance factors of the Enterobacteriaceae. J. Mol. Biol. **17**, 376 (1966).

Rubbo, S.D., A. Albert, and M. Maxwell: The influence of chemical compounds on antiseptic activity. I. A study of the monoaminoacridines. Brit. J. Exptl. Pathol. **23**, 69 (1942).

Rudzinska, M.A., and W. Trager: Phagotrophy and two new structures in the malaria parasite *Plasmodium berghei*. J. Biophys. Biochem. Cytol. **6**, 103 (1959).

Russell, H.K.: Eosinophilia caused by atebrin. U.S. Naval Med. Bull. **44**, 574 (1945).

Scatchard, G.: The attraction of proteins for small molecules and ions. Ann. N.Y. Acad. Sci. **51**, 660 (1949).

Schaffer, F.L.: Binding of proflavine by and photoinactivation of poliovirus propagated in the presence of dye. Virology **18**, 412 (1962).

Schellenberg, K.A., and G.R. Coatney: The influence of antimalarial drugs on nucleic acid synthesis in *Plasmodium gallinaceum* and *Plasmodium berghei*. Biochem. Pharmacol. **6**, 143 (1961).

Schueler, F.W., and W.F. Cantrell: Antagonism of the antimalarial action of chloroquine ferrihemate and an hypothesis for the mechanism of chloroquine resistance. J. Pharmacol. **143**, 278 (1964).

Seligman, M.L., and H.G. Mandel: Inhibition of growth and RNA biosynthesis of *Bacillus cereus* by quinacrine. J. Gen. Microbiol. **68**, 135 (1971).

Semmel, M., and M. Daune: Etudes des complexes de colorants basiques avec le RNA. Biochim. Biophys. Acta **145**, 561 (1967).

Sesnowitz-Horn, S., and E.A. Adelberg: Proflavin treatment of *Escherichia coli:* Generation of frameshift mutations. Cold Spring Harbor Symp. Quant. Biol. **33**, 393 (1968).

Setlow, R.B., and W.L. Carrier: Formation and destruction of pyrimidine dimers in polynucleotides by ultra-violet irradiation in the presence of proflavine. Nature **213**, 906 (1967).

Setlow, J.K., and R.B. Setlow: Contributions of dimers containing cytosine to ultra-violet inactivation of transforming DNA. Nature **213**, 907 (1967).

Sevag, M.G.: Prevention of the emergence of antibiotic-resistant strains of bacteria by atabrine. Arch. Biochem. Biophys. **108**, 85 (1964).

Sevag, M.G., and B. Ashton: Evolution and prevention of drug resistance. Nature **203**, 1323 (1964).

Sevag, M.G., and W.T. Drabble: Prevention of the emergence of drug resistant bacteria by polyamines. Biochem. Biophys. Res. Commun. **8**, 446 (1962).

Sherman, I.W., B.J. Mudd, and W. Trager: Chloroquine resistance and the nature of malarial pigment. Nature **208**, 691 (1965).

Sherman, F., and P.P. Slonimski: Respiration-deficient mutants of yeast. II. Biochemistry. Biochim. Biophys. Acta **90**, 1 (1964).

Silver, S.: Acriflavine resistance: A bacteriophage mutation affecting the uptake of dye by the infected bacterial cells. Proc. Natl. Acad. Sci. U.S. **53**, 24 (1965).

Silver, S.: Acridine sensitivity of bacteria T-2: A virus gene affecting cell permeability. J. Mol. Biol. **29**, 191 (1967).

SILVER, S., E. LEVINE, and P. SPIELMAN: Factors governing acridine uptake: Temperature, pH, surface chemistry. J. Bacteriol. **95**, 333 (1968).

SILVERMAN, M.: Antagonism of antibacterial action of atebrin. J. Biol. Chem. **172**, 849 (1948).

SILVERMAN, M., J. CEITHMAL, L. G. TALIAFERRO, and E. A. EVANS, JR.: The *in vitro* metabolism of *Plasmodium gallenaceum*. J. Infect. Diseases **75**, 212 (1944).

SILVERMAN, M., and E. A. EVANS, JR.: The effects of spermidine and other polyamines on the growth inhibition of *Escherichia coli* by atabrine. J. Biol. Chem. **15**, 265 (1943).

SILVERMAN, M., and E. A. EVANS, JR.: The effects of spermine, spermidine and other polyamines on the growth inhibition of *Escherichia coli* by atabrine. J. Biol. Chem. **154**, 521 (1944).

SINGER, J. A., and W. P. PURCELL: Huckel molecular orbital calculations for some antimalarial drugs and related molecules. J. Med. Pharm. Chem. **10**, 754 (1967).

SISSAKIAN, N. M., I. I. FILIPPOVICH, E. N. SVETAILO, and K. A. ALIYEV: On the protein-synthesizing system of chloroplasts. Biochim. Biophys. Acta **95**, 474 (1965).

SLONIMSKI, P., et B. EPHRUSSI: Action de l'acriflavine sur les levures V. Le systems des cytochromes des mutants ‹petite colonie›. Ann. Inst. Pasteur **77**, 47 (1949).

SLONIMSKI, P., et H. M. HIRSCH: Nouvelles données sur la constitution enzymatique du mutant ‹petite colonie› de Saccharomyces cerevisiae. Compt. Rend. **235**, 741 (1949).

SLONIMSKI, P. P., G. PERRODIN, and J. H. CROFT: Ethidium bromide induced mutation of yeast mitochondria: Complete transformation of cells into respiratory deficient non-chromosomal "petites". Biochem. Biophys. Res. Commun. **30**, 232 (1968).

SOFFER, R. L., and F. GROS: Effects of dinitrophenol and proflavine on information transfer mechanisms in *Escherichia coli*: A study *in vivo* and *in vitro*. Biochim. Biophys. Acta **87**, 423 (1964).

STEINER, R. F., and R. F. BEERS: Spectral changes accompanying the binding of acridine orange by polyadenylic acid. Science **127**, 335 (1958).

STEINER, R. F., and R. F. BEERS: Polynucleotides. V. Titration and spectrophotometric studies upon the interaction of synthetic polynucleotides with various dyes. Arch. Biochem. Biophys. **81**, 75 (1959).

STEINERT, M.: Specific loss of kinetoplastic DNA in Trypanosomatidae treated with ethidium bromide. Exptl. Cell Res. **55**, 248 (1969).

STEINERT, M., and S. VAN ASSEL: The loss of kinetoplast DNA in two species of Trypanosomatidae treated acriflavine. J. Cell Biol. **34**, 489 (1967).

STEINER, R. F., and R. F. BEERS, JR.: Spectral changes accompanying binding of acridine orange by polyadenylic acid. Science **127**, 335 (1958).

STEINER, R. F., and R. F. BEERS: Polynucleotides. V. Titration and spectrophotometric studies upon the interaction of synthetic polynucleotides with various dyes. Arch. Biochem. Biophys. **81**, 75 (1959).

STONE, A. L., and D. F. BRADLEY: Aggregation of cationic dyes on acid polysaccharides. I. Spectrophotometric titration with acridine orange and other metachromatic dyes. Biochim. Biophys. Acta **148**, 172 (1967).

STREISINGER, G., Y. OKADA, J. EMRICH, J. NEWTON, A. TSUGITA, E. TERZAGHI, and M. INOUYE: Frameshift mutations and the genetic code. Cold Spring Harbor Symp. Quant. Biol. **31**, 77 (1966).

SUGINO, Y.: Mutants of *Escherichia coli* sensitive to methylene blue and acridines. Genet. Res. Camb. **7**, 1 (1966).

SUSMAN, M., M. M. PIECHOWSKI, and D. A. RITCHIE: Studies on phage development. I. An acridine-sensitive clock. Virology **26**, 163 (1965).

SUTHERLAND, B. M., and J. C. SUTHERLAND: Mechanisms of inhibition of pyrimidine dimer formation in deoxyribonucleic acid by acridine dyes. Biophys. J. **9**, 292 (1969).

TAL, M., H. ROTEM, M. ALFASI, and R. A. BERG: The effect of acridine orange on the structure of ribosomes. Biopolymers **12**, 173 (1973).

THOMES, J. C., G. WEILL, and M. DAUNE: Fluorescence of Proflavine-DNA complexes: Heterogeneity of binding sites. Biopolymers **8**, 647 (1969).

THONARD-NEUMANN, E.: The treatment of natural malaria with atebrin in Columbia. Arch. Schiffs-Tropen-Hyg. **36**, 357 (1932).

TROPP, C., and W. WEISE: The excretion of atebrin in the urine and feces. Naunyn-Schmiedebergs Arch. Exptl. Pathol. Pharmak. **170**, 339 (1933).

TUBBS, R. K., W. E. DITMARS, and Q. VAN WINKLE: Heterogeneity of the interaction of DNA with acriflavine. J. Mol. Biol. **9**, 545 (1964).

UDENFRIEND, S., D. E. DUGGAN, B. M. VASTA, and B. B. BRODIE: A spectrophotofluorometric study of organic compounds of pharmacological interest. J. Pharmacol. Exptl. Therap. **120**, 26 (1957).

ULTMANN, J.E., A. GELLHORN, M. OSNOS, and E. HIRSCHBERG: Effect of quinacrine on neoplastic effusions and certain of their enzymes. Cancer **16**, 283 (1963).

VAN DYKE, K., C. LANTZ, and C. SZUSTKIEWICZ: Quinacrine: Mechanisms of antimalarial action. Science **169**, 492 (1970).

VAN DYKE, K., and C. SZUSTKIEWICZ: Apparent new modes of antimalarial action detected by inhibited incorporation of adenosine-8-H^3 into nucleic acids of *Plasmodium berghei*. Military Med. **134**, 1000 (1969).

VAN DYKE, K., C. SZUSTKIEWICZ, and C.H. LANTZ: Studies concerning the mechanism of action of antimalarial drugs: Inhibition of the incorporation of adenosine-8-H^3 into nucleic acids of *Plasmodium berghei*. Biochem. Pharmacol. **18**, 1417 (1969).

VAN DYKE, K., G.C. TREMBLAY, C.H. LANTZ, and C. SZUSTKIEWICZ: The source of purines and pyrimidines in *Plasmodium berghei*. Amer. J. Trop. Med. Hyg. **19**, 202 (1970).

WALLACE, R. A., A. N. KURTZ, and C. NIEMANN: Interaction of aromatic compounds with α-chymotrypsin. Biochemistry **2**, 824 (1963).

WALLACE, R. A., R. L. PETERSON, C. NIEMANN, and G. E. HEIN: Activation of chymotrypsin catalyted hydrolyses by 9-aminoacridine. Biochem. Biophys. Res. Commun. **23**, 246 (1966).

WARING, M.J.: Variations of the supercoils in closed circular DNA by binding of antibiotics and drugs. J. Mol. Biol. **54**, 247 (1970).

WARING, M. J.: In: Progress in Molecular and Subcellular Biology, Vol. 2, F. E. HAHN, ed. Berlin-Heidelberg-New York: Springer 1971.

WATANABE, T., and T. FUKASAWA: Episome-mediated transfer of drug resistance in enterobacteriaceae. II. Elimination of resistance factors with acridine dyes. J. Bacteriol. **81**, 679 (1961).

WEIL, R., and J. VINOGRAD: The cyclic helix and cyclic coil forms of polyoma viral DNA. Proc. Natl. Acad. Sci. U.S. **50**, 730 (1963).

WEINBERG, E.D., J.H. BILLMAN, and D. BORDERS: Lysis of *Bacillus subtilis* by amines, acridines and phenothiazines. Exptl. Cell Res. **15**, 625 (1958).

WEINSTEIN, I.B., and I.H. FINKELSTEIN: Proflavine inhibition of protein synthesis. J. Biol. Chem. **242**, 3757 (1967).

WEISBLUM, B., and P.L. DE HASETH: Quinacrine, a chromosome stain specific for deoxyadenylate-deoxythymidylate-rich regions in DNA. Proc. Soc. Natl. Acad. U.S. **69**, 629 (1972).

WERENNE, J., H. GROSJEAN, and H. CHANTRENNE: Effect of proflavine on the binding of isoleucine to transfer RNA. Biochim. Biophys. Acta **129**, 585 (1966).

WISELOGLE, F.Y.: Survey of antimalarial drugs, 1941–1945. Michigan: J.W. Edwards, Ann Arbor 1946.

WOESE, C., S. NAONO, R. SOFFER, and F. GROS: Studies on the breakdown of messenger RNA. Biochem. Biophys. Res. Commun. **11**, 435 (1963).

WOLFE, A.D.: In: Progress in Molecular and Subcellular Biology, Vol. 2, F.E. HAHN, ed. Berlin-Heidelberg-New York: Springer 1971.

WOLFE, A. D.: Unpublished observations. 1970.

WOLFE, A. D., R. G. ALLISON, and F. E. HAHN: Labilizing action of intercalating drugs and dyes on bacterial ribosomes. Biochemistry **11**, 1569 (1972).

WOLFE, A.D., T.M. COOK, and F.E. HAHN: Antibacterial Nitroacridine, Nitroakridin 3582: Effects on bacterial growth and macromolecular biosynthesis *in vivo*. J. Bacteriol. **108**, 320 (1971a).

WOLFE, A.D., T.M. COOK, and F.E. HAHN: Antibacterial Nitroacridine, Nitroakridine 3582: Binding to nucleic acids *in vitro* and effects on selected cell-free model systems of macromolecular biosynthesis. J. Bacteriol. **108**, 1026 (1971b).

WOLFE, A.D., and F.E. HAHN: Inhibition by atebrin (quinacrine) of model protein synthesis *in vitro*. Naturwissenschaften **59**, 277 (1972).

WRIGHT, C.I., and J.C. SABINE: The effect of atebrine on the oxygen consumption of the tissues. J. Biol. Chem. **155**, 315 (1944).

WRIGHT, C.I., and J.C. SABINE: Cholinesterases of human erythrocytes and plasma and their inhibition by antimalarial drugs. J. Pharmacol. Exptl. Therap. **93**, 230 (1948).

YAMAMOTO, N.: Photodynamic inactivation of bacteriophage and its inhibition. J. Bacteriol. **75**, 443 (1958).

YOSHIKAWA, M.: Selective enrichment of R-segregants as the main mechanism of curing of the R factor by acridine dyes. Genet. Res. Camb. **17**, 9 (1971).

YOSHIKAWA, M., and M. G. SEVAG: Sensitivity of *Escherichia coli* to atabrine conferred by R factor and its potential clinical significance. J. Bacteriol. **93**, 245 (1967).

YOTSUYANAGI, Y.: Etudes sur le chrondriome de la levure. II. Chondriomes des mutants a deficience respiratoire. J. Ultrastruct. Res. **7**, 141 (1962).

ZAMENHOF, P. J.: On the identity of two bacterial mutator genes: Effects of antimutagens. Mutation Res. **7**, 463 (1969).

ZELEZNICK, L. D., J. A. GRIM, and G. D. GRAY: Immunosuppression by compounds which complex with deoxyribonucleid acid. J. A. CRIM, Pharmacol. **18**, 1823 (1969).

Quinoxaline Antibiotics

Ken Katagiri, Tadashi Yoshida, and Kosaburo Sato

Quinoxaline antibiotic is a generic name given to a family of heterodetic cyclodepsipeptide antibiotics which contain a quinoxaline moiety in the molecule (KUROYA et al., 1961). These colorless antibiotics are derived from various species of Streptomycetes and are highly active against gram-positive bacteria. They also exhibit a cytotoxic effect on cultured cells. The antibiotics were shown to inhibit various experimental tumors (NITTA et al., 1955; ROSSOLIMO et al., 1959; KATAGIRI and SUGIURA, 1961) and to protect mice from viral infection (TSUNODA, 1962), although within a narrow concentration range.

Classification and Chemistry

Levomycin (CARTER et al., 1954) and actinoleukin (UEDA et al., 1954) were the first antibiotics in this family produced by streptomyces, and since then, a number of antibiotics have been reported, e.g. echinomycin (CORBAZ et al., 1957), antibiotics X-948 and X-1008 (BERGER et al., 1957), antibiotic 6270 (PREO-BRAZHENSKAYA et al., 1959), quinomycins (YOSHIDA et al., 1961) and triostins (SHOJI and KATAGIRI, 1961). In 1959, KELLER-SCHIERLEIN and PRELOG elucidated in detail the chemical structure of echinomycin (KELLER-SCHIERLEIN and PRELOG, 1957, KELLER-SCHIERLEIN et al., 1959), which showed two moles of quinoxaline-2-carboxylic acid are attached through amide bonds to the D-serine residues of an eight-membered cyclic depsipeptide, as shown in Fig. 1.

Subsequently, KATAGIRI and his collaborators established by chromatographic studies as well as by the physicochemical criteria (YOSHIDA et al., 1961; SHOJI and KATAGIRI, 1961), that quinoxaline antibiotics can be placed in two major series: quinomycins and triostins. Quinomycin and triostin only differ in the linkage between two N-methylcysteine residues in the peptide portion, and both are identical throughout the rest of the structure such as the quinoxaline moiety and the sequence of the amino acids in the peptide (Fig. 1). OTSUKA and SHOJI (1965b) determined the structural formula of triostin C which contains a N,N'-dimethylcystine residue instead of a six-membered dithian ring in the structure of quinomycin groups.

Both antibiotics are mixtures of several chemically related components which are synthesized simultaneously during the production process. These components proved to vary only at N-methylvaline sites in the peptide (OTSUKA and SHOJI, 1963). The synonomy and the amino acid compositions of various quinoxaline antibiotics are given in Table 1. Quinomycin-producing streptomyces produce three components, A, Bo, and C, under conditions generally employed, and

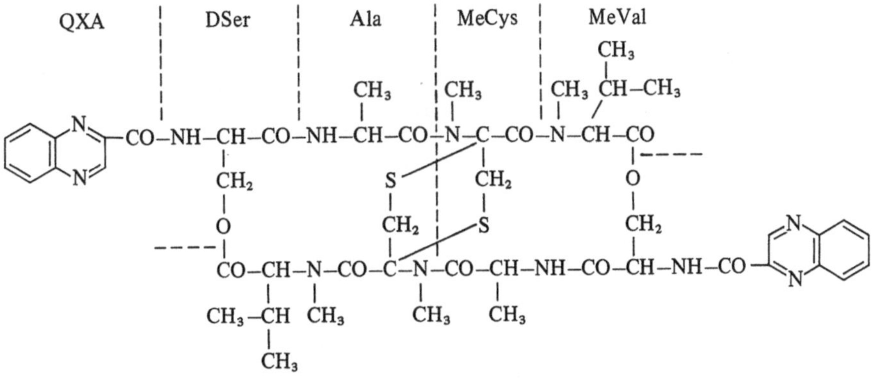

Echinomycin (Quinomycin A)

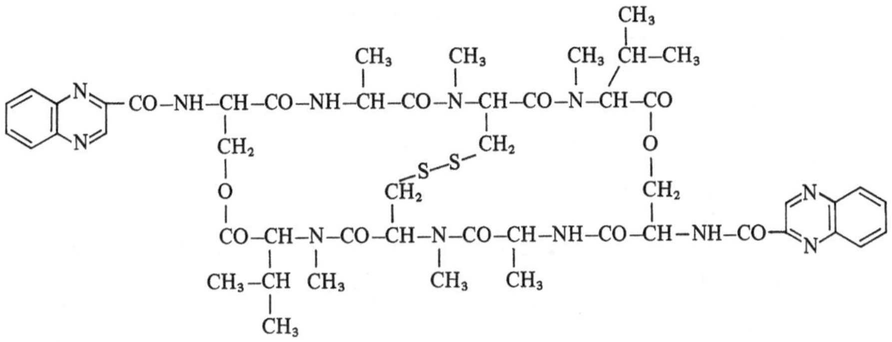

Triostin A

Fig. 1. Structure of quinoxaline antibiotics. QXY = quinoxaline 2-carboxylic acid, DSer = D-serine, Ala = L-alanine, MeCys = N-methyl-L-cysteine, MeVal = N-methyl-L-valine

the relative content of the three components in quinomycin depends upon the strains (YOSHIDA *et al.*, 1961). In addition, three quinomycins designated B, D, and E, were observed to be produced in the presence of DL-isoleucine in the culture medium (YOSHIDA and KATAGIRI, 1962, 1967). These components were found to contain one or two moles of N-methylalloisoleucine in place of N-methylvaline sites in the peptide (OTSUKA and SHOJI, 1965a; SHOJI *et al.*, 1965). As the replacement of the amino acids in the peptide occurs only at the N-methylvaline sites with N-methylalloisoleucine and N,γ-dimethylalloisoleucine, there can be as many as 12 components in both series of antibiotics. Ten components including six quinomycins: A, B, C, D, E, and Bo and four triostins: A, B, C, and Bo have been isolated so far and their structures have been determined (OTSUKA and SHOJI, 1966, 1967). Levomycin, actinoleukin, echinomycin, antibiotic X-948, and quinomycin A are chemically identical as described by several investigators (BERGER *et al.*, 1957; ISHIHARA *et al.*, 1958; KATAGIRI *et al.*, 1962). Although the structural formula of antibiotic 6270 has not been determined completely, this antibiotic was suggested to be identical to quinomycin C by chromatographic behavior (ROSSOLIMO *et al.*, 1959).

Table 1. *The quinoxaline antibiotics*

Antibiotic series	Component	Synonym	Amino acids in N-methylvaline sites	Culture condition	Reference
Quinomycin	A	Levomycin	MeVal, MeVal	normal	1
		Actinoleukin			2
		Echinomycin			3
		Antibiotic X-948			4
		Quinomycin A			5
	Bo	Quinomycin Bo	MeVal, Me$_2$aIle	normal	5
	C	Antibiotic 6270	Me$_2$aIle, Me$_2$aIle	normal	6
		Quinomycin C			5
	B	Quinomycin B	MeaIle, MeaIle	DL-isoleucine	7
	D	Quinomycin D	MeVal, MeaIle	DL-isoleucine	8
	E	Quinomycin E	MeaIle, Me$_2$aIle	DL-isoleucine	8
Triostin	A	Triostin A	MeVal, MeVal	normal	9
	Bo	Triostin Bo	MeVal, Me$_2$aIle	normal	10
	C	Triostin C	Me$_2$aIle, Me$_2$aIle	normal	9
	B	Triostin B	MeaIle, MeaIle	DL-isoleucine	11

MeVal = N-methyl-L-valine, MeaIle = N-methyl-alloisoleucine, Me$_2$aIle = N,γ-dimethyl-L-alloisoleucine.

References: 1. Carter *et al.* (1954). 2. Ueda *et al.* (1954). 3. Corbaz *et al.* (1957). 4. Berger *et al.* (1957). 5. Yoshida *et al.* (1961). 6. Preobrazhenskaya *et al.* (1959). 7. Yoshida and Katagiri (1962). 8. Yoshida and Katagiri (1967). 9. Shoji and Katagiri (1961). 10. Otsuka and Shoji (1966). 11. Otsuka and Shoji (1965a).

Table 2. *The producers of quinoxaline antibiotics*

Antibiotic series	Species	Strain No.	Component	Reference
Quinomycins	*S. flaveolus*		A	Ueda *et al.* (1954)
	S. aureus	Q-11A	A	Ueda *et al.* (1954)
	S. echinatus	8331	A	Corbaz *et al.* (1957)
	Species similar to *S. echinatus*	X-948	A	Berger *et al.* (1957)
	S. aureus	732	A, Bo, C	Kuroya *et al.* (1961)
	S. griseolus	3950	A, C	Ivanitskaya *et al.* (1961)
	S. flavochromogenes	6270	C	Ivanitskaya *et al.* (1961)
	S. lavendulae		A	Maksimova *et al.* (1965)
Triostins	*S. aureus*	s-2-210L	A, Bo, C	Kuroya *et al.* (1961)

The streptomyces species producing quinoxaline antibiotics are widely distributed in nature (Maksimova *et al.*, 1965). Table 2 gives a list of representative species and the antibiotic components produced in their cultures. Triostin is a unique antibiotic in this family because of its limited distribution as to producing organism; a particular strain belonging to *Streptomyces aureus* has been reported by KUROYA et al. (1961). On the contrary, quinomycin was found to be produced by a variety of species including even the actinomycin-producing strain of S. flaveolus (UMEZAWA et al., 1951).

On the basis of the biosynthetic studies on quinoxaline antibiotics (YOSHIDA and KATAGIRI, 1967, 1969), the possibility of substitution was indicated not only of amino acids in the peptide portion but also of the chromophore, the quinoxaline moiety. YOSHIDA and his collaborators (1968) examined a number of compounds chemically analogous to quinoxaline-2- carboxylic acid as a precursor, and the supplemention of quinaldinic acid to the culture of quinomycin-producing organism resulted in the synthesis of two new derivatives of quinomycin A, designated QN-quinomycin A and NX-quinomycin A. These were determined to be substituted quinomycin A molecule in which two and one residues respectively of quinoline had replaced the usual quinoxaline (YOSHIDA et al., 1968). It is important to note that such a replacement of the chromophore resulted in significant improvements in biological activity, as described later (Table 11). Similarly, QN-triostin A and NX-triostin A were isolated biosynthetically from the triostin-producing organism (YOSHIDA et al., 1970). Subsequently, KHAN and his group reported the production of an echinomycin analogue, quinazomycin, in which quinoxaline-2-carboxyl moiety is replaced by a quinazol-4-one-3-acetyl residue (KHAN et al., 1969).

Although none of the quinoxaline antibiotics have yet been chemically synthesized, attempts to improve their chemotherapeutic utilities as antitumor agents have been made by several investigators (KOPPEL et al., 1963; GERCHAKOV and SCHULTZ, 1969). They synthesized a number of N-(2-quinoxaloyl) amino acids or peptides, but none of the tested compounds displayed any antitumor activity.

Antimicrobial Activity

The antimicrobial activities of quinoxaline antibiotics are summarized in Table 3. Gram-positive bacteria, anaerobic bacteria and acid fast bacteria showed a relatively high degree of susceptibility, but gram-negative bacteria and fungi were generally insensitive. Very strong anti-mycoplasma activity was shown by these antibiotics (minimal inhibitory concentration is lower than 0.1 µg/ml) (KATAGIRI et al., 1966). Furthermore, moderate antiprotozoal activities were reported (UEDA et al., 1954, CORBAZ et al.,1957; KATAGIRI et al., 1957).

As triostin C showed high antibacterial activity in vitro and had a low toxicity to mice, its therapeutic effect on mice infected with Staphylococcus aureus (SMITH) was tested. No therapeutic activity was observed (SHOJI and KATAGIRI, 1961).

Resistance of Staphylococcus aureus 209P to the antibiotics increased more than 100-fold when the bacteria were subcultured repeatedly in media containing increased concentrations of the individual antibiotics. Cross-resistance tests among these antibotics were performed (KATAGIRI and MATSUURA, 1963) and the results are shown in Table 4. Each resistant strain showed a specific homologous resistance, but a very weak or almost negligible cross-resistance to the heterologous antibiotics. These results suggested that although these antibiotics were similar in chemical structure, they may have their own specific antibacterial modes of action.

Table 3. *Antimicrobial properties of quinoxaline antibiotics*
(YOSHIDA *et al.*, 1961; SHOJI *et al.*, 1961; KATAGIRI *et al.*, 1963)

Microorganism	Triostin	Quinomycin		
	C	A	B	C
Staphylococcus aureus (209P)	0.10	0.10	0.02	0.02
Bacillus subtilis (PCI219)	0.05	0.02	0.01	0.01
Bacillus anthracis	0.1	0.2	0.05	0.05
Sarcina lutea	0.05	0.02	0.005	0.005
Klebsiella pneumoniae	> 10.0	> 10.0	> 10.0	> 10.0
Escherichia coli	> 10.0	> 10.0	> 10.0	> 10.0
Salmonella typhosa	> 10.0	> 10.0	> 10.0	> 10.0
Shigella dysenteriae (Shiga)	> 10.0	1.0	2.0	2.0
Pseudomonas aeruginosa	> 10.0	> 10.0	> 10.0	> 10.0
Mycobacterium tuberculosis ($H_{37}RV$)	0.2	2.0	0.2	1.0
Diplococcus pneumoniae	0.02	0.05	0.05	0.05
Streptococcus heamolyticus	0.1	0.02	0.01	0.02
Corynebacterium diphtheriae	0.01	0.005	0.005	0.005
Clostridium chauvoei	0.001	0.01	0.001	0.001

Minimum inhibitory concentration, mcg/ml.

Table 4. *Cross-resistance among quinoxaline antibiotics* (KATAGIRI and MATSUURA, 1963)

Staph. aureus 209P	Antibiotic			
	QA	QB	QC	TC
QA Resistant	128	32	16	16
QC Resistant	2	16	128	16
TC Resistant	4	8	32	512

QA = quinomycin A, QB = quinomycin B, QC = quinomycin C, TC = triostin C.

Number: Resistance index $= \dfrac{\text{MIC of resistant strain}}{\text{MIC of original strain}}$

Toxicity

Data about the intraperitoneal LD_{50} of these antibiotics in mice are summarized in Table 5. As quinomycins showed a delayed toxicity, the results were also observed on the 5th day after injection. In spite of their similar physicochemical characteristics, the LD_{50} in mice (inbred Swiss albino, DS strain) showed a difference. LD_{50} values through various routes of administrations, have been determined with quinomycin A (Table 6). When administered intraperitoneally, the toxicity was higher compared to that shown through other routes of administration.

During the course of antitumor testing on these antibiotics, it was found that triostin C was ten times more toxic to the rat than to the mouse (KATAGIRI and SUGIURA, 1961). ROSSOLIMO *et al.* (1959) reported that antibiotics 6270, which are similar to quinomycin, caused atrophy of the spleen. Quinomycins, however, did not cause atrophy of the spleen in mice (KATAGIRI and SUGIURA, 1961).

Table 5. *Intraperitoneal acute toxicity in mice (mg/kg)* (YOSHIDA *et al.*, 1961; SHOJI *et al.*, 1961)

	QA	QB	QC	TC
LD$_{50}$ (2 days)	0.7	—	0.3	—
LD$_{50}$ (5 days)	0.4	0.04	0.03	100

Table 6. *Acute toxicity of quinomycin A in mice*

Route of administration	LD$_{50}$ (mg/kg)
i.p.	0.02
s.c.	0.28
i.v.	0.28

Observed on the 12th day after injection.

Cytotoxic Effect

HARADA *et al.* (1968) determined the cytotoxicity of quinoxaline antibiotics (quinomycin A and triostin C) in cultured cells JTC-13, in comparison with actinomycin B, by light and electron microscopic observations (after 4 hours of incubation with three dose levels ranging from 0.1 to 10 µg/ml). It was found that quinomycin A has a cytotoxicity close to that of actinomycin B, quantitatively. The cytotoxicity of triostin C was one-tenth that of quinomycin A.

By staining with acridine orange, a common pattern of cytotoxicity was observed among these antibiotics i.e. dispersion of the nucleoli and chromatin aggregation near the nuclear membrane. The ultrastructural changes consisted of coalescence of nucleolonema, the aggregation of the granular component and the decrease of amorphous materials in the nucleoli. Condensed granules were completely separated from the nucleolonema in the nucleus.

A difference in cytotoxic behavior was observed for the quinoxaline antibiotics, which showed a less alteration of cytoplasmic organellae than actinomycin.

These results suggested that quinoxaline antibiotics may act directly on nucleic acid synthesis. Thus, the mechanisms of action observed in bacteria (SATO *et al.*, 1967b) were also inferred to obtain in animal cell cultures.

Antitumor Activity

Among the quinoxaline antibiotics, actinoleukin (NITTA *et al.*, 1955) and antibiotics 6270 (ROSSOLIMO *et al.*, 1959), were reported to have antitumor activity.

KATAGIRI and SUGIURA (1961) examined the effect of quinoxaline antibiotics on the growth of 31 solid tumors, 2 ascites tumors and 1 virus leukemia. The results obtained with quinomycins A, B, C and triostin complex (a mixture of 40% A, 60% C with a trace of B) are shown in Table 7. The repeated daily injections of 0.01 mg/kg of quinomycin C had a destructive effect on Ehrlich ascites carcinoma, a marked inhibitory effect on sarcoma 180 ascites tumor,

Table 7. *Effect of the quinomycins and triostins on various mouse, rat, hamster, and chicken tumors* (KATAGIRI and SUGIURA, 1961)

Drug	Quinomycin A		Quinomycin B		Quinomycin C		Triostins	
	Dose[a]	Evaluation	Dose	Evaluation	Dose	Evaluation	Dose	Evaluation
Mouse tumors:								
Sarcoma 180 (solid)	0.05	—	0.01	—	0.01	—	25	—
Sarcoma 180 (solid)			0.005	—				
Sarcoma 180 (ascitic)	0.05	+	0.005	±	0.01	+ +	25	+
Sarcoma 180 (ascitic	0.025	+			0.005	+ +	12.5	±
Sarcoma T241	0.05	—	0.01	—	0.01	—	25	—
Sarcoma T241			0.005	—	0.005	—	12.5	—
Sarcoma MA387	0.05	—	0.005	—	0.01	—	25	—
Sarcoma MA387					0.005	—		
Ehrlich carcinoma (solid)	0.05	—	0.01	—	0.01	±	25	±
Ehrlich carcinoma (solid)					0.005	±		
Ehrlich carcinoma (ascitic)	0.05	+	0.01	+	0.01	+ + +	25	±
Ehrlich carcinoma (ascitic)			0.005	+	0.005	+ + +		
Bashford carcinoma 63	0.05	—	0.005	—	0.01	—	25	—
Bashford carcinoma 63					0.005	—		
Adenocarcinoma E 0771	0.05	—	0.01	—	0.01	—	25	±
Adenocarcinoma E 0771			0.005	—				
Miyono adenocarcinoma	0.05	—	0.005	+	0.005	±	25	+
Miyono adenocarcinoma							12.5	—
Carcinoma 1025	0.05	—	0.01	—	0.01	—	25	+
Carcinoma 1025			0.005	—				
Lewis bladder carcinoma	0.05	—	0.01	—	0.01	—	25	±
Lewis bladder carcinoma			0.005	—			12.5	—
Lewis lung carcinoma	0.05	—	0.005	—	0.01	—	25	±
Wagner osteogenic sarcoma	0.05	—	0.01	—	0.01	±	25	—
Wagner osteogenic sarcoma					0.005	±	12.5	—
Ridgway osteogenic sarcoma	0.05	—	0.01	—	0.01	—	25	+
Ridgway osteogenic sarcoma			0.005	—			12.5	+
Mecca lymphosarcoma	0.05	—	0.01	—	0.01	—	25	+ + +
Mecca lymphosarcoma							12.5	±
Harding-Passey melanoma	0.05	—	0.005	±	0.01	+	25	—
Harding-Passey melanoma					0.005	—		
Glioma 26	0.05	—	0.01	±	0.01	—	25	—
Friend virus leukemia	0.05	—	0.005	—	0.005	—	25	+
Friend virus leukemia							12.5	+
Friend virus leukemia (solid)	0.05	±	0.005	—	0.005	±	12.5	±
Leukemia L 4946	0.05	—	0.01	—	0.01	—	25	—
Leukemia L 4946					0.005	—		

[a] Daily dose (mg/kg).

Table 7 (continued)

Drug	Quinomycin A		Quinomycin B		Quinomycin C		Triostins	
	Dose[a]	Evaluation	Dose	Evaluation	Dose	Evaluation	Dose	Evalua: tion
Rat tumors:								
Flexner-Jobling carcinoma	0.025	−	0.005	−	0.005	−	1.0	−
Walker carcinosarcoma 256	0.025	±	0.01	−	0.005	−	1.0	−
Walker carcinosarcoma 256			0.005	−	0.005	−		
Jensen sarcoma	0.025	+	0.005	−	0.005	−	1.0	−
Iglesias sarcoma	0.025	−	0.005	−	0.005	−	1.0	−
Moore sarcoma No. 1	0.025	−	0.005	−	0.005	−	1.0	−
Murphy-Sturm lymphosarcoma	0.025	±	0.005	−	0.005	±	1.0	−
Murphy-Sturm lymphosarcoma	0.0125	±						
Iglesias ovarian tumor			0.005	−	0.005	−	1.0	
Iglesias adrenal tumor	0.025	−			0.0025	−	1.0	−
Babcock kidney tumor	0.025	−	0.01	±	0.005	±	1.0	−
Babcock kidney tumor			0.005	−				
Novikoff hepatoma	0.025	−	0.005	−	0.005	+	1.0	−
Hamster tumors:								
Crabb sarcoma	0.05	±	0.0025	−	0.005	−	1.0	−
Crabb sarcoma	0.025	−						
Fortner pancreatic tumor	0.05	±	0.005	−	0.005	−	1.0	−
Fortner pancreatic tumor	0.025	−						
Fortner small intestine tumor	0.025	±	0.005	−	0.005	±	1.0	−
Fortner small intestine tumor			0.0025	−				
Chicken tumor:								
Rous sarcoma	0.025	−	0.01	−	0.01	−	1.0	−

a moderate inhibitory effect on Harding-Passey melanoma and Novikoff hepatoma, and slight inhibitory effects on seven other tumors.

Quinomycin B showed very weak antitumor activity. In the case of rat and hamster tumors, quinomycin A had a definite inhibitory effect. It had a moderate inhibitory effect on Jensen sarcoma and slight inhibitory effects on six other tumors. The triostin complex had different antitumor activity. It was strongly active against Miyono adenocarcinoma, carcinoma 1025, Ridgway osteogenic sarcoma, Mecca lymphosarcoma and Friend virus leukemia. Perhaps, these antibiotics have different modes of antitumor activities.

To bring out more clearly the selective activity of the quinoxaline antibiotics, two ascites type tumors, Ehrlich carcinoma in mice and ascites hepatoma AH-130 in rats, were examined, and the total packed cell volume (TPCV) was employed as the criterion to evaluate the effect quantitatively (MATSUURA, 1965). The results are summarized in Tables 8, 9, and 10. It is clearly demonstrated that the results

Table 8. *Effect of quinoxaline antibiotics on Ehrlich ascites carcinoma in mice* (MATSUURA, 1965)

Antibiotic	Dose (mcg/kg/day)	No. of deaths	TPCV (at 10th day)	
			Tumor index	Effect
Quinomycin A	100	3/5	—	Toxic
	50	0/5	0.38	+
	25	0/5	0.77	—
	12.5	0/5	0.92	—
Quinomycin C	20	5/5	—	Toxic
	10	0/5	0.25	+ +
	5	0/5	0.27	+ +
	2.5	0/5	0.17	+ +
	1.25	0/5	0.69	±
	0.63	0/5	1.18	—
Triostin C	10000	4/5	—	+ + +
	5000	0/5	0.01	+ + +
	2500	0/5	0.07	+ + +
	1250	0/5	0.07	+ + +
	625	0/5	0.48	+
	313	0/5	0.52	±
	156	0/5	0.60	±
	78	0/5	0.79	—

Table 9. *Effect of quinoxaline antibiotics on ascites hepatoma AH-130 in rats* (MATSUURA, 1965)

Antibiotic	Dose (mcg/kg/day)	No. of deaths	TPCV (at 10th day)	
			Tumor index	Effect
Quinomycin A	50	3/4	—	Toxic
	25	0/4	0.01	+ + +
	12.5	0/4	0.04	+ + +
	6.25	0/4	0.21	+ +
	3.13	0/4	0.49	+
	1.56	0/4	0.73	±
	0.78	0/4	0.86	—
Quinomycin C	10	4/4	—	Toxic
	5	0/4	0.01	+ + +
	2.5	0/4	0.01	+ + +
	1.25	0/4	0.36	±
	0.63	0/4	0.39	—
Triostin C	1000	3/4	—	Toxic
	500	0/4	0.05	+ + +
	250	0/4	0.01	+ + +
	125	0/4	0.02	+ + +
	62.5	0/4	0.01	+ + +
	31.3	0/5	0.05	+ + +
	15.6	0/5	0.43	+
	7.8	0/5	0.68	±
	3.9	0/5	0.92	—

Table 10. *Chemotherapeutic index of quinoxaline antibiotics against two different ascites tumors*
(MATSUURA, 1965)

Antibiotic	Ehrlich	AH-130
Quinomycin A	1	8
Quinomycin C	4	2
Triostin C	8	32

are in good agreement with those of the tumor spectrum, as determined qualitatively.

Qinomycin A is active mainly against ascites hepatoma in rats, and quinomycin
C against Ehrlich ascites carcinoma in mice. Triostin C was the most active
among the antibiotics tested.

Antiviral Activity

TSUNODA (1962) reported that quinomycin had a prophylactic effect against
polio virus infections in mice at a daily dose of one half to one fourth of its
toxic dose. The mice infected with MEF 1 strain of type 2 polio virus were
treated with quinomycin complex by intraperitoneal injection for 5 days, starting
a day before the virus inoculation. Survival indices were calculated to be 4.02
for the dose of 0.1 µg/mouse/day, 2.29 for 0.05, and 1.63 for 0.025, respectively.
All antibiotics tested, quinomycin A, quinomycin B, and quinomycin C, showed
significant prophylactic effect, either by single intraperitoneal or subcutaneous
administration, against polio virus infection in mice. However, treatment with
the antibiotics starting on the day of virus inoculation or thereafter, was not
effective at all.

The effect of quinomycin A on the multiplication of bacteriophages was
also observed (SATO *et al.*, 1969). When *E. coli* infected with T2H, T4r or T3
phages were treated with 10 µg/ml of quinomycin A, the production of infective
phage particles was inhibited by 70% for T even phages, and by 14% for the
T3 phage. Since no inhibition of the host cell growth was seen at this concentration, a selective antiphage activity can be assumed. The antibiotic does
not inactivate free phages nor the adsorption of phages to host cells. Detailed
information on the mechanism of antiphage action of quinomycin A is reviewed
below.

Contribution of the Chromophore Moiety to Biological Activity

Although there is structural similarity among the quinoxaline antibiotics and
structure differences are restricted to the peptide portion, significant variations
in biological activity of different quinoxalines are observed. In addition, the
biosynthetic substitution of quinoxaline moiety by other residues has increased
the number of antibiotics belonging to this family and all these can provide
a useful means to elucidate the structural requirements for biological effects.

Table 11 lists some of the biological properties of NX-quinomycin A, QN-
quinomycin A, compared to those of the parent antibiotic, quinomycin A (YOS-
HIDA *et al.*, 1968). These new antibiotics remained virtually similar in activity

Table 11. *Biological properties of novel derivatives of quinomycin A* (YOSHIDA *et al.*, 1968)

System	Quinomycin A	NX-Quinomycin A	QN-Quinomycin A
Antibacterial activity (MIC, μg/ml):			
Staphylococcus aureus 209 P	0.2	0.2	0.2
Diplococcus pneumoniae I	0.05	0.02	0.05
Streptococcus pyogenes	0.1	0.05	0.1
Shigella dysenteriae	5.0	2.0	1.0
Mycobacterium tuberculosis $H_{37}RV$	2.0	1.0	0.5
Mycobacterium phlei	10.0	5.0	2.0
Mycobacterium sp. 607	50.0	5.0	2.0
Toxicity and antitumor activity:			
LD_{50} (20 days), mice, i.p. (mg/kg)	0.28	0.84	1.68
HeLaS$_3$ cells, (Inhib. % with 0.01 μg/ml)	96	87	76
Ehrlich ascites carcinoma in mice, i.p. (Effective dose, mg/kg)	0.04	0.2	0.4

against gram-positive bacteria, whereas they extensively enhanced the inhibitory effect on the acid fast bacteria, which could be correlated to the quinoline moiety in the molecule. This diverse effect can possibly be accounted for by an increased affinity of the drug for these microorganisms, since the quinoline-replacement causes an increase in the lipophilic nature of the compounds. In contrast, the derivatives show a considerable reduction in toxicity without loss of antitumor activity, compared to qinomycin A.

It would perhaps be worthwhile to extend this line of study to improve the chemotherapeutic index of these antibiotics. Some other quinomycin A analogues, quinazomycin and biquinazomycin (Fig. 2.), were prepared (KHAN *et al.*, 1969;

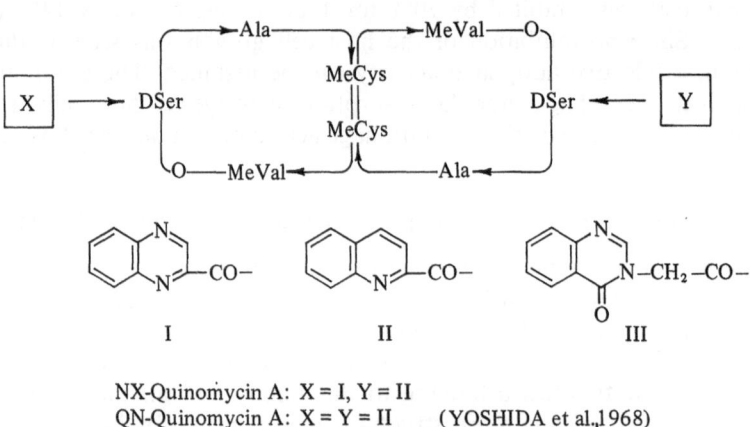

NX-Quinomycin A: X = I, Y = II
QN-Quinomycin A: X = Y = II   (YOSHIDA et al.,1968)
Quinazomycin: X = I, Y = III   (KHAN et al.,1969)
Biquinazomycin: X = Y = III   (ARIF et al.,1970)

Fig. 2. Structures of chromophore-substituted analogues of quinomycin A

ARIF et al., 1970), but nothing is known about their biological activities except the similarity to echinomycin about their activity against *S. aureus*.

Mechanism of Action

The effect of the antibiotics on macromolecular synthesis in susceptible bacteria and mammalian cells has been studied by several groups of investigators. In an early report, WARD et al. (1965) suggested that echinomycin inhibited DNA and RNA synthesis of bacteria and mammalian cells in culture. GAUZE et al. (1966) also indicated that antibiotic 6270 selectively suppressed RNA synthesis in both *B. subtilis* and Ehrlich ascites carcinoma cells. SATO et al. (1967b) arrived at a similar conclusion in that quinomycin A completely blocked RNA synthesis in *S. aureus* at a bacteriostatic concentration (0.5 µg/ml), while the synthesis of DNA and protein were affected to a lesser extent (Table 12). Also, in the protoplasts of *E. coli*, incorporation of ^{14}C-uracil into RNA fraction was inhibited with quinomycin A and quinomycin C (SATO et al., 1967a). It was inferred that the resistance of gram-negative bacteria to quinomycins may be a result of the impermeability of the cells for the antibiotics. In fact, the incorporation of ^{14}C-uracil into the RNA of protoplasts was strongly inhibited with quinomycin A at a concentration of 10 µg/ml, while quinomycin C at the same concentration showed a 50% inhibition. However, no inhibition of ^{14}C-uracil incorporation into the RNA faction was observed with quinomycins A and C in the intact cells of *E. coli*.

The interaction of nucleic acid with the antibiotics was studied by testing for template activity in RNA synthesis catalyzed by RNA polymerase and by testing for spectral differences. WARD et al. (1965) reported that echinomycin inhibited DNA-directed RNA synthesis primed with either calf thymus DNA or synthetic polydeoxyribonucleotides (Table 13). No inhibition was observed when polyribonucleotides served as the template for RNA polymerization. They also demonstrated that a complex is formed between echinomycin and DNA, but not between echinomycin and ribohomopolymers, by measuring the change in ultraviolet absorption spectra. From these findings, they concluded that echinomycin, like actinomycin, inhibits enzymic synthesis of RNA by binding to the

Table 12. *Effect of bacteriostatic concentration of QA on the synthesis of DNA, RNA, and protein in Staph. aureus* (SATO et al., 1967b)

QA (mcg/ml)	Length of incubation (hrs.)	Viable cells (per ml)	Relative increase in		
			DNA	RNA	Protein
0	0	2.2×10^8	1.00	1.00	1.00
0	1.5	7.6×10^8	4.52	4.53	6.00
0	3.0	1.1×10^9	7.15	5.67	7.00
0	4.5	2.1×10^9	9.10	6.33	10.00
0.5	1.5	2.2×10^8	2.03	1.02	1.53
0.5	3.0	2.5×10^8	2.98	0.91	1.38
0.5	4.5	2.3×10^8	3.71	0.94	1.61

Table 13. *Inhibition of DNA-primed RNA synthesis by echinomycin* (WARD et al., 1965)

Amount (mμMole/ml)	Inhibition of various DNA primers (%)							
	Calf-thymus DNA		Crab dAT	dAT	dGdc		dIdc	
	Drug/DNA	Inhibition (%)			GTP	CTP	GTP	CTP
27.5	0.21	100	98.0	97.2	41.8	29.8	13.7	25.4

template polydeoxyribonucleotides. They also speculated that a spatial organization of structures, which interact with echinomycin, is characteristic of the helical structure of DNA. The organization is common to all combinations of the base pairs in DNA, and is absent in RNA.

A similar observation with respect to the interaction with DNA was made by SATO et al. (1967 b). When quinomycin A was mixed with DNA, the absorption spectrum shifted from 325 mμ to 345 mμ (Fig. 3). Further evidence of the formation of quinomycin A-DNA complex was provided by analysis of the chromatographic profile of MAK column, to which a mixture of quinomycin A and nucleic acid extracted from *E. coli* with 90% phenol was applied. The fraction which contained DNA was accompanied with quinomycin A, whereas the fractions which contained t-RNA and ribosomal RNA were not.

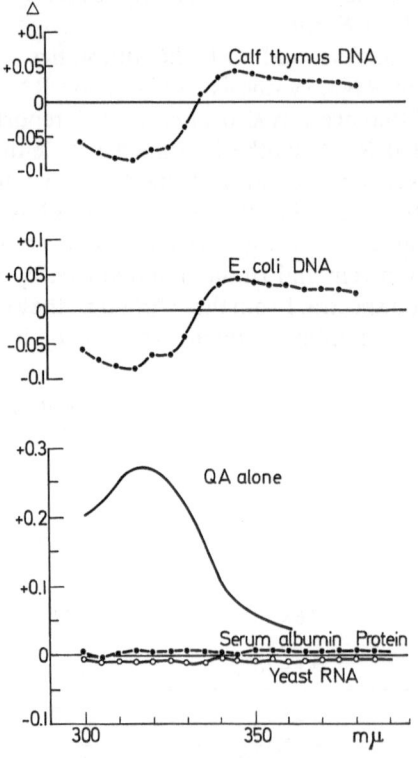

Fig. 3. Differential spectra of QA (SATO et al., 1967b)

Further, it was found that the binding of quinomycin A to DNA strongly reduced the antibacterial activity against *S. aureus*. When native DNA from various sources and some of their degradative products, including the heat-denatured form, and intermediaries between DNA and apurinic acid were used in the above experiment, the inhibitory effect of quinomycin A was reduced, but the hydrolysis of DNA resulted in an increase of the effect. Yeast RNA and bovine serum albumin failed to reduce the inhibition. In a parallel experiment, a distinct difference was observed between quinomycin A and actinomycin D. The antibacterial effect of actinomycin was partially reduced by the addition of oligonucleotides, which did not prevent inhibition of quinomycin A. A more marked difference was observed in the case when acid-insoluble polynucleotides were added. This reduced the inhibitory effect of actinomycin, but not that of quinomycin A. It was provisionally assumed that the binding of quinomycin A takes place only with highly polymerized DNA. However, it is still not known in what manner quinoxaline antibiotics form a complex preferentially with DNA. The interaction of quinoxaline antibiotics with DNA was also studied by testing the sensitivity of the complexes to a nucleolytic enzyme (KAGEYAMA *et al.*, 1970). The hyperchromic effect during the hydrolysis was measured in a reaction mixture containing 80 µg of native DNA, 1/150 millimole of $MgSO_4$, 7.5 µg of DNase I and various amounts of antibiotics, quinomycin A and triostin C. The antibiotics completely inhibited the digestion of DNA at a concentration of 10 mµmole/ml.

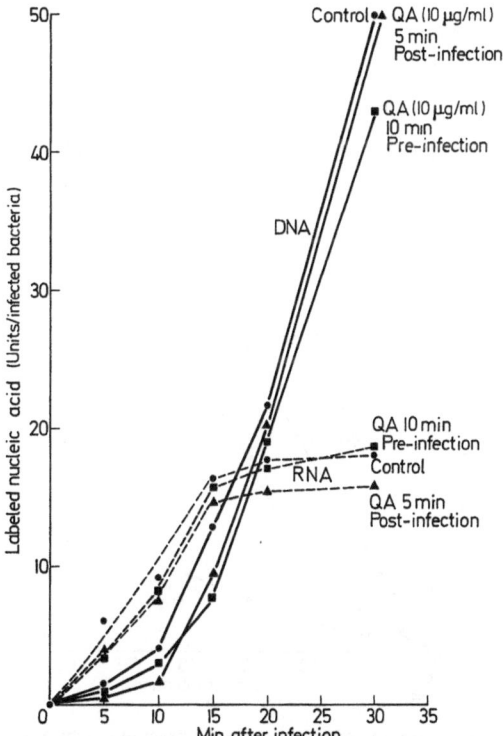

Fig. 4. Kinetics of the incorporation of ^{32}P into nucleic acids of infected bacteria (SATO *et al.*, 1969)

Fifty percent inhibition was observed when the molar ratios of drug to the nucleotide were, 1:16 for quinomycin A, 1:15 for triostin C, and 1:8 for actinomycin D. It was also shown that the ratio of binding sites for these antibiotics were 1:14 for quinomycin A, 1:16 for triostin C, and 1:10–20 for actinomycin, for a 100% change in the spectrum by spectrophotometric titration. This means that DNA lost its sensitivity to DNase by 50% when the binding sites were filled with the antibiotics. It was also concluded that these antibiotics did not act directly on DNase but rather on DNA, since the extent of inhibition was closely dependent on the drug/DNA ratio, but very little on the drug/enzyme ratio.

Other interesting biological effects have been observed with quinoxaline antibiotics. Sato *et al.* (1969) investigated the mechanism of the inhibitory effect of quinomycin A on the multiplication of phage T2. The infected cells were treated with 10 µg/ml of quinomycin A at 10 min before or 5 min after the infection, and the synthesis of phage DNA and RNA (m-RNA) was followed for 30 min, using radioisotopes. As shown in Fig. 4, phage DNA and m-RNA were synthesized in the presence of quinomycin A at almost the same rate as that in the absence of the antibiotic, even though the number of infective phage particles was reduced by 90%. Protein synthesis in the infected cells treated with quinomycin A(10 µg/ml), as well as phage DNA synthesis, were examined by the addition of specific antiserum or by high speed centrifugation. The results, shown in Fig. 5, indicate that the synthesis of ^{35}S-labelled protein precipitated

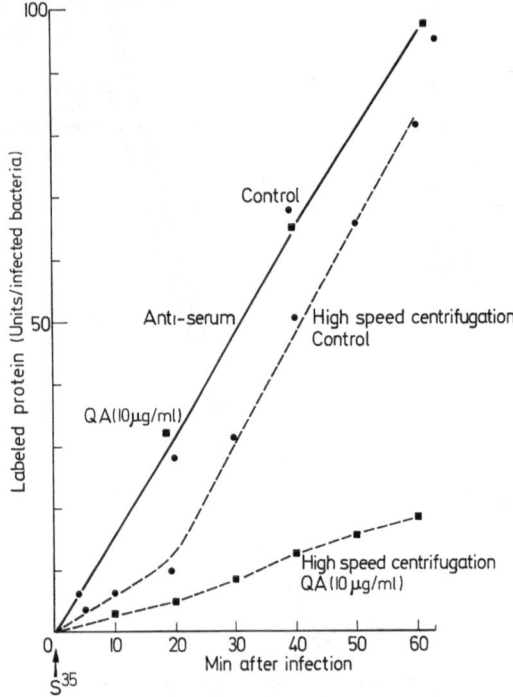

Fig. 5. Kinetics of the incorporation of ^{35}S into protein precipitable by anti-T2H serum and sedimentable by high speed centrifugation (Sato *et al.*, 1969)

with anti-T2 serum, was not affected. However, the formation of protein sedimentable by high speed centrifugation at 20,000 rpm for 60 min was strongly inhibited
as a reflection of the inhibition of infective particle formation. In the sedimentation
analysis of the phage DNA, more than 70% of ^{14}C-labelled-DNA in the untreated
lysate was precipitated with the addition of anti-T2 serum or by the centrifugation,
whereas less than 18% of labelled-DNA was precipitated in the lysate treated
with quinomycin A. It was also confirmed that the synthesis of tail fiber protein
in the experimentals was not impaired, by examining the formation of serum
blocking power. From these results, it was concluded that quinomycin A inhibited
phage development at a step during the maturation process. It can be assumed
that the phage DNA associated with quinomycin A may fail to assemble with
the head proteins, allowing accumulation of the naked DNA and phage antigenic
proteins, resulting in inhibition of formation of the infective particles.

References

ARIF, A. J., C. SINGH, A. P. BHADURI, C. M. GUPTA, A. W. KHAN, and M. M. DHAR: Actinomycetes
    studies. Part II. Cell-free synthesis of echinomycin and an echinomycin analogue. Indian J. Bio-
    chem. **7**, 193 (1970).
BERGER, J., E. R. LaSALA, W. E. SCOTT, B. R. MELTSNER, L. H. STERNBACH, S. KAISER, S. TEITEL,
    E. MACH, and M. W. GOLDBERG: Antibiotic X-948 and X-1008. Experientia **13**, 434 (1957).
CARTER, H. E., C. P. SCHAFFNER, and D. GOTTLIEB: Levomycin. I. Isolation and chemical studies.
    Arch. Biochem. Biophys. **53**, 282 (1954).
CORBAZ, R., L. ETTLINGER, E. GAUMANN, W. KELLER-SCHIERLEIN, F. KRADOLFER, L. NEIPP, V.
    PRELOG, P. REUSSER u. H. ZÄHNER: 23. Stoffwechselprodukte von Actinomyceten. Echinomycin.
    Helv. Chim. Acta **40**, 199 (1957).
GAUZE, G. G., YU. V. DUDNIK, N. P. LOSHGAREVA, and I. B. ZBARSKY: Inhibition of RNA synthesis
    by antibiotic 6270 from echinomycin group in bacterial and tissue cells. Antibiotiki **11**, 426
    (1966).
GERCHAKOV, S., and H. P. SCHULTZ: Quinoxaline studies. XIV. Potential Anticancer Agents. Some
    quinoxaline amino acid and dipeptide derivatives related to quinoxaline antibiotics. J. Med.
    Chem. **12**, 141 (1969).
HARADA, Y., N. SUNAGAWA, and K. KATAGIRI: Segregation of the nucleolar materials produced
    by quinoxaline antibiotics in JTC-13 cells. Comparison of effects among 4-nitro-quinoline 1-oxide,
    actinomycin B, and quinoxaline antibiotics. Gann **59**, 513 (1968).
ISHIHARA, S., R. UTAHARA, M. SUZUKI, Y. OKAMI, and H. UMEZAWA: Studies on actinoleukin,
    relation to echinomycin and levomycin. J. Antibiotics (Tokyo), Ser. A **11**, 160 (1958).
IVANITSKAYA, L. P., E. B. KRUGLYAK, T. S. MAKSIMOVA, and T. P. PREOBRAZHENSKAYA: The production
    of echinomycin-like substances by various species of actinomycetes. Antibiotiki **6**, 393 (1961).
KAGEYAMA, M., M. HASEGAWA, A. INAGAKI, and F. EGAMI: Interaction of antibiotics with deoxyri-
    bonucleic acid I. Sensitivity of the complexes to nucleolytic enzymes. J. Biochem. **67**, 549 (1970).
KATAGIRI, K., H. ENDO, M. TADA, and H. NIKAIDO: Biological characteristics of phenazine derivatives.
    III. Anti-mycoplasma effect. Ann. Rept. Shionogi Res. Lab. **16**, 58 (1966).
KATAGIRI, K., and S. MATSUURA: Studies on the antitumor activity of quinoxaline antibiotics. J.
    Antibiotics (Tokyo), Ser. B **16**, 122 (1963).
KATAGIRI, K., S. OKAMOTO, K. SATO, N. SHIMAOKA, K. TAWARA, and S. SASAKI: Studies on actinomy-
    cin-type antibiotics. II. Screening on antitumor activity. Ann. Rept. Shionogi Res. Lab. **7**, 191
    (1957).
KATAGIRI, K., J. SHOJI, and T. YOSHIDA: Identity of levomycin and quinomycin A. J. Antibiotics
    (Tokyo), Ser. A **15**, 273 (1962).
KATAGIRI, K., and K. SUGIURA: Antitumor action of the quinoxaline antibiotics. Antimicrobial
    Agents Chemotherapy **1961**, 162.
KELLER-SCHIERLEIN, W., M. LJ. MIHAILOVIC u. V. PRELOG: 26. Stoffwechselprodukte von Actinomy-
    ceten. Über die Konstitution von Echinomycin. Helv. Chim. Acta **42**, 305 (1959).

KELLER-SCHIERLEIN, W., u. V. PRELOG: 24. Stoffwechselprodukte von Actinomyceten. Hydrolysepro-
dukte des Echinomycins: D-Serin, L-Alanin und Chinoxaline carbonsäure-(2). Helv. Chim. Acta
40, 205 (1957).

KHAN, A. W., A. P. BHADURI, C. M. GUPTA, and M. M. DHAR: Actinomycetes studies. Part I. Microbial synthesis of quinazomycin, an echinomycin analogue containing one quinazol-4-one-3-acetyl
residue. Indian J. Biochem. **6**, 220 (1969).

KOPPEL, H. C., I. L. HONIGBERG, R. H. SPRINGER, and C. C. CHENG: Synthetic approaches to quinoxaline antibiotics. Synthesis of biquinoxaloyl derivatives. J. Org. Chem. **28**, 1119 (1963).

KUROYA, M., N. ISHIDA, K. KATAGIRI, J. SHOJI, T. YOSHIDA, M. MAYAMA, K. SATO, S. MATSUURA,
Y. NIINOMI, and O. SHIRATORI: Studies on quinoxaline antibiotics. I. General properties and
the producing strains. J. Antibiotics (Tokyo), Ser. A **14**, 324 (1961).

MAKSIMOVA, T. S., I. N. KOVSHAROVA, and V. V. PROSHLYAKOVA: Early identification of echinomycin
antibiotics and systematic position of their producers. Antibiotiki **10**, 298 (1965).

MATSUURA, S.: Studies on quinoxaline antibiotics IV. Selective antitumor activity of each quinoxaline
antibiotic. J. Antibiotics (Tokyo), Ser. A **18**, 43 (1965).

NITTA, K., T. TAKEUCHI, T. YAMAMOTO, and H. UMEZAWA: Studies on the effects of actinomycin,
cycloheximide and other known antibiotics on Ehrlich carcinoma of mice. J. Antibiotics (Tokyo),
Ser. A **8**, 120 (1955).

OTSUKA, H., and J. SHOJI: Studies on quinoxaline antibiotics. V. Degradative studies on quinoxaline
antibiotics. J. Antibiotics (Tokyo), Ser. A **16**, 52 (1963).

OTSUKA, H., and J. SHOJI: Configuration of the N-methylisoleucine, a constituent amino acid of
triostin B and quinomycin B. J. Antibiotics (Tokyo), Ser. A **18**, 134 (1965a).

OTSUKA, H., and J. SHOJI: The structure of Triostin C. Tetrahedron **21**, 2931 (1965b).

OTSUKA, H., and J. SHOJI: Isolation and structural study on minor components of quinoxaline
antibiotics. J. Antibiotics (Tokyo), Ser. A **19**, 128 (1966).

OTSUKA, H., and J. SHOJI: Structural studies on the minor components of quinoxaline antibiotics.
Tetrahedron **23**, 1535 (1967).

PREOBRAZHENSKAYA, T. P., L. P. IVANITSKAYA, E. G. TOROPOVA, G. F. GAUZE, N. P. ILICHEVA, G. V.
GAVRILINA, and K. K. IVANOV: Cultural conditions for the formation of the cancer inhibiting
antibiotic 6270. Puti i Metody Izyskan. Protivorak. Antibiotikov, Tr. Sympoziuma, Moscow
1958, 101 (1959). [Chem. Abstr. **54**, 25028e and g (1960)].

ROSSOLIMO, O. K., M. S. STANISLAVSKAYA, and G. N. LEPESHKINA: Experimental study of the antitumor
substance of the new antibiotic 6270. Antibiotiki **4**, 54 (1959).

SATO, K., Y. NIINOMI, K. KATAGIRI, A. MATSUKAGE, and T. MINAGAWA: Prevention of phage
multiplication by quinomycin A. Biochim. Biophys. Acta **174**, 230 (1969).

SATO, K., O. SHIRATORI, and K. KATAGIRI: The mode of action of quinoxaline antibiotics: Interaction
of quinomycin A with DNA. J. Antibiotics (Tokyo), Ser. A **20**, 270 (1967b).

SATO, K., T. YOSHIDA, and K. KATAGIRI: Inhibition of RNA synthesis in *E. coli* protoplasts by
quinomycins A and C. J. Antibiotics (Tokyo), Ser. A **20**, 188 (1967a).

SHOJI, J., and K. KATAGIRI: Studies on quinoxaline antibiotics III. New antibiotics, triostin A,
B, and C. J. Antibiotics (Tokyo), Ser. A **14**, 335 (1961).

SHOJI, J., K. TORI, and H. OTSUKA: Configuration of N,β-dimethylleucine, a constituent amino
acid of triostin C. J. Org. Chem. **30**, 2772 (1965).

TSUNODA, A.: Chemoprophylaxis of poliomyelitis in mice with quinomycin. J. Antibiotics (Tokyo),
Ser. A **15**, 60 (1962).

UEDA, M., Y. TANIGAWA, Y. OKAMI, and H. UMEZAWA: A new toxic antibiotic, Actinoleukin,
produced by a streptomycete. J. Antibiotics (Tokyo), Ser. A **7**, 125 (1954).

UMEZAWA, H., K. MAEDA, and Y. OKAMI: On the various souces of actinomycin. J. Antibiotics
(Tokyo), Ser. A **4**, 335 (1951).

WARD, D. C., E. REICH, and I. H. GOLDBERG: Base specificity in the interaction of polynucleotides
with antibiotic drug. Science **149**, 1259 (1965).

YOSHIDA, T., and K. KATAGIRI: Isolation of quinomycin B. J. Antibiotics (Tokyo), Ser. A **15**,
272 (1962).

YOSHIDA, T., and K. KATAGIRI: Influence of isoleucine upon quinomycin biosynthesis by *streptomyces*
sp. 732. J. Bacteriol. **93**, 1327 (1967).

YOSHIDA, T., and K. KATAGIRI: Biosynthesis of the quinoxaline antibiotic, Triostin, by *streptomyces*
s-2-210L. Biochemistry **8**, 2645 (1969).

YOSHIDA, T., K. KATAGIRI, and S. YOKOZAWA: Studies on quinoxaline antibiotics. II. Isolation and properties of quinomycin A, B, and C. J. Antibiotics (Tokyo), Ser. A **14**, 330 (1961).

YOSHIDA, T., Y. KIMURA, and K. KATAGIRI: Novel quinomycins. Biosynthetic replacement of the chromophores. J. Antibiotics (Tokyo), **21**, 465 (1968).

YOSHIDA, T., Y. KIMURA, and K. KATAGIRI: The biosynthesis of the quinoxaline antibiotic, triostin, by *streptomyces* s-2-210L. On the role of quinoxaline-2-carboxylic acid. Progress in Antimicrobial and Anticancer Chemotherapy, vol. II, p. 1160.: Univ. of Tokyo Press 1970.

Rifamycins and Other Ansamycins

W. Wehrli and M. Staehelin

The various actions of the rifamycins have been reviewed recently by the authors (WEHRLI and STAEHELIN, 1971). The antimicrobial, pharmacological and clinical aspects of rifampicin have also been summarized by BINDA et al. (1971); KONOPKA et al. (1972) and in the journal Drugs (Editorial staff, 1971). In the present report therefore, early data are dicussed only briefly. Detailed references can be found in the foregoing papers.

The rifamycins are ansa compounds, i. e. compounds containing an aromatic ring system spanned by a long aliphatic bridge (Fig. 1). They were discovered in 1957 by SENSI et al. (1960) and their structure was elucidated in 1963 by PRELOG and coworkers.

The rifamycins might easily have escaped detection altogether, since rifamycin B, the compound produced by *Streptomyces mediterranei* sp. n., has no antibacterial activity. However, it is readily oxidized to the active compound, rifamycin S, which inhibits the growth of Gram-positive bacteria at concentrations as low as 0.0025 µg/ml.

Fig. 1 shows the chemical relationship between rifamycin B and rifamycin S, a naphthoquinone derivative formed by oxidation and hydrolysis; reduction of rifamycin S yields the naphthohydroquinone derivative rifamycin SV. Many other rifamycin derivatives have been synthesized by substitutions in positions 3 and 4. The antibacterial activity of some of these derivatives is considerably improved as compared to that of rifamycin S and rifamycin SV, and some have the added advantage of being active orally, in contrast to rifamycin S, which must be administered parenterally. One of the compounds most active when

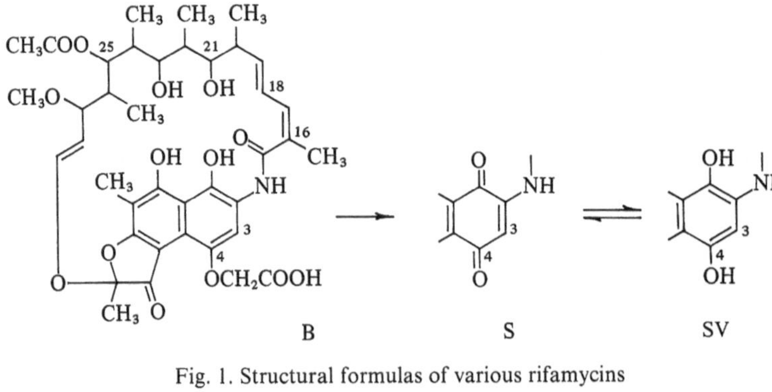

Fig. 1. Structural formulas of various rifamycins

given orally is rifampicin (Fig. 2), a rifamycin SV derivative having a (4-methyl-piperazinyl)-iminomethyl group in position 3. Rifampicin has a wide antibacterial spectrum, being particularly active at low concentrations against mycobacteria and Gram-positive organisms. Gram-negative bacteria are in general less sensitive, although the growth of Haemophilus and Neisseria is inhibited by concentrations of less than 1 µg/ml. Rifampicin is now in widespread clinical use and has proved especially effective in the treatment of tuberculosis, as well as various other infectious diseases.

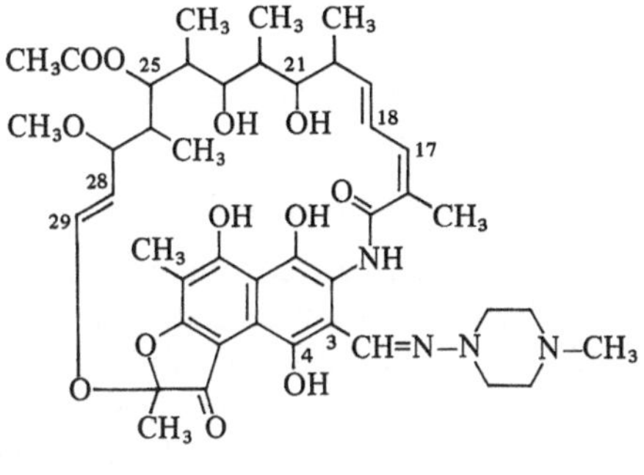

Fig. 2. Structural formula of rifampicin, in the U. S. rifampin. Trade name of Ciba-Geigy: Rimactane; trade name of Lepetit: Rifadin

The Various Modes of Action

Among the various actions of the rifamycins four different types can be clearly distinguished:

I. Action on transcription in bacteria.

Rifamycins inhibit the synthesis of RNA by inactivating the DNA-dependent RNA polymerase. This effect occurs at extremely low concentrations of the antibiotic and is highly specific. Other enzymes, such as DNA-dependent DNA polymerase (Kornberg enzyme), are not affected.

II. Action of rifampicin on RNA synthesis in eukaryotic cells.

Whereas bacterial RNA polymerase is inhibited at concentrations as low as 0.01 µg/ml, RNA polymerases in eukaryotic cells are not affected. Some enzymes in mitochondria and chloroplasts were found to be partially inhibited by concentrations from 1 µg/ml upwards. Thus 100–10,000-fold higher concentrations of rifampicin were required to elicit an effect similar to that observed with bacterial extracts. It cannot be stressed too strongly that these are two entirely different phenomena and should therefore be considered separately.

III. Effects of certain lipophilic rifamycin derivatives on DNA polymerases from oncogenic viruses and other DNA and RNA polymerases of bacterial and mammalian origin.

These actions are only exerted by rifamycins with lipophilic side chains and not by rifampicin or unsubstituted rifamycins. As in the case of eukaryotes, inhibition occurs at concentrations greater than 1 µg/ml. The inhibitory effects appear to be unspecific, since a large variety of different enzymes are affected.

Furthermore, inhibition can be completely counteracted by DNA and proteins such as bovine serum albumin (Wehrli, to be published).

IV. Effects on DNA viruses and larger infectious agents belonging to the class of Chlamydozoaceae.

Again, only certain rifamycin derivatives, such as rifampicin, inhibit the growth of these organisms. The underlying mechanism of action is poorly understood at present, but it does not seem to be related to RNA or DNA synthesis.

Each of the foregoing types of action is discussed separately below.

I. Effects on Bacteria

A. Rifamycins Inhibit RNA Polymerase

Most of the studies have been carried out with rifampicin. As far as the mechanism of action of rifamycins is concerned, what is true of rifampicin also holds good for most of the other derivatives, and rifampicin can therefore be taken as a representative model of this whole class of compounds. Rifampicin was first shown by Hartmann et al. (1967) to have a specific inhibitory effect on RNA polymerase from E. coli. This enzyme polymerizes ribonucleoside triphosphates on a DNA template and catalyses the process of transcription (Chamberlin, 1970). RNA polymerases from a large variety of bacteria besides E. coli have also been found to be sensitive to rifampicin. Bacterial RNA polymerase is inhibited by a number of toxic antibiotics, such as actinomycin and mitomycin. These antibiotics interfere with the action of the enzyme by interacting with the DNA template. In contrast, rifampicin interacts specifically with the enzyme itself, which it inactivates at very low concentrations (~ 0.01 µg/ml). As will become apparent below, the effective concentration depends on the amount of enzyme used in the assay.

The corresponding mammalian enzymes are not inhibited by concentrations up to 10,000-fold higher. The above-mentioned toxic antibiotics, in contrast, inhibit mammalian as well as bacterial enzymes at roughly equal concentrations, because they do not interact with the enzyme, but with the DNA template.

B. Rifampicin is Bound to the Enzyme

Using radioactive rifampicin, Wehrli et al. (1968a) were able to demonstrate that rifampicin forms a complex with the enzyme. One molecule of antibiotic is bound to each enzyme monomer (Wehrli and Staehelin, 1970). Binding takes place rapidly and the enzyme-antibiotic complex is very stable. The half life of the complex was found to be about 8 minutes at 37° and 8 hours at 0° (Wyss and Wehrli, 1973). The stability of the enzyme-antibiotic complex depends on various other factors besides temperature: the addition of organic solvents to the complex leads to a decrease in stability, whereas other compounds such as DNA and triphosphates increase its stability (Handschin and Wehrli, in preparation).

C. Nature of Rifampicin Resistance in Mutants

Mutants resistant to rifampicin have been isolated from a variety of micro-organisms. In *E. coli* the mutation rate is of the order of 10^{-8} and the conversion is apparently due to a one-step mutation. RNA polymerase has been prepared from rifamycin-resistant mutants of *E. coli* and *S. aureus*. In both cases it was found that the enzyme differs from the corresponding enzyme of the sensitive strain, in that it was no longer affected by rifampicin and, in addition, could no longer bind the drug (WEHRLI *et al.*, 1968a; WEHRLI *et al.*, 1968b; WHITE and LANCINI, 1971). Rifampicin resistance was, therefore, caused by an alteration in the structure of the enzyme, i.e. by replacement of one single amino acid. Rifampicin-resistant mutants have actually been used as a means of locating the RNA polymerase gene in the chromosome of *E. coli* and other micro-organisms.

Resistance to rifampicin is not an all-or-nothing effect. Mutants can be selected which are resistant to various concentrations of rifampicin. As the resistance of the bacteria against rifampicin increases, the sensitivity of the corresponding RNA polymerase decreases, and the enzyme-antibiotic complex becomes less stable. Hence, the "fit" between enzyme and antibiotic depends on the location and the nature of the amino acid substitution.

D. Location of the Rifampicin Binding Site in RNA Polymerase

The RNA polymerase molecule consists of several subunits, which are listed in the following table:

Subunit	Molecular weight	No. of subunits	
		In core enzyme	In holo enzyme
α	40,000	2	2
β	145,000	1	1
β'	160,000	1	1
σ	85,000	—	1

The following evidence lends support to the view that rifampicin binds to the β subunit:

1. The core enzyme still binds rifampicin, thus σ is not involved in the binding of the antibiotic (WEHRLI *et al.*, 1970).

2. RNA polymerase containing a modified β subunit has been found in a rifampicin-resistant strain of *E. coli* (RABUSSAY and ZILLIG, 1969).

3. Reconstitution experiments with subunits from rifampicin-sensitive and -resistant strains have shown that RNA polymerase containing α and β' subunits from a resistant strain and the β subunit from a sensitive strain remains sensitive to rifampicin, while, conversely, an RNA polymerase containing α and β' subunits

from a sensitive strain and the β subunit from a resistant strain remains rifampicin-resistant (HEIL and ZILLIG, 1970).

E. What Functions of the Enzyme Are Affected?

The various steps in the process of transcription can be classified as follows:
a) binding of the enzyme to DNA
b) binding of the first nucleoside triphosphate to the enzyme-DNA complex ("chain initiation")
c) assembly of more nucleotides to form a polyribonucleotide ("chain elongation")
d) release of the completed RNA chain from the template ("chain termination").

If rifampicin is added to an enzyme already synthesizing RNA chains, its activity is not immediately inhibited, but continues until all partly formed chains have been completed. The initiation of new chains, however, is stopped immediately. Rifampicin, therefore, does not interfere with chain elongation, but specifically inhibits DNA binding or chain initiation (HARTMANN et al., 1967; UMEZAWA et al., 1968).

The initial binding of the enzyme to DNA is not affected (UMEZAWA et al., 1968). Natural templates, such as viral DNA, however, seem to form a variety of specific complexes with the enzyme. HINKLE and CHAMBERLIN (1970); ZILLIG et al. (1970) and BAUTZ and BAUTZ (1970) have provided evidence showing that at low temperatures the interaction of enzyme with DNA is relatively weak, and the subsequent polymerization of nucleotides can still be prevented by rifampicin. If the temperature is raised above a critical value, the interaction between the enzyme and DNA is intensified and apparently a new sort of DNA-enzyme complex is created, whereby the enzyme seems to interact with a few specific promotor sites on the DNA. In this form the enzyme is resistant to rifampicin. Thus, it appears that when bound to specific promotor sites of intact DNA above a critical temperature or when engaged in an elongation complex, the enzyme is resistant to rifampicin. With synthetic polynucleotides as templates, rifampicin affects either DNA binding or the RNA chain initiation process, depending on the rates of formation and the stability of the various complexes under the given experimental conditions.

F. Binding of the Antibiotic to Various Enzyme-DNA Complexes

Theoretically, there are two possible explanations for the inability of rifampicin to inhibit the enzyme under certain conditions:
a) it may no longer be bound to the enzyme, and
b) it may still be bound but, after binding does not affect the action of the enzyme.

Both situations seem to exist during RNA synthesis. An initiation complex containing enzyme, DNA and purine triphosphates is protected against rifampicin inhibition, because it forms practically no complex with the drug. On the other hand, during elongation, when the enzyme is also not inhibited by rifampicin, the antibiotic can bind to the enzyme complex although at a much slower rate. Further experiments are necessary to clarify the exact behaviour of rifampicin during RNA-synthesis.

G. Structure-Activity Relationships

The rifamycin molecule has been extensively modified chemically. Most of the modifications consist of substitutions of the naphthoquinone chromophore in position 3 and/or 4. Whereas these modifications can have a pronounced effect on the *in vitro* antibacterial activity and especially on the *in vivo* action of the drug, there is little difference between the effects of various rifamycin derivatives exerted on isolated RNA polymerases (WEHRLI and STAEHELIN, 1969). They all inhibit the enzyme at about the same low concentrations. Differences in their *in vitro* antibacterial activity are therefore probably due to limitations of the penetration of the antibiotic into the bacterium, and differences in the *in vivo* activity to different pharmacokinetic behaviour of various derivatives in the animal. On the other hand, most chemical modifications of the ansa ring reduce the capacity of the substance to inhibit enzyme activity, to form a stable complex with the enzyme, and to affect bacterial growth (WEHRLI and STAEHELIN, 1969). Some gradual changes, such as successive hydrogenation of the three double bonds lead to a gradual decrease in both enzyme inhibition and complex stability.

It can be concluded that these changes alter the stereochemistry of the ansa ring step by step, in such a way as to impair the fit of the molecule with the acceptor site on the enzyme. Thus, the situation is analogous to that found with the various resistant enzymes, where substitution with different amino acids has a comparable effect. However, not all minor modifications of the ansa ring lead to a gradual loss of activity. Rifamycin YSV, for instance, differs from rifamycin SV only in having a keto group in position 21 instead of a hydroxyl group, and an additional hydroxyl group in position 20 (Fig. 3). This derivative was found to be totally inactive. It is therefore evident that certain parts of the aliphatic bridge play a crucial role in the interactions between the antibiotic and RNA polymerase.

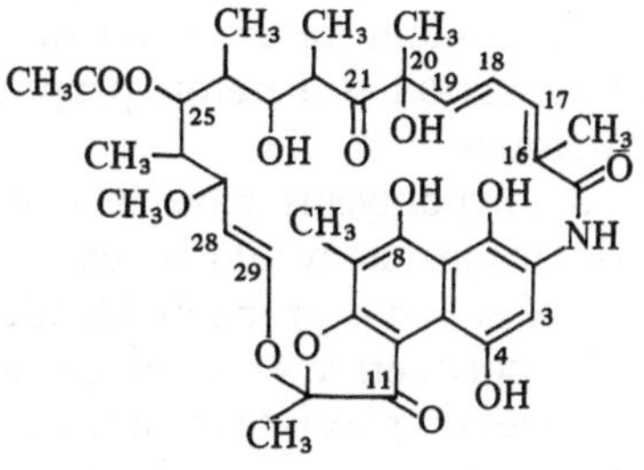

Fig. 3. Structural formula of rifamycin YSV

H. Effects of Rifampicin on Bacterial RNA Polymerases
Other than E. coli

RNA polymerases isolated from a large variety of bacteria were found to be inhibited by rifampicin in a way similar to the E. coli enzyme (see Wehrli and Staehelin, 1971; and the more recent papers by White et al., 1971; Johnson et al., 1971; Avila et al., 1971; Mišoň and Trnka, 1972).

A very interesting case is the RNA polymerase of an extremely halophilic bacterium, Halobacterium cutirubrum. This enzyme has a completely different structure than the RNA polymerase of other bacteria, such as E. coli, B. subtilis and Pseudomonas putida. It consists of only two subunits with a very low molecular weight (18,000 daltons), but it is also sensitive to rifampicin. Rifampicin binding apparently occurs with only one of the subunits, since inhibition by rifampicin could be overcome by the addition of an excess of β-subunits but not of α-subunits (Louis and Fitt, 1972).

Thus it seems that even a drastic reduction in the molecular weight of the β subunit does not abolish the binding of rifampicin and its influence on chain initiation. In fact, one wonders what special tasks the large subunits in bacteria such as E. coli have to fulfil. In this context, it is worth noting that in experiments on the spore formation of B. subtilis, mutants have been isolated which, in a single mutational event, acquire resistance to rifampicin and lose their ability to sporulate (Losick et al., 1970; Sonenshein and Losick, 1970) or form spores with an altered morphology (Doi et al., 1970).

In both cases the vegetative growth is not changed. In normal spores, one β subunit has a molecular weight of only 110,000 as against 155,000 in a vegetative cell. Therefore, it is argued that in drug-resistant strains of B. subtilis, the β subunit must be changed in such a way that it cannot be cleaved to the low-molecular-weight form, so that sporulation does not take place. Thus, the β subunit of RNA polymerase plays an important part in differentiating growth from sporulation.

J. Other Ansamycins

Three groups of antibiotics that are chemically very similar to the rifamycins have been described: the streptovaricins (Wang et al., 1971; Rinehart, Jr., 1972), the tolypomycins (Kamiya et al., 1969; Shibata et al., 1971; Kishi et al., 1972) and geldanamycin (De Boer et al., 1970; Sasaki et al., 1970). Like the rifamycins, they all contain an aromatic ring system spanned by an aliphatic bridge (Fig. 4). V. Prelog has therefore proposed the term ansamycins for the entire group of antibiotics.

The streptovaricins and tolypomycins have biochemical properties closely resembling those of the rifamycins, as would be expected from their chemical similarity; their mechanism of action is probably identical to that of rifampicin. However, they inhibit RNA polymerase to a somewhat lesser extent than the rifamycins, because a less stable complex is formed with the enzyme. Geldanamycin only inhibits bacterial growth at rather high concentrations (De Boer et al., 1970). RNA polymerase of E. coli is not inhibited by the drug at concentrations

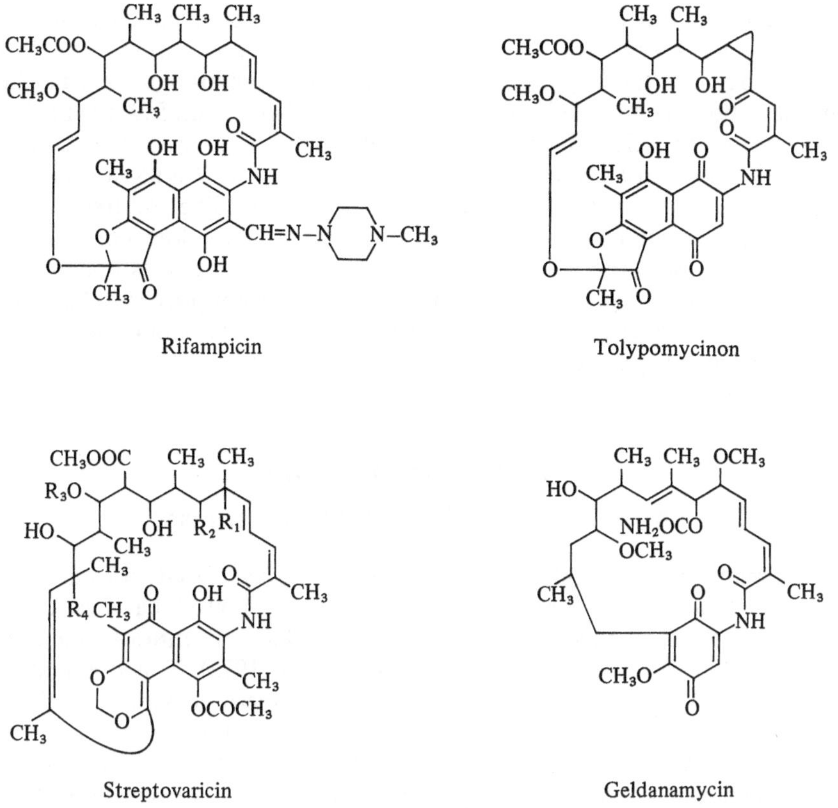

Rifampicin

Tolypomycinon

Streptovaricin

Geldanamycin

Fig. 4. Structural formulas of various ansamycins. The streptovaricins have the following structures:

	R_1	R_2	R_3	R_4
A:	OH	OH	OCOCH$_3$	OH
B:	H	OH	OCOCH$_3$	OH
C:	H	OH	H	OH
D:	H	OH	H	H
E:	H	=O	H	OH
G:	OH	OH	H	OH

up to 20 µg/ml (WEHRLI, unpublished). This seems to fit in with the structure-activity relationship found with rifamycins, from which it is evident that the hydroxyl group in position 21 must be intact to exhibit good activity. Geldanamycin has an —OCONH$_2$ group in the corresponding position.

K. Rifamycin as a Tool in Biochemical Studies

Owing to the specific action of rifamycins on bacterial RNA polymerase, this class of antibiotics has been extensively used in the study of RNA biosynthesis and metabolism. Furthermore, they are of great value in studying the significance of bacterial RNA synthesis in the biochemistry of bacterial viruses. We do not intend to describe the application of the drug for such studies in detail here. Some have been mentioned in our earlier review (WEHRLI and STAEHELIN, 1971), and some newer, selected papers are listed in the references (SIRBASKU and BUCHANAN, 1971; CHEN et al., 1971; BRUTLAG et al., 1971; WICKNER et al., 1972; AUSTIN et al., 1971; MATZURA et al., 1971). We would like to stress that the rifamycins, e.g. rifampicin, act specifically and at very low concentrations ($\ll$ 1 µg/ml) on bacterial RNA polymerase only. All other effects obtained with concentrations over 1 µg/ml (see section II to IV) have not proved to be specific, and thus inhibitory effects obtained under such conditions should not be interpreted as a sign of specific inhibition of RNA synthesis.

II. Effects on Eukaryotes

A. Actions on Nuclear RNA Polymerases

As already mentioned, the rifamycins have a very low toxicity for mammalian organisms. Examination of nuclear RNA polymerases showed that these enzymes were not affected by rifamycins (WEHRLI et al., 1968c; KEDINGER et al., 1970; FURTH et al., 1970; VOIGT et al., 1970; ASHBURNER, 1972). This observation is in agreement with the studies on the subunits of these enzymes, which seem to differ extensively from the bacterial RNA polymerase (WEAVER et al., 1971; GISSINGER and CHAMBON, 1972; KEDINGER and CHAMBON, 1972). The sensitivity of mammalian RNA polymerases to certain lipophilic rifamycin derivatives is discussed in Section III.

Nuclear RNA polymerases from plants, fungi and protozoa are also resistant to the antibiotic (WINTERSBERGER and WINTERSBERGER, 1970; RODRIQUEZ-LOPEZ et al., 1970; SURZYCKI, 1969; BYFIELD et al., 1970; STRAIN et al., 1971). However, in these lower eukaryotes some sensitive enzymes seem to exist. MONDAL et al. (1972) have found an RNA polymerase in the chromatin of coconut nuclei which is inhibited by rifampicin. This inhibition, however, can be counteracted by a protein factor found in the same organism. Furthermore, concentrations of rifampicin about 100 times higher are needed to obtain inhibition comparable to that obtained with bacterial RNA polymerase. Recently MEDOFF et al. (1972) reported that rifampicin together with amphotericin B inhibits the growth of yeast cells, whereas each antibiotic alone has no effect in the concentrations tested. The mechanism of action is not clear.

B. RNA Polymerases from Subcellular Organelles

Protein synthesis in subcellular organelles (i.e. mitochondria and chloroplasts) from eukaryotic cells has several features in common with bacterial systems, e.g. smaller ribosomes and sensitivity to chloramphenicol. This has prompted many authors to test transcription in these systems for rifamycin sensitivity.

The results obtained with well-characterized DNA-dependent RNA polymerases from mitochondria (WINTERSBERGER, 1970; WINTERSBERGER and WINTERSBERGER, 1970; TSAI et al., 1971; REID and PARSONS, 1971; WINTERSBERGER, 1972) and chloroplasts (BOTTOMLEY et al., 1971a; BOTTOMLEY et al., 1971b; POLYA and JAGENDORF, 1971) indicate that rifampicin does not inhibit these enzymes. However, some contradictory results, especially with intact mitochondria and chloroplasts, have been obtained.

One of the causes of confusion is that the term "rifampicin sensitivity" has been used in a very liberal way. For instance, in several reports "rifampicin sensitivity" was claimed on the grounds that partial inhibition was observed at concentrations of 40–250 µg/ml (HORGEN and GRIFFIN, 1971a and 1971b; SURZYCKI and ROCHAIX, 1971; SCRAGG, 1971). It is quite clear that "inhibitory effects" obtained at such high concentrations cannot be taken as evidence for a specific inhibition of RNA synthesis. It must always be remembered that bacterial RNA polymerase is inhibited by concentrations in the order of 0.01 µg/ml.

The difficulties encountered with crude enzyme preparations are best demonstrated by the results described by REID and PARSONS (1971) with mitochondrial RNA polymerase from rat liver. These authors find extensive inhibition with intact mitochondria and crude enzyme. However, highly purified enzyme is not inhibited consistently by concentrations of rifampicin as high as 10 µg/ml. This is quite a clear proof that inhibition of RNA synthesis observed in crude extracts is not due to a specific inhibition of the enzyme.

Similar results have been obtained with RNA polymerase from maize chloroplasts (BOTTOMLEY et al., 1971a). The pure enzyme was not affected by rifamycin SV concentrations of up to 200 µg/ml, whereas the crude enzyme was partially inhibited. This inhibition could be traced to unfavorable salt conditions (personal communication).

Neurospora crassa might be an exception to the rule that RNA polymerases from mitochondria and chloroplasts are not sensitive to rifampicin. KÜNTZEL and SCHÄFER (1971) have purified a mitochondrial RNA polymerase from this organism and the enzyme seems to be inhibited by 6 µg/ml of rifampicin. Unfortunately, no experiments have been done at lower concentrations of rifampicin and a comparison of sensitivity of this enzyme with that of the bacterial RNA polymerase could not be done. It is important to obtain this information since WINTERSBERGER has recently (1972) reported that mitochondrial RNA polymerase from Neurospora crassa is resistant to 20 µg/ml of rifampicin.

The data presented by KÜNTZEL and SCHÄFER (1971) do not warrant the conclusion that their enzyme is bacteria-like. Its primitive structure would rather suggest a resemblance with the enzyme found in bacteriophage T 7.

In summary, neither in mitochondria nor in chloroplasts has an RNA polymerase been isolated which is inhibited by concentrations of rifampicin as low as those required to affect bacterial RNA polymerase. In fact, most enzymes examined are not inhibited by 10,000-fold higher concentrations of the drug. Thus, on this basis, mitochondrial and chloroplastic RNA polymerases are not bacteria-like enzymes.

III. Effects of Certain Lipophilic Rifamycin Derivatives on DNA Polymerases from Oncogenic Virus and on Other DNA and RNA Polymerases from Bacterial and Mammalian Origin

Oncogenic RNA viruses contain an RNA-dependent DNA polymerase (reverse transcriptase) (GALLO, 1971; TEMIN, 1972). This enzyme seems to be required both for the proliferation of the virus and the transformation of its

Table 1. *Comparison of the action of lipophilic rifamycin derivatives on various nucleic acid polymerases. Rifampicin, which only acts on bacterial RNA polymerase is given as comparison. Chemically the lipophilic rifamycin derivatives differ from rifampicin only by their substituent in position 3, as indicated. Compound No. 3 is in addition hydrogenated in position 16, 17, 18, 19, 28, 29*

Compound No.	Partial formula	Concentration of antibiotic (µg/ml) needed for 40–60% enzyme inhibition				
		RNA polymerase E. coli ETH 2018	RNA polymerase E. coli Rifampicin resistant	DNA polymerase (Kornberg enzyme) M. lyso-deiktikus	Reverse transcriptase AMV	DNA polymerases calf thymus
1	3—N⟨ ⟩CH₃ CH₃ CH₃	0.2	20	60	70	70
2	3—N⟨ ⟩CH(CH₃)₂	0.1	100	60	70	100
3	3—N⟨ ⟩CH(CH₃)₂ 16, 17, 18, 19, 28, 29- Hexahydro-	6	15	40	60	120
4	3—CH=N—N⟨C(CH₃)₂ / C(CH₃)₂⟩N—CH₂—C₆H₅ (AF/ABDP)	0.1	200	90	100	200
5	3—CH=N—N⟨ ⟩N—CH₃ (Rifampicin)	0.02	∅	∅	∅	∅

host cell. Since RNA viruses might be involved in some cancers, such as the human breast cancer (AXEL et al., 1972; SARKAR and MOORE, 1972), inhibitors of reverse transcriptase could be valuable anticancer agents. It was thus not surprising that the reports on the inhibition of these enzymes by some lipophilic rifamycin derivatives, such as rifamycin AF/ABDP and rifamycin AF/013 (GALLO et al., 1970; GURGO et al., 1971) produced a flood of subsequent papers all supporting the idea of a specific inhibition of the reverse transcriptase by lipophilic rifamycins (e.g. GREEN, 1972; GREEN et al., 1972; TING et al., 1972; TEITZ, 1971; SMITH et al., 1972). However, some authors mentioned that not only the viral enzyme, but also DNA polymerases of normal cells were inhibited by this class of rifamycins (ROSS et al., 1971; SMITH et al., 1972; GREEN et al., 1972). Furthermore, several authors have shown that lipophilic rifamycins also inhibit mammalian RNA polymerase (BUTTERWORTH et al., 1971; MEILHAC et al., 1972; ONISHI and MURAMATSU, 1972).

We have compared the action of some lipophilic rifamycin derivatives on various nucleic acid polymerases, namely reverse transcriptase from AMV, DNA polymerase from calf thymus and *M. lysodeiktikus* (Kornberg enzyme) and RNA polymerase from *E. coli* resistant to rifampicin. The data presented in Table 1 clearly show that all these enzymes are inhibited by the lipophilic rifamycins to a similar extent. Furthermore, modifications in the ansa ring which significantly influence the effect on bacterial RNA polymerase (WEHRLI et al., 1969), do not influence the inhibition of these enzymes (compound No. 3 in Table 1). Thus, the inhibitory effect of lipophilic rifamycins is very unspecific and the structure-activity relationship is different from that observed in bacteria. This has recently been confirmed by the observation that both DNA and bovine serum albumin can completely counteract the inhibition of all the above-mentioned enzymes (WEHRLI, to be published).

Thus, it may be concluded that no rifamycin derivative known so far has a specific inhibitory effect on reverse transcriptase or any other DNA polymerase. Furthermore, the inhibition of mammalian RNA polymerase and ribonuclease H (hybridase) (MÖLLING et al., 1971; SEKERIS and ROEWEKAMP, 1972) clearly confirms the unspecific action of these derivatives.

IV. Effects on DNA Viruses and Larger Infectious Agents Such as Chlamydozoaceae

A. Effect on DNA Viruses

Most studies about the effect of rifamycins on DNA viruses have been made with vaccinia virus (for earlier literature, see WEHRLI and STAEHELIN, 1971). It has been found that some derivatives, such as rifampicin, inhibit the growth of this virus. There is no doubt, however, that this inhibition is not due to a block in RNA synthesis, as was found in bacteria (PENNINGTON and FOLLETT, 1971; SZILAGYI and PENNINGTON, 1971), but apparently the assembly of immature virus particles is affected (PENNINGTON and FOLLET, 1971; GRIMLEY and MOSS, 1971). As in the case of the RNA viruses, very large concentrations of antibiotic are needed to inhibit the virus growth. Thus, clinical application is certainly not possible.

B. Effect on Trachoma Agent

The infectious elementary bodies of trachoma agent belong to the Chlamydo-zoaceae which are parasites of mammalian cells and are considered unusually small bacterial cells. As in the case of vaccinia virus, only certain rifamycin derivatives, e.g. rifampicin at very high concentrations, affect the growth of trachoma agent. The mechanism of action is not known, but again RNA synthesis is not involved (Becker, 1971).

Summary and Conclusion

Rifamycins have been named "wonder drugs" because of their effects on a large variety of organisms, such as bacteria, eukaryotes and viruses. Unfortunately, however, many of these effects are not at all specific, but seem to be artifacts due to experimental conditions. It cannot be stressed strongly enough that not all inhibitory effects of rifamycins observed at extremely high concentrations result from specific inhibition of either nucleic acid synthesis or another defined biochemical reaction.

The fact is that rifamycins affect bacterial RNA synthesis in a highly specific manner by forming a complex with the bacterial DNA-dependent RNA polymer-ase. Since one molecule of rifamycin binds to and inactivates one molecule of enzyme, extremely low concentrations of antibiotic are required to inhibit the enzyme. In contrast, all inhibitory effects on eukaryotic RNA polymerases, be they of nuclear, mitochondrial or chloroplastic origin, are observed only at 100-10,000-fold higher concentrations. Very often the concentrations are so high (> 40 µg/ml) that impurities in the preparation or degradation of the antibiotic could be responsible for the observed inhibitions. In fact, this would explain the often contradictory findings reported by different laboratories.

As to the effect on DNA viruses such as vaccinia, it has been shown clearly that inhibition of virus growth is not due to a block in RNA synthesis, but that maturation of the virus might be affected.

Finally, the inhibition of reverse transcriptase by lipophilic rifamycin derivatives is completely non-specific, since a large variety of nucleic acid polymerases, as well as other enzymes, are affected in a similar way. In fact, these rifamycin derivatives bind to almost any kind of protein and even to nucleic acids.

In summary, the rifamycins are a very powerful and selective tool for inhibiting bacterial RNA synthesis; however, most other effects are either non-specific or inconclusive and much further work is needed to elucidate their significance.

References

Ashburner, M.: Ecdysone induction of puffing in polytene chromosomes of Drosophila melanogaster. Effects of inhibitors of RNA synthesis. Exptl. Cell Res. **71**, 433 (1972).

Austin, S.J., I.P.B. Tittawella, R.S. Hayward, and J.G. Scaife: Amber mutations of *Escherichia coli* RNA polymerase. Nature New Biol. **232**, 133 (1971).

Avila, J., J.M. Hermoso, E. Viñuela, and M. Salas: Purification and properties of DNA-dependent RNA polymerase from *Bacillus subtilis* vegetative cells. Eur. J. Biochem. **21**, 526 (1971).

Axel, R., J. Schlom, and S. Spiegelman: Presence in human breast cancer of RNA homologous to mouse mammary tumour virus RNA. Nature **235**, 32 (1972).

BAUTZ, E.K.F., and F.A. BAUTZ: Studies on the function of the RNA polymerase σ factor in promoter selection. Cold Spring Harbor Symp. Quant. Biol. **35**, 227 (1970).

BECKER, Y.: Antitrachoma activity of rifamycin B and 8-0-acetylrifamycin S. Nature **231**, 115 (1971).

BINDA, G., E. DOMENICHINI, A. GOTTARDI, B. ORLANDI, E. ORTELLI, B. PACINI, and G. FOWST: Rifampicin, a general review. Arzneimittel-Forsch. **21**, 1907 (1971).

BOTTOMLEY, W., H.J. SMITH, and L. BOGORAD: RNA polymerases of maize: Partial purification and properties of the chloroplast enzyme. Proc. Natl. Acad. Sci. U.S. **68**, 2412 (1971a).

BOTTOMLEY, W., D. SPENCER, A. M. WHEELER, and P.R. WHITFELD: The effect of a range of RNA polymerase inhibitors on RNA synthesis in higher plant chloroplasts and nuclei. Arch. Biochem. Biophys. **143**, 269 (1971b).

BRUTLAG, D., R. SCHEKMAN, and A. KORNBERG: A possible role for RNA polymerase in the initiation of M 13 DNA synthesis. Proc. Natl. Acad. Sci. U.S. **68**, 2826 (1971).

BUTTERWORTH, P.H.W., R.F. COX, and C.J. CHESTERON: Transcription of mammalian chromatin by mammalian DNA-dependent RNA polymerases. Eur. J. Biochem. **23**, 229 (1971).

BYFIELD, J.E., Y.C. LEE, and L.R. BENNET: Similarity of tetrahymena and mammalian RNA polymerases based on rifampicin resistance. Biochim. Biophys. Acta **204**, 610 (1970).

CHAMBERLIN, M.: Transcription 1970: a summary. Cold Spring Harbor Symp. Quant. Biol. **35**, 851 (1970).

CHEN, B., B. DE CROMBRUGGHE, W.B. ANDERSON, M.E. GOTTESMANN, and I. PASTAN: On the mechanism of action of lac repressor. Nature New Biol. **233**, 67 (1971).

DE BOER, C., P.A. MEULMAN, R.J. WNUK, and D.H. PETERSON: Geldanamycin, a new antibiotic. J. Antibiotics (Tokyo) **23**, 442 (1970).

DOI, R.H., L.R. BROWN, G. RODGERS, and Y. HSU: *B. subtilis* mutant altered in spore morphology and in RNA polymerase activity. Proc. Natl. Acad. Sci. U.S. **66**, 404 (1970).

Editorial Staff: Rifampicin: a review. Drugs **1**, 354 (1971).

FURTH, J.J., A. NICHOLSON, and G.E. AUSTIN: The enzymatic synthesis of RNA in animal tissue. III. Further purification of soluble RNA polymerase from lymphoid tissue and some general properties of the enzyme. Biochim. Biophys. Acta **213**, 124 (1970).

GALLO, R.C.: Reverse transcriptase, the DNA polymerase of oncogenic RNA viruses. Nature **234**, 194 (1971).

GALLO, R.C., S.S. YANG, and R.C. TING: RNA-dependent DNA polymerase of human acute leukaemic cells. Nature **228**, 927 (1970).

GISSINGER, F., and P. CHAMBON: Animal DNA-dependent RNA polymerases. 2. Purification of calf thymus AI enzyme. Eur. J. Biochem. **28**, 277 (1972).

GREEN, M.: Molecular basis for the attack on cancer. Proc. Natl. Acad. Sci. U.S. **69**, 1036 (1972).

GREEN, M., J. BRAGDON, and A. RANKIN: 3-Cyclic amine derivatives of rifamycin: Strong inhibitors of the DNA polymerase activity of RNA tumor viruses. Proc. Natl. Acad. Sci. U.S. **69**, 1294 (1972).

GRIMLEY, P.M., and B. MOSS: Similar effect of rifampin and other rifamycin derivatives on vaccinia virus morphogenesis. J. Virol. **8**, 225 (1971).

GURGO, C., R.K. RAY, L. THIRY, and M. GREEN: Inhibitors of the RNA and DNA-dependent polymerase activities of RNA tumor viruses. Nature New Biol. **229**, 111 (1971).

HARTMANN, G., K.O. HONIKEL, F. KNÜSEL and J. NÜESCH: The specific inhibition of the DNA-directed RNA synthesis by rifamycin. Biochim. Biophys. Acta **145**, 843 (1967).

HEIL, A., and W. ZILLIG: Reconstitution of bacterial DNA-dependent RNA polymerase from isolated subunits as a tool for the elucidation of the role of the subunits in transcription. FEBS Letters **11**, 165 (1970).

HINKLE, D.C., and M. CHAMBERLIN: The role of sigma subunit in template site selection by *E. coli* RNA polymerase. Cold Spring Harbor Symp. Quant. Biol. **35**, 65 (1970).

HORGEN, P.A., and D.H. GRIFFIN: Specific inhibitors of the three RNA polymerases from the aquatic fungus *Blastocladiella emersonii*. Proc. Natl. Acad. Sci. U.S. **68**, 338 (1971a).

HORGEN, P.A., and D.H. GRIFFIN: RNA polymerase III of *Blastocladiella emersonii* is mitochondrial. Nature New Biol. **234**, 17 (1971b).

JOHNSON, J.C., M. DEBACKER, and J.A. BOEZI: Deoxyribonucleic acid-dependent ribonucleic acid polymerase of *Pseudomonas putida*. J. Biol. Chem. **246**, 1222 (1971).

KAMIYA, K., T. SUGINO, Y. WADA, M. NISHIKAWA, and T. KISHI: The X-ray analysis of tolypomycinone tri-m-bromobenzoate. Experientia **25**, 901 (1969).

Kedinger, C., and P. Chambon: Animal DNA-dependent RNA polymerases. 3. Purification of calf thymus BI and BII enzymes. Eur. J. Biochem. **28**, 283 (1972).

Kedinger, C., M. Gniazdowski, J.L. Mandel, Jr., F. Gissinger, and P. Chambon: α-Amanitin: a specific inhibitor of one of two DNA-dependent RNA polymerase activities from calf thymus. Biochem. Biophys. Res. Commun. **38**, 165 (1970).

Kishi, T., H. Yamana, M. Muroi, S. Harada, M. Asai, T. Hasegawa, and K. Mizuno: Tolypomycin, a new antibiotic. III. Isolation and characterization of tolypomycin Y. J. Antibiotics (Tokyo) **25**, 11 (1972).

Konopka, E.A., F. Kradolfer, and J. Gelzer: Review of experimental antimycobacterial studies with rifampicin. Zbl. Bakteriol. Parasitenk. Abt. Ia. Ref. **228**, 1 (1972).

Küntzel, H., and K. P. Schäfer: Mitochondrial RNA polymerase from *Neurospora crassa*. Nature New Biol. **231**, 265 (1971).

Losick, R., R.G. Shorenstein, and A.L. Sonenshein: Structural alteration of RNA polymerase during sporulation. Nature **227**, 910 (1970).

Louis, G., and P.S. Fitt: The role of *Halobacterium cutirubrum* deoxyribonucleic acid-dependent ribonucleic acid polymerase subunits in initiation and polymerization. Biochem. J. **127**, 81 (1972).

Matzura, H., S. Molin, and O. Maaløe: Sequential biosynthesis of the β and β' subunits of the DNA-dependent RNA polymerase from *Escherichia coli*. J. Mol. Biol. **59**, 17 (1971).

Medoff, G., G.S. Kobayashi, C.N. Kwan, D. Schlessinger, and P. Venkov: Potentiation of rifampicin and 5-fluorocytosine as antifungal antibiotics by amphotericin B. Proc. Natl. Acad. Sci. U.S. **69**, 196 (1972).

Meilhac, M., Z. Tysper, and P. Chambon: Animal DNA-dependent RNA polymerases. Eur. J. Biochem. **28**, 291 (1972).

Mišoň, P., and L. Trnka: The effect of rifampicin on DNA-dependent RNA polymerase from mycobacteria. Collection Czech. Chem. Commun. **37**, 1049 (1972).

Mölling, K., D.P. Bolognesi, H. Bauer, W. Büsen, H.W. Plassmann, and P. Hausen: Association of viral reverse transcriptase with an enzyme degrading the RNA moiety of RNA-DNA hybrids. Nature New Biol. **234**, 240 (1971).

Mondal, H., A. Ganguly, A. Das, R.K. Mandal, and B.B. Biswas: Ribonucleic acid polymerase from eukaryotic cells. Effects of factors and rifampicin on the activity of RNA polymerase from chromatin of coconut nuclei. Eur. J. Biochem. **28**, 143 (1972).

Onishi, T., and M. Muramatsu: Inhibition by derivatives of rifamycin of soluble ribonucleic acid polymerase from rat liver. Biochem. J. **128**, 1361 (1972).

Pennington, T.H., and E.A.C. Follett: Inhibition of poxvirus maturation by rifamycin derivatives and related compounds. J. Virol. **7**, 821 (1971).

Polya, G.M., and A.T. Jagendorf: Wheat leaf RNA polymerases. I. Partial purification and characterization of nuclear, chloroplast and soluble DNA-dependent enzymes. Arch. Biochem. Biophys. **146**, 635 (1971).

Prelog, V.: Constitution of rifamycins. Pure Appl. Chem. **7**, 551 (1963).

Rabussay, D., and W. Zillig: A rifampicin resistant RNA polymerase from *E. coli* altered in the β subunit. FEBS Letters **5**, 104 (1969).

Reid, B.D., and P. Parsons: Partial purification of mitochondrial RNA polymerase from rat liver. Proc. Natl. Acad. Sci. U.S. **68**, 2830 (1971).

Rinehart, K.L., Jr.: Antibiotics with Ansa rings. Accounts of Chemical Res. **5**, 57 (1972).

Rodriquez-Lopez, M., M.L. Muñoz, and D. Vazquez: The effects of the rifamycin antibiotics on algae. FEBS Letters **9**, 171 (1970).

Ross, J., E.M. Scolnick, G.J. Todaro, and S.A. Aaronson: Separation of murine cellular and murine leukaemia virus DNA polymerases. Nature New Biol. **231**, 163 (1971).

Sarkar, N.H., and D.H. Moore: On the possibility of a human breast cancer virus. Nature **236**, 103 (1972).

Sasaki, K., K.L. Rinehart, Jr., G. Slomp, M.F. Grostic, and E.C. Olson: Geldanamycin. I. Structure assignment. J. Am. Chem. Soc. **92**, 7591 (1970).

Scragg, A.H.: Mitochondrial DNA-directed RNA polymerase from *Saccharomyces cerevisiae* mitochondria. Biochem. Biophys. Res. Commun. **45**, 701 (1971).

Sekeris, C.E., and W. Roewekamp: Inhibitory effects of rifampicin and some derivatives on ribonuclease H (hybridase) from rat liver. FEBS Letters **23**, 34 (1972).

SENSI, P., A.M. GRECO, and R. BALLOTTA: Rifomycins. I. Isolation and properties of rifomycin B and rifomycin complex. Antibiot. Ann. **1959–60**, 262.

SHIBATA, M., T. HASEGAWA, and E. HIGASHIDE: Tolypomycin, a new antibiotic. I. *Streptomyces tolypophorus* nov. sp., a new antibiotic, tolypomycin-producer. J. Antibiotics (Tokyo) **24**, 810 (1971).

SIRBASKU, D. A., and J. M. BUCHANAN: Patterns of ribonucleic acid synthesis in T5-infected *Escherichia coli*. V. Formation of stable, discrete, degradation products during turnover of phage-specific ribonucleic acid. J. Biol. Chem. **246**, 1665 (1971).

SMITH, R. G., J. WHANG-PENG, R. C. GALLO, P. LEVINE, and R. C. TING: Selective toxicity of rifamycin derivatives for leukaemic human leucocytes. Nature New Biol. **236**, 166 (1972).

SONENSHEIN, A. L., and R. LOSICK: RNA polymerase mutants blocked in sporulation. Nature **227**, 906 (1970).

STRAIN, G. C., K. P. MULLINIX, and L. BOGORAD: RNA polymerases of maize: nuclear RNA polymerases. Proc. Natl. Acad. Sci. U. S. **68**, 2647 (1971).

SURZYCKI, S. J.: Genetic functions of the chloroplast of chlamydomonas reinhardi: effect of rifampin on chloroplast DNA-dependent RNA polymerase. Proc. Natl. Acad. Sci. U. S. **63**, 1327 (1969).

SURZYCKI, S. J., and J. D. ROCHAIX: Transcriptional mapping of ribosomal RNA genes of the chloroplast and nucleus of Chlamydomonas reinhardi. J. Mol. Biol. **62**, 89 (1971).

SZILAGYI, J. F., and T. H. PENNINGTON: Effect of rifamycins and related antibiotics on the deoxyribonucleic acid-dependent ribonucleic acid polymerase of vaccinia virus particles. J. Virol. **8**, 133 (1971).

TEITZ, Y.: RNA dependent DNA polymerase in C type particles from normal rat thymus cultures. Nature New Biol. **232**, 250 (1971).

TEMIN, H. M.: The RNA tumor viruses—background and foreground. Proc. Natl. Acad. Sci. U. S. **69**, 1016 (1972).

TING, R. C., S. S. YANG, and R. C. GALLO: Reverse transcriptase, RNA tumour virus transformation and derivatives of rifamycins. Nature New Biol. **236**, 163 (1972).

TSAI, M., G. MICHAELIS, and R. S. CRIDDLE: DNA-dependent RNA polymerase from yeast mitochondria. Proc. Natl. Acad. Sci. U. S. **68**, 473 (1971).

UMEZAWA, H., S. MIZUNO, H. YAMAZAKI, and K. NITTA: Inhibition of DNA-dependent RNA synthesis by rifamycins. J. Antibiotics (Tokyo) **21**, 234 (1968).

VOIGT, H.-P., R. KAUFMANN, and H. MATTHAEI: Solubilized DNA-dependent RNA polymerase from human placenta: a Mn^{2+}-dependent enzyme. FEBS Letters **10**, 257 (1970).

WANG, A. H.-J., I. C. PAUL, K. L. RINEHART, JR., and F. J. ANTOSZ: Chemistry of streptovaricins. IX. X-ray crystallographic structure of a streptovaricin C derivative. J. Am. Chem. Soc. **93**, 6275 (1971).

WEAVER, R. F., S. P. BLATTI, and W. J. RUTTER: Molecular structures of DNA-dependent RNA polymerases (II) from calf thymus and rat liver. Proc. Natl. Acad. Sci. U. S. **68**, 2994 (1971).

WEHRLI, W., F. KNÜSEL, K. SCHMID, and M. STAEHELIN: Interaction of rifamycin with bacterial RNA polymerase. Proc. Natl. Acad. Sci. U. S. **61**, 667 (1968a).

WEHRLI, W., F. KNÜSEL, and M. STAEHELIN: Action of rifamycin on RNA polymerase from sensitive and resistant bacteria. Biochem. Biophys. Res. Commun. **32**, 284 (1968b).

WEHRLI, W., J. NÜESCH, F. KNÜSEL, and M. STAEHELIN: Action of rifamycins on RNA polymerase. Biochim. Biophys. Acta **157**, 215 (1968c).

WEHRLI, W., and M. STAEHELIN: The rifamycins-relation of chemical structure and action on RNA polymerase. Biochim. Biophys. Acta **182**, 24 (1969).

WEHRLI, W., and M. STAEHELIN: Interaction of rifamycin with RNA polymerase. In: Proc. 1st. Int. Lepetit Colloq. RNA Polymerase Transcript, p. 65. Amsterdam-London: North-Holland Publ. Co. 1970.

WEHRLI, W., and M. STAEHELIN: Actions of the rifamycins. Bacteriol. Rev. **35**, 290 (1971).

WHITE, R. J., and G. C. LANCINI: Uptake and binding of ³H-rifampicin by *Escherichia coli* and *Staphylococcus aureus*. Biochim. Biophys. Acta **240**, 429 (1971).

WHITE, R. J., G. C. LANCINI, and L. G. SILVESTRI: Mechanism of action of rifampin on *Mycobacterium smegmatis*. J. Bacteriol. **108**, 737 (1971).

WICKNER, W., D. BRUTLAG, R. SCHEKMAN, and A. KORNBERG: RNA synthesis initiates *in vitro* conversion of M 13 DNA to Its replicative form. Proc. Natl. Acad. Sci. U. S. **69**, 965 (1972).

Wintersberger, E.: DNA-dependent RNA polymerase from mitochondria of a cytoplasmic "petite" mutant of yeast. Biochem. Biophys. Res. Commun. **40**, 1179 (1970).

Wintersberger, E.: Isolation of a distinct rifampicin-resistant RNA polymerase from mitochondria of yeast, neurospora and liver. Biochem. Biophys. Res. Commun. **48**, 1287 (1972).

Wintersberger, E., and U. Wintersberger: Rifamycin insensitivity of RNA synthesis in yeast. FEBS Letters **6**, 58 (1970).

Wyss, E., and W. Wehrli: Kinetic studies of the interaction between *E. coli* RNA polymerase and rifampicin. Experientia **29**, 760 (1973).

Zillig, W., K. Zechel, D. Rabussay, M. Schachner, V.S. Sethi, P. Palm, A. Heil, and W. Seifert: On the role of different subunits of DNA-dependent RNA polymerase from *E. coli* in the transcription process. Cold Spring Harbor Symp. Quant. Biol. **35**, 47 (1970).

Sibiromycin

G. F. Gause

Antitumor antibiotic sibiromycin is a glycoside of a new animosugar sibiros-amine. The aglycone of this antibiotics is a derivative of 1,4-benzdiasepine (BRAZH-NIKOVA *et al.*, 1970; MESENTSEV *et al.*, 1971; MESENTSEV *et al.*, 1973).

Sibirosamine is 4-methylamino-4,6-didesoxy-3-C-methyl-hexopyranose:

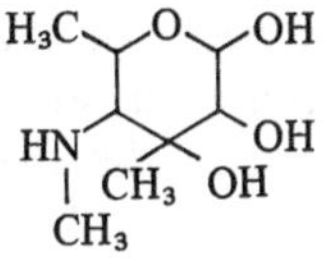

Fig. 1

which differs from all known natural aminosugars by its branching carbon chain. The aglycone of sibiromycin possesses the following structure:

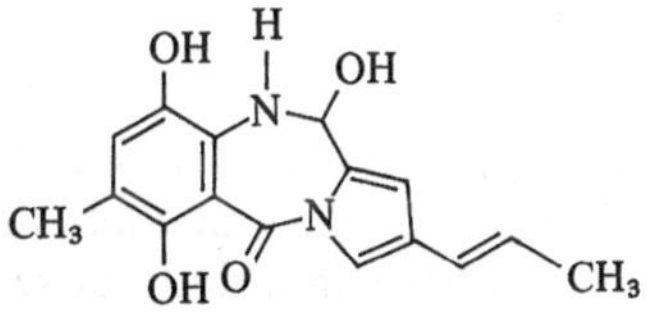

Fig. 2

Sibiromycin, produced by an actinomycete, *Streptosporangium sibiricum* (GAUSE *et al.*, 1969) was first detected by its capacity to inhibit the multiplication of suspended ascites tumor cells. For example, the culture liquid of *Streptosporangium sibiricum*, diluted 100 times, inhibited the multiplication of cells of reticulo-endothelial sarcoma of mice (strain RAB-1) as well as multiplication of cells of mouse lymphoadenoma (strain NK/Ly). In the culture liquid of the producing strain, sibiromycin is present in low concentrations and inhibits only the growth of tumor cells but not that of bacteria. However, when sibiromycin was later concentrated it was observed that at higher concentrations the antibiotic also inhibited the growth of various bacteria, notably that of *Bacillus mycoides*. This microorganism, therefore, was used for the biological assay of sibiromycin at early stages of research, before a chemical method for determining concentrations of sibiromycin was worked out (KONSTANTINOVA *et al.*, 1970).

Sibiromycin inhibits growth of *Bacillus mycoides* and *Bacillus subtilis* (0.3 mcg/ml), *Staphylococcus aureus* (1 mcg/ml) and *Escherichia coli* (20 mcg/ml) (GAUSE *et al.*, 1970). Antitumor action of sibiromycin against six different transplantable tumors of mice was studied in detail by SHORIN and ROSSOLIMO (1970). The activity of the antibiotic was most pronounced in the treatment of mice with inoculated squamous praegastric cancer cells (strain OG-5) and after two injections of sibiromycin in maximal tolerated doses the tumors disappeared completely. The development of lymphosarcoma (strain Lyo-1) was inhibited 90–97 percent by the maximum tolerated dose. The growth of ascitic forms of tumors and the solid sarcoma 180 was inhibited by 62–66 percent. Sibiromycin induced atrophic changes in the spleen, and the weight of the latter in treated animals was lower than in the control mice. The decrease in the weight of spleen was found to be dose-dependent: higher the dose, the lower was the weight.

LD_{50} of sibiromycin administered as a single dose to mice intravenously, intraperitoneally, subcutaneously and orally was found to be 58, 32, 84 and 459 mcg/kg respectively (Ilyushina and Goldberg, 1970). Maximal tolerated dose in dogs, inducing no impairment of the heart, kidneys and liver function on repeated daily intravenous injections for 30 days, was 0.37 mcg/kg. When the injections were made once a week for seven weeks, the corresponding dose was 2 mcg/kg. Sibiromycin exhibited a late toxic effect. Animals treated with the antibiotic in lethal doses died in the first twenty days and injuries to the myocardium and kidneys were found to be the cause of death. Acute toxicity experiments on cats showed that toxic doses do not significantly change the arterial blood pressure and the antibiotic is relatively without effect on the vegetative nervous system and neuromuscular transmission.

Vertogradova and Kunrat (1971) studied the effect of sibiromycin on peripheral blood, as well as its distribution and excretion in animals. When administered to dogs repeatedly in daily doses of 1.5, 1.0, and 0.75 mcg/kg, a markedly pronounced lympho- and thrombocytopenia as well as a slight decrease in the counts of leucocytes, granulocytes and erythrocytes was observed. Administered to dogs 30 times in a daily dose of 0.37 mcg/kg, once a week as single doses of 2.0 and 1.5 mcg/kg, and to rabbits in a daily dose of 2.5 mcg/kg, sibiromycin did not change the pattern of the peripheral blood. No changes were observed in the permeability of the skin capillaries in rabbits treated with the antibiotic in doses of 5 and 2.5 mcg/kg. After a single intravenous administration to rabbits, the antibiotic was detected in the blood for 15 to 30 minutes. Small amounts of sibiromycin were detected in all organs of the animals. About 35 to 40 percent of the injected dose of sibiromycin was excreted with the urine. Data on the histological examination of organs of dogs treated with sibiromycin has been published by Muraveiskaya (1971). Sibiromycin inhibits antibody production in mice and rabbits. This inhibition is associated with a decrease in the number of antibody-producing cells in the lymphoid tissue (Shapovalova, 1972).

Mode of action studies have shown that sibiromycin interacts specifically with DNA (Gause and Dudnik, 1971). In the cultures of *Staphylococcus aureus* and *Bacillus subtilis*, the antibiotic selectively inhibits the synthesis of DNA. Sibiromycin induces phage production in the lysogenic strains of *Micrococcus lysodeikticus* and *Escherichia coli*. The antibiotic complexes with isolated DNA, but does not interact with RNA, bases of nucleic acids, ribonucleosides, ribonucleotides, and deoxyribonucleotides. Heat denatured DNA binds lesser quantities of sibiromycin than the native DNA and DNA denatured by formaldehyde did not bind sibiromycin at all (Gause and Dudnik, 1972). The absorption spectrum of the antibiotic is changed on adding DNA, and these spectral changes were used for kinetic measurements. Magnesium ions prevent the binding of sibiromycin to DNA. The complexing of the antibiotic with DNA increases the melting temperature of the latter, but does not affect the viscosity of its solutions. It was observed that DNA with a high GC content binds more antibiotic than DNA with a low GC content.

Since the antibiotic complexes with native DNA even *in vitro*, it is reasonable to expect that the template activity of DNA would be decreased leading to inhibition of the synthesis of nucleic acids in bacterial cells. That such is the case has been demonstrated in several experiments and sibiromycin effectively inhibits the template activity of DNA in the system of DNA-dependent RNA polymerase of *E. coli* (GAUSE et al., 1972).

Sibiromycin selectively inhibits elongation of the molecules of RNA, as can be seen from the data shown in Table 1. DNA from phage T2 was used as the template in these experiments. Sibiromycin in the concentration 0.3 mcg/ml inhibits incorporation of ^{14}C-UMP from ^{14}C-UTP by 40 percent, but does not inhibit incorporation of γ-^{32}P-ATP. It follows that elongation of the RNA molecules is selectively vulnerable to the effect of sibiromycin while their initiation is affected less.

Table 1. *Effect of sibiromycin upon initiation and elongation of RNA chains in RNA polymerase system of E. coli.* (From GAUSE et al., 1972)

	^{14}C-UMP incorporation from ^{14}C-UTP	γ-^{32}P-ATP incorporation
Control	100	100
Sibiromycin 0.1 mcg/ml	102	110
Sibiromycin 0.3 mcg/ml	60	99
Sibiromycin 1.0 mcg/ml	35	87
Sibiromycin 3.0 mcg/ml	22	38

Since sibiromycin inhibits the growth of RNA chains in cell-free systems, it is possible to conclude that the antibiotic binds to the DNA template and thus stops or prevents the movement of the RNA polymerase molecule along its template. Sibiromycin is highly efficient in inhibiting DNA-dependent RNA synthesis which is inhibited by 50 percent when the molar ratio of antibiotic/DNA nucleotide is approximately 1/1300. Because sibiromycin does not selectively inhibit initiation of RNA molecules, it is reasonable to conclude that the high efficiency of action of this antibiotic is due to its high binding affinity to DNA. When added even in small concentration, the entire amount is found to be firmly bound to DNA.

Sibiromycin-DNA complex is very stable as follows from tests of its template activity in the DNA polymerase reaction (GAUSE et al., 1972). Although sodium dodecylsulfate completely blocks interaction of sibiromycin with DNA, it does not induce any dissociation of the preformed sibiromycin-DNA complex (DUDNIK and NETYKSA, 1972). Also, urea, formamide, dimethylformamide and ethanol are unable to bring about dissociation of the sibiromycin-DNA complex. All this would indicate that sibiromycin is covalently bound to the DNA.

DUDNIK and NETYKSA (1972) observed that sibiromycin competes with actinomycin D and olivomycin A for the binding sites on DNA. Sibiromycin effectively binds to DNA previously saturated with actinomycin or olivomycin, and almost completely displaces these antibiotics from their complexes with DNA. Conversely, both actinomycin and olivomycin practically fail to interact with DNA previously saturated with sibiromycin. This one-sided nature of the competition can be explained by a high stability of the sibiromycin-DNA complex.

DUDNIK et al. (1971 b) noticed considerable increase of antibacterial effect of sibiromycin in systems with impaired excision repair of DNA. Addition of trypaflavine to the plating medium significantly increased the bactericidal effect of sibiromycin on E. coli B. It was also noted that mutants (uvr⁻) of E. coli AB-1157 with impaired repair of DNA lesions induced by ultraviolet radiation or bifunctional alkylating agents had increased sensitivity to the bactericidal effect of sibiromycin.

Regarding the effect of sibiromycin upon animal cells, it is of considerable interest that in Ehrlich ascitic tumor cells the synthesis of both DNA and RNA is very sensitive to this antibiotic. Sibiromycin at a concentration of 0.04 mcg/ml suppresses the synthesis of DNA and of RNA approximately by 30–40 percent and at 4 mcg/ml completely stops the synthesis of both nucleic acids (GAUSE et al., 1972). In other words, while sibiromycin has a selective action on the DNA synthesis in bacteria, in Ehrlich ascitic tumor cells, both DNA and RNA syntheses are inhibited, almost to the same degree. Also in developing sea urchin embryos the synthesis of both DNA and RNA is equally sensitive to sibiromycin (GAUSE and FATKULLINA, 1971). In sharp contrast to the synthesis of nucleic acids, incorporation of labeled amino acids into proteins of the Ehrlich ascitic tumor cells is practically unaffected by sibiromycin and even the high concentrations of the antibiotic (4 mcg/ml), which completely stop the incorporation of labeled precursors into nucleic acids, have no effect on protein synthesis. In subcellular systems of isolated rat liver nuclei and mitochondria, sibiromycin effectively inhibits the synthesis of DNA. Numerous experiments, however, indicate that the inhibitory effect of sibiromycin upon animal cells results from an impairment of transcription and replication of DNA, as in the case of bacteria. Biological activity of sibiromycin in both cases may be ascribed to the formation of a stable complex with DNA template (G.G. GAUSE et al., 1972).

During studies on chemical structure of the antibiotic, a sulfur containing derivative with 5.5% sulfur content, was prepared (BRAZHNIKOVA et al., 1970). This derivative is more stable than sibiromycin and is also biologically active. This compound complexes in vitro with DNA in a way similar to the parent antibiotic with one important difference. Whereas the complex formation between sibiromycin and DNA is complete within 10–15 minutes, with the sulfur containing derivative the reaction continues for 2–3 hours. Studies with the labeled (^{35}S) derivative have indicated that during interaction with DNA, the ^{35}S is split off the molecule of the antibiotic and in fact the removal of this sulfur is a prerequisite for the interaction with DNA (DUDNIK, 1971 a). This also explains the increased time of interaction of the sulfur containing derivative with DNA, as compared to sibiromycin.

References

BRAZHNIKOVA, M.G., I.N. KOVSHAROVA, N.V. KONSTANTINOVA, A.S. MESENTSEV, V.V. PROSHLYA-KOVA, and I.V. TOLSTYKH: Chemical studies on sibiromycin, an antitumor antibiotic. Antibiotiki 15, 297 (1970).

DUDNIK, Y.V., V.L. KARPOV, and E.M. NETYKSA: Sulfur containing derivative of sibiromycin: Removal of sulfur on interaction with DNA. Antibiotiki 16, 6 (1971a).

DUDNIK, Y.V., and E.M. NETYKSA: Interaction of sibiromycin with DNA: Displacement of actinomycin D and olivomycin A from complexes with DNA. Antibiotiki 17, 44 (1972).

DUDNIK, Y.V., E.M. NETYKSA, and O.Y. VARIK: Increased antibacterial effect of sibiromycin and bruneomycin in systems with impaired reparation of DNA. Antibiotiki 16, 487 (1971b).

GAUSE, G.F., and Y.V. DUDNIK: Interaction of antitumor antibiotics with DNA: Studies on sibiromycin. Progr. Mol. Subcell. Biol. 2, 33 (1971).

GAUSE, G.F., and Y.V. DUDNIK: Mechanism of action of antitumor antibiotic sibiromycin. Proceed. 7th Intern. Congr. Chemotherapy. Prague 2, 87 (1972).

GAUSE, G.G., Y.V. DUDNIK, and S.M. DOLGILEVICH: Inhibition of synthesis of nucleic acids by antitumor antibiotic sibiromycin. Antibiotiki 17, 413 (1972).

GAUSE, G.F., Y.V. DUDNIK, E.M. NETYKSA, A.V. LAIKO, and G.G. GAUSE: Mechanism of action of sibiromycin. Antibiotiki 15, 867 (1970).

GAUSE, G.G., and L.G. FATKULLINA: The action of sibiromycin upon early cleavage of sea urchin embryos. Ontogenes (Moscow) 2, 648 (1971).

GAUSE, G.F., T.P. PREOBRAZHENSKAIA, L.P. IVANITSKAYA, and M.A. SVESHNIKOVA: Production of antibiotic sibiromycin by *Streptosporangium sibiricum*. Antibiotiki 14, 963 (1969).

ILYUSHINA, N.G., and L.E. GOLDBERG: Pharmacological studies on sibiromycin. Antibiotiki 15, 612 (1970).

KONSTANTINOVA, N.V., I.V. TOLSTYKH, and M.G. BRAZHNIKOVA: Quantitative chemical assay of antitumor antibiotic sibiromycin in preparations and culture liquids. Antibiotiki 15, 304 (1970).

MESENTSEV, A.S., V.V. KULYAEVA, L.M. RUBASHEVA, M.G. BRAZHNIKOVA, O.S. ANISIMOVA, T.V. VLASOVA, and Y.N. SHEINKER: The structure of sibirosamine, a new aminosugar from antibiotic sibiromycin. Khimia Prirodnych Soedinenii 5, 650 (1971).

MESENTSEV, A.S., L.M. RUBASHEVA, V.V. KULYAEVA, M.G. BRAZHNIKOVA, O.S. ANISIMOVA, T.V. VLASOVA, and Y.N. SHEINKER: The structure of a product of acid hydrolysis of sibiromycin. Khimia Prirodnych Soedinenii 7, 234 (1973).

MURAVEISKAYA, V.S.: Histological examination of organs of dogs treated with sibiromycin. Antibiotiki 16, 810 (1971).

SHAPOVALOVA, S.P.: The effect of sibiromycin on immunological reactivity of the organism. Antibiotiki 17, 270 (1972).

SHORIN, V.A., and O.K. ROSSOLIMO: Experimental studies on antitumor activity of sibiromycin. Antibiotiki 15, 300 (1970).

VERTOGRADOVA, T.P., and I.A. KUNRAT: Effect of sibiromycin on peripheral blood, its distribution and excretion in animals. Antibiotiki 16, 316 (1971).

Thiaxanthenones: Miracil D and Hycanthone

Erich Hirschberg

This review focuses on two synthetic compounds, the most widely studied in a large series of congeners and derivatives, which have been employed extensively in the treatment of human schistosomiasis, have attracted modest interest in experimental cancer chemotherapy, and have become useful probes in the examination of some aspects of molecular biology.

Beginning in 1932, a reliable animal screening test was developed for schistosomiasis or bilharzia by investigators at the Farbenfabriken Bayer (KIKUTH and GOENNERT, 1948). The availability of this test provided a stimulus for the synthesis and evaluation of many hundreds of compounds with potential schistosomicidal activity. Among these were a number of xanthenones and thiaxanthenones, including those later named the Miracil series. Miracil D (1-diethylaminoethylamino-4-methyl-10-thiaxanthenone, Lucanthone, Nilodin, Tixanthenone, NSC 14574) was prepared by MAUSS in December 1939 (MAUSS, 1948) and found to have pronounced biological activity in animals and man (BLAIR, 1958; HAWKING and ROSS, 1948; KIKUTH and GOENNERT, 1948, 1949; KIKUTH et al., 1946).

Several additional series of xanthenones and thiaxanthenones, as well as compounds representing or resembling portions of the active parent structure, were synthesized in Germany and the United States (e. g. ARCHER and SUTER, 1952; BLANZ and FRENCH, 1963; ELSLAGER et al., 1961; GOENNERT, 1961) and, indeed, the systematic search for agents endowed with a more favorable therapeutic index continues to this day.

In 1965, the research group at Sterling-Winthrop Research Institute headed by ARCHER reported the identification of Hycanthone (1-diethylaminoethylamino-4-hydroxymethyl-10-thiaxanthenone, Etrenol, NSC 142982) as a major metabolite of Miracil D and this derivative was shown to have considerably greater schistosomicidal activity in several test systems (ARCHER, 1968; ROSI et al., 1965; ROSI et al., 1967). Evaluation of therapeutic efficacy, toxicological and pharmacological properties, and mechanism of action of this derivative is still underway.

From a quantitative point of view, the major emphasis has been on the usefulness of these compounds in the treatment of schistosomiasis. However, some of the effects observed in the initial pharmacological evaluation of Miracil D suggested its examination for anticancer activity. Following brief positive reports by HACKMANN et al. (1949) and PIRWITZ et al. (1949), more extensive evaluation, including structure-activity studies, was carried out in several laboratories (BLANZ and FRENCH, 1963; HIRSCHBERG et al., 1959; LEITER, 1962). Miracil D inhibited the growth of a wide spectrum of experimental tumors, but could not be introduced into clinical anticancer testing because of unacceptable toxicity. The initial reports on Hycanthone (ADAMSON, 1971; HIRSCHBERG et al., 1968 b; SIEBER et al., 1973) indicate that it also has carcinostatic activity in experimental tumors but its examination in cancer patients must await clarification of its toxic and mutagenic potential.

As will be detailed in the following pages, Miracil D and Hycanthone have been studied in a wide variety of other biological systems ranging from *Escherichia coli* to the fruit fly and from HeLa cells to regenerating rat liver. Extensive knowledge has accumulated on their chemical and physical properties, on physiological disposition and metabolic conversions, on generalized and specific toxicities, on structure-activity relationships, and on mechanisms of action. Thirty-five years and hundreds of publications after the original synthesis of Miracil D, it appears to this reviewer that while many facts about this thiaxanthenone and its closely related derivative, Hycanthone, are definite and well established, there remains ample room for further vigorous investigation.

Physical and Chemical Properties

The structural formulae of Miracil D and Hycanthone are shown in Fig. 1. Details on the organic synthesis of the former were given by ARCHER and SUTER (1952), HAMMICK and MUNRO (1952), MAUSS (1948), and SHARP (1951). NABIH and ELSHEIKH (1965a, b) studied the chemical reactivity of the 4-methyl group. The microbiological oxidation of Miracil D to Hycanthone by *Aspergillus sclerotiorum* was described by ROSI et al. (1965, 1967) and confirmed by SALLAM et al. (1971) who showed that several other molds isolated from local Egyptian habitats also carried out this conversion to a significant extent.

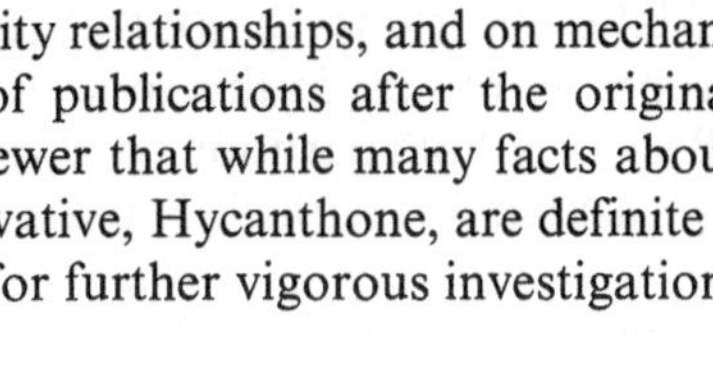

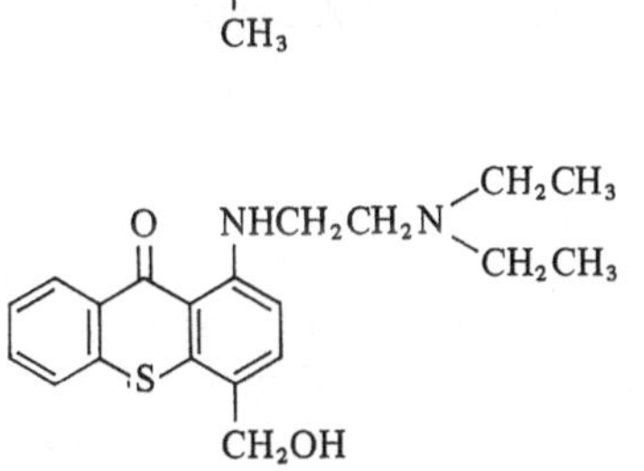

Fig. 1. Structural formulae of Miracil D and Hycanthone

Miracil D is a bright yellow crystalline powder, bitter to the taste, melting at 64–65° as the free base and at 191–197° as the hydrochloride (BLAIR, 1958; GOODMAN and GILMAN, 1970; HAWKING and ROSS, 1948; KUHNERT-BRANDSTAETTER et al., 1964; MAUSS, 1948). Solubility data in various solvents have been presented or summarized by several of the authors just cited and by ANONYMOUS (1963). Absorption spectra and related information may be found in several publications (Anonymous, 1963; Weinstein *et al.,* 1965; Zilversmit, 1970). Colorimetric assays for Miracil D in biological fluids were developed by Latner *et al.* (1947), Newsome and Robinson (1960b), and Strufe (1963) and applied to clinical or experimental investigations (Blair, 1958; Brindle *et al.,* 1964; Hawking and Ross, 1948). A polarographic approach to the determination of the compound was suggested by Munro (1960).

Scholtan and Goennert (1956) reported a tendency of xanthenones and thiaxanthenones, including Miracil D, to aggregate in aqueous solutions and to form micelles and colloidal suspensions. Scholtan (1960) provided further evidence on the colloidal properties of salts in this series including sedimentation and diffusion coefficients. Basicities, lipid partition coefficients, surface activities, and protein affinities were measured for Miracil D and a few homologous compounds (Munro, 1961). Factors affecting the exchange rate of Miracil D by Amberlite IR-120 were examined by Hussein *et al.* (1968 a, b).

Two recent papers by Zilversmit (1970, 1971) provide extensive comparative data on Miracil D and its biologically inactive N^{11}-methyl derivative with regard to electronic spectra, vibrational spectra, and ionization constants as a contribution to an understanding of the molecular mechanism of action (cf. below).

Since the discovery of Hycanthone occurred only a few years ago, it is not surprising that there have been only a few reports in this area so far. Most of the information has been provided by Rosi *et al.* (1965, 1967), including a melting point of 101–103° for the free base, the ultraviolet absorption spectrum, the nuclear magnetic resonance spectrum, thin layer and column chromatographic behavior, and the observation of exquisite sensitivity to acid. Chromatographic resolution of mixtures of Miracil D and Hycanthone with several solvents was also shown by Sallam *et al.* (1971). Ultraviolet absorption spectra at different pH values were presented for Hycanthone in connection with two recent biochemical studies (El-Sewedy *et al.,* 1972; Illingworth and Waring, unpublished).

Biological Activities

Treatment of Schistosomiasis. In the past twenty-five years, several hundred publications have been concerned with the evaluation of Miracil D and, more recently, Hycanthone in animals experimentally infected with *Schistosoma mansoni* or *S. haematobium* and in patients with this disease entity. These have ranged from single case reports in various, often exotic parts of the world through more substantial laboratory or clinical studies to extensive and intensive reviews and literature summaries. A detailed consideration of this large body of evidence on the first oral non-antimonial schistosomicide and its generally much more active close relative is beyond the scope of the present discussion. The following selected citations include key reviews, a few references of recognized historical interest, and some very recent papers dealing with Hycanthone: AMA Council on Drugs (1971), Archer (1968), Berberian *et al.* (1967a), Blair (1958), Elslager (1966, 1967), Foster *et al.* (1971), Foster and Richards (1970), Goennert (1955, 1961), Goennert and Koelling (1962), Goodman and Gilman (1970), Hoffer and Rachlin (1972), Jewsbury (1972), Jordan and Webbe (1969), Kikuth and Goennert (1948, 1949), Laemmler (1967, 1968), Lee (1972), Most

(1972), NEWSOME (1962), PELLEGRINO and KATZ (1968, 1969), ROSI et al. (1965), STANDEN (1963), SURREY and YARINSKY (1968, 1969).

Experimental Cancer Chemotherapy. In contrast to the abundance of reviews and summaries on the use of Miracil D and Hycanthone in the treatment of bilharzia, no attempt had been made to bring together at one place the gradually accumulating evidence of carcinostatic activity of these two compounds. Tables 1 and 2 represent such an initial summary and readily document the fact that

Table 1. *Anticancer activity of Miracil D*

Tumor	Activity	Literature citations
in vitro		
Glioblastoma multiforme	+	HIRSCHBERG (1958), HIRSCHBERG et al. (1959)
HeLa cells	+	BASES (1970), BASES and MENDEZ (1969), BRAMBILLA et al. (1966), JACQUEZ (1962)
KB cells	+	BRAMBILLA et al. (1966), MORASCA (1967), MORASCA and RAINISIO (1966), THAYER et al. (1971)
Leukemia L1210 cells	+	THAYER et al. (1971)
Embryonated eggs		
A-42	−	HARRIS (1962)
HAd No. 1	−	HARRIS (1962)
HS No. 1	+	HARRIS (1962)
Conditioned mice		
HEp No. 1	±	MERKER et al. (1962)
Conditioned rats		
HEp No. 1	−	TELLER (1962)
HS No. 1	−	TELLER (1962)
Hamster cheek pouch		
A-42	±	WOOLLEY (1962)
HEp No. 1	−	WOOLLEY (1962)
HS No. 1	−	WOOLLEY (1962)
Mouse		
Bashford carcinoma 63	±	SUGIURA (1962)
Carcinoma 1025	−	SUGIURA (1962)
C3HBA mammary tumor	+	MORAN and WOOLLEY (1962)
Ehrlich carcinoma	±	SUGIURA (1962)
Ehrlich carcinoma	−	HACKMANN et al. (1949)
Ehrlich carcinoma, ascites	±	HIRSCHBERG et al. (1959), SUGIURA (1962)
Ehrlich carcinoma, ascites	−	PIRWITZ et al. (1949)
Friend virus leukemia	±	SUGIURA (1962)
Glioma 26	±	SUGIURA (1962)
Glioma 26	−	HIRSCHBERG et al. (1959)
Harding-Passey melanoma	−	SUGIURA (1962)
Leukemia B82	±	BURCHENAL et al. (1963)
Leukemia L1210, ascites	+	BLANZ and FRENCH (1963), BURCHENAL et al. (1963), GRISWOLD et al. (1963), HIRSCHBERG (1963), HIRSCHBERG et al. (1964), HIRSCHBERG et al. (1959), PRESCOTT (1968), THAYER et al. (1971)

Table 1 (continued)

Tumor	Activity	Literature citations
Leukemia L1210, ascites	−	Hutchison et al. (1962a)
Leukemia L1210, Amethopterin-R	−	Hutchison et al. (1962a)
Leukemia L1210, 6-mercaptopurine-R	±	Hutchison et al. (1962a)
Leukemia L1210, azaserine-R	−	Hutchison et al. (1962a)
Leukemia L1210, 5-fluorouracil-R	−	Hutchison et al. (1962a)
Leukemia L4946	±	Sugiura (1962)
Leukemia P815	+	Burchenal et al. (1963)
Leukemia P1081	−	Burchenal et al. (1963)
Lewis bladder carcinoma	−	Sugiura (1962)
Lewis lung carcinoma	−	Sugiura (1962)
Mammary adenocarcinoma Eo771	+	Hirschberg et al. (1959)
Mammary adenocarcinoma Eo771	±	Sugiura (1962)
Mammary carcinoma 755	+	Blanz and French (1963), Griswold et al. (1963), Hirschberg (1963), Hirschberg et al. (1959), Prescott (1968)
Mecca lymphosarcoma	−	Sugiura (1962)
Miyono adenocarcinoma	±	Sugiura (1962)
Nelson tumor	+	Bross and Tarnowski (1962), Tarnowski and Bross (1962)
Nelson tumor, ascites	−	Tarnowski and Bross (1962)
Ridgway osteogenic sarcoma	±	Sugiura (1962)
Sarcoma 180	+	Hirschberg (1963), Hirschberg et al. (1959), Prescott (1968)
Sarcoma 180	−	Blanz and French (1963), Griswold et al. (1963)
Sarcoma M387	−	Sugiura (1962)
Sarcoma T241	−	Sugiura (1962)
Wagner osteogenic sarcoma	±	Sugiura (1962)

Rat		
Benzpyrene sarcoma	+	Hackmann et al. (1949)
Flexner-Jobling carcinoma	−	Sugiura (1962), Sugiura et al. (1972)
Iglesias ovarian tumor	+	Sugiura et al. (1972)
Iglesias sarcoma	±	Sugiura et al. (1972)
Jensen sarcoma	±	Sugiura et al. (1972)
Jensen sarcoma	−	Sugiura et al. (1962)
Jensen sarcoma, mitomycin C-R	−	Sugiura et al. (1972)
Methylcholanthrene-induced mammary carcinoma	+	Davis et al. (1966), Gropper and Shimkin (1967)
Moore sarcoma No. 1	−	Sugiura et al. (1972)
Murphy-Sturm lymphosarcoma	−	Sugiura (1962), Sugiura et al. (1972)
Walker carcinosarcoma 256	+	Hackmann et al. (1949)
Walker carcinosarcoma 256	−	Garattini and Palma (1961), Sugiura (1962), Sugiura et al. (1972)

Hamster		
Crabb sarcoma	−	Sugiura (1962), Sugiura et al. (1971)
Fortner pancreatic tumor	−	Sugiura (1962), Sugiura et al. (1971)
Fortner small bowel tumor	−	Sugiura (1962), Sugiura et al. (1971)

Note: +, ±, and − indicate the investigators' evaluation as active, borderline, and inactive. For details, the original publications should be consulted.

Table 2. *Anticancer activity of Hycanthone*

Tumor	Activity	Literature citations
in vitro		
HeLa cells	+	WITTNER *et al.* (1971)
Leukemia L1210 cells	+	ADAMSON (1971); SIEBER *et al.* (1973)
Novikoff hepatoma cells	+	SIEBER *et al.* (1973)
Walker carcinosarcoma 256 cells	+	SIEBER *et al.* (1973)
Mouse		
Leukemia 5178Y	+	SIEBER *et al.* (1973)
Leukemia L1210, ascites	+	ADAMSON (1971, 1972); HIRSCHBERG *et al.* (1968b); SIEBER *et al.* (1973); WEINSTEIN and HIRSCHBERG (1971)
Leukemia P388	+	ADAMSON (1971, 1972); SIEBER *et al.* (1973)
Mast cell tumor P815	+	SIEBER *et al.* (1973)
Rat		
Walker carcinosarcoma 256	+	SIEBER *et al.* (1973)

Note: + indicates the investigators' evaluation as active. For details, the original publications should be consulted.

both thiaxanthenones are effective against a wide variety of transplanted neoplasms in several animal species and *in vitro*. Miracil D causes significant inhibition of the growth of about half of the tumors examined, with no discernible pattern of response (cf. HIRSCHBERG, 1963) among tumors or host species. Hycanthone, in a more fragmentary series of test systems, has been found effective against every tumor to-date. In the few instances where both compounds have been tested against the same experimental tumor, their actions appear to be equivalent.

Antibacterial Activity. The first relevant study was that of SOBELL and ARNOLD (1955), who showed that 13 of 16 xanthones and 35 of 63 thiaxanthones tested were toxic to *Streptococcus faecalis* in the presence of limiting amounts of folic acid. In only one instance was the toxicity reversed by thymidine. Several close congeners of Miracil D were among these compounds. In an extensive study of the effects of 16 carcinostatic drugs on the growth of the same organism, HUTCHISON *et al.* (1962b) showed that on a medium devoid of purines and uracil but containing minimal folic acid, Miracil D was a moderately active inhibitor and that its effect on strains resistant to Amethopterin, 6-mercaptopurine, 8-azaguanine, or 5-fluorouracil was comparable to that on the parent strain.

In an investigation of biochemical changes accompanying impaired respiration in *Staphylococcus aureus* (GAUSE *et al.*, 1961) it was found that respiratory-deficient mutants were twenty times more sensitive to Miracil D than the parent strain.

The effect of this compound on *Bacillus subtilis* has been studied in some detail, in connection with the examination of the drug's mechanism of action (cf. below). Inhibition of growth has been observed uniformly at low concentration of Miracil D (HAIDLE *et al.*, 1970; HAIDLE *et al.*, 1967; HIRSCHBERG *et al.*, 1968c; MANDEL and BRINKLEY, 1967; WEINSTEIN *et al.*, 1965; WEINSTEIN and HIRSCH-

Berg, 1971). *Bacillus megaterium* exhibits comparable sensitivity (Mandel and Brinkley, 1967), whereas for *Escherichia coli* the 50 percent inhibitory concentration is two to three times higher (Cramer and Sinsheimer, 1972; Mandel and Brinkley, 1967; Weinstein *et al.*, 1965, 1967; Weinstein and Hirschberg, 1971).

Antiviral activity. Miracil D interferes with bacteriophage MS2 infection in *Escherichia coli* (Cramer and Sinsheimer, 1971, 1972) and both this compound and Hycanthone are mutagenic in a bacteriophage T4-*Escherichia coli* test system (Hartman *et al.*, 1971).

In an extensive series of experiments, Freeman *et al.* (1965a, b) developed methods for assessing the responses of cell-virus systems in culture to threshold concentrations of compounds which would control viral propagation and/or tumor cell induction and replication without jeopardy of host cell homeostasis. Two oncogenic viruses (Rous sarcoma and polyoma) and two lytic viruses (encephalomyocarditis and vaccinia) were included; the first and third are RNA viruses, the second and fourth DNA viruses. When 29 biologically active compounds were compared in terms of the cell-virus ratio, i.e. the ratio of the minimum amount of compound causing threshold injury to cells to the minimum causing significant reduction in viral plaques or foci, Miracil D was found to have ratios of 0.1–10 for three of the cell-virus systems and 0.01–1 for the fourth. This finding suggests moderate but not pronounced specificity for processes related to viral replication under these circumstances.

Inhibitory Effects in Plants, Fungi, and Protozoa. The grouping of these entries under a single subheading clearly has no taxonomic implications but simply attests to the paucity of data in these biological systems. The only reports of studies in a plant appear to be those of Weyland (1948) and Weyland *et al.* (1949) who mentioned that Miracil D blocked the initiation of mitosis in the roots of *Allium cepa* and that this was followed by gradual degeneration of the nuclei.

As an incidental finding in the studies of Rosi *et al.* (1967) on the microbiological conversion of Miracil D to Hycanthone, it was reported that the former was much more inhibitory than the latter to the growth of *Aspergillus sclerotiorum*, the fungus which had been found most effective in carrying out this conversion. The operational problem was circumvented by adding the substrate gradually to the fermentation medium.

Willard and Kodras (1967) tested 170 compounds *in vitro* against rumen protozoa for possible control of bloat and found Miracil D to be one of the most active. In a study directed toward molecular biology, rather than veterinary medicine, Hirschberg *et al.* (1968a) studied the effect of a series of thiaxanthenones, including Miracil D and Hycanthone, on proliferation and aggregation of the cellular slime mold, *Dictyostelium discoideum*. In its life cycle, aggregation of individual myxamoebae into a multicellular organism is the first morphogenetic event which takes place after the completion of the logarithmic growth phase. Miracil D inhibited proliferation at a concentration one-tenth that needed to inhibit aggregation; Hycanthone was ten and four times, respectively, less effective than Miracil D against these two stages of the life cycle.

Effects on Flatworms, Roundworms, Tapeworms, and Snails. The well established usefulness of Miracil D in schistosomiasis led logically to its examination in model systems for other trematode infections, such as fascioliasis, dicrocoeliasis, and opisthorchiasis. However, this schistosomicide, along with most other unrelated therapeutic agents for this disease, had no effect on rats or rabbits infected with *Fasciola hepatica* or hamsters infected with *Dicrocoelium dendriticum* or with *Opisthorchis felineus* (LAEMMLER, 1968).

Favorable therapeutic activity leading to 90–100 percent cures has been reported for Miracil D in a large series of Congolese adults infested with the roundworm *Strongyloides stercoralis* (GILLET et al., 1955). In experiments with mice infected with the tapeworm *Echinococcus multilocularis,* growth of vegetatively propagated cysts was inhibited fairly well by multiple injections of Miracil D (LUBINSKY, 1969). Another murine tapeworm, *Hymenolepsis nana,* was exposed *in vitro* to two dozen chemotherapeutic compounds of widely varying type; Miracil D was among those with pronounced inhibitory activity (SEN and HAWKING, 1960).

The last of this potpourri of biological test systems is the snail *Australorbis glabratus,* which is of particular interest because of its role as the intermediate host in the life cycle of *Schistosoma mansoni.* WARREN (1967) reported that Miracil D was the most potent of the drugs tested in suppressing molluscan schistosomiasis and preventing completely the growth and egg-laying capacity of the snail. More recently, YARINSKY and FREELE (1970) carried out a direct comparison of the molluscicidal and molluscan antischistosomal properties of Miracil D and Hycanthone. The former was more lethal to the snails and more toxic to the schistosome-infected mollusk than the latter, but Hycanthone was more effective in delaying and diminishing the shedding of cercariae. At nonlethal concentrations, neither drug affected growth significantly or interfered permanently with the reproductive ability of the snails.

Physiological Disposition and Metabolic Conversions

Preliminary data on blood levels of Miracil D following administration of single doses to rats and rabbits were obtained by LATNER et al. (1947) during the development of their colorimetric method of analysis. The highest values were obtained in the first 2 hours. Employing this assay procedure, HAWKING and ROSS (1948) carried out an extensive investigation on absorption, tissue distribution, and excretion in animals and human volunteers. In a monkey given the drug by stomach tube, peak blood levels were reached within several hours and maintained for about 20 hours. In man, absorption of a single oral dose from the alimentary canal was rapid, peak blood levels were reached in about 2 hours, over 90 percent of the drug was degraded in the body, and less than one-tenth was excreted in the urine. No appreciable quantities accumulated in the body (see also BLAIR, 1958; GOODMAN and GILMAN, 1970). HALAWANI et al. (1949) developed a dosage regimen for patients of two daily doses over an 8-day period which maintained a suitably high blood level to approximate the concentration of Miracil D shown to be toxic to schistosomes *in vitro.*

Tissue distribution studies were carried out by HAWKING and ROSS (1948) and by STRUFE (1963) in monkeys. The former showed that in animals dying from prolonged overdosage, concentrations were highest in lung and kidney, intermediate in heart, liver, and muscle, and lowest in brain. The latter reported that 7 days following a single oral dose the adrenal glands contained the highest levels, followed by kidney, liver, and lung, and at a great distance by spleen, pancreas, musculature, and brain.

Two recent reports from the Sterling-Winthrop group (Hernandez et al., 1971; Yarinsky et al., 1970) provide the first information about the physiological disposition of Hycanthone in several animal species. In rats and rhesus monkeys given moderate single intramuscular doses of randomly tritiated drug, peak blood and tissue concentrations were reached 30–60 minutes later. The highest levels were found in the liver, spleen, kidneys, and adrenals but these decreased rapidly and significantly in the next two to three days. In mice infected with *Schistosoma mansoni,* the administration of a large dose of the drug by the same route was followed by the attainment of peak plasma levels within 30 minutes; uptake into the parasite worms was highest at 2–6 hours when blood levels in the host were already declining.

In the elucidation of the therapeutic effectiveness and mechanism of action of any drug, the initial assumption usually is that the compound exerts its actions without undergoing alterations in structure. Even in some of the earliest studies with Miracil D, however, there were suggestions that this thiaxanthenone required metabolic conversion to an active form or forms. Bueding et al., (1947) found that the administration of Miracil D to schistosome-infected mice caused a decrease in the glycolytic rate of the flukes but that addition of the drug *in vitro* did not. Hawking and Ross (1948) reported the presence of considerable quantities of a yellow degradation product along with Miracil D in ether extracts of tissues of treated monkeys. Newsome and Robinson (1960a) in a more substantive contribution to this question showed that whereas a concentration of 10 mcg/ml of the drug did not kill *Schistosoma mansoni* kept in serum media for 20 days, the flukes were damaged by serum from treated patients although it contained only 5–6 mcg/ml of Miracil D and derivatives. Eleven chromatographic fractions were obtained from the urine of treated patients but none was active *in vitro*.

The next approach to this problem was reported by the group of investigators at Farbenfabriken Bayer who had been associated with the fundamental development of these schistosomicides from the beginning (Goennert, 1961; Goennert and Koelling, 1962; Strufe, 1963). It was postulated that enzymatic reactions might cleave the middle ring of Miracil D either at the 9- or 10-position, leading to derivatives shown in the top portion of Fig. 2. The absence of any schistosomicidal activity in these compounds or their derivatives eliminated this pathway as having biological significance.

In a systematic study of urinary excretion patterns following *in vivo* administration (Strufe, 1963), however, it was discovered that the previously demonstrated differences in the therapeutic efficacy of Miracil D against schistosomiasis in several animal species could be traced to differences in the relative importance of the metabolic conversions to which it was subject when different types of vertebrates served as the host for the same strain of the trematodal parasite. Urines were collected and separated into neutral, acid, and basic portions and a number of derivatives were identified following paper chromatography. In man, Miracil D sulfoxide (cf. bottom portion of Fig. 2) was the predominant metabolic product and exhibited schistosomicidal activity equivalent to the parent

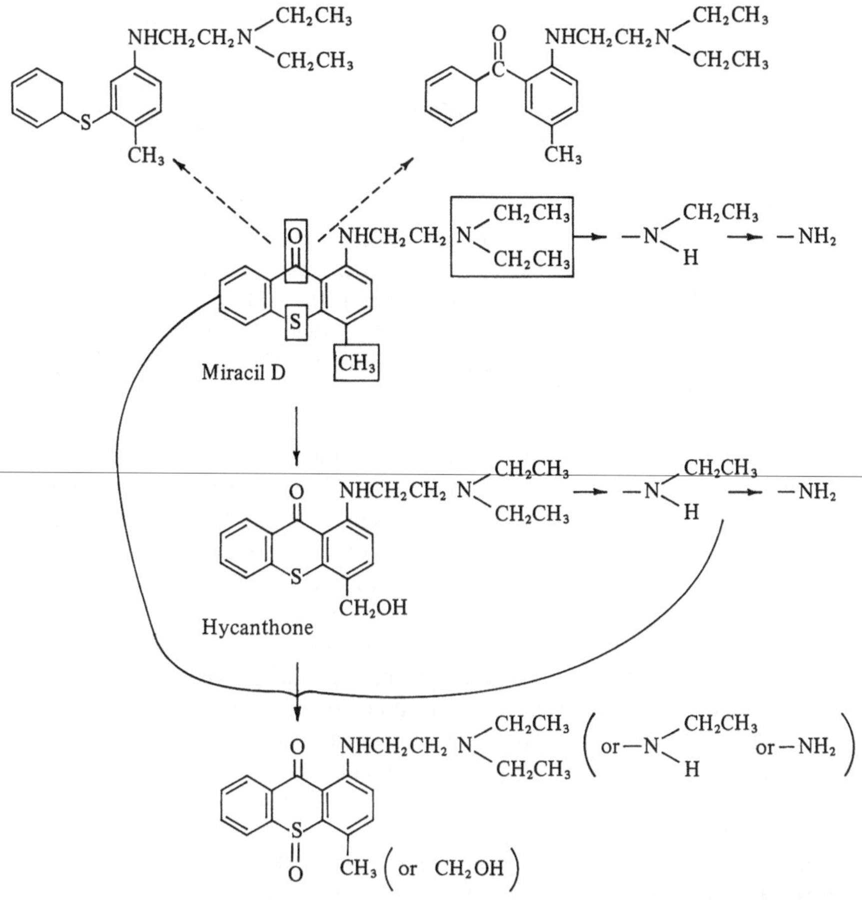

Fig. 2. Postulated and demonstrated metabolic conversions of Miracil D and Hycanthone

compound in infected mice (though not in hamsters—cf. ARCHER, 1967). There was some evidence that in human urine this derivative was present as a chromopeptide. In the rhesus monkey, Miracil D sulfoxide was only one, though the most active, of several derivatives excreted in approximately equal amounts. The principal metabolic product in the mouse was the biologically inactive sulfone. The excretory patterns in dogs and cattle resembled those in man; in sheep and hogs, an additional product was seen in significant amounts which was neither the sulfoxide nor the sulfone.

In blissful ignorance of these developments in the chemotherapy of schistosomiasis, HIRSCHBERG et al. (1964) were studying the inhibitory effect of Miracil D on mouse leukemia L1210. When it was injected simultaneously with non-inhibitory doses of β-diethylaminoethyl diphenylpropylacetate HCl·(SKF-525A), the carcinostatic activity of the thiaxanthenone was completely abolished. In view of the well established role of SKF-525A as an inhibitor of the microsomal enzyme systems involved in metabolic transformations of pharmacologically active compounds (e. g. THORP *et al.*, 1960), this finding suggested that tumor inhibition required metabolic activation.

Shortly thereafter, ROSI *et al.* (1965) published their brief account of the discovery of Hycanthone, which was followed soon by an extensive report (ROSI *et al.*, 1967) and the first of a number of reviews (ARCHER, 1968; BOND, 1969). The addition of Miracil D (4-CH_3) to fermentation media of *Aspergillus sclerotiorum* led to its rapid oxidation to Hycanthone (4-CH_2OH) as well as the corresponding aldehyde (4-CHO) and acid (4-COOH). In mice, Hycanthone was three times more effective against *Schistosoma mansoni* infections than Miracil D and in hamsters there was a nine-fold difference in the ED50. Extensive chemical, physical, and biological evidence was cited persuasively to buttress the conclusion that Hycanthone was indeed the active metabolite of Miracil D for schistosomiasis. The failure of STRUFE (1963) and other investigators to find it earlier was attributed to the great instability of Hycanthone to acid and to the finding that it was excreted almost entirely as a glucuronide.

Armed with this new information, ROSI *et al.* (1967) reinvestigated the metabolism of Miracil D in the monkey in more detail. In addition to Hycanthone glucuronide and Miracil D sulfoxide, the urines of monkeys treated with the parent compound also contained derivatives of Miracil D and Hycanthone which had lost one or both of the two ethyl groups on the terminal nitrogen of the sidechain (cf. middle portion of Fig. 2) and the sulfoxide derivatives of several of these metabolic products. The desethyl analogs and sulfoxides of both Miracil D and Hycanthone were less active in infected mice or hamsters than either parent compound but did exhibit activity (BERBERIAN *et al.*, 1967b). In the study of the metabolism of tritiated Hycanthone cited earlier (HERNANDEZ *et al.*, 1971) it was found that rapid conversion occurred to the sulfoxide in the rat and to the desethyl analog in the monkey.

In a simpler and more orderly universe, this might well have been the end of this particular story. However, when a number of these metabolites of Miracil D were compared as inhibitors of tumor cell or bacterial growth and in several *in vitro* test systems, it became apparent that the metabolic transformations favoring schistosomicidal activity were clearly not the same as those leading to the derivative(s) most active against mouse neoplasms or the processes of macromolecular synthesis (HIRSCHBERG *et al.*, 1968b; WEINSTEIN and HIRSCHBERG, 1971). Of the three demonstrated major pathways in mammalian organisms, *i.e.* hydroxylation of the 4-methyl group, desethylation of the terminal sidechain nitrogen, and sulfoxidation, the first appears crucial for schistosomicidal activity, whereas the second and third seem to be compatible with but not especially favorable for this type of biological action. In contrast, the antibacterial and carcinostatic activities and DNA-complexing capacity of Miracil D and Hycanthone are equivalent (see also HIRSCHBERG and WEINSTEIN, 1971) and, incidentally, the effect of Hycanthone against mouse leukemia L1210 is abolished by SKF-525A. Sidechain desethylation appears to increase activity in these test systems, but sulfoxidation leads to inactive derivatives. Further work is required for the definitive identification of the proximate carcinostatic metabolite of Miracil D.

Mammalian Toxicity, Mutagenicity, Carcinogenicity, and Teratogenicity

General Toxicity in Animals. Most of the available information was obtained during the extensive initial evaluations of Miracil D by German and British investigators (BLAIR, 1958; HAWKING and ROSS, 1948; KIKUTH and GOENNERT, 1948, 1949; WOOD, 1947). Maximum tolerated doses by various routes of administration were established in a half dozen animal species and pharmacological and histological evidence of toxic side effects was accumulated. A recurrent theme was the occurrence of gastrointestinal symptoms such as vomiting, anorexia, diarrhea, particularly in cats and monkeys; the appearance of fatty infiltration and degeneration of the liver, kidney, and heart in several species including rats, rabbits, and cats; and the observation of convulsions in rabbits, cats, and monkeys.

More recently, GARATTINI and PALMA (1961) included Miracil D in an elegant study designed to evaluate the specificity of anticancer drugs by comparing their effects at one-third, one-sixth, and one-twelfth the LD50 on a wide variety of tissue and function parameters in intact, partially hepatectomized, and tumorbearing rats. The toxicity of Miracil D was significantly greater in partially hepatectomized than in intact rats. The drug had an appreciable inhibitory effect on liver regeneration and on spleen weight but, as noted in Table 1, was ineffective against Walker carcinosarcoma 256.

In an investigation which provided a promising lead for a therapeutic application of Miracil D, BASES (1970) reported that this thiaxanthenone enhanced the damage caused in HeLa cells *in vitro* by exposure to X-rays. In a related study (BASES, 1972) it was shown that sublethal doses of Miracil D induced substantial mortality when administered to sublethally irradiated mice, in a manner reminiscent of the wellknown role of Actinomycin D as an adjuvant in clinical radiotherapy (e.g. D'ANGIO, 1962).

An analogous series of systematic investigations on the general animal toxicity of Hycanthone has not yet appeared in print in the open scientific literature.

General Toxicity in Man. The principal toxic reactions to Miracil D in human subjects have been gastrointestinal symptoms such as nausea, vomiting, anorexia, and epigastric pain (BLAIR, 1968; DAVIS, 1961; DAVIS *et al.,* 1965; EINHORN *et al.,* 1962; GOODMAN and GILMAN, 1970; HAWKING and ROSS, 1948); effects on the central nervous system (BLAIR, 1958; DAVIS, 1961; DAVIS *et al.,* 1965; EINHORN *et al.,* 1962; GELFAND, 1964; GOODMAN and GILMAN, 1970); and cardiovascular effects (ASLAMAZOV and MIKHAILOV, 1963; GERMINIANI *et al.,* 1964; GOODMAN and GILMAN, 1970; HAWKING and ROSS, 1948; NOR-EL-DIN and DAWOOD, 1950). Less consistent reports have appeared with regard to hepatic and renal damage (GOODMAN and GILMAN, 1970; HAWKING and ROSS, 1948; NOR-EL-DIN and DAWOOD, 1950). The nature of the salt of Miracil D administered had a definite effect on toxicity; the use of resinates, for example, in place of the usual hydrochloride caused a significant diminution in the severity of the side effects (DAVIS, 1961; DAVIS *et al.,* 1965). An intriguing finding which does not appear to have been followed up extensively or explained in terms of mechanism was the report of KUX (1956) that in a series of 27 patients Miracil

D decreased the frequency and severity of asthmatic attacks, presumably because of an effect on the neurovegetative system.

Gastrointestinal side effects have also been observed fairly extensively with Hycanthone (AMA Council on Drugs, 1971; COOK and JORDAN, 1971; KATZ et al., 1968; KATZ et al., 1969) and there have been reports of dizziness and headache (AMA Council on Drugs, 1971; KATZ et al., 1968, 1969), EKG changes (AMA Council on Drugs, 1971; KATZ et al., 1968; SALGADO et al., 1968), and minor other toxic reactions (DA CUNHA et al., 1971; DA SILVA et al., 1971). Of serious concern has been the recent discovery of hepatotoxicity leading to acute hepatitis and hepatocellular injury (FARID et al., 1972) and even, in four cases, to fatal liver necrosis (GANE, 1971).

Mutagenicity. The first indication that these thiaxanthenones might exhibit mutagenic effects preceded the clearcut demonstration of their interactions with DNA (see below) by a decade or more. LÜERS (1955) showed that Miracil D, fed to 1-5 day-old male imagoes of a wild strain of *Drosophila melanogaster*, was weakly mutagenic, especially in the early stages of spermatogenesis, and caused a significant number of chromosome mutations. OBE (1969, 1970) reported the induction of achromatic lesions and chromatid breaks when human leukocytes were exposed to Miracil D *in vitro*. A recent investigation (U, 1972) demonstrated that feeding of this drug to *Drosophila melanogaster* enhanced the frequency of X-ray-induced chromosome loss in male germ cells, particularly in spermatid and spermatocyte stages. This finding correlates with the data of BASES (1970, 1972) mentioned earlier.

When Miracil D and Hycanthone were compared in a test system consisting of T4 bacteriophage mutants during growth of *Escherichia coli* K-12 (HARTMAN et al., 1971), both compounds increased the reversion frequencies of two frameshift mutations in the rll region. In another test system studied by the same investigators, however, Hycanthone but not Miracil D effectively elicited reversions of two frameshift mutations in histidine-requiring tester strains of *Salmonella* (cf. also AMES, 1971). Significant and disturbingly high mutagenicity of Hycanthone has also been observed in a mammalian cell system involving forward mutations at the thymidine kinase locus of heterozygous mutants of L5178Y mouse lymphoma cells in cell culture; Hycanthone, though not as effective as X-irradiation or an alkylating agent, stimulated the mutation rate in this system 450-fold when administered at a reasonably biological concentration of 10^{-4} M (CLIVE et al., 1972a; CLIVE et al., 1972b).

When Hycanthone was added to a culture of phytohemagglutinin-stimulated human lymphocytes 4 hours prior to their harvest after 48 hours of incubation, there was a marked increase in chromosomal aberrations including chromosomal elongations, stickiness, chromatid erosion, chromatid exchanges and breaks. If the drug was present throughout the 48-hour incubation period, no significant effects were seen (SIEBER et al., 1973).

In a system involving an intact mammalian host rather than *in vitro* measurements, GREEN et al. (1973) demonstrated a significant increase in chromosomal abnormalities in the bone marrow of rats given intraperitoneal injections of Hycanthone. A significant linear relationship was observed between arithmetic dose and the proportion of cells affected.

It may be concluded that sufficient evidence of mutagenic potency has been accumulated, particularly for Hycanthone, to call for caution in the institution and evaluation of large-scale human trials.

Teratogenicity and Carcinogenicity. In a comparative study of several experimental anticancer agents, KARNOFSKY and LACON (1962) found that Miracil D caused slight feather inhibition when injected into the yolk sac of 4-day-old chick embryos. In a more extensive investigation, MOORE (1972) gave a single intramuscular injection of Hycanthone over a concentration range of 10–50 mg/kg to pregnant mice on day 7, the stage of gestation coinciding with the period of extensive organogenesis. Fetal mortality increased with increasing dosage, reaching 45% at 50 mg/kg. At 35 and 50 mg/kg, fetal malformations occurred in more than three-quarters of the litters and the percentage of abnormal live fetuses per litter rose concomitantly. Exencephaly, hydrocephaly, microphthalmia, fusion and branching of ribs, and cranial bone malformation were among the frequent abnormalities induced by this course of treatment. The embryotoxic effect of Hycanthone in this animal species in terms of resorptions and fetal deaths was confirmed by SIEBER *et al.* (1973), although these authors reported no significant fetal malformations when the drug was injected subcutaneously into mice on days 6-11 of gestation at concentrations of 12.5–50 mg/kg. It remains to be seen whether the difference in the route of administration in these two studies is significant.

These findings have stimulated interest in the teratology of Hycanthone and related compounds to the point where the National Institute of Environmental Health Sciences recently invited contract proposals (RFP-NIEHS-73-7) for an examination in depth of this potentially hazardous effect of these agents.

The state of knowledge about the potential carcinogenicity of Hycanthone is even more primitive. HETRICK and KOS (1973) showed that exposure to a combination of Rauscher virus and known potent chemical carcinogens transformed rat embryo cells *in vitro* whereas neither had this effect by itself. Hycanthone and Miracil D caused transformation in infected cells and implantation of these cells into newborn rats produced sarcomas. Preliminary results by HAESE *et al.* (1973) suggested that administration of Hycanthone to mice infected with *Schistosoma mansoni* led to higher incidence of gross masses and hyperplastic changes in the liver and to the appearance of a few hepatomas, but the significance of these rather specialized test systems for conclusions concerning carcinogenicity in general remains to be determined.

Mechanisms of Action

Of the more than three hundred publications dealing with Miracil D and Hycanthone which emerged from a reasonably thorough literature search, about one half have been included in this review. No more than a third of these have been concerned with one aspect or another of the mechanisms of action of these drugs. Evidence ranging from the descriptive to the molecular has been obtained in more than a dozen biological systems and in various tests *in vitro*. While much of the information which is available has the consistent ring of universality regardless of the part of nature from which it has been derived, there are some significant differences and the last word is very far from being at hand. To this reviewer, it appears particularly ironic that our knowledge is most fragmentary in that field which has been studied the longest and is of greatest and most urgent interest from the point of view of therapy, the elucidation of the mechanism of action of these agents in schistosomiasis.

Schistosomiasis. To a large extent, this may be due to the enormous complexity of the life cycle of this fluke and of its association with its victims (WRIGHT, 1967; see also JORDAN and WEBBE, 1969). There are three species commonly parasitic to man, *Schistosoma mansoni, Schistosoma haematobium,* and *Schistosoma japonicum.* Eggs produced by female worms pass through the wall of the fine blood vessels where they are deposited and then through the wall of the bladder or intestine. During their passage through the tissues, the embryos within the eggs develop and the resultant larvae are excreted with the urine or feces ready to hatch. If the larvae fall into fresh water, hatching occurs and the ciliated larvae or miracidia emerge to swim actively about until they make contact with their intermediate hosts, the aquatic snails *Biomphalaria, Australorbis,* or *Tropicorbis.* The miracidia attach themselves to the surface of the snail's body and enter through the skin, whereupon they develop into mother sporocysts within which balls of cells are budded off from the epithelial lining. The germ-balls develop into young daughter sporocysts which migrate to the snail's digestive gland where they grow and produce germ-balls which develop into the final larval stage or cercariae. The latter may be male or female and have the rudiments of the adult organs. They enter the definitive host, man, through the skin, reach the lymphatic vessels, the circulatory system, and eventually the liver where they mature before migrating through the portal system into the mesenteric or vesicle veins to begin oviposition.

Following two decades or more of biochemical and physiological studies of schistosomes in general, it is accepted (BUEDING, 1949, 1959; PELLEGRINO and KATZ, 1968, 1969; SENFT, 1965, 1969) that carbohydrates are a crucially important source of energy and a pool of amino acid precursors and that protein synthesis and breakdown are also of central concern. With specific regard to Miracil D and Hycanthone, however, information is scanty. BUEDING *et al.* (1947) found a decrease in the glycolytic rate in schistosomes isolated from infected mice following treatment with Miracil D, but not when the drug was added *in vitro* to a suspension of flukes. Oxygen uptake was unaffected in either instance and in the test tube experiments there was no effect on nucleoprotein metabolism or proteolytic activity. In their descriptions of gross and pathological effects of Miracil D on *Schistosoma mansoni* in the mouse (GOENNERT, 1947, 1955; KIKUTH and GOENNERT, 1948; cf. also BLAIR, 1948), the original group investigating this "new chemotherapeutic agent" concentrated on the loss of motility, the rather specific damage to the gonads and yolk cells of the flukes, and the resultant interference with egg production. It was suggested that Miracil D specifically blocked the initiation of mitosis but permitted the mitotic process, once begun, to be completed.

As mentioned, extensive work has been done on the effect of structural changes in the molecule on schistosomicidal activity, but the results have shed no direct light on the mechanism of action of these agents.

In the first studies on Hycanthone, ROGERS and BUEDING (1970) described a brief biphasic alteration in glycogen levels and changes in the motor activity of schistosomes obtained from mice or hamsters following a single intramuscular injection. Upon more prolonged observation, depletion of glycogen stores, damage to the female reproductive system, and, eventually, a hepatic shift were seen. In a more extensive paper, these authors demonstrated the appearance of stable resistance to Hycanthone in flukes surviving this treatment; when they resumed production of viable eggs after a period of 6–12 months, these gave rise to schistosomes which did not respond to Hycanthone, Miracil D, or a third antischistosomal drug (ROGERS and BUEDING, 1971). These effects were discussed in relation to the mutagenic potency of Hycanthone detailed earlier and the therapeutic implications were emphasized.

In contrast to ten compounds among 420 subjected to a preliminary screen for inhibitors of schistosome hemoglobin protease, neither Miracil D nor Hycanthone were active at the highest concentration tested (ZUSSMAN and BAUMAN, 1971).

It is little wonder, then, that even the most recent publications and reviews include specific statements to the effect that the mechanism of action of these drugs against the schistosome is not known, established, or elucidated (GOODMAN and GILMAN, 1970; JORDAN and WEBBE, 1969; SENFT, 1969; SIEBER et al., 1973).

Bacterial Systems. Considerably more information has been obtained in bacterial systems, particularly *Bacillus subtilis* and *Escherichia coli,* but it points in a different direction. In contrast to the complex structure, life cycle, and *in vivo* peregrinations of the liver fluke, these organisms lend themselves readily to biochemical examination in the presence of growth-inhibitory or lethal concentrations of the drugs. When Miracil D was added to a culture of *Bacillus subtilis* at a concentration sufficient to cause cessation of growth, there was a prompt and virtually complete inhibition of RNA synthesis, measured in terms of incorporation of uracil-2-^{14}C or uridine-2-^{14}C, but no inhibition at all of the incorporation of thymidine-^{14}C into DNA in the first 20 minutes. Protein synthesis from leucine-^{14}C or arginine-^{14}C was inhibited less strongly and less quickly, suggesting that this inhibition was a consequence of the interruption of RNA synthesis (WEINSTEIN et al., 1965).

MANDEL and collaborators further explored the effects of Miracil D on these bacterial cells (HAIDLE et al., 1967; 1970; MANDEL and BRINKLEY, 1967). They confirmed the early and progressive growth-inhibitory effect of the drug but found that DNA synthesis remained unaffected only up to 20 minutes of exposure but then ceased. At this stage, ghostlike cells were observed under phase-contrast microscopy. When cells were prelabeled with thymidine-^{3}H and exposed to the drug, labeled DNA was released into the medium. Transformation analyses with donor DNA from cells grown in the presence of drug showed a decrease with time in the ratio of origin markers to terminus markers. These findings, coupled with the electron microscopic demonstration of striking cytotoxic effects of Miracil D on the nucleoid and mesosomes, suggested that reinitiation of chromosome replication might be the specific point of attack of this inhibitor.

This bacterial system was included in an investigation of structure-activity relationships designed to explore the molecular mechanism of action of Miracil

D (HIRSCHBERG et al., 1968c). Several dozen derivatives were examined for growth-inhibitory potency, which in most instances proved to be related to their ability to interact with DNA (see below). Hycanthone and Miracil D had equivalent bacteriostatic activity (HIRSCHBERG and WEINSTEIN, 1971).

At slightly higher concentrations, Miracil D interfered with the growth of *Escherichia coli* as well (MANDEL and BRINKLEY, 1967; WEINSTEIN et al., 1965). It was therefore of interest to study the mechanism of action of the drug in this organism, in view of its extensive use in investigations in molecular biology and the inability of actinomycin D to penetrate its cell wall under standard conditions. WEINSTEIN et al. (1967) found that here too RNA synthesis was completely blocked from the beginning of exposure of a culture to the drug, but, in contrast to *Bacillus subtilis*, there also was an immediate though less pronounced inhibition of DNA synthesis. The effect on total protein synthesis was delayed and only moderate, but Miracil D caused complete cessation of the induction of β-galactosidase activity in the presence of the inducer methyl-β-D-thiogalactoside, proving that this process required *de novo* RNA synthesis. All these toxic effects were prevented by simultaneous addition of spermine to the culture; this normally occurring polyamine not only interfered with the cellular uptake of Miracil D by *Escherichia coli* but also with the interaction of the drug with DNA.

In an imaginative application of this tool to the elucidation of the mechanisms of viral replication, CRAMER and SINSHEIMER (1971, 1972) examined the detailed course of bacteriophage MS2 infection of *Escherichia coli* in the presence of concentrations of Miracil D which inhibited host cell growth and macromolecular synthesis. When the drug was added prior to phage infection, the phage adsorbed to the cells but penetration of viral RNA was inhibited. Penetration was achieved without further viral development by infection in the presence of chloramphenicol; if this antibiotic was then washed out and Miracil D was added up to 20 minutes after infection, a second viral function was inhibited and the yield of progeny phage was decreased. Synthesis of both double-stranded and single-stranded virus-specific RNA was limited, but the viral RNA species which were produced were the same as in the absence of drug. It was suggested that Miracil D may cause the production of an unstable viral RNA synthetase.

Mammalian Systems. Applying the information gained in bacterial systems to HeLa cells grown in suspension cultures, BASES and MENDEZ (1969) showed that the incorporation of precursors into total DNA and total RNA was inhibited by Miracil D to the same extent, but when the various species were separated by sucrose gradient centrifugation it became apparent that ribosomal RNA synthesis was blocked completely by the drug while messenger RNA synthesis was undisturbed. Protein synthesis was not affected at these concentrations. The effects of Miracil D were completely reversible if exposure was limited to 2 hours. The selective effect of the drug on ribosomal RNA synthesis was utilized in a study of the action of X-ray on these same cells *in vitro* to show that radiation did not inhibit messenger RNA synthesis (BASES et al., 1970).

When HeLa cells were exposed to the drug following an initial sublethal dose of X-irradiation, there was significant enhancement of damage to their colony-forming ability. Cells surviving irradiation gradually regained normal resistance to drug treatment, a finding which suggested that Miracil D inhibited a postradiation repair process (BASES, 1970). Potentiation of X-ray damage by this thiaxanthenone has also been reported in terms of lethality in mice (BASES, 1972) and chromosome loss in the fruit fly (U, 1973) and may have therapeutic implications (VARGAS, 1963).

Another group of investigators at Albert Einstein College of Medicine (WITTNER et al., 1971) examined the effects of Hycanthone on HeLa cells in analogous suspension cultures. Plating efficiency was reduced and exponential cell growth inhibited at concentrations at least an order of magnitude lower than with Miracil D (BASES and MENDEZ, 1969). The Hycanthone effect was selective for RNA synthesis, again providing a contrast with the cited study for Miracil D, where DNA and RNA synthesis were equally affected. However, the newer derivative as well as the parent compound specifically inhibited ribosomal RNA synthesis and its effects were also readily reversible. Hycanthone markedly delayed the nuclear and cytoplasmic processing of newly formed ribosomal precursor RNA.

For leukemia L1210, some of the evidence has been accumulated in the tumor-bearing mouse and some in test tube experiments in which the tumor cells were suspended in an appropriate medium. Under the latter circumstances, Miracil D caused a concentration-dependent inhibition of uridine-^{14}C incorporation into RNA and an even more pronounced inhibition of the incorporation of thymidine-^{14}C or -^{3}H into DNA (HIRSCHBERG et al., 1966; WILSON et al., 1972a). In further contrast to the bacterial systems, the inhibitory effect on RNA synthesis was not counteracted by spermine (WEINSTEIN and HIRSCHBERG, 1971). The effects of Hycanthone and Miracil D were equivalent (HIRSCHBERG et al., 1968b; HIRSCHBERG and WEINSTEIN, 1971). The desethyl derivatives of Miracil D were at least as if not more active than the parent compound (WILSON et al., 1972a).

In vivo, BLANZ and FRENCH (1963) and HIRSCHBERG et al. (1959) carried out extensive structure-activity investigations, using prolongation of survival time of tumor-bearing mice as the end point for leukemia L1210 and including several solid transplantable tumors as well. The results were consistent with the data obtained on interaction with DNA in vitro (see below) and confirmed the expectation generally held in cancer chemotherapy that DNA replication and/or transcription is the primary target in tumor inhibition.

Additional evidence of a different sort was assembled in prevention and cross-resistance experiments (HIRSCHBERG et al., 1964). At nontoxic levels, adenine counteracted the carcinostatic effect of Miracil D but pteroylglutamic acid and citrovorum factor were only partially effective and pyridoxine and glutamic acid had no activity. A Miracil D-resistant subline of leukemia L1210 was established and proved to have significant cross-resistance to 6-thioguanine and Cytoxan but not to other selected anticancer agents including 6-mercaptopurine and 8-azaguanine. Results with Amethopterin were equivocal.

In a recent extension of these investigations to other mammalian test systems, LEA et al. (1972) and REY (1973) examined the effects of Miracil D and Hycanthone in regenerating liver and of the former drug in several transplantable hepatomas in the rat. The two active drugs and a desethyl derivative of Miracil D caused a pronounced and prolonged inhibition of DNA synthesis and a less marked inhibition of RNA synthesis in 24-hour regenerating liver, whereas a non-DNA-complexing derivative was inactive. Administration of Miracil D 16 hours before measurement of DNA synthesis had a marked inhibitory effect on three slowly growing hepatomas but little or no effect on three rapidly growing representatives of this tumor spectrum, suggesting that sensitivity of nucleic acid synthesis to Miracil D may be lost with a decrease in the differentiation of hepatic tissues.

In more recent data from the same laboratory (Lea et al., 1973) it was demonstrated that both Miracil D and Hycanthone caused a pronounced reduction in amino acid incorporation into different histone fractions, but not in the synthesis of cytoplasmic and nuclear non-histone proteins, in 24-hour regenerating rat liver. Phosphorylation of lysine-rich f1 histones was also decreased significantly, but there was no change in the incorporation of phosphate into the cytoplasmic acid-insoluble fraction, f2b histones, combined f2a and f3 histones, and nuclear non-histone proteins. It was proposed that there may be coordinated inhibition of DNA and histone synthesis by these drugs.

In phytohemagglutin-stimulated human lymphocytes harvested from the blood of healthy donors, DNA and RNA synthesis was inhibited to an equivalent extent by Hycanthone at concentrations which had no effect on protein synthesis, though of course the level of the drug could be raised to the point where all macromolecular synthesis ceased entirely (Sieber et al., 1973).

If one compares the variegated effects of these two thiaxanthenones on these biological systems in terms of mechanisms of action, it is apparent that what little is known about their action on schistosomes has only the most peripheral relationship to the synthesis and metabolism of the nucleic acids, whereas it is just this area which provides the most cogent explanations for the effects on bacterial and mammalian cells. Among the latter, there is a wide range of selectivities, from those where RNA synthesis is the exclusive or primary target to those where DNA replication is consistently inhibited more significantly than DNA transcription. Protein synthesis is general is little affected, although pronounced effects on specific protein-synthetic mechanisms (e.g. β-galactosidase induction, histone synthesis and phosphorylation) have been detailed. Where comparative data are available, there has been little difference in effectiveness between Miracil D and Hycanthone.

In Vitro Systems. The final section of this review is concerned with those studies, representing the largest group in the publications on mechanisms of action, in which one or both drugs have been examined directly in the test tube. Here, again, the predominant focus has been the effect of these agents on the structure and function of DNA.

The only exceptions to this generalization are contained in two recent publications on Hycanthone. In view of the recognized association between vesical schistosomiasis and bladder cancer and the role of kynurenine metabolites as bladder carcinogens, El Sewedy et al., (1972) studied the effect of this schistosomicide on kynurenine metabolism in homogenates of mouse liver. Both kynureninase and kynurenine transaminase were inhibited, thus interfering with the conversion of kynurenine to anthranilic and kynurenic acids. Although these two enzymes are vitamin B_6-dependent, the inhibition did not appear to involve pyridoxal phosphate since increasing coenzyme concentrations did not counteract the inhibition. It was suggested that inactivation of the sulfhydryl groups of these enzymes may be involved.

One of the sections of the repeatedly cited paper by SIEBER *et al.* (1973) summarized experiments on the inhibitory effect of Hycanthone on the drug-metabolizing enzyme, aldehyde oxidase; it was shown that the K_i was 8×10^{-6} M, indicating that this thiaxanthenone was a highly effective inhibitor and suggesting that it should be evaluated in combination with carcinostatic agents such as Amethopterin which are catabolized oxidatively by this enzyme.

The first direct indication of an *in vitro* effect of Miracil D on nucleic acid structure or function was provided by LANG *et al.* (1950) who reported in a brief note that it caused a concentration-dependent increase in the viscosity of high molecular weight thymus nucleic acid, formed precipitates with ribo- and deoxyribonucleotides, and diminished the binding of the nucleic acid to purified globin, but had no effect on the enzymatic depolymerization, dephosphorylation, and deamination of deoxyribonucleotides.

Beginning in 1965, a series of measurements of the physical interaction of Miracil D and, later; Hycanthone with DNA were initiated with a variety of techniques which contributed to our present understanding of this process. Some of this information has been reviewed recently (NEWTON, 1970; WEINSTEIN and HIRSCHBERG, 1971). The first of these was an examination of the effects of nucleic acids on the absorption spectra of the drugs. WEINSTEIN *et al.* (1965) showed that at neutral pH Miracil D had absorption maxima at 330 and 442 mµ and that both native and denatured DNA caused a shift in the maxima to 337 and 450 mµ and decrease of absorption at the original maxima. The same effects, though to a smaller extent, were caused by the addition of sRNA or synthetic polynucleotides, but mononucleotides of adenine, guanine, cytosine, or uridine were inactive.

Analogous results have been obtained for Hycanthone by ILLINGWORTH and WARING (unpublished experiments), who also presented data on the difference spectra of the drug with mixtures of various homopolymers and obtained additional information from binding curves.

Another approach has been the determination of the effect of both compounds on the heat stability of calf thymus DNA. When assayed in buffer of low ionic strength, the heat denaturation profile of DNA reached 50% of total hyperchromicity (T_m) at 56°. This was raised to 71° in the presence of 6×10^{-6} M Miracil D and to 79° at twice that concentration. In standard high ionic strength saline citrate, this effect was not observed. Hycanthone was equivalent to Miracil D and extensive structure-activity studies were carried out with this simple and informative technique (HIRSCHBERG and WEINSTEIN, 1971; HIRSCHBERG *et al.*, 1968 b, c; WEINSTEIN *et al.*, 1965).

In buffer of low ionic strength, but not in standard saline-citrate, the interaction between Miracil D, Hycanthone, or their biologically active congeners caused an appreciable increase in the relative viscosity of DNA. This increase was prevented by the addition of low concentrations of spermine, intermediate concentrations of Mg^{++}, or high concentrations of Na^+. The structure-activity relationships for this effect correlated well with those for the increased heat stability of DNA (CARCHMAN et al., 1969; HIRSCHBERG and WEINSTEIN, 1971; HIRSCHBERG et al., 1968b).

In a subcellular system, Miracil D inhibited the incorporation of ATP-^{14}C into RNA in a system containing purified *Escherichia coli* RNA polymerase and a limiting amount of DNA as primer. The addition of excess DNA, but not of excess enzyme, reversed the inhibition, indicating that the interference by the drug in this process constituted additional evidence for its interaction with DNA. In the presence of spermine, RNA synthesis in this subcellular system was more resistant than in its absence to inhibition by Miracil D (WEINSTEIN et al., 1967). Structure-activity relationships with regard to this parameter of DNA complexing (HEBBORN et al., 1968; HIRSCHBERG et al., 1968c) once again correlated adequately with the effects on thermal denaturation of DNA. In recent additional studies, it was demonstrated that Miracil D inhibited the template properties of both polydeoxy- and polyribonucleotides for *Micrococcus luteus* RNA polymerase, the inhibition being most pronounced for those polynucleotides containing deoxyribose and adenine (WILSON et al., 1972a, b).

In a series of elegant papers, WARING (1970a, b, 1971, 1972, 1973) studied the relation between binding of drugs and the supercoiling of a closed circular duplex DNA, the replicative form of bacteriophage ϕX174 DNA. Local uncoiling of the double helix is required for the insertion of a molecule of drug in line with LERMAN's (1961) model for intercalation, which has proved a very useful and informative concept. This leads to an increase in the average number of base pairs per turn, which with closed circular DNA results in supercoiling of the circles and, in consequence, predictable changes in the sedimentation coefficient. In order to be able to intercalate, the drug molecule should possess a planar aromatic ring system capable of strong hydrophobic interaction with the DNA base pairs. The data obtained in these studies provide persuasive evidence that Hycanthone and Miracil D interact with DNA by intercalation.

Twelve compounds which are known or assumed to bind to duplex DNA by intercalation, accelerated the degradation of *Escherichia coli* ribosomes at 52°; Miracil D had weak but demonstrable activity in this test system (WOLFE et al., 1972).

An additional bit of evidence which may be mentioned is the fact that preliminary flow dichroism studies indicated that the Miracil D chromophore is oriented perpendicular to the long axis of the DNA, a feature consistent with intercalation (WEINSTEIN and HIRSCHBERG, 1971).

The application of physicochemical techniques to a comparison of Miracil D and its biologically inactive congener bearing a methyl group on the proximal sidechain nitrogen (ZILVERSMIT, 1970, 1971) has suggested that coplanarity of the thiaxanthenone ring system and the sidechain may be required for biological activity and that changes in physical properties concomitant with steric inhibition of resonance may explain why Miracil D is active and its N-methyl derivative is not. A diminished capacity of the latter to form intermolecular hydrogen bonds does not appear to be a factor.

The structure-activity relationships in these various test systems, excluding the effects of drugs on *Schistosoma mansoni*, have a high degree of internal consistency, as pointed out repeatedly in these pages. The presence of a hydrogen on the proximal nitrogen of the sidechain appears to be essential for optimum interaction with DNA, although very recent data indicate that indazole derivatives of Miracil D or Hycanthone, in which this nitrogen becomes part of a fourth ring leading to a substituent on the 9-position, are excellent DNA intercalators (WARING, 1973; CAMPANA and HIRSCHBERG, unpublished observations). In contrast, the length of the sidechain and the nature of the substituents on the terminal nitrogen have little influence on the ability of the drug to complex with DNA. Several molecular models for this interaction have been postulated (CARCHMAN et al., 1969; HIRSCHBERG et al., 1968c; WEINSTEIN and HIRSCHBERG, 1971) but a final decision among them would appear premature. The most recent formulation visualizes intercalation of the ring system between consecutive base pairs of DNA, with the terminal nitrogen extending outward to the periphery of the DNA helix to interact with the DNA backbone. If such a formulation leads to the conception of more sophisticated and more definitive experiments, it will have served its purpose.

Acknowledgments. It is a pleasure to thank Dr. SYDNEY ARCHER of Sterling-Winthrop Research Institute, Dr. ERNEST BUEDING of Johns Hopkins University, Dr. EDWARD F. ELSLAGER of Parke, Davis and Co. Research Laboratories, Dr. J.A.R. MEAD of the National Cancer Institute, and Dr. I. BERNARD WEINSTEIN of Columbia University College of Physicians and Surgeons for stimulating discussions and aid in assembling the pertinent literature. Dr. MICHAEL WARING of the University of Cambridge generously provided copies of material prior to its publication.

References

ADAMSON, R.H.: Antileukaemic activity of hycanthone. Lancet **1971-II**, 1206.
ADAMSON, R.H.: Hydroxyguanidine and hycanthone—two new antitumor drugs—rationale, evaluation, and activity (Abstract). Proc. Am. Assoc. Cancer Res. **13**, 7 (1972).
AMA Council on Drugs: Hycanthone mesylate. In: AMA drug evaluations, 1st edition, 1971, p. 477.
AMES, B.N.: The detection of chemical mutagens with enteric bacteria. In: Chemical mutagens. Principles and methods for their detection, HOLLAENDER, A., ed., p. 267. New York: Plenum Press 1971.
ANONYMOUS: Physical properties of standard reference list of agents. Cancer Res. **23** (4), Pt. 2, 259 (1963).
ARCHER, S.: Award address. Biological activity: A medicinal chemist's view. In: Annual report in medicinal chemistry—1967, CAIN, C.K., ed., p. VII. New York: Academic Press 1968.
ARCHER, S., and C.M. SUTER: The preparation of some 1-alkylamino- and dialkylamino-thiaxanthenones. J. Am. Chem. Soc. **74**, 4296 (1952).
ASLAMAZOV, E.G., and A.A. MIKHAILOV: [Toxic effect of trivalent antimony and miracil D preparations on cardiovascular system during therapy of schistosomiasis patients]. Soviet Med. **27**, 65 (1963).
BASES, R.: Enhancement of x-ray damage in HeLa cells by exposure to lucanthone (miracil D) following radiation. Cancer Res. **30**, 2007 (1970).
BASES, R.: Combined lethal effects of lucanthone (miracil D) and total-body irradiation in mice. Radiology **102**, 193 (1972).
BASES, R., and F. MENDEZ: Reversible inhibition of ribosomal RNA synthesis in HeLa by lucanthone (miracil D) with continued synthesis of DNA-like RNA. J. Cell Physiol. **74**, 283 (1969).
BASES, R., F. MENDEZ, and R. NICOLINO: Inhibition of ribosomal RNA synthesis in x-irradiated HeLa cells. Radiation Res. **44**, 456 (1970).
BERBERIAN, D.A., H. FREELE, D. ROSI, E.W. DENNIS, and S. ARCHER: A comparison of oral and parenteral activity of hycanthone and lucanthone in experimental infections with *Schistosoma mansoni*. Am. J. Trop. Med. Hyg. **16**, 487 (1967a).

BERBERIAN, D.A., H. FREELE, D. ROSI, E.W. DENNIS, and S. ARCHER: Schistosomicidal activity
    of lucanthone hydrochloride, hycanthone and their metabolites in mice and hamsters. J. Parasitol.
    53, 306 (1967b).
BLAIR, D.M.: Lucanthone hydrochloride. A review. Bull. World Health Organ. **18**, 989 (1958).
BLANZ, E.J., Jr., and F.A. FRENCH: A systematic investigation of thiaxanthen-9-ones and analogs
    as potential antitumor agents. J. Med. Pharm. Chem. **6**, 185 (1963).
BOND, H.W.: Chemistry of schistosomicides. Ann. N.Y. Acad. Sci. **160**, 519 (1969).
BRAMBILLA, G., L. BALDINI e G. GALLI: Osservazioni sull'attività di diversi chemioterapici antineoplas-
    tici sullo sviluppo di culture di cellule HeLa e KB. Boll. Soc. Ital. Biol. Sper. **42**, 1679 (1966).
BRINDLE, S.D., E. HIRSCHBERG, and P.R. GROSS: Uptake of miracil D by cells of sensitive and
    miracil D-resistant lines of mouse leukemia L1210. Cancer Res. **24**, 1738 (1964).
BROSS, I.D.J., and G.S. TARNOWSKI: A new approach to differential toxicity. Cancer Res. **22 (1)**,
    Pt. 2, 46 (1962).
BUEDING, E.: Metabolism of parasitic helminths. Physiol. Rev. **29**, 195 (1949).
BUEDING, E.: Mechanisms of action of schistosomicidal agents. J. Pharm. Pharmacol. **11**, 385 (1959).
BUEDING, E., A. HIGASHI, L. PETERS, and A.D. VALK: Some observations on the action of miracil
    (1-[β-diethyl-aminoethylamino]-4-methylthioxanthone hydrochloride) against *Schistosoma man-
    soni* (Abstract). Federation Proc. **6**, 313 (1947).
BURCHENAL, J.H., V.C. GREGG, J.R. PURPLE, and W. KREIS: The use of mouse leukemias in the
    screening of drugs. Proc. 3rd Internat. Congress Chemother., Stuttgart, 1963, p. 932.
CARCHMAN, R.A., E. HIRSCHBERG, and I.B. WEINSTEIN: Miracil D: Effect on the viscosity of DNA.
    Biochim. Biophys. Acta **179**, 158 (1969).
CLIVE, D., W.G. FLAMM, and M.R. MACHESKO: Mutagenicity of hycanthone in mammalian cells.
    Mutation Res. **14**, 262 (1972a).
CLIVE, D., W.G. FLAMM, M.R. MACHESKO, and N.J. BERNHEIM: A mutational assay system using
    the thymidine kinase locus in mouse lymphoma cells. Mutation Res. **16**, 77 (1972b).
COOK, J.A., and P. JORDAN: Clinical trial of hycanthone in schistosomiasis mansoni in St. Lucia.
    Am. J. Trop. Med. Hyg. **20**, 84 (1971).
CRAMER, J.H., and R.L. SINSHEIMER: Replication of bacteriophage MS2. X. Phage-specific ribonucleo-
    protein particles found in MS2-infected *Escherichia coli*. J. Mol. Biol. **62**, 189 (1971).
CRAMER, J.H., and R.L. SINSHEIMER: Use of miracil D to suppress bacterial ribonucleic acid and
    protein synthesis during bacteriophage MS2 infection. J. Virol. **9**, 189 (1972).
DA CUNHA, A.S., D.G. DE CARVALHO, J.N. SANTOS CAMBRAIA y J.R. CANCADO: Manifestacoes
    de intolerancia ao hycanthone no tratamento da esquistossomose mansoni. Rev. Inst. Med.
    Trop. Sao Paulo **13**, 213 (1971).
D'ANGIO, G.J.: Clinical and biologic studies of actinomycin D and roentgen irradiation. Am. J.
    Roentgenol. **87**, 106 (1962).
DA SILVA, L.C., M.E. CAMARGO, S. HOSHINO, J. GUNJI, J.D. LOPES, D.F. CHAMONE, J.O. MARTINEZ,
    R.G. FERRI, W. ROTHSTEIN, G.R. DA SILVA, A.C. CENEVIVA, and E.J. CARDOSO: [Serum antibody
    and eosinophil changes after treatment of human Manson's schistosomiasis with niridazole or
    hycanthone]. Rev. Inst. Med. Trop. Sao Paulo **13**, 121 (1971).
DAVIS, A.: Lucanthone resinates in schistosomiasis. Lancet **1961-I**, 201.
DAVIS, A., J. NEWSOME, J.R. HENDERSON, J. KETTLE, and L.F. WIGGINS: Resinates of lucanthone
    in the treatment of schistosomiasis. E. African Med. J. **42**, 10 (1965); Chem. Abstr. **63**, 1104c
    (1965).
DAVIS, A.P., M. GRUENSTEIN, and M.B. SHIMKIN: Evaluation of chemotherapeutic agents in mammary
    carcinoma induced by 3-methylcholanthrene in Wistar rats. Cancer Res. **26 (2)**, Pt. 2, 1 (1966).
EINHORN, A., A. FRITSCH, K.G. DWORK, and H.B. SHOOKHOFF: *Schistosoma mansoni* infection
    in children. Treatment with lucanthone hydrochloride (Nilodin). Am. J. Diseases Children **104**,
    30 (1962).
EL-SEWEDY, S.M., G.A. ABDEL-TAWAB, M.H. ABDEL-DAIM, and M.F. EL-SAWY: Studies with trypto-
    phan metabolites *in vitro*. V. Effect of the methanesulphonate derivative of hycanthone (etrenol)
    and lead acetate on kynureninase and kynurenine transaminase of normal mouse liver. Biochem.
    Pharmacol. **21**, 379 (1972).
ELSLAGER, E.F.: Human antiparasitic agents. In: Annual report in medicinal chemistry—1965, CAIN,
    C.K., ed., p.136. New York: Academic Press 1966.
ELSLAGER, E.F.: Human antiparasitic agents. In: Annual report in medicinal chemistry—1966, CAIN,
    C.K., ed., p. 131. New York: Academic Press 1967.

ELSLAGER, E.F., J.F. CAVALLA, W.D. CLOSSON, and D.F. WORTH: Synthetic schistosomicides. II. 1-[(Dialkylaminoalkyl)-methylamino]-4-methyl-10-thiaxanthenones. J. Org. Chem. **26**, 2837 (1961).

FARID, Z., J.H. SMITH, S. BASSILY, and H.A. SPARKS: Hepatotoxicity after treatment of schistosomiasis with hycanthone. Brit. Med. J. **1972**, 88.

FOSTER, R., B.L. CHEETHAM, E.T. MESMER, and D.F. KING: Comparative studies of the action of mirasan, lucanthone, hycanthone, and niridazole against *Schistosoma mansoni* in mice. Ann. Trop. Med. Parasitol. **65**, 45 (1971).

FOSTER, R., and H.C. RICHARDS: Recent advances toward the chemotherapeutic control of schistosomiasis. Bull. Chim. Therap. **4**, 293 (1970).

FREEMAN, G., A. KUEHN, and I. SULTANIAN: Response of tumorigenic viruses and of cells to biologically active compounds. I. Methods for determining response and application of methods. Cancer Res. **25 (10)**, Pt. 2, 1609 (1965a).

FREEMAN, G., A. KUEHN, and I. SULTANIAN: Tumorigenic virus and cell responses to biologically active compounds. II. Pharmacologic specificity in cell-virus relationships. Ann. N.Y. Acad. Sci. **130**, 330 (1965b).

GANE, N.F.C.: Hycanthone: Contraindications to its use. Central African J. Med. **17**, 108 (1971).

GARATTINI, S., and V. PALMA: An attempt to evaluate specific effects of antitumoral drugs. Cancer Chemotherapy Rept. **13**, 9 (1961).

GAUSE, G.F., G.V. KOCHETKOVA, and G.B. VLADIMIROVA: Biochemical changes accompanying impaired respiration in staphylocci. Nature **190**, 978 (1961).

GELFAND, M.: Temporary mental confusion in the African. Central African J. Med. **10**, 125 (1964).

GERMINIANI, H., F.S. DE LACERDA Jr., M.C. BARANSKI, H.G. MEDINA, M. BACILA, and C.C. MOTA: [Electrocardiographic changes in schistosomiasis patients submitting to treatment with a thiaxanthone derivative]. Rev. Inst. Med. Trop. Sao Paulo **6**, 123 (1964).

GILLET, J., R.M. DE SMET et P. NANNAN: L'action thérapeutique favorable d'un dérivé du thioxanthone (Nilodin) dans la strongyloidose. Ann. Soc. Belge Méd. Trop. **35**, 499 (1955).

GOENNERT, R.: Zur Frage des Wirkungsmechanismus von Miracil. Naturwissenschaften **34**, 347 (1947).

GOENNERT, R.: Schistosomiasis-Studien. III. Über die Einwirkungen von Miracil D auf *Schistosoma mansoni* im Mäuseversuch und die Verteilung des Pigmentes in der Wirtsleber. Z. Tropenmed. Parasitol. **6**, 257 (1955).

GOENNERT, R.: The structure-activity relationship in several schistosomicidal compounds. Bull. World Health Organ. **25**, 702 (1961).

GOENNERT, R., and H. KOELLING: Thiaxanthones and related compounds in experimental schistosomiasis. In: Drugs, parasites, and hosts, GOODWIN, L.G., and R.H. NIMMO-SMITH, eds., p. 29. London: J. & A. Churchill, Ltd. 1962.

GOODMAN, L.S., and A. GILMAN: Lucanthone. In: The pharmacological basis of therapeutics, 4th ed., p. 1073. New York: MacMillan Co. 1970.

GREEN, S., F.M. SAURO, and M.S. LEGATOR: Cytogenetic effects of hycanthone in the rat. Mutation Res. **17**, 239 (1973).

GRISWOLD, D.P., W.R. LASTER, Jr., M.Y. SNOW, F.M. SCHABEL, Jr., and H.E. SKIPPER: Experimental evaluation of potential anticancer agents. XII. Quantitative drug response of SA 180, CA 755, and leukemia L1210 systems to a "standard list" of "active" and "inactive agents". Cancer Res. **23 (4)**, Pt. 2, 271 (1963).

GROPPER, L., and M.B. SHIMKIN: Combination therapy of 3-methylcholanthrene-induced mammary carcinoma in rats: Effect of chemotherapy, ovariectomy, and food restriction. Cancer Res. **27**, 26 (1967).

HACKMANN, C., R. GOENNERT u. H. MAUSS: Untersuchung über die tumorhemmende Wirkung des Miracils. Naturwissenschaften **36**, 29 (1949).

HAESE, W.H., D.L. SMITH, and E. BUEDING.: Hycanthone-induced hepatic changes in mice infected with *Schistosoma mansoni* J. Pharmacol. Exptl. Therap. **186**, 430 (1973).

HAIDLE, C.W., B.R. BRINKLEY, and M. MANDEL: Effect of miracil D on marker frequency ratio and cytotoxicity in *Bacillus subtilis*. J. Bacteriol. **102**, 835 (1970).

HAIDLE, C., M.J. TEVETHIA, and M. MANDEL: Effect of miracil D on chromosome completion in *Bacillus subtilis* (Abstract). Bacteriol. Proc. 1967, 57.

HALAWANI, A., A. HAFEZ, J. NEWSOME, and S.G. COWPER: Miracil D: Effect on *B. mansoni in vitro* and in the treatment of urinary bilharziasis. J. Egypt. Med. Assoc. **32**, 29 (1949).

HAMMICK, D. L., and D. C. MUNRO: New synthesis of 1-amino-4-methylthioxanthone and of miracil D. J. Chem. Soc. **1952**, 1077.

HARRIS, J. J.: The effect of NSC survey compounds on human tumors in embryonated eggs. Cancer Res. **22** (1), Pt. 2, 1 (1962).

HARTMAN, P. E., K. LEVINE, Z. HARTMAN, and H. BERGER: Hycanthone: A frameshift mutagen. Science **172**, 1058 (1971).

HAWKING, F., and W. F. ROSS: Miracil D, its toxicology, absorption, and excretion in animals and human volunteers. Brit. J. Pharmacol. Chemother. **3**, 167 (1948).

HEBBORN, P., T. J. BARDOS, Z. F. CHMIELEWICZ, and D. J. TRIGGLE: Miracil D analogs: DNA-dependent RNA polymerase inhibitory activity and *in vivo* biological activity (Abstract). Proc. Am. Assoc. Cancer Res. **9**, 29 (1968).

HERNANDEZ, P., E. W. DENNIS, and A. FARAH: Metabolism of the schistosomicidal agent hycanthone by rats and rhesus monkeys. Bull. World Health Organ. **45**, 27 (1971).

HETRICK, F. M., and W. L. KOS: Transformation of Rauscher virus-infected cell cultures after treatment with hycanthone and lucanthone. J. Pharmacol. Exptl. Therap. **186**, 425 (1973).

HIRSCHBERG, E.: Tissue culture in cancer chemotherapy screening. Cancer Res. **18**, 869 (1958).

HIRSCHBERG, E.: Patterns of response of animal tumors to anticancer agents. Cancer Res. **23 (5)**, Pt. 2, 521 (1963).

HIRSCHBERG, E., S. D. BRINDLE, and G. SEMENTE: Development and properties of mouse leukemia L1210 resistant to miracil D. Cancer Res. **24**, 1733 (1964).

HIRSCHBERG, E., C. CECCARINI, M. OSNOS, and R. CARCHMAN: Effects of inhibitors of nucleic acid and protein synthesis on growth and aggregation of the cellular slime mold *Dictyostelium discoideum*. Proc. Natl. Acad. Sci. U.S. **61**, 316 (1968a).

HIRSCHBERG, E., A. GELLHORN, M. R. MURRAY, and E. F. ELSLAGER: Effects of miracil D, amodiaquin, and a series of other 10-thiaxanthenones and 4-aminoquinolines against a variety of experimental tumors *in vitro* and *in vivo*. J. Natl. Cancer Inst. **22**, 567 (1959).

HIRSCHBERG, E., and I. B. WEINSTEIN: Comparative ability of hycanthone and miracil D to interact with DNA. Science **174**, 1147 (1971).

HIRSCHBERG, E., I. B. WEINSTEIN, and R. CARCHMAN: Miracil D: An inhibitor of deoxyribonucleic and ribonucleic acid synthesis in mouse leukemia L1210 ascites (Abstract). Abstr. 9th Internat. Cancer Congress, Tokyo, 1966, p. 347.

HIRSCHBERG, E., I. B. WEINSTEIN, R. CARCHMAN, and S. ARCHER: Search for the carcinostatic metabolite of miracil D (Abstract). Proc. Am. Assoc. Cancer Res. **9**, 30 (1968b).

HIRSCHBERG, E., I. B. WEINSTEIN, N. GERSTEN, E. MARNER, T. FINKELSTEIN, and R. CARCHMAN: Structure-activity studies on the mechanism of action of miracil D. Cancer Res. **28**, 601 (1968c).

HOFFER, M., and A. I. RACHLIN: Antiparasitic agents. In: Annual report in medicinal chemistry, HEINZELMAN, R. V., ed., vol. 7, p. 145. New York: Academic Press 1972.

HUSSEIN, A. M., A. A. KASSEM, A. SINA, and A. A. BADAWY: Factors affecting exchange rate of lucanthone by amberlite IR-120 resin. I. Effect of solvent. Bull. Fac. Pharm. Cairo Univ. **7**, 73 (1968a); Chem. Abstr. **73**, No. 38493h (1970).

HUSSEIN, A. M., A. A. KASSEM, A. SINA, and A. A. BADAWY: Factors affecting exchange rate of lucanthone by amberlite IR-120. II. Effect of concentration and solution volume. Bull. Fac. Pharm. Cairo Univ. **7**, 81 (1968b); Chem. Abstr. **73**, No. 38492g (1970).

HUTCHISON, D. J., D. L. ROBINSON, D. MARTIN, O. L. ITTENSOHN, and J. DILLENBERG: Effects of selected cancer chemotherapeutic drugs on the survival times of mice with L-1210 leukemia: Relative responses of antimetabolite resistant strains. Cancer Res. **22 (1)**, Pt. 2, 57 (1962a).

HUTCHISON, D. J., W. V. ZUCKER, and E. A. BRAND: Effects of selected cancer chemotherapeutic drugs on the growth of *Streptococcus faecalis*: Relative responses of antimetabolite resistant strains. Cancer Res. **22 (1)**, Pt. 2, 73 (1962b).

JACQUEZ, J. A.: Tissue culture screening. Cancer Res. **22 (1)**, Pt. 2, 81 (1962).

JEWSBURY, J. M.: Experimental chemoprophylaxis against schistosomiasis. I. Introduction and rationale. Ann. Trop. Med. Parasitol. **66**, 409 (1972).

JORDAN, P., and G. WEBBE: Human schistosomiasis. London: William Heinemann Medical Books, Ltd. 1969.

KARNOFSKY, D. A., and C. R. LACON: Survey of cancer chemotherapy service center compounds for teratogenic effect in the chick embryo. Cancer Res. **22 (1)**, Pt. 2, 84 (1962).

KATZ, N., J. PELLEGRINO, M. T. FERREIRA, C. A. OLIVEIRA, and C. B. DIAS: Preliminary clinical trials with hycanthone, a new antischistosomal agent. Am. J. Trop. Med. Hyg. **17**, 743 (1968).

KATZ, N., J. PELLEGRINO, and C. A. OLIVEIRA: Further clinical trials with hycanthone, a new antischistosomal agent. Am. J. Trop. Med. Hyg. **18**, 924 (1969).

KIKUTH, W., and R. GOENNERT: Experimental studies on the therapy of schistosomiasis. Ann. Trop. Med. Parasitol. **42**, 256 (1948).

KIKUTH, W. u. R. GOENNERT: Experimentelle Untersuchungen und Erfahrungen mit dem neuen Schistosomiasismittel Miracil. Z. Tropenmed. Parasitol. **1**, 234 (1949).

KIKUTH, W., R. GOENNERT u. H. MAUSS: Miracil, ein neues Chemotherapeuticum gegen Darmbilharziose. Naturwissenschaften **33**, 253 (1946); Chem. Abstr. **43**, 9251 i (1949).

KUHNERT-BRANDSTAETTER, M., A. KOFLER u. H. C. RHI: Beitrag zur mikroskopischen Charakterisierung und Identifizierung von Arzneimitteln. Sci. Pharm. **32**, 308 (1964).

KUX, P.: Die Wirkung von Miracil D bei Asthma bronchiale. Medizinische **5**, 230 (1956).

LAEMMLER, G.: Chemotherapeutic activity of drugs in experimental schistosomiasis and other trematode-infections. Ann. Soc. Belge Méd. Trop. **47**, 215 (1967).

LAEMMLER, G.: Chemotherapy of trematode infections. In: Advances in chemotherapy, GOLDIN, A., F. HAWKING, and R. J. SCHNITZER, eds., vol. 3, p. 153. New York: Academic Press 1968.

LANG, K., G. SIEBERT u. W. LOENNECKE: Über biochemische Wirkungen von Miracil D. Klin. Wochschr. **28**, 104 (1950).

LATNER, A. L., R. V. COXON, and E. J. KING: Measurement of the concentration of miracil in biological fluids. Trans. Roy. Soc. Trop. Med. Hyg. **41**, 133 (1947).

LEA, M. A., F. L. KHALIL, and M. I. REY: Inhibition of histone synthesis and phosphorylation of lysine-rich histones in regenerating liver by miracil D and hycanthone (Abstract). Federation Proc. **32**, 588 Abs (1973).

LEA, M. A., S. MILLER, I. MACKAUF, E. HIRSCHBERG, and H. P. MORRIS: Action of miracil D and related compounds on DNA and RNA synthesis in regenerating liver and hepatomas. Internat. J. Cancer **9**, 484 (1972).

LEE, H. G.: Aspects of the effect of thioxanthone on *Schistosoma mansoni* in mice and *in vitro*. Bull. World Health Organ. **46**, 397 (1972).

LEITER, J.: Cancer chemotherapy screening data. XIII. Cancer Res. **22** (1), Pt. 2, 1 (1962).

LERMAN, L. S.: Structural considerations in the interaction of DNA and acridines. J. Mol. Biol. **3**, 18 (1961).

LUBINSKY, G.: Attempts at chemotherapy of *Echinococcus multilocularis* infections in rodents. Can. J. Zool. **47**, 1001 (1969); Chem. Abstr. **71**, No. 100154n (1969).

LÜERS, H.: Biologisch-genetische Untersuchungen über die Wirkung eines Thioxanthonderivates an *Drosophila melanogaster*. Z. Induktive Abstammungs-Vererbungslehre **87**, 93 (1955).

MANDEL, M., and B. R. BRINKLEY: Effect of miracil D on the ultrastructure of *Bacillus subtilis* and other bacteria (Abstract). Bacteriol. Proc. **1967**, 57.

MAUSS, H.: Über basisch substituierte Xanthon- und Thioxanthon-Abkömmlinge; Miracil, ein neues Chemotherapeuticum. Chem. Ber. **81**, 19 (1948).

MERKER, P. C., P. ANIDO, J. SARINO, and G. W. WOOLLEY: A study of human epidermoid carcinoma (H. Ep. No. 3) growing in conditioned Swiss mice. III. Chemotherapy with selected chemicals and observations on diet, food intake and drug toxicities. Cancer Res. **22** (1), Pt. 2, 9 (1962).

MOORE, J. A.: Teratogenicity of hycanthone in mice. Nature **239**, 107 (1972).

MORAN, M. R., and G. W. WOOLLEY: Chemotherapy of a transplantable mammary tumor in strain C3H mice. Cancer Res. **22** (1), Pt. 2, 87 (1962).

MORASCA, L.: Duration of cytoxicity in blood after treatment with antitumoral agents. In: Fifth International Congress of chemotherapy. Proceedings, SPITZY, K. H., ed., vol. 3, p. 383; Chem. Abstr. **70**, No. 18563r (1969). Vienna: Verlag Wien. Med. Akad. 1967.

MORASCA, L., and C. RAINISIO: A method to perfuse tissue culture with an *in vivo* system: Reliability of the method for metabolic studies. In: Cancer chemotherapy. Proceeding of Takeda International Conference, Osaka 1966, GOLDIN, A., ed., p. 141; Chem. Abstr. **70**, No. 36187a (1969). Tokyo: Maruzen Co., Ltd. 1967.

MOST, H.: Drug therapy. Treatment of common parasitic infections of man encountered in the United States. New Engl. J. Med. **287**, 495 and 698 (1972).

MUNRO, D. C.: Polarography of 1-amino-4-methylthioxanthone and of miracil D. J. Chem. Soc. **1960**, 4579.

MUNRO, D. C.: Physicochemical properties of some chemotherapeutic thioxanthones. J. Chem. Soc. **1961**, 5381.

NABIH, I., and M. ELSHEIKH: Chlorination and condensation reactions at the 4-methyl group of lucanthone and oxalucanthone. J. Pharm. Sci. **54**, 1672 (1965a).

NABIH, I., and M. ELSHEIKH: Mannich reaction at the 4-methyl group in schistosomicidal agents, lucanthone and oxalucanthone. J. Pharm. Sci. **54**, 1821 (1965b).

NEWSOME, J.: The search for non-antimonial schistosomicides. In: Bilharziasis, WOLSTENHOLME, G. E. W., and M. O'CONNOR, eds., Ciba Foundation Symposium, p. 310. Boston: Little, Brown & Co. 1962.

NEWSOME, J., and D. L. H. ROBINSON: Preliminary observations on metabolites of lucanthone. Trans. Roy. Soc. Trop. Med. Hyg. **54**, 582 (1960a).

NEWSOME, J., and D.L.H. ROBINSON: The estimation of lucanthone in plasma and serum. Trans. Roy. Soc. Trop. Med. Hyg. **54**, 454 (1960b).

NEWTON, B.A.: Chemotherapeutic compounds affecting DNA structure and function. In: Advances in pharmacology and chemotherapy, GARATTINI, S., F. HAWKING, A. GOLDIN, and I.J. KOPIN, eds., vol. 8, p. 149. New York: Academic Press 1970.

NOR EL DIN, G., and M. DAWOOD: Observations on the side-effects of miracil D. J. Egypt. Med. Assoc. **33**, 688 (1950).

OBE, G.: Zur Wirkung von Miracil auf menschliche Leukozytenchromosomen *in vitro*. Mol. Gen. Genet. **103**, 326 (1969).

OBE, G.: Die Verteilung Miracil D-induzierter achromatischer Läsionen und Chromatidbrüche auf den Chromosomen menschlicher Leukozyten. Z. Naturforsch. **25b**, 115 (1970).

PELLEGRINO, J., and N. KATZ: Experimental chemotherapy of schistosomiasis mansoni. In: Advances in parasitology, DAWES, B., ed., vol. 6, p. 233. New York: Academic Press 1968.

PELLEGRINO, J., and N. KATZ: Laboratory evaluation of antischistosomal agents. Ann. N.Y. Acad. Sci. **160**, 429 (1969).

PIRWITZ, J., B. SCHWAER u. K.A. SEYFARTH: Das Ehrlich-Asziteskarzinom der Maus als Testobjekt zytostatischer Substanzen: Lost, Urethan, Colchicin, Folinsäure, Miracil. Mikroskopie (Vienna), Sonderbd. **3**, 29 (1949).

PRESCOTT, B.: Some new salts of lucanthone as potential anticancer agents. J. Med. Pharm. Chem. **11**, 156 (1968).

REY, M.: Studies on the inhibition of DNA and histone synthesis in regenerating liver and hepatomas. M.S. Thesis, Graduate School of Biomedical Sciences, College of Medicine and Dentistry of New Jersey, Newark, New Jersey, 1973.

ROGERS, S.H., and E. BUEDING: The effects of hycanthone on *Schistosoma mansoni* in mice and hamsters (Abstract). Proc. 2nd Internat. Congress Parasitol. J. Parasitol. **56**, 288 (1970).

ROGERS, S. H., and E. BUEDING: Hycanthone resistance: Development in *Schistosoma mansoni*. Science **172**, 1057 (1971).

ROSI, D., G. PERUZZOTTI, E.W. DENNIS, D.A. BERBERIAN, H. FREELE, and S. ARCHER: A new, active metabolite of miracil D. Nature **208**, 1005 (1965).

ROSI, D., G. PERUZZOTTI, E.W. DENNIS, D.A. BERBERIAN, H. FREELE, B.F. TULLAR, and S. ARCHER: Hycanthone, a new active metabolite of lucanthone. J. Med. Pharm. Chem. **10**, 867 (1967).

SALGADO, J.A., C. VELOSO, C.A. OLIVEIRA, D.A.F. CHAMONE, M.S. LEMOS, N. KATZ y J. PELLEGRINO: Alteracoes electrocardiograficas observadas em pacientes com esquistossomose mansoni tratados com um derivado hidroximetilico do Miracil D (Hycanthone). Rev. Inst. Med. Trop. Sao Paulo **10**, 312 (1968).

SALLAM, L.A.R., A. EL-REFAI, and H.A. SHOEB: Microbiological transformation of lucanthone. Indian J. Exptl. Biol. **9**, 276 (1971).

SCHOLTAN, W.: Kolloidchemische Eigenschaften von Salzen basisch substituierter Xanthon- und Thioxanthonderivate. Kolloid-Z. **170**, 19 (1960).

SCHOLTAN, W., and R. GOENNERT: The relation between weight of micelles and the biological activity of xanthones and thioxanthones effective against bilharziasis. Med. Chem. Abhandl. Med.-Chem. Forschungsstätten Farbenfabriken Bayer **5**, 314 (1956); Chem. Abstr. **55**, 8766e (1961).

SEN, A.B., and F. HAWKING: The action of drugs *in vitro* on tapeworm (*Hymenolepsis nana*). Ann. Biochem. Exptl. Med. (Calcutta) **20**, Suppl., 547 (1960).

SENFT, A.W.: Recent developments in the understanding of amino-acid and protein metabolism by *Schistosoma mansoni in vitro*. Ann. Trop. Med. Parasitol. **59**, 164 (1965).

SENFT, A.W.: Considerations of schistosome physiology in the search for antibilharziasis drugs. Ann. N.Y. Acad. Sci. **160**, 571 (1969).

SHARP, T. M.: New synthesis of lucanthone (miracil D, nilodin). J. Chem. Soc. **1951**, 2961.

SIEBER, S. M., J. WHANG-PENG, D. G. JOHNS, and R. H. ADAMSON: Effects of hycanthone on rapidly proliferating cells. Biochem. Pharmacol. **22**, 1253 (1973).

SOBELL, S. D., and A. ARNOLD: Inhibition studies with *Streptococcus faecalis*. Xanthones and thiaxanthones. Proc. Soc. Exptl. Biol. Med. **90**, 594 (1955).

STANDEN, O. D.: Chemotherapy of helminthic infections. In: Experimental chemotherapy, SCHNITZER, R. J., and F. HAWKING, eds., vol. 1, p. 701. New York: Academic Press 1963.

STRUFE, R.: Stoffwechsel-Untersuchungen mit Miracil D. Med. Chem. Abhandl. Med.-Chem. Forschungsstätten Farbenfabriken Bayer **7**, 337 (1963).

SUGIURA, K.: Studies in a spectrum of mouse, rat and hamster tumors. Cancer Res. **22 (1)**, Pt. 2, 93 (1962).

SUGIURA, K., F. A. SCHMID, G. F. BROWN, and R. GONZÁLES: Chemotherapy of hamster tumors. Cancer Chemother. Rept. **2 (1)**, Pt. 2, 141 (1971).

SUGIURA, K., F. A. SCHMID, M. M. SCHMID, and G. F. BROWN: Effects of compounds on a spectrum of rat tumors. Cancer Chemother. Rept. **3 (1)**, Pt. 2, 231 (1972).

SURREY, A. R., and A. YARINSKY: Human antiparasitic agents. In: Annual report in medicinal chemistry—1967, CAIN, C. K., ed., p. 126. New York: Academic Press 1968.

SURREY, A. R., and A. YARINSKY: Human antiparasitic agents. In: Annual report in medicinal chemistry—1968, CAIN, C. K., ed., p. 126. New York: Academic Press 1969.

TARNOWSKI, G. S., and I. D. J. BROSS: Titration of the effects of a group of selected chemicals on the growth of ascites and solid forms of the Nelson mouse tumor. Cancer Res. **22 (1)**, Pt. 2, 136 (1962).

TELLER, M. N.: Chemotherapy of transplantable human tumors in the rat. Cancer Res. **22 (1)**, Pt. 2, 25 (1962).

THAYER, P. S., P. HIMMELFARB, and G. L. WATTS: Cytotoxicity assays with L1210 cells *in vitro*: Comparison with L1210 *in vivo* and KB cells *in vitro*. Cancer Chemother. Rept. **2 (1)**, Pt. 2, 1 (1971).

THORP, J. M., E. W. HURST, and A. R. MARTIN: The effect of β-diethylaminoethyl diphenylpropylacetate hydrochloride (SKF 525-A) on the therapeutic action of a variety of antibacterial, antiviral, or antiprotozoal drugs. J. Med. Pharm. Chem. **2**, 15 (1960).

U, R.: Enhancing effect of lucanthone (miracil D) on the frequency of x-ray-induced chromosome loss in *Drosophila* male germ cells. Mutation Res. **14**, 315 (1972).

VARGAS, M.: Carcinoma del endometrio. Tratamiento. Rev. Obstet. Ginecol. (Caracas) **23**, 385 (1963).

WARING, M. J.: Drugs and DNA: Uncoiling of the DNA double helix as evidence of intercalation. Humangenetik **9**, 234 (1970).

WARING, M.: Variation of the supercoils in closed circular DNA by binding of antibiotics and drugs: Evidence for molecular models involving intercalation. J. Mol. Biol. **54**, 247 (1970).

WARING: M.: Binding of drugs to supercoiled circular DNA: Evidence for and against intercalation. In: Progress in molecular and subcellular biology, HAHN, F. E., ed., vol. 2, p. 216. Berlin-Heidelberg-New York: Springer 1971.

WARING, M. J.: Inhibitors of nucleic acid synthesis. In: The molecular basis of antibiotic action, GALE, E. F., E. CUNDLIFFE, P. E. REYNOLDS, M. H. RICHMOND, and M. J. WARING, eds., p. 173. London: Wiley 1972.

WARING, M. J.: Interaction of indazole analogs of lucanthone and hycanthone with closed circular duplex deoxyribonucleic acid. J. Pharmacol. Exptl. Therap. **186**, 385 (1973).

WARREN, K. S.: Treatment of molluscan schistosomiasis mansoni with antibiotics, nonantibiotic metabolic inhibitors, molluscicides, and antischistosomal agents. Trans. Roy. Soc. Trop. Med. Hyg. **61**, 368 (1967); Chem. Abstr. **67**, No. 51541 t (1967).

WEINSTEIN, I. B., R. CARCHMAN, E. MARNER, and E. HIRSCHBERG: Miracil D: Effects on nucleic acid synthesis, protein synthesis, and enzyme induction in *Escherichia coli*. Biochim. Biophys. Acta **142**, 440 (1967).

WEINSTEIN, I. B., R. CHERNOFF, I. FINKELSTEIN, and E. HIRSCHBERG: Miracil D: An inhibitor of ribonucleic acid synthesis in *Bacillus subtilis*. Mol. Pharmacol. **1**, 297 (1965).

WEINSTEIN, I. B., and E. HIRSCHBERG: Mode of action of miracil D. In: Progress in molecular and subcellular biology, HAHN, F. E., ed., vol. 2, p. 232. Berlin-Heidelberg-New York: Springer 1971.

Weyland, H.: Untersuchungen über die Beeinflussung der Pflanze durch chemische Substanzen und ihre Bedeutung für die Beurteilung gewisser medizinischer Fragen. Z. Krebsforsch. **56**, 148 (1948).

Weyland, H., R. Weyland, K.W. Cremer, J. Reinert u. H.U. Koecke: Die Beeinflussung der Pflanze durch chemische Substanzen. Biol. Zentralblatt **68**, 140 (1949).

Willard, F.L., and R. Kodras: Survey of chemical compounds tested *in vitro* against rumen protozoa for possible control of bloat. Appl. Microbiol. **15**, 1014 (1967).

Wilson, R., J. Church, R. Bodner, N. Sylvain, and E. Hirschberg: The effect of miracil D and analogs upon nucleic acid synthesis in leukemia L1210 cells (Abstract). Proc. Am. Assoc. Cancer Res. **13**, 76 (1972a).

Wilson, R.G., J.A. Church, and N.P. Sylvain: The effect of miracil D upon the template properties of polynucleotides for a bacterial RNA polymerase. Biochim. Biophys. Acta **277**, 564 (1972b).

Wittner, M., H. Tanowitz, and R.M. Rosenbaum. Studies with the schistosomacide hycanthone: Inhibition of macromolecular synthesis and its reversal. Exptl. Mol. Pathol. **14**, 124 (1971).

Wolfe, A.D., R.G. Allison, and F.E. Hahn: Labilizing action of intercalating drugs and dyes on bacterial ribosomes. Biochemistry **11**, 1569 (1972).

Wood, D.R.: Observations on the pharmacology of miracil, a new chemotherapeutic agent for schistosomiasis. Quart. J. Pharm. Pharmacol. **20**, 31 (1947).

Woolley, G.W.: Chemotherapy of transplantable human tumors in the hamster. Cancer Res. **22** **(1)**, Pt. 2, 34 (1962).

Wright, C.A.: The schistosome life-cycle. In: Bilharziasis, Mostofi, K., ed., p. 3. Berlin-Heidelberg-New York: Springer 1967.

Yarinsky, A., and H. Freele: A comparison of molluscicidal and mollusc inhibitory activity of hycanthone and lucanthone and the effect of the drugs on the development of *Schistosoma mansoni* in the snail intermediate host, *Australobris glabratus*. J. Trop. Med. Hyg. **73**, 23 (1970).

Yarinsky, A., P. Hernandez, and E.W. Dennis: The uptake of tritiated hycanthone by male and female *Schistosoma mansoni* worms and distribution of the drug in plasma and whole blood of mice following a single intramuscular injection. Bull. World Health Organ. **42**, 445 (1970).

Zilversmit, R.: Thioxanthones. Structural differences between lucanthone and its N-methyl derivative. Mol. Pharmacol. **6**, 172 (1970).

Zilversmit, R.: Thioxanthones. II. Studies on the hydrogen-bonding capacity of lucanthone. Mol. Pharmacol. **7**, 674 (1971).

Zussman, R.A., and P.M. Bauman: Schistosome hemoglobin protease; search for inhibitors. J. Parasitol. **57**, 233 (1971).

Addendum

Note added in galley proof, April 1974: In the past several months, much pertinent additional information has become available in the published literature. In particular, the significance of mutagenic and other toxic effects of Hycanthone relative to its therapeutic potential continues to be a subject of vigorous investigation and dispute. The reader may wish to consult the following references to gain an updated view of the topics discussed in this chapter:

Archer, S., and A. Yarinsky: Recent developments in the chemotherapy of schistosomiasis. In: Progress in drug research, Jucker, E., ed., Vol. 16, p. 12, Basel: Birkhäuser Verlag, 1972.

Bueding, E., J. Fisher, and J. Bruce: The antischistosomal activity of a chloroindazole analog of hycanthone in mice infected with *Schistosoma mansoni*. J. Pharmacol. Exptl. Therap. **186**, 402 (1973).

Chou, T.C.T., J.L. Bennett, C. Pert, and E. Bueding: Effect of hycanthone and of two of its structural analogs on levels and uptake of 5-hydroxytryptamine in *Schistosoma mansoni*. J. Pharmacol. Exptl. Therap. **186**, 408 (1973).

Clive, D., W.G. Flamm, and J.B. Patterson: Specific-locus mutational assay systems for mouse lymphoma cells. In: Chemical mutagens. Principles and methods for their detection, Hollander, A., ed., Vol. 3, p. 79, New York: Plenum Press, 1973.

Farber, E.: Carcinogenesis—cellular evolution as a unifying thread: Presidential address. Cancer Res. **33**, 2537 (1973).

FRIEDHEIM, E. A. H.: Chemotherapy of schistosomiasis. In: International encyclopedia of pharmacology and therapeutics. Section 64. Chemotherapy of helminthiasis, CAVIER, R., ed., Vol. I, p. 29, Oxford: Pergamon Press, 1973.

GENEROSO, W. M., and G. E. COSGROVE: Total reproductive capacity in female mice: Chemical effects and their analysis. In: chemical mutagens. Principles and methods for their detection, HOLLANDER, A., ed., Vol. 3, p. 241, New York: Plenum Press, 1973.

GREEN, S., J. V. CARR, F. M. SAURO, and M. S. LEGATOR: Effects of hycanthone on spermatogonial cells, deoxyribonucleic acid synthesis in bone marrow and dominant lethality in rats. J. Pharmacol. Exptl. Therap.186, 437 (1973).

HARTMAN, P. E., H. BERGER, and Z. HARTMAN: Comparison of hycanthone ("etrenol"), some hycanthone analogs, myxin and 4-nitroquinoline-1-oxide as frameshift mutagens. J. Pharmacol. Exptl. Therap. 186, 390 (1973).

JANSMA, W. B., P. B. HULBERT, and E. BUEDING: Induction of hycanthone resistance in Schistosoma mansoni (Abstract). Federation Proc. 33, 556 (1974).

LAIDLAW, G. M., J. C. COLLINS, S. ARCHER, D. ROSI, and J. W. SCHULENBERG: The synthesis of hycanthone. J. Org. Chem. 38. 1743 (1973).

LEA, M. A., F. L. KHALIL, and M. I. REY: Action of miracil D and related compounds on histone synthesis and phosphorylation in regenerating liver. Chem. Biol. Interactions 7, 367 (1973).

LEGATOR, M. S. and W. G. FLAMM: Environmental mutagenesis and repair. Ann. Rev. Biochem. 42, 683 (1973).

LUCIER, G. W., O. S. MCDANIEL, J. R. BEND, and E. FAEDER: Effects of hycanthone and two of its chlorinated analogs on hepatic microsomes. J. Pharmacol. Exptl. Therap. 186, 416 (1973).

MEADOWS, M. G., S. K. QUAH, and R. C. VON BORSTEL: Mutagenic action of hycanthone and IA-4 on yeast. J. Pharmacol. Exptl. Therap. 186, 444 (1973).

OBE, G.: Action of hycanthone on human chromosomes in leukocyte cultures. Mutation Res. 21, 287 (1973).

ROSENKRANZ, H. D.: Miracil D: Inhibition of deoxyribonucleic acid polymerase-deficient Escherichia coli. Antimicrobial Agents Chemother. 3, 530 (1973).

SOURO, F. M., and S. GREEN: In vivo cytogenetic evaluation of chloroindazole thioxanthene IA-4 (a hycanthone analog) and niridazole in rat bone marrow. J. Pharmacol. Exptl. Therap. 186, 399 (1973).

SENFT, A. W., and G. R. HILLMAN: Effect of hycanthone, niridazole, and antimony tartrate on schistosome motility. Am. J. Trop. Med. Hyg. 22, 734 (1973).

SMITH, R. H.: Lack of mutagenicity of hycanthone in Habrobracon serinopae females. Mutation Res. 21, 279 (1973).

STRAUS, D. S., P. E. HARTMAN, and Z. HARTMAN: Actions of mutagens on Salmonella: Molecular mutagenesis. In: Molecular and environmental aspects of mutagenesis, 6th Rochester International Conference on Environmental Toxicity, MILLER, M. W., ed., Springfield: CC Thomas, 1974.

TURNER, S., R. BASES, A. PEARLMAN, M. NOBLER, and B. KABAKOW: Lucanthone: An adjuvant in clinical radiation therapy (Abstract). Proc. Am. Assoc. Cancer Res. 15, 30 (1974).

YARINSKY, A., H. P. DROBECK, H. FREELE, J. WILAND, and K. I. GUMAER: An 18-month study of the parasitologic and tumorigenic effects of hycanthone in Schistosoma mansoni-infected and noninfected mice. Toxicol. Appl. Pharmacol. 27, 169 (1974).

Trimethoprim and Pyrimethamine

James J. Burchall

Trimethoprim

Trimethoprim (2,4-diamino-5-(3'4'5'-trimethoxybenzyl)-pyrimidine is the most active and selective antibacterial of a series of inhibitors of dihydrofolate reductase synthesized by HITCHINGS and his colleagues during the mid-1950's (HITCHINGS and BUSHBY, 1961; ROTH et al., 1962; FALCO et al., 1951). For the treatment of human disease, trimethoprim is combined with a sulfonamide, sulfamethoxazole (Gantanol, Roche) and marketed under the names Septrin, Eusaprim, Wellcome; and Bactrim, Roche. The structure of trimethoprim is shown in Fig. 1.

It is a weak base with a pKa of about 7.3, slightly soluble in water (0.4 mg/ml), and forms stable salts with a variety of acids most of which have low solubility in water. The acetate and lactate are the most water soluble; solubility of the lactate salt being 25 mg of base/ml water. Trimethoprim is stable to autoclaving at 120°C for 20 minutes.

Trimethoprim possesses a high activity against a variety of Gram-positive and Gram-negative bacteria. Activity against *N. gonorrhoeae* is variable and apparently strain dependent. For all practical purposes, the drug is ineffective against *Ps. aeruginosa* and *M. tuberculosis;* in both cases their respective M.I.C. values exceed 100 µg/ml. Trimethoprim is also ineffective against *T. pallidum.* There are a number of protozoal and fungal diseases which, on the basis of incomplete laboratory and clinical data, may be susceptible to trimethoprim or the combination of trimethoprim and sulfonamide. Both sensitive and chloroquine-resistant strains of *P. falciparum* malaria have been successfully treated by trimethoprim alone or in combination with sulfametopyrazine (MARTIN and ARNOLD, 1967, 1968). Successful treatment of *P. vivax* with trimethoprim-sulfalene has been reported (MARTIN and ARNOLD, 1969). Activity against toxoplas-

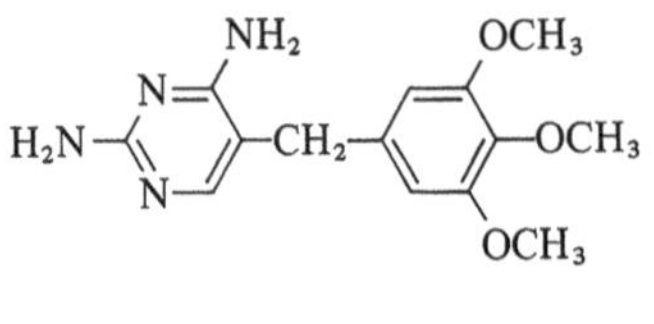

2,4-diamino-5-(3', 4', 5'-trimethoxybenzyl)
pyrimidine

Fig. 1. Trimethoprim

mosis has also been reported (MOSSNER, 1969). Success was reported in the treatment of several disseminated fungal diseases such as Nocardiosis (BAIKIE et al., 1970; BLACK and McNELLIS, 1970), Histoplasmosis (MACLEOD, 1970), and South American Blastomycosis (FERREIRA LOPES and ARMOND, 1968).

It should be emphasized that the ability of trimethoprim to inhibit various microorganisms is strongly dependent on composition of the medium in which the test is performed (BUSHBY and HITCHINGS, 1968; DARRELL et al., 1968). This is not unexpected in view of the mechanism of action of trimethoprim (see below). In brief, it appears that certain media may be the source of exogenous metabolites whose normal de novo synthesis is blocked by trimethoprim. When these metabolites are taken up by the cell and utilized by "salvage" pathways known to be present, the inhibition caused by trimethoprim could be bypassed in a noncompetitive manner. KOCH and BURCHALL (1971) showed that in minimal medium the inhibition caused by trimethoprim could be partially reversed by the addition of thymidine, ribonucleosides, amino acids, and vitamins. No reversal occurred when thymidine was absent. In a number of commercially prepared media, the inhibitory activity of trimethoprim correlated inversely with the amount of thymidine present.

In laboratories concerned with the determination of trimethoprim and/or sulfonamide sensitivity of microorganisms, lysed horse blood is frequently added to the test media (WATERWORTH, 1969) and this procedure appears to inactivate substances which interfere with the activity of these drugs (HARPER and CAWSTON, 1945; WALKER et al., 1947). GLAZ (1967) investigated the active substance in horse blood and found it to be a protein. Recent investigation in the Wellcome Laboratories (FERONE et al., 1972) has shown that Harper-Cawston factor is thymidine phosphorylase. The enzyme acts by cleaving thymidine to thymine, the latter compound being only 1/200 as active as thymidine in reversing the action of trimethoprim. The resistance of Pseudomonas aeruginosa to trimethoprim is probably due to the inability of the drug to enter the bacterial cell since the reductase of this organism is virtually indistinguishable from that of E. coli (see below). Similar problems may exist with M. tuberculosis, but it has been observed that this organism possesses a unique reductase which is inherently less susceptible to trimethoprim.

Resistance to trimethoprim may be acquired by successive passage of E. coli and S. aureus in increasing concentrations of the drug (DARRELL et al., 1968). This resistance can be due to a variety of factors including changes in permeability of the organism, alterations in the quantity or structure of the reductase present or other enzymatic changes in the folate pathway. Resistant microorganisms isolated from rich medium are usually thymidine-requiring and lack thymidylate synthetase, an enzyme which acts as a drain on the cellular supply of tetrahydrofolate (BERTINO and STACEY, 1966). Some microorganisms resistant to trimethoprim have been isolated from patients receiving the drug (DARRELL et al., 1968; WATERWORTH, 1969; MAY and DAVIES, 1972; LACEY et al., 1972), but to date clinical resistance has not been a factor in the use of the drug. Recently reports have appeared of the presence of R (resistance) factors in E. coli and other enteric organisms (FLEMING et al., 1972; DATTA and HEDGES, 1971, 1972). Plasmids of the W compatibility group confer high resistance (> 1000 µg/ml minimum

Table 1. *In vitro inhibitory concentrations of trimethoprim*

	Organism	Minimum inhibitory concentration (µg/ml)
St. pyogenes	CN10	0.1
St. pyogenes	S3640	0.3
St. faecalis	CN478	0.1
Staph. aureus	CN491	0.3
Vibrio comma	CN2005	0.3
Ery. rhusiopathiae	CN904	3.0
Past. boviseptica	CN1066	0.1
Monilia albicans	CN1863	> 100
Mycobacterium 607	S3254	3.0
Serra. marcescens	CN2398	0.1
Kleb. pneumoniae	CN3632	1.0
Kleb. aerogenes	CN2298	1.0
Sal. typhosa	CN512	0.1
Esch. coli	CN314	0.1
Shig. flexneri	CN6007	0.1
Shig. dysenteriae	CN1513	0.1
Entero. aerogenes	2200/86	0.3
Entero. cloacae	2200/87	1.0
Citro. freundii	2200/77	0.3
Pr. vulgaris	CN329	3.0
Pr. mirabilis	S2409	3.0
Pr. rettgeri	2200/135	3.0
Ps. aeruginosa	CN200	> 100
Ps. aeruginosa	D60	> 100

Medium: Wellcotest agar CM48 plus 7% lysed horse blood.
The activity of trimethoprim was determined in Wellcome nutrient agar containing 7.5% horse blood except for *M. tuberculosis* for which Peizer and Schecter medium was used. The inoculum was a 1 mm loopful of a 10^{-2} dilution of an 18-hr. dilution of broth cultures of all strains except those of *Streptococcus pyogenes* which was used undiluted.

inhibitory concentration) on the recipient organisms. The biochemical mechanism of the R^+ induced resistance is unknown.

The mechanism of action of trimethoprim is based on its ability to selectively interfere with the formation of folic acid cofactors in the target cell. A schematic outline of the pathway together with the loci of inhibition of sulfonamide and trimethoprim is shown in Fig. 2. A microorganism such as *E. coli* is able to synthesize dihydrofolate *de novo* from a dihydropteridine, p-ABA, and glutamic acid, but is unable to assimilate folate from its growth medium. Man cannot make folate and must obtain it from his diet. Sulfonamide is a close structural analog of p-aminobenzoic acid and acts as a competitive inhibitor of the enzyme that forms dihydropteroic acid. Since this pathway is absent in man, he is unaffected by the drug.

In both microorganism and man, tetrahydrofolate (FAH_4) is required to carry out reactions leading to nucleic acids and protein. For this reason the reduction of dihydrofolate to tetrahydrofolate is an attractive locus for inhibition of bacteria provided that inhibition of the same reaction could be avoided in the host. The enzyme which catalyses this reaction is dihydrofolate reductase.

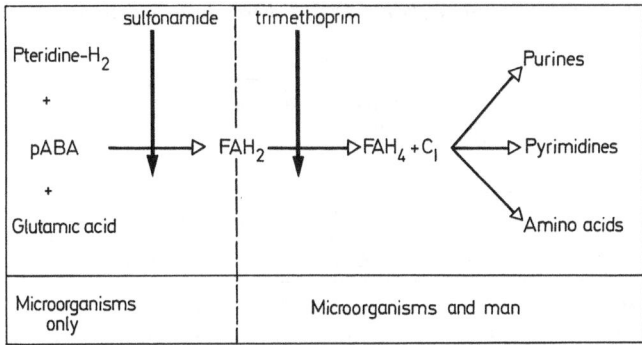

Fig. 2. The folic acid pathway: Loci of trimethoprim and sulfonamide inhibition

During the 1950's Hitchings and his group at The Wellcome Research Laboratories synthesized a number of compounds that showed selective activity against plasmodia and bacteria. These compounds acted as "anti-folic" agents (HITCHINGS *et al.*, 1948, 1950a, 1950b). The nature of this activity was clarified by the finding that folinic acid bypassed the effects of these drugs whereas folate did not. This suggested that the inhibitors interferred with some essential step in the conversion of folic acid to an active form. A number of these compounds are shown in Table 2 and the data show that by varying the substitutents on the pyrimidine and phenyl rings it is possible to obtain selective action against certain organisms. For example, addition of a methyl group to the pyrimidine-6 position decreased the amount of drug required for the inhibition of *Plasmodium gallinaceum* and increased it for the inhibition of *Staphylococcus aureus*. Manipulation of host toxicity was also possible as evidenced by the rise in the LD_{50} when the drug administered to mice contained 3',4'-dimethoxyl groups on the phenyl ring.

Concentrations required for 50% inhibition of bacterial and mammalian dihydrofolate reductases by a variety of compounds including trimethoprim are shown in Table 3. The right-hand column gives the ratio of these two activities which is a measure of the potential selective toxicity of the compound, i.e. larger

Table 2. *Toxicity of 5-benzylpyrimidines for various species*

Phenyl substituent	Pyrimidine 6-position	*P. gallinaceum*[a] (mg/kg)	*S. aureus*[a] (mg/l)	Mouse[b] (mg/kg)
4-H	H	100	32	40
4-H	CH_3	5	128	60
4-Cl	H	100	8	80
4-Cl	CH_3	1.5	64	30
3',4'-$(OCH_3)_2$	H	100	< 1	> 500
3',4'-$(OCH_3)_2$	CH_3	10	30	> 100

[a] Dose required for minimum inhibition.
[b] Dose required for LD_{50}.

Table 3. *Binding of 2,4-diamino-5-benzylpyrimidines to dihydrofolate reductase of E. coli and rat liver*

R_1	R_2	R_3	Molar conc. for 50% inhibition		Rat liver: $E. coli$
			$E. coli$	Rat liver	
H	H	H	3.4×10^{-6}	1.5×10^{-4}	44
H	CH_3	H	2.3×10^{-6}	1.0×10^{-4}	44
H	Cl	H	2.1×10^{-6}	9.0×10^{-5}	43
H	OH	H	1.8×10^{-6}	1.6×10^{-4}	89
H	OCH_3	H	1.1×10^{-6}	1.7×10^{-4}	155
OCH_3	H	H	4.9×10^{-7}	4.3×10^{-5}	88
OCH_3	OCH_3	H	1.0×10^{-7}	7.0×10^{-5}	700
OCH_3	OCH_3	OCH_3	5.0×10^{-9}	2.6×10^{-4}	52,000
OCH_3	OCH_3	Cl	4.0×10^{-8}	1.0×10^{-4}	2,500
OCH_3	OH	OCH_3	1.1×10^{-8}	1.0×10^{-4}	9,100

this ratio, the more potent the compound is as a relative inhibitor of *E. coli*. The compound substituted with hydrogen in the R^1, R^2, and R^3 positions is a weak inhibitor of both the *E. coli* and rat liver reductases. A major contribution to selectivity is made by the addition of a methoxyl group in the para or meta positions. This group causes increased binding to the *E. coli* reductase without a concomitant increase in activity against the rat liver enzyme. The substitution of methoxyl groups in both the R^1 and R^2 positions results in a significant further increase in selective activity. When methoxyl groups are present in all the three positions, a 50% inhibition of the enzyme from *E. coli* is obtained at 5×10^{-9} M whereas the quantity of compound required to inhibit the reductase of rat liver is increased to 2.4×10^{-4} M (BURCHALL and HITCHINGS, 1965).

The enzymatic basis for the selectivity of trimethoprim can be seen in Table 4 which gives the concentration $\times 10^{-8}$ molar needed for 50% inhibition of the reductase from *E. coli*, *S. aureus*, *P. vulgaris*, and from human and rat liver. The failure of amethopterin (compound No. 1) to distinguish between the enzymes from host and parasite is seen in the first column of data. The compound binds extremely tightly to reductases from all the species listed. Compound No. 3 is toxic for mammalian species, but ineffective as an antibacterial. Trimethoprim, compound No. 2, is a potent inhibitor of the bacterial reductases and relatively nontoxic to mammals. Table 4 also contains other interesting examples which illustrate the ability of inhibitors to detect small differences between and among bacterial and mammalian reductases (BURCHALL and HITCHINGS, 1965). Similar observations were made by BAKER (1964).

The chemical nature of differences among the reductases is unknown, although recent progress in obtaining pure reductase from various species promises solution of this problem in the reasonably near future (GUNDERSEN *et al.*, 1972; POE *et al.*, 1972; ERICKSON and MATHEWS, 1972). It seems likely that reductases vary

Table 4. *Inhibitor binding to bacterial and mammalian dihydrofolate reductases*

Source of enzyme	Inhibitory concentration[a]				
	Compound number				
	1	2	3	4	5
E. coli	0.6	0.5	65,000	50	2
S. aureus	0.1	1.5	50,000	4	7
P. vulgaris	0.5	0.5	10,000	50	1
Human liver	0.2	30,000	55	95	24
Rat liver	0.2	26,000	14	46	26

[a] Concentration of drug ($M \times 10^{-8}$) needed to reduce enzyme activity to 50% of control value (no drug) in the presence of 5×10^{-5} M dihydrofolate.

No. 1. Amethopterin: 2,4-diamino-N^{10}-methylpteroylglutamate.

No. 2. 2,4-diamino-5-(3′,4′,5′-trimethoxybenzyl)pyrimidine (trimethoprim).

No. 3. 1-(p-butylphenyl)-1,2-dihydro-2,2-dimethyl-4,6-diamino-5-triazine.

No. 4. 2,4-diamino-6-butylpyrido(2,3-d)pyrimidine.

No. 5. 2,4-diamino-5-methyl-6-butylpyrido(2,3-d)pyrimidine.

little in the region of their catalytic site in view of their common requirement for amino acid sequences capable of binding both dihydrofolate and TPNH and for carrying out the reduction of the substrate. Although bacterial and mammalian reductases may be distinguished from each other kinetically, there is remarkably little variation in K_m, V_{max}, and pH-activity relationships within each group (HITCHINGS and BURCHALL, 1965). On the other hand, region outside of the catalytic site may be permitted a greater degree of variation since the main function of this region is likely to be limited to maintenance of the overall conformation of the enzyme. HITCHINGS (1969) and BAKER (1967) have suggested that inhibitors showing species specificity may bind both within and partially outside the active center. The diamino-pyrimidine moiety presumably binds in the same area as the amino-hydroxy-pyrimidine portion of dihydrofolate. The remainder of the inhibitor may bind to a region outside the catalytic site, and BAKER (1967) suggests that hydrophobic binding may be the principal force involved in some cases.

The physiological consequences of the blockade of dihydrofolate reductase have been extensively investigated. The result is cessation of synthesis of purines, pyrimidines (most importantly thymine), several amino acids (methionine, glycine, and histidine [BLAKELEY, 1969]), pantothenate (TELLER and SNELL, 1969), and N-formylmethionyl-transfer ribonucleic acid (f-met-tRNA$_f$ [EISENSTADT and LENGYEL, 1966; ADAMS and CAPECCHI, 1966; CAPECCHI, 1966]). These findings are reasonable on the basis of the known pathways of tetrahydrofolate utilization as well as the results of studies in which reversal of the inhibition of the antifolates was overcome by end products of the folate pathway (HITCHINGS and BURCHALL, 1965).

A number of observations emphasize the critical importance of thymine deprivation as a consequence of inhibition of trimethoprim. DALE and GREENBERG (1972) studied the effect of trimethoprim on the incorporation of labeled thymine into DNA and showed that the inhibition caused by low levels of trimethoprim could be reversed by the addition of thymine in the presence of deoxyguanosine or when 5-fluorodeoxyuridine (an inhibitor of thymidylate synthetase) was added at the same time as was trimethoprim. The ease of selection of trimethoprim-resistant mutants which lack thymidylate synthetase (BERTINO and STACEY, 1966) also emphasizes the importance of this enzyme.

The importance of the N-formyl group of f-met-tRNA$_f$ is not as well understood. In spite of the presence of systems capable of catalyzing formylation, there is evidence that the formylated derivative (f-met-tRNA) is not essential for cell growth. Some of these studies have been carried out in *Streptococcus* which cannot synthesize its own folates. RABINOWITZ et al. (SAMUEL et al., 1970) have shown that *S. faecalis R* grown on media free of folates but containing serine, methionine, thymine, adenine, and guanine were not affected by the presence of trimethoprim. Extracts of cells grown on folate-free medium do not contain tetrahydrofolate, and therefore formylated methionyl-tRNA$_f$ is not synthesized, although the enzymes to do so are present. PINE et al. (1969) have observed that extracts of *Streptococcus faecium*, grown in medium supplemented with end products of folate metabolism, can grow and initiate protein synthesis in the absence of folates and in the presence of trimethoprim. HARVEY (1972a) has carried out similar studies in *E. coli*, an organism which can make its own folates. These studies have shown that sustained exponential growth of *E. coli* can occur in the presence of trimethoprim and the low molecular-weight products of folate metabolism, under conditions where formylation of met-tRNA$_f$ is undetectable.

The effect of trimethoprim on other enzymes of the folate pathway has been examined (DEV, 1972). No inhibition of the following enzymes in extracts of *E. coli* was observed at concentrations of trimethoprim $\geq 1 \times 10^{-4}$ M; thymidylate synthetase; serine hydroxymethyl transferase; 5,10 methylene tetrahydrofolate dehydrogenase; and 5,10 methenyl tetrahydrofolate cyclohydrolase. These observations favor the view that dihydrofolate reductase is the sole locus of action of trimethoprim.

Because trimethoprim and sulfonamide are inhibitors of sequential steps in the biosynthesis and utilization of folates, it might be expected that their combined use could produce inhibitory effects greater than the sum of their individual effects. In view of potential importance of "synergistic" effects in chemotherapy, there exists only a small body of theoretical knowledge which would allow *a priori* prediction of the degree of potentiation, if any, between two or more drugs. For a general discussion of the mathematical arguments involved, the reader is referred to the volume of WEBB (1963) and to the work of GARRETT (1958).

Synergism in the folate pathway has been discussed by ELION et al. (1954). Recently HARVEY (1972b) has reviewed this work and attempted to formulate a comprehensive theory which predicts pathway configurations that are amenable to synergistic inhibition. The models described predict that trimethoprim and sulfonamides act synergistically, and experimental data are given to support this conclusion. A theoretical and experimental study by SEYDEL et al. (1972) also supports the conclusion that the combination is potentiating. Evidence is presented that although sulfonamides act only bacteriostatically even at high concentrations, trimethoprim alone can cause a bactericidal effect if a concentration of ≥ 1 µmol/liter is exceeded. Combinations of trimethoprim and sulfonamide at concentrations where both drugs are acting merely bacteriostatically lead to effects considerably greater than would be expected from simple additivity.

BUSHBY and HITCHINGS (1968) have summarized a number of experiments conducted *in vitro* and *in vivo* that demonstrate potentiation for the combination of the two drugs. BUSHBY has added further documentation to these arguments (BUSHBY, 1969). In studies with *N. gonorrhoeae* he showed that although this organism is relatively resistant to trimethoprim, it is sensitive to the combination of trimethoprim and sulfonamide. The optimum ratio of trimethoprim to sulfonamide required to produce the greatest inhibition was 1:20 (T:S) (BÖHNI, 1969). In a series of experiments, mice infected with *H. influenzae, D. pneumoniae, E. coli,* or *N. meningitidis* were treated with each drug alone and in combination. The enhancing effect of the combinations was seen best with *H. influenzae,* an organism that is very susceptible to trimethoprim.

A number of studies have been conducted on the pharmacokinetics and renal handling of trimethoprim. SCHWARTZ and ZIEGLER (1969) have calculated the half-lifetime of trimethoprim as 10 hours in the plasma (sulfamethoxazole has a half-life of 9–11 hours). When 2 mg of C^{14}-labeled trimethoprim was administered to dogs or humans, the largest portion of the administered dose was found in the urine as unchanged drug. Conjugates such as glucuronides and sulfates accounted for only 1 to 7% of the radioactivity. In contrast, the urine of the rat contained much larger proportions of conjugates and of other water-soluble metabolites. Approximately 44% of the drug is bound to plasma protein.

BUSHBY and HITCHINGS (1968) noted that a 7 mg dose of trimethoprim given to a 20-gram mouse produced blood levels of the drug in 15 minutes and that significant amounts of activity were still present after 18 hours. The drug has a strong tendency to concentrate in tissues, and 15 minutes after administration the drug concentration was tenfold higher in kidney and lungs than in serum. SHARPSTONE (1969) estimated that the renal clearance of trimethoprim ranges from 19–148 ml/min in normal subjects. The rate of excretion increases steeply as the pH of the urine is lowered.

In mice the acute oral LD_{50} is greater than 2,000 mg/kg, and the acute *i.v.* LD_{50} is 200 mg/kg. In chronic toxicity tests, depression of hemopoesis was noted in dogs dosed 6 days a week with 135 mg/kg and in one of 6 monkeys dosed daily with 300 mg/kg. Studies of the acute toxicity of combinations of trimethoprim and sulfonamide have been reported by UDALL (1969) who noted the results of studies by P. J. FRASER on toxicity of trimethoprim to the fetus. Dosing pregnant rats with trimethoprim (500 mg/kg) on any single day between 8 and 16 had no effect on the fetus. A single dose of 2,000 mg/kg was lethal to most fetuses when given on the eighth or ninth day and caused a very high incidence of malformations when given on the tenth, eleventh, or twelfth day. After the sixteenth day, doses up to 133 mg/kg/day were not teratogenic, but those over 200 mg/kg/day were. Protection from the effects of 2,000 mg/kg trimethoprim on the eleventh day of pregnancy was provided by leucovorin (8 mg/rat). In rabbits which were fed a diet high in folates, no teratogenic effects were seen.

Whitman (1969) has reported the effects of prolonged administration of trimethoprim and sulfisoxazole in man. When trimethoprim was given at oral doses of 1,000 mg/day, the first detectable changes in folate metabolism were found after 7–10 days. The clinical significance of these changes were not apparent. When trimethoprim (200 mg/day) was given, no changes in folate metabolism were found after 13 weeks of administration.

For further information on the combination of trimethoprim and sulmethoxazole and its clinical effects, the reader is referred to the excellent article of Garrod (1971) and to articles in Postgraduate Medical J. Suppl. **45** regarding clinical uses of the combinations that are not reviewed in this paper.

Pyrimethamine

Pyrimethamine (2,4-diamino-5-p-chlorophenyl-6-ethylpyrimidine) is a potent antimalarial that was discovered by Hitchings and his colleagues (Hitchings *et al.*, 1952a, 1952b; Russell and Hitchings, 1951) in their studies on the biological activity of the 2,4-diaminopyrimidine series. The drug is issued under the brand name "Daraprim", Wellcome. The structure of pyrimethamine is shown in Fig. 3.

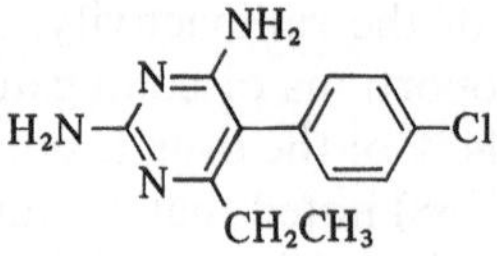

2,4-diamino-5-p-chlorophenyl-6-
ethylpyrimidine

Fig. 3. Pyrimethamine

Pyrimethamine base is slightly soluble in water, and alcohol, and moderately soluble in aqueous acid solution. Soluble salts include the isethionate, acetate, and lactate, while the hydrochloride and sulfate are less soluble. The ultraviolet spectrum of the compound has been the object of several studies (Maggiolo and Russell, 1951; Russell, 1954).

The antibacterial spectrum of pyrimethamine is given in Table 5 and although the compound has some activity against *Streptococcus pyogenes,* it is not an effective antibacterial agent. Pyrimethamine possesses high activity as a suppressive (prophylactic) agent in man and animals against malaria as an inhibitor of the development of plasmodial gametocytes in the mosquito (Hitchings, 1960).

Pyrimethamine is the outcome of extensive investigations on agents which inhibit selectively the synthesis of nucleic acid (see Trimethoprim, this chapter). Falco *et al.* (1951) established the basic structural requirements of the pyrimidine series necessary for antimalarial activity and these may be summarized as follows: 1. the 2,4-diamino groups are essential for activity and their substitution is detri-

Table 5. *In vitro inhibitory concentrations of pyrimethamine*

	Organism	Minimum inhibitory concentration (µg/ml)
St. pyogenes	CN10	0.3
St. pyogenes	S3640	0.3
St. faecalis	CN478	3.0
Staph. aureus	CN491	100
Vibrio comma	CN2005	30
Ery. rhusiopathiae	CN904	100
Past. boviseptica	CN1066	3.0
Monilia albicans	CN1863	>100
Mycobacterium 607	S3254	10
Serr. marcescens	CN2398	3.0
Kleb. pneumoniae	CN3632	>100
Kleb. aerogenes	CN2298	100
Sal. typhosa	CN512	30
Esch. coli	CN314	100
Shig. flexneri	CN6007	100
Shig. dysenteriae	CN1513	30
Entero. aerogenes	2200/86	100
Entero. cloacae	2200/87	>100
Citro. freundii	2200/77	100
Pr. vulgaris	CN329	>100
Pr. mirabilis	S2409	>100
Pr. rettgeri	2200/135	>100
Ps. aeruginosa	CN200	>100
Ps. aeruginosa	D60	>100

Medium: Wellcotest agar CM48 plus 7% lysed horse blood.

mental, and (2) the 5-phenyl group is superior to the 5-benzyl and 5-phenoxy substituents. The effects of substituents on the benzene nucleus of the phenylpyrimidines have also been studied and are summarized by HITCHINGS (1952c) as follows: 1. an electronegative group in the para position of the benzene ring enchances activity for both *P. gallinaceum* and *P. berghei*, and 2. an electron-donating ring fails to enhance activity and, especially when *P. berghei* is considered, may greatly diminish it. FALCO *et al.* (1951) have studied the effect of substitution in the 6-position of the pyrimidine on antimalarial activity. The 6-methyl compound has fivefold higher activity than the unsubstituted congener. The 6-unsubstituted 5-p-chlorophenyl-2,4-diaminopyrimidine is active against *P. berghei*, but the addition of an ethyl group at the 6-position produced the most potent derivative. With other species of plasmodia, the 6-methyl group is the most active. The strong antimalarial enhancing effect of the 6-methyl group against *P. gallinaceum* may be seen from the data presented in Table 2.

Experimental therapy of malaria in animals has suggested the main areas of activity of the drug (FALCO *et al.*, 1951; ROLLO, 1952a; GOODWIN, 1949). The minimum effective dose (MED) of pyrimethamine in the mouse and chick is approximately 40 µg/kg. In sporozoite-induced infections with *P. gallinaceum*, the exoerythrocytic stages of the parasite were cleared by pyrimethamine, but less success was achieved with short-term treatment of sporozoite-induced infections with *P. cynomolgi* in the monkey. SCHMIDT and GENTHER (1953) showed that daily doses of 0.02 mg/kg or weekly doses of 0.15 mg/kg of pyrimethamine suppressed sporozoite-induced infections, but relapse occurred when the drug was stopped. In acute infections of malaria, the infection was cured slowly.

Studies on the activity of the drug in humans were carried out by COATNEY *et al.* (1953) employing *P. vivax* (Chesson strain) and by COVELL *et al.* (1953) with *P. falciparum* infections. A variety of field trials finally led to the widespread use of the drug as an antimalarial agent (HITCHINGS, 1960).

It was observed by FALCO *et al.* (1949) that the diaminopyrimidines were related to a possible configuration of chloroguanide. On this basis the activity of pyrimethamine as an antimalarial was rationalized. A number of later studies have conclusively demonstrated that plasmodia possess a more or less complete folic acid pathway and that pyrimethamine interferes with the formation of tetrahydrofolate in this pathway. The indirect evidence for this may be summarized as follows:

1. known inhibitors of folate synthesis and interconversions are effective antimalarials (HILL, 1963; ROLLO, 1955).

2. *p*-ABA and folic acid reverse the effects of some inhibitors (ROLLO, 1955; THURSTON, 1954; JACOBS, 1964).

3. survival and growth of plasmodia increases in the presence of leucovorin or folic acid (GLENN and MANWELL, 1956; TRAGER, 1958).

Direct studies by FERONE and HITCHINGS (1966) provided evidence to support these early speculations. Malarial parasites (whole or disrupted) are able to convert folic to folinic acid. However, dihydrofolate was readily converted to folinic in the presence of NADPH. The synthesis of folinate was inhibited by pyrimethamine and suggested the presence of a plasmodial dihydrofolate reductase. Evidence for the presence of this enzyme in plasmodia and confirmation of its role as the site of pyrimethamine inhibition was provided by FERONE *et al.* (1969). The enzyme is clearly distinguished from reductases of bacterial and mammalian sources by its high molecular weight (MW = 190,000 ± 10%). Unlike bacterial enzymes, the reductase is stimulated by high concentrations of KCl and urea. Its Michaelis constant (K_m) for NADPH is 1.3 µM. The most striking feature of the enzyme is its inhibition by pyrimethamine. The reductase is inhibited 50% by approximately 5×10^{-10} M pyrimethamine. The inhibition appears stoichiometric when enzyme and drug are preincubated and shows competitive kinetics in the absence of preincubation. In view of the fact that pyrimethamine binds to the reductase even more tightly than does methotrexate and shows a 1,000-fold greater binding to the parasite over the host enzyme, it is concluded that the plasmodial reductase is the locus of action of pyrimethamine.

A summary of the ability of pyrimethamine to bind to reductases from various species is shown in Table 6. Bacteria and fungi show only moderate sensitivity to the drug. Mammalian liver enzymes are somewhat more sensitive. Guinea pig is the most sensitive to pyrimethamine and the enzyme also shows increased affinity to methotrexate as compared to other reductases in this group. By comparison, the affinity of pyrimethamine for the *P. berghei* reductase is extremely high.

Table 6. *Binding of pyrimethamine to dihydrofolate reductase of various species*

Enzyme source	50% Inhibitory concentration ($\times 10^{-8}$ M)
E. coli	250
S. aureus	300
P. vulgaris	150
Sacch. cerevisiae	1,100
Rat liver	70
Guinea pig liver	3
Rabbit liver	50
Human liver	180
P. berghei	0.05

It is relatively easy to develop pyrimethamine-resistant strains of plasmodia in the laboratory (BISHOP and BIRKETT, 1947; WILLIAMSON et al., 1947). ROLLO (1952 b) was able to increase levels of resistance rather rapidly when large numbers of organisms were passed and the mechanism of this resistance is not clear. In bacteria, resistance to diaminopyrimidines can be due to a variety of causes (BURCHALL, unpublished observations), and the type of resistant organism isolated depends largely on the selection procedure used. The most common mechanisms of resistance to diaminopyrimidines include alteration in the structure and/or quantity of dihydrofolate reductase produced, loss of thymidylate synthetase, and inability of the drug to penetrate the cell. In the best defined case to date, FERONE (1970) has demonstrated marked differences in dihydrofolate reductases from sensitive and resistant strains of *P. berghei*. The K_m value for dihydrofolate is 12.4-fold higher for the enzyme from the resistant strain. The reductase from the resistant strain (Pb/WLTM/50–63) is only slightly stimulated by KCl, whereas a threefold stimulation is observed with the wild strain. The inhibition constants (K_i) for pyrimethamine and two other antifolates are higher for the enzyme from the resistant strain. It was proposed that strain Pb/WLTM/50–63 is resistant owing to an increase in enzyme content and changes in the structure of the enzyme protein that result in decreased binding.

DIGGINS et al. (1970) also reported on the properties of a dihydrofolate reductase from a pyrimethamine-resistant strain of *P. berghei*. The enzyme showed an altered Michaelis constant (K_m) for dihydrofolate, an increased maximum velocity (V_{max}) and a thirtyfold decrease in affinity for pyrimethamine.

Increases in resistance to pyrimethamine in the field were observed early (CLYDE and SHUTE, 1954). The resistant strains isolated were frequently found to be cross-resistant to chloroguanide (SCHMIDT and GENTHER, 1953; THURSTON, 1954; HERNANDEZ et al., 1953; JONES, 1953). This trend toward increasing resistance has continued and is discussed at length in the following reviews and articles [PETERS, 1967; PETERS, 1969; THOMPSON, 1967; WHO Tech. Report Series 324 (1966); 357 (1967); WHO Tech. Reports 226 (1961); 375 (1967)]. Attempts to overcome the resistance problem by combination of pyrimethamine with a sulfonamide are discussed below.

Pyrimethamine has relatively long half-lifetime in the human (greater than 24 hours) (Smith and Ihrig, 1957). No major variations in individual tissue levels of the drug have been noted (Lorz et al., 1951). Hitchings (1960) has summarized most of the information on the toxicology of pyrimethamine. Massive doses produce emesis and acute convulsions (Goodwin, 1952). Humans have shown the same response to overdoses (Gunther, 1954). In general, pyrimethamine, when administered in sufficient doses, causes effects that resemble those of dietary folate deficiency or deficiency owing to antifolates such as aminopterin (Hamilton et al., 1954). As with other antifolates, the toxicity of the compound may be reversed by administration of leucovorin (Frenkel and Hitchings, 1957).

The advantages of combining pyrimethamine with a sulfonamide for the treatment of protozoal diseases are twofold. 1. In folate synthesizing organisms of this type, the simultaneous administration of both drugs causes a double blockade in the folate pathway. 2. The existence of two loci of inhibition requires that a potentially resistant organism be able to overcome inhibition at both sites. Regarding the first point, Hurly (1959) has clearly demonstrated that the activity of pyrimethamine is potentiated about tenfold in the presence of sulfonamide. A combination of pyrimethamine and sulfadiazine appears to be the drug of choice in the treatment of toxoplasmosis (Eyles, 1953). It is likely that the introduction of a combined drug regimen is beneficial in an area where no resistance to either drug has appeared. If resistance to either drug already exists, the outcome of combined therapy is less predictable. Much depends on the physiology of the individual strain. It is possible that resistance to sulfonamides can occur even though overall synthesis of folates is decreased somewhat by the drug. While no therapeutic value could be derived from this situation when only sulfonamide or pyrimethamine were used alone, the subtherapeutic effect of a sulfonamide could substantially increase the activity of pyrimethamine used concurrently.

References

Adams, J. M., and M. R. Capecchi; N-formylmethionyl-sRNA as the initiator of protein synthesis. Proc. Natl. Acad. Sci. U.S. 55, 147–155 (1966).

Baikie, A. G., C. B. Macdonald, and G. R. Mundy: Systemic nocardiosis treated with trimethoprim and sulfamethoxazole. Lancet 1970-II, 261.

Baker, B. R.: Design of active-site-directed irreversible enzyme inhibitors. New York: Wiley, 1967.

Baker, B. R., and Beng-Thong Ho: Differential inhibition of dihydrofolic reductase from different species. J. Pharm. Soc. 53, 1137–1138 (1964).

Bertino, J. B., and K. A. Stacey: A suggested mechanism for the selective procedure for isolating thymine-requiring mutants of Escherichia coli. Biochem. J. 101, 32C–33C (1966).

Bishop, A., and B. Birkett: Acquired resistance to paludrine in Plasmodium gallinaceum. Nature 159, 884–885 (1947).

Black, W. A., and D. A. McNellis: Sensitivity of Nocardia to trimethoprim and sulfonamides "in vitro". J. Clin. Pathol. 23, 423–426 (1970).

Blakeley, R. L.: The biochemistry of folic acid and related pteridines. New York: John Wiley & Sons, Inc. 1969.

Böhni, E.: Chemotherapeutic activity of the combination of trimethoprim and sulfamethoxazole in infections of mice. Postgrad. Med. J. 45 (Suppl.), 18–21 (1969).

Burchall, J. J., and G. H. Hitchings: Inhibitor binding analysis of dihydrofolate reductases from various species. Mol. Pharmacol. 1, 126–136 (1965).

BUSHBY, S. R. M.: Combined antibacterial action *in vitro* of trimethoprim and sulphonamides. Postgrad. Med. J. **45** (Suppl.), 10–18 (1969).

BUSHBY, S. R. M., and G. H. HITCHINGS: Trimethoprim, a sulphonamide potentiator. Brit. J. Pharmacol. **33**, 72–90 (1968).

CAPECCHI, M. R.: Initiation of *E. coli* proteins. Proc. Natl. Acad. Sci. U.S. **55**, 1517–1524 (1966).

CLYDE, D. F., and G. T. SHUTE: Resistance of East African varieties of *Plasmodium falciparum* to pyrimethamine. Trans. Roy. Soc. Trop. Med. Hyg. **48**, 495–500 (1954).

COATNEY, G. R., A. V. MYATT, T. HERNANDEZ, G. M. JEFFEREY, and W. C. COOPER: Studies in human malaria. XXXII. The protective and therapeutic effects of Pyrimethamine (Daraprim) against Chesson strain vivax malaria. Am. J. Trop. Med. Hyg. **2**, 777–787 (1953).

COVELL, G., P. G. SHUTE, and M. MARYON: Pyrimethamine (Daraprim) as a prophylactic agent against a West African strain of *P. falciparum*. Brit. Med. J. **1953**, 1081–1083.

DALE, B. A., and G. R. GREENBERG: Effect of the folic acid analogue, trimethoprim, on growth, macromolecular synthesis and incorporation of exogenous thymine in *Escherichia coli*. J. Bacteriol. **110**, 905–916 (1972).

DARRELL, J. H., L. P. GARROD, and P. M. WATERWORTH: Trimethoprim: laboratory and clinical studies. J. Clin. Pathol. **21**, 202–209 (1968).

DATTA, N., and R. W. HEDGES: Compatibility groups among fi-R factors. Nature **234**, 222–223 (1971).

DATTA, N., and R. W. HEDGES: Trimethoprim resistance conferred by W *plasmids* in enterobacteriaceae. J. Gen. Microbiol. **72**, 349 (1972).

DEV., I. K.: Personal Communication 1972.

DIGGENS, S. M., W. E. GUTTERIDGE, and P. I. TRIGG: Altered dihydrofolate reductase associated with a pyrimethamine-resistant *Plasmodium berghei* produced in a single step. Nature **228**, 579–580 (1970).

EISENSTADT, J., and P. LENGYEL: Formylmethionyl-t-RNA dependence of amino acid incorporation in extracts of trimethoprim-treated *Escherichia coli*. Science **154**, 524–527 (1966).

ELION, G. B., S. SINGER, and G. H. HITCHINGS: Antagonists of nucleic acid derivatives. VIII. Synergism in combinations of biochemically related antimetabolites. J. Biol. Chem. **208**, 477–488 (1954).

ERICKSON, J. S., and C. K. MATHEWS: T₄ bacteriophage-specific dihydrofolate reductase: Purification to homogeneity by affinity chromatography. Biochem. Biophys. Res. Commun. **43**, 1164–1170 (1972).

EYLES, D. E.: The present status of the chemotherapy of toxoplasmosis. Am. J. Trop. Med. Hyg. **2**, 429–444 (1953).

FALCO, E. A., L. G. GOODWIN, G. H. HITCHINGS, I. M. ROLLO, and P. B. RUSSELL: 2,4-diaminopyrimidines—a new series of antimalarials. Brit. J. Pharmacol. **6**, 185–200 (1951).

FALCO, E. A., G. H. HITCHINGS, P. B. RUSSELL, and H. VANDERWERFF: Antimalarials as antagonists of purines and pteroylglutamic acid. Nature **164**, 107–108 (1949).

FERONE, R.: Dihydrofolate reductase from pyrimethamine-resistant *Plasmodium berghei*. J. Biol. Chem. **245**, 850–854 (1970).

FERONE, R., J. J. BURCHALL, and G. H. HITCHINGS: *Plasmodium berghei* dihydrofolate reductase. Isolation, properties, and inhibition by antifolates. Mol. Pharmacol. **5**, 49–59 (1969).

FERONE, R., S. R. M. BUSHBY, J. J. BURCHALL, W. B. MOORE, and D. SMITH: Identification and purification of the antifolate potentiating substance in lysed horse blood (Harper-Cawston factor) (1972). (In Preparation.)

FERONE, R., and G. H. HITCHINGS: Folate cofactor biosynthesis by *Plasmodium berghei*. Comparison of folate and dihydrofolate as substrates. J. Protozool. **13**, 504–506 (1966).

FERREIRA LOPES, C., and S. ARMOND: Therapeutic trial in sulfonamide-resistant cases of South American blastomycosis. Hospital (Rio de Janeiro) **73**, 1245–1255 (1968).

FLEMING, M. P., N. DATTA, and R. N. GRÜNEBERG: Trimethoprim resistance determined by R factors. Brit. Med. J. **1972**, 726–728.

FRENKEL, J. K., and G. H. HITCHINGS: Relative reversal by vitamins (p-aminobenzoic, folic and folinic acids) of the effects of sulfadiazine and pyrimethamine on toxoplasma, mouse and man. Antibiot. & Chemoherapy **7**, 630–638 (1957).

GARRETT, E. R.: Classification and evaluation of combined antibiotic activity. Antibiot. & Chemotherapy **8**, 8–20 (1958).

GARROD, L. P.: Trimethoprim-Sulfamethoxazole. Drugs 1, 7–53 (1971).

GLAZ, E. T.: Partial purification of the Harper-Cawston/horse red blood/factor enhancing sulfonamide activity. Proc. Fifth Int. Cong. Chemotherapy 1, 211–214 (1967).

GLENN, S., and R. D. MANWELL: Further studies on the cultivation of the avian malaria parasites: II. The effects of heterologous sera and added metabolites on growth and reproduction *in vitro*. Exptl. Parasitol. 5, 22–33 (1956).

GOODWIN, L. G.: Response of *Plasmodium berghei* to antimalarial drugs. Nature 164, 1133 (1949).

GOODWIN, L. G.: Daraprim—clinical trials and pharmacology. Trans. Roy. Soc. Trop. Med. Hyg. 46, 485–495 (1952).

GUNDERSEN, L. E., R. B. DUNLAP, N. G. L. HARDING, J. H. FREISHEIM, F. OTTING, and F. M. HUENNE- KENS: Dihydrofolate reductase from amethopterin-resistant *Lactobacillus casei*. Biochemistry 11, 1018–1023 (1972).

GUNTHER, C. E.: A fatal overdose of "Daraprim". Med. J. Aust. 2, 970 (1954).

HAMILTON, L., F. S. PHILIPS, S. S. STERNBERG, D. A. CLARKE, and G. H. HITCHINGS: Hematological effects of certain 2,4-diaminopyrimidines, antagonists of folic acid metabolism. Blood 9, 1062–1081 (1954).

HARPER, G. J., and W. C. CAWSTON: *In vitro* determination of sulfonamide sensitivity of bacteria. J. Pathol. Bacteriol. 57, 59–66 (1945).

HARVEY, R. J.: Growth and initiation of protein synthesis in *Escherichia coli* in the presence of trimethoprim J. Bacteriol. 114, 309–322 (1973).

HARVEY, R. J.: Multiple inhibition in complex metabolic pathways (1972 b). Submitted to Molec. Pharmacol.

HERNANDEZ, T., A. V. MYATT, G. R. COATNEY, and G. M. JEFFEREY: Studies in human malaria. XXXIV. Acquired resistance to pyrimethamine (Daraprim) by the Chesson strain of *Plasmodium vivax*. Am. J. Trop. Med. Hyg. 2, 797–804 (1953).

HILL, J.: Chemotherapy of malaria: Part 2. The antimalarial drugs. In: Experimental chemotherapy (R. J. SCHNITZER and F. HAWKING, eds.), vol. 1, p. 513–601. New York: Academic Press 1963.

HITCHINGS, G. H.: Daraprim as an antagonist of folic and folinic acid. Trans. Roy. Soc. Trop. Med. Hyg. 46, 467–473 (1952 c).

HITCHINGS, G. H.: Pyrimethamine; the use of an antimetabolite in the chemotherapy of malaria and other infections. Clin. Pharmacol. Therap. 1, 570–589 (1960).

HITCHINGS, G. H.: Specific differences among dihydrofolate reductases as a basis for chemotherapy. Postgrad. Med. J. 45, Suppl. 7–10 (1969).

HITCHINGS, G. H., and J. J. BURCHALL: Inhibition of folate biosynthesis and function as a basis for chemotherapy. Advan. Enzymol. 27, 417–468 (1965).

HITCHINGS, G. H., and S. R. M. BUSHBY: 5-Benzyl-2,4-diaminopyrimidines, a new class of systemic antibacterial agents. Fifth Int. Congress Biochem., 165, Moscow (1961).

HITCHINGS, G. H., G. B. ELION, E. A. FALCO, P. B. RUSSELL, M. B. SHERWOOD, and H. VANDERWERFF: Antagonists of nucleic acid derivatives. I. The *Lactobacillus casei* model. J. Biol. Chem. 183, 1–9 (1950 a).

HITCHINGS, G. H., G. B. ELION, E. A. FALCO, P. B. RUSSELL, and H. VANDERWERFF: Studies on analogs of purines and pyrimidines. Ann. N. Y. Acad. Sci. 52, 1318–1335 (1950 b).

HITCHINGS, G. H., G. B. ELION, H. VANDERWERFF, and E. A. FALCO: Pyrimidine derivatives as antago- nists of pteroylglutamic acid. J. Biol. Chem. 174, 765–766 (1948).

HITCHINGS, G. H., E. A. FALCO, G. B. ELION, S. SINGER, G. B. WARING, D. I. HUTCHISON, and J. H. BURCHENAL: 2,4-Diaminopyrimidines as antagonists of folic acid and folinic acid. Arch. Biochem. Biophys. 40, 479–481 (1952 a).

HITCHINGS, G. H., E. A. FALCO, H. VANDERWERFF, P. B. RUSSELL, and G. B. ELION: Antagonists of nucleic acid derivatives. VII. 2,4-diaminopyrimidines. J. Biol. Chem. 199, 43–56 (1952 b).

HURLY, M. G. D.: Potentiation of pyrimethamine by sulfadiazine in human malaria. Trans. Roy. Soc. Trop. Med. Hyg. 53, 412–413 (1959).

JACOBS, R. L.: Role of p-aminobenzoic acid in *Plasmodium berghei* infection in the mouse. Exptl. Parasitol. 15, 213–225 (1964).

JONES, S. A.: Experiment to determine if a Proguanil-resistant strain of *P. falciparum* would respond to large doses of pyrimethamine. Brit. Med. J. 1953, 977.

KOCH, A. E., and J. J. BURCHALL: Reversal of the antimicrobial activity of trimethoprim by thymidine in commercially prepared media. Appl. Microbiol. 22, 812–817 (1971).

LABEK, G., u. E. WIEDMEN: Empfindlichkeit menschlicher Krankheitserreger gegen das Kombinations-Therapeutikum Sulfamethoxazol-Trimethoprim *in vitro*. Schweiz. Med. Wochschr. **101**, 1385 (1971).

LACEY, R. W., W. A. GILLESPIE, D. M. BRUTEN, and E. L. LEWIS: Trimethoprim-resistant coliforms. Lancet **1972-I**, 409–410.

LORZ, D. C., P. L. GRAHAM, and G. H. HITCHINGS: Distribution of 2,4-diaminopyrimidine antimalarial in the tissues of the dog. Federation Proc. **10**, 320 (1951).

MACLEOD, W. M.: The treatment of Nocardiosis and Histoplasmosis. Lancet **1970-II**, 363.

MAGGIOLO, A., and P. B. RUSSELL: The ultra-violet absorption spectra of phenylpyrimidines. J. Chem. Soc. **1951**, 3297–3300.

MARTIN, D. C., and J. D. ARNOLD: Trimethoprim in therapy of acute attacks of malaria. J. Clin. Pharmacol. **7**, 336–341 (1967).

MARTIN, D. C., and J. D. ARNOLD: Treatment of acute falciparum malaria with sulfalene and trimethoprim. J. Am. Med. Assoc. **203**, 476–480 (1968).

MARTIN, D. C., and J. D. ARNOLD: Trimethoprim and sulfalene therapy of *Plasmodium vivax*. J. Clin. Pharmacol. **9**, 155–159 (1969).

MAY, J. R., and DAVIES, J.: Resistance of *Haemophilus influenzae* to trimethoprim. Brit. Med. J. **1972**, 376–377.

MOSSNER, G.: Clinical results with the combined preparation sulfamethoxazole and trimethoprim. Proc. Sixth Int. Congress of Chemotherapy **1**, 966–970 (1969).

PETERS, W.: A review of recent studies on chemotherapy and drug resistance in malaria parasites of birds and animals. Trop. Diseases Bull. **64**, 1145–1175 (1967).

PETERS, W.: Drug resistance in malaria—a perspective. Trans. Roy. Soc. Trop. Med. Hyg. **63**, 25–45 (1969).

PINE, M. J., B. GORDON, and S. S. SARIMO: Protein initiation without folate in *Streptococcus faecium*. Biochim. Biophys. Acta **179**, 439–447 (1969).

POE, M., N. J. GREENFIELD, J. M. HIRSHFIELD, M. N. WILLIAMS, and K. HOOGSTEEN: Dihydrofolate reductase. Purification and characterization of the enzyme from an Amethopterin-resistant mutant of *Escherichia coli*. Biochemistry **11**, 1023–1030 (1972).

ROLLO, I. M.: Daraprim: experimental chemotherapy. Trans. Roy. Soc. Trop. Med. Hyg. **46**, 474–484 (1952a).

ROLLO, I. M.: "Daraprim" resistance in experimental malarial infections. Nature **170**, 415 (1952b).

ROLLO, I. M.: The mode of action of sulfonamides, proguanil and pyrimethamine of *Plasmodium gallinaceum*. Brit. J. Pharmacol. **10**, 208–214 (1955).

ROTH, B., E. A. FALCO, and G. H. HITCHINGS: 5-Benzyl-2,4-diaminopyrimidines as antibacterial agents. I. Synthesis and antibacterial activity *in vitro*. J. Med. Pharm.Chem. **5**, 1103–1123 (1962).

RUSSELL, P. B.: The effects of substitution on the ultra-violet absorption spectra of phenylpyrimidines. J. Chem. Soc. **1951**, 2951–2955.

RUSSELL, P. B., and G. H. HITCHINGS: 2,4-Diaminopyrimidines as antimalarials. III. 5-Aryl derivatives. J. Am. Chem. Soc. **73**, 3763–3770 (1951).

SAMUEL, C. E., L. D'ARI, and J. C. RABINOWITZ: Evidence against the folate-mediated formylation of formyl-accepting methionyl transfer ribonucleic acid in *Streptococcus faecalis* R. J. Biol. Chem. **245**, 5115–5121 (1970).

SCHMIDT, L. H., and C. S. GENTHER: The antimalarial properties of 2,4-diamino-5-p-chlorophenyl-6-ethylpyrimidine (Daraprim). J. Pharmacol. Exptl. Therap. **107**, 61–91 (1953).

SCHWARTZ, D. E., and W. H. ZIEGLER: Assay and pharmacokinetics of trimethoprim in man and animals. Postgrad. Med. J. **45**, Suppl. 32–37 (1969).

SEYDEL, J. K., E. WEMPE, G. H. MILLER, and L. MILLER: Kinetics and mechanisms of action of trimethoprim and sulfonamides, alone or in combination, upon *Escherichia coli*. Chemotherapy **17**, 217–258 (1972).

SHARPSTONE, P.: The renal handling of trimethoprim and sulphamethoxazole in man. Postgrad. Med. J. **45**, Suppl. 38–42 (1969).

SMITH, C. C., and J. IHRIG: The pharmacological basis for the prolonged antimalarial activity of pyrimethamine. Am. J. Trop. Med. Hyg. **6**, 50–57 (1957).

TELLER, J. H., and E. E. SNELL: Tetrahydrofolate-dependent formation of ketopantoate. Am. Chem. Soc. 158th Nat. Meeting Abst. (Biol.) 40 (1969).

THOMPSON, P. E.: Parasite chemotherapy. Ann. Rev. Pharmacol. **7**, 77–100 (1967).

Thurston, J.P.: The chemotherapy of *Plasmodium berghei*. II. Antagonism of the action of drugs. Parasitology **44**, 99–110 (1954).

Trager, W.: Folinic acid and non-dialyzable materials in the nutrition of malaria parasites. J. Exptl. Med. **108**, 753–771 (1958).

Udall, V.: Toxicology of sulphonamide-trimethoprim combinations. Postgrad. Med. J. **45**, Suppl. 42–45 (1969).

Walker, N., R. Philip, M.M. Smyth, and J. W. McLeod: Observations on the prevention of bacterial growth by sulfonamides with special reference to the Harper-Cawston effect. J. Pathol. Bacteriól. **59**, 631–645 (1947).

Waterworth, P.M.: Practical aspects of testing sensitivity to trimethoprim and sulfonamide. Postgrad. Med. J. **45**, Suppl. 21–27 (1969).

Webb, J.L.: Enzyme and metabolic inhibitors, vol. 1. New York: Academic Press 1963.

Whitman, E.N.: Effects in man of prolonged administration of trimethoprim and sulfisoxazole. Postgrad. Med. J. **45**, Suppl. 46–51 (1969).

Williamson, J., D.S. Bertram, and E.M. Lourie.: Effects of Paludrine and other antimalarials. Nature **159**, 885–886 (1947).

World Health Organization: Tech. Reports on Chemotherapy of Malaria No. 226 (1961); No. 375 (1967).

World Health Organization: Tech. Rep. Ser. 324 (1966); 357 (1967); 382 (1968).

II. Interference with Protein Biosynthesis

Althiomycin

Sidney Pestka

The isolation of althiomycin was described by YAMAGUCHI *et al.*, (1957) from a strain of *Streptomyces*. The antibiotic was found to inhibit the growth of Gram-positive and Gram-negative bacteria (Table 1). An apparently similar antibiotic was isolated independently at the Upjohn Company from culture no. 116a of an unspecified mold (EBLE and WHITFIELD as quoted in CRAM *et al.*, 1963). Isolation of the antibiotic matamycin from *Streptomyces matensis* was described by MARGALITH *et al.*, 1959. These three preparations have been considered by UMEZAWA (1967) to be probably independent isolates of the same antibiotic. However, the slight differences in antibacterial spectrum of althiomycin and matamycin (Table 1) suggests that the antibiotics although similar may not be identical. Alternatively, differences in sensitivity of *E. coli* to althiomycin (A) and matamycin (B) may possibly reflect differences in the *E. coli* strains (Table 1). The antibiotic appears to have a low toxicity, for mice tolerated intraperitoneal injections of 720 mg/kg.

Althiomycin is crystallized as white needles. It is soluble in ethyl cellosolve, dioxane, and pyridine; slightly soluble in acetone, methanol, butanol, and ethyl acetate; and relatively insoluble in water, ether, and benzene (YAMAGUCHI *et al.*, 1957; SENSI *et al.*, 1959). Its solubility in water is about 0.4 to 0.5 mg/ml. In 0.3 N NaOH, its ultraviolet spectrum shows maxima at 235 mμ ($E_{1\,cm}^{1\%}=611$) and at 300–305 mμ ($E_{1\,cm}^{1\%}=317$). In 0.03 N HCl, a maximum at 222 mμ ($E_{1\,cm}^{1\%}=810$) is seen (YAMAGUCHI *et al.*, 1957). Elementary analysis was consistent with the formula $C_{27}H_{28}O_{10}N_8S_3$ (CRAM *et al.*, 1963). Its molecular weight is

Table 1

Test organism	Minimal inhibitory concentration (μg/ml)	
	a	b
D. pneumoniae Type III	2.5	
M. pyogenes var aureus	6	5
B. anthracis	0.7	
B. subtilis	6	5
E. coli	0.7	50
Sh. dysenteriae	3	
S. typhosa	0.7	
Pr. vulgaris	>30	100
Ps. aeruginosa	>30	>100
S. cerevisiae	>30	

[a] Data from YAMAGUCHI *et al.* (1957).
[b] Data from MARGALITH *et al.* (1959).

about 708 (Cram et al., 1963). Slight differences in elementary composition obtained by the three groups may also be suggestive of differences in primary structure. Some differences in the published infrared spectra of matamycin (Sensi et al., 1959) and althiomycin (Yamaguchi et al., 1957) are also apparent as well as small differences in the ultraviolet absorption spectrum. The presence of about 14% sulfur and a thiazole ring suggests some resemblance to the micrococcins and the thiostrepton group of antibiotics (see Chapters Micrococcin and Micrococcin P and The Thiostrepton group of Antibiotics in this volume).

The antibiotic is not stable at pH values lower than 5 or higher than 7 in aqueous media. It is stable in organic solvents. Very little is known of the structure of the antibiotic. Cram et al. (1963 and 1964) evaluated some structural features of the antibiotic. On treatment with acetic anhydride and pyridine, acetyl-althiomycin is formed. Its structure is shown in Fig. 1. Although the three isolates of althiomycin are considered by Umezawa (1967) probably to represent isolates of an identical antibiotic, there appear to be differences between the isolates. Yamaguchi et al. (1957) report the antibiotic yields a positive ninhydrin, whereas Cram et al. (1963) report a negative reaction with ninhydrin. Whereas Escherichia coli are sensitive to 0.7 µg/ml of althiomycin (Yamaguchi et al., 1957) and 116a (Whitfield, personal communication), E. coli is not as sensitive to matamycin. Conceivably, there may be subtle differences between the antibiotic isolates. Definitive structural characterization of the various isolates should provide the answer.

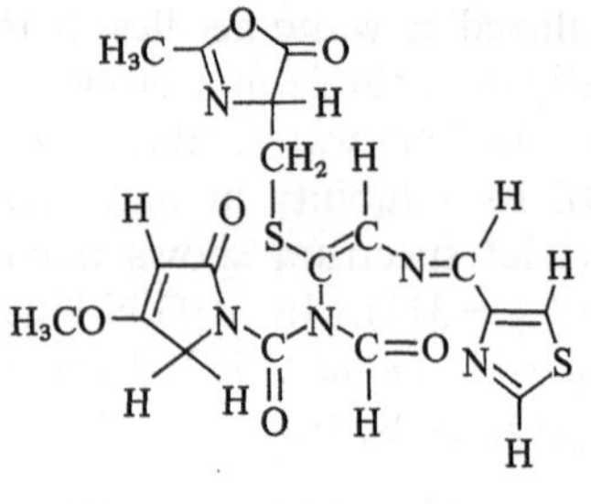

Fig. 1. Acetylalthiomycin

Mode of Action

Fujimoto et al. (1970) reported that althiomycin exhibited a strong inhibition of protein synthesis in intact cells of Escherichia coli. DNA synthesis was slightly inhibited and RNA synthesis was stimulated. Althiomycin inhibited incorporation of amino acids into protein in a cell-free system using endogenous RNA templates. It also inhibited poly U and poly C dependent incorporation of phenylalanine and proline, respectively; but lysine incorporation dependent on poly A was remarkably resistant to inhibition by the antibiotic. Protein synthesis in intact reticulocytes and reticulocyte lysates was resistant to inhibition by althiomycin.

Althiomycin inhibited N-acetyl-phenylalanyl-puromycin synthesis (Fujimoto et al., 1970). A 57% inhibition was observed at 1.4×10^{-5} M althiomycin. It should be noted that high concentrations of K^+ (0.4 M or greater) or of Mg^{++}

(0.04 M and greater) seem to prevent the inhibitory effect of althiomycin on N-acetyl-phenylalanyl-puromycin synthesis (PESTKA, unpublished observations). Thus, in the presence of 0.4 M K$^+$ and 0.04 M Mg^{++} PESTKA and BROT (1971) did not observe a significant inhibition of N-acetyl-phenylalanyl-puromycin formation by althiomycin.

Althiomycin also inhibited peptide bond synthesis on polyribosomes as measured by peptidyl-[^{3}H]puromycin synthesis (PESTKA, 1972). It can be seen that 10^{-5} M althiomycin produced a 65% inhibition of peptidyl-puromycin synthesis (Table 2). Althiomycin inhibited peptidyl-puromycin synthesis non-competitively with respect to puromycin as a substrate. The interaction coefficient for inhibition of peptidyl-puromycin synthesis was greater than one, indicating that several althiomycin molecules were involved in the inhibitory process.

Table 2. *Effect of althiomycin on peptidyl-[^{3}H]puromycin synthesis*

Althiomycin concentration	Percentage of control
0	100
10^{-6} M	104
3×10^{-6} M	109
10^{-5} M	35
3×10^{-5} M	20

Data taken from PESTKA (1972).

That the ribosome was the target of the inhibition of protein synthesis is indicated by several experiments. The inhibition of N-acetyl-phenylalanyl-puromycin synthesis, a ribosome function, was indicative of a ribosomal focus of action. Also, the antibiotic inhibited oligophenylalanine synthesis in the presence or absence of elongation factor G and GTP (PESTKA and BROT, 1971). This inhibition of both enzymic and non-enzymic translocation (PESTKA, 1973) further implicated the ribosome as its primary site of action. Furthermore, since althiomycin inhibited N-acetyl-phenylalanyl-puromycin synthesis with isolated 50 S subunits, it appears that the antibiotic can be classified as a 50 S inhibitor (PESTKA, unpublished observations).

Other intermediate reactions of protein synthesis which were tested were not inhibited by the antibiotic. Thus, aminoacyl-tRNA synthesis and Phe-tRNA binding to ribosomes were not significantly inhibited by the antibiotic (FUJIMOTO et al., 1970; PESTKA and BROT, 1971). The antibiotic did not inhibit C-A-C-C-A (Phe) binding to ribosomes, but it did inhibit transfer of nascent peptides from peptidyl-tRNA to C-A-C-C-A (Phe) (PESTKA and BROT, 1971). It did not inhibit elongation factor G and ribosome dependent hydrolysis of GTP. Although it inhibited peptidyl-puromycin synthesis, it did not stabilize polyribosomes as well as thiostrepton, siomycin A or micrococcin (PESTKA and HINTIKKA, 1971).

Since althiomycin inhibits the ribosomal function of the peptidyl transferase, it may be useful in obtaining new types of mutants of ribosomal proteins. So far, however, genetic studies with the antibiotic have not been reported nor have mutants resistant to the antibiotic been described.

Summary

Althiomycin is a sulfur containing antibiotic which is inhibitory to both Gram-positive and Gram-negative bacteria. It inhibits protein synthesis in bacteria, but not in mammalian cells. Its specific site of inhibition is the 50S subunit of the ribosome, where the antibiotic blocks transpeptidation.

References

CRAM, D.J., O. THEANDER, H. JAGER, and M.K. STANFIELD: Mold metabolites. IX. Contribution to the elucidation of the structure of althiomycin. J. Am. Chem. Soc. **85**, 1430–1437 (1963).

CRAM, D.J., O. THEANDER, H. JAGER, and M.K. STANFIELD: Beiträge zur Strukturaufklärung des Althiomycins. Angew. Chem. **76**, 793 (1964).

FUJIMOTO, H., T. KINOSHITA, H. SUZUKI, and H. UMEZAWA: Studies on the mode of action of althiomycin. J. Antibiotics (Tokyo) **23**, 271–275 (1970).

MARGALITH, P., G. BERETTA, and M.T. TIMBAL: Matamycin, a new antibiotic. I. Biological studies. Antibiot. & Chemotherapy **9**, 71–75 (1959).

PESTKA, S.: Studies on transfer ribonucleic acid-ribosome complexes. XIX. Effect of antibiotics on peptidyl-puromycin synthesis on polyribosomes from *Escherichia coli.* J. Biol. Chem. **247**, 4669–4678 (1972).

PESTKA, S.: Methods in Enzymology **27**, in press (1973).

PESTKA, S., and N. BROT: Studies on the formation of transfer ribonucleic acid-ribosome complexes. XV. Effect of antibiotics on steps of bacterial protein synthesis: some new ribosomal inhibitors of translocation. J. Biol. Chem. **246**, 7715–7722 (1971).

PESTKA, S., and H. HINTIKKA: Studies on the formation of ribonucleic acid-ribosome complexes. XVI. Effect of ribosomal translocation inhibitors on polyribosomes. J. Biol. Chem. **247**, 7723–7730 (1971).

SENSI, P., R. BALLOTTA, and G.G. GALLO: Matamycin, a new antibiotic. II. Isolation and characterisation. Antibiot. & Chemotherapy **9**, 76–80 (1959).

UMEZAWA, H. (Ed.): Index of antibiotics from actinomycetes: University Park Press, State College Pa. 1967.

YAMAGUCHI, H., Y. NAKAYAMA, K. TAKEDA, K. TAWARA, K. MAEDA, T. TAKEUCHI, and H. UMEZAWA: A new antibiotic, althiomycin. J. Antibiotics (Tokyo) **10** A, 195–200 (1957).

Aurintricarboxylic Acid
a Non-Antibiotic Organic Molecule
that Inhibits Protein Synthesis

David Apirion and Dennis Dohner

Introduction

One of the goals of the scientist who works with antibiotics is to evolve tailor-made molecules to combat infectious diseases. Thus far, this goal has not been advanced much beyond the stage of modifying some chemical groups in naturally occurring antibiotic molecules (COGHILL et al., 1949). Another avenue that might lead to the same goal could be the screening of known chemicals for antimicrobial action.

By screening chemicals for potential use as antimicrobial or antiviral agents, GROLLMAN and his colleagues found that aurintricarboxylic acid (ATA, Fig. 1), a triphenylmethane dye, could fit the requirements of such a molecule. (GROLLMAN and STEWART, 1968; GROLLMAN, 1968).

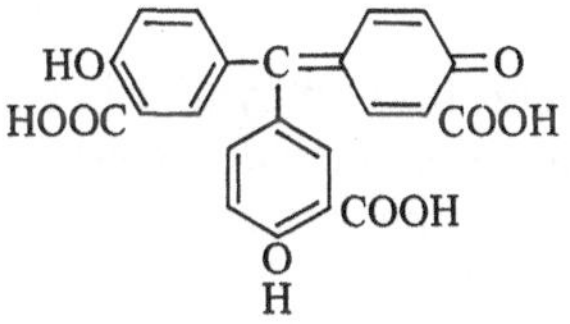

Fig. 1. Aurintricarboxylic Acid (ATA)

They showed that this chemical, ATA, inhibits protein synthesis in various cell-free systems, and there was a number of indications that this inhibition is specific for the binding of the mRNA to the ribosome. Since none of the familiar antibiotics is known to interfere with this step in protein synthesis, these experiments raised the possibility, as GROLLMAN (1968) pointed out, that ATA might become an anti-viral agent. This requires that the drug would differentially inhibit the attachment of a viral mRNA to the ribosome as opposed to the attachment of the cellular mesengers to ribosomes.

At that time there were some suggestions that interferon might act by specifically blocking synthesis directed by viral messengers as compared to cellular messengers, and that this specificity resided in the messenger ribosome binding step (JOKLIK and MERIGAN, 1966; LEVY and CARTER, 1968; MARCUS and SALB, 1966). The experiments carried out by GROLLMAN (1968) showing that ATA inhibits protein synthesis directed by mRNA from a bacteriophage (f2) much more strongly than endogenous globin synthesis, which is directed by an endogenous messenger from a mammalian cell, were rather encouraging in that direction, and the expectation that a chemical like ATA could be an antiviral agent was not unreasonable.

The spectrum of antimicrobial activity of the chemical cannot be discussed, since ATA is not known to penetrate cells. Grollman (1968) tested the uptake of radioactive ATA by He La cells, green monkey kidney cells, rabbit reticulocytes and a variety of bacteria, but in no case was uptake observed. *Escherichia coli* for instance can grow in medium containing as much as 1% (0.024 M) of ATA (Siegelman and Apirion unpublished observations). In our laboratory Siegelman isolated mutants of *Escherichia coli* that became sensitive to ATA, but analysis of polyribosome metabolism did not reveal a simple pattern for this chemical action and the results did not allow any obvious conclusion leading to an understanding of the mechanism of action of this chemical *in vivo*. Later studies by Lund and Kjelgaard (1972) with an *E. coli* strain permeable to ATA revealed that like some other inhibitors ATA blocked protein synthesis and RNA accumulation, while guanosine tetraphosphate, ppGpp, accumulated. Therefore the forthcoming discussion will by necessity concentrate mainly on the effects of ATA on protein synthesis *in vitro*.

We will try to show that ATA can differentially block initiation of protein synthesis as opposed to elongation of the polypeptide chain, but that its action is not restricted to, or directed against any specific step in protein synthesis. This review covers mostly material that appeared up to the summer of 1972. It is possible that important contributions were overlooked. We wish to apologize should this have happened.

A brief review on ATA and protein synthesis was written by Pestka (1971).

The studies with ATA can be somewhat arbitrarily divided into three categories: those which deal with its mechanism of action, those that are relevant to its mechanism of action and those which take for granted that its mechanism of action is inhibition of binding of mRNA to ribosomes. Herein we deal with studies that fall mainly into the first two categories.

When Grollman (1968) initiated a search for chemicals that affect protein synthesis, he found that several triphenylmethane dyes (including ATA) were active in inhibiting protein synthesis in cell-free systems. Further analysis indicated that ATA is a novel inhibitor of translation of mRNA (Grollman and Stewart, 1968). Although at first sight the literature on ATA seems to be beset with contradictions, examination of the data reveals a rather remarkable consistency, and experiments from different laboratories tend to reinforce the idea that ATA can be used under special circumstances as a differential inhibitor of initiation of protein synthesis. This is true in all systems tested. Inhibitory effects, however, are not exclusive to the initiation steps of protein synthesis.

Many systems have been studied, including protein synthesizing systems from *E. coli,* yeast, wheat embryo, ascites cells, rabbit reticulocytes, and rat liver with varied messengers, endogenous mRNA and exogenous mRNA such as polyuridylic acid, R17, f2, MS2, tobacco mosaic virus RNA, T4 mRNA, globin mRNA, and Rauscher leukemia virus RNA. (T4 mRNA: Wilhelm and Hasel-

KORN, 1970; RAUSCHER leukemia virus, WANG et al., 1972. These references are given here since they are not mentioned again in this review.)

Early Studies with ATA

Some of the biological properties of ATA had been studied before this compound became of interest to people working on protein synthesis. LINDENBAUM et al., (1952) studied the reversal of Be^{+2} poisoning of rats by ATA. They suggested that this reversal is related to chelation of Be^{+2} by ATA. They also noted that ATA at high concentrations would inhibit rat alkaline phosphatase. SCHUBERT and LINDENBAUM (1954) found that $1177\ \mu M$ of ATA were required for 50% inhibition of plasma alkaline phosphatase. This is a high level of ATA, compared with the concentrations sufficient for the inhibition of protein synthesis (see Table 1). LINDENBAUM and SCHUBERT (1956), studied the binding of ATA and various other aromatic compounds, many of which were salicylic acid derivatives, to bovine serum albumin (BSA), and found that ATA had the highest binding constant among those compounds studied, $(k_B = 48,000)$. *ATA therefore has two interesting properties: it can chelate divalent cations, and it can bind tightly to protein.*

Table 1. *Inhibition of protein synthesis by ATA*

System	Messenger RNA	ATA concent. (μM)	% of control (no ATA)	Reference
E. coli	endogenous	10	88	GROLLMAN (1968)
E. coli	Poly (U)	10	41	WEISSBACH and BROT (1970)
E. coli	Poly (U)	10	59	TAL et al. (1972)
E. coli	Poly (U)	10	49	GROLLMAN (1968)
E. coli	f2-RNA	10	13	GROLLMAN (1968)
E. coli	R17-RNA	20	50	SIEGELMAN and APIRION (1971 b)
Rabbit:				
Reticulocytes	endogenous	10	92	GROLLMAN (1968)
Reticulocytes	endogenous	10	75	MATHEWS (1971)
Reticulocytes	Poly (U)	10	10	MATHEWS (1971)
Reticulocytes	Poly (U)	10	8	GROLLMAN (1968)
Ascites	endogenous	50	98	MATHEWS (1971)
Ascites	Poly (U)	3.6	22	ROBERTS and COLEMAN (1971)
Ascites	Poly (U)	10	28	MATHEWS (1971)
Ascites	EMC-RNA	10	86	MATHEWS (1971)

ATA in Protein Synthesis

GROLLMAN and STEWART (1968) attempted to examine model compounds that could inhibit a step unique to viral replication, such as the attachment of bacteriophage RNA to the ribosome. They found that ATA at a concentration of $50\ \mu M$ prevented binding of f2 RNA to E. coli ribosomes in a cell-free system. Using $100\ \mu M$ ATA, they found that ATA prevented the formation of the ribosome-Q_β RNA-fMet-tRNA$_f$ complex. ATA was effective only when added to the ribosomes before the RNA. They also studied f2 RNA stimulated polypeptide synthesis and found that ATA was again inhibitory at low concentration; 100% inhibition was acheived at about 10 µM ATA, and 50% inhibition at about 3.6 µM. When the Mg^{+2} concentration was increased from 7 to 10 µM, inhibition increased from 50 to 85%. This high level of inhibition was not changed when Mg^{+2} was increased from 10 to 13 mM. In other systems (see for instance Siegelman and Apirion, 1971a and b; Leader, 1972), raising the level of Mg^{+2} did not decrease the level of inhibition by ATA as would be expected if ATA were acting by chelating Mg^{+2}. Therefore this suggests that the mechanism of action of ATA is not chelating of magnesium and lowering of its effective concentration. When increased concentrations of ribosomes were used (Grollman and Stewart, 1968) the inhibitory effect of ATA was decreased. However, increasing the concentration of the messenger, f2 RNA, had no effect on inhibition by ATA. Therefore, ATA was probably exerting its effect on the ribosomes (or on other proteins which were associated with ribosomes) and was binding in a manner which was noncompetitive with the binding of the f2 RNA.

To find out which step in protein synthesis is affected by ATA, Grollman and Stewart (1968) added ATA before and after the messenger (f2 RNA) was added. When 50 µM ATA was added after the RNA, it inhibited protein synthesis less than if added before the RNA. It was also noted that endogenous protein synthesis was much less sensitive to ATA, requiring about ten times as much ATA to obtain 50% inhibition as for protein synthesis directed by exogenous RNA. Therefore, it appears that initiation is affected preferentially over elongation.

Grollman (1968) compared endogenous, f2-directed, and poly(U)-directed polypeptide synthesis using E. coli constituents, and found that the f2-directed synthesis was most sensitive, poly(U)-directed synthesis being slightly less sensitive, and endogenous synthesis being relatively insensitive to inhibition by ATA. Using reticulocyte extracts he found that poly(U)-directed polypeptide synthesis was more sensitive to inhibition by ATA than endogenous polypeptide synthesis. These results do not seem to be consistent with each other but see further discussion.

In view of these results, a number of laboratories began using ATA as a specific inhibitor of initiation of protein synthesis using both prokaryotic and eukaryotic in vitro systems. Marcus et al. (1970) studied the effects of ATA on protein synthesis in wheat embryo in cell-free systems. Their findings were similar to those reported by Grollman's laboratory. In their case, synthesis directed by tobacco mosaic virus RNA was almost completely inhibited by 25 µM ATA, while endogenous protein synthesis was inhibited only slightly (about 20%) by 50 µM ATA. They also observed that if ATA was added before protein synthesis had begun, its inhibitory effect was much stronger than if it was added later.

Stewart et al. (1971) found that ATA (70 µM) could completely inhibit protein synthesis directed by f2 RNA, but, if ATA was added after synthesis had begun, inhibition started only after a delay. This suggests that protein chains that have already been started were completed, but new initiations were prevented. Similar experiments were carried out independently by Siegelman and Apirion

(1971 b) using R17 bacteriophage RNA. They found that 70 μM ATA completely inhibited protein synthesis if added before the R17 RNA, while adding the ATA 5 minutes after synthesis started allowed the synthesis to continue for about 3 minutes at a similar rate, after which synthesis was completely inhibited.

MATHEWS (1971) showed that ATA had similar effects on protein synthesis in rabbit reticulocyte and Krebs II ascites cell-free systems; he found that poly(U)-directed synthesis was most sensitive to ATA, encephalomyocarditis viral RNA-directed synthesis was much less sensitive, and endogenous synthesis was least sensitive.

Inhibition of Various Steps in Protein Synthesis by ATA

The effect of ATA on other reactions besides initiation of protein synthesis was also tested. WIESSBACH and BROT (1970) looked at the effect of ATA on Ts (elongation factor) catalyzed ^{3}H-GDP exchange with Tu-GDP. They found that ATA inhibited this reaction. When they testet poly(U)-directed polypeptide synthesis in an *E. coli* cell-free system, purified to require addition of elongation factors Ts, Tu, and G, they found that inhibition by ATA was similar to that found by other workers.

SIEGELMAN and APIRION (1971 a, b), studied the effect of ATA on several steps in cell-free protein synthesis, using *E. coli* extracts and poly(U) or R17 RNA as messengers. As mentioned earlier, using R17 RNA-directed polypeptide synthesis, they found that the addition of ATA after synthesis had started gave inhibition after a delay, suggesting that initiation was somehow more sensitive to ATA than elongation. They found that poly(U)-directed protein synthesis was inhibited by ATA at higher concentrations than R17 RNA-directed protein synthesis. Poly(U) and R17 RNA binding to the 30S ribosomal subunits was inhibited by ATA. G-factor specific ribosome-dependent GTPase was inhibited

Table 2. *Inhibition of binding of mRNA to ribosomes by ATA*

System	Messenger RNA	Ribosome concent. (μM)	ATA concent. (μM)	% of control (no ATA)	Reference
E. coli	Poly (U)	0.25 (70S)	30	53	TAL et al. (1972)
E. coli	Poly (U)	0.33 S1 protein	30	58	TAL et al. (1972)
E. coli	R17-RNA	1.2 (70S)	100	64	SIEGELMAN and APIRION (1971 b)
Rabbit reticulocytes	Globin mRNA (ribo-nucleo-protein)	?	200	0	LEBLEU et al. (1970)

Ribosome concentrations were calculated from the following assumptions:
1. $^1/_3$ of M. W. of ribosome is protein, $^2/_3$ is RNA; 2. I A_{260} unit of ribosomes is 60 μg; 3. *E. coli* ribosomes: 50S MW 1.5×10^6, 30S MW 0.9×10^6, eukaryotic ribosomes: 60S MW 2.4×10^6, 40S MW 1.0×10^6.

Table 3. *Inhibition of binding of tRNA to ribosomes by ATA*

System	tRNA and mRNA	Ribosome concent. (µM)		ATA concent. (µM)	% of control (No ATA)	Reference
E. coli	fmet-tRNA Poly (A, U, G)	1.8	70S	200	76	Siegelman and Apirion (1971 b)
Rat-liver	phe-tRNA Poly (U)	0.036	40S	1	22	Leader (1972)
Rat-liver	phe-tRNA Poly (U)	0.036	40S	10	4	Leader (1972)
Rat-liver	phe-tRNA Poly (U)	0.036	40S+60S	1	17	Leader (1972)
Rat-liver	phe-tRNA Poly (U)	0.036	40S+60S	10	4	Leader (1972)

at relatively high concentrations of ATA (50% inhibition at about 100 µM). The binding of fMet-tRNA$_f$ to the 30S subunit-poly (A, U, G) complex was tested and found to be sensitive to ATA inhibition (see Table 3). The binding of the 50S subunit to the 30S-R17 RNA complex was also found to be sensitive to ATA inhibition (about 40% inhibition at 100 µM ATA).

Roberts and Coleman (1971) examined polypeptide synthesis, directed by poly(U) in an Ehrlich ascites cell-free system, and found that it was inhibited by ATA (see Table 1). They also washed protein off ribosomes with NH_4Cl, which was found to bind poly(U). This binding was sensitive to inhibition by ATA as was the binding of poly(U) by whole ribosomes. They observed that both the 40S and 60S subunits would bind poly(U), though the 40S subunit was 7 times as active in poly(U) binding as the 60S. Both kinds of binding to poly(U) were inhibited by ATA, but the binding of poly(U) by the 40S was about 3 times as sensitive as the binding of poly(U) by the 60S. Similary Huang and Grollman (1972) showed that ATA would bind to 40S and 60S subunits of rabbit reticulocytes. The 40S peak bound five to ten times as much ATA per A_{260} unit as the 60S peak.

In *Escherichia coli*, Tal et al. (1972) showed that poly(U) binds to the S1 ribosomal protein (from the 30S subunit) and this binding is inhibited by ATA. When Roberts and Coleman (1971) investigated the binding of poly(U) by various proteins and the effect of ATA on those bindings, they found that bovine serum albumin, shown previously (Lindenbaum and Schubert, 1956) to bind ATA strongly, had essentially no binding affinity for poly(U). These results suggest that the specificity of ATA, if it exists, is not necessarily its binding to nucleic acid binding sites. Leader (1972) studied the effect of ATA on phe-tRNA binding to rat liver ribosomes. He found that ATA inhibited the binding of phe-tRNA to 40S and to 40S plus 60S subunits. When protein factor M1 was added to the 40S binding reaction an increased binding was observed, and ATA still inhibited the reaction but to a lesser extent. The same phenomenon occurred when elongation factor T-I was added to the 40S + 60S binding reaction. However, when BSA was added, the inhibitory effect of ATA was reduced, suggesting that the effective concentration of ATA was decreased by the binding of ATA to BSA. (BSA does not interfere with the binding of tRNA to ribosomes.) They also checked the effect of varying the Mg^{+2} concentration on the inhibitory effect of ATA, and found that with an increase of Mg^{+2} concentration the inhibition of phe-tRNA binding was also increased, again indicating as in the experiments of GROLLMAN and STEWART (1968) that ATA does not affect protein synthesis because of its chelating capacity.

TAL et al. (1972) purified a poly(U)-binding protein from E. coli ribosomes. They compared poly(U) binding by whole ribosomes, the purified binding protein, and poly(U)-directed phenylalanine incorporation, and found that similar concentrations of the ribosomes and the binding protein were inhibited to a similar extent by similar concentrations of ATA (Table 2). A number of experimental points concerning inhibition of protein synthesis or particular steps in the process are summarized in Tables 1–3.

Discussion

The effects of ATA on protein synthesis seem to be similar in the various systems studied. As mentioned previously the conclusion that emerges is that initiation of protein synthesis is more sensitive to inhibition by ATA than polypeptide chain elongation. It is quite remarkable that this conclusion can be derived from all the systems investigated, which include bacteria, yeast, ascites cells, rat liver cells, etc., and a variety of messenger RNAs. It is also interesting that equivalent reactions are inhibited more or less to the same extent by ATA (see Tables 1–3). However there are some differences. Using the E. coli system a number of investigators found that poly(U)-directed polyphenylalanine synthesis is less sensitive to inhibition by ATA than protein synthesis directed by phage RNA (R17 or f2), while using the ascites system MATHEWS (1971) showed that the poly(U)-directed polyphenylalanine synthesis is more sensitive than EMC viral RNA-directed protein synthesis, (see Table 1).

It is not clear to what extent the differences observed in different systems reflect a significant or a trivial difference in the mode tf action of ATA in the different systems. This problem will be further discussed below. Again we would like to emphasize that there are not serious disagreements between results obtained in different laboratories using similar materials.

Mode of action of ATA. From what is known about ATA one might imagine that *in vivo* or *in vitro* ATA could have two modes of actions a) acting as a chelator and interfering with various reactions by lowering the concentration of necessary cations and b) by binding to proteins and interfering with their function. Since we deal here with protein synthesis both these assumptions could be valid, but there are considerations which suggest that ATA does not inhibit protein synthesis by chelating Mg^{+2}. Considering that the lowest Mg^{+2} concentration usually used in the reactions is around 5 mM while the highest ATA concentrations used were 0.4 mM, but usually only around 0.05 mM, Mg^{+2} should not be limiting. Also in all experiments where ATA inhibition was tested against different Mg^{+2} concentrations, the changes in Mg^{+2} concentrations did not affect the inhibition by ATA in a manner expected from a chelator (see

Grollman and Stewart, 1968; Siegelman and Apirion, 1971a, b; Huang and Grollman, 1972; Leader, 1972). Since, at present, protein synthesis can be carried out only in a relatively narrow range of Mg^{+2} concentrations reversal experiments are limited to these concentrations.

Therefore, the above considerations suggest that ATA exerts its effect by binding to proteins. Since in almost all steps of protein synthesis there are interactions between nucleic acids and proteins, it is possible that ATA could bind to nucleic acids and, thus, affect protein synthesis. There are at least two experiments that suggest that ATA does not inhibit polypeptide synthesis in this manner. Grollman and Stewart (1968) used 5 µM of ATA (about 2 µg/ml), and varied the f2 RNA concentration from about 30 to 500 µg/ml and found that inhibition of polypeptide synthesis (about 50%) remained constant. In another experiment, Huang and Grollman (1972) showed that varying the concentration of poly(U) from 0 to 1.2 mg/ml did not change the level of binding of ATA to reticulocyte ribosomes, using ATA concentrations of 10, 20 and 80 µM. Therefore, the effect of ATA on protein synthesis is most likely caused by its binding to proteins that are involved in the process and hence, interfering with their functions.

We have mentioned that ATA binds tightly to bovine serum albumin (Lindenbaum and Schubert; 1956) while poly(U) does not bind well to BSA (see Roberts and Coleman, 1971). It is not clear what common parameter(s) distinguish proteins that can bind ATA. Perhaps some proteins that can interact with nucleic acids can interact also with ATA. That this could be the case, i.e. ATA binds to macromolecules that bind nucleic acids, is to some extent suggested by the experiments of Roberts and Coleman (1971) who showed that by and large the better the binding of poly(U) to a macromolecule, the lower the concentration of ATA sufficient to inhibit this binding. These experiments also suggest that the binding of ATA is not highly specific and that it can bind to ribosomes washed with either sucrose or NH_4Cl as well as to 40S and 60S subunits, pH 5-precipitated enzymes and high-speed supernatant. However ATA binds to 40S ribosomes much better than to the 60S ribosome (Huang and Grollman, 1972). Therefore one should expect that many steps in protein synthesis would be inhibited by ATA. Indeed, this is the case and many steps are inhibited (see Tables 1–3); they include charging of tRNA by amino acids (Siegelman, unpublished observations), various steps in the initiation process, binding of tRNA to the ribosomes as well as further elongation steps. It is worth noting that thus far not a single step tested has been shown to be specifically sensitive to inhibition by ATA, while there are a few steps that show a remarkable resistance to inhibition by ATA. These include peptide bond formation as measured by the fragment reaction in which CACCA-leuAc bound to ribosomes is transfered to puromycin (Battaner and Vazquez, 1971) as well as the binding of a tRNA fragment CACCA-phe to the ribosome (Pestka, 1969a, b; Battaner and Vazquez, 1971). It is interesting that the binding of the whole phe-tRNA molecule is inhibited by ATA (Leader, (1972).

In view of the above, how is the almost universal phenomenon of initiation being inhibited preferentially to elongation to be accounted for? First we would like to describe some of the experiments on which this conclusion is based. Endogenous protein synthesis, in all systems tested, is less sensitive to inhibition by ATA than protein synthesis directed by an exogenous messenger which necessitates initiation (see Table 1). In other experiments, it was shown that concentrations of ATA, added at the begining of the reaction, which completely inhibited protein synthesis, directed by R17 or f2, would not affect the reaction for a few minutes at least if added later after initiation had taken place, (SIEGELMAN and APIRION, 1971b; STEWART et al., (1971); HUANG and GROLLMAN 1972). The inhibition of the reaction after a few minutes is probably due to new initiations which are inhibited by ATA. These new initiations could be due to recycling of ribosomes that had participated in the reaction, or could be due to the recruitment of new ribosomes that had not been involved previously in protein synthesis.

There are at least two possible explanations of why initiation is more sensitive to inhibition by ATA than elongation. First, initiation might involve a larger number of ATA sensitive steps than elongation (SIEGELMAN and APIRION, 1971b), or during initiation the sites that can bind ATA are more available to binding with ATA than the same or different sites during elongation after the ribosome-messenger RNA-tRNA-polypeptide complex are formed. We prefer the second possibility. However these alternatives are not mutually exclusive.

There are two reasons for our preference. Since it is likely that most of the interference caused by ATA in protein synthesis is a result of ATA binding to ribosomes, it is assumed that before initiation more of the ribosome surface is exposed to the environment than during elongation, and therefore the ATA sensitive sites are more accessible prior to initiation. Such a possibility gains credence from the experiments performed by HUANG and GROLLMAN (1972), who showed that two to three times as much ATA bound to 40S than to 80S rabbit reticulocyte ribosomes. Secondly, there is the experiment performed by HSU (1971) that showed that when the initiation complex is formed but elongation is prevented, protein chain elongation becomes almost as sensitive to ATA inhibition as chain initiation.

To some extent the mode of action of an inhibitor can be reasoned out from the shape of the dosage-response-curve. The inhibition curve of protein synthesis by increasing levels of ATA is not exponential. We constructed such curves from data taken from a number of articles, (see for instance GROLLMAN and STEWART, 1968; MATHEWS, 1971; ROBERTS and COLEMAN, 1971). Since these curves are not exponential their shape suggests that it is not a single target which is affected by ATA. Where a sufficient range of concentrations were employed to measure ATA effects, two components of inhibition are clearly revealed. In the first part at low concentrations of ATA the reaction is very sensitive to ATA while in the second part it is rather insensitive. Only one of these cases was sufficiently analyzed to imply that in the first part inhibition was due to the effects(s) of ATA on initiation while in the second part inhibition was due to the effect(s) of ATA on elongation. Such an experiment was carried out by MATHEWS (1971) with a rabbit reticulocyte cell-free system synthesizing globin. He showed that the sensitive component was completely eliminated if NaF or poly(U) were added to the system. These two molecules inhibit initiation of new globin chains. The first component was completely inhibited with about 40 µM of ATA. Further addition of ATA up to about 130 µM had no effect. Further increase of ATA concentrations, however, increased the inhibition.

As expected, the inhibition by ATA of total protein synthesis is not the sum of the inhibitions of the single steps, (see for instance SIEGELMAN and APIRION, 1971b; TAL et al., 1972). This is logical since the steps do not operate in concurrent pathways, but in a single pathway. Therefore the inhibition should be the product of all steps accesible to inhibition.

As we mentioned earlier, ATA binds to ribosomes, (see ROBERTS and COLEMAN, 1971; HUANG and GROLLMAN, 1972). This is further indicated by the fact that ribosomes change their sedimentation rate in the presence of ATA. However, it is a little disturbing that GROLLMAN and STEWART (1968) reported a change in E. coli ribosomes from 70S to 76S, while LEBLEU et al. (1971), and HUANG and GROLLMAN (1972), reported a change in reticulocyte ribosomes from 80S to 60S. It should, however, be mentioned that with a rather more purified preparation of ATA, HUANG and GROLLMAN (1972), claim that this change in ribosome sedimentation is not observed.

Is ATA the Active Principle? Since ATA apparently binds to the ribosome it is interesting that changing levels of ribosomes affect the inhibition of protein synthesis by ATA. Unfortunately, there are very few data on this important point. In an experiment performed by GROLLMAN and STEWART (1968), keeping ATA at 9 µM and changing the ribosome concentration from 0.15 to 0.54 µM, reduced the inhibition of f2 RNA stimulated polypeptide synthesis from almost 100% to about 10%. Since the molar ratio of ATA to ribosomes was high there are two ways to explain this phenomenon. Perhaps a contaminant molecule rather than ATA was the inhibitor. Alternatively, each ribosome could bind a large number of ATA molecules and the affinity of ATA for the ribosome could be rather low. We prefer the second alternative. Indeed, there are a number of reasons to think that the second alternative is more likely. First of all, ATA from different commercial sources gives similar inhibitions, (see for instance MATHEWS, 1971). Moreover since there are at least 50 different proteins in the ribosome complex, many of which can probably interact with ATA, the assumption that a large number of ATA molecules can bind to a single ribosome is not far-fetched. Finally, in the experiments carried out by HUANG and GROLLMAN (1972) a very rough calculation from their membrane filter binding assay data suggests that about 100 molecules of ATA can bind to a single ribosome. When binding was tested in a sucrose gradient, about 10 molecules of ATA were bound to a 40S ribosome, (HUANG and GROLLMAN, 1972, Fig. 9). However, caution is needed since a crucial assumption in these calculations is that the tritiated ATA was pure. That the affinity of ATA for ribosomes is rather low is indicated by an experiment carried out by BATTANER and VAZQUEZ (1972). When ribosomes were pretreated with ATA, centrifuged, resuspended in buffer free of ATA and then used to assay poly(U)-directed protein synthesis, inhibition by ATA was much reduced as compared to the assay carried out in presence of ATA.

ATA Inhibits RNA Synthesis. BLUMENTHAL (unpublished observations) showed that QB replicase, Escherichia coli RNA polymerase, and coliphage T7 RNA polymerase are all Inhibited by ATA (about 5 µM). In each case, when ATA was added prior to initiation, inhibition was observed but not when it was added after initiation. Again, as in the case of protein synthesis, BLUMENTHAL found that changing the concentrations of Q_β replicase from 5×10^{-3} to 5×10^{-2} µM while keeping ATA concentration at $2\,\mu M$ reversed the inhibition from 100 to 0%. Such data might suggest that a contaminant rather than ATA is the inhibitor since with both concentrations of enzymes ATA seems to be in excess (1 to 400 and 1 to 40). However for the reasons mentioned above, we prefer the possibility that this phenomenon is due to the binding of many molecules of ATA to a single protein, or due to low affinity of the binding of ATA to some proteins (or both).

Some Odd Phenomena. There are, however, at least two phenomena related to ATA which might be quite puzzling. The first is the increase in protein synthesis at low levels of ATA observed by MATHEWS (1971) and by LEADER (1972). Perhaps this could be due to the fact that ATA, being a trifunctional molecule, where the functional groups are placed at some distance from each other, could bind to three ribosomes or other components of protein synthesis and bring them together, thus increasing the effective concentration of some component(s). Alternatively, it could be inhibiting ribonucleases. If ATA can bind to proteins that bind to nucleic acids, as we argued before, the second possibility is quite attractive.

Another phenomenon which seems peculiar is the decreased inhibition of polyphenylalanine synthesis when a natural initiation codon is added to the poly(U) molecule, (see MATHEWS, 1971). This could perhaps be due to the fact that with poly(U), initiation has to start with phe-tRNA, which could be a much more demanding reaction and leave the ribosomal sites exposed to ATA for much longer than when initiation can start efficiently with the natural initiator molecule fMet-tRNA$_f$. The fact that at higher Mg^{+2} concentrations, that probably distort the ribosome configuration and ease illegitimate initiation, this difference between $AUGU_n$ $AUUU_n$ and poly(U), was abolished gives some credence to this hypothesis. Differences in Mg^{+2} concentrations could also explain why poly(U)-directed protein synthesis is more sensitive to ATA inhibition in systems derived from eukaryotic cells than in the systems derived from *E. coli*. In the eukaryotic systems the Mg^{+2} concentrations used were relatively low (5 to 10 mM) compared to the concentrations used in the *E. coli* system (615 to 20mM). The higher Mg^{+2} concentrations used with *E. coli* extracts could have eased distorted initiation by phe-tRNA and made it less sensitive to inhibition by ATA as compared to initiation by phe-tRNA in the mammalian systems at low Mg^{+2} concentrations.

ATA as an Antibiotic. ATA as an inhibitor of protein synthesis can be compared with other antibiotics. It is clear from such a comparison that while other antibiotics usually inhibit a specific reaction, ATA as we mentioned above, interferes with a number of steps in protein synthesis. It is interesting to compare ATA with fusidic acid. Both of these drugs inhibit ribosome dependent GTPase activity.

HUANG and GROLLMAN (1972) showed for instance, that fusidic acid maintained about the same inhibitory effect whether the reaction was assayed with ribosomes only, TF-II or the complete set of components necessary for protein synthesis present. ATA on the other hand was most inhibitory when just ribosomes or TF-II were assayed and became less inhibitory when the complete set of components for protein synthesis were present. This again can be explained by decreased accessibility of ATA to its binding sites when the complete set of components were present.

Certain antibiotics affect single sites since one-step mutations can lead to high level resistance. This for instance is the case with streptomycin and spectinomycin. However, not all antibiotics behave in this way, and resistance to high levels of chloramphenicol can be achieved only by accumulating a number of mutations (Cavalli and Maccacaro, 1952). It is not known whether the large number of mutations that lead to resistance to chloramphenicol are expressed in components of the protein synthetic machinery. However, even in this case it is likely that the sites of interactions of chloramphenicol in the cell are less numerous than those of ATA.

The foregoing considerations suggest that while a molecule such as ATA is too non-specific to be used as an antibiotic, it may be possible that a molecule with a higher specificity could be designed and used as an antibiotic. It might have the advantage that bacteria would not be able to become resistant to it by mutations, since, too many such mutations would have to accumulate in order to endow the cell with such a resistance. Perhaps this needs for too many mutational steps to adapt to such an "antibiotic", selected against such molecules, since their existence could affect the ecological balance too drastically.

It is worth mentioning that ATA can indeed be used as a tool to indicate that a specific reaction is related to initiation rather than to elongation, but we must emphasize that since a number of factors can affect the activity of ATA, such as the concentration of ribosomes or non-ribosomal proteins, it is mandatory to show in each case that ATA was used at concentrations that in the given system affect initiation rather than elongation. This could be achieved for instance by finding a level of ATA that abolishes the reaction when it is added in the begining of synthesis but does not affect its rate for at least a certain length of time if added after initation took place.

Finally, even though protein synthesis in eukaryotes and prokaryotes is similar, differences do exist: e.g. the size of the ribosomes, the ease of dissociation of the ribosomes into subunits (prokaryotic ribosomes are easier to dissociate), initiation of prokaryotic protein synthesis with fmet-tRNA$_f$ rather than with met-tRNA$_f$ and differences in the elongation factors. Therefore it should be possible, in principle, to design molecules that would inhibit prokaryotic but not eukaryotic protein synthesis and therefore could have the potential of being used as effective "antibiotics". However, at present, it is not clear that there are sufficient differences between virus-directed protein synthesis and endogenous cellular protein synthesis to warrant such an attempt to design anti-viral agents aimed at distinguishing viral from host directed protein synthesis.

References

Battaner, E., and D. Vazquez: Inhibitors of protein synthesis by ribosomes of the 80-S type. Biochim. Biophys. Acta **254**, 316 (1971).

Cavalli, L.L., and G.A. Maccacaro: Polygenic inheritance of drug-resistance in the bacterium, *Escherichia coli.* Heredity **6**, 311 (1952).

Coghill, R.D., F.H. Stodola, and J.L. Wachtel: Chemical modifications of natural penicillins. In: *The Chemistry of Penicillin,* H.T. Clarke, J.R. Johnson, and R. Robinson, eds. p. 680. Princeton, N.J.: Princeton University Press 1949.

GROLLMAN, A.P.: Inhibition of messenger ribonucleic acid attachment to ribosomes. II. Proposed mechanism for the design of novel antiviral agents. Antimicrobial Agents and Chemotherapy 1968. American Society for Microbiology. G.L. Hobby, ed. p. 36.

GROLLMAN, A.P., and M.L. STEWART: Inhibition of the attachment of messenger ribonucleic acid to ribosomes. Proc. Natl. Acad. Sci. U.S. **61**, 719 (1968).

HSU, W.T.: Translation of an RNA viral message *in vitro;* one-step polypeptide chain elongation. Biochem. Biophys. Res. Commun. **42**, 405 (1971).

HUANG, M.T., and A.P. GROLLMAN: Effects of aurintricarboxylic acid on ribosomes and the biosynthesis of globin in rabbit reticulocytes. Mol. Pharm. **8**, 111 (1972).

JOKLIK, W.K., and T.C. MERIGAN: Concerning the Mechanism of action of interferon. Proc. Natl. Acad. Sci. U.S. **56**, 558 (1966).

LEADER, D.P.: Aurintricarboxylic acid inhibition of the binding of phenyl-alanyl-tRNA to rat liver ribosomal subunits. FEBS Letters **22**, 245 (1972).

LEBLEU, B., G. MARBAIX, J. WÉRENNE, A. BURNY, and G. HUEZ: Effect of aurintricarboxylic acid and of NaF on the binding of globin messenger RNA to reticulocyte 40S ribosomal subunits. Biochem. Biophys. Res. Commun. **40**, 731 (1970).

LEVY, H.B., and W.A. CARTER: Molecular basis of the action of interferon. J. Mol. Biol. **31**, 561 (1968).

LINDENBAUM, A., and J. SCHUBERT: Binding of organic anions by serum albumin. J. Phys. Chem. **60**, 1663 (1956).

LINDENBAUM, A., M.R. WHITE, and J. SCHUBERT: Effect of aurintricarboxylic acid on beryllium inhibition of alkaline phosphatase. J. Biol. Chem. **196**, 273 (1952).

LUND, E., and N.O. KJELDGAARD: Metabolism of guanosine tetraphosphate in *E. coli.* Eur. J. Biochem. **28**, 316 (1972).

MARCUS, A., J.D. BEWLEY, and D.P. WEEKS: Aurintricarboxylic acid and initiation factors of wheat embryo. Science **167**, 1735 (1970).

MARCUS, P.I., and J.M. SALB: Molecular basis of interferon action: Inhibition of viral RNA translation. Virology **30**, 502 (1966).

MATHEWS, M.B.: Mammalian chain initiation: the effect of aurintricarboxylic acid. FEBS letters **15**, 201 (1971).

PESTKA, S.: Translocation, aminoacyl-oligonucleotides, and antibiotic action. Cold Spring Harbor Symp. Quant. Biol. **34**, 395 (1969a).

PESTKA, S.: Studies on the formation of transfer ribonucleic acid-ribosome complexes, XI. Antibiotic effects on phenylalanyl-oligonucleotide binding to ribosomes. Proc. Natl. Acad. Sci. U.S. **64**, 709 (1969b).

PESTKA, S.: Inhibitors of ribosome function. Ann. Rev. Microbiol. **25**, 487 (1971).

ROBERTS, W.K., and W.H. COLEMAN: Polyuridylic acid binding by protein from Ehrlich ascites cell ribosomes and its inhibition by aurintricarboxylic acid. Biochemistry **10**, 4304 (1971).

SCHUBERT, J., and A. LJNDENBAUM: Studies on the mechanism of protection by aurintricarboxylic acid in beryllium poisoning. II. Equilibria involving alkaline phosphatase. J. Biol. Chem. **208**, 359 (1954).

SIEGELMAN, F.L., and D. APIRION: Inhibition of polyuridylic acid-directed protein synthesis by aurintricarboxylate in extracts of *Escherichia coli.* J. Bacteriol. **105**, 451 (1971a).

SIEGELMAN, F.L., and D. APIRION: Aurintricarboxylic acid, a perferential inhibitor of protein synthesis. J. Bacteriol. **105**, 902 (1971b).

STEWART, M.L., A.P. GROLLMAN, and M.T. HUANG: Aurintricarboxylic acid: inhibitor of initiation of protein synthesis. Proc. Natl. Acad. Sci. U.S. **68**, 97 (1971).

TAL, M., M. AVIRAM, A. KANAREK, and A. WEISS: Polyuridylic acid binding and translating by *Escherichia coli* ribosomes: stimulation by protein I, inhibition by aurintricarboxylic acid. Biochim. Biophys. Acta **281**, 381 (1972).

WANG, C.S., R.B. NASO, and R.B. ARLINGHAUS: Factor dependent binding of Rauscher leukemia virus RNA to ribosomes. Biochem. Biophys. Res. Commun. **47**, 1290 (1972).

WEISSBACH, H., and N. BROT: Inhibition of transfer factor Ts by aurintricarboxylic acid. Biochem. Biophys. Res. Commun. **39**, 1194 (1970).

WILHELM, J.M., and R. HASELKORN: The chain growth rate of T4 lysozyme *in vitro.* Proc. Natl. Acad. Sci. U.S. **65**, 388 (1970).

Aminoglycoside Antibiotics

Nobuo Tanaka

Some aminoglycosides of microbial origin are listed in Table 1. Recently derivatives of these antibiotics, which are effective against resistant organisms,

Table 1. *Aminoglycoside antibiotics of microbial origin*

I. Aminoglycosides without cyclitol or aminocyclitol	
1. Monosaccharides	Nojirimycin
	3-Amino-3-deoxy-D-glucose
	Streptozotocin
2. Disaccharides	Trehalosamine
	Mannosylglucosaminide
II. Aminoglycosides with cyclitol	
1. Disaccharide	Kasugamycin
2. Oligosaccharide	Validamycin
III. Aminoglycosides with aminocyclitol	
1. Disaccharides	Neamine (Neomycin A)
	Paromamine
	6-Amino-6-deoxy-D-glucosyldeoxystreptamine
	3-Amino-3-deoxy-D-glucosyldeoxystreptamine
	Hybrimycins A_3 and B_3
2. Oligosaccharides	
a) Streptomycin group	Streptomycin
	Mannosidostreptomycin (Streptomycin B)
	Hydroxystreptomycin
	Dihydrostreptomycin
	Glebomycin (Bluensomycin)
b) Kanamycin group	Kanamycin (A), B and C
	NK-1001, NK-1012-1
	Tobramycin (Nebramycin factor 6)
c) Gentamicin group	Gentamicins C_1, C_{1a} and C_2, and A
	Sisomicin
d) Spectinomycin (Actinospectacin)	
e) Neomycin group	Neomycins B and C (Streptothricin BII and BI)
	Hybrimycins A_1, A_2, B_1 and B_2
	Paromomycins I and II (Zygomycins A_1 and A_2)
	Lividomycins A and B
	Mannosylparomomycin
	Ribostamycin
	Butirosins A and B
f) Destomycin group	Destomycins A and B
	Hygromycin B
	A-396-I

have been developed; some of them are described below. Except these, there are many aminosugar-containing antibiotics, which are usually classified to other groups of antibiotics: macrolide, lincomycin, streptothricin, etc.

Of aminoglycosides, the mechanism of action has been studied most extensively with streptomycin, which are described in another chapter. Here are summarized the investigations concerning aminoglycosides except the streptomycin group: streptomycin, mannosidostreptomycin, hydroxystreptomycin, dihydrostreptomycin, bluensomycin etc.

Streptozotocin (VAVRA *et al.*, 1960; HERR, *et al.*, 1960; HERR *et al.*, 1967) produced by *Streptomyces achromogenes* strongly inhibits both Gram-positive and negative bacteria, and growth of animal tumors. It is a derivative of glucosamine (Fig. 1). Nojirimycin (ISHIDA *et al.*, 1967; ISHIDA *et al.*, 1967; INOUE *et al.*, 1966) produced by *S. nojiriensis, S. roseochromogenes, S. lavendulae* and other *Streptomycetes* is a labile substance; the structure is D-glucopiperidnose (5-amino-5-deoxy-D-glucopyranose) (Fig. 1). The antibiotic shows a weak antimicrobial activity, except that it strongly inhibits *Sarcina lutea* and a drug-resistant strain of *Shigella flexneri;* it interferes with α- and β-glucosidase reactions. 3-Amino-3-deoxy-D-glucose, a constituent of kanamycin, is also produced by *Bacillus subtilis;* it exhibits a weak antibacterial activity (UMEZAWA *et al.*, 1967c).

α-D-Glucopyranosido-2-deoxy-2-amino-α-D-glucopyranoside obtained from *Streptomyces lavendulae* is called trehalosamine; it exerts a weak activity against some bacteria and fungi, and the antimicrobial effect is antagonized by salicin, esculin and trehalose (ARCAMONE and BIZIOLI, 1957). α-D-Mannosyl-α-D-glucosaminide is produced by *Streptomyces virginiae* and shows a weak activity against *Mycobacteria* and other organisms. The antimicrobial activity is comparable with trehalosamine (URAMOTO *et al.*, 1967).

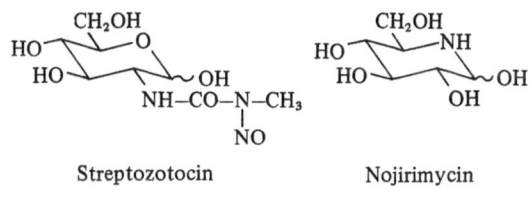

Streptozotocin   Nojirimycin

Fig. 1. The structure of aminoglycosides. a) Monosaccharides

Kasugamycin, produced by *Streptomyces kasugaensis,* is a cyclitol aminoglycoside with a side chain of amidinocarboxylic acid (Fig. 2). It inhibits a variety of microorganisms, including *Pseudomonas aeruginosa* and *Piricularia oryzae,* and is widely used to control rice blast disease caused by *P. oryzae* (UMEZAWA *et al.*, 1965; SUHARA *et al.*, 1966). Validamycin A is a main component of the validamycin complex which is used to control sheath blight in rice plants and produced by *Streptomyces hygroscopicus var. limoneus.* The structure in Fig. 2 is assigned for validamycin A (SHOJI and NAKAGAWA, 1970; HORII and KAMEDA, 1972).

Kanamycin is an aminoglycoside, consisting of deoxystreptamine, 6-amino-6-deoxy-D-glucose, and 3-amino-3-deoxy-D-glucose (Fig. 3). The absolute crystal

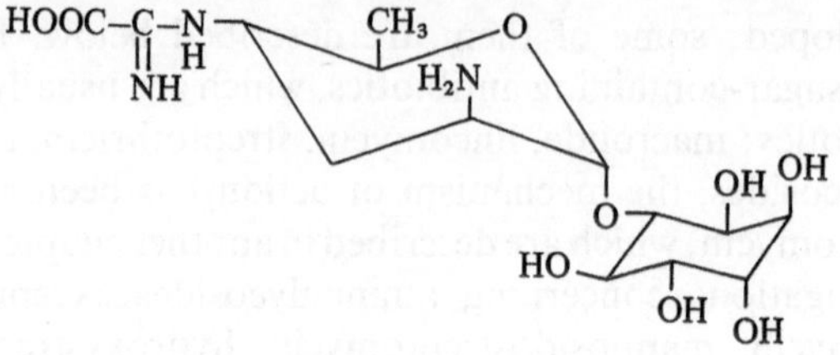

Kasugamycin

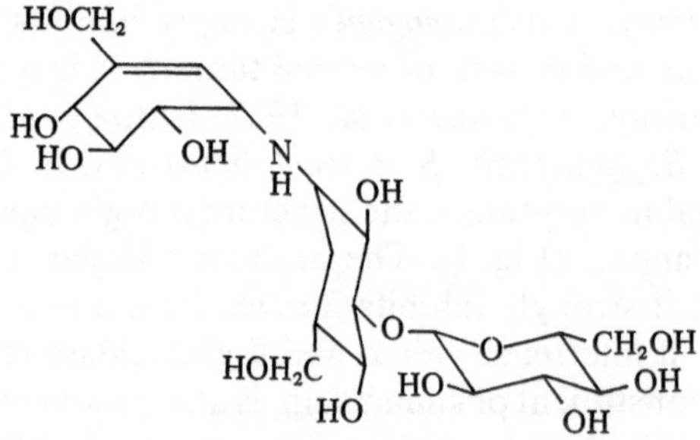

Validamycin A

Fig. 2. The structure of aminoglycosides. b) Cyclitol antibiotics

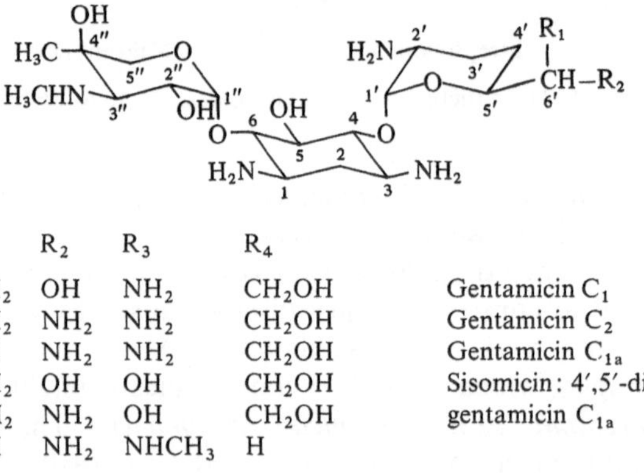

	R$_1$	R$_2$	R$_3$	R$_4$		R$_1$	R$_2$
Kanamycin	NH$_2$	OH	NH$_2$	CH$_2$OH	Gentamicin C$_1$	CH$_3$	NHCH$_3$
Kanamycin B	NH$_2$	NH$_2$	NH$_2$	CH$_2$OH	Gentamicin C$_2$	CH$_3$	NH$_2$
Kanamycin C	OH	NH$_2$	NH$_2$	CH$_2$OH	Gentamicin C$_{1a}$	H	NH$_2$
NK-1001	NH$_2$	OH	OH	CH$_2$OH	Sisomicin: 4',5'-didehydro-		
NK-1012-1	NH$_2$	NH$_2$	OH	CH$_2$OH	gentamicin C$_{1a}$		
Gentamicin A	OH	NH$_2$	NHCH$_3$	H			
Tobramycin:							
3'-deoxy-kanamycin B							

Fig. 3. The structure of aminoglycosides. c) Kanamycin and gentamicin groups

structure of kanamycin has been established (KOYAMA *et al.*, 1968). The antibiotic is obtained from *Streptomyces kanamyceticus*, which produces two minor components: kanamycins B and C. Kanamycin is sometimes described as kanamycin A to distinguish it from the minor components. Kanamycin exhibits a strong antimicrobial activity against a variety of aerobic bacteria, including *Mycobacteria, Staphylococci*, and *Enterobacteria*. Kanamycin B shows about five times higher activity against *Staphylococci* than kanamycin, but less activity against *Mycobacteria*. Kanamycin C exhibits almost the same activity as kanamycin except one-tenth activity against *Mycobacteria*. 6''-Deoxykanamycin shows antibacterial activity similar to kanamycin. Kanamycin has a low degree of toxicity, and is widely used for treatment of staphylococcal infection, urinary and intestinal infections, and tuberculosis. Kanamycin B is more toxic than kanamycin, and is also clinically used (cf. a review by H. UMEZAWA, 1964). NK-1001, NK-1012-1, 6-amino-6-deoxy-D-glucosyl-deoxystreptamine, 3-amino-3-deoxy-D-glucosyl-streptamine, neamine, paromamine, and deoxystreptamine are produced by various mutants of *S. kanamyceticus*. The former three antibiotics as well as neamine and paromamine exhibit fairly good antimicrobial activity (MURASE, *et al.*, 1970). Nebramycins are produced by *Streptomyces tenebrarius;* and the factor 6 (tobramycin) is identified with 3'-deoxykanamycin B (KOCH and RHOADES, 1971). Gentamicins A, C_1, C_{1a} and C_2 are produced by *Micromonospora purpura* or *M. echinospora* (WEINSTEIN *et al.*, 1963); sisomicin is produced by *Micromonospora inyoensis* (WEINSTEIN *et al.*, 1970). Their structures are similar to that of kanamycin (Fig. 3) (COOPER *et al.*, 1969; MAEHR and SCHAFFNER, 1970; COOPER *et al.*, 1971; REIMANN *et al.*, 1971). Gentamicin C_1, C_{1a} and C_2, and sisomicin as well as tobramycin (DEL BENE and FARRAR, 1972) show a strong antimicrobial activity against a variety of microorganisms, including *Pseudomonas aeruginosa* and some of kanamycin-neomycin-resistant bacteria as described below. Gentamicin C complex is clinically used, particularly for pseudomonas infections.

Spectinomycin (actinospectacin), produced by *Streptomyces spectabilis* and by *S. flavopersicus*, is a disaccharide, containing actinamine, a stereoisomer of N,N'-dimethylstreptamine (MASON *et al.*, 1961; OLIVER *et al.*, 1961; HOEKSEMA *et al.*, 1962). The antibiotic exhibits antibacterial activity of broad spectrum; is clinically used, especially for gonorrhea. (Fig. 4).

Neamine, neomycins B and C are produced by *Streptomyces fradiae* and by *S. albogriseolus*. Neomycins B and C show higher antibacterial activity than neamine, and are clinically used. Paromamine, paromomycins I and II are produced by *Streptomyces rimosus*. Paromomycins I and II exhibit higher activity than paromamine, and are clinically used. The structures in Fig. 5 are assigned for neomycins and paromomycins. Catenulin produced *by S. catenulae*, zygomycins produced by *S. pulveraceus*, and amminosidin produced by *S. chrestomycedicus* are identical with paromomycin (RINEHART *et al.*, 1962; HICHENS and RINEHART, 1963). There is a monography on neomycins and related antibiotics by RINEHART (1964). Paromomycin and neomycin are more toxic than kanamycin. Lividomycins A and B, and mannosylparomomycin are produced by *S. lividus*. (MORI *et al.*, 1971; ODA *et al.*, 1971). Lividomycins (Fig. 5) show a strong antibacterial activity against a variety of microorganisms, including *Mycobacterium tuberculosis*. Ribostamycin, obtained from *S. ribosidificus*, shows antibacterial activity

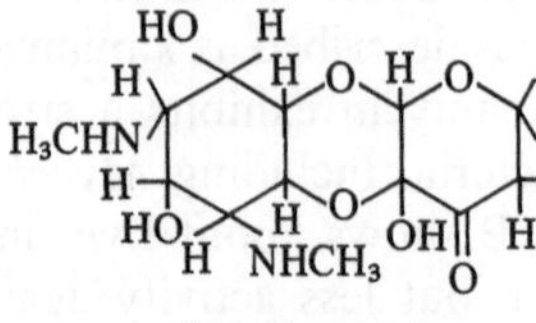

Spectinomycin

Fig. 4

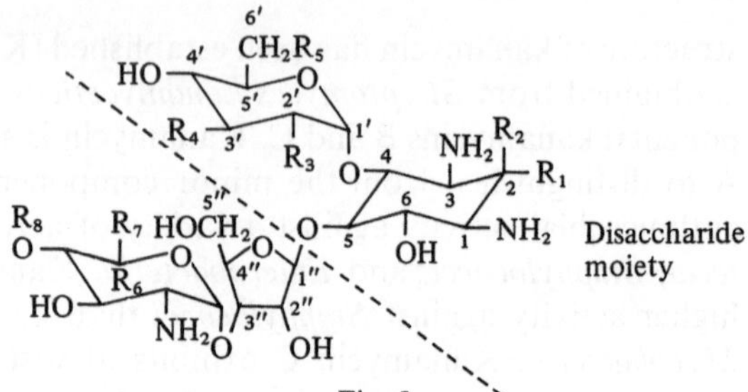

Disaccharide moiety

Fig. 5

Fig. 4. The structure of aminoglycosides. d) Spectinomycin

Fig. 5. The structure of aminoglycosides. e) Neomycin group I

	R_1	R_2	R_3	R_4	R_5	R_6	R_7	R_8
Neomycin B	H	H	NH_2	OH	NH_2	CH_2NH_2	H	H
Neomycin C	H	H	NH_2	OH	NH_2	H	CH_2NH_2	H
Hybrimycin A_1	OH	H	NH_2	OH	NH_2	CH_2NH_2	H	H
Hybrimycin A_2	OH	H	NH_2	OH	NH_2	H	CH_2NH_2	H
Hybrimycin B_1	H	OH	NH_2	OH	NH_2	CH_2NH_2	H	H
Hybrimycin B_2	H	OH	NH_2	OH	NH_2	H	CH_2NH_2	H
Paromomycin I	H	H	NH_2	OH	OH	CH_2NH_2	H	H
Paromomycin II	H	H	NH_2	OH	OH	H	CH_2NH_2	H
Mannosyl-paromomycin	H	H	NH_2	OH	OH	CH_2NH_2	H	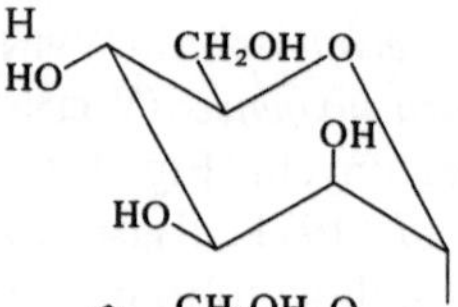
Lividomycin A	H	H	NH_2	H	OH	CH_2NH	H	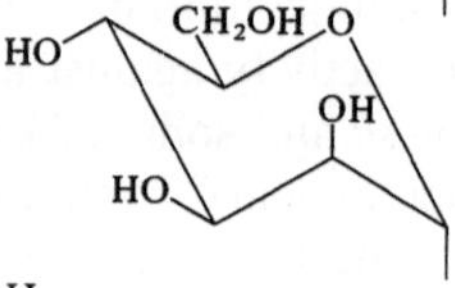
Lividomycin B	H	H	NH_2	H	OH	CH_2NH_2	H	H

Disaccharides:

	R_1	R_2	R_3	R_4	R_5
Neamine	H	H	NH_2	OH	NH_2
Paromamine	H	H	NH_2	OH	OH
6-Amino-6-deoxy-D-glucosyl-deoxystreptamine	H	H	OH	OH	NH_2

of broad spectrum and is clinically used (SHOMURA et al., 1970; AKITA et al., 1970). The structure of ribostamycin is similar to neomycin. The former is lacking in 2,6-diaminohexose of the neomycin molecule (Fig. 6). Butirosins A and B (WOO et al., 1971) are produced by *Bacillus circulans*. Butirosin B has the structure, in which 4-amino-2-hydroxylbutyryl group is attached to the N^1 position of deoxystreptamine moiety of ribostamycin. Butirosin A is a isomer of B, and has xylose instead of ribose (Fig. 6). Butirosin is less toxic than neomycin and shows a good antibacterial activity, including *Pseudomonas aeruginosa*, which is resistant to neomycin, ribostamycin and kanamycin.

SHIER et al., (1969, 1970) have prepared a series of semisynthetic neomycin analogs, hybrimycins, by addition of streptamine or 2-epistreptamine to the culture of a mutant of *Streptomyces fradiae,* which needs 2-deoxystreptamine to produce neomycins. The 2-deoxystreptamine residue of neomycins is replaced by streptamine or 2-epistreptamine in hybrimycins. Hybrimycin A_1 is the streptamine analog of neomycin B, and hybrimycin B_1 is the 2-epistreptamine analog of neomycin B; a similar relationship exists between neomycin C, hybrimycin A_2 or B_2, and between neamine and hybrimycin A_3 or B_3 (Fig. 5). The antimicrobial activity of hybrimycins is similar to that of neomycins.

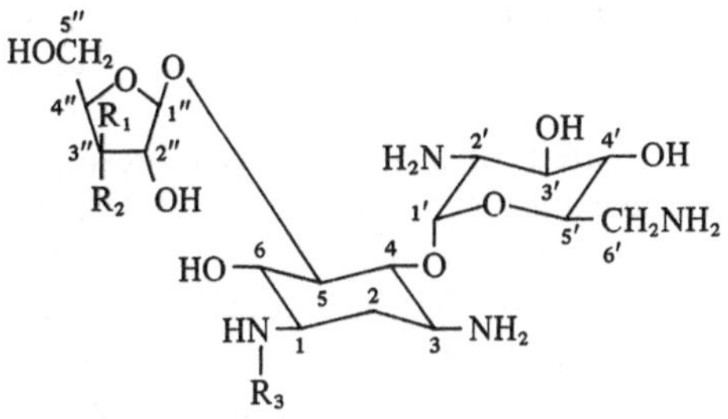

Fig. 6

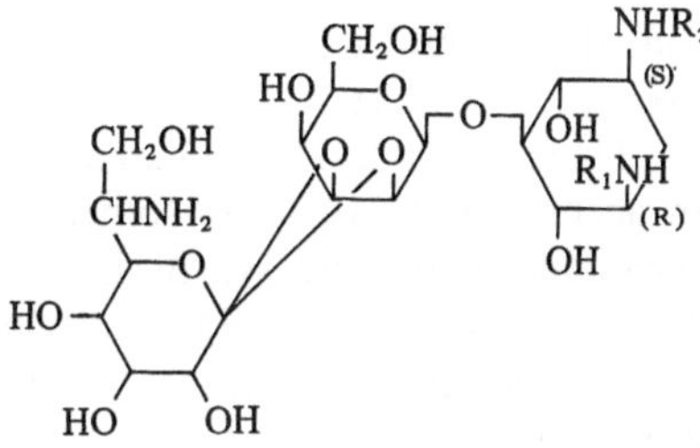

Fig. 7

Fig. 6. The structure of aminoglycosides. f) Neomycin group II

	R_1	R_2	R_3
Ribostamycin	H	OH	H
Butirosin A	OH	H	—CO—CH(OH)—CH$_2$—CH$_2$NH$_2$
Butirosin B	H	OH	—CO—CH(OH)—CH$_2$—CH$_2$NH$_2$

Fig. 7. The structure of aminoglycosides. g) Destomycin group

	R_1	R_2
Destomycin A	H	CH$_3$
Hygromycin B	CH$_3$	H
A-396-I	H	H

Destomycin A produced by *Streptomyces rimofaciens* (KONDO et al., 1965; KONDO et al., 1966), hygromycin B produced by *S. hygroscopicus* (NEUSS et al., 1970) and A-396-I produced by *Streptoverticillium eurocidicus* (SHOJI et al., 1970; SHOJI and NAKAGAWA, 1970) are aminoglycosides of another group, consisting of 6-amino-6-deoxyheptose, D-talose, and 2-deoxystreptamine with or without a methyl group (Fig. 7). Destomycin A and hygromycin B are used as antihelminthics in pigs and chicken.

Mechanism of Action

Of aminoglycoside antibiotics, the mechanism of action has been studied most precisely with streptomycin, which is described in another chapter. Here are summarized the investigations on antibiotics of kasugamycin, kanamycin, gentamicin, spectinomycin, neomycin, and destomycin groups (Table 1). The mechanism of action of these aminoglycosides is similar to that of streptomycin. However the details are different. The mechanism of action of the aminoglycosides will be discussed in this chapter in particular emphasis of the differences from that of streptomycin.

The in vivo Effects

The bactericidal action of streptomycin and kanamycin is reversed by chloramphenicol, erythromycin, mikamycin A, blasticidin S or tetracycline, but enhanced by puromycin (Yamaki and Tanaka, 1963; White and White, 1964). This implies that the normal ribosomal cycle must be in operation for aminoglycosides to exert their bactericidal effects.

The polysome level decreases in cells treated with streptomycin, resulting in accumulation of monosomes, that are bearing mRNA and fMet-tRNA but incapable of protein synthesis either *in vivo* or *in vitro* (Luzzatto et al., 1968, 1969a, b; Kogut and Prizant, 1970). They suggest that streptomycin may act on peptide chain initiation. Unlike streptomycin, neomycin and spectinomycin as well as chloramphenicol and tetracycline accumulate polysomes. In the presence of these antibiotics polysome formation continues and pulse-labelled mRNA still joins to ribosomes at rates comparable to that in untreated cells, suggesting that the antibiotics uncouple polysome formation from peptide bond formation (Gurgo et al., 1969).

Most of the evidence up to date favors the assumption that the lethal effect of aminoglycosides is the result of irreversible binding of the drugs to ribosomes and the subsequent interference with some phases of protein synthesis: initiation, elongation, and termination.

Interaction of Aminoglycosides with Ribosomes

Aminoglycosides interact with the 30S ribosomal subunit. Streptomycin or dihydrostreptomycin tightly bind to either the 70S ribosome or the 30S subunit of streptomycin-sensitive *E. coli* cells in a molar ratio of 1:1, and do not significantly bind to streptomycin-resistant or -dependent ribosomes (Leon and Brock, 1967; Kaji and Tanaka, 1968; Bollen et al., 1969b; Yamada et al., 1970; Vogel et al., 1970; Chang and Flaks, 1972). In contrast to streptomycin, neomycin, kanamycin, and gentamicin interact with more than one site on the ribosome (Davies and Davis, 1968).

The binding of dihydrostreptomycin is markedly enhanced by neomycin or paromomycin. The reason for the stimulation is not clear, although it seems likely that neomycin or paromomycin cause a conformational change of ribosomes and affect the binding of streptomycin to its normal site (Yamada et al., 1970; Chang and Flaks, 1972). Bound dihydrostreptomycin can be readily exchanged by the streptomycin group aminoglycosides (streptomycin, dihydrostreptomycin, hydroxystreptomycin, bluensomycin, and mannosidostreptomycin); but not by fragments of streptomycin (methylstreptobiosaminide, didesamidinostreptomycin, streptidine, and deoxystreptamine) and other aminoglycosides (spectinomycin, kanamycin and neamine) (CHANG and FLAKS, 1972). YAMADA et al. (1970) reported that the presence of streptomycin has no effect on the binding of dihydrostreptomycin to ribosomes, although CHANG and FLAKS (1972) observed that streptomycin can readily exchange bound dihydrostreptomycin. The presence of kanamycin or spectinomycin has no or little effect on the binding of dihydrostreptomycin, suggesting that the binding site of these antibiotics is not related to that of streptomycin.

In contrast to streptomycin, the binding of di^{3}H-spectinomycin is reversible: spectinomycin can be removed from ribosomes completely by dialysis for 2 hours. The binding of di^{3}H-spectinomycin to 30S spectinomycin-sensitive ribosomes is much slower than the streptomycin binding. The binding of streptomycin to ribosomes is complete in less than 2 min. at 37° C, whereas it takes 20 min. to reach a plateau for spectinomycin binding under the same conditions. The difference in ability of di^{3}H-spectinomycin to bind to spectinomycin-sensitive and -resistant ribosomes can only be shown at low drug concentrations. Neomycin, kanamycin or streptomycin do not affect its binding to ribosomes, suggesting that the binding site of these antibiotics is different from that of spectinomycin (BOLLEN et al., 1969b).

The ribosomes washed with 1 M NH$_4$Cl are more sensitive to aminoglycosides (kanamycin, gentamicin and streptomycin) than unwashed ribosomes (OKUYAMA et al., 1972). The binding of streptomycin to ribosomes is markedly reduced in the presence of added initiation factors (YAMADA et al., 1970).

Ribosomal Components Determining the Sensitivity and Resistance to Aminoglycosides

OZAKI et al. (1970) have identified a specific protein (P 10) of the 30S ribosomal subunit to be controlled by the str locus in E. coli. The reconstitution study has shown that P10 is responsible for the streptomycin sensitivity. The 30S particle reconstituted with P10 from streptomycin-resistant mutant and all other proteins from streptomycin-sensitive strain is resistant to streptomycin. Conversely, 30S particle reconstituted using P10 from the streptomycin-sensitive strain and all other proteins from the streptomycin-resistant mutant is sensitive to the antibiotic. Particles containing P10 from the streptomycin-resistant mutant or those not containing P10 bind little or no streptomycin or dihydrostreptomycin. Protein P10 itself, however, does not bind dihydrostreptomycin. Therefore, it appears that some structure involving both P10 and other ribosomal components is essential for binding of dihydrostreptomycin; and the mutation or alteration in P10 which accompanies mutation from streptomycin-sensitivity to streptomycin-resistance abolishes the binding ability of this structure.

The core particle obtained by treatment of the 30S ribosomal subunit with 5 M CsCl (16S RNA plus 15 proteins) does not bind streptomycin (TANAKA and KAJI, 1968). Six proteins can be removed from the 30S subunit with trypsin without affecting its ability to bind streptomycin. A decline in the ability of the 30S subunit to bind streptomycin is correlated with the removal of either one, or both, of two proteins (P8 and P11), neither one of which is the gene product of the *str* locus (CHANG and FLAKS, 1970). Protein P 10 may not be actually involved in the binding itself but may order the proper configuration of other proteins, possibly P8 and/or P11, or possibly the rRNA (BISWAS and GORINI, 1972). Alternatively, P10 binds streptomycin, but other proteins, particularly P8 and/or P11, participate in the binding (CHANG AND FLAKS, 1970).

By similar techniques as in the case of streptomycin, the sensitivity and resistance of other aminoglycosides have been determined to reside in the components of the 30S ribosomal subunit. The kanamycin sensitivity and resistance is localized to protein P10 (MASUKAWA *et al.*, 1968; MASUKAWA, 1969). Protein P4, distinctly different from P10, is responsible for the spectinomycin sensitivity and resistance (BOLLEN *et al.*, 1969a; FUNATSU *et al.*, 1972).

The kasugamycin sensitivity resides in 16S RNA but not in proteins of the 30S subunit. The resistance is attributed to non-methylation of 16S RNA. The 16S RNA from the resistant mutant differs from the sensitive in that it lacks the dimethylation of two adjacent adenine residues near 3′ end of the molecule (HELSER *et al.*, 1971). The kasugamycin-sensitive strain contains an RNA methylase which is capable of methylating the 16S RNA of the 30S core particle from the resistant mutant. The 21S core particle is a substrate for the methylase but the isolated 16S RNA is not a substrate. The methylation converts the resistant 16S RNA into sensitive. This methylase is lacking or inactive in the kasugamycin-resistant mutant. The *ksg A* locus seems to code for an RNA methylase (HELSER *et al.*, 1972).

Genetic Loci Controlling
the Sensitivity, Resistance or Dependency to Aminoglycosides

The sensitivity, resistance and dependency to streptomycin are determined by multiple alleles of a single locus at minute 64 in *E. coli* K12 (HASHIMOTO, 1960). The ribosomal mutations in *E. coli* resistance to spectinomycin and other aminoglycosides are located in a cluster near *str A* gene. They may form an operon with *ery* and *fus* loci (NOMURA and ENGBAECK, 1972). However, The *ksg* (kasugamycin) gene is located at a distance from this region, near the leucine region (minute 1) (SPARLING, 1968).

Effects of Aminoglycosides on Initiation Complex Formation

The *in vitro* 30S initiation complex formation is affected by various aminoglycosides; and the effect of aminoglycosides is highly dependent upon the cistron of mRNA. Using poly AUG as a messenger, the 30S initiation complex formation is inhibited by kasugamycin but not by streptomycin, kanamycin or gentamicin (OKUYAMA *et al.*, 1971; OKUYAMA *et al.*, 1972). The 30S initiation complex formation on f2 phage RNA is inhibited by kasugamycin, kanamycin or gentamicin; but not by streptomycin. The fMet-tRNA-stimulated binding of f2 RNA to the 30S ribosomal subunit is blocked by kasugamycin; but not by kanamycin and gentamicin. Kasugamycin inhibits initiation of translation of maturation protein cistron more markedly than that of coat protein cistron. Kanamycin and gentamicin block initiation of coat protein synthesis, but do not significantly affect initiation of maturation protein synthesis. Streptomycin blocks both protein syntheses at the same level.

Kanamycin and gentamicin inhibit initiation of translation of coat protein cistron, which form a base-pairing hairpin loop near initiation codon AUG and requires initiation factor IF3; but do not significantly affect initiation of translation of maturation protein cistron, which do not form a hairpin loop near initiation condon. This is in accordance with the results that kanamycin and gentamicin do not significantly affect the 30S initiation complex formation on poly AUG.

Kasugamycin preferentially blocks initiation of translation of maturation protein cistron, which lacks in a hairpin loop at the initiation site. It is also in accordance with the results that kasugamycin inhibits the 30S initiation complex formation on poly AUG (OKUYAMA and TANAKA, 1972).

KOZAK and NATHANS (1972) have observed that kasugamycin exerts differential inhibition of coliphage MS2 protein synthesis *in vivo* and *in vitro*, with coat protein synthesis being more resistant than maturation protein synthesis.

The reason for differential effects of aminoglycosides on cistron-specific initiation of protein synthesis remain to be determined. One possibility is that separate subclasses of 30S ribosomal subunits, differing in their sensitivity to aminoglycosides, initiate each phage cistron. An alternative explanation is that aminoglycosides differentially affect the ability of a single class of ribosome to bind to the three initiation sites on phage RNA.

Breakdown of Initiation Complex
by Streptomycin and Other Aminoglycosides

In vitro streptomycin does not affect the primary assembly step during which initiator tRNA is positioned on a 30S-ribosome-mRNA complex but causes release of bound fMet-tRNA from the preformed 70S initiation complex. Both ribosomal subunits and GTP hydrolysis are required for this reaction, indicating that fMet-tRNA cannot be released until bound to the puromycin-reactive (P) site on the sensitive ribosomes. Other aminoglycosides (paromomycin, neomycin and gentamicin) do not induce fMet-tRNA release, showing that streptomycin specifically induces distortion of the P site (LELONG et al., 1971; MODOLELL and DAVIS, 1970; ZAGORSKA, et al., 1971). The phenomenon may be related to the fact that streptomycin causes polysome breakdown and the other aminoglycosides accumulate polysomes.

OKUYAMA et al. (1972) have observed that kanamycin and gentamicin as well as streptomycin induce release of fMet-tRNA from the preformed 70S initiation complex. The results are different from those of LELONG et al. (1971). The discrepancy remains to be elucidated. However, since the sensitivity of washed ribosomes to the aminoglycosides is significantly different from that of unwashed ribosomes, the discrepancy may be due to the state of ribosomes employed.

The Inhibition by Aminoglycosides of Ribosomal Dissociation

Aminoglycosides (streptomycin, neomycin, kanamycin and spectinomycin) increase the ribosomal resistance to dissociation by heat or low Mg^{++} concentration (Herzog, 1964; Leon and Brock, 1967; Suzuki et al., 1968; Wolfe and Hahn, 1968).

The ribosomal dissociation produced in vitro by dissociation factor (DF) is antagonized by streptomycin, kanamycin, neomycin, paromomycin and gentamicin; but not by kasugamycin and spectinomycin. The resistance to DF seems to be caused by the direct interaction with ribosomes rather than by stabilization of complexing with fMet-tRNA or peptidyl-tRNA (García-Patrone et al., 1971; Herzog et al., 1971; Wallace et al., 1972).

Codon Misreading or Miscoding Induced by Aminoglycosides

As described above, the aminoglycosides affect peptide chain initiation by blocking the 30S initiation complex formation, by inducing breakdown of the 70S initiation complex and/or by inhibiting ribosomal dissociation. However, aminoglycosides with a deoxystreptamine or streptamine residue in the molecule cause codon misreading or miscoding, indicating that they also interfere with peptide elongation process in protein synthesis.

Kanamycin, neomycin and paromomycin as well as streptomycin and bluensomycin have been shown to disturb the fidelity of translation of the genetic code in vivo (Gorini and Kataja, 1965); and a similar effect has been observed in vitro with these antibiotics as well as hygromycin B and gentamicin (Davies et al., 1965).

Likover and Kurland (1967) have found that streptomycin has no miscoding effects in a system with highly purified ribosomes and supernatant fractions. Translation errors are introduced by addition of a small amounts of denatured DNA as a cofactor. This observation complicates the interpretation of miscoding caused by aminoglycosides. However, the evidence has been accumulated that in the presence of aminoglycosides miscoded amino acids are incorporated into protein in vivo: for instance, Bissel (1965) has observed that E. coli cells produce altered β-galactosidase in the presence of neomycin.

Using synthetic polynucleotides as templates, Davies et al. (1966) have revealed that the coding changes induced by neomycin are more pronounced than those induced by streptomycin. Streptomycin causes the misreading of the 5′-terminal and internal bases of a codon, of a single base at one time, and of pyrimidines more frequently than purines. Neomycin, in addition to the streptomycin coding errors, provokes the misreading of the 3′-terminal base and sometimes of two bases of a codon at one time.

Kanamycin induces substantial miscoding with cell-free extracts from kanamycin-sensitive E. coli; and the grade of miscoding is much reduced in extracts from kanamycin-resistant mutant. (Tanaka et al., 1967b). The miscoding spectrum of kanamycin resembles that of streptomycin (Davies et al., 1965; Tanaka et al., 1967b).

Neomycin, kanamycin, gentamicin and hygromycin B produce higher levels of miscoding than streptomycin and do not show a simple stoichiometric drug-to-ribosome ratio for this effect. These antibiotics show a 2- to 5-fold increase in level of miscoding as drug concentration is raised from 10^{-6} to 10^{-4} M, suggesting that each antibiotic may interact with more than one site. It is consistent with the absence of one-step mutants with high level resistance. Little increase in miscoding over the same concentration range is seen with streptomycin for which one-step high level resistant mutants are known. As described above, streptomycin acts on a single site of ribosome (DAVIES and DAVIS, 1968).

The binding of Ile-, Leu- and Ser-tRNA to ribosomes is enhanced by streptomycin at low concentration where the streptomycin:ribosome ratio is 1:1. The stimulation is observed with sensitive ribosomes but, only slightly with resistant ribosomes. The effects on aminoacyl-tRNA binding may be related to the miscoding (PESTKA et al., 1965; KAJI and KAJI, 1965; PESTKA, 1966). HATFIELD (1966) has shown that streptomycin not only inhibits the binding of Phe-tRNA to ribosomes with ^{3}H-UUC and -UUUC, but also interferes with the binding of these templates.

MCCARTHY and HOLLAND (1965, 1966) have found that denatured DNA, rRNA and tRNA can function as direct templates for protein synthesis in the presence of neomycin in extracts from E. coli. MASUKAWA and TANAKA (1967) have demonstrated that the amino acid incorporation into protein by DNA is markedly increased in the presence of neomycin and kanamycin. Less stimulation is observed with streptomycin; and kasugamycin does not significantly affect the amino acid incorporation by DNA. Kanamycin, neomycin or streptomycin stimulate the binding of aminoacyl-tRNA without affecting the attachment of DNA to ribosomes. Ribosomes obtained from drug-resistant mutants are resistant to these effects. PRICE and ROTTMAN (1970), and DUNLAP et al. (1971) have shown that amino acid incorporation dependent upon oligo-or polynucleotides containing 2'-O-methyl analogs is stimulated by neomycin.

All the results concerning the miscoding activity indicate that streptomycin may distort the A site of the 30S ribosomal subunit and interferes with aminoacyl-tRNA recognition (DAVIS, 1969). Streptomycin also seems to induce the P site distortion (MODOLELL and DAVIS, 1970).

The Structure of Antibiotics Required for Miscoding Activity

All the aminoglycosides which induce miscoding contain a deoxystreptamine or streptamine residue (streptomycin, kanamycin, neomycin, paromomycin, gentamicin and hygromycin B); and those, lacking in such a residue, do not induce miscoding (kasugamycin and spectinomycin). The chemical structure responsible for the miscoding activity of aminoglycosides has not yet been fully undertood, but some evidences for elucidating the structure-miscoding activity relationship have been accumulated.

TANAKA et al. (1967a) have proposed a hypothesis that the deoxystreptamine or streptamine moiety is responsible for the miscoding activity of aminoglycosides (the misreading of RNA and the reading of DNA templates in vitro). Since kasugamycin fails to induce miscoding (TANAKA *et al.*, 1965; TANAKA *et al.*, 1966c; TANAKA *et al.*, 1966b), they have studied the structural basis of kanamycin for miscoding activity and found that, of the three components of kanamycin, 2-deoxystreptamine has a weak miscoding activity but 6-amino-6-deoxy-D-glucose and 3-amino-3-deoxy-D-glucose lack miscoding activity (TANAKA *et al.*, 1967a) (see Fig. 3).

Paromamine and neamine also exhibit a miscoding activity *in vitro*; and the potency is lower than that of kanamycin but higher than that of deoxystreptamine. N-methyldeoxystreptamine shows less activity than deoxystreptamine. The activity of streptamine is at the same level or slightly less than that of deoxystreptamine. Actinamine lacks miscoding activity (see Fig. 8). The streochemistry of 2-position and free amino groups at 1 and 3 positions of aminocyclitol is important

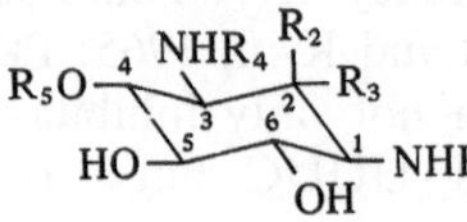

	R_1	R_2	R_3	R_4	R_5
Deoxystreptamine	H	H	H	H	H
Streptamine	H	OH	H	H	H
N-Methyl-deoxystreptamine	CH_3	H	H	H	H
Actinamine	CH_3	H	OH	CH_3	H
Paromamine	H	H	H	H	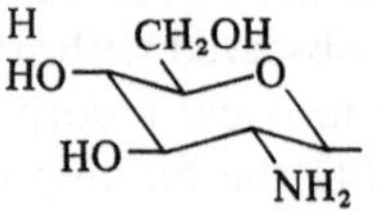
Neamine	H	H	H	H	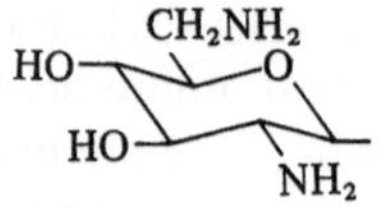

Fig. 8. 2-Deoxystreptamine and related compounds

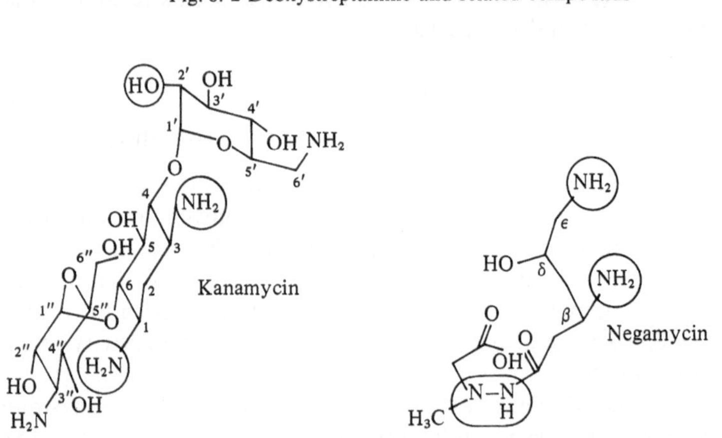

Fig. 9. The three-dimensional structure of kanamycin and negamycin (UEHARA *et al.*, 1972)

for miscoding activity (MASUKAWA and TANAKA, 1968). A marked loss of miscoding activity is noted in the neomycin-hybrimycin group, if the deoxystreptamine residue is replaced by an *epi*-streptamine moiety. This also supports the importance of the streochemistry of 2-position of aminocyclitol (DAVIES, 1970).

Basic peptide antibiotics, such as bottromycin A_2 and viomycin, do not induce miscoding, although they act on ribosomes and inhibit protein synthesis (TANAKA *et al.*, 1966; LIN and TANAKA, 1968; TANAKA and IGUSA, 1968). This suggests that the basic property of antibiotics is not enough to cause miscoding. However, recently negamycin, a basic peptide antibiotic with a unique structure, has been found to cause miscoding. (MIZUNO *et al.*, 1970; UEHARA *et al.*, 1972). A possible similarity in the three-dimensional structure between negamycin and the aminocyclitol antibiotics is revealed by use of CKP models (American Society of Biological Chemist, Inc.). As illustrated in Fig. 9, the positions of the hydrazide and β-amino groups in negamycin coincide with the 1- and 3-amino groups in the 2-deoxystreptamine moiety, and the ε-amino group can be placed in a position similar to that of the 2'-hydroxyl group in kanamycin. In effect, the basic group in a possible conformation of negamycin can be superimposed upon those of paromamine. For this conformation of negamycin, the existence of carbonyl and N-methyl groups and the stereochemistry of the β-amino and δ-hydroxyl groups are critical (UEHARA *et al.*, 1972). In support of this view, an antipode of negamycin synthesized from 3-amino-3-deoxy-D-glucose has been shown to exhibit much weaker miscoding activity than negamycin (SHIBAHARA *et al.*, 1972). The studies on negamycin seem to support the assumption that the 2-deoxystreptamine residue is important for the miscoding activity of aminocyclitol antibiotics.

Inhibition by Streptomycin of Peptide Chain Termination

Streptomycin has been reported to block peptide chain termination, probably by interfering with the interaction of release factor and ribosomes (CASKEY, 1971).

Mechanism of Resistance

Two major classes of resistance to aminoglycoside antibiotics have been reported: one is caused by the change of the 30S ribosomal subunit which is genetically controlled by the bacterial chromosome, and another is due to the inactivating enzyme production which is genetically carried by plasmid. The former is found in most of laboratory-developed mutants and in some of clinical isolates (see above). However, most of clinical isolates of drug-resistant *Enterobacteria* and *Staphylococci* as well as *Pseudomonas aeruginosa* show the latter form of resistance. Since the ribosomal change is described above, the production by bacteria of drug-inactivating enzymes will be discussed here.

OKAMOTO and SUZUKI (1965) have observed that *E. coli*, carrying R factor, produces chloramphenicol acetyltransferase. The enzyme inactivates the antibiotic by forming mono- and diacetates (OKAMOTO and SUZUKI, 1967; SHAW, 1967). The organism also produces other enzymes inactivating kanamycin and streptomycin respectively. Kanamycin acetyltransferase has been isolated from the same

Table 2. *Inactivation of aminoglycoside antibiotics by enzymes obtained from strains of E. coli carrying R factor, drug-resistant Staphylococcus aureus and/or Pseudomonas aeruginosa*

Enzymes	Strains	Substrates	Site of action	Inactivated products
kanamycin phospho- transferase (I)	E. coli P. aeruginosa S. aureus	kanamycins A, B and C gentamicin A, ribostamycin neomycins, paromomycins lividomycin A	3'-OH 5''-OH	3'-phosphate 5''-phosphate
kanamycin phospho- transferase (II)	E. coli	the above antibiotics butirosin A	3'-OH	3'-phosphate
streptomycin phospho- transferase	E. coli P. aeruginosa	streptomycin dihydrostreptomycin	3''-OH	3''-phosphate
kanamycin adenyl- transferase	E. coli	kanamycins A, B and C, DKB gentamicin C_1, tobramycin	2''-OH	2''-adenylate
streptomycin adenyl- transferase	E. coli	streptomycin	3''-OH	3''-adenylate
kanamycin acetyl- transferase	E. coli P. aeruginosa	kanamycins A and B, DKB gentamicins C_{1a} and C_2 tobramycin, ribostamycin neomycins	6'-NH_2	6'-acetate
gentamicin acetyl- transferase	P. aeruginosa	gentamicins C_1, C_{1a} and C_2	3-NH_2	3-acetate

DKB: 3',4'-dideoxykanamycin B.

organism. Since then, a number of enzymes, which inactivate aminoglycoside antibiotics, have been obtained from a variety of strains of *E. coli* carrying R factor, drug-resistant *Staphylococci,* and *Pseudomonas aeruginosa* (Table 2). The mechanism of resistance seems to be common in all the three groups of bacteria, producing the same sorts of inactivating enzymes (Fig. 10).

Phosphorylation of Aminoglycosides

Kanamycin acetyltransferase, which makes kanamycin 6'-acetate of kanamycin, acts on kanamycins A and B, but not on kanamycin C and paromamine (OKANISHI *et al.,* 1967; UMEZAWA *et al.,* 1967b). Another kanamycin-inactivating enzyme has been found in *E. coli* carrying a different type of R factor, which shows more broad resistance to aminoglycosides including kanamycin C. The enzyme converts kanamycin or paromamine to kanamycin 3'-phosphate or paromamine 3'-phosphate (UMEZAWA *et al.,* 1967a; OKANISHI *et al.,* 1968; KONDO *et al.,* 1968). Kanamycin phosphotransferase is also demonstrated in *Pseudomonas aeruginosa* strains that do not appear to be controlled by a transferable resistant factor (UMEZAWA *et al.,* 1968a; DOI *et al.,* 1968b), and in *Staphylococcus aureus*

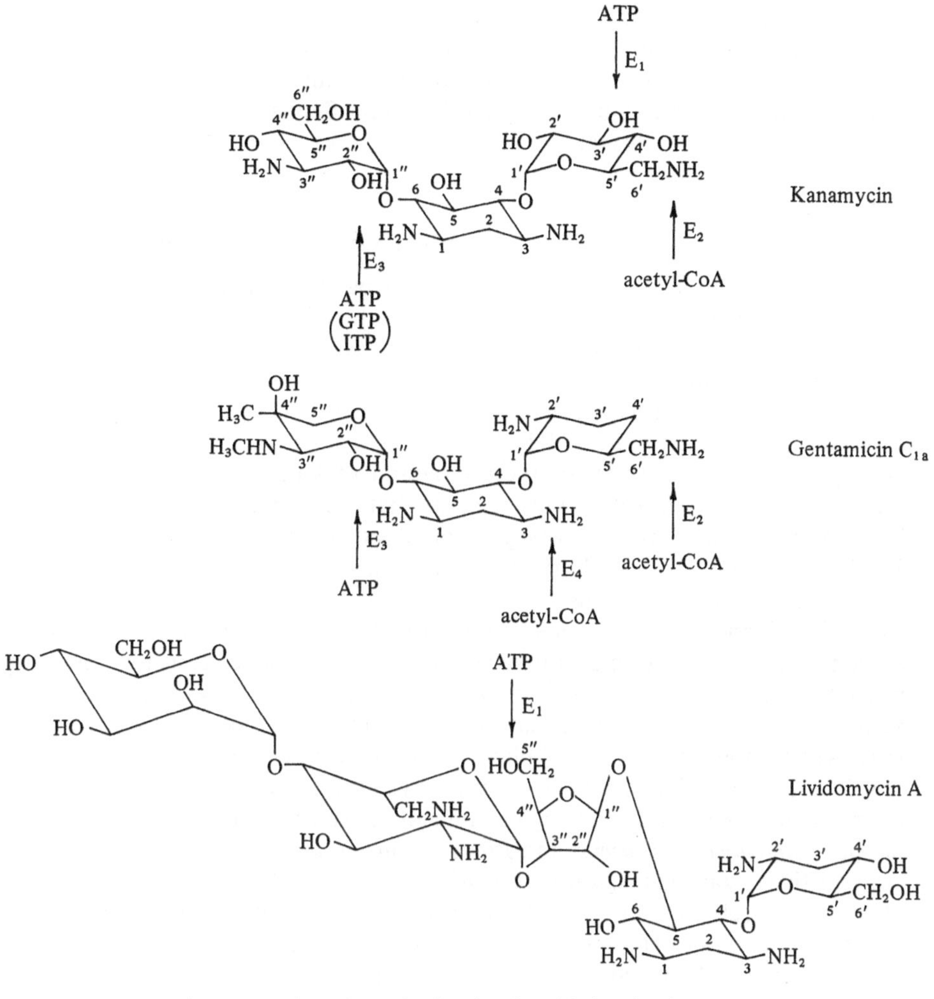

Fig. 10. The sites of attack of aminoglycoside inactivating enzymes:

E_1: kanamycin phosphotransferase
E_2: kanamycin acetyltransferase
E_3: kanamycin adenyltransferase
E_4: gentamicin acetyltransferase

strains carrying drug-resistant plasmid (DOI *et al.*, 1968a). Kanamycin phospho-transferase reacts with a number of aminoglycosides, containing 3'-hydroxyl group which is attacked by the enzyme, such as kanamycins A, B and C, neomycins, paromomycins and gentamicin A (DAVIES *et al.*, 1971). Gentamicins C_1, C_{1a} and C_2, 3'-deoxy-kanamycin A, 3',4'-dideoxy-kanamycin B (DKB), and tobramycin, do not serve as substrates for kanamycin phosphotransferase, since all lack a hydroxyl group on 3'-carbon. Thus kanamycin phosphotransferase is thought to be the major mechanism of resistance to kanamycin, neomycin and paromomycin of many clinically significant organisms: *Pseudomonas aeruginosa, Staphylococcus aureus,* and *Enterobacteriaceae.*

Kanamycin phosphotransferase is extracted from *Pseudomonas aeruginosa,* and purified by ammonium sulfate fractionation, Sephadex G-100 and DEAE Sephadex A-50 chromatography. The enzyme catalyzes the reaction, in which ATP reacts with kanamycin on an equimolar basis yielding kanamycin 3′-phosphate and ADP.

$$\text{Kanamycin} + \text{ATP} \rightarrow \text{Kanamycin 3′-phosphate} + \text{ADP}$$

Mg^{++} is required for the reaction and can be substituted by Mn^{++}, Zn^{++} or Co^{++}. The enzyme is protected from heat denaturation by kanamycin. Substrates possessing the specific structures of 6-amino-6-deoxy-α-D-glucopyranosyl, 2,6-diamino-2,6-dideoxy-α-D-glucopyranosyl or 2-amino-2-deoxy-α-D-glucopyranosyl 2-deoxystreptamine are required for the reaction. The studies on substrates and inhibitors suggest that 6′-NH_2 (or 6′-OH), 2′-OH (or 2′-NH_2), 1-NH_2 and 3-NH_2 groups are essential for the interaction of kanamycin A, B or C with the enzyme (Doi et al., 1969). This suggests that the derivatives effective against kanamycin-resistant organisms may be obtained by modifying these groups or 3′-OH group which is phosphorylated by the enzyme (Fig. 11).

3′-O-methylkanamycin and 3′-deoxykanamycin have been chemically synthesized. The former shows a weak antimicrobial activity, but the latter exhibits a strong activity comparable with kanamycin and is effective against kanamycin-resistant organisms which produce kanamycin phosphotransferase. The results seem to confirm the conclusion that the kanamycin resistance of these organisms is caused by the production of phosphotransferase (Umezawa et al., 1971b). DKB (3′,4′-dideoxykanamycin B) is chemically synthesized. It inhibits kanamycin-resistant organisms producing kanamycin phosphotransferase *(Enterobacteriaceae, Staphylococcus aureus,* and *Pseudomonas aeruginosa)* as well as kanamycin-sensitive organisms (Umezawa et al., 1971a). DKB has been shown to be effective clinically against both kanamycin-resistant and -sensitive organisms. DKB exhibits lower MIC (minimal growth-inhibitory concentration) than gentamicin C against *Pseudomonas aeruginosa.* 3′,4′-Dideoxyribostamycin has been

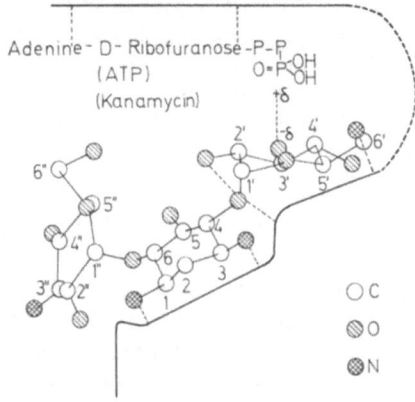

Fig. 11. Scheme for the reaction at the catalytic center of kanamycin phosphotransferase (Doi et al., 1969)

also synthesized and found to be effective against *Pseudomonas aeruginosa* and other organisms which produce kanamycin phosphotransferase (UMEZAWA *et al.*, 1972 b).

Another type of kanamycin phosphotransferase has been recently isolated from a strain of *E. coli* carrying R factor. This enzyme (II) is differentiated from kanamycin phosphotransferase (I) because it phosphorylates the 3'-hydroxyl group of butirosin A which is not affected by kanamycin phosphotransferase (I). It attacks kanamycin, neamine and ribostamycin as well as butirosin A (YAGISAWA *et al.*, 1972 c).

Lividomycin phosphotransferase has been obtained from *Pseudomonas aeruginosa* and *E. coli* carrying R factor. The enzyme is identical with kanamycin phosphotransferase I (UMEZAWA *et al.*, 1973), and converts lividomycin A to lividomycin A 5''-phosphate in the presence of ATP (KOBAYASHI *et al.*, 1972; KONDO *et al.*, 1972; YAMAMOTO *et al.*, 1972 a). 5''-Deoxylividomycin A, 5''-amino-5''-deoxylividomycin A and 5''-deoxylividomycin B have been synthesized and found to be effective against lividomycin-resistant organisms, although their antibacterial activity is not high (YAMAMOTO *et al.*, 1972 b; UMEZAWA *et al.*, 1972 c).

Streptomycin phosphotransferase, which produces streptomycin 3''-phosphate of streptomycin and ATP, is found in R factor-carrying *E. coli* and *Pseudomonas aeruginosa* (OZANNE *et al.*, 1969; KAWABE *et al.*, 1971; NAGANAWA *et al.*, 1971 a).

Adenylation of Aminoglycosides

Kanamycin adenyltransferase has been isolated from DKB-resistant *E. coli* carrying R factor. This enzyme catalyzes the reaction, in which kanamycin or DKB react with ATP, forming kanamycin or DKB 2''-adenylate (YAGISAWA *et al.*, 1971; NAGANAWA *et al.*, 1971 b). The enzyme can react with GTP or ITP instead of ATP, resulting in formation of 2''-guanylate or 2''-inosinate (YAGISAWA *et al.*, 1972 a). Gentamicin C_1 is also adenylated by this enzyme (BENVENISTE and DAVIES, 1971 a). Kanamycins B and C as well as tobramycin are also substrates for this enzyme, while aminoglycosides with pentose (neomycin, lividomycin and butirosin) are not susceptible.

Streptomycin adenyltransferase is found in *E. coli* carrying R factor. The antibiotic is converted to streptomycin 3''-adenylate by the enzyme (UMEZAWA *et al.*, 1968 b; TAKASAWA *et al.*, 1968).

Acetylation of Aminoglycosides

Kanamycin acetyltransferase has been isolated from *E. coli* carrying R factor and *Pseudomonas aeruginosa*. The enzyme reacts with kanamycin and acetyl-CoA, yielding kanamycin 6'-acetate (OKANISHI *et al.*, 1967; UMEZAWA *et al.*, 1967 b). Kanamycins A and B, DKB, gentamicins C_{1a} and C_2, tobramycin, ribostamycin and neomycins are substrates for the enzyme; but kanamycin C, gentamicins A and C_1, and paromomycins are not susceptible (BENVENISTE and DAVIES, 1971; YAGISAWA *et al.*, 1972 b; YAMAMOTO *et al.*, 1972 c).

6'-N-methylkanamycin and 3',4'-dideoxy-6'-N-methylkanamycin B have been synthesized and shown to be active against resistant organisms possessing 6'-N-acetylating enzyme (UMEZAWA *et al.*, 1972 a).

Gentamicin acetyltransferase, an enzyme acetylating gentamicins but not kanamycins, has been isolated from strains of *Pseudomonas aeruginosa*. Its site of action is the 3-NH_2 group of 2-deoxystreptamine (BRZEZINSKA *et al., 1972*). Those aminoglycosides which lack a 4'-hydroxyl group such as gentamicins C_1, C_{1a}, and C_2 are good substrates for the enzyme; but kanamycins A and B, tobramycin and gentamicin A are poor ones.

N^1-Acyl Aminocyclitol Antibiotics

The structure of butirosins has been elucidated: butirosin B is N^1-(L-γ-amino-α-hydroxybutyryl)ribostamycin and butirosin A contains xylose instead of ribose (Fig. 6). Butirosins A and B exhibit higher antimicrobial activity than ribostamycin, and are active against *Pseudomonas aeruginosa* and other drug-resistant organisms (WOO *et al., 1971*). It indicates that N^1-acylation enhances the antibacterial activity, particularly against drug-resistant organisms possibly by decreasing the affinity for inactivating enzymes. N^1-(L-γ-amino-α-hydroxybutyryl) kanamycin (BB-K8) has been synthesized and found to exhibit antibacterial activity generally equal to kanamycin. BB-K8 is active against kanamycin- and/or gentamicin-resistant organisms, including *Pseudomonas* strains, which produce kanamycin-phosphotransferase, -adenyltransferase and/or gentamicin acetyltransferase (KAWAGUCHI *et al., 1972*; PRINCE *et al., 1972*).

References

AKITA, E., T. TSURUOKA, N. EZAKI, and T. NIIDA: Studies on antibiotic SF-733, a new antibiotic. II. Chemical structure of antibiotic SF-733. J. Antibiotics (Tokyo) **23**, 173 (1970).

ARCAMONE, F., and F. BIZIOLI: Isolation and constitution of trehalosamine, a new aminosugar from a streptomycetes. Gazz. Chim. Ital. **87**, 896 (1957).

BENVENISTE, R., and J. DAVIES: R-Factor mediated gentamicin resistance: a new enzyme which modifies aminoglycosidic antibiotics. FEBS Letters **14**, 293 (1971 a).

BENVENISTE, R., and J. DAVIES: Enzymatic acetylation of aminoglycoside antibiotics by *Escherichia coli* carrying an R factor. Biochemistry **10**, 1787 (1971 b).

BISSEL, D. M.: Formation of an altered enzyme by *Escherichia coli* in the presence of neomycin. J. Mol. Biol. **14**, 619 (1965).

BISWAS, D. K., and L. GORINI: The attachment site of streptomycin to the 30S ribosomal subunit. Proc. Natl. Acad. Sci. U.S. **69**, 2141 (1972).

BODLEY, J. W., and E. M. DAVIE: A study of the mechanism of ambiguous amino acid coding by poly U: the nature of the products. J. Mol. Biol. **18**, 344 (1966).

BOLLEN, A., J. DAVIES, M. OZAKI, and S. MIZUSHIMA: Identification of the ribosomal protein conferring sensitivity to the antibiotic spectinomycin in *Escherichia coli*. Science **165**, 85 (1969a).

BOLLEN, A., T. HELSER, T. YAMADA, and J. DAVIES: Altered ribosomes in antibiotic-resistant mutants of *E. coli*. Cold Spring Harbor Symp. Quant. Biol. **34**, 95 (1969b).

BRZEZINSKA, M., R. BENVENISTE, J. DAVIES, P. J. L. DANIELS, and J. WEINSTEIN: Gentamicin resistance in strains of *Pseudomonas aeruginosa* mediated by enzymatic N-acetylation of the deoxystreptamine moiety. Biochemistry **11**, 761 (1972).

CASKEY, C. T.: In: Molecular mechanism of antibiotic action on protein biosynthesis and membranes. Amsterdam: Elsevier Pub. Co. 1972.

CHANG, F. N., and J. G. FLAKS: Topography of the *Escherichia coli* 30S ribosomal subunit and streptomycin binding. Proc. Natl. Acad. Sci. U.S. **67**, 1321 (1970).

CHANG, F. N., and J. G. FLAKS: The binding of dihydrostreptomycin to *E. coli* ribosomes: Characteristics and equilibrium of the reaction. Antimicr. Ag. Chemoth. **2**, 294 (1972).

CHANG, F. N., and J. G. FLAKS: The binding of dihydrostreptomycin to *E. coli* ribosomes: Kinetics of the reaction. Antimicr. Ag. Chemoth. **2**, 308 (1972).

COOPER, D. J., R. S. JARET, and H. REIMANN: Structure of sisomicin, a novel unsaturated aminoglycoside antibiotic from *Micromonospora inyoensis*. Chem. Commun. 285 (1971).

COOPER, D. J., H. M. MARIGLIANO, M. D. YUDIS, and T. TRAUBEL: Recent developments in the chemistry of gentamicin. J. Infect. Diseases 119, 342 (1969).

DAVIES, J.: Structure-activity relationships among the aminoglycoside antibiotics: Comparison of the neomycins and hybrimycins. Biochim. Biophys. Acta 222, 674 (1970).

DAVIES, J., M. BRZEZINSKA, and R. BENVENISTE: R factors: Biochemical mechanisms of resistance to aminoglycoside antibiotics. Ann. N. Y. Acad. Sci. 182, 226 (1971).

DAVIES, J., and B. D. DAVIS: Misreading of RNA codewords induced by aminoglycoside antibiotics: the effect of drug concentration. J. Biol. Chem. 243, 3312 (1968).

DAVIES, J., W. GILBERT, and L. GORINI: Streptomycin, suppression and the code. Proc. Natl. Acad. Sci. U. S. 51, 883 (1964).

DAVIES, J., L. GORINI, and B. D. DAVIS: Misreading of RNA codewords induced by aminoglycoside antibiotics. Mol. Pharmacol. 1, 93 (1965).

DAVIS, B. D.: Use of antibiotics in the study of ribosome action. Antimicr. Ag. Chemoth. 1969, 11.

DEL BENE, V. E., and W. E. FARRAR, JR.: Tobramycin: *In vitro* activity and comparison with kanamycin and gentamicin. Antimicr. Ag. Chemoth. 1, 340 (1972).

DOI, O., S. KONDO, N. TANAKA, and H. UMEZAWA: Phosphorylating enzyme from *Pseudomonas aeruginosa*. J. Antibiotics (Tokyo) 22, 273 (1969).

DOI, O., M. MIYAMOTO, N. TANAKA, and H. UMEZAWA: Inactivation and phosphorylation of kanamycin by drug-resistant *Staphylococcus aureus*. Appl. Microbiol. 16, 1282 (1968a).

DOI, O., M. OGURA, N. TANAKA, and H. UMEZAWA: Inactivation of kanamycin, neomycin and streptomycin by enzymes obtained in cells of *Pseudomonas aeruginosa*. Appl. Microbiol. 16, 1276 (1968b).

DUNLAP, B. E., K. H. FRIDERICI, and F. ROTTMAN: 2'-O-Methyl polynucleotides as templates for cell-free amino acid incorporation. Biochemistry 10, 2581 (1971).

FUNATSU, G., K. NIERHAUS, and B. WITTMANN-LIEBOLD: Ribosomal proteins. XXII. Studies on the altered protein S5 from a spectinomycin-resistant mutant of *Escherichia coli*. J. Mol. Biol. 64, 201 (1972).

GARCÍA-PATRONE, M., C. A. PERAZZOLO, F. BARALLE, N. S. GONZÁLEZ, and I. D. ALGRANATI: Studies on dissociation factor of bacterial ribosomes: Effect of antibiotics. Biochim. Biophys. Acta 246, 291 (1971).

GORINI, L., and E. KATAJA: Phenotypic repair by streptomycin of defective genotypes in *E. coli*. Proc. Natl. Acad. Sci. U. S. 51, 487 (1964).

GURGO, C., D. APIRION, and D. SCHLESSINGER: Polyribosome metabolism in *Escherichia coli* treated with chloramphenicol, neomycin, spectinomycin or tetracycline. J. Mol. Biol. 45, 205 (1969).

HASHIMOTO, K.: Streptomycin resistance in *Escherichia coli* analyzed by transduction. Genetics 45, 49 (1960).

HATFIELD, D.: Oligonucleotide-ribosome-AA-sRNA interactions. Cold Spring Harbor Symp. Quant. Biol. 34, 619 (1966).

HELSER, T. L., J. E. DAVIES, and J. E. DAHLBERG: Change in methylation of 16S ribosomal RNA associated with mutation to kasugamycin resistance in *Escherichia coli*. Nature New Biol. 233, 12 (1971).

HELSER, T. L., J. E. DAVIES, and J. E. DAHLBERG: Mechanism of kasugamycin resistance in *Escherichia coli*. Nature New Biol. 235, 6 (1972).

HERR, R. R., T. E. EBLE, M. E. BERGY, and H. K. JANKE: Isolation and characterization of streptozotocin. Antibiot. Ann. 1959/1960, 236 (1960).

HERR, R. R., H. K. JANKE, and A. D. ARGOUDELIS: The structure of streptozotocin. J. Am. Chem. Soc. 89, 4808 (1967).

HERZOG, A.: An effect of streptomycin on the dissociation of *Escherichia coli* 70S ribosomes. Biochem. Biophys. Res. Commun. 15, 172 (1964).

HERZOG, A., A. GHYSEN, and A. BOLLEN: Sensitivity and resistance to streptomycin in relation with factor-mediated dissociation of ribosomes. FEBS Letters 15, 291 (1971).

HICHENS, M., and K. L. RINEHART: Chemistry of the neomycins. XII. The absolute configuration of deoxystreptamine in the neomycins, paromomycins and kanamycins. J. Am. Chem. Soc. 85, 1547 (1963).

Hoeksema, H., A. D. Argoudelis, and P. F. Wiley: Chemistry of actinospectacin II. The structure of actinospectacin. J. Am. Chem. Soc. **84**, 3212 (1962).

Holland, J. J., C. A. Buck, and B. J. MaCarthy: Stimulation of protein synthesis *in vitro* by partially degraded ribosomal ribonucleic acid and transfer ribonucleic acid. Biochemistry **5**, 358 (1966).

Horii, S., and Y. Kameda: Structure of the antibiotic validamycin A. J. C. S. Chem. Commun. 747 (1972).

Inoue, S., T. Tsuruoka, and T. Niida: The structure of nojirimycin, a piperidinose sugar antibiotic. J. Antibiotics (Tokyo), Ser. A **19**, 288 (1966).

Ishida, N., K. Kumagai, T. Niida, K. Hamamoto, and T. Shomura: Nojirimycin, a new antibiotic. I. Taxonomy and fermentation. J. Antibiotics (Tokyo), Ser. A **20**, 62 (1967).

Ishida, N., K. Kumagai, T. Niida, T. Tsuruoka, and H. Yumoto: Nojirimycin, a new antibiotic. II. Isolation, characterization and biological activity. J. Antibiotics (Tokyo), Ser. A **20**, 66 (1967).

Kaji, H., and A. Kaji: Specific binding of sRNA to ribosomes: effect of streptomycin. Proc. Natl. Acad. Sci. U. S. **54**, 213 (1965).

Kaji, H., and Y. Tanaka: Binding of dihydrostreptomycin to ribosomal subunits. J. Mol. Biol. **32**, 221 (1968).

Kawabe, H., F. Kobayashi, M. Yamaguchi, R. Utahara, and S. Mitsuhashi: 3″-Phosphoryldihydrostreptomycin produced by the inactivating enzyme of *Pseudomonas aeruginosa*. J. Antibiotics (Tokyo), **24**, 651 (1971).

Kawaguchi, H., T. Naito, S. Nakagawa, and K. Fujisawa: BB-K8, a new semisynthetic aminoglycoside antibiotic. J. Antibiotics (Tokyo) **25**, 695 (1972).

Kobayashi, F., M. Yamaguchi, and S. Mitsuhashi: Activity of lividomycin against *Pseudomonas aeruginosa*: Its inactivation by phosphorylation induced by resistant strain. Antimicr. Ag. Chemoth. **1**, 17 (1972).

Koch, K. F., and J. A. Rhodes: Structure of nebramycin factor 6, a new aminoglycosidic antibiotic. Antimicr. Ag. Chemoth. **1970**, 313 (1971).

Kogut, M., and E. Prizant: Effects of dihydrostreptomycin treatment *in vivo* on the ribosome cycle in *Escherichia coli*. FEBS Letters **12**, 17 (1970).

Kondo, S., E. Akita, and M. Koike: The structure of destomycin A. J. Antibiotics (Tokyo), Ser. A **19**, 139 (1966).

Kondo, S., M. Okanishi, R. Utahara, K. Maeda, and H. Umezawa: Isolation of kanamycin and paromamine inactivated by *E. coli* carrying R factor. J. Antibiotics (Tokyo) **21**, 22 (1968).

Kondo, S., M. Sezaki, M. Koike, M. Shimura, E. Akita, K. Satoh, and T. Hara: Destomycins A and B, two new antibiotics produced by a *Streptomyces*. J. Antibiotics (Tokyo), Ser. A **18**, 38 (1965).

Kondo, S., H. Yamamoto, H. Naganawa, H. Umezawa, and S. Mitsuhashi: Isolation and characterization of lividomycin A inactivated by *Pseudomonas aeruginosa* and *Escherichia coli* carrying R factor. J. Antibiotics (Tokyo) **25**, 483 (1972).

Koyama, G., Y. Iitaka, K. Maeda, and H. Umezawa: The crystal structure of kanamycin. Tetrahedron Letters **1968**, 1875.

Kozak, M. and D. Nathans: Differential inhibition of coliphage MS2 protein synthesis by ribosome-directed antibiotics. J. Mol. Biol. **70**, 41 (1972).

Kreider, G., and B. L. Brownstein: A mutation suppressing streptomycin dependence II: An altered protein on the 30S ribosomal subunit. J. Mol. Biol. **61**, 135 (1971).

Kreider, G., and B. L. Brownstein: Ribosomal proteins involved in the suppression of streptomycin dependence in *Escherichia coli*. J. Bacteriol. **109**, 780 (1972).

Lelong, J. C., M. A. Cousin, D. Gros, M. Grunberg-Manago, and F. Gros: Streptomycin induced release of fMet-tRNA from the ribosomal initiation complex. Biochem. Biophys. Res. Commun. **42**, 530 (1971).

Lennette, E. T., and D. Apirion: The level of f-Met-tRNA on ribosomes from streptomycin treated cells. Biochem. Biophys. Res. Commun. **41**, 804 (1970).

Leon, S. A., and T. D. Brock: Effect of streptomycin and neomycin on physical properties of the ribosome. J. Mol. Biol. **24**, 391 (1967).

Likover, T. E. and C. G. Kurland: The contribution of DNA to translation errors induced by streptomycin *in vitro*. Proc. Natl. Acad. Sci. U. S. **58**, 2385 (1967).

Likover, T. E., and C. G. Kurland: Ribosomes from a streptomycin-dependent strain of *Escherichia coli*. J. Mol. Biol. **25**, 497 (1967).

LIN, Y., and N. TANAKA: Mechanism of bottromycin action in polypeptide biosynthesis. J. Biochem. (Tokyo) **63**, 1 (1968).

LUZZATTO, L., D. APIRION, and D. SCHLESSINGER: Mechanism of action of streptomycin in *E. coli*: interruption of the ribosome cycle at the initiation of protein synthesis. Proc. Natl. Acad. Sci. U.S. **60**, 873 (1968).

LUZZATTO, L., D. APIRION, and D. SCHLESSINGER: Streptomycin action: Greater inhibition of *Escherichia coli* ribosome function with exogenous than with endogenous messenger ribonucleic acid. J. Bacteriol. **99**, 206 (1969a).

LUZZATTO, L., D. APIRION, and D. SCHLESSINGER: Polyribosome depletion and blockage of the ribosome cycle by streptomycin in *Escherichia coli*. J. Mol. Biol. **42**, 315 (1969b).

MACHIYAMA, N.: Mechanism of action of lividomycin, a new aminoglycosidic antibiotic. J. Antibiotics (Tokyo) **24**, 706 (1971).

MAEHR, H., and C.P. SCHAFFNER: Chemistry of the gentamicins. II. Stereochemistry and synthesis of gentosamine. Total structure of gentamicin A. J. Am. Chem. Soc. **92**, 1697 (1970).

MASON, D.J., A. DIETZ, and R.M. SMITH: Actinospectacin, a new antibiotic. I. Discovery and biological properties. Antibiot. & Chemotherapy **11**, 118 (1961).

MASUKAWA, H.: Localization of sensitivity to kanamycin and streptomycin in 30S ribosomal proteins of *Escherichia coli*. J. Antibiotics (Tokyo) **22**, 612 (1969).

MASUKAWA, H., and N. TANAKA: Stimulation by aminoglycosidic antibiotics of DNA-directed protein synthesis. J. Biochem. (Tokyo) **62**, 202 (1966).

MASUKAWA, H., and N. TANAKA: Miscoding activity of aminosugars. J. Antibiotics (Tokyo), Ser. A **21**, 70 (1968).

MASUKAWA, H., N. TANAKA, and H. UMEZAWA: Localization of kanamycin sensitivity in the 23S core of 30S ribosomes of *E. oli*. J. Antibiotics (Tokyo), Ser. A **21**, 517 (1968).

MCCARTHY, B.J., and J.J. HOLLAND: Denatured DNA as a direct template for *in vitro* protein synthesis. Proc. Natl. Acad. Sci. U.S. **54**, 880 (1965).

MITSUHASHI, S., F. KOBAYASHI, and M. YAMAGUCHI: Enzymatic inactivation of gentamicin C components by cell-free extract from *Pseudomonas aeruginosa*. J. Antibiotics (Tokyo) **24**, 400 (1971).

MIZUNO, S., K. NITTA, and H. UMEZAWA: Mechanism of action of negamycin in *E. coli* K12. II. Miscoding activity in polypeptide synthesis directed by synthetic polynucleotide. J. Antibiotics (Tokyo) **23**, 589 (1970).

MODOLELL, J., and B.D. DAVIS: Breakdown by streptomycin of initiation complexes formed on ribosomes of *Escherichia coli*. Proc. Natl. Acad. Sci. U.S. **67**, 1148 (1970).

MORI, T., T. ICHIYANAGI, H. KONDO, K. TOKUNAGA, T. ODA, and K. MUNAKATA: Studies on new antibiotic lividomycins. II. Isolation and characterization of lividomycins A, B and other aminoglycosidic antibiotics produced by *Streptomyces lividus*. J. Antibiotics (Tokyo) **24**, 339 (1971).

MURASE, M., T. ITO, S. FUNATSU, and H. UMEZAWA: Studies on kanamycin related compounds produced during fermentation by mutants of *Streptomyces kanamyceticus*. Isolation and properties. Progr. Antimicr. Anticancer Chemoth. (Univ. Tokyo Press) **2**, 1098 (1970).

NAGANAWA, H., S. KONDO, K. MAEDA, and H. UMEZAWA: Structure determination of enzymatically phosphorylated products of aminoglycoside antibiotic by proton magnetic resonance. J. Antibiotics (Tokyo) **24**, 823 (1971a).

NAGANAWA, H., M. YAGISAWA, S. KONDO, T. TAKEUCHI, and H. UMEZAWA: The structure determination of an enzymatic inactivation product of 3',4'-dideoxykanamycin B. J. Antibiotics (Tokyo) **24**, 913 (1971b).

NEUSS, N., K.F. KOCH, B.B. MOLLOY, W. DAY, L.L. HUCKSTEP, D.E. DORMAN, and J.D. ROBERTS: Structure of hygromycin B, an antibiotic from *Streptomyces hygroscopicus*: the use of CMR spectra in structure determination. I. Helv. Chim. Acta **53**, 2314 (1970).

NOMURA, M., and F. ENGBAEK: Expression of ribosomal protein genes as analyzed by bacteriophage Mu-induced mutations. Proc. Natl. Acad. Sci. U.S. **69**, 1526 (1972).

ODA, T., T. MORI, Y. KYOTANI, and M. NAKAYAMA: Studies on new antibiotic lividomycins. IV. Structure of lividomycin A. J. Antibiotics (Tokyo) **24**, 511 (1971).

OKAMOTO, S., and Y. SUZUKI: Chloramphenicol-, dihydrostreptomycin-, and kanamycin-inactivating enzymes from multiple drug-resistant *Escherichia coli* carrying episome "R". Nature **208**, 1301 (1965).

OKANISHI, M., S. KONDO, Y. SUZUKI, S. OKAMOTO, and H. UMEZAWA: Studies on inactivation of kanamycin and resistance of *E. coli*. J. Antibiotics (Tokyo), Ser. A **20**, 132 (1967).

Okanishi, M., S. Kondo, R. Utahara, and H. Umezawa: Phosphorylation and inactivation of aminoglycosidic antibiotics by *E. coli* carrying R factor. J. Antibiotics (Tokyo) **21**, 13 (1968).

Okuyama, A., N. Machiyama, T. Kinoshita, and N. Tanaka: Inhibition by kasugamycin of initiation complex formation on 30S ribosomes. Biochem. Biophys. Res. Commun. **43**, 196 (1971).

Okuyama, A., and N. Tanaka: Differential effects of aminoglycosides on cistron-specific initiation of protein synthesis. Biochem. Biophys. Res. Commun. **49**, 951 (1972).

Okuyama, A., T. Watanabe, and N. Tanaka: Effects of aminoglycoside antibiotics on initiation of viral RNA-directed protein synthesis. J. Antibiotics (Tokyo) **25**, 212 (1972).

Old, D., and L. Gorini: Amino acid changes provoked by streptomycin in a polypeptide synthesized *in vitro*. Science **150**, 1290 (1965).

Oliver, T.J., A. Goldstein, R.R. Bower, J.C. Holper, and R.H. Otto: M-141, a new antibiotic. I. Antimicrobial properties, identity with actinospectacin, and production by *Streptomyces flavopersicus*, sp. n. Antimicr. Ag. Chemoth. 495 (1961).

Ozaki, M., S. Mizushima, and M. Nomura: Identification and functional characterization of the protein controlled by the streptomycin-resistant locus in *E. coli*. Nature **226**, 333 (1969).

Ozanne, B., R. Benveniste, D. Tipper, and J. Davies: Aminoglycoside antibiotics: inactivation by phosphorylation in *E. coli* carrying R factor. J. Bacteriol. **100**, 1144 (1969).

Pestka, S., R. Marchall, and M. Nirenberg: RNA codewords and protein synthesis. V. Effects of streptomycin on the formation of ribosome-sRNA complexes. Proc. Natl. Acad. Sci. U.S. **53**, 639 (1965).

Pestka, S.: Studies on the formation of transfer ribonucleic acid-ribosome complexes. I. The effect of streptomycin and ribosomal dissociation on ^{14}C-aminoacyl transfer ribonucleic acid binding to ribosomes. J. Biol. Chem. **241**, 367 (1966).

Petitpas-Dewandre, A., H. Barbason, and W.G. Verly: Affinite pour la streptomycine des ribosomes d'*Escherichia coli*. Europ. J. Biochem. **7**, 307 (1969).

Price, A.R., and F. Rottman: 2'-O-Methyloligoadenylates as templates for the binding of lysyl transfer ribonucleic acid to ribosomes. Biochemistry **9**, 4524 (1970).

Price, K.E., D.R. Chisholm, M. Misiek, F. Leitner, and Y.H. Tsai: Microbiological evaluation of BB-K8, a new semisynthetic aminoglycoside. J. Antibiotics (Tokyo) **25**, 709 (1972).

Reimann, H., R.S. Jaret, and D.J. Cooper: Sisomicin: Stereochemistry and attachment of the unsaturated sugar moiety. Chem. Commun. 924 (1971).

Rinehart, K.L.: The neomycins and related antibiotics. New York: John Wiley & Sons 1964.

Rinehart, K.L., M. Hichens, A.D. Argoudelis, W.S. Chilton, H.E. Carter, M.P. Georgiadis, C.P. Schaffner, and R.T. Schillings: Chemistry of the neomycins. X. Neomycins B and C. J. Am. Chem. Soc. **84**, 3218 (1962).

Schwartz, J.H.: An effect of streptomycin on the biosynthesis of the coat protein of coliphage of f2 by extracts of *E. coli*. Proc. Natl. Acad. Sci. U.S. **53**, 1133 (1965).

Shaw, W.V.: The enzymatic acetylation of chloramphenicol by extracts of R factor-resistant *Escherichia coli*. J. Biol. Chem. **242**, 687 (1967).

Sherman, M.I.: The role of ribosomal conformation in protein biosynthesis. Further studies with streptomycin. Europ. J. Biochem. **25**, 291 (1972).

Sherman, M.I., and M.V. Simpson: The role of ribosomal conformation in protein biosynthesis: the streptomycin-ribosome interaction. Proc. Natl. Acad. Sci. U.S. **64**, 1388 (1969).

Shibahara, S., S. Kondo, K. Maeda, H. Umezawa, and M. Ohno: The total synthesis of negamycin and the antipode. J. Am. Chem. Soc. **94**, 4353 (1972).

Shier, W.T., K.L. Rinehart, Jr., and D. Gottlieb: Preparation of four new antibiotics from a mutant of *Streptomyces fradiae*. Proc. Natl. Acad. Sci. U.S. **63**, 198 (1969).

Shier, W.T., K.L. Rinehart, Jr., and D. Gottlieb: Preparation of two new aminoglycoside antibiotics. J. Antibiotics (Tokyo) **23**, 51 (1970).

Shoji, J., S. Kozuki, M. Mayama, Y. Kawamura, and K. Matsumoto: Isolation of a new water-soluble basic antibiotic A-396-I. J. Antibiotics (Tokyo) **23**, 291 (1970).

Shoji, J., and Y. Nakagawa: Structural feature of antibiotic A-396-I. J. Antibiotics (Tokyo) **23**, 569 (1970).

Shomura, T., N. Ezaki, T. Tsuruoka, T. Niwa, E. Akita, and T. Niida: Studies on antibiotic SF-733, a new antibiotic. I. taxonomy, isolation and characterization. J. Antibiotics (Tokyo) **23**, 155 (1970).

Sparling, P.F.: Kasugamycin resistance: 30S ribosomal mutation with an unusual location on the *Escherichia coli* chromosome. Science **167**, 56 (1968).

SUHARA, Y., K. MAEDA, H. UMEZAWA, and M. OHNO: Chemical studies on kasugamycin. V. The structure of kasugamycin. Tetrahedron Letters 1966, 1239.

SUZUKI, Y., M. HORI, and H. UMEZAWA: Effect of antibiotics on magnesium ion removal from E. coli ribosome suspensions. J. Antibiotics (Tokyo) 21, 571 (1968).

SUZUKI, Y., and S. OKAMOTO: The enzymatic acetylation of chloramphenicol by the multiple drug-resistant Escherichia coli carrying R factor. J. Biol. Chem. 242, 4722 (1967).

TAKASAWA, S., R. UTAHARA, M. OKANISHI, K. MAEDA, and H. UMEZWA: Studies on adenylstreptomycin, a product of streptomycin inactivation by E. coli carrying R factor. J. Antibiotics (Tokyo) 21, 477 (1968).

TANAKA, N., and S. IGUSA: Effects of viomycin and polymyxin B on protein synthesis in vitro. J. Antibiotics (Tokyo) 21, 239 (1968).

TANAKA, N., H. MASUKAWA, and H. UMEZAWA: Structural basis of kanamycin for miscoding activity. Biochem. Biophys. Res. Commun. 26, 544 (1967a).

TANAKA, N., T. NISHIMURA, H. YAMAGUCHI, C. YAMAMOTO, Y. YOSHIDA, K. SASHIKATA, and H. Umezawa: Mechanism of action of kasugamycin. J. Antibiotics (Tokyo) 17, 140 (1965).

TANAKA, N., K. SASHIKATA, T. NISHIMURA, and H. UMEZAWA: Activity of ribosomes from kanamycin-resistant E. coli. Biochem. Biophys. Res. Commun. 16, 216 (1964).

TANAKA, N., K. SASHIKATA, and H. UMEZAWA: Antibiotic-sensitivity of ribosomes from kanamycin-resistant E. coli. J. Antibiotics, (Tokyo), Ser. A 20, 115 (1967b).

TANAKA, N., K. SASHIKATA, H. YAMAGUCHI, and H. UMEZAWA: Inhibition of protein synthesis by bottromycin A$_2$ and its hydrazide. J. Biochem. (Tokyo) 60, 405 (1966a).

TANAKA, N., H. YAMAGUCHI, and H. UMEZAWA: Mechanism of kasugamycin action on polypeptide synthesis. J. Biochem. (Tokyo) 60, 429 (1966b).

TANAKA, N., Y. YOSHIDA, K. SASHIKATA, H. YAMAGUCHI, and H. UMEZAWA: Inhibition of polypeptide synthesis by kasugamycin, an aminoglycoside antibiotic. J. Antibiotics (Tokyo) 19, 65 (1966c).

TANAKA, Y., and H. KAJI: The role of ribosomal protein for the binding of dihydrostreptomycin to ribosomes. Biochem. Biophys. Res. Commun. 32, 313 (1968).

TRAUB, P., K. HOSOKAWA, and M. NOMURA: Streptomycin sensitivity and the structural components of the 30S ribosomes of Escherichia coli. J. Mol. Biol. 19, 211 (1966).

UEHARA, Y., S. KONDO, H. UMEZAWA, K. SUZUKAKE, and M. HORI: Negamycin, a miscoding antibiotic with a unique structure. J. Antibiotics (Tokyo) 25, 685 (1972).

UMEZAWA, H.: Recent advances in chemistry and biochemistry of antibiotics. Microbial Chemistry Research Foundation (1964).

UMEZAWA, H., O. DOI, M. OGURA, S. KONDO, and N. TANAKA: Phosphorylation and inactivation of kanamycin by Pseudomonas aeruginosa. J. Antibiotics (Tokyo), Ser. A 21, 154 (1968a).

UMEZAWA, H., Y. NISHIMURA, T. TSUCHIYA, and S. UMEZAWA: Syntheses of 6'-N-methylkanamycin and 3',4'-dideoxy-6'-N-methylkanamycin B active against resistant strains having 6'-N-acetylating enzymes. J. Antibiotics (Tokyo) 25, 743 (1972a).

UMEZAWA, H., Y. OKAMI, T. HASHIMOTO, Y. SUHARA, M. HAMADA, and T. TAKEUCHI: A new antibiotic, kasugamycin. J. Antibiotics (Tokyo), Ser. A 18, 101 (1965).

UMEZAWA, H., M. OKANISHI, S. KONDO, K. HAMANA, R. UTAHARA, K. MAEDA, and S. MITSUHASHI: Phosphorylative inactivation of aminoglycosidic antibiotics by Escherichia coli carrying R factor. Science 157, 1559 (1967a).

UMEZAWA, H., M. OKANISHI, R. UTAHARA, K. MAEDA, and S. KONDO: Isolation and structure of kanamycin inactivated by a cell-free system of kanamycin-resistant E. coli. J. Antibiotics (Tokyo), Ser. A 20, 136 (1967b).

UMEZAWA, H., S. TAKASAWA, M. OKANISHI, and R. UTAHARA: Adenylstreptomycin, a product of streptomycin inactivated by E. coli carrying R factor. J. Antibiotics (Tokyo), Ser. A 21, 81 (1968b).

UMEZAWA, H., M. UEDA, K. MAEDA, K. YAGISHITA, S. KONDO, Y. OKAMI, R. UTAHARA, Y. OSATO, K. NITTA, and T. TAKEUCHI: Production and isolation of a new antibiotic, kanamycin. J. Antibiotics (Tokyo), Ser. A 10, 181 (1957).

UMEZAWA, H., S. UMEZAWA, T. TSUCHIYA, and Y. OKAZAKI: 3',4'-Dideoxykanamycin B active against kanamycin-resistant Escherichia coli and Pseudomonas aeruginosa. J. Antibiotics (Tokyo) 24, 485 (1971a).

UMEZAWA, H., H. YAMAMOTO, M. YAGISAWA, S. KONDO, T. TAKEUCHI, and Y. A. CHABBERT: Kanamycin phosphotransferase I: Mechanism of cross-resistance between kanamycin and lividomycin. J. Antibiotics (Tokyo) 26, 407 (1973).

Umezawa, S., T. Tsuchiya, R. Muto, Y. Nishimoto, and H. Umezawa: Synthesis of 3'-deoxykanamycin effective against kanamycin-resistant *Escherichia coli* and *Pseudomonas aeruginosa*. J. Antibiotics (Tokyo) **24**, 274 (1971b).

Umezawa, S., T. Tsuchiya, D. Ikeda, and H. Umezawa: Syntheses of 3',4'-dideoxy- and 3',4',5''-trideoxy-ribostamycin active against kanamycin-resistant *Escherichia coli* and *Pseudomonas aeruginosa*. J. Antibiotics (Tokyo) **25**, 613 (1972b).

Umezawa, S., K. Umino, S. Shibahara, M. Hamada, and S. Omoto: Fermentation of 3-amino-3-deoxy-D-glucose. J. Antibiotics (Tokyo), Ser. A **20**, 355 (1967c).

Umezawa, S., I. Watanabe, T. Tsuchiya, H. Umezawa, and M. Hamada: Synthesis of 5''-deoxylividomycin B. J. Antibiotics (Tokyo) **25**, 617 (1972c).

Uramoto, M., N. Otake, and H. Yonehara: Mannosyl glucosaminide, a new antibiotic. J. Antibiotics (Tokyo), Ser. A **20**, 236 (1967).

Vavra, J.J., C. Deboer, A. Dietz, L.J. Hannka, and W.T. Sokolski: Streptozotocin, a new antibacterial antibiotic. Antibiot. Ann. **1959/1960,** 230 (1960).

Vogel, Z., T. Vogel, A. Zamir, and D. Elson: Ribosome activation and the binding of dihydrostreptomycin: effect of polynucleotides and temperature on activation. J. Mol. Biol. **54**, 379 (1970).

Wallace, B.J., P.-C. Tai, and B.D. Davis: Effect of streptomycin on the response of *Escherichia coli* ribosomes to the dissociation factor. J. Mol. Biol. (in press).

Weinstein, W.J., G.M. Luedemann, E.M. Oden, G.H. Wagman, J.R. Rosselet, J.A. Marquez, C.T. Coniglio, W. Charney, H.L. Herzog, and J. Black: Gentamicin, a new antibiotic complex from *Micromonospora*. J. Med. Chem. **6**, 463 (1963).

Weinstein, M.J., J.A. Marquez, R.T. Testa, G.H. Wagman, E.M. Oden, and J.A. Waitz: Antibiotic 6640, a new *Micromonospora*-produced aminoglycoside antibiotic. J. Antibiotics (Tokyo) **23**, 551 (1970).

White, J.R., and H.L. White: Streptomycinoid antibiotics: synergism by puromycin. Science **146**, 772 (1964).

Wolfe, A.D., and F.E. Hahn: Stability of ribosomes from streptomycin exposed *Escherichia coli*. Biochem. Biophys. Res. Commun. **31**, 945 (1968).

Woo, P.W.K., H.W. Dion, and Q.R. Bartz: Butirosins A and B, aminoglycoside antibiotics. III. Structures. Tetrahedron Letters **28**, 2625 (1971).

Yagisawa, M., H. Naganawa, S. Kondo, M. Hamada, T. Takeuchi, and H. Umezawa: Adenyldideoxykanamycin B, a product of the inactivation of dideoxykanamycin B by *Escherichia coli* carrying R factor. J. Antibiotics (Tokyo) **24**, 911 (1971).

Yagisawa, M., H. Naganawa, S. Kondo, T. Takeuchi, and H. Umezawa: Inactivation of 3',4'-dedeoxykanamycin B by an enzyme solution of resistant *E. coli* and isolation of 3',4'-dideoxykanamycin B 2''-guanylate and 2''-inosinate. J. Antibiotics (Tokyo) **25**, 492 (1972a).

Yagisawa, M., H. Naganawa, S. Kondo, T. Takeuchi, and H. Umezawa: 6'-N-Acetylation of 3',4'-dideoxykanamycin B by an enzyme in a resistant strain of *Pseudomonas aeruginosa*. J. Antibiotics (Tokyo) **25**, 495 (1972b).

Yagisawa, M., H. Yamamoto, H. Naganawa, S. Kondo, T. Takeuchi, and H. Umezawa: A new enzyme in *Escherichia coli* carrying R-factor phosphorylating 3'-hydroxyl of butirosin A, kanamycin, neamine and ribostamycin. J. Antibiotics (Tokyo) **25**, 748 (1972c).

Yamada, T., K. Kvitek, and J. Davies: The binding of streptomycin and R-factor-inactivated streptomycin to ribosomes. Prog. Antimicr. Anticancer Chemoth. (Univ. Tokyo Press) **2**, 562 (1970).

Yamaki, H., and N. Tanaka: Effects of protein synthesis inhibitors on the lethal action of kanamycin and streptomycin. J. Antibiotics (Tokyo), Ser. A **16**, 222 (1963).

Yamamoto, H., S. Kondo, K. Maeda, and H. Umezawa: Synthesis of lividomycin A 5''-phosphate, an enzymatically inactivated lividomycin A. J. Antibiotics (Tokyo) **25**, 485 (1972a).

Yamamoto, H., S. Kondo, M. Maeda, and H. Umezawa: Syntheses of 5''-deoxylividomycin A and its amino derivative. J. Antibiotics (Tokyo) **25**, 487 (1972b).

Yamamoto, H., M. Yagisawa, H. Naganawa, S. Kondo, T. Takeuchi, and H. Umezawa: Kanamycin 6'-acetate and ribostamycin 6'-acetate, enzymatically inactivated products by *Pseudomonas aeruginosa*. J. Antibiotics (Tokyo) **25**, 746 (1972c).

Zagorska, L., J. Dondon, J.C. Lelong, F. Gros, and M. Grunberg-Manago: Decoding site of initiator transfer RNA. Biochemie **53**, 63 (1971).

Borrelidin

K. Poralla

Borrelidin was first isolated as a product of *Streptomyces rochei* in crystalline form in 1949 (BERGER *et al.*, 1949). It showed high activity against experimental infections of the spirochete Borrelia and was therefore called borrelidin.

The structure of borrelidin was elucidated by KELLER-SCHIERLEIN in 1966. It is a macrolide-type antibiotic without a sugar residue (Fig. 1). Further character-istics include a dienenitrile system, the regular distribution of the methyl groups, the OH group in β-position to the lactone group and an additional cyclopentane ring with a carboxyl group. Chemically it lies between the antifungal polyene antibiotics and the bacteriostatic macrolide antibiotics.

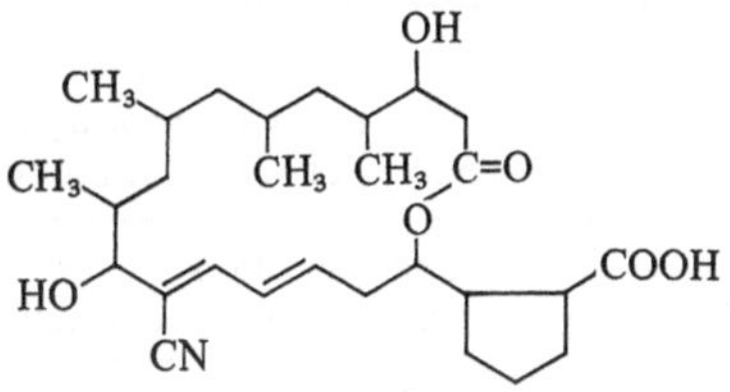

Fig. 1. Structure of borrelidin

Borrelidin is produced by a variety of Streptomyces species. These are *Strepto-myces rochei* (BERGER *et al.*, 1949), *Streptomyces griseus* (LUMB *et al.*, 1965; ECKARDT *et al.*, 1970) and *Streptomyces parvulus* (HÜTTER *et al.*, 1966). Borrelidin is active against a variety of organisms, including pro- and eucaryotes. It inhibits the growth of *Sarcina lutea* (BERGER *et al.*, 1949), some species of Corynebacterium (LUMB *et al., 1965*), *Bacillus subtilis* (HÜTTER *et al.*, 1966), *Bacillus polymyxa* (MONREAL and PAULUS, 1970), and *Escherichia coli* (NASS, *et al.*, 1969). The yeasts of the Candida spec. and *Saccharomyces cerevisiae* are also inhibited by borrelidin in concentrations of less than 10 µg/ml (PORALLA, unpublished). The minimal inhibitory concentration for *Bacillus subtilis* on minimal medium is 2.5 µg/ml (PORALLA, 1967). Although good activity was found in the plaque inhibition test against a variety of viruses (DICKINSON *et al.*, 1965; TONEW *et al.*, 1970), the cell division was also inhibited. Activity was also found against some tumors in mice (SUGIURA and SUGIURA, 1958).

Borrelidin is toxic (BERGER *et al.*, 1949); it causes severe skin irritations (LUMB *et al.* 1965) and for these reasons is excluded from chemotherapy.

It was found that borrelidin shows bigger inhibition zones on agar plates containing minimal medium (glucose and mineral salts) than plates containing complex medium (HÜTTER *et al.*, 1966). Since most tests on borrelidin are made using different complex media, this may explain why activity is found only against a few bacterial species. By a simple cross strip test it was found, that L-homoserine or L-threonine competitively antagonize the inhibition caused by borrelidin (PORALLA, 1967) and this should explain the lower or a lack of activity of borrelidin in complex media.

It has been shown with growth curves that the *Bacillus subtilis* growth is inhibited within a few minutes after the application of borrelidin. In the same temporal manner the inhibition can be counteracted by threonine (PORALLA, 1967).

Determined colorimetrically, macromolecular syntheses in the whole cells, of protein, RNA and DNA, are equally inhibited by bacteriostatic concentrations of borrelidin (PORALLA and ZÄHNER, 1968).

Since no other metabolites except threonine or homoserine antagonize the inhibition of borrelidin the threonyl-tRNA synthetases (ThRS) of *Escherichia coli* (HÜTTER *et al.*, 1966) and *Bacillus subtilis* (PORALLA and ZÄHNER, 1968) were tested in cell-free systems for inhibition. There is an effective and specific inhibition of ThRS *in vitro* (Table 1).

On exposing the growing cells of *E. coli* to borrelidin followed by extraction of the enzymes and testing the different aminoacyl-tRNA synthetases (NASS *et al.*, 1969), only ThRS was found to be inactivated (Table 2). The ThRS of

Table 1. *Effect of borrelidin on aminoacylation of tRNA with different amino acids in a cell-free system of Bacillus subtilis* (PORALLA and ZÄHNER, 1968)

Amino acid	Aminoacylation (nmole amino acid/sample)	
	− Borrelidin	+ Borrelidin
Thr	0.10	0.04
Ile	0.28	0.29
Val	0.06	0.06
Tyr	0.05	0.05
Ser	0.19	0.19
Pro	0.07	0.08
Glu	0.31	0.32

Table 2. *Specificity of the inactivation after treatment of whole cells of Escherichia coli with borrelidin and extraction of the enzymes* (NASS *et al.*, 1969)

Aminoacyl-tRNA synthetase	Enzyme activity	
	− Borrelidin	+ Borrelidin
threonyl-tRNA synthetase	0.024	0.002
methionyl-tRNA synthetase	0.030	0.031
arginyl-tRNA synthetase	0.119	0.117
phenylalanyl-tRNA synthetase	0.080	0.072
isoleucyl-tRNA synthetase	0.052	0.053
leucyl-tRNA synthetase	0.063	0.075
tyrosyl-tRNA synthetase	0.071	0.080
lysyl-tRNA synthetase	0.030	0.034
valyl-tRNA synthetase	0.168	0.192

Bacillus polymyxa (MONREAL and PAULUS, 1970), yeast (HÜTTER *et al.*, 1966), and from cytoplasm and chloroplasts of beans (BURKARD *et al.*, 1970) is also sensitive to borrelidin.

Experiments with the synthetase of *Bacillus polymyxa* show that whereas the pyrophosphate exchange reaction is not inhibited, the so-called second step, the transfer of threonine to tRNA is inhibited (MONREAL and PAULUS, 1970). The same authors have drawn attention to the structural similarity of borrelidin to the 3′-0-threonyl-adenosine-5′-phosphoryl moiety of threonyl-tRNA. This could be a reason for the lack of inhibition of the pyrophosphate exchange reaction. Further data on ThRS inhibition were obtained with a one hundred fold purified enzyme preparation from *Escherichia coli* K12. As shown in Fig. 2 the inhibition is noncompetitive with respect to threonine and the K_i for borrelidin is 3.0×10^{-9} M (PAETZ and NASS, 1973).

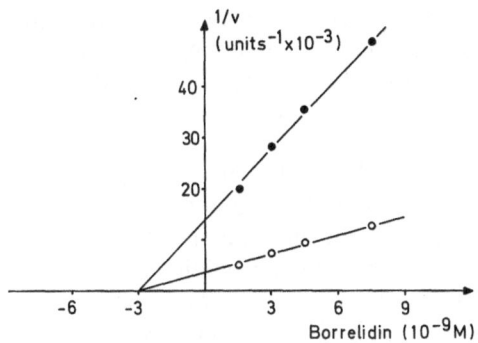

Fig. 2. Noncompetitive inhibition of threonyl-tRNA synthetase (ThRS) by borrelidin shown in a Dixon-plot. $\circ = 5 \times 10^{-6}$ M Thr, $\bullet = 2 \times 10^{-5}$ M Thr

The powerful and specific inhibitory action of borrelidin was used as a tool towards solving three problems in microbial metabolism. As a direct consequence of the inhibition of ThRS, protein synthesis is arrested. In addition by a yet not completely understood regulatory mechanism, RNA synthesis is also inhibited (PORALLA and ZÄHNER, 1968). This would appear to be similar to the inhibition of RNA synthesis in *Escherichia coli* mutants with a temperature sensitive aminoacyl-tRNA synthetase on shifting to the higher temperature (NEIDHARDT, 1966). The inhibition of the synthesis of stable RNA seems to be linked to the activity of every aminoacyl-tRNA synthetase.

A second problem could be examined with the aid of borrelidin. Is threonine the corepressor for the repression of its biosynthetic enzymes? After partially inhibiting the growth of *Escherichia coli,* threonine biosynthetic enzyme synthesis is specifically derepressed (Table 3).

Since even after the addition of threonine, thereby restoring growth, derepression was observed (NASS *et al.*, 1969), the amino acid itself could be excluded as the corepressor. Perhaps loaded tRNA is the corepressor. However, the nature of the corepressor has not been elucidated in the case of amino acid biosynthetic

Table 3. *Specific derepression of threonine biosynthetic enzymes after partial growth inhibition of Escherichia coli K12B by borrelidin* (Nass et al., 1969)

	Growth rate (generations/hr)	Specific activity of		
		Aspartyl kinase	Homoserine dehydrogenase	Threonine deaminase
− Borrelidin	0.64	0.052	3.0	0.054
+ Borrelidin	0.33	0.305	11.0	0.048

enzymes (Calvo and Fink, 1971). It is very probable that a similar derepression meachanism exists in eucaryotes. In yeast a two fold derepression of aspartyl kinase was found after addition of borrelidin (Nass and Hasenbank, 1970).

The specific mode of action of borrelidin was employed by Monreal and Paulus in 1970 to solve the question whether threonyl-tRNA was an intermediate in the biosynthesis of the threonine containing oligopeptide antibiotic, polymyxin B. They found no inhibition of polymyxin biosynthesis, in *Bacillus polymyxa*, even though both growth and protein synthesis were inhibited by borrelidin. Also the inhibition of the formation of threonyl-tRNA was observed *in vitro*. Therefore the possibility of threonyl-tRNA acting as an obligatory intermediate in the biosynthesis of polymyxin is ruled out.

Although borrelidin is chemically similar to the bacteriostatic macrolide antibiotics, no cross-resistance between borrelidin and erythromycin was found (Poralla, 1967). The mechanism of resistance was elucidated with borrelidin resistant mutants of *Escherichia coli* K12 (Paetz and Nass, 1973; Thomale and Nass, 1972). Out of the 15 resistant mutants tested, 9 showed a four to eight fold higher ThRS specific activity compared to the wild type. Two mutants were more thoroughly examined (Paetz and Nass, 1973) and the ThRS from these strains was purified about ninety fold. In Table 4 are listed the K_m and K_i values obtained on the enzymes isolated from the wild type and resistant mutants.

Bor Res 3 is a regulatory mutant which synthesizes about five times more ThRS than the wild type. In this the resistance is achieved by increasing the concentration of the target enzyme. On the other hand mutant Bor Res 2 has an structurally altered ThRS as shown by different K_m's and K_i and a different antibody neutralization curve.

Table 4. *Properties of ThRS from wild type and borrelidin resistant strains of Escherichia coli K12B* (Nass und Paetz, 1973)

Escherichia coli strain	K_m for		K_i for Borrelidin $(\times 10^{-9} \text{ M})$
	Threonine $(\times 10^{-5} \text{ M})$	ATP $(\times 10^{-4} \text{ M})$	
K12B	8.5	1.2	3.0
Bor Res 3	8.2	1.1	3.0
Bor Res 2	1.7	0.5	7.5

Borrelidin resistant strains of haploid *Saccharomyces cerevisiae* (S288C) were also isolated and biochemically and genetically characterized (NASS and PORALLA, to be published). Three unlinked dominant markers were found, two of which influence properties of the ThRS. One of these two lies very near to the gene of the first threonine biosynthetic enzyme, the aspartyl kinase, on chromosome 5 as shown by tetrad analysis.

Acknowledgements: I want to thank Dr. G. NASS for a critical reading of the article and Dr. H.S. HERTZ for help in writing a readable English.

References

BERGER, J., L. M. JAMPOLSKY, and M. W. GOLDBERG: Borrelidin, a new antibiotic with anti-borrelia activity and penicillin enhancement properties. Arch. Biochem. **22**, 476 (1949).

BURKARD, G., P. GUILLEMAUT, and J. H. WEIL: Comparative studies of tRNA's and the aminoacyl-tRNA synthetases from the cytoplasm and the chloroplasts of Phaseolus vulgaris. Biochim. Biophys. Acta **224**, 184 (1970).

CALVO, J. M., and G. R. FINK: Regulation of biosynthetic pathways in bacteria and fungi. Ann. Rev. Biochem. **40**, 943 (1971).

DICKINSON, L., A. J. GRIFFITHS, C. G. MASON, and R. F. N. MILLS: Anti-viral activity of two antibiotics isolated from a species of Streptomyces. Nature **206**, 265 (1965).

ECKARDT, K., W. FLECK, H. PRAUSER, E. TONEW u. P. ZÖPEL: Ein neues Verfahren zur Gewinnung des antiviralen Antibiotikums Borrelidin. Z. Allgem. Mikrobiol. **10**, 367 (1970).

HÜTTER, R., K. PORALLA, H. G. ZACHAU u. H. ZÄHNER: Über die Wirkungsweise von Borrelidin – Hemmung des Threonineinbaus in sRNA. Biochem. Z. **344**, 190 (1966).

KELLER-SCHIERLEIN, W.: Die Konstitution des Borrelidins. Experientia **22**, 355 (1966).

LUMB, M., P. E. MACEY, J. SPYVEE, J. M. WHITMARSH, and R. D. WRIGHT: Isolation of vivomycin and borrelidin, two antibiotics with anti-viral activity, from a species of Streptomyces (C2989). Nature **206**, 263 (1965).

MONREAL, J., and H. PAULUS: Nonparticipation of transfer RNA in the biosynthesis of polymyxin B. Biochim. Biophys. Acta **199**, 280 (1970).

NASS, G., and R. HASENBANK: Effect of borrelidin on the threonyl-tRNA synthetase activity and the regulation of threonine-biosynthetic enzymes in Saccharomyces cerevisiae. Molec. Gen. Genetics **108**, 28 (1970).

NASS, G., and K. PORALLA: Genetics and biochemistry of borrelidin resistant mutants of Saccharomyces with an altered threonyl-tRNA synthetase. To be published.

NASS, G., K. PORALLA, and H. ZÄHNER: Effect of the antibiotic borrelidin on the regulation of threonine biosynthetic enzymes in E. coli. Biochem. Biophys. Res. Commun. **34**, 84 (1969).

NEIDHARDT, F.C.: Roles of amino acid activating enzymes in cellular physiology. Bacteriol. Rev. **30**, 701 (1966).

PAETZ, W., and G. NASS: Biochemical and immunological characterization of threonyl-tRNA synthetase of two borrelidin resistant mutants of E. coli K12. Europ. J. Biochem. **35**, 331 (1973).

PORALLA, K.: Zur Wirkungsweise des makrolidartigen Antibiotikums Borrelidin und typischer Makrolid-Antibiotika. Ph. D. Thesis, University of Tübingen 1967.

PORALLA, K., u. H. ZÄHNER: Die Hemmung des Einbaus von Threonin in sRNS im zellfreien System und der Synthese von Protein und Nukleinsäuren in der Zelle durch das Antibiotikum Borrelidin. Arch. Mikrobiol. **61**, 143 (1968).

SUGIURA, K., and M. M. SUGIURA: Test of compounds against various mouse tumors. Cancer Res. **18** (Suppl.), 290 (1958).

THOMALE, J., u. G. NASS: Regulation der Bildung der Threonyl-tRNA-Synthetase in borrelidinresistenten E. coli Mutanten. Hoppe-Seylers Z. Physiol. Chem. **353**, 1572 (1972).

TONEW, E., K. ECKARDT, W. FLECK u. P. ZÖPEL: Die antivirale Wirkung des Streptomyceten-Antibiotikums IMET A 8136/A (Borrelidin) auf das Virus der klassischen Geflügelpest in vitro. Z. Allgem. Mikrobiol. **10**, 353 (1970).

Unusual abbreviation: ThRS = threonyl-tRNA synthetase.

Chloramphenicol

Sidney Pestka

A previous summary of chloramphenicol and its mode of action appeared in Vol I of the Mode of Action of Antibiotics. This review will chiefly concern the mode of action of chloramphenicol as currently understood and will concentrate on the work published since the previous volume. The previous review by HAHN (1967) has provided a summary of the work on chloramphenicol to that date. Other reviews published since that time also may be consulted for additional information (WEISBLUM and DAVIES, 1968; PESTKA, 1971).

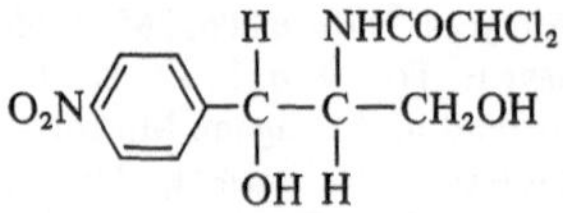

$$O_2N-\langle\bigcirc\rangle-\overset{\displaystyle H}{\underset{\displaystyle OH}{C}}-\overset{\displaystyle NHCOCHCl_2}{\underset{\displaystyle H}{C}}-CH_2OH$$

Fig. 1. Structure of chloramphenicol

Chloramphenicol is chiefly a bacteriostatic agent. It is a broad spectrum antibiotic inhibiting the growth of Gram-positive and -negative bacteria. Eucaryotic cells are generally resistant to the antibiotic although some inhibitions have been reported. The chemical structure of chloramphenicol (Fig. 1) is one of the simplest of the known antibiotics (REBSTOCK et al., 1949; CONTROULIS et al., 1949; DUNITZ, 1952). There are 4 stereoisomers, only one of which is active as an antibacterial agent, D(−)threo-chloramphenicol (MAXWELL and NICKEL, 1954). A variety of electronegative groups can be substituted for the aromatic nitro group without major loss of antimicrobial activity (HAHN et al., 1956; SHEMYAKIN, 1961). The dichloromethyl group can be replaced by vinyl, allenyl, ethynyl, cyclopropyl and isopropyl groups with retention of substantial antibacterial activity (RINGROSE and LAMBERT, 1973).

Inhibition of Protein Synthesis

Chloramphenicol inhibits protein synthesis (GALE and FOLKES, 1953; HAHN and WISSEMAN, 1951; WISSEMAN et al., 1954; MARMUR and SAZ, 1953; SORM and GRUNBERGER, 1954; NIRENBERG and MATTHAEI, 1961; NATHANS and LIPMANN, 1961; RENDI and OCHOA, 1962). It does not inhibit amino acid dependent pyrophosphate exchange (DeMoss and NOVELLI, 1956) nor formation of aminoacyl-tRNA (RENDI and OCHOA, 1962; LACKS and GROS, 1959) nor does it inhibit binding of poly A or poly UC to ribosomes (SPEYER *et al.*, 1963; KUCAN and LIPMANN, 1964) contrary to the report that chloramphenicol inhibited poly U binding to *E. coli* ribosomes (JARDETZKY and JULIAN, 1964). Chloramphenicol was found to inhibit incorporation of amino acids into protein in cell-free extracts from *E. coli* (RENDI and OCHOA, 1962; NIRENBERG and MATTHAEI, 1961; NATHANS and LIPMANN, 1961; TISSIÉRES *et al.*, 1960; LAMBORG and ZAMECNIK, 1960), but not in those from eucaryotic cells (RENDI, 1959; VON EHRENSTEIN and LIPMANN, 1961; ALLEN and SCHWEET, 1962; BORSOOK *et al.*, 1957; SO and DAVIE, 1963).

KUCAN and LIPMANN (1964) observed that the effect of chloramphenicol on polynucleotide-directed protein synthesis was a function of the template used. Polypeptide synthesis directed by poly U and poly UA templates was substantially more resistant to the action of chloramphenicol than polypeptide synthesis directed by poly A and poly UC templates. Similar observations have been reported by others (SPEYER *et al.*, 1963; VAZQUEZ, 1966c). It should be noted that protein synthesis directed by natural messengers such as phage f2RNA is markedly inhibited by chloramphenicol. The apparent paradox may partially be a reflection of the differential solubility and precipitability of the various polypeptides formed under the direction of the different templates. Small phenylalanine peptides of chain length 4 are substantially precipitable by cold trichloroacetic acid (PESTKA *et al.*, 1969). Lysine peptides or those containing a high proportion of lysine tend to be soluble in trichloroacetic acid used in the usual methods of assaying protein synthesis in cell-free extracts (GARDNER *et al.*, 1962). JULIAN (1965, 1966) examined in detail the distribution of the lysine peptides formed in the presence and absence of chloramphenicol under the direction of a poly A template. In the presence of chloramphenicol the distribution of peptides was shifted from long chain material to those of shorter chain lengths. Although 2×10^{-4} M chloramphenicol caused a 60% inhibition of lysine peptides precipitable by tungstate, it produced only a 23% inhibition of total peptide bond formation and no inhibition of the total number of lysine peptide chains formed. TERAOKA *et al.*, (1969) also reported that the synthesis of the smaller lysine peptides was found to be resistant to inhibition by chloramphenicol. In similar observations studying diphenylalanine and oligophenylalanine (phenylalanines of chain length 3 and greater) synthesis, PESTKA (1968, 1969a, 1969b, 1970b) found that although the rate of oligophenylalanine synthesis was inhibited by chloramphenicol, extent of formation of diphenylalanine was either unchanged or stimulated. COUTSO-GEORGOPOULOS (1971) has also reported similar observations. The formation of increased amounts of shorter chains is not inconsistent with inhibition of peptide bond formation by chloramphenicol, for chain-propagating systems in general show greater quantities of shorter chains in inhibited than in the uninhibited systems (FROST and PEARSON, 1953). Additionally, YUKIOKA and MORISAWA (1970) showed that increasing concentrations of elongation factor G and GTP could reduce the inhibition of poly U directed polyphenylalanine synthesis produced by chloramphenicol. Another possible contribution to the small inhibition of polyphenylalanine synthesis by chloramphenicol may be related to interaction of the amino acid with ribosomes. Puromycin analogs possess maximal puromycin-like activity when they contain constituent aromatic residues (NATHANS and NEIDLE, 1963; SYMONS *et al.*, 1969; WALLER, *et al.*, 1966; RYCHLÍK, *et al.*, 1969).

The aminoacyl-end of Phe-tRNA may have greater affinity for the ribosome than non-aromatic aminoacyl-tRNA˙; and thus chloramphenicol may inhibit peptide bond formation more in the case of aliphatic aminoacyl-tRNA. IRVINE and JULIAN (1970) reported that, in contrast to lysine peptides, chloramphenicol inhibits all sizes of proline peptides synthesized in the presence of a poly C template although large peptides were inhibited more than small ones.

Acetylation of both hydroxyl groups of chloramphenicol (1,3-diacetoxychloramphenicol) inactivates the molecule for *in vitro* activity against *E. coli* and eliminates its ability to inhibit protein synthesis in cell-free extracts (PIFFARETTI *et al.*, 1970). Acetylation of the 3-hydroxyl group only reduced the activity to about one-half of that of chloramphenicol in inhibiting protein synthesis in cell-free extracts of *E. coli*.

The Ribosome is the Site of Chloramphenicol Action

The resistance of protein synthesis by a combination of yeast ribosomes and *E. coli* supernatant, but not by *E. coli* ribosomes and yeast supernatant, to chloramphenicol early suggested that the site of action of chloramphenicol is on the ribosomes (SO and DAVIE, 1963). This conclusion is supported by studies of chloramphenicol binding to ribosomes and assays of various ribosome functions (see below).

Binding of Chloramphenicol to Ribosomes

Radioactive chloramphenicol is bound to purified ribosomes (VAZQUEZ 1963, 1964, 1966a; CHANG *et al.*, 1969; WOLFE and HAHN, 1965; DAS *et al.*, 1966; LESSARD and PESTKA, 1972b). Binding is relatively weak, and chloramphenicol can be removed by washing (VAZQUEZ, 1967; CANNON, 1968). Binding of chloramphenicol to intact *E. coli* was also weak (HURWITZ and BRAUN, 1967, 1968). The ability to remove chloramphenicol bound to ribosomes and to intact cells by washing accounts for the reversibility of the inhibitory effects on protein synthesis.

Binding of chloramphenicol to ribosomes requires K^+ or NH_4^+ and was localized to the 50S subunit (VAZQUEZ, 1966a; LESSARD and PESTKA, 1972b). In addition, binding of [^{14}C]chloramphenicol to *E. coli* ribosomes, studied by equilibrium dialysis, revealed two sites for chloramphenicol binding: a high affinity and a low affinity site, as shown in the Scatchard plot of Fig. 2 (LESSARD and PESTKA, 1972b). The binding site with the dissociation constant of about 2×10^{-6} M is saturated at about 0.6 mole of antibiotic per mole of ribosomes. The second ribosomal site appears to bind the molecule more loosely. The dissociation constant for this site is about 2×10^{-4} M and about 0.9 site per ribosome is available for this binding.

Thus equilibrium dialysis experiments indicate that chloramphenicol is bound to two classes of sites (Fig. 2): a high affinity and a low affinity site. Previously, FERNANDEZ-MUÑOZ *et al.* (1971) have reported 0.7 to 0.8 binding site per ribosome for chloramphenicol; their experiments, performed at lower chloramphenicol concentrations than the experiments described by the data of Fig. 2, did not

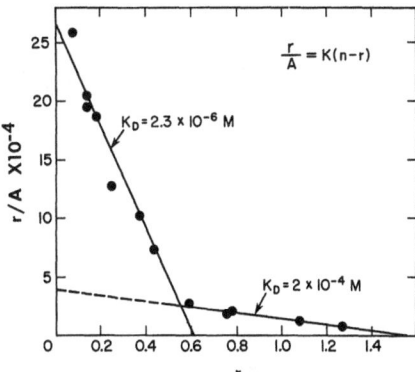

Fig. 2. Binding of [^{14}C]chloramphenicol to *Escherichia coli* ribosomes. The binding of [^{14}C]chloramphenicol to ribosomes was determined by equilibrium dialysis. The reaction mixtures were shaken at 7° for 18 hrs and contained the following components: 0.4 M KCl, 0.04 M MgCl$_2$, 0.05 M Tris-Cl, pH 7.2. The concentration of ribosomes employed was 1.8 to 2.2×10^{-5} M. The data are presented in the form a Scatchard plot according to the equation r/A = K (n − r), where r = moles of antibiotic bound per mole of ribosomes, n = the number of sites available per ribosome, A = concentration of free chloramphenicol, and K$_A$ = the association constant for the binding. The dissociation constant (1/k$_A$) for the binding of chloramphenicol under these conditions is estimated to be 2.3×10^{-6} M for the first site and 2×10^{-4} M for a second site. The data are from LESSARD and PESTKA (1972b)

demonstrate a low affinity binding site. The finding of two classes of inhibition of peptidyl-puromycin synthesis by chloramphenicol on polyribosomes (see below) is in accordance with the possibility of two binding sites for chloramphenicol on each ribosome. The value of 1.5 sites per ribosome for chloramphenicol (Fig. 2) is consistent with a total of two sites per ribosome. Also, CANNON (1968) observed the binding of two molecules of chloramphenicol to each ribosome.

Many antibiotics interfere with [^{14}C]chloramphenicol binding to ribosomes (WOLFE and HAHN, 1965; VAZQUEZ, 1966a, 1966b) and it has been presumed that those that do are also 50S inhibitors. The effect of a variety of antibiotics on the binding of [^{14}C]chloramphenicol to *B. megaterium* and *E. coli* ribosomes is shown in Tables 1 and 2. Puromycin, carbomycin, vernamycin A, lincomycin, gougerotin, and sparsomycin all appear to inhibit the binding of [^{14}C]chloramphenicol to *E. coli* ribosomes (Table 1). Although sparsomycin can inhibit chloramphenicol binding to ribosomes, it is not likely that both antibiotics are binding to identical sites. The dissociation constant for sparsomycin binding to polyribosomes is 1 to 2×10^{-7} M (PESTKA, 1972) and to ribosomes probably of the same order of magnitude (GOLDBERG and MITSUGI, 1967; PESTKA, 1970a). If sparsomycin were binding to the same sites as chloramphenicol, a 200-fold excess of sparsomycin over [^{14}C]chloramphenicol should have produced almost total inhibition of chloramphenicol binding. Since this was not noted (CHANG *et al.*, 1969; FERNANDEZ-MUÑOZ *et al.*, 1971; LESSARD and PESTKA, 1972b), it is possible that sparsomycin inhibits binding of chloramphenicol to only one of the two chloramphenicol binding sites or that the chloramphenicol and sparsomycin binding sites are not identical.

Table 1. *Effect of antibiotics on the binding of* [^{14}C]*chloramphenicol to E. coli ribosomes*

Antibiotic	nmoles added	Percent of [^{14}C]chloramphenicol bound in the absence of antibiotic	
		Expt. 1	Expt. 2
Chloramphenicol	100	1	8
	1000	2	5
	2500	—	—
Puromycin	100	95	80
	500	62	—
	1000	—	44
	2500	27	—
Sparsomycin	5	89	66
	50	75	43
	250	65	—
Gougerotin	100	—	84
	300	75	—
	3000	52	—
Lincomycin	10	—	81
	100	43	41
	1000	2	—
Amecitin	100	119	108
	1000	128	—
Blasticidin S	100	90	—
	1000	90	—
Carbomycin	1	—	81
	10	—	13
Vernamycin A	5	—	14
	50	—	12

Assays were performed by equilibrium dialysis. [^{14}C]Chloramphenicol concentration was 1.25×10^{-6} M (1.25 nmoles per ml) outside the dialysis bag. The nanomoles/ml of antibiotics added to the solution outside the dialysis bag are given in the Table. The values in the Table are reported as a percentage of [^{14}C]chloramphenicol bound in the absence of additional antibiotics. The data are taken from Lessard and Pestka (1972b).

Gougerotin in large excess inhibits chloramphenicol binding (Table 1); blasti-cidin S and amicetin probably do not inhibit chloramphenicol binding to *E. coli* ribosomes (Table 1), but do inhibit the binding to *B. stearothermophilus* ribosomes (Chang et al., 1969). Carbomycin and vernamycin A are strong inhibi-tors of chloramphenicol binding. Lincomycin inhibits chloramphenicol binding about 60 and 100% at 100- and 1000-fold, respectively, excess of lincomycin. Fernandez-Muñoz et al. (1971) reported a greater inhibition by lincomycin in their assays. The inhibition of chloramphenicol binding by puromycin (Table 2) is consistent with the ability of chloramphenicol to bind to the ribosomal acceptor site (Fernandez-Muñoz et al., 1971; Lessard and Pestka, 1972b). Even at very high puromycin concentrations (2.5×10^{-3} M) at least 25% of chloramphenicol remained bound. The inhibition of aminoacyl-oligonucleotide binding by chloramphenicol is also consistent with inhibition of binding of the acceptor end of aminoacyl-tRNA by this antibiotic (see below). Aminoglycosides, tetracyclines and other 30S inhibitors do not inhibit chloramphenicol binding to ribosomes although most 50S inhibitors do (Table 2). Thiostrepton is a 50S inhibitor which does not inhibit chloramphenicol binding to ribosomes (CHANG et al., 1969). Studies of this sort have led to a relatively consistent picture of the subunit localization of various inhibitors. It should be noted that although erythromycin is an excellent inhibitor of chloramphenicol binding, chloramphenicol does not inhibit erythromycin binding (TAUBMAN et al., 1966; TANAKA et al., 1966); and erythromycin bound to washed ribosomes can abolish the inhibitory effect of chloramphenicol on transpeptidation (TERAOKA, 1970).

Table 2. *Effect of antibiotics on chloramphenicol binding to ribosomes*

A. Antibiotics inhibiting chloramphenicol binding
   Macrolide antibiotics
      Angolamycin
      Carbomycin
      Erythromycin
      (Lancamycin)
      (Methymycin)
      (Oleandomycin)
      Spiramycin
   Streptogramins A
      Vernamycin A
      PA114A
   Streptogramins B
      Vernamycin B
      PA114B
   (Lincomycin)
   (Sparsomycin)
   (Puromycin)
   Celesticetin
   Aminonucleosides
      (Amicetin)
      (Gougerotin)
      (Blasticidin S)

B. Agents not inhibiting chloramphenicol binding
   Aminoglycosides
      Streptomycin
      Kanamycin
      Neomycin
   Bottromycin
   Fusidic Acid
   Tetracyclines
   Thiostrepton

Data from WOLFE and HAHN (1965), VAZQUEZ (1966a), CHANG et al. (1969) and LESSARD and PESTKA (1972b). Antibiotics in parentheses weakly interfere with chloramphenicol binding to ribosomes.

Inhibition of Peptide Bond Formation

Chloramphenicol inhibits peptide bond formation as measured by various assays. In intact bacteria, chloramphenicol inhibits peptidyl-puromycin formation (Nathans et al., 1962; Nathans, 1964). Traut and Monro (1964) showed that chloramphenicol inhibited polyphenylalanyl-puromycin formation. Synthesis of polylysyl-puromycin (Goldberg and Mitsugi, 1967; Rychlík, 1966; Coutso-Georgopoulos, 1967; Gottesman, 1967; Černá, et al., 1969), fMet-puromycin (Monro, 1967; Monro and Marcker, 1967) and acetyl-phenylalanyl-puromycin (Weissbach et al., 1968; Pestka, 1970a; Fico and Coutsogeorgopoulos, 1972) was also inhibited by chloramphenicol. With the use of ribosomes which catalyze the reaction of C-A-A-C-C-A(fMet) with puromycin Monro and Marcker (1967) and Monro and Vazquez (1967) showed that chloramphenicol inhibited this reaction; 50S subunits catalyze a similar reaction (Monro, 1967) which was also inhibited by chloramphenicol (Monro, 1969). In studying the effect of chloramphenicol on acetyl-phenylalanyl-puromycin synthesis, Pestka (1970a) showed that chloramphenicol acted as a competitive inhibitor of puromycin. Since puromycin is an analog of the aminoacyl-end of aminoacyl-tRNA, then chloramphenicol also may be. A similar suggestion was made by Coutsogeorgopoulos (1966). Additionally, after examination of the puromycin reaction in protoplasts of *Bacillus megaterium*, Cundliffe and McQuillen (1967) concluded that chloramphenicol inhibited peptidyl transfer. Weber and DeMoss (1969), on the other hand, have concluded that chloramphenicol does not directly block peptidyl transfer in protein synthesis, but inhibits the conversion of peptidyl-tRNA into the puromycin susceptible donor state.

The effects of chloramphenicol on the transfer of nascent chains to [^{3}H]puromycin in a cell-free extract containing native polyribosomes was examined (Pestka, 1972) to evaluate the effect of the antibiotic in a more nearly physiological cell-free system than ordinarily used with washed ribosomes and synthetic donors. When the kinetics of chloramphenicol inhibition of peptidyl-puromycin synthesis was evaluated by the double reciprocal plot (Fig. 3), a mixed type of competition

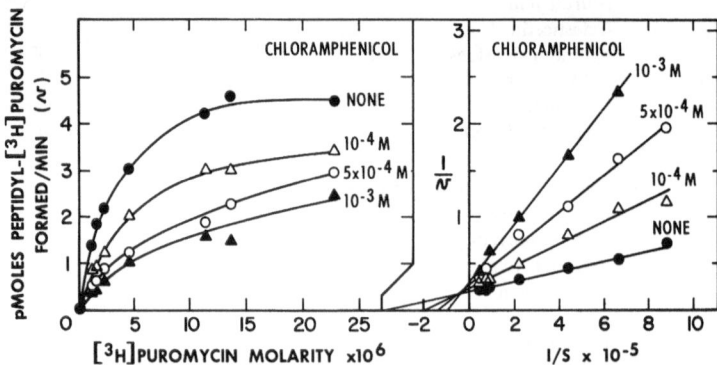

Fig. 3. Kinetics of chloramphenicol inhibition of peptidyl-[^{3}H]puromycin formation by *E. coli* polyribosomes. The concentration of [^{3}H]puromycin was varied as shown on the abscissa. A reciprocal plot of the data of the left panel is presented in the right panel. ●, no chloramphenicol; △, 10^{-4} M chloramphenicol; ○, 5×10^{-4} M chloramphenicol; ▲, 10^{-3} M chloramphenicol. The data are from Pestka (1972)

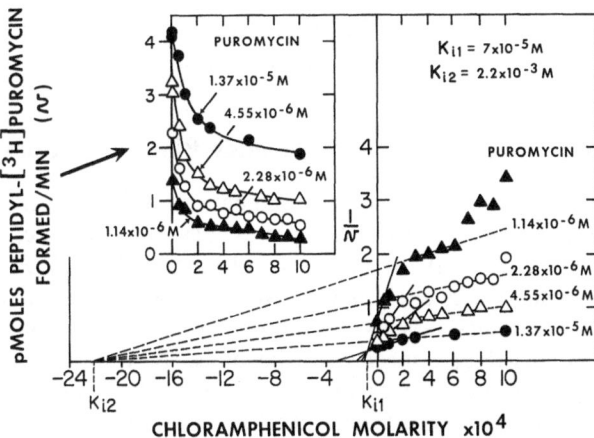

Fig. 4. Kinetics of chloramphenicol inhibition of peptidyl-[³H]puromycin formation. Chloramphenicol concentration was varied as indicated on the abscissa at four different puromycin concentrations. Polyribosomes were added last to start the reactions. The data of the left panel are plotted and calculated in the right panel according to Dixon (1953). ●, 1.37×10^{-5} M puromycin; △, 4.55×10^{-6} M puromycin; ○, 2.28×10^{-6} M puromycin; ▲, 1.14×10^{-6} M puromycin. The data are from Pestka (1972)

was apparent. Analysis of the inhibition by a Dixon plot (Fig. 4) suggested that two modes of inhibition were possible, competitive inhibition at low chloramphenicol concentrations ($K_{i1} = 7 \times 10^{-5}$ M) and a non-competitive type of inhibition at high concentrations of the antibiotic ($K_{i2} = 2.2 \times 10^{-3}$ M). Under different conditions, both types have been reported. Goldberg and Mitsugi (1967) reported a mixed type of inhibition of polylysyl-puromycin synthesis by chloramphenicol; Pestka (1970a) reported that chloramphenicol is a competitive inhibitor of puromycin in acetylphenylalanyl-puromycin synthesis. In studying the inhibition of the puromycin dependent release of nascent protein chains by chloramphenicol, Cannon (1968) noted that only about 40% of the reaction was inhibited by chloramphenicol. Although kinetic studies were not reported, this observation is similar to data of Fig. 4 where only partial inhibition of the reaction was noted even at high chloramphenicol concentrations.

The meaning of two modes of inhibition by chloramphenicol is not obvious. It may mean that there are two sites for chloramphenicol action or perhaps two classes of ribosomal states amenable to chloramphenicol inhibition. In any case, the inhibition of peptidyl-puromycin formation on polyribosomes by chloramphenicol appears to be asymmetric with respect to peptidyl-tRNA and puromycin. About half of the peptidyl-tRNA which is available for reaction with puromycin is in a state where chloramphenicol inhibits the reaction competitively with respect to puromycin; the other half of peptidyl-tRNA reacts with puromycin in such a way that the inhibition by chloramphenicol is noncompetitive. Inhibition of this latter reaction by chloramphenicol is only partial at very high chloramphenicol concentrations, whereas high chloramphenicol concentrations can almost completely inhibit the competitive reaction. The model of ribosome function previously described (Pestka, 1972) where each 50S site can alternately serve as donor and acceptor is consistent with these results.

Effect on Aminoacyl-Oligonucleotide Binding to Ribosomes

Isolated aminoacyl-oligonucleotide fragments of aminoacyl-tRNA are bound well to ribosomes in the absence of templates (Pestka, 1969b, c, d; Hishizawa et al., 1970; Pestka et al., 1970; Lessard and Pestka, 1972a). Binding of these aminoacyl-oligonucleotides such as C-C-A(Phe), C-C-A(Lys), C-C-A(Ser), and C-C-A(Leu) to ribosomes reflects the binding of the aminoacyl-termini of the respective aminoacyl-tRNA species to ribosomes. Binding of these aminoacyl-oligonucleotides to ribosomes was inhibited by chloramphenicol (Pestka, 1969b, c,; Pestka et al., 1970; Lessard and Pestka, 1972 b). The biologically active isomer, D($-$)threo-chloramphenicol, is most inhibitory; 50% inhibition is obtained at about 3×10^{-5} M (Fig. 5). The remaining three enantiomers exhibit relatively little effect on the binding of C-A-C-C-A(Phe) at concentrations as high as 10^{-3} M. At 10^{-3} M, L($+$)threo-chloramphenicol slightly inhibits C-A-C-C-A(Phe) binding.

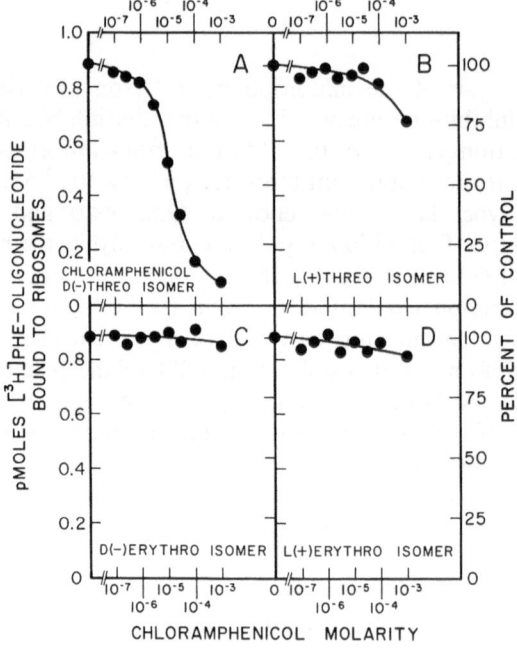

Fig. 5. A comparison of the effects of the isomers of chloramphenicol on the binding of C-A-C-C-A ([³H]Phe) to ribosomes. Each 0.050-ml reaction mixture contained the following components: 20% (v/v) ethanol; 0.05 M Tris-acetate, pH 7.2; 0.04 M MgCl₂; 0.40 M KCl; 0.06 M NH₄Cl; 4.1 A₂₆₀ units of ribosomes; 1.5 pmoles (0.02 A₂₆₀ unit) of C-A-C-C-A([³H]Phe). After incubating the reaction vessels at 24° for 20 min, the binding of C-A-C-C-A(Phe) to ribosomes was assayed by adsorbing ribosomes to cellulose nitrate filters. The concentrations of the isomers of chloramphenicol are given on the abscissa. The data are from Lessard and Pestka (1972b)

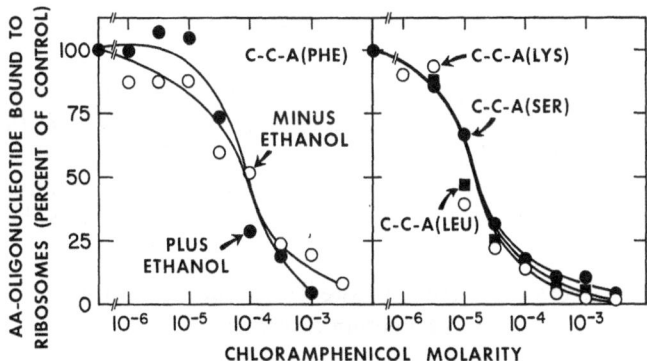

Fig. 6. The effect of chloramphenicol on the binding of C-C-A-(Phe), C-C-A(Ser), C-C-A(Leu), and C-C-A(Lys) to puromycin-treated ribosomes. Each 0.050-ml reaction mixture contained the following components: 0.05 M Tris-acetate, pH 7.2; 0.04 M $MgCl_2$; 0.4 M KCl; 0.08 M NH_4Cl; 4.0 A_{260} units of puromycin-treated ribosomes; 20% (v/v) ethanol where indicated; and 0.84 pmole (0.02 A_{260} unit) of C-C-A([³H]Phe), 7.1 pmoles (0.06 A_{260} unit) of C-C-A([³H]Ser), 9.6 pmoles (0.004 A_{260} unit) of C-C-A([³H]Lys), or 1.8 pmoles (0.001 A_{260} unit) of C-C-A([³H]Leu). In the absence of chloramphenicol, 0.26 and 0.41 pmole of C-C-A(Phe) were bound in the absence and presence of ethanol, respectively; and also in the absence of chloramphenicol 0.22 pmole of C-C-A(Ser), 1.6 pmoles of C-C-A(Lys), and 0.30 pmole of C-C-A(leu) were bound to ribosomes in the presence of ethanol. Reactions were incubated for 30 min at 24° and assays were performed as described in the legend to Fig. 5. Left panel: o, binding of C-C-A(Phe) in the absence of ethanol; •, binding of C-C-A(Phe) in the presence of 20% (v/v) ethanol. Right panel: o, binding of C-C-A(Lys) in the presence of 20% (v/v) ethanol; •, binding of C-C-A(Ser) in the presence of 20% (v/v) ethanol; ■, binding of C-C-A(Leu) in the presence of 20% (v/v) ethanol

To evaluate whether the amino acid side chains play a role in determining the binding characteristics of the aminoacyl-oligonucleotides in the presence of chloramphenicol, the effects of this antibiotic on the binding of several C-C-A(amino acid) species to ribosomes were compared (Fig. 6). To keep the oligonucleotide portion constant for comparison, the C-C-A(amino acid) derivatives were used (LESSARD and PESTKA, 1972b). Chloramphenicol appears to inhibit the binding of C-C-A(Phe) about equally in the presence and absence of ethanol. Fifty per cent inhibtion of the binding occurred at about 10^{-4} M antibiotic in both cases. The effect of chloramphenicol on the binding of C-C-A(Lys), C-C-A(Ser), and C-C-A(Leu) was examined only in the presence of 20% (v/v) ethanol since little binding of these compounds occurred in the absence of alcohol. The inhibition of the binding of all three fragments to ribosomes by chloramphenicol is essentially identical: 50% inhibition at 2×10^{-5} chloramphenicol. It is interesting to note that higher concentrations of chloramphenicol are required for 50% inhibition of C-C-A(Phe) binding than of 50% inhibition of the binding of the lysine, leucine, and serine fragments. This probably reflects the greater affinity of the Phe-oligonucleotides than the other fragments for ribosomes.

These results demonstrated directly that chloramphenicol inhibits the binding of the aminoacyl-end of aminoacyl-tRNA to ribosomes. Inhibition of functional attachment of the aminoacyl-end of aminoacyl-tRNA could account for inhibition of peptide bond formation. Conversely, C-A-C-C-A(Phe) can inhibit binding of chloramphenicol to ribosomes (YUKIOKA and MORISAWA, 1971). Also, CELMA *et al.* (1970) showed that chloramphenicol slightly stimulated binding of C-A-C-C-A(Ac-Leu) to ribosomes.

Furthermore, in the presence of deacylated tRNA and ethanol, 50S subunits from *E. coli* could bind C-A-C-C-A(Phe) (HISHIZAWA and PESTKA, 1971). Chloramphenicol (10^{-4} M) inhibits this binding of C-A-C-C-A(Phe) to 50S subunits 80%. This is an additional example of the interaction of chloramphenicol with isolated 50S subunits and inhibition of a reaction specific to this subunit.

Effects on Release or the Termination Reactions of Protein Synthesis

Genetic analysis indicated that the codons UAA, UAG, and UGA could act as termination signals (BRENNER, *et al.*, 1965; WEIGERT and GAREN, 1965; SAMBROOK *et al.*, 1967; ZIPSER, 1967). If by mutation such a codon appears in phase in the interior of a cistron, premature polypeptide chain termination occurs (SARABHAI *et al.*, 1964). With the use of RNA from a mutant of the bacteriophage R17 in which the seventh codon in the viral coat protein cistron has mutated from CAG to UAG, a cell-free assay for termination was devised (CAPECCHI, 1967a, b). In a cell-free amino acid incorporating system derived from *E. coli,* RNA from this mutant phage directs the synthesis of a small NH$_2$-terminal coat protein fragment, the hexapeptide fMet-Ala-Ser-Asn-Phe-Thr, which is released (ZINDER *et al.*, 1966). By isolating the hexapeptide-tRNA · ribosome · R17 mRNA complex, CAPECCHI (1967a, b) was able to use the release of hexapeptide as a specific assay for termination. With the use of this assay, CAPECCHI and KLEIN (1969) showed that chloramphenicol at a concentration of 5×10^{-4} M inhibited release about 65%.

An additional assay for termination has been the release of formylmethionine from fMet-tRNA bound to ribosomes (CASKEY *et al.*, 1968). Chloramphenicol (3×10^{-4} M) inhibited release of formylmethionine in this assay by about 20% (SCOLNICK *et al.*, 1968) although VOGEL *et al.*, (1969) reported greater than 90% inhibition of release by 10^{-4} M chloramphenicol. These latter workers correlated inhibition of peptidyl transferase activity (fMet-puromycin synthesis) with inhibition of release for chloramphenicol as well as other antibiotics. It should be noted, however, that inhibition of peptidyl transferase activity in model systems does not always correlate with inhibition of peptide bond synthesis in intact cells or on polyribosomes (PESTKA, 1972). It is thus probable that model systems for termination will not always reflect the mode of action of an agent in intact cells. For example, although lincomycin is a strong inhibitor of formylmethionyl-puromycin synthesis and termination reactions in model systems (VOGEL *et al.,* 1969) it is unlikely that it inhibits transpeptidation or termination in intact cells (CUNDLIFFE, 1969) or on native polyribosomes (PESTKA, 1972).

Effects on Polyribosomes and Ribosomes

Chloramphenicol has been shown to prevent polysome breakdown when protein synthesis is inhibited by it for short times of 5 min or less (DAS *et al.,* 1966; WEBER and DeMoss, 1966; FLESSEL, 1968; DRESDEN and HOAGLAND, 1967).

Chloramphenicol can prevent the breakdown of polysomes caused by puromycin or actinomycin (FLESSEL, 1968), and it can also block the degradation of mRNA observed in the presence of actinomycin (LEVINTHAL et al., 1964; GROS et al., 1964). Also, although chloramphenicol did not inhibit attachment of ribosomes to mRNA (DAS et al., 1966; DRESDEN and HOAGLAND, 1967), DRESDEN and HOAGLAND (1967) reported that chloramphenicol inhibited formation of polysomes in cells recovering from glucose starvation. This indicated that chloramphenicol inhibits ribosome movement along mRNA (or possibly initiation). On longer exposure of cells to chloramphenicol, polysome breakdown can occur (FLESSEL, 1968). Nevertheless, GURGO et al., (1969 a, b) have reported the continued entrance of mRNA into polysomes in the presence of 3×10^{-4} M chloramphenicol for 30 min; from this they inferred that movement of ribosomes on mRNA continues in cells treated with chloramphenicol. However, continual association of newly synthesized mRNA with ribosomes distorted by antibiotics may not imply translocation for a distorted ribosome may contain more than one mRNA. It is possible that ribosomal movement along mRNA occurs at a slow rate in the presence of chloramphenicol. This rate of movement is much slower than that occurring during protein synthesis and thus the apparent protection of polysomes during short incubations with chloramphenicol. On longer exposure of cells to chloramphenicol, polysome dissolution (FLESSEL, 1968) and entrance of newly synthesized mRNA into polysomes is observed (GURGO et al., 1969a, b). If polysomes continue to form at almost a normal rate as GURGO et al. (1969a, b) suggest, under conditions where protein synthesis is inhibited 95%, then inhibition of peptide bond formation by chloramphenicol must be uncoupled from ribosomal movement along the mRNA. Slow rate of protein synthesis may account for some of the entry of newly synthesized mRNA into polysomes. The results of WEBER and DeMoss (1966) and CAMERON and JULIAN (1968) indicate that peptidyl-tRNA remains attached to ribosomes on binding of chloramphenicol, for the polysome specific activity does not decrease shortly after chloramphenicol is added to growing cultures of E. coli, at least for several minutes. Perhaps, as suggested (GURGO et al., 1969a, b) peptidyl-tRNA is eventually released from ribosomes, which can then attach to new mRNA. KAEMPFER and MESELSON (1969) and KAEMPFER (1968) have shown that in the presence of 6×10^{-4} M chloramphenicol, ribosome subunit exchange and protein synthesis can occur at slow rates in cell-free extracts at about one-fifth the rate of control tubes without antibiotic. In the presence of fusidic acid, polyribosomes are fixed so that little mRNA enters polysomes (GURGO et al., 1969a, b). It thus was suggested that ribosomal movement is coupled to GTP hydrolysis and can occur in the absence of peptide bond formation. It is possible that chloramphenicol inhibits peptide bond formation by distorting the ribosome so that aminoacyl- and/or peptidyl-tRNA cannot bind appropriately to the peptidyl transferase, but that ribosomal movement can still occur. CAMERON and JULIAN (1968) as well as WEBER and DeMoss (1966) have indicated that chloramphenicol stimulates polysome formation; this was shown to be due to nonfunctional polysomes formed from the combination of 30 and 50S subunits with newly synthesized mRNA. Perhaps, the incorporation of mRNA into polysomes seen by GURGO et al. (1969a, b) is similarly due to stimulation of the formation of non-functional polysomal complexes.

In addition, chloramphenicol has been used to demonstrate the presence of messenger RNA-associated 30S ribosomal subunits as intermediary initiation complexes (Hori *et al.,* 1968). These complexes can be demonstrated in sucrose gradients of cell lysates prepared by rapid chilling of cells. Chloramphenicol, by stabilizing polysomes, inhibits the formation of these messenger RNA-30S complexes.

As indicated above when some *E. coli* strains are incubated in the presence of chloramphenicol, the cells fail to synthesize protein but continue to accumulate RNA. Following removal of the drug, the accumulated RNA's which exist in the form of ribonucleoprotein particles are converted into mature ribosomes when protein synthesis resumes (Nomura and Hosokawa, 1965; Horowitz and Hills, 1966). During recovery from chloramphenicol treatment preferential synthesis of ribosomal protein occurs over that of soluble protein (Aronson and Spiegelman, 1961 b; Davis and Sells, 1969). Early during recovery at least four proteins are preferentially synthesized and assembled into 50S ribosomal subunits; late in recovery the cells preferentially synthesize the complement to those ribosomal proteins synthesized early (Davis and Sells, 1969). In other studies, ribosomes from *E. coli* Q13 treated with chloramphenicol for a prolonged period were found to be deficient in protein content compared to control cells (Young and Nakada, 1971). Initiation factors F2 and F3 were either defective or substantially reduced in ribosomes from these chloramphenicol treated cells.

Protein Synthesis Resistant to Chloramphenicol

Some phage proteins are unusual in that they are synthesized in the presence of 30 µg of chloramphenicol per ml. Tessman (1966) observed the synthesis of gene products of phage S13 in the presence of 30 µg of chloramphenicol per ml. Sinsheimer, *et al.,* (1967) similarly observed the synthesis of a gene product of φX174 which could be synthesized in the presence of 30 µg of chloramphenicol per ml. These phage proteins, however, are not made, in 100 to 150 µg of chloramphenicol per ml (Tessman, 1966; Levine and Sinsheimer, 1968). Analogously, an *E. coli* protein required for the initiation of bacterial DNA replication also can be synthesized in the presence of lower concentrations of chloramphenicol (Lark, 1966; Lark and Lark, 1966). The chloramphenicol resistant protein of φX174 appears to be in the membrane fraction (Levine and Sinsheimer, 1969). Other bacterial and phage proteins have also been reported to be synthesized in the presence of chloramphenicol (Aronson and Spiegelman, 1961; Sinsheimer *et al.,* 1962). The protein which confers resistance to lysis from without by T_4 bacteriophage can be synthesized in the presence of 100 µg per ml of chloramphenicol (Peterson *et al.,* 1972). Additionally, an *E. coli* protein involved in lysogenization was reported to be formed in the presence of 100 µg/ml chloramphenicol (Naha, 1969).

It appears that the proteins synthesized in the presence of chloramphenicol are related to functions associated with the cell membrane. Thus, it is possible that bacterial protein synthesis on polyribosomes in or near the cell membrane may be relatively resistant to chloramphenicol. Alternatively, however, chloramphenicol resistance of particular proteins may conceivably be related to the composition of the messenger RNA template, for as noted above, inhibition of polypeptide synthesis with synthetic templates is dependent on the base composition of those templates. Possibly both considerations may be applicable.

Effect of Chloramphenicol on RNA Synthesis

The synthesis of stable RNA species in many strains of *E. coli* is sharply reduced by deprivation of a required amino acid (STENT and BRENNER, 1961; LAZZARINI and WINSLOW, 1970; PRIMAKOFF and BERG, 1970). The stringent response, as this control mechanism is called, depends upon function of the *rel* gene. Amino acid starvation causes a rapid accumulation of two unusual guanosine nucleotides, called magic spots I and II (MSI and MSII), in stringent (*rel$^+$*) but not in relaxed (*rel$^-$*) strains (CASHEL and GALLANT, 1969; CASHEL, 1969). CASHEL and GALLANT (1969) postulated that high intracellular concentrations of the MS compounds leads to the cessation of RNA accumulation and to the other characteristics of the stringent response such as inability to incorporate uracil from the medium and shrinkage of the nucleoside triphosphate pools. CASHEL and KALBACHER (1970) have identified MSI as a guanosine tetraphosphate (ppGpp, 5′ guanosine diphosphate 3′ or 2′ diphosphate) and MSII as a guanosine pentaphosphate (pppGpp). Relaxed mutants (rel$^-$) continue to accumulate stable RNA during amino acid starvation and there is no formation of these guanosine nucleotides (CASHEL, 1969). Chloramphenicol inhibits the synthesis of MSI and II in stringent cells and abolishes the stringent response (PARDEE and PRESTIDGE, 1956; GROS and GROS, 1956).

Thus, when bacterial cultures are inhibited by chloramphenicol, the cells accumulate considerable quantities of RNA (PARDEE *et al.,* 1957; NOMURA and WATSON, 1959; KURLAND and MAALOE, 1962). Study of the kinetics of RNA synthesis in inhibited cells indicate significant changes in the rates of formation of nucleic acids. The rate of RNA synthesis is almost instantaneously increased when the antibiotic is added to cultures growing slowly in minimal media, but increase in RNA synthesis is barely detectable in enriched medium (FRAENKEL and NEIDHARDT, 1961; KURLAND and MAALOE, 1962; MAALOE and KJELDGAARD, 1966). Detailed examination of the synthesis and stability of RNA in chloramphenicol-inhibited cultures of *E. coli* has been made by MIDGLEY and GRAY (1971). They concluded that accelerations in the rate of biosynthesis of ribosomal RNA by chloramphenicol in growth-limiting media are due primarily to an increase in the rate of initiation of new chains up to the rates existing in cultures growing rapidly in rich media. Thus, in poorer media, only a small fraction of the available DNA-dependent RNA polymerase molecules are active at any given instant, since the chain-initiation rate is limiting in these conditions. In cultures growing rapidly in enriched broth, antibiotic addition caused a rise of some 12% in the rate of incorporation of exogenous uracil into total RNA. This small acceleration was due entirely to the partial stabilization of the mRNA fraction, which accumulated as 14% of the RNA formed after the addition of chloramphenicol. In cultures growing more slowly in minimal media, chloramphenicol caused an immediate acceleration of two- to three-fold in the overall rate of RNA synthesis. Studies by DNA-RNA hybridization showed that the synthesis of mRNA was accelerated in harmony with the other affected species. However, just over half the mRNA formed after the addition of chloramphenicol quickly decayed to acid-soluble products, whereas the remainder was more stable and accumulated in the cells.

Bacterial Resistance to Chloramphenicol

The status of acquired resistance to chloramphenicol has not changed significantly in the past several years. There appear to be two major areas for the biochemical basis of chloramphenicol resistance. One is an acquired impermeability to the antibiotic. This has been shown to occur as a result of multistep mutations as well as episomal transfer (OKAMOTO and MIZUNO, 1962, 1964; HAHN, 1967). A second mode of acquired resistance results from the episomal transfer of chloramphenicol acetyl-transferase, the enzyme capable of acetylating chloramphenicol to form the inactive diacetyl derivative (OKAMOTO and SUZUKI, 1965; SHAW et al., 1970). The episome carrying the genes for chloramphenicol-resistance is capable of integrating into the E. coli chromosome (IYOBE et al., 1969, 1970).

In addition, the induction of chloramphenicol-resistance by compounds structurally related to chloramphenicol has been reported (MITSUHASHI et al., 1969; KONO et al., 1971). The populations of staphylococci exhibiting induced-resistance to chloramphenicol inactivated the antibiotic by acetylation (KONO et al., 1968). Thus, this phenomenon differs from inducible-resistance to erythromycin (WEISBLUM, 1971).

To date single step ribosomal mutations to chloramphenicol resistance have not been reported. Nevertheless, ČERNÁ and RYCHLÍK (1968) reported the isolation of an E. coli mutant resistant to erythromycin by cultivation of the sensitive E. coli B in enriched medium with increasing concentrations of erythromycin. Ribosomes from E. coli B, resistant to erythromycin, exhibited cross-resistance to chloramphenicol. A 50% inhibition of the puromycin reaction with ribosomes from the erythromycin-resistant strain required about 10^{-3} M chloramphenicol compared to 10^{-4} M chloramphenicol required for a similar inhibition of the reaction with ribosomes from the parent sensitive strain. In studies of neomycin and kanamycin resistant mutants of E. coli, APIRION and SCHLESSINGER (1968) observed that such mutants had decreased resistance to chloramphenicol. Such results probably reflect the pleiotropic manifestations of ribosomal mutations (PESTKA, 1971). Also, phenotypic suppression in E. coli by chloramphenicol has been reported (KIRSCHMANN and DAVIS, 1969). Chloramphenicol was able to replace streptomycin in supporting growth of conditionally streptomycin-dependent mutants, however, the mechanism of this effect has not been clarified.

Effect of Chloramphenicol on Eucaryotes, Mitochondria and Chloroplasts

Although the effect of chloramphenicol on eucaryotes is not as prominent as its effects on bacteria, a large number of observations have indicated that chloramphenicol can inhibit processes in the eucaryotes. In particular, protein synthesis in eucaryotic organelles is inhibited by chloramphenicol. Chloroplast ribosomes from tobacco leaves (ELLIS, 1969) and mitochondrial ribosomes from rat liver (FREEMAN, 1970b) and yeast (GORDON, LOWDON, and STEWART, 1972) show the same stereospecific inhibition of protein synthesis by chloramphenicol as do bacterial ribosomes; under the same conditions cytoplasmic ribosomes are unaffected by chloramphenicol. Protein synthesis by isolated mitochondria from mammalian liver (KROON, 1963; ASHWELL and WORK, 1968; CLARK-WALKER and LINNANE, 1966, 1967), HeLa cells (GALPER and DARNELL, 1971) and yeast (WINTERSBERGER, 1965; LINNANE, et al., 1968) is inhibited by chloramphenicol. During continuous infusion of chloramphenicol into pregnant rats, sufficient antibiotic accumulates in embryonic tissue to inhibit mammalian mitochondrial protein synthesis (OERTER and BASS, 1972).

Chloramphenicol has been shown to inhibit polypeptide synthesis with reticulocyte ribosomes when exogenous templates are used (WEISBERGER et al., 1963; ARMENTROUT and WEISBERGER, 1967 and 1968), particularly poly U. The effect is variable and strongly dependent on Mg^{++} concentration. If extracts are preincubated with poly U for 5 min prior to addition of chloramphenicol, the antibiotic has no inhibitory effect. So far this observation has not been confirmed by other laboratories. Chloramphenicol had no effect on protein synthesis by reticulocyte ribosomes with endogenous mRNA. Although this effect of chloramphenicol is observed at concentrations of 10^{-4} M or less, it is probably unrelated to the effects described on procaryotic ribosomes, and may be a reaction of chloramphenicol with exogenous templates or with ribosomal sites to which these exogenous templates bind. The effect of the other three enantiomers of chloramphenicol in this system has not been reported. The resemblance of chloramphenicol to uridylic acid (WEISBERGER et al., 1964) may possibly enable chloramphenicol to compete with uridylic residues on template codons for ribosomal sites. Additionally, WEISBERGER and WOLFE (1964) reported binding of chloramphenicol to reticulocyte 80S ribosomes. However, binding of chloramphenicol to eucaryotic 80S ribosomes has not been observed by other laboratories (VAZQUEZ, 1966a).

Although mammalian tissues are relatively insensitive to chloramphenicol, antibody synthesis appears to be an exception since it is inhibited by concentrations of chloramphenicol which are bacteriostatic (AMBROSE and COONS, 1963; WEISBERGER et al., 1964; SVEHAG, 1964; CRUCHAUD and COONS, 1964). In spleen, ribosomes bound to endoplasmic reticulum were sensitive to inhibition by 10^{-5} M chloramphenicol whereas free cytoplasmic ribosomes were relatively insensitive (TALAL and EXUM, 1966). Protein synthesis in intact reticulocytes was also inhibited by chloramphenicol at high concentrations (3×10^{-3} M). However, the inhibition occurred equally well with the bacteriostatically inactive L(+)threo-enantiomers and RNA synthesis as well as protein synthesis was inhibited (GODCHAUX and HERBERT, 1966). In contrast, even high concentrations of chloramphenicol (3×10^{-3} M) did not inhibit peptidyl-puromycin formation with rat liver polyribosomes (PESTKA, unpublished observations). This suggests that the mode of action of chloramphenicol in inhibiting protein synthesis in mammalian cells is significantly different from its mode of action in bacteria.

Effects on Respiration

Chloramphenicol and its isomers and analogs inhibit mitochondrial respiration. With the use of rat liver mitochondria, D-*threo*-chloramphenicol inhibited β-hydroxybutyrate oxidase 50% at 1.3 mM and succinoxidase at 3.5 mM. L-*threo*-chloramphenicol inhibited these 50% at 3.6 and 3.6 mM, respectively (FREEMAN, 1970a, b). FIRKIN and LINNANE (1968) have shown that at low concentrations of the antibiotic (20 µg/ml) there was selective inhibition of the synthesis of mitochondrial membrane-bound cytochromes, and this action of chloramphenicol appeared to result in the inhibition of synthesis of only a small proportion of the mitochondrial proteins. At comparatively high concentrations of chloramphenicol (100 to 150 µg/ml), the drug directly inhibits the respiration of isolated mitochondria as well as intact HeLa cells. The inhibition of mitochondrial respiration by chloramphenicol both *in vitro* and *in vivo* suggests that this phenomenon probably accounts for the immediate inhibition of HeLa cell growth by chloramphenicol at these high concentrations. Although FREEMAN and HALDAR (1968) have reported that chloramphenicol is a specific inhibitor of NADH oxidation, FIRKIN and LINNANE (1968) have found chloramphenicol to be a more general inhibitor of respiration in mammalian cells. In mouse ascites cells, HALDAR and FREEMAN (1968) have shown that both the D(−)-*threo*- and L(+)-*threo*-, and L-erythro-chloramphenicol isomers inhibit respiration at about 1 to 5 mM. Similarly, BALL and TUSTANOFF (1970) have demonstrated that glucose repression and chloramphenicol inhibition of respiration are additive effects.

Miscellaneous Observations

HAMBURGER (1966) has been able to produce chloramphenicol specific antibody by using chloramphenicol coupled to bovine gamma-globulin and to rabbit serum albumin. The reactivity of analogs of chloramphenicol with anti-chloramphenicol antibody indicated that the major antigenic sites are separate from the sites critical for antibiotic activity (HAMBURGER and DOUGLASS, 1969a). The presence in serum of antibody specific for chloramphenicol significantly inhibited the antibiotic activity in inhibiting growth of *E. coli* (HAMBURGER and DOUGLASS, 1969b). With the use of a radioimmune-assay, ORGEL and HAMBURGER (1971) developed a convenient method for detecting antichloramphenicol antibody in human serum.

Chloramphenicol suppressed virus production, but had no effect on malignant transformation as measured by focus formation produced by *Rous sarcoma* virus (RICHERT and HARE, 1972). Exposure of cells to 50 µg/ml of chloramphenicol for 5 days before and after addition of SV40 virus failed to inhibit synthesis of the tumor and virion antigens, but did inhibit the cytopathic effect of the virus (SABIN, 1966). Chloramphenicol was reported to inhibit regeneration of tissues in the newt (KOKOLIS et al., 1972a). KOKOLIS et al., (1972b) also reported that chloramphenicol reduced tumor growth in rats and prevented accumulation of tetrahydrobiopterin.

Lymphocyte transformation to blast cells induced by phytohemagglutinin was not inhibited by chloramphenicol in cultures (NASJLETI and SPENCER, 1968),

However, RINGROSE and LAMBERT (1973) reported that chloramphenicol inhibited induced RNA synthesis in phytohemagglutinin-stimulated lymphocytes. The mechanism of this inhibition has not been determined.

Chloramphenicol has been shown to inhibit active amino acid transport (GROSS and KING, 1969). The inhibition of amino acid transport was explained by assuming that some protein which is involved in carrier-mediated transport has a rapid turnover and its amount is reduced by chloramphenicol treatment so as to become rate limiting.

Although the primary action of chloramphenicol in inhibiting the growth of most microorganisms is probably a consequence of its effects on protein synthesis, the direct action in inhibiting the transfer of glucose to teichoic acid may be a contributory factor in some organisms. At 100 μg per ml of chloramphenicol some inhibition (10%) of techoic acid synthesis can be observed (STOW et al., 1971). Higher concentrations of the antibiotic produce proportionately greater inhibitions. The effect is apparently specific to glucose, and is not observed with the transfer of other residues from nucleotides; for example, glycerol phosphate, N-acetyl-glucosamine, and N-acetyl-galactosamine. The site of inhibition seems to be at the stage of transfer of glucose from nucleotide precursor to lipid carrier, and in the cases studied there was no inhibition of glucose transfer in syntheses where lipid carriers were not involved. The action of chloramphenicol on the synthesis of poly-glucosylglycerol phosphate in *Bacillus licheniformis* is not a secondary effect of inhibition of endogenous protein synthesis; the preparation of the enzyme included treatment with both deoxyribonuclease and ribonuclease, and higher concentrations of antibiotic were necessary than those normally required for inhibition of protein synthesis in cell-free systems. In addition, amino acids were absent from the incubation mixtures, and neither streptomycin nor puromycin were inhibitory.

References

ALLEN, E. H., and R.S. SCHWEET: Synthesis of hemoglobin in a cell-free system, I. Properties of the complete system. J. Biol. Chem. **237**, 760–767 (1962).

AMBROSE, C.T., and A.H. COONS: Studies on antibody production. VIII. The inhibitory effect of chloramphenicol on the synthesis of antibody in tissue culture. J. Exptl. Med. **117**, 1075–1088 (1963).

APIRION, D., and D. SCHLESSINGER: Coresistance to neomycin and kanamycin by mutations in an *Escherichia coli* Locus that affects ribosomes. J. Bacteriol. **96**, 768–776 (1968).

ARMENTROUT, S.A., and A.S. WEISBERGER: Inhibition of directed protein synthesis by chloramphenicol: Effect of magnesium concentration. Biochem. Biophys. Res. Commun. **26**, 712–716 (1967).

ARMENTROUT, S.A., and A.S. WEISBERGER: Ribonucleoprotein interaction with mammalian monosomes. Biochim. Biophys. Acta **161**, 180–187 (1968).

ARONSON, A.I., and S. SPIEGELMAN: Protein and ribonucleic acid synthesis in a cloramphenicol-inhibited system. Biochim. Biophys. Acta **53**, 70–84 (1961a).

ARONSON, A.I., and S. SPIEGELMAN: On the nature of the ribonucleic acid synthesized in the presence of chloramphenicol. Biochim. Biophys. Acta **53**, 84–95 (1961b).

ASHWELL, M.A., and T.S. WORK: Contrasting effects of cycloheximide on mitochondrial protein synthesis *in vivo* and *in vitro*. Biochem. Biophys. Res. Commun. **32**, 1006–1012 (1968).

BALL, A.J.S., and E.R. TUSTANOFF: Effect of D(−) and L(+)-*threo*-chloramphenicol on nucleotide and related respiratory activities in yeast undergoing metabolic repression and de-repression. Biochim. Biophys. Acta **199**, 476–489 (1970).

Borsook, H., E.H. Fischer, and G. Keighley: Factors affecting protein synthesis *in vitro* in rabbit reticulocytes. J. Biol. Chem. **229**, 1059–1070 (1957).

Brenner, S., A.O.W. Stretton, and S. Kaplan: Genetic code: The "nonsense" triplets for chain termination and their suppression. Nature **206**, 994–998 (1965).

Brock, T.D.: Chloramphenicol. Bacteriol. Rev. **25**, 32–48 (1961).

Brock, T.D.: Chloramphenicol. Experimental chemotheraphy, vol. III, p. 119–169. New York: Academic Press, 1964.

Cameron, H.J., and G.R. Julian: The effect of chloramphenicol on the polysome formation of starved stringent *Escherichia coli*. Biochim. Biophys. Acta **169**, 373–380 (1968).

Cannon, M.: The puromycin reaction and its inhibition by chloramphenicol. Eur. J. Biochem. **7**, 137–145 (1968).

Capecchi, M.R.: Polypeptide chain termination *in vitro*: Isolation of a release factor. Proc. Natl. Acad. Sci. U.S. **58**, 1144–1151 (1967a).

Capecchi, M.R.: A rapid assay for polypeptide chain termination. Biochem. Biophys. Res. Commun. **28**, 773–778 (1967b).

Capecchi, M.R., and H.A. Klein: Characterization of three proteins involved in polypeptide chain termination. Cold Spring Harbor Symp. Quant Biol. **34**, 469–477 (1969).

Cashel, M.: The control of ribonucleic acid synthesis in *E. coli* IV. Relevance of unusual phosphorylated compounds from amino acid-starved stringent strains. J. Biol. Chem. **244**, 3133–3141 (1969).

Cashel, M., and J. Gallant: Two compounds implicated in the function of the RC gene of *E. coli*. Nature **221**, 838–841 (1969).

Cashel, M., and B. Kalbacher: The control of ribonucleic acid synthesis in *E. coli* V. Characterization of a nucleotide associated with the stringent response. J. Biol. Chem. **245**, 2309–2318 (1970).

Caskey, T., R. Tompkins, E. Scolnick, T. Caryk, and M. Nirenberg: Sequential translation of trinucleotide codons for the initiation and termination of protein synthesis. Science **162**, 135–138 (1968).

Celma, M.L., R.E. Monro, and D. Vazquez: Substrate and antibiotic sites at the peptidyl transferase centre of *E. coli* ribosomes. FEBS Letters **6**, 273–277 (1970).

Černá, J., and I. Rychlík: Cross resistance of *Escherichia coli* B ribosomes to inhibition of the puromycin reaction by erythromycin, spiramycin and chloramphenicol. Biochim. Biophys. Acta **157**, 436–438 (1968).

Černá, J., I. Rychlík, and P. Pulkrábek: The effect of antibiotics on the coded binding of peptidyl-tRNA to the ribosome and on the transfer of the peptidyl residue to puromycin. Eur. J. Biochem. **9**, 27–35 (1969).

Chang, F.N., C. Siddhikol, and B. Weisblum: Subunit localization studies of antibiotic inhibitors of protein synthesis. Biochim. Biophys. Acta **186**, 396–398 (1969).

Clark-Walker, G.D., and A.W. Linnane: *In vivo* differentiation of yeast cytoplasmic and mitochondrial protein synthesis with antibiotics. Biochem. Biophys. Res. Commun. **25**, 8–13 (1966).

Clark-Walker, G.D., and A.W. Linnane: The biogenesis of mitochondria in *Saccharomyces cerevisiae*. A comparison between cytoplasmic respiratory-deficient mutant yeast and chloramphenicol-inhibited wild type cells. J. Cell Biol. **34**, 1–14 (1967).

Controulis, J., M.C. Rebstock, and H.M. Crooks: Chloramphenicol (Chloromycetin), V. Synthesis. J. Am. Chem. Soc. **71**, 2463–2468 (1949).

Coutsogeorgopoulos, C.: On the mechanism of action of chloramphenicol in protein synthesis. Biochim. Biophys. Acta **129**, 214–217 (1966).

Coutsogeorgopoulos, C.: Inhibitors of the reaction between puromycin and polylysyl-RNA in the presence of ribosomes. Biochem. Biophys. Res. Commun. **27**, 46–52 (1967).

Coutsogeorgopoulos, C.: Amino acylaminonucleoside inhibitors of protein synthesis II. Effect on oligophenylalanine formation. Biochim. Biophys. Acta **240**, 137–150; **247**, 632 (1971).

Cross, D.F.W., G.W. Kenner, R.C. Sheppard, and C.E. Stehr: Peptides. Part XIV. Thiazole amino-acids degradation products of thiostrepton. J. Chem. Soc. 2143–2159 (1963).

Cruchaud, A., and A.H. Coons: Studies on antibody production. XIII. The effect of chloramphenicol on priming in mice. J. Exptl. Med. **120**, 1061–1074 (1964).

Cundliffe, E., and K. McQuillen: Bacterial protein synthesis. The effects of antibiotics. J. Mol. Biol. **30**, 137–146 (1967).

Das, H.K., A. Goldstein, and L.C. Kanner: Inhibition by chloramphenicol of the growth of nascent protein chains in *Escherichia coli*. Mol. Pharmacol. **2**, 158–170 (1966).

DAVIS, F.C., and B.H. SELLS: Synthesis and assembly of ribosomal protein into 50S subunits during recovery from chloramphenicol treatment. J. Mol. Biol. **39**, 503–521 (1969).

DeMOSS, J.A., and G.D. NOVELLI: An amino acid dependent exchange between ^{32}P labeled inorganic pyrophosphate and ATP in microbial extracts. Biochim. Biophys. Acta **22**, 49–61 (1956).

DIXON, M.: The determination of enzyme inhibitor constants. Biochem. J. **55**, 170–171 (1953).

DRESDEN, M.H., and M.B. HOAGLAND: Polyribosomes of *Escherichia coli*. Breakdown during glucose starvation. J. Biol. Chem. **242**, 1065–1068 (1967).

DRESDEN, H.M., and M.B. HOAGLAND: Polyribosomes of *Escherichia coli*. Re-formation during recovery from glucose starvation. J. Biol. Chem. **242**, 1069–1073 (1967).

DUNITZ, J.D.: The crystal structure of chloramphenicol and bromamphenicol. J. Am. Chem. Soc. **74**, 995–999 (1952).

EHRENSTEIN, G. VON, and F. LIPMANN: Experiments on hemoglobin biosynthesis. Proc. Natl. Acad. Sci. U.S. **47**, 941–950 (1961).

ELLIS, R.J., Chloroplast ribosomes; stereospecificity of inhibition by chloramphenicol. Science **163**, 477–478 (1969).

FERNANDEZ-MUÑOZ, R., R.E. MONRO, R. TORRES-PINEDO, and D. VAZQUEZ: Substrate- and antibiotic-binding sites at the peptidyl-transferase centre of *E. coli* ribosomes. Studies on the chloramphenicol, lincomycin and erythromycin sites. Eur J. Biochem. **23**, 185–193 (1971).

FICO, R., and C. COUTSOGEORGOPOULOS: Peptidyl transferase. A new method for kinetic studies. Biochem. Biophys. Res. Commun. **47**, 645–651 (1972).

FIRKIN, F.C., and A.W. LINNANE: Differential effects of chloramphenicol on the growth and respiration of mammalian cells. Biochem. Biophys. Res. Commun. **32**, 398–402 (1968).

FIRKIN, F.C., and A.W. LINNANE: Biogenesis of mitochondria. VIII. The effect of chloramphenicol on regenerating rat liver. Exptl. Cell Res. **55**, 68–76 (1969).

FLESSEL, C.P.: Chloramphenicol protects polyribosomes. Biochem. Biophys. Res. Commun. **32**, 438–446 (1968).

FRAENKEL, D.G., and F.C. NEIDHARDT: Use of chloramphenicol to study control of RNA synthesis in bacteria. Biochim. Biophys. Acta **53**, 96–110 (1961).

FREEMAN, K.B.: Effects of chloramphenicol and its isomers and analogs on the mitochondrial respiratory chain. Can. J. Biochem. Physiol. **48**, 469–478 (1970a).

FREEMAN, K.B.: Inhibition of mitochondrial and bacterial protein synthesis by chloramphenicol. Can J. Biochem. Physiol. **48**, 479–485 (1970b).

FREEMAN, K.B., and D. HALDAR: The inhibition of mammalian mitochondrial NADH oxidation by chloramphenicol and its isomers and analogs. Can. J. Biochem Physiol. **46**, 1003–1008 (1968).

FROST, A.A., and R.G. PEARSON: Kinetics and mechanisms, 233 pp. New York: Wiley & Sons 1953.

GALE, E.F.: Mechanisms of antibiotic action. Pharmacol. Rev. **15**, 481–530 (1963).

GALE, E.F., and J.P. FOLKES: The assimilation of amino-acids by bacteria. 15. Actions of antibiotics on nucleic acid and protein synthesis in *Staphylococcus aureus*. Biochem. J. **53**, 493–498 (1953).

GALPER, J.B., and J.E. DARNELL: Mitochondrial protein synthesis in HeLa cells. J. Mol. Biol. **57**, 363–367 (1971).

GARDNER, R.S., A.J. WAHBA, C. BASILIO, R.S. MILLER, P. LENGYEL, and J.F. SPEYER: Synthetic polynucleotides and the amino acid code. VII. Proc. Natl. Acad. Sci. U.S. **48**, 2087–2094 (1962).

GODCHAUX, W., and E. HERBERT: The effect of chloramphenicol in intact erythroid cells. J. Mol. Biol. **21**, 537–553 (1966).

GOLDBERG, I.H.: Mode of action of antibiotics. II. Drugs affecting nucleic acid and protein synthesis. Am. J. Med. **39**, 722–752 (1965).

GOLDBERG, I.H., and K. MITSUGI: Inhibition by sparsomycin and other antibiotics of the puromycin-induced release of polypeptide from ribosomes. Biochemistry **6**, 383–391 (1967).

GORDON, P.A., M.J. LOWDON, and P.R. STEWART: Effects of chloramphenicol isomers and erythromycin on enzyme and lipid synthesis induced by oxygen in wild-type and petite yeast. J. Bacteriol. **110**, 504–510 (1972).

GOTTESMAN, M.: Reaction of ribosome-bound peptidyl transfer ribonucleic acid with aminoacyl transfer ribonucleic acid or puromycin. J. Biol. Chem. **242**, 5564–5571 (1967).

GROS, F., J. DUBERT, A. TISSIERES, S. BOURGEOIS, M. MICHELSON, R. SOFFER, and L. LEGAALT: Regulation of metabolic breakdown and synthesis of messenger RNA in bacteria. Cold Spring Harbor Symp. Quant. Biol. **28**, 299–313 (1964).

Gros, F., and F. Gros: Role des aminoacides dans la synthese des acides nucléiques chez *E. coli*. Biochim. Biophys. Acta **22**, 200–201 (1956).

Gross, W., and K. Ring: Effect of chloramphenicol on active amino acid transport. FEBS Letters **4**, 319–322 (1969).

Gurgo, C., D. Apirion, and D. Schlessinger: Effects of chloramphenicol and fusidic acid on polyribosome metabolism in *Escherichia coli*. FEBS Letters **3**, 34–36 (1969a).

Gurgo, C., D. Apirion, and D. Schlessinger: Polyribosome metabolism in *Escherichia coli* treated with chloramphenicol, neomycin, spectinomycin or tetracycline. J. Mol. Biol. **45**, 205–220 (1969b).

Guthrie, G.D., and J.M. Buchanan: Control of phage-induced enzymes in bacteria. Federation Proc. **25**, 864–873 (1966).

Hahn, F.E.: Chloramphenicol, antibiotics, vol. 1 (Gottlieb, D., Shaw, P.D., eds.) p. 308–330. Berlin-Heidelberg-New York: Springer 1967.

Hahn, F.E., J.E. Hayes, C.L. Wisseman, H.E. Hopps, and J.E. Smadel: Mode of action of chloramphenicol. VI. Relation between structure and activity in the chloramphenicol series. Antibiot. & Chemotherapy **6**, 531–543 (1956).

Hahn, F.E., and C.L. Wisseman: Inhibition of adaptive enzyme formation by antimicrobial agents. Proc. Soc. Exptl. Biol. Med. **76**, 533–535 (1951).

Haldar, D., and K.B. Freeman: The inhibition of protein synthesis and respiration in mouse ascites tumor cells by chloramphenicol and its isomers and analogs. Can. J. Biochem. Physiol. **46**, 1009–1017 (1968).

Hamburger, R.N.: Chloramphenicol-specific antibody. Science **152**, 203–204 (1966).

Hamburger, R.N., and J.H. Douglass: Chloramphenicol-specific antibody. II. Reactivity to analogues of chloramphenicol. Immunology **17**, 587–591 (1969a).

Hamburger, R.N., and J.H. Douglass: Chloramphenicol-specific antibody. IV. Neutralization of antibiotic effect on *Escherichia coli*. Immunology **17**, 599–602 (1969b).

Hishizawa, T., J.L. Lessard, and S. Pestka: Studies on the formation of transfer ribonucleic acid-ribosome complexes. XII. Phenylalanyl-oligonucleotide binding to *E. coli* ribosomes: Necessity for a free amino group. Proc. Natl. Acad. Sci. U.S. **66**, 523–530 (1970).

Hishizawa, T., and S. Pestka: Studies on the formation of transfer ribonucleic acid-ribosome complexes. XVII. The effect of tRNA on aminoacyl-oligonucleotide binding to ribosomes. Arch. Biochem. Biophys. **147**, 624–631 (1971).

Hori, M., and M. Rabinovitz: Polyribosomal changes during inhibition of rabbit hemoglobin synthesis by an isoleucine antagonist. Proc. Natl. Acad. Sci. U.S. **59**, 1349–1355 (1968).

Hori, M., J. Suzuki, and H. Umezawa: Messenger RNA-associated 30S ribosomal subunit: Extraction from *E. coli* and the effect of chloramphenicol on the content. J. Biochem. (Tokyo) **64**, 905–907 (1968).

Horowitz, J., and D.C. Hills: Evidence for the direct conversion of chloramphenicol particles into ribosomes in *Escherichia coli*. Biochim. Biophys. Acta **123**, 416–419 (1966).

Hurwitz, C., and C.B. Braun: Measurement of binding of chloramphenicol by intact cells. J. Bacteriol. **93**, 1671–1676 (1967).

Hurwitz, C., and C.B. Braun: Temperature-sensitivity of the weak bonds by which chloramphenicol is held in intact cells. Biochim. Biophys. Acta **157**, 392–403 (1968).

Irvin, J.D., and G.R. Julian: The distribution of ^{14}C-proline peptides synthesized *in vitro* directed by polycytidylic acid; the effect of chloramphenicol. FEBS Letters **8**, 129–132 (1970).

Iyobe, S., H. Hashimoto, and S. Mitsuhashi: Integration of chloramphenicol-resistance gene of an R factor on *Escherchia coli* chromosome. Japan. J. Microbiol. **13**, 225–232 (1969).

Iyobe, S., H. Hashimoto, and S. Mitsuhashi: Integration of chloramphenicol-resistance genes of an R factor into various sites of an *Escherichia coli* chromosome. Japan. J. Microbiol. **14**, 463–471 (1970).

Jardetzky, O.: Studies on the mechanism of action of chloramphenicol, I. The conformation of chloramphenicol in solution. J. Biol. Chem. **238**, 2498–2508 (1963).

Jardetzky, O., and G. Julian: Chloramphenicol inhibition of polyuridylic acid binding to *E. coli* ribosomes. Nature **201**, 397–398 (1964).

Julian, G.R.: [^{14}C]Lysine peptides synthesized in an *in vitro Escherichia coli* system in the presence of chloramphenicol. J. Mol. Biol. **12**, 9–16 (1965).

Julian, G.R.: Effect of chloramphenicol on synthesis of C^{14}-lysine peptides. Antimicrobial Agents Chemotherapy 1965, 992–1000 (1966).

KAEMPFER, R.: Ribosomal subunit exchange during protein synthesis. Proc. Natl. Acad. Sci. U.S. **61**, 106–113 (1968).

KAEMPFER, R., and M. MESELSON: Studies of ribosomal subunit exchange. Cold Spring Harbor Symp. Quant. Biol. **34**, 209–220 (1969).

KIRSCHMANN, C., and B.D. DAVIS: Phenotypic suppression in *Escherichia coli* by chloramphenicol and other reversible inhibitors of the ribosome. J. Bacteriol. **98**, 152–159 (1969).

KOKOLIS, N., N. MYLONAS, and I. ZIEGLER: Pteridine and riboflavin patterns during tail regeneration in *Triturus* species and the effects of chloramphenicol, isoxanthopterin and reserpine. Z. Naturforsch. **27b**, 285–291 (1972a).

KOKOLIS, N., N. MYLONAS, and I. ZIEGLER: Pteridine and riboflavin in tumor tissue and the effect of chloramphenicol and isoxanthopterin. Z. Naturforsch. **27b**, 292–295 (1972b).

KONO, M., K. OGAWA, and S. MITSUHASHI: Drug resistance of staphylococci. VI. Genetic determinant for chloramphenicol resistance. J. Bacteriol. **95**, 886–892 (1968).

KONO, M., K. O'HARA, M. NAGAWA, and S. MITSUHASHI: Drug resistance of staphylococci: Ability of chloramphenicol related compounds to induce chloramphenicol resistance in *Staphylococcus aureus*. Japan. J. Microbiol. **15**, 219–227 (1971).

KROON, A.M.: Protein synthesis in heart mitochondria. I. Amino acid incorporation into the protein of isolated beef-heart mitochondria and fractions derived from them by sonic oscillation. Biochim. Biophys. Acta **72**, 391–402 (1963).

KUCAN, Z., and F. LIPMANN: Differences in chloramphenicol sensitivity of cell-free amino acid polymerization systems. J. Biol. Chem. **239**, 516–520 (1964).

KURLAND, C.G.: The proteins of the bacterial ribosome. Protein synthesis: A series of advances, vol. 1 (McCONKEY, E., ed.), p. 179–228. New York: Marcel Dekker, 1971.

KURLAND, C.G., and O. MAALOE: Regulation of ribosomal and transfer RNA synthesis. J. Mol. Biol. **4**, 193–210 (1962).

LACKS, S., and F. GROS: A metabolic study of the RNA-amino acid complexes in *Escherichia coli*. J. Mol. Biol. **1**, 301–320 (1959).

LAMBORG, M.R., and P.C. ZAMECNIK: Amino acid incorporation into protein by extracts of *E. coli*. Biochim. Biophys. Acta **42**, 206–211 (1960).

LARK, K.G.: Regulation of chromosome replication and segregation in bacteria. Bacteriol. Rev. **30**, 3–32 (1966).

LARK, K.G. and C. LARK: Regulation of chromosome replication in *E. coli*: a comparison of the effects of phenethyl alcohol treatment with those of amino acid starvation. J. Mol. Biol. **20**, 9–19 (1966).

LAZZARINI, R.A., and R.M. WINSLOW: The regulation of RNA synthesis during growth rate transitions and amino acid deprivation in *E. coli*. Cold Spring Harbor Symp. Quant. Biol. **35**, 383–390 (1970).

LEMBACH, K.J., and J.M. BUCHANAN: The relationship of protein synthesis to early transcriptive events in bacteriophage T_4 Infected *Escherichia coli* B. J. Biol. Chem. **245**, 1575–1587 (1970).

LESSARD, J.L., and S. PESTKA: Studies on the formation of transfer ribonucleic acid-ribosome complexes. XXII. Binding of aminoacyl-oligonucleotides to ribosomes. J. Biol. Chem. **247**, 6901–6908 (1972a).

LESSARD, J.L., and S. PESTKA: Studies on the formation of transfer ribonucleic acid-ribosome complexes. XXIII. Chloramphenicol, aminoacyl-oligonucleotides, and *Escherichia coli* ribosomes. J. Biol. Chem. **247**, 6909–6912 (1972b).

LEVINE, A.J., and R.L. SINSHEIMER: The process of infection with bacteriophage φX174. XIX. Isolation and characterization of a chloramphenicol-resistant protein from φX-infected cells. J. Mol. Biol. **32**, 567–578 (1968).

LEVINE, A.J., and R.L. SINSHEIMER: The process of infection with bacterial phage φX174. XXVII. Synthesis of a viral-specific chloramphenicol-resistant protein in φX174 infected cells. J. Mol. Biol. **39**, 655–668 (1969).

LEVINTHAL, C., D.P. FAN, A. HIGA, and R.A. ZIMMERMAN: The decay and protection of messenger RNA in bacteria. Cold Spring Harbor Symp. Quant. Biol. **28**, 183–190 (1964).

LINNANE, A.W., A.J. LAMB, C. CHRISTODOLOU, and H.B. LUKINS: The biogenesis of mitochondria. VI. Biochemical basis of the resistance of *Saccharomyces cerevisiae* toward antibiotics which specifically inhibit mitochondrial protein synthesis. Proc. Natl. Acad. Sci. U.S. **59**, 1288–1293 (1968).

Maaloe, O., and N.O. Kjeldgaard: Control of macromolecular biosynthesis. New York: W.A. Benjamin, Inc. 1966.

Marmur, J., and A.K. Saz: The inhibition of adaptive enzyme formation in *Escherichia coli* by chloramphenicol. Antibiot. & Chemotherapy 3, 613–617 (1953).

Maxwell, R.E., and V.S. Nickel: The antibacterial activity of the isomers of chloramphenicol. Antibiot. & Chemotherapy 4, 289–295 (1954).

Midgley, J.E.M., and W.J.H. Gray: The control of ribonucleic acid synthesis in bacteria. The synthesis and stability of ribonucleic acid in chloramphenicol-inhibited cultures of *Escherichia coli*. Biochem. J. 122, 149–159 (1971).

Mitsuhashi, S., M. Kono, M. Sagawa, and H. Mori: Drug resistance of staphylococcus: X. Induction of chloramphenicol resistance by its derivatives. Japan. J. Microbiol. 13, 177–180 (1969).

Monro, R.E.: Catalysis of peptide bond formation by 50S ribosomal subunits from *Escherichia coli*. J. Mol. Biol. 26, 147–151 (1967).

Monro, R.E.: The peptidyl transferase activity of ribosomes. Cold Spring Harbor Symp. Quant. Biol. 34, 357–366 (1969).

Monro, R.E., and K.A. Marcker: Ribosome-catalysed reaction of puromycin with a formylmethionine-containing oligonucleotide. J. Mol. Biol. 25, 347–350 (1967).

Monro, R.E., and D. Vazquez. Ribosome-catalysed peptidyl transfer: Effects of some inhibitors of protein synthesis. J. Mol. Biol. 28, 161–165 (1967).

Naha, P.M.: A chloramphenicol resistant host protein involved in lysogenization. Biochem. Biophys. Res. Commun. 35, 920–925 (1969).

Nasjleti, C.E., and H.H. Spencer: The effects of chloramphenicol on mitosis of phytohemagglutinin stimulated human leukocytes. Exptl. Cell Res. 53, 11–17 (1968).

Nathans, D.: Puromycin inhibition of protein synthesis: Incorporation of puromycin into peptide chains. Proc. Natl. Acad. Sci. U.S. 51, 585–592 (1964).

Nathans, D., Ehrenstein, G. von, R. Monro, and F. Lipmann: Protein synthesis from aminoacyl-soluble ribonucleic acid. Federation Proc. 21, 127–135 (1962).

Nathans, D., and F. Lipmann: Amino acid transfer from aminoacyl-ribonucleic acids to protein on ribosomes of *Escherichia coli*. Proc. Natl. Acad. Sci. U.S. 47, 497–504 (1961).

Nathans, D., and A. Neidle: Structural requirements for puromycin inhibition of protein synthesis. Nature 197, 1076–1077 (1963).

Newton, B.A.: Mechanisms of antibiotic action. Ann. Rev. Microbiol. 19, 209–240 (1965).

Nirenberg, M.W., and J.H. Matthaei: The dependence of cell-free protein synthesis in *E. coli* upon naturally occurring or synthetic polyribonucleotides. Proc. Natl. Acad. Sci. U.S. 47, 1588–1602 (1961).

Nomura, M., and K. Hosokawa: Biosynthesis of ribosomes: Fate of chloramphenicol particles and of pulse-labeled RNA in *Escherichia coli*. J. Mol. Biol. 12, 242–265 (1965).

Nomura, M., and J.D. Watson: Ribonucleoprotein particles within chloromycetin-inhibited *Escherichia coli*. J. Mol. Biol. 1, 204–217 (1959).

Oerter, D., and R. Bass: Effect of chloramphenicol infusion on the rate of synthesis of cytochrome oxidase in mammalian embryonic tissue. Arch. Exptl. Pathol. Pharmakol. 272, 239–242 (1972).

Okamoto, S., and D. Mizuno: Inhibition by chloramphenicol of protein synthesis in the cell-free system of a chloramphenicol-resistant strain of *Escherichia coli*. Nature 195, 1022–1023 (1962).

Okamoto, S., and D. Mizuno: Mechanism of chloramphenicol and tetracycline resistance in *Escherichia coli*. J. Gen. Microbiol. 35, 125–133 (1964).

Okamoto, S., and Y. Suzuki: Chloramphenicol-, dihydrostreptomycin-, and kanamycin-inactivating enzymes from multiple drug-resistant *Escherichia coli* carrying episome 'R'. Nature 208, 1301–1303 (1965).

Orgel, H.A., and R.N. Hamburger: Chloramphenicol-specific antibody. IV. A method for the detection of anti-chloramphenicol antibody in human sera. Immunology 20, 233–239 (1971).

Pardee, A.B., K. Paigen, and L.S. Prestidge: A study of the ribonucleic acid of normal and chloromycetin-inhibited bacteria by zone electrophoresis. Biochim. Biophys. Acta 23, 162–173 (1957).

Pardee, A.B., and L.S. Prestidge: The dependence of nucleic acid synthesis on the presence of amino acids in *E. coli*. J. Bacteriol. 71, 677–683 (1956).

Pestka, S.: Studies on the formation of transfer ribonucleic acid-ribosome complexes. V. On the function of a soluble transfer factor in protein synthesis. Proc. Natl. Acad. Sci. U.S. 61, 726–733 (1968).

PESTKA, S.: Studies on the formation of transfer ribonucleic acid-ribosome complexes. VI. Oligopeptide synthesis and translocation on ribosomes in the presence and absence of souble transfer factors. J. Biol. Chem. **244**, 1533–1539 (1969a).

PESTKA, S.: Translocation, aminoacyl-oligonucleotides, and antibiotic action. Cold Spring Harbor Symp. Quant. Biol. **34**, 395–410 (1969b).

PESTKA, S.: Studies on the formation of transfer ribonucleic acid-ribosome complexes. X. Phenylalanyl-oligonucleotide binding to ribosomes and the mechanism of chloramphenicol action. Biochem. Biophys. Res. Commun. **36**, 589–595 (1969c).

PESTKA, S.: Studies on the formation of transfer ribonucleic acid-ribosome complexes. XI. Antibiotic effects on phenylalanyl-oligonucleotide binding to ribosomes. Proc. Natl. Acad. Sci. U.S. **64**, 709–714 (1969d).

PESTKA, S.: Studies on the formation of transfer ribonucleic acid-ribosome complexes. VIII. Survey of the effect of antibiotics on N-acetyl-phenylalanyl-puromycin formation: Possible mechanism of chloramphenicol action. Arch. Biochem. Biophys. **136**, 80–88 (1970a).

PESTKA, S.: Studies on the formation of transfer ribonucleic acid-ribosome complexes. IX. Effect of antibiotics on translocation and peptide bond formation. Arch. Biochem. Biophys. **136**, 89–96 (1970b).

PESTKA, S.: Inhibitors of ribosome function. Ann. Rev. Microbiol. **25**, 487–562 (1971).

PESTKA, S.: Studies on transfer ribonucleic acid-ribosome complexes. XIX. Effect of antibiotics on peptidyl-puromycin synthesis on polyribosomes from *Escherichia coli*. J. Biol. Chem. **247**, 4669–4678 (1972).

PESTKA, S., B.H. HECK, and E.M. SCOLNICK: A convenient assay for mono-, di, and oligophenylalanines. Anal. Biochem. **28**, 376–384 (1969).

PESTKA, S., T. HISHIZAWA, and J.L. LESSARD: Studies on the formation of transfer ribonucleic acid-ribosome complexes. XIII. Aminoacyl-oligonucleotide binding to ribosomes: Characteristics and requirements. J. Biol. Chem. **245**, 6208–6219 (1970).

PETERSON, R.F., P.S. COHEN and H.L. ENNIS: Properties of T_4 messenger RNA synthesized in the absence of protein synthesis. Virology **48**, 201–206 (1972).

PIFFARETTI, J.C., B. ALLET, and J.S. PITTON: Analogy between *in vivo* and *in vitro* biological effect of chloramphenicol and its acetylated derivatives. FEBS Letters **11**, 26–28 (1970).

PRIMAKOFF, P., and P. BERG: Stringent control of transcription of phage ϕ80psu$_3$. Cold Spring Harbor Symp. Quant. Biol. **35**, 391–396 (1970).

REBSTOCK, M.C., H.M. CROOKS, J. CONTROULIS, and Q.R. BARTZ: Chloramphenicol (chloromycetin). IV. Chemical studies. J. Am. Chem. Soc. **71**, 2458–2462 (1949).

RENDI, R.: The effect of chloramphenicol on the incorporation of labeled amino acids into proteins by isolated subcellular fractions from rat liver. Exptl. Cell Res. **18**, 187–189 (1959).

RENDI, R., and S. OCHOA: Effect of chloramphenicol on protein synthesis in cell-free preparation of *Escherichia coli*. J. Biol. Chem. **237**, 3711–3713 (1962).

RICHERT, N.J. and J.D. HARE: Distinctive effects of inhibitors of mitochondrial function on Rous sarcoma virus replication and malignant transformation. Biochem. Biophys. Res. Commun. **46**, 5–10 (1972).

RINGROSE, P.S., and R.W. LAMBERT: The action of novel chloramphenicol analogues on prokaryotic and eukaryotic systems. Biochim. Biophys. Acta, **299**, 374–384 (1973).

RYCHLÍK, I.: Release of lysine peptides by puromycin from polylysyl-transfer ribonucleic acid in the presence of ribosomes. Biochim. Biophys. Acta **114**, 425–427 (1966).

RYCHLÍK, I., J. CERNÁ, S. CHLADEK, J. ZEMLICKA, and Z. HALADOVA: Substrate specificity of ribosomal peptidyl transferase: 2'(3')-O-aminoacyl nucleosides as acceptors of the peptide chain on the amino acid site. J. Mol. Biol. **43**, 13–24 (1969).

SABIN, A.B.: Different effects of chloramphenicol, dactinomycin, and streptovitacin A on synthesis of tumor and virion antigens in SV40 virus-infected cells. Proc. Natl. Acad. Sci. U.S. **55**, 1141–1148 (1966).

SALSER, W., A. BOLLE, and R. EPSTEIN: Transcription during bacteriophage T_4 development: A demonstration that distant sub-classes of the "early" RNA appear at different times and that some are "turned off" at late times. J. Mol. Biol. **49**, 271–295 (1970).

SAMBROOK, J.F., D.P. FAN, and S. BRENNER: A strong suppressor specific for UGA. Nature **214**, 452–453 (1967).

SARABHAI, A.S., A.O.W. STRETTON, S. BRENNER, and A. BOLLE: Co-linearity of the gene with the polypeptide chain. Nature **201**, 13–17 (1964).

SCOLNICK, E., R. TOMPKINS, T. CASKEY, and M. NIRENBERG: Release factors differing in specificity for terminator codons. Proc. Natl. Acad. Sci. U.S. **61**, 768–774 (1968).

SHAW, W.V., D.W. BENTLEY, and L. SANDS: Mechanism of chloramphenicol resistance in *Staphylococcus epidermidis*. J. Bacteriol. **104**, 1095–1105 (1970).

SHEMYAKIN, M.N.: Khimia Antibiotikov 1, Moscow Acad. Sci. USSR (1961).

SINSHEIMER, R.L., C.A. HUTCHINSON, and B. LINDQVIST: Bacterial phage ØX174: viral functions. Molecular biology of viruses, ed. J.P. COLTER and W. PARANCHYCH, p. 175–192. New York: Academic Press 1967.

SINSHEIMER, R.L., B. STARMAN, C. NAGLER and S. GUTHRIE: The process of infection with bacteriophage ØX174. I. Evidence for a "replicative" form. J. Mol. Biol. **4**, 142–160 (1962).

SO, A.G., and E.W. DAVIE: The incorporation of amino acids into protein in a cell-free system from yeast. Biochemistry **2**, 132–136 (1963).

SORM, F., and D. GRUNBERGER: Inhibitory effect of chloramphenicol on the formation of some enzyme systems of *Escherichia coli*. Collection Czech. Chem. Commun. **19**, 167–173 (1954).

SPEYER, J.F., P. LENGYEL, C. BASILIO, A.J. WAHBA, R.S. GARDNER, and S. OCHOA: Synthetic polynucleotides and the amino acid code. Cold Spring Harbor Symp. Quant. Biol. **28**, 559–567 (1963).

STENT, G.S., and S. BRENNER: A genetic locus for the regulation of ribonucleic acid synthesis. Proc. Natl. Acad. Sci. U.S. **47**, 2005–2014 (1961).

STOW, M., B.J. STARKEY, I.C. HANCOCK and J. BADDILEY: Inhibition by chloramphenicol of glucose transfer in teichoic acid biosynthesis. Nature New Biol. **229**, 56–57 (1971).

SVEHAG, S.: Antibody formation *in vitro* by separated spleen cells. Inhibition by actinomycin or chloramphenicol. Science **146**, 659–661 (1964).

SYMONS, R.H., R.J. HARRIS, L.P. CLARKE, J.F. WHELDRAKE, and W.H. ELLIOTT. Structural requirements for inhibition of polyphenylalanine synthesis by aminoacyl and nucleotidyl analogues of puromycin. Biochim. Biophys. Acta **179**, 248–250 (1969).

TALAL, N., and E.D. EXUM: Two classes of spleen ribosomes with different sensitivities to chloramphenicol. Proc. Natl. Acad. Sci. U.S. **55**, 1288–1295 (1966).

TANAKA, K., H. TERAOKA, T. NAGIRA, and M. TAMAKI: [^{14}C]Erythromycin-ribosome complex formation and non-enzymatic binding of aminoacyl-transfer RNA to ribosome-messenger RNA complex. Biochim. Biophys. Acta **123**, 435–437 (1966).

TAUBMAN, S.B., N.R. JONES, F.E. YOUNG, and J.W. CORCORAN: Sensitivity and resistance to erythromycin in *Bacillus subtilis* 168: The ribosomal binding of erythromycin and chloramphenicol. Biochim. Biophys. Acta **123**, 438–440 (1966).

TERAOKA, H.: Reversal of the inhibitory action of chloramphenicol on the ribosomal peptidyl-transfer reaction by erythromycin. Biochim. Biophys. Acta **213**, 535–537 (1970).

TERAOKA, H., K. TANAKA, and M. TAMAKI: The comparative study on the effects of chloramphenicol, erythromycin and lincomycin on polylysine synthesis in an *Escherichia coli* cell-free system. Biochim. Biophys. Acta **174**, 776–778 (1969).

TESSMAN, E.S.: Mutants of bacteriophage S13 blocked in infectious DNA synthesis. J. Mol. Biol. **17**, 218–236 (1966).

TISSIÉRES, A., D. SCHLESSINGER, and F. GROS: Amino acid incorporation into proteins by *Escherichia coli* ribosomes. Proc. Natl. Acad. Sci. U.S. **46**, 1450–1463 (1960)

TRAUT, R.R., and R.E. MONRO: The puromycin reaction and its relation to protein synthesis. J. Mol. Biol. **10**, 63–72 (1964).

VAZQUEZ, D.: Antibiotics which affect protein synthesis: The uptake of ^{14}C-chloramphenicol by bacteria. Biochem. Biophys. Res. Commun. **12**, 409–413 (1963).

VAZQUEZ, D.: The binding of chloramphenicol by ribosomes from *Bacillus megaterium*. Biochem. Biophys. Res. Commun. **15**, 464–468 (1964).

VAZQUEZ, D.: Binding of chloramphenicol to ribosomes. The effect of a number of antibiotics. Biochim. Biophys. Acta **114**, 277–288 (1966a).

VAZQUEZ, D.: 16th Symp. Soc. Gen. Microbiol., p. 169–191 (1966b).

VAZQUEZ, D.: Antibiotics affecting chloramphenicol uptake by bacteria. Their effect on amino acid incorporation in a cell-free system. Biochim. Biophys. Acta **114**, 289–295 (1966c).

VAZQUEZ, D.: Inhibitors of protein synthesis at the ribosome level; studies on their site of action. Life Sci. **6**, 381–386 (1967).

VOGEL, Z., A. ZAMIR, and D. ELSON: The possible involvement of peptidyl transferase in the termination step of protein biosynthesis. Biochemistry **8**, 5161–5168 (1969).

WALLER, J.P., T. ERDÖS, F. LEMOINE, S. GUTTMAN, and E. SANDRIN: Inhibition of protein synthesis by aminoacyl 3' (2')-adenosine. Biochim. Biophys. Acta 119, 566–580 (1966).

WEBER, M.J., and J.A. DEMOSS: The inhibition by chloramphenicol of nascent protein formation in E. coli. Proc. Natl. Acad. Sci. U.S. 55, 1224–1230 (1966).

WEBER, M.J., and J.A. DEMOSS: Inhibition of the peptide bond synthesizing cycle by chloramphenicol. J. Bacteriol. 97, 1099–1105 (1969).

WEIGERT, M.G., and A. GAREN: Base composition of nonsense codons in E. coli. Nature 206, 992–994 (1965).

WEISBERGER, A.S., S. ARMENTROUT, and S. WOLFE.: Protein synthesis by reticulocyte ribosomes. I. Inhibition of polyuridylic acid-induced ribosomal protein synthesis by chloramphenicol. Proc. Natl. Acad. Sci. U.S. 50, 86–93 (1963).

WEISBERGER, A.S., T.M. DANIEL, and A. HOFFMAN: Suppression of antibody synthesis and prolongation of homograft survival by chloramphenicol. J. Exptl. Med. 120, 183–196 (1964).

WEISBERGER, A.S., and S. WOLFE: Effect of chloramphenicol on protein synthesis. Federation Proc. 23, 976–983 (1964).

WEISBLUM, B.: Macrolide resistance in Staphylococcus aureus. In: Drug action and drug resistance in bacteria. I. Macrolide Antibiotics (ed. by S. MITSUHASHI), p. 217–238. Baltimore: Univ. Park Press 1971.

WEISBLUM, B., and J. DAVIES: Antibiotic inhibitors of the bacterial ribosome. Bacteriol. Rev. 32, 493–528 (1968).

WEISSBACH, H., B. REDFIELD, and N. BROT: Studies on the reaction of N-acetyl-phenylalanyl-tRNA with puromycin. Arch. Biochem. Biophys. 127, 705–710 (1968).

WINTERSBERGER, E.: Proteinsynthese in isolierten Hefe-Mitochondrien. Biochem. Z. 341, 409–419 (1965).

WISSEMAN, C.L., J.E. SMADEL, F.E. HAHN, and H.E. HOPPS: Mode of action of chloramphenicol. I. Action of chloramphenicol on assimilation of ammonia and on synthesis of proteins and nucleic acids in Escherichia coli. J. Bacteriol. 67, 662–673 (1954).

WOLFE, A.D., and F.E. HAHN: Mode of action of chloramphenicol. IX. Effects of chloramphenicol upon a ribosomal amino acid polymerization system and its binding to bacterial ribosome. Biochim. Biophys. Acta 95, 146–155 (1965).

YOUNG, R.M., and D. NAKADA: Defective ribosomes in chloramphenicol-treated Escherichia coli. J. Mol. Biol. 57, 457–473 (1971).

YUKIOKA, M., and S. MORISAWA: Reversibility of chloramphenicol inhibition of the poly U directed polyphenylalanine synthesis by G factor and GTP. Biochem. Biophys. Res. Commun. 40, 1331–1339 (1970).

YUKIOKA, M., and S. MORISAWA: Enhancement of the phenylalanyl-oligonucleotide binding to the peptidyl recognition center of ribosomal peptidyl transferase and inhibition of the chloramphenicol binding to ribosomes. Biochim. Biophys. Acta 254, 304–315 (1971).

YUNIS, A.A. and G.R. BLOOMBERG: Chloramphenicol toxicity: Clinical features and pathogenesis. Prog. Hematol. 4, 138–159 (1964).

ZINDER, N.D., D.L. ENGELHARDT, and R.E. WEBSTER: Punctuation in the genetic code. Cold Spring Harbor Symp. Quant. Biol. 31, 251–256 (1966).

ZIPSER, D.: UGA: A third class of suppressible polar mutants. J. Mol. Biol. 29, 441–445 (1967).

The Erythromycins

Nancy L. Oleinick

Introduction

Erythromycin is an antibiotic which inhibits the growth of many Gram-positive and some Gram-negative organisms and is especially successful against group A streptococcal, staphylococcal, and pneumococcal infections. The bacteriostatic action of erythromycin is a result of the antibiotic's ability to selectively inhibit protein synthesis in the susceptible bacteria, but not in the host mammalian tissue.

Origin and Chemical Structure

The erythromycins are a group of structurally related antibiotics elaborated by strains of *Streptomyces erythreus*. A common feature to all erythromycins is an aglycone moiety (erythronolide) which is a highly substituted macrocyclic lactone ring containing one oxygen and 13 carbon atoms (Fig. 1; WILEY *et al.*, 1957a; WILEY *et al.*, 1957b; WILEY *et al.*, 1957c). A neutral and an amino-deoxy sugar are linked by glycosidic bonds to C-3 and C-5 of the ring, respectively (WILEY *et al.*, 1957a). In erythromycins A and B, the carbohydrate moieties are cladinose (WILEY and WEAVER, 1956) and desosamine (CLARK, 1953; FLYNN *et al.*, 1954), and in erythromycin C, mycarose (HOFHEINZ and GRISEBACH, 1962) and desosamine. The total absolute steric configurations of the erythromycins are known (Fig. 1; DJERASSI *et al.*, 1958; HARRIS *et al.*, 1965; CELMER, 1965 a, b, 1966, 1971; PERUN, 1971), and conformational models have been proposed (CELMER, 1971; PERUN, 1971).

The discovery of the erythromycins was reported in 1952 (McGUIRE *et al.*, 1952). Wild-type strains of *S. erythreus* produce all three erythromycins (A, B, and C) and possibly a fourth antimicrobial substance (MARTIN and GOLDSTEIN, 1970) in varying yields depending on strain differences and culture conditions. The erythronolide moiety is synthesized from propionate and 2-methylmalonate by a route which resembles that for fatty acid biosynthesis; some of the steps in the two pathways are identical (*e. g.*, CORCORAN *et al.*, 1960 a, b; GRISEBACH *et al.*, 1960 a, b; KANEDA *et al.*, 1962; CORCORAN, 1964; FRIEDMAN *et al.*, 1964; GRISEBACH and HOFHEINZ, 1964; WAWSZKIEWICZ and LYNEN, 1964; HUNG *et al.*, 1965; MARTIN and ROSENBROOK, 1967). The final steps in the biosynthesis of the erythromycins are not known with certainty, but introduction of putative intermediates and analysis of substances obtained from wild type strains and blocked mutants has suggested that first mycarose must be added to erythronolide b, followed by C-12 hydroxylation of the lactone ring, addition of desosamine,

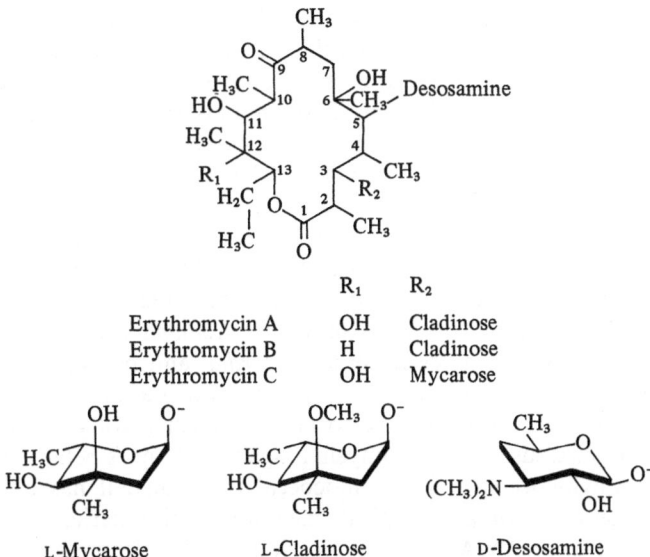

	R₁	R₂
Erythromycin A	OH	Cladinose
Erythromycin B	H	Cladinose
Erythromycin C	OH	Mycarose

L-Mycarose   L-Cladinose   D-Desosamine

Fig. 1. Structures of the erythromycins (from CELMER, 1971). The aglycone moiety of erythromycins A and C is (2R, 3S, 4S, 5R, 6R, 8R, 10R, 11R, 12S, 13R) erythronolide a; that of erythromycin B is (2R, 3S, 4S, 5R, 6R, 8R, 10R, 11S, 12R, 13R) erythronolide b

and methylation of mycarose to produce cladinose (HUNG et al., 1965; SPIZEK et al., 1965; MARTIN et al., 1966; CORCORAN and CHICK, 1966; MARTIN and ROSENBROOK, 1967; MARTIN and GOLDSTEIN, 1970; MCALPINE and CORCORAN, 1971). The only study done with cell-free systems has demonstrated that the methylation of erythromycin C by erythromycin C: S-adenosylmethionine trans-methylase is the final step in the formation of the A form of erythromycin (MCALPINE and CORCORAN, 1971).

Erythromycin A is the most common and widely used of the erythromycins both in research and in clinical practice. Unless otherwise noted, the simple term erythromycin hereafter will refer to the A form. The erythromycins are both structurally and functionally related to the other macrolide antibiotics (WOODWARD, 1957). Comparisons among the various macrolides are considered by VAZQUEZ in another chapter in this volume. In addition, other reviews on erythromycin have appeared (HAHN, 1967; WEISBLUM and DAVIES, 1968; PESTKA, 1971a; CORCORAN, 1971).

Toxicity and Antimicrobial Spectrum

Erythromycin is one of the safest of the antibiotics currently prescribed. The rare occurrences of serious adverse reactions are generally the result of patient hypersensitivity or unusually large doses (AMA Drug Evaluations, 1971). Since erythromycin acts by inhibiting protein synthesis in sensitive bacteria, it is significant that an accumulation of clinical data (HERRELL, 1958; MARTIN and WELLMAN, 1967) has revealed no side effects of erythromycin which can be related to an inhibition of protein synthesis in the host. Erythromycin is a basic substance which is slightly soluble in water and is acid-labile; however, for oral administration a variety of salts and esters are available which will satisfactorily survive stomach acid. Test animals show a rapid accumulation and sustained high level of the drug in liver, kidney, lung, and blood following intra-muscular administration (FUKAYA and KITAMOTO, 1970; OKUBO and FUJI-MOTO, 1970). An enzyme in rabbit liver microsomal preparations is capable of inactivating erythromycin by N-demethylation (MAO and TARDREW, 1965).

Erythromycin is highly effective against many Gram-positive and some Gram-negative organisms (Table 1) (FINLAND et al., 1952; WELCH et al., 1952; FUSILLO et al., 1953). Moderate sensitivity is shown by Hemophilus. Gram-negative coliform and enteric bacilli, mycobacteria, Nocardia, and pathogenic fungi are all highly resistant to the antibiotic as are yeasts, protozoa, and animal cells. Some atypical mycobacteria are sensitive to erythromycin (MOLAVI and WEINSTEIN, 1971). The principal use of erythromycin is against group A beta-hemolytic streptococcal, staphylococcal, and pneumococcal infections; it has also been used in the treatment of primary atypical pneumonia, diphtheria, erythrasma, and intestinal amebiasis (AMA Drug Evaluations, 1971). Representative examples

Table 1. *Erythromycin-sensitivities of selected bacteria*[a]

Organism	Inhibitory concentration (µg/ml)
Gram-positive	
Streptococcus Group A	0.01–0.1
Streptococcus Group B	0.2
Streptococcus Group D	0.2–6.3
Pneumococcus sp.	0.02–0.05
Staphylococcus albus	1.6
Staphylococcus aureus	0.4
Sarcina lutea	0.02
Bacillus cereus	0.4
Bacillus subtilis	0.5
Clostridium sp.	0.1–2.0
Corynebacterium diphtheriae	0.01–0.1
Actinomyces israeli	0.1
Mycobacterium kansasii	0.5–2.0
Mycobacterium tuberculosis	1–10
Gram-negative	
Neisseria meningitidis	0.1–0.4
Hemophilus influenzae	0.8–12.5
Escherichia coli	100
Aerobacter aerogenes	100– >200
Klebsiella pneumoniae	≧200
Salmonella salinatis	>200
Pseudomonas aeruginosa	>200

[a] From FINLAND et al. (1952); FUSILLO et al. (1953); HUDSON et al. (1956); GUY and CHAPMAN (1961); MOLAVI and WEINSTEIN (1971).

from the antimicrobial spectrum are given in Table 1 together with the inhibitory concentrations of erythromycin. The difference in sensitivity between Gram-positive and Gram-negative bacteria has been attributed to reduced uptake of erythromycin in the latter case (MAO and PUTTERMAN, 1968). The cell-free protein-synthesizing systems prepared from both types of bacteria were equally inhibited. However, intact Gram-positive bacteria were able to concentrate the drug from the medium and form a more stable antibiotic-ribosome complex. It was suggested that the barrier to uptake is the outer cell membrane of Gram-negative bacteria, since a highly resistant strain of *Proteus mirabilis* became 1000-fold more sensitive to erythromycin when converted to a stable L-form, lacking an intact cell wall (TAUBENECK, 1962). More recently, this suggestion has been confirmed by a study which shows that even with a highly sensitive Gram-positive organism like *B. subtilis* there is a major permeability barrier to erythromycin A (ARORA *et al.*, 1971). The accumulation of the antibiotic is sensitive to unknown factors in the growth medium, since nutrient broth (Difco) depresses the rate substantially as compared to a non-nutritive environment. The implications of this situation for the efficacy of erythromycin *in vivo* are obvious.

Erythromycin is normally bacteriostatic; however, at concentrations 25–1 000 fold greater than those necessary to produce bacteriostasis, the antibiotic can become bactericidal (GARROD and WATERWORTH, 1956; LUTZ *et al.*, 1957). Bacteria may mutate to resistance to erythromycin *in vivo* or upon serial transfer in culture in the presence of the drug (HAIGHT and FINLAND, 1952; FUSILLO *et al.*, 1953; CHABBERT, 1956; GARROD and WATERWORTH, 1956; JONES *et al.*, 1956; LUTZ *et al.*, 1957; GARROD, 1957). Resistant mutants are less able to accumulate extracellular erythromycin (TAUBMAN *et al.*, 1963; ARORA *et al.*, 1971) and form less stable antibiotic-ribosome complexes (discussed in detail below). Bacterial destruction of erythromycin has not been reported.

Erythromycin-resistant mutants are often found which are also resistant to other macrolide antibiotics (FINLAND *et al.*, 1952; FUSILLO *et al.*, 1953; COLEMAN *et al.*, 1953; GARROD and WATERWORTH, 1956; JONES *et al.*, 1956; AHMED, 1968; WILHELM *et al.*, 1968; SAITO *et al.*, 1970a), to lincomycin (SAITO *et al.*, 1970a) and chloramphenicol (ČERNÁ and RYCHLÍK, 1968; TANAKA *et al.*, 1972). However, such cross-resistance is not universal, since some erythromycin-resistant mutants remain sensitive to one or more of these other antibiotics (CHABBERT, 1956; JONES *et al.*, 1956; TAUBMAN *et al.*, 1966; WILHELM *et al.*, 1968; SAITO *et al.*, 1970a). Both synergism and antagonism between erythromycin and other antibiotics have been observed (COLEMAN *et al.*, 1953; CHABBERT, 1956; MATHIEU and FAGUET, 1958; OLEINICK *et al.*, 1968; WILHELM *et al.*, 1968; WEISBLUM, 1969).

The inhibition of bacterial growth by erythromycin is dependent upon environmental conditions. The minimum inhibitory concentration increases as the pH of the medium is lowered from 8.5 to 5.5 (GARROD and WATERWORTH, 1956). Accordingly, MAO and WIEGAND (1968) observed that inhibition by erythromycin of polyphenylalanine synthesis takes place preferentially at high pH; indeed, at pH 6.5 erythromycin stimulates this synthesis. Since the pKa of erythromycin is about 8.6, it is possible that the non-protonated form is the more active species.

Inhibitory activity is dependent upon the presence of at least one of the two sugars linked to the erythronolide moiety, the basic sugar conferring more activity than the neutral sugar (WILHELM et al., 1968). Maximal activity requires both sugars. Neither the free sugars nor the free lactone are active.

The Genetics of Bacterial Resistance to Erythromycin

The genetic locus which codes for resistance to erythromycin (the erythromycin locus) is linked to a gene for a 50S ribosomal protein in Bacillus subtilis (SMITH et al.,1969), and is clustered with other antibiotic markers for ribosomal proteins near the major group of genes for rRNA (GOLDTHWAITE et al.,1970). Segments of DNA which contain the erythromycin locus and the other ribosomal protein markers become distributed on columns of methylated albumin-kieselguhr in a manner different from that of the genes for rRNA, implying that rRNA genes neither are linked to nor code for these ribosomal proteins (SMITH et al., 1971). The erythromycin locus in Escherichia coli is near pro_2 (APIRION, 1967) or in the same region of the genetic map as the streptomycin locus (TAKATA et al., 1970).

Stable mutations producing erythromycin resistance in E. coli and B. subtilis result in a 50S ribosomal subunit containing one altered protein (TANAKA et al., 1968; HANSEN and CORCORAN, 1969), Mutants of E. coli which have an increased sensitivity to lincomycin and/or erythromycin (lir mutants) also differ in one 50S ribosomal protein (KREMBEL and APIRION, 1968). In several types of E. coli having different ribosomal protein compositions, resistance is always accompanied by the same altered protein, designated 50-8 (the eighth 50S-ribosomal protein eluted from a carboxymethyl-cellulose column) (DEKIO et al., 1970). Each erythromycin-resistant mutant has an altered 50-8 protein which can be distinguished from the wild-type and the other mutants by its elution characteristics (OTAKA et al., 1970). The altered chromatographic property is due to a difference in the amino acid sequence of one tryptic peptide of the 50-8 protein (OTAKA et al., 1971). The following phenotypes always co-transduced: cross-resistance to other macrolides, an altered 50-8 protein, and reduced affinity of the ribosomes for erythromycin and lower peptidyl transferase activity which could be restored to normal in vitro by increased levels of K^+ or NH_4^+ (TANAKA et al., 1971).

In Staphylococcus aureus, low levels of erythromycin can induce resistance to other macrolides, lincosamides, and streptogramin B-type antibiotics (WEAVER and PATTEE, 1964; WEISBLUM and DEMOHN, 1969; SAITO et al., 1970). The loci governing induction and resistance could not be separated (SAITO et al., 1971). Since the inducible change appears very rapidly (WEISBLUM et al., 1971), it probably is not the result of synthesis of a new ribosomal protein component, and no difference in ribosomal proteins has been detected in either induced or constitutively resistant cells (LAI and WEISBLUM, 1971).The optimal inducing concentration of erythromycin (10^{-8} M to 10^{-7} M) is lower than that necessary to inhibit protein synthesis (WEISBLUM et al., 1971). The induction process requires protein and RNA synthesis, but not DNA synthesis. It appears that a modification may be introduced by an erythromycin inducible enzyme, leading to methylation of a unique adenine residue in the 23S ribosomal RNA (LAI and WEISBLUM, 1971). The phenomenon of inducible resistance has been studied extensively by MITSUHASHI and coworkers and was recently reviewed (SAITO et al., 1971).

Inhibition of Protein Synthesis

In intact bacteria erythromycin acts by inhibiting protein synthesis without any direct effect on nucleic acid synthesis (BROCK and BROCK, 1959; TAUBMAN *et al.*, 1963). Amino acid incorporation into acid-insoluble material is susceptible to inhibition by erythromycin in cell-free preparations from *Bacillus subtilis* (TAUBMAN *et al.*, 1963), *Escherichia coli* (WOLFE and HAHN, 1964; VAZQUEZ, 1966b), and *Staphylococcus aureus* (MAO, 1967a; MAO and WIEGAND, 1968). The inhibition is exerted on a step in protein synthesis which comes after the activation of the amino acid and its transfer to tRNA (TAUBMAN *et al.*, 1964).

When the action of erythromycin on protein synthesis is studied with natural polysomes, the drug has very little or no inhibitory action on elongation of the nascent peptide chains (WILHELM, 1968; OLEINICK and CORCORAN, 1970). However, when the antibiotic is studied in simpler systems in which protein synthesis depends on added exogenous mRNAs, a very significant inhibition of protein synthesis is observed. The situation with the natural polysomes is best discussed later after a detailed consideration of the effect of erythromycin on peptide synthesis which depends on added polynucleotides. A great many studies of the latter sort have been done in which a variety of synthetic polynucleotides have been used as artificial mRNAs. Chief among these are poly A, poly C, and poly U, which direct the synthesis of polylysine, polyproline, and polyphenylalanine, respectively. The action of erythromycin on such systems has been extensively studied. VAZQUEZ (1966b) found that inhibition by erythromycin was greatest on proline incorporation, and less on lysine incorporation, while phenylalanine incorporation was least sensitive. In a cell-free system from *B. subtilis*, WILHELM and CORCORAN (1967) observed that polylysine synthesis was slightly more sensitive to erythromycin than polyproline synthesis, although polyphenylalanine synthesis again was the most resistant.

Erythromycin inhibits poly A-directed incorporation of lysine by inhibiting the synthesis of highly polymerized lysine peptides and by stimulating the accumulation of small lysine oligomers (TANAKA and TERAOKA, 1966, 1968; WILHELM, 1968; TERAOKA *et al.*, 1969; PULKRÁBEK *et al.*, 1970; MAO and ROBISHAW, 1971). Similar results were obtained with phenylalanine incorporation (WILHELM, 1968; MAO and ROBISHAW, 1971).

Binding of Erythromycin to Ribosomes

The inhibition of protein synthesis by erythromycin is due to its ability to bind to the 50S subunit of ribosomes from the target bacteria (TAUBMAN *et al.*, 1966; TANAKA and TERAOKA, 1966; OLEINICK and CORCORAN, 1969; MAO and PUTTERMAN, 1969; SAITO *et al.*, 1971). One molecule of erythromycin is bound to each 50S subunit at saturation (OLEINICK and CORCORAN, 1967, 1969; MAO, 1967b; FERNANDEZ-MUÑOZ *et al.*, 1971). Several additional molecules may be loosely bound at very high concentrations of antibiotic, but these are undoubtedly unrelated to the biological action (OLEINICK and CORCORAN, 1969). Erythromycin does not bind to cytosolic ribosomes from animal sources (MAO *et al.*, 1970).

Erythromycin and other macrolide antibiotics interfere with the binding of chloramphenicol to ribosomes from *B. megaterium* (VAZQUEZ, 1966a), *E. coli* (WOLFE and HAHN, 1965; FERNANDEZ-MUÑOZ et al., 1971; VOGEL et al., 1971a) and algae (RODRIGUEZ-LOPEZ and VAZQUEZ, 1968), and the ribosomal binding of lincomycin (CHANG and WEISBLUM, 1967; FERNANDEZ-MUÑOZ et al., 1971) and other macrolides (VAZQUEZ, 1967; AHMED, 1968; TAGO and NAGANO, 1970), suggesting a common site of interaction for these antibiotics. The affinity of this locus for erythromycin is higher than that for the other antibiotics (VAZQUEZ, 1966a) since erythromycin binding is not prevented by chloramphenicol (TAUB-MAN et al., 1966; OLEINICK et al., 1968) and may or may not be inhibited by lincomycin (TERAOKA et al., 1969; FERNANDEZ-MUÑOZ et al., 1971). On ribosomes from *B. subtilis*, however, erythromycin and chloramphenicol do not compete for a common binding site (OLEINICK et al., 1968), although erythromycin-binding is reduced by the presence of several other macrolides as well as by lincomycin (WILHELM et al., 1968). Finally, the erythromycin binding site is distinct from that which attracts tetracycline (OLEINICK and CORCORAN, 1969).

Analysis of the ribosomal binding of erythromycin under equilibrium conditions resulted in Scatchard plots which deviate from linearity (OLEINICK and CORCORAN, 1969; FERNANDEZ-MUÑOZ et al., 1971), suggesting a complex interaction between ribosome and antibiotic. The apparent affinity constant is $2.6 \times 10^7 M^{-1}$ for *B. subtilis* ribosomes (OLEINICK and CORCORAN, 1969), and $2.8 - 4.7 \times 10^5 M^{-1}$ for *S. aureus* ribosomes (MAO and PUTTERMAN, 1969). A recent reassessment of the binding to *E. coli* ribosomes produced a linear Scatchard plot and an apparent affinity constant of $0.92 \times 10^7 M^{-1}$ (FERNANDEZ-MUÑOZ and VAZQUEZ, 1973). The authors cited an impurity in their preparation of [N-methyl-^{14}C] erythromycin to explain the earlier non-linear plot. Since the preparation employed by OLEINICK and CORCORAN (1969) was chromatographically pure and still produced non-linear Scatchard plots, it will be necessary to reconfirm these data in *B. subtilis*.

The functional parts of the molecule involved in the interaction with ribosomes have been determined by assessing the ability of various macrolides and analogues to interfere with the binding of radioactively labeled erythromycin (WILHELM et al., 1968; MAO and PUTTERMAN, 1969). Both sugars are necessary for maximum competition, but some ability to compete is observed when only one sugar is present. Cross-resistance to most macrolides, but not to methymycin (which lacks the neutral sugar) suggests that the mutation may affect the region on the ribosome to which the neutral sugar binds (WILHELM et al., 1968). Functional groups which are important for complex formation include the 11- and 12-hydroxyl and 9-keto groups on the lactone, the 2-hydroxyl and 3-dimethyl-amino groups on desosamine, and the 3-methoxy group on cladinose (MAO and PUTTERMAN, 1969).

Chemical alterations of the ribosomal RNA or protein suggest that the intact polynucleotide and most ribosomal proteins are essential; the 42S "core" particles did not bind erythromycin (MAO and PUTTERMAN, 1969). However, BALLESTA et al. (1971) reported that β-cores (50S particles from which seven proteins had been removed) retained the ability to bind erythromycin, while γ-cores (deficient in seven more proteins) did not.

There is no covalent linkage between erythromycin and the ribosome, since dissociation proceeds readily upon dialysis, removal of K^+ or NH_4^+ ions, or introduction of organic solvents (OLEINICK and CORCORAN, 1969; MAO and PUTTERMAN, 1969). Ribosomal binding of erythromycin shows an absolute requirement for the presence of a monovalent cation, NH_4^+ or K^+ (OLEINICK and CORCORAN, 1969; MAO and PUTTERMAN, 1969; TERAOKA, 1970a). Na^+ or Li^+ will not substitute. Ribosomes can be reversibly inactivated and reactivated for erythromycin binding by incubation in a medium deficient or rich, respectively, in NH_4^+ or K^+ (TERAOKA, 1970a; VOGEL et al., 1971b).

Ribosomes from erythromycin-resistant mutants have a decreased affinity for erythromycin (OLEINICK and CORCORAN, 1969) and a lower capacity for in vitro peptide synthesis at concentrations of NH_4^+ or K^+ which are optimal for isogenic, sensitive strains (WILHELM and CORCORAN, 1967; OTAKA et al., 1970). Increased concentration of monovalent cations in the environment is sufficient by itself to increase the levels of erythromycin-binding, peptide synthesis, and peptidyl transferase activity to equal those in the sensitive strain (OTAKA et al., 1970). A plausible explanation is that the primary result of the mutation is a decreased affinity of the ribosomes or critical sites on them for NH_4^+ or K^+ and that the lowered activity for the above measured phenomena are secondary effects.

The requirement for high levels of NH_4^+ or K^+ is reminiscent of a similar requirement for protein synthesis and especially for the peptide bond-forming step (TRAUT and MONRO, 1964). This similarity and the higher levels of monovalent cation required by resistant strains of E. coli for both properties (erythromycin-binding and peptidyl transferase activity) have been considered as evidence for the binding of erythromycin to the peptidyl transferase "center" on the 50S subunit (MONRO et al., 1970; VOGEL et al., 1971b).

Erythromycin appears to have only a limited ability to bind to ribosomes which are actively engaged in protein synthesis. Crude ribosomal preparations bind less than one molecule of erythromycin per ribosome (OLEINICK and CORCORAN, 1970). If the associated nascent peptides are removed, either by dialysis against a buffer containing a low concentration of Mg^{++} or by treatment with puromycin, the maximum level (one mole equivalent) of antibiotic can be bound. Furthermore, erythromycin does not inhibit elongation of these nascent chains (OLEINICK and CORCORAN, 1970). TAI, WALLACE and DAVIS (personal communication) have found similarly that erythromycin is inactive against large polysomes which are devoid of initiating complexes. Finally, erythromycin can inhibit N-Ac-oligophenylalanine synthesis in a reconstituted cell-free system only if the antibiotic was bound previously to the free 50S subunit (OLEINICK and CORCORAN, 1970).

Sensitivity of Protein Synthesis
in Mitochondria and Chloroplasts to Erythromycin

Erythromycin inhibits the growth of the microbial eukaryotes Paramecium aurelia (ADOUTTE and BEISSON, 1970), Saccharomyces cerevisiae (LINNANE et al., 1968a), and Chlamydomonas reinhardi (SAGER and RAMANIS, 1970). Resistant mutations have been described, as have cross-resistance to other macrolides, lincomycin, and chloramphenicol (LINNANE et al., 1968a; THOMAS and WILKIE, 1968; reviewed by LINNANE and HASLAM, 1970). Genetic analyses indicate cytoplasmic inheritance of the resistance determinant in Paramecium (BEALE, 1969; ADOUTTE and BEISSON, 1970) and in some mutants of Chlamydomonas (SAGER and RAMANIS, 1970; METS and BOGORAD, 1971) and Saccharomyces (LINNANE et al., 1968a; THOMAS and WILKIE, 1968).

In these mutants, the resistance character appears to reside in the mitochondria or chloroplasts, making it a valuable genetic marker for extra-chromosomal linkage groups. Growth of yeast in the presence of erythromycin results in cells which are phenotypically similar to cells carrying the ϱ^-, cytoplasmic petite mutation; the oxygen-induced synthesis and/or assembly of cytochromes a, a_3, b and c on the inner mitochondrial membrane does not occur (LAMB et al., 1968). Both the ϱ^- and erythromycin-resistance determinants may be on the mitochondrial DNA, but they can be separated by recombination or reassortment during genetic crossing (LINNANE et al., 1968b; GINGOLD et al., 1969; LINNANE and HASLAM, 1970). Erythromycin resistance is linked to other non-Mendelian genes on chloroplast DNA (SAGER and RAMANIS, 1970).

Amino acid incorporation by isolated yeast mitochondria (LINNANE et al., 1968a; FIRKIN and LINNANE, 1969) and by chloroplast ribosomes from several plants (ELLIS and HARTLEY, 1971) is sensitive to erythromycin, although the synthesis of some proteins on mitochondrial ribosomes is more readily inhibited by the antibiotic than that of others (THOMAS and WILLIAMSON, 1971). The antibiotic binds well to the 52S subunit of chloroplast ribosomes from Chlamydomonas (METS and BOGORAD, 1971). Protein synthesis on cytosolic ribosomes from eukaryotes is universally insensitive to erythromycin (LAMB et al., 1968; MAO et al., 1970; BORST and GRIVELL, 1971). VILLA et al. (1972) have recently reported that the protein complement of mitochondrial and cytosolic ribosomes from yeast differ, and that a unique protein present in mitochondrial ribosomes from an erythromycin-resistant strain is absent in a sensitive strain.

In some erythromycin-resistant yeast mutants, resistance and cross-resistance to other macrolides and to lincomycin is also observed with isolated mitochondria even after the permeability of the mitochondria has been altered by freezing and thawing (LINNANE et al., 1968a). In another class of mutants, the intact cells are resistant to chloramphenicol, lincomycin, mikamycin, and carbomycin, but sensitive to erythromycin; amino acid incorporation by isolated mitochondria, however, is inhibited by all these antibiotics. It has been suggested (BUNN et al., 1970) that, in these mutants, resistance to erythromycin is the result of a structural alteration of the mitochondrial ribosomes (perhaps similar to that found by VILLA et al., 1972), while resistance to the other antibiotics is due to a permeability barrier in the mitochondrial membrane. Erythromycin, chloramphenicol, and cycloheximide (an antibiotic specific for cytosolic ribosomes) interfere with the oxygen-induced formation of several enzymes of yeast mitochondria, but none of these antibiotics is effective in a cytoplasmically determined erythromycin-resistant mutant (VARY et al., 1970). Thus, a complex interaction between the two protein synthetic systems, cytosolic and mitochondrial, is indicated.

Isolated mitochondria from mammalian cells (FIRKIN and LINNANE, 1969) and from insect flight muscle (WILLIAMS and BIRT, 1972), unlike those from wild-type yeast cells, are resistant to erythromycin. To explain this differential sensitivity, an evolutionary change in the nature of the mitochondrial ribosomes was invoked. However, mildly swollen rat liver mitochondria are susceptible to the drug (KROON and DEVRIES, 1971). Furthermore, BEATTIE (1971) reported that erythromycin can interfere with amino acid incorporation in mitochondria preincubated with the antibiotic and in digitonin-preparations of the inner membrane. Resolution of this controversy will require measurement of the affinity and inhibitory capacity of erythromycin during protein synthesis with mitochondrial ribosomes isolated from these diverse sources.

The Influence of Erythromycin on Binding
of Aminoacyl- and Peptidyl-tRNA to the Ribosomes

The non-enzymatic (T-factor-independent) binding of ^{14}C-lys-tRNA to ribosome-poly A complexes and of ^{14}C-phe-tRNA to ribosome-poly U complexes is insensitive to erythromycin (TANAKA et al., 1966); there is one report that enzymic (T-factor-dependent) binding can be weakly inhibited (HILL, 1969). Furthermore, VAZQUEZ and MONRO (1967) observed no effect of erythromycin on the binding of aminoacyl-tRNA to the 30S subunit, but a partial inhibition of the increase in binding of phe-tRNA (but not of lys-tRNA or pro-tRNA) which occurs when 50S subunits are added. Erythromycin does not interfere with either the binding of polylys-tRNA and N-Ac-phe-tRNA to the ribosome ČERNÁ et al., 1969) or the sparsomycin-induced stabilization of the binding of N-Ac-phe-tRNA (HERNER et al., 1969). In fact, a slight stimulation in binding of these substrates has been reported (ČERNÁ et al., 1969).

When various aminoacyl-tRNAs, either acetylated or unacetylated, are treated with Tl-ribonuclease, one of the fragments produced contains the penta- or hexa-nucleotide at the aminoacyl end of the molecule. These fragments can be isolated and used to measure either the binding of acyl-aminoacyl-oligonucleotide (e.g., CACCA-leu-Ac) to the donor (D, peptidyl, or P) site (MONRO et al., 1968) or the binding of aminoacyl-oligonucleotide fragments (e.g., CACCA-phe or CACCA-leu) to the acceptor (A or aminoacyl) site. In general, erythromycin enhances the binding of both kinds of substrates to their respective ribosomal sites, with the stronger effect exhibited on the CACCA-leu-Ac-binding to the donor site (PESTKA, 1969; CELMA et al., 1970, 1971; MONRO et al., 1970; ČERNÁ and RYCHLÍK, 1972). In contrast, on erythromycin-resistant E. coli ribosomes, erythromycin produces a weaker stimulation of CACCA-leu-Ac binding to the donor site and a much stronger stimulation of CACCA-phe binding to the acceptor site (ČERNÁ and RYCHLÍK, 1972). Furthermore, erythromycin completely abolishes the inhibition of binding of both types of substrates produced by spiramycin, carbomycin, and lincomycin, and reverses about 50% of the inhibition of substrate binding to the acceptor site caused by chloramphenicol, gougerotin, and puromycin (ČERNÁ and RYCHLÍK, 1972). These observations suggest that erythromycin interacts with the ribosome at a site removed from both the donor and acceptor sites, since a direct interaction at either site would compete with the corresponding substrate and therefore produce an inhibition.

The Action of Erythromycin
on Peptide Bond Synthesis and Translocation

The exact enzymatic reaction in polypeptide chain elongation which is inhibited by erythromycin remains obscure. Some evidence supports the notion that the reaction catalyzed by peptidyl transferase is the primary target, while other evidence favors the translocation reaction which promotes transfer of peptidyl-tRNA from the ribosomal acceptor site to the ribosomal donor site and is catalyzed by G-factor and GTP (cf. CORCORAN, 1971).

Peptidyl transferase is an integral part of the 50S ribosomal subunit (MONRO, 1967). This enzymatic activity has frequently been assayed by measuring the formation of a peptide bond between peptidyl- or aminoacyl-tRNA bound to the donor site and aminoacyl-tRNA or puromycin (the "puromycin reaction") in the acceptor site. Erythromycin produces varying effects on this enzymatic reaction depending upon the size and amino acid composition of the donor peptidyl residue, as well as the assay conditions. Erythromycin inhibits the "limited addition" of one lysine residue to oligolysyl-tRNA (JAYARAMAN and GOLDBERG, 1968) and the reaction of puromycin with polylysyl-tRNA (RYCHLÍK, 1966; TERAOKA and TANAKA, 1971b), but stimulates the puromycin reaction when N-Ac-phe-tRNA, CACCA-phe-Ac, or CACCA-leu-Ac are substrates (ČERNÁ et al., 1969; ČERNÁ et al., 1971; KUBOTA et al., 1972). The antibiotic also inhibits the release of peptides from the ribosome subsequent to formation of a covalent bond between the peptidyl moiety of the donor substrate and glycyl-ApC (RYCHLÍK et al., 1967).

When several aminoacyl-phe-tRNAs were tested as donors in the puromycin reaction, the synthesis of peptidyl puromycin was stimulated in each case except for phe-phe-puromycin formation which was inhibited (TANAKA et al., 1971). In a cell-free system from E. coli, OLEINICK and CORCORAN (1970) showed that N-Ac-diphe formation was not sensitive to erythromycin, while N-Ac-oligophe synthesis was markedly inhibited, provided that the antibiotic was bound to the 50S subunit before it combined with the initiation complex containing the 30S subunit, poly U, and N-Ac-phe-tRNA. This suggested that a step after the formation of the first peptide bond, possibly translocation, is blocked.

On the basis of a stimulation of the rate of formation of di- and tri-lysine and an inhibition of the synthesis of tetra- or longer lysine peptides, MAO (1971) proposed that the synthesis of the first peptide bond was the least sensitive to erythromycin. In agreement with this hypothesis, the synthesis of f-met-puromycin was found to be stimulated by erythromycin, while the synthesis of f-met-ala-puromycin was inhibited (KUBOTA et al., 1972). Many other macrolide antibiotics bind to the same site as does erythromycin and inhibit the formation of the first as well as of subsequent peptide bonds. Consequently, a common mechanism of action for all macrolides was suggested: Erythromycin may cause a conformational change in the ribosome which modifies the activity of peptidyl transferase, producing either a stimulation (of the formation of the first peptide bond) or an inhibition (of the formation of subsequent bonds and of overall synthesis) (MAO and ROBISHAW, 1971; TERAOKA and TANAKA, 1971b). It is noteworthy that the synthesis of peptidyl-puromycin on native polyribosomes (isolated with nascent peptidyl-tRNA) was not inhibited by erythromycin (PESTKA, 1972; 1973). Therefore, either peptidyl transferase is not sensitive to erythromycin on native polyribosomes or erythromycin is incapable of interacting with those ribosomes bearing nascent peptides (OLEINICK and CORCORAN, 1970; TAI, WALLACE and DAVIS, personal communication).

The peptidyl transferase reaction of 50S subunits can be assayed in 25–50% ethanol or methanol; in this case, only aminoacyl-oligonucleotides and puromycin are required (MONRO, 1969). This "fragment reaction" with f-met-hexanucleotide as substrate is not inhibited by erythromycin (MONRO and VAZQUEZ, 1967), and with other substrates, it may even be stimulated (MAO and ROBISHAW, 1972). Furthermore, on *E. coli* ribosomes, erythromycin reverses the inhibitory effect of chloramphenicol or lincomycin on the "fragment reaction" (MONRO *et al.,* 1970; VOGEL *et al.,* 1971 a), as well as the inhibition by chloramphenicol of N-Ac-phe-puromycin synthesis in the absence of alcohol (TERAOKA, 1970b). VOGEL *et al.* (1971 a) have proposed that the rate limiting step in the alcoholic medium is the interaction of the ribosome and f-met-tRNA, a reaction which is not affected by erythromycin. On the other hand, if the substrate is bound to the ribosome prior to the addition of alcohol, the new rate-limiting step (ribosome-puromycin interaction or peptidyl transferase) is stimulated by erythromycin.

MAO and ROBISHAW (1972) have examined a series of N-blocked and unblocked aminoacyl-tRNAs and dipeptidyl-tRNAs for their reaction with puromycin on salt-washed 50S subunits in the presence of methanol in a reaction similar to the "fragment reaction". Erythromycin stimulates the formation both of N-Ac-aminoacyl-puromycin and of aminoacyl-puromycin, and the dose-response curve coincides with the binding curve for erythromycin (OLEINICK and CORCORAN, 1969; WEISBLUM *et al.,* 1971). The extent of stimulation, however, depends on the particular substrate. Erythromycin can either stimulate or inhibit the formation of N-Ac-dipeptidyl-puromycins and dipeptidyl-puromycins in this system, again depending on the individual donor substrate (MAO and ROBISHAW, 1972). The effects of erythromycin are less pronounced when the substrate contains a hydrophobic (*e.g.,* leucine) or small (*e.g.,* glycine) amino acid side chain. These authors (MAO and ROBISHAW, 1972) concluded that erythromycin is an "effector" (positive or negative) of peptidyl transferase; when bound to the 50S subunit, it induces a conformational change in the ribosome which modifies the rate of this enzymatic reaction. The results suggest that donor substrates interact differentially with an erythromycin-modified site on the ribosome. It should be pointed out that the rate of the peptidyl transferase reaction under these conditions is very low and/or only a low percentage of the ribosomal subunits may be active. Consequently, it is difficult to relate the results of any studies in alcohol-containing media to the mechanism in the intact cell.

From the preceding discussion, it is clear that erythromycin can inhibit the peptidyl transferase reaction under certain conditions in cell-free systems. Further evidence in support of the susceptibility of peptide bond synthesis to erythromycin comes from a comparison of the requirements for ribosomal binding of erythromycin and for the peptidyl transferase activity. As noted earlier, both binding and transferase activity require the presence of a monovalent cation, K^+ or NH_4^+. Ribosomes from resistant mutants are deficient in both properties, and both can be restored to normal levels *in vitro* by an increase in the environmental K^+ or NH_4^+ (TERAOKA et al., 1970). At low K^+ concentration, erythromycin stimulates N-Ac-phe-puromycin synthesis, while at optimal K^+, there is no effect of the antibiotic (TERAOKA and TANAKA, 1971a). Therefore, although both the monovalent cation and erythromycin modify peptidyl transferase activity, K^+ is more effective.

Ribosomes may be reversibly inactivated merely by decreasing the K^+ or NH_4^+ concentration (VOGEL et al., 1971b) or by partially deproteinating the 50S subunit (BALLESTA et al., 1971). Upon reactivation or reconstitution, binding ability and peptidyl transferase activity are regained together, suggesting that erythromycin interacts either with the peptidyl transferase site or with a site that is altered along with peptidyl transferase. On *E. coli* ribosomes, the binding site for erythromycin overlaps the site for chloramphenicol (a known inhibitor of peptide-bond formation), while that for siomycin (an inhibitor of translocation), is at an independent site (MODOLELL et al., 1971). When ribosomes are isolated while still containing nascent peptidyl-tRNA, the requirement for monovalent cations for peptidyl-puromycin synthesis is much reduced and erythromycin is ineffective (PESTKA, 1973). However, the incorporation of ^3H-leucine into peptidyl-puromycin is inhibited, perhaps through an inhibition of translocation (PESTKA, 1971b).

SCHLESSINGER et al. (1969) reported that the formation of polyribosomes continues in the presence of erythromycin, despite a severe depression in protein synthesis. In this case, it appears that erythromycin uncouples translocation from peptide bond synthesis. However, in crude extracts of *E. coli*, KAEMPFER (1968) demonstrated the exchange of ribosomal subunits which is dependent upon ribosome movement along the mRNA, release from the mRNA, and mixing of the newly-released subunits with the pool of ribosomal subunits prior to reattachment to mRNA. Erythromycin partially blocks this exchange. Furthermore, at low concentrations of the antibiotic, the polyribosomes of *B. megaterium* spheroplasts are partially degraded, while at high concentrations, erythromycin preserves the integrity of the polysomes (CUNDLIFFE, 1969), an action consistent with a blockage of ribosome movement by the antibiotic.

Unlike fusidic acid and thiostrepton (known inhibitors of translocation), erythromycin does not block the G-factor- (nor the T-factor-) dependent GTPase activities (TANAKA et al., 1969; MAO and ROBISHAW, 1971). The G-factor- and ribosome-dependent GTPase is part of the translocation sequence, but a translocase inhibitor which acted through the ribosome and not through G-factor might be active on another partial reaction. For example, PESTKA and BROT (1971) showed that although erythromycin was only slightly inhibitory to oligophenylalanine synthesis in the presence of G-factor and GTP, it produced a large inhibition in the absence of G and GTP (during nonenzymic translocation). Fusidic acid allows one translocation event to occur, but prevents the dissociation of the G-factor-GDP complex from the single G- and T-dependent GTPase site (RICHMAN and BODLEY, 1972; MILLER, 1972). Therefore, attachment of the next amino acyl-tRNA is prevented and the nascent peptides are locked in the puromycin-reactive donor site (CUNDLIFFE, 1972). Erythromycin, on the other hand, completely inhibits the puromycin reaction by blocking the translocation of peptidyl-tRNA from the (puromycin-insensitive) acceptor site in spheroplasts (CUNDLIFFE, 1972) and in a cell-free system active in protein synthesis (CANNON and BURNS, 1971).

Erythromycin inhibits the cell-free synthesis of phenylalanyl-puromycin only under conditions permitting phenylalanyl-tRNA binding to both the peptidyl and aminoacyl sites (IGARASHI et al., 1969; IGARASHI et al., 1970). When only the peptidyl site is occupied, erythromycin does not interfere with the puromycin reaction. Furthermore, the antibiotic completely inhibits the G-factor-dependent release of deacylated tRNA^{phe} from the ribosomes (IGARASHI et al., 1969, 1970).

When chlortetracycline is added to spheroplasts, it prevents the binding of aminoacyl-tRNA to the ribosomal acceptor site and makes the aminoacyl-tRNA unavailable for translocation; after this treatment, erythromycin does not inhibit the puromycin reaction (CUNDLIFFE and McQUILLEN, 1967). Since chlortetracycline and erythromycin do not bind to the same site on isolated ribosomes (OLEINICK and CORCORAN, 1969), erythromycin appears to block translocation. One other interpretation of these results must be considered. Chlortetracycline prevents the exchange of ribosomal subunits which accompanies ribosome movement during active protein synthesis (KAEMPFER, 1968). Since erythromycin appears to interact preferentially with free 50S subunits and not directly with polyribosomes (OLEINICK and CORCORAN, 1970; TAI, WALLACE and DAVIS, personal communication), active movement, release of subunits, and re-initiation of protein synthesis may be essential for the binding of erythromycin to polyribosomes in intact cells. Consequently, erythromycin may be unable to interact with those polyribosomes bearing chlortetracycline and, for this reason, would not inhibit the puromycin reaction.

In conclusion, when erythromycin binds to the 50S subunit, it appears to alter the conformation of the ribosome such that modifications of both peptidyl transferase and translocase activities may be observed. In general, in the most purified systems reconstituted from washed ribosomes and synthetic substrates, peptidyl transferase appears to be modified, while in whole cells and in crude cell-free extracts, erythromycin appears to preferentially inhibit translocation. The apparent rate-limiting step in protein synthesis is profoundly affected by the assay conditions and furthermore, different macrolides might affect the different steps in individual ways (CORCORAN, 1971; see also VAZQUEZ, in this volume).

Summary

The erythromycins are a group of macrolide antibiotics each of which is composed of a 14-membered lactone ring and, glycosidically linked to the ring, a neutral and an amino-deoxy sugar. They are synthesized by Streptomyces erythreus from the lactone precursors, propionate and 2-methylmalonate, with subsequent attachment of the sugars. Erythromycin A, the most widely used form, is an extremely safe antibiotic, which is effective against Gram-positive and some Gram-negative organisms. The antibiotic is normally bacteriostatic, but at high concentrations it may be bactericidal.

Bacteria may mutate to resistance to erythromycin in vivo or upon serial transfer in culture in the presence of the drug. Cross-resistance is often, although not universally, found to other macrolide antibiotics, to lincomycin, and to chloramphenicol. Maximal inhibitory activity is dependent upon the presence of both sugars and a pH at which a significant concentration of the non-protonated form is present (pKa = 8.6). Resistant mutants have 50S ribosomal subunits which contain either one altered protein (in resistant *B. subtilis* or *E. coli*) or a uniquely methylated 23S rRNA (in inducibly- or constitutively-resistant *S. aureus*).

Erythromycin specifically inhibits protein synthesis in intact bacteria and in cell-free extracts. The inhibition is exerted on a step which comes after the activation of the amino acid and its transfer to tRNA. When erythromycin is directed against poly A-dependent lysine incorporation, highly polymerized lysine polymers are not produced, but the synthesis of small lysine peptides is stimulated.

The inhibition of protein synthesis by erythromycin is due to its ability to bind to the 50S subunit of susceptible bacteria. One molecule of erythromycin is bound at saturation to a site which, on the ribosomes from some bacteria, may also bind other macrolides, lincomycin, and chloramphenicol. The functional parts of the molecule involved in the interaction with ribosomes have been determined. Ribosomal binding of erythromycin shows an absolute requirements for the presence of a monovalent cation, NH_4^+ or K^+, as do overall protein synthesis and especially the peptide-bond forming step. This similarity and the higher levels of monovalent cation required by resistant strains for these properties have been considered as evidence for the binding of erythromycin to the peptidyl transferase "center" on the 50S subunit. Erythromycin has a limited ability to bind to ribosomes which are actively engaged in protein synthesis, and may have to bind before the 50S subunit enters a polyribosomal complex.

Protein synthesis on cytosolic ribosomes from eukaryotic cells is universally insensitive to erythromycin. Mitochondrial protein synthesis, on the other hand, is inhibited in several microbial eukaryotes and in swollen or preincubated mitochondria isolated from mammalian cells. In *Saccharomyces* and in *Chlamydomonas,* resistant mutants have been described. For some of these mutants, the resistance character appears to reside in the mitochondria or chloroplasts, making it a valuable genetic marker for extra-chromosomal linkage groups.

Erythromycin does not usually inhibit and in many cases may even stimulate the binding of aminoacyl- and peptidyl-tRNA to the ribosomes. This implies that the reaction which is the primary target of erythromycin is after ribosomal binding of these substrates. The peptidyl transferase reaction (the formation of a peptide bond between peptidyl- or aminoacyl-tRNA bound to the donor site and aminoacyl-tRNA or puromycin in the acceptor site) is sometimes inhibited, but often is stimulated by erythromycin, depending upon the size and amino acid composition of the donor peptidyl residue, as well as the assay conditions. Other macrolide antibiotics bind to the same site as does erythromycin and consistently produce an inhibition of peptidyl transferase. Consequently, it has been proposed that erythromycin also modifies this reaction. Ribosomes may be reversibly inactivated merely by decreasing the monovalent cation concentration or by partially deproteinating the 50S subunit. Upon reactivation or reconstitution, binding ability and peptidyl transferase activity are regained together, suggesting that erythromycin interacts either with the peptidyl transferase site or a related one.

In intact bacteria and in crude cell-free extracts, erythromycin blocks the movement of ribosomes along the mRNA which is required for the exchange of ribosomal subunits and preserves the integrity of the polysomes, consistent with an inhibition of the translocation reaction. Erythromycin does not inhibit the G-factor- and ribosome-dependent GTPase reaction as do some known inhibitors of translocation, but may inhibit translocation at another partial reaction. Erythromycin inhibits non-enzymic translocation and blocks peptidyl-tRNA in the (puromycin-insensitive) acceptor site. In the presence of chlortetracycline, which makes aminoacyl-tRNA unavailable for translocation, erythromycin does not inhibit the puromycin reaction. These results indicate either that erythromycin is an inhibitor of translocation or that it was unable to bind to polyribosomes which were previously blocked by chlortetracycline.

In conclusion, the primary target of erythromycin appears to be translocation in intact cells, but experiments in purified cell-free systems indicate that the peptidyl transferase reaction may also be affected.

Acknowledgements. I am indebted to Prof. JOHN W. CORCORAN for the opportunity to work in his laboratory and for his patient guidance of my studies of antibiotic action. I also wish to thank the following colleagues for critically reading this manuscript: Dr. ODDVAR F. NYGAARD, Dr. RONALD C. RUSTAD, Dr. HELEN H. EVANS, Dr. HENRY Z. SABLE, and Dr. JOSEPH ILAN. The preparation of this manuscript was supported by the U.S. Atomic Energy Commission (Contract No. W-31-109-ENG-78; Report No. C00-78-298).

Note Added in Proof

Since the submission of this manuscript, three reports have come to the author's attention, which deserve note here.

1. Nomura and Engbaeck [NOMURA, M., and F. ENGBAECK: Expression of ribosomal protein genes as analyzed by bacteriophage Mu-induced mutations. Proc. Nat. Acad. Sci. U. S. **69**, 1526 (1972)] have identified the genetic locus for erythromycin sensitivity as part of the "ribosomal protein operon", a single transcriptional unit which is located between argG and aroB on the *E. coli* chromosome and which contains genes for both 30S and 50S ribosomal proteins as well as for elongation factor EF G.

2. Recently, Tanaka *et al.* [TANAKA, S., T. OTAKA and A. KAJI: Further studies on the mechanism of erythromycin action. Biochim. Biophys. Acta **331**, 128 (1973)] presented a comprehensive model of erythromycin action on translocation. According to this model, translocation is composed of three steps: the evacuation of peptidyl-tRNA from the acceptor site, movement of peptidyl-tRNA toward the donor site, and positioning of peptidyl-tRNA in the donor site. They propose that erythromycin interferes with the third step. The model accounts for many of the discrepancies in the literature concerning the apparent varying sensitivities of peptide bond synthesis with different donor amino acyl-tRNAs by their differential ability to interact with peptidyl transferase even though improperly positioned at the donor site. Such a hypothesis certainly warrants further investigation.

3. Very recently, MAJER et al. [MAJER, J., J.R. MARTIN and J.W. CORCORAN: Erythromycin D. Isolation, properties and structure. Abstract No. 200, 14th Interscience Conference on Antimicrobial agents and Chemotherapy, San Francisco, (1974)] have isolated erythromycin D and its structure, as expected, consists of erythronolide b linked to desosamine and mycarose.

References

ADOUTTE, A., and J. BEISSON: Cytoplasmic inheritance of erythromycin resistant mutations in *Paramecium aurelia*. Mol. Gen. Genetics **108**, 70 (1970).

AHMED, A.: Altered ribosomes in spiramycin-resistant mutants of *Bacillus subtilis*. Biochim. Biophys. Acta **166**, 218 (1968).

AMA Drug Evaluations, 1st ed., 1971, American Medical Association.

APIRION, D.: Three genes that affect *Escherichia coli* ribosomes. J. Mol. Biol. **30**, 255 (1967).

ARORA, K.L., J. MAJER, M. CHEVALLIER, and J.W. CORCORAN: Comparative effects of erythromycin on gram-positive and negative bacteria in relation to its mode of action. Abstract No. 147. Interscience Conference on Antimicrobial Agents and Chemotherapy, Atlantic City, 1971.

BALLESTA, J.P.G., V. MONTEJO, and D. VAZQUEZ: Reconstitution of the 50S ribosome subunit. Localization of activities related to the peptidyl transferase centre. FEBS Letters **19**, 75 (1971).

BEALE, G.H.: A note on the inheritance of erythromycin-resistance in *Paramecium aurelia*. Genet. Res. **14**, 341 (1969).

BEATTIE, D.S.: The synthesis of mitochondrial proteins. Sub-Cell. Biochem. **1**, 1 (1971).

BORST, P., and L.A. GRIVELL: Mitochondrial ribosomes. FEBS Letters **13**, 73 (1971).

BROCK, T.D., and M.L. BROCK: Similarity in mode of action of chloramphenicol and erythromycin. Biochim. Biophys. Acta **33**, 274 (1959).

BUNN, C.L., C.H. MITCHELL, H.B. LUKINS, and A.W. LINNANE: Biogenesis of mitochondria, XVIII. A new class of cytoplasmically determined antibiotic resistant mutants in *Saccharomyces cerevisiae*. Proc. Natl. Acad. Sci. U.S. **67**, 1233 (1970).

CANNON, M., and K. BURNS: Modes of action of erythromycin and thiostrepton as inhibitors of protein synthesis. FEBS Letters **18**, 1 (1971).

CELMA, M.L., R.E. MONRO, and D. VAZQUEZ: Substrate and antibiotic binding sites at the peptidyl transferase centre of *E. coli* ribosomes. FEBS Letters **6**, 273 (1970).

CELMA, M.L., R.E. MONRO, and D. VAZQUEZ: Substrate and antibiotic binding sites at the peptidyl transferase centre of *E. coli* ribosomes: Binding of UACCA-Leu to 50S subunits. FEBS Letters **13**, 247 (1971).

CELMER, W.D.: Macrolide stereochemistry. II. Configurational assignments at certain centers in various macrolide antibiotics. J. Am. Chem. Soc. **87**, 1799 (1965a).

CELMER, W.D.: Macrolide stereochemistry. III. A configurational model for macrolide antibiotics. J. Am. Chem. Soc. **87**, 1801 (1965b).

CELMER, W.D.: Biogenetic, constitutional, and stereochemical unitary principles in macrolide antibiotics. Antimicrobial Agents Chemotherapy **1965**, 144.

CELMER, W.D.: Stereochemical problems in macrolide antibiotics. In: Symposium on antibiotics, p. 413, ed. by S. RAKHIT. London: Butterworths 1971.

ČERNÁ, J., J. JONÁK, and I. RYCHLÍK: Effects of macrolide antibiotics on the ribosomal peptidyl transferase in cell-free systems derived from *Escherichia coli* B and erythromycin-resistant mutant of *Escherichia coli* B. Biochim. Biophys. Acta **240**, 109 (1971).

ČERNÁ, J., and I. RYCHLÍK: Cross-resistance of *Escherichia coli* B ribosomes to inhibition of the puromycin reaction by erythromycin, spiramycin and chloramphenicol. Biochim. Biophys. Acta **157**, 436 (1968).

ČERNÁ, J., and I. RYCHLÍK: The effect of antibiotics on the substrate binding to the acceptor and donor site of ribosomal peptidyl transferase of an erythromycin-resistant mutant of *Escherichia coli*. Biochim. Biophys. Acta **287**, 292 (1972).

ČERNÁ, J., I. RYCHLÍK, and P. PULKRÁBEK: The effect of antibiotics on the coded binding of peptidyl-tRNA to the ribosome and on the transfer of the peptidyl residue to puromycin. Eur. J. Biochem. **9**, 27 (1969).

CHABBERT, Y.: Antagonisme *in vitro* entre l'erythromycine et la spiramycine. Ann. Inst. Pasteur **90**, 787 (1956).

CHANG, F. N., and B. WEISBLUM: The specificity of lincomycin binding to ribosomes. Biochemistry **6**, 836 (1967).

CLARK, R. K.: The chemistry of erythromycin. I. Acid degradation products. Antibiot. & Chemotherapy **3**, 663 (1953).

COLEMAN, V. R., J. B. GUNNISON, and E. JAWETZ: Participation of erythromycin and carbomycin in combined antibiotic action *in vitro*. Proc. Soc. Exptl. Biol. Med. **83**, 668 (1953).

CORCORAN, J. W.: The biosynthesis of erythromycin. Lloydia **27**, 1 (1964).

CORCORAN, J. W.: Erythromycin and the bacterial ribosome: A study of the mechanism of sensitivity and resistance to macrolide antibiotics and lincomycin in *Bacillus subtilis* 168. In: Drug action and drug resistance in bacteria: Macrolide antibiotics and lincomycin, p. 177, MITSUHASHI, S., ed. Tokyo, Japan: Univ. of Tokyo Press 1971.

CORCORAN, J. W., and M. CHICK: Biochemistry of the macrolide antibiotics. In: Biosynthesis of antibiotics, J. F. SNELL, ed., p. 159. New York: Academic Press 1966.

CORCORAN, J. W., T. KANEDA, and J. C. BUTTE: Propionate incorporation into a unique branched fatty acid. Federation Proc. **19**, 227 (1960a).

CORCORAN, J. W., T. KANEDA, and J. C. BUTTE: Actinomycete antibiotics. I. The biological incorporation of propionate into the macrocyclic lactone of erythromycin. J. Biol. Chem. **235**, PC 29, (1960b).

CUNDLIFFE, E.: Antibiotics and polyribosomes. II. Some effects of lincomycin, spiramycin, and streptogramin A *in vivo*. Biochemistry **8**, 2063 (1969).

CUNDLIFFE, E.: The mode of action of fusidic acid. Biochem. Biophys. Res. Commun. **46**, 1794 (1972).

CUNDLIFFE, E. and K. McQUILLEN: Bacterial protein synthesis: The effects of antibiotics. J. Mol. Biol. **30**, 137 (1967).

DEKIO, S., R. TAKATA, S. OSAWA, K. TANAKA, and M. TAMAKI: Genetic studies of the ribosomal proteins in *Escherichia coli*. IV. Pattern of the alteration of ribosomal protein components in mutants resistant to spectinomycin or erythromycin in different strains of *Escherichia coli*. Mol. Gen. Genetics **107**, 39 (1970).

DJERASSI, C., O. HALPERN, D. I. WILKINSON, and E. J. EISENBRAUN: Macrolide antibiotics. VIII. The absolute configuration of certain centers in neomethymycin, erythromycin, and related antibiotics. Tetrahedron **4**, 369 (1958).

ELLIS, R. J., and M. R. HARTLEY: Sites of synthesis of chloroplasts proteins. Nature New Biol. **233**, 193 (1971).

FERNANDEZ-MUÑOZ, R., R. E. MONRO, R. TORRES-PIÑEDO, and D. VAZQUEZ: Substrate- and antibiotic-binding sites at the peptidyl-transferase centre of *Escherichia coli* ribosomes. Studies on the chloramphenicol, lincomycin and erythromycin sites. Eur. J. Biochem. **23**, 185 (1971).

FERNANDEZ-MUÑOZ, R., and D. VAZQUEZ: Quantitative binding of ^{14}C-erythromycin to *E. coli* ribosomes. J. Antibiotics (Tokyo) **26**, 107 (1973).

FINLAND, M., C. WILCOX, S. S. WRIGHT, and E. M. PURCELL: Cross-resistance to antibiotics: Effect of exposures of bacteria to carbomycin or erythromycin *in vitro*. Proc. Soc. Exptl. Biol. Med. **81**, 725 (1952).

FIRKIN, F. C., and A. W. LINNANE: Phylogenetic differences in the sensitivity of mitochondrial protein synthesizing systems to antibiotics. FEBS Letters **2**, 330 (1969).

FLYNN, E. H., M. V. SIGAL, JR., P. F. WILEY, and K. GERZON: Erythromycin. I. Properties and degradation studies. J. Am. Chem. Soc. **76**, 3121 (1954).

FRIEDMAN, S. M., T. KANEDA, and J. W. CORCORAN: Antibiotic glycosides. V. A comparison of 2-methylmalonate and propionate as precursors of the C_{21} branched chain lactone in erythromycin. J. Biol. Chem. **239**, 2386 (1964).

FUKAYA, K., and O. KITAMOTO: Studies on tissue distribution of several new antibiotics. In: Progress in antimicrobial and anticancer chemotherapy (Proceedings of the 6th International Congress of Chemotherapy), vol. I, p. 503. Tokyo: University of Tokyo Press 1970.

FUSILLO, M. H., H. E. NOYES, E. J. PULASKI, and J. Y. S. TOM: Antimicrobial spectrum and cross-resistance studies of erythromycin and carbomycin. Antibiot. & Chemotherapy **3**, 581 (1953).

GARROD, L.P.: The erythromycin group of antibiotics. Brit. Med. J. **1957**, 57.

GARROD, L.P., and P.M. WATERWORTH: Behaviour *in vitro* of some new antistaphylococcal antibiotics. Brit. Med. J. **1956**, 61.

GINGOLD, E.B., G.W. SAUNDERS, H.B. LUKINS, and A.W. LINNANE: Biogenesis of mitochondria. X. Reassortment of the cytoplasmic genetic determinants for respiratory competence and erythromycin resistance in *Saccharomyces cerevisiae*. Genetics **62**, 735 (1969).

GOLDTHWAITE, C., D. DUBNAU, and I. SMITH: Genetic mapping of antibiotic resistance markers in *Bacillus subtilis*. Proc. Natl. Acad. Sci. U.S. **65**, 96 (1970).

GRISEBACH, H., H. ACHENBACH, and U.C. GRISEBACH: Zur Biogenese des Erythromycins. Naturwissenschaften **47**, 206 (1960a).

GRISEBACH, H., H. ACHENBACH u. W. HOFHEINZ: Untersuchungen zur Biogenese des Erythromycins. I. Mitt.: Der Aufbau des Lactonringes. Z. Naturforsch. **15b**, 560 (1960b).

GRISEBACH, H., and W. HOFHEINZ: Biosynthesis of the macrolide antibiotics. J. Roy. Inst. Chem. **88**, 332 (1964).

GUY, L.R., and J.S. CHAPMAN: Susceptibility *in vitro* of unclassified mycobacteria to commonly used antimicrobials. Am. Rev. Respirat. Diseases **84**, 746 (1961).

HAHN, F.E.: Erythromycin and oleandomycin. In: Antibiotics. I. Mode of action, D. Gottlieb and P.D. Shaw, eds., p. 378. Berlin-Heidelberg-New York: Springer 1967.

HAIGHT, T.H., and M. FINLAND: Resistance of bacteria to erythromycin. Proc. Soc. Exptl. Biol. Med. **81**, 183 (1952).

HANSEN, A., and J.W. CORCORAN: Alteration in a ribosomal protein in *Bacillus subtilis* 168 associated with sensitivity to erythromycin. Federation Proc. **28**, 725 (1969).

HARRIS, D.R., S.G. McGEACHIN, and H.H. MILLS: The structure and stereochemistry of erythromycin A. Tetrahedron Letters **11**, 679 (1965).

HERNER, A.E., I.H. GOLDBERG, and L.B. COHEN: Stabilization of N-acetylphenylalanyl transfer ribonucleic acid binding to ribosomes by sparsomycin. Biochemistry **8**, 1335 (1969).

HERRELL, W.E.: Hazards of antibiotic therapy. J. Am. Med. Assoc. **168**, 1875 (1958).

HILL, R.N.: The effects of antibiotics on the interaction of T-factor, aminoacyl-tRNA and ribosomes. J. Gen. Microbiol. **58**, viii (1969).

HOFHEINZ, W., u. H. GRISEBACH: X. Mitt.: Über das Vorkommen von L-Mycarose in Erythromycin C. Z. Naturforsch. **17b**, 852 (1962).

HUDSON, D.G., G.M. YOSHIHARA, and W.M.M. KIRBY: Spiramycin, clinical and laboratory studies. Arch. Internal Med. **97**, 57 (1956).

HUNG, P.P., C.L. MARKS, and P.L. TARDREW: The biosynthesis and metabolism of erythromycins by *Streptomyces erythreus*. J. Biol. Chem. **240**, 1322 (1965).

IGARASHI, K., H. ISHITSUKA, and A. KAJI: Comparative studies on the mechanism of action of lincomycin, streptomycin, and erythromycin. Biochem. Biophys. Res. Commun. **37**, 499 (1969).

IGARASHI, K., H. ISHITSUKA, Y. KURIKI, and A. KAJI: Use of antibiotics in studies of protein synthesis. Progr. in Antimicrob. and Anticancer Chemother. **2**, 445 (1970).

JAYARAMAN, J., and I.H. GOLDBERG: Localization of sparsomycin action to the peptide-bond-forming step. Biochemistry **7**, 418 (1968).

JONES, W.F., R.L. NICHOLS, and M. FINLAND: Development of resistance and cross-resistance *in vitro* to erythromycin, carbomycin, spiramycin, oleandomycin and streptogramin. Proc. Soc. Exptl. Biol. Med. **93**, 388 (1956).

KAEMPFER, R.: Ribosomal subunit exchange during protein synthesis. Proc. Natl. Acad. Sci. U.S. **61**, 106 (1968).

KANEDA, T., J.C. BUTTE, S.B. TAUBMAN, and J.W. CORCORAN. Actinomycete antibiotics. III. The biogenesis of erythronolide, the C_{21} branched lactone in erythromycin. J. Biol. Chem. **237**, 322 (1962).

KREMBEL, J., and D. APIRION: Changes in ribosomal proteins associated with mutants in a locus that affects *Escherichia coli* ribosomes. J. Mol. Biol. **33**, 363 (1968).

KROON, A.M., and H. DeVRIES: Mitochondriogenesis in animal cells: Studies with different inhibitors. In: Autonomy and biogenesis of mitochondria and chloroplasts, p. 318. North Holland 1971.

KUBOTA, K., A. OKUYAMA, and N. TANAKA: Differential effects of antibiotics on peptidyl transferase reactions. Biochem. Biophys. Res. Commun. **47**, 1196 (1972).

LAI, C.J., and B. WEISBLUM: Altered methylation of ribosomal RNA in an erythromycin-resistant strain of *Staphylococcus aureus*. Proc. Natl. Acad. Sci. U.S. **68**, 856 (1971).

LAMB, A.J., G.D. CLARK-WALKER, and A.W. LINNANE: The biogenesis of mitochondria. 4. The differentiation of mitochondrial and cytoplasmic protein synthesizing systems *in vitro* by antibiotics. Biochim. Biophys. Acta **161**, 415 (1968).

LINNANE, A.W., and J.M. HASLAM: The biogenesis of yeast mitochondria. In: *Current topics in cellular regulation*, vol. 2, p. 101, B.L. HORECKER and E.R. STADTMAN, eds. New York: Academic Press 1970.

LINNANE, A.W., A.J. LAMB, C. CHRISTODOULOU, and H.B. LUKINS. The biogenesis of mitochondria, VI. Biochemical basis of the resistance of *Saccharomyces cerevisiae* toward antibiotics which specifically inhibit mitochondrial protein synthesis. Proc. Natl. Acad. Sci. U.S. **59**, 1288 (1968a).

LINNANE, A.W., G.W. SAUNDERS, E.B. GINGOLD, and H.B. LUKINS: The biogenesis of mitochondria. V. Cytoplasmic inheritance of erythromycin resistance in *Saccharomyces cerevisiae*. Proc. Natl. Acad. Sci. U.S. **59**, 903 (1968b).

LUTZ, A., O. GROOTEN et J. HOFFERER: Evolution et modifications de la resistance des staphylocoques pathogenes à six anibiotiques usuels de 1950 à 1956. L'action comparée *in vitro* de l'erythromycine, de la magnamycine, de la spiramycine, de la novobiocine (albamycine) et de l'oléandomycine. Ann. Inst. Pasteur **92**, 778 (1957).

MAO, J.C.-H.: Protein synthesis in a cell-free extract from *Staphylococcus aureus*. J. Bacteriol. **94**, 80 (1967a).

MAO, J.C.-H.: The stoichiometry of erythromycin binding to ribosomal particles of *Staphylococcus aureus*. Biochem. Pharmacol. **16**, 2441 (1967b).

MAO, J.C.-H.: Mode of action of erythromycin. In: Drug action and drug resistance in bacteria, vol. 1, MITSUHASHI, S., ed. p. 153. Tokyo, Japan: Univ. of Tokyo Press 1971.

MAO, J.C.-H., and M. PUTTERMAN: Accumulation in gram-positive and gram-negative bacteria as a mechanism of resistance to erythromycin. J. Bacteriol. **95**, 1111 (1968).

MAO, J.C.-H., and M. PUTTERMAN: The intermolecular complex of erythromycin and ribosome. J. Mol. Biol. **44**, 347 (1969).

MAO, J.C.-H., M. PUTTERMAN, and R.G. WIEGAND: Biochemical basis for the selective toxicity of erythromycin. Biochem. Pharmacol. **19**, 391 (1970).

MAO, J.C.-H., and E.E. ROBISHAW: Effects of macrolides on peptide-bond formation and translocation. Biochemistry **10**, 2054 (1971).

MAO, J.C.-H., and E.E. ROBISHAW: Erythromycin-a peptidyl-transferase effector. Biochemistry **11**, 4864 (1972).

MAO, J.C.-H., and P.L. TARDREW: Demethylation of erythromycins by rabbit tissues *in vitro*. Biochem. Pharmacol. **14**, 1049 (1965).

MAO, J.C.-H., and R.G. WIEGAND: Mode of action of macrolides. Biochim. Biophys. Acta **157**, 404 (1968).

MARTIN, J.R., and A.W. GOLDSTEIN: Final steps in erythromycin biosynthesis. In: Progress in antimicrobial and anticancer chemotherapy (Proceedings of the 6th International Congress of Chemotherapy), vol. I, p. 199. Tokyo: University of Tokyo Press, 1970.

MARTIN, J.R., T.J. PERUN, and R.L. GIROLAMI: Studies on the biosynthesis of the erythromycins. I. Isolation and structure of an intermediate glycoside, 3-α-L-Mycarosylerythronolide B. Biochemistry **5**, 2852 (1966).

MARTIN, J.R., and W. ROSENBROOK: Studies on the biosynthesis of the erythromycins. II. Isolation and structure of a biosynthetic intermediate, 6-deoxyerythronolide B. Biochemistry **6**, 435 (1967).

MARTIN, W.J., and W.E. WELLMAN: Clinically useful antimicrobial agents: Untoward reactions. Post-grad. Med. **42**, 397 (1967).

MATHIEU, N., et M. FAGUET: Activité *in vitro* de la spiramycine en association avec la tétracycline, l'erythromycine, la pénicilline, la streptomycine sur la multiplication de *Staphylococcus aureus* etudiée au microbiophotomètre. Ann. Inst. Pasteur **94**, 69 (1958).

McALPINE, T.S., and J.W. CORCORAN: Enzymatic O-methylation of erythromycin C as the final step in the biogenesis of erythromycin A. Federation Proc. **30**, 1168 (1971).

McGUIRE, J.M., R.L. BUNCH, R.C. ANDERSON, H.E. BOAZ, E.H. FLYNN, H.M. POWELL, and J.W. SMITH: "Ilotycin", a new antibiotic. Antibiot. & Chemotherapy **2**, 281 (1952).

METS, L.J., and L. BOGORAD: Mendelian and uniparental alterations in erythromycin binding by plastid ribosomes. Science **174**, 707 (1971).

Miller, D. L.: Elongation factors EF Tu and EF G interact at related sites on ribosomes. Proc. Natl. Acad. Sci. U. S. **69**, 752 (1972).

Modolell, J., D. Vazquez, and R. E. Monro: Ribosomes, G-factor and siomycin. Nature **230**, 109 (1971).

Molavi, A., and L. Weinstein: *In vitro* activity of erythromycin against atypical mycobacteria. J. Infect. Diseases **123**, 216 (1971).

Monro, R. E.: Catalysis of peptide bond formation by 50S ribosomal subunits from *Escherichia coli*. J. Mol. Biol. **26**, 147 (1967).

Monro, R. E.: Protein synthesis: Uncoupling of polymerization from template control. Nature **223**, 903 (1969).

Monro, R. E., J. Černá, and K. A. Marcker: Ribosome-catalyzed peptidyl transfer: substrate specificity at the P-site. Proc. Natl. Acad. Sci. U. S. **61**, 1042 (1968).

Monro, R. E., R. Fernandez-Muñoz, M. L. Celma, A. Jimenez, E. Battaner, and D. Vazquez. Antibiotics acting on the peptidyl transferase center of ribosomes. In: Progress in antimicrobial and anticancer chemotherapy, vol. II, p. 473. Tokyo: University of Tokyo Press 1970.

Monro, R. E., and D. Vazquez: Ribosome-catalysed peptidyl transfer: Effects of some inhibitors of protein synthesis. J. Mol. Biol. **28**, 161 (1967).

Okubo, H., and Y. Fujimoto: Distribution of antibiotics in the body. In: Progress in antimicrobial and anticancer chemotherapy (Proceedings of the 6th International Congress of Chemotherapy), vol. I, p. 495. Tokyo: University of Tokyo Press 1970.

Oleinick, N. L., and J. W. Corcoran: Two types of erythromycin binding to ribosomes of *Bacillus subtilis*. Federation Proc. **26**, 285 (1967).

Oleinick, N. L., and J. W. Corcoran: Two types of binding of erythromycin to ribosomes from antibiotic-sensitive and -resistant *Bacillus subtilis* 168. J. Biol. Chem. **244**, 727 (1969).

Oleinick, N. L., and J. W. Corcoran: Evidence of a limited access of erythromycin A to functional polysomes and its action on bacterial translocation. In: Progress in antimicrobial and anticancer chemotherapy, vol. I, p. 202. Tokyo: University of Tokyo Press 1970.

Oleinick, N. L., J. M. Wilhelm, and J. W. Corcoran: Nonidentity of the site of action of erythromycin A and chloramphenicol on *Bacillus subtilis* ribosomes. Biochim. Biophys. Acta **155**, 290 (1968).

Otaka, E., T. Itoh, S. Osawa, K. Tanaka, and M. Tamaki: Peptide analyses of a protein component, 50-8, of 50S ribosomal subunit from erythromycin resistant mutants of *Escherichia coli* and *Escherichia freundii*. Mol. Gen. Genetics **114**, 14 (1971).

Otaka, E., H. Teraoka, M. Tamaki, K. Tanaka, and S. Osawa: Ribosomes from erythromycin-resistant mutants of *Escherichia coli* Q 13. J. Mol. Biol. **48**, 499 (1970).

Perun, T. J.: The chemistry and conformation of erythromycin. In: Drug action and drug resistance in bacteria: 1. Macrolide antibiotics and lincomycin, p. 123, ed. by S. Mitsuhashi. Tokyo: University of Tokyo Press, 1971.

Pestka, S.: Studies on the formation of transfer ribonucleic acid-ribosome complexes, XI. Antibiotic effects on phenylalanyl-oligonucleotide binding to ribosomes. Proc. Natl. Acad. Sci. U. S. **64**, 709 (1969).

Pestka, S.: Inhibitors of ribosome functions. Ann. Rev. Microbiol. **25**, 487 (1971a).

Pestka, S.: Ribosomal inhibitors of translocation and transpeptidation. Proc. symp. molecular mechanisms of antibiotic action on protein biosynthesis and membranes. (1971b) Granada, Spain, eds. E. Muñoz, F. Ferrandiz, D. Vazquez. Berlin-Heidelberg-New York: Springer 1971b.

Pestka, S.: Studies on transfer ribonucleic acid-ribosome complexes. XIX. Effect of antibiotics on peptidyl puromycin synthesis on polyribosomes from *Escherichia coli*. J. Biol. Chem. **247**, 4669 (1972).

Pestka, S.: Effect of antibiotics on peptidyl puromycin formation on polyribosomes and a model of ribosome function. Proc. of the VIIth International Congress of Chemotherapy, Prague, Czechoslovakia, 1973 (in press).

Pestka, S., and N. Brot: Studies on the formation of transfer ribonucleic acid-ribosome complexes. XV. Effect of antibiotics on steps of bacterial protein synthesis: Some new ribosomal inhibitors of translocation. J. Biol. Chem. **246**, 7715 (1971).

Pulkrábek, P., J. Černá, and I. Rychlík: Synthesis of tRNA-bound lysine peptides in the presence of puromycin or of antibiotics inhibiting ribosomal transpeptidation. Collection Czech. Chem. Commun. **35**, 2973 (1970).

RODRIGUEZ-LOPEZ, M., and D. VAZQUEZ: Comparative studies on cytoplasmic ribosomes from algae. Life Sci. **7**, 327 (1968).

RICHMAN, N., and J. W. BODLEY: Ribosomes cannot interact simultaneously with elongation factors EF Tu and EF G. Proc. Natl. Acad. Sci. U. S. **69**, 686 (1972).

RYCHLÍK, I.: Release of lysine peptides by puromycin from polylysyl-transfer ribonucleic acid in the presence of ribosomes. Biochim. Biophys. Acta **114**, 425 (1966).

RYCHLÍK, I., S. CHLÁDEK, and J. ŽEMLIČKA: Release of peptide chains from the polylysyl-tRNA-ribosome complex by cytidyl-(3′→5′)-2′(3′)-O-glycyladenosine. Biochim. Biophys. Acta **138**, 640 (1967).

SAGER, R., and Z. RAMANIS: A genetic map of non-Mendelian genes in *Chlamydomonas*. Proc. Natl. Acad. Sci. U. S. **65**, 593 (1970).

SAITO, T., M. OSHIMA, M. SHIMIZU, M. HASHIMOTO, and S. MITSUHASHI: Macrolide resistance in *Staphylococcus aureus*. In: Progress in antimicrobial and anticancer chemotherapy, vol. II, p. 572. Tokyo: University of Tokyo Press 1970.

SAITO, T., M. SHIMIZU, and S. MITSUHASHI: Macrolide resistance in *Staphylococci*. In: Drug action and drug resistance in bacteria: Macrolide antibiotics and lincomycin, p. 239, MITSUHASHI, S., ed. Tokyo, Japan: Univ. of Tokyo Press 1971.

SCHLESSINGER, D., C. GURGO, L. LUZZATTO, and D. APIRION: Polyribosome metabolism in growing and nongrowing *Escherichia coli*. Cold Spring Harbor Symp. Quant. Biol. **34**, 231 (1969).

SMITH, I., W. COLLI, and M. OISHI: Studies on the physical linkage of antibiotic resistance markers to ribosomal RNA genes in *Bacillus subtilis*. J. Mol. Biol. **62**, 111 (1971).

SMITH, I., C. GOLDTHWAITE, and D. DUBNAU: The genetics of ribosomes in *Bacillus subtilis*. Cold Spring Harbor Symp. Quant. Biol. **34**, 85 (1969).

SPIZEK, J., M. CHICK, and J. W. CORCORAN: Biogenetic relationship of the erythromycins and the lactone of erythromycin B. Antimicrobial Agents Chemotherapy **1965** 138.

TAGO, K., and M. NAGANO: Mechanism of inhibition of protein synthesis by leucomycin. In: Progress in antimicrobial and anticancer chemotherapy (Proceedings of the 6th International Congress of Chemotherapy), vol. I, p. 199. Tokyo: University of Tokyo Press 1970.

TAKATA, R., S. OSAWA, K. TANAKA, H. TERAOKA, and M. TAMAKI: Genetic studies of the ribosomal proteins in *Escherichia coli* V. Mapping of erythromycin resistance mutations which lead to alteration of a 50S ribosomal protein component. Mol. Gen. Genetics **109**, 123 (1970).

TANAKA, K., M. TAMAKI, R. TAKATA, and S. OSAWA: Low affinity for chloramphenicol of erythromycin resistant *Escherichia coli* ribosomes having an altered protein component. Biochem. Biophys. Res. Commun. **46**, 1979 (1972).

TANAKA, K., and H. TERAOKA: Binding of erythromycin to *Escherichia coli* ribosomes. Biochim. Biophys. Acta **114**, 204 (1966).

TANAKA, K., and H. TERAOKA: Effect of erythromycin on polylysine synthesis directed by polyadenylic acid in an *Escherichia coli* cell-free system. J. Biochem. (Tokyo) **64**, 635 (1968).

TANAKA, K., H. TERAOKA, T. NAGIRA, and M. TAMAKI: [^{14}C] Erythromycin-ribosome complex formation and non-enzymatic binding of aminoacyl-transfer RNA to ribosome-messenger RNA complex. Biochim. Biophys. Acta **123**, 435 (1966).

TANAKA, K., H. TERAOKA, and M. TAMAKI: Peptidyl puromycin synthesis: Effect of several antibiotics which act on 50S ribosomal subunits. FEBS Letters **13**, 65 (1971).

TANAKA, K., H. TERAOKA, M. TAMAKI, E. OTAKA, and S. OSAWA: Erythromycin-resistant mutant of *Escherichia coli* with altered ribosomal protein component. Science **162**, 576 (1968).

TANAKA, K., H. TERAOKA, M. TAMAKI, R. TAKATA, and S. OSAWA: Phenotypes represented by a mutational change in a 50S ribosomal protein component, 50-8, in *Escherichia coli*. Mol. Gen. Genetics **114**, 9 (1971).

TANAKA, N., T. KINOSHITA, and H. MASUKAWA: Mechanism of inhibition of protein synthesis by fusidic acid and related steroidal antibiotics. J. Biochem. **65**, 459 (1969).

TAUBENECK, U.: Susceptibility of *Proteus mirabilis* and its stable L-forms to erythromycin and other macrolides. Nature **196**, 195 (1962).

TAUBMAN, S. B., N. R. JONES, F. E. YOUNG, and J. W. CORCORAN: Sensitivity and resistance to erythromycin in *Bacillus subtilis* 168: the ribosomal binding of erythromycin and chloramphenicol. Biochim. Biophys. Acta **123**, 438 (1966).

TAUBMAN, S. B., A. G. SO, F. E. YOUNG, E. W. DAVIE, and J. W. CORCORAN: Effect of erythromycin on protein-biosynthesis in *Bacillus subtilis*. Antimicrobial Agents Chemotherapy **1963**, 395.

TAUBMAN, S.B., F.E. YOUNG, and J.W. CORCORAN: Antibiotic glycosides, IV. Studies on the mechanism of erythromycin resistance in *Bacillus subtilis*. Proc. Natl. Acad. Sci. U.S. **50**, 955 (1963).

TERAOKA, H.: A reversible change in the ability of *Escherichia coli* ribosomes to bind to erythromycin. J. Mol. Biol. **48**, 511 (1970a).

TERAOKA, H.: Reversal of the inhibitory action of chloramphenicol on the ribosomal peptidyl transfer reaction by erythromycin. Biochim. Biophys. Acta **213**, 535 (1970b).

TERAOKA, H., M. TAMAKI, and K. TANAKA: Peptidyl transferase activity of *Escherichia coli* ribosomes having an altered protein component in the 50S subunit. Biochem. Biophys. Res. Commun. **38**, 328 (1970).

TERAOKA, H., and K. TANAKA: An alteration in ribosome function caused by equimolar binding of erythromycin. Biochim. Biophys. Acta **232**, 509 (1971a).

TERAOKA, H., and K. TANAKA: Reaction of puromycin with N-acetylphenylalanyl-tRNA on ribosomes reassociated from *Escherichia coli* ribosomal subunits. In: Molecular mechanisms of antibiotic action on protein synthesis and membranes, Proceedings, Symposium, Granada, Spain, 1971b, Elsevier, Amsterdam.

TERAOKA, H., K. TANAKA, and M. TAMAKI: The comparative study on the effects of chloramphenicol, erythromycin, and lincomycin on polylysine synthesis in an *Escherichia coli* cell-free system. Biochim. Biophys. Acta **174**, 776 (1969).

THOMAS, D.L., and D. WILKIE: Inhibition of mitochondrial synthesis in yeast by erythromycin: cytoplasmic and nuclear factors controlling resistance. Genet. Res. **11**, 33 (1968).

THOMAS, D.Y., and D.H. WILLIAMSON: Products of mitochondrial protein synthesis in yeast. Nature New Biol. **233**, 196 (1971).

TRAUT, R.R., and R.E. MONRO: The puromycin reaction and its relation to protein synthesis. J. Mol. Biol. **10**, 63 (1964).

VARY, M.J., P.R. STEWART, and A.W. LINNANE: Biogenesis of mitochondria. XVII. The role of mitochondrial and cytoplasmic ribosomal protein synthesis in the oxygen-induced formation of yeast mitochondrial enzymes. Arch. Biochem. Biophys. **141**, 430 (1970).

VAZQUEZ, D.: Binding of chloramphenicol to ribosomes. The effect of a number of antibiotics. Biochim. Biophys. Acta **114**, 277 (1966a).

VAZQUEZ, D.: Antibiotics affecting chloramphenicol uptake by bacteria. Their effect on amino acid incorporation in a cell-free system. Biochim. Biophys. Acta **114**, 289 (1966b).

VAZQUEZ, D.: Binding to ribosomes and inhibitory effect on protein synthesis of the spiramycin antibiotics. Life Sci. **6**, 845 (1967).

VAZQUEZ, D., and R.E. MONRO: Effects of some inhibitors of protein synthesis on the binding of aminoacyl tRNA to ribosomal subunits. Biochim. Biophys. Acta **142**, 155 (1967).

VILLA, V.D., H. MORIMOTO, and H.O. HALVORSON: Mitochondrial and cytoplasmic ribosomal proteins in erythromycin resistant and sensitive yeast strains. Federation Proc. **31**, A456 (1972).

VOGEL, Z., T. VOGEL and D. ELSON: The effect of erythromycin on peptide bond formation and the termination reaction. FEBS Letters **15**, 249 (1971a).

VOGEL, Z., T. VOGEL, A. ZAMIR, and D. ELSON: Correlation between the peptidyl transferase activity of the 50S ribosomal subunit and the ability of the subunit to interact with antibiotics. J. Mol. Biol. **60**, 339 (1971b).

WAWSZKIEWICZ, E.J., and F. LYNEN: Propionyl-Co A dependent $H^{14}CO_3$ exchange into methyl-malonyl-Co A in extracts of *Streptomyces erythraeus*. Biochem. Z. **340**, 213 (1964).

WEAVER, J., and P.A. PATTEE: Inducible resistance to erythromycin in *Staphylococcus aureus*. J. Bacteriol. **88**, 574 (1964).

WEISBLUM, B.: Antibiotic inhibitors of protein synthesis which are antagonizable by erythromycin. Federation Proc. **28**, 466 (1969).

WEISBLUM, B., and J. DAVIES: Antibiotic inhibitors of the bacterial ribosome. Bacteriol. Rev. **32**, 493 (1968).

WEISBLUM, B., and V. DEMOHN: Erythromycin-inducible resistance in *Staphylococcus aureus*: Survey of antibiotic classes involved. J. Bacteriol. **98**, 447 (1969).

WEISBLUM, B., S. SIDDHIKOL, C.J. LAI, and V. DEMOHN: Erythromycin-inducible resistance in *Staphylococcus aureus*: requirements for induction. J. Bacteriol. **106**, 835 (1971).

WELCH, H., W. A. RANDALL, R. J. REEDY, and J. KRAMER: Bacterial spectrum of erythromycin, carbomycin, chloramphenicol, aureomycin, and terramycin. Antibiot. & Chemotherapy **2**, 693 (1952).

WILEY, P. F., R. GALE, C. W. PETTINGA, and K. GERZON: Erythromycin. XII. The isolation, properties and partial structure of erythromycin C. J. Am. Chem. Soc. **79**, 6074 (1957c).

WILEY, P. F., K. GERZON, E. H. FLYNN, M. W. SIGAL, JR., O. WEAVER, U. C. QUARCK, R. R. CHAUVETTE, and R. MONAHAN: Erythromycin X. Structure of erythromycin. J. Am. Chem. Soc. **79**, 6062 (1957a).

WILEY, P. F., M. W. SIGAL, JR., O. WEAVER, R. MONAHAN, and K. GERZON: Erythromycin. XI. Structure of erythromycin B. J. Am. Chem. Soc. **79**, 6070 (1957b).

WILEY, P. F., and O. WEAVER: Erythromycin. VII. The structure of cladinose. J. Am. Chem. Soc. **78**, 808 (1956).

WILHELM, J. M.: Antibiotic action and protein synthesis in *Bacillus subtilis*. Thesis, Case Western Reserve University, 1968.

WILHELM, J. M., and J. W. CORCORAN: Antibiotic glycosides. VI. Definition of the 50S ribosomal subunit of *Bacillus subtilis* 168 as a major determinant of sensitivity to erythromycin A. Biochemistry **6**, 2578 (1967).

WILHELM, J. M., N. L. OLEINICK, and J. W. CORCORAN: Interaction of antibiotics with ribosomes: Structure-function relationships and a possible common mechanism for the antibacterial action of the macrolides and lincomycin. Antimicrobial Agents Chemotherapy **1968**, 236.

WILLIAMS, K. L., and L. M. BIRT: Sensitivity to erythromycin of mitochondrial protein synthesis in isolated flight muscle mitochondria of the blowfly *Lucilia*. FEBS Letters **22**, 327 (1972).

WOLFE, A. D., and F. E. HAHN: Erythromycin: Mode of action. Science **143**, 1445 (1964).

WOLFE, A. D., and F. E. HAHN: Mode of action of chloramphenicol. IX. Effects of chloramphenicol upon a ribosomal amino acid polymerizing system and its binding to bacterial ribosome. Biochim. Biophys. Acta **95**, 146 (1965).

WOODWARD, R. B.: Struktur und biogenese der makrolide. Eine neue Klasse von Naturstoffen. Angew. Chem. **69**, 50 (1957).

Emetine and Related Alkaloids

Arthur P. Grollman and Zelda Jarkovsky

Emetine, the active principle of ipecac, is isolated from the roots of *Cephaelis ipecacuanha*, a small plant indigenous to the tropical rain forests of Brazil and countries immediately to the north. The medicinal properties of ipecac in the treatment of dysentery have long been known to the natives of Brazil (PURCHAS, 1625). The drug was introduced into Europe during the latter part of the 17th Century and sold by a Paris physician, HELVETIUS, as a secret remedy for dysentery. By 1688, ipecac had achieved such fame that the formula was purchased by Louis XIV and placed in the public domain (LLOYD, 1921; STAUB, 1927).

PELLETIER and MAGENDIE isolated a crude alkaloid from ipecac in 1817, which they designated "emetine;' much later, PAUL and COWNLEY showed this material to be a mixture consisting of emetine, cephaeline and psychotrine. PYMAN identified two additional components of ipecac: o-methylpsychotrine and emetamine; other alkaloids have subsequently been isolated (cf reviews by JANOT, 1953, and OPENSHAW, 1970).

During the 19th Century, the therapeutic value of ipecac was occasionally questioned. Although many physicians cited its effectiveness in the treatment of dysentery, others failed to consistently achieve favorable results. These divergent experiences probably reflect the failure of physicians of that time to distinguish between bacterial, amoebic and other causes of dysentery. It remained for VEDDER (1914) to show that ipecac was a powerful amoebicide but inactive against bacteria. VEDDER also found that the amoebicidal action of ipecac on free-living amoebae was due to the presence of the major constitutent alkaloid, emetine. ROGERS (1912) demonstrated the amoebicidal effect of emetine on the human pathogen, *Entamoeba histolytica,* and is credited with introducing the alkaloid as a specific remedy for amoebic dysentery and extraintestinal forms of amoebiasis.

Until recently, the therapeutic use of emetine was limited to amoebiasis and a few other protozoal infections. Its demonstrated effectiveness against certain nonspecific granulomatous disorders (GROLLMAN, 1965) drew attention to the cytotoxic action of the drug. Earlier reports of tumor regression after emetine therapy (LEWISOHN, 1918; VAN HOOSEN, 1919) appear to have been overlooked; presently emetine is again being tried in the therapy of certain forms of cancer (PANETTIERE and COLTMAN, 1971).

Chemistry

The structure of emetine is shown in Fig. 1. A number of related alkaloids have been isolated from the root of the ipecacuanha plant; these include cephae-

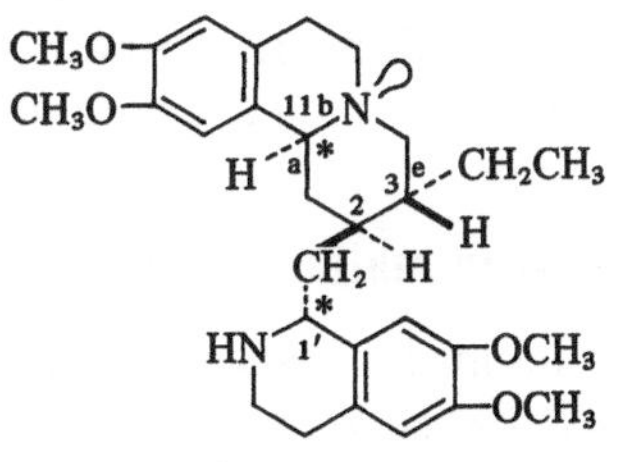

Fig. 1. Structural formula of emetine

line, emetamine, o-methylpsychotrine, psychotrine, protoemetine and "ipecac alkaloid A". Some of the same compounds occur in the seeds and bark of the Indian plant, *Alangium lamarckii* Thwaites; this plant contains other alkaloids related in structure to emetine, including alangicine (11-hydroxypsychotrine), desmethylpsychotrine, dihydroprotoemetine and ankorine (11-hydroxydihydro-protoemetine). A family of indole compounds: tubulosine, isotubulosine, demethyltubulosine, and alangimarckine (11-hydroxydeoxytubulosine) have also been isolated from *Alangium lamarckii*. Tubulosine (Fig. 2) resembles emetine in configuration and shares many of its biological properties. Structural formulae of the previously-mentioned alkaloids can be found in the review by OPENSHAW (1970).

The structure of emetine (Fig. 1) was primarily deduced from the work of BATTERSBY *et al.* (for reviews, see JANOT, 1953, SZANTAY, 1967, and OPENSHAW, 1970). Emetine possesses asymmetric centers at the 1', 2, 3 and 11b positions; thus, sixteen stereoisomers are possible. The relative and absolute configuration of the alkaloid has been elucidated largely through the work of VAN TAMELEN (1959) and BATTERSBY (1960).

The hexahydrobenzoquinolinozine ring system of emetine can assume several conformations; in the preferred form, the hydrogen atom at the 11b position is *trans* to the unshared pair of electrons on the nitrogen atom. This conformation is comparable to that of *trans*-decalin.

The total synthesis of emetine was first achieved by EVSTIGNEEVA and PREO-BRAZHENSKII (1958); further work by OPENSHAW *et al.* (reviewed by SZANTAY,

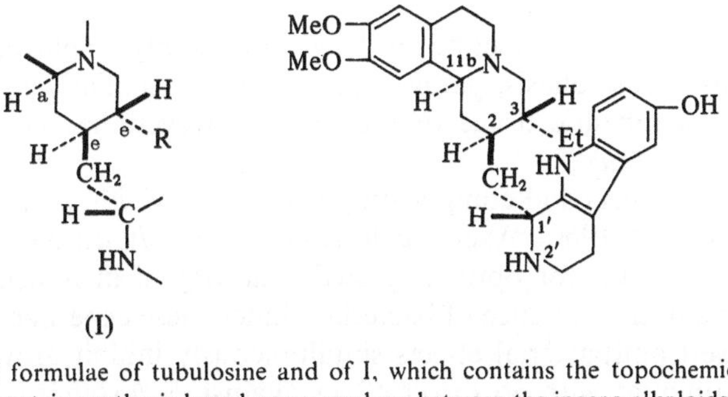

(I)

Fig. 2. Structural formulae of tubulosine and of I, which contains the topochemical requirements for inhibition of protein synthesis based on an analogy between the ipecac alkaloids and glutarimide antibiotics (GROLLMAN, 1966)

1967 and Openshaw, 1970) resulted in the synthesis of numerous stereoisomers and derivatives of emetine, some of which are biologically active.

Mammalian Toxicity

The distribution of emetine in the tissues of rats, dogs and rabbits has been reported (Gimble et al., 1948; Parmer and Cottrill, 1949; Davis et al., 1962). The highest concentrations of the drug were recorded in the lung, liver, spleen and kidney; moderate levels in cardiac tissue; and low amounts in muscle and brain. Emetine is slowly excreted by the kidney (Gimble et al. 1948); accumulation of the drug over a period of time may contribute to the toxicity observed in animals and man.

The clinical use of emetine is associated with cardiotoxicity, muscle weakness and gastrointestinal symptoms (Klatskin and Friedman, 1948). Cardiotoxicity represents a potentially serious side-effect; electrocardiographic abnormalities were observed in 53% of patients treated with therapeutic doses of emetine, and some form of cardiovascular manifestation (hypotension, tachycardia, etc.) occurred in 85% of the 93 patients studied. The literature on the cardiotoxic effects of emetine has been reviewed by Wenzel (1967).

Serial biopsies of the myocardium in dogs treated with emetine revealed selective damage to mitochondria (Pearce et al., 1971). In cells cultured from embryonic chick-heart, 1 μM emetine produced gross alterations in membrane structure (Watkins and Guess, 1968).

Muscle weakness following emetine therapy may reflect a drug-induced myopathy (Duane and Engel, 1970). In rats treated with emetine for one to two weeks, mitochondrial degeneration was observed in white and red skeletal muscle fibers; myofibrillar degeneration appeared only in the latter. Morphological or physiological changes were not found in the neural component of the motor unit.

The gastrointestinal toxicity of emetine includes diarrhea, nausea and vomiting. The emetic effect is exerted through a central medullary action and, partly, by a direct irritant effect on the gastric mucosa (Wang, 1965).

Therapeutic Properties

Antiparasitic Activity. Emetine has been successfully employed in such parasitic infections as fascioliasis, paragonimiasis, schistosomiasis and balantidiasis. The primary therapeutic usage of the drug, however, has been in infections caused by *E. histolytica*.

The multiple facets and complicating features of *E. histolytica* infections are well known (Faust, 1960). Axenic cultures of *E. histolytica* have been reported (Diamond, 1961) but, for optimal growth, the organism is usually grown on complex media in the presence of bacteria. Under these experimental conditions, drugs that exert antibacterial effects simultaneously inhibit growth of amoeba. Likewise, antibiotics with no direct effect against *E. histolytica* will effectively eradicate this parasite from infected animals (Woolfe, 1963). In the case of emetine and related ipecac alkaloids, the lack of demonstrable antibacterial effects indicates that the amoebicidal action of the alkaloid is directed against the amoeba.

Strains of *E. histolytica* that are relatively less sensitive to the effects of emetine (when tested *in vitro*) have been isolated. A genetic basis for this drug resistance has been established (ENTNER *et al.,* 1962; ENTNER and MOST, 1965). However, even these "resistant" strains are killed by therapeutic levels of emetine, and most treatment failures can be attributed to incorrect diagnosis or improper use of the drug.

Antiviral Activity. Emetine has been reported to be effective in the therapy of a number of viral diseases. VIDAL (1952) showed that patients with herpes zoster responded to the drug by a rapid disappearance of pain and drying of vesicles. HANISCH *et al.* (1966a) reported the use of emetine in opthalmic herpes zoster and epidemic keratoconjunctivitis. DEL PUERTO *et al.* (1968) described 600 children with virus hepatitis that had been treated with emetine; the clinical course of this disease in this group of patients was markedly shortened.

In experimental viral infections in mice, emetine was partially effective against Columbia SK virus but inactive against Coxsackie virus B1 and influenza A (GRUNBERG and PRINCE, 1966). GROLLMAN (1968) has reported inhibition by emetine of poliovirus replication in HeLa cells.

Antitumor Activity. LEWISOHN (1918) first described tumor regression following treatment with emetine. Shortly thereafter, VAN HOOSEN (1919) reported a series of 100 patients with advanced malignancies, a number of whom showed subjective and objective responses to emetine therapy. More recently, ABD-RABBO (1966, 1969) and WYBURN-MASON (1966) observed the effectiveness of dehydro-emetine in chronic granulocytic leukemia and other malignancies.

A systematic study of emetine in human malignancies has been initiated by the National Cancer Institute, and Phase I experience has been reported (PANETTIERE and COLTMAN, 1971). The toxic effects of emetine were generally minor and not dissimilar to those seen in patients with amoebiasis who are treated with this drug. The toxicity of emetine is unlike that of most other antitumor agents; it has no myelosuppressive effect. The drug may well prove useful when used in combination with other chemotherapeutic agents (PANETTIERE and COLTMAN, 1971).

The effectiveness of emetine and other ipecac alkaloids in experimental animal tumors has also been tested. ISAKA (1950) found that a single injection of emetine prolonged survival in the rat Yoshida sarcoma. In L1210 mouse leukemia, intermittent emetine therapy prolonged life span by 72%; in P388 mouse leukemia, by 20%; and in the B-16 melanoma, by 40% (JONDORF *et al.,* 1971). In this study, various ipecac alkaloids and synthetic derivatives of emetine were tested against L1210 leukemia. The structure-activity relationships correlate closely with the values reported for these compounds as inhibitors of protein synthesis (GROLLMAN, 1966; GROLLMAN and JARKOVSKY, 1974).

Effects on HeLa Cells

RNA, DNA and Protein Synthesis. The effects of emetine on nucleic acid and protein synthesis in nonsynchronized suspension cultures of HeLa cells have been studied by GROLLMAN (1968). The rate of protein synthesis is inhibited 50% by emetine at a concentration of 0.04 µM and by 99% at a concentration of 1.0 µM. There is a parallel effect on the synthesis of DNA, but inhibition, in this case, is incomplete. Inhibition of protein and DNA synthesis is not reversed by washing the cells and resuspending them in fresh media.

RNA synthesis was slightly stimulated at low concentrations of emetine but was partially inhibited by higher concentrations of drug. At much higher concentrations (1.0 mM), synthesis of ribosomal RNA is blocked (GILEAD and BECKER, 1971). This effect is reversible and may be related to a direct action of emetine on the DNA template.

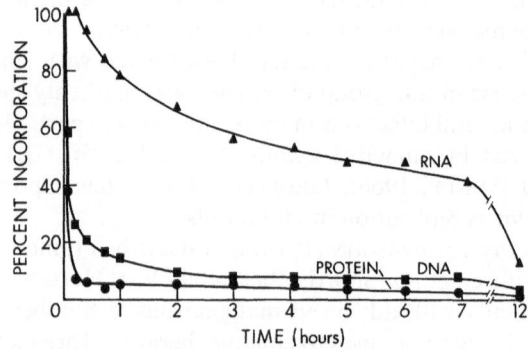

Fig. 3. Effect of emetine on RNA, DNA and protein synthesis in HeLa cells.

The effects of emetine on protein and DNA synthesis in HeLa cells are observed within several minutes after exposure to the alkaloid; the inhibitory effect on DNA synthesis was slightly less rapid than that on protein synthesis (GROLLMAN, 1968). The rate of RNA synthesis decreased over a period of hours after inhibition of protein synthesis was complete (Fig. 3). The observed inhibition of DNA synthesis is probably secondary to a primary effect on protein synthesis, as the latter process is required for concurrent synthesis of DNA in animal cells. Other inhibitors of protein synthesis in mammalian cells are known to simultaneously inhibit synthesis of DNA (cf GROLLMAN and HUANG, 1973).

Uptake of Ipecac Alkaloids. The rate of uptake of ^{3}H-emetine by intact HeLa cells is temperature-dependent and proportional to the concentration of alkaloid in the media (at concentrations less than 0.5 µM) (GROLLMAN, 1968). No differences in uptake between emetine and the biologically inactive epimer, isoemetine, were detected. Using an estimated value for intracellular volume, it was calculated that emetine and isoemetine are concentrated approximately 50-fold by HeLa cells. An uptake of one molecule of alkaloid per ribosome occurs at a concentration (0.2 µM) at which emetine inhibits protein biosynthesis by 90%. Maximum uptake at an emetine concentration of 5 µM is equivalent to approximately 30 molecules of alkaloid per ribosome.

Attachment of Nascent Peptide to Polyribosomes. Exposure of growing cultures of HeLa cells to emetine for two minutes results in a decrease in the number

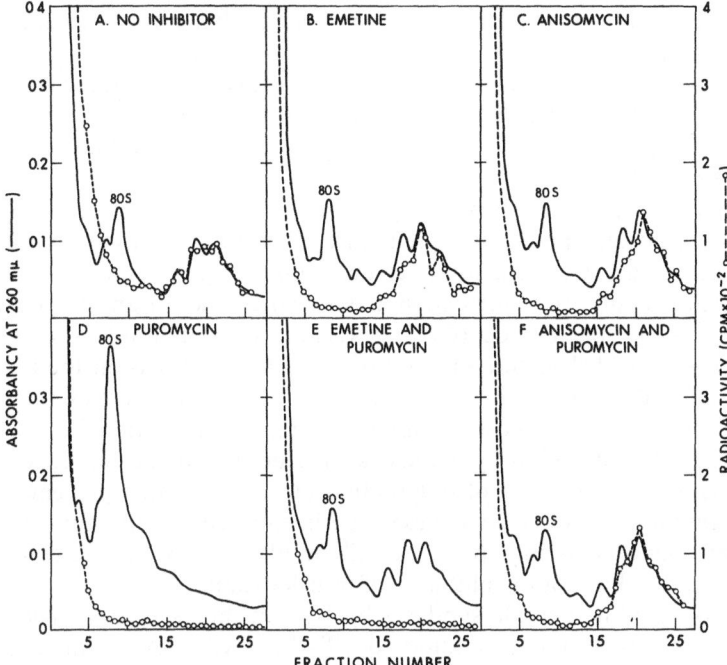

Fig. 4. Effect of puromycin on attachment of nascent peptide bound to HeLa cell polyribosomes in the presence of emetine and anisomycin (GROLLMAN and HUANG, 1973)

of single ribosomes and a concomitant increase in polyribosomes (Fig. 4B). In cells exposed to labeled amino acids for two minutes, nascent peptide was detected in the polyribosome region; if emetine was added at that time, nascent peptide remained attached to polyribosomes.

The effects of emetine can be distinguished from other inhibitors of protein synthesis, such as puromycin and anisomycin. Puromycin dissociates polyribosomes to single ribosomes and causes the release of incomplete polypeptides (Fig. 4D). The effects of puromycin are modified by prior exposure to emetine for 2 minutes; under these conditions, 85% of the bound nascent peptide is released from the polyribosome and polyribosomes fail to dissociate (Fig. 4E). Anisomycin is an inhibitor of peptide bond formation (GROLLMAN, 1967; GROLLMAN and HUANG, 1973); puromycin fails to release peptides from anisomycin-treated cells (Fig. 4F).

Protein Biosynthesis

Cytoplasmic Protein Synthesis. Amino acid incorporation into protein is rapidly inhibited by emetine in cell-free extracts (GROLLMAN, 1966); these kinetics, taken together with the observed immobilization of polyribosomes and blocked release of nascent peptides in intact cells, suggest that the drug inhibits elongation of peptide chains rather than chain initiation or termination.

Among inhibitors of chain elongation, agents that affect translocation can be distinguished from inhibitors of peptide bond formation and various other intermediate steps in this process. Several assays have been used for this purpose; in eukaryotes, measuring the release of nascent peptides by puromycin and the two-step assay described by MCKEEHAN and HARDESTY (1969) have proved particularly useful.

Puromycin reacts directly with peptidyl-tRNA located in the peptidyl (donor) site on the ribosome and causes premature release of peptidyl-puromycin from the polyribosome. This reaction does not require supernatant factors or GTP. Inhibitors of peptide bond formation in eukaryotes, such as anisomycin (GROLLMAN, 1967a) and sparsomycin (GOLDBERG et al., 1973) inhibit this reaction.

The assay of MCKEEHAN and HARDESTY (1969) is based on the assumption that peptidyl-tRNA will be primarily located at the peptidyl site on the ribosome following incubation of ribosomes in the presence of elongation factor II (EF-II) and GTP. Labeled aminoacyl-tRNA is then bound to the aminoacyl site on the ribosome in the presence of sodium fluoride, an agent that prevents initiation of new globin peptides. Under these experimental conditions, further incorporation of amino acids into peptides will be blocked by inhibitors of peptide bond formation but not by inhibitors of translocation.

Emetine inhibits peptide synthesis by 50% in the assay described but only if the drug is present prior to addition of EF-II and GTP. Assuming that nascent peptide is equally distributed between the amino-acyl and peptidyl sites on the ribosome, such partial inhibition would be anticipated in the case of an agent that either blocks translocation or movement of mRNA along the ribosome.

In the presence of emetine, puromycin releases 90% of the nascent peptide chains bound to the polyribosome (Fig. 4E), suggesting that the drug does not affect peptide bond formation. However, if translocation was completely inhibited by the drug, peptidyl-tRNA should accumulate in the aminoacyl site and be unable to react with puromycin. It would appear, therefore, that emetine affects some aspect of translocation that involves movement of mRNA along the ribosome.

Binding of emetine to the ribosome has not been demonstrated, despite the irreversible effect on protein synthesis and the fact that emetine of high specific activity is available (GROLLMAN, 1968).

Mitochondrial Protein Synthesis. PERLMAN and PENMAN (1970) and MAHLER et al., 1971) reported that mitochondrial protein synthesis is unaffected by concentrations of emetine that completely block protein synthesis on cytoplasmic ribosomes. These findings are consistent with the observation that emetine has no effect on protein synthesis in prokaryotes (GROLLMAN, 1966). Very high concentrations of emetine (0.1–1.0 mM) are reported to inhibit protein synthesis in isolated mitochondria (LEITMAN, 1971).

Protein Synthesis in Various Species. Emetine inhibits protein synthesis in certain animal cells (HeLa cells, rabbit reticulocytes), plants (gametophytes of *Anemia phylitidis)* and in cell-free extracts of yeast (*Saccharomyces cerevisiae)* (GROLLMAN, 1966). The drug is an effective inhibitor of protein synthesis in *E. histolytica* (ENTNER and GROLLMAN, 1973) and other protozoa.

Emetine does not penetrate the cell wall of yeast or bacteria. Protein synthesis in extracts prepared from *E. coli* (directed by endogenous or exogenous mRNA) is unaffected even by high concentrations of emetine. Thus, emetine has inhibited protein synthesis in all eukaryotes tested, while bacteria and other prokaryotes are unaffected by the drug. Trachoma agent, an obligate prokaryotic parasite of eukaryotes, is resistant to concentrations of emetine that suppress RNA and protein synthesis in the host cell (GILEAD and BECKER, 1971).

Emetine inhibits protein synthesis in experimental animals treated with the drug; however, microsomes isolated from emetine-treated rats show enhanced rates of amino acid incorporation when tested *in vitro* (JONDORF and SZAPARY, 1968; JONDORF et al., 1969). Protein (actomyosin) synthesis in extracts of cardiac muscle is inhibited by the alkaloid (BELLER, 1968).

Other Biochemical Processes

Oxidation and Respiration. The effects of emetine on oxidative processes have chiefly been studied in myocardial tissue. Using homogenates prepared from rat heart, DEITRICH and HEIM (1956) observed that emetine decreased oxygen consumption in the presence of glucose, pyruvate, fumurate and malate. Oxygen uptake increased in the presence of succinate. APPELT and HEIM (1964) found that high concentrations (3 mM) of emetine inhibited oxygen uptake, but this effect was not specific for emetine, and inhibition of endogenous respiration was not observed at 0.1 mM concentrations of the drug.

Rats treated with emetine over a period of days did not demonstrate impaired oxidative phosphorylation nor did 1 mM emetine inhibit this process in isolated mitochondria (APPELT and HEIM, 1964). Oxygen uptake was decreased in homogenates prepared from chronically-poisoned rats when tested in the presence of citrate, malate or α-ketoglutarate. Liver homogenates, however, prepared from the same animals, readily oxidized all of these substrates (APPELT and HEIM, 1965). It is pertinent to note that the livers of emetine-treated animals accumulate much higher concentrations of emetine than does the heart (GIMBLE et al., 1948).

In the isolated, perfused rat heart, addition of emetine to the perfusate depressed myocardial metabolism of pyruvate and glucose and decreased oxygen uptake and CO_2 formation (BRINK et al., 1969).

WATKINS and GUESS (1968) observed that the effects of emetine on chick-heart cells in culture could be reversed by NADH but not by NAD or other coenzymes. These authors concluded that emetine acts on the respiratory chain by affecting the enzymatic oxidation of substrates, other than succinate, that are mediated by NAD.

Carbohydrate Metabolism. There is a significant decrease in the content of liver glycogen in emetine-treated animals due to a decrease in the rate of glycogen synthesis (DIAMANT, 1958). A significant decrease in phosphorylase and aldolase activities in liver homogenates was also recorded in this study. JORDA et al. (1958) and HANISCH et al. (1966b) have reported indirect evidence that suggests an inhibitory effect of emetine on hyaluronidase.

Structure-Activity Relationships

The availability of many stereoisomers and derivatives of the ipecac alkaloids (OPENSHAW *et al.*, 1969; TEITEL and BROSSI 1966 and previous papers in these series) made it possible to demarcate precisely the structure-activity requirements for the inhibition of protein synthesis by members of this series (GROLLMAN, 1966; GROLLMAN and JARKOVSKY, 1974). The (R) configuration at C-1′ and the secondary nitrogen atom at the 2′ position are essential requirements for activity (Fig. 1 and Table 1). These conclusions are based, in part, on the inactivity of the epimer with the (S) configuration at C-1′ (isoemetine) and the loss of activity by unsaturation at the 1′-2′ position (o-methylpsychotrine) or by substitution of the secondary nitrogen (N-methylemetine). Unsaturation at the 2-3 position (dehydroemetine) destroys the asymmetry at carbons 2 and 3 without loss of biological activity, but further oxidation to 1,2,3,4,5,11b-trisdehydroemetine creates a positive charge at the tertiary nitrogen atom and results in inactivation of the compound. The presence of a cis-ethyl side chain results in optimal activity but this group is not essential since noremetine, 3-propylemetine and even desethylemetine retain some biological activity. If the ethyl side chain is present as the 2–3 trans-isomer, activity is markedly reduced. Similarly, inactivity of the 11b epimer of dehydroemetine most probably results from steric influences.

SIEGEL *et al.* (1966) have tested various glutarimide antibiotic isomers of cycloheximide for their capacity to inhibit protein synthesis. Replacement of the imide hydrogen of cycloheximide by a methyl group, esterification of the hydroxyl with acetate or conversion of the ketone to an oxime, results in great diminution or complete loss of biological activity, suggesting the keto, hydroxyl and imide groups are involved in a three-point attachment of the glutarimide antibiotics to their receptor. GROLLMAN (1966) has suggested that the essential positions for biological activity in the glutarimide antibiotics have corresponding positions in the ipecac alkaloids. In addition, there are topochemical resemblances between cycloheximide and the part of the emetine molecule containing the essential positions for biological activity (Fig. 5).

Interpretations of a necessarily tentative character define the possible interactions of the active ipecac alkaloids and glutarimide antibiotics with their biological receptor (enzyme). One postulated receptor site would presumably bind the secondary nitrogen of emetine and the hydroxyl of cycloheximide. The loss of activity which accompanies the replacement of hydrogens at these positions with methyl or acetyl groups is consistent with the view that these functional groups are involved in hydrogen-bonding to the receptor. Although an intramolecularly hydrogen-bonded conformation appears to be favored in cycloheximide, preferential bonding of the hydroxyl hydrogen to a receptor site is not precluded.

Additional testing will be required to define fully the receptor site corresponding to the tertiary nitrogen of emetine or the imide nitrogen of cycloheximide. The imide grouping of cycloheximide is rendered inactive by N-methylation. This observation implied that the hydrogen atom on the imide nitrogen is involved in bonding since the two adjacent carbonyl groups effectively prevent bonding through the π-electrons. If the tertiary nitrogen of emetine is hydrogen-bonded in a similar manner, the hydrogen atom must be supplied via the receptor. A

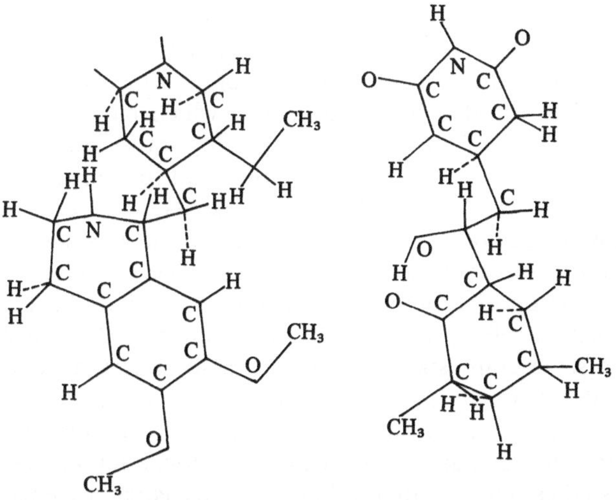

Fig. 5. Topological similarities between Dreiding models of cycloheximide (right) and a portion of the emetine molecule (left) (cf Fig. 1 and 2)

bifunctional site in the receptor that could protonate as well as accommodate a hydrogen atom, such as a hydroxyl or an imidazole group, would satisfy these requirements. Inactivity of trisdehydroemetine may result from repulsion by the positively charged nitrogen atom or from the effective removal of free electrons at this position. Steric effects also seem to be involved in the area of this nitrogen atom since reversal of configuration at the adjacent 11 b position ((+)-dehydro-isoemetine) creates significant hindrance at the lower face of the molecule and is associated with a corresponding decrease in biological activity (BROSSI *et al.*, 1962).

Although C-1′ of emetine and the asymmetric carbon of the side chain of cycloheximide may be involved in binding to the receptor, it is probable that they serve instead to fix the spatial position of the secondary nitrogen atom of emetine and the hydroxyl group of cycloheximide. The asymmetric carbon at C-6 of cycloheximide would fulfill a similar function since, in order for the quasi ring of cycloheximide to correspond to the ring of emetine, the configuration at C-6 must be such that the side chain is equatorial to the cyclohexanone ring (GROLLMAN, 1966).

Cycloheximide acts reversibly in the above biochemical reactions while the effects of emetine are irreversible. Among the glutarimide antibiotics, where extensive structure-activity relationships have been established, reversibility is associated with the presence of an oxygen atom at a position other than those involved in the inhibition of protein synthesis. The action of streptovitacin A and acetoxycycloheximide, which differ from cycloheximide only in having an equatorial hydroxyl or acetoxy substitution in place of the methyl group at the C-4 position, resembles emetine in being partially or totally irreversible. It appears that the property of irreversibility may be conferred by a secondary binding site which is not essential for inhibition of protein synthesis.

The usefulness of the foregoing analysis of structure-function relationships in drug design is illustrated by the degree to which biological activity can be predicted on the basis of structure in nonanalogous series. One example may be cited: a postulated structure common to the glutarimide antibiotics and ipecac alkaloids (I, Fig. 2), which is found in the indole alkaloid, tubulosine. Although biological activity had not been reported for tubulosine, it contained the topochemical requirements of structure I. Biochemical studies (GROLLMAN, 1967b) established that the action of tubulosine is: 1. species specific, being active against certain mammalian cells, protozoa and yeast but inactive against preparations of bacteria; 2. structurally specific, requiring a secondary nitrogen atom at the 2'-position and the (R) configuration at the 1'-carbon for activity; 3. selective, as RNA synthesis is unaffected at concentrations of tubulosine which totally inhibit protein synthesis; and 4. exerted during elongation of the peptide chain. Subsequent studies showed that tubulosine was equal to emetine in amoebicidal activity against several strains of *E. histolytica* (ENTNER and GROLLMAN, 1973).

Determination of precise structural requirements for inhibition of protein biosynthesis has implications for drug design (GROLLMAN, 1971; GROLLMAN et al., 1971a). The assumption that spatial position of the two nitrogen atoms of emetine was a major determinant of biological activity led to the preparation of a number of diamines, some of which have many times the amoebicidal activity of emetine *in vitro* (HALL et al., 1950; BERBERIAN et al., 1961). However, the stereochemistry of the carbon corresponding to C-1' of emetine was not considered in formulating the structure-activity relationships of these compounds, and their inactivity as inhibitors of protein synthesis is not surprising.

Mechanisms of Emetine Action and Toxicity

The profound metabolic disturbance created by inhibition of protein synthesis may well account for most of the therapeutic and toxic actions of emetine; this hypothesis will now be considered in light of the available data.

Amoebicide. As shown in Table 1, the amoebicidal properties of emetine and related ipecac alkaloids correlate with their capacity to inhibit protein synthesis in eukaryotes; this action may underly their amoebicidal activity. That inhibition of protein synthesis in *E. histolytica* represents a potent mechanism of amoebicidal action is supported by studies of direct-acting amoebicides (ENTNER and GROLLMAN, 1973). Cycloheximide, anisomycin, puromycin, tubulosine, acriflavine, diodoquin and tylocrebrine rapidly inhibit protein synthesis at the minimal concentration that kills amoebae. Such inhibition of protein synthesis could be secondary to effects on other metabolic pathways; however, several of the aforementioned drugs are known to act directly on protein synthesis in other eukaryotes (cf GROLLMAN and HUANG, 1973). Antibiotics, such as streptomycin, which inhibits protein synthesis in prokaryotes but not eukaryotes (cf PESTKA, 1971), have no direct amoebicidal activity against *E. histolytica*.

Table 1. *Specificity of ipecac alkaloids and other compounds as amebicides and as inhibitors of protein biosynthesis in rabbit reticulocytes* (ENTNER and GROLLMAN, 1973)

Inhibitor	Relative amebicidal activity	Concentration (μM) required for 50% inhibition of protein synthesis	
		Intact cells[a]	Lysates[b]
Ipecac Alkaloids			
(−)-Emetine	100	0.7	3
(−) Dehydroemetine	100	0.4	4
(−) Noremetine	20	20	30
(+) o-methylpsychotrine	1	300	800
(−) N-methylemetine	1	100	700
(−) Isoemetine	1	12	200
(±) Trisdehydroemetine	1	1,000	650

[a] Determined in intact reticulocytes.
[b] Determined in reticulocyte lysates.

Antitumor Agent. All inhibitors of protein synthesis in eukaryotes possess some degree of antitumor activity *in vivo.* Among the ipecac alkaloids, the structural requirements for inhibition of protein synthesis (GROLLMAN and JARKOVSKY, 1974) correspond to their relative cytotoxicity and to their effects on experimental leukemia in rats (JONDORF *et al.,* 1971). The chemotherapeutic usefulness of inhibitors of protein synthesis has always been limited by their cytotoxic action; nevertheless, despite relative toxicity, emetine has been successfully used for over three hundred years in the treatment of amoebiasis.

Emetic Properties. Emetine, in the form of ipecac, is widely used in the therapy of croup and for acute poisonings. The central action of the drug and its isomers on the medulla in dogs does not correspond to their ability to inhibit protein synthesis (S.C. WANG, personal communication) and must involve another cellular mechanism.

Antiviral. Inhibitors of protein synthesis in animal cells manifest antiviral activity in cell culture and, to some extent, in experimental infections in animals. The formation of enzymes utilized by the virus in synthesizing viral products is inhibited by these drugs; furthermore, protein synthesis is required for the concurrent synthesis of viral RNA. Although selective antiviral effects have been achieved in experimental animals and, possibly, in man, protein synthesis in the host cell is affected by the concentrations of drug required to inhibit virus replication. Accordingly, the therapeutic use of emetine as an antiviral agent may be associated with toxicity.

Cardiotoxicity. The most important adverse effect limiting the clinical use of emetine is its cardiotoxic action. At first glance, this observation is surprising, considering that turnover of protein in cardiac muscle is relatively slow. However, another cardiotoxin encountered in clinical practice, diphtheria toxin, is also a relatively specific inhibitor of protein biosynthesis in animal cells (GILL *et al.,* 1973). Like emetine, diphtheria toxin has been shown to act on one of the chain-elongation reactions (HONJO *et al.,* 1968).

There are several ways in which myocardial toxicity induced by emetine could be mediated through its effects on protein synthesis. The drug may interfere with the synthesis of a rapidly turning-over protein constituent of the cardiac cell membrane which functions as a carrier in membrane transport. Evidence for such proteins in cell membranes has been reviewed by OXENDER (1972). In addition, emetine might inhibit cellular oxidation by interfering with synthesis of enzymes involved in cellular respiration. Low concentrations of emetine do not interfere with substrate oxidation *in vitro,* but treatment of animals with low doses of emetine decreases oxygen uptake by the myocardium (APPELT and HEIM 1964; 1965). The observed inhibition (which could only be achieved at high concentrations *in vitro*) may be due to diminished activity of the enzymes involved in respiration. Alternatively, emetine intoxication might cause inhibition of protein synthesis and interfere with a slowly-synthesized contractile protein. The experiments of BELLER (1968) suggest that synthesis of actomyosin, although slow, is significantly impaired by low concentrations of emetine.

In rabbits, a structural abnormality induced by emetine is found in the mitochondria of the myocardium (PEARCE *et al.,* 1971). It seems unlikely that the low concentrations of drug reaching the myocardial cell (GIMBLE *et al.,* 1948) would directly affect mitochondrial protein synthesis, a process that appears to be relatively resistant to inhibition by emetine (PERLMAN and PENMAN, 1970; MAHLER *et al.,* 1971). However, some of the proteins required for mitochondrial function may be synthesized on cytoplasmic ribosomes.

Myopathy. The skeletal myopathy induced by emetine bears certain similarities to the cardiac myopathy produced by the alkaloid. Mitochondrial degeneration is confined to regions of decreased mitochondrial enzyme activity and was observed only in red muscle (DUANE and ENGEL, 1970). Further experimentation will be required before the mechanism of emetine-induced myopathy can be unequivocally attributed to inhibition of protein synthesis.

Summary and Conclusions

Emetine is the principal member of a family of alkaloids isolated from the root of the ipecacuanha plant. In the form of ipecac, emetine has been used therapeutically for over three centuries; presently, the drug is primarily employed in the treatment of amoebic dysentery but has also been used to treat other protozoal infections in man and as an antiviral and antitumor agent.

The toxic properties of emetine in animals and man include emetic effects, cardiotoxicity and a drug-induced myopathy. Selective lesions have been observed in mitochondria of muscle and heart tissue in emetine-treated animals.

Emetine and certain other ipecac alkaloids irreversibly inhibit cytoplasmic protein synthesis in eukaryotes. The drug acts by preventing movement of messenger RNA along the ribosome; thus, inhibiting peptide chain elongation. Inhibition of protein synthesis accounts for most of the toxic and therapeutic properties of emetine, however, higher concentrations of the drug affect other biochemical processes in animal cells, including oxygen uptake and biosynthesis of ribosomal RNA.

Acknowledgements. The experimental studies conducted by one of the authors of this chapter (A. P. G.) were supported by research grants from the National Institutes of Health, The American Cancer Society and Hoffman-La Roche.

References

ABD-RABBO, H.: Dehydroemetine in chronic leukemia. Lancet **1966-I**, 1161.

ABD-RABBO, H.: Chemotherapy of neoplasia (cancer) with dehydroemetine. J. Trop. Med. Hyg. **72**, 287 (1969).

APPELT, G.D., and H.C. HEIM: Effect of chronic poisoning by emetine on oxidative processes in rat heart: I. Effects on lipid metabolism and oxidative phosphorylation. J. Pharm. Sci. **53**, 1080 (1964).

APPELT, G.D., and H.C. HEIM: Effect of chronic poisoning by emetine on oxidative processes in rat heart: II. Effect on oxidation of citric acid cycle intermediates and nicotinamide adenine dinucleotide metabolism. J. Pharm. Sci. **54**, 1621 (1965).

BATTERSBY, A.R., R. BINKS, and T.P. EDWARDS: Ipecacuanha Alkaloids VI. The absolute stereo-chemistry at position 1 of emetine by chemical correlation with the natural amino acids. J. Chem. Soc. **1960**, 3474.

BELLER, B.M.: Observations on the mechanism of emetine poisoning of myocardial tissue. Circulation Res. **22**, 501 (1968).

BERBERIAN, D.A., R.G. SLIGHTER, and A.R. SURREY: *In vitro* and *in vivo* amebicidal activity of N, N′-bis (dichloroacetyl) diamines. Antibiot. & Chemotherapy **11**, 245 (1961).

BRINK, A.J., J.C.N. KOTZE, S.P. MULLER, and A. LOCHNER: The effect of emetine on metabolism and contractility of the isolated rat heart. J. Pharmacol. Exptl. Therap. **165**, 251 (1969).

BROSSI, A., M. BAUMANN, F. BURKHARDT, R. RICHLE u. J.R. FREY: Syntheseversuche in der Emetin-reihe. 9. Die absolute Konfiguration von (-)-2-Dehydro-Emetin. Helv. Chim. Acta **45**, 2219 (1962).

DAVIS, B., M.G. DODDS, and E.G. TOMICH: Spectrophotofluorometric determination of emetine in animal tissues. J. Pharm. Pharmacol. **14**, 249 (1962).

DEITRICH, R.A., and H.C. HEIM: The effect of emetine upon rat heart respiration. J. Am. Pharm. Assoc. **45**, 562 (1956).

DEL PUERTO, B.M., J.C. TATO, A. KOLTAN, O.M. BURES, P.R. DE CHIERI, A. GARCIA, T.I. ESCARAY y B. LORENZO: Hepatitis viral en el nino con especial referencia a su tratamiento con emetina. Prensa Med. Arg. **55**, 818 (1968).

DIAMANT, E.J.: Carbohydrate metabolism in emetine-poisoned rats. J. Pharmacol. Exptl. Therap. **122**, 465 (1958).

DIAMOND, L.S.: Axenic cultivation of *Entamoeba histolytica.* Science **134**, 336 (1961).

DUANE, D.D., and A.G. ENGEL: Emetine myopathy. Neurology **20**, 733 (1970).

ENTNER, N., L.A. EVANS, and C. GONZALEZ: Genetics of *Entamoeba histolytica:* Differences in drug sensitivity between Laredo and other strains of *Entamoeba histolytica.* J. Protozool. **9**, 466 (1962).

ENTNER, N., and A.P. GROLLMAN: Inhibition of protein synthesis: A mechanism of amebicide action of emetine and other structurally-related compounds. J. Protozool. **20**, 160 (1973).

ENTNER, N., and H. MOST: Genetics of *Entamoeba:* Characterization of two new parasitic strains which grow at room temperature (and at 37° C). J. Protozool. **12**, 10 (1965).

EVSTIGNEEVA, R.P., and N.A. PREOBRAZHENSKII: Synthesis of ipecacuanha alkaloids. Tetrahedron **4**, 223 (1958).

FAUST, E.C.: The multiple facets of *Entamoeba histolytica* infection. Intern. Rev. Trop. Med. **1**, 43 (1960).

GILEAD, Z., and Y. BECKER: Effect of emetine on ribonucleic acid biosynthesis in HeLa cells. Eur. J. Biochem. **23**, 143 (1971).

GILL, D.M., A.M. PAPPENHEIMER, JR., and T. UCHIDA: Diphtheria toxin, protein synthesis and the cell. Federation Proc. **32**, 1508 (1973).

GIMBLE, A.I., C. DAVISON, and P.K. SMITH: Studies on the toxicity, distribution and excretion of emetine. J. Pharmacol. Eptl. Therap. **94**, 431 (1948).

GOLDBERG, I.H., M.L. STEWART, M. AYUSO, and L. KAPPEN: On the mechanisms of inhibition of polypeptide synthesis by the antibiotics sparsomycin and pactamycin. Federation Proc. **32**, 1688 (1973).

GROLLMAN, A.I.: Emetine in the treatment of intra-abdominal and retroperitoneal nonspecific granulomas. Surg. Gynecol. Obstet. **120**, 792 (1965).

Grollman, A.P.: Structural basis for inhibition of protein synthesis by emetine and cycloheximide based on an analogy between ipecac alkaloids and glutarimide antibiotics. Proc. Natl. Acad. Sci. US **56**, 1867 (1966).

Grollman, A.P.: Mode of action of anisomycin. J. Biol. Chem. **242**, 3226 (1967a).

Grollman, A.P.: Structural basis for the inhibition of protein biosynthesis: Mode of action of tubulosine. Science **157**, 84 (1967b).

Grollman, A.P.: Inhibitors of protein biosynthesis. V. Effects of emetine on protein and nucleic acid biosynthesis in HeLa cells. J. Biol. Chem. **243**, 4089 (1968).

Grollman, A.P.: Inhibition of protein biosynthesis: Its significance in drug design. In: Medicinal chemistry, vol. II, Drug design (Ariens, E.J., ed.). New York: Academic Press 1971.

Grollman, A.P., and M.T. Huang: Inhibitors of protein synthesis in eukaryotes: tools in cell research. Federation Proc. **32**, 1673 (1973).

Grollman, A.P., and Z. Jarkovsky: Structure-activity relationships of the ipecac alkaloids. unpublished data.

Grollman, A.P., S. Rosen, and G. Hite: Potential inhibitors of protein synthesis. J. Med. Chem. **14**, 885 (1971a).

Grunberg, E., and H.N. Prince: Antiviral activity of emetine, 2-dehydroemetine and 2-dehydro-3-noremetine. Antimicrobial Agents Chemotherapy, 527 (1966).

Hall, D.M., S. Mahboob, and E.E. Turner: Structure and amebicidal activity Part I. Aliphatic diamines. J. Chem. Soc. **1950**, 1842.

Hanisch, J., K. Jarfas i T. Orban: O stosowaniu emetyny w niektorych chorobach oczu. Klinika Oczna **36**, 565 (1966a).

Hanisch, J., G. Vajda, and I. Bertha: The mode of action of emetine. Acta Chir. Acad. Sci. Hung. **7**, 51 (1966b).

Honjo, T., Y. Nishizuka, O. Hayaishi, and I. Kato: Diphtheria toxin-dependent adenoside diphosphate ribosylation of aminoacyl transferase II and inhibition of protein synthesis. J. Biol. Chem. **243**, 3553 (1968).

Isaka, H.: The effect of emetine hydrochloride upon the Yoshida sarcoma gann **41**, 165 (1950).

Janot, M.M.: The ipecac alkaloids In: The alkaloids, vol. III (Manske, R.H.F., ed.) New York: Academic Press 1953.

Jondorf, W.R., B.J. Abbott, N.H. Greenberg, and J.A.R. Mead: Increased lifespan of leukemic mice treated with drugs related to (−) emetine. Chemotherapy **16**, 109 (1971).

Jondorf, W.R., J.D. Drassner, R.K. Johnson, and H.H. Miller: Effect of various compounds related to emetine on hepatic protein synthesis in the rat. Arch. Biochem. Biophys. **131**, 163 (1969).

Jondorf, W.R., and D. Szapary: Enhanced protein synthesis at the liver microsomal level in emetine-pretreated rats. Arch. Biochem. Biophys. **126**, 892 (1968).

Jorda, V.V., J. Lenfeld u. L. Rothschild: Zur Frage der Wirkung des Emetins bei Herpes zoster. Z. Ges. Inn. Med. Ihre Grenzgebiete **13**, 71 (1958).

Klatskin, G., and H. Friedman: Emetine toxicity in man: Studies on the nature of early toxic manifestations, their relation to the dose level and their significance in determining safe dosage. Ann. Internal. Med. **28**, 892 (1948).

Lewisohn, R.: Action of emetine on malignant tumors. J. Am. Med. Assoc. **70**, 9 (1918).

Lloyd, J.U.: Ipecacuanha. In: Origin and history of all the pharmacopeial vegetable drugs, chemicals and preparations, vol. 1, p. 168. Cincinnati: Caxton Press. 1921.

Mahler, H.R., L.R. Jones, and W.J. Moore: Mitochondrial contribution to protein synthesis in cerebral cortex. Biochem. Biophys. Res. Commun. **42**, 384 (1971).

McKeehan, W., and B. Hardesty: The mechanism of cycloheximide inhibition of protein synthesis in rabbit reticulocytes. Biochem. Biophys. Res. Commun. **36**, 625 (1969).

Openshaw, H.T.: The ipecacuanha alkaloids. In: Chemistry of the alkaloids (Pelletier, S.W., ed.) New York: Reinhold Book Co. 1970.

Openshaw, H.T., N.C. Robson, and N. Whittaker: The synthesis of emetine and related compounds. Part X. The synthesis of emetine analogues including C(3)-bisnoremetine and C(3)-noremetine. Correlations of structure with amoebicidal activity. J. Chem. Soc. **1969**, 101.

Oxender, D.L.: Membrane transport. Ann. Rev. Biochem. **41**, 777 (1972).

Panettiere, F., and C.A. Coltman: Phase I experience with emetine hydrochloride (NSC 33669) as an antitumor agent. Cancer **27**, 835 (1971).

PARMER, L. G., and C. W. COTTRILL: Distribution of emetine in tissues. J. Lab. Clin. Med. **34**, 818 (1949).

PEARCE, M. B., R. T. BULLOCH, and M. L. MURPHY: Selective damage of myocardial mitochondria due to emetine hydrochloride. Arch. Pathol. **91**, 8 (1971).

PERLMAN, S., and S. PENMAN: Mitochondrial protein synthesis: Resistance to emetine and response to RNA synthesis inhibitors. Biochem. Biophys. Res. Commun. **40**, 941 (1970).

PESTKA, S.: Inhibitors of ribosome functions. Ann. Rev. Microbiol. **25**, 487 (1971).

PURCHAS, S.: His pilgrimes, vol. IV, London 1311, 1625–26.

ROGERS, L.: The rapid cure of amoebic dysentery and hepatitis by hypodermic injections of soluble salts of emetine. Brit. Med. J. **1912**, 1424.

SIEGEL, M. R., H. D. SISLER, and F. JOHNSON: Relationship of structure to fungitoxicity of cyclohexi-mide and related glutarimide derivatives. Biochem. Pharmacol. **15**, 1213 (1966).

STAUB, H.: En jahrhundert chemischer forschung uber ipecacuanha-alkaloide. Ph. D. Dissertation. Univ. of Zurich (1927).

SZANTAY, Cs.: Structure and synthesis of ipecac alkaloids. In: Recent developments in the chemistry of natural carbon compounds, vol. 2, p. 65. Budapest, Hungary: Akadémiai Kiadó 1967.

TEITEL, S., and A. BROSSI: Synthesis in the emetine series XIII: Structure and synthesis of psychotrine and 6'-O-methyl-7'-desmethylpsychotrine. J. Am. Chem. Soc. **88**, 4068 (1966).

VAN HOOSEN, B.: Emetin hydrochlorid in malignancy. Womens Med. J. **29**, 101 (1919).

VAN TAMELEN, E. E., P. E. ALDRICH, and J. B. HESTER, JR.: The stereochemistry of the ipecac alkaloids. J. Am. Chem. Soc. **81**, 6214 (1959).

VEDDER, E. B.: Origin and present status of the emetin treatment of amebic dysentery. J. Am. Med. Assoc. **62**, 501 (1914).

VIDAL, J.: Hospital **40**, 305 (1952).

WANG, S. C.: Emetic and antiemetic drugs. In: Physiological pharmacology, vol. II, p. 256 (ROOT, W. S. and F. G. HOFMAN, eds.). New York: Academic Press 1965.

WATKINS, W. D., and W. L. GUESS: Toxicity of emetine to isolated embryonic chick-heart cells. J. Pharm. Sci. **57**, 1968 (1968).

WENZEL, D. G.: Drug-induced cardiomyopathies. J. Pharm. Sci. **56**, 1209 (1967).

WOOLFE, G.: Chemotherapy of amebiasis. In: Experimental Chemotherapy I, 355 (SCHNITZER, R. S. and F. HAWKING, eds.). Academic Press N. Y., (1963).

WYBURN-MASON, R.: Dehydroemetine in chronic leukemia. Lancet **1966-I**, 1266.

Fusidic Acid

Nobuo Tanaka

Fusidic acid, $C_{31}H_{48}O_6$, was originally isolated from *Fusidium coccineum* (GODTFREDSEN *et al.*, 1962) and subsequently shown to be identical with ramycin, obtained from several *Cephalosporia* and a *Phycomycete* (VANDERHAEGHE *et al.*, 1965). It is an antibacterial terpenoid acid having protolanostane skeleton with the structure illustrated in Fig. 1 (GODTFREDSEN and VANGEDAL, 1962; ARIGONI *et al.*, 1963; GODTFREDSEN *et al.*, 1965).

Cephalosporin P produced by *Cephalosporium sp.* and helvolic acid produced by *Aspergillus fumigatus* are also steroidal antibiotics. Cephalosporin P consists of one major component P_1 and at least four minor components: P_2, P_3, P_4 and P_5. Fusidic acid, helvolic acid, and cephalosporin P_1 are three major antibacterial steroidal antibiotics and they are chemically and biosynthetically related. Each contains a tetracyclic ring with α, β-unsaturated carboxylic acid side chain and 16,21-cis oriented acetoxyl group on carbon 16 (Fig. 1).

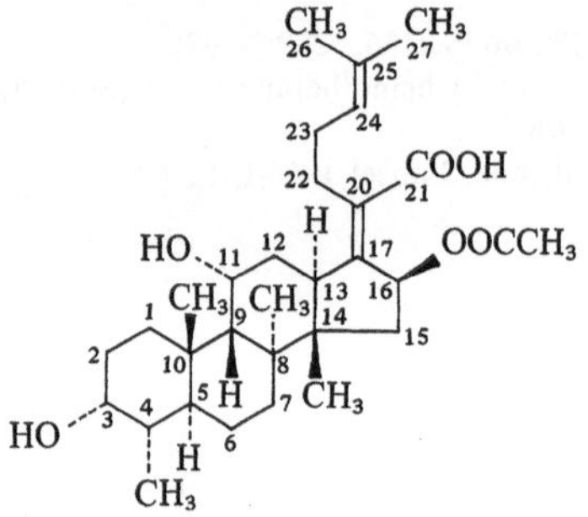

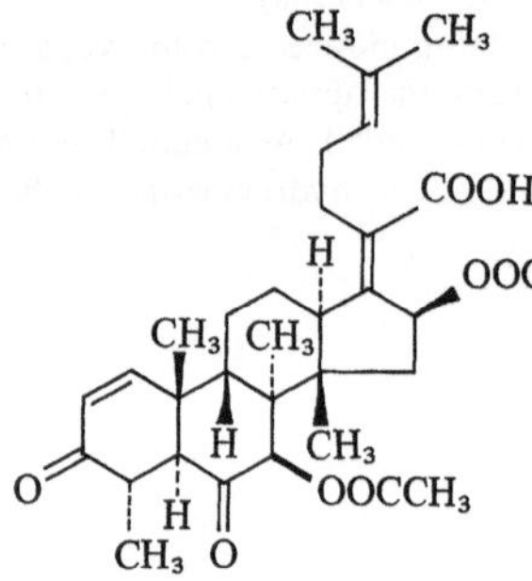

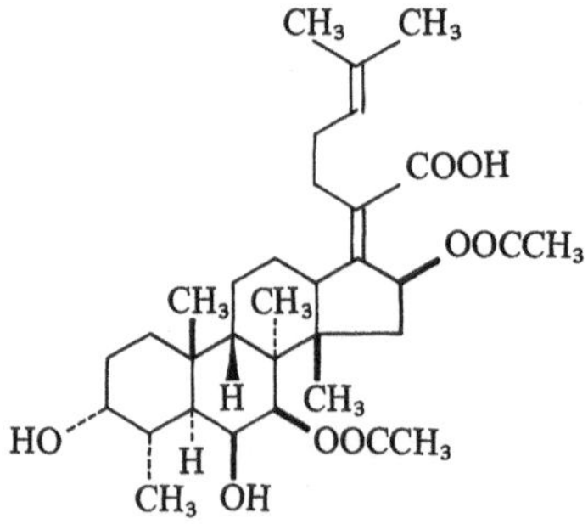

Fig. 1. The chemical structures of steroidal antibiotics. (I) Fusidic acid. (II) Helvolic acid. (III) Cephalosporin P_1. Helvolinic acid is 7-deacetyl-helvolic acid

Fusidic acid is active against Gram-positive bacteria but has no significant activity against the Gram-negative organisms and fungi. The minimal growth-inhibitory concentrations for *Staphylococcus aureus*, *Corynebacterium diphtheriae* and *Clostridium tetani* are in the range of 0.02–0.2 µg/ml, and that for *Mycobacterium tuberculosis* is 0.8 µg/ml. Cephalosporin P_1 and helvolic acid are about one-tenth as active as fusidic acid. *Styphylococcus aureus* rapidly acquires resistance to the three steroidal antibiotics, and there is a complete cross-resistance among them (GODTFREDSEN *et al.*, 1962; BARBER and WATERWORTH, 1962).

Fusidic acid exhibits a low toxicity, LD_{50} being 200 mg/kg (i.v., mice). It is used against Gram-positive infections by topical application or systemic administration. Oral administration of 500 mg to adult humans leads to highest blood level after an hour or two (mean 28 µg/ml). There is a slight excretion in the urine and the antibiotic is mainly excreted through the bile.

Structure-Activity Relationship

More than 50 derivatives of fusidic and helvolic acids have been examined for their antimicrobial activity (GODTFREDSEN *et al.*, 1966; JANSSEN and VANDER-HAEGHE, 1967). Whereas 24,25-Dihydro fusidic is as active as fusidic acid, the 17, 20, 24,25-tetrahydro derivative shows little activity. This would indicate that a 17,20-double bond is essential but a 24,25-double bond is dispensable for biological activity. The deacetyl lactone and 21-methyl derivatives lack activity, suggesting that the free carboxyl group at C-20 is also essential. Although the 16-acetoxyl group is dispensable, replacement of this with an hydroxyl decreases the antibacterial activity.

In summary, the presence of a free carboxyl group at C-20, the 17,20-double bond as well as its stereochemistry (16,21-cis) seem to be essential for activity. The 24,25-double bond and the terminal three carbon side chain are dispensable.

A similar structure-activity relationship has been demonstrated with respect to the capacity for inhibiting the function of elongation factor EF G in cell-free systems: polypeptide synthesis, ribosome-dependent GTPase activity, or stabilization of ribosome-EF G-GDP complex (Table 3, TANAKA *et al.*, 1968, 1969; BODLEY and GODTFREDSEN, 1972).

Inhibition of Protein Synthesis

Fusidic acid inhibits *in vivo* and *in vitro* protein synthesis in prokaryotic system (YAMAKI, 1965; HARVEY *et al.*, 1966) as well as in eukaryotic system (TANAKA *et al.*, 1969; MALKIN and LIPMANN, 1969; TANAKA *et al.*, 1970). The antibiotic neither affects the synthesis of aminoacyl tRNA nor the formation of aminoacyl-tRNA-messenger-ribosome complex but it inhibits amino acid transfer from aminoacyl-tRNA to protein on the ribosomes.

As shall be described below, it is well established that fusidic acid, in common with other related steroidal antibiotics, is a selective inhibitor of the elongation factor EF G (prokaryotic) or EF 2 (eukaryotic).

The Interaction of Fusidic Acid with Elongation Factor EF G. The ribosome-dependent GTPase activity of EF G is inhibited by the steroidal antibiotics,

Table 1. *Effect of antibiotics on ribosome-dependent GTPase activity of elongation factor EF G*

Antibiotics		GTP hydrolyzed	
(µM)		mµmoles	% of control
Control		4.75	100
Blasticidin S	50	4.66	98
Mikamycin B	120	4.94	104
Bottromycin A$_2$	120	4.37	92
Thiophenicol	300	4.42	93
Chloramphenicol	310	4.99	105
Erythromycin	70	4.97	105
Fusidic acid	18.5	1.47	31
Helvolinic acid	90	1.66	35

(TANAKA *et al.* (1968, 1969).

Table 2. *Effect of fusidic acid on ribosome-dependent GTPase activity of elongation factor EF G in the presence and absence of tRNA, phenylalanyl-tRNA or poly U*

Additions		GTP hydrolyzed (mµmoles)		
(µg/ml)		minus Fus[a]	plus Fus[a]	+ Fus/ − Fus[a]
None		5.16	1.85	0.36
tRNA	500	5.93	2.02	0.34
Phe-tRNA	500	5.75	2.08	0.36
Poly U	40	6.98	2.23	0.32
Phe-tRNA	500+poly U 40	8.25	2.89	0.35

TANAKA *et al.* (1969).
[a] Fus: fusidic acid 18.5 µM.

fusidic and helvolinic acids, but not by other antibiotics like blasticidin S, mikamy-cins A and B, bottromycin A$_2$, chloramphenicol, and erythromycin (Table 1) (TANAKA *et al.*, 1968, 1969). It is also inhibited by thiostrepton-siomycin-thiopep-tin group of antibiotics, inhibitors of the 50 S ribosome subunit function (WEISB-LUM and DEMOHN, 1970; PESTKA, 1970; TANAKA, *et al.*, 1970; MODOLELL *et al.*, 1971; BODLEY *et al.*, 1970; KINOSHITA *et al.*, 1971). The ribosome-dependent GTPase activity of EF G is stimulated by the addition of tRNA, phenylalanyl-tRNA, and poly U. Fusidic acid inhibits the GTP split reaction to the same extent in the presence and absence of these factors (Table 2)

Fusidic acid inhibits polyphenylalanine synthesis and ribosome-dependent GTPase activity of EF G to the same extent. As illustrated in Fig. 2, similar dose-inhibition curves are obtained for both the reactions. Approximately 65% inhibition is observed at a concentration of 18.5 µM fusidic acid. Various deriv-atives of fusidic acid on poly A-directed polylysine synthesis and on ribosome-dependent GTPase activity of EF G show parallel inhibitory effects (Table 3). The above data would indicate that the GTP hydrolysis reaction, inhibited by fusidic acid, is responsible for the protein synthesis inhibition (TANAKA *et al.*,

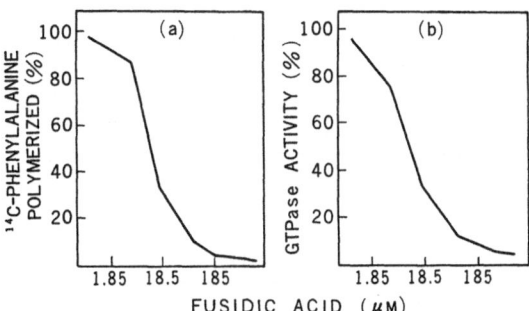

Fig. 2. Effect of fusidic acid on polyphenylalanine synthesis and on ribosome-dependent GTPase activity of elongation factor EF G. (TANAKA *et al.*, 1968, 1969)

Table 3. *Effect of various steroidal antibiotics on polylysine synthesis and on ribosome-dependent GTPase activity of elongation factor EF G*

Steroidal antibiotics (μM)		GTP hydrolyzed	Polylysine synthesized
Control		100	100
Fusidic acid	37	27	25
	185	9	8
Helvolinic acid	36	52	57
	180	22	40
Helvolic acid	36	20	32
	180	12	16
7-Propionylhelvolinic acid	37	12	15
	185	8	10
24,25-Dibromohelvolic acid	32	27	35
	160	13	22
3-Dihydrohelvolic acid-3-acetate	34	82	81
	170	52	48
Methylhelvolinate	37	101	97
	370	92	93

TANAKA *et al.* (1968, 1969).

1968, 1969). It has been further demonstrated by equilibrium dialysis that ^{3}H-labelled fusidic acid binds to EF G in a molar ratio of 1:1 (OKURA *et al.*, 1970, 1971).

Fusidic acid-resistant mutants of *E. coli* can be obtained by treatment with a nitrosoguanidine derivative in which the mutation has altered the EF G (KINOS-HITA *et al.*, 1968; TANAKA *et al.*, 1971). The GTPase activity exhibited with a combination of this EF G and ribosomes is resistant to the action of fusidic acid. If the ribosomes and EF G from sensitive cells and resistant mutants are mixed in various combinations, the effect of fusidic acid on the functioning of EF G-ribosome complex can be examined. The results obtained indicate that

Table 4. *Effect of fusidic acid on GTPase reaction and polyphenylalanine synthesis by ribosomes and EF G from sensitive and resistant cells of E. coli in various combinations*

Ribosomes	EF G	Inhibition by fusidic acid	
		GTP hydrolysis	poly-Phe synthesis
S	S	88%	88%
S	R	9	13
R	S	91	93
R	R	11	8

Kinoshita *et al.* (1968); Tanaka *et al.* (1968).

EF G *per se* is the site of action of fusidic acid, since the GTP split reaction and polyphenylalanine synthesis are inhibited only if the EF G from sensitive cells is involved in the assay system (Table 4). It has also been demonstrated that fusidic acid-resistance is localized in the EF G but not in EF T (Tocchini-Valentini *et al.*, 1969).

Elongation factor EF G from the resistant mutant shows a Km value of 3.5×10^{-4} M for GTP in the GTPase reaction, whereas Km of EF G from the sensitive cells is 7.0×10^{-5} M. The other physicochemical and immunological characteristics of the mutant EF G are the same as those of the sensitive EF G (Bernardi and Leder, 1970; Tanaka *et al.*, 1971).

The chromosomal location of a fusidic acid resistant marker (*fus*) has been studied by linkage in conjugation and transduction. It is located at minute 64.2, about 0.2 min. from the *strA* gene, lying near the *strA* and between the *strA* and *malA* loci (Tanaka *et al.*, 1971; Kuwano *et al.*, 1971).

Formation of a Stable Complex of Fusidic Acid-EF G-GDP-Ribosome. The ribosome and elongation factor EF G form a labile complex of EF G-GDP-ribosome in the presence of GTP or GDP (Brot *et al.*, 1969; Parmeggiani and Gottschalk, 1969). Fusidic acid stabilizes this ternary complex, which can be isolated by gel and millipore filtration, or by sucrose density gradient sedimentation.

$$\text{ribosome} + \text{EF G} + \text{GTP} \rightleftharpoons \underset{\text{(complex I)}}{\text{ribosome-EF G-GTP}} \tag{1}$$

$$\underset{\text{(complex I)}}{\text{ribosome-EF G-GTP}} \rightarrow \underset{\text{(complex II)}}{\text{ribosome-EF G-GDP} + \text{Pi}} \tag{2}$$

$$\text{ribosome-EF G-GDP} \underset{\text{Fus}}{\overset{}{\rightleftharpoons}} \text{ribosome} + \text{EF G} + \text{GDP} \tag{3}$$

Stoichiometric examination reveals a relatively stable ternary complex involving ribosomes, EF G, and GDP in a molar ratio of 1:1:1, assuming that the ribosomes are half active in this function. The rate and extent of GTP hydrolysis by a molar excess of ribosomes and EF G remain unaffected by fusidic acid. When GTP concentration level is increased to exceed the macromolecular components, there is a burst of fusidic acid-resistant hydrolysis equivalent to the molar amount of EF G when ribosomes are present in excess. With initial GTP concentrations as low as 10^{-8} M, essentially all of the nucleotide, as GDP, is in the form of a ternary complex with ribosomes and EF G, in the presence of fusidic acid. Furthermore, the binding of the reaction product, GDP, is stabilized by fusidic acid, and the preliminary binding of GDP prevents the subsequent binding of GTP. In order to account for the above observations, the following mechanism has been proposed for the uncoupled hydrolysis of GTP by the ribosome and EF G (BODLEY et al., 1969, 1970; BODLEY et al., 1970).

^{3}H-GTP binds to the 50S ribosomal subunit but not to the 30S subunit in the presence of EF G. EF G interacts with the 50S ribosomal subunit in the presence of GTP, forming EF G-nucleotide-50S complex, in which GTP is hydrolyzed to GDP and Pi. The complex of EF G and the 50S subunit with guanosine nucleotide is stabilized by fusidic acid (BODLEY and LIN, 1970).

^{3}H-Fusidic acid binds with elongation factor EF G in a molar ratio of 1:1 with an association constant 1.2×10^5 M^{-1}. The binding is strongly stimulated by ribosomes and GTP. Formation of fusidic acid-EF G-GDP-ribosome complex can be demonstrated by equilibrium dialysis and ultracentrifugal separation methods. Measurements of binding at equilibrium indicate a stoichiometric combination of the four substances in a molar ratio of 1:1:1:1, provided that half of the ribosomes employed are active in this function. The association constant of fusidic acid is 2.2×10^6 M^{-1}. Less binding of the antibiotic is observed when fusidic acid-resistant EF G is used. A significant binding of fusidic acid is seen when 70S ribosomes are replaced by the 50S ribosomal subunit or when GDP is used instead of GTP. Ki value for fusidic acid is 10^{-6} M in the GTPase reaction. This is in accordance with the association constant in the complex formation (OKURA et al., 1970, 1971).

GTP is hydrolyzed to GDP in the quartet complex: fusidic acid-EF G-nucleotide-ribosome. This indicates that fusidic acid inhibits the GTPase reaction by interfering with the dissociation of the complex but may not directly affect the hydrolysis itself (Fig. 3).

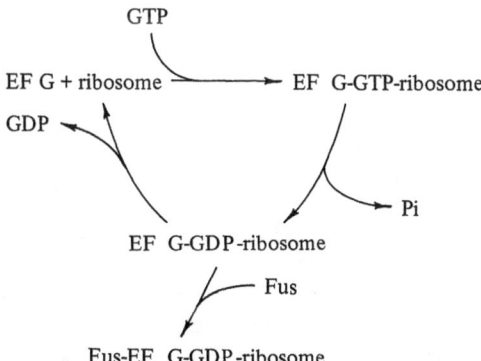

Fig. 3. Diagrammatic representation of mechanism of fusidic acid (Fus) inhibition of the GTPase reaction (OKURA et al., 1970, 1971)

Inhibition by Fusidic acid of Translocation. The movement of mRNA and
the simultaneous translocation of peptidyl-tRNA from the acceptor site to the
donor site on the ribosome are mediated by EF G and GTP (NISHIZUKA and
LIPMANN, 1966; HAENNI and LUCAS-LENARD, 1968; ERBE *et al.*, 1969). The effect
of fusidic acid on translocation has been examined for polyphenylalanyl-puromycin synthesis. The results, observed by sucrose density gradient centrifugation
analysis, are illustrated in Fig. 4. In the absence of EF G and GTP, the puromycin
reaction is not inhibited by fusidic acid. But the puomycin reaction enhanced
by EF G and GTP is inhibited by the antibiotic. Since polyphenylalanyl-tRNA
bound to the donor site is reactive to puromycin but polyphenylalanyl-tRNA
at the acceptor site is not reactive to puromycin, the above results indicate that
fusidic acid inhibits translocation of peptidyl-tRNA and mRNA from the acceptor
site to the donor site on the ribosome (TANAKA *et al.*, 1968, 1969). The inhibition
of translocation by fusidic acid has been confirmed further by the observations
that the antibiotic does not inhibit diPhe synthesis, but inhibits oligoPhe (Phe$_3 \geqq$)
synthesis or the formation of N-acetyl-diPhe-puromycin or N-acetyltriPhe-puro-
mycin (HAENNI and LUCAS-LENARD, 1968; PESTKA, 1968, 1969).

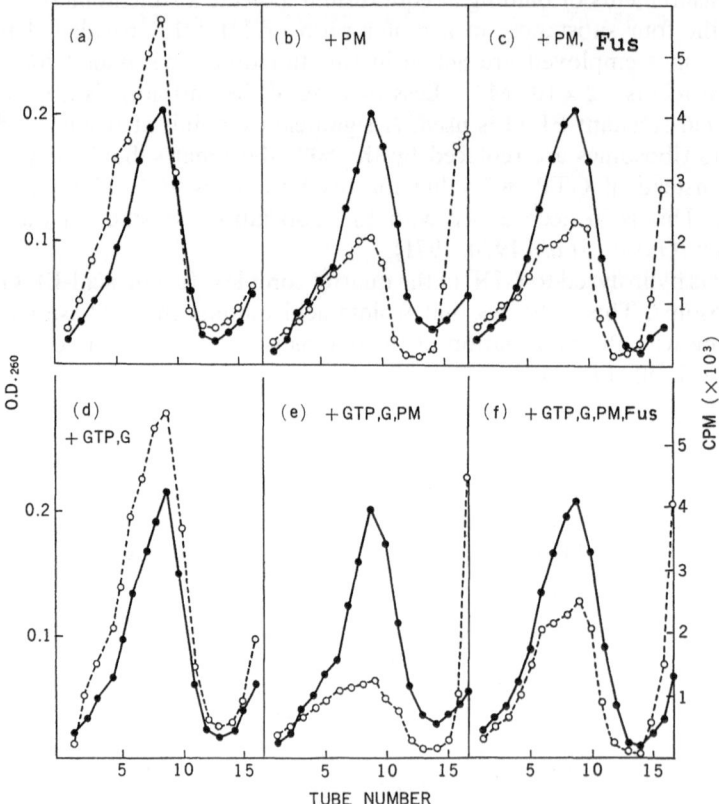

Fig. 4. Effect of fusidic acid (Fus) on puromycin-dependent release of peptide from the ribosomes
in the presence and absence of GTP and EF G. (TANAKA *et al.*, 1968, 1969). a Control. b With
puromycin 100 µM. c With puromycin 100 µM and fusidic acid 50 µM. d With EF G 15 µg/ml and
GTP 0.1 mM. e (d) + puromycin 100 µM. f (d) + puromycin 100 µM and fusidic acid 10 µM

Fusidic acid causes an interruption of polysome metabolism in *E. coli* cells. The ribosome cycle is apparently blocked and the pre-existing polysomes are preserved. This indicates that the antibiotic blocks the ribosomal movement along the mRNA strand (translocation) *in vivo* (GURGO et al., 1969).

The Inhibition of tRNA Release by Fusidic Acid. During protein synthesis, each time the peptide chain is elongated by the addition of one amino acid, free tRNA produced from peptidyl-tRNA is released from the donor site, simultaneously with translocation of mRNA and peptidyl-tRNA on the ribosome. The release of deacylated $tRNA^{Phe}$ (tRNA specific for phenylalanine) from the $tRNA^{Phe}$-poly U-ribosome complex is catalysed by EF G and GTP. Fusidic acid inhibits the release of $tRNA^{Phe}$, presumably by interacting with the EF G (KURIKI and KAJI, 1968; KAJI et al., 1969).

The EF G-catalyzed release of deacylated tRNA is inhibited by fusidic acid but not by bottromycin A_2. The ribosome-catalyzed translocations of peptidyl-tRNA, and mRNA which occur as the donor site is vacated, are inhibited by bottromycin A_2 but not by fusidic acid. This observation suggests a mechanism by which EF G primarily participates in the tRNA release and translocation *per se* is catalyzed by the 50S ribosomal subunit, which is the primary site of action of bottromycin A_2. The primary action of fusidic acid, an inhibitor of EF G, is the inhibition of tRNA release from the donor site and thus it prevents translocation (TANAKA et al., 1971). However, TANAKA and KAJI (1972) have observed that the presence of tRNA at the donor site is not necessary for the translocation by EF G and its inhibition by fusidic acid. This seems to support the assumption that fusidic acid inhibits translocation *per se*, resulting in blocking tRNA release from the donor site. The problem of whether the primary action of fusidic acid, i.e. the primary action of EF G, is tRNA release or translocation remains to be studied.

The Inhibition by Fusidic Acid of Aminoacyl-tRNA Binding to Ribosomes. The thiopeptin-siomycin-thiostrepton peptide antibiotics, which interact with the 50S ribosomal subunit, have been demonstrated to inhibit both EF Tu-associated function (fusidic acid-resistant GTPase and aminoacyl-tRNA binding) and EF G-associated function (fusidic acid-sensitive GTPase and translocation). This finding indicates that the antibiotic acts on a single site of the 50S ribosomal subunit, which participates in the interactions both with EF Tu and with EF G (KINOSHITA et al., 1971; MODOLELL et al., 1971).

The identity of the binding site of EF Tu-GTP-aminoacyl-tRNA (the Tu site) with the binding site of EF G (the G site) has been confirmed by further studies (RICHMAN and BODLEY, 1972; MILLER, 1972; WEISSBACH et al., 1972; RICHTER, 1972). A single or two acidic protein(s), required for the function of both EF G and EF Tu, have been isolated from the 50S ribosomal subunit and characterized (BROT et al., 1972; KISHA et al., 1972; SANDER et al., 1972). It has also been revealed that EF G, bound to the ribosome with fusidic acid, prevents the binding of EF Tu-GTP-aminoacyl-tRNA complex (CABRER et al., 1972).

In vivo, fusidic acid inhibits aminoacyl-tRNA binding more significantly than translocation (CUNDLIFFE, 1972). The antibiotic stabilizes polysomes and freezes peptidyl-tRNA in the donor site on the ribosome, indicating that it acts primarily as an inhibitor of aminoacyl-tRNA binding *in vivo* (Modolell and Davis, 1970; Pestka and Hintikka, 1971; Celma *et al.*, 1972).

The Effect of Fusidic Acid on Peptide Chain Initiation. Fusidic acid selectively inhibits at low concentration peptide chain elongation. The initiation complex (fMet-tRNA-f2 RNA-ribosome) formation is not significantly affected by fusidic acid (Tanaka *et al.*, 1971). However, at high concentration, fusidic acid interferes with peptide chain initiation as well as chain elongation. The antibiotic may act on the initiation factor IF 1 (Sala and Ciferri, 1970).

The Inhibition by Fusidic Acid of the Function of Eukaryotic Elongation Factor EF 2. Fusidic acid inhibits ribosome-dependent GTPase activity of EF 2 and translocation of mRNA and peptidyl-tRNA on the ribosomes in mammalian systems (Tanaka *et al.*, 1969, 1970; Malkin and Lipmann, 1969) and in yeast and algal systems (Tiboni and Ciferri, 1971). The ribosome-GDP-EF 2 complex is stabilized by fusidic acid in eukaryotic systems (Richter *et al.*, 1971). It has been reported that mitochondrial EF 2 from *Neurospora crassa* is resistant to fusidic acid (Grandi *et al.*, 1971) but mitochondrial EF 2 from yeast is sensitive to the antibiotic (Richter *et al.*, 1971).

Uncoupling of Oxidative Phosphorylation by Fusidic Acid

Fusidic, glycyrrhetic, and some related terpenoid acids uncouple oxidative phosphorylation: i.e. they inhibit the mitochondrial biosynthesis of ATP without interfering with mitochondrial respiration. Glycyrrhetic acid (100 μM) and poly-porenic acid A (200 μM) are fairly potent uncouplers of oxidative phosphorylation in rat liver mitochondria. A higher concentration of fusidic acid (600 μM) is needed for the same effect *in vitro*. Fusidic acid and related terpenoid acids appear to uncouple phosphorylation by interaction with key lysyl amino-groups participating in mitochondrial phosphorylation (Whitehouse *et al.*, 1967).

References

Arigioni, D., W. von Daehne, W.O. Godtfredsen, A. Marquet, and A. Melera: The location of the ring C hydroxyl group in fusidic acid. Experientia **19**, 521 (1963).

Barber, M., and P.M. Waterworth: Antibacterial activity *in vitro* of fucidin. Lancet **1962**, 931.

Bernardi, A., and P. Leder: Protein biosynthesis in *Escherichia coli*. Purification and characterization of a mutant G factor. J. Biol. Chem. **245**, 4263 (1970).

Bodley, J. W., and W. O. Godtfredsen: Studies on translocation. XI. Structure-function relationships of the fusidane-type antibiotics. Biochem. Biophys. Res. Commun. **46**, 871 (1972).

Bodley, J. W., and L. Lin: Interaction of *Escherichia coli* G factor with the 50S ribosomal subunit. Nature **227**, 60 (1970).

Bodley, J. W., L. Lin, and J. H. Highland: Studies on translocation. VI. Thiostrepton prevents the formation of a ribosome-G factor-guanine nucleotide complex. Biochem. Biophys. Res. Commun. **41**, 1406 (1970).

Bodley, J. W., F. J. Zieve, and L. Lin: Studies on translocation. IV. The hydrolysis of a single round of guanosine triphosphate in the presence of fusidic acid. J. Biol. Chem. **245**, 5662 (1970).

Bodley, J. W., F. J. Zieve, L. Lin, and S. T. Zieve: Formation of the ribosome-G factor-GDP complex in the presence of fusidic acid. Biochem. Biophys. Res. Commun. **37**, 437 (1969).

Bodley, J. W., F. J. Zieve, L. Lin, and S. T. Zieve: Studies on translocation. III. Conditions necessary for the formation and detection of a stable ribosome-G factor-guanosine diphosphate complex in the presence of fusidic acid. J. Biol. Chem. **245**, 5656 (1970).

BROT, N., C. SPEARS, and H. WEISSBACH: The formation of a complex containing ribosomes, transfer factor G and a guanosine nucleotide. Biochem. Biophys. Res. Commun. **34**, 843 (1969).

BROT, N., E. YAMASAKI, B. REDFIELD, and H. WEISSBACH: B. properties of an *E. coli* ribosomal protein required for the function of factor G. Arch. Biochem. Biophys. **148**, 148 (1972).

CABRER, B., D. VAZQUEZ, and J. MODOLELL: Inhibition by elongation factor EF G of aminoacyl-tRNA binding to ribosomes. Proc. Natl. Acad. Sci. U.S. **69**, 733 (1972).

CELMA, M.L., D. VAZQUEZ, and J. MODOLELL: Failure of fusidic acid and siomycin to block ribosomes in the pretranslocated state. Biochem. Biophys. Res. Commun. **48**, 1240 (1972).

CUNDLIFFE, E.: The mode of action of fusidic acid. Biochem. Biophys. Res. Commun. **46**, 1794 (1972).

ERBE, R.W., M.M. NAU, and P. LEDER: Translation and translocation of defined RNA messengers. J. Mol. Biol. **38**, 441 (1969).

GODTFREDSEN, W.O., W. VON DAEHNE, L. TYBRING, and S. VANGEDAL: Fusidic acid derivatives. I. Relationship between structure and antibacterial activity. J. Med. Chem. **9**, 15 (1966).

GODTFREDSEN, W.O., W. VON DAEHNE, S. VANGEDAL, M. MARQUET, D. ARIGONI, and A. MELERA: The stereochemistry of fusidic acid. Tetrahedron **21**, 3505 (1965).

GODTFREDSEN, W.O., S. JAHNSEN, H. LORCK, K. ROHOLT, and L. TYBRING: Fusidic acid, a new antibiotic. Nature **193**, 987 (1962).

GODTFREDSEN, W.O., and S. VANGEDAL: The structure of fusidic acid. Tetrahedron **18**, 1092 (1962).

GRANDI, M., A. HELMS, and H. KÜNTZEL: Fusidic acid resistance of mitochondrial G factor from *Neurospora crassa*. Biochem. Biophys. Res. Commun. **44**, 864 (1971).

GURGO, C., D. APIRION, and D. SCHLESSINGER: Effects of chloramphenicol and fusidic acid on polysome metabolism in *Escherichia coli*. FEBS Letters **3**, 34 (1969).

HAENNI, A.L., and J. LUCAS-LENARD: Stepwise synthesis of a tripeptide. Proc. Natl. Acad. Sci. U.S. **61**, 1363 (1968).

HARVEY, C.L., S.G. KNIGHT, and C.J. SIH: On the mode of action of fusidic acid. Biochemistry **5**, 3320 (1966).

JANSSEN, G., and H. VANDERHAEGHE: Modification of the side chain of fusidic acid (ramycin). J. Med. Chem. **10**, 205 (1967).

KAJI, A., K. IGARASHI, and H. ISHITSUKA: Interaction of tRNA with ribosomes — binding and release of tRNA. Cold Spring Harbor Symp. Quant. Biol. **34**, 167 (1969).

KINOSHITA, T., G. KAWANO, and N. TANAKA: Association of fusidic acid sensitivity with G factor in a protein-synthesizing system. Biochem. Biophys. Res. Commun. **33**, 769 (1968).

KINOSHITA, T., Y.-F. LIOU, and N. TANAKA: Inhibition by thiopeptin of ribosomal functions associated with T and G factors. Biochem. Biophys. Res. Commun. **44**, 859 (1971).

KISHA, K., W. MÖLLER, and G. STÖFFLER: Reconstitution of a GTPase activity by a 50S ribosomal protein from *E. coli*. Nature New Biol. **233**, 62 (1971).

KURIKI, Y., and A. KAJI: Factor and GTP dependent release of deacylated tRNA from 70S ribosomes. Proc. Natl. Acad. Sci. U.S. **61**, 1399 (1968).

KUWANO, M., D. SCHLESSINGER, G. RINALDI, L. FELICETTI, and G.P. TOCCHINI-VALENTINI: G factor mutants of *Escherichia coli*: Map location and properties. Biochem. Biophys. Res. Commun. **42**, 441 (1971).

MALKIN, M., and F. LIPMANN: Fusidic acid: Inhibition of factor T2 in reticulocyte protein synthesis. Science **164**, 71 (1969).

MILLER, D.L.: Elongation factors EF Tu and EF G interact at related sites on ribosomes. Proc. Natl. Acad. Sci. U.S. **69**, 752 (1972).

MODOLELL, J., B. CARBER, A. PARMEGGIANI, and D. VAZQUEZ: Inhibition by siomycin and thiostrepton of both aminoacyl-tRNA and factor G binding to ribosomes. Proc. Natl. Acad. Sci. U.S. **68**, 1796 (1971).

MODOLELL, J., and B.D. DAVIS: Significance of the effect of streptomycin and fusidic acid on polysome stability. Progr. Antimicr. Anticancer Chemoth. vol. II, p. 464. Tokyo: University of Tokyo Press 1970.

MODOLELL, J., D. VAZQUEZ, and R.E. MONRO: Ribosomes, G factor and siomycin. Nature New Biol. **230**, 109 (1971).

NISHIZUKA, Y., and F. LIPMANN: Comparison of guanosine triphosphate split and polypeptide synthesis with a purified *E. coli* system. Proc. Natl. Acad. Sci. U.S. **55**, 212 (1966).

NISHIZUKA, Y., and F. LIPMANN: The interrelationship between guanosine triphosphate and amino acid polymerization. Arch. Biochem. Biophys. **116**, 344 (1966).

Okura, A., T. Kinoshita, and N. Tanaka: Complex formation of fusidic acid with G factor, ribosome and guanosine nucleotide. Biochem. Biophys. Res. Commun. **41**, 1545 (1970).

Okura, A., T. Kinoshita, and N. Tanaka: Formation of fusidic acid-G factor-GDP-ribosome complex and the relationship to the inhibition of GTP hydrolysis. J. Antibiotics (Tokyo) **24**, 655 (1971).

Parmeggiani, A., and E. M. Gottschalk: Properties of the crystalline amino acid polymerization factors from *Escherichia coli*: binding of G to ribosomes. Biochem. Biophys. Res. Commun. **35**, 861 (1969).

Pestka, S.: Studies on the formation of transfer ribonucleic acid-ribosome complexes. V. On the function of a soluble transfer factor in protein synthesis. Proc. Natl. Acad. Sci. U.S. **61**, 726 (1968).

Pestka, S.: Studies on the formation of transfer ribonucleic acid-ribosome complexes. VI. Oligopeptide synthesis and translocation on ribosomes in the presence and absence of soluble transfer factors. J. Biol. Chem. **244**, 1533 (1969).

Pestka, S.: Studies on the formation of transfer ribonucleic acid-ribosome complexes. IX. Effect of antibiotics on translocation and peptide bond formation. Arch. Biochem. Biophys. **136**, 89 (1970).

Pestka, S.: Thiostreptone: A ribosomal inhibitor of translocation. Biochem. Biophys. Res. Commun. **40**, 667 (1970).

Pestka, S., and H. Hintikka: Studies on the formation of ribonucleic acid-ribosome complexes. XVI. Effect of ribosomal translocation inhibitors on polyribosomes. J. Biol. Chem. **246**, 7723 (1971).

Richman, N., and J. W. Bodley: Ribosomes cannot interact simultaneously with elongation factors EF Tu and EF G. Proc. Natl. Acad. Sci. U.S. **69**, 686 (1972).

Richter, D.: Inability of *E. coli* ribosomes to interact simultaneously with the bacterial elongation factors EF Tu and EF G, Biochem. Biophys. Res. Commun. **46**, 1850 (1972).

Richter, D., L. Lin, and J. W. Bodley: Studies on translocation. IX: The pattern of action of antibiotic translocation inhibitors in eukaryotic and prokaryotic systems. Arch. Biochem. Biophys. **147**, 186 (1971).

Sala, F., and O. Ciferri: Inhibition of peptide chain initiation in *E. coli* by fusidic acid. Biochim. Biophys. Acta **224**, 199 (1970).

Sander, G., R. C. Marsh, and A. Parmeggiani: Isolation and characterization of two acidic proteins from the 50S subunit required for GTPase activity of both EF G and T. Biochem. Biophys. Res. Commun. **47**, 866 (1972).

Tanaka, S., and A. Kaji: Does translocase (G-factor) require the presence of unesterified tRNA on the donor site for its action? -- The effect of fusidic acid. Biochem. Biophys. Res. Commun. **46**, 136 (1972).

Tanaka, N., G. Kawano, and T. Kinoshita: Chromosomal location of a fusidic acid resistant marker in *Escherichia coli*. Biochem. Biophys. Res. Commun. **42**, 564 (1971).

Tanaka, N., T. Kinoshita, and H. Masukawa: Mechanism of protein synthesis inhibition by fusidic acid and related antibiotics. Biochem. Biophys. Res. Commun. **30**, 278 (1968).

Tanaka, N., T. Kinoshita, and H. Masukawa: Mechanism of inhibition of protein synthesis by fusidic acid and related steroidal antibiotics. J. Biochem. (Tokyo) **65**, 459 (1969).

Tanaka, N., Y.-C. Lin, and A. Okuyama: Studies on translocation of fMet-tRNA and peptidyl-tRNA with antibiotics. Biochem. Biophys. Res. Commun. **44**, 477 (1971).

Tanaka, N., T. Nishimura, and T. Kinoshita: Inhibition by fusidic acid of transferase II in reticulocyte protein synthesis. J. Biochem. (Tokyo) **67**, 459 (1970).

Tanaka, N., T. Nishimura, T. Kinoshita, and H. Umezawa: The effect of fusidic acid on protein synthesis in a mammalian system. J. Antibiotics (Tokyo) **22**, 181 (1969).

Tanaka, K., S. Watanabe, H. Teraoka, and M. Tamaki: Effect of siomycin on protein synthesizing activity of *Escherichia coli* ribosomes. Biochem. Biophys. Res. Commun. **39**, 1189 (1970).

Tiboni, O., and O. Ciferri: Selective inhibition of the reactions catalyzed by ribosome-specific transfer factor G. FEBS Letters **19**, 174 (1971).

Tocchini-Valentini, G. P., L. Felicetti, and G. M. Rinaldi: Mutants of *Escherichia coli* blocked in protein synthesis: Mutants with an altered G factor. Cold Spring Harbor Symp. Quant. Biol. **34**, 463 (1969).

Vanderhaeghe, H., P. Van Dijck, and P. De Somer: Identity of ramycin with fusidic acid. Nature **205**, 710 (1965).

WEISBLUM, B., and V. DEMOHN: Thiostrepton, an inhibitor of 50S ribosome subunit function. J. Bacteriol. **101**, 1073 (1970).

WEISSBACH, H., B. REDFIELD, E. YAMASAKI, R.C. DAVIS, JR., S. PESTKA, and N. BROT: Studies on the ribosomal sites involved in factors Tu and G-dependent reactions. Arch. Biochem. Biophys. **149**, 110 (1972).

WHITEHOUSE, M.W., P.D.G. DEAN, and T.G. HALSALL: Uncoupling of oxidative phosphorylation by glycyrrhetic acid, fusidic acid and some related triterpenoid acids. J. Pharm. Pharmacol. **19**, 533 (1967).

YAMAKI, H.: Inhibition of protein synthesis by fusidic and helvolinic acids, steroidal antibiotics. J. Antibiotics (Tokyo). Ser. A **18**, 228 (1965).

Gougerotin

Munehiko Yukioka

Gougerotin is a water-soluble basic antibiotic obtained from culture filtrate of *Streptomyces gougerotii* No. 21544, isolated from a soil sample in Kyoto by workers at the Takeda Chemical Industries, Ltd., Osaka, Japan (KANZAKI *et al.*, 1962). Gougerotin is a broad spectrum but weak antibiotic and inhibits Gram-positive, Gram-negative bacteria and mycobacteria. Yeasts and fungi are not inhibited (Table 1). Gougerotin inhibits the growth of pseudorabies, vaccinia, Newcastle disease, fowl plague, and Western equine encephalomyelitis viruses (THIRY, 1968). The LD_{50} for mice is 57 mg/Kg by intravenous route (KANZAKI *et al.*, 1962).

Table 1. *Inhibitory levels of gougerotin* (KANZAKI *et al.*, 1962)

Organism	Detection limit (µg/ml) using agar streak method
E. coli	200
Prot. vulgaris	800
Staph. aureus 209 P	400
B. subtilis PCI 219	400
B. cereus	>800
Microc. flavus	40
Sarcina lutea	800
Ps. aeruginosa	800
B. brevis	200
Mycobacterium avium	800
M. avium streptomycin-fast	800
Mycobacterium 607	800
Mycobacterium smegmatis	>800
Mycobacterium phlei	800
M. tuberculosis H 37 Rv	100[a]
Pen. chrysogenum Q 176	>500
Sacc. cerevisiae	>500
Candida albicans	>500
Piricularia oryzae	>500
Gib. fujikuroi	>500
Phytophythora infestans	>500
Colleto, lagernarium	>500
Glomerella cingulata	>500
Alternaria kikuchiana	>500

[a] Serial broth dilution method.

Gougerotin was recently reviewed (Fox *et al.*, 1966; Korzybski *et al.*, 1967; Clark, 1967; Suhadolnik, 1970; Pestka, 1971; Fox and Watanabe, 1971).

Gougerotin is a 4-aminohexose pyrimidine nucleoside antibiotic. The first proposed structure for gougerotin, 1-(N-sarcosyl-1-cytosinyl)-3-D-serylamino-1,3-dideoxy-β-D-allopyranuronamide (Fig. 1 A) (Iwasaki, 1962), was reinvestigated and corrected to 1-(cytosinyl)-4-sarcosyl-D-serylamino-1,4-dideoxy-β-D-galacto-pyranurononamide (Fig. 1 B) (Fox *et al.*, 1964). Further, these structures were most recently revised by Fox *et al.* (1968). The correct structure assigned to gougerotin is 1-(cytosinyl)-4-sarcosyl-D-serylamino-1,4-dideoxy-β-D-glucopyr-anuronamide (Fig. 1 C).

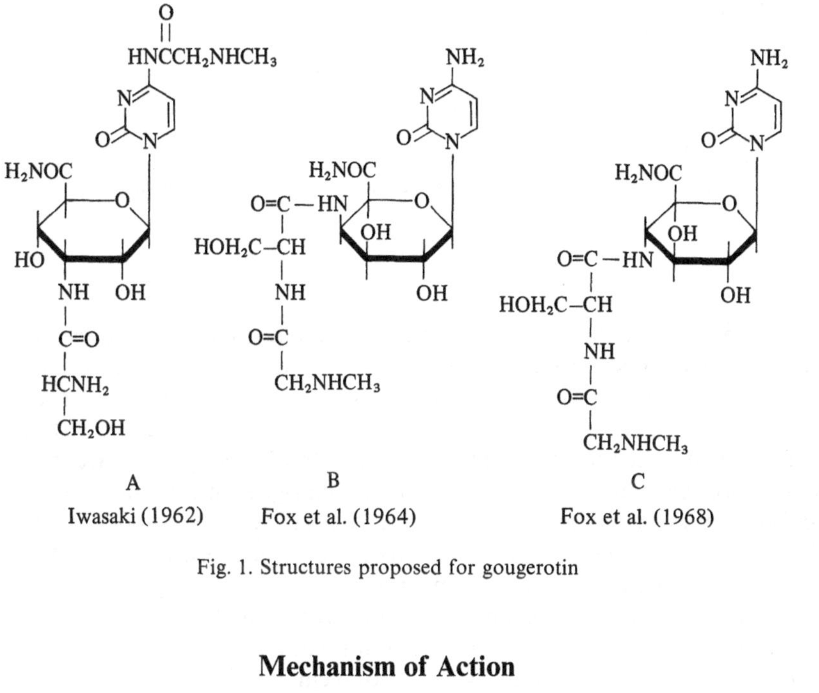

A	B	C
Iwasaki (1962)	Fox et al. (1964)	Fox et al. (1968)

Fig. 1. Structures proposed for gougerotin

Mechanism of Action

Gougerotin is a potent inhibitor of protein synthesis in prokaryotes and eukaryotes. Thus, the protein synthesis in cell-free systems obtained from *E. coli* (Clark and Gunther, 1963; Clark and Chang, 1965), rabbit reticulocyte (Casjens and Morris, 1965; Clark and Chang, 1965) and mouse liver (Sinohara and Sky-Peck, 1965) were inhibited by gougerotin. However, gougerotin does not inhibit either the formation of aminoacyl-RNA (Clark and Chang, 1965; Sinohara and Sky-Peck, 1965) or the binding of aminoacyl-tRNA into the aminoacyl-tRNA, messenger-RNA, ribosome complex (Clark and Chang, 1965; Pestka, 1970b). The release of completed protein from ribosomes was also not affected by gougerotin (Casjens and Morris, 1965). These lines of evidence would indicate that the gougerotin action is directed against the actual peptide formation in the ribosomes. More direct information concerning the mechanism of action of gougerotin came from its effect on the puromycin-dependent release of nascent peptides from ribosomes (puromycin reaction).

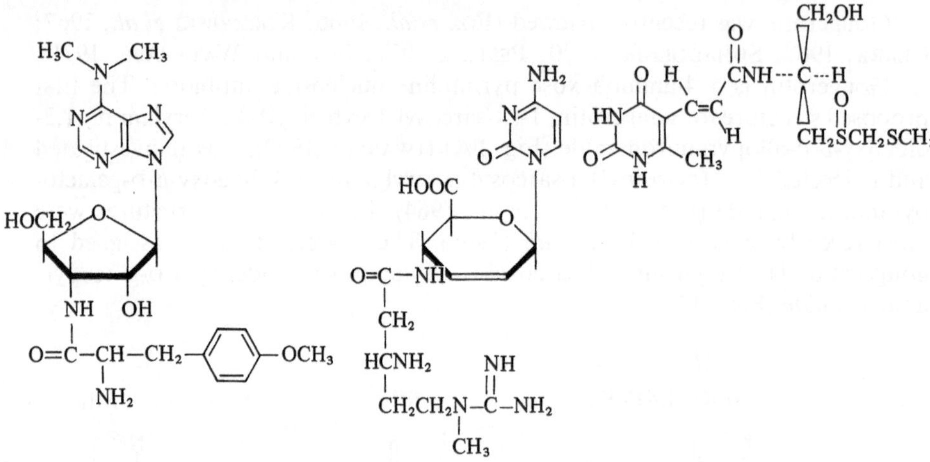

A   B   C

Puromycin   Blasticidin S   Sparsomycin

Fig. 2

Puromycin, as a functional analogue of aminoacyl-tRNA, releases polypeptide from peptidyl-tRNA bound to ribosomes by reacting with the ester bond of the peptidyl-tRNA (ALLEN and ZAMECNICK, 1962; NIRENBERG et al., 1962; NATHANS et al., 1963; GILBERT, 1963). Puromycin analogues with aromatic L-amino acid linked to the 3'-amino group of the 3'-deoxyribose are good inhibitors of protein synthesisyl. D-Amino acid analogues and nonaromatic L-amino acid analogues are inactive (NATHANS and NEIDLE, 1963). SINOHARA and SKY-PECK (1965) compared the first proposed structure of gougerotin (Fig. 1 A) with the structural analogues of puromycin and pointed out that gougerotin does not meet the structural requirements to inhibit protein synthesis in the same manner as puromycin. Actually, gougerotin did not catalyze the release of peptide from ribosomes, capable of reacting with puromycin, but it inhibited the puromycin-dependent release of polypeptides from prelabeled ribosomes (CASJENS and MORRIS, 1965; CLARK and CHANG, 1965). In this inhibition, it was reported that gougerotin acted in strict competition with puromycin (CASJENS and MORRIS, 1965). From these results, it was inferred that gougerotin acted as a nonfunctional analogue of aminoacyl-tRNA or puromycin (CLARK, 1967; COUTSOGEORGO-POULOS, 1967).

However, GOLDBERG and MITSUGI (1967) also studied the effect of gougerotin on the puromycin-induced release of polylysyl-puromycin from polylysyl-tRNA bound to ribosomes. Their kinetic data, different from the results of CASJENS and MORRIS (1965), indicate that the competition between gougerotin and puromycin belongs the "mixed" type inhibition (DIXON and WEBB, 1964). Gougerotin (the revised structure shown in Fig. 1 C), which contains a cytosine aglycon and a N-blocked dipeptide (sarcosyl-D-serine) in acylamino linkage to a 4-aminohexose, differs radically in structure from the amino acid bearing end of tRNA. Further, although gougerotin contains a dipeptide, it does not prevent the binding of peptidyl-tRNA to ribosomes (GOLDBERG and MITSUGI, 1967) nor competes with peptidyl-tRNA in the puromycin reaction (Fig. 4). From these results, it seems that gougerotin is neither a simple structural analogue of aminoacyl-tRNA or pepidyl-tRNA nor one of puromycin.

Table 2. *Effect of various antibiotics on reactions of polylysyl-tRNA with lysyl-tRNA or puromycin* (COUTSOGEORGOPOULOS, 1970)

Inhibitor (M)		C^{12}-polylysyl-tRNA C^{14}-lysyl-tRNA Inhibition (%)	C^{14}-polylysyl-tRNA puromycin Inhibition (%)
Chloramphenicol	10^{-4}	52	55
Blasticidin S	10^{-5}	57	52
Gougerotin	8×10^{-5}	45	48
Amicetin	4×10^{-5}	66	72
Sparsomycin	10^{-6}	76	69
Tetracycline	10^{-4}	80	50
Puromycin	8×10^{-6}	69	—

Several investigators have reported that gougerotin is a potent inhibitior of the formation of single peptide bonds (Table 2). Examples are 1. reaction of polylysyl-tRNA and lysyl-tRNA in the presence of poly A (GOTTESMAN, 1967; JAYARAMAN and GOLDBERG, 1968; COUTSOGEORGOPOULOS, 1970), 2. reaction of polylysyl-tRNA and puromycin in the presence of poly A (GOTTESMAN, 1967; GOLDBERG and MITSUGI, 1967; CERNA et al.,1971), and 3. reaction of N-acetyl-phenylalanyl-tRNA and puromycin in the presence or absence of poly U (PESTKA, 1970a; COUTSOGEORGOPOULOS, 1970; CERNA et al., 1971). All these reactions can take place on NH_4Cl washed ribosomes from E.coli, in the absence of supernatant factors and GTP, and in the presence of relatively high Mg^{++} concentrations (10–20 mM). On the basis of the observed inhibition of these systems, it has been concluded that gougerotin specifically inhibits the formation of the peptide bond *per se*, i.e. the reaction assigned to the peptidyl transferase that is an integral part of ribosomes (GOTTESMAN, 1967; MONRO and MARCKER, 1967; MONRO, 1967). This idea was confirmed by the inhibitory effect of gougerotin on the "fragment reaction", a reaction between N-acylated aminoacyl nucleotide fragments of tRNA and puromycin in the presence of ethanol (MONRO and MARCKER, 1967; MONRO, 1967; MONRO and VAZQUEZ, 1967; MONRO et al., 1968; MONRO et al., 1969; VAZQUEZ et al., 1969; CERNA et al., 1971). Ribosomal peptidyl transferase also catalyzes the formation of ester bonds as shown by the formation of N-formyl-methionyl-hydroxypuromycin (FAHNESTOCK, et al., 1970). Gougerotin inhibits this reaction.

The classification of gougerotin, along with other antibiotics such as chloramphenicol, blasticidin S, sparsomycin and amicetin, as specific inhibitors of peptidyl transfer reaction does not easily explain the formation of phenylalanine-containing *oligo*peptides from phenylalanyl-tRNA in the presence of these antibiotics, at concentrations at which *poly*-phenylalanine formation is inhibited (COUTSO-

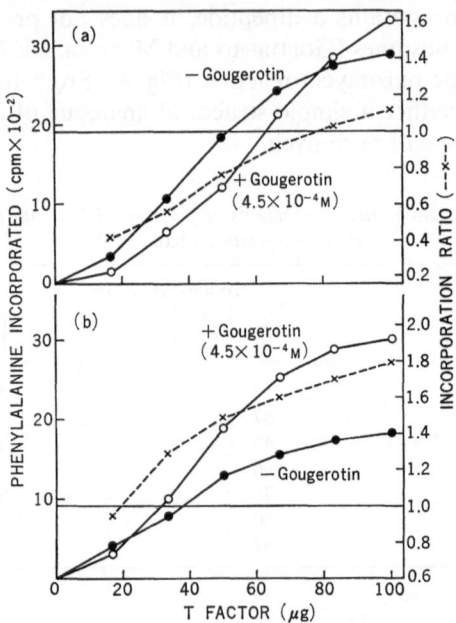

Fig. 3. Effect of T factor concentration on polyphenylalanine synthesis in the presence of gougerotin. The reaction was carried out in the presence (a) 160 mM ammonium acetate, b 160 mM ammonium acetate and 200 mM KCl. Incorporation ratio is the ratio of phenylalanine incorporated in the presence of gougerotin to that of in the absence of gougerotin. (From YUKIOKA and MORISAWA, 1969 b)

GEORGOPOULOS, 1969, 1970). Furthermore, gougerotin stimulated the formation of polyphenylalanine under certain conditions (YUKIOKA and MORISAWA, 1969 a, 1969 b) (Fig. 3). If the formation of the peptide bond *per se* were inhibited, then an inhibition of the formation of peptide chains of all sizes should be expected and a stimulation of polypeptide formation should not occur. The effect of gougerotin on the poly U-stimulated polyphenylalanine synthesis was greatly influenced by the concentration of monovalent ions (K^+ or NH_4^+) and by the amount of T factor or poly U in reaction mixtures (YUKIOKA and MORISAWA, 1969 a, 1969 b). COUTSOGEORGOPOULOS (1970) believes that the mechanism of initiation of polypeptide chains or the state of the ribosome-poly U-phenylalanyl-tRNA complex at the time of peptide chain initiation determine the extent of inhibitions of peptide formation by certain antibiotics, including gougerotin. Thus, it appears that gougerotin exerts additional actions on peptide formation besides its effect on peptidyl transferase.

Some of the above-mentioned results might be explained by following observations: 1. stimulation of the poly U-dependent binding of N-acetyl phenylalanyl-tRNA to ribosomes in the presence of ribosomal supernatant by gougerotin (HERNER et al., 1969) and 2. enhancement of the T factor-dependent binding of phenylalanyl-tRNA to ribosomes by gougerotin (YUKIOKA and MORISAWA, 1969 c). T factor stimulates the binding of phenylalanyl-tRNA to the donor site (D-site) as well as to the acceptor site (A-site) (KAJI et al., 1969). Ribosomal "wash" which contains the initiation factors stimulates the binding of N-acetyl-phenylalanyl-tRNA to the donor site (LUCAS-LENARD and HAENNI, 1968). Gougerotin might bring about, by its stimulating effect on chain initiation, the increase of the number of polyphenylalanine peptide chains or the stimulation of oligophenylalanine formation which have been observed under certain conditions (YUKIOKA and MORISAWA, 1969a, 1969b; COUTSOGEORGOPOULOS, 1970).

Recently, it was found that sparsomycin (Fig. 2C) markedly stimulated the binding of N-substituted aminoacyl nucleotide fragment of tRNA to isolated large subunits of ribosomes (50S or 60S) in the conditions of fragment reaction ("sparsomycin reaction") (MONRO *et al.*, 1969; VAZQUEZ *et al.*, 1969; JIMENEZ *et al.*, 1970a). This sparsomycin-induced complex is unreactive with puromycin or hydroxylamine (MONRO *et al.*, 1969; JIMENEZ *et al.*, 1970b). It was proposed that sparsomycin induced the formation of a nonreactive complex between the CCA-peptide moiety of the peptidyl donor substrates and the D-site of the peptidyl transferase (MONRO *et al.*, 1969). Gougerotin inhibited this sparsomycin reaction (Table 3) (MONRO *et al.*, 1969; VAZQUEZ *et al.*, 1969). But in the absence of sparsomycin, gougerotin stimulated the binding of CACCA-(AcLeu) by 40%. From this it was inferred that the mode of action of gougerotin may be related to that of sparsomycin in that they compete for the same site on the ribosomes. However, the inert complexes formed with gougerotin are less stable than the sparsomycin complex. The data in Table 3 strongly suggest that the action of chloramphenicol, in peptide bond formation, differs from the action of gougerotin.

Table 3. *Effect of various antibiotics on complex formation* (MONRO et al., 1969)

Addition	Final concentration mM	Percentage of control sparsomycin-stimulated binding
None	—	100
Chloramphenicol	1	3
Carbomycin	0.1	0
Spiramycin III	0.1	0
Streptogramin A	0.1	2
Lincomycin	1	5
Amicetin	1	27
Gougerotin	1	42

Meanwhile, the active site of peptidyl transferase is composed of two binding sites, the donor and the acceptor site, which specifically interact with the donor and acceptor substrate, respectively; the peptidyl transfer reaction which is catalyzed by peptidyl transferase on the ribosome could be subdivided into 3 steps: A. the binding of the aminoacyl adenyl terminus of aminoacyl-tRNA to A-site, B. the binding of the peptidyl adenyl terminus of peptidyl-tRNA to D-site, and C. the formation of the peptide bond subsequent to the correct alignment of the substrates listed under A. and B. at the proper site (peptidyl transfer step) (PESTKA, 1969b; COUTSOGEORGOPOULOS, 1970). The inhibitory action of gougerotin on peptidyl transfer reaction could affect one or more of these proc-

Table 4. *Effect of gougerotin and analogue (III) on the CACCA-(AcLeu) and CACCA-Phe binding to ribosomes* (Cerna et al., 1971)

	Concentration (M)	CACCA-(AcLeu) binding		CACCA-Phe binding	
		cpm	%	cpm	%
Control	—	345	100	2680	100
Gougerotin	10^{-4}	535	155	2168	81
	10^{-3}	565	164	1256	47
Analogue (III)	10^{-3}	357	104	2567	96
	2×10^{-3}	385	112	2243	80

esses. As mentioned above, gougerotin stimulates the formation of oligophenyla-lanine or increases the number of polyphenylalanines under certain conditions; it might be reasonable to assume that gougerotin is involved in, or interferes with, the binding steps (step A and B) rather than the actual peptide forming step (peptidyl transfer step). This notion was supported by recent experimental results reported by Pestka (1969a, 1969b), Celma et al. (1970), and Cerna et al., (1971). Thus, gougerotin at a concentration which inhibits the transfer of an acylaminoa-cyl group, increases the binding of donor substrate bound to the donor site, but decreases the binding of acceptor substrate at acceptor site (Table 4). As shown in Fig. 4 and Fig. 5, 4-aminohexose pyrimidine nucleotide antibiotics, gou-

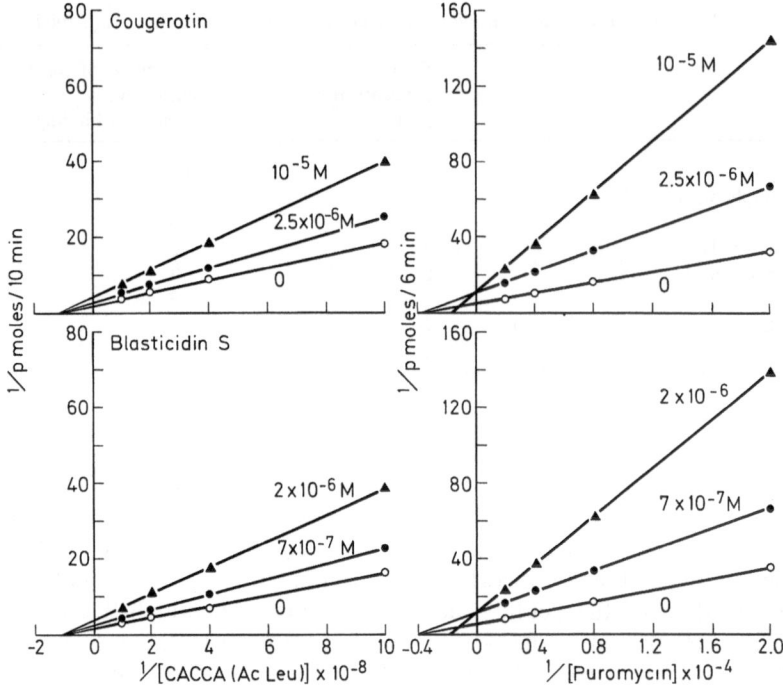

Fig. 4. Effect of gougerotin and blasticidin S on the rate of puromycin reaction as a function of the concentration of CACCA-(AcLeu) or puromycin. (From Yukioka and Morisawa, 1972)

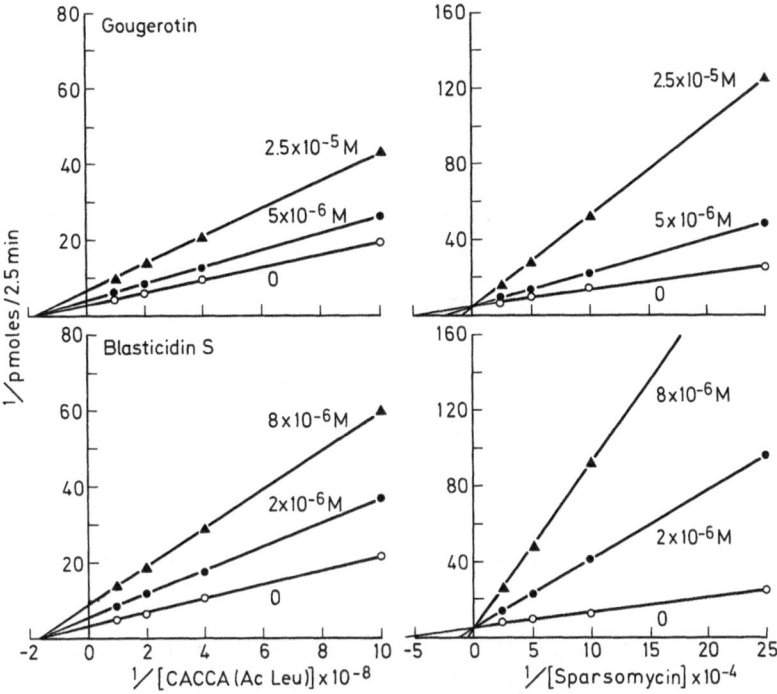

Fig. 5. Effect of gougerotin and blasticidin S on the rate of sparsomycin reaction as a function of the concentration of CACCA-(AcLeu) or sparsomycin. (From YUKIOKA and MORISAWA, 1972)

gerotin and blasticidin S (Fig. 2B), exhibited competitive inhibition with sparsomycin in the sparsomycin reaction, but showed mixed type inhibitions with puromycin in the puromycin reaction. In both reactions, they exhibited noncompetitive type inhibitions with peptidyl donor substrate (CACCA-(AcLeu)) (YUKIOKA and MORISAWA, 1972). These results suggest that gougerotin, blasticidin S and sparsomycin interact with one ribosomal site (sparsomycin site), which is not identical, but very close to, the puromycin locus. The acceptor site might be inclusive of sparsomycin site. Thus, gougerotin and blasticidin S inhibit the peptidyl transferase by competing with acceptor substrate at the acceptor site. The increase of binding of donor substrate to the donor site might be caused by a conformational change at the donor site which occurs when the acceptor site has been occupied by the acceptor substrate (CERNA et al., 1971).

CAPECCHI and KLEIN (1969) showed that gougerotin inhibited the factor R_1-mediated release of F-met-Ala-Ser-Asn-Phe-Thr from ribosomes. This inhibition might be due to gougerotin acting as a steric inhibitor by its interaction at the acceptor site of ribosomes.

Recently, three gougerotin analogues have been synthesized, 1-[3-(sarcosyl-D-seryl-)amino-3-deoxy-β-D-glucopyranosyl-]uracil (I), 1-[3-(sarcosyl-D-seryl-)amino-3-deoxy-β-D-glucopyranosyl-]cytosine (II) and 1-[4-(sarcosyl-D-seryl-)amino-4-deoxy-β-D-glucopyranosyl-] cytosine (III) (Fig. 6), and their effect on the ribosomal peptidyl transferase was reported (Table 4) (LICHTENTHALER et al.,

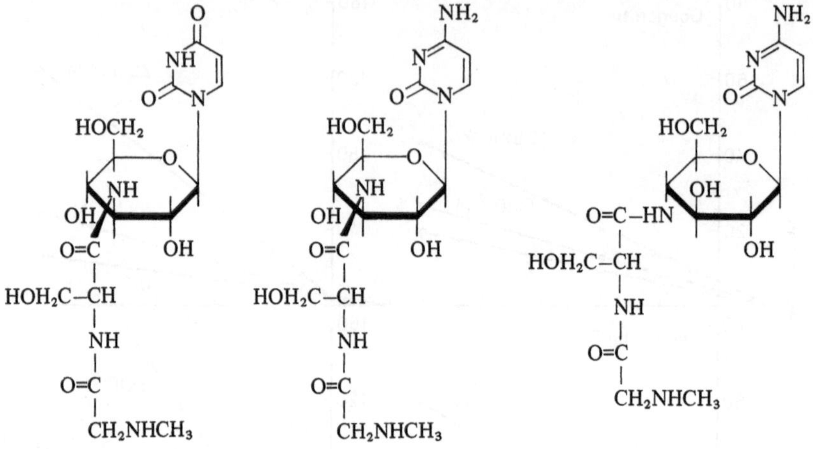

Fig. 6. Analogues of gougerotin. (From LICHTENTHALER et al., 1970, 1971)

1970; LICHTENTHALER et al., 1971; CERNA et al., 1971). In all of these analogues, the glucuronic acid amide of parent compound (gougerotin) is replaced by amino-glucose, i.e. the analogues contain a —CH$_2$OH group on 6' carbon instead of carboxamide —CONH$_2$ grouping. Although, analogue III was approximately one-tenth as active as gougerotin (Table 4), the analogues with the sarcosyl-D-seryl amide chain attached to the 3'-carbon (analogue I and II) did not show any inhibitory effect on peptidyl transferase. These result indicate the position of the sarcosyl-D-seryl amide residue on the glucose moiety is directly involved in the inhibitory mechanism of gougerotin.

In concluding, gougerotin along with other 4-aminohexose antibiotics has become an important tool in elucidating the complex mechanism of peptide bond formation on ribosomes.

References

ALLEN, D. W., and P. C. ZAMECNIK: The effect of puromycin on rabbit reticulocyte ribosomes. Biochim. Biophys. Acta 55, 865 (1962).

CAPECCHI, M. R., and H. A. KLEIN: Characterization of three proteins involved in polypeptide chain termination. Cold Spring Harbor Symp. Quant. Biol. 34, 469 (1969).

CASJENS, S. R., and A. J. MORRIS: The selective inhibition of protein assembly by gougerotin. Biochim. Biophys. Acta 108, 677 (1965).

CELMA, M. L., R. E. MONRO, and D. VAZQUEZ: Substrate and antibiotic binding sites at the peptidyl transferase centre of E. coli ribosomes. FEBS Letters 6, 273 (1970).

CERNA, J., F. W. LICHTENTHALER, and I. RYCHLIK: The effect of gougerotin analogues on ribosomal peptidyl transferase, FEBS Letters 14, 45 (1971).

CLARK, J. M., JR.: Gougerotin. In: Antibiotics, vol. 1, (eds. D. GOTTLIEB and P. D. SHAW) p. 278. Berlin-Heidelberg-New York: Springer 1967.

CLARK, J. M., JR., and A. Y. CHANG: Inhibitors of the transfer of amino acids from aminoacyl soluble ribonucleic acid to protein. J. Biol. Chem. 240, 4734 (1965).

CLARK, J.M., JR., and J.K. GUNTHER: Gougerotin, a specific inhibitor of protein synthesis. Biochim. Biophys. Acta **76**, 636 (1963).

COUTSOGEORGOPOULOS, C.: Amino acylaminonucleoside inhibitors of protein synthesis. The effect of amino acyl ribonucleic acid on the inhibition. Biochemistry **6**, 1704 (1967).

COUTSOGEORGOPOULOS, C: Formation of oligophenylalanine in the presence of certain inhibitors of protein synthesis. Federation Proc. **28**, 844 (1969).

COUTSOGEORGOPOULOS, C.: On the inhibition of ribosome-catalyzed peptide bond formation by certain antibiotics. Proc. 6th Intern. Congr. Chemotherapy, Tokyo, p. 482, 1970.

DIXON, M., and E.C. WEBB: Enzymes. London: Longmans, 1964.

FAHNESTOCK, S., H. NEUMANN, V. SHASHOUA, and A. RICH: Ribosome-catalyzed ester formation. Biochemistry **9**, 2477 (1970).

FOX, J.J., Y. KUWADA, and K.A. WATANABE: Nucleosides. LVI. Structure of nucleoside antibiotic, gougerotin. Tetrahedron Letters **1968**, 6029.

FOX, J.J., Y. KUWADA, K.A. WATANABE, T. UEDA, and E.B. WHIPPLE: Nucleosides. XXV. Chemistry of gougerotin. Antimicrobial Agents chemotherapy **1964**, 518.

FOX, J.J., and K.A. WATANABE: Studies directed towards the total synthesis of the nucleoside antibiotics, gougerotin and blasticidin S. Pure Appl. Chem. **28**, 475 (1971).

FOX, J.J., K.A. WATANABE, and A. BLOCH: Nucleoside antibiotics. Progr. Nucleic Acid Res. Mol. Biol. **5**, 251 (1966).

GILBERT, W.: Polypeptide synthesis in *Escherichia coli*. II. The polypeptide chain and sRNA. J. Mol. Biol. **6**, 389 (1963).

GOLDBERG, I.H., and K. MITSUGI: Inhibition by sparsomycin and other antibiotics of the puromycin-induced release of polypeptide from ribosomes. Biochemistry **6**, 383 (1967).

GOTTESMAN, M.E.: Reaction of ribosome-bound peptidyl transfer ribonucleic acid with aminoacyl transfer ribonucleic acid or puromycin. J. Biol. Chem. **242**, 5564 (1967).

HERNER, A.E., I.H. GOLDBERG, and L.B. Cohen: Stabilization of N-acetyl-phenylalanyl transfer ribonucleic acid binding to ribosomes by sparsomycin. Biochemistry **8**, 1335 (1969).

IWASAKI, H.: Studies on the structure of gougerotin. XI. Structure of gougerotin. Yakugaku Zasshi **82**, 1393 (1962).

JAYARAMAN, J., and I.H. GOLDBERG: Localization of sparsomycin action to the peptide-bond-forming step. Biochemistry **7**, 418 (1968).

JIMENEZ, A., R.E. MONRO, and D. VAZQUEZ: Interaction of Ac-Phe- tRNA with *E. coli* ribosomal subunits. 1. Sparsomycin-induced formation of a complex containing 50S and 30S subunits but not mRNA. FEBS Letters **7**, 103 (1970a).

JIMENEZ, A., R.E. MONRO, and D. VAZQUEZ: Interaction of Ac-Phe-tRNA with *E. coli* ribosomal subunits. 2. Resistance of the sparsomycin-induced complex to hydroxylamine action. FEBS Letters **7**, 109 (1970b).

KAJI, A., K. IGARASHI, and H. ISHITSUKA: Interaction of tRNA with ribosomes, Binding and release of tRNA. Cold Spring Harbor Symp. Quant. Biol. **34**, 167 (1969).

KANZAKI, T., E. HIGASHIDE, H. YAMAMOTO, M. SHIBATA, K. NAKAZAWA, H. IWASAKI, T. TAKEWAKA, and A. MIYAKE: Gougerotin, a new antibacterial antibiotic. J. Antibiotics (Tokyo), Ser. A **15**, 93 (1962).

KORZYBSKI, T., Z. KOWSZYK-GINDIFER, and W. KURYLOWICZ: Antibiotics, vol.1, Origin, nature and properties. New York: Pergamon Press 1967.

LICHTENTHALER, F.W., G. TRUMMLITZ, G. BAMBACH, and I. RYCHLIK: Nucleosides. XI. Synthesis of a biologically active gougerotin analogue. Angew. Chem. **10**, 334 (1971).

LICHTENTHALER, F.W., G. TRUMMLITZ, and P. EMIG: Nucleosides. X. Synthesis of dipeptidyl amino sugar nucleosides structurally related to gougerotin. Tetrahedron Letters **1970**, 2061.

LUCAS-LENARD, J., and A.L. HAENNI: Requirement of guanosine 5'-triphosphate for ribosomal binding of aminoacyl-sRNA. Proc. Natl. Acad. Sci. U.S. **59**, 554 (1968).

MONRO, R.E.: Catalysis of peptide bond formation by 50S ribosomal subunits from *Escherichia coli*. J. Mol. Biol. **26**, 147 (1967).

MONRO, R.E., M.L. CELMA, and D. Vazquez: Action of sparsomycin on ribosome-catalysed peptidyl transfer. Nature **222**, 356 (1969).

MONRO, R.E., J. CERNA, and K.A. MARCKER: Ribosome-catalyzed peptidyl transfer: Substrate specificity at the P-site. Proc. Natl. Acad. Sci. U.S. **61**, 1042 (1968).

MONRO, R.E., and K.A. MARCKER: Ribosome-catalyzed reaction of puromycin with a formylmethionine-containing oligonucleotide, J. Mol. Biol. **25**, 347 (1967).

Monro, R. E., T. Staehelin, M. L. Celma, and D. Vazquez: The peptidyl transferase activity of ribosomes. Cold Spring Harbor Symp. Quant. Biol. **34**, 357 (1969).

Monro, R. E., and D. Vazquez: Ribosome-catalysed peptidyl transfer: Effects of some inhibitors of protein synthesis. J. Mol. Biol. **28**, 161 (1967).

Nathans, D., J. E. Allende, T. W. Conway, G. I. Spyrides, and F. Lipmann: Protein synthesis from aminoacyl-sRNA's. Symposium on Information Macromolecules (H. J. Vogel, V. Bryson and J. O. Lampen, eds). Academic Press: New York 1963.

Nathans, D., and A. Neidle: Structural requirements for puromycin inhibition of protein synthesis. Nature **197**, 1076 (1963).

Nirenberg, M. W., J. H. Matthae, and O. W. Jones: An intermediate in the biosynthesis of poly-phenylalanine directed by synthetic template RNA. Proc. Natl. Acad. Sic. U. S. **48**, 104 (1962).

Pestka, S.: Studies on the formation of transfer ribonucleic acid-ribosome complexes, XI. Antibiotic effects on phenylalanyl-oligonucleotide binding to ribosomes. Proc. Natl. Acad. Sci. U.S. **64**, 709 (1969a).

Pestka, S.: Translocation, aminoacyl-oligonucleotides, and antibiotic action. Cold Spring Harbor Symp. Quant. Biol. **34**, 395 (1969b).

Pestka. S.: Studies on the formation of transfer ribonucleic acid-ribosome complexes, VIII. Survey of the effect of antibiotics on N-acetyl-phenylalanyl-puromycin formation: Possible mechanism of chloramphenicol action. Arch. Biochem. Biophys. **136**, 80 (1970a).

Pestka, S.: Studies on the formation of transfer ribonucleic cid-ribosome complexes, IX. Effect of antibiotics on translocation and peptide bond formation. Arch. Biochem. Biophys. **136**, 89 (1970b).

Pestka, S.: Inhibitors of ribosome functions. Ann. Rev. Biochem. **40**, 697 (1971).

Sinohara, H., and H. H. Sky-Peck: Effect of gougerotin on the protein synthesis in the mouse liver. Biochem. Biophys. Res. Commun. **18**, 98 (1965).

Suhadolnik, R. J.: Nucleoside antibiotics. New York: Wiley-Interscience 1970.

Thiry, L.: The action of sparsomycin and gougerotin on virus growth. J. Gen. Virol. **2**, 143 (1968).

Vazquez, D., E. Battaner, R. Neth, G. Heller, and R. E. Monro: The function of 80S ribosomal subunits and effects of some antibiotics. Cold Spring Harbor Symp. Quant. Biol. **34**, 369 (1969).

Yukioka, M., and S. Morisawa: Studies on the mechanism of action of gougerotin. I. Enhancement of polyphenylalanine synthesis by gougerotin. J. Biochem. (Tokyo) **66**, 225 (1969a).

Yukioka, M., and S. Morisawa: Studies on the mechanism of action of gougerotin. II. Effect of various factors on gougerotin action. J. Biochem. (Tokyo) **66**, 233 (1969b).

Yukioka, M., and S. Morisawa: Studies on the mechanism of action of gougerotin. III. Enhancement of enzymatic binding of phenylalanyl-tRNA to ribosomes by gougerotin. J Biochem. (Tokyo) **66**, 241 (1969c).

Yukioka, M., and Morisawa: Manuscript in preparation.

The Macrolide Antibiotics

D. Vazquez

Source, Chemical Structure, Properties and Inhibitory Spectra. The term macrolide has been applied to members of a group of structurally related antibiotics produced by species of streptomyces (Table 1). All macrolide antibiotics (WOOD-

Table 1. *Macrolide antibiotics*

Antibiotic complex (synonyms)	Individual components (synonyms)	Source, isolation and chemistry	
		Streptomyces or *Micromonospora* strain producer	References
	Angolamycin (Shincomycin A)	*S. eurythermus*	CORBAZ *et al.* (1955a); KINUMAKI and SUZUKI (1972)
Carbomycin (Magnamycin) (M-4209)	Carbomycin A Carbomycin B	*S. halstedii* *S. hygroscopicus* *S. albireticuli*	TANNER *et al.* (1952); PAGANA *et al.* (1953); HOCHSTEIN and MURAL (1954); MIYAKE *et al.* (1959); WOODWARD (1957); KUEHNE and BENSON (1965); WOODWARD *et al.* (1965)
	Chalcomycin (Bandamycin B)	*S. bikinensis*	COFFEY *et al.* (1963); WOO *et al.* (1964)
Erythromycin	Erythromycin A Erythromycin B Erythromycin C	*S. erythreus*	MCGUIRE *et al.* (1952); WELCH *et al.* (1952); WILEY *et al.* (1957a, b, c); HOFHEINZ and GRISEBACH (1962)
	Josamycin (Leucomycin A$_3$)	*S. narbonensis* var. *josamyceticus*	OSONO *et al.* (1967); OSONO and UMEZAWA (1971)
Kujimycin	Kujimycin A Kujimycin B (Lancamycin)	*S. spinichromogenes* var. *Kujimyceticus*	OMURA *et al.* (1969); OMURA *et al.* (1970b)
	Lancamycin	*S. violaceoniger*	GÄUMANN *et al.* (1960); KELLER-SCHIERLEIN and RONCARI (1964); EGAN and MARTIN (1970)
Leucomycin (Kitasamycin)	Leucomycin A$_1$ Leucomycin A$_3$ (Josamycin) Leucomycin A$_4$ Leucomycin A$_5$ Leucomycin A$_6$ Leucomycin A$_7$ Leucomycin A$_8$ Leucomycin A$_9$	*S. Kitasatoensis*	HATA *et al.* (1953); OMURA *et al.* (1967); OMURA *et al.* (1970a); TOJU and OMURA (1971)

Table 1 (continued)

Antibiotic complex (synonyms)	Individual components (synonyms)	Source, isolation and chemistry	
		Strptomyces or *Micromonospora* strain producer	References
Macrocin		*S. fradiae*	Hamill and Stark (1964)
Megalomycin	Megalomycin A	*M. megalomicea*	Weinstein *et al.*, 1969; Marquez *et al.*, 1969; Waitz *et al.*, 1969; Mallams *et al.*, 1969
	Methymycin	*S. eurocidicus*	Donin *et al.* (1953/54); Djerassi and Zderic (1956)
	Narbomycin	*S. narbonensis*	Corbaz *et al.* (1955b); Prelog *et al.* (1962)
	Neomethymycin	*S. venezuelae*	Djerassi and Halpern (1958)
	Neutramycin	*S. rimosus*	Mitscher and Kunstmann (1969)
	Niddamycin	*S. djakartensis*	Huber *et al.* (1962)
	Oleandomycin (Matromycin) (Romicil)	*S. antibioticus* *S. olivochromogenes*	Celmer *et al.* (1957/58); Hochstein *et al.* (1960)
	Picromycin	*S. felleus* *S. venezuelae*	Brockmann and Henkel (1950); Brockmann and Oster (1957); Anliker and Gubler (1957); Rickards *et al.* (1968); Muxfeldt *et al.* (1968)
	Relomycin	*S. hygroscopicus*	Whaley *et al.* (1963)
Spiramycin (Rovamycin) (Sequamycin) (Selectomycin) (Provamycin)	Spiramycin I Spiramycin II Spiramycin III	*S. ambofaciens*	Pinnert-Sindico *et al.* (1954/55); Paul and Tchelitcheff (1965); Kuehne and Benson (1965); Omura *et al.* (1969)
	Tylosin	*S. hygroscopicus* *S. fradiae*	McGuire *et al.* (1961); Morin *et al.* (1970)

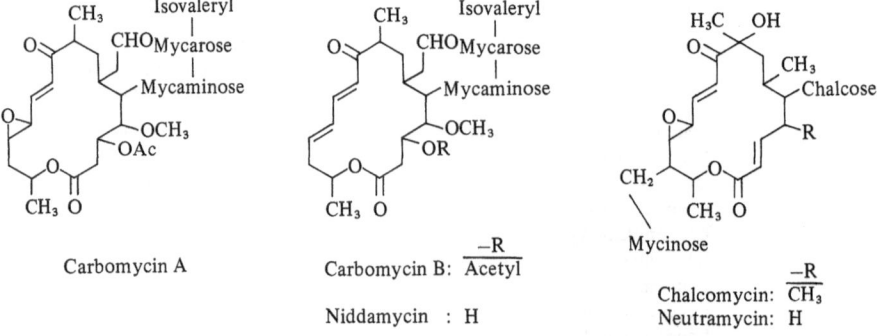

Carbomycin A

Carbomycin B: $\overline{\text{Acetyl}}$ −R

Niddamycin : H

Chalcomycin: $\overline{\text{CH}_3}$ −R

Neutramycin: H

Fig. 1. Continuation, see next page

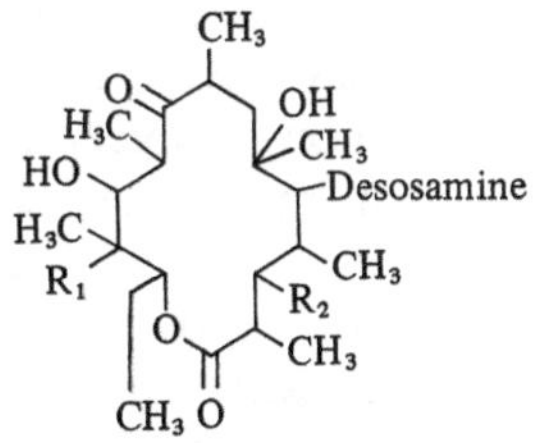

	−R₁	−R₂
Erythromycin A:	OH	Cladinose
Erythromycin B:	H	Cladinose
Erythromycin C:	OH	Mycarose

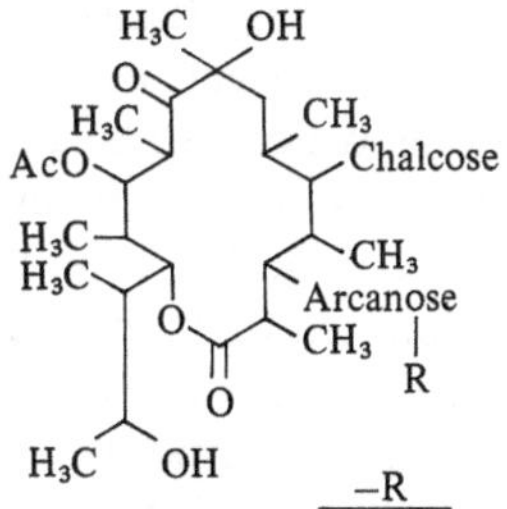

	−R
Lancamycin:	Acetyl
Kujimycin A:	H

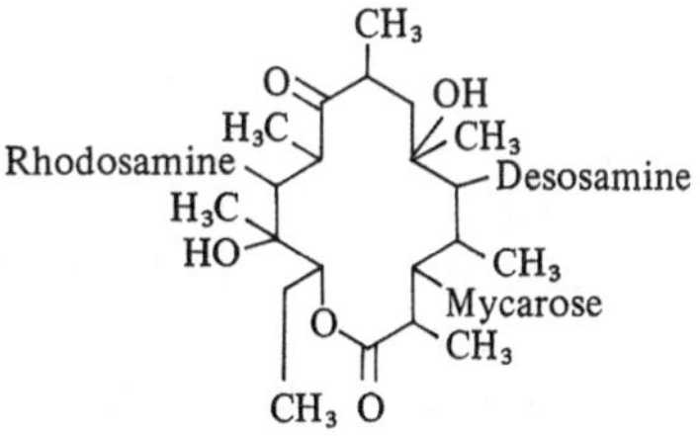

Megalomycin A

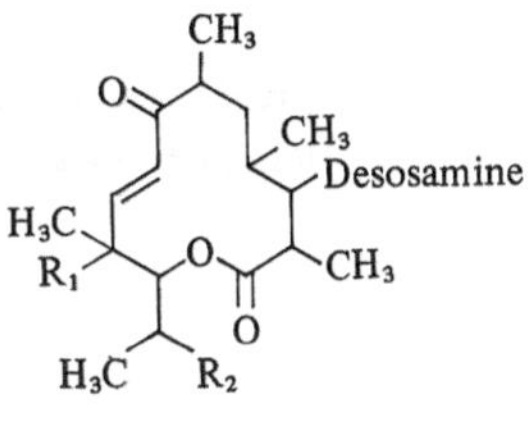

	−R₁	−R₂
Methymycin:	OH	H
Neomethymycin:	H	OH

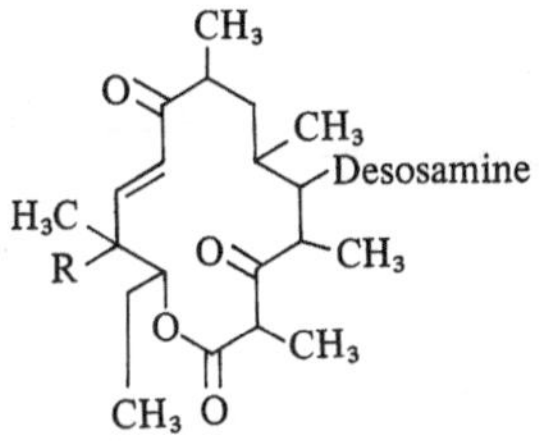

	−R
Narbomycin:	H
Picromycin :	OH

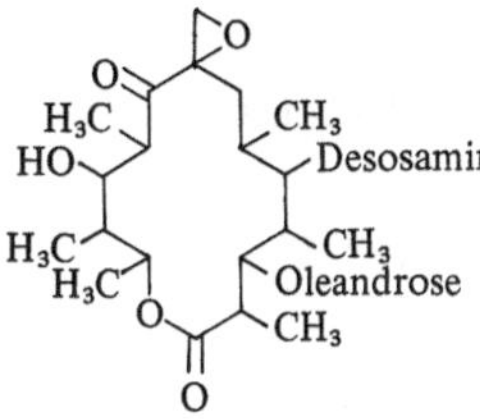

Oleandomycin

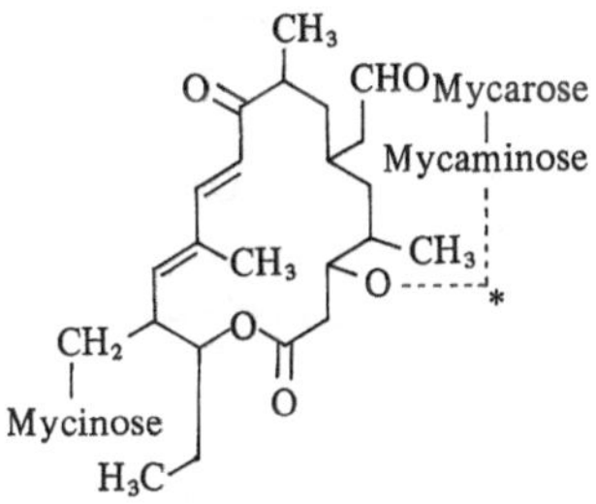

Tylosin

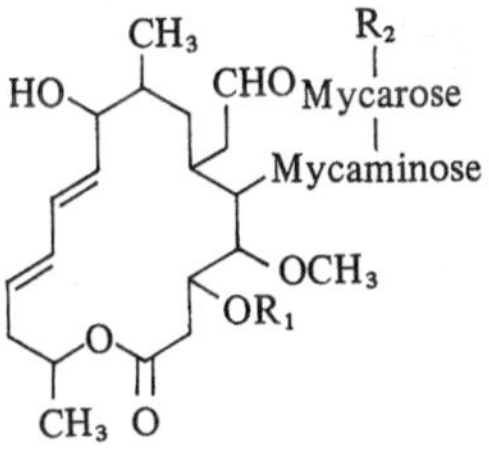

Leucomycins

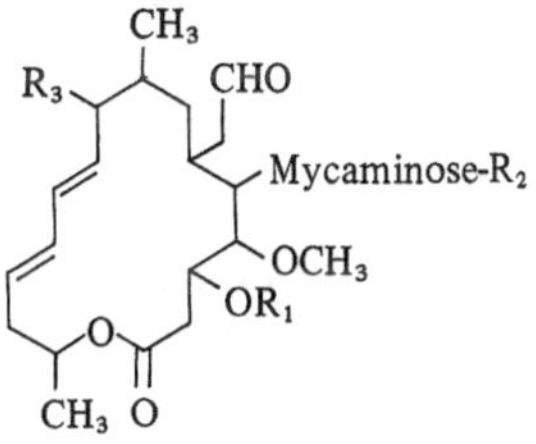

	−R₁	−R₂		−R₁	−R₂	−R₃
A₁:	H	Isovaleryl	Spiramycin I:	H	Mycarose	Forosamine
A₃:	Acetyl	Isovaleryl	Spiramycin II:	Acetyl	Mycarose	Forosamine
A₄:	Acetyl	n-Butyryl	Spiramycin III:	Propionyl	Mycarose	Forosamine
A₅:	H	n-Butyryl	Neospiramycin I:	H	H	Forosamine
A₆:	Acetyl	Propionyl				
A₇:	H	Propionyl	Neospiramycin II:	Acetyl	H	Forosamine
A₈:	Acetyl	Acetyl				
A₉:	H	Acetyl	Neospiramycin III:	Propionyl	H	Forosamine
			Forocidin I:	H	H	H
Leucomycin A₃			Forocidin II:	Acetyl	H	H
(synonym Josamycin).			Forocidin III:	Propionyl	H	H

Fig. 1. Structures of some macrolide antibiotics (taken from CELMER, 1971)
* Not completely established

WARD, 1957) contain a large lactone ring (aglycone of 12 to 22 atoms) with a few or no double bonds and no nitrogen atoms (Fig. 1); they have one or more sugars, linked to this aglycone, which can be amino sugars, non-nitrogenous sugars or both (Fig. 2). In the widest sense however, the term macrolide has been applied to all the antibiotics containing a large lactone ring; in this sense the polyene antibiotics and antibiotics of streptogramin A and B groups can be also termed macrolides. This article will be restricted to the "classical" macrolides of WOODWARD (1957) and related compounds discovered after 1957. These antibiotics will be considered in a comparative manner in this chapter.

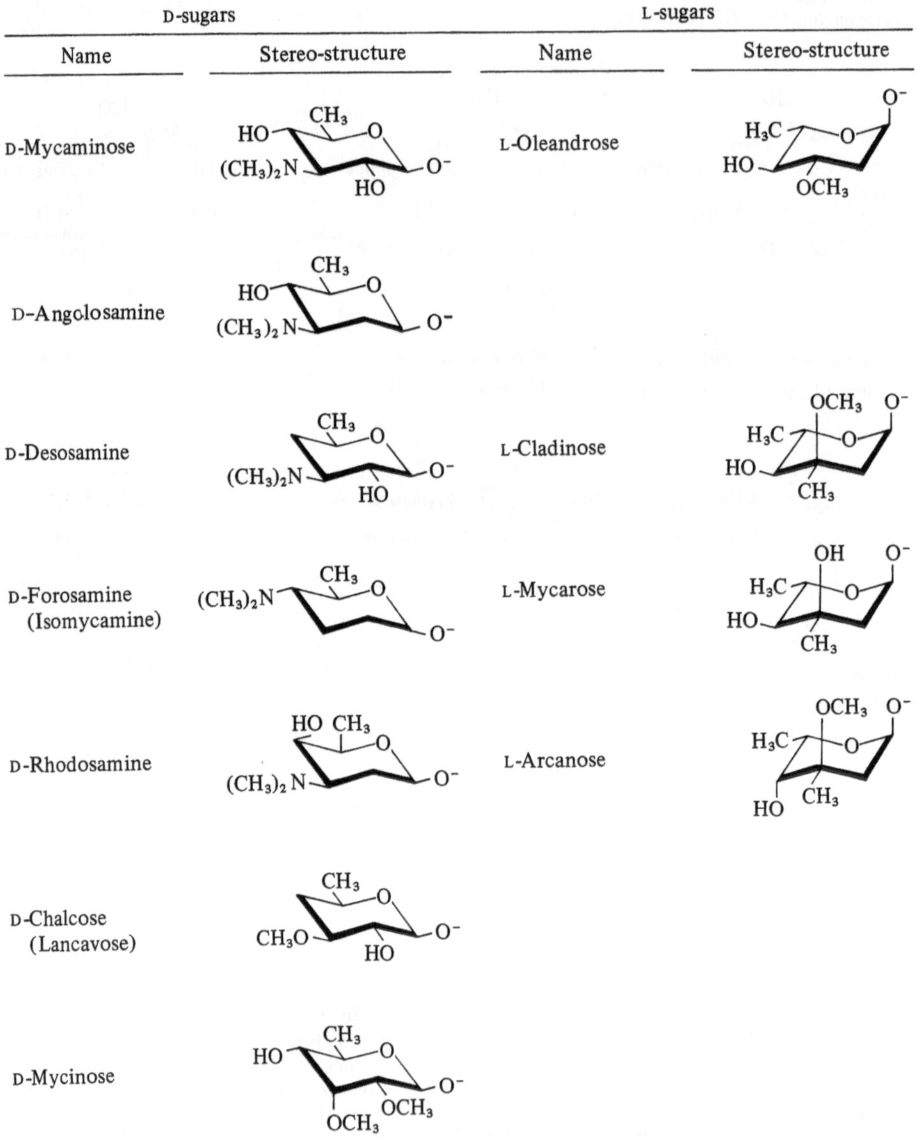

D-sugars		L-sugars	
Name	Stereo-structure	Name	Stereo-structure
D-Mycaminose		L-Oleandrose	
D-Angolosamine			
D-Desosamine		L-Cladinose	
D-Forosamine (Isomycamine)		L-Mycarose	
D-Rhodosamine		L-Arcanose	
D-Chalcose (Lancavose)			
D-Mycinose			

Fig. 2. Sugar moieties from macrolide antibiotics. (Taken from CELMER, 1971)

In 1950 BROCKMANN and HENKEL described the isolation of picromycin, the first known macrolide antibiotic. Since then, more than thirty antibiotics, classified as macrolides or suspected macrolides have been discovered. Most of the streptomyces species which produce macrolide antibiotics produce two or more of these compounds which are usually distinguished for purposes of nomenclature by the addition of letters or figures to the name of the original complex antibiotic, e.g. erythromycin A, B and C, spiramycin I, II and III. In the case of some macrolides their chemical structures are not yet known. Chemical structures of the aglycone and sugar moieties of the best known macrolides are shown in Figs. 1 and 2. There are a number of macrolide antibiotics derived from those produced by streptomyces such as neospiramycins I, II and III obtained by chemical removal of the mycarose moiety of spiramycins I, II and III respectively, the forocidins I, II and III obtained by chemical removal of the -isomycamine (synonym forosamine) moiety of the neospiramycins I, II and III respectively (PAUL and TCHELITCHEFF, 1965), the acetyl spiramycin and the triacetyl oleandomycin. The sugar configurations shown in Fig. 2 are those already established by different workers or predicted by CELMER (1965a, b, c, d, 1971) on the basis of his model for the macrolides. Total absolute configurations have been assigned for carbomycins, erythromycins, leucomycins, methymycin, oleandomycin, picromycin and spiramycins and there is also important information concerning the aglycones of chalcomycin, lancamycin, narbomycin and neomethymycin (CELMER, 1971; PERUN, 1971; reviews).

Some of the macrolide antibiotics, including oleandomycin and triacetyl oleandomycin and erythromycin, leucomycin, spiramycin and acetyl spiramycin complexes are used clinically and also as supplements in animal feeds. Tylosin is used in preservation of food and also as supplement in animal feeding.

The macrolide antibiotics have been considered as a homogeneous group of compounds. Their chemical compositions are related and their antibacterial spectra are similar. However some of them differ in certain chemical and physical properties; thus most of the macrolide antibiotics are basic substances but some are neutral compounds; some are fairly soluble in water whereas others are very insoluble in water but soluble in ethanol.

Both the aglycone and sugar moieties are required for the antibiotic activity of the macrolides. Although the sugars are not active themselves, differences in the sugar moiety linked to the same aglycone (for instance spiramycins, neospiramycins and forocidins) cause important differences in properties and activity. According to their sugar components, the macrolide antibiotics can be classified in different groups with some common characteristics (Table 2). This classification appears to be quite pertinent to the mode of action of the macrolide antibiotics and we will refer to it in this Chapter when considering the different antibiotics comparatively.

The macrolide antibiotics have very reduced or no activity in eucaryotic cells and their cell-free system. They are active mainly against Gram-positive bacteria and in general have a much reduced activity against Gram-negative organisms. This is shown in Table 3 which presents the antimicrobial spectra of macrolides representatives of the different groups included in Table 2 except erythromycin (see chapter by OLEINICK). *In vivo*, macrolide antibiotics of the

Table 2. *The macrolide group of antibiotics. Classification according to the sugar moieties linked to the aglycone ring*

Group	Characteristics	Members of the group	Sugars linked to the aglycone	
Spiramycin	a	Spiramycins	-mycaminose-mycarose	-isomycamine (synonym forosamine)
		Angolamycin (synonym Shincomycin A)	-angolosamine-mycarose	-mycinose
		Relomycin	-mycaminose-mycarose	-mycinose
		Tylosin	-mycaminose-mycarose	-mycinose
Carbomycin	b	Carbomycins	-mycaminose-mycarose-isovaleryl	
		Josamycin (synonym Leucomycin A₃)	-mycaminose-mycarose-isovaleryl	
		Leucomycins (synonyms Kitasamycins)	-mycaminose-mycarose-R	
		Niddamycin	-mycaminose-mycarose-isovaleryl	
Erythromycin	c	Erythromycins	-desosamine	-cladinose or -mycarose
		Neospiramycins	-mycaminose	-isomycamine
		Oleandomycin	-desosamine	-oleandrose
		Megalomycin A	-desosamine	-mycarose
Methymycin	d	Methymycin	-desosamine	
		Forocidins	-mycaminose	
		Narbomycin	-desosamine	
		Neomethymycin	-desosamine	
		Picromycin	-desosamine	
Lancamycin	e	Chalcomycin	-mycinose	-chalcose
		Neutramycin	-mycinose	-chalcose
		Lancamycin (synonym Kujimycin B)	-acetylarcanose	-chalcose (synonym lancavose)
		Kujimycin A	-arcanose	-chalcose

[a] Containing the -mycaminose-mycarose moiety and also another sugar linked to the aglycone.
[b] Containing the -mycaminose-mycarose-R moiety linked to the aglycone.
[c] Containing the moieties -amino sugar and either -amino sugar or -non nitrogenous sugar linked to the aglycone.
[d] Containing the moiety -amino sugar linked to the aglycone.
[e] Containing the moieties -non nitrogenous sugar and -non nitrogenous sugar linked to the aglycone.

The anglicized names lancamycin, picromycin and megalomycin are used rather than the names lankamycin, pikromycin and megalomicin which are sometimes found in the literature.

Table 3. *In vitro inhibitory concentrations of different macrolide antibiotics (μg/ml)*

Organism	Carbo-mycin	Spira-mycin	Methy-mycin	Lanca-mycin
Gram-positive bacteria:				
Bacillus megaterium	—	—	—	100
Bacillus subtilis	0.36	3	730	—
Staphylococcus aureus	0.22	1	40	100
Streptococcus pyogenes	0.08	0.6	10.5	100
Streptococcus faecalis	1.3	1	1 875	>100
Diplococcus pneumoniae	0.07	0.2	6.0–50	—
Corynebacterium diphtheriae	—	3	20	100
Mycobacterium sp. 607	—	23	—	—
Mycobacterium tuberculosis	<50	—	1 875	—
Gram-negative bacteria:				
Escherichia coli	100	31	1 875	>100
Aerobacter aerogenes	100	31	1 875	—
Neisseria catarrhalis	—	10	—	—
Klebsiella pneumoniae	3	33	3.2	100
Pseudomonas aeruginosa	60	>1 500	1 875	>100
Proteus vulgaris	100	>1 500	1 875	—
Yeasts:				
Candida albicans	100	—	—	>100
Protozoa:				
Endamoeba histolytica	>100		—	125

Data taken from CORBAZ et al. (1955a); GÄUMANN et al. (1960); DONIN et al. (1954); ENGLISH et al. (1952); PAGANO et al. (1953); PINNERT-SINDICO et al. (1954/55).

carbomycin, spiramycin and erythromycin groups are very effective in protecting mice against lethal infections of *Staphylococcus aureus*, *Diplococcus pneumoniae* and *Streptococcus pyogenes* whereas antibiotics of the methymycin and lancamycin groups have very small activity. Toxicity of macrolides in mice is very low and their LD_{50} ranges from 1–3 g/kg (ENGLISH et al., 1952; PINNERT-SINDICO et al., 1954/55; CORBAZ et al., 1955a; DONIN et al., 1953/54; GASTAL, 1958; MANIAR and EIDUS, 1961; GÄUMANN et al., 1960; OSONO and UMEZAWA, 1971). Anti-rickettsial and antiviral activities of carbomycin have been reported (PAGANO et al., 1953) and at high concentrations the antibiotic is active against protozoa (SENECA and IDES, 1953).

Macrolide antibiotics are bacteriostatic and their minimum growth inhibitory concentration does not depend significantly on the inoculum size (GARROD and WATERWORTH, 1956). However it has been found that antibiotics of the carbomycin and spiramycin groups at concentrations 4–10 times their minimum growth inhibitory concentration are bactericidal (LUTZ et al., 1957). Macrolide antibiotics antagonize and interfere with the bactericidal effect of penicillin (COLEMAN et al., 1953; MATHIEU and FAGUET, 1958).

Cross-resistance of bacteria to the different macrolide antibiotics has been frequently observed (FINLAND et al.,1952; FUSILLO et al., 1953; JONES et al., 1956; GARROD and WATERWORTH, 1956). However it has been observed in some cases that erythromycin-resistant mutants are sensitive to other macrolides and vice-versa (Hsie et al., 1955/56; Chabbert, 1956; Hudson et al., 1956). Weisblum and Demohn (1969) have studied a class of erythromycin-resistant mutants of S. aureus, which in normal conditions are sensitive to other macrolide antibiotics, but in the presence of erythromycin become resistant to them. This phenomenon of erythromycin-induced resistance is not fully understood but presumably indicates that the binding of erythromycin excludes, or at least disfavours the binding of the other macrolides to which the bacteria was sensitive.

   Protein Synthesis Inhibition. Inhibition of protein synthesis in intact bacteria by macrolide antibiotics has bee reported using erythromycin (Brock and Brock, 1959; Taubman et al., 1963), spiramycin III, neospiramycin III and forocidin III (Vazquez, 1967b), spiramycin (Ahmed, 1968a), chalcomycin (Jordan, 1963) and leucomycin (Tago and Nagano, 1970) and it is widely accepted that all the macrolide antibiotics are inhibitors of protein synthesis. Studies in bacterial cell-free systems have shown the inhibitory effect of a number of macrolides on amino acid incorporation directed by synthetic polynucleotides. These effects were initially observed with erythromycin (Wolfe and Hahn, 1964; Taubman et al., 1966) spiramycin, carbomycin, angolamycin, methymycin, oleandomycin, erythromycin and lancamycin (Vazquez, 1966b), leucomycin (Tago and Nagano, 1970), spiramycin III, neospiramycin III and forocidin III (Vazquez, 1967b) and chalcomycin, tylosin and niddamycin (Mao and Wiegand, 1968). Effects of the antibiotics in these systems are much dependent on the experimental conditions and it has been observed that their action on polyphenylalanine synthesis takes place preferentially at high pH (Mao and Wiegand, 1968). However it is generally concluded from different reports that macrolides of the carbomycin and spiramycin groups (Table 2) are good inhibitors of proline, lysine and phenylalanine incorporation directed respectively by poly C, poly A and poly U whereas macrolides of the erythromycin, methymycin and lancamycin groups (Table 2) are moderately active as inhibitors of proline and lysine incorporation directed respectively by poly C and poly A but have small or no activity on poly U-dependent phenylalanine incorporation (Vazquez, 1966b, 1967b; Mao and Wiegand, 1968; Wilhelm et al., 1968). These results suggest the importance of the -mycaminose-mycarose moiety on these differential effects since all the macrolides having this moiety do have a strong inhibitory effect with the different polynucleotides used as mRNA. On the other hand neospiramycins and forocidins are derived from spiramycins and have the same aglycone but differ in the sugar moiety (Table 2) and behave in the cell-free systems like erythromycin and methymycin respectively (Vazquez, 1966b, 1967b).

   The effects of the macrolide antibiotics have been extensively studied on the length of the poly A-dependent polylysine and poly U-dependent polyphenylalanine synthesised in the presence or the absence of EF G. The experimental conditions of these studies by various workers were quite different but, in general, there is good agreement on the results obtained. Macrolides of the carbomycin group strongly inhibit rate and total synthesis of all peptides whereas antibiotics of the spiramycin group cause a slight or no inhibition of dilysine formation but block synthesis of larger peptides and macrolides of the erythromycin group always enhance synthesis of dipeptide and even in some cases stimulate tri-

and tetrapeptides synthesis but in all cases block synthesis of larger peptides (Mao and Robishaw, 1971; Tanaka and Teraoka, 1968; Teraoka et al., 1969; Pulkrábek et al., 1970).

The macrolide antibiotics have also been found to block protein synthesis by mitochondria and chloroplasts ribosomes and have been very useful markers on genetic studies with mitochondria (Linnane et al., 1968; Gingold et al., 1969; Ellis and Hartley, 1971; Borst and Grivell, 1971). It has also been reported that erythromycin is active on yeast mitochondrial ribosomes but not on mammalian ribosomes (Firkin and Linnane, 1969; Davey et al., 1970). This very interesting finding should be confirmed by other workers.

Binding to Ribosomes of Macrolide Antibiotics. It has early been shown that antibiotics of the spiramycin, carbomycin, erythromycin, methymycin and lancamycin groups inhibit chloramphenicol binding to bacterial ribosomes (Vazquez, 1963, 1966a, Wolfe and Hahn, 1965). These results suggested that all macrolides act, like chloramphenicol, on the 50S subunit of bacterial ribosomes. Evidence for spiramycin III action on the 50S ribosomal subunit was also obtained by studying polyphenylalanine synthesising activity ribosomes reconstituted from subunits treated or untreated with the antibiotic (Vazquez, 1967a). Direct evidence for macrolide binding to bacterial ribosomes has been obtained with erythromycin (Taubman et al., 1966; Wilhelm et al., 1968; Mao and Putterman, 1969; Tanaka et al., 1968; Tanaka et al., 1966; Teraoka, 1970; Otaka et al., 1970; Wilhelm and Corcoran, 1967; Wilhelm et al., 1968; Oleinick and Corcoran, 1969; Fernandez-Muñoz et al., 1971; Teraoka and Tanaka, 1971), spiramycin III (Vazquez 1967b), spiramycin (Ahmed 1968a, b; Saito et al., 1971), and leucomycin (Tago and Nagano, 1970). In most of these studies specific interaction of the macrolide antibiotics with the 50S subunit was also demonstrated. Binding to ribosomes of radioactive macrolides requires K^+ or NH_4^+ and is inhibited by other non-radioactive macrolides but not by synthetic polynucleotides and the aglycone or sugar moieties of a number of macrolides. In the case of spiramycin III and erythromycin binding, high concentrations of antibiotics of the methymycin and lancamycin groups are required for inhibition probably due to a small affinity of these antibiotics for the ribosome (Vazquez, 1967b; Wilhelm et al., 1968; Fernandez-Muñoz et al., 1971). Similarly to which occurs with some other centers of the 50S ribosome subunit, activity of the binding site for erythromycin can be lowered or lost due to the method of preparation of ribosomes. Erythromycin binding activity can be recovered by heating the ribosome preparations between 20° C and 50° C in the presence of NH_4^+ or K^+ and it has been suggested that the process of activation involves a change in the ribosomal conformation (Teraoka, 1970; Vogel et al., 1971).

Ribosomes from Gram-positive bacteria bind erythromycin with a higher affinity than those from Gram-negative microorganisms. In both cases affinity is enhanced by the presence of ethanol but specific binding is complete in either case at one molecule of the antibiotic per ribosome (Wilhelm et al., 1968; Oleinick and Corcoran, 1969; Otaka et al., 1970; Fernandez-Muñoz et al., 1971). Erythromycin and spiramycin are known to have no effect on eucaryotic systems; in good agreement with these data it has been observed that these antibiotics do not bind to cytoplasmic ribosomes from green algae, rabbit reticulocyte and rat liver (MAO et al., 1970; RODRIGUEZ-LOPEZ and VAZQUEZ, 1968; RODRIGUEZ-LOPEZ et al., 1968).

Non-linear plots have been obtained in certain cases in studies on saturation of erythromycin binding to ribosomes (OLEINICK and CORCORAN, 1969; FERNANDEZ-MUÑOZ et al., 1972) which appear to be due to impurities in the preparations of the radioactive antibiotic since linear plots have been obtained by other workers using pure ^{14}C-erythromycin A (MAO and PUTTERMAN, 1969; FERNANDEZ-MUÑOZ and VAZQUEZ, 1973).

There is some disagreement on lincomycin and chloramphenicol effects on macrolide binding to ribosomes which might be due to the use of different experimental conditions and ribosomes from different bacterial strains. TERAOKA et al. (1969) found no effect by lincomycin on erythromycin binding to E. coli ribosomes and OLEINICK et al. (1968) found no effect by chloramphenicol on erythromycin binding to B. subtilis ribosomes. On the other hand WILHELM et al. (1968) found inhibitory effect by lincomycin on erythromycin binding to B. subtilis and FERNANDEZ-MUÑOZ et al. (1971) observed inhibition by lincomycin and chloramphenicol on erythromycin binding to E. coli ribosomes either in the presence or absence of ethanol. VAZQUEZ (1967b) has also observed inhibition by lincomycin and chloramphenicol on spiramycin III binding to B. subtilis ribosomes. A low affinity for chloramphenicol has been observed in the case of erythromycin-resistant ribosomes from a mutant of Escherichia coli (TANAKA et al., 1972). Similarly resistance to spiramycin and erythromycin was observed in the case of Escherichia coli mutants having ribosomes resistant to erythromycin (ČERNÁ and RYCHLIK, 1968; ČERNÁ et al., 1971). These results suggest that sites for chloramphenicol, spiramycin and erythromycin binding on the ribosome are very closely related or even overlap.

Specific Ribosomal Site of Action for the Macrolide Antibiotics. Ribosomes from B. subtilis and E. coli erythromycin-resistant mutants have a smaller affinity for the antibiotic than ribosomes from the parental sensitive strains (OLEINICK and CORCORAN, 1969; OTAKA et al., 1970). However it was observed that in the case of the E. coli mutants, the ribosomes are less active not only for erythromycin binding but also for catalyzing peptide synthesis and both activities are recovered by the ribosomes when pretreated with rather high concentrations of K^+/NH_4^+ (OTAKA et al., 1970). Ribosomal mutations leading to resistance to erythromycin have been shown to be located in the streptomycin region in B. subtilis and E. coli (DUBNAU et al., 1967; GOLDTHWAITE et al., 1970; TANAKA et al., 1970). Furthermore in the case of E. coli mutants it has been observed that at least the protein component L4 (also known as 50-8) of the 50S ribosomal subunit is distinct from that of the parental sensitive strain and the chemical nature of the alteration differs in different mutants (TANAKA et al., 1970; TANAKA et al., 1968; OTAKA et al., 1970; OTAKA et al., 1971). These erythromycin-resistant ribosomes have a low peptidyl transferase activity and are cross-resistant to tylosin, spiramycin, oleandomycin and leucomycin (TANAKA et al., 1971a). On these considerations it has been proposed that changes in the L4 protein in the erythromycin-resistant mutants cause an alteration in the entire conformation of the 50S subunit affecting not only the erythromycin binding site but also the peptidyl transferase centre. It is not known if erythromycin binds specifically to the L4 protein on the 50S ribosome subunit. In any case some very specific structural components of the subunit are required since γ-cores derived from 50S subunits contain the L4 protein (HOMANN and NIERHAUS, 1971) but do not bind erythromycin whereas β-cores prepared from γ-cores by addition of the split protein fraction are able to bind erythromycin (BALLESTA *et al.*, 1971). *E. coli* mutants with a higher sensitivity to erythromycin and lincomycin have been found and it was shown that these mutants have altered at least one of the proteins of the 50S subunit (APIRION, 1967; KREMBEL and APIRION, 1968).

A considerable reduction in spiramycin binding by ribosomes from resistant strains of *Staphylococcus aureus* when compared with ribosomes from sensitive strains (SHIMIZU *et al.*, 1970a, b) has been observed. *E. coli* mutants resistant to spiramycin have also been selected; one of these mutants contains an altered protein of the 50S subunit (protein known as L4 or 50-8) whereas no such change has been detected in any other ribosomal protein. Some of the spiramycin-resistant mutants, but not others, showed cross-resistance with erythromycin, tylosin and leucomycin (TANAKA *et al.*, 1971b).

Protein Synthesis Reaction(s) Affected by the Macrolide Antibiotics. It is well established from the above reports that the macrolide antibiotics act at the level of the 50S ribosome subunit and there is no evidence for these antibiotics having any effect on aminoacyl-tRNA formation or codon-anticodon interaction at the level of the 30S ribosome subunit. Macrolide antibiotics were found to have no inhibition on non-enzymic binding of either polylysyl- or Ac-Phe-tRNA to ribosomes (ČERNÁ *et al.*, 1969). However ENNIS and DUFFY (1972) found that macrolides of the carbomycin group inhibit non-enzymic binding of f-Met-tRNA whereas macrolides of the other groups rather enhance this binding.

Concerning the non-enzymic binding of aminoacyl-tRNA, it has been observed that macrolides do not affect binding by 30S subunits but antibiotics of the carbomycin and spiramycin groups inhibit to a certain extent the enhancement in poly U-directed Phe-tRNA binding which takes places when 50S subunits are added; macrolides of the other groups enhance or have no effect on this reaction (VAZQUEZ and MONRO, 1967). There is apparently some discrepancy about the effects of the macrolides on enzymic aminoacyl-tRNA binding since HILL (1969) has observed inhibition of Phe-tRNA binding by spiramycin in the presence of factor T and GTP whereas MAO and ROBISHAW (1971) did not find any effect by the macrolide antibiotics of the carbomycin, spiramycin and erythromycin groups on EF T- and EF G-dependent GTPase and on EF T-dependent Phe-tRNA binding in the presence of GMP-PCP. This GTP analogue was added to prevent any possible polymerization which might occur in the system for non-enzymic binding described-above (VAZQUEZ and MONRO, 1967) and also in the experiments of HILL (1969) if EF T was slightly contaminated with EF G.

Effects by macrolide antibiotics on peptide bond formation have been studied in several different experimental systems. Early work has shown that erythromycin blocks peptide bond formation by ribosomes when polylysyl-tRNA is used as a donor substrate and either puromycin (RYCHLIK, 1966) or lysyl-tRNA is used as an acceptor substrate (GOTTESMAN, 1964). However inhibition of peptide bond formation in these systems might be an indirect effect due to a blockade on

other reaction in protein synthesis. The fragment reaction assay for peptide bond
formation involves only the peptidyl transferase centre of the larger ribosome
subunit, either CACCA-Met-f or CACCA-Leu-Ac as a donor substrate and
puromycin as an acceptor substrate. Using this system it has been shown that
antibiotics of the carbomycin and spiramycin groups block peptide bond formation whereas macrolides of the erythromycin, methymycin and lancamycin groups
have no inhibitory effect (MONRO and VAZQUEZ, 1967). Many other different
experimental systems and donor substrates have been used to study the effects
of the macrolide antibiotics on peptide bond formation since it has been found
that the results obtained might change with the experimental conditions. Antibiotics of the carbomycin group have been found to block peptide bond formation
in most of the experimental systems for reaction with puromycin in which they
were tested, f. i. using as a donor substrate either polylysyl- or f-Met- or Ac-Phe-
tRNA or CACCA-Phe-Ac or CACCA-Leu-Ac (ČERNÁ et al., 1969; ČERNÁ et al.,
1971; MAO and ROBISHAW, 1971). Antibiotics of the spiramycin group have
been found to block peptide bond formation using as a donor substrate either
polylysyl- or Ac-Phe- or Gly-Phe- or Phe-Phe or Leu-Phe- or Val-Gly-Phe-tRNA
or CACCA-Phe-Ac or CACCA-Leu-Ac and as an acceptor substrate puromycin
(ČERNÁ et al., 1969; ČERNÁ et al., 1971; TANAKA et al., 1971); however antibiotics
of the spiramycin group are very poor inhibitors of the puromycin reaction
when f-Met-tRNA is used as a donor substrate (MAO and ROBISHAW, 1971).
Antibiotics of the erythromycin group are good inhibitors of the puromycin
reaction when polylysyl-tRNa or Phe-Phe-tRNA is used as a donor substrate
but not in the case of using either f-Met- or Ac-Phe- or Gly-Phe- or Leu-Phe-
or Val-Gly-Phe-tRNA as a donor substrate (ČERNÁ et al., 1969; ČERNÁ et al.,
1971; MAO and ROBISHAW, 1971; Tanaka et al., 1971). RYCHLIK and ČERNÁ
(1972) have studied the effect of erythromycin on the puromycin reaction using
as donor substrates specific polylysyl-tRNA and found a total inhibition by
the antibiotics when pentalysyl-tRNA or larger polypeptidyl-tRNA are used
as donor substrates but the inhibition was much smaller in the case of using
tetralysyl-tRNA as a donor substrate. The authors concluded that inhibition
of longer peptides by erythromycin is caused by interaction with peptide chains
longer than two or three amino acids. Extensive studies have been carried out
by MAO and ROBISHAW (1972) on the effect of erythromycin on the fragment
and puromycin reactions with a wide range of substrates and found inhibition
using some of them (Ac-Pro-Gly-, Ac-Phe-Gly-, Ac-Phe-Phe-, Ac-Phe-Leu-, Ac-
Phe-Pro- and Phe-Phe-tRNA) and stimulation using some others (Ac-Gly-Gly-
and Ac-Gly-Pro-tRNA). The authors concluded that hydrophilicity and length
of the peptidyl chain are important factors on the different results obtained
with erythromycin; they consider the antibiotic as an effector acting on the
donor site of the peptidyl transferase center affecting the velocity of the enzymic
reaction in a positive or negative way depending on the nature of the donor
substrate.

    In order to study substrate-interaction at the peptidyl transferase center, assays
have been developed to measure CACCA-Leu-Ac binding to the donor site and
either UACCA-Leu-Ac or CACCA-Leu or CACCA-Phe to the acceptor site.
Either no inhibitory effect or some stimulation was found in these systems by

antibiotics of the erythromycin and methymycin groups on substrate binding to either the donor or the acceptor sites of the peptidyl transferase center whereas antibiotics of the spiramycin and carbomycin groups inhibit binding of substrates to both sites (PESTKA, 1969; MONRO *et al.*, 1970; CELMA *et al.*, 1970, 1971; RYCHLIK and ČERNÁ, 1972). Antibiotics of the carbomycin and spiramycin groups, but not those from the erythromycin, methymycin and lancamycin groups also block the sparsomycin-induced binding of CACCA-Leu-Ac to the donor site of the peptidyl transferase center (MONRO *et al.*, 1969).

Other experiments have shown that erythromycin blocks "in vivo" release of nascent peptides by puromycin from ribosomes or *B. megaterium* protoplasts and this might be due to an effect of the macrolide on peptide bond formation or translocation. However erythromycin hardly inhibits this puromycin reaction in the presence of chlortetracycline and based on these considerations it was postulated that erythromycin might inhibit any reaction in the translocation step (CUNDLIFFE and McQUILLEN, 1967). This conclusion is also supported by the finding of IGARASHI *et al.* (1969) and TANAKA *et al.* (1973) that erythromycin inhibits the EF G-factor dependent release of tRNA from ribosomes which is coupled or precedes translocation of peptidyl-tRNA. It has also been shown that erythromycin preserves polysome breakdown (CUNDLIFFE, 1969) and this is compatible with the antibiotic acting as an inhibitor of either peptide bond formation or translocation. An effect of erythromycin in translocation is also compatible with the results obtained by CANNON and BURNS (1971) studying the effect of the antibiotic on the puromycin reaction under different conditions of incubation.

Although most of the data mentioned above suggest that macrolides of the erythromycin group cannot be considered as inhibitors of peptide bond formation there are a number of reports clearly showing that erythromycin blocks the inhibitory effect of chloamphenicol and lincomycin on the peptidyl transferase center (TERAOKA, 1970; CHANG and WEISBLUM, 1967; MONRO *et al.*, 1970,1971). These results and those mentioned above showing parallel changes in activity, inactivation and reactivation of peptidyl transferase and binding of erythromycin (OTAKA *et al.*, 1970; TANAKA *et al.*, 1971a) suggest that erythromycin site of action on the 50S subunit is very closely related to the peptidyl transferase center and the hypotheses mentioned above of RYCHLIK and ČERNÁ (1972) and MAO and ROBISHAW (1972) on the mode of action of erythromycin might be correct.

Spiramycin has been shown to cause polysome breakdown (CUNDLIFFE, 1969). This might be due either to a premature detachment of ribosomes from polysomes or an inhibition of the initiation step of polypeptide synthesis by the antibiotic. This second possibility could be correct since the antibiotic has been found to block substrate binding to the donor- (MONRO *et al.*, 1970; CELMA *et al.*, 1970) and acceptor-sites (PESTKA, 1969; CELMA *et al.*,1971; RYCHLIK and ČERNÁ, 1972) of the peptidyl transferase center. Certainly by blocking interaction with the donor site of the peptidyl transferase center, antibiotics of the carbomycin and spiramycin groups might block not only peptide bond formation but also initiation of polypeptide synthesis. PESTKA (1972) has reported that a number of macrolide antibiotics of the carbomycin, spiramycin and erythromycin groups which have been tested do not inhibit peptide bond formation.

Discussion

Summarizing the results mentioned above concerning the site and mode of action of the macrolide antibiotics, it can be said that there is ample experimental evidence that all of them are inhibitors of protein synthesis in bacterial, but not in eucaryotic systems, as a consequence of their interaction with the 50S ribosome subunit. Protein L 4 (also known as 50-8) from this subunit of *E. coli* ribosomes might be directly implicated in the interaction of the macrolide antibiotics since it is altered in the case of mutations leading to resistance to spiramycin and erythromycin.

There is a controversy concerning the reaction(s) in protein synthesis which might be affected by the macrolide antibiotics in intact bacteria. Most of the experimental evidence presented above using different systems suggests that antibiotics of the carbomycin and spiramycin groups act at the peptidyl transferase center blocking peptide bond formation and possibly initiation by preventing the attachment of the 3′ end of substrates to the donor and acceptor site of that center. Nevertheless even these antibiotics were found by PESTKA (1972) not to be inhibitors of peptide bond formation in a polysomal system.

Antibiotics of the methymycin and lancamycin groups have been found not to block peptide bond formation in the few experimental systems in which they have been tested.

Antibiotics of the erythromycin group are not inhibitors of peptide bond formation in most of the experimental systems devised and using certain donor substrates; however these antibiotics are known to act at or near the peptidyl transferase center either as an effector (MAO and ROBISHAW, 1972) or blocking peptide bond formation when the donor site has a peptidyl moiety of a certain length (RYCHLIK and ČERNÁ, 1972). These results appear to contradict the finding by PESTKA (1972) that antibiotics of the erythromycin group do not inhibit peptide bond formation by bacterial polysomes. Also results from other workers suggest that erythromycin might act on the complex step of translocation (CUNDLIFFE and MCQUILLEN, 1967; IGARASHI et al., 1969; TANAKA et al., 1973).

Although not having the same mechanism of action, it is known that most of the macrolide antibiotics block binding of chloramphenicol, lincomycin and spiramycin to the ribosome (VAZQUEZ, 1967b; FERNANDEZ-MUÑOZ et al., 1971). It is possible that the aglycone ring is responsible for this common effect of the macrolides whereas the sugar moieties might be responsible for the differential effect of the macrolides on peptide bond formation. In support of this hypothesis are the results obtained with spiramycin III and their derived antibiotics neospiramycin III and forocidin III which have the same aglycone but lack one and two of their sugars respectively (Table 2) (Fig. 1). Although having the same aglycone as spiramycin III, neospiramycin III and forocidin III behave in their mode of action similarly to other antibiotics of the erythromycin and methymycin group respectively. It is therefore highly probable that they bind at the same site as spiramycin III (and probably also all the other macrolides) and that it is only the extra sugar residue of spiramycin III (mainly the mycarose of the -mycaminose-mycarose moiety) which interferes with substrate binding at the peptidyl transferase center (MONRO et al., 1971). Therefore, there might be a common site in the ribosome for the aglycone moiety of the macrolide antibiotics, which is not on the peptidyltransferase centre. However the effects of the macrolide antibiotics are dependent on their sugar and aminosugar moieties which interact with other ribosomal sites.

References

AHMED, A.: Mechanism of inhibition of protein synthesis by spiramycin. Biochim. Biophys. Acta **166**, 205 (1968a).

AHMED, A.: Altered ribosomes in spiramycin-resistant mutants of *Bacillus subtilis*. Biochim. Biophys. Acta **166**, 218 (1968b).

ANLIKER, R., u. K. GUBLER: Stoffwechselprodukte von Actinomyceten. (6. Mitteilung.) Über die Konstitution des Pikromycins. I. Helv. Chim. Acta **40**, 119 (1957).

APIRION, D.: Three genes that affect *Escherichia coli* ribosomes. J. Mol. Biol. **30**, 255 (1967).

BALLESTA, J.P.G., V. MONTEJO, and D. VAZQUEZ: Reconstitution of the 50S ribosome subunit. Localization of activities related to the peptidyl transferase centre. FEBS Letters **19**, 75 (1971).

BORST, P., and L.A. GRIVELL: Mitochondrial ribosomes. FEBS Letters **13**, 73 (1971).

BROCK, T.D., and M.L. BROCK: Similarity in mode of action of chloramphenicol and erythromycin. Biochim. Biophys. Acta **33**, 274 (1959).

BROCKMANN, H., u. W. HENKEL: Pikromycin ein neues Antibiotikum aus Actinomyceten. Naturwissenschaften **37**, 138 (1950).

BROCKMANN, H., u. R. OSTER: Zur Konstitution des Pikromycins und Kromycins (Pikromycin, VI). Chem. Ber. **90**, 605 (1957).

CANNON, M., and K. BURNS: Modes of action of erythromycin and thiostrepton as inhibitors of protein synthesis. FEBS Letters **18**, 1 (1971).

CELMA, M.L., R.E. MONRO, and D. VAZQUEZ: Substrate and antibiotic binding sites at the peptidyl transferase centre of *E. coli* ribosomes. FEBS Letters **6**, 273 (1970).

CELMA, M.L., R.E. MONRO, and D. VAZQUEZ: Substrate and antibiotic binding sites at the peptidyl transferase centre of *E. coli* ribosomes: Binding of UACCA-Leu to 50S subunits. FEBS Letters **13**, 247 (1971).

CELMER, W.D.: Macrolide stereochemistry. I. The total absolute configuration of oleandomycin. J. Am. Chem. Soc. **87**, 1797 (1965a).

CELMER, W.D.: Macrolide stereochemistry. II. Configurational assignments at certain centers in various macrolide antibiotics. J. Am. Chem. Soc. **87**, 1799 (1965b).

CELMER, W.D.: Macrolide stereochemistry. III. A configurational model for macrolide antibiotics. J. Am. Chem. Soc. **87**, 1801 (1965c).

CELMER, W.D.: Basic stereochemical research topics in the macrolide antibiotics. In: Biogenesis of antibiotic substances, ed. by Z. VANĚK and Z. HOŠŤÁLĚK. New York and London: Academic Press (1965d).

CELMER, W.D.: Stereochemical problems in macrolide antibiotics. Pure Applied Chem. **28**, 413 (1971).

CELMER, W.D., H. ELS, and K. MURAI: Oleandomycin derivatives. Preparation and characterization. Antibiotics Ann. **1957/58**, 476.

ČERNÁ, J., J. JONAK, and I. RYCHLIK: Effects of macrolide antibiotics on the ribosomal peptidyl transferase in cell-free systems derived from *Escherichia coli* B and erythromycin-resistant mutant of *Escherichia coli* B. Biochim. Biophys. Acta **240**, 109 (1971).

ČERNÁ, J., and I. RYCHLIK: Cross-resistance of *Escherichia coli* B ribosomes to inhibition of the puromycin reaction by erythromycin, spiramycin and chloramphenicol. Biochim. Biophys. Acta **157**, 436 (1968).

ČERNÁ, J., I. RYCHLIK, and P. PULKRÁBEK: The effect of antibiotics on the coded binding of peptidyl-tRNA to the ribosome and on the transfer of the peptidyl residue to puromycin. Eur. J. Biochem. **9**, 27 (1969).

CHABBERT, Y.: Antagonisme *in vitro* entre l'erythromycine et la spiramycine. Ann. Inst. Pasteur **90**, 787 (1956).

CHANG, F.N., and B. WEISBLUM: The specificity of lincomycin binding to ribosomes. Biochemistry **6**, 836 (1967).

Coffey, G.L., L.E. Anderson, J.D. Douros, A.L. Erlandson, M.W. Fisher, R.J. Hans, R.F. Pittillo, D.K. Vogler, K.S. Weston, and J. Ehrlich: Chalcomycin, a new antibiotic: Biological studies. Can. J. Microbiol. **9**, 665 (1963).

Coleman, V.R., J.B. Gunnison, and E. Jawetz: Participation of erythromycin and carbomycin in combined antibiotic action *in vitro*. Proc. Soc. Exptl. Biol. Med. **83**, 668 (1953).

Corbaz, R., L. Ettlinger, E. Gäumann, W. Keller, F. Kradolfer, E. Kyburz, L. Neipp, V. Prelog, R. Reusser u. H. Zähner: Stoffwechselprodukte von Actinomyceten, Narbomycin. Helv. Chim. Acta **38**, 935 (1955b).

Corbaz, R., L. Ettlinger, E. Gäumann, W. Keller-Schierlein, L. Neipp, V. Prelog, P. Reusser u. H. Zähner: Stoffwechselprodukte von Actinomyceten. Angolamycin. Helv. Chim. Acta **38**, 1202 (1955a).

Cundliffe, E.: Antibiotics and polyribosomes. II. Some effects of lincomycin, spiramycin and streptogramin A *in vivo*. Biochemistry **8**, 2063 (1969).

Cundliffe, E., and K. McQuillen: Bacterial protein synthesis: The effects of antibiotics. J. Mol. Biol. **30**, 137 (1967).

Davey, P.J., J.M. Haslam, and A.W. Linnane: Biogenesis of mitochondria. 12. The effects of aminoglycoside antibiotics on the mitochondrial and cytoplasmic protein-synthesizing systems of *Saccharomyces cerevisiae*. Arch. Biochem. Biophys. **136**, 54 (1970).

Djerassi, C., and O. Halpern: Macrolide antibiotics. VII. The structure of neomethymycin. Tetrahedron **3**, 255 (1958).

Djerassi, C., and J.A. Zderic: The structure of the antibiotic methymycin. J. Am. Chem. Soc. **78**, 6390 (1956).

Donin, M.N., J. Pagano, J.D. Dutcher, and C.M. McKee: Methymycin, a new crystalline antibiotic. Antibiotics Ann. **1953/54**, 179.

Dubnau, D., C. Goldthwaite, I. Smith, and J. Marmur: Genetic mapping in *Bacillus subtilis*. J. Mol. Biol. **27**, 163 (1967).

Egan, R.S., and J.R. Martin: Structure of lankamycin. J. Am. Chem. Soc. **92**, 4129 (1970).

Ellis, R.J., and M.R. Hartley: Sites of synthesis of chloroplasts proteins. Nature New Biol. **233**, 193 (1971).

English, A.R., M.F. Field, S.R. Szendy, N.J. Tagliani, and R.A. Fitts: Magnamycin. I. *In vitro* studies. Antibiot. & Chemotherapy **2**, 678 (1952).

Ennis, H.L., and K.E. Duffy: Vernamycin A inhibits the non-enzymic binding of Fmet-tRNA to ribosomes. (Submited for publication.)

Feldman, L.I., I.K. Dill, Ch.E. Holmlund, H.A. Whaley, E.L. Patterson, and N. Bohonos: Microbiological transformation of macrolide antibiotics. Antimicrobial Agents Chemotherapy **1963**, 54.

Fernandez-Muñoz, R., R.E. Monro, and D. Vazquez: Binding of chloramphenicol, lincomycin and erythromycin to *E. coli* ribosomes. In: Molecular mechanisms of antibiotic action on protein synthesis and membranes, p. 207, ed. by E. Muñoz, F. Garcia-Ferrandiz and D. Vazquez. Amsterdam-London-New York: Elsevier 1972.

Fernandez-Muñoz, R., R.E. Monro, R. Torres-Pinedo, and D. Vazquez: Substrate- and antibiotic-binding sites at the peptidyl-transferase centre of *Escherichia coli* ribosomes. Studies on the chloramphenicol, lincomycin and erythromycin sites. Eur. J. Biochem. **23**, 185 (1971).

Fernandez-Muñoz, R., and D. Vazquez: Erythromycin binding to *E. coli* ribosomes. J. Antibiotics (Tokyo) **26**, 107 (1972).

Finland, M., C. Wilcox, S.S. Wright, and E.M. Purcell: Cross-resistance to antibiotics: Effect of exposures of bacteria to carbomycin or erythromycin *in vitro*. Proc. Soc. Exptl. Biol. Med. **81**, 725 (1952).

Firkin, F.C., and A.W. Linnane: Phylogenetic differences in the sensitivity of mitochondrial protein synthesizing systems to antibiotics. FEBS Letters **2**, 330 (1969).

Fusillo, M.H., H.E. Noyes, E.J. Pulaski, and J.Y.S. Tom: Antimicrobial spectrum and cross-resistance studies of erythromycin and carbomycin. Antibiot. & Chemotherapy **3**, 581 (1953).

Gäumann, E., R. Hütter, W. Keller-Schierlein, L. Neipp, V. Prelog u. H. Zähner: Stoffwechselprodukte von Actinomyceten. Lankamycin und Lankacidin. Helv. Chim. Acta **43**, 601 (1960).

Garrod, L.P., and P.M. Waterworth: Behaviour *in vitro* of some new antistaphylococcal antibiotics. Brit. Med. J. **1956**, 61.

Gastal, R.: Action de la spiramycine sur l'infection expérimentale de la souris par *H. pertussis*. Ann. Inst. Pasteur **94**, 636 (1958).

GINGOLD, E. B., G. W. SAUNDERS, H. B. LUKINS, and A. W. LINNANE: Biogenesis of mitochondria. X. Reassortment of the cytoplasmic genetic determinants for respiratory competence and erythromycin resistance in *Saccharomyces cerevisiae*. Genetics **62**, 735 (1969).

GOLDTHWAITE, C., D. DUBNAU, and I. SMITH: Genetic mapping of antibiotic resistance in markers *Bacillus subtilis*. Proc. Natl. Acad. Sci. U. S. **65**, 96 (1970).

GOTTESMAN, M.: Reaction of ribosome-bound peptidyl transfer ribonucleic acid with aminoacyl transfer ribonucleic acid or puromycin. J. Biol. Chem. **242**, 5564 (1967).

HAMILL, R. L., and W. M. STARK: Macrocin, a new antibiotic, and lactenocin, an active degradation product. J. Antibiotics (Tokyo) **17**, 133 (1964).

HATA, T., Y. SANO, N. OHKI, Y. YOKOYAMA, A. MATSUMAE, and S. ITO: Leucomycin, a new antibiotic. J. Antibiotics (Tokyo) **6**, 87 (1953).

HILL, R. N.: The effects of antibiotics on the interaction of T-factor, aminoacyl-tRNA and ribosomes. J. Gen. Microbiol. **58**, VIII (1969).

HOCHSTEIN, F. A., H. ELS, W. D. CELMER, B. L. SHAPIRO, and R. B. WOODWARD: The structure of oleandomycin. J. Am. Chem. Soc. **82**, 3225 (1960).

HOCHSTEIN, F. A., and K. MURAI: Magnamycin B, a second antibiotic from *Streptomyces halstedii*. J. Am. Chem. Soc. **76**, 5080 (1954).

HOFHEINZ, W., u. H. GRISEBACH: X. Mitt.: Über das Vorkommen von L-Mycarose in Erythromycin C. Z. Naturforsch. **17b**, 852 (1962).

HOMANN, H. E., and K. H. NIERHAUS: Ribosomal proteins: Protein compositions of biosynthetic precursors and artificial subparticles from ribosomal subunits in *Escherichia coli* K 12. Eur. J. Biochem. **20**, 249 (1971).

HSIE, J.-Y., R. KOTZ, and W. NUSSER: Analysis of cross-resistance to erythromycin and carbomycin in *Micrococcus pyogenes* var. *aureus*. Antibiotics Ann. **1955/56**, 773.

HUBER, G., K. H. WALLHAEUSSER, L. FRIES, A. STEIGLER u. H. WEIDENMUELLER: Niddamycin, ein neues Makrolid-Antibioticum. Arzneimittel-Forsch. **12**, 1191 (1962).

HUDSON, D. G., G. M. YOSHIHARA, and W. M. M. KIRBY: Spiramycin. Clinical and laboratory studies. Arch. Internal. Med. **97**, 57 (1956).

IGARASHI, K., H. ISHITSUKA, and A. KAJI: Comparative studies on the mechanism of action of lincomycin, streptomycin and erythromycin. Biochem. Biophys. Res. Commun. **37**, 499 (1969).

JONES, W. F., R. L. NICHOLS, and M. FINLAND: Development of resistance and cross-resistance *in vitro* to erythromycin, carbomycin, spiramycin, oleandomycin and streptogramin. Proc. Soc. Exptl. Biol. Med. **93**, 388 (1956).

JORDAN, D. C.: Effect of chalcomycin on protein synthesis by *Staphylococcus aureus*. Can. J. Microbiol. **9**, 129 (1963).

KELLER-SCHIERLEIN, W., u. G. RONCARI: Stoffwechselprodukte von Mirkoorganismen. 46. Mitteilung [1]. Die Konstitution des Lankamycins. Helv. Chim. Acta **47**, 78 (1964).

KINUMAKI, A., and M. SUZUKI: Proposed structure of angolamycin (shincomycin A) by mass spectrometry. J. Antibiotics (Tokyo) **25**, 480 (1972).

KREMBEL, J., and D. APIRION: Changes in ribosomal proteins associated with mutants in a locus that affects *Escherichia coli* ribosomes. J. Mol. Biol. **33**, 363 (1968).

KUEHNE, M. E., and B. W. BENSON: The structures of the spiramycins and magnamycin. J. Am. chem. Soc. **87**, 4660 (1965).

LINNANE, A. W., G. W. SAUNDERS, E. B. GINGOLD, and H. B. LUKINS: The biogenesis of mitochondria, V. Cytoplasmic inheritance of erythromycin resistance in *Saccharomyces cerevisiae*. Proc. Natl. Acad. Sci. U. S. **59**, 903 (1968).

LUTZ, A., O. GROOTTEN et J. HOFFERER: Evolution et modifications de la resistance des staphylocoques pathogènes à six antibiotiques usuels de 1950 à 1956. L'action comparée *in vitro* de l'erythromycine, de la magnamycine, de la spiramycine, de la novobiocine (albamycine) et de l'oléandomycine. Ann. Inst. Pasteur **92**, 778 (1957).

MALLAMS, A. K., R. S. JARET, and H. REIMANN: The megalomicins. II. The structure of megalomicin A. J. Am. Chem. Soc. **91**, 7506 (1969).

MANIAR, A. C., and L. EIDUS: Un des facteurs influençant l'action des antibiotiques. Ann. Inst. Pasteur **101**, 887 (1961).

MAO, J., and E. E. ROBISHAW: Erythromycin a peptidyl transferase effector. (Submited for publication.)

MAO, J. C. H., and M. PUTTERMAN: The intermolecular complex of erythromycin and ribosome. J. Mol. Biol. **44**, 347 (1969).

Mao, J.C.H., M. Putterman, and R.G. Wiegand: Biochemical basis for the selective toxicity of erythromycin. Biochem. Pharmacol. **19**, 391 (1970).

Mao, J.C.H., and E.E. Robishaw: Effects of macrolides on peptide bond formation and translocation. Biochemistry **10**, 2054 (1971).

Mao, J.C.H., and R.G. Wiegand: Mode of action of macrolides. Biochim. Biophys. Acta **157**, 404 (1968).

Marquez, J., A. Murawski, G.H. Wagman, R.S. Jaret, and H. Reimann: Isolation, purification and preliminary characterization of megalomicin. J. Antibiotics (Tokyo) **22**, 259 (1969).

Mathieu, N., et M. Faguet: Activité *in vitro* de la spiramycine en association avec la tétracycline, l'erythromycine, la pénicilline, la streptomycine sur la multiplication de *Staphylococcus aureus* étudiée au microbiophotomètre. Ann. Inst. Pasteur **94**, 69 (1958).

McGuire, J.M., W.S. Boniece, C.E. Higgens, M.M. Hoehn, W.M. Stark, J. Westhead, and R.N. Wolfe: Tylosin, a new antibiotic. I. Microbiological studies. Antibiot. & Chemotherapy **11**, 320 (1961).

McGuire, J.M., R.L. Bunch, R.C. Anderson, H.E. Boaz, E.H. Flynn, H. Powell, and J.E. Smith: Ilotycin, a new antibiotic. Antibiot. & Chemotherapy **2**, 281 (1952).

Mitscher, L.A., and M.P. Kunstmann: The structure of neutramycin. Experientia **25**, 12 (1969).

Miyake, A., H. Iwasaki, T. Takewata, M. Shibata, and K. Nakazawa: Production of tertiomycin A by *Streptomyces albireticuli*. J. Antibiotics (Tokyo), Ser. A **12**, 59 (1959).

Monro, R.E., M.L. Celma, and D. Vazquez: Action of sparsomycin on ribosome-catalysed peptidyl transfer. Nature **222**, 356 (1969).

Monro, R.E., R. Fernandez-Muñoz, M.L. Celma, A. Jimenez, and D. Vazquez: Antibiotics acting on the peptidyl transferase center of ribosomes. In: Progress in antimicrobial and anticancer chemotherapy, vol. II, p. 473. Tokyo: University of Tokyo 1970.

Monro, R.E., R. Fernandez-Muñoz, M.L. Celma, and D. Vazquez: Mode of action of the lincomycin and related antibiotics. In: Drug action and drug resistance in bacteria: Macrolide antibiotics and lincomycin, p. 305, ed. by S. Mitsuhashi. Tokyo: University of Tokyo Press 1971.

Monro, R.E., and D. Vazquez: Ribosome-catalysed peptidyl transfer: Effects of some inhibitors of protein synthesis. J. Mol. Biol. **28**, 161 (1967).

Morin, R.B., M. Gorman, R.L. Hamill, and P.V. De Marco: The structure of tylosin. Tetrahedron Letters **1970**, 4737.

Muxfeldt, H., S. Shrader, P. Hansen, and H. Brockmann: The structure of pikromycin. J. Am. Chem. Soc. **90**, 4748 (1968).

Oleinick, N.L., and J.W. Corcoran: Two types of binding of erythromycin to ribosomes from antibiotic-sensitive and -resistant *Bacillus subtilis* 168. J. Biol. Chem. **244** 727 (1969).

Oleinick, N.L., J.M. Wilhelm, and J.W. Corcoran: Non-identity of the site of action of erythromycin A and chloramphenicol on *Bacillus subtilis* ribosomes. Biochim. Biophys. Acta **155**, 290 (1968).

Omura, S., Y. Hironaka, and T. Hata: Chemistry of leucomycin. IX. Identification of leucomycin A$_3$ with josamycin. J. Antibiotics (Tokyo) **23**, 511 (1970a).

Omura, S., M. Katagiri, and T. Hata: The structures of leucomycin A$_4$A$_5$A$_6$A$_7$A$_8$ and A$_9$. J. Antibiotics (Tokyo), Ser. A **20**, 234 (1967).

Omura, S., A. Nakagawa, M. Otani, T. Hata, H. Ogura, and K. Furuhata: Structure of the spiramycins (Foromacidines) and their relationship with the leucomycins and carbomycins (Magnamycins). J. Am. Chem. Soc. **91**, 3401 (1969a).

Omura, S., S. Namiki, M. Shibata, T. Muro, H. Nakayoshi, and J. Sawada: Studies on the antibiotics from *Streptomyces spinichromogenes* var. Kujimyceticus. II. Isolation and characterization of Kujimycins A and B. J. Antibiotics (Tokyo) **22**, 500 (1969b).

Omura, S., S. Namiki, M. Shibata, T. Muro, and J. Sawada: Studies on the antibiotics from *Streptomyces spinichromogenes* var. Kujimyceticus. V. Some antimicrobial characteristics of Kujimycin A and Kujimycin B against macrolide resistant staphylococci. J. Antibiotics (Tokyo) **23**, 448 (1970b).

Osono, T., Y. Oka, S. Watanabe, Y. Numazaki, K. Moriyama, H. Ishida, K. Suzaki, Y. Okami, and H. Umezawa: A new antibiotic, josamycin. I. Isolation and physico-chemical characteristics. J. Antibiotics (Tokyo) **20**, 174 (1967).

Osono, T., and H. Umezawa: Josamycin, a new macrolide antibiotic of resistance non-inducing type. In: Drug action and drug resistance in bacteria. 1. Macrolide antibiotics and lincomycin, p. 41, ed. by S. Mitsuhashi. Tokyo: University of Tokyo 1971.

OTAKA, E., I. TAKUZI, S. OSAWA, K. TANAKA, and M. TAMAKI: Peptide analyses of a protein component, 50-8, of 50S ribosomal subunit from erythromycin resistant mutants of *Escherichia coli* and *Escherichia freundii*. Mol. Gen. Genetics **114**, 14 (1971).

OTAKA, E., H. TERAOKA, M. TAMAKI, K. TANAKA, and S. OSAWA: Ribosomes from erythromycin-resistant mutants of *Escherichia coli* Q 13. J. Mol. Biol. **48**, 499 (1970).

PAGANO, J. F., M. J. WEINSTEIN, and C. M. McKEE: An anti-rickettsial antibiotic from a streptomycete, M-4209. I. Biological characterizations. Antibiot. & Chemotherapy **3**, 899 (1953).

PAUL, R. et S. TCHELITCHEFF: Structure de la spiramycine. VI. Etablissement de la formule développée. Bull. Soc. Chim. France **1965**, 650.

PERUN, T. J.: The chemistry and conformation of erythromycin. In: Drug action and drug resistance in bacteria: 1. Macrolide antibiotics and lincomycin, p. 123, ed. by S. MITSUHASHI. Tokyo: University of Tokyo Press 1971.

PESTKA, S.: Studies on the formation of transfer ribonucleic acid-ribosome complexes. XI. Antibiotic effects on phenylalanyl-oligonucleotide binding to ribosomes. Proc. Natl. Acad. Sci. U.S. **64**, 709 (1969).

PESTKA, S.: Inhibitors of ribosome functions. In: Molecular mechanisms of antibiotic action on protein biosynthesis and membranes, ed. by E. MUÑOZ, F. GARCIA-FERRANDIZ and D. VAZQUEZ. Amsterdam-London-New York: Elsevier 1972.

PINNERT-SINDICO, S., L. NINET, J. PREUD'HOMME, and C. COSAR: A new antibiotic, spiramycin. Antibiotics Ann. **1954/55**, 724.

PRELOG, V., A. M. GOLD, G. TALBOT u. A. ZAMOJSKI: Stoffwechselprodukte von Actinomyceten. Über die Konstitution der Narbomycins. Helv. Chim. Acta **45**, 4 (1962).

PULKRÁBEK, P., J. ČERNÁ, and I. RYCHLIK: Synthesis of tRNA-bound lysine peptides in the presence of puromycin or of antibiotics inhibiting ribosomal transpeptidation. Collection Czech. Chem. Commun. **35**, 2973 (1970).

RICKARDS, R. W., R. M. SMITH, and J. MAJER: The structure of the macrolide antibiotic picromycin. Chem. Commun. **1968**, 1049.

RODRIGUEZ-LOPEZ, M., M. L. CELMA, R. FERNANDEZ-MUÑOZ, and D. VAZQUEZ: Studies on algae ribosomes. Inhibitors of protein synthesis in blue-green algae. Atti del VII Simposio Internazionale di Agrochimica su La sintesi biologica delle proteine, p. 63 (1968).

RODRIGUEZ-LOPEZ, M., and D. VAZQUEZ: Comparative studies on cytoplasmic ribosomes from algae. Life Sci. **7**, 327 (1968).

RYCHLIK, I.: Release of lysine peptides by puromycin from polylysyltransfer RNA in presence of ribosomes. Biochim. Biophys. Acta **114**, 425 (1966).

RYCHLIK, I., and J. ČERNÁ: Effect of macrolide antibiotics on ribosomal peptidyl transferase. Proc. 7th International Congress Chemotherapy p. 793 (1972).

SAITO, T., M. SHIMIZU, and S. MITSUHASHI: Macrolide resistance in Staphylococci. In: Drug action and drug resistance in bacteria: Macrolide antibiotics and lincomycin, p. 239, ed. by S. MITSUHASHI. Tokyo: University of Tokyo Press 1971.

SENECA, H., and D. IDES: The effect of magnamycin on protozoa and spermatozoa. Antibiot. & Chemotherapy **3**, 117 (1953).

SHIMIZU, M., T. SAITO, H. HASHIMOTO, and S. MITSUHASHI: Spiramycin resistance in *Staphylococcus aureus*. Decrease in spiramycin-accumulation and the ribosomal affinity of spiramycin in resistant Staphylococci. J. Antibiotics (Tokyo) **23**, 63 (1970a).

SHIMIZU, M., T. SAITO, and S. MITSUHASHI: Macrolide resistance in *Staphylococcus aureus*. Decrease of spiramycin-binding to 50S ribosomal subunit in macrolide resistant strains of Staphylococci. J. Antibiotics (Tokyo) **23**, 467 (1970b).

TAGO, K., and M. NAGANO: Mechanism of inhibition of protein synthesis by leucomycin. In: Progress in antimicrobial and anticancer chemotherapy (Proceedings of the 6th International Congress of Chemotherapy), vol. I, p. 199. Tokyo: University of Tokyo Press 1970.

TANAKA, S., T. OTAKA and A. KAJI: Further studies on the mechanism of erythromycin action. Biochim. Biophys. Acta **331**, 128 (1973).

TANAKA, K., M. TAMAKI, T. ITOH, E. OTAKA, and S. OSAWA: Ribosomes from spiramycin resistant mutants of *Escherichia coli* Q 13. Mol. Gen. Genetics **114**, 23 (1971b).

TANAKA, K., M. TAMAKI, R. TAKATA, and S. OSAWA: Low affinity for chloramphenicol of erythromycin resistant *Escherichia coli* ribosomes having an altered protein component. Biochem. Biophys. Res. Commun. **46**, 1979 (1972).

Tanaka, K., and H. Teraoka: Effect of erythromycin on polylysine synthesis directed by polyadenylic acid in an *Escherichia coli* cell-free system. J. Biochem. **64**, 635 (1968).

Tanaka, K., H. Teraoka, T. Nagira, and M. Tamaki: Formation of C^{14}-erythromycin-ribosome complex. J. Biochem. **59**, 632 (1966).

Tanaka, K., H. Teraoka, and M. Tamaki: Peptidyl puromycin synthesis: Effect of several antibiotics which act on 50S ribosomal subunits. FEBS Letters **13**, 65 (1971).

Tanaka, K., H. Teraoka, M. Tamaki, E. Otaka, and S. Osawa: Erythromycin-resistant mutant of *Escherichia coli* with altered ribosomal protein component. Science **162**, 576 (1968).

Tanaka, K., H. Teraoka, M. Tamaki, R. Tanaka, and S. Osawa: Phenotypes represented by a mutational change in a 50S ribosomal protein component, 50-8 in *Escherichia coli*. Mol. Gen. Genetics **114**, 9 (1971a).

Tanaka, R., S. Osawa, K. Tanaka, H. Teraoka, and M. Tamaki: Genetic studies of the ribosomal proteins in *Escherichia coli*. V. Mapping of erythromycin resistance mutations which lead to alteration of a 50S ribosomal protein component. Mol. Gen. Genetics **109**, 123 (1970).

Tanner, F.W., A.R. English, T.M. Lees, and J.B. Routien: Some properties of magnamycin, a new antibiotic. Antibiot. & Chemotherapy **2**, 441 (1952).

Taubman, S., N. Jones, F. Young, and J. Corcoran: Sensitivity and resistance to erythromycin in *Bacillus subtilis* 168: The ribosomal binding of erythromycin and chloramphenicol. Biochim. Biophys. Acta **123**, 438 (1966).

Taubman, S.B., F.E. Young, and J.W. Corcoran: Antibiotic glycosides. IV. Studies on the mechanism of erythromycin resistance in *Bacillus subtilis*. Proc. Natl. Acad. Sci. U.S. **50**, 955 (1963).

Teraoka, H.: A reversible change in the binding ability of *Escherichia coli* ribosomes to bind to erythromycin. J. Mol. Biol. **48**, 511 (1970).

Teraoka, H., and K. Tanaka: An alteration in ribosome function caused by equimolar binding of erythromycin. Biochim. Biophys. Acta **232**, 509 (1971).

Teraoka, H., K. Tanaka, and M. Tamaki: The comparative study on the effects of chloramphenicol, erythromycin und lincomycin on polylysine synthesis in an *Escherichia coli* cell-free system. Biochim. Biophys. Acta **174**, 776 (1969).

Toju, H., and S. Omura: Chemical and biological studies on leucomycins (Kitasamycins). In: Drug action and drug resistance in bacteria. 1. Macrolide antibiotics and lincomycin, p. 267, ed. by S. Mitsuhashi. Tokyo: University of Tokyo Press 1971.

Vazquez, D.: Antibiotics which affect protein synthesis: The uptake of ^{14}C-chloramphenicol by bacteria. Biochem. Biophys. Res. Commun. **12**, 409 (1963).

Vazquez, D.: Binding of chloramphenicol to ribosomes. The effect of a number of antibiotics. Biochim. Biophys. Acta **114**, 277 (1966a).

Vazquez, D.: Antibiotics affecting chloramphenicol uptake by bacteria. Their effect on amino acid incorporation in a cell-free system. Biochim. Biophys. Acta **114**, 289 (1966b).

Vazquez, D.: Inhibitors of protein synthesis at the ribosome level. Studies on their site of action. Life Sci. **6**, 381 (1967a).

Vazquez, D.: Binding to ribosomes and inhibitory effect on protein synthesis of the spiramycin antibiotics. Life Sci. **6**, 845 (1967b).

Vazquez, D., and R.E. Monro: Effects of some inhibitors of protein synthesis on the binding of aminoacyl tRNA to ribosomal subunits. Biochim. Biophys. Acta **142**, 155 (1967).

Vogel, Z., T. Vogel, A. Zamir, and D. Elson: Correlation between the peptidyl transferase activity of the 50S ribosomal subunit and the ability of the subunit to interact with antibiotics. J. Mol. Biol. **60**, 339 (1971).

Waitz, J.A., E.L. Moss, E.M. Oden, and M.J. Weinstein: Biological activity of megalomicin, a new Micromonospora-produced macrolide antibiotic complex. J. Antibiotics (Tokyo) **22**, 265 (1969).

Weinstein, M.J., G.H. Wagman, J.A. Marquez, R.T. Testa, E. Oden, and J.A. Waitz: Megalomicin, a new macrolide antibiotic complex produced by Micromonospora. J. Antibiotics (Tokyo) **22**, 253 (1969).

Weisblum, B., and V. Demohn: Erythromycin-inducible resistance in *Staphylococcus aureus*: Survey of antibiotic classes involved. J. Bacteriol. **98**, 447 (1969).

Welch, H., W.R. Randall, R.J. Reedy, and J. Kramer: Bacterial spectrum of erythromycin, carbomycin, chloramphenicol, aureomycin and terramycin. Antibiot. & Chemotherapy **2**, 693 (1952).

WHALEY, H. A, E. L. PATTERSON, A. C. DORNBUSH, E. J. BACKUS, and N. BOHONOS: Isolation and characterization of relomycin, a new antibiotic. Antimicrobial Agents Chemotherapy 1963, 45.

WILEY, P. F., R. GALE, C. W. PETTINGA, and K. GERZON: Erythromycin. XII. The isolation, properties and partial structure of erythromycin C. J. Am. Chem. Soc. 79, 6074 (1957a).

WILEY, P. F., K. GERZON, E. H. FLYNN, M. V. SIGAL, Jr., O. WEAVER, U. C. QUARCK, R. R. CHAUVETTE, and R. MONAHAN: Erythromycin. X. Structure of erythromycin. J. Am. Chem. Soc. 79, 6062 (1957b).

WILEY, P. F., M. V. SIGAL, Jr., O. WEAVER, R. MONAHAN, and K. GERZON: Erythromycin. XI. Structure of erythromycin B. J. Am. Chem. Soc. 79, 6070 (1957c).

WILHELM, J. M., and J. W. CORCORAN: Antibiotic glycosides. VI. Definition of the 50S ribosomal subunit of *Bacillus subtilis* 168 as a major determinant of sensitivity to erythromycin A. Biochemistry 6, 2578 (1967).

WILHELM, J. M., N. L. OLEINICK, and J. W. CORCORAN: Interaction of antibiotics with ribosomes: Structure-function relationships and a possible common mechanism for the antibacterial action of the macrolides and lincomycin. Antimicrobial Agents Chemotherapy 1967, 236.

WOLFE, A. D., and F. E. HAHN: Erythromycin: Mode of action. Science 143, 1445 (1964).

WOLFE, A. D., and F. E. HAHN: Mode of action of chloramphenicol. IX. Effects of chloramphenicol upon a ribosomal amino acid polymerizing system and its binding to bacterial ribosome. Biochim. Biophys. Acta 95, 146 (1965).

WOO, P. W. K., H. W. DION, and Q. R. BARTZ: The structure of chalcomycin. J. Am. Chem. Soc. 86, 2726 (1964).

WOODWARD, R. B.: Struktur und Biogenese der Makrolide. Eine neue Klasse von Naturstoffen. Angew. Chem. 69, 50 (1957).

WOODWARD, R. B., L. S. WEILER, and P. C. DUTTA: The structure of magnamycin. J. Am. Chem. Soc. 87, 4662 (1965).

Micrococcin and Micrococcin P

Sidney Pestka

Micrococcin was isolated from a species of *micrococcus* (No. 7218 of the National Collection of Type Cultures, London, England) by SU (1948a, b). This strain of *micrococcus* was isolated from sewage and showed a resemblance to *M. varians*. The organism is a Gram-positive facultative anaerobe. Media were developed to produce 40–60 µg of micrococcin per ml of culture (KELLY *et al.*, 1952). An identical antibiotic, later called micrococcin P, was isolated from *Bacillus pumilus* (No. 8738 of the National Collection of Industrial Bacteria, Aberdeen, Scotland) by FULLER (1955) (ABRAHAM *et al.*, 1956). The antibiotic was found to be inhibitory chiefly to Gram-positive microorganisms (Table 1). Sensitive strains readily become resistant to micrococcin as reported by SU (1948b). Its inhibition of growth of *Streptomyces* (Table 1) perhaps deserves further study.

Because of the sensitivity of cultured cells of *Mycobacterium tuberculosis* to micrococcin, its effectiveness against experimental tuberculosis was evaluated. Although the antibiotic inhibits the growth of the tubercle bacillus in culture, it was not effective against experimental tuberculosis in animals (MARKHAM *et al.*, 1951b; SANDERS *et al.*, 1951; HEATLEY *et al.*, 1952a). Since micrococcin is hardly

Table 1

Species	Minimal inhibitory concentration (µg/ml)
Staphylococcus albus	0.10
Streptococcus faecalis	0.025
Corynebacterium diphtheriae gravis	0.10
Bacillus subtilis	0.016
Bacillus anthracis	0.10
Clostridium sporogenes	6.25
Clostridium welchii	3.13
Mycobacterium tuberculosis	6.25
Streptomyces AI "Tai"	0.10
Bacterium coli	>1000
Pseudomonas pyocyanea	>1000
Salmonella typhi	>1000
Proteus vulgaris	>1000
Vibrio cholerae	500

Taken from the data of SU (1948b).

soluble in water, the effects of injections of suspensions of micrococcin were examined in conjunction with the studies on experimental tuberculosis. These studies determined that particulate micrococcin is taken up by the reticuloendothelial system (MARKHAM et al., 1951a).

The antibiotic is relatively non-toxic (HEATLEY et al., 1952b; FULLER, 1955). Mice received 1 gm/kgm/day for repeated subcutaneous injections without ill effects. The antibiotic appears to be chiefly excreted in the bile (MARKHAM et al., 1951a) substantially as intact antibiotic. It cannot be given orally, however.

The antibiotic has been assayed microbiologically (SU, 1948b) as well as by fluorescence (HEATLEY et al., 1952b). Ethanolic solutions of micrococcin have a strong blue or purple fluorescence under ultraviolet light (SU 1948b). For microbiological assay Bacillus subtilis was found satisfactory as a test organism. It is sensitive to micrococcin and shows relatively little tendency to become resistant to the antibiotic.

Sensitivity and Resistance

Sensitive strains of staphylococci, streptococci, tubercle bacilli and actinomycetes readily become resistant to micrococcin (SU, 1948b). Micrococcin binds to both sensitive and resistant strains of staphylococci, but antibacterial activity can be recovered only from resistant strains. The mechanism of resistance has not been described. With the naturally insensitive Escherichia coli, there seems to be no affinity between the cells and micrococcin. Resistance and sensitivity markers to micrococcin have been mapped in B. subtilis (DUBNAU et al., 1967). The micrococcin marker maps in the area adjacent to the streptomycin, spectinomycin, oleandomycin, kanamycin, neomycin, thiostrepton and erythromycin loci; all are between the cys A region and the markers for the ribosomal RNA and transfer RNA genes in B. subtilis.

Physical and Chemical Properties

Micrococcin and micrococcin P are odorless, tasteless, white compounds. It is soluble in ethanol, acetone, chloroform, propylene glycol, glacial acetic acid and pyridine; but is insoluble in ether, benzene, amyl acetate, and glycerol; it is only very slightly soluble in water. For present purposes, I shall refer to micrococcin and micrococcin P interchangeably as micrococcin unless otherwise stated since it is likely they are identical. Micrococcin can be autoclaved without loss of activity (SU, 1948b). No significant reduction of antibacterial activity was noted when micrococcin was incubated with pepsin or trypsin for 5 hrs. It stable in 0.5 N HCl for 10 min at 100° C. In contrast, 0.1 N NaOH destroyed 50 percent of the activity in 1 min or 75 percent in 5 min at 100° C; however, no destruction was noted in 0.1 N NaOH at room temperature for 1 hr. At 37° C at pH 9, it also seems to be stable. Samples of crude culture fluid seemed to maintain stable antibacterial titers at 0° and 37° for one year; alcoholic solutions of micrococcin at room temperature lost no activity after more than one year of storage. When kept dry and exposed to air the powder generally turns yellow, then orange and decreases in solubility and potency.

The molecular weight of micrococcin P was reported as 2290 (MIJOVIC and WALKER, 1960). However, the molecular weight may be closer to half the above; judging from the component fragments a molecular weight of 1119 appears to be a more reasonable estimate (JAMES WALKER, personal communication). It contains about 16% of sulfur (FULLER, 1955; ABRAHAM et al., 1956). The concentration of the antibiotic can be estimated by its absorption at 345 mμ in ethanolic solution where $E_{1\,cm}^{1\%}=177.5$ (HEATLEY et al., 1952b). A spectrum of purified micrococcin is given by SU (1948b). Both micrococcin and micrococcin P seem to be indistinguishable chromatographically and both consist essentially of two main fractions roughly in the ratio of 7 to 1 (JAMES WALKER, personal communication).

For use in studies in cell-free extracts, we have usually dissolved the antibiotic in 90% ethanol (v/v) to make a stock solution of the antibiotic at about 0.01 M. Further dilutions are made directly into water and any suspensions formed are quite stable. The stock solutions and dilutions are kept at $-20°$ C, despite the fact that micrococcin is a stable molecule. The concentration of solutions is determined by measuring absorbance at 345 mμ in ethanol (Table 2).

Table 2. *Summary of properties of micrococcin*

Molecular weight	2,290
$E_{1\,cm}^{1\%}$ at 345 mμ in ethanol	177.5
E at 345 mμ in ethanol	40,648
Stable in solution at room temperature. Blue-purple fluorescence under UV light	

See text for appropriate references.

Structure

The complete structure of micrococcin P has not yet been determined. However, the structures of degradation products have been reported in a series of papers on the chemistry of micrococcin P. In the first paper of the series (BROOKES et al., 1957) three groups of acid-degradation products were reported. By the second paper of the series (MIJOVIC and WALKER, 1960) some of the components of the acid hydrolytic fractions were determined. On acid hydrolysis three fractions, termed for convenience acid-insoluble, ether-soluble, and acid-soluble, were separated. The ether-soluble fraction contained 2-propionylthiazole-4-carboxylic acid and propionic acid. The acid-soluble fraction contained ammonia, two molecules of L-threonine, four of 2-aminopropan-1-ol (alaninol), aminoacetone, and two of (+)-2-(1-amino-2-methylpropyl)thiazole-4-carboxylic acid. The two main fractions of micrococcin P have been separated chromatographically; only the minor component comprising about 15% of the total yields aminoacetone (JAMES WALKER, personal communication). In the intact antibiotic it appears that the carboxyl groups of the two threonines are exposed. The 2-aminopropan-1-ol is related configurationally to D-alanine. The acid-insoluble fraction contained a material very difficult to characterize (BROOKS et al., 1957; BROOKES et al., 1960a; BROOKES et al., 1960b; MIJOVIC and WALKER, 1961; CLARK and WALKER, 1966). Proper elucidation of the structures required studies on the racemization of derivatives of thiazole-4-carboxylic acid (DEAN et al., 1961) as well as development of new techniques for their degradation (HALL and WALKER, 1966).

After treatment of the acid-insoluble fraction with methanol and sulphuric acid, dimethyl micrococcinate was isolated. The bis-4-bromoanilide derivative of dimethyl micrococcinate was crystallized and its molecular structure determined by X-ray crystal structure analysis (JAMES and WATSON, 1966). A summary of the fragments obtained from micrococcin as elucidated in these studies is given in Fig. 1. Some resemblances to the structure of thiostrepton can be noted: the presence of multiple thiazole rings and threonine.

The biosynthesis of micrococcinic acid (Fig. 1) probably occurs through ring-closure, dehydrogenation and oxidative deamination of a precursor peptide composed of four molecules of cysteine, one of α-aminobutyric acid, and one of α-aminoadipic acid (HALL et al., 1966). It is of interest that 2-propionylthiazole-4-carboxylic acid (Fig. 1) is isolated as an acid hydrolytic product of micrococcin and also is incorporated in the structure of micrococcinic acid. The propionyl residue in 2-propionylthiazole-4-carboxylic acid and in micrococcinic acid probably arises by oxidative deamination of a 1-aminopropyl group derived from α-aminobutyric acid in the primary peptide precursor, analogous to the conversion of bacitracin A into bacitracin F. In confirmation of the incorporation of cysteine into the antibiotic, it has been shown that the availability of cysteine is a limiting factor in the production of micrococcin P (BROOKES et al., 1960a).

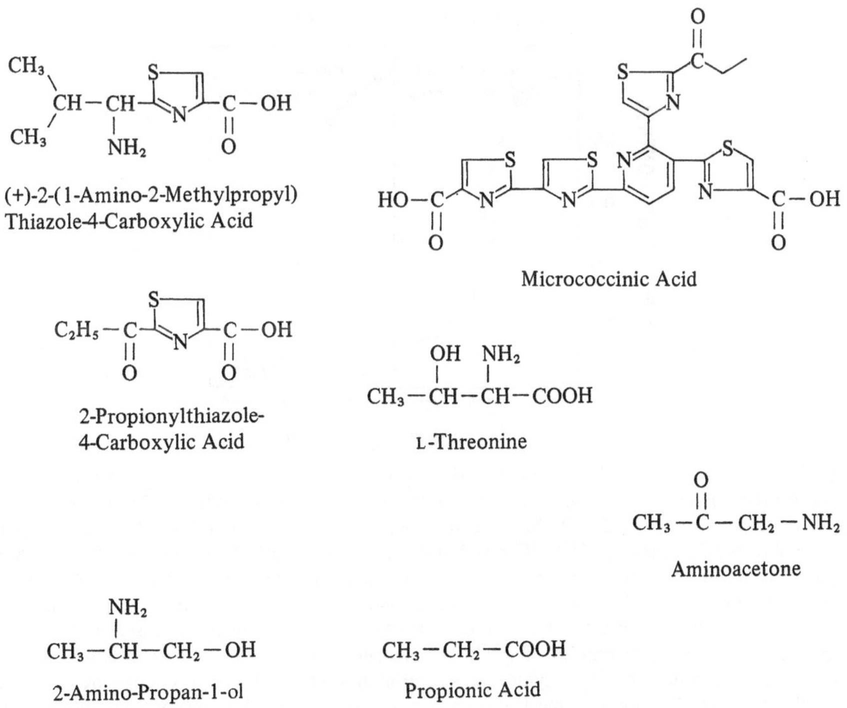

Fig. 1. The structures of the acid hydrolytic fragments of micrococcin are presented in the figure. Appropriate references to these structures are given in the text

Mode of Action

Micrococcin inhibits protein synthesis in cell-free extracts from *Escherichia coli*. Although *E. coli* is not sensitive to micrococcin (Table 1), cell-free extracts from this organism are (PESTKA and HINTIKKA, 1971). Studies with polyribosomes show that micrococcin stabilizes polyribosomes even after treatment with puromycin. In order to elucidate the site of action of micrococcin, the effect of the antibiotic on various steps in protein synthesis was examined. Binding of Phe-tRNA to ribosomes, acetyl-phenylalanyl-puromycin synthesis, and C-A-C-C-A(Phe) binding to ribosomes were not inhibited by micrococcin. Studies of non-enzymic and enzymic translocation indicated that micrococcin and micrococcin P (Fig. 2) inhibited both of these processes. The fact that non-enzymic translocation is inhibited as well as the enzymic reaction indicates that micrococcin functions through the ribosome in inhibiting these events rather than through the soluble elongation factor G (PESTKA and BROT, 1971; PESTKA, 1973). This conclusion is consistent with experiments mapping the markers for resistance and sensitivity to micrococcin. Genetic mapping in *Bacillus subtilis* indicates that the micrococcin locus is close to the loci of other markers for antibiotics which are closely associated with ribosomal function (DUBNAU *et al.*, 1967). In addition, studies on the GTPase activity of elongation factor G have shown that micrococcin has little or no effect on the hydrolysis of GTP associated with ribosomes and

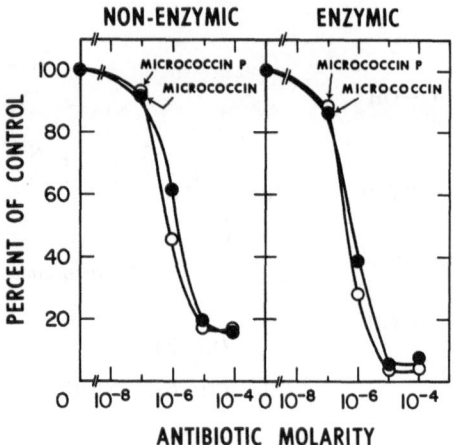

Fig. 2. Enzymic (right) and non-enzymic (left) oligopeptide synthesis as a function of antibiotic concentration. Each reaction mixture contained the following components: 0.2 M magnesium acetate, 0.05 M Tris-acetate, pH 7.2, 0.05 M potassium acetate, 2×10^{-4} M GTP, 0.2 A_{260} unit of poly(U), 1.9 µg of factor G, 5 pmoles of $[^{14}C]$Phe-tRNA (0.02 A_{260} unit; specific activity 367 mCi per mmole), and 0.4 A_{260} unit of ribosomes. Antibiotic concentration is given on the abscissa. Reactions were started by adding $[^{14}C]$Phe-tRNA to the reaction mixtures. Incubations were performed at 37° for 10 min. The data are expressed as a percentage of the control value in the absence of micrococcin. The control values in picomoles of $[^{14}C]$Phe- incorporated were as follows: 0.8 for micrococcin (●) and micrococcin P (○) for enzymic translocation and 0.6 pmoles for non-enzymic translocation. For the reactions for non-enzyme translocation, 0.5 A_{260} unit of ribosomes was present in each reaction mixture and GTP and elongation factor G were omitted. The incubations were performed at 37° for 60 min

factor G (PESTKA and BROT, 1971); and micrococcin has little or no effect on the binding of GTP to ribosomes. Furthermore, micrococcin does not inhibit peptide bond formation as measured by acetyl-phenylalanyl-puromycin synthesis or by peptidyl-[^{3}H]puromycin synthesis on bacterial polyribosomes (PESTKA, 1972). At the present time, it has not been determined whether micrococcin is an inhibitor of the 30 or 50S subunit and stable complexes of micrococcin with 30S or 50S subunits have not been able to be isolated so far (PESTKA, unpublished observations).

Thus, results indicate that micrococcin inhibits translocation through interference with ribosomal events associated with these processes. Unlike thiostrepton, micrococcin does not inhibit the GTPase activity associated with translocation and does not inhibit Tu dependent binding of Phe-tRNA to ribosomes (WEISSBACH and PESTKA, unpublished observations). Further studies are in progress to determine if micrococcin can inhibit tRNA release and other steps in translocation. Its use in studies on protein synthesis should not only elucidate its mechanism of action, but elucidate some of the steps of protein synthesis as well as details of ribosomal events.

Summary

The antibiotics micrococcin and micrococcin P appear to be identical. The antibiotic is inhibitory chiefly to Gram-positive microorganisms. Although the complete structure of micrococcin is not known, the structures of several of the component fragments have been determined.

The antibiotic inhibits protein synthesis in cell-free extracts from *Escherichia coli*. It appears to inhibit translocation through its interaction with the ribosome rather than through the soluble elongation factor G. It does not inhibit other steps in protein synthesis which have been measured such as aminoacyl-tRNA binding to ribosomes, transpeptidation, and GTP hydrolysis. As a consequence of its inhibition of translocation, polyribosomes are stabilized by micrococcin. Consistent with these results in cell-free extracts, genetic studies have determined that resistance and sensitivity to micrococcin map close to the genetic markers for other antibiotics which inhibit ribosomal function.

Acknowledgement. I am grateful to Dr. JAMES WALKER for his comments regarding this chapter and for the unpublished information relevant to the chemistry of micrococcin.

References

ABRAHAM, E.P., N.G. HEATLEY, P. BROOKES, A.T. FULLER, and J. WALKER: Probable identity of an antibiotic produced by a spore-bearing bacillus of the *B. pumilis* group with micrococcin. Nature **178**, 44 (1956).

BROOKES, P., R.J. CLARK, A.T. FULLER, M.P.V. MIJOVIC, and J. WALKER: Chemistry of micrococcin P. Part III. J. Chem. Soc. **1960a**, 916.

BROOKES, P., R.J. CLARK, B. MAJHOFER, M.P.V. MIJOVIC, and J. WALKER: Chemistry of micrococcin P. Part IV. A method for the structural study of thiazoles. J. Chem. Soc. **1960b**, 925.

BROOKES, J., A.T. FULLER, and J. WALKER: Chemistry of micrococcin P. Part I. J. Chem. Soc. **1957**, 689.

CLARK, R.J., and J. WALKER: Chemistry of micrococcin P. Part VII. Dimethyl micrococcinate and some synthetic pyridine-polythiazole carboxylic esters. J. Chem. Soc. **1966**, 1354.

DEAN, B.M., M.P.V. MIJOVIĆ, and J. WALKER: Chemistry of micrococcin P. Part VI. Racemisation of 2-(1-amino-2-methylpropyl)thiazole-4-carboxylic acid, and related studies. J. Chem. Soc. **1961**, 3394.

DUBNAU, D., C. GOLDTHWAITE, I. SMITH, and J. MARMUR: Genetic mapping in *Bacillus subtilis*. J. Mol. Biol. **27**, 163 (1967).

FULLER, A.T.: A new antibiotic of bacterial origin. Nature **175**, 722 (1955).

HALL, G.E., N. SHEPPARD, and J. WALKER: Chemistry of micrococcin P. Part X. Proton magnetic resonance spectrum of dimethyl micrococcinate, and the probable mode of biosynthesis of micrococcinic acid. J. Chem. Soc. **1966**, 1371.

HALL, G.E. and J. WALKER: Chemistry of micrococcin P. Part VIII. A method for the degradation of thiazole-4-carboxylic acids. J. Chem. Soc. **1966**, 1357.

HEATLEY, N.G., J.L. GOWANS, H.W. FLOREY, and A.G. SANDERS: The effect on experimental tuberculosis and other infections of a micrococcin-triton solution. Brit. J. Exptl. Pathol. **33**, 105 (1952a).

HEATLEY, N.G., B.K. KELLY, and N. SMITH: The assay of micrococcin, an almost insoluble antibiotic. J. Gen. Microbiol. **6**, 30 (1952b).

JAMES, M.N.G., and K.J. WATSON: Chemistry of micrococcin P. Part IX. The crystal and molecular structure of micrococcinic acid bis-4-bromoanilide. J. Chem. Soc. **1966**, 1361.

KELLY, B.K., G.A. MILLER, and C.W. HALE: Culture media for large-scale production of micrococcin. J. Gen. Microbiol. **6**, 41 (1952).

MARKHAM, N.P., N.G. HEATLEY, A.G. SANDERS, and H.W. FLOREY: The behavior *in vivo* of particulate micrococcin. Brit. J. Exptl. Pathol. **32**, 136 (1951a).

MARKHAM, N.P., A.Q. WELLS, N.G. HEATLEY, and H.W. FLOREY: The effect on experimental tuberculosis of the intravenous injection of micrococcin. Brit. J. Exptl. Pathol. **32**, 353 (1951b).

MIJOVIC, M.P.V., and J. WALKER: Chemistry of micrococcin P. Part II. J. Chem. Soc. **1960**, 909

MIJOVIC, M.P.V., and J. WALKER: Chemistry of micrococcin P. Part V. The infrared absorption spectra of thiazoles. J. Chem. soc. **1961**, 3381.

PESTKA, S.: Studies on transfer ribonucleic acid-ribosome complexes. XIX. Effect of antibiotics on peptidyl-puromycin synthesis on polyribosomes from *Escherichia coli*. J. Biol. Chem. **247**, 4669 (1972).

PESTKA, S.: Enzymic and non-enzymic translocation. Methods in Enzymology. In press (1973).

PESTKA, S., and N. BROT: Studies on the formation of ribonucleic acid-ribosome complexes. XV. Effect of antibiotics on steps of bacterial protein synthesis: Some new ribosomal inhibitors of translocation. J. Biol. Chem. **246**, 7715 (1971).

PESTKA, S. and H. HINTIKKA: Studies on the formation of ribonucleic acid-ribosome complexes. XVI Effect of ribosomal translocation inhibitors on polyribosomes. J. Biol. Chem. **246**, 7723 (1971).

SANDERS, A.G., H.W. FLOREY, and A.Q. WELLS: The behavior of intravenously injected particles of carbon and micrococcin in normal and tuberculous tissue. Brit. J. Exptl. Pathol. **32**, 452 (1951).

SU, T.L.: Antibiotic-producing organisms in faeces and sewage. Brit. J. Exptl. Pathol. **29**, 466 (1948a).

SU, T.L.: Micrococcin. An antibacterial substance formed by a strain of micrococcus. Brit. J. Exptl. Pathol. **29**, 473 (1948b).

Mikamycin

Nobuo Tanaka

The antibiotics of the mikamycin group exist as a complex of synergistic compounds (ARAI *et al.*, 1956; WATANABE, 1961), closely related to or identical in structure with ostreogrycin (EASTWOOD *et al.*, 1960; DELPIERRE *et al.*, 1966; KINGSTON *et al.*, 1966a; KINGSTON *et al.*, 1966b; COX and EASTWOOD, 1970), streptogramin (CHARNEY *et al.*, 1953; VERWAY *et al.*, 1958; VAZQUEZ, 1967), vernamycin (BODANSZKY and ONDETTI, 1964; SQUIBB, 1967), PA 114 = synergistin (CELMER and SOBIN, 1956; HOBBS and CELMER, 1960), pristinamycin (BENAZET *et al.*, 1962; PREUD'HOMME, 1968), and virginiamycin = staphylomycin (DE SOMER and VAN DIJCK, 1955; VANDERHAEGHE *et al.*, 1957; VANDERHAEGHE and PARMENTIER, 1960; VANDERHAEGHE *et al.*, 1971; COMPERNOLLE *et al.*, 1972; CROOY and DE NEYS, 1972). The antibiotic complex consists of two or more components belonging to two major groups, A and B. The antibiotics in group A exhibit a marked synergism with those in the group B in their antimicrobial activity against Gram-positive bacteria. The synergism is not so significantly observed with *Escherichia coli* and other Gram-negative organisms as with the Gram-positive organisms.

Patricin A and B (SQUIBB, 1968), doricin (BODANSZKY and SHEEHAN, 1964; CHARLES-SIGLER and GIL-AV, 1966) and viridogrisein (BARTZ *et al.*, 1955; HEINEMANN *et al.*, 1955; ARNOLD *et al.*, 1958; SHEEHAN *et al.*, 1958) are peptide antibiotics related to the group B antibiotic.

The chemical structures and the relationship of these antibiotics have been determined. These are presented in Figs. 1 and 2.

Mikamycin A, $C_{28}H_{35}N_3O_7$ (MW 526) is a macrocyclic lactone, containing pyrrolidine and oxazole rings, with the structure as shown in Fig. 1. It is identical

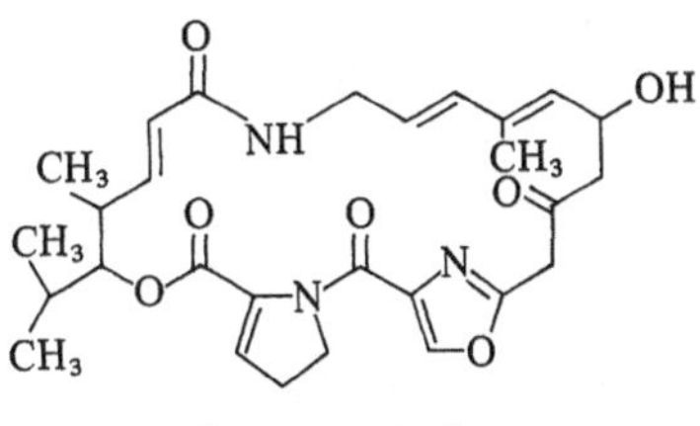

Ostreogrycin A
= Mikamycin A =
vernamycin A = PA 114 A1
= streptogramin A =
pristinamycin IIA = virginiamycin M

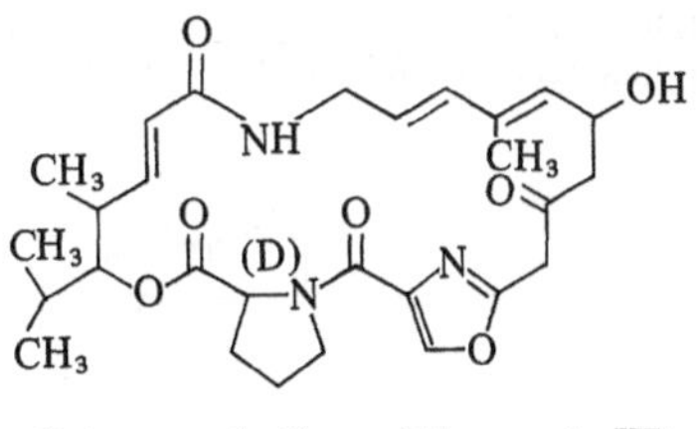

Ostreogrycin G = pristinamycin IIB
= virginiamycin M_2

Fig. 1. The structure of the group A antibiotic

with ostreogrycin A, pristinamycin IIA, streptogramin A, PA 114 A1, vernamycin A and virginiamycin M. Ostreogrycin G, pristinamycin IIB or virginiamycin M_2 has the structure with Δ-2,3-saturated.

The group B antibiotic is a depsipeptide (Fig. 2) and includes mikamycin B, streptogramin B, PA 114 B1, vernamycins Bα, Bβ, Bγ, Bδ and C, ostreogrycins B, B_1, B_2 and B_3, pristinamycins IA, IB and IC, virginiamycins S, S_2, S_3 and S_4, patricins A and B, doricin, viridogrisein, and etamycin. The structures and their interrelationships are shown in Fig. 2.

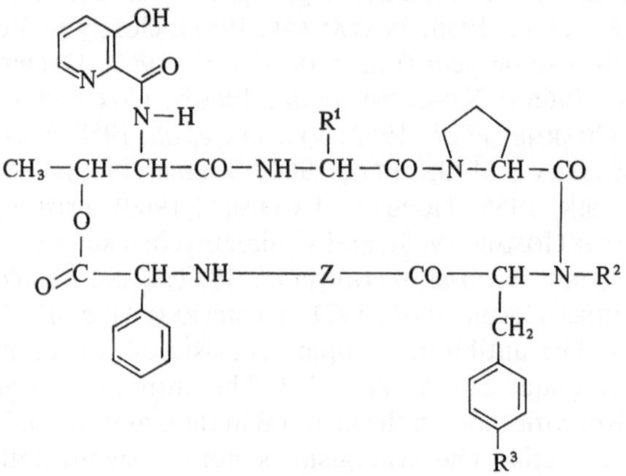

Mikamycin B = Streptogramin B = PA 114 B1 = Pristinamycin IA = Vernamycin Bα = Ostreogrycin B: $R^1 = C_2H_5$, $R^2 = CH_3$, $R^3 = N(CH_3)_2$, Z = 4-oxopipecolic acid.
Vernamycin Bγ = Ostreogrycin B_1 = Pristinamycin IC: $R^1 = CH_3$, $R^2 = CH_3$, $R^3 = N(CH_3)_2$, Z = 4-oxopipecolic acid.
Vernamycin Bβ = Ostreogrycin B_2 = Pristinamycin IB: $R^1 = C_2H_5$, $R^2 = CH_3$, $R^3 = NHCH_3$, Z = 4-oxopipecolic acid.
Vernamycin Bδ: $R^1 = CH_3$, $R^2 = CH_3$, $R^3 = NHCH_3$, Z = 4-oxopipecolic acid.
Ostreogrycin B_3: $R^1 = C_2H_5$, $R^2 = CH_3$, $R^3 = N(CH_3)_2$, Z = 3-hydroxy-4-oxopipecolic acid.
Vernamycin C = Doricin: $R^1 = C_2H_5$, $R^2 = CH_3$, $R^3 = N(CH_3)_2$, Z = aspartic acid.
Patricin A: $R^1 = C_2H_5$, $R^2 = CH_3$, $R^3 = H$, Z = proline.
Patricin B: $R^1 = C_2H_5$, $R^2 = CH_3$, $R^3 = H$, Z = pipecolic acid.
Virginiamycin S: $R^1 = C_2H_5$, $R^2 = CH_3$, $R^3 = H$, Z = 4-oxopipecolic acid.
Virginiamycin S_4: $R^1 = CH_3$, $R^2 = CH_3$, $R^3 = H$, Z = 4-oxopipecolic acid.
Virginiamycin S_2: $R^1 = C_2H_5$, $R^2 = H$, $R^3 = H$, Z = 4-hydroxypipecolic acid.
Virginiamycin S_3: $R^1 = C_2H_5$, $R^2 = CH_3$, $R^3 = H$, Z = 3-hydroxy-4-oxopipecolic acid.

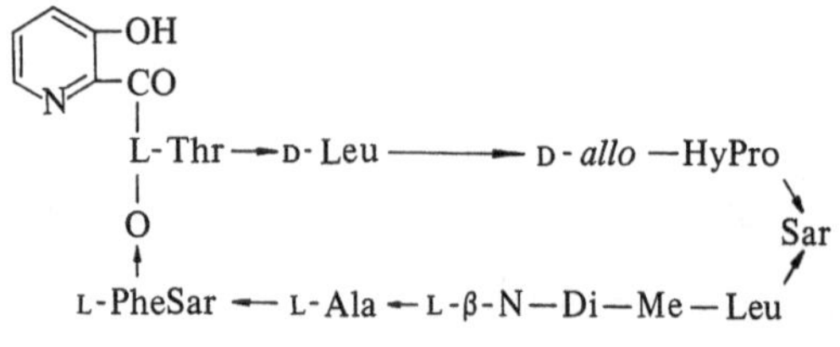

Viridogrisein = Etamycin

Fig. 2. The structure of the group B antibiotic

All the antibiotics of this group are produced by *Streptomycetes: S. gramino-faciens* (CHARNEY *et al.*, 1953), *S. virginiae* (DE SOMER and VAN DIJCK, 1955), *S. mitakaensis* (ARAI *et al.*, 1956), *S. olivaceus* (CELMER and SOBIN, 1956), *S. loidensis* (BODANSZKY and ONDETTI, 1964), *S. ostreogriseus* (BALL *et al.*, 1958), *S. pristinaespiralis* (BENAZET *et al.*, 1962), *S. griseus, S. lavendulae* (BARTZ *et al.*, 1955), etc.

Table 1. *Antimicrobial spectrum of mikamycins A and B*
(minimum inhibitory concentration µg/ml)

Test organisms	Mikamycin		Test organisms	Mikamycin	
	A	B		A	B
Staphylococcus aureus 209 P	2.0	10.0	*Br. melitensis*	20.0	10.0
Staphylococcus aureus 193	10.0	10.0	*H. pertussis*	0.4	10.0
M. flavus	2.0	4.0	*S. typhosa*	>80.0	>80.0
Sarcina lutea	0.2	2.0	*S. paratyphosa A*	40.0	>80.0
Str. pyogenes group A			*S. paratyphosa B*	>80.0	>80.0
B 930/24 (type 3)	4.0	4.0	*S. typhimurium*	>80.0	>80.0
J 17D (type 19)	4.0	4.0	*Sh. dysenteriae*	>80.0	>80.0
D. pneumoniae type I	16.0	1.0	*Sh. flexneri*	>80.0	>80.0
D. pneumoniae type II	16.0	2.0	*Sh. sonnei*	>80.0	>80.0
D. pneumoniae type III	16.0	1.0	*E. coli*	>80.0	>80.0
B. subtilis PCI 219	80.0	1.0	*Pr. vulgaris*	80.0	>80.0
B. agri	10.0	1.0	*K. pneumoniae* 602	>80.0	>80.0
B. cereus	20.0	1.0	*Ps. aeruginosa*	>80.0	>80.0
B. cereus var. *mycoides*	10.0	0.1	*Serratia marcescens*	>80.0	>80.0
B. megatherium APF	80.0	2.0	*Pen. chrysogenum*	>80.0	>80.0
B. megatherium 10778	40.0	4.0	*Asp. oryzae*	>80.0	>80.0
B. anthracis	10.0	2.0	*Asp. niger*	>80.0	>80.0
C. xerosis	>80.0	4.0	*Candida albicans*	>80.0	>80.0
Mycobacterium 607	>80.0	>80.0	*Torula utilis*	>80.0	>80.0
M. phlei	>80.0	>80.0	*Sacch. cerevisiae*	>80.0	>80.0
M. tuberculosis H 37Rv	20.0	>80.0	*Tetrahymena geleii*	>80.0	>80.0

TANAKA *et al.* (1959).

The mikamycin group antibiotics are active against Gram-positive organisms both *in vitro* and *in vivo*. The minimal growth-inhibitory concentrations of mikamycins A and B are listed in Table 1 (TANAKA *et al.*, 1959). All the other antibiotics of this group show similar antimicrobial activity. In an earlier review on the streptogramin family of antibiotics, VAZQUEZ cites very high minimum inhibitory concentrations of mikamycins implying that these are less active than other antibiotics of this group (VAZQUEZ, 1967). This is apparently not true. Mikamycins A and B protect bacteria against the bactericidal actions of streptomycin and kanamycin (YAMAKI and TANAKA, 1963). Mikamycins A and B exhibit cross-resistance with macrolide antibiotics (TANAKA *et al.*, 1961). As illustrated in Fig. 3, mikamycins A and B exhibit a marked synergism in combination (TANAKA *et al.*, 1959).

Mice can tolerate a daily intraperitoneal dose of 350 mg/kg body weight mikamycin (TANAKA *et al.*, 1959, 1961).

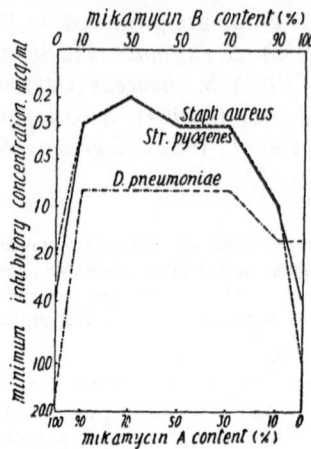

Fig. 3. Antimicrobial activity of mikamycins A and B in combination. (Tanaka *et al.*, 1959)

Mikamycin, a mixture of A and B, is mainly employed as an additive to animal feed (1–2 µg/g) for promoting growth of chicken and other poultry. Virginiamycin and mikamycin find a limited use against certain Gram-positive infections and are applied topically or administered orally.

Mechanism of Action

The Interaction of the Antibiotics A and B with Bacterial Ribosomes. The primary site of action of the mikamycin group antibiotics is localized in the ribosomes and they interfere with protein synthesis. The ^{14}C-chloramphenicol uptake by bacteria and its binding to their ribosomes are inhibited by the group A antibiotic. Since the site of action of chloramphenicol is known to be the 50S ribosomal subunit, these results suggest that the group A antibiotic may also act on the 50S ribosomal subunit (Vazquez, 1967). In the *E. coli,* chloramphenicol, lincomycin and erythromycin act on closely related sites on the 50S ribosomal subunit, and there is one binding site per ribosome for each drug. The binding of lincomycin, chloramphenicol, and erythromycin is inhibited by the antibiotics A and B (Fernandez-Muñoz *et al.,* 1971).

The *E. coli* 70S ribosomes are dissociated into the 50S and 30S subunits, treated with the antibiotics, dialyzed against antibiotic-free buffer and the functioning of these ribosomal subunits are then studied by reconstituting the antibiotic-treated and -untreated ribosomal subunits in all combinations. The group A antibiotic markedly inactivates the 50S subunit and is less effective in inhibiting the activity of the 30S subunit; i.e. the group A antibiotic interacts with the 50S subunit but not with the 30S subunit (Ennis, 1966).

The ^{3}H-labelled group A and B antibiotics bind specifically to the 50S ribosomal subunit (Ennis, 1971; Cocito and Kaji, 1971; Tanaka, N., unpublished results). Ennis (1971) reports that binding of the group A antibiotic (^{3}H-vernamycin) requires K^+ or NH_4^+, and Mg^{++}. It does not occur at 0°C but takes place rapidly at 37°C. At saturation, one mole of the antibiotic is bound per mole of the 70S ribosome. The antibiotic-ribosome complex formed in the presence of K^+ is stable. However, upon removal of the cation, the complex readily dissociates. The binding is enhanced by the group B antibiotic. Macrolide antibiotics, which interact with the 50S ribosomal subunit, can prevent the binding of the group A antibiotic. Although the group A antibiotic can inhibit chloramphenicol binding to the ribosomes (see above), the reverse is not observed.

In vitro Inhibition of Protein Synthesis by the Antibiotics A and B. The mikamycin family antibiotics, both groups A and B, inhibit protein synthesis in cell-free extracts of *E. coli* directed by a variety of synthetic polynucleotides and natural messengers. The type of polynucleotide used as messenger greatly affects the degree of inhibition of polypeptide synthesis observed. The group A antibiotic significantly inhibits all the messenger systems. However, the effects of the group B antibiotic on protein synthesis seems to be highly dependent upon the nucleotide sequence of the mRNA. As illustrated in Fig. 4, it is ineffective against poly U-directed synthesis of polyphenylalanine but inhibits the poly A- and poly C-directed polypeptide syntheses (YAMAGUCHI *et al.*, 1966; VAZQUEZ, 1966; ENNIS, 1970).

The degree of inhibition of polypeptide synthesis by the group A antibiotic decreases when it is introduced to the reaction mixture later, whereas that by the group B antibiotic shows a tendency to increase with the delay of addition (Fig. 5). The results suggest that the group A antibiotic primarily affects initiation or early phase of the polypeptide synthesis, while the site of action of the group B antibiotic may be somewhere after the phase attacked by the group A antibiotic

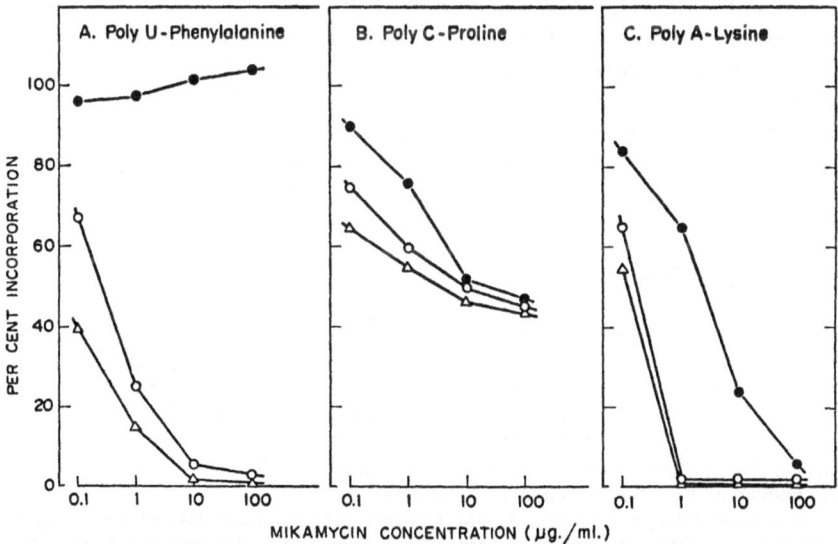

Fig. 4. Effect of mikamycin on polynucleotide-directed polypeptide synthesis in the *E. coli* system. Incorporation of C^{14}-phenylalanine (A), C^{14}-proline (B) and C^{14}-lysine (C) was measured in the presence of 4 µg of poly U, poly C and poly A, respectively, with the addition of the indicated concentrations of mikamycin A (—○—), mikamycin B (—●—) and mikamycin A-B complex (—△—). (YAMAGUCHI *et al.*, 1966)

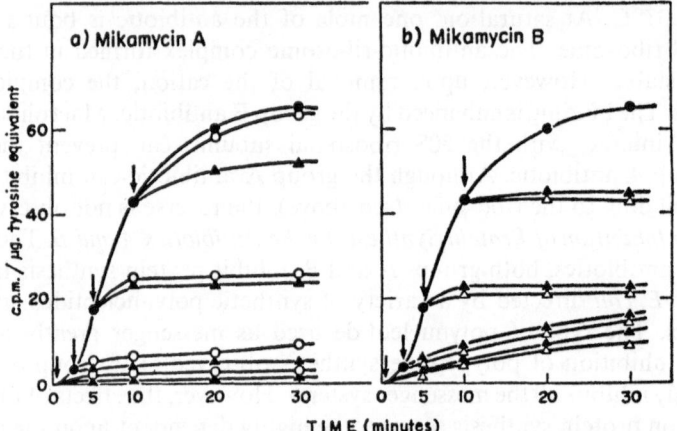

Fig. 5. Effects of the time of addition of mikamycins A and B on poly A-directed polylysine synthesis. △ 100 µg/ml of Mikamycin A or B, ▲ 10 µg/ml ○ 1 µg/ml ● 0 µg/ml (Yamaguchi and Tanaka, 1967)

(Yamaguchi and Tanaka, 1967). The non-enzymic binding of fMet-tRNA to 70S ribosomes or to the 50S subunit is inhibited by the group A antibiotic. It inhibits the binding of fMet-tRNA to a puromycin-reactive site (the donor site), which is not affected by tetracycline (Ennis, 1972). The group A antibiotic inhibits the 70S initiation complex formation on f2 phage RNA in the presence of initiation factors, but does not significantly affect the 30S initiation complex formation (Table 2). This indicates that the group A antibiotic does not affect the real initiation of protein synthesis, but blocks a subsequent step following chain initiation (Okuyama and Tanaka, unpublished data). The same conclusion has been arrived at in the *in vivo* experiments (Ennis, 1972).

The group A antibiotic seems to affect the peptidyl transferase center of ribosomes, since it inhibits the puromycin reaction on 70S ribosomes (Tanaka *et al.,* 1970; Ennis, 1970; Pestka, 1970) and the fragment reaction on the 50S

Table 2. *Effects of mikamycin A on initiation complex formation in vitro*

Antibiotic	Initiation complex formed on	
	30S subunit	70S ribosome
—	3.17[a] pmoles	1.56 pmoles
Mikamycin A 1.7×10^{-5} M	3.20 (—)	0.44 (72)[b]
1.7×10^{-6}	3.18 (—)	0.78 (50)

[a] The number represents pmoles of ^{14}C-fMet-tRNA bound.
[b] The number in the bracket represents % inhibition.

The reaction mixture contained: 50 mM Tris-HCl, pH 7.5, 60 mM NH$_4$Cl, 6 mM Mg(AcO)$_2$, 10 mM 2-mercaptoethanol, 2 mg/ml washed 70S ribosomes, 0.6 mg/ml initiation factor, 150 µg/ml f-^{14}C-Met-tRNA (300,000 cpm/mg), 1 mg/ml f2 RNA, 0.2 mM GTP, 0.1 ml in each tube. It was incubated at 37° C for 15 min. The radioactivity, collected on Millipore filter, was assayed with corrections for values without messenger. In the 30S experiment, the ribosomes and initiation factor were replaced by 0.6 mg/ml native 30S ribosomal subunit. The ribosomes and initiation factor were obtained from *E. coli* Q13. (Okuyama, A. and N. Tanaka, unpublished data.)

ribosomal subunit (MONRO and VAZQUEZ, 1967). However, the group A antibiotic produces little or no inhibition of peptidyl-puromycin synthesis on native poly-somes. It supports the above-mentioned assumption that the group A antibiotic interferes with peptide chain initiation (PESTKA and HINTIKKA, 1971; PESTKA, 1972).

The group B antibiotic affects peptidyl transferase center in a manner different from the group A antibiotic, since the former inhibits formation of lysyl- or polylysyl-puromycin (TANAKA et al., 1970; ENNIS, 1970) but stimulates N-acetyl-phenylalanyl-puromycin synthesis (PESTKA, 1970; KUBOTA et al., 1972). In a sys-tem using f2 phage RNA as a messenger, the group B antibiotic exhibits diverse effects on peptidyl transferase reactions. It stimulates fMet-puromycin synthesis but inhibits fMet-Ala-puromycin reaction (Table 3). The accumulation of fMet-Ala is observed in the presence of the group B antibiotic, while the polypeptide synthesis is suppressed. The group A antibiotic inhibits all the peptidyl transferase reactions. The results suggest that the first and second peptide bond formation may involve somewhat different mechanism (KUBOTA et al., 1972). It is in accord-ance with the previous observation that oligopeptide is formed when polypeptide synthesis is blocked by mikamycin B (YAMAGUCHI and TANAKA, 1967). Oligopep-tide synthesis is inhibited by the group A antibiotic (YAMAGUCHI and TANAKA, 1967; PESTKA, 1970). The group A antibiotic inhibits the binding of aminoacyl end of aminoacyl-tRNA (CACCA-Phe) to the ribosome or 50S ribosomal subunit, while the group B antibiotic stimulates the binding (PESTKA, 1969; HISHIZAWA and PESTKA, 1971). The binding of UACCA-Leu to the 50S subunit is also inhibited by the group A antibiotic (CELMA et al., 1971).

Table 3. *Effects of antibiotics on puromycin reactions*

Additions		fMet-Ala-PM	fMet-PM	Ac-Phe-PM
Mikamycin B	1.1×10^{-5} M	67	103	181
	4.5×10^{-5}	50	118	196
Mikamycin A	1.7×10^{-5}	36	76	2
	6.7×10^{-5}	23	55	
Spiramycin	1.2×10^{-5}	29	98	16
	2.4×10^{-4}	15	93	13
Erythromycin	1.4×10^{-4}	80	102	138
Blasticidin S	2.3×10^{-6}	28		43
	2.3×10^{-5}	18	76	17

The number represents % incorporation. PM: puromycin. (KUBOTA et al., 1972.)

Sparsomycin, an inhibitor of the peptidyl transfer reaction, induces formation of an inert complex between the CCA-peptide moiety of the peptidyl donor substrate and the peptidyl transferase center on the 50S ribosomal subunit. The sparsomycin-induced binding of N-acetylphenylalanyl-tRNA is inhibited by the group A antibiotic as well as other peptidyl transferase inhibitors (JIMENEZ et al., 1970).

Table 4. *Effects of antibiotics on the binding of phenylalanyl-tRNA to the ribosomes with poly U*

a

Antibiotics		Phenylalanyl-tRNA bound			
		EF T and GTP (+)		EF T and GTP (−)	
—		1274[a]	(100)	218	(100)
Mikamycin A	12 µM	647	(51)	218	(100)
	60	460	(36)	198	(91)
	240	310	(24)	173	(79)
—		1611	(100)	301	(100)
Tetracycline	50 µM	954	(56)		
	200	420	(26)	331	(110)

[a] cpm.　[b] Relativeradioactivity.

b

Antibiotics		EFT factor and GTP (—)					
		8 mM Mg^{++}		14 mM Mg^{++}		20 mM Mg^{++}	
—		250	(100)	288	(100)	431	(100)
Mikamycin A	120 µM	238	(95)	200	(70)	202	(46)
Tetracycline	200	226	(91)	168	(58)	285	(66)

Tanaka *et al.* (1967).

The group A antibiotic inhibits enzymic and non-enzymic binding of amino-acyl-tRNA to the acceptor site of ribosomes. As shown in Table 4, mikamycin A (the group A antibiotic) inhibits the binding of phenylalanyl-tRNA to the ribosomes with poly U in the presence of elongation factor EF T and GTP at 8 mM of Mg^{++} concentration. In the absence of EF T and GTP, the antibiotic causes up to 50% inhibition of the binding of phenylalanyl-tRNA at high Mg^{++} concentrations, but less inhibition is observed at low Mg^{++} concentrations. These results indicate that the binding of aminoacyl-tRNA to the acceptor site is preferentially inhibited by the group A antibiotic. The binding to the donor site may not be significantly affected by the antibiotic. The inhibition of the peptidyl transferase reaction is demonstrated at much lower concentration of the antibiotic. Therefore the main site of action of the group A antibiotic seems to be localized in a certain mechanism linked to the peptidyl transferase. There is a possibility that the group A antibiotic interacts with the acceptor site of the ribosome resulting in the steric hinderance of peptidyl transferase or *vice versa* (Tanaka *et al.*, 1970; Cocito and Kaji, 1971).

The Inhibition of Protein Synthesis in Vivo. At the minimal growth-inhibitory concentration, the group A and/or B antibiotics inhibit protein synthesis without affecting nucleic acid and cell wall syntheses. The accumulation of oligopeptide is observed with the group B antibiotic, but not with the group A antibiotic (see the review by Vazquez, 1967).

The group A antibiotic allows normal degradation of polysomes with a half-life of about one min at 37°C. The same phenomenon is observed with rifampicin, whereas puromycin promotes rapid degradation, complete within 15 sec. Amicetin, chloramphenicol, erythromycin, oleandomycin and fusidic acid stabilize polysomes. In the presence of the group A antibiotic, the polysomes break down almost exclusively to 70S ribosomes. Essentially no peptide is associated with the ribosomes by the antibiotic treatment of cells. Chloramphenicol, fusidic acid and amicetin, added prior to the group A antibiotic, prevent the usual breakdown of polysomes observed in the presence of the group A antibiotic. The *in vivo* effect of the group A antibiotic indicates that it inhibits initiation or early stage of protein synthesis (PESTKA and HINTIKKA, 1971; ENNIS, 1972).

The Mechanism of Synergism of the Group A and B antibiotics. The mixture of the antibiotics A and B shows much higher antimicrobial activity than each antibiotic alone. For Gram-positive cocci, which are more sensitive to the group A antibiotic than to the group B antibiotic, the latter enhances the antimicrobial activity of the former approximately 10 times (Fig. 3).

This synergism of the antibiotics A and B is also demonstrated in a cell-free system obtained from *E. coli* (ENNIS, 1970; TANAKA *et al.,* 1970). The extent of the synergistic activity, as well as that of the inhibition, is reversed with the increased concentrations of ribosomes and tRNA (YAMAGUCHI *et al.,* 1967). The data indicate that the synergism is not due to the heterogeneous sensitivity of the cell population to the antibiotics A and B. The synergism may occur at the ribosome level but not at the cell population level.

The binding of the group A antibiotic (^{3}H-vernamycin) to the ribosomes is enhanced by the group B antibiotic. This enhancement may possibly be the cause of the synergism (ENNIS, 1971). However, the precise mechanism of the synergism remains to be determined.

References

ARAI, M., S. NAKAMURA, Y. SAKAGAMI, K. FUKUHARA, and H. YONEHARA: A new antibiotic, mikamycin. J. Antibiotics (Tokyo), Ser. A **9**, 193 (1956).

ARNOLD, R. B., A. W. JOHNSON, and A. B. MAUGER: The structure of viridogrisein (etamycin). J. Chem. Soc. **1958**, 4466.

BALL, S., B. BOOTHROYD, K. A. LEES, A. H. RAPER, and E. L. SMITH: Preparation and Properties of an antibiotic complex E 219. Biochem. J. **68**, 24 P (1958).

BARTZ, Q. R., J. STANDIFORD, J. D. MOLD, D. W. JOHANNESSEN, A. RYDER, A. MARETZKI, and T. H. HASKELL: Griseoviridin and viridogrisein: isolation and characterization. Antibiotics Ann. **1954/1955**, 777 (1955).

BENAZET, F., C. COSAR, M. DUBOST, L. JULOU, et D. MANCY: Un nouvel antibiotique, la pristinamycine (7293 R. P.). Semaine Hop. Paris **38**, 13 (1962).

BODANSZKY, M., and M. A. ONDETTI: Structures of the vernamycin B group of antibiotics. Antimicrobial Agents Chemotherapy **1963**, 360 (1964).

BODANSZKY, M., and J. T. SHEEHAN: Structure of doricin, a peptide related to the vernamycin B group. Antimicrobial Agents Chemotherapy **1963**, 38 (1964).

CELMA, M. L., R. E. MONRO, and D. VAZQUEZ: Substrate and antibiotic binding sites at the peptidyl transferase centre of *E. coli* ribosomes: Binding of UACCA-Leu to 50S subunits. FEBS Letters **13**, 247 (1971).

CELMER, W. D., and B. A. SOBIN: The isolation of two synergistic antibiotics from a single fermentation source. Antibiotics Ann. **1955/1956**, 437.

CHARLES-SIGLER, R., and E. GIL-AV: Gas-chromatographic determination of the configuration of amino-acids in antibiotics of the vernamycin B group (doricin). Tetrahedron Letters 1966, 4231.

CHARNEY, J., W.P. FISHER, C. CURRAN, R.A. MACHLOWITZ, and A.A. TYTELL: Streptogramin, a new antibiotic. Antibiot. & Chemotherapy 3, 1283 (1953).

COCITO, C., and A. KAJI: Virginiamycin M—A specific inhibitor of the acceptor site of ribosomes. Biochemie 53, 763 (1971).

COMPERNOLLE, F., H. VANDERHAEGHE, and G. JANSSEN: Mass spectra of staphylomycin A components and related cyclodepsipeptide antibiotics. Organic Mass Spectrometry 6, 151 (1972).

COX, B.R., and F.W. EASTWOOD: The structure of the antibiotic ostreogrycin B_3. Chem. Commun. 1970, 1623.

CROOY, P., and R. DE NEYS: Virginiamycin: Nomenclature. J. Antibiotics (Tokyo) 25, 371 (1972).

DELPIERRE, G.R., F.W. EASTWOOD, G.E. GREAM, D.G.I. KINGSTON, P.S. SARIN, L. TODD, and D.H. WILLIAMS: The structure of ostreogrycin A. Tetrahedron Letters 1966, 369.

EASTWOOD, F.W., B.K. SNELL, and A. TODD: Antibiotics of the E129 (ostreogrycin) complex. I. The structure of E129 B. J. Chem. Soc. 1960, 2286.

ENNIS, H.L.: Inhibition of protein synthesis by polypeptide antibiotics. III. Ribosomal site of inhibition. Mol. Pharmacol. 2, 444 (1966).

ENNIS, H.L.: A synergistic antibiotic complex which selectively inhibits protein synthesis. Progr. Antimicr. Anticancer Chemoth. II. 489 (1970). University of Tokyo Press, Tokyo.

ENNIS, H.L.: Interaction of vernamycin A with Escherichia coli ribosomes. Biochemistry 10, 1265 (1971).

ENNIS, H.L.: Polysome metabolism in E. coli: Effect of antibiotics on polysome stability. Antimicrobial Agents Chemotherapy 1, 197 (1972).

ENNIS, H.L.: Polysome metabolism in E. coli: Amicetin, an antibiotic that stabilizes polysomes. Antimicrobial. Agents Chemotherapy 1, 204 (1972).

ENNIS, H.L., and DUFFY, K.E.: Vernamycin A inhibits the non-enzymatic binding of fMet-tRNA to ribosomes. Biochim. Biophys. Acta 281, 93 (1972).

FERNANDEZ-MUÑOZ, R., R.E. MONRO, R. TORRES-PINEDO, and D. VAZQUEZ: Substrate- and antibiotic-binding sites at the peptidyl-transferase centre of Escherichia coli ribosomes. Eur. J. Biochem. 23, 185 (1971).

HEINEMANN, B., A. GOUREVITCH, J. LEIN, D.L. JOHNSON, M.A. KAPLAN, D. VANAS, and I.R. HOOPER: Etamycin, a new antibiotic. Antibiotics Ann. 1954/1955, 728.

HISHIZAWA, T., and S. PESTKA: Studies on the formation of transfer ribonucleic acid-ribosome complexes. XVII. The effect of tRNA on aminoacyl-oligonucleotide binding to ribosomes. Arch. Biochem. Biophys. 147, 624 (1971).

HOBBS, D.C., and W.D. CELMER: Structure of the antibiotics PA 114 B-1 and PA 114 B-3. Nature 187, 598 (1960).

JIMENEZ, A., MONRO, R.E., and D. VAZQUEZ: Interaction of Ac-Phe-tRNA with E. coli ribosomal subunits. 1. Sparsomycin-induced formation of a complex containing 50S and 30S subunits but not mRNA. FEBS Letters 7, 103 (1970).

KINGSTON, D.G.I., P.S. SARIN, L. TODD, and D.H. WILLIAMS: Structure of ostreogrycin G. IV. J. Chem. Soc. 19C, 1856 (1966a).

KINGSTON, D.G.I., L. TODD, and D.H. WILLIAMS: Antibiotics of the ostreogrycin complex. III. The structure of ostreogrycin A. J. Chem. Soc. 19C, 1669 (1966b).

KUBOTA, K., OKUYAMA, A., and TANAKA, N.: Differential effects of antibiotics on peptidyl transferase reactions. Biochem. Biophys. Res. Commun. 47, 1196 (1972).

MONRO, R.E., M.L. CELMA, and D. VAZQUEZ: Action of sparsomycin on ribosome-catalysed peptidyl transfer. Nature 222, 356 (1969).

MONRO, R.E., R. FERNANDEZ-MUÑOZ, M.L. CELMA, A. JIMENEZ, E. BATTANER, and D. VAZQUEZ: Antibiotics acting on the peptidyl transferase centre of ribosomes. Progr. Antimicr. Anticancer Chemoth. II, 473 (1970). University of Tokyo Press, Tokyo.

MONRO, R.E., and D. VAZQUEZ: Ribosome-catalysed peptidyl transfer: Effects of some inhibitors of protein synthesis. J. Mol. Biol. 28, 161 (1967).

PESTKA, S.: Studies on the formation of transfer ribonucleic acid-ribosome complexes. V. On the function of a soluble transfer factor in protein synthesis. Proc. Natl. Acad. Sci. U.S. 61, 726 (1968).

PESTKA, S.: Studies on the formation of transfer ribonucleic acid-ribosome complexes. XI. Antibiotic effects on phenylalanyl-oligonucleotide binding to ribosomes. Proc. Natl. Acad. Sci. U.S. **64**, 709 (1969).

PESTKA, S.: Studies on the formation of transfer ribonucleic acid-ribosome complexes. VIII. Survey of the effect of antibiotics on N-acetyl-phenylalanyl-puromycin formation: possible mechanism of chloramphenicol action. Arch. Biochem. Biophys. **136**, 80 (1970).

PESTKA, S.: Studies on the formation of transfer ribonucleic acid-ribosome complexes. IX. Effect of antibiotics on translocation and peptide bond formation. Arch. Biochem. Biophys. **136**, 89 (1970).

PESTKA, S.: Studies on transfer ribonucleic acid-ribosome complexes. XIX. Effect of antibiotics on peptidyl puromycin synthesis on polyribosomes from *Escherichia coli.* J. Biol. Chem. **247**, 4669 (1972).

PESTKA, S., and H. HINTIKKA: Studies on the formation of ribonucleic acid-ribosome complexes. XVI. Effect of ribosomal translocation inhibitors on polyribosomes. J. Biol. Chem. **246**, 7723 (1971).

PREUD'HOMME, J., P. TARRIDEC et A. BELLOC: Isolement de la pristinamycine et étude de ses principales propriétés physico-chimiques. Rev. Med. Toulouse **9**, 619 (1968).

SHEEHAN, J.C., H.G. ZACHAU, and W.B. LAWSON: The structure of etamycin. J. Am. Chem. Soc. **80**, 3349 (1958).

SOMER, P. DE, and P. VAN DIJCK: A preliminary report on antibiotic number 899, a streptogramin-like substance. Antibiot. & Chemotherapy **5**, 632 (1955).

SQUIBB, E.R. & SONS: Vernamycin C. U.S.P.O. 3,299,047 (1967).

SQUIBB, E.R. & SONS: Patricin A & B and related compounds. U.S.P.O. 3,373,151 (1968).

TANAKA, N., T. KINOSHITA, Y. LIN, T. NISHIMURA, and H. UMEZAWA: Inhibition sites of several antibiotics in polypeptide synthesis. Progr. Antimicr. Anticancer Chemoth. II, 502 (1970). Univ. Tokyo Press.

TANAKA, N., N. MIYAIRI, T. NISHIMURA, and H. UMEZAWA: Activity of mikamycins, angustmycins and emimycin against antibiotic-resistant staphylococci. J. Antibiotics (Tokyo), Ser. A **14**, 18 (1961).

TANAKA, N., N. MIYAIRI, K. WATANABE, N. SHINJO, T. NISHIMURA, and H. UMEZAWA: Biological studies on mikamycin. II. Laboratory investigations of mikamycin A and mikamycin B.J. Antibiotics (Tokyo), Ser. A **12**, 290 (1959).

VANDERHAEGHE, H., G. JANSSEN, and F. COMPERNOLLE: The structure of minor components of virginiamycin S. Tetrahedron Letters **1971**, 2687.

VANDERHAEGHE, H., and G. PARMENTIER: The structure of factor S of staphylomycin. J. Am. Chem. Soc. **82**, 4414 (1960).

VANDERHAEGHE, H., P. VAN DIJCK, G. PARMENTIER, and P. DE SOMER: Isolation and properties of the components of staphylomycin. Antibiot. & Chemotherapy **7**, 606 (1957).

VAZQUEZ, D.: Antibiotics affecting chloramphenicol uptake by bacteria. Their effect on amino acid incorporation in a cell-free system. Biochim. Biophys. Acta **114**, 289 (1966).

VAZQUEZ, D.: The streptogramin family of antibiotics. Antibiotics I. Mechanism of action, ed. by GOTTLIEB, D. and P.D. SHAW, p. 387, 1967.

VAZQUEZ, D., and R.E. MONRO,: Effects of some inhibitors of protein synthesis on the binding of aminoacyl tRNA to ribosomal subunits. Biochim. Biophys. Acta **142**, 155 (1967).

VERWEY, W.F., M.K. WEST, and A.K. MILLER: Laboratory studies of streptogramin. Antibiot. & Chemotherapy **8**, 500 (1958).

WATANABE, K.: Studies on mikamycin. VII. Structure of mikamycin B. J. Antibiotics (Tokyo), Ser. A **14**, 14 (1961).

YAMAGUCHI, H., and N. TANAKA: Site of action of mikamycin A and B in polypeptide-synthesizing system. J. Biochem. (Tokyo) **61**, 18 (1967).

YAMAGUCHI, H., N. TANAKA, and H. UMEZAWA: Effects of biopolymers and magnesium on the mikamycin inhibition of polyphenylalanine synthesis and the synergistic action of mikamycins A and B. J. Antibiotics (Tokyo) Ser. A **20**, 41 (1967).

YAMAGUCHI, H., Y. YOSHIDA, and N. TANAKA: Inhibition by mikamycins of polypeptide synthesis directed by native messengers and synthetic polynucleotides. J. Biochem. (Tokyo) **60**, 246 (1966).

YAMAKI, H., and N. TANAKA: Effects of protein synthesis inhibitors on the lethal action of kanamycin and streptomycin. J. Antibiotics (Tokyo), Ser. A **16**, 222 (1963).

Pactamycin*

Irving H. Goldberg

Pactamycin (Fig. 1), an antibiotic isolated from the fermentation broth of *Streptomyces pactum* (BHUYAN *et al.*, 1961; ARGOUDELIS *et al.*, 1961; BRODASKY and LUMMIS, 1961), is active against a variety of Gram-positive and Gram-negative microorganisms and against several animal tumor lines in culture or *in vivo* (BHUYAN *et al.*, 1961; WHITE, 1962). When administered to animals or cancer patients, the drug is rapidly cleared from the blood (BHUYAN and JOHNSTON, 1963) and is rapidly degraded in cultures of *B. amylofaciens* (BOTH, *et al.*, 1971).

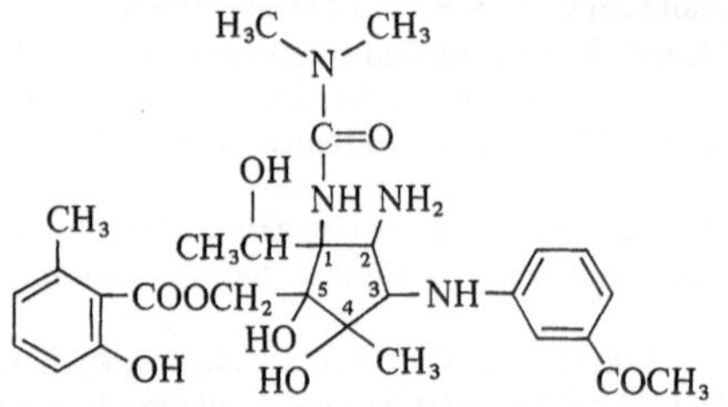

Fig. 1. Structure of pactamycin (WILEY *et al.*, 1970)

Pactamycin has been shown to act primarily by inhibiting protein synthesis in intact cells and in extracts of microorganisms and animals (YOUNG, 1966; COLOMBO, *et al.*, 1966; BHUYAN, 1967; CUNDLIFFE and MCQUILLEN, 1967; COHEN and GOLDBERG, 1967; MACDONALD and GOLDBERG, 1970; HARFORD and SUEOKA, 1970). The ribosomes have been implicated as the site of action of the antibiotic (FELICETTI *et al.*, 1966; BHUYAN, 1967). In common with other inhibitors of protein synthesis, pactamycin can stimulate the synthesis of RNA in microorganisms (KERSTEN *et al.*, 1967; EZEKIEL and ELKINS, 1968). Both ribosomal and transfer RNA of *B. subtilis* treated with pactamycin have been found to be deficient in methyl groups (KERSTEN *et al.*, 1968). Pactamycin resistance, unlike the resistance loci to several other antibiotic inhibitors of protein synthesis, maps on the origin of the *B. subtilis* chromosomes on the distal side of the reference marker *cys* A14 (HARFORD and SUEOKA, 1970).

Several lines of evidence show that at low concentrations, pactamycin selectively blocks polypeptide chain initiation. Thus, in extracts from *E. coli*, the initiation complex in the presence of pactamycin possessed an altered structure with

---

*This work was supported by U.S. Public Health Service Research Grant GM 12573 from the National Institutes of Health.

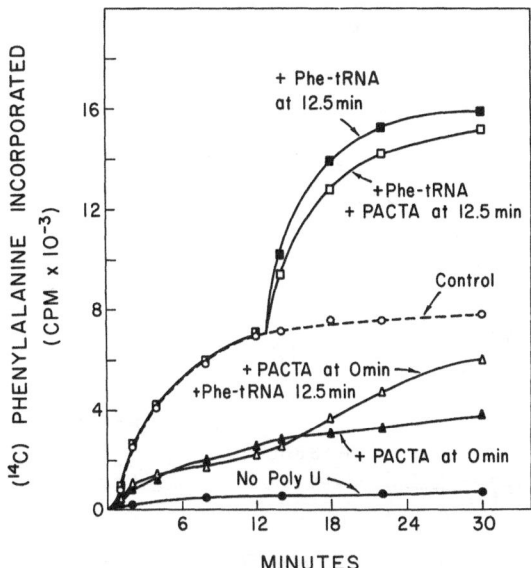

Fig. 2. Effect of the delayed addition of pactamycin on polyphenylalanine synthesis initiated by (^{12}C) N-acetylphenylalanyl-tRNA. Conditions were as described in Fig. 2, COHEN *et al.*, 1969 a, except that the reaction contained half the amount of (^{14}C) phenylalanyl-tRNA from the start of the reaction and the other half was added as indicated

decreased stability at low magnesium concentrations (COHEN and GOLDBERG, 1967; COHEN *et al.*, 1969 b; COHEN *et al.*, 1969 a). Further, in polyphenylalanine synthesis at low magnesium concentrations where N-acetylphenylalanine-tRNA served as the initiator-tRNA, pactamycin appeared to have a primary effect on polypeptide initiation. This is shown in Fig. 2 where the addition of pactamycin to a polypeptide-synthesizing reaction after initiation has been completed was found to have little effect on subsequent elongation in contrast to the situation where the antibiotic was present during initiation. Further, pactamycin does not affect peptide bond formation as such, since the puromycin reaction with prebound peptide or its equivalent in the P-site is not inhibited by this agent (COHEN *et al.*, 1969 b).

A strong support for a primary action of pactamycin as being on polypeptide chain initiation comes from experiments using intact cells and extracts from mammalian sources (MACDONALD and GOLDBERG, 1970; STEWART-BLAIR *et al.*, 1971; LODISH *et al.*, 1971; AYUSO and GOLDBERG, in press). As shown in Fig. 3, two cell-free globin-synthesizing systems derived from rabbit reticulocytes, which differ in their ability to start new peptide chains were compared for their sensitivity to different concentrations of pactamycin. Comparison of A and B reveals that polypeptide synthesis in the lysate system (B) in which over 70 percent of the globin made is due to the formation of new chains, is considerably more sensitive to inhibition by pactamycin than is the fractionated system (A) where less than 10 percent can be so accounted for. Further, inhibition of incorporation is seen after a lag of about 2 minutes during which time the chain elongation is completed

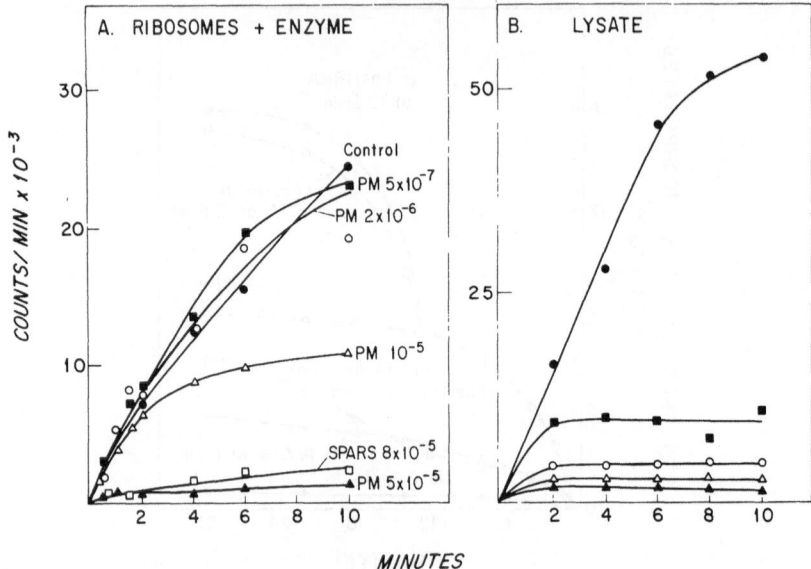

Fig. 3. Effect of pactamycin on protein synthesis in fractionated (A) and whole lysates (B) from rabbit reticulocytes. On the ordinate is plotted the incorporation of (^{14}C) valine into polypeptide. NH$_2$-terminal (^{14}C) valine analyses of the synthetized globin revealed that less than 10 percent of the globin made in (A) is due to the formation of new chains, whereas over 70 percent of that made in (B) is synthesized *de novo*. PM, pactamycin; SPARS, sparsomycin. See STEWART-BLAIR *et al.*, (1971) for details

(and is the time required to translate the globin message). COLOMBO *et al.*, 1966, found with intact reticulocytes, in lysate which are able to initiate new chains, that a low level of pactamycin (about 10^{-6} M, dependent on the amount of ribosomes) leads to the rapid (within 2 minutes) and sequential breakdown of larger polyribosomes to smaller polyribosomes and finally to single 80S ribosomes with the release of completed α and β globin chains (Fig. 4) (STEWART-BLAIR *et al.*, 1971). The latter have been identified by polyacrylamide gel electrophoresis. In contrast to aurintricarboxylic acid which causes the accumulation of ribosomal subunits in addition to 80S monosomes (STEWART-BLAIR *et al.*, 1971; HUANG and GROLLMAN, 1972), subunits are not produced by pactamycin. At higher concentrations of pactamycin (10^{-5} M), polyribosome breakdown is incomplete and globin release is slowed. These data suggest that at low concentrations of pactamycin, initiation of protein synthesis is affected much more than elongation or termination, but at the higher concentrations, elongation may be affected as well. By interfering primarily with the intiation step, pactamycin prevents the formation of new polyribosomes while allowing for read-off of mRNA and run-off ribosomes by the existing ones. Thus, antibiotic inhibitors of peptide-chain elongation such as sparsomycin, cycloheximide or fusidic acid, prevent the pactamycin-induced polyribosome breakdown. While polyribosome decay and radioactive peptide release is completely accomplished within 2 minutes of the addition of 3.5×10^{-6} M pactamycin, there appears to be a dissociation between these two parameters at the 30 second point (Fig. 5). The radioactivity

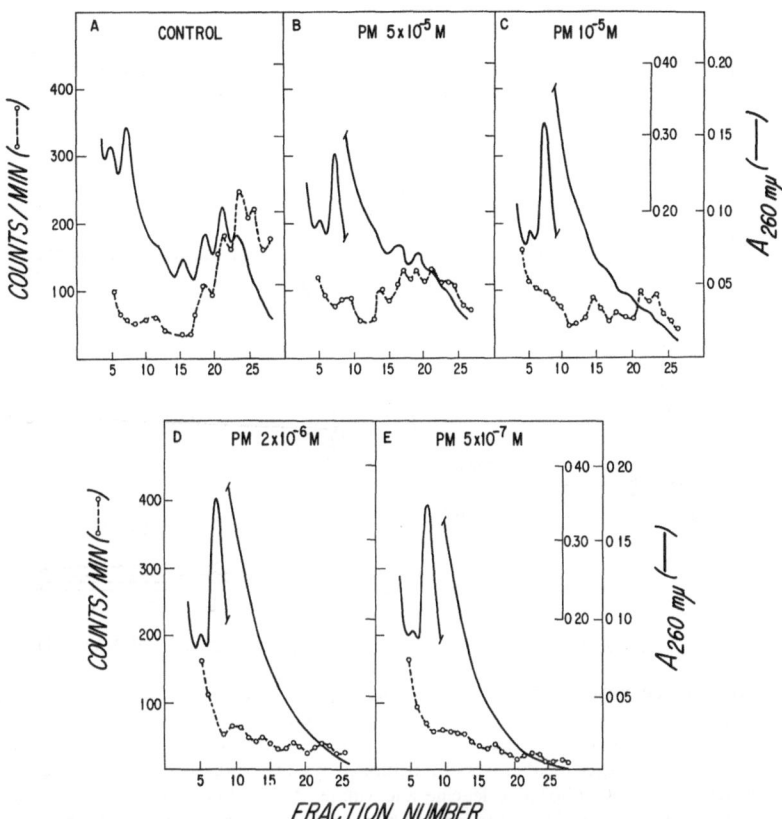

Fig. 4. Effect of pactamycin concentration on the breakdown of polyribosomes and release of peptide in rabbit reticulocyte lysates. Polyribosomes prelabelled with (^{14}C) amino acids in lysates were then incubated for 2.25 min with different pactamycin concentrations. An aliquot of the reaction was analyzed by sucrose density gradient centrifugation (from left to right: the main single peak at $A_{260m\mu}$ is at 80S (STEWART-BLAIR *et al.*, 1971)

which persists on the polyribosomes for the first 30 seconds results from the continued addition of radioactive amino acids onto elongating peptide, since the addition of excess unlabelled amino acids obliterates the lag in the radioactivity release from the polyribosomes. The continued incorporation of radioactive amino acids associated with elongation in the absence of initiation also accounts for the slightly more than two-fold increase in labelled protein, released into the soluble fraction in the presence of pactamycin. Further, as expected of an agent which predominately blocks polypeptide chain initiation, the patterns of release of radioactivity from polyribosomes and of accumulation in the supernatant fraction are virtually identical with those in the control when further incorporation of radioactivity is prevented by excess unlabelled amino acids. It should be noted that the fact that labelled nascent peptide is better maintained on the polyribosomes at 30 seconds than the polyribosomes themselves is further evidence against the premature release of incomplete polypeptide chains owing to pactamycin's action.

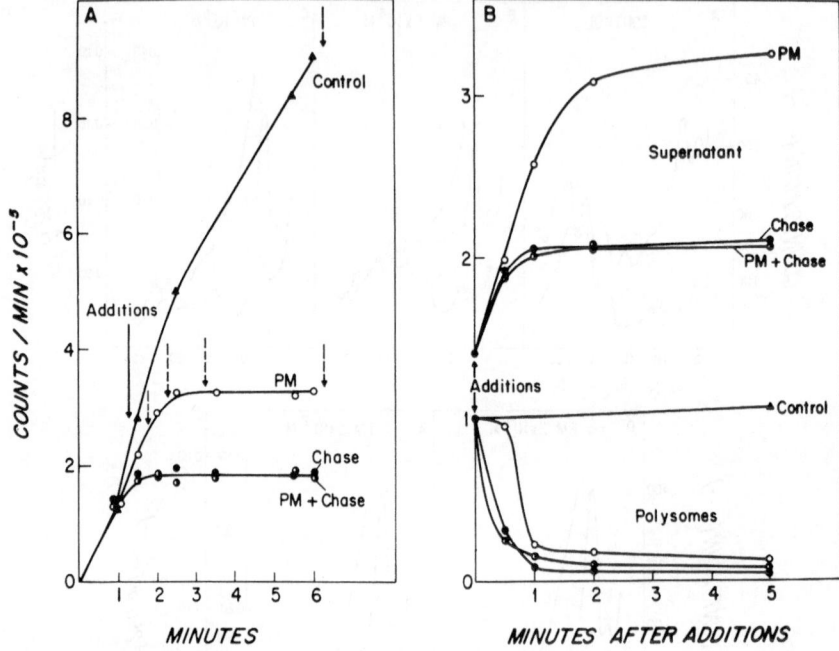

Fig. 5. Effect of pactamycin (3.5×10^{-6} M) on polypeptide synthesis and distribution as a function of time. At 1.25 min (additions, A and B), solutions containing pactamycin, pactamycin+chase (containing large excess of unlabelled amino acids), and a chase solution without pactamycin were added as indicated, to three incubations of reticulocyte lysates synthesizing polypeptide as measured by the incorporation of radioactive leucine and valine. At the times indicated by broken arrows in (A), aliquots were removed and fractionated by centrifugation into supernatant and polyribosome fractions (B). See STEWART-BLAIR et al., 1971 for experimental details

Table 1. *Effects of different antibiotics on the KCl wash-stimulated incorporation of amino acids into protein*

Antibiotic (M)		Percent Inhibition	
		− KCl wash	+ KCl wash
Pactamycin	1×10^{-6}	0	70
Chlortetracycline	1×10^{-4}	27	19
	2×10^{-4}	42	42
Sparsomycin	1×10^{-7}	50	37
	2×10^{-7}	71	61
Cycloheximide	2×10^{-6}	23	40

A fractionated system similar to that used in Fig. 3A, except that the ribosomes had been extracted with 0.5 KCl to remove initiation factors, was allowed to incorporate amino acids into protein for 2 min. when 380μg of KCl wash protein was added and incubation continued for an additional 8 min. When included, the antibiotic was present from zero time (AYUSO and GOLDBERG, in press).

Additional evidence supporting an action of pactamycin on initiation is provided by experiments in which polypeptide synthesis, stimulated by a preparation of initiation factors (KCl wash), is found to be much more sensitive to inhibition by the antibiotic than is synthesis in the absence of added initiation factors. In the former, new globin chains are being made, while in the latter, incorporation of radioactive amino acid represents mainly completion of existing nascent chains. This differential effect of pactamycin on initiating and non-initiating systems is not found with antibiotics such as chlortetracycline and sparsomycin which affect only elongation, but is found to some extent with cycloheximide (Table 1) which has been reported to inhibit preferentially peptide chain initiation at low concentrations.

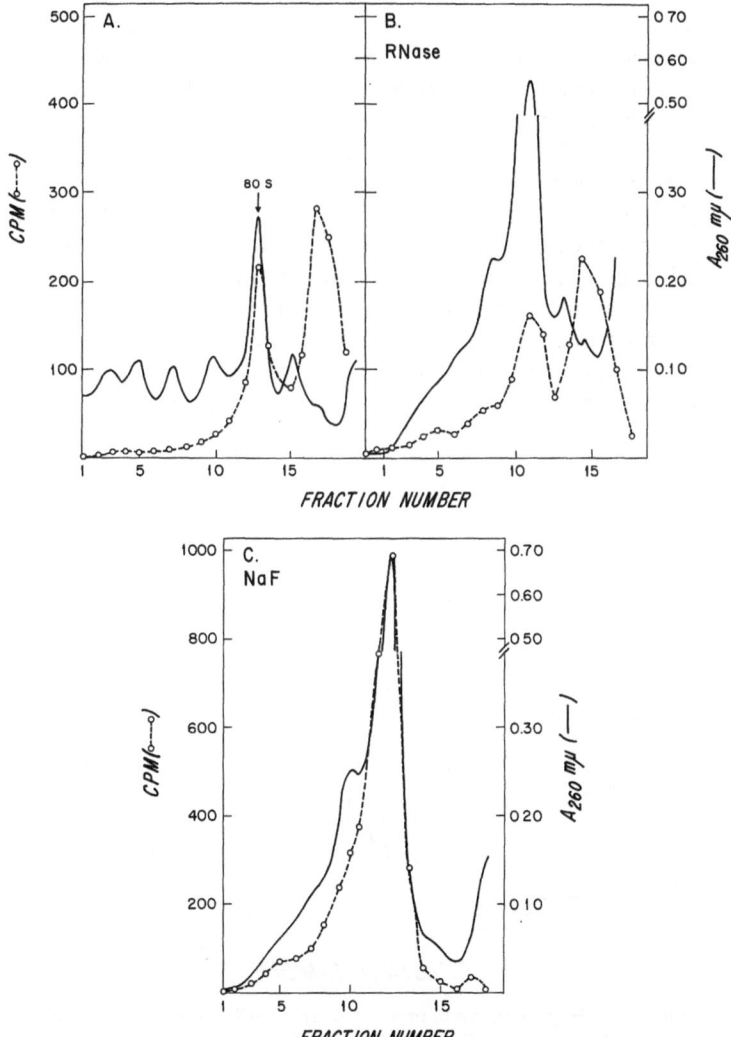

Fig. 6. Cold binding of (^{3}H) pactamycin to reticulocyte ribosomes as analyzed by sucrose density centrifugation (from right to left). See MACDONALD and GOLDBERG, 1970 for experimental details

Pactamycin binds rapidly to the smaller (30S) ribosomal subunit from bacteria (COHEN et al., 1979a) and also interacts with the smaller (40S) ribosomal subunit from mammalian cells (MACDONALD and GOLDBERG, 1970). At 0°C, pactamycin binds to the rabbit reticulocyte 40S ribosomal subunit, as well as to the 80S ribosome, but not to polyribosomes (Fig. 6A). The binding to 80S ribosomes is presumably by way of the smaller subunit, since there is no binding to the isolated larger subunit. Further, binding to the free smaller subunit has the highest specific activity. Single ribosomes derived from polyribosomes by RNase treatment appear not to bind pactamycin at 0°C (Fig. 6B). On the other hand, 80S ribosomes which are produced by NaF treatment of reticulocytes (by blocking initiation) and which, for the most part, are run-off ribosomes, and hence, free or mRNA (and likely other components involved in polypeptide synthesis) bind pactamycin readily (Fig. 6C). Similar results have been obtained in E. coli extracts where polyribosomes are produced in response to added f2 viral RNA as messenger (Fig. 7) (STEWART and GOLDBERG, in press). These data indicate that ribosomes which bear even a small fragment of mRNA are not able to bind pactamycin at 0°C. While this may be due to a direct blocking of the antibiotic binding site by the mRNA or to a conformational change induced in the ribosome by the mRNA such that this site is inaccessible to the antibiotic, it may rather be due to the presence of some other component(s) of the protein-synthesizing system which is present in the polyribosomes and prevents pactamycin binding.

There is an inverse relationship between the formation of f2 RNA-bearing ribosomal complexes engaged in polypeptide synthesis and the cold binding of pactamycin (STEWART and GOLDBERG, in press). Thus, pactamycin binding decreases as the amount of f2 RNA added to the reaction is increased (and polypep-

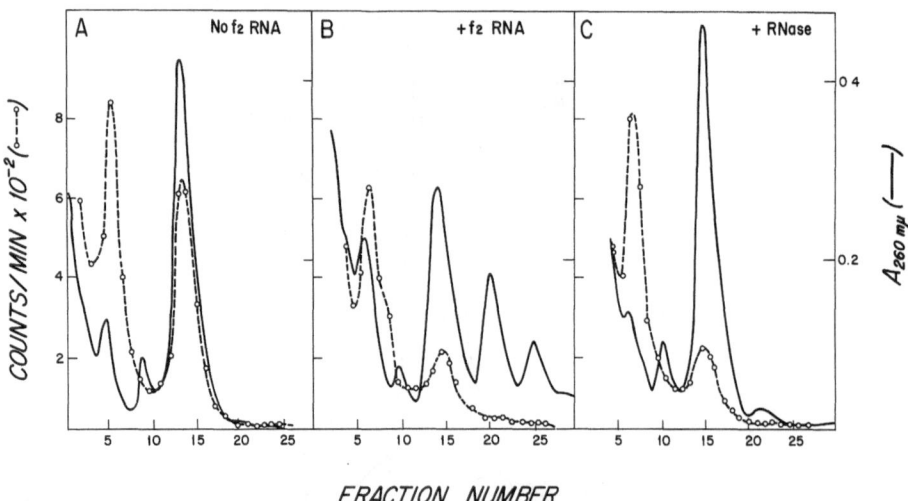

Fig. 7. Cold binding of (^{3}H) pactamycin to preformed mRNA-ribosome complexes in E. coli extracts. In (C), polyribosomes formed with f2 RNA were degraded to monosomes by treatment with ribonuclease before addition of (^{3}H) pactamycin at 0°C. Sucrose gradients were centrifuged from left to right (STEWART and GOLDBERG, in press)

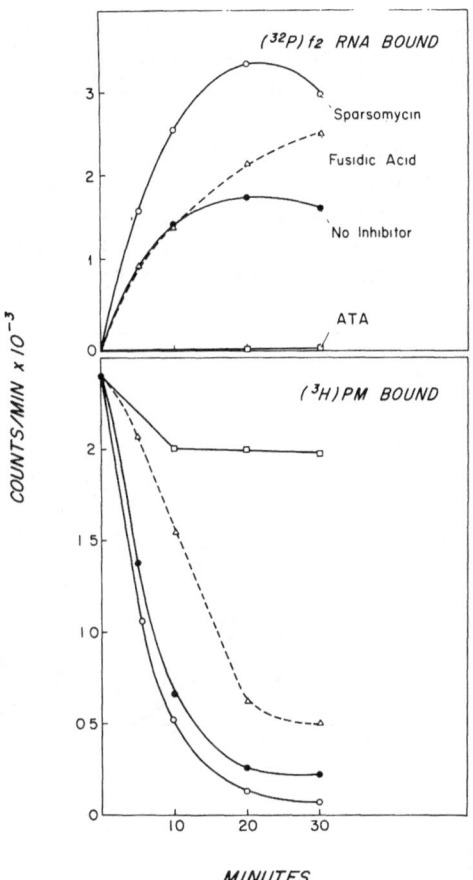

Fig. 8. Time course of binding of (^{32}P) f2 RNA and (^{3}H) pactamycin to *E. coli* ribosomes in the presence of polypeptide synthesis inhibitors. f2 RNA-promoted polypeptide synthesis was allowed to proceed at 35°C in the absence or in the presence of aurintricarboxylic acid (ATA, 6×10^{-5} M), fusidic acid (1×10^{-3} M), or sparsomycin (5×10^{-6} M) and then (^{3}H) pactamycin was added at 0°C. Binding of (^{3}H) pactamycin and (^{32}P) f2 RNA to ribosomes was measured on sucrose gradients (STEWART and GOLDBERG, in press)

tide synthesis thereby increased) and as the binding of f2 RNA to ribosomes increases with time of incubation (Fig. 8). It is not necessary, however, to have active chain elongation taking place on the f2 RNA-bearing ribosomes in order to block the subsequent cold binding of pactamycin. When elongation and polyribosome formation is prevented by the antibiotics sparsomycin or fusidic acid, f2 RNA still binds to single ribosomes (the apparent increase in bound f2 RNA is probably due to lack of translation-induced mRNA degradation so that the pieces of mRNA bound are longer) and pactamycin binding is blocked (Fig. 8). Further, as expected, aurintricarboxylic acid which prevents f2 RNA attachment permits maximal pactamycin binding.

From the foregoing, one would predict that conditions which lead to the release of ribosomes from mRNA would reinstate their ability to bind pactamycin. In the experiment shown in Fig. 9, puromycin was used to release nascent peptide and produce single ribosomes from polyribosomes, resulting in the return of pactamycin binding in the cold (Fig. 9C) to the extent found in the absence of f2 RNA (Fig. 9A). In the presence of puromycin, however, some single ribosomes are able to rebind mRNA and form a peak of more rapidly sedimenting bound (^{32}P) f2 RNA which is free of pactamycin. If rebinding of puromycin-released ribosomes to mRNA is completely prevented by the presence of aurintri-

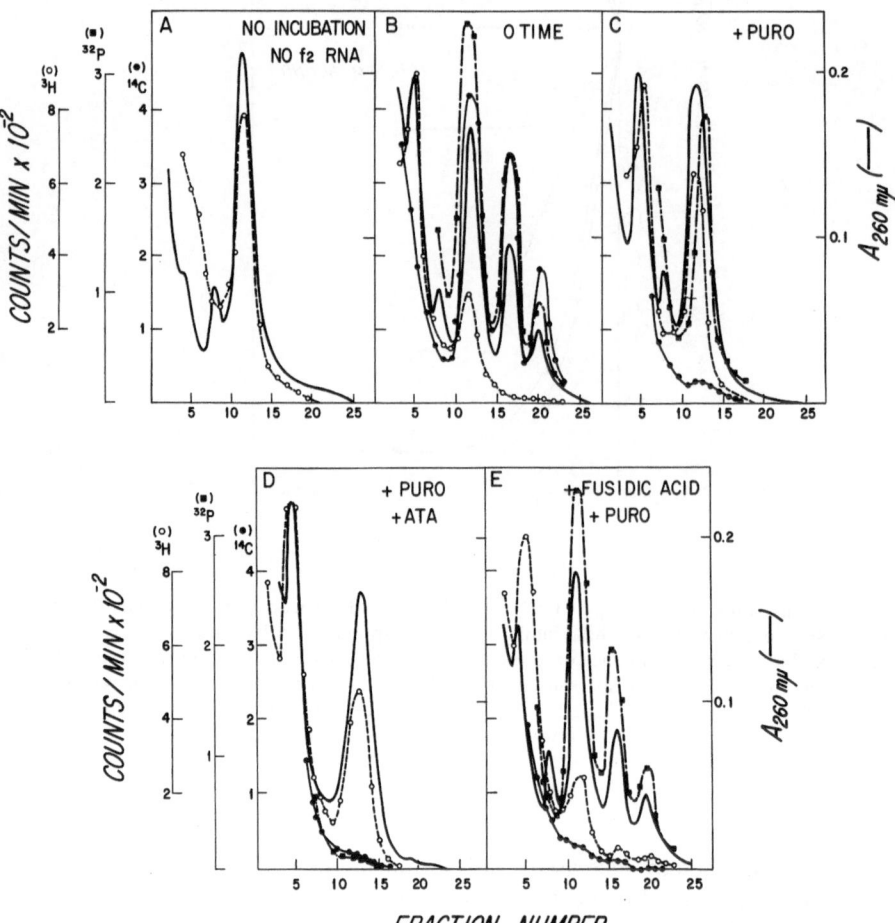

Fig. 9. Restoration of pactamycin binding (at 0°C) to ribosomes following puromycin-induced degradation of polyribosomes. Polypeptide synthesis [(^{14}C) valine incorporation], (^{32}P) f2 RNA binding and subsequent cold (^{3}H) pactamycin binding were followed in identical reactions in the presence or absence of puromycin (3×10^{-3} M), fusidic acid (10^{-3} M), and aurintricarboxylic acid (5×10^{-5} M) as indicated. After incubation at 35° for 15 min (zero time), two reactions (A and B) were cooled to 0°C and the indicated protein synthesis inhibitors were added to the other reactions (C, D and E) for an additional 4 min incubation at 35°C. Centrifugation on sucrose gradients was from left to right (STEWART and GOLDBERG, in press)

carboxylic acid, there is, however, no additional binding of pactamycin to these ribosomes (Fig. 9 D). On the other hand, if the puromycin-induced polyribosome disaggregation is blocked by fusidic acid, the cold binding of pactamycin remains very low even though almost all of the polypeptide has been released (Fig. 9 E). This effect is not due to a direct interference by fusidic acid of pactamycin binding to ribosomes.

Although pactamycin binds negligibly to mRNA-bearing monosomes and polyribosomes at 0°C, adding ^{3}H-pactamycin at 35° results in binding of the antibiotic to polyribosomes (Fig. 10). Unlike the rapid binding at 0°C to ribosomes lacking mRNA, binding of pactamycin to polyribosomes is slow, accumulative and sensitive to inhibitors of elongation such as sparsomycin or chloramphenicol.

These experiments suggest that the cold binding of ^{3}H-pactamycin may be used as a simple test for the presence of mRNA on ribosomes. It is possible, however, that cold pactamycin binding to the ribosome is not blocked by the mRNA *per se* but rather because the donor or acceptor site is occupied by initiator- or aminoacyl-tRNA (or factors). When pactamycin is present *during* active polyribosome formation and polypeptide synthesis (i.e. at elevated temperatures), it may compete during a round of peptide formation for binding to the appropriate site with one or the other types of acylated-tRNA or factor(s) (Fig. 10 C). When pactamycin is added after polyribosome and polypeptide formation (i.e. cold binding), however, these sites are filled and pactamycin cannot

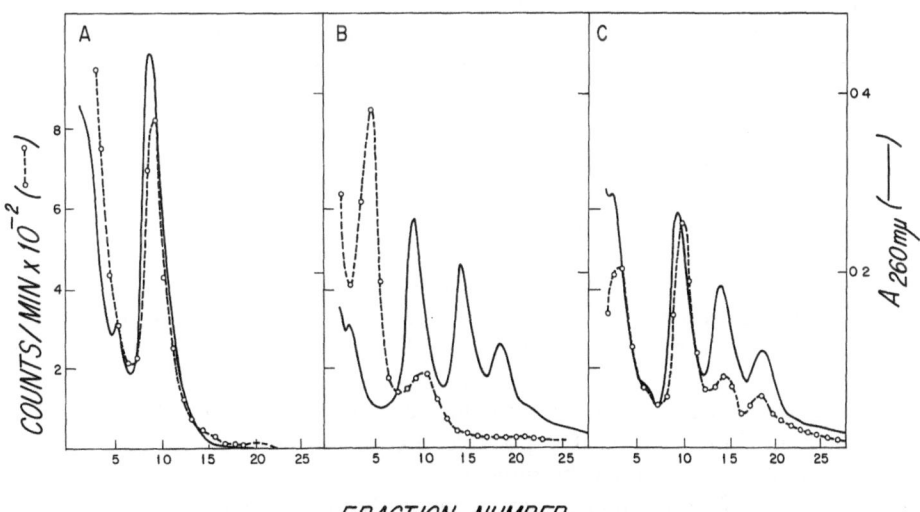

Fig. 10. (^{3}H) pactamycin binding to f2 RNA-bearing ribosomes at 0°C and 35°C. No f2 RNA is present in (A) and the reaction was always at 0°C. In (B) and (C), f2 RNA-promoted polyribosome formation was allowed to take place (15 min, 35°C). In (C), (^{3}H) pactamycin was present throughout the incubation at 35°C, whereas in (B), the labelled antibiotic was added at 0°C after completion of incubation. Sucrose gradient centrifugation was from left to right (STEWART and GOLDBERG, in press)

bind (Fig. 7 B). The experiment in which both puromycin and fusidic acid were used (Fig. 9E) is compatible with such a mechanism for even though peptide has been released by puromycin, the release of the deacylated tRNA from the donor site may be blocked by fusidic acid. It should be noted, furthermore, that binding to the smaller ribosomal unit is always of the highest specific activity and remains very high even when f2 RNA is added.

While such mechanisms may account for the binding of pactamycin to polyribosomes at elevated temperatures an perhaps be responsible for inhibitory effects on polypeptide elongation in both pro- and eucaryotes, results obtained in the reticulocyte system (to be described later) suggest an alternate explanation for the polyribosomal binding of pactamycin and the selectivity of inhibition of initiation. These data suggest that the polyribosomes may become labelled with (^{3}H) pactamycin by virtue of the presence of a pactamycin-inactivated 40S ribosomal subunit on the initiation site of the mRNA. Thus, labelling of the polyribosome would occur only at the elevated temperature since attachment of the 40S subunit bearing labelled pactamycin would not occur in the cold.

These data suggest that pactamycin binds preferentially to free smaller ribosomal subunits, (and possibly to run-off single ribosomes) before monosome or polysome formation and that this results in a relatively selective action on the initiation of polypeptide synthesis rather than on peptide chain elongation and release. In f2 RNA-promoted polypeptide synthesis, however, we have not been able as yet to obtain evidence that initiation is significantly more sensitive to pactamycin than is elongation. In this system, pactamycin (10^{-5} M) addition to ribosomes undergoing active protein synthesis results in the immediate cessation of further amino acid incorporation (GOLDBERG et al., in press); P.-C. TAI and B. J. WALLACE, personal communication) and even at concentrations of antibiotic less than is needed for a complete inhibition of synthesis, the polyribosomes do not decay with the release of peptide into the soluble fraction, unlike the situation in the eucaryotic system. A similar "freezing" action by pactamycin has also been found with E. coli polyribosomes which use endogenous mRNA (P.-C. TAI and B.J. WALLACE, personal communication). It should be recalled, however, that in the E. coli ribosomal system where an unnatural mRNA, poly U, was used, a selective effect of pactamycin on initiation could be demonstrated (COHEN et al., 1969 b, Fig. 2). Furthermore, we have found that the overall binding of N-formylmethionyl-tRNA$_f$ to E. coli ribosomes stimulated by f2 RNA is inhibited 50–60% by pactamycin at 6 mM Mg^{2+}. One wonders whether the fact that both the endogenous and viral RNA's are polycistronic in the E. coli system, but monocistronic in the reticulocyte system might account for the different findings. For instance, it is possible that given the relatively small size of the intracistronic ribosomal binding site, pactamycin by allowing formation of a defective initiation complex may, secondarily prevent elongation and release of peptide and dissociation of ribosomes from the 5'-adjacent cistron of a polycistronic message such as f2 RNA. Thus, it is conceivable that elongation and release of coat protein might be prevented by a ribosome occupying the initiation site for the RNA synthetase cistron of the f2 RNA. It is also possible that the "nativeness" of the mRNA is determining the sensitivity of initiation to pactamycin.

The accumulated observations suggest that in the reticulocyte system, an initiation complex is formed in the presence of pactamycin but is not functionally active. Experiments using ribosomes from Ehrlich Ascites cells by Drs. AYUSO, HIRCH and HENSHAW (personal communication) have shown that pactamycin does not prevent the binding of initiation factors to the 40 S subunit or the attachment of synthetic mRNA to the ribosomes. Further, the ability of a "dissociation factor" to convert reticulocyte 80S ribosomes into subunits is also not affected by the antibiotic. If the relative sensitivities of initiation and elongation to pactamycin inhibition in the mammalian system are not determined solely by the availability of a binding site on the ribosome for the antibiotic, then some step peculiar to initiation might be selectively blocked by pactamycin. Thus, during initiation, the transfer or conversion of the initiator-tRNA from a site (? pre-donor site) or form in which it is not able to form a peptide bond (or react with puromycin) to one in which it can, might possibly be prevented by pactamycin, resulting in a defective initiation complex. This "conversion" to puromycin reactivity, at least in the bacterial system, apparently does not involve movement of the initiator-tRNA through the acceptor site on its way to the donor site (LELONG et al., 1971; MODOLELL et al., 1971; THACH and THACH, 1971; BENNE and VOORMA, 1972).

Evidence has been obtained in eucaryotes that pactamycin does not block the formation of the initiation complex but at low magnesium concentrations interferes with the conversion of initiator methionyl-tRNA$_f$ into a puromycin-reactive form on ribosomes from wheat embryos (SEAL and MARCUS, 1972) and reticulocytes (KAPPEN et al., in press) (Table 2 and Fig. 11). The total binding of [^{35}S] Met-tRNA$_f$ to reticulocyte ribosomes bearing endogenous mRNA is

Table 2. *Effect of pactamycin and other agents on Met-tRNA$_f$ binding to reticulocyte ribosomes and Met-puromycin formation*

Experiment 1: [^{35}S] Met-tRNA$_f$ binding			Experiment 2: Met-puromycin formation		
Inhibitor (concentration)	cpm bound	Percent inhibition	Inhibitor (concentration)	Met- puromycin (cpm)	Percent inhibition
0	7823	—	0	3433	—
Pactamycin (2 μM)	7791	0	Pactamycin (1 μM)	766	78
Pactamycin (10 μM)	7190	8	Pactamycin (1 μM) at 7 min	2952	14
NaF (10 mM)	5602	28	Pactamycin (2 μM)	665	81
ATA (0.2 mM)	2347	70	Pactamycin (2 μM) at 7 min	2495	27
			Pactamycin (10 μM)	267	92
			NaF (10 mM)	852	75

Binding incubations (7 min, 23°) in a total volume of 0.1 ml contained 20 mM Tris-HCl, pH 7.5, 1 mM dithiothreitol, 0.5 mM GTP, 100 mM KCl, 2 mM MgCl$_2$, 2.2 A$_{260}$ units of 0.5 m KCl washed ribosomes, 0.17 A$_{260}$ units of [^{35}S] Met-tRNA (46,000 cpm) and 278 μg of ribosomal wash protein. For Met-puromycin synthesis, 0.5 mM puromycin was added to the binding incubation at 7 min and incubation was continued for 10 min at 30°. Pactamycin, NaF and aurintricarboxylic acid (ATA) were present from the start of the reaction unless otherwise indicated. (KAPPEN et al., in press).

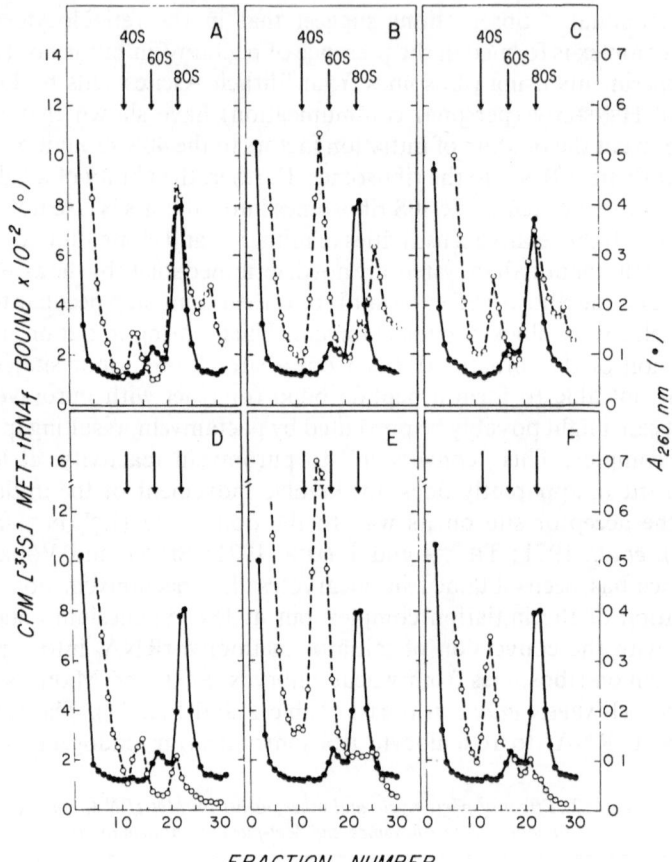

Fig. 11. Sucrose gradient centrifugation of [^{35}S] Met-tRNA$_f$ bound to washed reticulocyte ribosomes and effect of pactamycin. Reaction mixtures (0.1 ml) containing 20 mM Tris-HCl, pH 7.5, 1 mM dithiothreitol, 1 mM GTP, 100 mM KCl, 2 mM MgCl$_2$, 2.9 A$_{260}$ units of ribosomes, 400 μg of ribosomal wash protein, 0.39 A$_{260}$ units of [^{35}S]-Met-tRNA (61,600 cpm) were incubated for 14 min at 30° in the presence or absence of pactamycin (1 μM) and layered onto 15–30% sucrose gradients. After centrifugation, the gradients were fractionated and the fractions were diluted to 1 ml with the gradient buffer before absorbance at 260 nm was measured in a Zeiss spectrophotometer. Centrifugation was from left to right. A. Control without pactamycin. B. Pactamycin present from the beginning. C. Pactamycin added at 7 min. D, E. F. 0.5 mM puromycin was added at 7 min to incubations identical to A, B, C, respectively (KAPPEN et al., in press)

not significantly affected by pactamycin or NaF but is inhibited by aurintricarboxylic acid (Table 2). The synthesis of a functional initiation complex, as judged by its ability to form [^{35}S] Met-puromycin however, is impaired by both pactamycin and NaF provided that these agents were present during complex formation. If added after most of the active complex has been formed, pactamycin is not able to interfere significantly with the puromycin reaction (Table 2). These results are in agreement with those found with wheat embryo ribosomes and tobacco mosaic virus RNA as messenger (SEAL and MARCUS, 1972).

The formation of an active 80S initiation complex proceeds via a 40S ribosomal subunit-mRNA-Met-tRNA$_f$ intermediate to which a 60S ribosomal subunit joins (CRYSTAL *et al.*, 1971; Weeks *et al.*, 1972). The overall process requires initiation factors and GTP. The analysis of binding reactions similar to those described in Table 2 by sucrose density gradient centrifugation shows that the formation of the puromycin-reactive 80S initiation complex (Fig. 11A, D) is prevented by pactamycin (Fig. 11B, E) and that this is associated with a pronounced increase in the amount of [^{35}S] Met-tRNA$_f$ bound to the smaller initiation complex, which sediments slightly further down the gradient (about 50S) than the 40S ribosomal subunit. Pactamycin, furthermore, has altered the structure of the accumulated smaller initiation complex since it is considerably more resistant to degradation by pancreatic RNAse than is the normal intermediate in protein synthesis. If the addition of pactamycin is delayed until after most of the active 80S complex has been formed, there is only a slight inhibition of [^{35}S] Met-puromycin formation (Fig. 10C, F). It should be noted, however, that the small amount of [^{35}S] Met-tRNA$_f$ bound to the 80S initiation complex in the presence of pactamycin (Fig. 11B) is poorly releasable by puromycin (Fig. 11E). Pactamycin does not in itself cause the 80S complex to dissociate since the pattern of binding found in Fig. 11C is identical to a reaction lacking pactamycin (not shown) which was stopped at the time the antibiotic was added in Fig. 11C.

Although in the control incubation a small peak of [^{35}S] Met-tRNA$_f$ is found in an area on the gradient corresponding to the dimer region (Fig. 11A), in the presence of pactamycin the predominant fast-moving radioactive peak is neither in the monomer or the dimer region but inbetween (about 95S) (Fig. 11B). While the amount of this particle has been found to vary in different experiments for as yet unexplained reasons, its location 3 to 4 fractions earlier on the gradient than the dimer region is highly reproducible. This material is not produced when pactamycin is added late in the binding reaction (Fig. 11C). Its position on the sucrose gradient suggests that it represents a hybrid ("1^1/$_2$ mer") between a monosome (ribosome couple on a strand of mRNA) and a pactamycin-inactivated 40S ribosomal subunit bearing [^{35}S] Met-tRNA$_f$, located on the initiation site of the mRNA. Such ribosomal structures have been found under various conditions (HOERZ and MCCARTY, 1969; CRYSTAL *et al.*, 1971) especially in the presence of NaF (HOERZ and MCCARTY, 1969). Additional support for the notion that pactamycin produces initiation-inactivated 1^1/$_2$ mers to which the 60S ribosomal subunit fails to join comes from experiments using puromycin or pancreatic RNase. The addition of puromycin to the reaction after binding results in the disappearance of radioactivity (about 2500 cpm) from the putative 1^1/$_2$ mer region of the gradient and a concomitant increase (about 2500 cpm) in the amount of radioactivity found in the smaller initiation complex (Fig. 11E). This result is the expected one since puromycin should lead to the release of the uninvolved ribosomal couple from the mRNA while the inactivated ribosomal subunit with its [^{35}S] Met-tRNA$_f$ remains attached. By contrast, [^{35}S] methionine associated with ribosome monomers and dimers is released to the top of the gradient as [^{35}S] Met-puromycin (Fig. 11D, F). Further, RNase treatment of binding reactions results in the disappearance of the putative radioactive 1^1/$_2$ mer with little change or a small increase in the radioactivity associated with the 80S complex.

The sensitivity of Met-puromycin formation to pactamycin depends on the concentration of Mg^{2+} (Fig. 12). When the binding reaction takes place at higher Mg^{2+} concentrations, the pactamycin effect is considerably decreased. Sucrose gradients confirm that the formation of the functional 80S initiation complex at 5 mM Mg^{2+} is inhibited to the same small extent as is the synthesis of Met-puromycin. If the Mg^{2+} concentration is increased to 5 mM after binding in the presence of pactamycin is over, there is a partial reversal of the inhibition of the puromycin reaction and the formation of the 80S initiation complex.

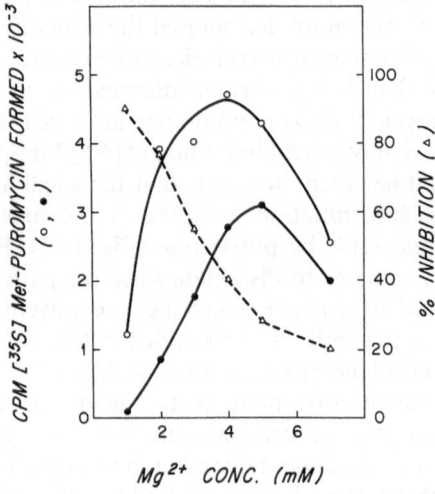

Fig. 12. Effect of magnesium concentration on pactamycin inhibition of Met-puromycin synthesis. The reaction conditions are similar to those described in the legend for Table 2 except that the concentration of the initiation factors was 400 µg and puromycin (0.5 mM) and pactamycin (1 µM) were present from the start of the reaction. ○, control; ●, with pactamycin; △, % inhibition by pactamycin (KAPPEN *et al.*, in press)

These experiments provide additional support for the concept that pactamycin produces a non-functional initiation complex with reticulocyte ribosomes bearing endogenous mRNA (STEWART-BLAIR *et al.*, 1971; GOLDBERG *et al.*, in press). A similar conclusion has been reached from experiments using wheat embryo ribosomes and tobacco mosaic virus RNA as messenger (SEAL and MARCUS, 1972). Our results on sucrose density gradients, however, differ from those reported by SEAL and MARCUS (1972) and WEEKS and BAXTER (1972) with wheat embryo ribosomes and viral mRNA. These workers report that pactamycin does not lead to the accumulation of the smaller ribosomal initiation complex. In the only sucrose gradient data actually presented from the wheat embryo system, the Mg^{2+} concentration in the binding reaction was relatively high, 3.6 mM (WEEKS *et al.*, 1972). It is not yet clear, however, whether the difference in results is due to differences in reaction conditions, or basic differences between reticulocyte and wheat embryo ribosomes, mRNA or initiation factors. The data from Fig. 11B, E are compatible with the formation of an inactive 80S complex in the presence of pactamycin in the reticulocyte system but, in contrast to that in the wheat embryo system, the altered larger initiation complex dissociates into its subunit components. Furthermore, in experiments in reticulocyte lysates monosomes and disomes bearing [^{35}S] Met-tRNA$_f$ accumulate in the presence of pactamycin to a much greater degree than smaller initiation complexes (unpublished data).

It is possible that as a consequence of the interaction of the antibiotic in the region of the initiator or donor site, the functional binding of aminoacyl-tRNA to the acceptor site is also interferred with. This is compatible with effects we have found on the binding of aminoacyl-tRNA to ribosomes by pactamycin (COHEN and GOLDBERG, 1967) and may account for the inhibition of peptide chain elongation found in the bacterial system and in the mammalian system especially at the higher antibiotic concentrations. Such a mechanism would also explain the findings of CUNDLIFFE (1971) in intact bacteria that suggest that pactamycin prevents access of aminoacyl-tRNA to the acceptor site.

Finally, pactamycin has proved to be a useful tool in the study of biological processes. Recently, several research groups have taken advantage of the ability of pactamycin to block the initiation of viral proteins in mammalian cells infected with virus to determine the gene order for the viral genome (SUMMERS and MAIZEL, 1971; TABER et al., 1971; BUTTERWORTH and RUECKERT, 1972). By labelling the virus-specific nascent proteins at different times after the addition of pactamycin and comparing the labelling pattern in the absence of the antibiotic, it has been possible to determine the order of synthesis of proteins for poliovirus and encephalomyocarditis virus.

References

ARGOUDELIS, A. D., H. K. JAHNKE, and J. A. FOX: Pactamycin, a new antitumor antibiotic. II. Isolation and characterization. Antimicrobial Agents Chemotherapy 1961, 191–197.

AYUSO, M., and I. H. GOLDBERG: Pactamycin inhibition of eucaryotic polypeptide synthesis dependent on added initiation factors. Biochim. Biophys. Acta in press.

BENNE, R., and H. O. VOORMA: Entry site of formylmethionyl-tRNA. FEBS Letters 20, 347–351 (1972).

BHUYAN, B. K.: Pactamycin, an antibiotic that inhibits protein synthesis. Biochem. Pharmacol. 16, 1411–1420 (1967).

BHUYAN, B. K., A. DIETZ, and C. G. SMITH: Pactamycin, a new antitumor antibiotic. I. Discovery and biological properties. Antimicrobial Agents Chemotherapy 1961, 184–190.

BHUYAN, B. K., and R. L. JOHNSTON: Metabolic studies of pactamycin. Biochem. Pharmacol. 12, 1001–1010 (1963).

BOTH, G. W., J. L. MCINNES, B. K. MAY, and W. H. ELLIOTT: Recovery of Bacillus amyloliquefaciens protein synthesis from inhibition by pactamycin. Biochem. Biophys. Res. Commun. 43, 1095–1101 (1971).

BRODASKY, T. F., and W. L. LUMMIS: Pactamycin, a new antitumor antibiotic. III. Spectrophotometric quantitative paper chromatographic assay. Antimicrobial Agents Chemotherapy 1961, 198–204.

BUTTERWORTH, B. E., and R. R. RUECKERT: The genetic map of encephalomyocarditis (EMC) virus as determined from studies with pactamycin. Federation Proc. 31, 407 (1972).

COHEN, L. B., and I. H. GOLDBERG: Inhibition of peptidyl-sRNA binding to ribosomes by pactamycin. Biochem. Biophys. Res. Commun. 29, 617–622 (1967).

COHEN, L. B., I. H. GOLDBERG, and A. E. HERNER: Inhibition by pactamycin of the initiation of protein synthesis. Effect on the 30S ribosomal subunit. Biochemistry 8, 1327–1335 (1969b).

COHEN, L. B., A. E. HERNER, and I. H. GOLDBERG: Inhibition by pactamycin of the initiation of protein synthesis. Binding of N-acetylphenylalanyl transfer ribonucleic acid and polyuridylic acid to ribosomes. Biochemistry 8, 1312–1326 (1969a).

COLOMBO, B., L. FELICETTI, and C. BAGLIONI: Inhibition of protein synthesis in reticulocytes by antibiotics. I. Effects on polysomes. Biochim. Biophys. Acta **119**, 109–119 (1966).

CRYSTAL, R.G., D.A. SHAFRITZ, P.M. PRICHARD, and W.F. ANDERSON: Initial dipeptide formation in hemoglobin biosynthesis. Proc. Natl. Acad. Sci. US **68**, 1810–1814 (1971).

CUNDLIFFE, E.: Bacterial protein synthesis: the effects of antibiotics upon the puromycin reaction *in vivo*. Symposium in Granada, June, 1971. MUNOZ, FERRANDIZ and VAZQUEZ, eds. Berlin-Heidelberg-New York: Springer 1971.

CUNDLIFFE, E., and K. MCQUILLEN. Bacterial protein synthesis: the effects of antibiotics. J. Mol. Biol. **30**, 137–146 (1967).

EZEKIEL, D.H., and B.N. ELKINS: The stimulation of ribonucleic acid synthesis by ribosome inhibitors in amino acid-starved *Escherichia coli*. Biochim. Biophys. Acta **166**, 466–474 (1968).

FELICETTI, L., B. COLOMBO, and C. BAGLIONI: Inhibition of protein synthesis in reticulocytes by antibiotics. II. The site of action of cycloheximide, streptovitacin A and pactamycin. Biochim. Biophys. Acta **119**, 120–129 (1966).

GOLDBERG, I.H., M.L. STEWART, M. AYUSO, and L.S. KAPPEN: On the mechanisms of inhibition of polypeptide synthesis by the antibiotics sparsomycin and pactamycin. Federation Proc., in press.

HARFORD, N., and R. SUEOKA: Chromosomal location of antibiotic resistance markers in *Bacillus subtilis*. J. Mol. Biol. **51**, 267–286 (1970).

HOERZ, W., and K.S. MCCARTY: Evidence for a proposed initiation complex for protein synthesis in reticulocyte polyribosome profiles. Proc. Natl. Acad. Sci. U.S. **63**, 1206–1213 (1969).

HUANG, M.-T., and A.P. GROLLMAN: Effects of aurintricarboxylic acid on ribosomes and the biosynthesis of globin in rabbit reticulocytes. Mol. Pharmacol. **8**, 111–127 (1972).

KAPPEN, L.S., S. SUZUKI, and I.H. GOLDBERG: Inhibition of reticulocyte peptide chain initiation by pactamycin: accumulation of inactive ribosomal initiation complexes. Proc. Natl. Acad. Sci. U.S. in press.

KERSTEN, H., P. CHANDRA, W. TANCK, W. WIEDEMHOVER, and W. KERSTEN: Effect of pactamycin on methylation of RNA and protein synthesis. Z. Physiol. Chem. **349**, 659–663 (1968).

KERSTEN, H., W. KERSTEN, B. EMMERICH, and P. CHANDRA: Studies on the mode of action of pactamycin. Z. Physiol. Chem. **348**, 1424–1430 (1967).

LELONG, J.C., M.A. COUSIN, D. GROS, M. GRUNBERG-MANAGO, and F. GROS: Streptomycin induced release of fMet-tRNA from the ribosomal initiation complex. Biochem. Biophys. Res. Commun. **42**, 530–537 (1971).

LODISH, H.F., D. HOUSMAN, and M. JACOBSEN: Initiation of hemoglobin synthesis. Specific inhibition by antibiotics and bacteriophage ribonucleic acid. Biochemistry **10**, 2348-2356 (1971).

MACDONALD, J.S., and I.H. GOLDBERG: An effect of pactamycin on the initiation of protein synthesis in reticulocytes. Biochem. Biophys. Res. Commun. **41**, 1–8 (1970).

MODOLELL, J., B. CABRER, A. PARMEGGIANI, and D. VAZQUEZ: Inhibition by siomycin and thiostrepton of both aminoacyl-tRNA and factor G binding to ribosomes. Proc. Natl. Acad. Sci. U.S. **68**, 1796–1800 (1971).

SEAL, S.N., and A. MARCUS: Reactivity of ribosomally bound methionyl-tRNA with puromycin and the locus of pactamycin inhibition of chain initiation. Biochem. Biophys. Res. Commun. **46**, 1895–1902 (1972).

STEWART, M.L. and I.H. GOLDBERG: Pactamycin binding to *E. coli* ribosomes: interference by formation of the protein synthesizing complex with f2 viral RNA. Biochim. Biophys. Acta in press.

STEWART-BLAIR, M.L., I.S. YANOWITZ, and I.H. GOLDBERG: Inhibition of synthesis of new globin chains in reticulocyte lysates by pactamycin. Biochemistry **10**, 4198–4206 (1971).

SUMMERS, D.F., and J.V. MAIZEL: Determination of the gene sequence of poliovirus with pactamycin. Proc. Natl. Acad. Sci. U.S. **68**, 2852 (1971).

TABER, R., D. REKOSH, and D. BALTIMORE: Effect of pactamycin on synthesis of poliovirus proteins: a method for genetic mapping. J. Virol. **8**, 395–401 (1971).

THACH, S.S., and R.E. THACH: Translocation of messenger RNA and "accommodation" of fMet-tRNA. Proc. Natl. Acad. Sci. U.S. **68**, 1791–1795 (1971).

WEEKS, D.P., and R. BAXTER: Specific inhibition of peptide-chain initiation by 2-(4-Methyl-2,6-dinitroanilino)-N-methylpropionamide. Biochemistry **11**, 3060–3064 (1972).

WEEKS, D.P., D.P.S. VERMA, S.N. SEAL, and A. MARCUS: Role of ribosomal subunits in eukaryotic protein chain initiation. Nature **236**, 167–168 (1972).

WHITE, F. R.: Pactamycin. Cancer Chemotherapy Rept. **24**, 75–78 (1962).

WILEY, P. F., H. K. JAHNKE, F. MCKELLAR, R. B. KELLY, and A. D. ARGOUDELIS: The structure of pactamycin. J. Org. Chem. **35**, 1420–1425 (1970).

YOUNG, C. W.: Inhibitory effects of acetoxycycloheximide, puromycin and pactamycin upon synthesis of protein and DNA in asynchronous populations of HeLa cells. Mol. Pharmacol. **2**, 50–55 (1966).

Primaquine

John G. Olenick

During the First World War, the shortage in supplies of the antimalarial drug quinine prompted the Germans to search for a synthetic substitute. Their chemical starting point was methylene blue, the only compound (with the exception of the cinchona alkaloids and arsenicals) then known to possess activity, albeit weak, against malaria (GUTTMANN and EHRLICH, 1891). Attempts were made to enhance the antimalarial properties of this dye: by changing the alkyl substituents attached to the extranuclear nitrogen atoms, it was found that the activity varied and by replacing one of the N-methyl constituents with a dialkylamino group, the potency was greatly increased (SCHULEMANN, 1932). Having discovered that the attachment of a basic side chain to the phenothiazine nucleus of methylene blue led to an enhancement of antimalarial activity, the effect of attaching various dialkylaminoalkyl groupings to other heterocyclic nuclei was investigated. Since the quinoline nucleus in quinine was thought to be important for activity, a large series of quinoline derivatives were synthesized. After testing for antimalarial efficacy using a model laboratory screen devised by RÖEHL (1926) in which canaries were infected with *Plasmodium relictum,* an 8(4-diethyl-amino-l-methylbutyl-amino)-6-methoxyquinoline, named pamaquine, was selected for practical clinical investigation (SCHULEMANN, 1932).

Pamaquine has had an unusually variegated chemotherapeutic history. This first synthetic antimalarial of clinical value was initially hailed with great enthusiasm but later found infrequent use owing to occasionally grave toxic side effects. During World War II. because of its remarkable effectiveness in preventing relapses of vivax malaria, pamaquine was reinvestigated and, in turn, became the starting point from which a large number of 8-aminoquinoline derivatives were synthesized and tested in the antimalarial research program of the United States (WISELOGLE, 1946) with a view to finding more potent but less toxic compounds. Primaquine emerged as the most satisfactory compound and is currently the drug of choice for the radically curative treatment (complete eradication of the parasites) of vivax malaria.

Primaquine ($C_{15}H_{21}N_3O$; mol.wt.259) differs solely from pamaquine in that the former has a terminal primary amino group attached to the 8-position of

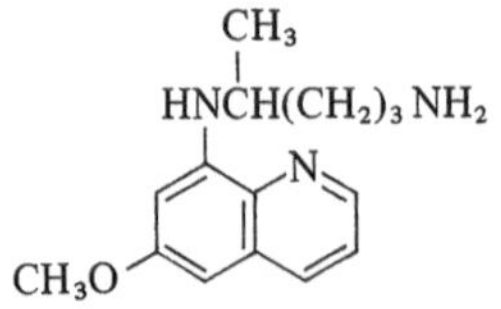

Fig. 1. Primaquine

the quinoline nucleus (see accompanying structural formula), whereas in the latter, the terminal function is a diethylamino group. The presence of a methoxyl or hydroxyl grouping at position 6 is essential for antimalarial efficacy; replacement by an ethoxyl or methyl markedly diminishes potency (FOURNEAU *et al.,* 1931 and 1933). Primaquine diphosphate is an orange-red crystalline powder having moderate solubility in water, is supplied in tablet form for oral administration and is never given by any other route. Undesirable gastrointestinal reactions of anorexia, nausea, vomiting, epigastric distress and abdominal cramps may occur. After large doses or in sensitive individuals, severe hematological reactions of leukopenia, hemolytic anemia and methemoglobinemia may be produced. The degree of toxicity appears to be directly related to the extent of substitution on the terminal amino nitrogen (SCHMIDT, 1951). Thus, pamaquine, having a dialkyl-substituted (tertiary) amino group, is more toxic than primaquine with a primary amino terminus; pentaquine and isopentaquine, which have secondary terminal amino functions, are intermediate in toxicity.

Acute hemolytic crises are most prevalent and marked in persons having erythrocytes genetically deficient in glucose-6-phosphate dehydrogenase (CARSON *et al.,* 1956; ALVING, *et al.,* 1958; TARLOV *et al.,* 1962), the enzyme that catalyzes the initial oxidative step of the pentose phosphate pathway of glucose metabolism. This pathway provides the only source for the generation of reduced nicotinamide adenine dinucleotide phosphate (NADPH), the coenzymic hydrogen carrier essential to the mature red cell for the maintenance of reducing potential (via glutathione and its respective reductase) to guard against oxidative damage. Deficiency in glucose-6-phosphate dehydrogenase (G-6-PD), attended by an incapacity to provide sufficient NADPH and coupled with a concomitantly diminished content of reduced glutathione (GSH), presumably makes the erythrocyte unusually predisposed to oxidative damage and hemolysis (CARSON and TARLOV, 1962).

Based on the hemolytic properties of primaquine, a mode of antimalarial action has been inferred in which primaquine or its metabolites may exert antimalarial activity by adversely affecting NADPH-linked reductive processes in cells containing exoerythrocytic schizonts (an asexual form in the life cycle of the malarial parasite). Consequently, these cells are unable to provide nutrients essential for rapid development of the parasites. Tissue schizonts, as compared to blood schizonts, may also have deficiencies involving enzymes or cofactors of the pentose phosphate pathway that render the tissue schizonts more vulnerable to oxidative damage (ALVING *et al.,* 1962). However, although frequently termed primaquine-sensitivity, hemolytic anemia in subjects with gene-linked erythrocyte abnormality can be caused by a plethoric variety of agents (MARKS and BANKS, 1965). Many pharmacologically different drugs, foods and chemicals, as well as viral and bacterial infections, have been implicated in provoking acute intravascular hemolysis in susceptible individuals. WITTELS (1970) has recently provided evidence which suggests that the mechanism of hemolysis may be unrelated to that of a disturbance in redox systems and may instead be the result of an alteration of red cell phospholipid metabolism. Such observations collectively tend to lessen the plausibility of the related causality between the mechanisms of the antimalarial and of the hemolytic effects of primaquine.

Systematic studies on the mode of action of primaquine have tended to lag behind the early recognition of this drug as the most effective agent available against the exoerythrocytic stages of malaria. One reason for this stems from the widely held but unsubstantiated belief that the antimalarial properties of primaquine and of other 8-aminoquinolines reside in intermediates formed during metabolic transformations of these drugs in the host (ALVING et al., 1962). The general in vitro insensitivity of test microorganisms to primaquine has tended to augment this belief.

Nevertheless, the mode of action of primaquine has been examined in the ciliate protozoan, Tetrahymena pyriformis (CONKLIN and CHOU, 1972), and in the bacterium, Bacillus megaterium (OLENICK and HAHN, 1972). Using T. pyriformis forced into synchronous cell division (not cell growth, see SCHERBAUM and ZEUTHEN, 1954) by several cycles of heat treatment, CONKLIN and CHOU (1972) attempted to show that primaquine directly blocks the uptake of precursors into DNA, RNA and protein which results indirectly in an inhibition of all categories of macromolecular biosynthesis. The data they obtained are not unambiguously interpretable and consequently lack conclusiveness. OLENICK and HAHN (1972) have found the primary effect of primaquine on B. megaterium to be a preferential inhibition of protein biosynthesis; the macromolecular synthesis of protein was immediately and completely halted, while DNA, RNA and cell wall synthesis continued for a brief time at normal or slightly less than normal rates. Macromolecular biosyntheses displaying this type of inhibitory pattern, when subject to the action of a drug, have come to be recognized as characteristic of a blockade of protein synthesis at a step subsequent to the formation of aminoacyl-sRNA (NEIDHARDT, 1964). Indeed, at concentrations of primaquine as high as 10^{-3}M, LANDEZ et al., (1969) were unable to show an effect on rat liver cell-free aminoacylation of sRNA. ROSKOSKI, and JASKUNAS (1972) have observed that, in a cell-free system from rat liver, primaquine inhibited the polyuridylic acid- and poly AGU-directed incorporation of phenylalanine, provided these polynucleotides were preincubated with the drug. In contrast, the endogenous amino acid incorporation, mediated by residual natural messenger RNA attached to liver ribosomes, was not affected by primaquine. This suggests that the inhibitions of in vitro polypeptide synthesis were produced by complexing of synthetic model messengers with the drug. MORRIS et al., (1970) have studied the binding of primaquine to various polynucleotides and have speculated that protein synthesis as well as the actions of nucleic acid polymerases may be affected by such drug-binding reactions.

Morphological changes evoked by primaquine have been studied by AIKAWA and BEAUDOIN (1969) in the exoerythrocytic forms of avian malarial parasites grown in tissue culture. The most pronounced change observed was a severe swelling of the mitochondria of the parasites, whereas those of the host cells were apparently unaltered. Subsequently, using techniques of autoradiography, AIKAWA and BEAUDOIN (1970) confirmed their previous results by showing the selective localization of primaquine within the mitochondrion of the parasite. These investigators considered the damage to mitochondria sufficiently severe to account for the probability that a similar action on the exoerythrocytic forms of human vivax and malariae malaria might explain the ability of primaquine to produce a radical cure of infections by these parasites. Similar deformations of mitochondria were demonstrated by HOWELLS et al., (1970) who investigated the in vivo (in the living animal) effects of primaquine on the morphology of erythrocytic forms of a rodent malarial parasite. In addition, they found that, unlike the exoerythrocytic forms, survivors of the primaquine treatment in the erythrocytic population showed a substantial increase in the amount and complexity of the mitochondrial membrane systems, suggesting to them that the parasite may overcome the effects of primaquine by actively increasing the biogenesis of mitochondrial organelles. It is of interest to note that a prominent morphological feature of primaquine-resistant strains of Plasmodium berghei is an increased number of mitochondrial membrane systems (HOWELLS et al., 1968).

OLENICK and HAHN (1972), in reporting the antibacterial action of primaquine to be a preferential inhibition of protein biosynthesis and bearing in mind that mitochondria contain a protein-synthesizing system which differs from its cytoplasmic counterpart but resembles the protein synthesizing machinery of bacteria (CIFERRI and PARISI, 1970), have speculated that the antimalarial action may be due to a specific inhibition of mitochondrial protein synthesis. It can be further speculated that the selective action of primaquine on the exoerythrocytic stages of infection may be due to a specific inhibition of mitochondrial protein synthesis in a parasitic form already deficient in its capacity for renewed biogenesis of mitochondria (as discussed above). AIKAWA and BEAUDOIN (1970) propose that the morphological changes observed in the mitochondria of malarial parasites may be the result of primaquine binding to the DNA of the mitochondrion or alternatively, may be due to a reduction of oxidative phosphorylation followed by oxidative damage.

Experimentally, the development of resistance to primaquine is a slow and gradual process that can be produced under extreme drug pressure in avian (BISHOP, 1967), rodent (PRAKASH et al., 1961; PETERS, 1966), simian (RAMAKRISH-NAN and PRAKASH, 1961) and, even in, human (ARNOLD et al., 1961) malarial parasites. Clinically, the low incidence of resistance to primaquine does not present a serious therapeutic problem.

References

AIKAWA, M., and R.L. BEAUDOIN: Morphological effects of 8-aminoquinolines on the exoerythrocytic stages of Plasmodium fallax. Military Med. **134**, 986 (1969).

AIKAWA, M., and R.L. BEAUDOIN: Plasmodium fallax: High-resolution autoradiograhy of exoerythrocytic stages treated with primaquine in vitro. Exptl. Parasitol. **27**, 454 (1970).

ALVING, A.S., R.W. KELLERMEYER, A. TARLOV, S. SCHRIER, and P.E. CARSON: Biochemical and genetic aspects of primaquine-sensitive hemolytic anemia. Ann. Internal Med. **49**, 240 (1958).

ALVING, A.S., R.D. POWELL, G.J. BREWER, and J.D. ARNOLD: Malaria, 8-aminoquinolines and haemolysis. In: L.G. GOODWIN and R.H. NIMMO-SMITH (eds.), Drugs, parasites and hosts, p. 83. London: J.&A. Churchill, Ltd. 1962.

ARNOLD, J., A.S. ALVING, C.B. CLAYMAN, and R.S. HOCHWALD: Induced primaquine resistance in vivax malaria. Trans. Roy. Soc. Trop. Med. Hyg. **55**, 345 (1962).

BISHOP, A.: Resistance to primaquine in Plasmodium gallinaceum, and the problem of resistance to quinoline compounds in malaria parasites. Parasitology **57**, 755 (1967).

CARSON, P.E., C.L. FLANAGAN, C.E. ICKES, and A.S. ALVING: Enzymatic deficiency in primaquine-sensitive erythrocytes. Science **124**, 484 (1956).

CARSON, P. E., and A. R. TARLOV: Biochemistry of hemolysis. Ann. Rev. Med. 13, 105 (1962).

CIFERRI, O., and B. PARISI: Ribosome specificity of protein synthesis in vitro. In: J. N. Davidson and W. E. COHN (eds.), Progress in nucleic acid research and molecular biology, vol. 10, p. 135. New York: Academic Press 1970.

CONKLIN, K. A., and S. C. CHOU: The effects of antimalarial drugs on uptake and incorporation of macromolecular precursors by Tetrahymena pyriformis. J. Pharmacol. Exptl. Therap. 180, 158 (1972).

FOURNEAU, E., J. TRÉFOUEL, MME. TRÉFOUEL, D. BOVET et G. BENOIT: Contribution à la chimiothérapie du paludisme: essais sur les calfats. Ann. Inst. Pasteur 46, 514 (1931).

FOURNEAU, E., J. TRÉFOUEL, MME. TRÉFOUEL, D. BOVET et G. BENOIT: Contribution à la chimiothérapie du paludisme: essais sur les calfats (deuxiéme mémoire). Ann. Inst. Pasteur 50, 731 (1933).

GUTTMANN, P., u. P. EHRLICH: Über die Wirkung des Methylenblau bei Malaria. Berlin. Klin. Wochenschr. 28, 953 (1891).

HOWELLS, R. E., W. PETERS, and J. FULLARD: The chemotherapy of rodent malaria. XIII. Fine structural changes observed in the erythrocytic stages of Plasmodium berghei berghei following exposure to primaquine and menoctone. Ann. Trop. Med. Parasitol. 64, 203 (1970).

HOWELLS, R. E., W. PETERS, and E. A. THOMAS: The chemotherapy of rodent malaria. IV. Host-parasite relationships, part 4: The relationship between haemozoin formation and host-cell age in chloroquine- and primaquine-resistant strains of Plasmodium berghei. Ann. Trop. Med. Parasitol. 62, 271 (1968).

LANDEZ, J. H., R. ROSKOSKI, JR., and G. L. COPPOC: Ethidium bromide and chloroquine inhibition of rat liver cell-free aminoacylation. Biochim. Biophys. Acta 195, 276 (1969).

MARKS, P. A., and J. BANKS: Drug-induced hemolytic anemias associated with glucose-6-phosphate dehydrogenase deficiency: A genetically heterogeneous trait. Ann. N. Y. Acad. Sci. 123, 198 (1965).

MORRIS, C. R., L. V. ANDREW, L. P. WHICHARD, and D. J. HOLBROOK, JR.: The binding of antimalarial aminoquinolines to nucleic acids and polynucleotides. Mol. Pharmacol. 6, 240 (1970).

NEIDHARDT, F. C.: The regulation of RNA synthesis in bacteria. In: J. N. DAVIDSON and W. E. COHN (eds.), Progress in nucleic acid research and molecular biology, vol. 3, p. 145. New York: Academic Press 1964.

OLENICK, J. G., and F. E. HAHN: Mode of action of primaquine: Preferential inhibition of protein biosynthesis in Bacillus megaterium. Antimicrobial Agents Chemotherapy 1, 259 (1972).

PETERS, W.: Drug responses of mepacrine- and primaquine-resistant strains of Plasmodium berghei Vincke & Lips, 1948. Ann. Trop. Med. Parasitol. 60, 25 (1966).

PRAKASH, S., A. K. CHAKRABARTI, and D. S. CHOUDHURY: Studies on Plasmodium berghei Vincke and Lips, 1948. XXXI. Selection of a primaquine resistant strain. Indian J. Malariol. 15, 115 (1961).

RAMAKRISHNAN, S. P., and S. PRAKASH: A note on the rapid selection of a primaquine-resistant strain of Plasmodium knowlesi in Macaca mulatta. Bull. Natl. Soc. Indian Malaria, Other Mosquito-Borne Dis. 9, 261 (1961).

RÖEHL, W.: Die Wirkung des Plasmochins auf die Vogelmalaria. Arch. Schiffs- u. Tropen-Hyg. 30, Beiheft, 311 (1926).

ROSKOSKI, R., JR., and S. R. JASKUNAS: Chloronique and primaquine inhibition of rat liver cell-free polynucleotide-dependent polypeptide synthesis. Biochem. Pharmacol. 21, 391 (1972).

SCHERBAUM, O., and E. ZEUTHEN: Induction of synchronous cell division in mass cultures of Tetrahymena priformis. Exptl. Cell Res. 6, 221 (1954).

SCHMIDT, L. H.: First symposium on chemical-biological correlation. The relations between chemical structure and toxicity among the 8-aminoquinolines. Natl. Research Council, Natl. Acad. Sci., Washington, D. C., Chem.-Biol. Coordination Center, Publ. No. 206, 181 (1951).

SCHULEMANN, W.: Synthetic antimalarial preparations. Proc. Roy. Soc. Med. 25, 897 (1932).

TARLOV, A. R., G. J. BREWER, P. E. CARSON, and A. S. ALVING: Primaquine sensitivity. Glucose-6-phosphate dehydrogenase deficiency. An inborn error of metabolism of medical and biological significance. Arch. Internal Med. 109, 209 (1962).

WISELOGLE, F. Y. (ed.): A survey of antimalarial drugs, 1941–1945. Ann Arbor, Mich.: J. W. Edwards, Inc. 1946.

WITTELS, B.: Modification of phospholipid metabolism in human red cells by primaquine. A possible mechanism in drug-induced hemolysis. Biochim. Biophys. Acta 210, 74 (1970).

The Streptogramin Family of Antibiotics

D. Vazquez

This contribution will be mainly concerned with advances in the subject for the last seven years since we have already reviewed this group of antibiotics in 1967 (VAZQUEZ, 1967a).

The antibiotics virginiamycin, ostreogrycin, synergistin, mikamycin, pristinamycin, and vernamycin are included in the streptogramin family. These antibiotics are mixtures of two or more active compounds which can be placed in two major groups A and B (Table 1). The antibiotics in the group A show a marked synergism with those in group B in their activity against Gram-positive bacteria and consequently all the complex antibiotics of this family have a markedly higher activity than the individual components. The antibiotics griseoviridin and viridogrisein are also included in Table 1 since they are produced by the same streptomyces as a complex mixture and viridogrisein has a chemical structure very similar to other members of group B. Viridogrisein was indeed found to have a synergistic effect with different antibiotics included in group A but mixtures of griseoviridin and viridogrisein have been found to be synergistic only in one particular strain of *Staphylococcus aureus* (MAGYAR *et al.*, 1962). On the other hand, griseoviridin is only tentatively included in group A since it differs from others in its chemical structure and has not been tested in mixture with antibiotics of group B other than viridogrisein. Etamycin, unlike all the other members of the family, is not obtained from an antibiotic complex but as a single component which is identical to viridogrisein. Patricins A and B are synthetic compounds of the streptogramin B group.

Antibiotics of the streptogramin family are not active in inhibiting eucaryotic organisms but are active in blocking growth of organellae and procaryotic organisms other than bacteria. Some complex antibiotics of the streptogramin family (mikamycin, virginiamycin and pristinamycin) are used clinically.

Antibiotics of Group A

The chemical structures of ostreogrycin A and griseoviridin are shown in Fig 1. Ostreogrycin A is identical with streptogramin A, virginiamycin M, vernamycin A, synergistin A-1, mikamycin A and pristinamycin II$_A$ (CROOY and DE NEYS, 1972). Since griseoviridin differs from the other compounds of group A in its chemical structure, and possibly in its mode of action and inhibitory spectra, we will consider it separately.

The antibiotics in group A are preferentially active against Gram-positive cocci (Table 2); their action is bacteriostatic on *Bacterium agri, Staphylococcus aureus* (VAZQUEZ, 1967a) and *Bacillus subtilis* (COCITO and FRASELLE, 1973).

Table 1

Antibiotic complex (synonyms)	Group A (synonyms)	Group B (synonyms)
Ostreogrycin (E 129)	Ostreogrycin A Ostreogrycin C Ostreogrycin D Ostreogrycin G Ostreogrycin Q	Ostreogrycin B Ostreogrycin B_1 Ostreogrycin B_2 Ostreogrycin B_3
Streptogramin	Streptogramin A	Streptogramin B
Virginiamycin (Staphylomycin) (Virgimycin)	Virginiamycin M or M_1 (Staphylomycin M_1) Virginiamycin M_2 (Staphylomycin M_{II})	Virginiamycin S or S_1 (Staphylomycin S) Virginiamycin S_2 Virginiamycin S_3 Virginiamycin S_4
Synergistin (PA 114)	Synergistin A-1 (PA 114 A-1)	Synergistin B-1 (PA 114 B-1) Synergistin B-3) (PA 114 B-3)
Mikamycin	Mikamycin A	Mikamycin B
Pristinamycin (Pyostacin) (7293 RP)	Pristinamycin II_A Pristinamycin II_B	Pristinamycin I_A Pristinamycin I_B Pristinamycin I_C
Vernamycin	Vernamycin A	Vernamycin B_α Vernamycin B_β Vernamycin B_γ Vernamycin B_δ Vernamycin C (Doricin)
	Griseoviridin	Viridogrisein
		Etamycin
		Patricin A Patricin B

The antibiotics ostreogrycin A, streptogramin A, virginiamycin M, synergistin A-1, mikamycin A, pristinamycin II_A and vernamycin A have the same chemical structure Viridogrisein and etamycin are identical compounds. (Data taken from Vazquez (1967a) and Crooy and Neys (1972).)

However, mixtures of these compounds with antibiotics of either group B or the tetracycline family are bactericidal against *Staphylococcus aureus* (Vazquez, 1967a) and *Bacillus subtilis* (Cocito and Fraselle, 1973).

Although wild-type *Escherichia coli* is naturally resistant to streptogramin A antibiotics, mutants have been obtained sensitive to these antibiotics; however these mutants have been found to be pleiotropic being also sensitive to certain antibiotics having a different target and to lysis by detergents (Ennis, 1971b). Conversely *Bacillus subtilis* mutants resistant to streptogramin A antibiotics have been described (Cocito and Fraselle, 1973).

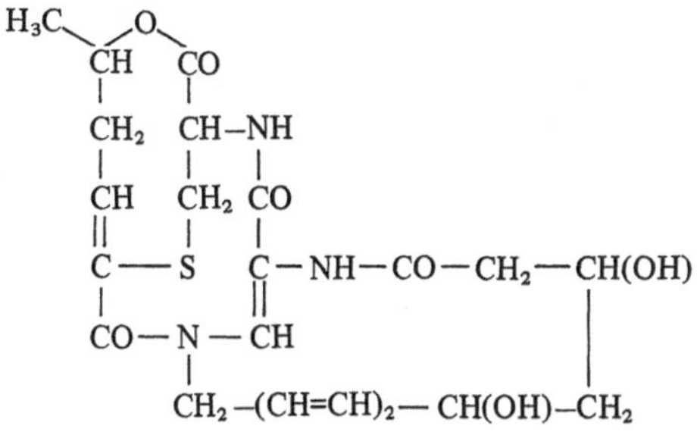

H₃C, O
CH CO
| |
CH₂ CH–NH
| | |
CH   CH₂ CO
|| | |
C — S   C–NH–CO–CH₂—CH(OH)
| ||
CO—N—CH
|
CH₂–(CH=CH)₂— CH(OH)–CH₂

Griseoviridin 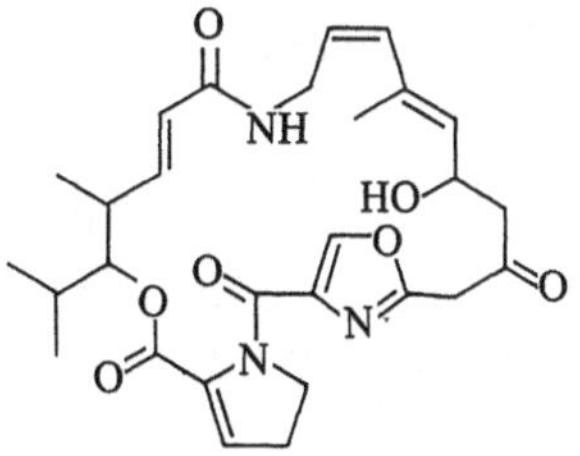 Ostreogrycin A

Fig. 1. Chemical structures of griseoviridin and antibiotics of the streptogramin A group

Table 2. *"In vitro" inhibitory concentration of the streptogramin antibiotics in group A and B*

Organism	Inhibitory concentration in µg/ml									
	Griseo-viridin	Virido-grisein (Eta-mycin)	Strepto-gramin		Syner-gistin		Virginia-mycin		Mika-mycin	
			A	B	A-1	B-1	M_1	S	A	B
Bacteria, Gram-positive:										
Bacillus megaterium	—	—	40.00	3.00	—	—	—	—	80.00	2.00
Bacillus subtilis	—	2.50	—	—	100.00	3.12	50.00	13.00	80.00	1.00
Staphylococcus aureus	2–100	0.31	6.00	10.00	0.78	6.25	5.00	125.00	2.00	10.00
Sarcina lutea	—	--	—	—	—	—	2.50	14.00	—	—
Streptococcus pyogenes	0.5	0.63	—.	—	0.19	50.00	—	—	0.40	4.00
Bacteria, Gram-negative:										
Escherichia coli	10	200.00	40.00	100.00	—	—	—	—	80.00	80.00
Hemophilus pertussis	0.1	5.00	—	—	—	—	—	—	0.40	10.00

Data taken from VAZQUEZ (1967a) and TANAKA et al. (1961).

In vivo studies have shown that these antibiotics have little activity in protecting mice against experimental infections with a virulent strain of *Staphylococcus aureus*. They are not toxic to mice when injected intraperitoneally at doses of 350 mg/kg daily (VAZQUEZ, 1967a) and there is no tendency of mikamycin A to accumulate in specific organs and tissues (WATANABE et al., 1970). At growth inhibitory concentrations there is no effect of these antibiotics on endogenous respiration, oxidation of ethanol or glucose or anaerobic fermentation of glucose by *Staphylococcus aureus*. Antibiotics in group A immediately block protein synthesis in Gram-positive bacteria without inhibiting either nucleic acid or cell wall synthesis or accumulation of low molecular weight constituents in the "pool".The increase in nucleic acid synthesis observed early after addition of group A antibiotics to bacterial cultures is due to the RNA synthesised and there is only a slight inhibition in the rate of DNA synthesised under these conditions. The increase in RNA is due to a higher synthesis of mRNA, rRNA and tRNA and to a lower turnover of mRNA whereas the rRNA synthesised in the presence of group A antibiotics is undermethylated and has higher turnover than normal rRNA. The effects of antibiotics of group A in the metabolism of nucleic acids are due in all cases to their interference with protein synthesis (VAZQUEZ, 1967a).

Formation of new ribosomes does not take place in *Bacillus subtilis* in the presence of these antibiotics but a ribosomal pattern similar to that of control cultures can be observed soon after removal of the inhibitors (COCITO, 1973a). An unusual 60S peak has been described in sucrose density gradients of lysates from *B. subtilis* treated with streptogramin A antibiotics; this peak might be due to the formation of labile monosomes in the presence of the antibiotics which are transformed in the 60S particles in the centrifugation procedure (COCITO, 1973b).

Antibiotics of group A are also inhibitors of protein synthesis by green algae chloroplasts (COCITO *et al.*, 1972). As expected, antibiotics of the group A, by inhibiting protein synthesis, prevent the formation of phage particles as observed in *Bacillus subtilis* infected with the virulent phage 2C (COCITO, 1969).

When an antibiotic of group A is added to growing cultures of *Euglena gracilis*, chlorophyll synthesis and chloroplast development is halted but no alteration of the mitochondria has been observed (COCITO *et al.*, 1972). The bleaching is not permanent and can be reversed by removal of the antibiotic. Protein synthesis is prevented in the chloroplasts of Euglena treated with the antibiotic and there is no new formation of chloroplast ribosomes and 16S ribosomal RNA (VAN PEL and COCITO, 1973).

The inhibitory effect of antibiotics of the group A on protein synthesis in intact bacteria has been confirmed by studies on amino acid incorporation in cell-free systems both from bacteria (VAZQUEZ, 1967a) and blue-green algae (RODRIGUEZ-LOPEZ *et al.*, 1970). On the other hand, antibiotics of the group A are not active on cell-free systems from human tonsils, yeast (VAZQUEZ *et al.*, 1969; NETH *et al.*, 1970), the protozoa *Crithidia oncopelti* (G.A.M. GROSS, personal communication) and rat liver (YAMAGUCHI *et al.*, 1966). In the bacterial cell-free systems, antibiotics of group A, unlike chloramphenicol, do not show a significant differential inhibition of proline, lysine and phenylalanine incorporation directed respectively by poly C, poly A and poly U (VAZQUEZ, 1967a). Inhibition of protein synthesis by group A antibiotics has also been confirmed in cell-free systems using f2 (ENNIS, 1970) and MS2 (COCITO and KAJI, 1971) phage RNA as mRNA.

Antibiotics of the streptogramin A group inhibit both uptake by bacteria and binding to ribosomes of ^{14}C-chloramphenicol. Binding of ^{14}C-chloramphenicol to ribosomes is readily reversible whereas ribosomes treated with group A antibiotics are unable to bind ^{14}C-chloramphenicol even after repeated washing with buffer. Since it is known that chloramphenicol interacts with the 50S ribosomal subunit these findings suggest that the site of action of group A antibiotics is also located in the 50S ribosomal subunit (VAZQUEZ, 1967a). This is in agreement with the findings that streptogramin A strongly inhibits binding of spiramycin I, lincomycin and erythromycin to *Escherichia coli* ribosomes since it is well known that these antibiotics also bind to the 50S ribosome subunit (VAZQUEZ, 1967b; RODRIGUEZ-LOPEZ et al., 1968; FERNANDEZ-MUÑOZ et al., 1971). Binding of group A antibiotics to the 50S ribosomal subunit and requirement of K^+ or NH_4^+ for this interaction has been confirmed by studying polyphenylalanine synthesising activity of ribosomes reconstituted after pretreatment of either 50S or 30S subunits with the antibiotic followed by dialysis or gel filtration prior to addition of the untreated complementary subunits to reconstitute the ribosomes (ENNIS, 1966; VAZQUEZ, 1967c). Direct studies on binding of ^{3}H-vernamycin A to bacterial ribosomes have confirmed the specificity of its interaction with the 50S subunit and the requirement of K^+ or NH_4^+ and further shown that the antibiotic binding is temperature dependent and that saturation is reached when one molecule of the antibiotic is bound per 50S ribosome subunit (ENNIS, 1971a).

Concerning the reaction inhibited by the streptogramin A antibiotics, it is clear that they do not affect aminoacyl-tRNA synthetases or any other reaction prior to formation of aminoacyl-tRNA. Streptogramin A antibiotics have been found to inhibit to a certain extent non-enzymic binding of Phe-tRNA directed by poly U to bacterial ribosomes (VAZQUEZ, 1967a; YAMAGUCHI and TANAKA, 1967; COCITO and KAJI, 1971). However group A antibiotics block very effectively EF T-dependent binding of either Phe-tRNA directed by poly U (HILL, 1969; MODOLELL et al., 1971a; COCITO and KAJI, 1971) or Ala-tRNA directed by R 17 phage RNA after the initiation complex was formed (MODOLELL et al., 1971a) but do not affect the GTP hydrolysis which is normally coupled to the enzymic binding of aminoacyl-tRNA (MODOLELL et al., 1971b). These results suggest that streptogramin A antibiotics block the acceptor site of aminoacyl-tRNA on the 50S ribosome subunit. On the other hand streptogramin A antibiotics have also been found to block non-enzymic binding of f-Met-tRNA to ribosomes and to 50S subunits to the donor site of the ribosome since the bound substrate is reactive with puromycin (ENNIS and DUFFY, 1972). This inhibitory action of streptogramin A antibiotics on binding of the initiator substrate might indicate that the antibiotics block initiation. This is also in agreement with the polysome breakdown which has been observed in studies with bacterial polysomes "in vivo" (CUNDLIFFE, 1969) and intact bacteria pretreated with streptogramin A antibiotics (ENNIS, 1972). However, apparently conflicting with these results, it has also been reported that treatment of sensitive bacteria with streptogramin A antibiotics prevents polysome breakdown (COCITO, 1971).

Streptogramin A antibiotics have been found to be strong inhibitors of peptide bond formation in the well resolved systems of the fragment reaction in which either CACCA-Met-f or CACCA-Leu-Ac was used as a donor substrate and puromycin as an acceptor substrate (MONRO and VAZQUEZ, 1967). This inhibitory effect by group A antibiotics on peptide bond formation has been confirmed by other workers using puromycin as an acceptor substrate and either Ac-Phe-tRNA (PESTKA, 1970) or polylysyl-tRNA (ENNIS, 1970) as a donor substrate. Further works have shown that streptogramin A compounds inhibit binding of the substrate CACCA-Leu-Ac to the donor site (CELMA et al., 1970) and the substrates CACCA-Phe (PESTKA, 1969) and CACCA-Leu to the acceptor site (CELMA et al., 1971) of the ribosomal peptidyl transferase centre. All this evidence supports that streptogramin A antibiotics, by binding to the peptidyl transferase center of the 50S subunit, prevent the subsequent binding of the 3′ terminal end of donor- and acceptor-substrates and consequently block peptide bond formation. This strong inhibitory effect of streptogramin A antibiotics on the binding of the 3′ terminal moiety of substrates to the donor and acceptor sites might well explain the reported polysome breakdown (CUNDLIFFE, 1969; ENNIS, 1970, 1972) and the inhibition above mentioned on binding of f-Met-tRNA (ENNIS and DUFFY, 1972), aminoacyl-tRNA (HILL, 1969; MODOLELL et al., 1971; COCITO and KAJI, 1971) The interaction of streptogramin A antibiotics with the peptidyl transferase center has been confirmed by other experimental evidence. Sparsomycin is known to block peptidyl transferase action by inducing formation of the complex sparsomycin-ribosome-donor substrate which is saturated when there is one molecule of the donor substrate bound per ribosome, strongly suggesting that there is one peptidyl transferase center in each 50S ribosome subunit. Formation of this sparsomycin-ribosome-substrate bound to the donor site is strongly inhibited by streptogramin A supporting the interaction of this antibiotic with the peptidyl transferase centre (MONRO et al., 1969; JIMENEZ et al., 1970).

However some other workers have found that under certain experimental conditions, streptogramin A antibiotics hardly affect peptide bond formation as observed studying the ribosomal catalysed reaction of puromycin with either Phe-tRNA prior and after the addition of EF G (COCITO and KAJI, 1971). This lack of inhibition might be due to differences between the Phe-tRNA and the normal donor substrate peptidyl-tRNA. Nevertheless, streptogramin A antibiotics do not cause either polysome breakdown or inhibition of puromycin-induced breakdown of polysomes "in vitro" (ENNIS, 1972) or inhibition of peptidyl-tRNA reaction with puromycin in a crude bacterial polysomal system (PESTKA, 1972). The lack of effect of the streptogramin A antibiotics in these systems might be due to lack of accessibility of the antibiotics to their ribosomal binding site under certain conditions in which the substrates are previously bound or actively synthesising proteins.

Griseoviridin

Perhaps griseoviridin might be considered as a streptogramin A antibiotic since it is an antibacterial compound produced simultaneously with viridogrisein (EHRLICH et al., 1955) and at least in one case it has been shown to be synergistic with this antibiotic of the streptogramin B group (MAGYAR et al., 1962). Although griseoviridin has a chemical structure very different from that of the streptogramin A antibiotics (Fig.1) (FALLONA et al., 1964), it has also been shown to inhibit bacterial protein synthesis in cell-free systems by blocking peptide bond formation using as an acceptor substrate puromycin and as a donor-substrate either Ac-Phe- or fMet- or polylysyl-tRNA or CACCA-Phe-Ac (PESTKA, 1972). However griseoviridin also inhibits poly U-directed polyphenylalanine synthesis in yeast and human tonsil cell-free systems, unlike streptogramin A antibiotics (L. SANCHEZ and L. CARRASCO, personal communication).

Antibiotics of Group B

The known chemical structures of the antibiotics included in the streptogramin B group are shown in Fig 2. All these consist of a polypeptide containing some unusual amino acids linked through the hydroxyl of threonine to form a macrocyclic lactone ring.

These antibiotics are mainly active against Gram-positive bacilli (Table 2) and are bacteriostatic for *Staphylococcus aureus, Bacterium agri* (VAZQUEZ, 1967a) and *Bacillus subtilis* (COCITO, 1973a). Antibiotics in the group B are not active on yeasts, fungi and protozoa.

Bacillus subtilis mutants resistant to streptogramin B antibiotics have been described (COCITO and FRASELLE, 1973). Certain erythromycin-resistant strains of *Staphylococcus aureus* have been shown to be sensitive to other macrolides and to streptogramin B antibiotics; however in these mutants exposure to low levels of erythromycin induces resistance to these antibiotics (WEISBLUM and DEMOHN, 1969).

In vivo studies have shown that most of the antibiotics in group B have little activity in protecting mice against *Staphylococcus aureus* and *Diplococcus pneumoniae* infections. Viridogrisein is an exception and when administered orally or parenterally prevents death of the mice which have been experimentally infected with *Haemophilus pertussis* or *Diplococcus pneumoniae*. Antibiotics in group B are toxic to mice only in high doses (2–3 g/kg) (VAZQUEZ, 1967a). There is no tendency for ^{3}H-mikamycin B to accumulate in specific tissues or organs of mice and most of the antibiotic is rapidly excreted in the urine and feces (WATANABE *et al.*, 1970).

Antibiotics in the group B have no effect on endogenous respiration, oxidation of ethanol or glucose, or anaerobic fermentation of glucose by *Staphylococcus aureus*. These antibiotics initially block the protein synthesis by intact bacteria and, with longer times of exposure, reduce to a lesser extent RNA, DNA and cell wall synthesis and the accumulation of low molecular weight constituents in the pool (VAZQUEZ, 1967a). *Bacillus subtilis* treated with B group antibiotics does not synthesize new ribosomes but upon removal of the antibiotic the normal ribosomal pattern is recovered (COCITO, 1973a). In *Bacillus subtilis* infected with DNA from phage 2C, antibiotics of the group B also block synthesis of DNA dependent on protein synthesis (COCITO, 1969).

The inhibitory effect of group B antibiotics on protein synthesis has been confirmed in cell-free systems. The extent of inhibition by the antibiotics in these systems is dependent on the polynucleotide used as the mRNA. Little or no inhibition is found in these systems by studying poly U-directed synthesis of polyphenylalanine whereas there is a good inhibition by these antibiotics in polylysine or polyproline synthesis directed by poly A and poly C respectively (VAZQUEZ, 1967a). Inhibition of protein synthesis by group B antibiotics has also been confirmed in cell-free systems using RNA from phages f2 (ENNIS, 1970) and MS2 as mRNA (COCITO and KAJI, 1971).

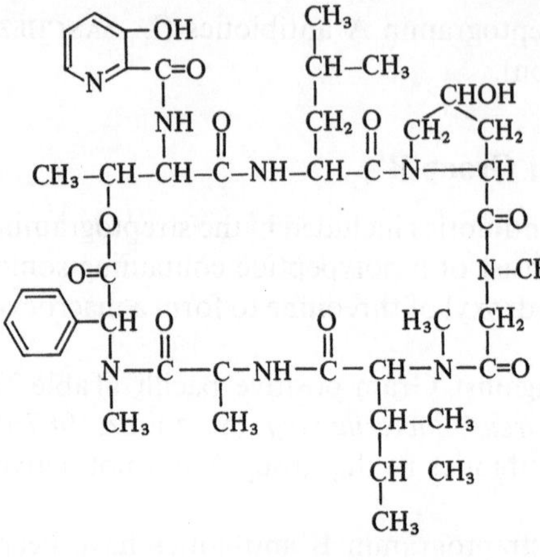

Viridogrisein

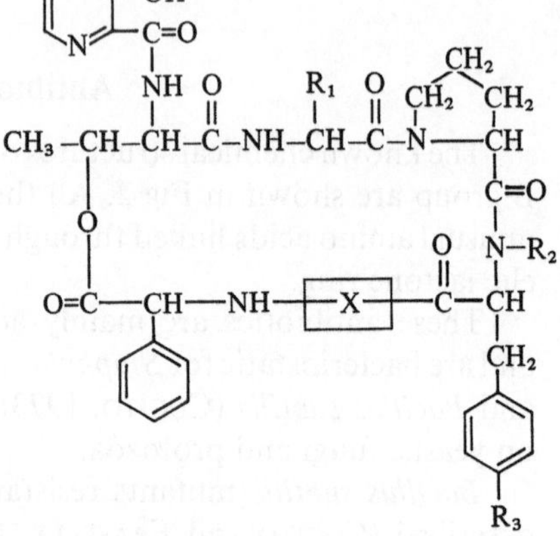

Viridogrisein
(synonym etamycin)

Other antibiotics of the streptogramin B group

	R_1	R_2	R_3	X
Ostreogrycin B Mikamycin B Streptogramin B Synergistin B-1 Vernamycin B$_\alpha$ Pristinamycin I$_A$	—CH$_2$—CH$_3$	—CH$_3$	—N(CH$_3$)$_2$	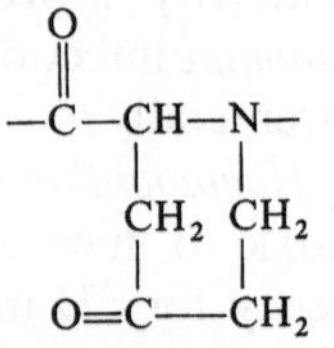 (4-oxopipecolic acid)
Ostreogrycin B$_1$ Pristinamycin I$_c$ Vernamycin B$_\gamma$	—CH$_3$	—CH$_3$	—N(CH$_3$)$_2$	4-oxopipecolic acid
Ostreogrycin B$_2$ Pristinamycin I$_B$ Vernamycin B$_\beta$	—CH$_2$—CH$_3$	—CH$_3$	—NH—CH$_3$	4-oxopipecolic acid
Ostreogrycin B$_3$	—CH$_2$—CH$_3$	—CH$_3$	—N—(CH$_3$)$_2$	3-hydroxy- 4-oxopipecolic acid
Patricin A	—CH$_2$—CH$_3$	—CH$_3$	—H	proline
Patricin B	—CH$_2$—CH$_3$	—CH$_3$	—H	pipecolic acid
Vernamycin B$_\delta$	—CH$_3$	—CH$_3$	—NH—CH$_3$	4-oxopipecolic acid
Vernamycin C	—CH$_2$—CH$_3$	—CH$_3$	—N(CH$_3$)$_2$	aspartic acid
Virginiamycin S	—CH$_2$—CH$_3$	—CH$_3$	—H	4-oxopipecolic acid
Virginiamycin S$_2$	—CH$_2$—CH$_3$	—H	—H	4-hydroxypipecolic acid
Virginiamycin S$_3$	—CH$_2$—CH$_3$	—CH$_3$	—H	3-hydroxy- 4-oxopipecolic acid
Virginiamycin S$_4$	—CH$_3$	—CH$_3$	—H	4-oxopipecolic acid

Fig. 2. Chemical structures of antibiotics of the streptogramin B group

There is ample experimental evidence supporting the interaction of streptogramin B antibiotics with the 50S ribosomal subunit. They have been found to inhibit ^{14}C-chloramphenicol uptake by *Bacillus megaterium* and *Staphylococcus aureus* (VAZQUEZ 1967a). Furthermore, in cell-free systems they inhibit the binding to bacterial ribosomes of ^{14}C-spiramycin I (VAZQUEZ, 1967b) and ^{14}C-chloramphenicol (CHANG *et al.*, 1969; MONRO *et al.*, 1971; FERNÁNDEZ-MUÑOZ *et al.*, 1971) and both spiramycin and chloramphenicol are known to interact with the 50S ribosome subunit. Streptogramin B antibiotics have also been found to enhance interaction of the streptogramin A antibiotics with the ribosome in cell-free systems also suggesting the ribosome as their site of action (ENNIS, 1971). Evidence in favor of the 50S ribosome subunit being the site of action of the streptogramin B antibiotics is also supported by studies on polyphenylalanine synthesising activity of reconstituted ribosomes after pretreatment of either 50S or 30S subunits with synergistin B followed by dialysis against buffer to remove unbound antibiotic prior to addition of the untreated complementary subunits to reconstitute the ribosome (ENNIS, 1966). Association of ^{3}H-dihydrovirginiamycin S with bacterial ribosomes and polysomes has also been reported (COCITO, 1971).

Streptogramin B antibiotics do not affect either aminoacyl-tRNA synthetases or any reaction prior to the formation of aminoacyl-tRNA or aminoacyl-tRNA binding to ribosomes (VAZQUEZ, 1967a). Although elongation might be the most probable target for the B group antibiotics, it has been shown that at least in this process of elongation they do not affect the step of peptide bond formation catalysed by the peptidyl transferase activity of the 50S ribosome subunit (MONRO and VAZQUEZ, 1967). By exclusion, translocation from the A- to the P-site of the growing polypeptide chain has been already postulated as the specific step which might be blocked by group B antibiotics (SPIRIN, 1968). Elongation factor G-dependent GTPase activity has been found not to be affected by antibiotics of group B (MODOLELL, personal communication) but this finding does not totally exclude the possibility of streptogramin B antibiotics blocking translocation since pederine and diphtheria toxin have been found to block the translocation step by eucaryotic ribosomes without affecting elongation factor 2 (the counterpart of elongation factor G in eucaryotic systems) dependent GTPase activity (CARRASCO and VAZQUEZ, 1972). Contradictory reports have appeared concerning the effect of the streptogramin B antibiotics on *Bacillus subtilis* polysomes. ENNIS (1972) has observed an increased polysome breakdown in the presence of streptogramin B antibiotics whereas COCITO (1971) found a prevention of polysome decay by these antibiotics.

Complex Antibiotics of the Streptogramin Family.
The Synergistic Effect of Mixtures of Antibiotics
in Groups A and B

It would seem obvious that the mixture of two or more compounds will affect metabolic processes which are inhibited by the individual antibiotics. However in the case of the synergistic complex of streptogramins, the effect of the

Table 3. *"In vitro" inhibitory concentrations of the complex antibiotics of the streptogramin family*

Organism	Inhibitory concentrations in µg/ml				
	Strepto-gramin	Syner-gistin	Staphylo-mycin	Mika-mycin	Pristina-mycin
Bacteria, Gram-positive:					
Bacillus megaterium	2.00	—	—	6.40	—
Bacillus subtilis	—	0.78	1.00	3.20	0.70
Staphylococcus aureus	0.60	0.19	0.20	0.40	0.20
Sarcina lutea	—	—	0.10	0.10	—
Streptococcus pyogenes	0.05	0.08	0.07	—	0.10
Streptococcus faecalis	1.49	0.39	0.50	—	0.20
Diplococcus pneumoniae	0.25	3.12	0.07	0.60	0.15
Corynebacterium diphtheriae	0.04	0.39	—	0.10	0.02
Mycobacterium sp. 607	11.00	6.25	—	28.00	—
Mycobacterium tuberculosis	5.00	—	20.00	20.00	—
Bacteria, Gram-negative:					
Salmonella typhosa	11.80	100.00	—	160.00	—
Escherichia coli	40.00	100.00	—	160.00	50.00
Aerobacter aerogenes	—	100.00	100.00	—	250.00
Hemophilus pertussis	0.04	3.12	—	—	—
Neisseria gonorrheae	—	3.12	—	—	0.20
Pseudomonas aeruginosa	50.00	100.00	—	160.00	250.00
Yeasts:					
Saccharomyces cerevisiae	85.00	—	—	160.00	—
Candida albicans	—	100.00	100.00	160.00	—
Fungi:					
Aspergillus niger	85.00	—	—	—	—
Aspergillus oryzae	—	—	—	160.00	—
Protozoa:					
Trichomonas vaginalis	490.00	—	—	—	—
Trichophyton sulfureum	—	100.00	—	—	—

Data taken from VAZQUEZ (1967a) and TANAKA *et al.* (1958).

separate compounds are negligible when tested at the concentrations in which they occur in the minimal growth inhibitory concentration of the complex (VAZQUEZ, 1967a). This allows the study of the mode of action of synergistic mixtures without interference by the individual compounds.

The inhibitory spectra of various complexes are shown in Table 3; the antibiotics have no significant inhibitory action on fungi, yeasts, protozoa and most Gram-negative bacteria. Mixtures of A and B antibiotics are bactericidal on growing *Staphylococcus aureus*. This killing effect of the synergistic mixtures can be prevented by incubation in the presence of chloramphenicol or erythromycin but is not prevented by using a medium in which glutamic acid has been omitted and is even enhanced in the presence of chlorotetracycline or oxytetracycline. These findings suggest that the effect of chloramphenicol and erythromycin cannot be explained only on the basis that they inhibit bacterial growth (VAZQUEZ,

1967a). No bactericidal effect can be observed when resistant *Bacillus subtilis* is incubated with the synergistic mixture. The pattern of protein synthesis inhibition in different *B. subtilis* mutants correlates with the killing effect of the synergistic mixtures. Furthermore, B group antibiotics prevent the inhibitory effect of A antibiotics in mutants resistant to the B antibiotics (COCITO and FRASELLE, 1973).

The streptogramin complexes are active when administered orally, subcutaneously or intraperitoneally in protecting mice against infections by *Staphylococcus aureus, Streptococcus pyogenes* and *Diplococcus pneumoniae*. The toxicity of the antibiotics to mice is small when administered orally (2–2.5 g/kg) but higher when administered intraperitoneally (450 mg/kg) (VAZQUEZ, 1967a).

As described earlier, the streptogramin complexes have no effect on bacterial endogenous respiration, oxidation of ethanol or glucose, or anaerobic fermentation of glucose (VAZQUEZ, 1967a). However the mikamycin complex has been reported to inhibit respiration as well as protein synthesis by rat liver mitochondria and it was proposed that this might be due to ribosome integration into the membrane in these organellae (DIXON *et al.*, 1971).

The streptogramin complexes inhibit bacterial growth by blocking protein synthesis; unlike chloramphenicol the inhibitory effect on protein synthesis is not readily reversible, and removal of the streptogramin is followed by only a slight increase in the rate of protein synthesis. The antibiotics do not significantly affect accumulation in the "pool" of low molecular weight constituents or the synthesis of nucleic acid and cell-wall material. The inhibitory effect of the antibiotics on protein synthesis can be detected within 2 minutes of addition. In the presence of the complex antibiotics there is an increase in nucleic acid content which is due to RNA. The synthesis of RNA is unaffected by minimun growth inhibitory concentrations of the streptogramin complexes but is reduced in the presence of higher concentrations (VAZQUEZ, 1967a).

There is an irreversible block of ribosome assembly in the presence of the synergistic mixture in *B. subtilis* and a 60S particle is accumulated (COCITO, 1973a and b). The synergistic mixture of A and B compound causes permanent bleaching of *Euglena gracilis* preventing chlorophyll synthesis and chloroplast development (COCITO *et al.*, 1972) by irreversibly blocking chloroplasts ribosome assembly and synthesis of new 16S ribosomal RNA (VAN PEL and COCITO, 1973).

In bacterial cell-free systems the "streptogramins" inhibit amino acid incorporation in a manner very similar to antibiotics of the group A (VAZQUEZ, 1967a). No synergism could be demonstrated in most cases between antibiotics of the groups A and B in well defined cell-free systems (VAZQUEZ, 1967a) but ENNIS (1965) has reported synergism between antibiotics in groups A and B when they are present at certain critical concentrations. In studies on the binding of ^{3}H-vernamycin A, it has been found that vernamycin B at low concentrations of K$^+$ (55 mM) stimulates about 20% the binding of ^{3}H-vernamycin A to bacterial ribosomes, but this stimulation was not found at higher concentrations of K$^+$. However it was also found that the binding of ^{3}H-vernamycin A appears to be of a different nature and stronger in the presence of the component B; whereas ^{3}H-vernamycin A normally bound to ribosomes can be washed with buffer without K$^+$, it remains bound to the ribosomes when the binding took place in the presence of vernamycin B (Ennis, 1971 a). A stronger binding by component A in the presence of component B might explain the synergism of the mixture of the A and B components. These results are in agreement with the previous report showing that a dialyzable, low molecular weight, heat stable constituent of the soluble bacterial fraction is required for the synergism of the components A and B (Vazquez, 1967a). This required constituent might be a critical concentration of either K^+ or other cation normally present in the bacterial supernatant fraction.

As both streptogramin A and B act on the ribosome and the inhibitory effect of the streptogramin complex appears to occur at the ribosomal level, it seems probable that the synergism of the streptogramin antibiotics in Grampositive bacteria is due to a) an enhancement by streptogramin B of the binding of streptogramin A to the ribosomes and b) a stronger binding of this antibiotic to the ribosome in the presence of components B.

Discussion

Antibiotics in group A are inhibitors of protein synthesis. The accumulation of RNA in the presence of the antibiotics is not a direct effect of the antibiotics. An increase in the rate of RNA accumulation has also been observed in bacteria treated with chloramphenicol and it is thought to be a secondary effect following inhibition of protein synthesis. The RNA which accumulates in bacteria treated with chloramphenicol is mainly ribosomal RNA (accumulated due to an alteration of the control mechanism), mRNA (protected from decay when protein synthesis is blocked) and also tRNA (see the review by Pestka in this book). It is likely that a similar situation exists when protein synthesis in inhibited by the streptogramin A antibiotics.

The binding of antibiotics of group A appears to be located on the peptidyl transferase center of the 50S ribosome subunit of 70S ribosomes. These antibiotics block binding of the CCA-peptidyl and CCA-aminoacyl moiety of the peptidyl- and aminoacyl-tRNA substrates respectively to the P- and A-sites of the ribosomal peptidyl transferase center. As a consequence, streptogramin A antibiotics prevent not only peptide bond formation but also f-Met-tRNA and aminoacyl-tRNA binding and so block the initiation and the elongation phases in protein synthesis. However, under certain experimental conditions, streptogramin A antibiotics do not block all these reactions perhaps due to a lack of accessibility of these antibiotics to their binding site on the ribosome.

Antibiotics in group B are also inhibitors of protein synthesis. The inhibitory effect of these antibiotics on protein synthesis is located on the 50S subunit of the bacterial ribosomes. By exclusion of any effect of the B antibiotics on reactions at the ribosome level which are well resolved and studied, it has been postulated that these compounds block the complex translocation reaction but this has still to be demonstrated.

The overall effects of the "streptogramins" are very similar to those of antibiotics in the group A. The results reported above suggest that the synergism of A and B antibiotics can be explained by an enhancement by component B of the binding of component A to ribosomes. Furthermore, in the presence of antibiotics of the B group, affinity of the A group components for the ribosome is increased and this might account for the bactericidal and more permanent effects observed when the synergistic mixtures are used.

References

CARRASCO, L., and D. VAZQUEZ: Survey of inhibitors in different steps of protein synthesis by mammalian ribosomes. J. Antibiotics (Tokyo) **25**, 732 (1972).

CELMA, M.L., R.E. MONRO, and D. VAZQUEZ: Substrate and antibiotic binding sites at the peptidyl transferase centre of *Escherichia coli* ribosomes. FEBS Letters **6**, 273 (1970).

CELMA, M.L., R.E. MONRO, and D. VAZQUEZ: Substrate and antibiotic binding sites at the peptidyl transferase centre of *Escherichia coli* ribosomes: Binding of UACCA-Leu to 50S subunits. FEBS Letters **13**, 247 (1971).

CHANG, F.N., C. SIDDHIKOL, and B. WEISBLUM: Subunit localization studies of antibiotic inhibitors of protein synthesis. Biochim. Biophys. Acta **186**, 396 (1969).

COCITO, C.: The action of virginiamycin on nucleic acid and protein synthesis in *Bacillus subtilis* infected with bacteriophage 2C. J. Gen. Microbiol. **57**, 195 (1969).

COCITO, C.: Formation and decay of polyribosomes and ribosomes during the inhibition of protein synthesis and recovery. Biochimie **53**, 987 (1971).

COCITO, C.: Formation of ribosomal particles in virginiamycin sensitive and resistant mutants of *B. subtilis*. Biochimie **55**, 153 (1973 a).

COCITO, C.: The ribosomal cycle in bacteria treated with an inhibitor of protein synthesis. Biochimie, **55**, 309. (1973 b).

COCITO, C.G., R. BRONCHART, and B. VAN PEL: Phenotypic and genotypic changes induced in eukaryotic cells by protein inhibitors. Biochem. Biophys. Res. Commun. **46**, 1688 (1972).

COCITO, C., and G. FRASELLE: The properties of virginiamycin-resistant mutants of *B. subtilis*. J. Gen. Microbiol **76**, 115 (1973).

COCITO, C., and A. KAJI: Virginiamycin M, a specific inhibitor of the acceptor site of ribosomes. Biochimie **53**, 763 (1971).

CROOY, P., and R. DE NEYS: Virginiamycin: Nomenclature. J. Antibiotics (Tokyo) **25**, 371 (1972).

CUNDLIFFE, E.: Antibiotics and polyribosomes. II. Some effects of lincomycin, spiramycin and streptogramin A *in vivo*. Biochemistry **8**, 2063 (1969).

DIXON, H., G.M. KELLERMAN, C.H. MITCHELL, N.H. TOWERS, and A.W. LINNANE: Mikamycin, an inhibitor of both mitochondrial protein synthesis and respiration. Biochem. Biophys. Res. Commun. **43**, 780 (1971).

EHRLICH, J., G.L. COFFEY, M.W. FISHER, M.M. GALBRAITH, M.P. KNUDSEN, R.W. SARBER, A.S. SCHLINGMAN, R.M. SMITH, and J.K. WESTON: Griseoviridin and viridogrisein/Biological studies. Antibiotics Ann. **1954/55**, 790.

ENNIS, H.L.: Inhibition of protein synthesis by polypeptide antibiotics. II. *In vitro* protein synthesis. J. Bacteriol. **90**, 1109 (1965).

ENNIS, H.L.: Inhibition of protein synthesis by polypeptide antibiotics III. Ribosomal site of inhibition. Mol. Pharmacol. **2**, 444 (1966).

ENNIS, H.L.: Synergistin: A synergistic antibiotic complex which selectively inhibits protein synthesis. Progress in Antimicrobial and Anticancer Chemotherapy, vol. II, p. 489. University of Tokyo Press (1970).

ENNIS, H.L.: Interaction of vernamycin A with *Escherichia coli* ribosomes. Biochemistry **10**, 1265 (1971 a).

ENNIS, H.L.: Mutants of *Escherichia coli* sensitive to antibiotics. J. Bacteriol. **107**, 486 (1971 b).

ENNIS, H.L.: Polysome metabolism in *Escherichia coli*: Effect of antibiotics on polysome stability. Antimicrobial Agents Chemotherapy **1**, 197 (1972).

ENNIS, H.L., and K.E. DUFFY: Vernamycin A inhibits the non-enzymatic binding of Fmet-tRNA to ribosomes. Biochim. Biophys. Acta **281**, 93 (1972).

FALLONA, C., P. DE MAYO, and A. STOESSL: Mold metabolites: 3: The structure of the griseoviridin salts. Can. J. Chem. **42**, 394 (1964).

FERNANDEZ-MUÑOZ, R., R.E. MONRO, R. TORRES-PINEDO, and D. VAZQUEZ: Substrate- and antibiotic-binding sites at the peptidyl-transferase centre of *Escherichia coli* ribosomes. Studies on the chloramphenicol, lincomycin and erythromycin sites. Eur. J. Biochem. **23**, 185 (1971).

534 D. VAZQUEZ:

HILL, R. N.: Inhibitors of bacterial protein synthesis and the interaction of T-factor, aminoacyl-tRNA and ribosomes. FEBS Abstracts, p. 218 (1969).

JIMENEZ, A., R.E. MONRO, and D. VAZQUEZ: Interaction of Ac-Phe-tRNA with *Escherichia coli* ribosomal subunits. 1. Sparsomycin-induced formation of a complex containing 50S and 30S subunits but not mRNA.FEBS Letters 7, 103 (1970).

MAGYAR, K., J. STVERTECZKY, and I. HORVÁTH: The synergism of viridogrisein and griseoviridin. Acta Microbiol. Acad. Sci. Hung. 9, 247 (1962).

MODOLELL, J., B. CABRER, A. PARMEGGIANI, and D. VAZQUEZ: Inhibition by siomycin and thiostrepton of both aminoacyl-tRNA and factor G binding to ribosomes. Proc. Natl. Acad. Sci. U.S. 68, 1796 (1971a).

MODOLELL, J., D. VAZQUEZ, and R.E. MONRO: Ribosomes, G-factor and siomycin. Nature New Biol. 230, 109 (1971b).

MONRO, R.E., M.L. CELMA, and D. VAZQUEZ: Action of sparsomycin on ribosome-catalysed peptidyl transfer. Nature 222, 356 (1969).

MONRO, R.E., and D. VAZQUEZ: Ribosome-catalysed peptidyl transfer: Effects of some inhibitors of protein synthesis. J. Mol. Biol. 28, 161 (1967).

NETH, R., R.E. MONRO, G. HELLER, E. BATTANER, and D. VAZQUEZ: Catalysis of peptidyl transfer by human tonsil ribosomes and effects of some antibiotics. FEBS Letters 6, 198 (1970).

PESTKA, S.: Studies on the formation of transfer ribonucleic acid-ribosome complexes, XI. Antibiotic effects on phenylalanyl-oligonucleotide binding to ribosomes. Proc. Natl. Acad. Sci. U.S. 64, 709 (1969).

PESTKA, S.: Studies on the formation of transfer ribonucleic acid-ribosomes complexes. VIII. Survey of the effect of antibiotics on N-acetyl-phenylalanyl-puromycin formation: Possible mechanism of chloramphenicol action. Arch. Biochem. Biophys. 136, 80 (1970).

PESTKA, S.: Inhibitors of ribosome functions. In: Molecular mechanisms of antibiotic action on protein synthesis and membranes (Granada, 1971), ed. E. MUÑOZ, F. GARCÍA-FERRANDIZ and D. VAZQUEZ, p. 160. Amsterdam: Elsevier 1972.

RODRIGUEZ-LOPEZ, M., M.L. CELMA, R. FERNANDEZ-MUÑOZ e D. VAZQUEZ: Atti VII Simposio Internazionale di Agrochimica su: La sintesi biologica delle protein, p. 63 (1968).

RODRIGUEZ-LOPEZ, M.: M.L. MUÑOZ, and D. VAZQUEZ: The effects of the rifamycin antibiotics on algae. FEBS Letters 9, 171 (1970).

SPIRIN, A.S.: How does the ribosome work? A hypothesis based on the two subunit construction of the ribosome. Currents in Modern Biology. 2, 115 (1968).

TANAKA, N., N. MIYAIRI, T. NISHIMURA, and H. UMEZAWA: Activity of mikamycin, angustamycins and emimycin against antibiotic-resistant staphylococci. J. Antibiotics (Tokyo) 14, 18 (1961).

TANAKA, N., N. SHINJO, N. MIYAIRI, and H. UMEZAWA: Biological studies on mikamycin. J. Antibiotics (Tokyo) 11, 127 (1958).

VAN PEL, B., and C. COCITO: Formation of chloroplast ribosomes and ribosomal RNA in *Euglena gracilis* incubated with protein inhibitors. Exptl. Cell Res. 78, 103 (1973).

VAZQUEZ, D.: The streptogramin family of antibiotics. In: Antibiotics, vol. I, Mechanism of action, eds. D. GOTTLIEB and P.D. SHAW, p. 387. Berlin-Heidelberg-New York: Springer 1967a.

VAZQUEZ, D.: Binding to ribosomes and inhibitory effect on protein synthesis of the spiramycin antibiotics. Life Sci. 6, 845 (1967b).

VAZQUEZ, D.: Inhibitors of protein synthesis at the ribosome level. Studies on their site of action. Life Sci. 6, 381 (1967c).

VAZQUEZ, D., E. BATTANER, R. NETH., G. HELLER, and R.E. MONRO: The function of 80S ribosomal subunits and effects of some antibiotics. Cold Spring Harbor Symp. Quant. Biol. 34, 369 (1969).

WATANABE, K., K. YONEZAWA, T. KOMAI, and T. TAKEUCHI: Biological studies on mikamycin. V. Absorption and excretion of tritiated mikamycin A and B. J. Antibiotics (Tokyo) 23, 394 (1970).

WEISBLUM, B., and V. DEMOHN: Erythromycin-inducible resistance in *Staphylococcus aureus:* Survey of antibiotic classes involved. J. Bacteriol. 98, 447 (1969).

YAMAGUCHI, H., and N. TANAKA: Site of action of mikamycins A and B in polypeptide synthesising systems. J. Biochem. (Tokyo) 61, 18 (1967).

YAMAGUCHI, H., Y. YOSHIDA, and N. TANAKA: Inhibition by mikamycins of polypeptide synthesis directed by native messengers and synthetic polynucleotides. J. Biochem. (Tokyo) 60, 246 (1966).

Streptomycin, Dihydrostreptomycin, and the Gentamicins *

David Schlessinger and Gerald Medoff

History and Chemical Structure

Streptomycin

Streptomycin was isolated from *Streptomyces griseus* by SCHATZ *et al.* in 1944 and was the first nontoxic broad spectrum antibiotic discovered which was also effective against the tubercle bacillus. The chemical structure is N-methyl-L-glucosaminidostreptoside-streptidine and the antibiotic is a highly polar organic base with a large number of hydrophilic and functional groups (Fig. 1).

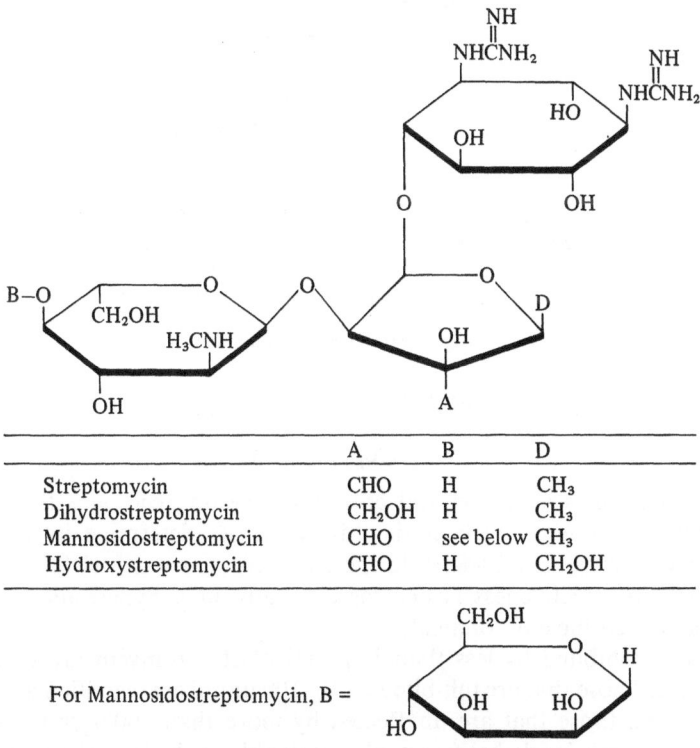

	A	B	D
Streptomycin	CHO	H	CH_3
Dihydrostreptomycin	CH_2OH	H	CH_3
Mannosidostreptomycin	CHO	see below	CH_3
Hydroxystreptomycin	CHO	H	CH_2OH

For Mannosidostreptomycin, B =

Fig. 1. Structure of streptomycins

* Supported in part by grants CA 12021 from the P.H.S., GB 23052 from the NSF and a grant from the John A. Hartford Foundation, Inc.

Dihydrostreptomycin was first produced in 1947 by the catalytic hydrogenation of streptomycin. It has about the same degree of antibacterial activity as
streptomycin, but it is less effective against some gram-negative microorganisms.
Because it has a higher risk of irreversible deafness, and its effectiveness is no
greater that that of streptomycin, dihydrostreptomycin is no longer used clinically.

Gentamicin

Gentamicin is a broad spectrum antibiotic derived from *Micromonospora
purpurea* of the family Actinomycetales. It was first described in 1963 (WEINSTEIN
et al., 1963) and isolated, purified and characterized by ROSSELET and colleagues
(ROSSELET *et al.*, 1963). Gentamicin consists of three closely related components,
gentamicins C_1, C_2 and C_{1a} with very similar molecular weights (COOPER *et al.*,
1969) (Fig. 2). In many respects it is more similar to kanamycin and neomycin
than streptomycin and is covered in this chapter mainly for comparison.

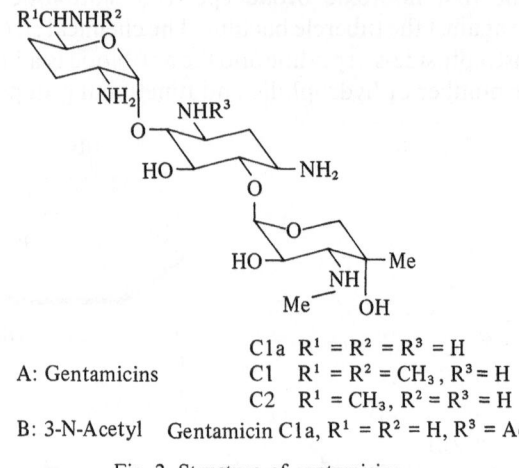

A: Gentamicins

C1a	$R^1 = R^2 = R^3 = H$
C1	$R^1 = R^2 = CH_3$, $R^3 = H$
C2	$R^1 = CH_3$, $R^2 = R^3 = H$

B: 3-N-Acetyl   Gentamicin C1a, $R^1 = R^2 = H$, $R^3 = Ac$

Fig. 2. Structure of gentamicins

Antimicrobial Spectrum and Medicinal Use

Streptomycin

Streptomycin is a bactericidal antibiotic. Resting cells are less susceptible
to it than are multiplying bacteria. Many factors determine its antimicrobial
effectiveness: there is a 20–80 fold increase in potency at pH 8.0 as compared
to pH 6.0; its effectiveness is decreased greatly in a hypertonic solution and
also in an anaerobic environment.

Bacteria inhibited by less than 10 µg/ml of streptomycin are considered to
be sensitive. Those that are inhibited by 10–100 µg/ml are classified as moderately
sensitive; and those that are unaffected by more than 100 µg/ml are resistant.
Because the peak level that is readily attainable without toxicity in the blood
of patients is 15–20 µg/ml, only the first group can be called sensitive from
a clinical point of view. In this group of bacteria are *Brucella, Erysipelothrix,
Listeria, Nocardia, Shigella, Tularensis,* the plague organism (*Yersinia pestis*)

and many but not all strains of tubercle bacilli. Bacteria exhibiting a wide variation in susceptibility include the *Pneumococcus*, the Typhoid bacillus and other *Salmonella, Escherichia coli, Hemophilus influenza*, the *Gonococcus* and *Meningococcus, Proteus vulgaris, Staphylococcus aureus* and *albus, Streptococci* and *Vibrio comma*. Anaerobic bacteria, fungi, parasites and all viruses are totally resistant to streptomycin (WEINSTEIN, 1971).

Streptomycin is used clinically only for a very limited number of infections. It is used alone for the treatment of plague and tularemia and occasionally for brucellosis. Its usual use is in combination with other antibiotics. The most frequent infections for which streptomycin is given with other antibiotics are tuberculosis (INH), brucellosis (tetracycline), Group D enterococcal infections of the heart valve (penicillin) and *Listeria monocytogenes* infection (penicillin). The effect of penicillin and streptomycin on the latter two bacteria illustrates the phenomenon of antibiotic synergism, in which the potency of the streptomycin is enhanced by the penicillin, which alters cell permeability so that an increased amount of streptomycin penetrates the cell envelope and binds to the ribosome (ZIMMERMANN *et al.*, 1971).

Gentamicin

The spectrum of gentamicin is wider than that of any other clinically available antimicrobial agent. It is highly effective against *Pseudomonas, E. coli, Klebsiella, Enterobacteria, Proteus* and *Serratia*, as well as other less common gram-negative bacteria. Recent studies have also shown it to be effective against both penicillinase and nonpenicillinase producing *Staphylococci*. It is also a highly active inhibitor of growth of the tubercle bacillus *in vitro*. Although its activity against gram-negative anaerobic bacteria is greater than that of streptomycin, it is still not a satisfactory antibiotic against these bacteria. The drug is ineffective against the *Pneumococcus* and the *Streptococcus*, although, as with streptomycin, there is synergism with penicillin against the Group D enterococci and *Listeria*. Gentamicin has no effect on the growth of rickettsia, fungi, viruses and protozoa, (SNELLING *et al.*, 1971). In most studies, 95% of clinically isolated *Pseudomonas aeruginosa* are inhibited by less than 6.0 µg/ml of gentamicin. Achievable blood levels are in the range of 8–12 µg/ml (RIFF and JACKSON, 1971). On the basis of laboratory sensitivity tests and its clinical performance to date, gentamicin has become the preferred antibiotic for life threatening gram-negative infection in general and Pseudomonas infection in particular.

Mammalian Toxicity

Streptomycin

The most important toxic effects of streptomycin are dependent on the quantity of drug administered and the duration of treatment (WEINSTEIN, 1971). Nearly 75% of patients given 2 grams of streptomycin a day for 60–120 days manifest some vestibular abnormality. Disturbances of hearing are less common but do occur in an appreciable number of patients. The basis for the 8th nerve toxicity is not known. Streptomycin may also produce a dysfunction of the optic nerve and rarely a peripheral neuritis. Abnormalities of kidney functions may also develop from prolonged or high dose streptomycin treatment.

Gentamicin

The most important toxic effect of gentamicin involves the eighth cranial nerve. Ototoxicity appears in about 2% of patients treated with gentamicin. Vestibular function is impaired to a greater extent than is hearing. As with streptomycin, the risk of ototoxicity is highest in the presence of renal insufficiency. It can be avoided in most patients by keeping blood levels below 12 µg/ml. Electron microscopic studies of the vestibular end organs from monkeys that received large doses of gentamicin have demonstrated that the type I hair cells, both in the cristae and maculae, were more vulnerable to the effects of gentamicin than were the type II hair cells (Igarashi et al., 1971). These cells showed a progressive deterioration with a disappearance of the ribosomes, pyknosis of the nuclei, edema of the cytoplasm, and finally disintegration of the cells (Wersall et al., 1969). The reason for this differential sensitivity to the drug is unknown but it suggests the possibility of important differences in the nutrition and metabolism of these two groups of sensory cells.

Gentamicin also has a nephrotoxicity which is dose dependent and increased in incidence in patients with abnormalities in renal function.

Mode and Mechanism of Action

Streptomycin

In their review of the effects of aminoglycosides, Jacoby and Gorini (1967) suggested that "consideration of the number of different hypothesis which have advanced to explain streptomycin action suggests caution in assuming that the mechanism of streptomycin lethality is now known. Undoubtedly, continued study of streptomycin action will reveal further surprises". In the following pages, this *caveat* remains; what is possible, on the basis of the information gleaned in the five years since that earlier review, is a partial characterization of the results of interaction of streptomycin with ribosomes. These are summarized below, with a further discussion of the probable relationship of these interactions with lethality.

I. *The Structural Basis of Streptomycin Effects.* The continuing fascination of the complexity of streptomycin action on cells is clearly based on its direct interaction with ribosomes. The reaction is strong, specific, and greatly depressed in the major class of genetically resistant strains.

The earliest detectable effects of streptomycin are a relatively rapid uptake of drug, with some attendant effects on membrane function, including K^+ efflux from the cell (Anand and Davis, 1960; Dubin and Davis, 1961); once the antibiotic enters the cell, the major binding sites in the cell are the 30S ribosomes (see review by Jacoby and Gorini, 1967).

Since streptomycin inhibits polypeptide synthesis at concentrations that are equivalent to about one molecule per ribosome, high binding affinities to ribosomes were early inferred. In agreement with this notion, LEON and BROCK (1967) for 70 S, and KAJI and TANAKA (1968) for 30 S ribosomes, reported binding of one or at most several molecules per ribosome. In those studies, resistant ribosomes did not bind the antibiotic, and others (WOLFGANG and LAWRENCE, 1967; PETITPAS-DEWANDRE et al., 1969) specified that sensitive ribosomes show a greater affinity than do dependent ribosomes, with resistant ribosomes showing much lower, if any, binding affinity.

More detailed binding studies have been carried out with dihydrostreptomycin by CHANG and FLAKS (1972a, b). Above concentrations of 10^{-5} M dihydrostreptomycin, weaker binding of larger amounts of drug is observed; but at concentrations of drug up to 1×10^{-5} M, more than 95% of washed 70 S ribosomes (obtained as products of "runoff" from messenger RNA in absence of excess initiation factors) bound exactly one molecule per particle. The Kdiss for the bound complex at 25° was 9.4×10^{-8} M. No significant binding was observed with ribosomes from streptomycin-resistant or dependent strains. 50 S subunits displayed no binding; however, 30 S subunits bound with a Kdiss about 10-fold lower than 70 S particles (1.0×10^{-6}). In other words, the 50 S ribosome seems to render binding to the 30 S subunit more stable.

In extensive work on the nature of the binding of streptomycin to ribosomes, OZAKI et al. (1969) showed that when 30 S ribosomes were reconstituted from isolated RNA and proteins, a particular ribosomal protein (S12, in the terminology of WITTMAN et al., 1971) showed an especially great effect on the capacity of the ribosomes to bind ^{3}H-dihydrostreptomycin, as well as to determine other responses to the antibiotic (see part II below). When the protein S12 was derived from a resistant ribosome, the reconstituted ribosome showed only 3 to 9% of binding capacity of reconstituted ribosomes containing protein S12 from a sensitive strain. The results were indifferent to the source of the RNA or of other proteins (whether they were derived from "sensitive" or "resistant" ribosomes).

The changes in the S12 protein have now been specified and consist of amino acid replacements at only a few sites (FUNATSU and WITTMAN, 1972), consistent with the genetic mapping data of BRECKENRIDGE and GORINI (1970).

A somewhat different approach to the study of the attachment site of streptomycin to the 30 S ribosomal subunit has been taken by BISWAS and GORINI (1972). They have presented evidence that one or a few binding sites for streptomycin are actually present in the 16 S-RNA isolated from each 30 S subunit, whether from sensitive or resistant strains; they suggest a model in which the binding site(s) are relatively available in sensitive ribosomes, but become masked in the resistant ribosomes, presumably by a different spatial arrangement conferred by the altered S12 protein. At any event, there seems no doubt that the tightness of the interaction of streptomycin with the 30 S ribosome is the source of its direct effects on protein synthesis and directly or indirectly of its lethality.

II. *Functional Consequences of Streptomycin Binding.* The consequences of streptomycin binding to ribosomes are profound, and are one major reason for the sustained interest in its mechanism of action.

A. Effects *in vitro*: When streptomycin binds to ribosomes, every measurable phase of protein synthesis is modified. Some of these effects (such as 5 and 6) require rather higher concentrations of antibiotic, but all can be observed at levels normally used in studies *in vivo*. The changes analyzed *in vitro* include the following:

1. Inhibition of initiation complex formation and its destabilization. Luzzatto *et al.* (1968) presented evidence that formation of a 70S initiation complex could apparently proceed in cell extracts in presence of fmet-tRNA, but that the binding of the second aminoacyl tRNA was severely inhibited by streptomycin, and that translation of phage RNA was also nearly totally blocked, with the ribosome presumably never able to progress very far, if at all, along the messenger RNA.

With the studies of Lelong *et al.* (1971), and of Modolell and Davis (1970), the mechanism and severity of the interference with initiation has become clearer. These authors showed that the partial 70S initiation complex as formed in presence of streptomycin is unusually unstable, tending to lose bound fmet-tRNA at least 10 times faster than in any comparable process in absence of streptomycin.

Streptomycin, in further studies of these authors, seemed to antagonize the binding and effective function of the initiation factor preparations; probably streptomycin competes with factors for binding to the 30S subunit (Lelong *et al.*, 1972) and inhibits the shift of 70S particles to 30S and 50S particles, usually seen in presence of the partially purified "dissociation factor" (Wallace *et al.*, 1973) so that many ribosomes are observed stopped short of reinitiation as free 70S particles (see Section B2 below).

2. Stabilization of 70S ribosomes. Even the general equilibrium of 30S, 50S and 70S ribosomes that lack initiation factors is shifted toward 70S particles by dihydrostreptomycin (Chang and Flaks, 1972b; see also part B2 below).

3. Inhibition during translation. Translation of synthetic ribopolymers is, in many cases, inhibited by streptomycin (see review of Gorini and Jacoby, 1967). Binding of aminoacyl tRNA is inhibited (Pestka *et al.*, 1966) and at least in the case of RNA from phage R17, the ribosome peptidyl-tRNA complex has been reported to break down in incubation mixtures with a half-time of about five minutes (Modolell and Davis, 1969).

4. Stimulation of G factor-dependent GTPase. Novogrodsky (1972) presented data that the "uncoupled" ribosome-dependent GTPase activity of the G translation factor with 70S ribosomes is augmented by streptomycin.

5. Inhibition of termination. In an *in vitro* model system for polypeptide chain termination, streptomycin inhibited release-factor function (Caskey *et al.*, 1971).

6. Inhibition of ribosome release. A factor that seems to function in the release of ribosomes from mRNA has been characterized by Ogawa *et al.* (1972); this function is inhibited by streptomycin.

7. Increased miscoding. Starting with the observations by Davies *et al.* (1964), there have been many reports that streptomycin induces increased miscoding of artificial polymers and natural messenger RNA. In the most widely accepted view, the ribosome is distorted so that false pairing of codon and tRNA anticodon occurs.

8. Inhibition of inactivation and reactivation of ribosomes. Several measurable functions of ribosomes, including binding of aminoacyl tRNA can be inactivated by various treatments—for example, upon exposure to low concentrations of Mg^{2+}; the ribosomes can be reactivated by heating in appropriate buffers. These phenomena have been specified and carefully explored by ZAMIR et al. (1971). In a recent extension of the work, MISKIN and ZAMIR (1972) report that dihydro-streptomycin both retards the inactivation of ribosomes and inhibits the reactivation process. These results suggest that streptomycin may inhibit conformational changes in ribosomes (CHUANG and SIMPSON, 1971) that are important in protein synthesis. SHERMAN et al. (1969) showed that even at catalytic levels, streptomycin affects the bulk of 70S ribosomes enough to cause a detectable conformational change, measurable by hydrogen exchange assays.

B. *Reversal of effects of streptomycin:* As might be expected, the effects of streptomycin can be counteracted or modified by mutations to resistance, by removal of the S12 protein from the ribosomes, or by changes in the ribosome brought about by antibiotics or by appropriate mutations.

The most obvious of these is the change to the resistant phenotype, which permits normal initiation and function of the ribosomes, and is the standard control in the studies reported above—in which, for example, streptomycin-induced miscoding is greatly restricted (DAVIES et al., 1964) and there is almost no effect of dihydrostreptomycin on ribosome inactivation or reactivation (MISKIN and ZAMIR, 1972). No binding,—no effect.

Resistant ribosomes are nevertheless modified, so that they are generally deficient in miscoding and show differences in ribosome stability, subunit association, etc. In the extreme, the studies of OZAKI et al. (1969) show that ribosomes which completely lack the S12 protein required for dihydrostreptomycin binding are competent to make polyphenylanine, though deficient in translation of phage messenger RNA. More subtle modifications of the ribosome can modify the effects of streptomycin or of the resistant mutation, as in the specific reversal of miscoding restriction by ram mutations (ROSSET and GORINI, 1969).

C. *Physiological consequences of streptomycin in vivo:* Many of the *in vitro* effects of streptomycin are evinced in whole cells, including effects on protein synthesis and on accompanying miscoding.

1. Miscoding. Since the initial observations of miscoding *in vivo* by GORINI and KATAJA (1964), the study of such effects has been greatly extended. For the most part, these studies have intrinsic interest for an understanding of the role of the ribosome in translation of the genetic code (very likely miscoding *per se* plays no role in the major killing mechanism of streptomycin—see below).

Evidence was initially accumulated for suppression of both missense and nonsense mutation by streptomycin, in keeping with its action in increasing miscoding in cell-free extracts.

Again, just as with miscoding *in vitro*, mutation—or specific treatment of cells—can act to counteract or reverse the effects of streptomycin. Thus, mutations to streptomycin resistance generally restrict phenotypic suppression *in vivo* (GORINI et al., 1966), while second mutations in ribosomes often restore suppressibility; similar results have been obtained with both nonsense and missense suppression (KUWANO et al., 1969; APIRION and SCHLESSINGER, 1969). The most comprehensive analysis was made possible by the analysis of a special class of mutations by ROSSET and GORINI (1969). These lesions, in a ribosomal protein different from the *str* A protein, were named *ram* and extensively analyzed: they increase the efficiency of suppression to counteract the restriction by *str* A mutations. Models have been inferred from these studies for the role of the ribosome in the readout of the genetic code (BISWAS and GORINI, 1972).

2. Effects on the ribosome cycle and coordination with effects *in vitro:* the critical block at "free" ribosomes.

Just as the miscoding effects of streptomycin on ribosomes are evinced both *in vitro* and *in vivo*, one might anticipate that the concomitant effects of streptomycin observed in extracts also occur in intact cells. With the range of effects tabulated above, one is then led to ask which, if any, of these effects dominates ribosome function *in vivo*—i.e., is any effect of streptomycin the most critical physiologically? In this section, we consider some kinetic effects and physiological experiments; in part III below we consider separately the question of lethality.

Generally speaking, one expects the ribosomes in actively growing cells to be distributed as translating ribosomes, ribosomes involved in 30S and 70S initiation complex formation, and ribosomes that are just finishing translation. If an antibiotic were to inhibit selectively one or another specific step, then all the ribosomes would be expected to stop accordingly and, for example, all the ribosomes might accumulate as translating ribosomes. But there are additional assumptions involved: 1. The ribosomes, containing antibiotic, must behave like a ribosome "frozen" in something close to one of its normal conformational states; 2. The ribosome must undergo no other "abnormal" transformations after its blockage. Although both of these assumptions fail, at least in part, with streptomycin as the antibiotic of choice, nevertheless some inferences can be drawn.

When moderate concentrations of streptomycin are added to cells, protein synthesis is rapidly, but not completely, inhibited. At a rate that depends on the strain and dose used, polyribosomes progressively break down and 70S particles accumulate (LUZZATTO *et al.*, 1969; MODOLELL and DAVIS, 1969; KOGUT and PRIZANT, 1970). These particles could be detected to a lesser extent as inactive ribosomes in polyribosomes and then in accumulating amounts as 70S particles. They were characterized by a deficit in their incorporation of radioactive amino acids and of newly-labeled [3]HmRNA (LUZZATTO *et al.*, 1969; WALLACE *et al.*, 1972) (though whether they contain pre-existing mRNA is in doubt; see below); along with the patent block of initiation on natural mRNA *in vitro*, these particles were the source of the suggestions that the primary block of streptomycin is somewhere at initiation of protein synthesis.

As KOGUT and PRIZANT (1972) point out, 70S ribosomes accumulate in many conditions, such as energy starvation or decreased growth rates; why are the ribosomes observed in streptomycin-treated cells "special"? *In vivo*, their specialness can be defined as above by their deficit in uptake of newly-labeled mRNA; other antibiotics that inhibit protein synthesis at least as extensively (for example, chloramphenicol and spectinomycin) produce no class of 70S ribosomes with such a selective incompetence; with those antibiotics, all ribosomes continue to join to new mRNA with a comparable frequency (GURGO *et al.*, 1969; see also section III below).

In extracts, the specialness of 70S ribosomes from streptomycin treated cells was first pointed out by HERZOG (1964) who showed that the ribosomes show an increased stability against dissociation by low levels of Mg^{2+} ions.

The exact nature of the streptomycin 70S ribosomes is not yet clear, and it may differ somewhat for different strains (as do many features of the rate and extent of streptomycin action and ribosome stability). Thus, the ribosomes isolated by LUZZATTO et al. (1969) were reported to contain mRNA, while KOGUT and PRIZANT (1972) and WALLACE et al. (1973) find that the ribosomes have many properties of 30S–50S couples.

Even the special stability of 70S ribosomes from streptomycin-treated cells (HERZOG, 1964; WOLFE and HAHN, 1968; WALLACE et al., 1973) is not always seen (KOGUT and PRIZANT, 1972). However, as HERZOG (1964) and CHANG and FLAKS (1972b) point out, the formation of "stuck 70S" in vitro by addition of streptomycin to ribosomes has not been easy, though the recent work does show some unexpected stability of treated ribosomes upon dilution and WALLACE and DAVIS (1973) report the formation of what they consider to be such ribosomes in an appropriate in vitro system. Perhaps the variations in properties of extracted 70S particles reflect strain differences (WALLACE and DAVIS, 1973), as well as possible differential losses of ligands like GTP, IF3, or polyamines.

In this regard, the work on the ribosome cycle in extracts is ahead of the measurements on extracted ribosomes; the interference with initiation, detailed above, is clear and the difference in the findings of various laboratories with extracted 70S particles may reflects only differential losses of ligands.

In addition to the obvious correlation of detailed work on the inhibition of initiation, a number of other experiments in vitro strongly support the inference that initiation is the critical locus of streptomycin action. Thus, LUZZATTO et al. (1969) presented evidence, and WALLACE and DAVIS (personal communication) confirmed that the translation of endogenous mRNA is less inhibited than is that of added exogenous mRNA. Also consistent is the finding of KAEMPFER (1969) that in a protein-synthesizing system in extracts, ribosomes undergo one round of subunit exchange (presumably when they are released from mRNA) and then become inert to further subunit exchange. Finally, consistent though it is in no way predicted or required by the other results, is the observation by OZAKI et al. (1969) that 30S subunits reconstituted without protein S12 fail in the initiation of protein synthesis.

3. Streptomycin dependence. Changes in the same ribosomal protein can lead to resistance or dependence (BIRGE and KURLAND, 1969) and ribosomes from dependent strains starved of streptomycin show very poor capacity for protein synthesis (GADÓ and HORVÁTH, 1963). The most recent work tends to confirm the possibility of a lesion that affects ribosomes in absence of streptomycin in a way analogous to the presence of streptomycin in sensitive strains. Thus, ZITOMER and FLAKS (1972) provide evidence that the 30S subunit from dependent cells is relatively rapidly inactivated in absence of the drug; while LELONG et al. (personal communication, 1972) have shown that the ribosomes deprived of streptomycin still respond to poly U, but fail at initiation of protein synthesis on natural mRNA; once again, the failure is beyond the initial binding of fmet-tRNA.

An interesting alternative model was suggested by Jacoby and Gorini (1967) based on the correlation between killing and misreading in various drug dependent strains. In this case the notion of excessive mistranslation, here in the absence of the required drug, is retained as a possible cause of lethality.

III. *The question of lethality.* How might the effects of streptomycin produce lethality? Since measurements of lethality are made by observations of colony formation 48 hours after the event discussed above, and the response of various strains differs somewhat, all the observations discussed might be irrelevant or only preliminary to the onset of true killing events. Such a line of reasoning is followed, for example, in the experiments of Stern, Barner and Cohen (1966), who propose an explanation for lethality based on a later synthesis of "odd" RNA, possibly related to changes in polyamine function. Furthermore, even bacteriostatic antibiotics can often become bacteriocidal on prolonged incubation, so that death represents a kind of epiphenomenon (as Gorini has put it, "we tend to think of even our own death as an epiphenomenon"). Finally, we are faced in the following discussion with the sobering basis for prediction that the given "explanation" is completely different from those offered by Jacoby and Gorini (1967) and earlier authors. One can construct the following order of events following addition of streptomycin to sensitive cells:

1. First, great but not complete inhibition of protein synthesis.

2. Polyribosomes are depleted, with the release of substantially inert "70S ribosomes". (This might happen either by slow completion of polypeptides, or by breakdown of inhibited polyribosomes, or by both mechanisms.)

3. The 70S ribosomes have difficulty forming an initiation complex, at least in part because the binding and function of IF3 is inhibited. They usually jam before or at the stage of initiation; unstable complexes can be formed that lose f met-tRNA with a half-life of about five minutes (Modolell and Davis, 1970; Lelong et al., 1971).

4. At a slow rate, many ribosomes do succeed in forming small peptides before they are released from mRNA. This supports the continued production of some partially functional polyribosomes (Wallace and Davis, 1973) comparable to the case of chloramphenicol treatment (Cremer and Schlessinger, 1972).

In this "construction", the blockage of ribosomes at initiation (complete or almost complete) is sufficient to explain the dominance of sensitivity in protein synthesis in diploid strains containing one sensitive and one streptomycin resistant allele. A version of this model that emphasizes measurements of the dynamics of the system in several strains is detailed by Wallace and Davis (1973). Lethality as colony-formers could then depend on whether the ribosomes containing streptomycin are irreversibly inhibited or are provoked to irreversible denaturation in presence of the drug. At this point, the "structural" and "functional" analysis of streptomycin action (parts I and II above) merge. In a S/R diploid, the fate of the cell would depend on its residual capacity to provide mRNA that could be rescued by unblocked resistant ribosomes when the drug is withdrawn [cf. Gorini (1969) with Sparling et al., 1968; and Wallace and Davis, 1973].

The only point that seems specifically important in this "construction" is that not just any interaction with ribosomes will produce lethality; the drug must be exposed to ribosomes when they are free of mRNA. This permits one to understand the protection against streptomycin killing by chloramphenicol (JAWETZ et al., 1951), and the promotion of killing by puromycin (WHITE and WHITE, 1964); the one protects polyribosomes, the other promotes their disassociation. (These observations alone originally prompted WHITE and WHITE to this type of model.)

This general view gains some credence from a comparison of streptomycin action with that of other antibiotics. Thus, chloramphenicol produces similar effects on the ribosome cycle, except for the selective block at initiation; the effects are all reversible (bacteriostatic). Among the aminoglycosides, one finds that spectinomycin, which shows reversible effects predominantly at chain elongation (GURGO et al., 1969) is bacteriostatic; again, the more severe bactericidal aminoglycosides (neomycin, kanamycin) provoke the accumulation of inert 70S ribosomes. As for gentamicin, little is known about the molecular basis of its action, but its bactericidal effect suggests that the mechanism of killing is again similar to that of streptomycin (OKUYAMA et al., 1971).

Structure-Activity Relationships

While miscoding is generally discounted as a cause of lethality, the levels of misreading produced by the various antibiotics have proven useful as a guide to their relative antimicrobial effectiveness. For all of them the effects are heavily dose-dependent and each antibiotic gives a curve with a characteristic shape (DAVIES, 1967). The concentration-activity curves correlate reasonably well with the in vivo activity of the drugs. Because drug activity in this system is reflected in changes in the shape of the curves, this type of experiment provides a very sensitive assay for structure-activity relationships, eliminating many problems of such studies in vivo.

Streptomycin

Very little structural modification without substantial loss of activity is possible. The modifications which cause little or no change in activity are: reduction of the aldehyde to a primary alcohol (dihydrostreptomycin), linkage of mannose residue to the 4′ OH of the N-methyl-L-glucosamine residue (mannidostreptomycin), or removal of one amidino residue (bluensomycin).

Substantial loss of activity is accompanied by removal of both amidino residues, blocking of the 3′ OH on the N-methyl-L-glucosamine residue, or by attachment of alkyl side chains through the aldehyde moiety. None of the hydrolysis products of streptomycin have any antibiotic activity.

Gentamicin

The gentamicin complex induces the highest levels of misreading of all the aminoglycosides, although neomycin is a more potent antibiotic (DAVIES et al., 1969). There is very little difference in activity between the complex and the pure components C_1, C_1a, and C_2 and probably in this group of antibiotics, the structural features responsible for drug activity are identical. There is some evidence that as with streptomycin, the primary site of gentamicin action is subsequent to the 30S initiation complex formation. (Okuyama *et al.*, 1971).

Resistance

Streptomycin

Besides the one-step mutants to high resistance by change in the 30S ribosome, resistance based on a relative impermeability of the cell envelope to streptomycin also occurs; perhaps the best example of this is the Group D enterococcus. Resistance in this organism can be overcome with the use of penicillin, which increases the uptake of streptomycin into the cell (Zimmermann *et al.*, 1971).

The most important mechanism of resistance operative in streptomycin resistant strains found in clinical infection appears to be through enzymes that inactivate the antibiotic by adenylation (Yamada *et al.*, 1968). A distinguishing characteristic of these strains is that the genetic loci determining the inactivating enzymes are usually carried on extrachromosomal elements (Davies, 1971). In enterobacteria, these R factors are the principle cause of antibiotic resistant bacteria in clinical situations.

Resistance to streptomycin in bacteria that carry an R factor has been shown to be due to enzymatic activities, a streptomycin phosphotransferase and a streptomycin adenylate synthetase (Davies *et al.*, 1971; Umezawa *et al.*, 1968; Harwood and Smith, 1969). It should not be assumed that only R factor strains can inactivate aminoglycoside antibiotics by enzymatic esterification. Certain strains of pseudomonas, and staphylococci not carrying R factors, have an enzyme system which can phosphorylate the neomycin-kanamycin-streptomycin group of antibiotics, and the presence of this enzyme alone is responsible for resistance to these drugs (Dori *et al.*, 1968).

Gentamicin

The general sensitivity of gram-negative bacteria to gentamicin has been confirmed, but with a few disquieting notes. About 95 % of strains of *Pseudomonas aeruginosa* isolated from patients are sensitive to gentamicin, but highly resistant strains have emerged, particularly where gentamicin has been used topically on a closed ward population, as in burn units (Shulman *et al.*, 1971). These strains have caused local epidemics and in those instances in which the organisms have been studied, the resistance has been transferable, presumably by an episomal mechanism. The two mechanisms by which gentamicin has been inactivated by resistant organisms are by acetylation of the purpurosamine ring of gentamicin 1 a by *E. coli* (Benveniste and Davies, 1971), and by acetylation of the deoxystreptamine ring by *Pseudomonas* (Brzezinska *et al.*, 1972). Attempts to determine if the latter two mechanisms are chromosomal or episomal have thus far been unsuccessful.

It is of interest that the only well characterized natural isolates are not resistant as a consequence of a change in the ribosome. Possibly resistant ribosomes are changed in a way that would give cells a strong selective disadvantage in natural environments; alternatively, R factors may be sufficiently prevalent to make the rare mutants to streptomycin resistance a more unusual mode of natural defense against the drug.

Acknowledgement. We wish to thank colleagues, including Dr's B.D. DAVIS, J. DAVIES, J.G. FLAKS, L. GORINI, H. KAJI and G. JACOBY who kindly contributed information in advance of publication, and helped us correct some of the inevitable misstatements.

References

ANAND, N., and B.D. DAVIS: Damage by streptomycin to the cell membrane of *Escherichia coli.* Nature **185**, 22 (1960).

APIRION, D., and D. SCHLESSINGER: The effect of ribosome alterations on ribosome function, and on expression of ribosome and non-ribosome mutations. In: Ciba Foundation symposium on mutation as cellular process, ed. by G. E. W. WOLSTENHOLME and MAEVE O'CONNER, p. 155–167. London: Churchill Ltd. 1969.

BENVENISTE, R., and J. DAVIES: Enzymatic acetylation of aminoglycoside antibiotics by *Escherichia coli* carrying an R factor. Biochemistry **10**, 1787 (1971).

BIRGE, E.A., and C.G. KURLAND: Altered ribosomal protein in streptomycin-dependent *Escherichia coli.* Science **166**, 1282 (1969).

BISWAS, D.K., and L. GORINI: The attachment site of streptomycin to the 30S ribosomal subunit. Proc. Natl. Acad. Sci. U.S. **69**, 2141 (1972).

BISWAS, D.K., and L. GORINI: Restriction, de-restriction and mistranslation in missence suppression. Ribosomal discrimination of transfer RNA's. J. Mol. Biol. **64**, 119 (1972).

BRECKENRIDGE, L., and L. GORINI: Genetic analysis of streptomycin resistance in *Escherichia coli.* Genetics **65**, 9 (1970).

BRZEZINSKA, M., R. BENVENISTE, J. DAVIES, P.J.L. DANIELS, and J. WEINSTEIN: Gentamicin resistance in strains of *Pseudomonas aeruginosa* mediated by enzymatic N-acetylation of the deoxystreptamine moiety. Biochemistry **11**, 761 (1972).

CASKEY, C.T., A. L. BEAUDET, E. M. SCOLNICK, and M. ROSMAN: Hydrolysis of fMet-tRNA by peptidyl transferase. Proc. Natl. Acad. Sci. U.S. **68**, 3163 (1971).

CHANG, F. N., and J. G. FLAKS: The binding of dihydrostreptomycin to *E. coli* ribosomes: characteristics and equilibrium of the reaction. Antimicrobial Agents Chemotherapy **2**, 294 (1972a).

CHANG, F. N., and J. G. FLAKS: The binding of dihydrostreptomycin to *E. coli* ribosomes: kinetics of the reaction. Antimicrobial Agents Chemotherapy **2**, 308 (1972b).

CHUANG, D., and M.V. SIMPSON: A translocation-associated ribosomal conformational change detected by hydrogen exchange and sedimentation velocity. Proc. Natl. Acad. Sci. U.S. **68**, 1474 (1971).

COOPER, D.J., H.M. MARIGLIANO, M.D. YUDIS, and T. TRAUBEL: Recent developments in the chemistry of gentamicin. J. Infect. Diseases **119**, 342 (1969).

CREMER, K., and D. SCHLESSINGER: Polypeptide formation and polyribosomes in antibiotic-inhibited *Escherichia coli.* In preparation (1972).

DAVIES, J.: Structure-activity relationships among the aminoglycoside antibiotics. Antimicrobial Agents Chemotherapy **1967**, 297.

DAVIES, J.: Bacterial resistance to aminoglycoside antibiotics. J. Infect. Diseases **124**, 7 (1971).

DAVIES, J., R. BENVENISTE, K. KVITEK, B. OZANNE, and T. YAMADA: Aminoglycosides: biologic effects of molecular manipulation. J. Infect. Diseases **119** (Supplement), 351 (1969).

DAVIES, J., M. BRZEZINSKA, and R. BENVENISTE: R Factors: Biochemical mechanisms of resistance to aminoglycoside antibiotics. Ann. N.Y. Acad. Sci. **182**, 226 (1971).

DAVIES, J., W. GILBERT, and L. GORINI: Streptomycin, suppression, and the code. Proc. Natl. Acad. Sci. U.S. **51**, 883 (1964).

548　　　　　　　　　　　　　　　D. Schlessinger and G. Medoff:

Doi, O., M. Ogura, N. Tanaka, and H. Umezawa: Inactivition of kanamycin, neomycin, and streptomycin by enzymes obtained in cells of *Pseudomonas aeruginosa*. Appl. Microbiol. **16**, 1276 (1968).

Dubin, D.T., and B.D. Davies: The effect of streptomycin on potassium flux in *Escherichia coli*. Biochim. Biophys. Acta **52**, 400 (1961).

Dubin, D.T., R. Hancock, and B.D. Davis: The sequence of some effects of streptomycin in *Escherichia coli*. Biochim. Biophys. Acta **74**, 476 (1963).

Funatsu, G., and H.G. Wittmann: Ribosomal proteins XXXIII: Location of amino-acid replacements in protein S12 isolated from *Escherichia coli* mutants resistant to streptomycin. J. Mol. Biol. **68**, 547 (1972).

Gadó, I., and I. Horváth: The effect of methanol on the growth of a streptomycin-dependent strain of *Escherichia coli* in streptomycin-free media. Life Sci. **2**, 741 (1963).

Gorini, L.: Discussion in: Cold Spring Harbor Symp. Quant. Biol. **34**, 241 (1969).

Gorini, L., G.A. Jacoby, and L. Breckenridge: Ribosomal ambiguity: Cold Spring Harbor Symp. Quant. Biol. **31**, 657 (1966).

Gorini, L., and E. Kataja: Phenotypic repair by streptomycin of defective genotypes in *E. coli*. Proc. Natl. Acad. Sci. U.S. **51**, 487 (1964).

Gurgo, D., D. Apirion, and D. Schlessinger: Polyribosome metabolism in *Escherichia coli* treated with chloramphenicol, neomycin, spectinomycin or tetracycline. J. Mol. Biol. **45**, 205 (1969).

Harwood, J.H., and D.H. Smith: Resistance Factor – mediated streptomycin resistance. J. Bacteriol. **97**, 1262 (1969).

Herzog, A.: An effect of streptomycin on the dissociation of *Escherichia coli* 70S ribosomes. Biochem. Biophys. Res. Commun. **15**, 172 (1964).

Herzog, A., A. Ghysen, and A. Bollen: Sensitivity and resistance to streptomycin, relation with factor-mediated dissociation of ribosomes. FEBS Letters **15**, 291 (1971).

Igarashi, M., P.G. Lundquist, B.R. Alford, and H. Miyata: Experimental ototoxicity of gentamicin in squirrel monkeys. J. Infect. Diseases **124**, (Supplement) 114 (1971).

Jacoby, G.A., and L. Gorini: The effect of streptomycin and other aminoglycoside antibiotics on protein synthesis. In: Antibiotics, vol I, Mechanism of action, ed. by D. Gottlieb and P.D. Shaw, p. 726–747. Berlin-Heidelberg-New York: Springer 1967.

Jawetz, E., J.B. Gunnison, and R.S. Speck: Studies on antibiotic synergism and antagonism; the interference of aureomycin, chloramphenicol, and terramycin with the action of streptomycin, Am. J. Med. Sci. **222**, 404 (1951).

Kaji, H., and Y. Tanaka: Binding of dihydrostreptomycin to ribosomal subunits. J. Mol. Biol. **32**, 221 (1968).

Kogut, M., and E. Prizant: Effects of dihydrostreptomycin treatment *in vivo* on ribosome cycle in *Escherichia coli*. FEBS Letters **12**, 17 (1970).

Kuwano, M., H. Endo, and Y. Ohnishi: Mutations to spectinomycin resistance which alleviate the restriction of an amber suppressor by streptomycin resistance. J. Bacteriol. **97**, 940 (1969).

Lelong, J.C., M.A. Cousin, D. Gros, M. Grunberg-Manago, and F. Gros: Streptomycin induced release of fMet-tRNA from the ribosomal initiation complex. Biochem. Biophys. Res. Commun. **42**, 530 (1971).

Lelong, J.C., M.A. Cousin, F. Gros, R. Miskin, Z. Vogel, Y. Groner, and M. Revel: Protection of *Escherichia coli* ribosomes against streptomycin by purified initiation factors. Eur. J. Biochem. **27**, 174 (1972).

Leon, S.A., and T.D. Brock: Effect of streptomycin and neomycin on physical properties of the ribosome. J. Mol. Biol. **24**, 391 (1967).

Luzzatto, L., D. Apirion, and D. Schlessinger: Mechanism of action of streptomycin in *E. coli*: interruption of the ribosome cycle at the initiation of protein synthesis. Proc. Natl. Acad. Sci. U.S. **60**, 873 (1968).

Luzzatto, L., D. Apirion, and D. Schlessinger: Polyribosome depletion and blockage of the ribosome cycle by streptomycin in *Escherichia coli*. J. Mol. Biol. **42**, 315 (1969a).

Luzzatto, L., D. Apirion, and D. Schlessinger: Streptomycin action: greater inhibition of *Escherichia coli* ribosome function with exogenous than with endogenous messenger ribonucleic acid. J. Bacteriol. **99**, 206 (1969b).

Miskin, R., and A. Zamir: Effect of streptomycin on ribosome inter-conversion, a possible basis for the action of the antibiotic. Nature New Biol. **238**, 78 (1972).

MODOLELL, J., and B. D. DAVIS: Rapid inhibition of polypeptide chain extension by streptomycin. Proc. Natl. Acad. Sci. U.S. **61**, 1279 (1968).

MODOLELL, J., and B. D. DAVIS: Breakdown by streptomycin of initiation complexes formed on ribosomes of *Escherichia coli*. Proc. Natl. Acad. Sci. U.S. **67**, 1148 (1970).

NOVOGRADSKY, A.: Effect of streptomycin on G-factor dependent ribosomal GTPase activity. Biochem. Biophys. Res. Commun. In press (1972).

OGAWA, K., A. HIRASHIMA, and A. KAJI: Factor dependent release of ribosomes at the termination signal. In preparation (1972).

OKUYAMA, A., N. MACHIYAMA, T. KINOSHITA, and N. TANAKA: Inhibition by kasugamycin of initiation complex formation on 30S ribosomes. Biochem. Biophys. Res. Commun. **43**, 196 (1971).

OZAKI, M., S. MIZUSHIMA, and M. NOMURA: Identification and functional characterization of the protein controlled by the streptomycin-resistant locus in *E. coli*. Nature **222**, 333 (1969).

PESTKA, S., R. MARSHALL, and M. NIRENBERG: RNA codewords and protein synthesis. V. Effect of streptomycin on the formation of ribosome-sRNA complexes. Proc. Natl. Acad. Sci. U.S. **53**, 639 (1965).

PETITPAS-DEWANDRE, A., H. BARBASON et W. G. VERLY: Affinite pour la streptomycine des ribosomes *Escherichia coli*. Eur. J. Biochem. **7**, 307 (1969).

RIFF, L. J., and G. G. JACKSON: Pharmacology of gentamicin in man. J. Infect. Diseases **124**, (Supplement) 98 (1971).

ROSSELET, J. P., J. MARQUEZ, E. MESECK, A. MURAWSKI, A. HAMDAN, C. JOYNER, R. SCHMIDT, D. MIGLIORE, and H. L. HERZOG: Isolation, purification, and characterization of gentamicin. Antimicrobial Agents Chemotherapy **1963**, 14.

ROSSET, R., and L. GORINI: A ribosomal ambiguity mutation. J. Mol. Biol. **39**, 95 (1969).

SCHATZ, A., E. BUGIE, and S. A. WAKSMAN: Streptomycin, a substance exhibiting antibiotic activity against Gram-positive and Gram-negative bacteria. Proc. Soc. Exptl. Biol. Med. **55**, 66 (1944).

SCHATZ, A., and S. A. WAKSMAN: Effect of streptomycin and other antibiotic substances upon *Mycobacterium tuberculosis* and related organisms. Proc. Soc. Exptl. Biol. Med. **57**, 244 (1944).

SHERMAN, M. I., D. SCHUENG, and M. SIMPSON: Streptomycin and ribosome conformation changes. Cold Spring Harbor Symp. Quant. Biol. **34**, 109 (1969).

SHULMAN, J. A., P. M. TERRY, and C. E. HOUGH: Colonization with gentamicin-resistant *Pseudomonas aeruginosa*, Pyocine type 5, in a burn unit. J. Infect. Diseases **124**, (Supplement) 18 (1971).

SNELLING, C. F. T., A. R. RONALD, C. Y. CATES, and W. C. FORSYTHE: Resistance of gram-negative bacilli to gentamicin. J. Infect. Diseases **124**, (Supplement), 264 (1971).

SPARLING, P. F., J. MODOLELL, Y. TAKEDA, and B. D. DAVIS: Ribosomes from *Escherichia coli* merodiploids heterozygous for resistance to streptomycin and spectinomycin. J. Mol. Biol. **37**, 407 (1968).

STERN, J. L., H. D. BARNER, and S. S. COHEN: The lethality of streptomycin and the stimulation of RNA synthesis in the absence of protein synthesis. J. Mol. Biol. **17**, 188 (1966).

TANAKA, N.: Biochemical studies on gentamicin resistance. J. Antibiotics (Tokyo) **23**, 469 (1970).

UMEZAWA, H., S. TAKASAWA, M. OKANISHI, and R. UTAHARA: Adenylyl-streptomycin, a product of streptomycin inactivated by *E. coli* carrying R Factor. J. Antibiotics (Tokyo) **21**, 81 (1968).

WALLACE, B. J., and B. D. DAVIS: Cyclic blockade of initiation sites by streptomycin damaged ribosomes in *Escherichia coli*: An explanation for dominance of sensitivity. J. Mol. Biol. (1973), in press.

WALLACE, B. J., P. C. TAI, and B. D. DAVIS: Effects of streptomycin on the response of *Escherichia coli* ribosomes to the dissociation factor. J. Mol. Biol. (1973), in press.

WEINSTEIN, L.: Streptomycin. In: L. S. GOODMAN and A. GILMAN, ed., The pharmacological basis of therapeutics, chap. 58. London: The Macmillan Co. 1971.

WEINSTEIN, M. J., G. M. LUEDEMANN, E. M. ODEN, and G. H. WAGMAN: Gentamicin, a new broad spectrum antibiotic complex. Antimicrobial Agents Chemotherapy **1963**, 1.

WERSALL, J., P. G. LUNDQUIST, and B. BJORKROTH: Ototoxicity of gentamicin. J. Infect. Diseases **119**, 410 (1969).

WHITE, J. R., and H. L. WHITE: Streptomycinoid antibiotics: synergism by puromycin. Science **146**, 772 (1964).

WITTMANN, H. G., G. STOFFLER, I. HINDENNACH, C. G. KURLAND, E. A. BIRGE, L. RANDALL-HAZELBAUER, M. NOMURA, E. KALTSCHMIDT, S. MIZUSHIMA, R. R. TRAUT, and T. A. BICKLE: Correlation of 30S ribosomal proteins of *Escherichia coli* isolated in different laboratories. Mol. Gen. Genet. **111**, 327 (1971).

Wolfe, A. D., and F. E. Hahn: Stability of ribosomes from streptomycin-exposed *Escherichia coli*. Biochem. Biophys. Res. Commun. **31**, 945 (1968).

Wolfgang, R. W., and N. L. Lawrence: Binding of streptomycin by ribosomes of sensitive, resistant and dependent *Bacillus megaterium*. J. Mol. Biol. **29**, 531 (1967).

Yamada, T., D. Tipper, and J. Davies: Enzymatic inactivation of streptomycin by R-factor resistant *Escherichia coli*. Nature **219**, 288 (1968).

Zamir, A., R. Miskin, and D. Elson: Inactivation and reactivation of ribosomal subunits: amino acyl-transfer RNA binding activity of the 30S subunit of *Escherichia coli*. J. Mol. Biol. **60**, 347 (1971).

Zimmermann, R. A., R. C. Moellering, Jr., and A. N. Weinberg: Mechanism of resistance to antibiotic synergism in enterococci. J. Bacteriol. **105**, 873 (1971).

Zitomer, R. S., and J. G. Flaks: The nature of streptomycin dependency. Federation Proc. **31**, 456A, abst. No. 1353 (1972).

The Thiostrepton Group of Antibiotics

Sidney Pestka and James W. Bodley

The antibiotics thiostrepton, siomycin, A-59, multhiomycin, thiopeptin, and sporangiomycin have been included in the thiostrepton group on the basis of similarities in antibacterial activity, mode of action, chemical and physical properties, and structure. However, since studies on the mode of action of some of these antibiotics are only preliminary and since the entire structure of thiostrepton only has been reported, this grouping may be premature. Nevertheless, as will become apparent in this chapter these antibiotics exhibit sufficient similarities to warrant inclusion in a single category at this time.

The thiostrepton group of antibiotics consists of several sulfur-containing peptide antibiotics with similar chemical and biological properties. Although the structure of only one has been reported it is likely they will probably have many structural features in common. They also have some structural and biological properties in common with the micrococcins (see Chapter Micrococcin and Micrococcin P, in this volume). Some effects of this group of antibiotics on protein synthesis have been reviewed (PESTKA, 1971, 1973 a, b).

Isolation and Preparation of the Antibiotics

Thiostrepton was isolated from the fermentation broth inoculated with *Streptomyces azureus* (PAGANO *et al.*, 1956). The antibiotic bryamycin was isolated independently from *Streptomyces hawaiiensis* (CRON *et al.*, 1956). It appears that the antibiotics thiostrepton and bryamycin (or thiactin) are identical (BODANSZKY, *et al.*, 1963). Because of its low solubility, at least 95 percent of the antibiotic remains with the mycelium when this is separated from the fermentation broth. The antibiotic can be extracted from the mycelial cake by extraction into chloroform (VANDEPUTTE and DUTCHER, 1956). The thiostrepton can then be further purified by repeated crystallization and column chromatography over alumina. The presence of metal cations in the fermentation broth stimulated the production of thiostrepton (PLATT and FRAZIER, 1962). Particulary effective were Fe^{++}, CO^{++}, and Mn^{++}; their stimulatory effects were additive. In the presence of all three cations at 1 mM up to 980 mg of thiostrepton per liter could be produced.

Siomycin is produced by *Sreptomyces sioyaensis* (NISHIMURA *et al.*, 1961). Thiopeptin B was discovered in the fermentation broth of a species subsequently named *Streptomyces tateyamensis* (MIYAIRI *et al.*, 1970). These antibiotics are similar to thiostrepton in chemical, physical and biological properties. Multhiomycin, produced by *Streptomyces antibioticus,* also has some similarities to the above antibiotics (TANAKA *et al.*, 1970c). Further studies should clarify whether multhiomycin belongs to the thiostrepton group. The antibiotic A-59 (KONDO et al., 1961) has been considered identical to siomycin (Umezawa, 1967), but this is yet uncertain. An antibiotic isolated from Actinomyces (*Planomonospora parontospora*), sporangiomycin, also appears to be related to thiostrepton (Thie-mann et al., 1968).

Chemical and Physical Properties

Thiostrepton is a relatively stable molecule. Antibiotic activity remained stable in peptone broth for at least 4 days at 37°; and no appreciable activity was lost by incubation at room temperature in intestinal juice, gastric juice, blood plasma or water (Pagano et al., 1956). The antibiotic melts with decomposition over the range 246 to 256° C (Vandeputte and Dutcher, 1956). It is soluble in chloroform, dioxane, pyridine, glacial acetic acid, dimethylformamide and dimethylsulfoxide. It is insouble in water, lower alcohols and nonpolar organic solvents such as hexane or benzene. It is not dissolved in dilute aqueous acid or base. The ultraviolet spectrum presents no maxima, but shows characteristic shoulders at 225, 250 and 280 mμ with $E_1^{1\%}$ equal to 520, 380 and 225, respectively. A strong biuret reaction indicates the presence of peptide linkages; and hydrolysis in 6 N HCl results in products which react with ninhydrin. Thiostrepton itself does not give a positive ninhydrin reaction. The antibiotic appears to be a modified peptide containing a large proportion of sulfur held in thiazole rings (Vandeputte and Dutcher, 1956; Bodanszky et al., 1964). Sulfur comprises about 7.4 percent of the antibiotic by weight. The products of acid hydrolysis included a number of derivatives of thiazole-4-carboxylic acid as well as pyruvic acid, several amino acids, and 4-(α-hydroxyethyl)-8-hydroxyquinaldic acid (Bodanszky et al., 1964; Cross et al., 1963; Kenner et al., 1960; Barton et al., 1966). The molecular weight of thiostrepton has been estimated to be 1650 (Bodanszky et al., 1964) or 1613 (Cross et al., 1963). Other reports dealing with the structures of some of the degradation products have been published (Bodanszky et al., 1960; Drey et al., 1961; Bodanszky et al., 1962; Bodanszky et al., 1970).

Siomycin A contains 8.4 percent sulfur and melts with decomposition at approximately 255 to 260° C (Nishimura et al., 1961). Its solubility is similar to that of thiostrepton. As with thiostrepton, the biuret test is positive; the ninhydrin reaction is negative with intact siomycin A, but its acid hydrolysate gives a strong ninhydrin reaction. The ultraviolet spectrum shows no maxima and is similar to that of thiostrepton. At 250 mμ the molar extinction coefficient is about 5×10^4 based on a molecular weight of 1712 (Ebata et al., 1969a). The specific rotations of thiostrepton and siomycin A differ significantly as do their chromatographic migrations in methanol:acetic acid:water (25:3:72). Sio-mycin shows an Rf of 0.12 to 0.14, whereas thiostrepton migrates faster, its characteristic Rf being 0.38 to 0.40. Siomycin was found to consist of one major (A) and two minor components (B and C). Siomycin B is derived from siomycin A during storage whereas siomycin C is a natural product of *Streptomyces sioyaensis* (Ebata et al., 1969a). The content of siomycins A, B, and C in the usual crude siomycin preparations is 98%, 1.2–2%, and 0.4–1%, respectively. Siomycins A, B, and C have approximately similar antibacterial spectra. Products of acid hydrolysis of siomycin A included amino acids, pyruvic acid, several derivatives of thiazole-4-carboxylic acid and 4-(α-hydroxyethyl)-8-hydroxyquinaldic acid similar to the acid degradation products of thiostrepton (EBATA et al., 1969 b). Thiostreptine was also isolated as a hydrolytic product. Preliminary studies on the sequence and association of the components in siomycin A have also been reported (EBATA et al., 1969 c). Monochloroacetyl, Monoiodoacetyl, and tetraacetyl siomycin A have been prepared for X-ray analysis of crystalline siomycin A (EBATA et al., 1969 d). Although the tetraacetyl siomycin A had no detectable antimicrobial activity, the monoacetylated derivatives had about 1 to 2% of the activity of unsubstituted siomycin A. Additionally, soluble derivatives of siomycin A have been prepared by reaction with thiolcarboxylic acids (thioglycolic or thiomalic acid). The sodium salts of the derivatives were remarkably soluble in water, whereas siomycin A is remarkably insoluble (EBATA et al., 1969 c). These soluble derivatives exhibited antibacterial spectra similar to the parent siomycin A, but they were approximately one-fifth as active as the parent compound. From chemical and physical properties antibiotic A-59 appears identical to siomycin A (KONDO et al., 1961).

Thiopeptin B, the main component of thiopeptin antibiotics, is a faint yellowish crystalline material having a decomposition point between 219 and 222° C. Its molecular weight was estimated to be 1782 to 1942. Elementary analysis showed a content of 10.8% sulfur (MIYAIRI et al., 1970). The ultraviolet spectrum of thiopeptin B is distinctly different from that of thiostrepton and siomycin A, showing a broad almost flat shoulder from 230 to 250 mμ and minor shoulders at 295 mμ and 305 mμ. At 240 mμ its absoption $E_{1\,cm}^{1\%}$ is about 430. Its chromatographic migration in several solvents is also distinctly different from those of siomycin A and thiostrepton (MIYAIRI et al., 1970). Its solubility characteristics are similar to thiostrepton and siomycin A; and like the other two antibiotics it exhibits a negative ninhydrin reaction, but its acid hydrolysate is strongly ninhydrin positive. Thiopeptin B is stable on treatment at 60° C for 1 hour at pH values in the range of 2 to 8. It exhibits optical activity.

Multhiomycin is a yellow crystalline compound which melts at greater than 300° C. In contrast to the other antibiotics of this group it shows little or no optical activity. It is soluble in dimethylformamide, pyridine, dimethylsulfoxide; slightly soluble in ethyl acetate, methanol, ethanol, dioxane; and insoluble in water and glacial acetic acid. Ultraviolet and visible spectra show maxima at 328 mμ ($E_{1\,cm}^{1\%}$, 220) and a shoulder at 420 mμ ($E_{1\,cm}^{1\%}$, 20) in neutral or acid methanol; and maxima at 292 mμ ($E_{1\,cm}^{1\%}$, 255) and at 406 mμ ($E_{1\,cm}^{1\%}$, 132) in alkaline methanol. Its molecular weight was estimated to be 1043 (TANAKA et al., 1970). Sulfur content was determined to be 15.1%. It shows a negative ninhydrin test, but its acid hydrolysate is ninhydrin positive.

Sporangiomycin is a white, crystalline material melting between 260 and 262° C. It is insoluble in water, acetone, dichloromethane, methanol, and higher alcohols; it is slightly soluble in chloroform; and it is soluble in dimethylformamide, dimethylsulfoxide, and glacial acetic acid. About 10.5% of the antibiotic consists of sulphur (THIEMANN et al., 1968). The molecular weight of sporangiomycin was estimated to be 1 800 ± 100. The chromatographic behavior of sporangiomycin indicates that it is distinct from thiostrepton. As most of the thiostrepton antibiotics, sporangiomycin yields a positive biuret reaction, but the ninhydrin reaction is positive only after acid hydrolysis. In methanol or aqueous solutions, sporangiomycin does not show any characteristic ultraviolet absorption spectrum; however, in 6 N sulfuric acid, it shows an absorption peak at 318 mμ. Table 1 summarizes the molecular weights and selected ultraviolet absorption values for these antibiotics.

Table 1. *Molecular weights and ultraviolet absorption of the thiostrepton antibiotics*

Antibiotic	Molecular weight	$E_{1cm}^{1\%}$ at λ		ε
Thiostrepton	1613–1650	520	225 mμ	8.4×10^4
Siomycin A	1712–1740	292	250 mμ	5×10^4
Thiopeptin B	1782–1942	430	240 mμ	7.7×10^4
Multhiomycin	1043	292	255	3.0×10^4
A-59	2000	340	245	6.8×10^4
Sporangiomycin	1700–1900	–	–	–

The molar extinction coefficient (ε) is based on the lower value of the molecular weight given in the Table. References for the values presented in this table are given in the text.

Structure

Of these antibiotics, the complete structure of thiostrepton only has been reported (Anderson *et al.*, 1970). Although the structure given in Fig. 1 has been derived from X-ray crystallographic techniques, there may be some doubt about the nature of the five terminal atoms of the long side chain. Intramolecular forces stabilize the main part of the molecule, but the linear part extends into disordered solvent which provides little to constrain the atoms of the chain. In this region, therefore, bond lengths, angles, and atom type cannot be determined at all precisely and the final five atom fragment has been characterized from the chemistry as much as from the crystallography.

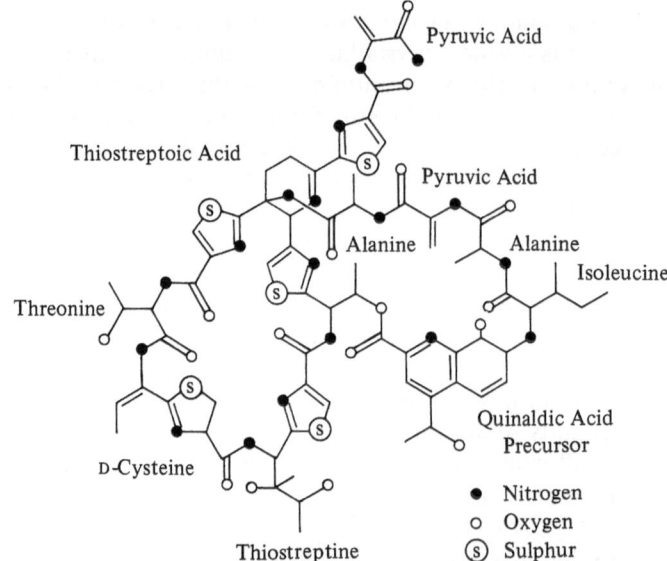

Fig. 1. Structure of thiostrepton (taken from the paper of Anderson *et al.*, 1970)

Effects on Eucaryotes

Siomycin A showed little or no inhibitory activity on hemoglobin synthesis in rabbit reticulocytes or on protein synthesis in cell-free extracts from reticulocytes (TANAKA et al., 1970a). However, thiostrepton at a concentration of 0.07 µg/ml produced a 50% inhibition of multiplication of mouse fibroblasts (L-cells) in suspension culture (PERLMAN et al., 1959). In addition, thiostrepton had some inhibitory effect on the growth of one species of alga, but not on another (PERLMAN, 1965): thiostrepton inhibited the growth of Scenedesmus obliquus, but not Chlorella pyrenoidosa. Sporangiomycin had little or no effect on polypeptide synthesis in cell-free extracts from Saccharomyces cerevisiae (PIRALI et al., 1972).

Antimicrobial Properties

Table 2 presents the minimal inhibitory concentrations of the various thiostrepton antibiotics against several test organisms. The data are a compilation

Table 2. *Minimal inhibitory concentrations of antibiotics against various microorganisms*

Test Organisms	Sio-mycin (µg/ml)	Thio-strep-ton (µg/ml)	Thio-peptin B (µg/ml)	Multhio-mycin (µg/ml)	A-59 (µg/ml)	Sporangio-mycin (µg/ml)
Staphylococcus aureus 209P	0.05	0.1	0.12	0.01	<0.02	0.02–0.08
B. subtilis	0.05	0.1	0.25	0.2	<0.02	0.04
S. lutea	0.05	0.05	0.03	0.006	0.04	–
B. anthracis	0.2	0.2	–	–	–	–
Diplococcus pneumoniae, type I	0.005	0.01	0.02	–	–	–
Streptococcus hemolyticus	0.01	0.02	0.05	–	–	–
Streptococcus faecalis	–	–	–	–	–	0.04
Corynebacterium diphtheriae	0.005	0.01	–	–	–	–
Corynebacterium xerosis	–	–	0.25	0.1	–	–
Mycobacterium tuberculosis var. hominis, H37Rv	2.0	5.0	–	–	–	–
Mycobacterium phlei	2.0	2.0	50	>100	5.0	0.6
Mycobacterium avium	20.0	20.0	50	–	–	–
Mycobacterium 607	10.0	>20.0	–	>100	5.0	–
Candida albicans	>20.0	>20.0	>128	–	>40	>100
Escherichia coli	>20.0	>20.0	>128	>100	>40	>100
Salmonella typhosa	>20.0	>20.0	–	–	>40	–
Salmonella paratyphi A	>20.0	>20.0	–	–	>40	–
Shigella dysenteriae	>20.0	>20.0	–	–	>40	–
Shigella paradysenteriae, Ohara	>20.0	>20.0	–	–	–	–

Minimal inhibitory concentration determined by the agar plate dilution method. References are given in the text.

from many sources (Pagano et al., 1956; Nishimura et al., 1961; Miyairi et al., 1970; Tanaka et al., 1970; Thiemann et al., 1968; Kondo et al., 1961). In general, Gram-positive bacteria are very sensitive to the antibiotics. Gram-negative organisms and yeasts are naturally resistant to these agents. Growth of sensitive organisms often was inhibited by extremely low concentrations of thiostrepton (Kelly et al., 1959a, b; Kutscher et al., 1958; Kutscher et al, 1959a). As will be pointed out later, these antibiotics are unable to enter gram-negative cells, but they strongly act in cell-free extracts derived from these cells. They are not effective against eucaryotic cells or cell-free extracts derived from eucaryotic cells. In this latter case the antibiotics appear unable to interact with eucaryotic ribosomes (Richter et al., 1971).

Antibiotic Resistance

Thiostrepton resistant mutants of *Bacillus subtilis* appear to be genetically linked to other ribosomal markers such as streptomycin resistance (Dubnau et al., 1967; Goldthwaite and Smith, 1972; Smith and Smith, 1973).

Mutants resistant to thiostrepton show cross-resistance to siomycin and conversely siomycin-resistant mutants are resistant to thiostrepton (Nishimura et al., 1961). Furthermore, it is of interest that thiostrepton-resistant mutants show cross-resistance to micrococcin; but micrococcin-resistant mutants of *B. subtilis* do not demonstrate cross-resistance to thiostrepton (Goldthwaite and Smith, 1972). This might be explained by the binding sites of these antibiotics on ribosomes. Micrococcin is a smaller molecule than thiostrepton and appears to bind to ribosomes with less affinity than does thiostrepton judging from the inhibition of protein synthesis as a function of antibiotic concentration (Pestka and Brot, 1971). It is thus possible that the micrococcin binding site overlaps that of thiostrepton which includes the micrococcin site. Ribosomal mutants which no longer bind thiostrepton would no longer bind micrococcin. Changes in ribosomal structure in micrococcin-resistant mutants preventing binding of micrococcin may not sufficiently decrease the affinity of the very tight binding thiostrepton. Studies of binding of these antibiotics to ribosomes of appropriate mutant strains should allow examination of these possibilities directly.

It is noteworthy that although Gram-negative bacteria are generally resistant to this group of antibiotics, cell-free extracts from these bacteria are sensitive. With the use of appropriate mutants, it may be possible to obtain ribosomes from *E. coli*, resistant to this antibiotic group. In fact, protein synthesis in an actinomycin sensitive strain of *E. coli* (AS-19) was found to be sensitive to high concentrations of thiostrepton (Elsebet Lund, personal communication; Smith and Pestka, unpublished studies).

Resistance to thiostrepton developed readily (Pagano et al., 1956; Kelly et al., 1960). Since no cross-resistance with penicillin, erythromycin, carbomycin, streptomycin, neomycin, chloramphenicol, novobiocin, and tetracycline was observed, the thiostrepton binding site on the ribosome is distinct from the sites binding these other antibiotics.

In Vivo Studies and Clinical Aspects

Although the Gram-positive bacteria are highly sensitive to these antibiotics, the *in vivo* action and clinical use of these antibiotics have been limited by their insolubility. Subcutaneous injection of thiostrepton into mice protected them from streptococcal infection (STEINBERG *et al.*, 1956). Siomycin A injection protected mice from pneumococcal infection (NISHIMURA *et al.*, 1961). Generally, these agents were effective in protecting mice at levels in the range of about 1 mg/kgm/day. However, they were not effective systemically through oral administration. These antibiotics are fairly non-toxic. Mice were found to have an LD_{50} of 100 mg/kg for siomycin A on intravenous injection and were able to tolerate at least 0.5 gm/kg of thiopeptin B intraperitoneally without toxic symptoms (NISHIMURA *et al.*, 1961; MIYAIRI, 1970). In the case of sporangiomycin subcutaneous injection of 1 gm/kg did not produce any toxic symptoms in mice (THIEMANN *et al.*, 1968).

Clinical studies of thiostrepton in humans after topical oral administration were found to produce no toxic side effects (KUTSCHER *et al.*, 1959b). After 10 mg of thiostrepton in a troche was dissolved in the mouth, significant salivary levels of thiostrepton beyond the minimal inhibitory concentrations for streptococcal and staphylococcal growth were found (BERMAN *et al.*, 1959). Although thiostrepton is very insoluble, a solubilized form of thiostrepton was developed (KUTSCHER *et al.*, 1961). This consists of thiostrepton desoxycholate. The thiostrepton desoxycholate can be supplied as a lyophilized powder which produces a very fine colloidal dispersion upon reconstitution in water. This colloidal suspension approaches a solution in clarity and yields suspensions of thiostrepton up to 10 mg/ml.

It has been suggested that thiostrepton might be useful in the treatment of some staphylococcal infections (WISE, 1958). Thiostrepton ointments and sprays, however, were ineffective in preventing or eradicating nasal carrier states for staphylococci (LEPPER *et al.*, 1963). Although thiostrepton was not recommended for use as an intestinal antiseptic prior to colon surgery, effective intestinal antisepsis was observed in combination with neomycin (COHN and LONGACRE, 1957; COHN, 1958, 1960). In any case, thiostrepton or other members of this group so far have not found a place in clinical medicine. Thiostrepton, however, is used in veterinary medicine in the treatment of bovine mastitis (LEVIN *et al.*, 1960).

Mode of Action*

Inhibition of Protein Synthesis

Many antibiotics function by interfering with some aspects of macro-molecular formation. Moreover, these processes and the effects of drugs on them are relatively easy to investigate with the use of radioactive precursors. On the basis of such experiments it was apparent that thiostrepton (WEISBLUM and DEMOHN, 1970a),

---

* Ensuing discussion of the mode of action of these antibiotics will follow more or less historical lines. For a general review of protein synthesis see HASELKORN and ROTHMAN-DENES (1973).

siomycin (Tanaka et al., 1970a), multhiomycin (Tanaka et al., 1970d) and sporangiomycin (Pirali et al., 1972) all rapidly and completely stop the polymerization of amino acids in cells while producing neither an immediate nor profound effect on either DNA or RNA synthesis. The studies of Weisblum and Demohn (1970a) on the effect of thiostrepton on [^{14}C]thymidine and [^{14}C]leucine polymerization (measures of DNA and protein synthesis, respectively) are shown in Figs. 2 and 3. On the basis of these and similar findings it was concluded that the inhibition of protein synthesis constituted the primary if not sole cause of the cessation of bacterial growth produced by these antibiotics.

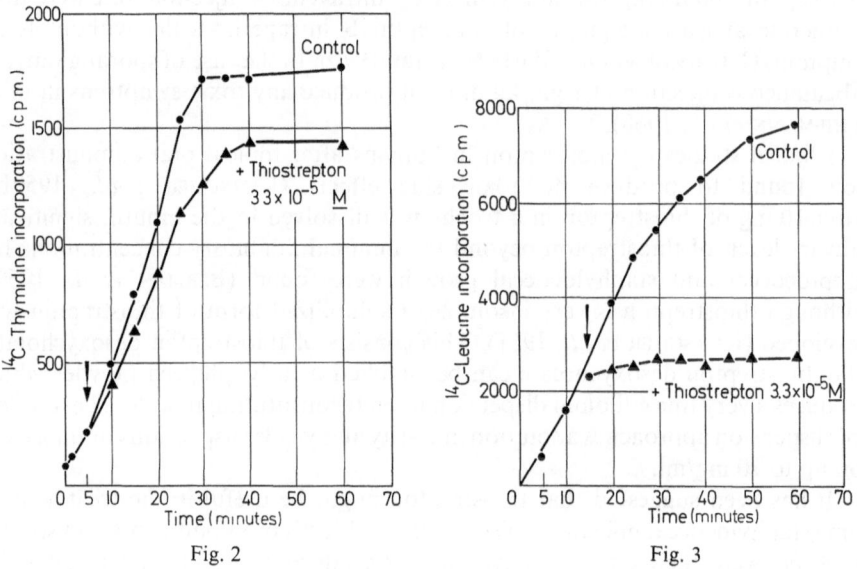

Fig. 2   Fig. 3

Fig. 2. Effect of thiostrepton on incorporation of [^{14}C]thymidine by B. megaterium. Data taken from Weisblum and Demohn (1970a)

Fig. 3. Effect of thiostrepton on incorporation of [^{14}C]leucine by B. megaterium. Data taken from Weisblum and Demohn (1970a)

Inhibition of Protein Chain Elongation

The synthesis by E. coli components of polyphenylalanine in response to the synthetic messenger poly U is a convenient and widely used manifestation of the reactions involved in protein chain growth. While this reaction does not embody all of the steps necessary for the complete assembly of proteins, many antibiotics which inhibit protein synthesis interfere with some aspect of chain growth (elongation) and hence inhibit polyphenylalanine formation. After observing the in vivo inhibition of protein synthesis by thiostrepton in B. megaterium, it was, therefore, natural for Weisblum and Demohn (1970a) to examine the effect of the antibiotic on polyphenylalanine formation in an E. coli cell-free system. A portion of their data is shown in Table 3 (line A) and it is clear that this reaction is inhibited more than 90%. Therefore, thiostrepton (as well

Table 3. *Effects of thiostrepton on polyphenylalanine synthesis by cell-free extracts of E. coli*

Experimental Condition	Percent inhibition by thiostrepton
A. Unfractionated S-30	94
B. S-30 dialyzed after exposure to thiostrepton	95
C. Ribosomes and S-100 derived from thiostrepton-treated S-30:	
S-100	0
ribosomes	96
D. S-30 dialyzed after exposure to thiostrepton and assayed in the presence of added:	
30S subunits	91
50S subunits	38

Data taken from Table 1 of WEISBLUM and DEMOHN (1970a).

as siomycin A and multhiomycin) can be classified as an inhibitor of protein chain elongation.

At the same time WEISBLUM and DEMOHN made a number of other important observations concerning the mode of thiostrepton action. First of all they found that the effect on the *E. coli* components was irreversible in that after exposure to the antibiotic, activity could not be restored by dialysis (Table 3, line B).

The 50S Ribosomal Subunit as Site of Action

It then was of interest to determine which of the many macromolecules, necessary for protein elongation, was the target of thiostrepton. High-speed centrifugation of the crude polymerizing system (S-30, 30,000 xg supernatant fraction) readily separates ribosomes from "soluble" macromolecules (S-100, 100,000 xg supernatant fraction), both of which are necessary for polymerization. When a thiostrepton-treated S-30 is fractionated in such a way (Table 3, line C), only the ribosomes are inactive.

Similarly, ribosomes can be fractionated into two subunits with sedimentation coefficients of 50S and 30S. In order to determine if only one of these was being inactivated by thiostrepton, WEISBLUM and DEMOHN attempted to restore polymerizing activity to a dialyzed, thiostrepton-inactivated S-30. As can be seen in Table 3 (line D), this activity is specifically restored by the addition of 50S subunits which therefore must contain the site of thiostrepton action. Similar observations with siomycin were reported by TANAKA et al. (1970a) and TANAKA et al. (1970b); and with sporangiomycin by PIRALI et al. (1972). Additionally, since thiostrepton inhibited non-enzymic translocation, a ribosomal function, it was clear that it inhibited the ribosome rather than some supernatant factor (PESTKA, 1970). This was confirmed also by GOLDTHWAITE and SMITH (1972) by studying protein synthesis in cell-free systems from thiostrepton-sensitive and -resistant strains. When ribosomes from thiostrepton-resistant *B. subtilis* strains were present, the cell-free extracts were resistant to thiostrepton no matter what the source of the supernatant fraction.

Transpeptidation

Protein chain elongation involves several discrete reactions and the ability of the ribosome to participate in each of these can be assessed in a variety of ways. The formation of the peptide bond (transpeptidation) is one of the simplest of these reactions to measure for it requires no non-ribosomal components beyond proper substrates and no energy input beyond that present in the ester linkage by which the amino acid (or nascent peptide) is attached to tRNA. One of the simplest manifestations of transpeptidation is the reaction between a ribosome-bound (at the donor site) peptidyl-tRNA (or peptidyl-tRNA analogue such as N-acetyl-phenylalanyl-tRNA) and the antibiotic puromycin which serves as an analogue of aminoacyl-tRNA and can accept the carboxyl terminus of the nascent peptide in the transfer reaction. Pestka (1970) demonstrated that thiostrepton is without effect on this model reaction and these data are shown in Table 4.

Table 4. *Effect of thiostrepton on N-acetyl-phenylalanyl-puromycin synthesis*

Thiostrepton	N-Acetyl-Phe-Puro (pmoles)	Percent
Absent	0.47	100
Present	0.44	94

Data from Table 2 of Pestka (1970).

A more physiological representation of peptidyl transfer is provided by the reaction of puromycin with the nascent peptides of polysomes derived from growing cells and this reaction, too, as seen in Table 5, is unaffected by levels of thiostrepton which completely inhibit overall protein chain growth. Similar negative results employing a number of model reactions of peptidyl transfer have also been obtained with siomycin A (Tanaka et al., 1970a, b), multhiomycin (Tanaka et al., 1971a and b), and sporangiomycin (Pirali et al., 1972). The binding of acceptor substrates such as C-A-C-C-A(Phe) to ribosomes is related to transpeptidation. Thiostrepton had no effect on the binding of C-A-C-C-A(Phe) to 50S subunits (Hishizawa and Pestka, 1971). Therefore, so far as can be determined, these antibiotics have no direct effect on this reaction.

Table 5. *Effects of antibiotics on peptidyl-puromycin formation by E. coli polysomes*

Antibiotic	Percent of control		
	10^{-6} M	10^{-5} M	10^{-4} M
Sparsomycin[a]	44	16	1
Thiostrepton	97	106	105
Siomycin A	107	105	104

[a]Sparsomycin is a known inhibitor of peptidyl transferase. Data taken from Table 1 of Pestka (1972).

Translocation

As a result of transpeptidation, peptidyl-tRNA is unable to react with puromycin and indeed is visualized to occupy the physical region on the ribosome (the acceptor site) with which puromycin interacts. In order for protein elongation to proceed, the peptidyl-tRNA must be repositioned in the donor site and the ribosome moved along the message so as to be able to "read" the codon corresponding to the next amino acid. (An alternative view is presented by PESTKA, 1972.) This reaction, known as enzymic translocation, normally requires the participation of the non-ribosomal protein, elongation factor G (EF-G), and GTP which in the course of the reaction is hydrolyzed to GDP and Pi. Non-enzymic translocation (PESTKA, 1969) can also occur in the absence of EF-G and GTP, however, this reaction is non-physiological and substantially less efficient than its enzymic counterpart.

The interactions of translocation can be followed in a variety of ways in attempting to determine if an antibiotic interferes with this process. Perhaps the simplest way is to examine the effect of the antibiotic on the hydrolysis of GTP mediated by EF-G and the ribosome. As shown by PESTKA (1970) and as seen in Table 6 this process is essentially abolished by the presence of thiostrepton. Siomycin also inhibits GTP hydrolysis (WATANABE and TANAKA, 1971) as does sporangiomycin (TIBONI and CIFERRI, 1971). Potential inhibitors of translocation may also be examined with respect to their effect on the movement of peptidyl-tRNA from the ribosomal A site to the P site as brought about by the action of EF-G and GTP. Acetyl-phenylalanyl-tRNA in the presence of poly U and under appropriate conditions binds to both ribosomal sites, but in the absence of EF-G and GTP only that which binds in the P site is capable of reacting with puromycin to form acetyl-phenylalanyl-puromycin. As demonstrated by TANAKA et al. (1970b) and shown in Table 7, siomycin A is without

Table 6. *Effect of thiostrepton on uncoupled GTP hydrolysis*

	GTP hydrolyzed (pmoles)	Percent
Control	183	100
Plus thiostrepton	15	8

Data taken from Table 3 of PESTKA (1970).

Table 7. *Effect of siomycin A on the factor G-dependent puromycin reaction*

EF-G	N-Acetyl-^{14}C-phenylalanyl-puromycin (cpm)	
	Control	Plus Siomycin A
Absent	2880	3168
Present	6173	4393

Data taken from Tables 3 and 4 of TANAKA et al. (1970b).

effect on the puromycin reaction which occurs in the absence of EF-G and GTP (as already shown it does not inhibit transpeptidation), but greatly suppresses the increment in this reaction which is brought about by these components. These results again identify thiostrepton as a translocation inhibitor.

Normally, as has been indicated, translocation as defined by the movement of peptidyl-tRNA from the A to the P site, requires the participation of EF-G and GTP. However, under appropriate conditions the ribosome alone can be induced to perform this transition (non-enzymic translocation). Since this process proceeds in the absence of EF-G it presumably reflects some aspect of the ribosomal component of the reaction. If translocation is prohibited, amino acid polymerization stops with the production of dipeptide, but when non-enzymic translocation is allowed, oligopeptides are formed. Although the oligopeptides contain more than two amino acids under most conditions, they are substantially shorter than those produced in the presence of EF-G and GTP. Consequently, the formation of oligopeptides in the absence of EF-G can be taken as a measure of some portion of the ribosomal contribution to translocation and as shown by Pestka (1968, 1969, 1970, and 1973 a) this reaction as well as that which occurs in the presence of EF-G is inhibited effectively by thiostrepton (Fig. 4). Also, Pestka and Hintikka (1971) demonstrated that thiostrepton and siomycin freeze ribosomes in polyribosomes. Thus, these antibiotics inhibit ribosomal movement along the messenger RNA template, another feature of the translocation process.

From what has been described, it can be concluded that these antibiotics interact and probably bind tightly to the 50S ribosomal subunit and in so doing destroy its ability to participate in translocation either in the presence or absence of EF-G.

Prior to the observations described above only one other class of antibiotics was known to inhibit protein synthesis by interfering with translocation, the steroid-like antibiotics best typified by fusidic acid. However, other ribosomal inhibitors had been implicated as possible inhibitors of translocation (see Pestka, 1971, for a review). While it was originally thought that fusidic acid inhibited

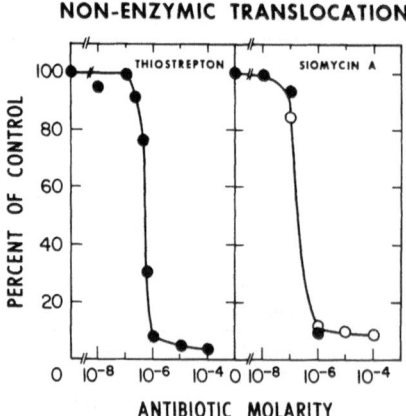

Fig. 4. Inhibition of non-enzymic translocation by thiostrepton and siomycin A. The data are taken from Pestka (1970 and 1973)

GTP hydrolysis directly, as indeed it does when assayed as described above for thiostrepton, it was later determined that it does so indirectly and by a very unusual mechanism. In the usual assay for ribosomal-dependent GTP hydrolysis thousands of equivalents of the nucleotide are hydrolyzed. However, in the presence of fusidic acid the ribosome and EF-G are able to hydrolyze *one* equivalent of GTP. Fusidic acid bound to the resulting ribosome · EF-G · GDP complex prevents its dissociation and hence prevents the *second* and subsequent rounds of GTP hydrolysis. It was therefore pertinent to inquire if thiostrepton inhibited the translocation-associated hydrolysis of GTP by the same or different mechanism as fusidic acid. The simplest test of this possibility was to determine if the fusidic acid-stabilized complex of the ribosome · EF-G · GDP could be formed on thiostrepton-treated ribosomes. The results of this experiment were reported by BODLEY et al. (1970), by WEISBLUM and DEMOHN (1970b), and later by MODOLELL et al. (1971b) and PESTKA and BROT (1971). As summarized in Fig. 5 (open circles) prior treatment of ribosomes with approximately one equivalent of thiostrepton completely prevented the formation of the fusidic acid-stabilized complex.

It had already been deomonstrated that EF-G and GDP bind to the 50S ribosomal subunit. Since it appeared that thiostrepton also binds to this subunit and in so doing prevented the binding of EF-G and GDP, it was appropriate to determine if the binding of these components to the ribosome was mutually exclusive. The experiment illustrated by the closed circles in Fig. 5 demonstrates that thiostrepton, if previously bound to the ribosome, can prevent the binding of EF-G and GDP. However, if these components are first bound to the ribosome, thiostrepton cannot displace them from it (Fig. 5, closed circles). These and other experiments lead to the conclusion that the binding of thiostrepton on the one hand and EF-G and GDP on the other to the 50S subunit were mutually exclusive (HIGHLAND et al., 1971).

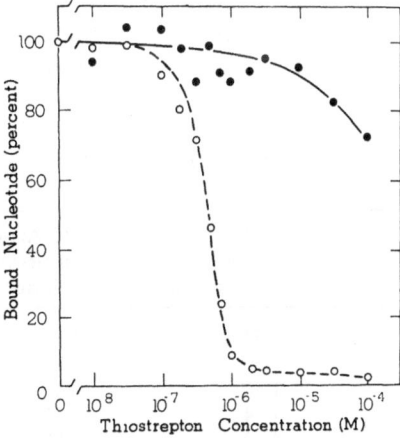

Fig. 5. Effect of thiostrepton on the formation and stability of the ribosome · EF-G · GDP complex. The data are taken from Fig. 1 of HIGHLAND et al. (1971)

Aminoacyl-tRNA Binding

The binding of aminoacyl-tRNA to the post-translocational complex of the ribosome, mRNA, and nascent peptidyl-tRNA represents, when viewed in the above sequence, the third and final step necessary to complete a single cycle of protein chain elongation. This binding reaction requires the participation of another non-ribosomal protein, EF-Tu, and GTP which in the course of the reaction is hydrolyzed to GDP and Pi. As a consequence, this reaction can be observed *in vitro* by following either or both the hydrolysis of GTP or the binding of aminoacyl-tRNA to the ribosome. In contrast to GTP hydrolysis by EF-G, however, EF-Tu-mediated hydrolysis is generally closely coupled to aminoacyl-tRNA binding.

After both thiostrepton and siomycin were found to be inhibitors of transloca-tion, it was determined that they also inhibit enzymic aminoacyl-tRNA binding. This is illustrated by the experiments of MODOLELL *et al.* (1971a) who used a protein synthesizing system programmed with the natural messenger, R17 RNA. Moreover, they began protein synthesis with a natural initiating system in which the first amino acid, N-formylmethionyl-tRNA (the universal initiator amino acid in *E. coli)* is bound to the ribosome in a GTP-dependent reaction which requires the participation of initiation factors (non-ribosomal proteins specifically involved in starting the synthesis of protein chains). This reaction precedes the formation of the first peptide bond and, although it resembles the reaction cata-lyzed by EF-Tu, it is completely independent of it. They reasoned that in such a system a specific inhibitor of translocation should arrest protein synthesis *after* the binding of the second amino acid (alanine is the second amino acid specified by the messenger) and the formation of the first peptide bond (generating in this case formyl-methionyl-alanine). A portion of their data is shown in Table 8. As seen in the control experiment, both F-Met-tRNA and Ala-tRNA bind to the ribosome in the absence of any inhibitor. (They demonstrated independently that peptide bond formation had occurred). Moreover, an inhibitor of transloca-tion, fusidic acid, had as expected, no effect on this system, while an inhibitor of EF-Tu-mediated aminoacyl-tRNA binding, tetracycline, completely and speci-fically blocked Ala-tRNA binding. Surprisingly, both thiostrepton and siomycin mimicked the effect of tetracycline. The results shown in Table 9 demonstrate

Table 8. *Effects of antibiotics on formyl-methionyl-tRNA and alanyl-tRNA binding to the ribosome and R17 RNA*

Additions	F-^{3}H-Met-tRNA pmoles/A_{260} unit ribosomes	^{14}C-Ala-tRNA pmoles/A_{260} unit ribosomes
None	1.32	1.36
Fusidic acid	1.30	1.35
Tetracycline	1.46	0.21
Thiostrepton	1.74	0.14
Siomycin A	1.69	0.16

The data are taken from Table 3 of MODOLELL *et al.* (1971a).

Table 9. *The effect of antibiotics on the EF-T-dependent binding of phenylalanyl-tRNA to the ribosome and the attendant hydrolysis of GTP*

Addition	^{14}C-Phe-tRNA bound (pmoles/A_{260} ribosomes)	γ-^{32}P-GTP hydrolyzed (pmoles/A_{260} ribosomes)
None	1.42	1.36
Thiostrepton	0.07	0.00
Siomycin	0.13	0.00
Fusidic acid	1.11	1.07
Tetracycline	0.27	0.87

The data are taken from Table 4 of MODOLELL *et al.* (1971a).

directly that thiostrepton and siomycin inhibit EF-Tu function. In this case MODO-LELL *et al.* (1971a) examined the EF-Tu-mediated binding of Phe-tRNA to the ribosome-poly U complex as well as the concomitant GTP hydrolysis. Both were inhibited by thiostrepton and siomycin. By contrast, fusidic acid inhibited neither reaction and tetracycline depressed only the binding reaction but had little effect on hydrolysis. Similar results were simultaneously reported by KINOS-HITA *et al.* (1971) with thiopeptin and siomycin. Subsequently, PIRALI *et al.* (1972) reported the inhibition of enzymic aminoacyl-tRNA binding to ribosomes by sporangiomycin.

As a result of these and other experiments as well as those described above dealing with translocation it became clear that thiostrepton and related antibiotics inhibited *both* the aminoacyl-tRNA binding and translocation reactions. These findings, in view of the fact that the thiostrepton-like antibiotics function by binding to the ribosome, led both MODOLELL *et al.* (1971) and KINOSHITA *et al.* (1971) to suggest that perhaps the two elongation factors Tu and G interact alternately with the same physical region of the 50S ribosomal subunit in the course of each cycle of elongation. Similar conclusions were reported by WATANABE and TANAKA (1971), WATANABE (1972), and WEISSBACH *et al.* (1972). This suggestion led to activity in a number of laboratories to test the possibility that, despite the fact that the ribosome is large with respect to the size of the elongation factors, it cannot interact with them simultaneously. This suspicion was confirmed (RICHMAN and BODLEY, 1972; CABRER *et al.,* 1972; MILLER, 1972; RICHTER, 1972) and led to the suggestion that EF-G, EF-Tu and the thiostrepton-like antibiotics interact in a mutually exclusive manner with the same or overlapping physical region of the 50S subunit. While the observations upon which this conclusion is based are undoubtedly correct the conclusion itself is probably somewhat naive for reasons that will become more apparent later.

Initiation of Protein Synthesis

The unique events involved in the initiation of protein synthesis (the binding of the ribosome to the initiator codon and the binding of the initiator tRNA, formyl-methionyl-tRNA, to this complex) bear at least superficial resemblance to the aminoacyl-tRNA binding and translocation steps of the elongation cycle.

These similarities include the involvement of non-ribosomal proteins and GTP and have led to speculation that initiation may be analogous to these two steps of elongation except employing factors which are specific for the initiator tRNA, F-Met-tRNA.

Antibiotic sensitivity provides one means of defining the relationship between initiation and elongation. The thiostrepton-like antibiotics are of particular importance in this regard because they inhibit both factor-requiring elongation steps. As mentioned earlier, experiments indicated that thiostrepton and siomycin had no effect on the initiation factor-dependent binding of F-Met-tRNA to the ribosome and R17 RNA. Moreover, the bound F-Met-tRNA was fully reactive with puromycin, a circumstance which is generally taken to indicate that initiation has been completed normally. A more recent report, however, has demonstrated that thiostrepton under some conditions can at least partially prevent the IF2 dependent hydrolysis of GTP without inhibiting F-Met-tRNA binding (Grun-berg-Manago et al., 1972). More thorough investigation of this situation, however, revealed (Lockwood and Maitra, in press; Hershey, personal communication) that under conditions where initiation proceeds most efficiently, thiostrepton is without effect on the IF2 dependent GTPase. Only under extreme conditions (e. g. with a large molar excess of initiation factors over ribosomes) does the antibiotic partially inhibit the reaction.

In light of these most recent observations it is reasonable to conclude that these antibiotics probably do not affect the initiation process and that the locus of inhibition is not the ribosomal IF2 GTPase site. Under some conditions the initiation factor IF2 can be forced to interact with the portion of the ribosome which normally interacts with the elongation factors Tu and G and this abnormal interaction is prevented by thiostrepton.

Experiments with Bacterial Protoplasts

Investigations of the mechanistic basis of complex biological processes in cell-free extracts are always in need of confirmation by experimentation in intact cells. This is particularly true when dealing with complex inhibitory agents such as the thiostrepton antibiotics which might be expected to exhibit pleiotropic effects.

Cundliffe (1971) used intact protoplasts for testing the site of action of antibiotic inhibitors of protein synthesis. In this technique B. megaterium cells are first steady-state labeled with ^{32}P to facilitate the detection of ribosomes and then made permeable to antibiotics by conversion to protoplasts. Following this, nascent protein chains are labeled with a [^{3}H]amino acid and after a brief time the antibiotic under investigation is added. If the antibiotic inhibited some step of elongation, as opposed to initiation or termination, labeling of protein would cease immediately and the labeled nascent chains would remain on the ribosome. This was the case with thiostrepton (see Fig. 6A). If the test antibiotic inhibits either translocation or the peptidyl transferase reaction, then the labeled nascent peptides will remain on the ribosome after the addition of puromycin. If, however, the test antibiotic inhibits aminoacyl-tRNA binding and thus leaves the nascent peptide in the P site with the ribosomal A site vacant, the peptide

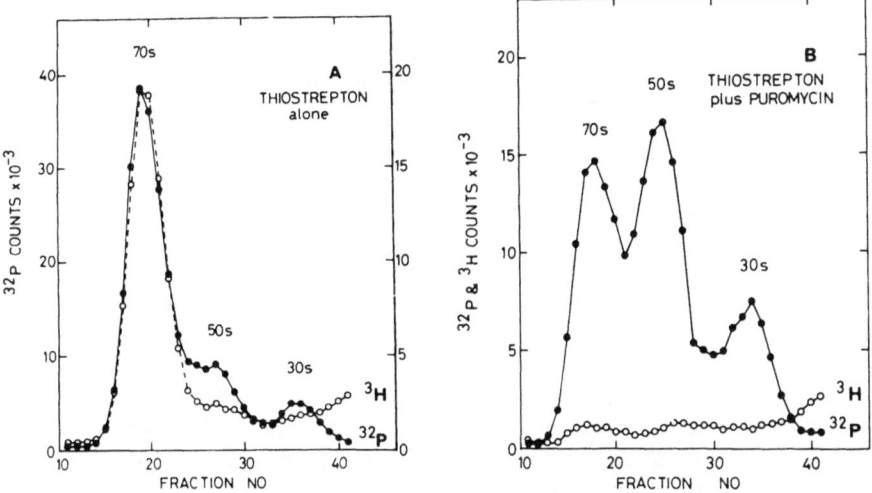

Fig. 6. The effect of thiostrepton on nascent peptide chains and their ability to react with puromycin.
The data are taken from Fig. 1 of CUNDLIFFE (1971)

will react with puromycin and be released from the ribosome. This latter result
was obtained with thiostrepton (Fig. 6 B) and thus it appears to inhibit aminoacyl-
tRNA binding. Similar experiments were performed by CANNON and BURNS
(1971) using an *E. coli* cell-free system.

The fact that in the intact cell thiostrepton selectively inhibits the ribosome
prior to aminoacyl-tRNA binding rather than prior to translocation presumably
results from the relative rate and extent of inhibition of these individual reactions
or the predominant state of the ribosome during the ribosomal epicycle (PESTKA,
1971). If, as is likely, aminoacyl-tRNA binding is the limiting (slow) reaction
of the elongation process, then inhibition of this reaction by these antibiotics
would be anticipated.

Binding of the Antibiotics to the 50S Ribosomal Subunit

On the basis of the type of indirect experiments cited earlier it is reasonable
to conclude that thiostrepton binds very tightly to the 50S ribosomal subunit
in a molar ratio of about 1:1. Approximately one molar equivalent of the
antibiotic is necessary and adequate to irreversibly inactivate the ribosome and
prolonged incubation with less than this amount of antibiotic leads only to
partial inactivation (HIGHLAND *et al.*, 1971). Chemical modification or covalent
attachment of the antibiotic to the ribosome probably does not occur because
thiostrepton can be recovered intact (it will inhibit the ribosome) by extraction
of antibiotic-inactivated ribosomes with organic solvents which denature the
ribosome (N. RICHMAN and J. W. BODLEY, unpublished observations).

SOPORI and LENGYEL (1972) have directly demonstrated the binding of thiostrepton to the ribosome. They prepared radioactive thiostrepton by fermentation
and as shown in Table 10 were successful in demonstrating its specific and equimolar binding to the 50S subunit. Thiostrepton bound just as effectively to a protein-

Table 10. *Binding of thiostrepton to various ribosomal components*

Ribosomal component	Moles thiostrepton bound per mole component
30S	0.16
50S	0.98
50S core	1.09
50S core + split proteins	1.05

The data are taken from Table 4 of Sopori and Lengyel (1972).

deficient 50S particle lacking proteins designated L7 and L12. Non-enzymic trans-location occurs on ribosomes deficient in L7 and L12. The fact that thiostrepton inhibits non-enzymic translocation on these ribosomal particles also indicates that thiostrepton binds to ribosomes deficient in L7 and L12 (Pestka, 1973a).

Analogous experiments with radioactive sporangiomycin revealed its binding to bacterial 70S ribosomes, but not to ribosomes from *S. cerevisiae* (Pirali et al., 1972); 50S, but not 30S, bacterial subunits also bound sporangiomycin. When 50S subunits are treated with 2.1 M LiCl, ribosomal core particles and split proteins result. These core particles do not bind sporangiomycin; however, after reconstitution of the 50S particle by addition of the split protein fraction, sporangiomycin binding was restored (Pirali et al., 1972).

The bacterial ribosome is composed of 3 molecules of RNA and 55 different proteins, approximately two-thirds of which comprise the 50S subunit. One of the most challenging problems facing molecular biologists today is to assign structural and functional roles to these individual components, a task which will certainly occupy them for a number of years. Since thiostrepton inhibits two very central ribosomal functions, it is particularly important to define its ribosomal site of binding.

A beginning to the analysis of the ribosome structure-function has been made. For example, it has been known for several years that the presence of proteins L7 and L12 on the 50S subunit were necessary for its interaction with EF-Tu and EF-G. Recently, through the use of antibodies prepared against the individual ribosomal proteins, Highland et al. (1973) have presented evidence which indicates that contact between EF-G and the ribosome is limited to the ribosomal proteins L7 and L12. Moreover, recent preliminary evidence (J.H. Highland, personal communication) suggests that the antibodies to ribosomal proteins L7 and L12 do not inhibit the binding of thiostrepton to the ribosome. In light of arguments advanced earlier concerning the relationship between the ribosomal sites of interaction of EF-G and thiostrepton a complete analysis of the binding of the antibiotic to the ribosome is clearly in order.

Summary

Thiostrepton, siomycin, multhiomycin, thiopeptin, A-59, and sporangiomycin comprise the thiostrepton group of antibiotics at present. These antibiotics are sulfur-containing peptide antibiotics. They generally inhibit Gram-positive, but not Gram-negative or eucaryotic organisms. Cell-free extracts from Gram-negative bacteria, however, are sensitive to inhibition by this group of antibacterial agents. In appropriate intact cells and cell-free extracts, these compounds inhibit protein synthesis chiefly and directly. They bind to the larger ribosomal subunit (50S) and consequently block several reactions in which this subunit is intimately involved. Thus, they inhibit aminoacyl-tRNA binding and translocation reactions, which require the elongation factors Tu and G of *Escherichia coli;* and they inhibit the hydrolysis of GTP associated with these reactions. In addition, under certain conditions some reactions related to the initiation events of protein synthesis can be inhibited. Bacterial mutants resistant to these antibiotics have been described and appear to have altered ribosomes.

References

ANDERSON, B., D.C. HODGKIN, and M.A. VISWAMITRA: The structure of thiostrepton. Nature **225**, 233–235 (1970).

BARTON, M.A., G.W. KENNER, and R.P. SHEPPARD: Peptides. Part XXIII. Experiments on the oxidation of thiostrepton. J. Chem. Soc. **1966**, 2115–2119.

BERMAN, C., F. BEUBE, A.H. KUTSCHER, E.V. ZEGARELLI, and I.B. STERN: Antibiotic levels in saliva obtained with a new antibiotic: thiostrepton troche studies. Oral Surg. Oral Med. Oral Pathol. **12**, 315–319 (1959).

BODANSZKY, M., J. ALICINO, C.A. BIRKHIMER, and N.J. WILLIAMS: Degradation of thiostrepton. The structure of thiostreptine. J. Am. Chem. Soc. **84**, 2003–2004 (1962).

BODANSZKY, M., J.D. DUTCHER, and N.J. WILLIAMS: The establishment of the identity of thiostrepton with thiactin (bryamycin). J. Antibiotics (Tokyo) **16**, 76–79 (1963).

BODANSZKY, M., J. FRIED, J.T. SHEEHAN, N.J. WILLIAMS, J. ALICINO, A.I. COHEN, B.T. KEELER, and C.A. BIRKHIMER: Thiostrepton. Degradation products and structural features. J. Am. Chem. Soc. **86**, 2478–2490 (1964).

BODANSZKY, M., J.A. SCOZZIE, and I. MURAMATSU: Dehydroalanine residues in thiostrepton. J. Antibiotics (Tokyo) **23**, 9–12 (1970).

BODANSZKY, M., J.T. SHEEHAN, J. FRIED, N.J. WILLIAMS, and C.A. BIRKHIMER: Degradation of thiostrepton. Thiostreptoic acid. J. Am. Chem. Soc. **82**, 4747–4748 (1960).

BODLEY, J.W., L. LIN, and J.H. HIGHLAND: Studies on translocation. VI Thiostrepton prevents the formation of a ribosome-G factor-guanine nucleotide complex. Biochem. Biophys. Res. Commun. **41**, 1406–1411 (1970).

BROT, N., W.P. TATE, C.T. CASKEY, and H. WEISSBACH: The requirement for ribosomal proteins L7 and L12 in peptide chain termination. Proc. Nat. Acad. Sci. US **71**, 89–92 (1974).

CABRER, B., D. VAZQUEZ, and J. MODOLELL: Inhibition by elongation factor EF-G of amino acyl-tRNA binding to ribosomes. Proc. Natl. Acad. Sci. U.S. **69**, 733–736 (1972).

CANNON, M., and K. BURNS: Modes of action of erythromycin and thiostrepton as inhibitors of protein synthesis. FEBS Letters **18**, 1–5 (1971).

COHN, I., JR.: Antibiotics for colon surgery. Gastroenterology **35**, 583–591 (1958).

COHN, I., JR.: Choice of agents for preoperative preparation of the colon. Southern Med. J. **53**, 881–884 (1960).

COHN, I., JR., and A.B. LONGACRE: Preoperative sterilization of the colon: Comparison of various antibacterial agents. III. Antibiotics Ann. **1957–1958**. N.Y., Med. Encyclopedia, 635–650 (1957).

CRON, M.J., D.F. WHITEHEAD, I.R. HOOPER, B. HEINEMANN, and J. LEIN: Bryamycin, a new antibiotic. Antibiot. & Chemotherapy **6**, 63–67 (1956).

CROSS, D.F.W., G.W. KENNER, R.C. SHEPPARD, and C.E. STEHR: Peptides. Part XIV. Thiazoleamino-acids, degradation products of thiostrepton. J. Chem. Soc. **1963**, 2143–2150.

CUNDLIFFE, E.: The mode of action of thiostrepton *in vivo*. Biochem. Biophys. Res. Commun. **44**, 912–917 (1971).

DREY, C.N.C., G.W. KENNER, H.D. LAW, R.C. SHEPPARD, M. BODANSZKY, J. FRIED, N.J. WILLIAMS, and J.T. SHEEHAN: Degradation of thiostrepton. Derivatives of 8-hydroxyquinoline. J. Am. Chem. Soc. **83**, 3906–3908 (1961).

570   S. Pestka and J.W. Bodley:

Dubnau, D., C. Goldthwaite, I. Smith, and J. Marmur: Genetic mapping in *Bacillus subtilis*. J. Mol. Biol. **27**, 163–185 (1967).

Ebata, M., K. Miyazaki, and H. Otsuka: Studies on siomycin. I. Physicochemical properties of siomycins A, B and C. J. Antibiotics (Tokyo) **22**, 364–368 (1969a).

Ebata, M., K. Miyazaki, and H. Otsuka: Studies on siomycin. II. The composition and degradation products of siomycin A. J. Antibiotics (Tokyo) **22**, 423–433 (1969b).

Ebata, M., K. Miyazaki, and H. Otsuka: Studies on siomycin. III. Structural features of siomycin A. J. Antibiotics (Tokyo) **22**, 434–441 (1969c).

Ebata, M., K. Miyazaki, and H. Otsuka: Studies on siomycin. IV. Acyl derivatives of siomycin A. J. Antibiotics (Tokyo) **22**, 506–507 (1969d).

Ebata, M., K. Miyazaki, and H. Otsuka: Studies on siomycin. V. Derivatives of siomycin A prepared with thiolcarboxylic acid. J. Antibiotics (Tokyo) **22**, 451–456 (1969e).

Goldthwaite, C., and I. Smith: Physiological characterization of antibiotic resistant mutants of *Bacillus subtilis*. Mol. Gen. Genet. **114**, 190–204 (1972).

Grunberg-Manago, M., J. Dondon, and M. Graffe: Inhibition by thiostrepton of the IF-2-dependent ribosomal GTP-ase. FEBS Letters **22**, 217–221 (1972).

Haselkorn, R., and L.B. Rothman-Denes: Protein synthesis. Ann. Rev. Biochem. **42**, 397–438 (1973).

Highland, J.H., J.W. Bodley, J. Gordon, R. Hasenbank, and G. Stöffler: Identity of the ribosomal proteins involved in the interaction with elongation factor G. Proc. Natl. Acad. Sci. U.S. **70**, 147–150 (1973).

Highland, J.H., L. Lin, and J.W. Bodley: Protection of ribosomes from thiostrepton inactivation by the binding of G factor and guanosine diphosphate. Biochemistry **10**, 4404–4409 (1971).

Hishizawa, T., and S. Pestka: Studies on the formation of transfer ribonucleic acid —ribosome complexes. XVII. The effect of tRNA on aminoacyl-oligonucleotide binding to ribosomes. Arch. Biochem. Biophys. **147**, 624–631 (1971).

Kelly, J., A.H. Kutscher, and F. Tuoti: Thiostrepton, a new antibiotic: tube dilution sensitivity studies. Oral Surg. Oral Med. Oral Pathol. **12**, 1334–1339 (1959a).

Kelly, J., A.H. Kutscher, and F. Tuoti: Activity of a new antibiotic, thiostrepton, against microorganisms and the pattern of resistance developed. Antibiot. & Chemotherapy **10**, 78–83 (1960).

Kelly, J., W.C. Schraft, A.H. Kutscher, and F. Tuoti: Antibacterial spectrum of a new antibiotic, thiostrepton: disc sensitivity studies. Antibiot. & Chemotherapy **9**, 87–89 (1959b).

Kenner, G.W., R.C. Sheppard, and C.E. Stehr: Synthesis of thiazole amino-acids derived from natural peptides. Tetrahedron Letters **1960**, 23–26.

Kinoshita, T., Y.-F. Liou, and N. Tanaka: Inhibition by thiopeptin of ribosomal functions associated with T and G factors. Biochem. Biophys. Res. Commun. **44**, 859–863 (1971).

Kondo, S., E. Akita, J.M.J. Sakamoto, M. Ogasawara, T. Niida, and T. Hatakeyama: Studies on a new antibiotic produced by streptomyces SP. A-59 J. Antibiotics (Tokyo) **14**, 194–198 (1961).

Kutscher, A.H., L. Seguin, J.B. Campbell, and M.L. Gochenour: Thiostrepton, a new antibiotic: disk sensitivity studies. II. Antibiot. & Chemotherapy **11**, 340–344 (1961).

Kutscher, A.H., L. Seguin, R.M. Rankow, and J.D. Piro: Thiostrepton, a new antibiotic: disc sensitivity studies. Antibiot. & Chemotherapy **8**, 576–583 (1958).

Kutscher, A.H., L. Seguin, E.V. Zegarelli, and J.D. Piro: Antimicrobial activity of thiostrepton: tube dilution studies. J. Am. Dental Assoc. **59**, 715–720 (1959a).

Kutscher, A.H., E.V. Zegarelli, R.M. Rankow, J. Mercadante, and J.D. Piro: Clinical laboratory studies on a new topical antibiotic: thiostrepton. Oral Surg. Oral Med. Oral Pathol. **12**, 967-974 (1959b).

Lepper, M.H., H.F. Dowling, G.G. Jackson, H.W. Spies, and J. Norsen: Effect of antibiotic nasal ointments on carrier states in patients and on the antibiotic pattern of organisms from personnel caring for these patients. Antimicrobial Agents Chemotherapy **1962**, 140–149.

Levin, J.D., H. Stander, and J.F. Pagano: An agar diffusion microbiological assay for thiostrepton. Antibiot. & Chemotherapy **10**, 422–429 (1960).

Lockwood, A.H., and U. Maitra: Relationship between the ribosomal sites involved in initiation and elongation of polypeptide chains. J. Biol. Chem. **249**, 346–352 (1974).

Mazumder, R.: Effect of thiostrepton on recycling of *E. coli* initiation factor 2. Proc. Nat. Acad. Sci. US **70**, 1939–1942 (1973).

MILLER, D. L.: Elongation factors EF-Tu and EF-G interact at related sites on the ribosome. Proc. Natl. Acad. Sci. U. S. **69**, 752–755 (1972).

MIYAIRI, N., T. MIYOSHI, H. AOKI, M. KOHSAKA, H. IKUSHIMA, K. KUNUGITA, H. SAKAI, and H. IMANAKA: Studies on thiopeptin antibiotics. I. Characteristics of thiopeptin B. J. Antibiotics (Tokyo) **23**, 113–119 (1970).

MODOLELL, J., B. CABRER, A. PARMEGGIANI, and D. VAZQUEZ: Inhibition by siomycin and thiostrepton of both amino acyl-tRNA and factor G binding to ribosomes. Proc. Natl. Acad. Sci. U. S. **68**, 1796–1800 (1971 a).

MODOLELL, J., D. VAZQUEZ, and R. E. MONRO: Ribosomes, G-factor and siomycin. Nature New Biol. **230**, 109–112 (1971 b).

NISHIMURA, H., S. OKAMOTO, N. MAYAMA, H. OHTSUKA, K. NAKAJIMA, K. TAWARA, M. SHIMOHIRA, and N. SHIMAOKA: Siomycin, a new thiostrepton-like antibiotic. J. Antibiotics (Tokyo) **14**, 255–263 (1961).

PAGANO, J. F., M. J. WEINSTEIN, H. A. STOUT, and R. DONOVICK: Thiostrepton, a new antibiotic. I. *In vitro* studies. Antibiotics Ann. **1955–1956**, 554–559.

PERLMAN, D.: Antibiotic inhibition of algal growth. Antimicrobial Agents Chemotherapy **1964**, 114–119.

PERLMAN, D., N. A. GUIFFRE, P. W. JACKSON, and F. E. GIARDINELLO: Effects of antibiotics on multiplication of L cells in suspension culture. Proc. Soc. Exptl. Biol. Med. **102**, 290–292 (1959).

PESTKA, S.: Studies on the formation of transfer ribonucleic acid-ribosome complexes. V. On the function of a soluble transfer factor in protein synthesis. Proc. Natl. Acad. Sci. U. S. **61**, 726–733 (1968).

PESTKA, S.: Studies on the formation of transfer ribonucleic acid-ribosome complexes. VI. Oligopeptide synthesis and translocation on ribosomes in the presence and absence of soluble transfer factors. J. Biol. Chem. **244**, 1533–1539 (1969).

PESTKA, S.: Thiostrepton: a ribosomal inhibitor of translocation. Biochem. Biophys. Res. Commun. **40**, 667–674 (1970).

PESTKA, S.: Inhibitors of ribosomal functions. Ann. Rev. Microbiol. **25**, 487–562 (1971).

PESTKA, S.: Studies on transfer ribonucleic acid-ribosome complexes. XIX. Effect of antibiotics on peptidyl puromycin synthesis on polyribosomes from *Escherichia coli*. J. Biol. Chem. **247**, 4669–4678 (1972).

PESTKA, S.: Assay for non-enzymic and enzymic translocation with *E. coli* ribosomes. Methods Enzymol. **25**, in press (1973 a).

PESTKA, S.: The use of inhibitors in studies of protein synthesis. Methods Enzymol. **25**, (in press) (1973 b).

PESTKA, S., and N. BROT: Studies on the formation of transfer ribonucleic acid-ribosome complexes. XV. Effect of antibiotics on steps of bacterial protein synthesis: some new ribosomal inhibitors of translocation. J. Biol. Chem. **246**, 7715–7722 (1971).

PESTKA, S., and H. HINTIKKA: Studies on the formation of ribonucleic acid-ribosome complexes. XVI. Effect of ribosomal translocation inhibitors on polyribosomes. J. Biol. Chem. **246**, 7723–7730 (1971).

PIRALI, G., G. C. LANCINI, B. PARISI, and F. SALA: Interaction of sporangiomycin with the bacterial ribosome. J. Antibiotics (Tokyo) **25**, 561–568 (1972).

PLATT, T. B., and W. R. FRAZIER: Effect of metal cations on production of thiostrepton. Antimicrobial Agents Chemotherapy **1961**, 205–211.

RICHMAN, N., and J. W. BODLEY: Ribosomes cannot interact simultaneously with elongation factors EF-Tu and EF-G. Proc. Natl. Acad. Sci. U. S. **69**, 686–689 (1972).

RICHTER, D.: Inability of *E. coli* Ribosomes to interact simultaneously with the bacterial elongation factors EF-Tu and EF-G. Biochem. Biophys. Res. Commun. **46**, 1850–1856 (1972).

RICHTER, D., L. LIN, and J. W. BODLEY: Studies on translocation. IX. The pattern of action of antibiotic translocation inhibitors in eukaryotic and prokaryotic systems. Arch. Biochem. Biophys. **147**, 186–191 (1971).

SMITH, I., and H. SMITH: Location of the SP02 attachment site and the bryamycin resistance marker on the *Bacillus subtilis* chromosome. J. Bacteriol., **114**, 1138–1142 (1973).

SOPORI, M. L., and P. LENGYEL: Components of the 50S ribosomal subunit involved in GTP cleavage. Biochem. Biophys. Res. Commun. **46**, 238–244 (1972).

STEINBERG, B. A., W. P. JAMBOR, and L. O. SUYDAM: Thiostrepton, a new antibiotic. III. *In vivo* studies. Antibiotics Ann. **1955–1956**. N. Y., Med. Encyclopedia, 1956, p. 562–565.

Tanaka, K., S. Watanabe, and M. Tamaki: Mode of action of siomycin. J. Antibiotics (Tokyo) 23, 13–19 (1970a).

Tanaka, K., S. Watanabe, H. Teraoka, and M. Tamaki: Effect of siomycin on protein synthesizing activity of Escherichia coli ribosomes. Biochem. Biophys. Res. Commun. 39, 1189–1193 (1970b).

Tanaka, T., T. Endo, A. Shimazu, R. Yoshida, Y. Suzuki, N. Otake, and H. Yonehara: A new antibiotic, multhiomycin. J. Antibiotics (Tokyo) 23, 231–237 (1970c).

Tanaka, T., K. Sakaguchi, and H. Yonehara: On the mode of action of multhiomycin, I. Effects of multhiomycin on macromolecular synthesis. J. Antibiotics (Tokyo) 23, 401–407 (1970d).

Tanaka, T., K. Sakaguchi, and H. Yonehara: On the mode of action of multhiomycin. II. Effects of multhiomycin on phe-tRNA binding to ribosomes and on other steps in protein synthesis. J. Antibiotics (Tokyo) 24, 537–542 (1971a).

Tanaka, T., K. Sakaguchi, and H. Yonehara: Inhibition by multhiomycin of T factor and GTP-dependent binding of phenylalanyl-tRNA to ribosomes and GTP hydrolysis associated with it. J. Biochem. (Tokyo) 69, 1127–1130 (1971b).

Thiemann, J.E., C. Coronelli, H. Pagani, G. Beretta, G. Tamoni, and V. Arioli: Antibiotic production by new form-genera of the Actinomycetales. I. Sporangiomycin, an antibacterial agent isolated from Planomonospora Parontospora Var. Antibiotica Var. Nov. J. Antibiotics (Tokyo) 21, 525–531 (1968).

Tiboni, O., and O. Ciferri: Selective inhibition of the reactions catalyzed by ribosome-specific transfer factors G. FEBS Letters 19, 174–179 (1971).

Umezawa, H.: Index of antibiotics from actinomycetes. Pennsylvania: University Park Press, State College, 1967.

Vandeputte, J., and J.D. Dutcher: Thiostrepton, a new antibiotic. II. Isolation and chemical characterization. Antibiotics Ann. 1955–1956. N.Y., Med. Encyclopedia, 1956, p. 560–561.

Watanabe, S.: Interaction of siomycin with the acceptor site of Escherichia coli ribosomes. J. Mol. Biol. 67, 443–457 (1972).

Watanabe, S., and K. Tanaka: Effect of siomycin on the acceptor site of Escherichia coli ribosomes. Biochem. Biophys. Res. Commun. 45, 728–734 (1971).

Watanabe, S., and K. Tanaka: Effect of siomycin on the G factor dependent GTP hydrolysis by Escherichia coli ribosomes. FEBS Letters 13, 267–268 (1971).

Weisblum, B., and V. Demohn: Thiostrepton, an inhibitor of 50S ribosome subunit function. J. Bacteriol. 101, 1073–1075 (1970a).

Weisblum, B., and V. Demohn: Inhibition by thiostrepton of the formation of a ribosome-bound guanine nucleotide complex. FEBS Letters 11, 149–152 (1970b).

Weissbach, H., B. Redfield, E. Yamasaki, R.C. Davis, Jr., S. Pestka, and N. Brot: Studies on the ribosomal sites involved in factors Tu and G-dependent reactions. Arch. Biochem. Biophys. 149, 110–117 (1972).

Wise, R.I.: Principles of management of staphylococcic infections. J. Am. Med. Assoc. 166, 1178–1182 (1958).

Note Added in Proof

Gordon and Highland (submitted for publication) have recently examined in detail the interaction between radiolabeled thiostrepton and the ribosome. They have confirmed the observation by Sopori and Lengyel (1972) that the antibiotic binds specifically to the prokaryotic 50S subunit in a molar ratio of ca. 1:1 and also shown that this binding is inhibited by the binding of EF-G. Moreover, although thiostrepton did not appear to bind to the individual components of the subunit, they found (Highland, Howard, Ochsner, Stöffler, Hasenbank and Gordon, manuscript in preparation) a unique involvement of protein L11 in this binding. Selective removal of this protein destroys the ability of the ribosome to bind thiostrepton and this activity can be restored by the rebinding of protein L11 to a ribosomal core containing only 8 proteins. Finally, after rebinding L11 to this simplified ribosomal core, antibody to protein L11 and no other ribosomal protein inhibits thiostrepton binding. Hence it is reasonable to conclude that thiostrepton actually binds to protein L11 on the ribosome.

It has now become clear that thiostrepton does inhibit some aspects of the interaction of the ribosome with IF-2 (MAZUMDER, 1973, and LOCKWOOD and MAITRA, 1974), but the exact nature of this inhibitory effect remains somewhat unclear. In addition it has recently been found that the binding of thiostrepton to the ribosome inhibits the termination reactions catalyzed by the Release Factors (BROT et al., 1974). Thus it now appears that the binding of thiostrepton alters ability of the ribosomes to interact with all nonribosomal factors which utilize GTP.

III. Interference with Cell Wall/Membrane Biosynthesis, Specific Enzyme Systems and Those in Which Mode of Action Not Known with Certainty

Berberine

Fred E. Hahn and Jennie Ciak

I. Introduction

Berberine is a yellow alkaloid which occurs in numerous plants (JEFFS,1967). It was first isolated from *Xanthoxylon cava* in 1826 by CHEVALLIER and PELLETAN. The chemical structure of berberine is shown in Fig. 1. The compound exists as an equilibrium mixture of three tautomeric forms (GADAMER, 1905). Structure Ia which contains a quarternary nitrogen, forms salts with mineral acids such as HCl and H_2SO_4 through elimination of one molecule of water; these salts are correctly designated as berberinium compounds. Berberinium nitrate is poorly soluble, and the addition of nitric acid to aqueous solutions of the alkaloid precipitates the nitrate. The quarternary base has an extraordinary high pK_α of 15.23 in water (SHIH-CHIEH CH'OU and LIANG-CHUN CH'A, 1965).

Extracts of berberine-containing plants are traditional folk remedies, especially in Hindu medicine (*Berberis aristata Linn*) but also in southern parts of the United States (*Hydrastis canadensis*). The pharmacological and therapeutic properties of berberine have been reviewed by SHIDEMAN (1950), and more recent pharmacological studies have been published by several authors (UCHIZUMI, 1957; TUROVA *et al.* 1962).

The mammalian toxicity of berberine depends upon the experimental animal (mouse, rat, guinea pig, rabbit); the minimal lethal doses varied from 27.5 mg/kg to 250 mg/kg upon intravenous injection (UCHIZUMI, 1957; SHIDEMAN, 1950). Intraperitoneal or subcutaneous administration did not greatly modify the toxicity of the alkaloid. Oral administration to cats of 100 mg/kg soon caused extensive vomiting followed by death 8–10 days later (TUROVA *et al.*,1962).

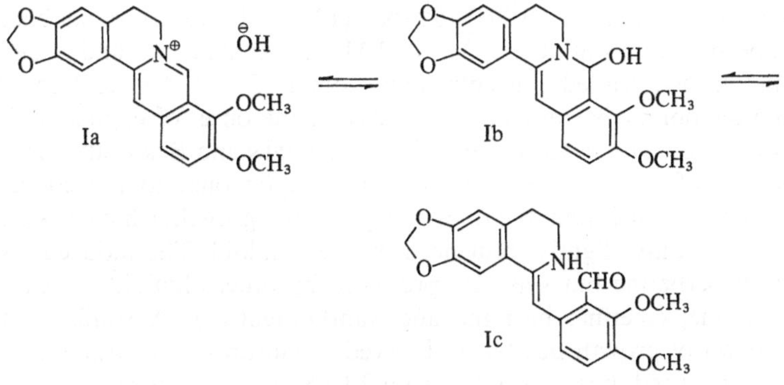

Fig. 1. Structure of berberine in its three tautomeric forms

II. Berberine as a Therapeutic Agent

The only practical therapeutic use of berberine is in the treatment of cutaneous Leishmaniasis ("oriental sore") by direct injection into the local lesions (Varma, 1927). The alkaloid inhibits the growth of *Leishmania tropica in vitro* at concentrations as low as 3×10^{-5} M (Das Gupta and Dikshit, 1929). At $\sim 10^{-7}$ M, berberine immobilized the cercariae of *Schistosoma mansoni* within 30 minutes (Cushing, 1957).

Berberinium sulfate at $> 10^{-3}$ M inhibits the trophozoites of *Entamoeba histolytica* after a 24 hour lag and shows effects on hepatic amebiasis in hamsters (Subbaiah and Amin, 1967). The use of the alkaloid in experimental cholera infections (Dutta and Panse, 1962) and in two human cholera epidemics (Lahiri and Dutta, 1967) cannot be ascribed unambiguously to the growth inhibitory effect of berberine on *Vibrio cholerae* since "experimental cholera" induced in infant rabbits by soluble toxic materials from cultures of *V. cholerae* was also susceptible to treatment with the alkaloid with survival or significantly increased survival times of the experimental animals (Dutta *et al.*,1972). This points to a pharmacological antidysenteric effect of berberine on the intestinal tract. *V. cholerae*, growing in berberine-containing media, fail to produce toxins (Modak *et al.*, 1970a). While the daily administration of 30 mg berberine for 10 consecutive days to healthy infants produced little change in the intestinal flora except for a disappearance of staphylococci (Homma *et al.*, 1961), a treatment of 50 cases of gastroenteritis in children showed that berberine was a good antidiarrheal agent (Sharda, 1970).

Although antimalarial effects have been ascribed to berberine, there exists no conclusive experimental data to substantiate such claims.

III. Antibacterial Effects of Berberine

Data on antibacterial effects of the alkaloid are widely scattered throughout the literature and only a few systematic screening studies have been reported (Sado, 1947; Foley *et al.*, 1958; Amin *et al.*,1969). Data in Table 1 have been selected from Amin *et al.* (1969). Stickl (1928) obseved a rapid bactericidal effect of berberinium sulfate at 2.3×10^{-3} M for staphylococci and *Bacillus anthracis*. In contrast, graded concentrations from 1.15×10^{-5} M to 1.15×10^{-4} M produced temporary bacteriostasis by delaying the onset of exponential growth of *Staphylococcus aureus* for up to 20 hours (Lambin and Bernard, 1955). While the duration of this drug-induced lag was proportional to the concentration of berberine in the medium, the rates of exponential growth, which was ultimately attained, were only slightly influenced by the alkaloid. The induced lag is not specific for berberine but was also produced by other alkaloids (cocaine, morphine, quinine, emetine, harmine and xanthofagarine). A similar pattern of delayed onset of growth has been observed in cultures of *V. cholerae* in which berberine produced increases in lag from 2 to 8 hours at concentrations ranging from 6×10^{-5} M to 1.1×10^{-4} M (Modak *et al.*, 1970a).

Table 1. *Antibacterial activity of berberinium sulfate*[a]

Organism	Minimal growth inhibitory concentration in µg/ml
Corynebacterium diphtheriae	6.2
Staphylococcus aureus	6.2–50.0
Streptococcus pyogenes	12.5
Shigella boydii	12.5
Vibrio cholerae	12.5–25.0
Vibrio cholerae El Tor	50.0
Klebsiella pneumoniae	25.0
Bacillus subtilis	25.0
Escherichia coli	50.0–> 100.0
Staphylococcus albus	50.0
Pseudomonas sp.	> 100.0
Salmonella sp.	> 100.0
Proteus sp.	> 100.0

[a] Data selected from AMIN *et al.* (1969). The minimal inhibitory concentrations can be converted into molarities using a calculated molecular weight of berberinium sulfate of 433 daltons.

Berberine is assimilated by "several species of bacteria" which are not inhibited by the alkaloid's hydrochloride at 6×10^{-5} M (calculated for the anydrous salt) (GRAY and LACHANCE, 1956, 1957). This requires active bacterial metabolism in complete growth medium while washed, so-called "resting", cells in saline show minimal absorption of berberine.

One of us (J.C.) has carried out preliminary mode-of-action studies on berberinium chloride in a strain of *Diplococcus pneumoniae* with the following (unpublished) results. At 5×10^{-4} M, the chloride (turbidimetrically) inhibited growth in liquid medium partially, while complete growth inhibition was observed at 1.5×10^{-3} M. The lower concentration caused a decline in the number of viable bacteria by three decadic logarithms during the first 60 min of exposure and the higher concentration produced such a decline by five decadic logarithms which was of first order with time. Incorporations of ^{14}C-thymidine or ^{14}C-phenylalanine into bacterial polymers were immediately and completely arrested at 10^{-3} M berberinium chloride while the rate of incorporation of ^{14}C-uridine was reduced to 10 per cent of that of a drug-free culture and was linear for a 60 min experimental period.

The limited conclusions which can be drawn from these observations are that berberinium chloride had a very rapid bactericidal effect on the test organism and that biochemical measurements beyond the first 15 min of drug exposure described *post mortem* events. We have no indication that the bactericidal effect of berberine is a result of unbalanced growth.

Growth of *Escherichia coli* from small inocula was inhibited by berberinium chloride at 7.7×10^{-4} M. Larger inocula required an increase in drug concentration to $> 10^{-3}$ M in order to produce complete growth inhibition. Microscopic examination of these bacteria which had grown in subinhibitory concentrations of berberinium chloride showed enlargement of the cells and many pleomorphic and filamentous forms. Others have reported that the concentrations of berberine required to inhibit growth of individual strains of *E. coli* are subject to considerable variation (Johnson *et al.*, 1952).

Similar studies have been reported (Modak *et al.*, 1970a) in *V. cholerae* and its El Tor variant. At 7×10^{-5} M berberine, which caused delay in growth for 8 hours, there occurred, during this induced lag, partial inhibitions of incorporations of ^{14}C-labelled adenine, guanine, phenylalanine and leucine into bacterial polymers. When growth resumed, a sudden burst in protein biosynthesis and a significant increase in the lipid fraction, both in relation to bacterial density as the determined variable, were observed.

V. cholerae which had been grown in uniformly ^{14}C-labelled glucose lost considerable radioactivity into the experimental medium during the first 40 min of treatment with 1.4×10^{-4} M berberine (Modak *et al.*, 1970a). The authors of that work interpret this observation as a loss of material from the metabolic pool of the bacteria but have not considered the possibility that the release of radioactivity might signal an autolytic breakdown of preformed bacterial polymers. Fractionation of berberine-exposed *V. cholerae* followed by fluorometric determinations of berberine showed that, in balance, 75 per cent of the alkaloid was recovered in the lipid fraction.

Amin *et al.* (1969) studied the mode of action of berberine in *V. cholerae, E. coli* and *S. aureus*. They found, like one of these authors (J. C.), that the minimal inhibitory concentration of berberine (as the sulfate) increased with increases in the size of the inoculum. The optimal pH for the antibacterial action of berberinium sulfate was 8.0. It was confirmed that the alkaloid is temporarily bacteriostatic for *S. aureus* at concentrations of 8×10^{-5} M and 1.15×10^{-4} M, followed by onset of exponential growth. At the same concentration range, berberinium sulfate had no influence on DNA biosynthesis in *V. cholerae* [in contrast to results of Modak *et al.* (1970a)] but inhibited protein biosynthesis and actually produced a 50 per cent decrease in the amount of global RNA of *V. cholerae* during the first 60 minutes of exposure to the alkaloid. This agrees with our interpretation of data of Modak *et al.*, (1970a) that the loss of ^{14}C from prelabelled *V. cholerae* signals some form of autolysis of preformed bacterial polymers rather than merely a loss of material from the metabolic pool.

IV. Berberine and DNA

Of greater interest than the fragmentary information on the mode of antibacterial action of berberine are the effects of the alkaloid on nonchromosomal genetic entities in microorganisms. Like certain aminoacridines and ethidium bromide, berberine converts *Saccharomyces cerevisiae* to respiration-deficient mitochondrial mutants which grow on plates in small ("petite") colonies (Meisel and Sokolova, 1959). The alkaloid eliminates, with low frequency, the determinants of resistance to kanamycin and chloramphenicol from an R-factor, harbored by *E. coli* RS-2 (Hahn and Ciak, 1971). When this R-factor was transferred into *Salmonella typhimurium* LT-2, berberinium chloride at 10^{-4} M eliminated the resistance determinants to kanamycin (83%), streptomycin (70%), chloramphenicol (68%) and ampicillin (62%) (Hahn and Ciak, unpublished data).

Since the petite conversion in yeast is based on a rapid block of the synthesis and transcription of yeast mitochondrial DNA (MAHLER *et al.,* 1971) and the elimination of resistance determinants of bacterial R-factors is a group property of DNA-complexing compounds which may inhibit, selectively, the autonomous replication of episomal DNA (HAHN and CIAK, 1971), it is logical to assume that berberine produces its effect on non-chromosomal genetic entities in such a manner.

This hypothesis is supported by a body of information concerning the formation of a complex of DNA with berberine. DNA and, to a lesser extent, RNA, alter the absorption spectrum of berberine (MORTHLAND *et al.,* 1954). DNA causes both a decrease in light absorption and a bathochromic shift in the two absorption bands of berberine above 300 nm (KLÍMEK AND HNILICA, 1959). In decreasing order of activity, heat-denatured and rapidly cooled DNA, heat-denatured but slowly cooled DNA, ribosomal RNA, native DNA and transfer RNA enhance the fluorescence of berberinium sulfate at 540 nm; these effects are directly proportional to the nucleic acid concentration (given as phosphate equivalents) (YAMAGISHI, 1962).

Berberine cosediments with DNA; the yellow alkaloid accumulates below the hypersharp boundary of calf-thymus DNA in analytical ultracentrifugation and the phenomenon can be visually observed through the viewport of the instrument (KREY and HAHN, 1969). Such observations remain the most simple and direct proof of the presence of a DNA-ligand complex. The DNA-berberine complex can be flow-oriented and the linear dichroism of the oriented complex has been measured (KREY and HAHN, 1969). The signs and magnitudes of the dichroic phenomenon were identical for the DNA bases (measured at 259 nm) and for the chromophore of berberine (measured at 350 nm). This means that the chromophoric ring system of berberine is oriented in a highly regular manner relative to the planes of the base pairs in DNA. Provided that the transition moment of berberine's absorption peak at ~350 nm lies in the plane of the chromophore, the result of the dichroism measurement would indicate that this plane is arranged coplanarely to the component bases of the DNA double helix. This suggests the possibility of intercalation binding in agreement with berberine's effect of yeast mitochondria (MEISEL and SOKOLOVA, 1959) and with its ability to eliminate resistance determinants from bacterial R-factors (HAHN and CIAK, 1971 and unpublished). Indeed, the viscosity of the DNA-berberine complex is significantly higher than that of the same uncomplexed DNA preparation (HAHN and KREY, 1971). Viscosity enhancement of DNA is one of the hydrodynamic criteria of intercalation binding (LERMAN, 1964).

Berberine stabilizes DNA to strand separation by heat (KREY and HAHN, 1969). The thermal denaturation profile ("melting curve") of DNA is shifted by the alkaloid to higher temperatures and is rendered more steep than that of DNA alone. This indicates that the strand separation process is more cooperative when the alkaloid is complexed with DNA. The hyperchromic shift of the complex at 260 nm is larger than the standard 40 per cent increment in DNA's absorbance. Since the absorption spectrum of free berberine has a maximum at 263 nm, the larger hyperchromicity of the complex points to a contribution of liberated berberine to the overall absorbance at 260 nm.

The bathochromic shifts in the absorption spectrum of the alkaloid, upon complexing with DNA (KLÍMEK and HNILICA,1959; KREY and HAHN, 1969), suggests that single alkaloid molecules become bound to DNA and, in this condition, exhibit a true monomer spectrum (MICHAELIS, 1947). Upon progressive dilution, solutions of berberinium hydrochloride showed a comparable redshift in absorption bands (KREY and HAHN, 1969). Evidently, free berberine has a tendency to self-aggregation.

Berberine was among 17 DNA-complexing compounds which were tested (KREY and HAHN, 1971) for their ability to displace methyl green from its stable complex with DNA. The first order rate constant of this reaction and its endpoint of 34 per cent methyl green displaced were among the lowest values measured and placed berberine in the vicinity of actinomycin D as concerns the rate and endpoint of the methyl green displacement reaction. The result could suggest either a low affinity of the alkaloid for native DNA or a low stoichiometry of binding to DNA, or both.

This then was studied in detail by spectrophotometric and spectrophotofluorometric titrations of berberinium chloride with graded concentrations of native calf thymus DNA (HAHN and KREY, 1971). The data from these titrations were converted into non-linear adsorption isotherms ("Scatchard plots") from which stoichiometries and apparent association constants were determined. A strong binding reaction had a stoichiometry of one berberine molecule per 5 DNA base pairs and an apparent association constant of the order of 10^6 M^{-1}, while, by a weaker process, one alkaloid molecule was bound per 2 base pairs with an apparent association constant of the order of 10^5 M^{-1}. The standard evaluation of non-linear adsorption isotherms in terms of two underlying binding processes, involving two categories of binding sites, is based upon the assumption that DNA binding sites for ligands are independent, i.e. that upon partial binding the remaining free sites are not modified or influenced by the extent of binding which already has occurred. This unproved assumption in the interpretation of binding curves is not of critical importance in the consideration of berberine's binding to DNA since the stoichiometry of the weaker process did not exceed one alkaloid molecule per two base pairs, i.e. the stoichiometric limit for intercalation binding.

V. Discussion

The bodies of information on the antibacterial actions of berberine and on the non-chromosomal genetic effects of the alkaloid and its binding to DNA are unconnected and appear to be unrelated. The elimination of the yeast mitochondrial genetic system and of resistance determinants from bacterial R-factors are consistent with biophysical data on the formation and properties of a DNA-berberine complex and might be interpreted to mean that berberine can act as a DNA template poison to inhibit the replication of non-chromosomal DNA of mitochondria and of plasmids, i.e. of autonomous genetic systems. Effects of berberine upon cell-free nucleic acid polymerase reactions have not been reported.

The antibacterial effects of berberine are, however, not results of inhibitions of chromosomal DNA replication. While in the bactericidal action of the alkaloid, all biosyntheses that were studied (J.C.) failed, berberinium sulfate, acting on *V. cholerae*, produced only a slight or no inhibition of DNA biosynthesis. The loss of RNA from berberine-treated *V. cholerae* and the attending complete inhibition of protein biosynthesis rather point to the possibility of a direct effect of the alkaloid on ribosomes. Indeed, in *in vitro* experiments on the labilizing effect of intercalative drugs on *E. coli* ribosomes (WOLFE et al., 1972) it was discovered that berberinium chloride accelerated the disassembly of ribosomes under thermal stress. It is not impossible that the alkaloid renders ribosomal RNA susceptible to hydrolysis by ribonuclease which is found in association with isolated bacterial ribosomes, but *ad hoc* experiments on the fate of ribosomes in berberine-treated bacteria have not been reported. The few studies of berberine on individual enzyme reactions have shed no light on the mechanism of antibacterial action of the alkaloid (KUWANO and YAMAUCHI, 1960 a and b; KUWANO et al., 1961).

In view of observations that berberine associates preferentially with the lipid fraction (membranes?) of *V. cholerae*, inhibits the incorporation of ^{14}C-glycine into that fraction, alters its fatty acid composition in bacteria which are grown in the presence of the alkaloid (MODAK et al.,1970b) and inhibits the formation of cholera toxins, the possibility must finally be considered that berberine acts as a membrane poison; this would explain the uniform failure of macromolecular biosyntheses in experiments in which the alkaloid was bactericidal for *D. pneumoniae*. At the time of this writing, the mode and mechanism of berberine's antibacterial actions remain unknown and would appear an interesting field for further study.

References

AMIN, A.H., T.V.SUBBAIAH, and K.M.ABBASI: Berberine sulfate: Antimicrobial activity, bioassay, and mode of action. Can. J. Microbiol. **15**, 1067 (1969).

CUSHING, E.C.: Apparent specific inhibitive action of certain oxytoxic and spasmogenic drugs and substances against cercariae of *Schistosoma mansoni*. Military Med. **121**, 17 (1957).

Das GUPTA, B.M., and B.B.DIKSHIT: Berberine in treatment of oriental sore. Indian Med. Gaz. **64**, 67 (1929).

DUTTA, N.K., P.H.MARKER, and N.R.RAO: Berberine in toxin-induced experimental cholera. Brit. J. Pharmacol. **44**, 153 (1972).

DUTTA, N.K., and M.V.PANSE: Usefulness of berberine (an alkaloid from *Berberis aristata*) in the treatment of cholera (experimental). Indian J. Med. Res. **50**, 732 (1962).

FOLEY, G.E., R.E. McCARTHY, V.M. BINNS, E.E. SNELL, B.M. GUIRARD, G.W. KIDDER, V.C. DEWEY, and P.S. THAYER: A comparative study of the use of microorganisms in the screening of potential antitumor agents. Ann. N.Y. Acad. Sci. **76**, 413 (1958).

GADAMER, J.: Über das Berberin. Arch. Pharm. **243**, 31 (1905).

GRAY, P.H.H., and R.A. LACHANCE: Assimilation of berberine by bacteria. Nature **177**, 1182 (1956).

GRAY, P.H.H., and R.A. LACHANCE: Assimilation of berberine and chelidoxanthine by bacteria. Plant Soil **8**, 354 (1957).

HAHN, F.E., and J. CIAK: Elimination of bacterial episomes by DNA-complexing compounds. Ann. N.Y. Acad. Sci. **182**, 295 (1971).

HAHN, F.E., and A.K. KREY: Interaction of alkaloids with DNA. Progr. Molec. Subcell. Biol. **2**, 134 (1971).

HOMMA, N., M. KONO, H. KADOHIRA, S. YOSHIHARA, and S. MASUDA: Influence of berberine chloride on the intestinal flora of infants. Arzneimittel-Forsch. **11**, 450 (1961).

JEFFS, P.W.: The protoberberine alkaloids. *In*: The alkaloids, IX, R.H.F. MANSKE (ed.), p. 41 ff. Academic Press 1967.

JOHNSON, C. C., G. JOHNSON, and C. F. POE: Toxicity of alkaloids to certain bacteria. II. Berberine, physostigmine and sanguinarine. Acta Pharmacol. Toxicol. **8**, 71 (1952).

KLÍMEK, M., and L. HNILICA: The influence of deoxyribonucleic acid on ultraviolet and visible light absorption of berberine. Arch. Biochem. Biophys. **81**, 105 (1959).

KREY, A. K., and F. E. HAHN: Berberine: Complex with DNA. Science **166**, 755 (1969).

KREY, A. K., and F. E. HAHN: Methyl green-DNA complex: Displacement of dye by DNA-binding substances. Proc. First Eur. Biophys. Congr. **1**, 223 (1971).

KUWANO, S., and K. YAMAUCHI: Effect of berberine on tyrosine decarboxylase activity of *Streptococcus faecalis*. Chem. Pharm. Bull. (Tokyo) **8**, 491 (1960a).

KUWANO, S., and K. YAMAUCHI: Competition of berberine with pyridoxal phosphate in the tryptophanase system of *Escherichia coli*. Chem. Pharm. Bull. (Tokyo) **8**, 497 (1960b).

KUWANO, S., K. YAMAUCHI, and T. G. BAK: Competition of berberine with pyridoxal phosphate in the tryptophanase system of *Escherichia coli*. Chem. Pharm. Bull. (Tokyo) **9**, 651 (1961).

LAHIRI, S. C., and N. K. DUTTA: Berberine and chloramphenicol in the treatment of cholera and severe diarrhoea. J. Indian Med. Assoc. **48**, 1 (1967).

LAMBIN, S., and J. BERNARD: Sur les modalités d'action des substances alcaloïdiques. Compt. Rend. Soc. Biol. **149**, 492 (1955).

LERMAN, L.: Acridine mutagens and DNA structure. J. Cell. Comp. Physiol. **64**, Suppl. 1, 1 (1964).

MAHLER, H. R., B. D. MEHROTRA, and P. S. PERLMAN: Formation of yeast mitochondria. V. Ethidium bromide as a probe for the functions of mitochondrial DNA. Progr. Molec. Subcell. Biol. **2**, 274 (1971).

MEISEL, M. N., and T. S. SOKOLOVA: Inherited cytoplasmic changes induced in yeast by acriflavine and berberine. Dokl. Akad. Nauk SSSR **131**, 436 (1959).

MICHAELIS, L.: The nature of the interaction of nucleic acids and nuclei with basic dyestuffs. Cold Spring Harbor Symp. Quant. Biol. **12**, 131 (1947).

MODAK, S., M. J. MODAK, and A. VENKATARAMAN: Mechanism of action of berberine on *Vibrio cholerae* and *Vibrio cholerae* biotype eltor. Indian J. Med. Res. **58**, 1510 (1970a).

MODAK, M. J., S. M. MODAK, and A. VENKATARAMAN: Effect of berberine on the fatty acid composition of *Vibrio cholerae* and *Vibro cholerae* biotype eltor. Indian J. Med. Res. **58**, 1523 (1970b).

MORTHLAND, F. W., P. P. H. DE BRUYN, and N. H. SMITH: Spectrophotometric studies on the interaction of nucleic acids with aminoacridines and other basic dyes. Expt. Cell Res. **7**, 201 (1954).

SADO, S.: Pharmacological study of berberine and its derivatives II. J. Pharm. Soc. Japan **67**, 166 (1947).

SHARDA, D. C.: Berberine in the treatment of diarrhoea in infancy and childhood. J. Indian Med. Ass. **54**, 22 (1970).

SHIDEMAN, F. E.: A review of the pharmacology and therapeutics of hydrastis and its alkaloids, hydrastine, berberine and canadine. Bull. Natl. Formulary Committee **18**, 3 (1950).

SHIH-CHIEH CH'OU, and LIANG-CHUN CH'A: Physical and chemical properties of berberine as related to the biological effects. I. Acid dissociation constants. Yao Hsüeh T'ung Pao, **12**, 57 (1965) through Chem. Abstr. **64**, 11027g (1966).

STICKL, O.: Die bactericide Wirkung der Extrakte und Alkaloide des Schöllkrautes (Chelidonium maius) auf grampositive pathogene Mikroorganismen. Z. Hyg. **108**, 567 (1928).

SUBBAIAH, T. V., and A. H. AMIN: Effect of berberine sulphate on *Entamoeba histolytica*. Nature **215**, 527 (1967).

TUROVA, A. D., A. I. LESKOV, and V. I. BICHEVINA: Berberine. Lekarstv. Sredstva iz Rast. **1962**, 303.

UCHIZUMI, S.: Pharmacological action of berberine. Nippon Yakurigaku Zasshi **53**, 63 (1957).

VARMA, R. L.: Berberine sulphate in oriental sore. Indian Med. Gaz. **62**, 84 (1927).

WOLFE, A. D., R. G. ALLISON, and F. E. HAHN: Labilizing action af intercalating drugs and dyes on bacterial ribosomes. Biochemistry **11**, 1569 (1972).

YAMAGISHI, H.: Interaction between nucleic acids and berberine sulfate. J. Cell Biol. **15**, 589 (1962).

Boromycin

W. Pache

Boromycin is the first antibiotic as well as the first natural product found to contain the trace element boron (HÜTTER et al., 1967). Its structure was elucidated by degradation studies which yielded D-valine, boric acid and a polyhydroxy compound, and finally by X-ray analysis of the anion obtained from boromycin by splitting the D-valine (DUNITZ et al., 1971). The complete structure is given in Fig. 1.

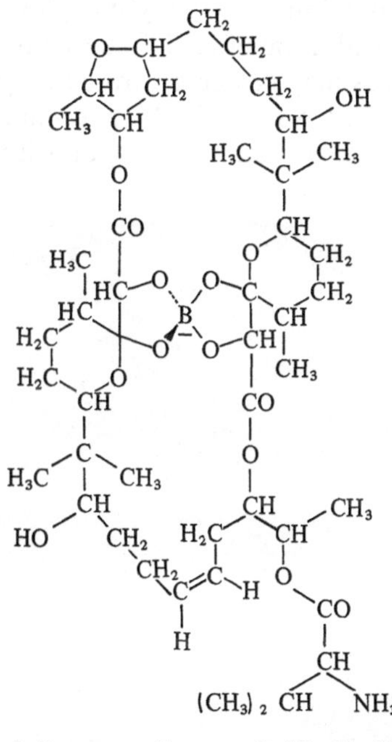

Fig. 1. Structure of boromycin ($C_{45}H_{74}BNO_{15}$)

Boromycin is produced by a streptomycete of the species *Streptomyces antibioticus* (WAKSMAN and WOODRUFF). Because of its lipophilic character the antibiotic can be extracted from the mycelium with organic solvents; only trace amounts are found in the culture filtrate. Purification is achieved by countercurrent distribution and chromatography on silica gel (HÜTTER et al., 1967).

Boromycin inhibits gram-positive bacteria, certain fungi and protozoae, whereas gram-negative bacteria are not sensitive to the antibiotic (HÜTTER et al., 1967). The resistance of gram-negative organisms seems to be due to an impeded access since halobacteria and mycoplasmas, gram-negative organisms lacking a complex cell envelope, are sensitive to boromycin (PACHE, unpublished results). The mammalian toxicity was found to be 180 mg/kg p.o. (LD_{50}) (HÜTTER et al., 1967). Considering the lipophilic character of the antibiotic it is possible that it is poorly absorbed from the intestine and the actual toxic dose is much lower. No medical use of the antibiotic has been reported so far.

At low concentrations boromycin increases branching in the hyphae of the fungus Botrytis cinerea, very similar to the alterations caused by the scopamycins on this organism (HÜTTER et al., 1965).

The mode of action of boromycin was studied using Bacillus subtilis (PACHE and ZÄHNER, 1969b). Inhibition of growth takes place immediately after addition of the antibiotic, and lysis is observed when the dose is slightly higher than the minimum inhibitory concentration. Lysis is also observed when the antibiotic is added to cells grown up to the stationary phase. With log-phase cells, protein, RNA and DNA syntheses are stopped simultaneously. Action of boromycin is prevented or reversed by adding a variety of lipids, lipoproteins (PACHE and ZÄHNER, 1969a) and by the addition of KCl. Potassium ions cannot be substituted by other alkali metal ions as may be demonstrated on plates containing counter-current gradients of KCl and NaCl (Fig. 2). The inhibition zone in the region of high NaCl/KCl ratio is wider than in the region with high ratio of KCl/NaCl.

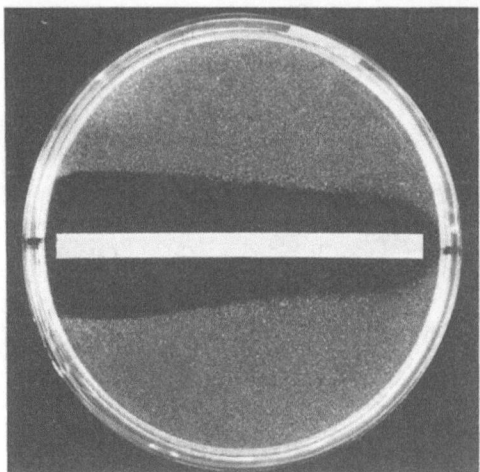

Fig. 2. Inhibition zone of boromycin on a test plate with countercurrent gradients of K^+ and Na^+. Plate prepared by pouring agar containing 0.5 m KCl and agar with 0.5 m NaCl after adjusting the dish on a slope and covering these layers with a layer seeded with spores of B. subtilis. Concentration of K^+ increases from left to right, Na^+ from right to left

The antagonising effect of lipids and lipoproteins on the action of boromycin indicated that the antibiotic attacks the cytoplasmic membrane. Investigating the accumulation of boromycin in different cell fractions by measuring the boron content a marked accumulation of the antibiotic in the membrane fraction could be demonstrated (PACHE and ZÄHNER, 1969b). The reversing effect of potassium ions gave a clue to the mechanism of action of boromycin, since a similar effect was reported for the action of ion-transporting antibiotics on *Streptococcus fecalis* (HAROLD and BAARDA, 1967; HAROLD and BAARDA, 1968). The action of these potassium complexing antibiotics is explained by the carrier function they perform within the cytoplasmic membrane and the resulting breakdown of the permeability barrier for potassium ions, which are accumulated by the bacterial cell. Addition of potassium ions to the medium prevents loss of K^+ from the cell and therefore blocks the action of the antibiotic.

For boromycin the ability to transport ions across artificial membranes could be demonstrated using bulk membranes and lecithin bilayers (black films), although only a low specifity for potassium ions was observed in these systems (PACHE and ZÄHNER, 1969 b; PACHE, unpublished results). No interaction of boromycin with phospholipids was found using calorimetric and spin label techniques (PACHE and CHAPMAN, 1972), and it can therefore be concluded that the action of boromycin is due to its carrier properties.

X-ray analysis of boromycin (DUNITZ *et al.*, 1971) confirmed that its structure is in accordance with the postulated mechanism of action of the antibiotic. The overall shape of the anion obtained by cleavage of D-valine is spherical with a lipophilic surface and a cleft lined with oxygen atoms. In the anion the cation is housed in the cleft coordinated by 6–8 oxygen atoms, whereas in boromycin itself the amino group of D-valine is accomodated within the cleft and this region is blocked by the lipophilic part of the amino acid.

The presence of a polar center covered in a lipophilic sphere is exactly the requirement for ion-transporting antibiotics, as was first shown for nonactin (KILBOURN *et al.*, 1967). Since boromycin and the anion obtained share the same rigid conformation (DUNITZ *et al.*, 1971) it is likely that the anion mediates ion-transport too, and actually antibiotic action could be demonstrated for the degradation product (PACHE and ZÄHNER, 1969 b).

References

DUNITZ, J. D., D. M. HAWLEY, D. MIKLOS, D. N. J. WHITE, YU. BERLIN, R. MARUSIC, and V. PRELOG: Structure of boromycin. Helv. Chim. Acta **54**, 1709 (1971).

HAROLD, F. M., and J. F. BAARDA: Gramicidin, valinomycin and cation permeability of Streptococcus fecalis. J. Bacteriol. **94**, 53 (1967).

HAROLD, F. M., and J. F. BAARDA: Effect of nigericin and monactin on cation permeability of Streptococcus fecalis and metabolic capacities of potassium-depleted cells. J. Bacteriol. **95**, 816 (1968).

HÜTTER, R., W. KELLER-SCHIERLEIN, F. KNÜSEL, V. PRELOG, G. C. RODGERS, JR., P. SUTER, G. VOGEL, W. VOSER, and H. ZÄHNER: Boromycin. Helv. Chim. Acta **50**, 1533 (1967).

HÜTTER, R., W. KELLER-SCHIERLEIN, J. NÜESCH, and H. ZÄHNER: Scopamycine Arch. Mikrobiol. **51**, 1 (1965).

KILBOURN, B. T., J. D. DUNITZ, L. A. R. PIODA, and W. SIMON: Structure of the K^+-complex with nonactin, a macrotetrolide antibiotic possessing highly specific K^+-transport properties. J. Mol. Biol. **30**, 559 (1967).

PACHE, W., and D. CHAPMAN: Interaction of antibiotics with membranes. Chlorothricin. Biochim. Biophys. Acta **255**, 348 (1972).

PACHE, W., u. H. ZÄHNER: Bindung von Antibiotica an Lipoprotein. Arch. Mikrobiol. **66**, 281 (1969 a).

PACHE, W., and H. ZÄHNER: Studies on the mechanism of action of boromycin. Arch. Mikrobiol. **67**, 156 (1969 b).

Colicins 1972

Louis W. Wendt

Introduction

Colicins are bactericidal proteins synthesized by certain strains of enteric bacteria and active against other strains of the same, or related, bacterial species. The well-studied colicins fall into three groups (Table 1) as defined by their predominant physiological effects: 1. inhibition of protein synthesis; 2. degradation of DNA; and 3. general blockage of energy dependent cellular functions. Colicins of all three classes exert their effects via poorly understood processes involving the cell membrane. Colicins are not antibiotics in the usual sense of the word, since they apparently require specific membrane proteins by means of which they adsorb to, and initiate their attack upon sensitive cells. No effects of colicins upon the cells of unrelated species (including mammalian cells) have been reported, nor have they been reported to attack artifical lipid bilayer membranes. Inclusion of a review of colicins in a volume such as this thus depends on the light their study can shed upon the action of agents of broader spectrum and upon basic mechanisms of attack upon the particular cellular functions affected.

Table 1. *Classes of colicins*

Colicin	Approximate molecular weight	Processes affected
E2	60,000[a]	DNA synthesis: DNA degraded Cell division: blocked, filaments form Prophage replication: induced
E3 D	60,000[a] 92,000[b]	Protein synthesis; 30S ribosome inactivated Protein synthesis; —
A E1 K Ia; Ib Q[f]	— 55,000[c] 45,000[d] 80,000; 78,000[e] —	Membrane functions; energy production blocked* Membrane functions; energy production blocked* Membrane functions; energy production blocked* Membrane functions; energy production blocked* Membrane functions; — —

– Signifies not reported.

* A variety of other effects are listed in Table 3.

[a] HERSCHMAN and HELINSKI (1967a)

[b] TIMMIS (1972).

[c] SCHWARTZ and HELINSKI (1971).

[d] JESAITIS (1970).

[e] KONISKY and RICHARDS (1970).
  KONISKY (1972).

[f] CAVARD et al. (1968).

The information on colicins has been reviewed several times (FREDERICQ, 1957; REEVES, 1965; NOMURA, 1967a and 1967b), including a summary review previously published in this series (NOMURA, 1967a). Discussions of the relationship of colicins to other bacteriocins appear elsewhere and will not be repeated here (REEVES, 1965; HAMON, 1964). This review focuses primarily on recent information on the mechanism of action of colicins, and the relationship of this information to basic problems of membrane structure and function. An attempt has been made to stitch together the information on colicin action into a fabric sufficiently tight and detailed that the major problems and relationships are brought into sharp focus. A **Medline** survey of colicins for the years 1970–1972 produced 140 references. (NOMURA's 1967 review in which most of the references were drawn from the decade 1957–67 listed a total of 178 references). Considering the rapidly accumulating information, and the complexity of some of the phenomena encountered (e. g., the extreme pleiotropism of some of the colicin tolerant mutations—Section B), this review must be regarded as a status report and not as definitive. Some familiarity with phage and bacterial genetics is a useful prerequisite (or co-requisite) for the reading of this discussion. HAYES' classic text is highly recommended (HAYES, 1968).

A. Cellular Synthesis and Origin

Although the name colicins does not indicate it, these agents are produced by and are active upon several members of the family *Enterobacteriaceae* including *Shigella, Salmonella* and *Escherichia* species. Within the cell, colicin synthesis is directed by genes carried on extra-chromosomal elements called colicinogenic factors (col factors), representatives of the general class of genetic elements known as plasmids, which exist and replicate independently of the host cell chromosome. Where col factor DNA's have been isolated they are found to be of molecular weights ranging from $4–5 \times 10^6$ (for the E. colicins; BAZARAL and HELINSKI, 1968) to about 10^8 (col V2; CLOWES, 1970). Two size classes seem apparent at the present time, since, except for the E col factors, the others whose molecular weights have been measured fall in the range from 6×10^7 to 10^8 (CLOWES, 1972, Table 3). Thus, even the smallest col factors are large enough to contain 25–30 genes. The functions of nearly all of these genes are unknown. Some must be involved with DNA replication and its control, others may be concerned with plasmid establishment and maintenance at sites associated with cell membrane (NOVICK, 1969). Just how, or indeed if, the colicin molecules themselves function in relation to these other plasmid activities remains to be determined.

Colicin production is increased in cells which have been treated with mitomycin C or irradiated with ultraviolet light (HERSCHMAN and HELINSKI, 1967b). By analogy with the induction of temperate bacteriophage this could be taken to indicate that colicin synthesis is coupled to colicin factor replication. However, no increase in colicin factor DNA following mitomycin C treatment has been observed with cells colicinogenic for colicin E2 (HARDY and MEYNELL, 1972) or for colicin I (KONISKY, personal communication), even though the colicin titer increased in these same experiments.

It has frequently been speculated, and there is indeed some evidence, that colicins may be evolutionarily related to defective bacteriophage (e. g. see FREDE-RICQ, 1957; NOMURA, 1967), but a detailed comparison (FIELDS, 1969) of abortive T-phage infection and colicin E1 and K action led to the conclusion that the inhibitions of cellular function observed were related only in that the effects of phages and colicins both involved membrane alterations. It was considered probable that the mechanisms were quite different (FIELDS, 1969). REEVES (1965) has suggested fertility recognition sites as an alternative evolutionary origin. This question is one of many stemming from the larger topic of the origin of phages, both temperate and virulent, and as such is far beyond the scope of this discussion. In addition to the reviews already cited the reader is referred again to the monograph by HAYES (1968) for the background it provides to this topic. At a more restricted level it is possible that colicins will be found to be derived from proteins which function in relation to replication, or membrane site maintenance, of the corresponding col factor. Consistent with this proposal is the fact that four different colicins have been found to exist as protein—lipocar-bohydrate complexes identical to the somatic 0 antigen of the organism which produced them (GOEBEL, 1962; HUTTON and GOEBEL, 1963; BARRY et al., 1963; NUSKE et al., 1957). Although this has not been found for other colicins (HERSCH-MAN and HELINSKI, 1967a) the inference remains that the bacterial surface is the normal cellular location of colicins. This location and proposed function are compatible with a fertility recognition site origin.

B. Mechanisms of Colicin Resistance

Sensitive bacterial cells may acquire resistance to colicins either by genetic changes, or by particular chemical and physical treatments which produce changes in the bacterial cell surface. The genetic changes include: 1. harboring a colicinoge-nic factor that produces an homologous colicin ("immunity"); 2. mutations that alter the cell wall so as to prevent colicin adsorption (resistance); and 3. mutations that alter the cell membrane and block the action of adsorbed colicin (tolerance). Treatments which render the cells colicin resistant include exposure to dinitrophe-nol, plasmolysis, conversion to sphereoplasts, and low temperature.

1. Immunity and Immunity Breakdown

The colicin-immune state is formally analogous to the lysogenic immune state found with temperate phage such as lambda, since in both cases cells carrying the genetic information for the production of a lethal agent (phage or colicin) are immune to attack from the outside by the same agent. In both cases, immunity is relative, i.e., dependent upon the concentration of attacking colicin molecules, or phages. At high colicin/cell ratios (20—100) immunity "breaks down" and the whole spectrum of effects of the particular colicin, including lethality, is exhibited (LEVISOHN et al., 1968). Just what this means is not clear. In the temperate phage system—as, for example, with lambda phage—lysogenic cells were found to undergo immunity breakdown at multiplicities of 25—30 (JACOB et al., 1962); this can be accounted for by the titration of the lambda repressor substance by the injected genome of the superinfecting phage. When sufficient genomes are present to account for all cellular repressor molecules, additional genomes escape repression and phage multiplication ensues. Thus colicin immunity breakdown also may be due to the titration of an "immunity substance".

As an alternative to this mechanism LEVISOHN et al. (1968) suggested that immunity led to an "alteration in the process whereby the initial local effect of a colicin molecule at the receptor is transmitted to all of the biochemical targets". This explanation was predicated on Nomura's model of colicin action (NOMURA, 1964) which involves the transmission of colicin effects through the membrane by an unspecified process. NOMURA's model, however, requires modification because of recent results bearing on the nature of the assumed transmission process (Section C, below). It is sufficient to observe here that the two proposed mechanisms are not necessarily distinct—an immunity substance could act via specific local effects within the cell membrane. Finally, as pointed out by LEVISOHN et al., (1968), the titration of an immunity substance by excess colicin would be expected to lead to a multi-hit killing of the immune strain, with the number of hits corresponding approximately to the ratio of colicin molecules/cell required to produce immunity breakdown. In fact this was not observed. A 2–3 hit curve (rather than a 20–100 hit curve) was obtained when the multiplicity required to kill a colicin immune strain was expressed in units determined by titering the colicin on the sensitive parental strain. As a partial explanation of this result it was suggested that the immune cells may be heterologous with respect to the number of immunity substances (LEVISOHN et al., 1968).

An activity, inhibitory for colicin E3 action in vitro, has been detected in crude colicin E3 preparations (BOWMAN et al., 1971). This activity is not found in similar preparations made from colicin-sensitive cells, and may therefore be the hypothetical immunity substance. Whatever the mechanism of immunity, it is not always completely specific for a given colicin, since cells colicinogenic for colicin E2 have been observed to be approximately eight times less sensitive to colicin E3 (NOMURA, 1964).

2. Resistance Due to Mutation

a) Resistance and Tolerance. In the context of colicins, resistance and tolerance refer to distinct ways in which bacterial cells may become insensitive to colicin by mutation. The distinction between the two types of mutationally produced resistance was made independently by CLOWES (1965) and by NOMURA (1964) who both isolated mutants resistant to colicins which were then shown to adsorb colicin normally. This type of resistance, now termed tolerance, has also been distinguished from resistance which alters the cell wall and prevents adsorption in studies of subcellular membrane vesicle preparations. Membrane vesicles derived from tolerant cells adsorb colicin but are not sensitive to colicin action, whereas membrane vesicles derived from resistant cells (which do not adsorb colicin) are colicin-sensitive (BHATTACHARYYA et al., 1970). Removal of the cell wall during vesicle preparation removes the site involved in superficial resistance, and the full machinery for colicin response and the site of tolerance mutations seem clearly linked to the cell membrane. One can argue, as SABET and SCHNAITMAN

(1971) have done that some cell wall material is still associated with the vesicles, but if so, it has clearly lost the altered structure needed to prevent adsorption. It seems more reasonable to regard the operation of making the vesicles as a separation of cell wall from membrane so that material remaining with the vesicles is considered membrane rather than wall.

b) Tolerant Mutants. The results summarized above point to a role for the cell wall in permitting access to specific membrane sites at which colicins adsorb and initiate their lethal and physiological effects. Such sites can be identified with the sites revealed in electron microscopic studies (BAYER, 1968) as locations at the base of adsorbed phage where the membrane continues to adhere to the wall in plasmolysed cells, since phages T1, T6, and BF23 apparently share adsorption sites with colicins M, K, and the E group colicins respectively (FREDERICQ, 1953, 1956; NOMURA, 1967b). However, these results shed no light on the specific membrane sites altered by the tolerance mutations, if indeed they are specific.

Cells which are tolerant to colicin apparently exhibit a tolerance breakdown (NAGEL DE ZWAIG and LURIA, 1967, Fig. 1; NOMURA and MAEDA, 1965, pp. 235–236) which may be similar to the immunity breakdown described above. Detailed studies of this phenomenon have not been reported.

Genetic studies of colicin tolerance have defined nine classes of colicin tolerant mutants (HILL and HOLLAND, 1967; NAGEL DE ZWAIG and LURIA, 1967; NOMURA and WITTEN, 1967; REEVES, 1966; Table 2). These classes are based on the pattern of sensitivity and resistance (tolerance) to colicins K, E1, E2 and E3. Although several other colicins (including A, B, C, D, H, I, and V) have been used in two of these studies, the unique tolerance classes can be defined on the basis of the mutants' response to colicin K and the E group colicins only (Table 2). The genetic loci at which mutations can occur to give rise to these phenotypes, where known, have been designated *tol A, B, tol C, tol D, tol P,* and *cet B, C* (BERNSTEIN et al.,1972; BURMAN and NORDSTROM, 1971; HOLLAND et al., 1970; TAYLOR, 1970; WHITNEY, 1971). The *tol A, B* loci correspond to the phenotypic classes tol II, tol III, and tol IV (BERNSTEIN et al., 1972; NAGEL DE ZWAIG and LURIA, 1967; NOMURA and WITTEN, 1967) and the *tol C* locus corresponds to the class tol VIII (NAGEL DE ZWAIG and LURIA, 1967; WHITNEY, 1971). *Tol D* mutants resemble the tol III and tol IV classes, but have been incompletely characterized and could well belong to a new class. Their map location has been placed at a site distinct from *tol A, B* (BURMAN and NORDSTROM, 1971). *Tol P* mutants are partially tolerant mutants occurring in a distinct cistron at the *tol A, B* locus (BERNSTEIN et al., 1972). The *cet B, C* locus is the name now given to a locus previously called *ref II,* and corresponds to the phenotypic class tol VII (or ref II; Table 2).

The spectrum of tolerant mutants so far described is intriguing and begs for a detailed interpretation. A few of the more provocative cases are pointed out here. Mutants of type tol II are tolerant to colicins A, E1, E2, E3, and K, but sensitive to colicins B, D, H, and I, and V (NAGEL DE ZWAIG and LURIA, 1967; NOMURA and WITTEN, 1967). (Colicins A, E1, K and I have similar effects, as do D and E3. Colicin E2 has yet different effects. See Table 1). There have been no mutants so far clearly shown to be tolerant to only one (singly tolerant) of the above array of colicins, although *tol C* mutants, characterized as tolerant

Table 2. *Phenotypic classes of tolerant mutants*

Designation(s)	Resistance pattern			
	E1	E2	E3	K
tol II (tol IId)[a,b]	r	r	r	r
tol V[c]	r	r	r	s
tol III (ref V, VI and VII)[b]	s	r	r	r
tol VI	r	r	s	s
tol IV, ref IV (tol IVt)	s	r	r	s
ref III	s	s	r	r
tol VIII, ref I[b]	r	s	s	s
tol VII (ref II, tol VIIt)	s	r	s	s
tol I[c]	s	s	s	r

The basic classification is derived from the resistance pattern exhibited at 37C, or at both 30 C and 40 C. Designations referring to mutants in which this pattern is altered at either 30 C, or 40 C, are in parentheses. The tol designations are used here as the primary designations since they apply to the greatest number of basic classes. The ref designations refer to a series of mutants isolated by HILL and HOLLAND (1967) which they called refractory, and several of which correspond to particular tol classes in their resistance pattern as shown here.

[a] Tol II types are also resistant to colicin A (NAGEL DE ZWAIG and LURIA, 1967), and probably colicin C (REEVES, 1966).

[b] These classes have been tested against, and found sensitive to, colicins B, D, H, I, and V.

[c] Mutants also exist with these same phenotypes but which are also resistant to specific bacteriophage. These are cell wall receptor mutants (e. g. see Section 2b, or BHATTACHARYYA *et al.*, 1970).

only to colicin E1, apparently show only slightly decreased sensitivity to colicins A and K (NAGEL DE ZWAIG and LURIA, 1967). Possible candidates for singly tolerant mutations (again restricting the discussion to the array above) are among the tol I and tol VII types which have been identified as tolerant to colicins K and E2 respectively, but which have apparently not been tested for tolerance to A, B, D, H, I, or V (NOMURA and WITTEN, 1967). Although the E group colicins share the same adsorption site, mutants have been reported which are tolerant to E1 and E2, but not E3, and to E2 and E3, but not E1 (tol IV and tol VI; NOMURA and WITTEN, 1967; BURMAN and NORDSTROM, 1971).

A further peculiar property of tolerance mutations is their extreme pleiotropism. Examples of this pleiotropy are found in the multiple tolerances described above, and in effects which several of these mutants have on a variety of other characteristics related to the cell membrane, or surface. The *tol A, B* mutants were originally described as fragile, hypersensitive to ethylene diamine tetraacetate, and deoxycholate (NAGEL DE ZWAIG and LURIA, 1967) and have now been reported to also be sensitive to vancomycin, bacitracin, and sodium dodecyl sulfate (SDS), and to show an increased efficiency of plating of lysis-defective mutants of bacteriophage lambda (BERNSTEIN *et al.*, 1972). *Tol C* mutants show increased sensitivity to acridine dyes, methylene blue, deoxycholate, erythromycin and lincomycin, and resistance to vancomycin and bacitracin (NAGEL DE ZWAIG and LURIA, 1967; BERNSTEIN *et al.* 1972; WHITNEY, 1971). HOLLAND and his coworkers have isolated mutants, specifically tolerant to colicin E2 (*cet B, C* mutants), which show increased sensitivity to ultraviolet light and deoxycholate,

defective cell division, abortive growth of bacteriophage lambda, and recombination deficiency (HOLLAND, et al., 1970). It might be expected from these results
that selection for resistance to at least some of the large variety of antibiotics
would also uncover colicin tolerant mutants, perhaps of new classes. The tol
D mutants, which were selected for ampicillin resistance and subsequently found
to be tolerant to colicins E2 and E3, and sensitive to deoxycholate (BURMAN
and NORDSTROM, 1971), confirm this expectation.

These pleiotropic effects can be explained in general if the products of the
tol genes are proteins which play an organizational role and interact with the
components of other functional systems related to the membrane. Two laboratories have, in fact, reported changes in membrane proteins related to colicin tolerance using SDS gel electrophoresis analysis of membranes derived from mutant
cells – a mutant with an extensive deletion including the tol A, B locus has
been found to be missing several membrane proteins (ONODERA et al., 1970),
one protein band was found to be absent from a tol C deletion mutant (ROLFE
and ONODERA, 1971), and a cet B mutant was found to exhibit a new protein
band (HOLLAND and TUCKETT, 1972). There is no evidence at present for alterations in the lipid components of the membrane in any of these mutants. Indeed,
the specificity of the various classes with regard to the array of colicins to which
they are resistant (Table 2) seems difficult to reconcile with such a possibility,
since changes in membrane lipids might be expected to produce generally resistant
cells. In contrast, treatments exist which appear to block colicin action via general
changes in the cell membrane, and which may, therefore, act nonspecifically.

3. Treatments Which Produce Resistance

The first potentially general inhibitor of the action of adsorbed colicin reported
was dinitrophenol (REYNOLDS and REEVES, 1963; NOMURA, 1964). It is now known
that dinitrophenol inhibits the action of colicins E1, E2, E3 and K, and that
other compounds with similar modes of action (for example p-trifluoro-methoxycarbonylcyanide phenylhydrazone: FCCP) prevent the lethal action of these same
colicins (NOMURA and MAEDA, 1965; PLATE and LURIA, 1972). Although the
primary mode of action of agents such as dinitrophenol and FCCP is now better
explained as a result of the conduction of protons into cells, rather than the
uncoupling of oxidative phosphorylation (see e. g. HAROLD, 1972), what is crucial
here is that they block cellular energy production. The implication seems clear
that a membrane with its energy producing functions intact is required for colicin
action, even for those colicins whose ultimate targets are DNA and ribosomes.

The action of colicins E1, E2 and K has been reported to be prevented
or retarded at temperatures between 0 and 37 C (WENDT, 1970a: LUSK and
NELSON, 1972; PLATE and LURIA, 1972). This probably results from changes
in the fluidity of membrane lipids (WENDT, 1970a; SINGER and NICOLSON,1972),
either by indirect effects upon the energy generating capacity of the membrane,
and/or by a direct dependence of some step in colicin action on this fluidity.
It is significant that at intermediate temperatures, colicin K action, as monitored
by potassium loss, is delayed rather than prevented (WENDT, 1970a). The length
of the delay is dependent on colicin multiplicity as well as on temperature, being

shorter at higher multiplicities and temperatures. This can be explained if an energy-requiring step in colicin action which is retarded at low temperatures occurs more rapidly at higher multiplicities, or if some process not requiring cellular energy (for example diffusion of colicins within the membrane—see SIN-GER and NICOLSON, (1972); Section C. 3., below) is slowed by a decrease in lipid fluidity.

Blockage of colicin action has also been reported to occur upon plasmolysis of sensitive cells (for colicins K and E2; BEPPU and ARIMA, 1967), and with spheroplast formation (for colicin K, E2, and E3; NOMURA and MAEDA, 1965). It may be that these two conditions differ in the way they affect colicin action even though both involve changes in the relationship between cell wall and membrane, and both involve exposure to conditions of high osmolarity. The protection afforded by plasmolysis seems less complete than that due to "spheroplasting" since comparable experiments (similar colicin multiplicity and time scale) with colicin E2 show degradation of DNA and killing of plasmolyzed cells at the solute concentration conferring the maximal protection (BEPPU and ARIMA, 1967), but no degradation of DNA within spheroplasts (NOMURA and MAEDA, 1965). However, this difference may depend on the particular colicin studied—breakdown of the resistance of spheroplasts to colicin E3 is observed at high multiplicities (NOMURA and MAEDA, 1965).

The effect of plasmolysis is not due to a complete separation of the membrane from the wall since, as noted earlier, many of the membrane-wall junctions on which colicins exert their action are retained. A change within the membrane itself seems more likely to be responsible for the resistance. A general hypothesis which could account for the apparent greater resistance of spheroplasts would be that a similar change in membrane structure, inhibitory for colicin action, is produced by the physical deformations which accompany plasmolysis and "spheroplasting", but that such changes are more extensive in spheroplasts. However, ALMENDINGER and HAGER (1972) have implicated endonuclease I in colicin E2 action and in the resistance of spheroplasts to colicin E2, in part by showing that osmotically shocked cells lose this periplasmic enzyme and become relatively insensitive to the colicin (see also Section C. 2. a.). They also state that, under the conditions used by NOMURA and MAEDA, spheroplast formation releases essentially all the endonuclease I, and point out that this can account for the colicin E2 resistance of spheroplasts. In addition it raises the possibility that changes in the location, or function, of specific surface proteins rather than general membrane changes may underlie the colicin resistance developed by plasmolysis and the other conditions discussed here.

C. Mechanisms of Colicin Action

1. General Processes in Colicin Action

One effect common to the action of various colicins is single hit killing. Experiments studying the loss of cell viability with time at a given colicin concentration, or the fraction of cells killed within the same time period at different colicin concentrations, generate survival curves which at low multiplicities are negative exponential functions passing through the origin. This result permits the conclusion that a single colicin molecule may, with a certain probability, cause the death of a sensitive cell (NOMURA, 1967b). In the case of a colicin such as E3 which has multiple targets within the cell some means must exist for transmitting the effects of single molecule to at least a large fraction of these targets. One can argue from the fact that the response to other colicins such as K (WENDT, 1970a) and E2 (NOMURA, 1963; HOLLAND and HOLLAND, 1970), is more rapid at high multiplicities, that colicins generally have multiple targets, and that the effects of all colicins must be capable of being transmitted from a single molecule to many targets. The most economical hypothesis would be that the mechanism responsible for this effect is similar for all colicins.

The existence of mutants tolerant to several colicins, and the general inhibitors of colicin action, indicate that many colicins also share in common a step, or steps, in their initial attack on cells. Further definition of these shared early steps comes from experiments in which the action of colicins (including most of those which have been studied intensively: E1, E2, E3, K and I; NOMURA, 1967b; LEVISOHN et al. 1968; PLATE and LURIA, 1972) can be halted by treatment with trypsin. If a certain stage in colicin action has not yet been reached the cell may be rescued from the lethal action of colicin. Parallel studies of the inhibition of physiological functions and the effect of trypsin indicate that those cells in which the inhibition of physiological function has been initiated are identical with those which cannot be trypsin rescued (PLATE and LURIA, 1972). At sufficiently low temperature, or in the presence of proton conductors, the cells appear to remain in the trypsin-rescuable state indefinitely (WENDT, 1970a; REYNOLDS and REEVES, 1963). Studies of the transition from the rescuable condition to the non-rescuable condition are stated to show that, as was expected, it is this transition which is blocked, or retarded, by low temperatures and proton conductors (PLATE and LURIA, 1972).

Finally, in spite of the differences between them, it should be noted that low temperature, plasmolysis, and spheroplasting all produce resistance which, like that owing to immunity and tolerance, may be overcome by increasing colicin multiplicity. If this were a reflection of effects on a process common to the action of various colicins one might expect that delays in colicin action would occur in immune, tolerant or plasmolyzed cells, or spheroplasts, as they do at low temperature. For colicin E2, at least, a delay in DNA degradation has been observed in plasmolyzed cells, and killing data are consistent with an effect of plasmolysis in delaying the transition to the trypsin insensitive state (BEPPU and ARIMA, 1967).

2. Mechanism of Action of Specific Colicins

a) Colicin E2. Colicin E2 has effects on DNA synthesis and cell division in *Escherichia coli.* The isolation of mutants in which cell division is blocked but DNA synthesis is unaffected (BEPPU et al., 1972) suggests that these processes may be independently influenced by the colicin in normal cells. We shall therefore discuss the two effects separately.

Early observations on colicin E2 action showed that all macromolecular syntheses in the cells could be affected (NOMURA, 1963; REEVES, 1965), but that the effect on DNA synthesis seemed to be primary since it could be blocked as early as two minutes after colicin addition (REEVES, 1965). Phage T4 and T5 DNA syntheses are not blocked (NOMURA, 1963; REEVES, 1965), hence this action of the colicin is not directed at all classes of intra-cellular DNA. Phage lambda replication is induced in lysogenic cells at low colicin multiplicities, suggesting some effect related to lambda excision and/or to the regulatory mechanisms for DNA synthesis involved in prophage induction (ENDO et al., 1963; NOMURA, 1963). The extent of induction is maximal at a multiplicity of one and declines exponentially with increasing multiplicity (NOMURA, 1963). The growth of lambda in lysogens induced by a temperature shift is also inhibited by the colicin (HULL and REEVES, 1971). The inhibition of phage multiplication and of cellular DNA synthesis may be explained by the findings that E2 provokes the degradation of both cellular DNA (NOMURA, 1963; HOLLAND, 1967), and replicating or non-replicating lamda DNA (HULL and REEVES, 1971). RINGROSE (1970) has studied the breakdown of cellular DNA in detail and shown that it proceeds in three stages: 1. single-strand scission, 2. limited endonucleolytic cleavage of double strands and 3. exonucleolytic degradation to acid soluble fragments. The breakdown of lambda DNA follows a similar pattern (HULL and REEVES, 1971). Cellular DNA breakdown has been found to be accelerated by thymine deprivation (BEPPU and ARIMA, 1971).

Analysis of colicin E2 action on DNA has now focused on the question of the particular nuclease activities involved in the degradation. So far endonuclease I has been clearly implicated (ALMENDINGER and HAGER, 1972), but its specific role is unknown. The existence of the colicin E2-tolerant, pleiotropic mutants which can show recombination deficiency (HOLLAND et al., 1970) suggests that recombination nucleases may also be involved. Of these, the ATP-dependent activity related to the rec B, rec C genes (BARBOUR and CLARK, 1970) has been ruled out (HOLLAND and HOLLAND, 1970). It remains possible that colicin E2 will itself be found to exhibit nucleolytic activity under appropriate conditions.

The formation of filaments by cells challenged with colicin E2 has been observed by several authors (BEPPU and ARIMA, 1971; HOLLAND, 1968; RINGROSE, 1970). Since filament formation is a common effect seen with agents that disrupt DNA synthesis it is possible that this blockage of cell divison resulted from the inhibition of DNA synthesis in colicin E2-treated cells. However, BEPPU et al., (1972) have isolated a mutant which cannot form colonies in the presence of colicin E2, but recovers colony-forming ability when exposed to trypsin even after more than two hours of exposure to the colicin. Cell division in the mutant is inhibited by colicin, causing the formation of filaments, but DNA synthesis proceeds at the same rate as in untreated cells. DNA degradation and induction of lambda prophage are not observed. Clearly, interference with DNA synthesis is not necessary for the inhibition of cell division produced by colicin E2. The fact that the colicin-challenged cells remain in the trypsin-rescuable stage implies that the inhibition of cell division is a function of this stage in the action of the colicin, and suggests that it is the transition to the next stage, in which DNA synthesis is attacked, which is blocked in this mutant.

Colicin E2 activity has also been studied *in vitro* (BEPPU and ARIMA, 1972) where it has been reported to provoke the relase of DNA from a DNA-membrane complex derived from either sensitive or tolerant cells. The reaction requires the presence of ATP or other nucleotide triphosphates, ADP, or inorganic pyrophosphate. Interestingly, it is extremely temperature sensitive, and does not occur below 20° C.

b) Colicin E3. In intact cells colicin E3 affects predominantly protein synthesis. Isolation of ribosomes from E3 treated cells showed that the ribosomes were intact but unable to support protein synthesis. The alteration could be monitored by messenger-stimulated transfer RNA binding and was traced to the 30S subunit (KONISKY and NOMURA, 1967). It has now been shown that E3 can act on ribosomes *in vitro*, cleaving a fragment of about fifty nucleotides from the 16S ribosomal RNA (BOON, 1971; BOWMAN *et al.*, 1971; SENIOR and HOLLAND, 1971). The same fragment is detected *in vivo*. It is not formed if purified ribosomal RNAs, in particular the 16S RNA, are incubated with colicin. Nucleolytic activity of colicin E3 preparations has not been reported, and known ribonucleases are stated to be less specific in their action than the agent responsible for the 16S RNA cleavage (BOWMAN *et al.*, 1971). It appears that colicin E3 itself exhibits a ribonucleolytic activity which may require a native ribosomal structure, or that it mobilizes such an activity which is present in ribosomal preparations or associated with ribosomes. This action of colicin E3 in the apparent absence of cellular membrane material has raised the possibility that colicins act from within the cell rather than via the membrane. This question is considered in the final section on common processes in colicin action, where these findings with colicin E3 are discussed in relation to the large body of data linking colicin action and the cell membrane which we have already surveyed.

c) Energy Metabolism Colicins. Although the action of colicins E2 and E3 can be traced, at least in part, to a specific attack on DNA or ribosomes, the action of the other main class of colicins—that represented by colicins A, E1, K, I and Q—is less understood. These colicins attack the basic functions of the cell membrane itself. Most of the effects they are known to produce are summarized in Table 3 which also shows the particular effects that have been reported for given colicins. The overlapping of the arrays of colicins showing the same effect impels the conclusion that all the effects can be produced by each of these colicins, and raises several inter-related questions:

1. What are the similarities and differences among these various colicin molecules? 2. What basic system in the membrane is inactivated by each of them? and 3. How do their pathways of attack differ, and at what point do they converge to generate the characteristic array of effects? Partial answers can be provided for questions 1. and 2.

The chemical properties, molecular weights and amino acid composition data show that colicins E1, K and I are distinct molecules. E1 and I may be more closely related to one another than to either A or K (SCHWARTZ and HELINSKI, 1971; JESAITIS, 1970; DANDEU and BARBU, 1968; KUNUGITA and MATSUHASHI, 1970). The original classification of colicins by FREDERICQ (1948) was based, in part on the ability of each colicin to select a unique class of resistant mutants, and this implies differences somewhere in the pathway of action of all five of these colicins. It may well be that each colicin interacts with structurally distinct sites within the membrane.

The general system attacked by these colicins is responsible for the generation of cellular energy, and the data available (Table 3) suggest that the common effect is the dissipation of a membrane state, rather than the disruption of ion gradients or a depletion of ATP. The argument stands on three pillars: 1. The large variety of energy-dependent processes blocked indicates that virtually all cellular energy supplies are affected [glycolysis continues at a reduced rate but the energy derived here is apparently not available, or insufficient to support the other processes (FIELDS and LURIA, 1969 b)]; 2. The cessation of energy-dependent functions is not due to leakage of ATP since the membrane does not become generally leaky, and the decline of the ATP level is not a necessary condition for the other effects; 3. The most rapid events provoked by challenge with these colicins (ANS fluorescence, K^+/Na^+ exchange, and the fall in ATP level) take place concurrently, suggesting they are interrelated. Yet, the exit of potassium can occur in the absence of a drop in ATP (FEINGOLD, 1970), and ANS fluorescence

Table 3. *The variety of effects of energy metabolism colicins*

Colicin(s)	Effects	References[a]
E1	ANS[b] fluorescence increase	3
K	K^+/Na^+ exchange	8
E1, K	Fall in ATP	3, 5, 6, 8
	Inhibition of energy dependent functions	
	Transport	
K	K^+	13, 14
E1, K	Mg^{2+}	10
E1, K	Thiomethyl galactoside	6
E1, K	Proline	1
A	Leucine	11
	Macromolecular synthesis	
E1, K, I	DNA	9, 12
A, E1, K, I	RNA	3, 9, 11, 12
E1, K, I	Protein	3, 9, 12
E1, K	Polysaccharide	7
A, E1, K	Motility	6, 11, 15
	Efflux, or altered permeability	7
E1, K	Pyruvate	7
E1, K	Glycolytic intermediates	10
A, E1, K, Q	K^+	2, 4, 5, 13, 14, 15
E1, K	Mg^{2+}; Co^{2+}	10
E1	α-methyl glucoside (partial)	6
K	"Phosphorylated constituents"	2
A, E1, K, Q	Shift in phospholipid composition	2

[a] 1. BHATTACHARYYA *et al.* (1970). 2. CAVARD *et al.* (1968). 3. CRAMER and PHILLIPS (1970). 4. DANDEU *et al.* (1969). 5. FEINGOLD (1970). 6. FIELDS and LURIA (1969a). 7. FIELDS and LURIA (1969b). 8. HIRATA *et al.* (1969). 9. LEVISOHN *et al.* (1968). 10. LUSK and NELSON (1972). 11. NAGEL DE ZWAIG (1969). 12. NOMURA (1963). 13. NOMURA and MAEDA (1965). 14. WENDT (1970). 15. WENDT (unpublished).

[b] 8-anilino-1-naphthalenesulfonate.

shifts have been observed under conditions (REEVES et al., 1972; KABACK, 1972) where ion gradients should be disrupted. This experimental dissociation of these early effects leaves the ANS fluorescence as the likely indicator of the actual primary events. The data supporting these statements and some information regarding the significance of the fluorescence data are summarized below.

Evidence for general integrity of the membranes attacked by these colicins is of two kinds. First, gross lysis is not observed, and although indications of changes in the membrane (including shifts in phospholipid composition, and a decrease in optical density—CAVARD et al., 1968; Table 3) have been reported, they are very slow to appear compared to the rapid shifts in ANS fluorescence, monovalent cation content, and ATP level. Furthermore, similar changes are provoked by colicin E3 which is unrelated to this group in its mode of action. Second, there is no solid evidence that any small molecules (including intermediary metabolites) can enter or leave the cells via a non-specific mechanism. Although the ATP level falls, the residual ATP remains within the cell. The loss of internal glycolytic intermediates, and accumulated substrates or analogues such as proline and thiomethyl galactoside, can be explained as occurring by means of specific transport systems which facilitate diffusion toward equilibrium when the energy supply is cut off (WINKLER and WILSON, 1966). There is no increase in permeability to external small molecules such as orthonitrophenyl galactoside (FIELDS and LURIA, 1969a) or acridine dyes (WENDT, unpublished). Finally, although a stimulation of cobalt and magnesium exchange indicative of increased permeability is produced by treatment with colicins of this class, it begins after the inhibition of energy-dependent functions (LUSK and NELSON, 1972).

FEINGOLD (1970) has shown that the fall in ATP can be prevented by means of an inhibitor (dicyclohexylcarbodiimide: DCCD) known to inhibit bacterial ATPases (FEINGOLD, 1970; HAROLD, 1972), and that potassium loss and the inhibitons of RNA and protein synthesis still occur. In fact, the cellular ATP level increased appreciably when DCCD treated cells were challenged with colicin. The potassium loss measured by Feingold is indicative of the K^+/Na^+ exchange process (HIRATA et al., 1969; WENDT, 1970a; MOGELSON and WENDT, unpublished). Colicin treatment does not cause protons to move into the cells and thus K^+/H^+ exchange is not involved in the action of these colicins (FEINGOLD, 1970; WENDT, unpublished).

One is led to infer that the K^+/Na^+ exchange, or the ANS fluorescence, might more closely reflect the primary action of these colicins. While the basis of ANS fluorescence during interactions with membranes is incompletely understood (BRAND and GOHLKE, 1972), fluorescence changes can occur in response to energization of mitochondria and this is, at least partially, due to changes in the membrane binding of the dye (AZZI, 1969; BROCKLEHURST et al., 1970). REEVES et al., (1972) and KABACK (1972) have described experiments showing that energization of membrane vesicles from E. coli by addition of D-lactate leads to a rapid decrease of ANS fluorescence. Since treatment with electron transport inhibitors which act before the point of coupling of D-lactate to the electron transport chain inhibit the fluorescence decrease, and those which act after this point do not, the fluorescence change seems associated with events occurring during electron transfer (REEVES et al., 1972). Vesicles treated with phospholipase, or extracted with acetone, are still capable of exhibiting fluorescence shifts, although they are unable to retain transported solutes (summarized in KABACK, 1972). It is highly unlikely that acetone-treated vesicles can maintain an ion gradient; hence, the ANS fluorescence change in the vesicles appears to be independent of ion fluxes. This implies that the colicin-provoked fluorescence shift may also be independent of ion movements.

We are thus led to a view of energy metabolism colicin action in which the primary effect is mirrored in the ANS fluorescence increase. This increase is associated with a loss of cellular energy production just as a fluorescence decrease is associated with energization of the vesicles by D-lactate. Presumably the change in ANS fluorescence is correlated with alterations in the structural state of the membrane, such as those observed during energization and de-energization of mitochondria (PENNISTON et al., 1968; WILLIAMS et al., 1970).

3. Common Processes and Membrane Models

A fuller explantation of colicin action, especially the early, and common, events awaits a more complete understanding of membrane structure in E. coli. Conversely the information concerning these events must be accounted for by any membrane model. In this context it is worth considering once again the processes common to colicin action and other information, having implications for the initial stages of colicin attack.

The basic question here is how these diverse colicins reach, or exert their effects upon, their ultimate targets. The facts that colicins kill with a single hit, and that all well-studied colicins go through a transition from trypsin-sensitivity to insensitivity which can be blocked by metabolic inhibitors and low temperature, and which culminates in the initiation of their lethal action, indicate that these early steps are similar for all colicins. The loss of sensitivity to trypsin clearly indicates that colicins may penetrate the membrane, and this is supported by the action of colicin E3 on ribosomes in vitro. Since the bulk of labelled colicin E2 remains with the membrane fraction (MAEDA and NOMURA, 1966), a mechanism in which the colicin molecules maintain a membrane association is favored over the penetration of colicins to a 'free' state in the cellular interior.

Two basic mechanisms can account for the action of a single colicin molecule on multiple targets: the colicin could remain localized and exert its effects by the transmission of an altered state within the membrane; or the colicin could be mobile within the membrane. The former possibility has been favored by NOMURA (1964), and an explicit model formulated by CHANGEUX et al. (1967) was applied to explain this aspect of colicin action as due to a domino-like propagation of conformational changes via cooperative interactions between membrane subunits (CHANGEUX and THIERY, 1967). In the case of energy metabolism colicins there is evidence which argues for structural changes as already discussed, and which has been related to a cooperative effect within the membrane (WENDT, 1970a and b). These phenomena appear to be subsumed by the more detailed model proposed for mitochondria by GREEN and JI (1972). However, if the assumption of a unitary mechanism in the early stages of colicin action is maintained, then the action of colicin E3 directly on ribosome preparations can be construed as evidence for the movement of a single colicin molecule from its site of penetration to each of many targets. Such movement can be explained on the basis of the fluid mosaic membrane model as diffusion of the colicin within the fluid phase of the membrane (SINGER and NICOLSON, 1972). This explanation does not assume that colicins dissolve themselves in the membrane, but rather colicins would be bound to specific membrane proteins which are themselves mobile in the membrane. SINGER and NICOLSON (1972) thus predicted that lowered temperatures, which decrease membrane fluidity, might affect the kinetics of the ANS fluorescence shift produced by colicin E1 treatment. Although effects of temperature on ANS fluorescence kinetics have not been reported, the delays in the initiation of potassium loss, and in the transition from trypsin sensitivity to insensitivity, could represent slower diffusion in agreement with this prediction. The fluid mosaic model is general enough to accomodate cooperative phenomena; it is philosophically preferable, since, parsimoniously, it does not introduce new physical entities; and, not least, it is testable. It will be well worth entertaining further.

Acknowledgements. An early version of this manuscript was read and criticized by H. SCRIBNER, S. SILVER and A. SUMMERS. J. KONISKY contributed information on the relationship between colicin factor replication and colicin production. L. WENDT was supported by Public Health Service research grant AIO8062.

References

ALMENDINGER, R., and L.P. HAGER: Role for endonuclease I in the transmission process of colicin E2. Nature New Biol. **235**, 199 (1972).

AZZI, A.: Redistribution of the electrical charge of the mitochondrial membrane during energy conservation. Biochem. Biophys. Res. Commun. **37**, 254 (1969).

BARBOUR, S.D., and A.J. CLARK: Biochemical and genetic studies of recombination proficiency in *Escherichia coli* I. Enzymatic activity associated with *rec B* and *rec C* genes. Proc. Natl. Acad. Sci. U.S. **65**, 955 (1970).

BARRY, G.T., D. EVERHART, and M. GRAHAM: Colicin A. Nature **198**, 211 (1963).

BAYER, M.E.: The adsorption of bacteriophages to adhesions between wall and membrane of *Escherichia coli*. J. Virol. **2**, 346 (1968).

BAZARAL, M., and D.R. HELINSKI: Circular DNA forms of colicinogenic factors E1, E2, and E3 from *Escherichia coli*. J. Mol. Biol. **36**, 185 (1968).

BEPPU, T., and K. ARIMA: Protection of *Escherichia coli* from the lethal effect of colicins by high osmotic pressure. J. Bacteriol. **93**, 80 (1967).

BEPPU, T., and K. ARIMA: Properties of the colicin E2 induced degradation of deoxyribonucleic acid in *Escherichia coli*. J. Biochem. **70**, 263 (1971).

BEPPU, T., and K. ARIMA: Dissociating activity of purified colicin E2 on the isolated DNA-membrane complex of *Escherichia coli*. Biochim. Biophys. Acta **262**, 453 (1972).

BEPPU, T., K. KAWABATA, and K. ARIMA: Specific inhibition of cell division by colicin E2 without degradation of deoxyribonucleic acid in a new colicin sensitivity mutant of *Escherichia coli*. J. Bacteriol. **110**, 485 (1972).

BERNSTEIN, A., B. ROLFE, and K. ONODERA: Pleiotropic properties and genetic organization of the *tol A,B* locus of *Escherichia coli* K-12. J. Bacteriol. **112**, 74 (1972).

BHATTACHARYYA, P., L. WENDT, E. WHITNEY, and S. SILVER: Colicin-tolerant mutants of *Escherichia coli*: Resistance of membranes to colicin E1. Science **168**, 998 (1970).

BOON, T.: Inactivation of ribosomes *in vitro* by colicin E3. Proc. Natl. Acad. Sci. U.S. **68**, 2421, (1971).

BOWMAN, C.M., J. SIDIKARO, and M. NOMURA: Specific inactivation of ribosomes by colicin E3 *in vitro* and mechanism of immunity in colicinogenic cells. Nature New Biol. **234**, 133 (1971).

BRAND, L., and J.R. GOHLKE: Flourescence probes for structure. Ann. Rev. Biochem. **41**, 843 (1972).

BROCKLEHURST, J.R., R.B. FREEDMAN, D.J. HANCOCK, and G.K. RADDA: Membrane studies with polarity-dependent and excimer-forming flourescent probes. Biochem. J. **116**, 721 (1970).

BURMAN, L.G., and K. NORDSTRÖM: Colicin tolerance induced by ampicillin, or mutation to ampicillin resistance, in a strain of *Escherichia coli* K-12. J. Bacteriol. **106**, 1 (1971).

CAVARD, D., C. RAMPINI, E. BARBU, and J. POLONOVSKI: Activité phospholipasique et autre modifications du métabolisme des phospholipides consécutives a l'action des colicines sur *E. coli*. Bull. Soc. Chim. Biol. **50**, 1455 (1968).

CHANGEUX, J.P., and J. THIERY: On the mode of action of colicins: a model of regulation at the membrane level. J. Theoret. Biol. **17**, 315 (1967).

CHANGEUX, J.P., J. THIERY, Y. TUNG, and C. KITTEL: On the cooperativity of biological membranes. Proc. Natl. Acad. Sci. U.S. **57**, 335 (1967).

CLOWES, R.C.: Transmission and elimination of colicin factors and some aspects or immunity to colicin E1 in *Escherichia coli*. Zentr. Bakteriol. Parasitenk. Abt. I Orig. **196**, 152 (1965).

CLOWES, R.C.: Molecular weight and number of copies of various sex factor DNA molecules in *Escherichia coli*. In: A. PEREZ MIRAVETE and D. PELAEZ (eds.), Xth International Congress for Microbiology, Mexico, 1970, 58 p. (Cd-1).

CLOWES, R.C.: Molecular structure of bacterial plasmids. Bacteriol. Rev. **36**, 361 (1972).

CRAMER, W.A., and S.K. PHILLIPS: Response of an *Escherichia coli*-bound flourescent probe to colicin E1. J. Bacteriol. **104**, 819 (1970).

DANDEU, P., and E. BARBU: Etude comparee de quelques colicines. Compt. Rend. **266**, 634 (1968).

DANDEU, P., A. BILLAULT, and E. BARBU: Action des colicines sur la vitesse de sortie du potassium intracellulaire. Compt. Rend. **269**, 2044 (1969).

ENDO, H., T. KAMIYA, and M. ISHIZAWA: Lamda phage induction by colicin E2. Biochim. Biophys. Res. Comun. **11**, 477 (1963).

FEINGOLD, D.S.: The mechanism of colicin E1 action. J. Membrane Biol. **3**, 372 (1970).

FIELDS, K.L.: Comparison of the action of colicins E1 and K on *Escherichia coli* with the effects of abortive infection by virulent bacteriophage. J. Bacteriol. **97**, 78 (1969).

FIELDS, K.L., and S.E. LURIA: Effects of colicins E1 and K on transport systems. J. Bacteriol. **97**, 57 (1969a).

FIELDS, K.L., and S.E. LURIA: Effects of colicins E1 and K on cellular metabolism. J. Bacteriol. **97**, 64 (1969b).

FREDERICQ, P.: Action antibiotiques reciproques chez les Enterobacteriaceae. Rev. Belge Pathol. Med. Exptl. **29**, (Suppl. 4), 1 (1948).

FREDERICQ, P.: Colicines et bacteriophages. Ann. Inst. Pasteur **84**, 294 (1953).

FREDERICQ, P.: Resistance et immunité aux colicines. Sceances Soc. Biol. Filiales **150**, 1514 (1956).

FREDERICQ, P.: Colicins. Ann. Rev. Microbiol. **11**, 7 (1957).

GOEBEL, W.F.: The chromatographic fractionation of colicin K. Proc. Natl. Acad. Sci. U.S. **48**, 214 (1962).

GREEN, D.E., and S. JI: The electromechanical model of mitochondrial structure and function. In: The molecular basis of electron transport. New York and London: Academic Press, 1972.

HAMON, Y.: Les Bacteriocines. Ann. Inst. Pasteur **107**, 18 (1964).

HARDY, K.G., and G.G. MEYNELL: "Induction" of colicin factor E2-P9 by mitomycin C. J. Bacteriol. **112**, 1007 (1972).

HAROLD, F.M.: Conservation and transformation of energy by bacterial membranes. Bacteriol. Rev. **36**, 172 (1972).

HAYES, W.: The genetics of bacteria and their viruses, 2nd edition. New York: John Wiley & Sons. 1968.

HERSCHMAN, H.R., and D.R. HELINSKI: Purification and characterization of colicin E2 and colicin E3. J. Biol. Chem. **242**, 5360 (1967a).

HERSCHMAN, H.R., and D.R. HELINSKI: Comparative study of the events associated with colicin induction. J. Bacteriol. **94**, 691 (1967b).

HILL, C., and I.B. HOLLAND: Genetic basis of colicin E susceptibility in *Escherichia coli* I. Isolation and properties of refractory mutants and the preliminary mapping of their mutations. J. Bacteriol. **94**, 677 (1967).

HIRATA, H., S. FUKUI, and S. ISHIKAWA: Initial events caused by colicin K infection—cation movement and depletion of ATP pool. J. Biochem. **65**, 843 (1969).

HOLLAND, E. M., and I. B. HOLLAND: Induction of DNA breakdown and inhibition of cell division by colicin E2. Nature of some early steps in the process and properties of the E2 specific nuclease system. J. Gen. Microbiol. **64**, 223 (1970).

HOLLAND, I. B.: The properties of UV sensitive mutants of *Escherichia coli* which are also refractory to colicin E2. Molec. Gen. Genetics **100**, 242 (1967).

HOLLAND, I. B.: Properties of *Escherichia coli* K-12 mutants which show conditional refractivity to colicin E2. J. Mol. Biol. **31**, 267 (1968).

HOLLAND, I. B., E. J. TRELFALL, E. M. HOLLAND, V. DARBY, and A. C. R. SAMSON: Mutants of *Escherichia coli* with altered surface properties which are refractory to colicin E2, sensitive to ultraviolet light, and which can also show recombination deficiency, abortive growth of bacteriophage lamda and filament formation. J. Gen. Microbiol. **62**, 371 (1970).

HOLLAND, I. B., and S. TUCKETT: Study of envelope proteins in *E. coli cet* and *rec A* mutants by SDS acrylamide gel electrophoresis. J. Supramol. Struct. **1**, 77 (1972).

HULL, R. R., and P. REEVES: Sensitivity of intracellular bacteriophage lamda to colicin CA42-E2. J. Virol. **8**, 355 (1971).

HUTTON, J. J., and W. F. GOEBEL: The isolation of colicin V and a study of its immunological properties. J. Gen. Physiol. **45**, (Suppl.), 125 (1963).

JACOB, F., R. SUSSMAN, and J. MONOD: Sur la nature du répresseur assurant l'immunité des bateries lysogenes. Compt. Rend. **254**, 4212 (1962).

JESAITIS, M. A.: The nature of colicin K from *Proteus mirabilis*. J. Exptl. Med. **131**, 1016 (1970).

KABACK, H. R.: Transport across isolated bacterial cytoplasmic membranes. Biochim. Biophys. Acta **265**, 367 (1972).

KONISKY, J.: Characterization of colicin Ia and colicin Ib. Chemical studies of protein structure. J. Biol. Chem. **247**, 3750 (1972).

KONISKY, J., and M. NOMURA: Interactions of colicins with bacterial cells II. Specific alteration of *Escherichia coli* ribosomes induced by colicin E3 *in vivo*. J. Mol. Biol. **26**, 181 (1967).

KONISKY, J., and F. M. RICHARDS: Characterization of colicin Ia and colicin Ib. Purification and some physical properties. J. Biol. Chem. **245**, 2972 (1970).

KUNUGITA, K., and M. MATSUHASHI: Purification and properties of colicin K. J. Bacteriol. **104**, 1017 (1970).

LEVISOHN, R., J. KONISKY, and M. NOMURA: Interaction of colicins with bacterial cells IV. Immunity breakdown studied with colicins Ia and Ib. J. Bacteriol. **96**, 811 (1968).

LUSK, J. E., and D. L. NELSON: Effects of colicins E1 and K on permeability to magnesium and cobaltous ions. J. Bacteriol. **12**, 148 (1972).

MAEDA, A., and M. NOMURA: Interaction of colicins with bacterial cells. I. Studies with radioactive colicins. J. Bacteriol. **91**, 685 (1966).

NAGEL DE ZWAIG, R.: Mode of action of colicin A. J. Bacteriol. **99**, 913 (1969).

NAGEL DE ZWAIG, R., and S. E. LURIA: Genetics and physiology of colicin-tolerant mutants of *Escherichia coli*. J. Bacteriol. **94**, 1112 (1967).

NOMURA, M.: Mode of action of colicins. Cold Spring Harbor Symp. Quant. Biol. **28**, 315 (1963).

NOMURA, M.: Mechanism of action of colicins. Proc. Natl. Acad. Sci. U. S. **52**, 1514 (1964).

NOMURA, M.: Colicins. In: Antibiotics, vol. I. Mechanism of action. Berlin-Heidelberg-New York: Springer 1967a.

NOMURA, M.: Colicins and related bacteriocins. Ann. Rev. Microbiol. **21**, 257 (1967b).

NOMURA, M., and A. MAEDA: Mechanism of action of colicins. Zentr. Bakteriol. Parasitenk. Abt. I, Orig. **196**, 216 (1965).

NOMURA, M., and C. WITTEN: Interaction of colicins with bacterial cells III. Colicin-tolerant mutations in *Escherichia coli*. J. Bacteriol. **94**, 1093 (1967).

NOVICK, R. P.: Extrachromosomal inheritance in bacteria. Bacteriol. Rev. **33**, 210 (1969).

NUSKE, R., G. HOSEL, H. VENNER u. H. ZINNER: Über ein Colicin aus *Escherichia coli* SG 710. Biochem. Z. **346**, (1957).

ONODERA, K., A. BERNSTEIN, and B. ROLFE: Demonstration of missing membrane proteins in deletion mutants of *E. coli* K-12. Biochem. Biophys. Res. Comun. **39**, 969 (1970).

PENNISTON, J. T., R. HARRIS, J. ASAI, and D. E. GREEN: The conformational basis of energy transformations in membrane systems I. Conformational changes in mitochondria. Proc. Natl. Acad. Sci. U. S. **59**, 624 (1968).

PLATE, C. A., and S. E. LURIA: Stages in colicin K action as revealed by the action of trypsin. Proc. Natl. Acad. Sci. U. S. **69**, 2030 (1972).

REEVES, J.P., F.J. LOMBARDI, and H.R. KABACK: Mechanisms of active transport in isolated bacterial membrane vesicles VII. Fluorescence of 1-anilino-8-naphthalene-sulfonate during D-lactate oxidation by membrane vesicles from *Escherichia coli*. J. Biol. Chem. **247**, 6204 (1972).

REEVES, P.: The bacteriocins. Bacteriol. Rev. **29**, 24 (1965).

REEVES, P.: Mutants resistant to colicin CA42-E2. Cross-resistance and genetic mapping of a special class of mutants. Australian J. Exptl. Biol. Med. Sci. **44**, 301 (1966).

REYNOLDS, B.L., and P.R. REEVES: Some observations on the mode of action of colicin F. Biochem. Biophys. Res. Commun. **11**, 140 (1963).

RINGROSE, P.: Sedimentation analysis of DNA degradation products resulting from the action of colicin E2 on *Escherichia coli*. Biochim. Biophys. Acta **213**, 320 (1970).

ROLFE, B., and K. ONODERA: Demonstration of missing membrane proteins in a colicin-tolerant mutant of *Escherichia coli* K-12. Biochem. Biophys. Res. Commun. **44**, 767 (1971).

SABET, S.F., and C.A. SCHNAITMAN: Localization and solubilization of colicin receptors. J. Bacteriol. **108**, 422 (1971).

SCHWARTZ, S.A., and D.R. HELINSKI: Purification and characterization of colicin E1. J. Biol. Chem. **246**, 6318 (1971).

SENIOR, B.W., and I.B. HOLLAND: Effect of colicin E3 upon the 30S ribosomal sub-unit of *Escherichia coli*. Proc. Natl. Acad. Sci. U.S. **68**, 959 (1971).

SINGER, S.J., and G.L. NICOLSON: The fluid mosaic model of the structure of cell membranes. Science **175**, 720 (1972).

TAYLOR, A.L.: The current linkage map of *Escherichia coli*. Bacteriol. Rev. **34**, 155 (1970).

TIMMIS, K.: Purification and characterization of colicin D. J. Bacteriol. **109**, 12 (1972).

WENDT, L.: Mechanism of colicin action; early events. J. Bacteriol. **104**, 1236 (1970a).

WENDT, L.: The action of irehdiamine A on *Escherichia coli*: inhibition of the membrane response at low temperatures. Biochem. Biophys. Res. Commun. **40**, 489 (1970b).

WHITNEY, E.N.: The tol C locus in *Escherichia coli* K-12. Genetics **67**, 39 (1971).

WILLIAMS, C.H., W.J. VAIL, R.A. HARRIS, M. CALDWELL, E. VALDIVIA, and D.E. GREEN: The conformational basis of energy transduction in membrane systems VIII. Configurational changes of mitochondria *in situ* and *in vitro*. Bioenergetics **1**, 147 (1970).

WINKLER, H.H., and T.H. WILSON: The role of energy coupling in the transport of β-galactosides by *Escherichia coli*. J. Biol. Chem. **241**, 2200 (1966).

Griseofulvin

Floyd M. Huber

In 1946, BRIAN, CURTIS and HEMMING isolated a substance from cultures of *Penicillium janczewskii* Zal which caused abnormal curling of fungal hyphae. This material was later shown to be identical to griseofulvin (GROVE and McGOWAN, 1947; BRIAN *et al.*, 1949). The latter compund was first isolated in 1939 by OXFORD, RAISTRICK and SIMONART. The structure of griseofulvin, as proposed by GROVE *et al.* (1952) is shown in Fig. 1.

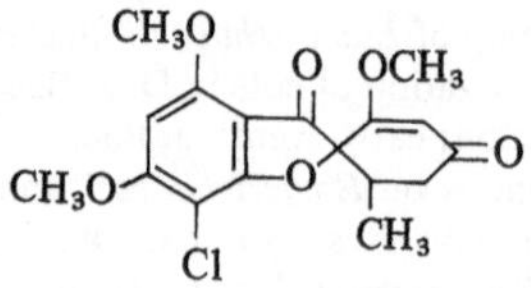

Fig. 1. Griseofulvin

The bromo- and dechloro-analogs of griseofulvin have also been isolated from fungal cultures (MACMILLAN, 1951, 1954). Fully deuterated griseofulvin has been reported to be synthesized when the producing organism is grown in deuterium oxide (NONA *et al.*, 1968).

Several recent reviews concerning griseofulvin have been published. OSMENT (1969) presented a general summary of griseofulvin knowledge with a brief discussion of the mode of action. In contrast, reviews by LAMPEN *et al.* (1965) and BENT and MOORE (1966) dealt primarily with the mechanism of action of griseofulvin. The latter review not only encompassed prior work, but also included much previously unpublished data and information derived from personal communications.

Biological Spectrum

Although griseofulvin inhibits the growth of fungi with chitinous cell walls, it has no effect on fungi with cellulose cell walls (BRIAN *et al.*, 1949). Examples of organisms sensitive to griseofulvin are presented in Table 1. In the microbial world, bacteria, yeasts, yeast protoplasts, and many fungi are not inhibited by this antibiotic (BRIAN *et al.*, 1949; EL-NAKEEB *et al.*, 1965c; SHOCKMAN and LAMPEN, 1962).

Griseofulvin has been reported to inhibit germination of seeds and root growth of certain plants (BRIAN *et al.*, 1949; BRIAN *et al.*, 1951; WRIGHT 1951).

Table 1. *Sensitivity of fungi to griseofulvin*

Organism	Percent inhibition	Griseofulvin (μg/ml)	Reference
Alternaria tenuis	39[a]	10	EL-NAKEEB, *et al.* (1965c)
Glomerella cingulata	65[a]	10	EL-NAKEEB, *et al.* (1965c)
Microsporum gypseum	97[a]	5	EL-NAKEEB, *et al.* (1965c)
Trichophyton interdigitalis	95[a]	10	EL-NAKEEB, *et al.* (1965c)
Trichophyton mentagrophytes	95[a]	5	EL-NAKEEB, *et al.* (1965c)
Armillaria mellea	50[b]	1	BRIAN, *et al.* (1949)
Botrytis allii	50[b]	1	BRIAN, *et al.* (1949)
Diaporthe perniciosa	50[b]	1	BRIAN, *et al.* (1949)
Phoma betae	50[b]	1	BRIAN, *et al.* (1949)
Botrytis cinerea	90[a]	20	HUBER and GOTTLIEB (1968)

[a] Based on dry weight.
[b] Based on radial growth.

In human subjects oral administration of griseofulvin produces no side effects (ROTH *et al.*, 1959b; BURGOON *et al.*, 1960). However, in tissue cultures at concentrations of 100 μg/ml, griseofulvin inhibits the growth of both HeLa and skin cell lines (DENNIS *et al.*, 1960). According to MUNTONI and LODDO (1964), griseofulvin causes budding of human amnion and monkey kidney cells. With respect to other animal systems, griseofulvin causes inhibition of spermatogenesis in mice and necrosis of the seminiferous epithelium in rats (SCHWARZ and LOUTZENHISER, 1960; PAGET and WALPOLE, 1960). BARICH *et al.* (1960) have demonstrated that excessively large doses of oral griseofulvin promote tumors in mice treated with methyl-cholanthrene; however, in rats the antibiotic was found not to be carcinogenic (PAGET and ALCOCK, 1960).

Resistance to Griseofulvin

Trichophyton quinckeanum has been reported to develop resistance to griseofulvin both *in vitro* and *in vivo* (VIDMAR-CVJETANOVIC *et al.*, 1966). Resistance was shown to correlate with reduced pathogenicity. Two reports by LENHART (1968, 1969) described the development of several *Microsporum gypseum* strains possessing griseofulvin resistance. A spontaneous mutation frequency of 10^{-8} was reported for resistance to griseofulvin; however, this frequency was increased ten-fold by ultraviolet radiation. The tolerance of all mutants to griseofulvin was approximately 100 times greater than in the wild strain.

Mutagenic Effects

LENHART (1969) treated *Nannizzia incurvata* and *M. gypseum* with griseofulvin and isolated eighty-four mutants with significantly altered growth rates. Approximately ten percent of these mutants had macromorphological changes. To prove stability, the cultures were transferred twenty-six times over a two-year period. None of the mutants were found to have greater resistance to griseofulvin than the parent strains Further examination of these strains revealed that they sectored on solid media and that these sectors were not like the parent type. This investigator suggested that the mutagenic effect of griseofulvin was at the level of extranuclear factors.

Reversal of Griseofulvin Inhibition

Early reports by McNall (1960a, b) indicated that several nucleotides and bases could partially reverse griseofulvin inhibition. Combinations of adenylic acid and guanylic acid have been reported to reverse griseofulvin inhibition of *M. gypseum* (El-Nakeeb *et al.,* 1965c). However, the same investigators and others (Huber and Gottlieb, 1968) reported that many nucleic acids, nucleotides, nucleosides, purines and pyrimidines failed to reverse the inhibition of fungal growth caused by griseofulvin.

Effect of Griseofulvin on Growth and Morphology

Only fungal hyphae in contact with griseofulvin are altered by the antibiotic and therefore aerial hyphae are uneffected (Aytoun *et al.,* 1960). Brian *et al.* (1946) demonstrated that neither spore germination nor germination time are altered by the antibiotic. Low concentrations of griseofulvin caused germtubes to curl, hyphae to curl and have shortened internodes, and excessive branching and unusual swellings (Brian *et al.,* 1946; Brian, 1949; Srivastava and Vora, 1961). At high concentrations of the antibiotic, hyphae lost direction of growth, became larger in diameter and finally ruptured (Aytoun, 1956; Larpent, 1963; Brian, 1949). Drawings, photographs and terminology associated with these morphological abnormalities in fungi have been published by Brian *et al.* (1946); Napier *et al.* (1956), and Roth *et al.* (1959b).

Griseofulvin has also been reported to cause morphological changes in biological systems other than of fungal origin. Griseofulvin treatment of larvae of the mosquito *Aedes atropalpus* produced gross anatomical changes in the cuticle, detachment of somatic muscles from the integument, and prolongation of the molting cycle (Anderson, 1966). Curvature of the setae associated with administration of griseofulvin to these larvae resembles the morphological effects produced by this antibiotic on fungi.

Using the protozoan *Stentor coeruleus,* Margulis *et al.* (1969) demonstrated that griseofulvin caused twisting and folding of the cells. Although sub-lethal concentrations of griseofulvin arrested cell division in *Stentor,* the cells continued to swim actively for weeks.

Cytological Effects

Dermatophytes exposed to griseofulvin have abnormalities only in the growing region (Thyagarajan *et al.,* 1963). Although more nuclei were present in the hyphal tips, they were of abnormal size and shape. Extremely large nuclei were found, suggesting that these structures did not separate during division.

Bent and Moore (1966) treated *Botrytis allii* and *M. gypseum* with griseofulvin and demonstrated by electron microscopy that unusually large and irregularly shaped nuclei plus many smaller nuclei had been generated by this treatment. Other notable changes included a general breakdown of organelle membranes and an increase in vacuoles and fat droplets.

Griseofulvin was demonstrated to possess very limited antimitotic activity in hamsters, but was significantly active in both rats and swine (MUNTONI, 1965). PAGET and WALPOLE (1958, 1960) found that intravenous injections of griseofulvin in rats caused cessation of mitosis in metaphase. These arrested cells exhibited disorientation and scattering of chromosomes throughout the cytoplasm. Polarization and electron microscopy were applied to the study of griseofulvin effects on mitosis in unfertilized oocytes of the marine annelid *Pectinaria gouldi* (MALAWISTA *et al.*, 1968). The antibiotic decreased spindle size and birefringence; however, the system was capable of quick recovery. Ultrastructural changes associated with griseofulvin resembled those produced by vinblastine and colcemid. The authors suggested that interference with gelated structures other than spindles may be implicated in the fungistatic activity of griseofulvin.

Similar antimitotic activity has been observed in both bean roots and onion when treated with griseofulvin (DEYSON, 1964 a, b; PAGET and WALPOLE, 1960).

Effect of Griseofulvin on Enzyme Systems

Griseofulvin has been found to have no effect on the respiration of numerous fungi (BRIAN, 1949; HUBER and GOTTLIEB, 1968; LARSEN and DEMIS, 1963; ROTH *et al.*, 1959 b). Although there is no effect on total respiration, MALE and HOLUBAR (1967 a, b) have reported that griseofulvin treatment reduced activities of succinic dehydrogenase and alkaline phosphatase in a group of dermatophytes. Administration of griseofulvin to six epidermatophytes produced a marked increase in alkaline phosphatase and a reduction in esterase activity (KNOTH *et al.*, 1967). Although the phosphorus uptake of *M. canis* has been reported to be interrupted for short periods, griseofulvin did not effect the proteolytic activity of the culture until 15 days had elapsed (ZIEGLER, 1963).

Effect on Cell Walls

Since griseofulvin inhibits only those fungi possessing chitin in the cell wall, it was possible that organisms producing this antibiotic would not have chitinous cell walls. However, APPLEGARTH and BOZOIAN (1967) demonstrated the presence of chitin in cell walls of *Penicillium patulum*, a griseofulvin-synthesizing organism.

When ISAAC and MILTON (1967) adapted a *Verticillium* species to grown on griseofulvin-containing media, they observed that the isolate so adapted was stable with respect to glucosamine content when the griseofulvin concentration was increased. Non-adapted mycelium demonstrated a rise in glucosamine under these conditions. In cell wall preparations from griseofulvin treated and control cultures of *T. mentagrophytes,* it was found by EVELEIGH and KNIGHT (1965) that the content of total carbohydrate, amino sugars, protein, and phosphate were similar.

Cell wall thickening resulted from griseofulvin treatment of *B. allii*, but electron microscopy revealed no alteration in the random fibrillar network of the wall. The fibril length appeared to decrease with larger concentrations of the antibiotic. The cell wall changes noted in *Botrytis* were not produced in *Microsporum gypseum* under identical conditions (BENT and MOORE, 1966).

Effect on Synthesis and Composition of Cellular Components

When *M. gypseum* was exposed to griseofulvin and analyses made on a culture volume basis, protein, RNA, and DNA were found to be much decreased (EL-NAKEEB et al., 1965c). A similar approach by HUBER and GOTTLIEB (1968) was used in testing the effect of griseofulvin on *B. cinerea*. In the latter study the data were related on a dry weight basis and it was observed that only a slight decrease in RNA occurred. The most significant result was that treatment with the antibiotic caused an increase in DNA content and that this increase was proportional to griseofulvin concentration. The difference in the data between the two investigations were resolved when the results from *M. gypseum* were recalculated on a dry weight basis (HUBER, 1967).

The only effect of griseofulvin on the low molecular weight mycelial constituents of *B. allii* was generation of a three-fold increase in glutamine content. Protein, RNA, and DNA were reduced by 20–30 percent (BENT and MOORE, 1966). When *Botrytis* spores were exposed to griseofulvin, the protein content, RNA, DNA, hexosamine polymers, and dry weight were inhibited by 43, 56, 52, 19, and 32 percent, respectively, over a 12-hour germination period. Using spores of *M. gypseum,* BARASH et al. (1967) demonstrated that griseofulvin inhibited the incorporation of tritiated uridine into RNA only after germination. Thymidine-H^3 incorporation was similarly affected, but no change in tritiated leucine incorporation into protein was apparent. The incorporation data of EL-NAKEEB et al. (1965c), when recalculated to reflect uptake of the radioactive substrate per milligram dry weight of cells, indicates that the uptake of uridine was severely depressed and the incorporation of thymidine greatly increased (HUBER, 1967). Radioactive amino acids were rapidly incorporated in both treated and untreated cells.

The incorporation of labeled glucose and glycine into a RNA fraction of *B. cinerea* was not inhibited by griseofulvin (HUBER and GOTTLIEB, 1968). When the incorporation of labeled aspartic acid into the RNA fraction was measured, the incorporation rate was greatly depressed in the presence of griseofulvin. The same nucleic acid precursors were also examined with respect to incorporation into a DNA fraction and an increased rate of uptake was observed. With respect to the incorporation of labeled substrates into a protein fraction, no difference was observed between those cells subjected to griseofulvin and the control population. Recalculation of the data of EL-NAKEEB et al. (1965c), indicates that the synthesis of protein was stimulated in the presence of griseofulvin (HUBER, 1967).

Binding of Griseofulvin to Cellular Components

EL-NAKEEB and LAMPEN (1964a) reported that griseofulvin forms complexes with ribonucleic acid isolated from an organism highly susceptible to the antibiotic. These complexes were stable to dialysis, molecular sieve chromatography, and density gradient centrifugation. Similar preparations from insensitive fungi have not shown a high affinity for griseofulvin (EL-NAKEEB and LAMPEN, 1964a; HUBER and GOTTLIEB, 1968). Further studies by LAMPEN et al. (1965), indicated that the labeled griseofulvin taken up by a susceptible fungus was bound covalently to a nucleic acid fraction. Continued fractionation revealed the label to be attached to adenylic acid residues of RNA. In a similar line of investigation, M. gypseum ribosomes bound griseofulvin at 4° C and this process was stimulated by small volumes of a 105,000 × g supernatant fraction (EL-NAKEEB and LAMPEN, 1964b). Incubation of the ribosomes at 30° C caused the release of nucleotides and griseofulvin. When labeled griseofulvin was added to broken cells of B. cinerea and this preparation subjected to fractionation by centrifugation, most of the activity was associated with a 500 × g pellet (HUBER and GOTTLIEB, 1968).

Uptake of Griseofulvin by Fungi

The uptake of griseofulvin into susceptible fungal cells is an energy-requiring process. EL-NAKEEB and LAMPEN (1965a, b) using M. gypseum and radioactive griseofulvin found that the uptake of the antibiotic was dependent on concentration, temperature, pH, and an energy source such as glucose. This uptake could be inhibited by sodium azide, 2,4-dinitrophenol and p-fluorophenylalanine. In another publication the same authors reported that insensitive fungi and yeasts did not bind appreciable amounts of the antibiotic (EL-NAKEEB and LAMPEN, 1965d).

Summary

The effects produced by exposure of fungi to griseofulvin vary among organisms with respect to the parameter in question. Morphological and cytological changes are noted in most sensitive organisms, including cell-wall thickening, changes in size, shape, and number of nuclei, and breakdown in organelle membranes. Cell division and conformation of mitotic figures are often affected by the antibiotic. Griseofulvin also produces changes in the synthetic capability for nucleic acid formation. As yet, the exact mechanism of action of griseofulvin has not been elucidated. However, the foregoing observations coupled with the report that the compound causes mutation does point to interference with the cell division process.

References

ANDERSON, J.F.: Anomalous development of the cuticle of mosquitoes induced by griseofulvin. J. Econ. Entomol. **59**, 1476 (1966).

APPLEGARTH, D.A., and G. BOZOIAN: Presence of chitin in the cell wall of a griseofulvin-producing species of Penicillium. J. Bacteriol. **94**, 1787 (1967).

AYTOUN, R.S.C.: The effects of griseofulvin on certain phytopathogenic fungi. Ann. Botany (London) **20**, 297 (1956).

AYTOUN, R.S.C., A.H. CAMPBELL, E.J. NAPIER, and D.A.L. SEILER: Mycological aspects of the action of griseofulvin against dermatophytes. Arch. Dermatol. **81**, 650 (1960).

BARASH, I., L.M. CONWAY, and D.H. HOWARD: Carbon catabolism and synthesis of macromolecules during spore germination of Microsporum gypseum. J. Bacteriol. **93**, 656 (1967).

BARICH, L. L., T. NAKAI, J. SCHWARZ, and D. J. BARICH: Tumour promoting effect of excessively large doses of oral griseofulvin on tumours induced in mice by methylcholanthrene. Nature **187**, 335 (1960).

BENT, K. J., and R. H. MOORE: Mode of action of griseofulvin. Symp. Soc. Gen. Microbiol. **16**, 82 (1966).

BRIAN, P. W.: Studies on the biological activity of griseofulvin. Ann. Botany (London) **13**, 59 (1949).

BRIAN, P. W., P. J. CURTIS, and H. G. HEMMING: A substance causing abnormal development of fungal hyphae produced by *Penicillium janczewskii* ZAL. I. Biological assay, production and isolation of "curling factor". Trans. Brit. Myco. Soc. **29**, 173 (1946).

BRIAN, P. W., P. J. CURTIS, and H. G. HEMMING: A substance causing abnormal development of fungal hyphae produced by *Penicillium janczewskii* ZAL. III. Identity of "curling factor" with griseofulvin. Trans. Brit Myco. Soc. **32**, 20 (1949).

BRIAN, P. W., J. M. WRIGHT, J. STUBBS, and A. M. WAY: Uptake of antibiotic metabolites of soil microorganisms by plants. Nature **167**, 347 (1951).

BURGOON, C. F., J. H. GRAHAM, R. J. KEIPER, F. URBACH, J. S. BURGOON, and E. B. HELWIG: Histopathologic evaluation of griseofulvin in *Microsporum audouini* infections. Arch. Dermatol. **81**, 724 (1960).

DENNIS, D. J., M. J. DAVIS, and J. C. CAMPBELL: The effects of griseofulvin on epithelial cells in tissue culture. J. Invest. Dermatol. **34**, 99 (1960).

DEYSON, G.: Antimitotic properties of griseofulvin. Ann. Pharm. Franc. **22**, 17 (1964a)

DEYSON, G.: The influence of griseofulvin on the antimitotic properties of colchicine. Ann. Pharm. Franc. **22**, 89 (1964b).

EL-NAKEEB, M. A., and J. O. LAMPEN: Formation of complexes of griseofulvin and nucleic acids of fungi and its relation to griseofulvin sensitivity. Biochem. J. **92**, 59P (1964a).

EL-NAKEEB, M. A., and J. O. LAMPEN: Binding of tritiated-griseofulvin by ribosomes and RNA of *Microsporum gypseum*. Bacteriol. Proc. **1964**, P60 (1964b).

EL-NAKEEB, M. A., and J. O. LAMPEN: Uptake of griseofulvin by the sensitive dermatophyte, *Microsporum gypseum*. J. Bacteriol. **89**, 564 (1965a).

EL-NAKEEB, M. A., and J. O. LAMPEN: Distribution of griseofulvin taken up by *Microsporum gypseum*: Complexes of the antibiotic with cell constituents. J. Bacteriol. **89**, 1075 (1965b).

EL-NAKEEB, M. A., and J. O. LAMPEN: Uptake of H^3-griseofulvin by microorganisms and its correlation with sensitivity to griseofulvin J. Gen. Microbiol. **39**, 285 (1965d).

EL-NAKEEB, M. A., W. L. MCLELLAN and J. O. LAMPEN: Antibiotic action of griseofulvin on dermatophytes. J. Bacteriol. **89**, 557 (1965c).

EVELIGH, D. E., and S. G. KNIGHT: The effect of griseofulvin on the cell wall composition of *Trichophyton mentagrophytes*. Bacteriol. Proc. **1965**, G82, p. 27.

GROVE, J. F., J. MACMILLAN, T. P. C. MULHOLLAND, and M. A. T. ROGERS: Griseofulvin. Part IV. Structure. J. Chem. Soc. **1952**, 3977.

GROVE, J. F., and J. C. MCGOWAN: Identity of griseofulvin and "curling factor". Nature **160**, 574 (1947).

HUBER, F. M.: Antibiotics, vol. I, p. 181. Berlin-Heidelberg-New York: Springer 1967.

HUBER, F. M., and D. GOTTLIEB: The mechanism of action of griseofulvin. Can. J. Microbiol. **14**, 111 (1968).

ISAAC, I., and J. M. MILTON: Response of *Verticillium* species to griseofulvin. J. Gen. Microbiol. **46**, 273 (1967).

KNOTH W., K. IRISAWA, and R. C. KNOTH-BORN: Enzyme-histochemical investigations on human pathogenic fungi after exposure to specific antimycotics. Mykosen **10**, 1 (1967).

LAMPEN, J. O., W. L. MCLELLAN, and M. A. EL-NAKEEB: Antibiotics and fungal physiology. Antimicrobial Agents Chemotherapy **1965**, 1006.

LARPENT, J. P.: Comparative action of griseofulvin and colchicine on growth and branching of the young thallus of *Saprolegnia monoica*. Compt. Rend. **157**, 2219 (Chem. Abstr. **61**, 2407f) (1963).

LARSEN, W. G., and D. J. DEMIS: Metabolic studies of the effect of griseofulvin and candicidin on fungi. J. Invest. Dermatol. **41**, 335 (1963).

LENHART, K.: Spontaneous mutants of *Microsporum gypseum* resistant to griseofulvin. Mycopathol. Mycol. Appl. **36**, 150 (1968).

LENHART, K.: Griseofulvin-resistant mutants in dermatophytes. I. Frequency of spontaneous and ultraviolet-induced mutants. Mykosen **12**, 655 (1969).

LENHART, K.: Mutagenic effect of griseofulvin. Mykosen **12**, 687 (1969).

MACMILLAN, J.: Griseofulvin. Part 9. Isolation of the bromoanalogue from *Penicillium griseofulvum* and *Penicillium nigricans*. J. Chem. Soc. **1954**, 2585.

MACMILLAN, J.: Dechlorogriseofulvin--a metabolic product of *Penicillium griseofulvum* DIECK and *Penicillium janczewskii* ZAL. Chem. & Ind. (London) **1951**, 179.

MALAWISTA, S. E., H. SATO, and K. G. BENSCH: Vinblastine and griseofulvin reversibly disrupt the living mitotic spindle. Science **160**, 770 (1968).

MALE, O., and K. HOLUBAR: Enzyme histochemical studies on the mode of action of some antimycotic drugs. Int. Congr. Chemother. Proc. 5th, **2**, 73 (1967a)

MALE, O., and K. HOLUBAR: Enzyme histochemistry of dermatophytes. II. Alteration of enzymatic reactions due to griseofulvin. Arh. Klin. Exptl. Dermatol. **227**, 973 (1967b).

MARGULIS, L., J. A. NEVIACKAS, and S. BANERJEE: Cilia regeneration in *Stentor*: Inhibition, delay and abnormalities induced by griseofulvin. J. Protozool. **16**, 660 (1969).

MCNALL, E. G.: Biochemical studies on the metabolism of griseofulvin. Arch. Dermatol. **81**, 657 (1960a).

MCNALL, E. G.: Metabolic studies on griseofulvin and its mechanism of action. Antibiotics Ann. **1959/60**, 674 (1960b).

MUNTONI, S.: Analogous effects of griseofulvin and colchicine on the mitosis. Boll. Soc. Ital. Biol. Sper. **41**, 813 (1965).

MUNTONI, S., and B. LODDO: Griseofulvin-induced budding in cell cultures. Arch. Intern. Pharmacodyn. **151**, 365 (Chem. Abstr. **62**, 5755g) (1964).

NAPIER, E. J., D. I. TURNER, and A. RHODES: The *in vitro* action of griseofulvin against pathogenic fungi of plants. Ann Botany (London) **20**, 461 (1956).

NONA, D. A., M. I. BLAKE, H. L. CRESPI, and J. J. KATZ: Effect of deuterium oxide on the culturing of *Penicillium janczewskii*. III. Antifungal activity of fully deuterated griseofulvin. J. Pharm. Sci. **57**, 1993 (1968).

OSMENT, L. S.: The many effects of griseofulvin. Alabama J. Med. Sci. **6**, 392 (1969).

OXFORD, A. E., H. RAISTRICK, and P. SIMONART: Studies on the biochemistry of microorganisms. 60. Griseofulvin, $C_{17}H_{17}O_6Cl$, a metabolic product of *Penicillium griseofulvum* DIECK. Biochem. J. **33**, 240 (1939).

PAGET, G. E., and S. J. ALCOCK: Griseofulvin and colchicine: lack of carcinogenic action. Nature **188**, 867 (1960).

PAGET, G. E., and A. L. WALPOLE: Some cytological effects of griseofulvin. Nature **182**, 1320 (1958).

PAGET, G. E., and A. L. WALPOLE: The experimental toxicology of griseofulvin. Arch. Dermatol. **81**, 750 (1960).

ROTH, F. J.: Griseofulvin: Ann. N. Y. Acad. Sci. **89**, 81, 750 (1960).

ROTH, F. J., B. SALLMAN, and H. BLANK: *In vitro* studies of the antifungal antibiotic griseofulvin. J. Invest. Dermatol. **33**, 403 (1959b).

SCHWARZ, J., and J. K. LOUTZENHISER: Laboratory experiences with griseofulvin. J. Invest. Dermatol. **34**, 295 (1960).

SHOCKMAN, G. D., and J. O. LAMPEN: Inhibition by antibiotics of the growth of bacterial and yeast protoplasts. J. Bacteriol. **84**, 508 (1962).

SRIVASTAVA, O. P., and V. C. VORA: Effect of griseofulvin on dermatophytes including locally isolated strains of *Trichophyton rubrum* and on *Microsporum canis* grown in keratin. J. Sci. Ind. Res. (India) **20C**, 163 (1961).

THYAGARAJAN, T. R., O. P. SRIVASTAVA, and V. C. VORA: Some cytological observations on the effect of griseofulvin on dermatophytes. Naturwissenschaften **50**, 524 (1963).

VIDMAR-CVJETANOVIC, B. P. GAUDIN, H. LOZERON, and W. JADASSOHN: Induced griseofulvin resistance of *Achorion* (Trichophyton) *quinckeanum in vitro* and *in vivo*. Experientia **22**, 737 (1966).

WRIGHT, J. M.: Phytotoxic effects of some antibiotics. Ann. Botany (London) **15**, 493 (1951).

ZIEGLER, H.: Effect of griseofulvin on *Microsporum canis*. IV. Z. Allgem. Mikrobiol. **3**, 211 (1963).

Irehdiamine and Malouetine

Simon Silver, Louis Wendt, and Pinakilal Bhattacharyya

Irehdiamine and malouetine are both alkaloids and can be considered "antibiotics" only in a broad sense of the term. They are two among a large number of steroidal alkaloids isolated from plants of the family *Apocyanaceae* by GOUTAREL and his collaborators at the Institut de Chimie des Substances Naturelles du C.N.R.S., Gif-sur-Yvette, France. GOUTAREL (1964) published a monograph on the isolation, identification and partial syntheses of many of these steroidal compounds. BISSET (1958, 1961) has considered pharmacological aspects of the alkaloids and taxonomy of their sources.

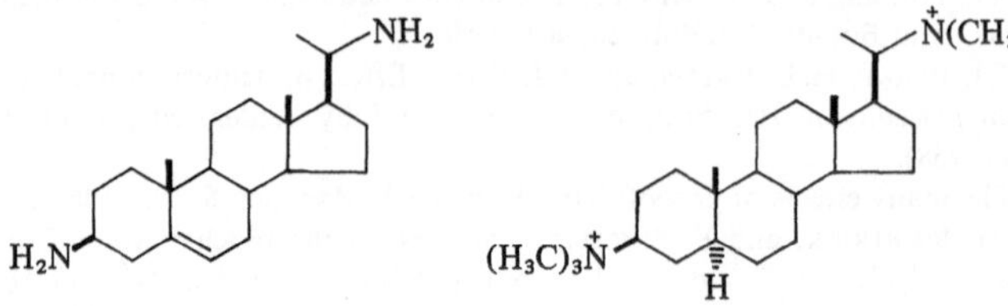

Irehdiamine A (IDA)   Malouetine (MAL)

Fig. 1

Irehdiamine (IDA) is the common name given to the primary diamine pregn-5-ene-3β, 20α-diamine (Fig. 1) by TRUONG-HO *et al.*, (1963) who isolated this steroidal diamine from leaves of the West African Ireh rubber tree *Funtumia elastica*. In addition to confirming the structure of irehdiamine by partial synthesis, GOUTAREL *et al.*, (1967) also prepared a series of secondary diamine (bis-mono-methyl-amino substituted IDA), tertiary diamine (bis-dimethyl-amino substituted IDA) and quaternary diamine (bis-trimethyl-amino substituted IDA) analogues for use in determining structural specificity of the action of IDA (see below).

Malouetine (MAL) is a steroidal diamine isolated from a related plant *Malouetia bequaertiana*, also of West African origin (JANOT *et al.*, 1960). Malouetine is 3β, 20α-bis-(trimethylammonium)-5α-pregnane (Fig. 1) and differs from irehdiamine in being a quaternary diamine and having the B-ring of the steroidal backbone fully saturated. The genus *Malouetia* also includes the plant producing the well-known Venezuelan toxin *guachamacá* (BISSET, 1961).

Effects on Bacterial Growth and Viability

MAHLER and BAYLOR (1967) reported that irehdiamine and malouetine are growth inhibitory for *Escherichia coli* and in addition are inhibitory for the growth of bacteriophages T2 and T4 within *E. coli*. While lower concentrations of irehdiamine are bacteriostatic, higher concentrations are bacteriocidal (MAHLER and BAYLOR, 1967; SILVER *et al.*, 1970). Malouetine is less potent than irehdiamine under comparable conditions (MAHLER and BAYLOR, 1967). A careful consideration of *E. coli* killing by irehdiamine showed single-hit kinetics with multiply-nucleated cells, suggesting that the nucleus (or DNA) is not the primary lethal target. However, the finding that irehdiamine and malouetine are mutagenic for bacteriophages T2 and T4 strongly suggested an interaction with the viral DNA. The mode of action will be discussed in detail below. The sensitivity of bacterial cells to these steroidal diamines appears to be dependent upon the ionic strength of the growth medium, with more dilute media favoring toxicity (MAHLER and BAYLOR, 1967). Magnesium ions can protect the cells from the action of steroidal diamines (SILVER and LEVINE, 1968b). Detailed studies have been carried out with *E. coli;* however, the steroidal diamines affect similarly Gram-positive bacteria such as *Bacillus subtilis* and some fungi, as well as animal cells in culture (MAHLER and BAYLOR, 1967; SILVER *et al.*, 1970; GOUTAREL, personal communication).

On short exposures with low concentrations of steroidal diamines, the *E. coli* cells remained viable indicating little if any loss of large macromolecules; there was no discernible cell lysis or change in appearance observed by phase contrast microscopy.

Effects on Phage Replication-Mutagenesis

Irehdiamine effectively inhibits the reproduction of bacteriophages T2 and T4 (MAHLER and BAYLOR, 1967). When 1.2×10^{-4} M irehdiamine was added at the time of infection, phage yield was reduced 10^4-fold. The extent of inhibition increased steeply (i.e. cooperatively) over a narrow range of concentrations and depended strongly on low ionic strength and low magnesium concentration in the growth media. Malouetine was less effective in reducing phage growth than irehdiamine (MAHLER and BAYLOR, 1967). Lower temperatures favored resistance to the action of steroidal diamines. Growth of the small RNA-containing virus MS2 was somewhat more resistant to irehdiamine inhibition than was growth of the large DNA-containing phage T2.

With higher concentrations of irehdiamine (10^{-4} M), ability to produce phage progeny was lost progressively and with single hit kinetics. The progeny burst size (phage yield per infected cell) was reduced by irehdiamine much more rapidly that plaque forming ability, i.e. the ability to produce any phage at all, leading to the conclusion that irehdiamine could interfere with the biosynthetic processes of phage multiplication without totally inactivating the infected cell. MAHLER and BAYLOR (1967) studied the inhibition in detail including measurements of DNA replication as well as RNA and protein synthesis, processes which were inhibited to greater or lesser extents. Genetic recombination did not appear to be inhibited although comparisons were difficult because of the greatly reduced progeny yield in the presence of irehdiamine or malouetine.

A most striking result of the experiments by MAHLER and BAYLOR (1967) was the finding that these steroidal diamines are mutagenic agents, irehdiamine being more potent than malouetine. A variety of mutations were produced and although no large deletions were obtained, some of the steroidal diamine-induced mutations could be reverted with base analogue mutagens and the steroidal diamines themselves could induce reversions in both base analogue-and proflavine (addition-deletion)-induced r_{II} mutants. Therefore, the mutagenic action of the steroidal diamines provided presumptive evidence for effects on DNA synthesis but the reversion pattern did not fit into the simple patterns sometimes found with chemical mutagens.

Mammalian Toxicity

Although observations on human cells in culture (SILVER et al., 1970) should make one cautious, irehdiamine has not been found to be toxic to animals (KHUONG-HUU-LAINÉ et al., 1965). The bitter bark of the Ireh rubber tree has been used as a remedy for piles (BISSET, 1958). Malouetine, on the other hand, is highly toxic. QUEVAUVILLER and LAINÉ (1960) found malouetine to have a curare-like effect on several types of animals (LD_{50} of 0,5 mg/kg with mice) and a potency in this respect only three-times less than tubocurare. The stereospecificity is such that the 3β, 20β-, 3α, 20α-, and 3α, 20β-analogues of malouetine (3β, 20α-) are equally toxic (KHUONG-HUU-LAINÉ and PINTO-SCOGNAMIGLIO, 1964). In this context it is of interest that tubocurare and the synthetic curare-substitute gallamine triethiodide are without effect on bacterial cells (SILVER et al., 1970). It was considered likely (KHUONG-HUU-LAINÉ and PINTO-SCOGNA-MIGLIO, 1964) that malouetine is either identical to or very similar in structure to guachamachá, the curare-like agent isolated from another species of the genus Malouetia of South American origin and which appears prominently in nineteenth century literature on curare-like materials (e. g. ERNST, 1870; SCHIFFER, 1882; BISSET, 1958; KHUONG-HUU-LAINÉ et al., 1965).

Biological Effects and Mode of Action

Two primary targets have been proposed for the bacteriocidal effects of IDA and malouetine: the cellular DNA and the cell membrane. The results of both series of studies were summarized in the "Polyamines" monograph of the New York Academy of Sciences (MAHLER and GREEN, 1970; SILVER et al., 1970). Since it seems now clear that the initial effects of the drugs are at the level of cell membrane (although nucleic acid interactions may occur subsequently), it would be logical to describe the cell membrane data first, followed by an attempt to relate these to the effects reported on nucleic acids.

A. Effects on Membrane Permeability

In media of low ionic strength and low magnesium, the uptake of small molecules (K^+, Mg^{2+}, thiomethylgalactoside and proline) by E. coli cells was inhibited strongly by steroidal diamines (SILVER and LEVINE, 1968a and b; SILVER et al., 1970; WENDT, 1970). Transport in isolated membrane "ghosts" from E. coli was also susceptible to the action of steroidal diamines and both proline

(SILVER *et al.*, 1970) and valinomycin-dependent potassium accumulations (BHATTACHARYYA *et al.*, 1971; BHATTACHARYYA, unpublished) were inhibited. There appeared to be a lack of specificity towards the type of membrane since cation accumulation and retention by mammalian cells in tissue culture was also affected by the steroidal diamines (SILVER *et al.*, 1970). Further, since both accumulation by active transport and retention of small molecules were affected, a nonspecific attack by the steroidal diamines on the membrane permeability properties is indicated rather than this being on some more specific membrane attribute or component. The breakdown in permeability barriers was further evidenced by the observation that acriflavine, to which *E. coli* cells are normally impermeable, was rapidly taken up by the cells and bound in the presence of the steroidal diamines (SILVER *et al.*, 1970). After short exposures followed by removal of irehdiamine, the cells recovered their previous impermeable state (SILVER and LEVINE, 1968*b*) by a process that did not require protein synthesis. Irehdiamine appears to act as a lipid soluble (steroidal) positively charged (primary diamine) molecule that inserts or dissolves into the membrane lipid of the cell, distorting the basic structure and leading to a lessening of the essential impermeability to small molecules. However, it is not simply the presence of amino groups that leads to these effects of IDA, since the primary diamine IDA was much more potent than was the secondary (bis-mono-methyl IDA) diamine and the tertiary and quaternary diamines were still less effective (SILVER and LEVINE, 1968*a*; SILVER *et al.*, 1970). The pK$_a$ for both amino groups of IDA is about 9.6 and therefore both irehdiamine and the quaternary diamine malouetine are divalent cations at physiological pH (MAHLER and GREEN, 1970). Related monovalent monoamines were also effective in altering cellular permeability: 3α-amino-5α-androstane and 3α-amino-5α-pregnane had about the equivalent efficacy to irehdiamine as far as has been tested (SILVER *et al.*, 1970). The shape of the steroidal backbone seems less important, however, since both malouetine and quaternary diamine bis-trimethyl IDA were about equally effective in causing membrane permeability changes (SILVER and LEVINE, 1968*a*; SILVER *et al.*, 1970). Furthermore, 5,6-dihydro-irehdiamine, which is the primary diamine with the same ring structure as malouetine, was indistinguishable from irehdiamine in terms of potency of its effects on membrane permeability (WENDT, 1970). Although the shape of the steroidal ring structure did not affect potency, the steroidal structure itself seemed to be required. Acriflavine, a primary diamine with about equivalent spacing to IDA between the amino groups, but with a system of three aromatic six-membered rings is known to be an anti-bacterial agent; however, acriflavine did *not* affect membrane permeability under these conditions (SILVER *et al.*, 1970). Furthermore, adamantane-1,3-diamine, a fully aliphatic compound with compact six-carbon rings that place the amino groups about half the distance one from the other as found with acriflavine or the steroidal diamines, was without effect on membrane permeability (unpublished data). Some feature of the steroid size or shape is required for attack on the membranes, but the currently available data are insufficient to define the structure-activity relationship more precisely.

WENDT (1970) studied the relationship between irehdiamine effects on membrane permeability and binding of the radioactive irehdiamine to bacterial cells.

The binding was found to be both concentration and temperature dependent. At low concentrations of irehdiamine, the amount bound was linearly related to the amount added. However, in the concentration range (10^{-5} to 10^{-4} M IDA) where effects on membrane permeability are apparent, cooperative binding, i.e. a sigmoidal relationship to concentration, was found. There was a comparable sigmoidal relationship between the concentration of IDA added and the *rate* of potassium loss (a measure of leakage) (SILVER *et al.*, 1970; WENDT, 1970) and the initial rate of potassium loss was directly proportional to the amount of IDA bound. The degree of cooperativity decreased at lower temperatures (WENDT, 1970) suggesting that the comparatively less fluid lipoprotein structure of the cell membrane at lower temperatures offers resistance to "dissolve in" the steroidal diamines or vice versa.

B. Effects on Nucleic Acids

The results of an extensive series of studies about the effects of steroidal diamines on nucleic acid secondary structure (MAHLER *et al.*, 1966; GOUTAREL *et al.*, 1967; MAHLER *et al.*, 1968) have been systematically reviewed and brought together by MAHLER and GREEN (1970). The diamines interact with DNA forming two complexes of different conformational structures depending upon concentration of the ligands and ionic strength of the medium.

Complex I ("stabilized"), at low steroid/nucleic acid ratios, is a deformed, somewhat expanded helix. This complex saturates at about 1 ligand bound per 2 base pairs and has been monitored by its effects on the physical and optical properties of the DNA including a) higher thermal denaturation temperature, b) changes in the UV absorption spectrum with resultant hyperchromicity and a "red-shift", c) altered optical rotatory dispersion and circular dichroism, again in the UV wavelength range, d) an increase in sedimentation coefficient, and e) a decrease in the intrinsic viscosity of the DNA solution (MAHLER and GREEN, 1970). WARING (1970) and WARING and CHISHOLM (1972) measured the changes in sedimentation velocity of supercoiled circular DNA and determined that the DNA in Complex I extends slightly with an unwinding angle of 5° per steroidal diamine bound. The stoichiometry of Complex I appears to be independent of base composition of the DNA, at least between the limits of 30% and 72% guanine plus cytosine (G+C). However, the incremental stabilization varied inversely with the base composition. Under a particular set of ionic conditions, the observed increase in the thermal midpoint ΔTm was 50°C for the DNA with 30% G+C, 46.5°C for DNA with 41% G+C, 40.5°C for DNA with 50% G+C and 35°C for DNA with 72% G+C. The extent of formation of Complex I was dependent upon the ionic strength of the solvent. At low ionic strength, complex formation was virtually quantitative, while increasing the ionic strength resulted in the dissociation of the complex. The properties of DNA in the dissociated mixture were indistinguishable from those of an undisturbed solution of DNA at the same ionic strength (MAHLER and GREEN, 1970). Thus, the complex formation seemed to be completely reversible. The relative ability to form the Complex I, as measured by increase in the thermal transition midpoint at $r = $ [amine]/[DNA-P] $= 0.25$ was irehdiamine $\geq$ malouetine $>$ Mg^{2+}. With the sub-

stituted steroidal diamines, the ability to stabilize DNA varied with the nature
and the position of the amino groups: although all 3,20-diamino steroids had
effects, at $r = 0.25$ the order of effectiveness within a series with an identical
steroid backbone was di $4° \geq$ dil° > di2° $\approx$ di3°. The efficacy series for forming
Complex I with regard to conformation of the amino groups was β, $\alpha > \beta$, $\beta > \alpha$,
$\alpha > \alpha$, β for the most effective (di4°) members of both series, preg-5-ene (bis-
trimethyl IDA) and pregane (malouetine). Thus the order for effectiveness in
forming Complex I with DNA was very different from the order for membrane
destabilization (above).

Complex II ("destabilizing") occurs at bound ligand to nucleotide pair ratios
above 1:2 and appears complete at a ratio of one steroidal diamine per DNA
nucleotide. The formation and properties of Complex II were followed using
the same physical-chemical parameters as used in studying the Complex I formation; however, the effects are generally in the opposite direction. In Complex
II the DNA becomes more sensitive to thermal denaturation, UV absorption
decreases, and there was decreased optical rotatory dispersion (MAHLER and
GREEN, 1970). The order of effectiveness of steroidal amines in forming Complex
II was essentially the opposite of that for Complex I formation: primary diamines
were more effective than secondary diamines and tertiary and quaternary diamines
were essentially without effect. Monoamines worked almost as well as diamines
(MAHLER and GREEN, 1970). As with Complex I, the formation and maintenance
of Complex II required low ionic strength conditions; the complex was completely
dissociated upon adding high salt (MAHLER and GREEN, 1970). The sum of the
physical chemical measurements of Mahler and coworkers lead to a picture
of Complex II with the steroidal diamine associated in short stacks or micelles
along one groove of the DNA double helix, bound electrostatically between
the amino groups and the DNA phosphates. As more steroidal diamines join
the micelles, the DNA molecules partially unfold exposing single stranded regions
and these regions grow more readily as they lengthen—i.e. the denaturation
shows cooperative kinetics with regard to the size of the micelles.

Most studies have been carried out with native DNA. However, similar effects
of steroidal diamines on the secondary structure of synthetic polynucleotides
occur (MAHLER and GREEN, 1970; LEFRESNE et al., 1967; GABBAY et al., 1970).
GABBAY et al. (1970) also measured cooperative destabilization of the DNA structure by steroidal diamines at high ligand/DNA ratios but proposed from thermal
denaturation and nuclear magnetic resonance studies that the steroidal diamines
destabilize the nucleic acids by stabilizing the single-stranded regions and preventing quick renaturation following transient and localized formation of single-
stranded regions.

Mode of Biological Action

More than one primary target for the steroidal diamines has been proposed.
The proponents of the alternative targets (cell membrane or nucleic acids) have
studied the effects of the steroidal diamines on *their* proposed targets in detail
(MAHLER and GREEN, 1970; SILVER et al. 1970). The key to the question of
primary target seems to lie in the structure-activity relationships, which were

Table 1. *Efficacy of steroidal diamines on biological functions*

Process or property affected	Efficacy sequence	Reference
Stabilization of DNA to heat (Complex I)	di4° $\geq$ di1° > di2° $\approx$ di3° (monoamines ineffective)	MAHLER and GREEN (1970) and MAHLER et al. (1966)
Labilization of DNA to heat (Complex II)	di1° > di2° $\gg$ di3° $\approx$ di4° (monoamines effective)	MAHLER and GREEN (1970) and GABBAY et al. (1970)
Potassium leakage and transport inhibition	di1° > di2° > di3° > di4° monoamines $\approx$ IDA	SILVER and LEVINE (1968a); SILVER et al. (1970)
Galactoside leakage and transport inhibition	di1° > di2°; di3° and di4° ineffective; IDA > MAL	SILVER and LEVINE (1968a)
Acriflavine permeability	IDA > MAL	SILVER et al. (1970)
Phage functions:		
burst size	IDA > MAL	MAHLER and BAYLOR (1967)
recombination frequency	no effect detected	MAHLER and BAYLOR (1967)
mutagenesis	IDA > MAL	MAHLER and BAYLOR (1967)
Cell viability	not determined	

described above and are summarized in Table 1. However, structure-activity relationships for bacterial killing or bacteriophage inhibition have not been as thoroughly studied as the relationships for DNA interactions or membrane permeability effects.

The antimicrobial activity of irehdiamine and malouetine can be understood in terms of the primary target being the cell membrane and the primary effect being a relatively non-specific loss of the normal membrane permeability barrier function. Small molecules (ions, amino acids, sugars) would first leak out from the cells. Next, the loss of small molecules and/or direct membrane association would greatly decrease general cellular metabolism, including energy metabolism, active transport, and nucleic acid and protein synthesis. This would in itself be lethal since energy metabolism and synthetic activity would be required for any recovery. With the membrane barrier destroyed and the ionic strength including that of magnesium reduced, steroidal diamines would now be able to interact electrostatically with the cellular (or bacteriophage) DNA. Irehdiamine is a stronger mutagen than malouetine (MAHLER and BAYLOR, 1967), although both amines appear to be equally effective in converting DNA to Complex I *in vitro*. It is this first complex that leads to profound alterations in DNA conformation, resulting in an unwinding of covalently closed circular DNA similar in geometry to that brought about by mutagenic intercalating dyes such as ethidium bromide (WARING, 1970; WARING and CHISHOLM, 1972). The steroidal diamines must first reduce the ionic strength in the cellular cytoplasm and then associate with the DNA leading to mistakes in DNA replication—i. e. mutations, or alternatively, the loss of essential cations and small molecular weight cofactors or the direct association of phage DNA replication to the cell membrane (there is wide-spread evidence for both cation requirements and membrane association) leads to mistakes in DNA replication. A similar sequence of considerations has been suggested for the detailed mechanism of mutagenesis by ethidium bromide (MAHLER and PERLMAN, 1972). Some details of this proposed mode of action remain open but the overall picture seems to be consistent with existing knowledge about the steroidal diamines.

Acknowledgements. The authors' research on steroidal diamines has been supported by research grant AI-08062 from the National Institute of Allergy and Infectious Diseases. Drs. HENRY MAHLER and MARTHA BAYLOR provided encouragement and advice. Irehdiamine and malouetine are not currently commercial products. Dr. ROBERT GOUTAREL and his colleagues at the Institut de Chimie des Substances Naturelles, C.N.R.S., 91 Gif-sur-Yvette, France, have responded generously and supplied these steroidal amines.

References

BHATTACHARYYA, P., W. EPSTEIN, and S. SILVER: Valinomycin-induced uptake of potassium in membrane vesicles from *Escherichia coli*. Proc. Natl. Acad. Sci. U.S. **68**, 1488 (1971).

BISSET, N.G.: The occurrence of alkaloids in the Apocyanaceae. Ann. Bogor. **3**, 105 (1958).

BISSET, N.G.: The occurrence of alkaloids in the Apocynaceae. Part II. A review of recent developments. Ann. Bogor. **4**, 65 (1961).

ERNST, M.A.: On the Guachamacan, a poisonous plant growing in the Ilanos (plains) of Venezuela. Trans. Bot. Soc. Edinburgh **10**, 448 (1870).

GABBAY, E.J., and R. GLASER: Topography of nucleic acid helices in solutions. Proton magnetic resonance studies of the interaction specificities of steroidal amines with nucleic acid systems. Biochemistry **10**, 1665 (1971).

GABBAY, E.J., R. GLASER, and B.L. GAFFNEY: Interaction specificity of nucleic acids. Ann. N.Y. Acad. Sci. **171**, 810 (1970).

GOUTAREL, R.: Les alcaloïdes steroïdiques des Apocynacées. Paris: Hermann 1964.

GOUTAREL, R., H.R. MAHLER, G. GREEN, Q. KHUONG-HUU, A. CAVÉ, C. CONREUR, F.-X. JARREAU et J. HANNART: Alcaloïdes stéroïdiques. LXXI. Synthèse des quatre diamino-3,20 prégnène-5: irehdiamine-A et stéréoisomères. Étude comparée de leurs interactions avec *ADN d'Escherichia coli*. Bull. Soc. Chim. France **1967**, 4575.

JANOT, M.-M., F. LAINÉ, et R. GOUTAREL: Alcaloïdes stéroïdiques. V: Alcaloïdes du *Malouetia bequaertiana* E. Woodson (Apocynacées): la funtuphyllamine B et la malouétine. Ann. Pharm. Franc. **18**, 673 (1960).

KHUONG-HUU-LAINÉ, F., N.G. BISSET et R. GOUTAREL: Alcaloïdes stéroïdiques. XXXIX. Alcaloïdes du *Malouetia bequaertiana* Woods. Mise au point sur le genre *Malouetia* (Apocynacées) et sur une drogue curarisante du Venezuela, le *Guachamacá*. Ann. Pharm. Franc. **23**, 395 (1965).

KHUONG HUU-LAINÉ, F., et W. PINTO-SCOGNAMIGLIO: Activité curarisante du dichlorure de 3β-20α bistrimethylammonium 5α-prégnane (malouétine) et de ses stéréoisomères. Arch. Intern. Pharmacodyn. **147**, 209 (1964).

LEFRESNE, P., J.M. SAUCIER, and C. PAOLETTI: Structural changes of polyriboinosinic acid induced by a steroidal diamine, irehdiamine A. Biochem. Biophys. Res. Commun. **29**, 216 (1967).

MAHLER, H.R., and M.B. BAYLOR: Effects of steroidal diamines on DNA duplication and mutagenesis. Proc. Natl. Acad. Sci. U.S. **58**, 256 (1967).

MAHLER, H.R., R. GOUTAREL, Q. KHUONG-HUU, and M. TRUONG-HO: Nucleic acid interactions. VI. Effects of steroidal diamines. Biochemistry **5**, 2177 (1966).

MAHLER, H.R., and G. GREEN: Interaction of steroidal diamines with DNA and polynucleotides. Ann. N.Y. Acad. Sci. **171**, 783 (1970).

MAHLER, H.R., G. GREEN., R. GOUTAREL, and Q. KHUONG-HUU: Nucleic acid-small molecule interactions. VII. Further characterization of DNA diamino steroid complexes. Biochemistry **7**, 1568 (1968).

MAHLER, H.R., and P.S. PERLMAN: Mitochondrial membranes and mutagenesis by ethidium bromide. J. Supramol. Struc. **1**, 105 (1972).

Quevauviller, A., et F. Lainé: Sur la toxicité et le pouvoir curarisant du chlorure de malouétine. Ann. Pharm. Franc. **18**, 678 (1960).

Schiffer, J.: Physiological Society, Berlin, report of meeting. Nature **25**, 620 (1882).

Silver, S., and E. Levine: Action of steroidal diamines on active transport and permeability properties of *Escherichia coli.* J. Bact. **96**, 338 (1968a).

Silver, S., and E. Levine: Reversible alterations in membrane permeability of *Escherichia coli* induced by a steroidal diamine, irehdiamine A. Biochem. Biophys. Res. Commun. **31**, 743 (1968b).

Silver, S., L. Wendt, P. Bhattacharyya, and R. S. Beauchamp: Effects of polyamines on membrane permeability. Ann. N. Y. Acad. Sci. **171**, 838 (1970).

Truong-Ho, M., Q. Khuong-Huu et R. Goutarel: Alcaloïdes stéroïdiques XV. Les irehdiamines A et B, alcaloïdes du *Funtumia elastica* (Preuss) Stapf. Bull. Soc. Chim. France **1963**, 594.

Waring, M.: Variation of the supercoils in closed circular DNA by binding of antibiotics and drugs: evidence for molecular models involving intercalation. J. Mol. Biol. **54**, 247 (1970).

Waring, M. J., and J. W. Chisholm: Uncoiling of bacteriophage PM2 DNA by binding of steroidal diamines. Biochim. Biophys. Acta **262**, 18 (1972).

Wendt, L. W.: The action of irehdiamine A on *Escherichia coli:* inhibition of the membrane response at low temperatures. Biochem. Biophys. Res. Commun. **40**, 489 (1970).

Isonicotinic Acid Hydrazide*

C. R. Krishna Murti

1. Introduction

Isonicotinic acid hydrazide (INH), though known as early as 1912 (MEYER and MALLY, 1912), was rediscovered in the early 1950's almost simultaneously by three independent groups of investigators (Fox, 1952a; BERNSTEIN et al., 1952; OFFE et al., 1952). The advent of INH has been justifiably hailed as a very significant milestone in the history of the conquest of tuberculosis. In combination with streptomycin and given in ambulatory drug regimens, INH has made it possible now to treat large numbers of tubercular patients without recourse either to hospitalize them or to keep them in sanatoria for prolonged periods (McDERMOTT, 1960; RUSSELL and MIDDLEBROOK, 1961; BARRY, 1964; Tuberculosis Chemotherapy Centre, Madras, 1970).

The specific inhibitory action of INH on the tubercle bacillus has stimulated numerous studies designed for gaining insight into the molecular mechanism by which it destroys the pathogen. These investigations have indeed uncovered a number of biochemical effects that occur when susceptible mycobacteria are exposed to INH (GOLDMAN, 1954; ROBSON and SULLIVAN, 1963; WINDER, 1964; NESTLER, 1966; RAMAKRISHNAN et al., 1972; LEWIS and SHEPHERD, 1970).

INH is a double-edged weapon. While it has a powerful tuberculostatic action, it also exerts certain pronounced effects on the central nervous system of many mammalian species including man. The drug produces symptoms of neuritis associated with pyridoxine deficiency in laboratory animals (LEVY et al., 1967). An attempt is made in the following pages to identify some of the meaningful approaches that have been pursued in the elucidation of the mode of action of INH in the above dual role.

2. Chemistry of INH

The parent compound of INH (I) is by a strange coincidence closely related to the vitamin nicotinamide (II) which itself shows a weak tuberculostatic action (CHLORINE, 1945). During a study of pyridine carboxylic acids, Fox (1952b)

I   II

$$\text{III} \quad \xrightarrow{} \quad \text{IV} \qquad \text{V} \qquad \text{VI} \quad \xrightarrow{NH_2NH_2} \quad \text{I}$$

COOH / NH₂ (III) → COOCH₃ / NH₂ (IV)   CHO (V)   COOCH₃ (VI)   CO—NH—NH₂ (I)

found that 3-aminoisonicotinic acid (III) and its methyl ester (IV) were tuberculo-static but had no vitamin activity. When, with the objective of preparing isonico-tinaldehyde (V), the methyl ester of isonicotinic acid (VI) was treated with hydra-zine, INH was obtained as the intermediate on the pathway of Mac-Fadyen and Steven synthesis. The presence of isoniazid-like substances has been reported in the exudate of the date palm, *Phoenix sylvestris*, but its identity with INH has not been established (BHAGWAT and SOHONIE, 1955).

The highly specific growth inhibitory action against the tubercle bacillus and the relatively low host toxicity of INH have prompted extensive structure-activity relationship studies and preparation of a large number of analogues, derivatives and substitution compounds (FOX and GIBAS, 1952, 1953, 1955; McMILLAN *et al.*, 1953; ISLER *et al.*, 1955; LEWIS and SHEPHERD, 1970). Of the many reported in the literature, a few show high activity by virtue of their *in vivo* transformation into INH, but none has so far excelled INH in efficacy and low toxicity. The pyridine nucleus, the —CO.NH.N< moiety as well as the basicity of the terminal nitrogen of the hydrazide moiety are all indispensable for growth inhibitory activity (BERNSTEIN *et al.*, 1953; CYMERMAN-CRAIG *et al.*, 1955). The best known is the isopropyl derivative, isoproniazid (VII) which is also a useful CNS-active drug. Some other interesting derivatives are bromoisonicotinic acid hydrazide (VIII) and bromoisonicotinyl thioamide (IX) which are reported to be less toxic than INH (PALAT *et al.*, 1967). Some relevant information on activity as affected by substitutions in INH is summarized in Table 1.

CO—NHNHCH(CH₃)CH₃ (VII)   CO—NHNH₂ / Br (VIII)   CSNH₂ / Br (IX)

3. Antimicrobial Spectrum of INH

3.1. Activity against Pathogenic Bacteria, Fungi and Protozoa

Unlike streptomycin with its broad spectrum of activity, INH is unusually specific for mycobacteria (Table 2). The relatively high concentrations (600 µg/ml) required to inhibit Gram-positive and Gram-negative bacteria as compared to the very low concentrations (0.05–0.1 µg/ml) sufficient to inhibit *Mycobacterium tuberculosis* indicate different modes of action on these species of microorganisms.

Table 1. *Structure-activity relationship in INH*

$CO \cdot \overset{1}{N}H\overset{2}{N}HH$	$CO \cdot NH \cdot N <$ Essential part Terminal N must be basic

Substitution:	*Results:*
2 Monoalkyl (C_1 through C_8 and sugar alcohol) derivatives (—$CO \cdot NH \cdot NHR$)	Inactive. Branched-chain alkyl (Isopropyl) highly active
2 Cycloalkyl or arylalkyl (—$CO \cdot NH \cdot NHR$)	Highly active but less than INH
2 Monoaryl or heteroaryl, 1-monoalkyl (—$CO \cdot NR \cdot NHR$)	Inactive
2,2-Dialkyl (—$CO \cdot NH \cdot NR_2$)	Highly active but decreased with bulk of R.
Trialkyl (—$CO \cdot N$—$R \cdot NR_2$)	Completely inactive
Substitution of pyridine nucleus by benzene, piperidine, or a thiazole	Inactive
Addition of halogen to ring, o-monobromo	Very active against H37Rv
Addition of halogen to ring plus substitution of hydrazine	Most effective against INH-resistant strains of H37Rv

Even among the mycobacteria, *M. tuberculosis* is the most susceptible to INH (PANSY *et al.*, 1952). The minimum inhibitory concentration required to suppress the growth of pathogenic fungi is far too high to make INH a useful fungistatic or fungicidal agent. It does not affect the growth of protozoal parasites.

3.2. Action of INH on the Tubercle Bacillus

Depending upon components of the medium and duration of contact with the bacillus, complete inhibition of growth of the human virulent strain, *M. tuberculosis var. hominis H37Rv*, can be achieved with concentrations as low as 25 nanograms but the usual inhibitory dose range *in vitro* is $50-250$ nanograms (see Tables 2 and 3) (ARONSON *et al.*, 1952; BERNSTEIN *et al.*, 1952; GRUNBERG *et al.*, 1952; STEENKEN and WOLINSKY, 1952; ARMSTRONG, 1960). There is apparently very little action on resting bacteria. Susceptible bacteria have to multiply at least once before the inhibitory effect manifests itself (BARCLAY *et al.*, 1953a; MACKANESS and SMITH, 1953; HOBBY and LENERT, 1957). The validity of this generally held opinion has, however, been questioned (SCHAEFER, 1954; YOUATT, 1969). In media that support vigorous growth, INH exerts, additionally, a bactericidal effect (MIDDLEBROOK, 1952; SCHAEFER, 1954), and the drug is effective against H37Rv strains which are resistant to streptomycin or p-aminosalicylic acid (KARLSON and FELDMAN, 1952). The amount of drug bound to the cells determines the extent of growth inhibition: microorganisms that are not susceptible do not bind the drug (YOUATT, 1958c; BEGGS and JENNE, 1969). Furthermore, INH exhibits a high activity against both intracellular and extracellular mycobacteria at concentrations attained under *in vivo* conditions (Table 4). The drug

Table 2. *Antimicrobial spectrum of INH*

Test organism	M.I.C. (µg/ml)
Micrococcus pyogenes var. aureus	> 600
Streptococcus faecalis	> 600
Streptococcus pyogenes	> 600
Klebsiella pneumoniae	> 600
Escherichia coli	> 600
Aerobacter aerogenes	> 600
Salmonella typhosa	> 600
Salmonella schottmulleri	> 600
Shigella dysenteriae	> 600
Shigella sonnei	> 600
Pseudomonas aeruginosa	> 600
Proteus vulgaris	> 600
Cornyebacterium diphtheriae	> 600
Diplococcus pneumoniae Type 2	600
Diplococcus pneumoniae Type 3	> 600
Lactobacillus acidophillus	600
Brucella abortus	> 600
Bacillus subtilis	> 600
Mycobacterium smegmatis	0.5
Mycobacterium tuberculosis var. bovis strain BCG	0.04
Mycobacterium tuberculosis var. bovis strain Ravenel	0.06
Mycobacterium tuberculosis var. hominis strain H37Rv	0.03
Candida albicans	100
Saccharomyces cerevisiae	100
Ceralostomella ulmi	50
Aspergillus fumigatus	100
Aspergillus niger	100
Pencillium notatum	25
Microsporum audouni	100
Microsporum canis	100
Epidermophyton floccosum	100
Cryptococcus neoformans	400
Trichophyton mentagrophytes ATCC 95–23	100
Trichophyton mentagrophytes Squibb 1920	100
Trichophyton tonsurans	100
Trichophyton rubrum	100
Trichophyton schoenleini	100
Trichophyton violaceum	100
Trichophyton flaviforme var. discoides	100
Trichophyton flaviforme var. album	100
Trichophyton flaviforme var. ocharacea	100

Data from PANSKY *et al.* (1952).

Entamoeba histolytica	500
Leishmania donovani	not active
L. tropica	not active
T. cruzi	not active
T. vaginalis	not active
T. foetus	not active

Data from SENECA and IDES (1953).

Table 3. *In vitro anti-tubercular activity of INH in different media*

Growth media	M.I.C. (μg/ml)
Tween-albumin liquid medium	0.050
Proskauer and Beck synthetic medium	0.025
Proskauer and Beck synthetic medium with 10% horse serum	0.050
Oleic acid albumin medium	0.250

Seven day culture of H37Rv in Tween-albumin used for inoculation. 0.1 ml for liquid medium and one double loopful for plates of solid medium. Incubation at 37° for 14 days in liquid media and 21 days in solid media.

Data from STEENKEN and WOLINSKY (1952).

Table 4. *Effect of INH on intracellular mycobacteria*

Growth medium of H37Rv	Minimal concentration to completely inhibit growth (μg/ml)	
	Streptomycin	INH
Tween-albumin liquid medium	0.5	0.05
Phagocytes *in vitro*	80–100	0.05
Rabbit macrophages	25.0	0.05

Based on data of SUTER (1952); MACKANESS (1952) and MACKANESS and SMITH (1952).

is equally effective on all species and strains of tubercle bacilli of mammals or birds.

3.3. Cytological Changes Produced by INH

INH brings about the loss of acid-fastness in mycobacteria (MIDDLEBROOK, 1952; SCHAEFER, 1954; BARCLAY et al., 1954, 1956; GRUMBACH, 1966) and this loss is correlated with the sterilization of the bacteria. Electron microscopic studies on mycobacteria, exposed to INH, reveal little change in length or girth but tapering rods are seen; internal structure is lost within 24 hours and bacilli become intensely electron dense (BRIEGER et al., 1953; GUPTA and VISWANATHAN, 1956). Swelling of the bacteria followed by retraction of protoplasm and eventual intracellular plasmolysis (BRAUNSTEINER et al., 1953); elongation of cells and increases in electron density at low concentrations of the drug have also been reported (BRINGMANN, 1951). There is need to repeat these studies at higher optical resolution and by scanning devices now available in order to gain more precise insight into the cytological changes initiated either directly by INH or indirectly on account of its primary action.

3.4. In vivo Activity of INH

Experimental tuberculosis induced in mice (BERNSTEIN et al., 1952; GRUNBERG et al., 1952) guinea pigs or rabbits (STEENKEN and WOLINSKY, 1952; KARLSON and FELDMAN, 1953) lends itself to treatment with INH. Monkeys treated with INH are not only cured of clinical tuberculosis but also become tuberculin-negative (SCHMIDT, 1956). The very first reports of its effectiveness in human patients (ROBITZEK and SELIKOFF, 1952; GRUNBERG et al., 1952) revealed that INH is a powerful chemotherapeutic agent against both pulmonary and non-pulmonary tuberculoses and generates a very rapid improvement in patients. Significant serum levels of antimicrobially active INH are readily achieved in humans following oral or intravenous administration and the drug readily diffuses into cavitary contents, pleural fluid and cerebrospinal fluid (BARCLAY et al., 1953b). Rapid sterilization of tubercle bacilli is achieved in human patients treated with INH particularly in conjunction with streptomycin or p-aminosalicylic acid. INH finds prophylactic use toward preventing primary infection of children from developing into clinical tuberculosis (RUSSELL et al., 1956).

4. Resistance of Tubercle Bacilli to INH

In vitro resistance develops in mycobacteria grown in increasing concentrations of INH; ordinarily there is no cross-resistance between INH and streptomycin. At any given period, over 60% of the cell population in growing cultures is INH susceptible and 40% resistant. However, when cells multiply rapidly, the population susceptible to INH increases and, in contrast, the population resistant to INH predominates in the static phase of growth (BIEHL, 1956, 1957; BARCLAY and WINBERG, 1964).

Resistant strains have been isolated from monkeys undergoing INH therapy for experimental tuberculosis (SCHMIDT, 1958). When used singly in humans, resistance develops rapidly and usually within a few weeks of therapy, the pathogens isolated from infected tissue or sputum are mainly the resistant variants (MIDDLEBROOK, 1952; McDERMOTT, 1960). The development of resistance in vitro to INH follows a "one step" pattern with a mutation rate of about 1 to 3×10^{-6} per bacterium per generation (SZYBALSKI and BRYSON, 1952). INH dependent strains of mycobacteria have also been isolated (BYRSON et al., 1953).

5. Criteria for an Effective Tuberculostat

Three principal criteria are needed for an ideally effective tuberculostat:

i) high toxicity for the pathogen and a low toxicity for the host to conform to the dogma of pharmacological specificity of action;

ii) ability to circumvent barriers interposed between the tubercle bacillus and the host tissue, particularly, the cell walls of macrophages, epitheloid and giant cells; and

iii) since even in the most propitious circumstances the drug has to be used for prolonged periods, it should be slow in producing resistant variants.

INH meets to a large extent the above exacting specifications. It is highly active against the pathogen in small doses, is relatively non-toxic and well-tolerated by the host. It has the ability to reach the target organism even when the latter seeks refuge within nodules in the host and above all, it has a low rate of conferring resistance to the susceptible organism as compared to other drugs. Any acceptable mechanism of action of INH has, therefore, to bear some relation to the above three criteria. Studies on the uptake and transformation of the drug by the tubercle bacillus, the effect of the drug on the metabolic activity of the bacillus especially on specific enzymes and the fate of the drug in the host in relation to various toxic manifestations are, therefore, relevant to the present discussion.

6. Uptake of INH by Mycobacteria

6.1. Uptake and Binding by M. tuberculosis

Susceptible strains of *M. tuberculosis*, exposed to INH-^{14}C (carbonyl), accumulate considerable amouts of radioactivity while resistant bacteria under identical conditions associate less radioactivity. Some typical results of such uptake studies with the H37Rv and a strain resistant up to $20-60$ µg INH/ml are represented in Fig. 1. It is evident that whereas niacinamide and nicotinic acid are readily taken up by the susceptible and the resistant strains alike, there is a very marked difference between the extents of uptake of INH by the two strains. The small difference between the uptakes of isonicotinic acid-^{14}C by the two strains was due to contamination of the acid with INH-^{14}C. That the uptake is mediated by an active process is inferred from the fact that heated cells show very much reduced levels of radioactivity. The amount of INH bound to susceptible cells is dependent on its concentration in the medium and the relationship can be represented graphically as a Freundlich's equation indicative of adsorption of the drug on specific binding sites of the bacillus.

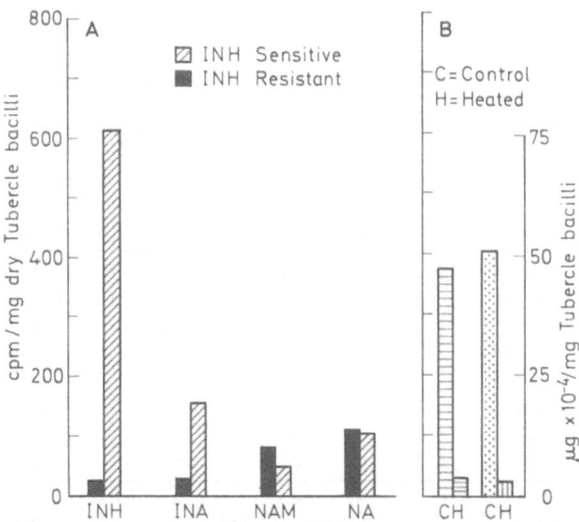

Fig. 1. A. Uptake of INH-^{14}C and related metabolites by INH sensitive and INH resistant *M. tuberculosis*. INA = Isonicotinic acid; NAM = Nicotinamide; NA = Nicotinic acid. B. Effect of heat on uptake and binding of INH-^{14}C by *M. tuberculosis*: Bars on left = radioactivity; bars on right = INH bound. From data of BARCLAY *et al.* (1954)

6.2. Factors Controlling Uptake of INH

Experimental studies (BARCLAY et al., 1953b, 1954, 1956; YOUATT, 1958c; YOUATT, 1960a; EDA, 1963; TSUKAMURA et al., 1963; WIMPENNY, 1967a; JUTTE, 1967) on the uptake of radioactive INH by different strains of M. tuberculosis, when critically evaluated (YOUATT, 1969), indicate the following features:

a) Surface cultures show increasing ability to take up INH throughout the logarithmic phase;

b) The presence of metabolizable substrates is not essential for uptake;

c) The binding of INH to growing cells and to washed suspensions is mediated by an identical mechanism;

d) Under both conditions, sodium cyanide and dinitrophenol inhibit the uptake suggesting the involvement of an active transport mechanism (perhaps a permease);

e) Uptake is dependent on external concentration of INH;

f) Uptake is dependent on the supply of oxygen and is inhibited under anaerobic conditions;

g) Cyanacetic hydrazide, benzoic acid hydrazide, p-, o- and m-aminobenzoic acid hydrazides, nicotinic and picolinic acid hydrazides inhibit the uptake and also antagonize the anti-tubercular action of INH;

h) Uptake is stimulated by sulphydryl reagents; and

i) the activation energy of binding is of the same order as that of an enzyme.

To these factors must be added the permeability barrier of the cell envelope. That the INH-resistant mutants can vary in their ability to take up INH and that drug resistance is perhaps associated with a permeability barrier to the drug are suggested by the data presented in Table 5.

Table 5. Uptake of INH by sensitive and resistant strains of M. tuberculosis and the effect of surface active compounds

15 days old cultures were inoculated into Youmans and Karlson medium containing $0.05 \mu Ci$ INH-^{14}C/ml and incubated for 40 hours. The organisms were then washed with Youmans and Karlson medium containing $10 \mu g$ unlabelled INH/ml and finally with distilled water.

Treatment	M. tuberculosis H37Rv		Mutant inh.-γ-4		Mutant inh.-γ-6	
	Counts[a] 5 min 10 mg dry wt.	Relative % of uptake INH-^{14}C	Counts[a] 5 min 10 mg dry wt.	Relative % of uptake INH-^{14}C	Counts[a] 5 min 10 mg dry wt.	Relative % of uptake INH-^{14}C
None	420	100	70	17	140	33
EDTA 5×10^{-3} M	360	85	110	26	220	52
DDS 0.01%	440	105	120	29	230	55

[a] Represent total counts minus background (110 counts/5 min); EDTA = Ethylenediaminetetraacetate; DDS = Sodium dodecyl sulphate.

Growth of mutant inh.-4 is inhibited by $1 \mu g/ml$ INH when cells are grown in presence of 5×10^{-3} M EDTA or 0.01% DDS.

From data of SRIPRAKASH and RAMAKRISHNAN (1970).

7. Fate of INH Taken up by Tubercle Bacillus

7.1. Metabolic Products

The radioactivity taken up by susceptible mycobacteria exposed to INH-^{14}C can be recovered from the cells as water soluble compounds. Among the identified metabolites of INH are: isonicotinic acid (X) and 4-pyridyl methanol (XI) (BOONE et al., 1957; YOUATT, 1958a, 1960b, 1961). Part of the radioactivity accumulated by the cells and recovered as hydrazones are presumably artifacts of the extraction and identification procedure. The metabolic pathway may be visualized to follow the sequence:

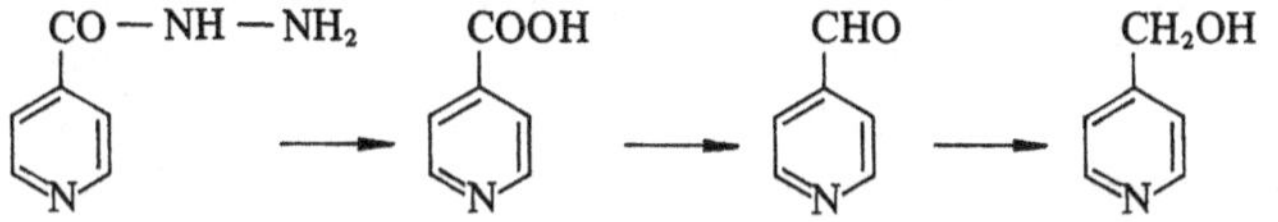

4-Pyridyl aldehyde (V) cannot be recovered from culture fluids. The intermediate formation of V is, however, surmised from the fact that whole cells and extracts of BCG are able to metabolize exogenously added V approximately a thousand times faster than INH; it may not be possible to trap the aldehyde (YOUATT, 1962).

7.2. Enzymes Mediating Metabolism of INH

Culture filtrates and extracts of mycobacteria are known to inactivate INH (YOUMANS and YOUMANS, 1955). The enzymic machinery catalyzing the above reactions has not been studied in detail. *Mycobacterium avium* shows an enzyme activity designated as "hydrazidase" which converts INH to isonicotinic acid and hydrazine (TOIDA, 1962). On the basis of distinct differences in the response to chelating agents, cyanide and azide, this enzyme seems to be unrelated to the activity present in *M. tuberculosis*. Tests made with cell extracts do not reveal hydrazine or ammonia and the fate of the hydrazide nitrogen remains unknown. Neither isonicotinic acid (X) nor 4-pyridyl methanol (XI) have any anti-tubercular activity. Possibly heavy metal ions such as Mn^{2+}, present in crude bacterial extracts, bring about a non-enzymic cleavage of INH to isonicotinic acid.

7.3. Inter-Relation of Uptake and Metabolism

The processes that mediate the uptake and metabolism of INH in mycobacteria are presumably identical (MILLER and ROESSLER, 1956; WIMPENNY, 1967a; YOUATT, 1960b). Both are inhibited by anaerobic conditions and by other hydrazides which do not possess any anti-tubercular activity. The uptake, binding and metabolism of INH are indispensable prerequisites for growth inhibition. The INH binding mechanism has many properties similar to those of catalase or peroxidase (WIMPENNY, 1967a). Since, however, the two well-characterized

$$\text{XII} \xrightarrow{\quad} \text{XIII} \xrightarrow{\quad} \text{XIV}$$

XII   XIII   XIV

metabolites do not exert any effect on the tubercle bacillus, the question remains to be answered as to whether there are other as yet unidentified metabolites of INH responsible for the specific molecular reaction or reactions that lead to growth inhibition of the bacillus.

Isonicotinic acid and 4-pyridyl methanol can be recovered from media in which susceptible cells are exposed to the drug under well-defined experimental conditions. INH-susceptible and-resistant strains exhibit significant differences in the uptake and metabolism of INH (see Table 6). The resistant organism takes up INH at a very slow rate and produces only isonicotinic acid but metabolizes other hydrazides such as nicotinic acid hydrazide (XII) which is converted into nicotinic acid (XIII) and 3-pyridyl methanol (XIV).

Table 6. *Enzymic breakdown of INH in mycobacteria*

	Hydrazidase $R \cdot CO \cdot NH \cdot NH_2 + H_2O \rightarrow R \cdot COOH + NH_2NH_2$	Dehydrogenase complex $R \cdot COOH \rightleftharpoons R \cdot CHO \rightleftharpoons R \cdot CH_2OH$
INH Susceptible Mycobacteria	Specific for INH	Specific for isonicotinic acid and possibly does not act on nicotinic acid
INH Resistant Mycobacteria	Non-specific Acts both on INH and NH	Does not act on isonicotinic but acts on nicotinic acid

NH = Nicotinic acid hydrazide

7.4. INH Resistance and Loss of Virulence

Mycobacteria are generally catalase-positive whereas INH-resistant strains of *M. tuberculosis* do not show catalase or peroxidase activities (Middlebrook, 1954). All highly INH-resistant catalase-negative mutants recovered from patients show a markedly reduced ability to initiate progressive tuberculosis in experimental animals (Middlebrook and Cohn, 1953; Mitchison, 1954; Morse *et al.*, 1954; Stewart, 1954). This direct relationship between INH susceptibility, pathogenicity and catalase activity of *M. tuberculosis* (Maher *et al.*, 1957) has led to an attractive hypothesis that the bactericidal effect of INH is ultimately produced by a breakdown product or free radicals arising out of interaction between INH, catalase or peroxidase (Winder, 1960; Tirunarayanan and Vischer, 1957, 1959).

The *in vitro* antibacterial action of INH against *M. tuberculosis* is reversed competitively by haemin and the MIC of INH required to inhibit the bacilli progressively increases as concentration of exogenously added haemin increases (Fisher, 1954a, 1954b). Evidence has also been adduced to show that haemin and inorganic iron can function as growth factors for certain strains of INH-resistant *M. tuberculosis* (KNOX, 1955; KNOX et al., 1956). It is not known as to whether porphyrin synthesis is inhibited by INH and the mechanism that regulates the activity of catalase is thereby disturbed. Furthermore, resistant strains isolated from a mixed population of cells, previously unexposed to INH, have been shown to be catalase-negative, indicating that loss of catalase and the loss of virulence in resistant strains are independent mutational events. In spite of extensive studies aimed at correlating the loss of virulence with INH resistance (FISHER, 1954a, 1954b; COHN et al., 1954a; COHN et al., 1954b; MIDDLEBROOK et al., 1954; MITCHISON, 1954; TOMPSETT, 1954; OESTREICHER, 1955; STIEF, 1956; WOLINSKY et al., 1956; SCHMIDT, 1958; WINDER, 1964), an acceptable explanation of the mechanism of action of INH has not emerged so far.

8. Metabolic Changes Produced in Mycobacteria by INH

8.1. Gross Changes

The gross changes in morphology resulting from the action of INH on *M. tuberculosis* and manifest as loss of acid fastness and lipid structures (FEINGOLD, 1963) are accompanied by alterations in the metabolic activity of the pathogen. These events are schematically represented in Fig. 2 to indicate the complexity of the biochemical effects that result from the action of INH. Large amounts of soluble compounds can be detected in the culture medium after one hour

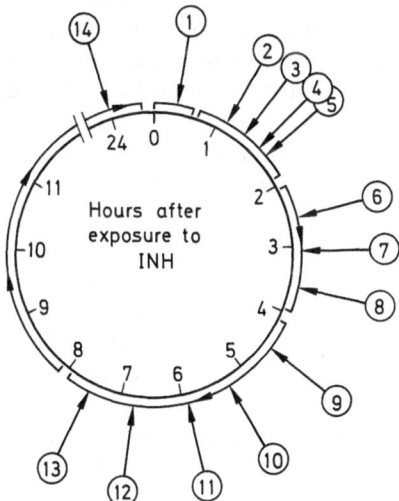

Fig. 2. Time sequence of cellular events in Mycobacteria on exposure to INH. Events shown as numbers 1–14 in circles are: (1) Uptake of INH. (2) Biotransformation of INH. (3) Production of pigments. (4) Accumulation of soluble phosphate esters. (5) Accumulation of soluble carbohydrates. (6) Gradual decrease of acid-fastness. (7) Decreased synthesis of DNA. (8) Accumulation of 260 nm absorbing materials. (9) Loss of acid fastness and viability (?) (10) Decreased synthesis of RNA. (11) Decrease in alkali soluble polyphosphates. (12) Loss of viability (?) (13) Decreased protein synthesis. (14) Cessation of growth. Based mostly on Table 1 of YOUATT (1969)

of exposure to INH. Within six hours, both the uptake of INH and the accumulation of soluble compounds reach peak values. Incorporation of P^{32} into nucleic acids decreases rapidly and within 24 hours protein synthesis comes to a halt.

The fatty acid composition changes and acids of chain lengths greater than 16 carbon atoms decrease in amount. Synthesis of mycolic acid and of the dimycolate ester of trehalose (cord factor) are also suppressed (WINDER and COLLINS 1970; TSUKAMURA and TSUKAMURA, 1963). Incorporation by mycobacteria of acetate-^{14}C (KOCH-WESER et al., 1955; EBINA et al., 1961; WIMPENNY 1967b; WINDER and COLLINS, 1970) and glycerol-^{14}C (BRENNAN et al., 1970) into ether-soluble bound lipids such as mycolic acids and phospholipids is inhibited. Synthesis of phospholipids by M. avium measured by incorporation of P^{32} is also inhibited (SENGHVI and SUBRAHMANYAM, 1967).

8.2. Effect of INH on the Biosynthesis of Macromolecules

In the presence of INH, DNA synthesis measured by incorporation of thymidine-^{3}H in sensitive M. tuberculosis decreases and is completely abolished within 2 hours. RNA synthesis followed by incorporation of uridine-^{3}H stops about 2 hours later, whereas protein synthesis measured by incorporation of valine-1-^{14}C is affected only after 8 hours. INH does not interfere with the uptake of thymidine-^{3}H by the bacterium nor do nucleic acid precursors decrease the bactericidal effect of INH. Although the effect of INH on DNA synthesis is irreversible, DNA is not degraded as evident from the fact that bacteria grown for several generations in medium containing thymidine-^{3}H when exposed to INH do not release any TCA soluble radioactivity.

Interference with the initiation of DNA synthesis has been suggested as a mode of action of INH from these results (MCCLATCHY, 1971; MCCLATCHY and SMITH, 1968). The inhibition of the synthesis of nucleic acids in M. tuberculosis is accompanied by loss of acid fastness and viability (GANGADHARAM et al., 1963; WIMPENNY, 1967; TSUKAMURA and MIZUNO, 1962). INH inhibits incorporation of amino acids-^{14}C into protein of mycobacteria but not uniformly (TSUKAMURA and TSUKAMURA, 1963; TSUKAMURA et al., 1964). In cell-free systems INH inhibits incorporation of amino acids into proteins whereas phenylhydrazine and acetyl hydrazine do not (TOMCSANYI, 1965). Inhibition of protein synthesis exerted by the drug on M. tuberculosis (GANGADHARAM et al., 1963; TSUKAMURA and MIZUNO, 1962; TSUKAMURA and TSUKAMURA, 1963) also appears to be an indirect effect. In the absence of precise knowledge concerning the mean generation time and the factors that regulate macromolecular synthesis in mycobacteria, it is difficult to fit the above experimental results, interesting as they are, into a general scheme of the mechanism of action of INH at the molecular level.

8.3. Accumulation of Soluble Carbohydrates and Phosphate Compounds

An accumulation of carbohydrates has been observed in cultures shaken with INH for one to six hours (WINDER, 1964; WINDER et al., 1967) but a similar phenomenon is seen with higher concentrations of nicotinic and picolinic acid hydrazides as well as with other anti-tubercular agents.

8.4. Accumulation of Phosphate Esters

Submerged cultures of INH-sensitive BCG when exposed to INH, accumulate several sugar phosphates (WINDER et al., 1967), and surface cultures accumulate both orthophosphate and non-nucleotide phosphates soluble in perchloric acid (YOUATT, 1965). No such accumulation occurs with INH-resistant mycobacteria (YOUATT, 1965; WINDER, 1964; WINDER et al., 1967). Since, however, ethambutol produces similar changes, the accumulation of phosphates is not a specific effect of the primary action of INH.

8.5. Accumulation of Pigments

It has been suggested by YOUATT (1969) that pigment production by M. tuberculosis is not only closely related to the uptake of INH but may even be a manifestation of the same reaction. Pigments are not produced by resistant bacteria or atypical mycobacteria nor do p-aminosalicylic acid, streptomycin, pyrizinamide, ethambutol, ethionamide or pyridine-4-ethionamide induce pigment production in sensitive M. tuberculosis. INH seems to be very specific in stimulating the formation of pigment since none of its metabolites, viz. isonicotinic acid, hydrazine, hydrogen peroxide, 4-pyridyl-methanol, or 4-pyridine aldehyde can mediate this reaction.

The production of pigment in the presence of INH is inhibited competitively by hydrazides which are not anti-tubercular but which reverse the growth inhibitory action of INH on M. tuberculosis. Pigments are not formed under anaerobic conditions. The colorless precursor of this pigment is formed in stationary culture or in washed cell suspensions but the pigment itself appears only in the presence of a metabolizable substrate. The Km value of the reaction is 2×10^{-4} M INH (YOUATT, 1961; YOUATT and THAM, 1969a, 1969b, 1969c). NAD and NADP can serve as cofactors in this reaction but NMN gives only 50% of the activity. Two different types of single step INH mutants of M. tuberculosis do not mediate this reaction whereas an INH sensitive strain does (GAYATHRI DEVI et al., 1972).

It is postulated that INH binds to specific sites on sensitive strains either through the ring nitrogen or through the hydrazide group. The INH receptor could be an enzyme which, on structural alteration, releases a toxic factor which then interferes with some essential metabolic process. The identity of the pigment with this hypothetical toxic metabolite has not been established. The chemical nature of the pigment has also not been elucidated.

9. INH and Metabolism of NAD

9.1. INH Analogue of NAD

As concerns an anti-metabolite action, the effect of INH on enzymes dependent on NAD and pyridoxal phosphate is very significant and has been extensively investigated. In experimental tuberculosis there is a fall in blood levels of NAD and NADP and a return to normal values after discontinuation of INH treatment (PATIALA, 1954). The levels of nicotinamide nucleotides in mycobacteria decrease on exposure to INH (Winder and Collins, 1968) and exogenously added NAD readily reverses the growth-inhibitory action of INH in drug sensitive mycobacteria (Sriprakash and Ramakrishnan, 1969).

Enzymes, prepared from mammalian sources, that cleave NAD also catalyze an exchange reaction by which the niacinamide moiety can be replaced to form NAD analogues (Zatman et al., 1954) as shown in Fig. 3. An INH analogue of NAD can be prepared in this manner. The analogue does not function as electron acceptor in dehydrogenase reactions in which NAD participates nor does it inhibit the reduction of NAD or NADP by their specific dehydrogenases (Goldman, 1954).

On the basis of formation of a similar NAD analogue of isonicotinic acid, the major metabolite of INH, it has been proposed that INH inferes primarily with the metabolism of NAD in mycobacteria (Kruger-Thiemer, 1958). Although INH or isonicotinic acid analogues of NAD have not been isolated from mycobacteria exposed to INH, the possibility exists that the drug or its main metabolite can displace niacinamide and disturb the rates of enzyme reactions mediated

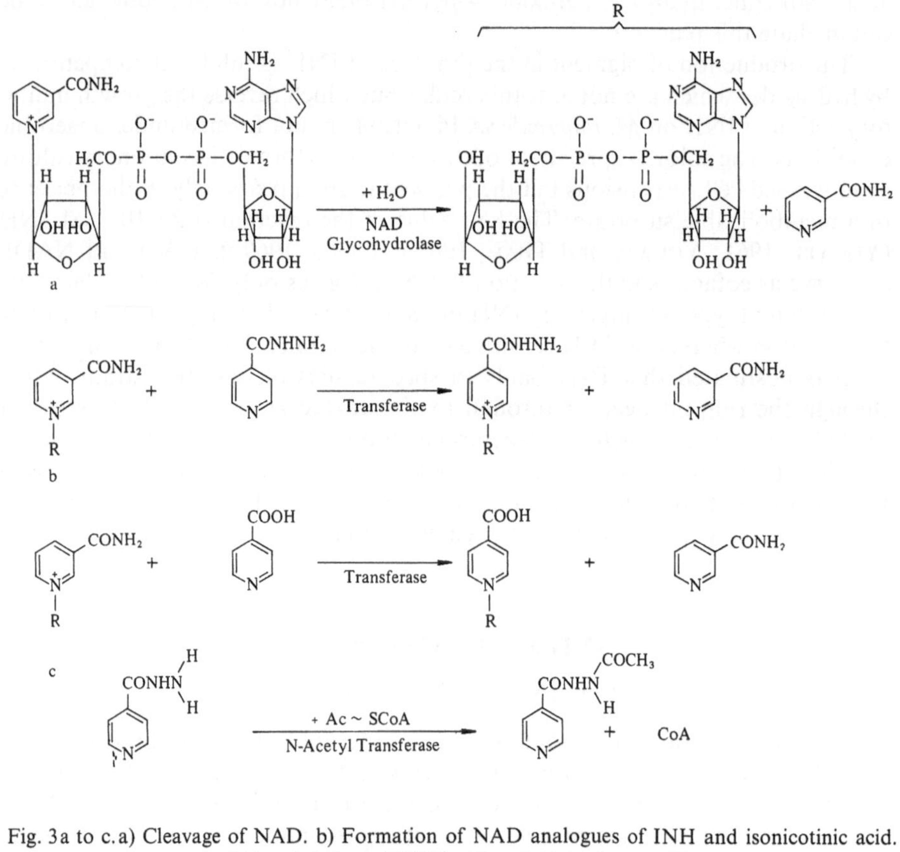

Fig. 3a to c.a) Cleavage of NAD. b) Formation of NAD analogues of INH and isonicotinic acid. c) N-acetylation of INH by N-acetyl transferase of liver

by NAD and NADP dependent dehydrogenases. Earlier reports of the effect of INH on oxidative metabolism of mycobacteria (MEADOW and KNOX, 1956) and on its dehydrogenases (SCHAEFER, 1960) could be attributed to decrease in NAD concentration, resulting from drug action.

9.2. Effect of INH on NAD Synthesis

Human strains of *M. tuberculosis* are known to produce excessive amounts of nicotinic acid, a phenomenon used as a test to distinguish them from other mycobacteria (KONNO, 1956; KONNO *et al.*, 1958). Besides, nicotinic acid-^{14}C or nicotinamide-^{14}C are not incorporated directly into NMN or NAD by extracts of mycobacteria (GOPINATHAN *et al.*, 1963; KONNO *et al.*, 1965b). Quinolinate transphosphoribosylase, an enzyme present in mycobacteria, can however, convert quinolinate directly into NMN without the intermediate formation of nicotinic acid (KONNO *et al.*, 1965a). When ATP is limiting the NMN formed by the enzyme is degraded to release nicotinic acid (NAKAMURA *et al.*, 1964). A regulatory role for the enzyme in the NAD metabolism of mycobacteria is implicit in these apparently diverse reactions.

$$
\begin{array}{ccc}
\text{C} & \text{C-COOH} & \text{COOH} \\
| \quad + & | & \longrightarrow \quad \text{COOH} \\
\text{C} & \text{C-COOH} & \\
& \text{N} & \\
& \text{H}_2 & \\
\text{Glycerol} & \text{Aspartate} & \text{XV}
\end{array}
$$

It is not known as to whether the condensation of the three carbon unit from glycerol with aspartic acid to give rise to quinolinic acid (XV), the pathway followed by mycobacteria, is inhibited by INH, or whether INH can interfere with the possible allosteric effects of ATP on the regulatory enzyme quinolinate transphosphoribosylase. Relatively high concentrations of INH inhibit NAD synthesis by cell-free extracts of both INH-sensitive and -resistant H 37 Rv (SRIPRA-KASH and RAMAKRISHNAN, 1969).

9.3. Degradation of NAD and Effect of INH on the Inhibitor of NADase

The degradation of NAD or NADP is brought about by the enzyme NAD glycohydrolase which cleaves the nicotinamide riboside bond of the oxidized form of the coenzyme (Fig. 3). NAD glycohydrolase activity in mycobacteria is associated with a heat labile protein inhibitor (KERN and NATALE, 1958; TOIDA, 1963; GOPINATHAN *et al.*, 1963; GOPINATHAN *et al.*, 1964). Partially purified NAD glycohydrolases from *M. butyricum* or from *M. tuberculosis H37Rv* do not, however, catalyze the NAD-INH exchange reaction mediated by the mammalian enzymes.

The inhibitor of NAD glycohydrolase of H37Rv has been purified; the complex formed between the enzyme and the purified inhibitor does not dissociate under physiological conditions but INH can inactivate the inhibitor and release the NADase (Gopinathan et al., 1966). A highly purified NAD hydrolase from *M. butyricum* with a molecular weight of 39,000 daltons and its specific proteinaceous inhibitor with a molecular weight of 26,000 daltons form complexes as evident from Sephadex G-100 chromatography, sucrose gradient centrifugation and cellulose acetate electrophoresis (Ogasawara et al., 1970). It is not known whether this complex formation is affected by INH. The drug combines with the heat labile inhibitor of NAD glycohydrolase present in *M. tuberculosis*. The inhibitor present in an INH resistant strain of H37Rv obtained by multiple step mutation is, however, not inactivated by INH (Bekierkunst and Bricker, 1967).

9.4. Permeability Barrier on INH Mutants and Inactivation of NADase Inhibitor

Independent single-step mutants, isolated by exposure to the lowest inhibitory concentrations of INH, contain inhibitors of NAD glycohydrolase which are either sensitive or insensitive to INH. The uptake of INH-^{14}C is very much decreased in both types of the above one step mutants as compared to the parent INH sensitive strain. However, when the permeability barrier to the drug is removed by the use of surface active agents, INH becomes lethal to the mutants and decreases the NAD concentration just as in the INH sensitive parent strain (Sriprakash and Ramakrishnan, 1970).

A direct correlation thus exists between the lethality of INH and the lowering of intracellular concentrations of NAD. DNA ligase of bacteria depends on NAD as a cofactor (Olivera, 1971) and since DNA ligase is essential for elongation of polydeoxynucleotide chains formed in DNA synthesis, the unavailability of NAD for this vital reaction may lead to inhibition of DNA synthesis. The

Table 7. *Effect of INH on NAD metabolism in H37Rv*

Reaction studied	INH-sensitive parent strain H37Rv (NCTS 4716)	INH-resistant single step mutant Type I	INH-resistant single step mutant Type II
Permeability to INH	Permeable	Not permeable	Not permeable
Synthesis of NAD	Inhibited	Inhibited	Inhibited
Binding to NADase inhibitor	Bound and activity annulled	Bound and activity annulled	Not bound
Depletion of Intracellular concentration of NAD	Depleted	Depleted after permeability barrier to INH was removed	Depleted after permeability barrier to INH was removed
NAD hydrolysis	Activated	Activated	Not activated
Peroxidative activity	Yes. Associated with NADase inhibitor	No activity in the NADase inhibitor	No activity in the NADase inhibitor

Data from Sriprakash and Ramakrishnan (1969).

early effect of INH in inhibiting DNA synthesis (McCLATCHY, 1971) but not RNA and protein synthesis could thus be attributed to the primary effect of INH in decreasing intracellular NAD concentrations. The proteinaceous inhibitor of the mutants does not bind INH as has been shown for the corresponding inhibitor from the INH sensitive strain nor does INH activate directly the NAD glycohydrolase. Our current knowledge about the effect of INH on NAD metabolism of *M. tuberculosis* is summarized in Table 7.

10. INH and Its Effect on Structural Organization of Mycobacteria

10.1. Damage to Outer Envelope and Leakage of Materials

One of the very first effects of the lethal action of INH on susceptible *M. tuberculosis* is to eliminate the acid fast staining reaction of mycobacteria. The components responsible for this reaction have not been unequivocally characterized but are assumed to be present on the cell envelope. It has been suggested that INH can interfere with the synthesis of a lipid component accompanied by unbalanced production of other compounds and thereby lead to permeability changes of the bacterium (SCHAEFER, 1960). The obvious result of this would be the release into the medium of carbohydrates, amino acids, phosphate esters and 260 nm absorbing substances before cessation of growth takes place. The carbohydrates released are heterogenous judged from their elution profile from Sephadex. They are presumably precursor units or degradation products of the lipopolysaccharide present in the cell envelope or wall of *M. tuberculosis*.

INH might disorganize the outer layer of the cell wall of mycobacteria by cross linking with the peptidoglycan moiety of the inner wall. The outer layer is known to consist predominantly of heteropolysaccharides made up of arabinose, glucose, mannose and a little galactose. he inner layer contains peptidoglycan. Lipopolysaccharides are formed by the linking of mycolic acids with the outer layer polysaccharides (WINDER *et al.*, 1967; WINDER and ROONEY, 1970).

10.2. Inhibition of Synthesis of Mycolic Acid

INH in bacteriostatic concentrations inhibits the synthesis of mycolic acid in sensitive strains but not in the resistant strains (WINDER and COLLINS, 1970). The cell envelope formed now is defective and perhaps permits exit of essential structural constituents resulting ultimately in the loss of acid fastness and death of the organism. The defective envelope is also presumably responsible for the leakage of amino acids and a variety of other cytoplasmic constituents as recorded in literature. Inhibitory action on mycolic acid synthesis is indeed very specific for the drug sensitive strain. The mechanism of biosynthesis of mycolic acids and the enzymes involved have not been elucidated. It is also not clear as to whether INH *per se* or one of its as yet unidentified metabolites brings about the inhibitory effect. The mechanism proposed on the basis of these results is represented diagrammatically in Fig. 4.

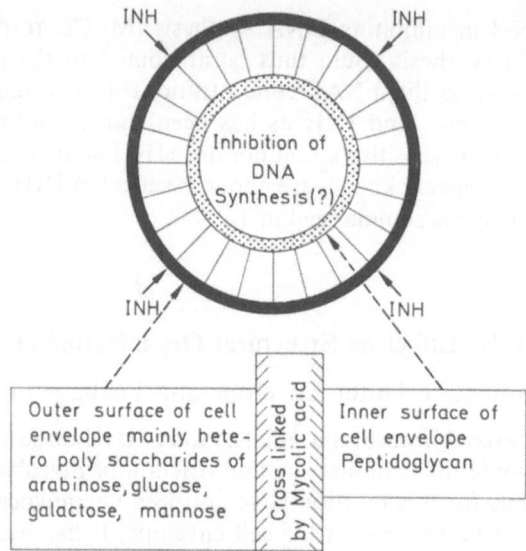

Fig. 4. Hypothetical scheme of action of INH on cell envelope of mycobacteria

11. Metal Chelation Hypothesis of Mode of Action of INH

One of the most widely advocated theories of the mode of action of INH has postulated that the bacteriostatic effect is related to its ability to complex certain essential heavy metals such as Cu and Fe. The inhibitory effects of INH on hepatic catalase (MIDDLEBROOK, 1954; ARORA and KRISHNA MURTI, 1960) on the succinoxidase system of pigeon breast muscle (ARORA and KRISHNA MURTI, 1954) or on the organic nitroreductase of gram-negative bacteria (ARORA et al., 1959) are presumably related to sequestering of essential metal ion moieties from the enzymes. These experiments have not been repeated with highly purified enzymes. However, tuberculostatic activity of the drug is not a necessary consequence of its chelating ability. Thus, nicotinic acid hydrazide is as effective as INH as a chelating agent, but is not tuberculostatic. Picolinic acid hydrazide with about 1 000–1,000,000 times more chelating ability than INH is only 1/80th as active as INH as a tuberculostat. The derivative 1-isonicotinyl-1-isopropyl hydrazine is very strongly anti-tubercular but has no chelating activity at all (ALBERT, 1956).

12. Biochemical Pharmacology of INH

12.1. Absorption and Distribution of INH in the Host

After oral administration, INH is well absorbed and its level in plasma attains a peak in about two hours. The drug is distributed throughout the body water and substantial concentrations are found in the cerebrospinal fluid and pleural effusions. The drug also passes through the placental barrier and into milk (RUBIN et al., 1952; BENSON et al., 1952; ELMENDORF et al., 1952; RUBIN and BURKE,

1953 a, 1953 b). Using INH, labelled with ^{14}C in the carbonyl position, it has been shown that the drug penetrates into caseous material, reaches lymph nodes and is localized mainly in liver, skin, lungs, brain and kidneys (ROTH and MANTHEI, 1952; BARCLAY et al., 1953 b).

12.2. Metabolism of INH in the Host

In the rhesus monkey and in man, the main excretion product is 1-isonicotinyl-2-acetyl hydrazine (HUGHES, 1953). All species studied excrete part of INH unchanged in the urine; the percentage of unchanged INH excreted varies with species and from individual to individual in *Homo sapiens*. The two major excretion products in man are: acetyl isoniazid and isonicotinic acid; minor metabolites are: isonicotinyl glycine, isonicotinyl hydrazones and traces of N-methyl isoniazid. Various metabolites of the drug have also been characterized in experimental animals (HUGHES et al., 1955). The dog is unique in that it excretes mainly isonicotinic acid because of its weak hepatic acetylating system.

12.3. Relation of INH Metabolism to Toxicity

Three mechanisms operate the degradation of INH in man. The first and the preferred one involves acetylation of the terminal nitrogen of the hydrazine moiety; the second is mediated through the cleavage of hydrazide to give rise to isonicotinic acid or a closely related derivative, and the third brings about degradation in as yet uncharacterized manner. One of the main toxic symptoms encountered with INH viz. peripheral neuritis and peripheral neuropathies (OEST-REICHER et al., 1954) is closely associated with an apparent defect in the acetylation of INH and in the third pattern of degradation (HUGHES et al., 1954).

The rate and speed of acetylation and thus the elimination of the drug depend largely on the activity of hepatic INH acetylase. The drug is acetylated relatively slowly in Whites and Negroes and most rapidly in Eskimos while the Japanese and other Orientals show an intermediate rate of acetylation. Acetylation of INH is controlled genetically and slow acetylation is due to a recessive homozygous autosome whereas the "rapid inactivators" can belong to any one of the two genotypes—heterozygous or homozygous dominants (ARMSTRONG and PEART, 1960; SUNAHARA et al., 1961). The far-reaching implications of the genetic control of INH metabolism in man in the development of polyneuritis with long term treatment and in the emergence of INH resistant population of bacilli are obvious and require more world-wide surveys of patients undergoing chemotherapeutic treatment with INH (EVANS et al., 1960; EVANS et al., 1961; RUSSELL and MIDDLEBROOK, 1961; TURPIN et al., 1969).

N-Acetyl transferases have been obtained in purified form from the liver of man, monkey and the rat (SCHLOOT et al., 1969). Barbiturates can mitigate the toxicity of INH (PAN et al., 1952). It is now well known that barbiturates alter the ratio of smooth to rough forms of endoplasmic reticulum and thereby stimulate drug metabolizing enzymes. If liver N-acetyl transferase is assumed to be a constitutive regulatory enzyme, genetic control of its activity can be considered to be exercised through a molecular modulation mechanism as is now known for a number of allosteric enzymes.

Table 8. *Inhibition by INH of some pyridoxal phosphate dependent enzymes*

Reaction studied	Source of enzyme	Nature of inhibition	Reference
Amino-transferase:			
Aspartic: 2-oxoglutarate aminotransferase	*M. tuberculosis* BCG INH sensitive	Not studied	Sakai (1954)
Aspartic, asparagine, glycine, phenylalanine, leucine, valine or alanine: 2-oxoglutarate aminotransferase	Mycobacteria	Not studied	Ito and Sugano (1954)
Different aminotransferases	Living mycobacteria	Not studied	Vodicka and DeLong (1956)
Different amino acids:2-oxoglutarate aminotransferase	*M. tuberculosis* BCG INH sensitive and resistant strains	Inhibition in both strains relieved by pyridoxal phosphate	Youatt (1958b)
Leucine:2-oxoglutarate aminotransferase	Cell-free extracts of *Vibrio cholerae*	Not studied	Saxena *et al.* (1956)
Alanine or aspartate:2-oxoglutarate aminotransferase	Cell-free extracts of *Pasteurella pestis*	Not studied	Saxena *et al.* (1957)
Aspartate, leucine, isoleucine and valine:2:oxoglutarate aminotransferase	Cell-free extracts of *M. tuberculosis* H37Rv, BCG and avian strains	Not studied	Saxena and Arora (1959)
Aspartate:2-oxoglutarate aminotransferase	Cell-free extracts of *Salmonella typhosa*	Reversed by pyridoxal phosphate	Saxena and Rastogi (1961)
Alanine:2-oxoglutarate aminotransferase	Purified enzyme from pig heart muscle	Not studied	Hicks and Cymerman-Craig (1957)
Alanine:2-oxoglutarate aminotransferase	Purified enzyme from equine red blood cells	Not reversed by pyridoxal phosphate	Balasaraswathi and Krishna Murti (1967)
Asparate:2-oxoglutarate aminotransferase	Purified enzymes from human and rabbit red blood cells	Not reversed by pyridoxal phosphate	Balasaraswathi and Krishna Murti (1973)
Tryptophanase:			
Production of indole	Whole cells of *Escherichia coli*	Not studied	Yoneda *et al.* (1952)
Production of indole	Whole cells and cell-free extracts of *E. coli*	Reversed by yeast extract	Arora and Krishna Murti (1955)
Amino acid decarboxylases:			
Arginine	*E. coli* cells	Reversed by pyridoxal phosphate	Yoneda and Asano (1953)

Table 8 (continued)

Reaction studied	Source of enzyme	Nature of inhibition	Reference
Arginine	*Bacterium cadaveris*	Not studied	HOARE (1956)
Glutamic acid	*M. tuberculosis* BCG cell-free extracts	Not studied	YAMAMOTO (1955)
Glutamic acid	Guinea pig brain	Not studied	DAVISON (1956)
Dihydroxyphenylalanine	Guinea pig kidney	Not studied	DAVISON (1956)
Cysteinesulfinic acid	Guinea pig liver	Not studied	DAVISON (1956)
Aspartic acid	*Desulphovibrio desulfuricans*	Not studied	CATTENEO-LACOMBE and SENEZ (1956)
Glutamic acid	Guinea pig kidney	INH treated animals not suppressed	CANAL and GARATTINI (1957)
Dihydroxyphenylalanine	Guinea pig brain	INH treated animals suppressed	CANAL and GARATTINI (1957)
Cysteic acid	Guinea pig liver	INH treated animals suppressed	CANAL and GARATTINI (1957)
Desulphydration: Cysteine	*M. tuberculosis*	Not studied	YAMAMOTO (1955)
Pyridoxal phosphate phosphatase	Goat Brain	INH inhibits	BEGUM and BACHHAWAT (1960)

12.4. Pyridoxine and the Toxicity of INH

The peripheral neuritis produced by INH show symptoms that resemble those of pyridoxine deficiency. Patients receiving large doses of INH excrete abnormal amounts of pyridoxine and its metabolites, and symptoms of neuritis are reversed by the administration of pyridoxine (BIEHL and VILTER, 1954; CARLSON et al., 1956; RAJTAR-LEONTIEW, 1970; MORALES and LINCOLN, 1957). Pyridoxine does not, however, interfere with the *in vivo* anti-tubercular activity of INH (UNGAR et al., 1954; GRUNBERG and BLENCOME, 1955; CROWLE and RIEMENSNIDER, 1960), and it has been suggested that anti-tubercular activity and the toxic action reside in different parts of the drug molecule. Hydrazine derivatives, in general, elicit reactions on CNS activity and the toxicity of INH is due to the hydrazine moiety.

Pyridoxal reverses the *in vitro* anti-tubercular activity of INH (LICHSTEIN, 1955; POPE, 1956; BEGGS and JENNE, 1967) and the possibility of a structural antagonism between INH and pyridoxal (BOONE and WOODWARD, 1953) exists, and therefore the effect of INH has been studied on a number of pyridoxal phosphate dependent enzymes involved in the intermediary metabolism of amino acids. Some examples of such interaction are summarized in Table 8 and would appear to be of greater relevance to the host toxicity of the drug rather than to its mode of anti-tubercular action.

Experimental B_6-hypervitaminosis induced by INH, is associated with the following early signs: diminished excretion of 4-pyridoxamic acid in urine, fall in α and β serum globulins, a rise of albumin, fall in activity of aspartate amino transferase, increase in lipoproteins, fatty degeneration of liver, kidneys and fatty infiltration of aortic walls and all these are corrected by administering pyridoxine (Karkalitskii et al., 1970). Electron microscopy is being increasingly used to map the subcellular changes in neuron cells of man in INH neuropathy (Ochoa, 1970) and such studies may in the near future throw more light on the molecular mechanism of neurotoxicity associated with INH.

12.5. INH as a Cancer Inducing Agent

Among the potential toxic effects of INH are the reports on its carcinogenic action (Roe and Lancaster, 1964; Biancifiori and Severi, 1966) in mice. In human patients no correlation has been observed between the high incidence of cancer in patients treated with INH (Campbell and Guilfoyle, 1970). In contrast to these, there is also a report on the cytostatic effect of INH particularly on Walker carcinoma (Gross, 1956). The hydrazide moiety is implicated in this action.

13. Conclusions and Future Projection

There is a voluminous literature on the mechanism of action of INH but no single unified concept based on unequivocal molecular terms has emerged. A number of attractive hypotheses have been vigorously advocated and their protagonists have marshalled the support of interesting experimental findings but the primary action of the drug is still unknown. As things stand today, it can at best be said that in the susceptible mycobacteria, INH presumably binds to certain receptor sites and by an as yet unknown mechanism, causes, the disruption of the cell envelope. The ensuing structural alteration produces a number of secondary effects and the eventual death of the cell. No attempts have been made so far to identify the receptor site. With the availability of a large number of derivatives of INH it should be possible to design meaningful experiments on the binding mechanism to specific sites particularly on the plasma membrane of the tubercle bacillus. Obviously a lot more has to be learnt about the nature of the cell envelope and the biosynthesis of its constituent units, above all about the mechanisms that regulate the biosynthesis of these units and their assembly to form the cell envelope of the bacterium.

The genetics of mycobacteria is also a relatively new subject, and the application of the elegant techniques of phage transduction using mycophages has given very promising results in attempts to map biosynthetic pathways in mycobacteria. With a greater exploitation of the knowledge gained by genetic studies one can look forward hopefully that the hitherto mysterious primary action of INH on M. tuberculosis will be uncovered soon and explained in clear molecular language.

Acknowledgement. This is communication No 1797 from CDRI, Lucknow. The author is grateful to Dr. O.P. Shukla, Scientist, Central Drug Research Institute, Lucknow, for very stimulating discussions on the subject.

References

ALBERT, A.: Mode of action of isoniazid. Nature **177**, 525 (1956).

ARMSTRONG, A. R.: Time-concentration relationship of isoniazid with tubercle bacilli *in vitro*. Am. Rev. Respirat. Diseases **81**, 498 (1960).

ARMSTRONG, A. R., and H. E. PEART: A comparison between the behaviour of Eskimos and Non-Eskimos to the administration of isoniazid. Am. Rev. Respirat. Diseases **81**, 588 (1960).

ARONSON, J. D., S. L. EHRLICH, and W. FLAGG: Effects of isonicotinic acid derivatives on tubercle bacilli. Proc. Soc. Exptl. Biol. Med. **80**, 259 (1952).

ARORA, K. L., and C. R. KRISHNA MURTI: Enzyme inhibition studies in relation to drug action. Part VI. Action of certain anti-bacterial agents on the succinoxidase system. J. Sci. Ind. Res. (India) **13A**, 482 (1954).

ARORA, K. L., and C. R. KRISHNA MURTI: Studies on enzyme inhibition in relation to drug action. VII. Action of certain anti-bacterial agents on tryptophanase. J. Sci. Ind. Res. (India) **14C**, 6 (1955).

ARORA, K. L., and C. R. KRISHNA MURTI: Enzyme inhibition studies in relation to drug action. Part IX. Action of certain anti-bacterial agents on catalases. J. Sci. Ind. Res. (India) **19C**, 103 (1960).

ARORA, K. L., C. R. KRISHNA MURTI, and D. L. SHRIVASTAVA: Studies in the enzyme make-up of *Vibrio cholerae*. Part XIV. Organic nitroreductase. J. Sci. Ind. Res. (India) **18C**, 105 (1959).

BALASARASWATHI, K., and C. R. KRISHNA MURTI: Partial purification and properties of L-alanine: 2-oxoglutarate aminotransferase of equine red blood cells. Indian J. Biochem. **4**, 22 (1967).

BALASARASWATHI, K., and C. R. KRISHNA MURTI: L-aspartate: 2-oxoglutarate aminotransferases of human and rabbit blood cells. Indian J. Biochem. **10**, 3 (1973).

BARCLAY, W. R., R. H. EBERT, and D. KOCH-WESER: Mode of action of isoniazid. Am. Rev. Tuberc. **67**, 490 (1953a).

BARCLAY, W. R., R. H. EBERT, G. V. LEROY, R. W. MANTHEI, and L. J. ROTH: Distribution and excretion of radioactive isoniazide in tuberculous patients. J. Am. Med. Assoc. **151**, 1384 (1953b).

BARCLAY, W. R., D. KOCH-WESER, and R. H. EBERT: Mode of action of isoniazid. Part II. Am. Rev. Tuberc. **70**, 784 (1954).

BARCLAY, W. R., D. KOCH-WESER, and R. H. EBERT: Mode of action of isoniazid. Am. Rev. Tuberc. Pulmonary Diseases **74**, Suppl. 109 (1956).

BARCLAY, W. R., and E. WINBERG: Bactericidal effect of isoniazid as a function of time. Am. Rev. Respirat. Diseases **90**, 749 (1964).

BARRY, V. C.: In: Chemotherapy of tuberculosis, BARRY, V. C. (ed.) p. 46. London: Butterworths 1964.

BARTMANN, K., H. COPER u. R. JUTTE: Die NAD (P)-Glykohydrolase in Tuberkulosebakterien: Ein Beitrag zum Wirkungsmechanismus des INH. Arch. Exptl. Pathol. Pharmakol. **257**, 8 (1967).

BEGGS, W. H., and J. W. JENNE: Mechanism for the pyridoxal neutralization of INH action on *M. tuberculosis*. J. Bacteriol. **94**, 793 (1967).

BEGGS, W. H., and J. W. JENNE: Isoniazid uptake and growth inhibition of *M. tuberculosis* in relation to time and concentration of pulsed drug exposures. Tubercle **50**, 377 (1969).

BEGUM, A., and B. K. BACHHAWAT: Role of isonicotinic acid hydrazide in the inactivation of pyridoxal phosphate. Ann. Biochem. Exptl. Med. (Calcutta) **20**, 143 (1960).

BEKIERKUNST, A.: Nicotinamide-adenine dinucleotide in tubercle bacilli exposed to isoniazid. Science **152**, 525 (1966).

BEKIERKUNST, A., and A. BRICKER: Studies on the mode of action of isoniazid on mycobacteria. Arch. Biochem. Biophys. **122**, 385 (1967).

BENSON, W. M., P. L. STEFKO, and M. D. ROE: Pharmacologic and toxicologic observations on hydrazine derivatives of isonicotinic acid. Am. Rev. Tuberc. **65**, 376 (1952).

BERNSTEIN, J., W. P. JAMBOR, W. A. LOTT, F. PANSY, B. A. STEINBERG, and H. L. YALE: Chemotherapy of experimental tuberculosis. VI. Derivatives of isoniazid. Am. Rev. Tuberc. **67**, 354 (1953).

BERNSTEIN, J., W. A. LOTT, B. A. STEINBERG, and H. L. YALE: Chemotherapy of experimental tuberculosis. V. Isoniazid and related compounds. Am. Rev. Tuberc. **65**, 357 (1952).

BHAGWAT, R. V., and K. SOHONIE: Isoniazid like substances in *Neera* from Date-Palm: *Phoenix sylvestris*. Sci. Cult. (Calcutta) **21**, 265 (1955).

Biancifiori, C., and L. Severi: The relation of isoniazid and allied compounds to carcinogensis in some species of small laboratory animals. Brit. J. Cancer **20**, 528 (1966).

Biehl, J.P.: The role of the dose and the metabolic fate of isoniazid in the emergence of isoniazid resistance. In: Transactions Conference Chemotherapy of Tuberculosis (St. Louis) **15**, 279 (1956).

Biehl, J.P.: Emergence of drug resistance as related to the dosage and metabolism of isoniazid. In: Transactions Conference Chemotherapy of Tuberculosis (St. Louis) **16**, 108 (1957).

Biehl, J.P. and R.W. Vilter: Effect of isoniazid on vitamin B_6 metabolism; its possible significance in producing isoniazid neuritis. Proc. Soc. Exptl. Biol. Med. **85**, 389 (1954).

Bonicke, R., u. W. Reif: Enzymatische Inaktivierung von Isonicotinsäure-hydrazid im menschlichen und tierischen Organismus. Arch. Exptl. Pathol. Pharmakol. **220**, 321 (1953).

Boone, I.U., V. Strang, and B. Rogers: Effect of pyridoxal on the uptake of ^{14}C activity from labelled isoniazid by *M. tuberculosis*. Am. Rev. Tuberc. Pulmonary Diseases, **76**, 568 (1957).

Boone, I.U., and K.T. Woodward: Relationship of pyridoxine and its derivatives to the mechanism of action of isoniazid. Proc. Soc. Exptl. Biol. Med. **84**, 292 (1953).

Braunsteiner, H., F. Milczoch, and W. Zischa: Schweiz. Z. Tuberk. **10**, 91 (1953) cit. by Winder (1964) in: Chemotherapy of tuberculosis, Barry (ed.): Butterworths.

Brennan, P.J., S.A. Rooney, and F.G. Winder: The lipids of *M. tuberculosis* BCG: fractionation, composition, turnover and effects of isoniazid. Irish J. Med. Sci. **3**, 371 (1970).

Brieger, E.M., V.E. Cosslett, and A.M. Glauert: Action of antibiotics on avian tubercle bacilli studied with electron microscope. Nature **171**, 211 (1953).

Bringmann, G.: Light and electron microscope studies on the nature of the granules of *Corynebacterium diphtheriae*. Zbl. Bakteriol. Parasitenk. Abt. I. Orig. **156**, 493 (1951).

Bryson, V., H. Deiches, and W. Szybalski: Isoniazid dependent strains of *M. ranae*. Am. Rev. Tuberc. **68**, 631 (1953).

Campell, A.H., and P. Guilfoyle: Pulmonary tuberculosis, INH and cancer. Brit. J. Diseases Chest. **64**, 141 (1970).

Canal, N., and S. Garattini: Inhibition of glutamic acid decarboxylase by INH. Arzneimittel-Forsch. **7**, 158 (1957).

Carlson, H.B., E.M. Anthony, W.F. Russell, Jr., and G. Middlebrook: Prophylaxis of isoniazid neuropathy with pyridoxine. New Engl. J. Med. **255**, 118 (1956).

Catteneo-Lacombe, J., and J.C. Senez: Inhibition of aspartic acid decarboxylase by INH. Compt. Rend. Soc. Biol. **150**, 748 (1956).

Chorine, V.: Action of nicotinamide on bacilli of the genus *Mycobacterium*. Compt. Rend. **220**, 150 (1945).

Cohn, M.L., U. Oda, C. Kovitz, and G. Middlebrook: Studies on isoniazid and tubercle bacilli. II. The growth requirements, catalase activities and pathogenic properties of isoniazid resistant mutants. Am. Rev. Tuberc. **70**, 641 (1954a).

Cohn, M.L., U. Oda, C. Kovitz, and G. Middlebrook: Studies on isoniazid and tubercle bacilli. I. The isolation of isoniazid resistant mutants *in vitro*. Am. Rev. Tuberc. **70**, 465 (1954b).

Crowle, A.J., and D.K. Riemensnider: Lack of any antagonism between pyridoxine and INH in the chemotherapy of acutely or chronically tuberculous mice. Tubercle **41**, 450 (1960).

Cymerman-Craig, J., D. Willis, S.D. Rubbo, and J. Edgar: Mode of action of isonicotinylhydrazide. Nature **176**, 34 (1955).

Davison, A.N.: The mechanism of inhibition of decarboxylases by INH. Liver cysteine sulphinic acid decarboxylase. Biochim. Biophys. Acta **19**, 131 (1956).

Ebina, T., K. Munakata, and M. Motomiya: Effect of isoniazid on the incorporation of acetate-1-^{14}C into fatty acids of mycobacteria. Compt. Rend. Soc. Biol. **155**, 1190 (1961).

Eda, T.: Studies on the action of INH on tubercle bacilli and the mechanism of isoniazid resistance in mycobacteria. Amer. Rev. Respirat. Diseases **88**, 590 (1963).

Elmendorf, D.F., W.U. Cawthon, C. Muschenheim, and W. McDermott: The absorption, distribution, excretion and short-term toxicity of isonicotinic acid hydrazide in man. Am Rev. Tuberc. **65**, 429 (1952).

Evans, D.A.P., K.A. Manley, and V.A. McKusick: Genetic control of isoniazid metabolism in man. Brit. Med. J. **1960**, 285.

Evans, D.A.P., P.B. Storey, and V.A. McKusick: Further observations on the determination of the isoniazid inactivator phenotype. Bull. Johns Hopkins Hosp. **108**, 60 (1961).

FEINGOLD, D.S.: Antimicrobial chemotherapeutic agents: The nature of their action and selective toxicity. New Engl. J. Med. **269**, 900 (1963).

FISHER, M.W.: The antagonism of the tuberculostatic action of isoniazid by hemin. Am. Rev. Tuberc. **69**, 469 (1954a).

FISHER, M.W.: Hemin as a growth factor for certain isoniazid resistant strains of *M. tuberculosis*. Am. Rev. Tuberc. **69**, 797 (1954b).

Fox, H.H.: The chemical approach to the control of tuberculosis. Science **116**, (1952a).

Fox, H.H.: Synthetic tuberculostats: I. Pyridine carboxylic acid derivatives. II. Amino and hydroxy pyridine carboxylic acid derivatives. III. Isonicotinaldehyde, thiosemicarbazone and some related compounds. J. Org. Chem. **17**, 542, 547, 555 (1952b).

Fox, H.H., and J.T. GIBAS: Synthetic tuberculostats. IV. Pyridine carboxylic acid hydrazides and benzoic acid hydrazides. J. Org. Chem. **17**, 1653 (1952).

Fox, H.H., and J.T. GIBAS: Synthetic tuberculostats. VII. Monoalkyl derivatives of isonicotinylhydrazine. J. Org. Chem. **18**, 994 (1953).

Fox, H.H., and J.T. GIBAS: Synthetic tuberculostats. IX. Dialkyl derivatives of isonicotinylhydrazine. J. Org. Chem. **20**, 60 (1955).

Fox, H.H., and J.T. GIBAS: Synthetic tuberculostats. XI. Trialkyl and other derivatives of isonicotinylhydrazine. J. Org. Chem. **21**, 356 (1955).

GANGADHARAM, P.R.J., F.M. HAROLD, and W.B. SCHAEFER: Selective inhibition of nucleic acid synthesis in *M. tuberculosis* by INH. Nature **198**, 712 (1963).

GAYATHRI DEVI, B., T. RAMAKRISHNAN, and K.P. GOPINATHAN: Enzymatic interaction between INH and NAD. Biochem. J. **128**, 63 P (1972).

GOLDMAN, D.S.: On the mechanism of action of INH, cited in WINDER (1964), J. Am. Chem. Soc. **76**, 2841 (1954).

GOLDMAN, D.S.: Enzyme systems in mycobacteria. A review. Adv. Tuberc. Res. **11**, 1 (1961).

GOPINATHAN, K.P., T. RAMAKRISHNAN, and C.S. VAIDYANATHAN: Purification and properties of an inhibitor for nicotinamide-adenine-dinucleotidase from *M. tuberculosis H37Rv*. Arch. Biochem. Biophys. **113**, 376 (1966).

GOPINATHAN, K.P., M. SIRSI, and T. RAMAKRISHNAN: Nicotinamide-adenine-dinucleotides of *M. tuberculosis H37Rv*. Biochem. J. **87**, 444 (1963).

GOPINATHAN, K.P., M. SIRSI, and C.S. VAIDYANATHAN: Nicotinamide-adenine-dinucleotide glycohydrolase of *M. tuberculosis H37Rv*. Biochem. J. **91**, 277 (1964).

GROSS, W.: Cytostatic effect of INH. Klin. Wochschr. **34**, 495 (1956).

GRUMBACH, F.: Antituberculous activity in mice of ethambutol associated with isoniazid and ethionamide. Ann. Inst. Pasteur **110**, 69 (1966).

GRUNBERG, E., and W. BLENCOME: The influence of pyridoxine on the *in vivo* antituberculous activity of isoniazid. Am. Rev. Tuberc. Pulmonary Diseases **71**, 898 (1955).

GRUNBERG, E., R.J. SCHNITZER, B. LEIWANT, I.L. D'ASCENSIO, and E. TITSWORTH: Studies on the activity of hydrazine derivatives of isonicotinic acid in the experimental tuberculosis of mice. Quart. Bull. Sea View Hosp. **13**, 3 (1952).

GUPTA, K.C., and R. VISWANATHAN: Electron microscopic and phase contrast studies of effects of PASA, INH and viomycin on tubercle bacilli. Am. Rev. Tuberc. Pulmonary Diseases **73**, 296 (1956).

HAMMOND, E.C., I.J. SELIKOFF, and E.H. ROBITZEK: Isoniazid therapy in relation to later occurrence of cancer in adults and infants. Brit. Med. J. **1961**, 792.

HICKS, R.M., and J. CYMERMAN-CRAIG: Inhibition of alanine α-KG transaminase of pig heart muscle by cyclic hydrazides. Biochem. J. **67**, 353 (1957).

HOARE, D.S.: The progressive reaction of INH with two bacterial amino acid decarboxylases. Biochim. Biophys. Acta **19**, 141 (1956).

HOBBY, G.L., and T.F. LENERT: The *in vitro* action of antituberculous agents against multiplying and non-multiplying microbial cells. Am. Rev. Tuberc. Pulmonary Diseases **76**, 1031 (1957).

HUGHES, H.B.: On the metabolic fate of isoniazid. J. Pharmacol. Exptl. Therap. **109**, 444 (1953).

HUGHES, H.B., J.P. BIEHL, A.P. JONES, and L.H. SCHMIDT: Metabolism of isoniazid in man as related to the occurrence of peripheral neuritis. Am. Rev. Tuberc. **70**, 266 (1954).

HUGHES, H.B., L.H. SCHMIDT, and J.P. BIEHL: The metabolism of isoniazid: Its implications in therapeutic use. In: Transactions of the 14th Conference on the Chemotherapy of Tuberculosis. Atlanta, Ga., Feb. 7–10, **14**, 217 (1955).

Isler, O., H. Gutmann, O. Straub, B. Fust, E. Bohni u. A. Studer: Chemotherapie der experimentellen Tuberkulose. II. Kernsubstituierte INH. Helv. Chim. Acta. **38**, 1033 (1955).

Ito, F., and T. Sugano: Transaminases of avian tubercle bacilli. Kekkaku **29**, 368 (1954).

Jutte, R.: Dissertation. Pharmakologisches Institut der Freien Universität, Berlin, 1967.

Karkalitskii, I. M., G. V. Karkalitskaya, E. M. Ashikhmina, N. D. Kovrizhnykh, G. P. Tuzova, G. F. Plotnikova, and M. P. Berdnikov: Characteristics of biochemical shifts during experimental B_6-hypovitaminosis. Vopr. Pitaniya **29**, 23 (1970); Biol. Abstr. **51**, 82670 (1970).

Karlson, A. G., and W. H. Feldman: Isoniazid in experimental tuberculosis of guinea pigs infected with tubercle bacilli resistant to streptomycin and paraamino salicylic acid. Am. Rev. Tuberc. **66**, 477 (1952).

Karlson, A. G., and W. H. Feldman: The effect of combined therapy with INH and streptomycin on experimental tuberculosis of guinea pigs. Am. Rev. Tuberc. **68**, 575 (1953).

Kern, M., and R. Natale: A diphosphopyridine nucleotidase and its protein inhibitor from *Mycobacterium butyricum*. J. Biol. Chem. **231**, 41 (1958).

Knox, R.: Haemin and isoniazid resistance of *M. tuberculosis*. J. Gen. Microbiol. **12**, 191 (1955).

Knox, R., P. Meadow, and A. R. H. Worssam: The relationship between the catalase activity, H_2O_2 sensitivity and INH resistance of mycobacteria. Am. Rev. Tuberc. Pulmonary Diseases **73**, 726 (1956).

Koch-Weser, D., B. J. Tricou, W. R. Barclay, and R. H. Ebert: The use of ^{14}C labelled compounds in tuberculosis research: In: Peaceful uses of atomic energy. Proc. Int. Conf. of the Peaceful Uses of Atomic Energy, Geneva, 1955, vol. 10, p. 469 United Nations, N. Y., 1956.

Konno, K.: New chemical method to differentiate human-type tubercle bacilli from other mycobacteria. Science **124**, 985 (1956).

Konno, K., R. Kurzmann, K. T. Bird, and A. Sbarra: Differentiation of human tubercle bacilli from atypical acid-fast bacilli. I. Niacin production of human tubercle bacilli and atypical acid-fast bacilli. Am. Rev. Tuberc. Pulmonary Diseases **77**, 669 (1958).

Konno, K., K. Oizumi, and S. Oka: Biosynthesis of niacin ribonucleotide from quinolinic acid by mycobacteria. Nature **205**, 874 (1965a).

Konno, K., K. Oizumi, Y. Shimizu, and S. Oka: Niacin metabolism in mycobacteria: Quinolinic acid as a precursor of niacin ribonucleotide and the differences in niacin biosynthesis among various species of mycobacteria. Am. Rev. Respirat. Diseases **91**, 383 (1965b).

Kruger-Thiemer, E.: Isonicotinic acid hypothesis of the antituberculosis action of isoniazid. Am. Rev. Tuberc. Pulmonary Diseases **77**, 364 (1958).

Levy, L., L. J. Higgins, and T. N. Burbridge: INH-induced vitamin B_6 deficiency. Am. Rev. Respirat. Diseases **96**, 910 (1967).

Lewis, A., and Shepherd, R. G.: In: Medicinal chemistry, Alfred Burger (ed.), p. 456. New York: Wiley Interscience, 1970.

Lichstein, H.: Mechanism of competitive action of INH and Vitamin B_6. Proc. Soc. Exptl. Biol. Med. **88**, 519 (1955).

Lincoln, E. M., and E. M. Sewell: In: Tuberculosis in children, p. 61. McGraw-Hill 1963.

Mackaness, G. B.: The action of drugs on intracellular tubercle bacilli. J. Pathol. Bacteriol. **64**, 429 (1952).

Mackaness, G. B., and N. Smith: The action of INH on intracellular tubercle bacilli. Am. Rev. Tuberc. **66**, 125 (1952).

Mackaness, G. B., and G. N. Smith: The bactericidal action of isoniazid, streptomycin and terramycin on extracellular and intracellular tubercle bacilli. Am. Rev. Tuberc. **67**, 322 (1953).

Maher, J. R., J. F. Speyer, and M. Levine: Mode of action of isoniazid. Am. Rev. Tuberc. Pulmonary Diseases **75**, 517 (1957).

Maher, J. R., J. F. Speyer, and M. Levine: The role of trace metals in the inhibition of bovine liver catalase by INH. Am. Rev. Tuberc. Pulmonary Diseases **77**, 501 (1958).

McClatchy, J. K.: Mechanism of action of INH on *M. bovis* strain BCG. Infect. Immun. **3**, 530 (1971); Biol. Abstr. **52**, 70860 (1971).

McClatchy, J. K., and J. A. Smith: Inhibition of DNA synthesis in *M. tuberculosis* BCG by INH. In: Pulmonary Disease Research Conference, Veterans Administration Department of Medicine and Surgery, Washington D. C. (1968).

McDermott, W.: Antimicrobial therapy of pulmonary tuberculosis. Bull. World Health Organ. **23**, 427 (1960).

McMILLAN, F., F. LEONARD, R. I. MELTZER, and J. A. KING: Antitubercular substances. II. Substitution products of isonicotinic hydrazide. J. Am. Pharm. Assoc. **42**, 457 (1953).

MEADOW, P., and R. KNOX: The effect of isonicotinic acid hydrazide on the oxidative metabolism of *M. tuberculosis var. bovis BCG*. J. Gen. Microbiol. **14**, 414 (1956).

MEYER, H., and J. MALLY: Hydrazine derivatives of pyridine carboxylic acids. Monatsh. **33**, 393 (1912); Chem. Abstr. **6**, 2073 (1912).

MIDDLEBROOK, G.: Sterilization of tubercle bacilli by INH and the incidence of variants resistant to drug *in vitro*. Am. Rev. Tuberc. **65**, 765 (1952).

MIDDLEBROOK, G.: INH resistance and catalase activity of tubercle bacilli. Am. Rev. Tuberc. **69**, 471 (1954).

MIDDLEBROOK, G.: Isoniazid-resistance and catalase activity of tubercle bacilli: A preliminary report. Am. Rev. Tuberc. **70**, 922 (1954).

MIDDLEBROOK, G., and M. L. COHN: Some observations on the pathogenicity of isoniazid-resistant variants of tubercle bacilli. Science **118**, 297 (1953).

MIDDLEBROOK, G., M. L. COHN, and W. B. SHAEFER: Studies on isoniazid and tubercle bacilli. III. The isolation, drug susceptibility and catalase-testing of tubercle bacilli from isoniazid-treated patients. Am. Rev. Tuberc. **70**, 852 (1954).

MILLER, I. L., and W. C. ROESSLER: Growth of *M. tuberculosis* in liquid media. Am. Rev. Tuberc. Pulmonary Diseases **73**, 716 (1956).

MITCHISON, D. A.: Tubercle bacilli resistant to isoniazid: virulence and response to treatment with isoniazid in guinea pig. Brit. Med. J. **1954**, 128.

MORALES, S. M., and E. M. LINCOLN: The effect of isoniazid therapy on pyridoxine metabolism in children. Am. Rev. Tuberc. Pulmonary Diseases **75**, 594 (1957).

MORSE, W. C., O. L WEISER, D. M. KUHNS, M. FUSILLO, M. D. DAIL, and J. R. EVANS: Study of the virulence of isoniazid-resistant tubercle bacilli in guinea pigs and mice: a preliminary report. Am. Rev. Tuberc. **69**, 464 (1954).

NAKAMURA, S., Y NISHIZUKA, and O. HAYAISHI: Regulation of nicotinamide adenine dinucleotide biosynthesis by adenosine triphosphate. J. Biol. Chem. **239**, 2717 (1964).

NESTLER, H. J.: Chemische Untersuchungen zum Wirkungsmechanismus des Isoniazids. Arzneimittel.-Forsch. **16**, 1442 (1966).

OCHOA, J.: INH neuropathy in man: Quantitative electron microscope study. Brain **93**, 831 (1970).

OESTREICHER, R.: Observations on the pathogenicity of isoniazid-resistant mutants of tubercle bacilli for tuberculous patients. Am. Rev. Tuberc. Pulmonary Diseases **71**, 390 (1955).

OESTREICHER, R., S. H. DRESSLER, and G. MIDDLEBROOK: Peripheral neuritis in tuberculous patients treated with isoniazid. Am. Rev. Tuberc. **70**, 504 (1954).

OFFE, H. A., W. SIEFKEN, and G. DOMAGK: The tuberculostatic activity of hydrazine derivatives from pyridine carboxylic acids and carbonyl compounds. Z. Naturforsch. **7b**, 462 (1952).

OGASAWARA, N., N. SUZUKI, M. YOSHINO, and Y. KOTAKE: Complex formation between NADase and protein inhibitor from *M. butyricum*. Fed. Eur. Biochem. Soc. Letters **6**, 337 (1970).

OLIVERA, B. M.: The DNA joining enzyme from *Escherichia coli*. In: SIDNEY P. COLOWICK and NATHAN O. KAPLAN, Methods in enzymology, L. GROSSMAN and K. MOLDAVE (ed.), vol. XXI, p. 311. New York: Academic Press, 1971.

PALAT, K., L. NOVACEK, and M. CELADNIK: Antituberculotics. VII. Some ring-substituted isonicotinic acid derivatives. Collect. Czech. Chem. Commun. **32**, 1191 (1967).

P'AN, S. Y., L. MARKAROGLU, and J. REILLY: The effects of barbiturate on the toxicity of isoniazid. Am. Rev. Tuberc. **66**, 100 (1952).

PANSY, F., H. STANDER, and R. DONOVICK: *In vitro* Studies on isonicotinic acid hydrazide. Am. Rev. Tuberc. **65**, 761 (1952).

PATIALA, J.: The amount of pyridine nucleotides (Co I & II) in blood in experimental tuberculosis before and during INH treatments. Am. Rev. Tuberc. **70**, 453 (1954).

PETERS, J. H., K. S. MILLER, and P. BROWN: Studies on the metabolic basis for the genetically determined capacities for isoniazid inactivation in man. J. Pharmacol. Exptl. Therap. **150**, 298 (1965).

POPE, H.: The neutralization of INH activity in *M. tuberculosis* by certain metabolites. Am. Rev. Tuberc. Pulmonary Diseases **73**, 735 (1956).

PREZ, R. D., and I. U. BOONE: Metabolism of ^{14}C-isoniazid in humans. Am. Rev. Respirat. Diseases **84**, 42 (1961).

Rajtar-Leontiew, Z.: The action of isoniazid on pyridoxine metabolism in children. Ann. Pediat. Semaine Hop. **17**, 414 (1970); Biol. Abstr. **52**, 93280 (1971).

Ramakrishnan, T., P. Suryanarayana Murthy, and K. P. Gopinathan: Intermediary metabolism of mycobacteria. Bacteriol. Rev. **36**, 65 (1972).

Robitzek, E. H., and I. J. Selikoff: Hydrazine derivatives of isonicotinic acid (Rimifon, Marsilid) in the treatment of active progressive caseous-pneumonic tuberculosis: a preliminary report. Am. Rev. Tuberc. **65**, 402 (1952).

Robson, J. M., and Sullivan, F. M.: Antituberculosis drugs. Pharmacol. Rev. **15**, 169 (1963).

Roe, F. J., E. Boyland, and A. Haddow: Chemotherapy of tuberculosis. Brit. Med. J. **1965**, 1550.

Roe, F. J. C., and M. C. Lancaster: Natural, metallic and other substances as carcinogens. Brit. Med. Bull. **20**, 127 (1964).

Roth, L. J., and R. W. Manthei: The distribution of ^{14}C labelled isonicotinic acid hydrazide in normal mice. Proc. Soc. Exptl. Biol. Med. **81**, 566 (1952).

Rubin, B., and J. C. Burke: Absorption, distribution and excretion of isoniazid in the dog. J. Pharmacol. Exptl. Therap. **107**, 219 (1953a).

Rubin, B., and J. C. Burke: Further observations on the pharmacology of isoniazid. Am. Rev. Tuberc. **67**, 644 (1953b).

Rubin, B., G. L. Hassert, B. G. H. Thomas, and J. C. Burke: Pharmacology of isonicotinic acid hydrazide. Am. Rev. Tuberc. **65**, 392 (1952).

Russell, W. F., Jr., S. H. Dressler, and G. Middlebrook: Chemotherapy of tuberculosis. Advan. Intern. Med. **8**, 221 (1956).

Russell, W. F., Jr., and G. Middlebrook: Chemotherapy of tuberculosis. Springfield: Charles C. Thomas, Publ. 1961.

Sakai, J.: The mechanism of bacteriostatic action of INH. The influence of INH on transaminase of BCG. Kekkaku **27**, 161 (1954).

Saxena, K. C., and K. L. Arora: Studies in *M. tuberculosis:* transaminase activity. J. Sci. Ind. Res. (India) **18 C**, 237 (1959).

Saxena, K. C., C. R. Krishna Murti, and D. L. Shrivastava: Studies on the enzyme make-up of *Vibrio cholerae:* Part X. Transaminase systems. J. Sci. Ind. Res. (India) **15 C**, 101 (1956).

Saxena, K. C., and M. K. Rastogi: Transamination in *Salmonella typhosa.* J. Sci. Ind. Res. (India) **20 C**, 287 (1961).

Saxena, K. C., P. Sagar, S. C. Agarwala, and D. L. Shrivastava: Studies in the enzyme make-up of *Pasteurella pestis.* Part IV. Transamination reactions in virulent and avirulent strains. Indian. J. Med. Res. **45**, 161 (1957).

Schaefer, W. B.: The effect of INH on growing and resting tubercle bacilli. Am. Rev. Tuberc. **69**, 125 (1954).

Schaefer, W. B.: Effect of INH on the dehydrogenase activity of *M. tuberculosis.* J. Bacteriol. **79**, 236 (1960).

Schloot, W., T. Franz-Jurgen, H. Blaesner, and H. Werner-Goedde: N-Acetyl transferase and serotonin metabolism in man and other species. I. Hoppe-Seylers Z. Physiol. Chem. **350**, 1353 (1969).

Schmidt, L. H.: Some observations on the utility of simian pulmonary tuberculosis in defining the therapeutic potentialities of isoniazid. Am. Rev. Tuberc. Pulmonary Diseases **74**, Suppl. 138 (1956).

Schmidt, L. H.: The emergence of isoniazid-sensitive bacilli in monkeys inoculated with isoniazid-resistant strains. In: Transaction of the 17th Conference on the Chemotherapy of Tuberculosis, Memphis, Tenn., Feb. 3–6, 1958, **17**, p. 264.

Seneca, H., and D. Ides: *In vitro* action of isonicotinic acid hydrazide on protozoa. Antibiot. & Chemotherapy **3**, 241 (1953).

Senghvi, D. R., and D. Subrahmanyam: Effect of certain anti-tubercular drugs on the metabolism of phospholipids of mycobacteria. Symp. Chemotherapy of Tuberculosis. Indian J. Chest. Dis., (1967) p. 18.

Sriprakash, K. S., and T. Ramakrishnan: Comparative study of nicotinamide adenine dinucleotide nucleotidase from *M. tuberculosis H37Rv* and pig brain. Effect of INH on the enzyme inhibitor complex. Indian J. Biochem. **3**, 211 (1966).

Sriprakash, K. S., and T. Ramakrishnan: Isoniazid and nicotinamide adenine dinucleotide synthesis in *M. tuberculosis.* Indian J. Biochem. **6**, 50 (1969).

SRIPRAKASH, K. S., and T. RAMAKRISHNAN: INH-resistant mutants of *M. tuberculosis H37Rv:* Uptake of INH and properties of NADase inhibitor. J. Gen. Microbiol. **60**, 125 (1970).

STEENKEN, W., and E. WOLINSKY: Anti-tuberculous properties of hydrazines of isonicotinic acid. Am. Rev. Tuberc. **65**, 365 (1952).

STEWART, S. M.: Virulance of tubercle bacilli recovered from patients treated with isoniazid. Am. Rev. Tuberc. **69**, 641 (1954).

STIEF, M. E.: Guinea pig virulence and catalase activity of INH-resistant tubercle bacilli. Am. J. Med. Technol. **22**, 265 (1956).

SUNAHARA, S., M. URANO, and M. OGAWA: Genetical and geographic studies on isoniazid inactivation. Science **134**, 1530 (1961).

SUTER, E.: Multiplication of tubercle bacilli within phagocytes cultivated *in vitro* and effect of streptomycin and isonicotinic acid hydrazide. Am. Rev. Tuberc. **65**, 775 (1952).

SZYBALSKI, W., and V. BRYSON: Bacterial resistance studies with derivatives of isoniazid. Am. Rev. Tuberc. **65**, 768 (1952).

TIRUNARAYANAN, M. O., and W. A. VISCHER: Relationship of isoniazid to the metabolism of mycobacterial catalase and peroxidase. Am. Rev. Tuberc. Pulmonary Diseases **75**, 62 (1957).

TIRUNARAYANAN, M. O., and W. A. VISCHER: Inactivation of isoniazid by peroxidase. Nature **183**, 681 (1959).

TOIDA, I.: Isoniazid-hydrolyzing enzyme of mycobacteria. Am. Rev. Respirat. Diseases **85**, 720 (1962).

TOIDA, I.: Nicotinamide-adenine-dinucleotide nucleosidase of *Mycobacterium butyricum*. Acta Chem. Scand. **17**, 161S (1963).

TOMCSANYI, A.: The effect of isoniazid on the incorporation of amino acids into protein by a soluble system of mycobacteria. Am. Rev. Respirat. Diseases **92**, 119 (1965).

TOMPSETT, R.: Quantitative observations on the pattern of emergence of resistance to isoniazid. Am. Rev. Tuberc. **70**, 91 (1954).

TSUKAMURA, M., and S. MIZUNO: Mode of action of isoniazid viewed from isotope incorporation studies. Kekkaku **37**, 29 (1962).

TSUKAMURA, M., S. MIZUNO, and S. TSUKAMURA: Effect of isoniazid on the incorporation of some radioactive substances into the cellular fractions of a strain of mycobacterium. Japan. J. Microbiol. **8**, 105 (1964).

TSUKAMURA, M., and S. TSUKAMURA: Isotopic studies on the effect of isoniazid on protein synthesis of mycobacteria. Japan. J. Tuberc. **11**, 14 (1963).

TSUKAMURA, M., S. TSUKAMURA, and E. NAKANO: The uptake of INH by mycobacteria and its relation to INH susceptibility. Am. Rev. Respirat. Diseases **87**, 269 (1963).

Tuberculosis Chemotherapy Centre, Madras: A controlled comparison of a twice-weekly and three once-weekly regimens in the initial treatment of pulmonary tuberculosis. Bull. World Health Organ. **43**, 143 (1970).

TURPIN, R., E. BERGOGNE-BEREZIN, B. CAILLE, and D. SALMON-BONNEREAU: The interest of INH metabolism in genetic studies: Anomalies discovered in trisomic patients. Ann. Med. Interne. **120**, 243 (1969); Biol. Abstr. **51**, 81889 (1970).

UNGAR, J., E. G. TOMICH, K. R. PARKIN, and P. W. MUGGLETON: Effect of pyridoxine on the action of isoniazid. Lancet **1954-II**, 220.

VODICKA, Z., and V. DELONG: Effect of isonicotinyl hydrazide on transaminases of living mycobacteria. Rozhledy Tuberk. **16**, 198 (1956).

WILLET, H. P.: The production of lysine from diaminopimelic acid by cell-free extracts of *M. tuberculosis*. Am. Rev. Respirat. Diseases **81**, 653 (1960).

WIMPENNY, J. W. T.: The uptake and fate of INH in *M. tuberculosis var bovis BCG*. J. Gen. Microbiol. **47**, 389 (1967a).

WIMPENNY, J. W. T.: Effect of INH on biosynthesis in *M. tuberculosis var bovis BCG*. J. Gen. Microbiol. **47**, 379 (1967b).

WINDER, F.: Catalase and peroxidase in mycobacteria. Possible relationship to the mode of action of INH. Am. Rev. Respirat. Diseases **81**, 68 (1960).

WINDER, F.: Early changes induced by INH in the composition of *M. tuberculosis*. Biochim. Biophys. Acta **82**, 210 (1964).

WINDER, F.: In: Chemotherapy of tuberculosis, V. C. BARRY (ed.), p. 111. London: Butterworths 1964.

Winder, F.G., P.J. Brennan, and I. McDonnell: Effects of isoniazid on the composition of mycobacteria with particular reference to soluble carbohydrates and related substances. Biochem. J. **104**, 385 (1967).

Winder, F.G., and P. Collins: The effect of isoniazid on nicotinamide dinucleotide levels in *M. bovis*. strain BCG. Am. Rev. Respirat. Diseases **97**, 719 (1968).

Winder, F.G., and P.B. Collins: Inhibition by INH of synthesis of mycolic acids in *M. tuberculosis*. J. Gen. Microbiol. **63**, 41 (1970).

Winder, F.G., and S.A. Rooney: Effects of isoniazid on the triglycerides of BCG. Am. Rev. Respirat. Diseases **97**, 938 (1968).

Winder, F.G., and S.A. Rooney: The effects of INH on the carbohydrates of *M. tuberculosis* BCG. Biochem. J. **117**, 355 (1970).

Wolinsky, E., M.M. Smith, and W. Steenken, Jr.: INH susceptibility, catalase activity and guinea pig virulence of recently isolated cultures of tubercle bacilli. Am. Rev. Tuberc. Pulmonary Diseases **73**, 768 (1956).

Yamamoto, M.: Mechanism of bacteriostatic action of isonicotinic acid hydrazide (IV). Influence of INH on L-cysteine disulfhydrase of BCG. Kekkaku **30**, 252 (1955).

Yoneda, M., and M. Asano: Competitive action of isonicotinic acid hydrazide and pyridoxal in the amino acid decarboxylation of *Escherichia coli*. Science **117**, 277 (1953).

Yoneda, M., N. Kato, and M. Okajima: Competitive action of isonicotinic acid hydrazide and vitamin B_6 in the formation of indole by *E. coli*. Nature **170**, 803 (1952).

Youatt, J.: Metabolism of isoniazid by *M. tuberculosis* BCG with reference to current theories of the mode of action. Am. Rev. Tuberc. Pulmonary Diseases **78**, 806 (1958a).

Youatt, J.: The action of INH on the transaminases of *M. tuberculosis (BCG)*. Biochem. J. **68**, 193 (1958b).

Youatt, J.: The uptake of INH by washed cell suspensions of mycobacteria and other organisms. Australian J. Exptl. Biol. Med. Sci. **36**, 223 (1958c).

Youatt, J.: The metabolism of isoniazid and other hydrazides by mycobacteria. Australian J. Exptl. Biol. Med. Sci. **38**, 245 (1960a).

Youatt, J.: The uptake of INH and related compounds by mycobacteria. Australian J. Exptl. Biol. Med. Sci., **38**, 331 (1960b).

Youatt, J.: The uptake of INH and related compounds by mycobacteria. Australian J. Exptl. Med. Sci. **39**, 93 (1961).

Youatt, J.: The metabolism of isoniazid and pyridine aldehydes by mycobacteria. Australian J. Exptl. Biol. Med. Sci. **40**, 191 (1962).

Youatt, J.: Changes in the phosphate content of mycobacteria produced by INH and ethambutol. Australian J. Exptl. Biol. Med. Sci. **43**, 305 (1965).

Youatt, J.: A review of the action of isoniazid. Am. Rev. Respirat. Diseases **99**, 729 (1969).

Youatt, J., and S.H. Tham: An enzyme system of *M. tuberculosis* that reacts specifically with isoniazid. I. Am. Rev. Respirat. Diseases **100**, 25 (1969a).

Youatt, J., and S.H. Tham: An enzyme system of *M. tuberculosis* that reacts specifically with INH. II. Correlation of this reaction with the binding and metabolism of INH. Am. Rev. Respirat. Diseases **100**, 31 (1969b).

Youatt, J., and S.H. Tham: Radioactive content of *M. tuberculosis* after exposure to ^{14}C-isoniazid. Am. Rev. Respirat. Diseases **100**, 77 (1969c).

Youmans, A.S., and G.P. Youmans: The inactivation of isoniazid by filtrates and extracts of mycobacteria. Am. Rev. Tuberc. Pulmonary Diseases **72**, 196 (1955).

Zatman, L.J., S.P. Colowick, N.O. Kaplan, and M.M. Ciotti: The action of nicotinamide and isonicotinic acid hydrazide on DPNase. Bull. Johns Hopkins Hosp. **91**, 211 (1952).

Zatman, L.J., N.O. Kaplan, S.P. Colowick, and M.M. Ciotti: Effect of isonicotinic acid hydrazide on diphosphopyridine nucleotidases. J. Biol. Chem. **209**, 453 (1954).

Sideromycins

F. Knüsel and W. Zimmermann

I. Introduction

In the first summary report on the sideromycins (NÜESCH and KNÜSEL, 1967) a detailed account was given of the chemical structure of these ferritrihydroxamate antibiotics, of their biological activity, as well as of their toxicological and pharmacological properties. In this review reference was also made to the relationship between the sideromycins and the sideramines, which are structurally akin to the sideromycins, whose antibiotic effect they antagonise (ZÄHNER et al., 1962). At that time, knowledge of the biochemical properties and mode of action of the sideromycins was still only fragmentary. In experiments carried out with the haemin-heterotrophic mutant S. aureus JT 52, the haemin requirement of which is attributable to an inability to incorporate iron into protoporphyrin, it was found that sideromycins do not intervene in the biosynthesis of haemin. In the light of results obtained in experiments with Brevibacterium flavum, the possibility that the sideromycin antibiotics might exert an influence on carbohydrate metabolism was also discussed.

The following review is concerned chiefly with studies on the mechanism of action of the sideromycins and the manner in which they permeate into their site of action inside the cell, with the antagonism of this permeation by sideramines, and with the problem of sideromycin resistance.

No reference will be made to the role which the sideramines or sideromycins possibly play in microbial iron metabolism, because this topic has already been dealt with in other papers. NEILANDS (1967) has examined in detail the possible biological functions of the naturally occurring hydroxamic acids, and these same problems have been discussed at length in a paper by SNOW (1970) on mycobactins. In experiments undertaken with Neurospora crassa, PADMANABAN et al. (1968) showed that two sideramines produced by this micro-organism are capable of promoting the incorporation of iron into protoporphyrin. Additional work on problems connected with the formation and biosynthesis of sideramines has been carried out by MÜLLER (1968) and CRUEGER (1968).

II. Results

a) Influence of Sideromycins on the Incorporation
of Low-Molecular Substances into Intact Bacterial Cells

In an initial series of experiments we studied the influence exerted by the sideromycin antibiotic A-22,765 on the incorporation of amino-acids and uracil into intact cells of *S. aureus* SG 511 (KNÜSEL *et al.*, 1967). The incorporation of ^{14}C-leucine was immediately and completely blocked by A-22,765 in a concentration of 2 mcg./ml. (Fig. 1); identical results were obtained with ^{14}C-phenylalanine, ^{14}C-isoleucine, and ^{14}C-valine. Similarly, the incorporation of amino acids into *S. aureus* SG 511 was inhibited by ferrimycin A_1, The same phenomenon also being observed in the case of *B. subtilis* ATCC 6633 in response to the four sideromycins tested, i.e. albomycin, danomycin, ferrimycin A_1, and A-22,765; in each instance the degree of inhibition was dependent upon concentration.

When, in experiments performed in the same manner, ferrioxamine B was employed simultaneously in addition to A-22,765, inhibition of the amino-acid incorporation was counteracted (Fig. 2). By raising the concentration of ferrioxamine B, while leaving the quantity of A-22,765 unchanged, it was found possible

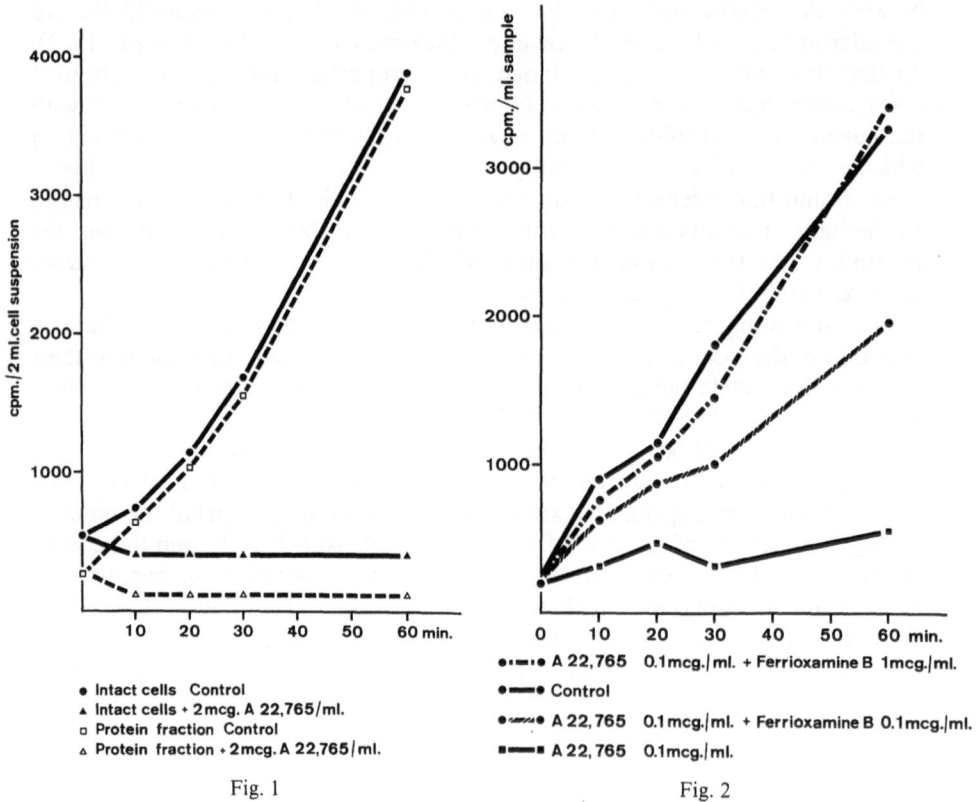

Fig. 1

Fig. 2

Fig. 1. Influence exerted by antibiotic A-22,765 on the incorporation of ^{14}C-leucine into intact cells and protein fraction of *S. aureus* SG 511

Fig. 2. Influence exerted by antibiotic A-22,765 and ferrioxamine B on the incorporation of ^{14}C-leucine into intact cells of *S. aureus* SG 511

to enhance this counteraction; at a sideramine:sideromycin weight ratio of 10:1, inhibition of amino-acid incorporation was completely abolished.

In sideromycin-resistant cells of *S. aureus* SG 511, on the other hand, even concentrations of A-22,765 as high as 50 mcg./ml. failed to produce any inhibition of amino-acid incorporation.

A-22,765 also inhibited the incorporation of ^{14}C-uracil (Fig. 3), although in this case inhibition did not set in until cell division had begun in the control culture; in contrast to the inhibition of amino-acid incorporation, no inhibition of uracil incorporation was observed during the lag phase.

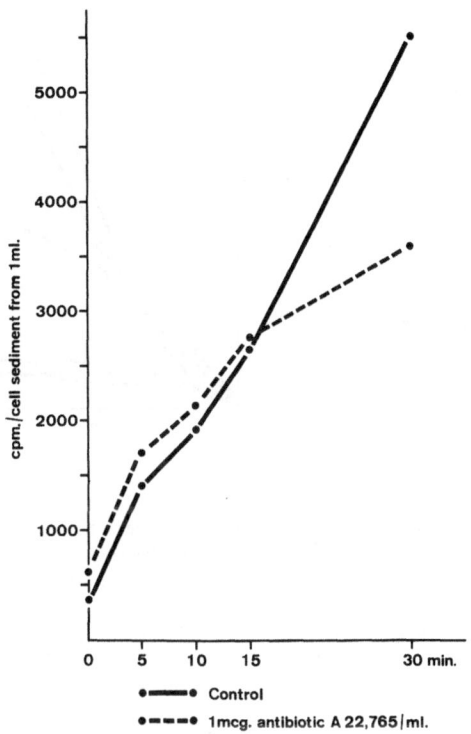

Fig. 3. Influence exerted by antibiotic A-22,765 on the incorporation of ^{14}C-uracil into intact cells of *B. subtilis* ATCC 6633

b) Influence of Sideromycins on Poly-U-Directed Protein Synthesis in a Cell-Free System of *S. aureus*

The results outlined above prompted us to undertake a further series of experiments designed to shed light on the influence exerted by sideromycin antibiotics upon protein synthesis in a cell-free system of *S. aureus* SG 511 (KNÜSEL *et al.*, 1969). The methods adopted in these experiments were basically the same as described by MAO (1967). From Fig. 4 it can be seen that danomycin (TSUKIURA *et al.*, 1964) inhibits the poly-U-directed incorporation of ^{14}C-phenylalanine in the S-30 fraction of sideromycin-sensitive cells of *S. aureus* SG 511; here, too,

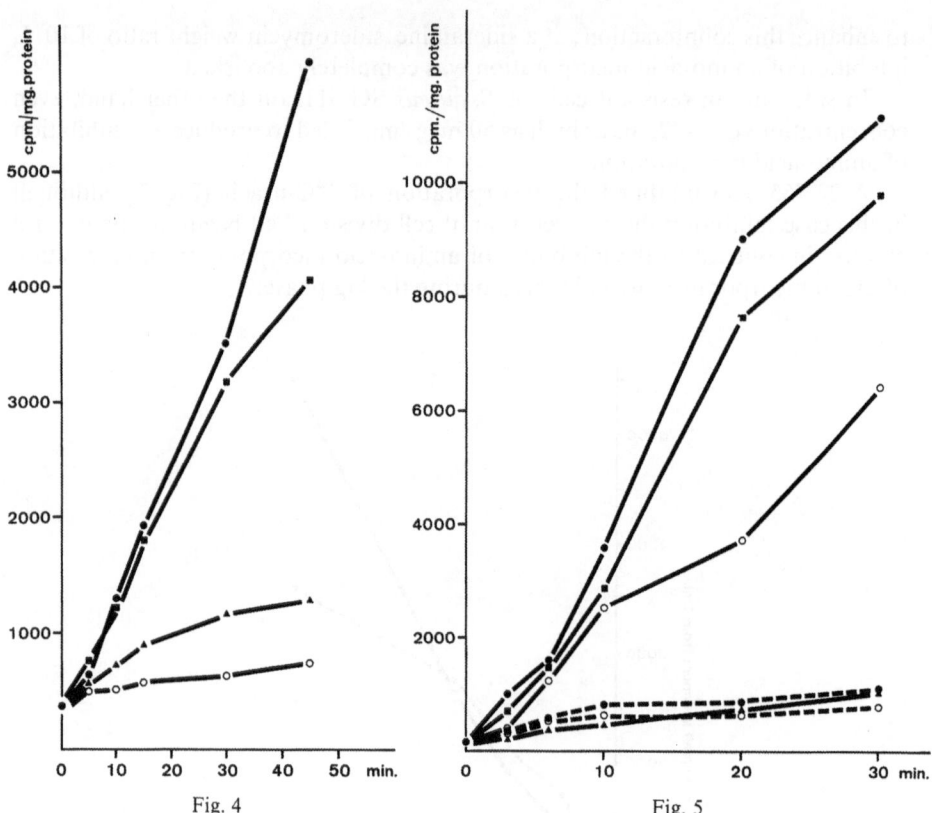

Fig. 4. Inhibition of poly U-directed polyphenylalanine synthesis by danomycin; S-30 fractions of danomycin-sensitive cells of *S. aureus* SG 511. Symbols: ●——● controls, without danomycin; ■——■ danomycin, 0.01 mcg/ml; ▲——▲ danomycin, 1 mcg/ml; ○——○ danomycin, 100 mcg/ml

Fig. 5. Inhibition of poly U-directed polyphenylalanine synthesis by the antibiotic A-22,765; S-30 fractions of A-22,765-resistant cells of *S. aureus* SG 511. Symbols: ●——● controls without A-22,765; ■——■ A-22,765, 0.1 mcg/ml; ○——○ A-22,765, 1 mcg/ml; ▲——▲ A-22,765, 10 mcg/ml; ●----● A-22,765, 10 mcg/ml, ferrioxamine B, 10 mcg/ml; ○----○ A-22,765, 10 mcg/ml, ferrioxamine B, 100 mcg/ml

the degree of inhibition was dependent on the concentration employed. In similarly conducted experiments, the same results were also obtained with A-22,765. The latter is probably identical with danomycin; it is at all events impossible to distinguish between the two substances by paper chromatographic analysis (NÜESCH and KNÜSEL, 1967). On the other hand, little is yet known about their chemical structure (KELLER-SCHIERLEIN *et al.*, 1964).

In S-30 fractions prepared from A-22,765-resistant cultures of *S. aureus* SG 511, the incorporation of ^{14}C-phenylalanine was also inhibited by A-22,765 (Fig. 5).

In contrast to the results obtained in experiments with intact cells, however, ferrioxamine B was in this system no longer able to counteract the inhibition

produced by danomycin or A-22,765, regardless of whether the S-30 fractions
had been prepared from sensitive or from resistant cultures. The findings observed
with S-30 fractions from A-22,765-resistant cells in response to the simultaneous
use of A-22,765 and ferrioxamine B are also presented in Fig. 5.

The results which danomycin and A-22,765 yielded do not tally with those
obtained by VORISEK and GRÜNBERGER (1966) with albomycin in a cell-free system
of *E. coli*. For this reason, we extended the scope of our experiments on the
influence exerted by sideromycin antibiotics upon protein synthesis in a cell-free
system of *S. aureus* by also studying the effect of albomycin, albomycin A_1, and
ferrimycin A_1. These three antibiotics differ from A-22,765 and danomycin not only
chemically but also with respect to their biological stability (NÜESCH and KNÜSEL,
1967). The results of these series of experiments are outlined in Fig. 6. Although
in intact cells of *S. aureus* both albomycin as well as ferrimycin A_1 strongly inhibit
the incorporation of amino acids in concentrations as low as 1 mcg./ml., albomycin and albomycin A_1 failed to exert any inhibitory influence on the incorporation
of phenylalanine in S-30 fractions from these *S. aureus* cells; in response to
ferrimycin, the incorporation of phenylalanine was in this case even strongly

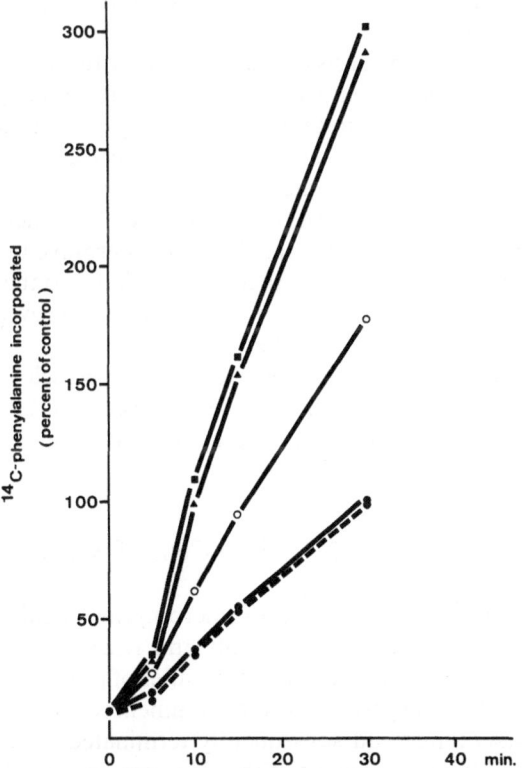

Fig. 6. Influence of ferrimycin A_1, albomycin, and albomycin A_1 on poly U-directed polyphenylalanine
synthesis; S-30 fraction of sensitive cells of *S. aureus* SG 511. Symbols: ●——● controls, with
no additions; ○——○ ferrimycin A_1, 1 mcg/ml; ▲——▲ ferrimycin A_1, 10 mcg/ml; ■——■ ferrimycin A_1, 100 mcg/ml; ●----● albomycin (albomycin A_1), 100 mcg/ml

enhanced, and in the system as employed by us neither ferrioxamine B nor ferrichrysin (Zähner *et al.*, 1962) displayed any antagonistic effect against the activity of these three antibiotics.

In contrast to the results described in the preceding section, which show that all the sideromycins studied clearly and consistently inhibit the incorporation of amino acids into sensitive intact cells of *S. aureus* SG 511 and that this inhibition can be completely abolished by the simultaneous addition of ferrioxamine, the experiments on protein synthesis in cell-free systems yielded the following four groups of results:

1. Concentration-dependent inhibition by danomycin and A-22,765 only.

2. No influence whatsoever exerted by albomycin.

3. Increase in the incorporation of ^{14}C-phenylalanine in response to ferrimycin A_1.

4. In no instance was ferrioxamine B capable of abolishing the effect produced by the sideromycin antibiotics.

In view of the appreciable differences in these results, it is still not fully clear whether the primary effect of all sideromycins does in fact consist in the inhibition of protein synthesis; the results obtained with S-30 fractions would if anything seem to suggest that the various sideromycins might also have differing mechanisms of action.

Since—particularly on the basis of their behaviour towards the sideramines— the sideromycins have been characterised as a homogenous group, and since, on the other hand, the sideramines no longer display de-inhibitory activity when added to a cell-free system, we considered it quite conceivable that the sideramines might act, not by cancelling out the effect of the sideromycins by competing with the latter at their site of action, but by antagonising their permeation into the interior of the cell. If this were so, then the sideromycins would constitute a homogeneous group only insofar as their permeation into the cell is susceptible to inhibition by sideramines; in their mode of action, however, they might well be completely different from one another.

To test this hypothesis, the experiments to be described below were undertaken, in which we studied the permeation of sideromycin antibiotics into cells of *S. aureus* SG 511 and the influence exerted by ferrioxamine B on this permeation process.

c) Permeation of A-22,765 into Cells of *S. aureus* SG 511

In all our studies on the permeation of sideromycins or sideramines, exponentially growing cultures were centrifuged off and resuspended in fresh medium in one tenth of the original volume. After 5 minutes of pre-incubation in a shaking water-bath at 37° C, the experiments were commenced by adding the respective sideromycin or sideramine and subsequently terminated by rapidly cooling the cultures. Quantitative determination of the sideromycins or sideramines was perfomed using a procedure based on measurement of turbidity (Oberzill, 1967). A detailed description of the methods employed has already been published (Zimmermann, 1970).

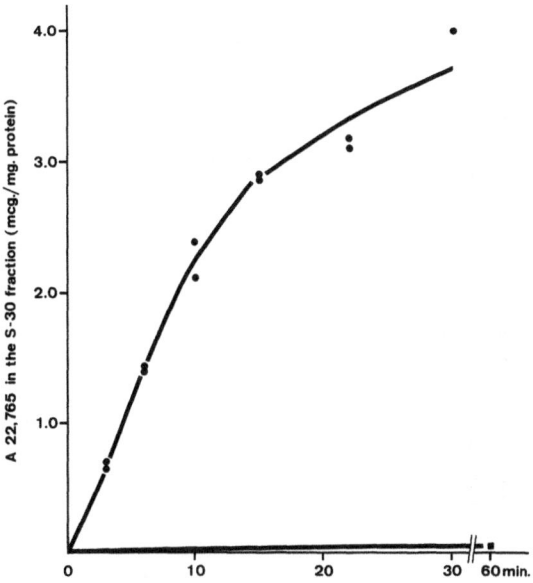

Fig. 7. Permeation of A-22,765 in cells of *S. aureus* SG 511. ●———● A-22,765-sensitive cells; ■———■
A-22,765-resistant cells

The permeation of sensitive cells of *S. aureus* SG 511 by A-22,765 in relation
to the time elapsing is illustrated in Fig. 7, from which it can also be seen that,
in similarly performed experiments with resistant cells, only very little A-22,765
could be detected in the S-30 fractions. After an incubation period lasting
60 minutes, the quantity measured was equivalent to only about 1% of that
found after only 30 minutes in the case of sensitive cells (ZIMMERMANN and
KNÜSEL, 1969). Following paper chromatography of the S-30 fraction, a bio-
autogram, in which a comparison was made with an A-22,765 solution chromatographed in parallel revealed that the only inhibitory substance present within
the cells was indeed A-22,765.

In further experiments carried out along the same lines we determined, not
the amount of sideromycin contained in the S-30 fraction, but the quantity still
present in the medium in which A-22,765-sensitive or A-22,765-resistant cells
had been cultured. Fig. 8 shows that in the case of A-22,765-resistant cultures
this quantity was practically the same as at the start of incubation, whereas
in sensitive cultures the concentration of A-22,765 detectable outside the cells
rapidly decreased. The uptake of A-22,765 by sensitive cells of *S. aureus* SG 511
cannot occur simply by a process of diffusion, because the amount within the
cells very quickly exceeds that present in the medium. Consequently, *S. aureus*
SG 511 must be capable of taking up A-22,765 from the surrounding medium
by an active transport mechanism, i.e. against a concentration gradient. The
existence of such an energy-dependent active mechanism for the transport of
A-22,765 was confirmed by the results of further experiments in which incubation

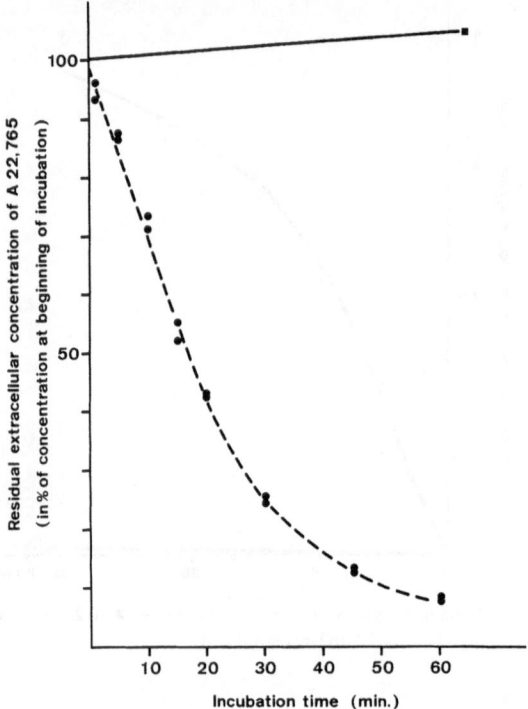

Fig. 8. A-22,765 permeation into cells of *S. aureus* SG 511; residual extracellular concentration of A-22,765. The values are expressed in percentages of the concentration of A-22,765 present at the start of the experiment. ●-----● A-22,765-sensitive cells; ■——■ A-22,765-resistant cells

was effected at 2° C or in a buffer solution in which, though having no source of energy, the cells are nevertheless still able to survive (Tables 1 and 2). In both instances, practically all of the A-22,765 remained in the medium, only a very small decrease being detectable in the number of surviving cells after an incubation period of 60 minutes. In contrast to these findings, in the control cultures which had been treated in the usual manner, 90% or more of the A-22,765 had been taken up into the cells, while at the same time the number of cells still capable of forming colonies dropped to 0.1–0.5% of the initial value.

The permeation experiments undertaken with *S. aureus* SG 511 and the sidero-mycin antibiotic A-22,765 were repeated using other strains, and were also carried out with *S. aureus* SG 511 and albomycin.

Sensitive and resistant cells of *B. subtilis* ATCC 6633 behaved similarly to corresponding cells of *S. aureus* with respect to permeation by A-22,765. In cultures of *E. coli* B, against which the antibiotic A-22,765 generally proves to be inactive, the entire quantity of sideromycin employed remained in the medium. Since, in a cell-free system of this micro-organism, poly U-directed protein synthesis is inhibited by A-22,765 in the same way as in the case of *S. aureus* SG 511, it would appear reasonable to assume that the resistance which *E. coli* displays towards A-22,765 can be ascribed to the fact that it is impermeable to this

Table 1. *Permeation of A-22,765 into A-22,765-sensitive cells of S. aureus SG 511 during incubation at 37° and at 2° C*

Culture No.	Residual extracellular concentration of A-22,765		Number of surviving cells after a 60-min. incubation period		
	mcg./ml.	in % of quantity present at start of experiment	before start of experiment cell count/ml.	cell count/ml.	in % of controls Nos. 5 and 10
1	0.58	5.8	3.2×10^9	1.2×10^6	0.02
2	0.62	6.2	(mean value	1.5×10^6	0.03
3	0.60	6.0	for all	1.3×10^6	0.03
4	0.60	6.0	10 cultures)	2.6×10^6	0.05
5	—	—		5.0×10^9	100
Mean value Nos. 1–4	0.60	6.0		1.7×10^6	0.03
6	9.9	99		3.0×10^9	83
7	9.8	98		3.8×10^9	105
8	9.7	97		2.8×10^9	78
9	9.9	99		3.1×10^9	86
10	—	—		3.6×10^9	100
Mean value Nos. 6–9	9.8	98		3.2×10^9	89

Incubation temperature of Cultures Nos. 1–5: 37° C.
Incubation temperature of Cultures Nos. 6–10: 2° C.
The control cultures were Cultures Nos. 5 and 10.
Concentration of A-22,765 at the start of the experiment: 10 mcg./ml.
Duration of incubation: 60 min.

antibiotic. Experiments in which we studied the permeation of albomycin into cells of *S. aureus* SG 511 fully confirmed the findings that had already been obtained previously with the same micro-organism and A-22,765, i.e.: uptake of the antibiotic into the interior of the cell against a concentration gradient, and virtually complete impermeability in the case of resistant cells.

d) Permeation of Ferrioxamine B into Cells of *S. aureus* SG 511

Curiously enough, ferrioxamine B, which antagonises the antibiotic effect of A-22,765, is hardly taken up at all by cells of *S. aureus* SG 511. Following addition of ferrioxamine B to the medium in a concentration of 100 mcg./ml, less than 0.2 mcg. ferrioxamine B per mg. protein could be detected in the S-30 fraction of each of 4 parallel cultures after an incubation period of 60 minutes. The theoretical possibility that the sideramine molecule may become inactivated inside the cell was excluded by determining the quantity still present in the medium. It was found that, in the case both of A-22,765-sensitive and of A-22,765-resistant cultures, practically all the ferrioxamine B remained outside the cells (ZIMMERMANN, 1970).

Table 2. *Permeation of A-22,765 into sensitive resting cells of S. aureus SG 511*

| Culture No. | Residual extracellular concentration of A-22,765 | | Number of surviving cells after a 60-min. incubation period | | |
	mcg./ml.	in % of quantity present at start of experiment	before start of experiment cell count/ml.	cell count/ml.	in % of controls Nos. 5 and 10
1	1.0	10	2.5×10^9	3.6×10^5	0.01
2	0.9	9	(mean value	5.2×10^5	0.02
3	1.0	10	for cultures	4.9×10^5	0.02
4	1.0	10	Nos. 1–5)	6.3×10^5	0.02
5	—	—		3.0×10^9	100
mean value Nos. 1–4	1.0	10		5.0×10^5	0.02
6	10.4	104	2.1×10^9	1.8×10^9	82
7	10.3	103	(mean value	1.8×10^9	82
8	10.5	105	for cultures	1.3×10^9	59
9	10.5	105	Nos. 6–10	1.5×10^9	68
10	—	—		2.2×10^9	100
mean value Nos. 6–9	10.4	104		1.6×10^9	73

Cultures Nos. 1–4: proliferating cells.
Cultures Nos. 6–9: resting cells.
Cultures Nos. 5 and 10: corresponding control cultures.
Concentration of A-22,765 at the start of the experiment: 10 mcg./ml.
Duration of incubation: 60 min.

e) Antagonism between Sideromycins and Sideramines

The experiments already described showed that sideromycins are very rapidly accumulated in the interior of sensitive cells; in contrast to this finding, the amount of ferrioxamine B that had permeated into the cells over the same period of time was extremely small. Since, on the other hand, Tables 1 and 2 clearly reveal that A-22,765 has to penetrate into the cell in order to produce an inhibitory action, it was obviously of interest to investigate the influence which ferrioxamine B exerts on permeation by A-22,765. All the results obtained thus far had in fact indicated that the sideramines do not antagonise the primary effect displayed by the sideromycins, but that they are capable of influencing permeation by sideromycins into the interior of the cell. Cultures of *S. aureus* SG 511 were therefore simultaneously incubated with unvarying quantities of A-22,765 and increasing concentrations of ferrioxamine B. The concentration of the sideromycin present in the S-30 fraction was then determined after the small quantity of ferrioxamine B which had also permeated had been separated off by paper chromatography. The A-22,765 remained almost entirely at the spot where it had been applied, whereas the ferrioxamine B ran on down the paper.

From Fig. 9 it is evident that ferrioxamine B inhibits the permeation of A-22,765 into sensitive cells of *S. aureus* and that the degree of inhibition increases as the concentration rises. Table 3 shows the results of a similar experiment in which the quantity of sideromycin remaining in the medium was determined, as well as the number of surviving micro-organisms present after a 60-minute incubation period.

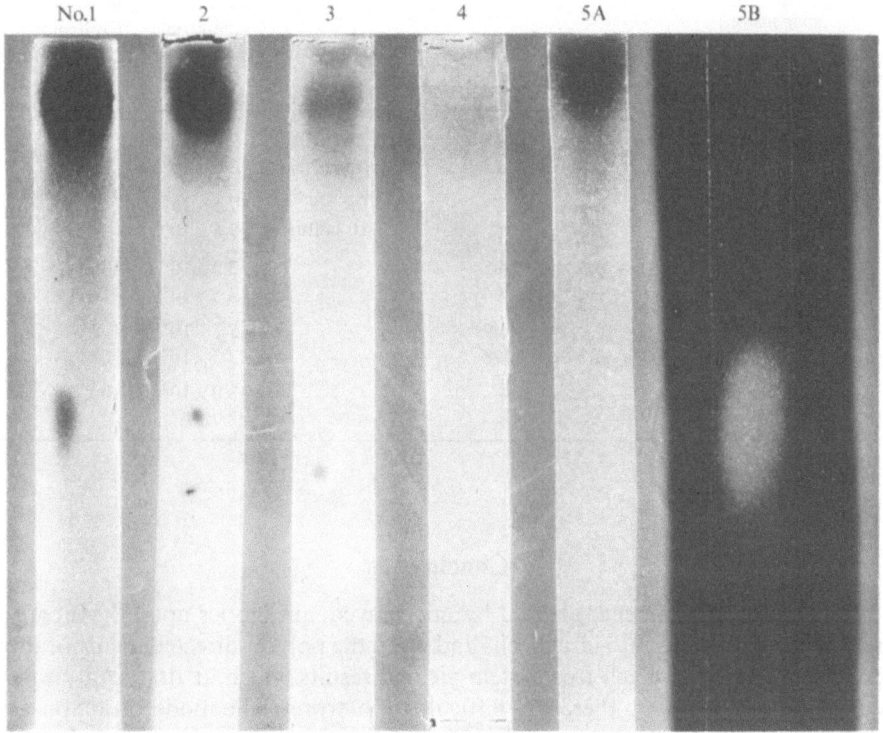

Fig. 9. Inhibition of A-22,765 permeation by ferrioxamine B

Bio-autogram of the chromatographed S-30 fractions; A-22,765-sensitive cultures of *S. aureus* SG 511; determination of A-22,765.

No.	Quantity of siderochrome present at start of incubation (mcg./ml. cell suspension)	
	A-22,765	Ferrioxamine B
1	10	0
2	10	1
3	10	10
4	10	100

No. 5 = control, containing a mixture of 10 mcg. A-22,765 and 50 mcg. ferrioxamine B/ml. BHI. 5A: determination of A-22,765; 5B: determination of ferrioxamine B

Table 3. *Inhibition of A-22,765 permeation by ferrioxamine B in A-22,765-sensitive cultures of S. aureus SG 511. Determination of the quantity of A-22,765 still present in the medium after an incubation period of 60 minutes*

Cul-ture No.	Quantity of sidero-chrome present at start of experiment (mcg./ml. cell suspension)		Residual extra-cellular concentration of A-22,765		Number of surviving cells		
			mcg./ml.	in % of quantity present at start of experiment	before incubation cell count/ml.	after a 60-min. incubation period	
	A-22,765	Ferri-oxamine B				cell count/ml.	in % of Controls Nos. 9 and 10
1	10	0	1.7	17	2.3×10^9	2.6×10^6	0.09
2	10	0	1.5	15	(mean value	3.0×10^6	0.10
3	10	1	5.3	53	for all	7.1×10^6	0.24
4	10	1	5.2	52	10 cultures)	9.6×10^6	0.32
5	10	10	9.5	95		3.0×10^9	100
6	10	10	9.4	94		3.3×10^9	110
7	10	100	9.9	99		3.1×10^9	103
8	10	100	9.8	98		2.5×10^9	83
9	—	—	—	—		3.0×10^9	100
10	—	—	—	—			

Conclusions

Studies on the influence exerted by sideromycin antibiotics upon the incorporation of amino acids into intact cells and upon the poly-U-directed incorporation of phenylalanine in a cell-free system yielded results which at first sight appear contradictory and were therefore difficult to interpret. The mode of action and the site of action of the sideromycins have not yet been determined in full detail. Conclusive proof that danomycin and A-22,765 primarily inhibit protein synthesis has yet to be furnished: However, the working hypothesis established on the basis of the present results is strongly supported by the findings obtained in the experiments on permeation by sideromycins and sideramines and on the antagonism existing between these two classes of compounds. Reliable evidence is now available to substantiate the following statements:

1. The antibiotic A-22,765 is very rapidly taken up by sideromycin-sensitive cells of *S. aureus,* in which it accumulates intracellularly against a concentration gradient. Experiments with A-22,765 and *B. subtilis,* as well as with albomycin and *E. coli,* yielded identical findings.

2. In sideromycin-resistant mutants this active transport mechanism is either absent or impaired.

3. In contrast to A-22,765, ferrioxamine B is taken up into the interior of the cell to only a very small extent. When A-22,765 and ferrioxamine B are simultaneously present in the medium, the sideramine inhibits permeation by the sideromycin antibiotic.

No answer has yet been found to the question why ferrioxamine B, though capable of inhibiting permeation by A-22,765, is itself taken up into the cells to only a very limited degree. That a clear and unequivocal answer to this question cannot at present be given is understandable in view of the rudimentary nature of our existing knowledge of active transport mechanisms. PARDEE (1968) has described a model in which the substrate to be transported is, in an initial step, specifically identified and bound by a corresponding structural element in the cell membrane, the substrate then being transported from the outer to the inner surface of the membrane. Here, in an energy-consuming process, the substrate becomes detached, whereupon the system reverts to its original status. Given such a mechanism, it is conceivable that, by virtue of the sideramine denominator common to them both (cf. Figs. 10 and 11), the structurally related sideromycins and sideramines are identified and bound by the cell membrane, but that the subsequent step required to enable the substrate to penetrate into the cell interior is dependent on that portion of the A-22,765 molecule which is of non-sideramine structure.

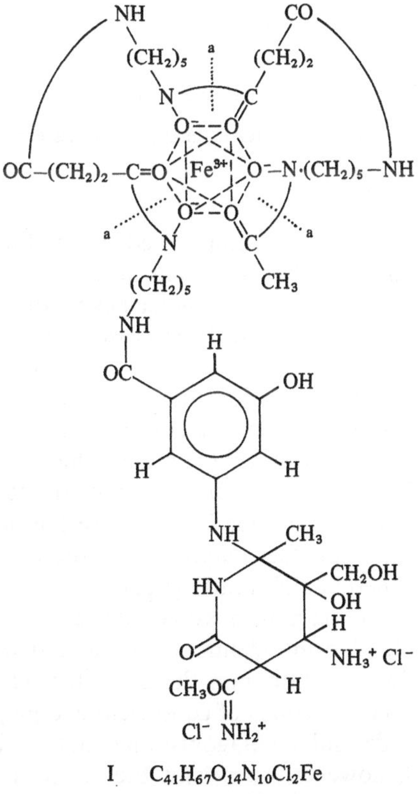

I $C_{41}H_{67}O_{14}N_{10}Cl_2Fe$

Fig. 10. Structure of ferrimycin A_1

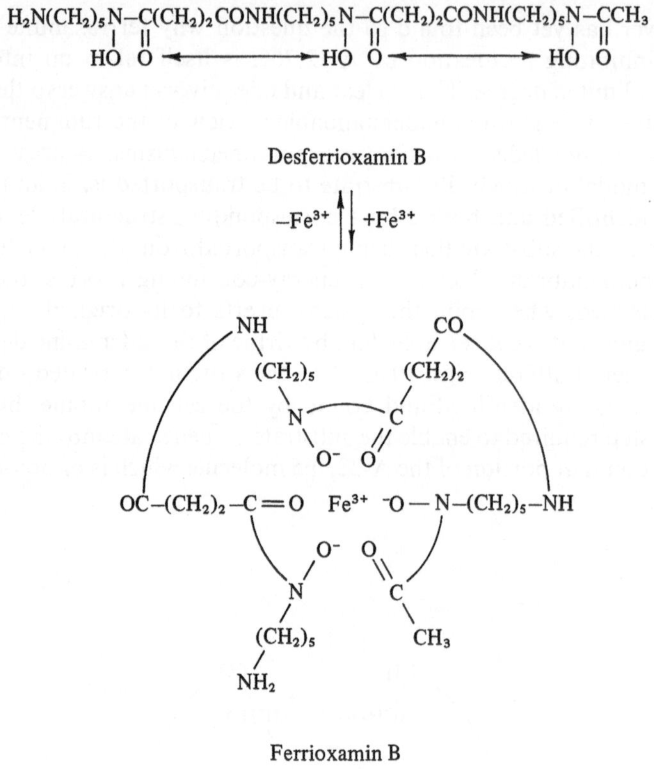

$$H_2N(CH_2)_5\,N-C(CH_2)_2\,CONH(CH_2)_5\,N-C(CH_2)_2CONH(CH_2)_5N-CCH_3$$

$$\begin{array}{ccc} | \;\; || & | \;\; || & | \;\; || \\ HO \;\; O \longleftarrow\!\!\!\longrightarrow HO \;\; O \longleftarrow\!\!\!\longrightarrow & HO \;\; O \end{array}$$

Desferrioxamin B

$$-Fe^{3+}\;\Big\updownarrow\;+Fe^{3+}$$

Ferrioxamin B

Fig. 11. Structure of ferrioxamine B and desferrioxamine B

The whole problem is further complicated by the fact that the antibotic A-22,765 is not a substance of absolute chemical purity and also that the spatial configuration of the two siderochrome molecules involved—a factor of cardinal importance in connection with all such processes—is not known. The ferri ion is in principle able to produce eight octahedral, diastereomeric complexes with a thread-shaped trihydroxamic acid molecule, and in addition each of these complexes may occur in the form of the corresponding enantiomeric compound (BICKEL et al., 1960). In the case of ferrioxamine B the number of possible configurations is reduced, because the chains which bind the hydroxamate residues consist in each instance of only nine links. These are not sufficient to connect two hydroxamate groups together diagonally in the octahedron; consequently, three of the diastereomeric configurations can be disregarded. It is completely unknown which of the remaining possible configurations do indeed occur in nature and are capable of being identified by those structures of sensitive cells that are responsible for transport into the cell interior; it is even possible that there exist several spatial structures of equivalent energy potential.

The sideromycin/sideramine antagonism has not yet been clarified in detail. The results obtained, however, do permit the conclusion that ferrioxamine B inhibits the activity of A-22,765 by reducing sideromycin permeation; but this has no bearing on the antibiotic effect of A-22,765 as such, because it is very probable that the latter effect is not exerted until the non-sideramine portion of the compound has reached the interior of the cell. Since this non-sideramine portion varies from one sideromycin to another, it may well be that these antibiotics have differing mechanisms of action, as is also suggested by the finding that several of the sideromycins—A-22,765, danomycin, ferrimycin, and albomycin—differ with regard to their effects on protein synthesis in a cell-free system. The sideromycins thus would constitute a distinctive group only insofar as their uptake into sensitive cells can be inhibited by sideramines. The fact that the activity of the sideromycins, on the one hand, and sideromycin/sideramine antagonism, on the other, are mutually independent would also explain why in cell-free systems ferrioxamine B is no longer able to block the action of A-22,765.

References

BICKEL, H., G. E. HALL, W. KELLER-SCHIERLEIN, V. PRELOG, E. VISCHER u. A. WETTSTEIN: Stoffwechselprodukte von Actinomyceten. 27. Mitt. Über die Konstitution von FerrioxaminB. Helv. Chim. Acta **43**, 2129 (1960).

CRUEGER, W.: Untersuchungen zur Sideramin-Biogenese. Diss. Universität Tübingen (1968).

KELLER-SCHIERLEIN, W., V. PRELOG u. H. ZÄHNER: Siderochrome. Fortschr. Chem. Org. Naturstoffe **22**, 279 (1964).

KNÜSEL, F., J. NÜESCH, M. SCHERRER u. B. SCHIESS: Der Einfluß von Siderochromen auf die Inkorporation niedermolekularer Substanzen in Ganzzellen von Bakterien. Pathol. Microbiol. **30**, 900 (1967).

KNÜSEL, F., B. SCHIESS, and W. ZIMMERMANN: The influence exerted by sideromycins on poly-U-directed incorporation of phenylalanine in the S-30 fraction of Staphylococcus aureus. Arch. Mikrobiol. **68**, 99 (1969).

MAO, J.-H.: Protein synthesis in a cell-free extract from Staphylococcus aureus. J. Bacteriol. **94**, 80 (1967).

MÜLLER, A.: Siderochrome aus Mikroorganismen. Diss. Universität Tübingen (1968).

NEILANDS, J. B.: Hydroxamic acids in nature. Science **156**, 1443 (1967).

NÜESCH, J., and F. KNÜSEL: Sideromycins. In: Antibiotics, vol. I. Berlin-Heidelberg-New York: Springer 1967.

OBERZILL: Mikrobiologische Analytik. Nürnberg: H. Carl 1967.

PADMANABAN, G., S. MUTHU KRISHNAN, K. N. SUBRAMANIAN, and P. S. SARMA: The in vivo iron donor for haem synthesis in Neurospora crassa. Indian J. Biochem. **5**, 153 (1968).

PARDEE, A. B.: Membrane transport proteins. Science **162**, 632 (1968).

SNOW, G. A.: Mycobactins: Iron chelating growth factors from mycobacteria. Bacteriol. Rev. **34**, 99 (1970).

TSUKIURA, H., M. OKANISHI, T. OHMORI, H. KOSHIYAMA, T. MIYAKI, H. KITAZIMA, and H. KAWAGUCHI: Danomycin, a new antibiotic. J. Antibiotics (Tokyo), Ser. A. **17**, 39 (1964).

VORISEK, H., and D. GRÜNBERGER: A study on the mechanism of action of albomycin (grisein) with regard to the origin of resistance. IX. International Congress for Microbiology, Moscow Abstracts of Papers 19 (1966)

ZÄHNER, H., E. BACHMANN, R. HÜTTER u. J. NÜESCH: Sideramine, eisenhaltige Wachstumsfaktoren aus Mikroorganismen. Pathol. Microbiol. **25**, 708 (1962).

ZIMMERMANN, W.: Sideromycin-Resistenz und Sideromycin/Sideramin-Antagonismus bei Staphylococcus aureus. Diss. Universität Zürich (1970).

ZIMMERMANN, W., and F. KNÜSEL: Permeability of Staphylococcus aureus to the sideromycin antibiotic A-22,765. Arch. Mikrobiol. **68**, 107 (1969).

Sulfonamides and Sulfones

Nitya Anand

The discovery of the antibacterial activity of sulfonamides[1] in 1930's marked the beginning of the era of modern antibacterial drug research. Subsequent observation of variations in the activities of these compounds by structural changes brought into sharp focus the power of molecular modification in drug development. The elucidation of the relationship between sulfanilamide and p-aminobenzoic acid provided one long sought-after mechanistic basis for a biochemical approach to chemotherapy. Much of the subsequent work in chemotherapy, and drug research in general, has been based on consideration of these facts.

The availability of a variety of sulfonamides with widely differing absorption and excretion rates has greatly increased their value in therapeutics. This, coupled with their ease of administration, wide spectrum of antimicrobial activity, non-interference with host defence mechanisms and relative freedom from the problem of superinfection, are responsible for their wide use in clinical practice even four decades after their introduction. The use of sulfonamides and sulfones now extends from the treatment of acute and chronic Gram-positive and Gram-negative bacterial infections, through leprosy, malaria, trachoma, nocardiosis, coccidiosis to toxoplasmosis.

Development of Sulfonamides and Sulfones

The synthesis of Prontosil (*1*) by MIETZSCH and KLARER (1934), the epoch-making report of its marked antibacterial activity in infected mice by DOMAGK (1935), followed by the suggestion of TRÈFOUËL et al. (1935) that the activity may be due to sulfanilamide (*2*), formed *in vivo*, and finally the isolation of *2* from the urine of patients under treatment with *1* by FULLER (1937) started intense activity in this field. Numerous derivatives of sulfanilamide were soon synthesized to improve upon its activity and tested against various bacterial, protozoal and viral infections. Sulfapyridine (MAY and BAKER 693) (*3*) reported

(1) Prontosil   (2) Sulfanilamide

---

[1] The general term sulfonamides has been used for derivatives of p-aminobenzenesulfonamide (sulfanilamide) while specific compounds have been described as substituted sulfanilamides. Similarly, the term sulfone has been used for all derivatives of 4,4'-diaminodiphenylsulfone (DDS) while specific compounds have been described as substituted diaminodiphenylsulfones.

in 1938, was one of the earliest new sulfonamides[2] to be used in clinical practice for treatment of pneumonia and remained the drug of choice till it was replaced by sulfathiazole (*4*), which possessed a higher therapeutic index. Sulfathiazole

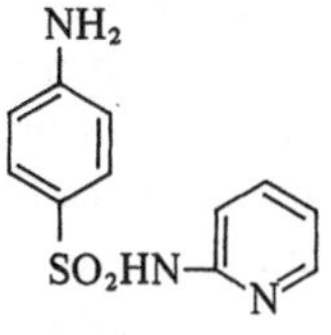

(3) Sulfapyridine 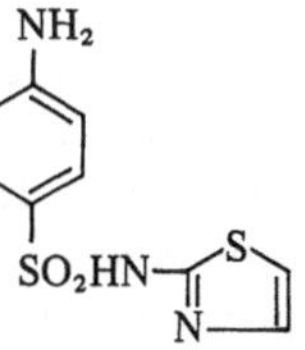   (4) Sulfathiazole

was in turn replaced by sulfadiazine (*5*), which has retained a preeminent position among the sulfonamides ever since. Two methylated derivatives of sulfadiazine, sulfamerazine (*6*) and sulfamethazine (*7*) were introduced into therapy later.

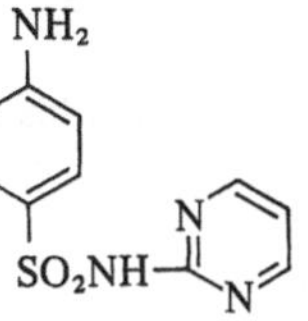

(5) Sulfadiazine 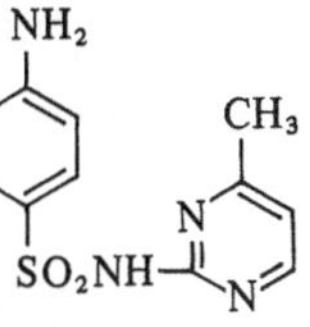   (6) Sulfamerazine

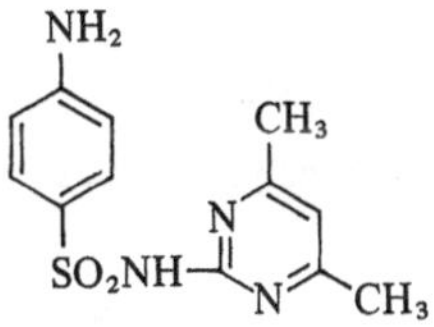

(7) Sulfamethazine

Two other developments took place during this period and had marked effect on future events in chemotherapy. 1. The standardization of a simple method for the quantitative determination of sulfonamides in body tissues and fluids (MARSHAL, 1937; BRATTON and MARSHAL, 1939) permitted a more precise determination of absorption, excretion, disposition and conjugation of these drugs, thus providing a rational basis for the formulation of proper dosage regimens. 2. FILDES' (1940) theory of metabolite antagonism, to explain the observation of WOODS (1940) of the competitive reversal of sulfonamide action by *p*-aminobenzoic acid (PAB), was the second and more important development. While competitive blocking of PAB utilization in bacteria leads to folic acid deficiency, animals cannot utilize PAB, need preformed folic acid, and are thus not affected. Even though Ehrlich's concept of the relationship between the affinity of a dye for a parasite and its antimicrobial activity proved irrelevant to the discovery of sulfonamides as antibacterials, sulfonamides did prove to be the "magic bullet"

---

[2] For references to compounds *3–14* see NORTHEY, 1948.

of Ehrlich attacking the parasite selectively! The concept of metabolite antagonism has been the basis of numerous and intensive studies in the field of chemotherapy.

Meanwhile the synthesis and study of sulfonamides continued with unabated vigor. About twenty-five sulfonamides are now used in clinical practice. They vary widely in their absorption, distribution and excretion. Some remain largely unabsorbed in the gastrointestinal tract after oral administration and produce changes only in the local bacterial flora and, even when used in high doses, their toxic effects are only minor. Important members of this class are sulfaguanidine (*8*), succinylsulfathiazole (*9*) and phthalylsulfathiazole (*10*). These drugs are used extensively for pre-surgical sterilization of the gut and to a lesser extent for the treatment of bacillary dysentery.

(8) Sulfaguanidine   (9) Succinylsulfathiazole

(10) Phthalylsulfathiazole

(11) Sulfisomidine   (12) Sulfamethizole

(13) Sulfisoxazole   (14) Sulfacetamide

Another group is characterized by rapid absorption, high solubility and rapid excretion, mainly in the unaltered form. This includes sulfisomidine (*11*), sulfamethizole (*12*), sulfisoxazole (*13*) and sulfacetamide (*14*). These sulfonamides, on account of their high solubility in urine, do not pose any serious danger of crystalluria. They are, therefore, widely used in the treatment of urinary tract infections. Sulfacetamide sodium salt solution, being non-irritant, is also widely used for treatment of local infections of the eye.

A major advance in sulfonamide therapy took place when it was observed that some sulfonamides were rapidly absorbed but slowly excreted resulting in maintenance of adequate blood level for almost 24 hrs after a single oral dose. Sulfamethoxypyridazine (*15*) was the first such long-acting[3] sulfonamide (NICHOLS et al., 1956), having a half-life[4] of 37 hrs. Sulfadimethoxine (*16*) (FUST and BOHNI, 1959), sulfaphenazole (*17*) (TRIPOD et al., 1960), sulfamethoxazole (*18*) (FUST and BOHNI, 1962), sulfamoxole (*19*) (DEININGER and GUTBROD, 1960) and sulformethoxine (*20*) (REBER et al., 1964) have since been added. Because of the long half-life of these sulfonamides, they require less frequent administration, and are thus particularly useful for chronic conditions and for prophylaxis.

(15) Sulfamethoxypyridazine   (16) Sulfadimethoxine

(17) Sulfaphenazole   (18) Sulfamethoxazole

(19) Sulfamoxole   (20) Sulformethoxine

---

[3] Long-acting sulfonamides should be differentiated from the depot form. In the latter case the drug is formulated in such a way that it tends to remain at the site of administration, from where it is gradually released into the blood or tissues. The total quantity of the drug is thus not made available immediately, in contrast to the long-acting preparations.

[4] Half-life is the the period of time needed for a given blood level of the drug to be reduced by half.

These studies led to a proper appraisal of pharmacokinetic concepts in sulfonamide therapy. It was recognized that even some of the earlier sulfonamides such
as sulfamerazine and sulfadiazine were long-acting.

A further development was the use of two or more sulfonamides in combination, which made it possible to lower the dosage of the individual components,
thereby reducing significantly the tendency to crystalluria, the major toxic hazard
of earlier sulfonamides (LEHR, 1945; YOW, 1955). Two of the successful combinations used are a "triple sulfa" consisting of sulfadiazine, sulfamerazine and sulfamethazine and a "di-sulfa", a combination of sulfisomidine and sulfadiazine.

Sulfonamides are also used in combination with other antibacterial agents.
In attempts to synergize the action of sulfonamides and to avoid the development
of resistance to them, the most logical approach seemed to combine them with
those agents which would block the same metabolic pathway as was done by
sulfonamides, but at different sites. The recognition that combinations of sulfonamides and antifolic drugs were synergistic, first recorded by GREENBERG (1949),
laid the basis for the use of dihydrofolate reductase inhibitors, such as pyrimethamine (21) and trimethoprim (22), in combination with sulfonamides in chemotherapy (see HITCHINGS and BURCHALL, 1965). This combination therapy has added
a new dimension to the place of sulfonamides in chemotherapy (see GARROD
et al., 1969).

(21)   Pyrimethamine   (22) Trimethoprim

Another major breakthrough in this field came with the demonstration by
RIST (1939) and FELDMAN et al. (1942) that 4,4'-diaminodiphenylsulfone (DDS)
(23) and promin (24) (disodium 4,4'-diaminodiphenylsulfone-N,N'-didextrosesul-
fonate) could effectively control experimental tuberculosis. Although DDS and
promin proved disappointing in the therapy of human tuberculosis, the interest
aroused in the possibility of treatment of mycobacterial infections with sulfones
led to the demonstration that promin exerted a favourable effect in rat leprosy

(23)   4,4'-Diaminodiphenyl-
         Sulfone (DDS)

(24) Promin

(COWDRY and RUANGSIRI, 1941). This was soon followed by successful treatment of leprosy patients first with promin and later with DDS itself; DDS has since then remained the drug of choice for leprosy (see BROWNE, 1969). The promise shown by acedapsone (N,N'-diacetyl-4,4'-diaminodiphenylsulfone, DADDS) (25) as a repository antimalarial (*vide infra*) led to its successful use in lepromatous leprosy (SHEPARD, 1967; SHEPARD *et al.*, 1968). A number of long-acting sulfonamides, such as sulfamethoxine have also been found useful in the treatment of leprosy (LANGUILLON, 1964; SCHNEIDER and LANGUILLON, 1963).

The antimalarial activity of sulfonamides and sulfones, noticed quite early (HILL and GOODWIN, 1937; COGGESHAL *et al.*, 1941), did not arouse much interest until ARCHIBALD and ROSS (1960), based on their observation of the lower prevalence of malaria in leprosy patients, showed that DDS could clear the blood of trophozoites of both *P. falciparum* and *malariae*. It was soon found that DDS potentiated the action of pyrimethamine and a combination of the two drugs markedly delayed the development of resistance (RAMAKRISHNAN *et al.*, 1962; RAMAKRISHNAN *et al.*, 1963; BASU *et al.*, 1964). A further advance in this therapy took place with the demonstration of the repository effect of DADDS; in the host it is slowly hydrolysed to release DDS (ELSLAGER and WORTH, 1965; GLAZKO *et al.*, 1968). DDS, its repository forms such as 25 and 26 and long-acting sulfonamides in combination with dihydrofolate reductase inhibitors now occupy an important place in the prophylaxis and treatment of falciparum malaria, particularly that caused by chloroquin-resistant strains (see ELSLAGER, 1969; RICHARDS, 1970).

(25) Acedapsone   (26) PSBA

The observation of the activity of sulfonamides (SABIN and WARREN, 1941) and sulfones (BIOCCA, 1943) against experimental toxoplasmosis, and the synergization of this action by pyrimethamine (EYLES, 1956; EYLES and COLEMAN, 1955) has led to the wide use of this combination in human toxoplasmosis (WERNER, 1970).

Sulfonamides have been shown to be effective against Eimeria infection in chickens (LEVINE, 1939, 1940; JOYNER *et al.*, 1963) and are now commonly used, alone or preferably in combination with pyrimethamine (KENDALL and JOYNER, 1958), for the treatment of coccidiosis.

McCALLUM and FINDLAY (1938) showed that the "large virus" of lymphogranuloma venereum in mice was susceptible to sulfonamides. Later, other chlamydiae were also shown to be inhibited by sulfonamides, which led to their successful clinical trial in trachoma (FORSTER and McGIBONY, 1944). Since then, sulfonamides, including long-acting ones, have been largely used in the treatment of trachoma (JAWETZ, 1969; TARIZZO, 1972).

Sulfonamides were also found to have marked *in vivo* activity against *Nocardia asteroides* and are largely used for the treatment of systemic nocardiosis (STRAUSS et al., 1951; CONNAR et al., 1951).

Mode of Action

Sulfonamides are one of the few groups of drugs whose mode of action is known at the enzyme level. Though some of the finer points about the nature of the enzymic antagonism have yet to be settled, the theory that sulfonamides inhibit the enzymes involved in the condensation of 2-amino-4-oxo-6-hydroxy-methyldihydropteridine pyrophosphate (27) with PAB is unlikely to be seriously questioned.

The presence of substances antagonizing the bacteriostatic action of sulfona-mides was observed in peptones by LOCKWOOD (1938), in bacteria by STAMP (1939) and GREEN (1940) and in yeast extract by WOODS (1940); the latter suggested that the antagonism was caused by PAB. This antagonistic action of PAB was confirmed *in vitro* experiments by WOODS (1940) and later *in vivo* by SELBIE (1940) and FINDLAY (1940). PAB was ultimately isolated from yeast extract, as its benzoyl derivative by RUBBO and GILLESPIE (1940) and as the free compound by BLANCHARD (1941). WOODS (1940) suggested that sulfanilamide through its similarity of structure interfered with the utilization of PAB in enzyme systems necessary to the growth of bacteria. Based on these facts, a more general and clear enunciation of the theory of metabolite antagonism to explain the action of chemotherapeutic agents was given by FILDES (1940) in his classical paper "A rational approach to research in chemotherapy".

Further studies showed that inhibition of growth by sulfonamides in simple media can be reversed not only competitively by PAB, but also non-competitively by a number of compounds not related to PAB such as methionine, serine, glycine, adenine, guanine and thymine (see HENRY, 1943). The relationship to purines was uncovered by the finding that sulfonamide-inhibited cultures accumu-lated 4-amino-5-imidazolecarboxamide ribotide (STETTEN and FOX, 1945), a com-pound later shown to be a precursor of purine biosynthesis (SHIVE et al., 1947; GOTS, 1953).

With simultaneous developments taking place in the field of bacterial metabo-lism and nutrition, these isolated facts could be fitted into a pattern. The purifica-tion and determination of the structure of folic acid (ANGIER et al., 1946; MOWAT et al., 1946) revealed the presence of PAB as one integral part. Following this, the suggestion was made that the molecule is formed by the condensation of PAB or *p*-aminobenzoylglutamic acid (PABG) with a pteridine (TSCHESCHE, 1947) and that sulfonamides compete in this condensation. Soon the structure of the active coenzymic form of folic acid as leucovorin (folinic acid, citrovorum factor) was established, and its involvement in biosynthetic steps where 1-carbon units are added was elucidated (see WELCH and NICHOL, 1952; FRIEDKIN, 1963). The amino acids, purines and pyrimidines which are able to replace or spare PAB are precisely those the formation of which needs 1-carbon addition as catalyzed by folic acid.

Direct evidence of inhibition of folic acid synthesis by sulfonamides was soon obtained by studies on bacterial cultures. It was already known that a number of organisms could use PAB and folic acid as alternative essential growth factors (see WOODS, 1954). LAMPEN and JONES (1946, 1947) found that the growth of some strains of S. faecalis, L. arabinosus and L. plantarum in media containing PAB was inhibited competitively by sulfonamides, and folic acid caused a non-competitive type of reversal of this inhibition, suggestive of its being a product of the inhibited reaction. MILLER (1944), and MILLER et al. (1947) using growing cultures of E. coli showed inhibition of folic acid synthesis by sulfonamides. A similar inhibition of folic acid synthesis by sulfonamides and its competitive reversal by PAB was shown in non-growing suspensions of L. plantarum (NIMMO-SMITH et al., 1948), S. aureus, a PAB-requiring mutant of E. coli and its parent wild strain (LASCELLES and WOODS, 1952).

The elucidation of the enzymic synthesis of dihydropteroate and dihydrofolate (Fig. 1) in cell-free extracts by SHIOTA (1959), SHIOTA and DISRAELY (1961) and SHIOTA et al. (1964) using enzymes mainly from L. plantarum and Veillonella, and by BROWN et al. (1961) and WEISMAN and BROWN (1964) working with an enzyme system from E. coli, made it possible to examine the action of sulfonamides at the enzyme level. BROWN (1962) demonstrated that the synthesis of dihydropteroate from PAB is sensitive to inhibition by a number of sulfonamides. BROWN (1962) with the E. coli system and SHIOTA et al. (1964) with the Veillonella system found that the relation between a sulfonamide and PAB remained strictly competitive as long as the two compounds were either added simultaneously or PAB was added first. If the enzyme and sulfonamide were preincubated, the subsequent addition of PAB failed to reverse the inhibition. BROWN showed that the enzyme was not irreversibly inactivated, and suggested the possibility of incorporation of sulfonamide in place of PAB. Using S^{35}-labelled sulfanilic acid in this enzyme system, he obtained evidence of incorporation, but the nature of the products formed was not elucidated. BROWN (1962, 1971) suggested that the product formed by sulfonamide incorporation may be the dihydropteroate analog (34) or its fully aromatic form. Metabolic incorporation of antimetabolite

(34)

is not unknown; however, additional work needs to be carried out with sulfonamides to obtain definitive proof of this incorporation and to show whether the folic acid analogs, thus formed, have inhibitory action. In intact organisms there is no evidence of metabolic incorporation.

BROWN (1962) observed that the enzymic synthesis was many times more sensitive to inhibition by sulfonamides than bacterial growth, suggestive of im-

peded permeability to sulfonamides of the intact organisms as compared to PAB.
The more potent inhibitors of folate biosynthesis were, in general, also better
growth inhibitors. Differences in the response of various organisms to sulfonamides may be due to differences in the quantitative ability of the individually
distinct isoenzymes to produce folic acid from PAB in the presence of sulfonamides
(HOTCHKISS and EVANS, 1960).

This enzyme can also use PABG as substrate to form dihydrofolate directly;
the utilization of both substrates is competitively inhibited by sulfonamides
(SHIOTA et al., 1964; ORTIZ and HOTCHKISS, 1966). On the basis of non-additive
and competitive utilization of these two substrates it appears that it is one single
enzyme which utilizes either PAB or PABG (SHIOTA et al., 1969; ORTIZ, 1970).
SHIOTA et al. (1969) suggested that the enzyme may be allosteric, although competitive antagonism is not normally associated with allosterism.

RICHEY and BROWN (1969) have purified this enzyme from E. coli, virtually
free from pyrophosphatase, and have proposed that it may be called dihydropteroate synthetase. ORTIZ (1970) has questioned the use of this name, because all
preparations of the enzyme can use PABG also as substrate to yield dihydrofolate
directly. PAB, however, is utilized more efficiently.

This enzyme has been shown to be present in a number of bacteria (SHIOTA
et al., 1964; ORTIZ, 1970; JONES and WILLIAMS, 1968), yeasts (JAENICKE and
CHAN, 1960) and plants (MITSUDA and SUZUKI, 1968; IWAI and OKINAKA, 1968).

Thus, sulfonamides, by competing for the enzyme site for PAB, inhibit the
biosynthesis of tetrahydrofolate, which is involved in 1-carbon transferase action.
This would then prevent or slow down the formation of a number of raw materials
of protein, DNA and RNA biosynthesis, thereby affecting a number of synthetic
processes of the organism concurrently.

On consideration of the formal electronic charges on sulfanilamide
(FOERNZLER and MARTIN, 1967) and PAB (PULLMAN and PULLMAN, 1963), determined by HMO, MORIGUCHI and WADA (1968) have proposed that on the enzyme
there are two binding sites, located 6.7–7 A° apart, one specific for 4-NH_2 group,
and the other non-specific where the acidic group of PAB or sulfonamide binds.

Sulfonamides inhibit only growing organisms and their bacteriostatis is preceded by a lag phase. The lag phase before bacteriostatic action sets in can
now be explained as due to stored PAB/folic acid, and its duration is dependent
on the quantity stored.

Animal cells require preformed folic acid and so they are unaffected by sulfonamides. It is not clearly resolved as to how in spite of the presence of folic acid
in blood and tissues, sulfonamides exert their bacteriostatic action. This may
be due to the fact that folic acid in animal tissues normally occurs linked to
polyglutamate conjugates or to proteins, and in this form, it cannot be used
by bacteria. Moreover, the concentration of PAB and other products of the
action of folic acid coenzymes, which are able to reverse the action of sulfonamides, may be too low in tissues to prevent action of sulfonamides when used
in adequate dosage.

The mechanism of action of DDS (and other sulfones) is probably similar
to that of sulfonamides, since the action is antagonised by PAB in mycobacteria
(DONOVICK et al., 1952; BROWNLEE et al., 1948) as in other bacteria (TSUKAMURA,

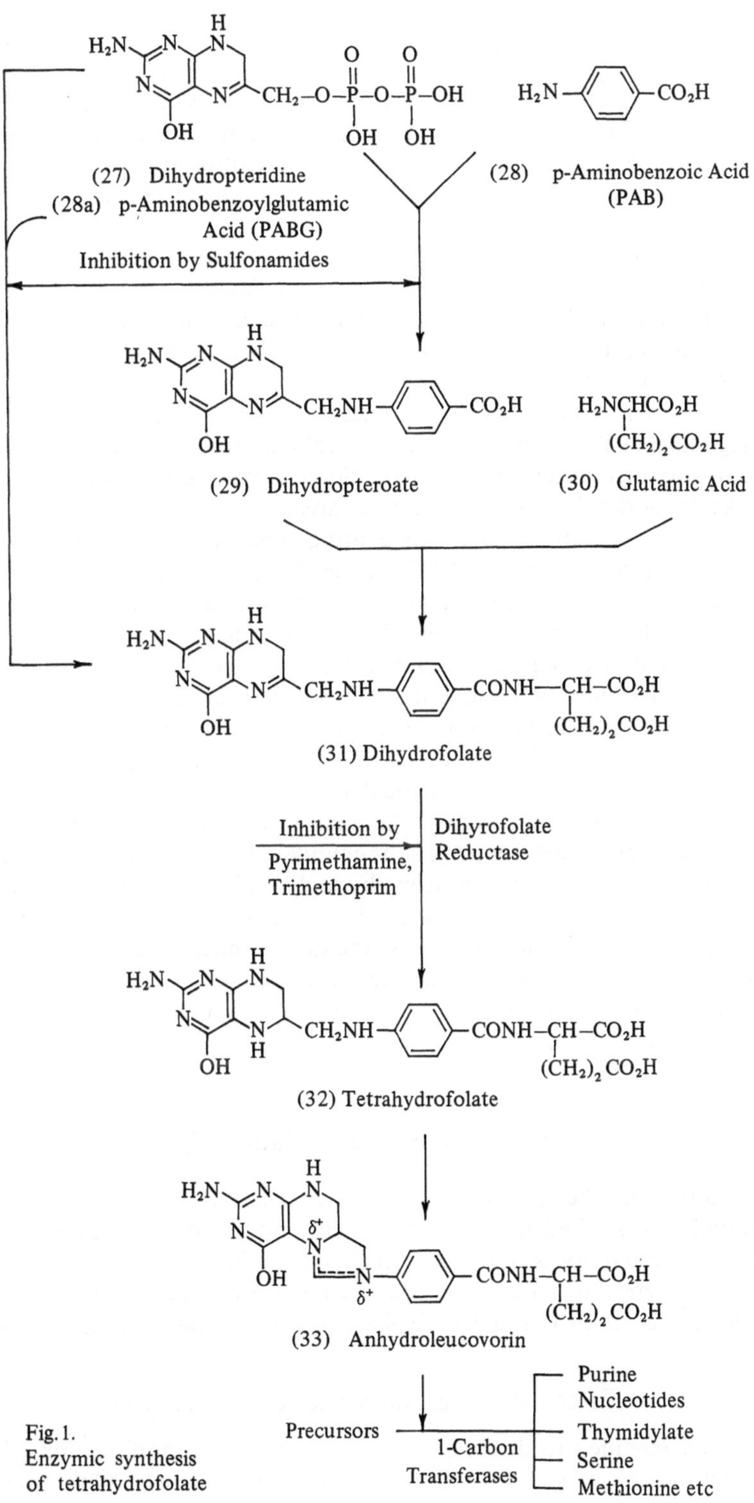

Fig. 1.
Enzymic synthesis
of tetrahydrofolate

(27) Dihydropteridine
(28a) p-Aminobenzoylglutamic
Acid (PABG)
Inhibition by Sulfonamides

(28) p-Aminobenzoic Acid
(PAB)

(29) Dihydropteroate

(30) Glutamic Acid

(31) Dihydrofolate

Inhibition by
Pyrimethamine,
Trimethoprim

Dihyrofolate
Reductase

(32) Tetrahydrofolate

(33) Anhydroleucovorin

Precursors — 1-Carbon Transferases —
Purine Nucleotides
Thymidylate
Serine
Methionine etc

1955). The action of DDS like that of sulfonamides, against *P. gallinaceum* and other malarial parasites is inhibited competitively by PAB and non-competitively by folic acid (BISHOP, 1963; MAIER and RILEY, 1942; SEELER *et al.,* 1943). CENEDELLA and JARRELL (1970) have suggested a new mechanism for the antimalarial action of DDS involving inhibition of glucose utilisation by the intraerythrocytic parasite; this inhibition was shown to be antagonized by raising glucose concentration of the medium.

Similarly, in the case of chlamydia it has been shown that the sulfonamide-sensitive members of this group such as trachoma-inclusion conjunctivitis viruses, have a folic acid metabolism similar to that of bacteria, and the action of sulfonamides is competitively antagonised by PAB (MORGAN, 1948; MOULDER, 1962; KURNOSOVA and LENKEVICH, 1964).

Other Actions. The action of sulfonamides is not, however, restricted only to antagonism of PAB, and they have other weak to strong effects (side effects) not related to their antibacterial activity. Some of these side effects have provided useful leads for developing drugs in other areas.

The clinical acidosis and alkaline urine observed following sulfanilamide administration (SOUTHWORTH, 1937), the carbonic anhydrase inhibiting activity of sulfonamides (MANN and KEILIN, 1940) and the demonstration of high concentration of this enzyme in kidney (DAVENPORT and WILHELMI, 1941) suggested a diuretic potential for sulfonamides. SCHWARTZ (1949) showed that sulfanilamide did indeed have diuretic action, though not of practical significance. Pursuit of this lead proved rewarding and was the beginning of the era of modern diuretics.

That some of the sulfonamides produce hypoglycemia as a side-effect was noticed quite early. The marked hypoglycemic effect of 5-isopropyl-2-sulfanilamido-1,3,4-thiodiazole observed in humans by JANBON *et al.* (1942) led to a search for an oral hypoglycemic in sulfonamides and culminated in the discovery of sulfonyl ureas, which possess strong hypoglycemic activity without much antibacterial action (FRANKE and FUCHS, 1955). This discovery opened a new chapter in the treatment of diabetes. Similarly, the development of probenecid (MILLER, 1952) followed by sulfinpyrazole as uricosuric agents was a result of the observation that some sulfonamides, particularly sulfamyl derivatives of PAB, had penicillin sparing effect by decreasing its renal clearance.

Structure-Activity Relationship

In interpreting the results of SAR studies it is important to ensure that the compounds are studied in the same system and act by the same mechanism, and that their activity is not due to their metabolic products. For the present discussion, therefore, only those compounds are considered whose activity is counteracted by PAB (termed PAB mechanism or sulfanilamide type activity).

Effect of Structural Variation on Activity

It is not intended to discuss in detail the effects of modification of various structural features of sulfanilamide on its pattern of activity. Only some of the broad conclusions and generalizations that have emerged from the vast amount of work carried out in this field, are discussed. The earlier studies in this area have been covered in the now classical review by NORTHEY (1948), and subsequent developments have been reviewed by SHEPARD (1970).

The Sulfanilamide Part. In the sulfanilamide part, the following features are more or less inviolate for compounds showing good activity: a) the 4-amino group, b) the benzene ring with 1,4-disubstitution and c) the singly substituted SO_2NH_2 group. Some compounds have been reported with variations in this part, but in almost every case it has been found that either they are biodegraded to compounds which fulfil these criteria or they act by a non-PAB mechanism.

N^1-*Substituents.* Substitution at N^1 has proved very fruitful, and most of the clinically useful sulfonamides are N^1-substituted derivatives.

N^1-Alkyl and carbo-aromatic substitution in sulfanilamides leads in general to lowering or loss of activity; the only exceptions are N^1-methyl derivatives which are almost as active as sulfanilamide.

N^1-Acyl substitution leads to compounds with improved activity and useful biological properties. On account of their significantly high acidity, the compounds form highly soluble and neutral sodium salts, which offer useful pharmaceutical properties for administration. Sulfacetamide (*13*) is more active than sulfanilamide and as active as sulfadiazine against some organisms *in vivo*, and is widely used clinically. The N^1-higher acyl derivatives are deacetylated very substantially in the body in contrast to sulfacetamide, and are not useful clinically. The N^1-acetyl derivatives of N^1-sulfanilamido-heterocycles are inactive *in vitro*, but have the advantage of being tasteless as compared to the bitter taste of the parent compound; these are quantitatively decomposed in the intestine to the sulfanilamido-heterocycles and find use in liquid medications (FLAKE *et al.*, 1954).

Another group of active N^1-substituted sulfanilamides includes the sulfanilyl ureas, thioureas (and their lower-akyl ethers) and guanidine which are equal to or more active than, sulfanilamide, both *in vitro* and *in vivo*; the corresponding amidines and imidates have very little activity.

The most fruitful variation has been the introduction of N^1-heterocycle residues and most of the clinically used sulfonamides belong to this group. The heterocycle substituents can vary widely and the compounds still possess activity. Compounds with wider antibacterial spectrum, reduced toxicity and more favorable pharmaceutical properties have been obtained and these have helped to provide drugs for a wide variety of infectious conditions.

In the five membered heterocycles, the most active compounds belong to thiazoles, oxazoles, isoxazoles, 1,3,4-thiadiazoles and pyrazoles. The azoles, bearing an NH in the ring, have little activity, while N-aryl and N-alkyl compounds are active and some of them have been found to be of clinical interest. The corresponding hetero or benzo annellates are less active. Most of the clinically used 5-membered heterocyclic compounds (except sulfaphenazole) possess one or two CH_3 groups; this would obviously increase their hydrophobicity and thus impart favorable properties for protein binding and perhaps for folate-enzyme binding. Position of substituents on the hetero ring has also a significant effect on the activity pattern of the resultant compounds.

Among the six membered heterocycles, although sulfapyridine was the first
N^1-heterocyclic compound to achieve clinical success, mainly diazines (pyrimidines, pyridazines and pyrazines) have been found to possess useful activity.
Of these, derivatives of pyrimidines have received the most attention and have
provided a large number of clinically useful agents. Sulfonamido linkage at both
2- and 4-positions gives active compounds. The *in vivo* activity and antibacterial
spectrum of sulfadiazine (2-sulfapyrimidine) is still almost unsurpassed, although
some other agents may be preferred clinically on account of specific pharmacokinetic considerations. The corresponding 4- and 5-sulfapyrimidines are much
less active.

Substituents in the hetero ring seem to have a marked effect on the activity
of these compounds. The most common substituents introduced are lower alkyl,
alkoxy and halogens; these substituents seem to affect not so much the intrinsic
antibacterial activity of the compounds, but rather the pharmacokinetic properties
by altering the lipophilic character and pattern of metabolism.

Sulfones. Structure-activity relationship studies of sulfones have posed some
problems. Their major use is in the treatment of leprosy. Only recently (SHEPARD,
1960) a suitable experimental model for leprosy has become available, but even
this method is not amenable to easy and quick screening of a large number
of compounds. Clinical evaluation, therefore, continues to remain the only reliable
method of evaluation of antileprosy compounds. This rules out comparison of
a large number of compounds for purposes of SAR studies. Most of the conclusions about SAR of sulfones are therefore based on antituberculosis testing.

As in the case of sulfonamides, a *p*-aminobenzenesulfonyl residue as such,
or carrying groups, which can generate it *in vivo* seems essential for activity
of sulfones. The corresponding 4,4'-diaminodiphenylsulfoxide showed significant
antileprosy activity in a clinical trial (BUU-HOI, 1954) but was found to be more
toxic (BROWN and DAVEY, 1961). Its activity may be due to its *in vivo* oxidation
to DDS, as this sulfoxide and its N-methyl derivatives have been shown to
undergo *in vivo* oxidation to sulfones (ELLARD, 1966; KHOSLA *et al.*, 1955).

Compounds in which one benzene ring has been replaced by a 2- or 4-pyridyl,
2- or 5-thiazolyl or 8-quinolyl residue, preferably carrying an amino group, possess
significant activity, but replacement of both benzene rings by these heterocycles
leads to inactive compounds. None of these mono-heterocyclic sulfones, however,
offer any marked advantage over DDS.

A number of mono-N-alkyl sulfones (SMITH *et al.*, 1949; BAKER *et al.*, 1950;
ANAND *et al.*, 1953; ANAND *et al.*, 1954) and their corresponding N^1-galactosyl
derivatives have been prepared (ANAND *et al.*, 1953). Some of the lower alkyl
derivatives have been shown to possess significant antituberculosis activity in
guinea-pigs and mice (GUPTA and CHAKRAVARTI, 1955; GUPTA and MATHUR,
1957). N-Ethyl-4,4'-diaminodiphenylsulfone was also successfully tested in human
leprosy by CHANDY (1954) at the Mission to the Lepers Leprasorium, Faizabad,
India, and found to show lower incidence of side reactions. Its overall advantage
over DDS, however, was only marginal. A number of 2-, 3- and 2,2'- and 3,3'-
methoxy- and hydroxy-4-alkylamino-4'-aminodiphenylsulfones have also been
synthesised (LINNELL and STENLAKE, 1950; AMSTUTZ, 1950; VYAS *et al.*, 1954,
1955; TANDON *et al.*, 1958) and found to be less active than DDS in experimental

tuberculosis of guinea-pigs (GUPTA and MATHUR, 1958). None of its derivatives has thus shown any distinct superiority over DDS; DADDS, which is used instead of DDS, acts by slow release of DDS.

Physicochemical Properties and Biological Activity

Studies to find a correlation between the physicochemical properties and biological activity of sulfonamides have been actively pursued almost since their discovery. These studies do point to a relationship between the electronic charge, pKa and lipophilicity of sulfonamides and their biological activity. However, no generally applicable quantitative relationship has yet emerged which may partly be due to the inherent difficulties of the problem. A number of these studies have been concerned with N^1-phenyl-sulfanilamides (sulfanilides) and N^1-benzoyl-sulfanilamides, on account of the ease of studying the effect of substituents in these two series. But these sulfonamides have low activity, are of minor interest as drugs and some of them act by non-PAB mechanism—results from such studies are thus of limited value. Most of these investigations have been carried out in *in vitro* and in cell-free systems, and were restricted to one or two organisms, on account of ease of quantitation in such systems. Obviously, extrapolation of the results of such studies to *in vivo* systems, where many additional factors are involved, would be of limited utility. It would be necessary to consider separately the parameters related to the physicochemical action of the compounds and those that are more concerned with pharmacokinetic properties.

Fox and ROSE (1942), SCHMELKES (1943) and COWLES (1942) suggested that a relationship exists between the ability of sulfonamides to ionize as acids and their antibacterial activity. They noticed that the ratio of PAB to the ionized drug at a concentration where growth was just restored was more or less constant, whereas the ratio in terms of total concentration of drug varied widely. This suggested that the ionic form of the drug was the active agent. COWLES (1942) and BRUECKNER (1943) found that the activity increased with increase in pH of the medium only up to the point at which ionization was about 50% and then decreased. COWLES postulated that the sulfonamides penetrate the bacterial cell in the unionized form, but once inside the cell, the bacteriostatic action is due to the ionic form; the half dissociated state appeared the best compromise between transport and activity.

BELL and ROBLIN (1942) observed a similar relationship between the pKa's and *in vitro* antibacterial activity against. *E. coli* for an extensive series of sulfonamides. They found that the plot of log l/MIC against pKa gave a parabolic curve, and the highest points of this curve lay between pKa 6 to 7.4; the maximal activity was thus attained in compounds whose pKa approximated the physiological pH. Using this relationship BELL and ROBLIN were able to predict the *in vitro* efficacy of sulfonamides. The pKa of most of the more active sulfonamides discovered since then and particularly of the long-acting ones falls in this range (RIEDER, 1963).

BELL and ROBLIN (1942) wove WOODS and FILDE's hypothesis about the structural similarity of the metabolite and its antagonist into the observed facts of ionization. They emphasized the need for the sulfonyl group of active sulfonamides to polarize for them to resemble as closely as possible the geometrical[5] and electronic characteristics of the *p*-aminobenzoate ion, which led them to postulate that "the more negative the SO_2 group of N^1-substituted sulfanilamides the more bacteriostatic the compound will be".

The inconsistency of the activity of sulfaguanidine and of ring N-methyl sulfanilamido-heterocycles with the ionization theory is resolved by considering the availability of electron pairs, regardless of the anionic charge, in response to the electrophilic center of the enzyme, as of critical importance. The more favoured resonance forms of these molecules would fulfill this requirement. The activity of DDS can similarly be explained when related to the resonance forms having a high electron availability at position *p*- to NH_2 group (see SHEPHERD, 1970).

CAMMARATA and ALLEN (1967) and SEYDEL (1968) have cited examples of active sulfonamides whose pKa's lie outside the limits cited by BELL and ROBLIN. CAMMARATA *et al.* suggested that the data could be interpreted better as a linear relationship by considering the compounds in terms of homologous series. They plotted Hammett sigma values for a series of substituted N^1-benzoylsulfanila-mides against log 1/MIC and showed that the *m*- and *p*-substituted compounds fitted into one straight line, while the *o*-substituted compounds formed another straight line with opposite slope; the break in the line at the transition point was assumed to be due to a change in the mechanism of action.

FUJITA and HANSCH (1967) in a multiparameter linear free-energy approach correlated the pKa, π values, Hammett sigma values and protein binding of a series of sulfanilides and N^1-heterocyclic-sulfanilamides with their MIC data, obtained from the literature, for Gram-positive and Gram-negative organisms, and devised suitable equations for the relationship by regression analysis. In the case of sulfanilides it was observed that suitable separate equations could be devised for the two groups of bacteria, provided the *m*- and *p*-substituted derivatives were treated separately. The authors obtained two equations, correlat-ing log 1/MIC with pKa and π for the N^1-heterocyclic-sulfanilamides by employ-ing linear free-energy approach. Keeping the lipophilicity of the substituents constant the plot of log 1/MIC against pKa in both the cases corresponded to two straight lines with opposite slopes. They indicated that in this class of sulfonamides, a pKa of 7 and $\pi = 1.67$ (isobutyl alcohol-H_2O) was essential for optimum activity.

YAMAZAKI *et al.* (1970), in a study of the relationship between antibacterial activity and pKa of fourteen N^1-heterocyclic sulfanilamides, considered separately the activity of compounds in terms of ionized and unionized forms and total concentration in the culture medium, and obtained separate equations for each of these. They found that whereas the correlation between pKa and activity is linear for ionized and unionized states, it is parabolic when total concentration is considered. The two straight lines are of opposite slopes, intersecting each other as pointed out by FUJITA *et al.* (1967), but the authors suggested that

---

[5] Recent crystallographic studies of sulfanilamide and PAB have shown close structural similari-ties in the bond lengths and angles of these two molecules (ALLEAUME and DECAP, 1968; LAI and MARSH, 1967).

the point of intersection corresponds to the pH of the culture medium. They found the pKa for optimal activity to be between 6.61–7.4.

MORIGUCHI and KANENIWA (1969), on the assumption that enzyme-substrate complexes are generally charge-transfer complexes formed under the influence of the local field in the living body, found a parabolic relationship between the activity of sulfonamides and their charge-transfer transition energy or ionization potential.

BELL and ROBLIN (1942) laid emphasis on polarisability of the SO_2 group and related the negative charge of this group to the *in vitro* bacteriostatic activity. SEYDEL et al. (1960) using IR spectroscopy, and SCHNAARE and MARTIN (1965) and FOERNZLER and MARTIN (1967) using molecular orbital calculations could not find any evidence for this correlation. They have, therefore, in their studies attached greater importance to the electronic charge on 1-NH. MARTIN et al. have calculated the MO characteristics of a large number of N^1-substituted sulfonamides, and have found that the electron density for the 4-NH_2, S and O atoms were essentially constant, while that for 1-NH varied. They found a correlation between the pKa, electron density and MIC of the sulfonamides against *E. coli*; this relationship was more significant when the compounds were separated into classes depending upon the nature of N^1-substituent.

SEYDEL and his associates have confined their studies to sulfanilides and N^1-(3-pyridyl)sulfanilamides, and extrapolated the electron density on the 1-NH group from a study of infra-red and nmr data and Hammett sigma values of the parent anilines used in the study, and correlated the data with the *in vitro* MIC against *E. coli*, determined in a manner which would give precise self-consistent results. The anilines were used for studying the infra-red spectra because non-polar solvents could be employed which would give more valid data, while this was not possible with sulfanilides on account of low solubility in such solvents. SEYDEL (1966, 1968) and GARRETT et al. (1969) found an approximately linear relationship between bacteriostatic activity, Hammett sigma values and electron density on the 1-NH of a group of *m*- and *p*-substituted sulfanilides and have emphasised the possibility of predicting the *in vitro* antibacterial activity of sulfanilides by use of this relationship. Later SEYDEL (1971b) included in this study 3-sulfapyridines, carried out regression analysis of the data and obtained a very acceptable correlation coefficient. In a more recent study MILLER et al. (1972) extended this investigation to a cell-free folate synthesizing system, and correlated the inhibitory activity of these compounds on this enzyme system and on the intact organisms to their pKa, Hammett sigma, chemical shift and π values. From a comparison of the linear free-energy relationships obtained in the two systems, they suggested that the observed parabolic dependency of the antibacterial activity indicates that it is not the extracellular ionic concentration which governs the potency of the sulfonamides but rather the intracellular ionic concentration, which in turn is limited by the permeation of unionized compounds, thus supporting COWLES (1942) and BRUECKNER'S (1942) postulates. They concluded that the lipophilic factors are not important in the cell-free system or for *in vitro* antibacterial activity when permeability is not limited by ionisation.

It does appear, therefore, that in sulfonamides for good activity there has to be an optimal balance between dissociation and hydrophobic character, which would allow their transport across cellular membranes and interaction of the ionised species with the active site of the enzyme.

Protein Binding. The binding of sulfonamides and sulfones to blood proteins has been a subject of considerable interest (see SHEPHERD and LEWIS, 1967). The important characteristics of protein binding have been studied experimentally using equilibrium dialysis (DAVIS, 1942; KLOTZ and WALKER, 1948), ultrafiltration (RIEDER, 1963) and spectrophotometry (MORIGUCHI et al., 1968). The binding in plasma occurs almost entirely to the albumin fraction. The binding is weak (4–5 Kcal) and easily reversible. Binding affinity and capacity of the proteins vary with the structure of the sulfonamides as also with the animal species and the physiological status of the animal. Binding appears to be predominantly hydrophobic in nature, with ionic binding playing a less significant role (IRMSCHER et al., 1966; FUJITA, 1972). Therefore, structural features which favour binding are the same that increase lipophilicity such as presence of alkyl, alkoxy, aryl or heteroaryl groups (SHANNON, 1943; RIEDER, 1966; SCHOLTAN, 1964, 1968). The N^4-acetyl derivatives are more hydrophobic and are more strongly bound to proteins. Introduction of OH or NH_2 groups decreases protein binding and glucuronidation almost abolishes it (IRMSCHER et al., 1966). The binding sites are believed to be the basic centers of arginine, lysine and histidine. These sites are not unique for sulfonamide binding, and various other drugs displace them or are displaced by them, which has an important bearing on the toxicity of sulfonamides.

The implications of protein binding for chemotherapeutic activity are not yet clearly understood. The factors favoring protein binding are also those which would favour transfer across cell membranes, tubular reabsorption and increased binding to the enzyme proteins. As yet no simple relationship has been found between protein binding and the half-life of drugs (Table 1). Protein binding abolishes the antibacterial activity of sulfonamides and also their propensity to N^4-acetylation. It is, therefore, the unbound fraction of the drug in the plasma which is important for activity, toxicity and metabolism. Protein binding does seem to modulate drug availability. However, a high degree of protein binding is frequently paralleled by high bacteriostatic activity. MORIGUCHI et al. (1968) observed a parabolic relationship between protein binding and *in vitro* bacteriostatic activity for a series of sulfonamides and explained it thus that too high an affinity between sulfonamides and proteins would prevent them from reaching their target sites in bacteria, and too low an affinity would not enable them to bind effectively with enzyme proteins to cause bacteriostasis, assuming that affinity for enzyme proteins is paralleled by affinity to bacterial proteins.

Absorption, Metabolism and Excretion

Most of the sulfonamides are well absorbed, mainly from the small intestine, slightly from the large intestine and insignificantly from the stomach. Absorption through the intestinal wall occurs *via* the unionized form, in proportion to their lipid solubility and molecular size. In rate and extent of absorption most sulfonamides behave similarly within the pKa range of 4.5–10.5. N^1-Acetyl- and N^1-sulfanilyl-sulfanilamide derivatives and sulfonamides bearing additional acidic or basic groups are poorly absorbed. Sulfonamides are quite evenly distributed throughout the body tissues. Those now available, offer a wide range of solubility and transport characteristics which permit their access to almost any site in the body, and thus greatly add to their usefulness as antibacterial agents.

Metabolism of sulfonamides involves mainly N^4-acetylation, to a lesser extent N^1-glucuronidation, and to a very small degree C-hydroxylation of phenyl and heterocyclic rings and of alkyl substituents, and O-, N^4-, N^1- and ring N-dealkylation. The structural requirements for each metabolic path are quite specific, and variation of the substituents markedly influences the metabolic fate (Table 1).

Excretion of sulfonamides occurs chiefly through the kidneys. The free, non-protein-bound drugs and their metabolic products are ultrafiltered in the glomeruli and then partly reabsorbed. Tubular secretion also plays an important role in the excretion of sulfonamides and their metabolites. Molecular characteristics of the compounds have marked effect on these processes and determine the rate of excretion. The renal clearance rate of the metabolites is generally higher than that of the intact drugs.

Half-life. The consideration of the half-life of sulfonamides is of great importance as the dosage regimen must be related to it. The dose schedule is a function of the molar activity and pharmacokinetic parameters; KRUGER-THIEMER (1966) has developed a mathematical model for the relationship between these parameters and evolved a computer programme for their calculation (see SEYDEL, 1971a).

The half-lives of different sulfonamides in clinical use vary widely[from 2.5 to 150 hr (Table 1)] and show marked differences in different animal species too. The factors which are responsible for differing rates of excretion of different sulfonamides are complex and not yet clearly understood. RIEDER (1963) correlated the pKa, liposolubility, surface activity and protein binding of a group of 21 sulfonamides with their half-life in man. They found that the long-acting sulfonamides were in general more lipid-soluble than the short-acting compounds, but no relationship could be established and other factors such as protein-binding, tubular secretion and tubular reabsorption seem to affect this. No direct relationship between the extent of protein binding and the half-life has been observed. In the case of compounds with comparable protein binding properties an increased lipid solubility is accompanied by an increased half-life (STRULLER, 1968). In general, the structural features which contribute to longer half-life are the same which increase liposolubility such as lower alkyl, alkoxy, and halogen groups; the methoxy group seems particularly effective in increasing the half-life. The orientation of these groups in the N^1-substituent is also very important and there are some subtle structural features affecting the excretion rate which are at present difficult to explain. In 2-sulfapyrimidines, a 4-CH_3 increases the half-life, 4,6-$(CH_3)_2$ reduces the half-life to less than one-half, the corresponding methoxy derivatives have a much longer half-life, while both 5-CH_3 and 5-OCH_3 prolong the half-life to the same extent. Similarly, in 4-sulfapyrimidines, the 2,6-$(CH_3)_2$ derivative is short-acting, 2,6-$(OCH_3)_2$ is long-acting, while the isomeric 5,6-$(OCH_3)_2$ is the most persistent sulfonamide known.

Sulfones. DDS after oral administration is well-absorbed and evenly distributed in almost all the body tissues. It is excreted mainly through the kidneys. Less than 5 per cent is excreted unchanged, very little N-acetylation takes place,

Table 1[a]

Name[b]	Common generic/proprietary name	pKa
1	2	3
Short-acting		
1. 2-Sulfa-5-methyl-1,3,4-thiadiazole	Sulfamethizole/Lucosil	5.5
2. 2-Sulfathiazole	Sulfathiazole/Cibazol	7.1
3. 5-Sulfa-3,4-dimethyl-isoxazole	Sulfisoxazole/Gantrisin	5.0
4. 2-Sulfa-4,6-dimethyl pyrimidine	Sulfamethazine/Diazil	7.4
5. 4-Sulfa-2,6-dimethyl pyrimidine	Sulfisomidine/Elkosin	7.4
6. 2-Sulfa-5-ethyl-1,3,4-thiadiazole	Sulfaethidole/Globucid	5.6
7. N^1-Acetylsulfanilamide	Sulfacetamide/Albucid	5.4
8. 2-Sulfapyridine	Sulfapyridine/Dagenan	8.4
9. Sulfanilamide	Sulfanilamide	10.5
Medium-Acting		
10. 5-Sulfa-1-phenylpyrazole	Sulfaphenazole/Orisul	5.9
11. 3-Sulfa-5-methylisoxazole	Sulfamethoxazole/Gantanol	6.0
12. 2-Sulfa-4,5-dimethyloxazole	Sulfamoxole/Sulfuno	7.4
13. 2-Sulfapyrimidine	Sulfadiazine/Debenal	6.4
Long-Acting		
14. 2-Sulfa-4-methylpyrimidine	Sulfamerazine/Debenal-M	7.0
15. 2-Sulfa-5-methyl-pyrimidine	Sulfamethyldiazine/Pallidin	6.7
16. 3-Sulfa-6-methoxy-pyridazine	Sulfamethoxypyridazine/Lederkyn	7.2
17. 4-Sulfa-2,6-dimethoxy-pyrimidine	Sulfadimethoxine/Madribon	6.1
18. 2-Sulfa-3-methoxypyrazine	Sulfalene/Kelfizina	6.1
19. 4-Sulfa-5,6-dimethoxy-pyrimidine	Sulformethoxine/Fanasil	6.1

[a] Data drawn from RIEDER, 1963; WILLIAMS and PARK, 1964; SCHOLTAN, 1964; KRÜGER-THIEMER
[b] Sulfa stands for *p*-aminobenzenesulfonamide (sulfanilamide) radical.
[c] Determined by partition between ethylene dichloride and sodium phosphate buffer (RIEDER,
[d] Figures in parenthesis refer to % N'-glucuronide in urine and the balance is mainly the

and most of it is present as the mono-N-glucuronide (BUSHBY and WOIWOOD, 1955, 1956; ELLARD, 1966). Acedapsone (DADDS) after intramuscular injection is very slowly absorbed and deacetylated. After a single injection of 300 mg in humans the peak plasma level is reached in 6 days, a plateau level of 0.03 µg/ml is maintained for about 60 days, and the level drops slowly thereafter, with a half-life of about 42.6 days (DDS has a half-life of about 20 hours) (GLAZKO *et al.*, 1968).

Antimicrobial Activity and Present Status in Therapeutics

Following the initial dramatic results obtained with sulfonamides in the treatment of streptococcal infections, studies with these drugs were extended to other bacteria, protozoa and viruses. It was found that many Gram-positive and Gram-

Plasma half-life (hrs)	Protein binding: % free in blood	Lipid[c] solubility (%)	*In vitro* effect against *E. coli* (μmol/l)	% N^4-Acetyl metabolite in urine[d] (Man)
4	5	6	7	8
2.5	22			5
4	22	15.3	1.6	30 (40)
5.0	8	4.8	2.15	16 (2,2'-glucuronide)
7	18	82.6	1.7	60
7.5	10	19.0	1.5	4 (0)
7	1.5	6.2	2.0	(30)
7	82	2.0	2.3	5
9	70		4.8	30
9	88	10.5	128	
11	1.5	69	1.0	20 (80) of 2'-glucuronide
11	30	20.5	0.8	60 (14)
11	22	41.4	4.0	
17	55	25.4	1.0	25
24	25	62.4	0.95	
32	10	69.6	1.0	
37	10	70.4	1.0	50 (15)
40	0.5	78.7	0.8	15 (70)
65	30		1.85	65
150	5		0.8	55 (2)

and BÜNGER, 1961; STRULLER, 1968; BRIDGES *et al.*, 1969; SHEPHERD, 1970; UEDA *et al.*, 1972.

1963).
unchanged drug.

negative cocci and bacilli, mycobacteria, some viruses and protozoa are highly susceptible to the action of the sulfonamides. Differences in the antibacterial spectra of individual sulfonamides are quantitative rather than qualitative. Table 2 gives broadly the antimicrobial spectrum of sulfonamides and sulfones.

With the introduction of penicillin in 1945 followed by the discovery of other powerful antibiotics, the sulfonamides were overshadowed for sometime. Very soon, however, in view of the problems encountered with antibiotics such as emergence of resistance strains, superinfection and allergic reactions, interest in sulfonamides was revived, and now they have an established and important place in therapeutics. Combination therapy with dihydrofolate reductase inhibitors has greatly increased their usefulness (see GARROD *et al.*). Clinical indications for sulfonamides and antibiotics have been carefully reassessed and their relative merits reasonably well charted out (WEINSTEIN *et al.*, 1960).

Table 2

Gram-positive	Gram-negative	Others
	Highly susceptible	
Streptococcus pyogenes	*Neisseiria meningitidis*	*Coccidia*
Staphylococcus aureus	*N. gonorrhoeae*	*Plasmodium malariae*
Streptococcus pneumoniae	*Haemophilus influenzae*	*Actinomyces bovis*
Bacillus anthracis	*H. ducreri*	*Nocardia asteroides*
Mycobacterium tuberculosis	*Pasteurella pestis*	*Toxoplasma*
M. leprae	*Shigella sonnei*	*Trachoma virus*
Listeria monocytogenes	*S. dysenteriae*	*Lymphogranuloma-psitta-*
Corynebacterium diphtheriae	*Salmonella paratyphi*	*cosis viruses*
	Vibrio cholerae	*Lymphocytic chorio-*
	Escherichia coli	*meningitis virus*
	Weakly sensitive	
Clostridium welchii	*Donovania granulomatis*	
Streptococcus viridans	*E. coli*	
	Proteus vulgaris	
	Pseudomonos aeruginosa	
	S. typhosa	
	S. paratyphi (some strains)	
	Shigella sonnei (some strains)	
	N. gonorrhoeae (some strains)	
	Klebsiella pneumoniae	
	Brucella abortus	
	Aerobacter species	

Sulfonamides differ from antibiotics such as penicillin in that their action is largely bacteriostatic and sulfonamide-initiated recovery from bacterial infections requires active phagocytosis. It is, therefore, important to maintain bacteriostatic concentration of sulfonamides for a period of time sufficient for the normal defence mechanism of the host to become active against the bacteria.

Sulfonamides continue to remain the drugs of choice in the treatment of chancroid, urinary tract infections, lymphogranuloma venereum, trachoma, inclusion conjunctivitis and nocardiosis. They are recommended for meningococcal meningitis when organisms are sensitive to them, and in haemophilus influenzae meningitis as an adjunct to streptomycin therapy. Sulfonamides continue to occupy an important place in the treatment of erysipelas, especially in patients allergic to penicillin. Sulfadiazine is recommended for the prophylaxis of rheumatic fever as an alternative to penicillin. In combination with penicillin, sulfonamides are recommended for the treatment of otitis media in children. In combination with pyrimethamine/trimethoprim they are recommended for toxoplasmosis and chloroquine-resistant falciparum malaria. (Report of the National Academy of Science National Research Council Committee to study the efficacy of drugs, quoted in J. Amer. Pharmaceutical Association, NS9, 535–36.) The combination of sulfamethoxazole with trimethoprim (which have similar half-life) has emerged as a useful treatment for salmonellosis (JAFARY and BURKE, 1970) and chronic bronchitis (MALLET and MUSSELWHITE, 1970). Sylicylazosulfapyridine has been found effective in the treatment of ulcerative colitis; combined with steroids it is considered the treatment of choice for chronic, intermittent ulcerative recto-colitis (MARTINEK, 1970).

DDS remains the treatment of choice for all forms of leprosy (BROWNE, 1970). It has been demonstrated recently that *M. leprae* is sensitive to very low doses of DDS (SHEPARD *et al.*, 1966), and that a weekly dose of 1–5 mg per adult may suffice to maintain the clinical and bacteriological improvements in patients (see BROWNE, 1969). These findings and the possibility of the use of acedapsone as a repository drug are likely to further enhance the usefulness of DDS in the treatment of leprosy.

Drug Resistance

Emergence of drug-resistant strains is one of the serious handicaps of sulfona-mide therapy. The development of resistance is considered to arise owing either to overproduction of PAB (LANDY *et al.*, 1943), altered sensitivity of enzymes involved in sulfonamide action (WOLF and HOTCHKISS, 1963; PATO and BROWN, 1963) or altered permeability of the organisms to sulfonamides (PATO *et al.*, 1963). Different sulfonamides show cross-resistance but there is no cross-resistance to other antibacterials.

One of the important aspects of drug resistance in sulfonamides which has come to light during the last decade is the transfer of multiple drug resistance from resistant to susceptible strains of different species of Enterobacteriaceae. It was first noticed by OCHIAI *et al.* (1959) (quoted by WATANABE, 1963) that multiple drug resistance involving streptomycin, chloramphenicol, tetracyclin and sulfonamides could be transferred between *Shigella* and *E. coli* in mixed cultiva-tion. Subsequent studies have established that the transfer of resistant factors (R-factors) is carried out through cell to cell contact (conjugation) by autono-mously replicating extra-chromosomal genetic particles called episomes (now preferred to be termed plasmids, HAYES, 1969). This transfer can take place both *in vitro* as well as in the alimentary tract. Drug resistance acquired in this manner can be transferred to other sensitive strains indefinitely.

It has been proposed that the R-factors consist of reversible covalently linked units that separately harbor either transfer (RTF) or resistance functions (R-determinants) (WATANABE and FUKASAWA, 1961a), and the two may be trans-ferred together or independently (ANDERSON, 1969; COHEN and MILLER, 1970a). COHEN and MILLER (1970b) have separated the R-factor DNA from chromosomal DNA of *Proteus* and *E. coli*, and have shown them to be closed circular DNA molecules. They also obtained circular segregants which have the capacity for transfer (RTF) but do not carry drug-resistance markers.

This transferred multiple drug resistance has been differentiated from classical drug resistance acquired by spontaneous mutation. WATANABE and FUKASAWA (1960, 1961b) reported a strain of *Shigella* with multiple drug resistance which, on treatment with acridines, gave a strain resistant to sulfonamides only. This resistance was, however, not transmissible. They thus concluded that the original multiple drug resistant *Shigella* strain possessed two forms of sulfonamide resist-ance, one chromosomal and the other cytoplasmic (extrachromosomal). Present evidence indicates that the sulfonamide resistance involved in multiple drug resistance is due to reduced permeability of the cells to sulfonamides.

The increased incidence of R-factors, conferring resistance to particular drugs, has paralleled the extensive use of the corresponding drugs. This fact indicates the importance of selection by the drugs of R-factor carrying drug-resistant bacteria. This is a serious problem, since such transfer can take place to almost every genus of Enterobacteriaceae, by non-pathogenic bacteria such as *E. coli*. Individuals who ingest *E. coli* carrying this resistance factor can become carriers of drug-resistant intestinal flora. If such individuals are infected with sensitive strains of Salmonella, Shigella etc., these pathogenic organisms could easily acquire resistance by transfer of R-factors carrying intestinal flora.

Sulfones. In spite of the extensive use of sulfones in the mass treatment of leprosy over the last 25 years, remarkably few cases of sulfone-resistance have so far been reported (see BROWN, 1969). Resistance may be developed in a stepwise manner (MORRISON, 1968) and partial resistance may occur (PEARSON et al., 1968). BROWN has suggested that the resistance may be related to low-dosage treatment.

Adverse Side-Reactions

A large number of adverse reactions to sulfonamides involving almost every system of the body have been reported in the literature. Some of these are trivial and can be easily managed. Others, however, may be serious and even fatal, thus necessitating a close surveillance of the patient and avoidance of indiscriminate use of these drugs.

Crystalluria was one of the earliest serious toxic reactions observed with sulfonamides. The significant point is the amount of drug and metabolite excreted per unit time in relation to urine flow rate and to the therapeutic dosage of the particular sulfonamide. The extent of crystalluria thus varies for different sulfonamides depending upon their concentration in the urine and their solubility. A short-acting sulfonamide requiring high dosage needs to be much more soluble in order to permit a higher rate of excretion of the drug. With the development of more soluble sulfonamides and the reduction in dosage of the earlier sulfonamides, due to the realisation that they had long half-lives, the incidence of such reactions is certainly less but the potential danger still exists.

The most serious side effects of sulfonamides are their actions on the bone marrow. Various types of blood dyscrasias have been reported, including leukopenia, agranulocytosis, aplastic anaemia, thrombocytopenia and pancytopenia. All these are serious complications and need close surveillance.

A large number of drugs are known to be bound to albumin, and sulfonamides can displace those with lesser binding affinity, e. g., warfarin sodium, tetracyclines, penicillins and, therefore, when used in combination with other drugs, a close watch should be kept for side reactions. Similarly, sulfonamides share chemical structures with other compounds like sulfonyl-ureas, thiazides and some uricosuric agents and can potentiate their action.

Sulfonamides have a higher affinity for albumin than the unconjugated bilirubin and, therefore, by displacing it from albumin can aggravate physiological hyperbilirubinemia in neonates, particularly in premature deliveries, and can thus precipitate the occurrence of kernicterus. Lack of development of acetyl transferase system particularly in premature neonates contributes to higher blood levels of sulfonamides and adds to the risk of precipitation of kernicterus.

Both topical and systemic administration of sulfonamides can lead to hypersensitivity reactions. Cutaneous manifestations include urticaria, exfoliative dermatitis, photosensitization, erythema nodosum and in its most severe form erythema multiforme-exudativum (Stevens-Johnson syndrome). The latter is considered to be a particularly serious hazard of long-acting sulfonamides.

Acknowledgement. I am very thankful to Drs. ARUNA V. RAO (Miss.), N.K. KAPOOR and S.S. AGARWAL for their sparing the time to read and check the sections on "Structure—activity relationship", "Mode of action" and "Antimicrobial activity and present status in therapeutics", respectively.

References

ALLEAUME, P.M., et J. DeCAP: Affinement Tridimensionnel du Sulfanilamide. Acta Cryst. **B24**, 214 (1968).

AMSTUTZ, E.D.: Studies in the sulfone series VII. The preparation of 2,8-diaminophenoxathiin-5-dioxide and bis-(2-hydroxy-4-aminophenyl)sulfone. J. Am. Chem. Soc. **72**, 3420 (1950).

ANAND, N., G.N. VYAS, and M.L. DHAR: Studies in potential antimycobacterial agents: Part I. Synthesis of *p*-alkylamino-*p*′-(galactosylamino)-diphenylsulfones. J. Sci. Ind. Res. (India) **12B**, 353 (1953).

ANAND, N., P.S. WADIA, and M.L. DHAR: Studies in potential antimycobacterial agents: Part III. Synthesis of *p*-(ω-hydroxy-alkylamino, ω-alkoxyalkylamino & haloethylamino)-*p*′-aminodiphenyl-sulfones. J. Sci. Ind. Res. (India) **13B**, 260 (1954).

ANDERSON, E.S.: Ecology and epidemiology of transferable drug resistance. Bacterial episomes & plasmids. A Ciba Foundation Symposium, ed. by G.W.W. WOLSTENHOLME and MAEVE O'CONNOR, p. 102. London: J. & A. Churchill Ltd. 1969.

ANGIER, R.B., J.H. BOOTHE, B.L. HUTCHINGS, J.H. MOWAT, J. SEMB, E.L.R. STOKSTAD, Y. SUBBAROW, C.W. WALLER, D.B. COSULICH, M.J. FAHRENBACH, M.E. HULTQUIST, E. KUH, E.H. NORTHEY, D.R. SEEGER, J.P. SICKLESS, and J.M. SMITH, JR.: The structure and synthesis of the liver *L. casei* factor. Science **103**, 667 (1946).

ARCHIBALD, H.M., and C.M.A. ROSS: Preliminary report on the effect of diaminodiphenylsulphone on malaria in northern Nigeria. J. Trop. Med. Hyg. **63**, 25 (1960).

BAKER, B.R., M.V. QUERRY, and A.F. KADISH: Sulfones II. Derivatives of 4,4′-diaminodiphenylsulfone. J. Org. Chem. **15**, 402 (1950).

BASU, P.C., N.N. SINGH, and N. SINGH: Potentiation of activity of diphenylsulfone and pyrimethamine against *P. gallinaceum* and *P. cynomolgi bastianellii.* Bull. World Health Organ. **31**, 699 (1964).

BELL, P.H., and R.O. ROBLIN, JR.: Studies in chemotherapy. VII. A theory of the relation of structure to activity of sulfanilamide type compounds. J. Am. Chem. Soc. **64**, 2905 (1942).

BIOCCA, E.: Quimioterapia sulfonica da toxoplasmose. Arquiv. Biol. (Sao Paulo) **7**, 27 (1943).

BISHOP, A.: Some recent developments in the problem of drug resistance in malaria. Parasitology **53**, 10p (1963).

BLANCHARD, K.C.: The isolation of *p*-aminobenzoic acid from yeast. J. Biol. Chem. **140**, 919 (1941).

BRATTON, A.C., and E.K. MARSHALL, JR.: A new coupling component for sulphanilamide determination. J. Biol. Chem. **128**, 537 (1939).

BRIDGES, J.W., S.R. WALKER, and R.T. WILLIAMS: Species difference in the metabolism of sulfasomidine and sulfamethomidine. Biochem. J. **111**, 173 (1969).

BROWN, G.M.: The biosynthesis of folic acid. Inhibition by sulfonamides. J. Biol. Chem. **237**, 536 (1962).

BROWN, G.M.: The biosynthesis of pteridines. Advan. Enzymol. **35**, 35 (1971).

BROWN, G.M., R.A. WEISMAN, and D.A. MOLNAR: The biosynthesis of folic acid: I. Substrate and cofactor requirements for enzymatic synthesis by cell-free extracts of *Escherichia coli.* J. Biol. Chem. **236**, 2539 (1961).

Browne, S.G.: The evaluation of present antileprosy compounds. Advan. Pharmacol. Chemother. 7, 211 (1969).

Browne, S.G.: Dapsone-resistant *M. leprae* in a patient receiving dapsone in low doses. Intern. J. Leprosy 37, 296 (1969).

Browne, S.G., and T.F. Davey: Diaminodiphenylsulphoxide in the treatment of leprosy: A definitive report on expanded trials. Leprosy Rev. 32, 194 (1961).

Brownlee, G., A.F. Green, and M. Woodbine: Sulphetrone: A chemotherapeutic agent for tuberculosis: Pharmacology and chemotherapy. Brit. J. Pharmacol. 3, 15 (1948).

Brueckner, A.H.: The effect of pH on sulphonamide activity. Yale J. Biol. Med. 15, 813 (1943).

Bushby, S.R.M., and A.J. Woiwood: Excretion products of 4:4'-diamino-diphenylsulfone. Am. Rev. Tuberc. Pulmonary Diseases 72, 123 (1955).

Bushby, S.R.M., and A.J. Woiwood: The identification of the major diazotizable metabolite of 4:4'-diaminodiphenylsulphone in rabbit urine. Biochem. J. 63, 406 (1956).

Buu-Hoi, N.P.: The selection of drug for chemotherapy in leprosy. Intern. J. Leprosy 22, 16 (1954).

Cammarata, A., and R.C. Allen: Observations concerning the correlation of *in vitro* sulfonamide activity with pKa and Hammett values. J. Pharm. Sci. 56, 640 (1967).

Cenedella, R.J., and J.J. Jarrell: Suggested new mechanisms of antimalarial action for DDS involving inhibition of glucose utilization by the intraerythrocytic parasite. Am. J. Trop. Med. Hyg. 19, 592 (1970).

Coggeshall, L.T., J. Maier, and C.A. Best: The effectiveness of two new types of chemotherapeutic agents in malaria: sodium p,p'- diaminodiphenylsulphone-N,N'-di dextrosesulfonate (Promin) and 2-sulfanilamidopyrimidine (sulfadiazine). J. Am. Med. Assoc. 177, 1077 (1941).

Cohen, S.N., and C.A. Miller: Non-chromosomal antibiotic resistance in bacteria. III Isolation of the discrete transfer unit of the R-factor RI. Proc. Natl. Acad. Sci. U.S. 67, 510 (1970a).

Cohen, S.N., and C.A. Miller: Non-chromosomal antibiotic resistance in bacteria. II Molecular nature of R-factors isolated from *Proteus mirabilis* and *E. coli*. J. Mol. Biol. 50, 671 (1970b).

Connar, R.G., T.B. Ferguson, W.C. Sealy, and N.F. Conant: Report of a single case with recovery. J. Thorac. Surg. 22, 424 (1951).

Cowdry, E.V., and C. Ruangsiri: Influence of promin, starch and heptaldehyde in experimental leprosy in rats. Arch. Pathol. 32, 632 (1941).

Cowles, P.B.: Ionisation and the bacteriostatic action of sulfonamides. Yale J. Biol. Med. 14, 599 (1942).

Davenport, H.W., and A.E. Wilhelmi: Renal carbonic anhydrase. Proc. Soc. Exptl. Biol. Med. 48, 53 (1941).

Davis, B.D.: Binding of sulfonamide drugs by plasma proteins. Science 95, 78 (1942).

Deininger, R., and H. Gutbrod: The pharmacodynamic action of the new sulfonamide, 2-(p-amino-benzene-sulfonamido)-4,5-dimethyl-oxazole. Arzneimittel-Forsch. 10, 612 (1960).

Domagk, G.: Ein Beitrag zur Chemotherapie der bakteriellen Infektionen. Deut. Med. Wochschr. 61, 250 (1935).

Donovick, R., A. Bayan, and D. Hamre: The reversal of the activity of antituberculous compounds *in vitro*. Am. Rev. Tuberc. 66, 219 (1952).

Ellard, G.A.: Absorption, metabolism and excretion of Di(p-aminophenyl)sulphone (Dapsone) and Di(p-aminophenyl)sulphoxide in man. Brit. J. Pharmacol. 26, 212 (1966).

Elslager, E.F.: Progress in malaria chemotherapy. Part I. Repository antimalarial drugs. Progr. Drug. Res. Vol. 13, 170 (1969).

Elslager, E.F., and D.F. Worth: Repository antimalarial drugs: N,N'-diacetyl-4,4'-diamino-diphenylsulfone and related 4-acylaminodiphenylsulfones. Nature 206, 630 (1965).

Eyles, D.E.: Newer knowledge of the chemotherapy of toxoplasmosis. Ann. N. Y. Acad. Sci. 64, 252 (1956).

Eyles, D.E., and N. Coleman: An evaluation of the curative effects of pyrimethamine and sulfadiazine, alone and in combination, on experimental mouse toxoplasmosis. Antibiot. & Chemotherapy 5, 529 (1955).

Feldman, W.H., H.C. Hinshaw, and H.E. Moses: Promin in experimental tuberculosis. Am. Rev. Tuberc. 45, 303 (1942).

Fildes, P.: Rational approach to research in chemotherapy. Lancet 1940-I, 955.

Findlay, G.M.: The action of sulfanilamide on the virus of lymphogranuloma venereum. Brit. J. Exptl. Pathol. 21, 356 (1940).

FLAKE, R. E., J. GRIFFIN, E. TOWNSEND, and E. M. YOW: Studies of the absorption, distribution and excretion of acetyl sulfisoxazole, an insoluble sulfonamide appearing as sulfisoxazole in the blood. J. Lab. Clin. Med. **44**, 582 (1954).

FOERNZLER, E. C., and A. N. MARTIN: Molecular orbital calculations on sulfonamide molecules. J. Pharm. Sci. **56**, 608 (1967).

FORSTER, W. G., and J. R. McGIBONY: Trachoma. Am. J. Ophthalmol. **27 C**, 1107 (1944).

FOX, C. L. JR., and H. M. ROSE: Ionisation of sulfonamides. Proc. Soc. Exptl. Biol. Med. **50**, 142 (1942).

FRANKE, H., und J. FUCHS: Ein neues antidiabetisches Prinzip (Ergebnisse klinischer Untersuchungen). Deut. Med. Wochschr. **80**, 1449 (1955).

FRIEDKIN, M.: Enzymatic aspects of folic acid. Ann. Rev. Biochem. **32**, 185 (1963).

FUJITA, T.: Hydrophobic bonding of sulfonamide drugs with serum albumin. J. Med. Chem. **15**, 1049 (1972).

FUJITA, T., and C. HANSCH: Analysis of the structure-activity relationship of the sulfonamide drugs using substituent constants. J. Med. Chem. **10**, 991 (1967).

FULLER, A. T.: Is p-aminobenzenesulfonamide an active agent in prontosil therapy? Lancet **1937-II**, 194.

FUST, B., and E. BOHNI: Tolerance and antibacterial properties of 2,4-dimethoxy-6-sulfanilamido-1,3-diazine (Madribon) and some other sulfonamides. Antibiot. Med. Clin. Therapy **6**, 3 (1959).

FUST, B., and E. BOHNI: Vergleichende experimentelle Untersuchungen mit 5-Methy 1-3-sulfanilamido-isoxazol und anderen Sulfanilamiden und Antibiotica. Schweiz. med. Wochschr. **92**, 1599 (1962).

GARRETT, E. R., J. B. MIELCK, J. K. SEYDEL, and H. J. KESSLER: Kinetics and mechanisms of action of drugs on microorganisms. VIII. Quantification and prediction of the biological activities of *m*- and *p*-substituted N'-phenylsulfanilamides by microbial kinetics. J. Med. Chem. **12**, 740 (1969).

GARROD, L. P., D. G. JAMES, and A. A. G. LEWIS: The synergy of trimethoprim and sulfonamides. Proc. of Conf. at Royal Coll. Physicians, May 1969, Postgrad. Med. J. (Supplement) **45**, 1 (1969).

GLAZKO, A. J., W. A. DILL, R. G. MONTALBO, and E. L. HOLMES: A new analytical procedure for dapsone: Application to blood level and urinary-excretion studies in normal men. Am. J. Trop. Med. Hyg. **17** (3), 465 (1968).

GOTS, J. S.: Occurrence of 4-amino-5-imidazolecarboxamide as a pentose derivative. Nature **172**, 256 (1953).

GREEN, H. N.: The mode of action of sulfanilamide. Brit. J. Exptl. Pathol. **21**, 38 (1940).

GREENBERG, J.: The antimalarial activity of 2,4-diamino-6,7-diphenylpterin; its potentiation by sulphadiazine and inhibition by pteroylglutamic acid. J. Pharmacol. Exptl. Therap. **97**, 484 (1949).

GUPTA, S. K., and R. N. CHAKRAVARTI: The therapeutic activity of some sulphones and sulphoxides in experimental tuberculosis of guineapigs. Brit. J. Pharmacol. **10**, 113 (1955).

GUPTA, S. K., and I. S. MATHUR: Therapeutic activity of S. N. 44 (*p*-ethylamino-*p'*-aminodiphenylsulphone) and S. N. 47 (*p*-isobutylamino-*p'*-aminodiphenylsulphone) in experimental tuberculosis of mice. J. Sci. Ind. Res. (India) **16 C**, 192 (1957).

GUPTA, S. K., and I. S. MATHUR: The evaluation of some nuclear and side-chain substituted sulfones in experimental tuberculosis of guineapigs. Arch. Intern. Pharmacodyn. **64**, 354 (1958).

HAYS, W.: What are episomes and plasmids? Bacterial episomes and plasmids. A Ciba Foundation Symposium, ed. by G. W. W. WOSTENHOLME & MAEVE O'CONNOR, p. 4. London: J. & A. Churchill Ltd. 1969.

HENRY, R. J.: The mode of action of sulphonamides. Bacteriol. Rev. **7**, 175 (1943).

HILL, R. A., and H. M. GOODWIN, JR.: 5th Med. J. Nashville **30**, 1170 (1937) (quoted by Richards, 1970).

HITCHINGS, G. H., and J. J. BURCHALL: Inhibition of folate biosynthesis and function as a basis for chemotherapy. Advanc. Enzymol. **27**, 417 (1965).

HOTCHKISS, R. D., and A. H. EVANS: Fine structure of a genetically modified enzyme as revealed by relative affinities for modified substrate. Federation Proc. **19**, 912 (1960).

IRMSCHER, K., D. GABE, K. JAHNKE UND W. SCHOLTAN: Untersuchungen zur Serumeiweißbindung und zur renalen Elimination von isomeren Sulfonamiden (5-Sulfanilamido-3-athyl-1,2,4-thiodiazol und 5-Sulfanilamido-2-athyl-1,3,4-thiodiazol). Arzneimittel-Forsch. **16** (8), 1019 (1966).

Iwai, K., and O. Okinaka: The biosynthesis of folic acid compounds in plants. II Some properties of dihydropteroate-synthesizing enzyme in pea seedlings. J. Vitaminol. (Kyoto) **14**, 170 (1968).

Jaenicke, L., und P.C. Chan: Die Biosynthese der Folsäure. Angew. Chem. **72**, 752 (1960).

Jafary, M.H., and G.J. Burke: Antibiotics and salmonella excretors. Brit. Med. J. **1970**, 605, also see editorial review **3**, 297 (1970).

Janbon, M., J. Chaptal, A. Vedel, and J. Schaap: Accidents hypoglycemiques granes par un sulfamidothiazol (1e VK57 OU 2254 RP). Montpellier. Med. **21–22**, 441 (1942).

Jawetz, E.: Chemotherapy of chlamydial infections. Advanc. Pharmacol. Chemother. **7**, 253 (1969).

Jones, L.P., and F.D. Williams: Rhizopterin biosynthesis in *Staphylococcus epidermidis*. Can. J. Microbiol. **14**, 933 (1968).

Joyner, L.P., S.F.M. Davies, and S.B. Kendall: Chemotherapy of coccidiosis. In: Experimental chemotherapy (ed. R.J. Schnitzer, and F. Hawking), p. 445. New York: Academic Press, 1963.

Kendall, S.B.: A comparison of the efficacy of sulphamezathine and sulphaquinozaline in the control of experimentally induced caecal coccidiosis in chicks. Vet. Record **62**, 381 (1950).

Kendall, S.B., and L.P. Joyner: Potentiation of the coccidiostatic effects of sulphadimidine by five different folic acid. antagonists. Vet. Record **70**, 632 (1958).

Khosla, M.C., P.S. Wadia, N. Anand, and M.L. Dhar: Studies in potential antimycobacterial agents: Part VIII — Metabolism and mode of excretion of *p*-ethylamino-*p′*-aminodiphenylsulfone and *p*-methylamino-*p′*-aminodiphenyl-sulphoxide. J. Sci. Ind. Res. (India) **14C**, 152 (1955).

Klotz, I.M., and F.M. Walker: The binding of some sulfonamides by bovine serum albumin. J. Am. Chem. Soc. **70**, 943 (1948).

Krüger-Thiemer, E.: Die Lösung pharmakologischer Probleme durch Rechenautomaten. Arzneimittel-Forsch. **16**, 1431 (1966).

Krüger-Thiemer, E., und P. Bünger: Kumulation und Toxizität bei falscher Dosierung von Sulfanilamiden. Arzneimittel-Forsch. **11**, 867 (1961).

Kurnosova, L.M., and M.M. Lenkevich: Mode of action of sulfonamides on trachoma virus. Acta Virol. **8**, 350 (1964).

Lai, T.F., and R.E. Marsh: The crystal structure of *p*-aminobenzoic acid. Acta Cryst. **22**, 855 (1967).

Lampen, J.O., and M.J. Jones: The antagonism of sulfonamide inhibition of certain lactobacilli and enterococci by pteroylglutamic acid and related compounds. J. Biol. Chem. **166**, 435 (1946).

Lampen, J.O., and M.J. Jones: The growth-promoting and antisulfonamide activity of *p*-aminobenzoic acid, pteroylglutamic acid and related compounds for *L. arabinosus* and *Strept. plantarum*. J. Biol. Chem. **170**, 133 (1947).

Landy, M., N.W. Larkun, E.J. Oswald, and F. Streightoff: Increased synthesis of *p*-aminobenzoic acid associated with the development of resistance in *Staph. aureus*. Science **97**, 265 (1943).

Languillon, J.: La sulfamidotherapie dans le lepre. Sulfamethoxy-pyridazine, acetylsulfamethoxy-pyridazine, sulfadimethoxine, acetylsulfadimethoxine, R04-4393. Med. Trop. **24**, 522 (1964).

Lascelles, J., and D.D. Woods: The synthesis of "folic acid" by *Bacterium coli* and *Staphylococcus aureus* and its inhibition by sulphonamides. Brit. J. Exptl. Pathol. **33**, 288 (1952).

Lehr, D.: Inhibition of drug precipitation in the urinary tract by the use of sulfonamides mixtures. I. Sulfathiazole-sulfadiazine mixture. Proc. Soc. Exptl. Biol. Med. **58**, 11 (1945).

Levine, P.P.: The effect of sulfanilamide on the course of experimental avian coccidiosis. Cornell Vet. **29**, 309 (1939).

Levine, P.P.: The effect of sulfapyridine on experimental avian coccidiosis. J. Parasitol. **26**, 233 (1940).

Linnell, W.H., and J.B. Stenlake: Nuclear derivatives of 4,4′-diaminodiphenylsulfone. J. Pharm. Pharmacol. **2**, 937 (1950).

Lockwood, J.S.: Action of sulfanilamide and its applications to surgical infections. J. Am. Med. Assoc. **111**, 2259 (1938).

Maier, J., and E. Riley: Inhibition of antimalarial action of sulfonamides by *p*-aminobenzoic acid. Proc. Soc. Exptl. Biol. Med. **50**, 152 (1942).

Mallett, E., and D. Musselwhite: The use of septrin in the treatment of upper respiratory tract infections. Practitioner **205**, 807 (1970).

Mann, T., and D. Keilin: Sulfanilamide as a specific inhibitor of carbonic anhydrase. Nature **146**, 164 (1940).

MARSHALL, E.K. (JR.): Determination of sulfanilamide in blood and urine. J. Biol. Chem. **122**, 263 (1937).

MARTINEK, K.: Wein. Med. Wochenschr. **120**, 278 (1970).

McCALLUM, F.O., and G.M. FINDLAY: Chemotherapeutic experiments on the virus of lymphogranuloma inguinale in the mouse. Lancet **1938-II**, 136.

MIETZSCH, F., und J. KLARER: Herstellung von Azoverbindungen. Deutsches Reichspatent 607537 (1934).

MILLER, A.K.: Folic acid and biotin synthesis by sulphonamide-sensitive and sulphonamide-resistant strains of Escherichia coli. Proc. Soc. Exptl. Biol. Med. **57**, 151 (1944).

MILLER, A.K., P. BRUNO, and R.M. BERGLUND: The effect of sulfathiazole on the in vitro synthesis of certain vitamins by Escherichia coli. J. Bacteriol. **54**, 9 (1947).

MILLER, C.S.: Dialkylsulfamyl, benzoic acid. U.S. Patent, 2,608,507, Aug. 1952.

MILLER, G.H., P.H. DOUKAS, and J.K. SEYDEL: Sulfonamide structure-activity relationship in a cell-free system. Correlation of inhibition of folate synthesis with antibacterial activity and physicochemical parameters. J. Med. Chem. **15**, 700 (1972).

MITSUDA, H., and Y. SUZUKI: Pteridines in plants. III. Biogenesis of folic acid in green leaves; inhibitors acting on the biosynthetic pathway for the formation of dihydropteroic acid from guanylic acid. J. Vitaminol. (Kyoto) **14**, 106 (1968).

MORGAN, H.R.: Studies on the relationship of pteroylglutamic acid to the growth of Psittacosis virus (strain 6BC). J. Exptl. Med. **88**, 285 (1948).

MORIGUCHI, I., and N. KANENIWA: Spectroscopic studies on molecular interactions. IV. Charge-transfer properties and antibacterial activity of sulfonamides. Chem. Pharm. Bull. (Tokyo) **17**, 2554 (1969).

MORIGUCHI, I., and S. WADA: Protein bindings. IV. Relations of an index for electronic structure to binding constant with serum albumin and bacteriostatic activities of sulfonamides. Chem. Pharm. Bull. (Tokyo) **16**, 734 (1968).

MORIGUCHI, I., S. WADA, and T. NISHIZAWA: Protein bindings. III. Binding of sulfonamides to bovine serum albumin. Chem. Pharm. Bull. (Tokyo) **16**, 601 (1968).

MORRISON, N.E.: Sulfone resistant states. Intern. J. Leprosy **36**, Abst. No. 199 (1968).

MOULDER, J.W.: The biochemistry of intracellular parasitism. p. 105. Chicago: University of Chicago Press 1962.

MOWAT, J.H., J.H. BOOTHE, B.L. HUTCHINGS, E.L.R. STOKSTAD, C.W. WALLER, R.B. ANGIER, J. SEMB, D.B. COSULICH, and Y. SUBBAROW: Structure and synthesis of the pteridine degradation products of the fermentation L. casei factor. Ann. N.Y. Acad. Sci. **98**, 279 (1946).

NICHOLS, R.L., W.F. JONES, and M. FINLAND: Sulfamethoxypyridazine: preliminary observations on absorption and excretion, a new long-acting antibacterial sulfonamide. Proc. Soc. Exptl. Biol. Med. **92**, 637 (1956).

NIMMO-SMITH, R.H., J. LASCELLES, and D.D. WOODS: Synthesis of folic acid by streptobacterium plantarum and its inhibition by sulfonamides. Brit. J. Exptl. Pathol. **29**, 264 (1948).

NORTHEY, E.H.: The sulfonamides and allied compounds. New York: Reinhold 1948.

ORTIZ, P.J.: Dihydrofolate and dihydropteroate synthesis by partially purified enzymes from wild-type and sulfonamide-resistant pneumococcus. Biochemistry **9**, 355 (1970).

ORTIZ, P.J., and R.D. HOTCHKISS: The enzymatic synthesis of dihydrofolate and dihydropteroate in cell-free preparations from wild-type and sulfonamide-resistant pneumococcus. Biochemistry **5**, 67 (1966).

PATO, M.L., and G.M. BROWN: Mechanisms of resistance of Escherichia coli to sulfonamides. Arch. Biochem. Biophys. **103**, 443 (1963).

PEARSON, J.M.H., J.H.S. PETIT, and R.J.W. REES: Studies on sulfone resistance in leprosy.3. A case of partial resistance. Intern. J. Leprosy **36**, 171 (1968).

PULLMAN, B., and A. PULLMAN: Quantum biochemistry, p. 108. New York: Interscience Publishers 1963.

RAMAKRISHNAN, S.P., P.C. BASU, H. SINGH, and N. SINGH: Studies on the toxicity and action of diaminodiphenylsulfone (DDS) in avian and simian malaria. Bull. World Health Organ. **27**, 213 (1962).

RAMAKRISHNAN, S.P., P.C. BASU, H. SINGH, and B.L. WATTAL: A study on the joint action of diaminodiphenylsulfone (DDS) and pyrimethamine in the sporogony cycle of Plasmodium gallinaceum: Potentiation of the sporontocidal activity of pyrimethamine by DDS. Indian J. Malariol. **17**, 141 (1963).

Reber, M., G. Rutishauser und H. Tholen: Clearance-Untersuchungen am Menschen mit Sulfa-methoxazol und Sulforthodimothoxin, 3rd int. Congr. Chemother., Stuttgart 1963, vol. 1, p. 648 (eds. H.P. Kuemmerle and Preziosi). Stuttgart: Thieme 1964.

Richards, W.H.G.: The combined action of pyrimidines and sulfonamides or sulfones in the che-motherapy of malaria and other protozoal infections. Advanc. Pharmacol. Chemother. 8, 121 (1970).

Richey, D.P., and G.M. Brown: The biosynthesis of folic acid. IX. Purification and properties of the enzymes required for the formation of dihydropteroic acid. J. Biol. Chem. 244, 1582 (1969).

Rieder, J.: Physikalisch-chemische und biologische Untersuchungen an Sulfonamiden. Arzneimit-tel-Forsch. 13, 81, 89, 95 (1963).

Rist, N.: Action du p-aminophenylsulfamide et de la p,p'-diamino-diphenylsulfone sur la culture des bacilles tuberculeux des mammiferes et des oiseaux. Compt. Rend. Soc. Biol. 130, 972 (1939).

Rubbo, S.D., and J.M. Gillepsie: p-Aminobenzoic acid as a bacterial growth factor. Nature 146, 838 (1940).

Sabin, A.B., and J. Warren: Therapeutic effect of the sulfonamides on infection by an intracellular protozoan (Toxoplasma). J. Bacteriol. 41, 80 (1941).

Schmelkes, F.C.: The mechanism of sulfonamide potentiation through ionisation and oxidation. J. Bacteriol. 45, 67 (1943).

Schnaare, R.S., and A.N. Martin: Quantum chemistry in drug design. J. Pharm. Sci. 54, 1707 (1965).

Schneider, J., and J. Languillon: The retard-action sulfonamides in the treatment of leprosy and the efficient action of acetyl-sulfamethoxypyridazine. Abstr. 8th Intern. Cong. Leprology, Rio de Janeiro, 1963, p. 36 (1964).

Scholtan, W.: Die Bindung der Sulfonamide an Eiweißkörper. Arzneimittel-Forsch. 14, 348 (1964).

Scholtan, W.: Die hydrophobe Bindung der Pharmaka an Human-Albumin und Ribonucleinsäure. Arzneimittel-Forsch. 18, 505 (1968).

Schwartz, W.B.: The effect of sulfanilamide on salt and water excretion in congestive heart failure. New Engl. J. Med. 240, 173 (1949).

Seeler, A.O., O. Graessle, and E.D. Dusenbery: The effect of PAB on the chemotherapeutic activity of sulfonamides in lympho-granuloma venereum and in duck malaria. J. Bacteriol. 45, 205 (1943).

Selbie, F.R.: The inhibition of the action of sulfanilamide in mice by p-aminobenzoic acid. Brit. J. Exptl. Pathol. 21, 90 (1940).

Seydel, J.K.: Prediction of in vitro activity of sulfonamides, using Hammett constants or spectropho-tometric data of the basic amines for calculation. Mol. Pharmacol. 2, 259 (1966).

Seydel, J.K.: Molecular basis for the action of chemotherapeutic drugs, structure-activity studies on sulfonamides. In: Physico-chemical aspects of drug action (ed. E.J. Ariens), p. 169. Pergamon Press 1968.

Seydel, J.K.: Physicochemical approaches to the rational development of new drugs. In: Drug design (ed. E.J. Ariens), p. 343. Vol. 1 Academic Press 1971a.

Seydel, J.K.: Prediction of the in vitro activity of sulfonamides synthesised from simple amines by use of electronic data obtained from the simple amines. J. Med. Chem. 14, 724 (1971b).

Seydel, J.K., E. Kruger-Thiemer, and E. Wempe: Relation between antibacterial activity and 1R of sulfanilamide. Z. Naturforsch. 15b, 628 (1960).

Shannon, J.A.: The relationship between chemical structure and physiological disposition of a series of substances allied to sulfanilamide. Ann. N.Y. Acad. Sci. 44, 455 (1943).

Shiota, T.: Enzymic synthesis of folic acid-like compounds by cell-free extracts of loctobacillus ara-binosus. Arch. Biochem. Biophys. 80, 155 (1959).

Shiota, T., C.M. Baugh, R. Jackson, and R. Dillard: The enzymatic synthesis of hydroxymethyldi-hydrofolate. Biochemistry 8, 5022 (1969).

Shiota, T., and M.M. Disraely: The enzymic synthesis of dihydrofolate from 2-amino-4-hydroxy-6-hydroxymethyl dihydropteridine and p-aminobenzoyl glutamate by extracts of Lactobacillus plan-tarum. Biochim. Biophys. Acta 52, 467 (1961).

Shiota, T., M.N. Disraely, and M.P. McCann: The enzymatic synthesis of folate-like compounds from hydroxymethyldihydropteridine pyrophosphate. J. Biol. Chem. 239, 2259 (1964).

Shive, W., W.C. Ackerman, M. Gordon, M.E. Getzendaner, and R.E. Eakin: 5(4)-Amino-4-(5)-imidazole-carboxamide, a precursor of purines. J. Am. Chem. Soc. 69, 725 (1947).

SHEPARD, C.C.: The experimental disease that follows the injection of human leprosy bacilli into the foot-pads of mice. J. Exptl. Med. **112**, 445 (1960).

SHEPARD, C.C., D.H. McRAE, and J.A. HABAS: Sensitivity of *M. leprae* to low levels of 4,4'-diamino-diphenylsulfone. Proc. Soc. Exptl. Biol. Med. **122**, 893 (1966).

SHEPARD, C.C., J.G. TOLENTINO, and D.A. McRAE: The therapeutic effect of 4,4-diacetyldiaminodi-phenylsulfone (DADDS) in leprosy. Am. J. Trop. Med. Hyg. **17**, 192 (1967).

SHEPHERD, R.G.: Sulfanilamides and other *p*-amino benzoic acid antagonists. In: Medicinal chemistry (ed. A. BURGER), P. 255. Vol. 1: Wiley-Interscience 1970.

SHEPHERD, R.G., and A. LEWIS: Synthesis of antibacterial agents. Ann. Rep. Med. Chem. **1966**, 114.

SMITH, M.I., M.J. JACKSON, and H. BAUER: Evaluation of sulfones and streptomycin in experimental tuberculosis. Ann. N.Y. Acad. Sci. **52**, 704 (1949).

SOUTHWORTH, H.: Acidosis associated with the administration of sulfanilamide. Proc. Soc. Exptl. Biol. Med. **36**, 58 (1937).

STAMP, T.C.: Bacteriostatic action of sulphanilamide *in vitro*. Lancet **1939-II**, 10.

STETTEN, M.R., and C.L. FOX: An amine formed by bacteria during sulphonamide bacteriostasis. J. Biol. Chem. **161**, 333 (1945).

STOCKSTAD, E.L.R.: The isolation of a nucleotide essential for the growth of *L. casei*. J. Biol. Chem. **139**, 475 (1941).

STRAUSS, A.M., A.M. KLIGGMAN, and D.M. PILLSBURY: The chemotherapy of actinomycosis and nocardiosis. Am. Rev. Tuberc. **63**, 441 (1951).

STRULLER, TH.: Long-acting and short-acting sulfonamides. Recent developments. Antibiot. Chemotherapia **14**, 179 (1968).

TANDON, J.K., G.N. VYAS, and N. ANAND: Studies in potential antimycobacterial agents. Part. XIV. Synthesis of *p*-alkylamino-*p*'-amino-o,m and m'-methoxy and hydroxy and m,m'-dimethoxy and dihydroxy-diphenylsulfones. J. Sci. Ind. Res. (India) **17B**, 192 (1958).

TARIZZO, M.L.: Chemotherapy of trachoma. Chronicle World Health Organ. **26**, 99 (1972).

TREFOUEL, J., MME J. TREFOUEL, F. NITTI, and D. BOVET: Activite du p-aminophenylsulfamide sur les infections streptococciques experimentales de la souris et du lapin. Compt. Rend. Soc. Biol. **120**, 756 (1935).

TRIPOD, J., L. NEIPP, W. PADOWTZ, and W. SACKMANN: Relations experimentales entre l'action curative et les taux sanguins de sulfamides, en particulier du sulfaphenazol. Antibiot. Chemotherapia **8**, 17 (1960).

TSCHESCHE, R.: A new explanation of the mode of action of the sulfonamides. Z. Naturforsch. **26b**, 10 (1947).

TSUKAMURA, M.: The *in vitro* tuberculostatic action of sulfonamides. Chemotherapy (Japan) **3**, 187 (1955).

UEDA, M., K. ORITA, and T. KOIZUMI: Studies on metabolism of drug XIII. Quantitative separation of metabolites in human urine after oral administration of sulfisomezole and sulfaphenazole. Chem. Pharm. Bull. (Tokyo) **20**, 2047 (1972).

VYAS, G.N., N. ANAND, and M.L. DHAR: Studies in potential antimycobacterial agents: Part IV— Synthesis of *p*-alkylamino-*p*'-amino-o'-hydroxydiphenyl sulfones. J. Sci. Ind. Res. (India) **13B**, 270 (1954).

VYAS, G.N., N. ANAND, and M.L. DHAR: Studies in potential antimycobacterial agents: Part X— Synthesis of *p*-alkylamino-*p*'-amino-o,o'-dihydroxydiphenyl sulphones. J. Sci. Ind. Res. (India) **14C**, 218 (1955).

WATANABE, T.: Infective heredity of multiple drug resistance in bacteria. Bacteriol. Rev. **27**, 87 (1963).

WATANABE, T., and T. FUKASAWA: Episomic resistance factors in *Enterobacteriaceae*. III. Elimination of resistance factors by treatment with acridine dyes. Med. Biol. (Tokyo) **56**, 71 (1960).

WATANABE, T., and T. FUKASAWA: Episome-mediated transfer of drug resistance in *Enterobacteria-ceae*. III. Transduction of resistance factors. J. Bacteriol. **82**, 202 (1961a).

WATANABE, T., and T. FUKASAWA: Episome-mediated transfer of drug resistance in *Entero-bacteriaceae*. II. Elimination of resistance factors with acridine dyes. J. Bacteriol. **81**, 679 (1961b).

WEINSTEIN, L., M.A. MADOFF, and C.M. SAMET: The sulfonamides. New Engl. J. Med. **263**, 793, 842, 900 (1960).

WEISMAN, R., and G. M. BROWN: The biosynthesis of folic acid. V. Characteristics of the enzyme system that catalyzes the synthesis of dihydropteroic acid. J. Biol. Chem. **239**, 326 (1964).

WELCH, A. D., and C. A. NICHOL: Water-soluble vitamins concerned with one- and two-C atom intermediates. Ann. Rev. Biochem. **21**, 633 (1952).

WERNER, A.: Tratemiento de la toxoplasmosis. Bol. Chileno Parasitol. **25**, 65 (1970).

WILLIAMS, R. T., and D. V. PARK: Metabolic fate of drugs. Ann. Rev. Pharmacol. **4**, 85 (1964).

WOLF, B., and R. D. HOTCHKISS: Genetically modified folic acid synthesizing enzymes of pneumococcus. Biochemistry **2**, 145 (1963).

WOODS, D. D.: Relation of p-aminobenzoic acid to mechanism of action of sulphanilamide. Brit. J. Exptl. Path. **21**, 74 (1940).

WOODS, D. D.: Metabolic relations between p-aminobenzoic and folic acids in micro-organisms. In: Chemistry and biology of pteridines, CIBA Foundation Symposium, p. 220. Boston, Mass.: Little Brown, 1954.

YAMAZAKI, M., N. KAKEYA, T. MORISHITA, A. KAMADA, and A. AOKI: Biological activity of drugs X. Relation of structure to the bacteriostatic activity of sulfonamides (1). Chem. Pharm. Bull. (Tokyo) **18**, 702 (1970).

YOW, E. M.: A re-evaluation of sulfonamide therapy. Ann. Intern. Med. **43**, 323 (1955).

Suramin

John G. Olenick

Suramin, the first nonmetallic drug ushered into the era of chemotherapy, is a complex carbamide used prophylactically to prevent, and to treat the early stage of, trypanosome-caused African sleeping sickness. It is also used in the treatment of onchocerciasis, a condition produced by filarial worm infection.

The drug was developed as an outgrowth of studies following the initial observation by EHRLICH and SHIGA (1904) that trypan red, a member of the Congo Red series of dyes, can cure experimentally induced trypanosomiasis in mice. NICOLLE and MESNIL (1906) systematically tested a very large range of cotton dyes and found two new trypanocidal dyes (trypan blue and afridol violet) which were both related to, but more active than, trypan red. Highly colored dyes with strong staining properties obviously have disadvantages as therapeutic agents, and chemists from the firm of Bayer in Germany sought to prepare active colorless derivatives of these dyes. The successful end-product of this work was suramin which was produced in 1916 and was introduced into therapy in Germany under the name "Bayer 205". For wartime and post-war political reasons, the chemical structure was kept secret until it was finally elucidated by the French workers, FOURNEAU, TRÉFOUEL, TRÉFOUEL and VALLÉE in 1924. Details of the historical evolution of suramin from the azo dyes have been interestingly related in a review by WILLIAMSON (1970).

Suramin sodium ($C_{51}H_{34}N_6O_{23}S_6Na_6$; mol.wt. 1 429) is a slightly pink-colored powder, highly soluble in water and extremely hygroscopic; structurally, as shown in the accompanying formula, it is the symmetrical urea of the sodium salt of m-aminobenzoyl-m-amino-p-methylbenzoyl-1-naphthylamino-4,6,8-tri-sulfonic acid and has variously been named Bayer 205, Germanin, Fourneau 309, Belganyl, Moranyl, Antrypol, Naphuride, Naganin and Naganol. The complex chemical structure is highly specific in that minor modifications, such as the removal of the two methyl groups, result in the loss of trypanocidal activity

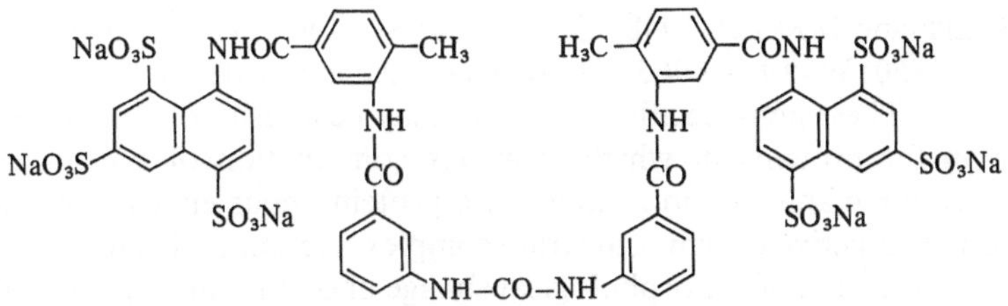

Fig. 1. Suramin

(FOURNEAU *et al.*, 1924). Drastic changes in the molecule as, for example, the substitution of the naphthalene trisulfonate portions by various heterocyclic systems, also produce inactive analogues (ADAMS *et al.*, 1956). Marked persistence (for as long as 3 months) of suramin in the blood stream is due to its firm combination with plasma proteins and is the basis of its prophylactic action (DEWEY and WORMALL, 1946; WILSON and WORMALL, 1949). However, structural features conferring persistence are much less specific than, and not related to, those required for therapeutic activity and can be met by any symmetrical high molecular weight compound containing polysulfonic acid end-groups (SPINKS, 1948). Poor absorption from the gastrointestinal tract and inability to enter the red cell and to penetrate the cerebrospinal fluid are properties of suramin that may be explained by its strongly electronegative nature and its impassability through collodion membranes.

Suramin must be administered parenterally and is usually given intravenously because intramuscular or subcutaneous injections cause severe local irritation and necrosis. A variety of adverse reactions may ensue after suramin administration but the effects are not likely to be harmful in well-nourished individuals. In a small percentage of patients the most serious reaction is immediately manifested by nausea, vomiting, shock and loss of consciousness. Delayed-in onsets are side effects consisting of sensory disturbances (photophobia, paresthesia or hyperesthesia), papular eruptions, lacrimation or palpebral edema and, still later, an irritant action on the kidney in which albuminuria, hematuria or cylindruria may occur. In view of the known teratogenic activity of certain azo dyes (GILLMAN *et al.*, 1948; WILSON, 1955) chemically related to suramin coupled with the finding that this drug, when administered to mice, can prevent or interrupt pregnancy (WHITTEN, 1956), one possible toxic effect in man that bears further investigation is chemical teratogenesis.

The trypanocidal action of suramin, as measured by a reduction of parasitemia, is exerted in the infected animal only after a prolonged latent period (HAWKING and SEN, 1960); exposure of isolated trypanosomes to suramin renders them non-infective but a likewise equally protracted contact of drug with parasite is required for lethal action (HAWKING, 1939). Initiating mode of action studies, early investigators (TOWN *et al.*, 1950; WILLS and WORMALL, 1950) tested a number of different enzymes isolated from various sources for their possible inactivation by suramin in an attempt to find specific enzyme inhibition and to correlate the specificity with trypanocidal activity. This screening of enzymes has been of little value in pinpointing the mode of action of suramin for, of the numerous enzymes reported to be inhibited by low concentrations of the drug (BEILER and MARTIN, 1948; QUASTEL, 1931; TOWN *et al.*, 1950; WILLS and WORMALL, 1950; WARING, 1965), not one has been implicated in the anti-trypanosomal action. These non-specific inhibitions may be a result of the strong affinity of suramin for proteins in which vicarious combination of the anionic drug with free cationic amino acid residues of a proteinaceous enzyme precludes the formation of an active enzyme-substrate complex. Action on bacteria is confined to the demonstration of a morphological change in cellular arrangement in which bacteria that are normally associated as single cells or as pairs can be induced to form long thread-like chains (not to be identified as filament formation owing to unbalanced growth) when grown in the presence of suramin (LOMINSKI *et al.,* 1958; LOMINSKI and GRAY, 1961). This chaining of cells was attributed to the inhibition of a lysozyme-like enzyme that was elaborated into the culture medium and is presumably involved in the lytic separation of cells. Suramin has no virucidal action but, when present throughout the initial phase of infectivity, the adsorption of bacteriophage specific to *Streptococcus lactis* ML₃ is prevented (REITER and ORAM, 1962). REITER and ORAM (1962) have suggested that the attachment was blocked because of the inactivation of a phage lysin that may function in the initial phase of adsorption.

Since EHRLICH (1913) propounded his theory of "chemorezeptors" there has been a tendency to conceptualize the action of drugs as taking place in two stages: a preliminary specific binding to some cellular component (chemoreceptor) followed by subsequent interaction to produce a characteristic biological effect. One such cellular entity that may be considered to be the first-stage receptor or target site for the action of certain drugs is the lysosome. This subcellular cytoplasmic structure is generally a single-membrane-bound vesicle containing various hydrolytic enzymes (DUVE *et al.,* 1955) and is capable of rapid and selective uptake of a variety of drugs (ALLISON and YOUNG, 1964; ALLISON, 1968).

The basophilic inclusion bodies (volutin or chemotherapy granules) produced in the cytoplasm of trypanosomes as a cytological sequela of suramin treatment of the infected animal host (ORMEROD, 1951; ORMEROD, 1958) were later found by ORMEROD and SHAW (1963) to be naturally occurring granules, some of which were associated with drug uptake. After ALLISON and YOUNG (1964) had demonstrated that lysosomes of mammalian cells in culture rapidly concentrate certain trypanocidal drugs, WILLIAMSON and MACADAM (1965) examined the *in vivo* (in the living animal) effect of suramin, as well as other trypanocides, on the ultrastructure of *Trypanosoma rhodesiense* and found evidence suggesting that the basophilic inclusion bodies might be drug-filled lysosomes. The lysosomes, although remaining structurally intact, appeared to be emptied of particulate contents while the ribosomes were progressively depleted and ultimately disappeared from the cytoplasm. These morphological changes may be considered non-specific insofar as they were caused by all the representative trypanocidal drugs tested, however, the damage to ribosomes was more extensive with suramin than with any other drug (MACADAM and WILLIAMSON, 1964). The suggestion by MACADAM and WILLIAMSON (1964) that the disappearance of ribosomes might be due to enzymatic hydrolysis by a ribonuclease released from drug-damaged lysosomes was gleaned from preliminary experiments which seemed to indicate that suramin treatment evoked increased ribonuclease activity in *T. rhodesiense* preparations. The significant increases found in alkaline ribonuclease activities of total homogenates of rat liver and kidney after intraperitoneal administration of trypan blue and other azo dyes structurally related to suramin (RABINOVITCH *et al.,* 1961) lent support to this tentative explanation. What appeared at first to be a valuable lead into the mode of action of suramin was later depreciated by WILLIAMSON and MACADAM (1965) who, upon re-examining their drug-treated trypanosomes for ribonuclease activation by using a direct spectrophotometric assay and optimizing conditions for preservation of lysosomes, discovered that an active alkaline ribonuclease was indeed present but now an increase of acid

or alkaline ribonuclease could not be detected. It is difficult to assess the significance of this negative, as well as the previous positive, result since no data
nor methodological details were presented in either of the reports. A study by
WEISSMAN *et al.,* (1969) on the stabilization of lysosomes by certain cationic
trypanocides (stilbamidines) in which, as an anionic control, suramin was tested
and found not to be a stabilizer (no mention of a labilizing action) serves to
point out that further investigation is needed on the interaction of suramin with,
and possible release and activation of a sequestered and latent ribonuclease from,
lysosomes.

In the laboratory, trypanosomes in rats and mice can be made resistant to
the action of suramin by repeated exposure to subeffective doses; after repeated
passage through non-treated animals, the resistance is gradually lost in about
6 months (HAWKING, 1963). In the field, suramin resistance is less common
and not clinically significant. When trypanosomes are made resistant to suramin,
they become hypersensitive to puromycin (WILLIAMSON, 1966). Against the normal
parent strain, *T. rhodesiense,* suramin and puromycin together produce a strong
potentiating synergistic effect (WILLIAMSON, 1966). The significance of this inter-
linked hypersensitivity-synergism is not apparent. Perhaps the potentiating effect
may be due to puromycin-induced breakdown of polysomes (HARDESTY *et al.,*
1963; BURKA and MARKS, 1964) rendering the concomitantly released monosomes
or ribosomal sub-units more susceptible to hydrolytic attack (by a suramin-
liberated lysosomal ribonuclease?).

References

ADAMS, A., J.N. ASHLEY, and H. BADER: A search for new trypanocides. Part III. Some analogues
of suramin. J. Chem. Soc. **1956**, 3739.

ALLISON, A.C.: The role of lysosomes in the action of drugs and hormones. Advanc. Chemother.
3, 253 (1968).

ALLISON, A.C., and M.R. YOUNG: Uptake of dyes and drugs by living cells in culture. Life Sci.
3, 1407 (1964).

BEILER, J.M., and G.J. MARTIN: Inhibition of hyaluronidase action by derivatives of hesperidin.
J. Biol. Chem. **174**, 31 (1948).

BURKA, E.R., and P.A. MARKS: Protein synthesis in erythroid cells. II. Polyribosome function
in intact reticulocytes. J. Mol. Biol. **9**, 439 (1964).

DEWEY, H.M., and A. WORMALL: Studies on suramin (antrypol: Bayer 205). 5. The combination
of the drug with the plasma and other proteins. Biochem. J. **40**, 119 (1946).

DUVE, C. DE, B.C. PRESSMAN, R. GIANETTO, R. WATTIAUX, and F. APPELMANS: Tissue fractionation
studies. 6. Intracellular distribution patterns of enzymes in rat-liver tissue. Biochem. J. **60**, 604
(1955).

EHRLICH, P.: Chemotherapeutics: Scientific principles, methods and results. Lancet **1913-II**, 445.

EHRLICH, P., u. K. SHIGA: Farbentherapeutische Versuche bei Trypanosomenerkrankung. Berlin.
Klin. Wochenschr. **41**, 329; 362 (1904).

FOURNEAU, E., J. TRÉFOUEL, Mme. TRÉFOUEL et J. VALLÉE: Recherches de chimiotherapie dans
la série du 205 Bayer. Ann. Inst. Pasteur **38**, 81 (1924).

GILLMAN, J., C. GILBERT, T. GILLMAN, and I. SPENCE: A preliminary report on hydrocephalus, spina
bifida and other congenital anomalies in the rat produced by trypan blue. S. African J. Med.
Sci. **13**, 47 (1948).

HARDESTY, B., R. MILLER, and R. SCHWEET: Polyribosome breakdown and hemoglobin synthesis.
Proc. Natl. Acad. Sci. U.S. **50**, 924 (1963).

HAWKING, F.: The mode of action of germanin (Bayer 205). Ann. Trop. Med. Parasitol. **33**, 13 (1939).

HAWKING, F.: Chemotherapy of trypanosomiasis. In: R.J. SCHNITZER and F. HAWKING (eds.), Exptl. chemotherapy, vol. I, p. 169. New York: Adademic Press 1963.

HAWKING, F., and A.B. SEN: The trypanocidal action of homidium, quinapyramine and suramin. Brit. J. Pharmacol. **15**, 567 (1960).

LOMINSKI, I., J. CAMERON, and G. WYLLIE: Chaining and unchaining *Streptococcus faecalis*—a hypothesis of the mechanism of bacterial cell separation. Nature **181**, 1477 (1958).

LOMINSKI, I., and S. GRAY: Inhibition of lysozyme by suramin. Nature **192**, 683 (1961).

MACADAM, R.F., and J. WILLIAMSON: The effect of some trypanosomicidal drugs on the fine structure of *Trypanosoma rhodesiense*. Trans. Roy. Soc. Trop. Med. Hyg. **58**, 7 (1964).

NICOLLE, M., et F. MESNIL: Traitement des trypanosomiases par les 'couleurs de benzidine.' Premiere partie. Etude chimique. Ann. Inst. Pasteur **20**, 417 (1906).

ORMEROD, W.E.: A study of basophilic inclusion bodies produced by chemotherapeutic agents in trypanosomes. Brit. J. Pharmacol. **6**, 334 (1951).

ORMEROD, W.E.: A comparative study of cytoplasmic inclusions (volutin granules) in different species of trypanosomes. J. Gen. Microbiol. **19**, 271 (1958).

ORMEROD, W.E., and J.J. SHAW: A study of granules and other changes in phase-contrast appearance produced by chemotherapeutic agents in trypanosomes. Brit. J. Pharmacol. **21**, 259 (1963).

QUASTEL, J.H.: Trypanocidal action and toxicity to enzymes. Biochem. J. **25**, 1121 (1931).

RABINOVITCH, M., R. BRENTANI, S. FERREIRA, N. FAUSTO, and T. MAACK: Alkaline ribonuclease activity increase in rat kidney cortex and liver after trypan blue and other azo dyes administration. J. Biophys. Biochem. Cytol. **10**, 105 (1961).

REITER, B., and J.D. ORAM: Inhibition of a streptococcal bacteriophage by suramin. Nature **193**, 651 (1962).

SPINKS, A.; The persistence in the blood stream of some compounds related to suramin. Biochem. J. **42**, 109 (1948).

TOWN, B.W., E.D. WILLS, E.J. WILSON, and A. WORMALL: Studies on suramin. 8. The action of the drug on enzymes and some other proteins. General considerations. Biochem. J. **47**, 149 (1950).

WARING, M.J.: The effects of antimicrobial agents on ribonucleic acid polymerase. Mol. Pharmacol. **1**, 1 (1965).

WEISSMAN, G., P. DAVIES, and K. KRAKAUER: Stabilization of lysosomes by stilbamidines. Federation Proc. **28**, 265 (1969).

WHITTEN, W.K.: Physiological control of population growth. Nature **178**, 992 (1956).

WILLIAMSON, J.: Trypanocidal drug action: Some observations on synergism. Trans. Roy. Soc. Trop. Med. Hyg. **60**, 121 (1966).

WILLIAMSON, J.: Review of chemotherapeutic and chemoprophylactic agents. In: H.W. MULLIGAN (ed), The African trypanosomiases, p. 128. New York: Wiley 1970.

WILLIAMSON, J., and R.F. MACADAM: Effects of trypanocidal drugs on the fine structure of *Trypanosoma rhodesiense*. Trans. Roy. Soc. Trop. Med. Hyg. **59**, 367 (1965).

WILLS, E.D., and A. WORMALL: Studies on suramin. 9. The action of the drug on some enzymes. Biochem. J. **47**, 158 (1950).

WILSON, E.J., and A. WORMALL: Studies on suramin (antrypol, Bayer 205). VII. Further observations on the combination of the drug with proteins. Biochem. J. **45**, 224 (1949).

WILSON, J.G.: Teratogenic activity of several azo dyes chemically related to trypan blue. Anat. Record **123**, 313 (1955).

Vancomycin

D.C. Jordan and P.E. Reynolds

Vancomycin, the first of a series of chemically related antibiotics, was isolated in the Eli Lilly Company laboratories from a *Streptomyces* species found in soil obtained in Borneo and India (MCCORMICK *et al.*, 1956). This organism, which is also found in domestic soils, has been described by PITTENGER and BRIGHAM (1956), who named it *Streptomyces orientalis*. The hydrochloride of this antibiotic bears the trade name Vancocin and was introduced into clinical practice in 1958.

The chemical properties of vancomycin hydrochloride have been reported by MCCORMICK *et al.* (1956), NISHIMURA *et al.* (1957) and HIGGINS *et al* (1958). It is a white, acid-stable solid, very soluble in water, moderately soluble in aqueous methanol and insoluble in organic solvents except polar ones such as dimethyl sulfoxide, dimethyl formamide and dimethyl acetamide. It is resistant to a variety of hydrolytic enzymes and although it is reasonably stable in 2% glycine buffers within the pH range 3 to 5 (providing the temperature is kept low, 5° C) it shows marked instability at 37° C.

Vancomycin is a complex amphoteric glycopeptide chemically related to the ristocetins, the ristomycins and the actinoidins. A tentative structure has been proposed for ristomycin A (Fig. 1) but the structures of the other antibiotics

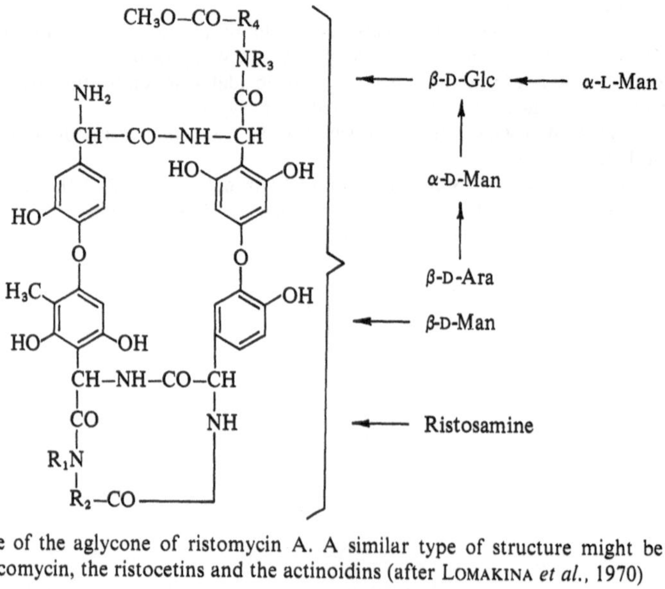

Fig. 1. Structure of the aglycone of ristomycin A. A similar type of structure might be found in vancomycin, the ristocetins and the actinoidins (after LOMAKINA *et al.*, 1970)

have not been determined. Elemental analysis of vancomycin reveals the presence of 8.7% nitrogen, 16–17% carbohydrate and 4.3% organically-bound chlorine. The ultraviolet absorption spectrum shows a maximum at 282 nm in acidic solution, shifting to 305 nm in basic solution, indicative of phenolic chromophores. Infrared absorption spectra indicate the presence of hydroxyl or amino, amide and aromatic groups. Electrometric and spectrophotometric titrations show vancomycin to contain groups with pK values of about 2.9, 7.2, 8.6, 9.6, 10.5 and 11.7, the last four being phenolic (NIETO and PERKINS, 1971 a).

Russian workers (LOMAKINA et al., 1970 b) on the basis of potentiometric titrations, reported that vancomycin had a molecular weight of approximately 1 600. The physico-chemical properties of vancomycin also have been examined in detail by NIETO and PERKINS (1971 a): on the basis of titration curves, organic chlorine and iodine (for iodinated vancomycin) analyses, and combination with synthetic peptides they suggested a M.Wt. of 1 700–1 800 and decided that the value of 3 200–3 500 obtained in earlier investigations (MCCORMICK et al., 1956; MARSHALL, 1965) was due to association of vancomycin molecules at high concentrations.

The chelating properties of vancomycin have been employed in a purification method (MARSHALL, 1965) based on the formation of a copper complex. Removal of a glucose residue leads to the formation of an aglucovancomycin which still retains about 75% of the biological activity of the unaltered antibiotic. Upon mild hydrolysis vancomycin is converted into a crystalline substance (CDP-I) with the tentative formula $C_{38}H_{185}Cl_2N_{10}O_{32-33}$ (M.Wt.=about 1840). The yield of CDP-I corresponds to 93% of the initial weight of the vancomycin molecule. Upon refluxing for 1 minute in 0.6 N HCl, CDP-I is converted into CDP-II and glucose (Fig. 2). Table 1 summarizes the current data on the structural components of vancomycin. MARSHALL (1965) suggests that this antibiotic may be made up of repeating units, perhaps linked through the glucose and amino acid units. Of some interest was his finding that acylation of the N-methylleucine residue results in a marked decrease (about 25%) in the antibacterial activity of the vancomycin molecule and suggests that this amino acid plays a role in the antibiotic's mechanism of action.

BEST et al. (1969), BEST and MCCONNELL (1970) and NIETO and PERKINS (1971 a) found that vancomycin can be resolved by gel filtration into three distinct fractions. These fractions, designated CM-1, CM-2 and CM-3 in order of elution from the gel, comprised approximately 10, 15 and 75% of the native complex respectively (BEST et al., 1969). CM-3 was seven times more active biologically

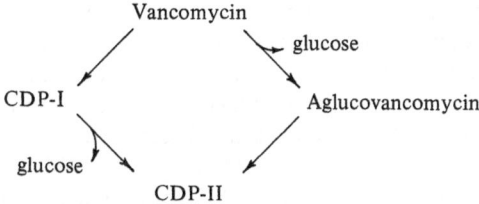

Fig. 2. Relationship of products formed from vancomycin degradation

Table 1. *Components of the vancomycin molecule*

Component	Probable number of residues
Glucose	1
Aspartic acid	1
D-N-methylleucine	1
HO—⟨benzene ring with Cl⟩—CH_2—	2
⟨benzene ring with OH⟩—CH_2—	2
$CH_3\,NH_2$ / OH / HO—O—CH_3 (sugar ring) a	1

a The 3-methyl-4-ketohexanoic acid found in earlier work (Marshall, 1965) probably arises from this desoxyamino sugar (reported by Lomakina *et al.*, 1968) during acid treatment. Recently Weringa *et al.* (1972) have identified this amino sugar as 3-amino-2,3,6-trideoxy-3-C-methyl-*lyxo*-hexopentose, and an absolute configuration has been assigned by Smith *et al.* (1972) who refer to it as vancosamine. It is present also in the related antibiotics ristocetin, ristomycin and actinoidin.

than CM-2 and fourteen times more active than CM-1. All the fractions contained 2 moles of glucose per mole of vancomycin and although aspartic acid was present in all three fractions, CM-1 had only 50% as much of this compound as CM-2 and CM-3. The results obtained suggested to the authors that CM-1 was only one-half as large as the other two fractions. All fractions were capable of binding to cell walls of *Bacillus subtilis*, although CM-3 and native vancomycin bound to a greater extent than CM-1, CM-2 or aglucovancomycin.

Vancomycin is a narrow-spectrum antibiotic active against Gram-positive bacteria and some spirochetes (McCormick *et al.*, 1956; Fairbrother and Williams, 1956; Ziegler *et al.*, 1956) and its biological activity is shown in Table 2. The outstanding potency of this antibiotic against staphylococci has been reported by a large number of workers, including Waisbren and Strelitzer (1958), Schneierson and Amsterdam (1957), Kirby and Divelbiss (1957), Daikos *et al.* (1960) and Jordan and Inniss (1962). Of particular significance is the fact that the development of staphylococcal resistance and cross-resistance to vancomycin is considerably less than that observed with most other antibiotics (Ziegler *et al.*, 1956; Garrod and Waterworth, 1956; Griffith and Peck, 1956; Geraci *et al.*, 1957; Kirby, 1963). Microbiological assays for vancomycin have been described by Kavanagh (1963).

Table 2. *Bacteriological activity of vancomycin*

Organism	Minimum inhibitory concentration (μg/ml)
Corynebacterium diphtheriae	0.80
Diplococcus pneumoniae	0.29
Leptospira pomona	1.00
Sarcina lutea	0.4 – 1.6
Staphylococcus aureus	0.15– > 10
Streptococcus faecalis	0.31– 2.5
Streptococcus pyogenes	0.15– 2.5

Microorganisms insensitive to 100 μg/ml include filamentous fungi, yeast, *Mycobacterium tuberculosis, Pseudomonas aeruginosa, Proteus vulgaris, Escherichia coli* and most *Shigella* and *Salmonella* species.

Both acute and chronic toxicity tests in mice, rats, guinea pigs, dogs and monkeys show vancomycin to have a low order of toxicity with little or no effect on respiration, blood pressure, electrocardiogram, urinary flow, intestinal motility or the isolated ileum of experimental animals (ANDERSON *et al.*, 1957; LEE and ANDERSON, 1962). Intravenous acute toxicity tests in mice indicate an LD$_{50}$ of 400–500 mg/kg while daily oral doses of 1 000 mg/kg and subcutaneous doses of 100 mg/kg are tolerated and show no abnormalities upon necropsy. Tissue culture tests show little toxicity in comparison with other clinically-used antibiotics (McCORMICK *et al.*, 1956).

Since there is little absorption of vancomycin from the gastrointestinal tract (LEE *et al.*, 1957) and since intramuscular injection causes mild to moderate pain, clinical administration of this antibiotic is limited to the intravenous route. In man intravenous doses of 50–100 mg yield serum levels of 0.5–2.0 μg/ml within 2–6 hours with high urine concentrations in 24 hours. Doses of this size, given every 6–8 hours for periods up to 7 days are well tolerated (GRIFFITH and PECK, 1956). However, side effects of vancomycin therapy can occur (KIRBY and DIVELBISS, 1957) and include local tissue irritation, drug fever and renal irritation. In the presence of impaired renal function, unnecessarily high blood levels of vancomycin (about 100 μg/ml) may damage the eighth cranial nerve and cause tinnitus or deafness (GERACI *et al.*, 1958) making it necessary to institute a system of serial serum assays to ensure keeping the serum level at a safe therapeutic dose of 10 μg/ml (DUTTON and ELMES, 1959).

Vancomycin is used clinically mainly for severe staphylococcal infections, especially when the responsible organism is resistant to the more commonly used antibiotics or when a bactericidal drug is desirable, as with patients whose normal defense mechanisms are defective. Its importance has been emphasized since the removal of ristocetin from the market because of limited sales. Good clinical responses to vancomycin therapy have been observed in cases of pneumococcal pneumonia, streptococcal pharyngitis and erysipelas (GRIFFITH and PECK, 1956) and it would appear to be the antibiotic of choice in the treatment of staphylococcal ileocolitis, in which case it should be administered orally (GERACI *et al.*, 1957). Further clinical data have been presented by Kirby *et al.* (1959); Ehrenkranz 1959); Louria *et al.* (1961) and Kirby (1963).

Non-clinical uses of vancomycin include its employment for the prevention of certain plant diseases (Mehta *et al.*, 1959; Boyle and Price, 1963) and its incorporation into media for the selective isolation of *Veillonella* (Rogosa *et al.*, 1958), *Fusobacterium*, and *Leptospira* (McCarthy and Snyder, 1963). Its use cannot be recommended for contamination control during culture of *Bacillus popilliae* and *B. lentimorbus* because sublethal concentrations elicit morphological changes in these bacteria (Pridham *et al.*, 1965).

Mode of Action

Ziegler *et al.* (1956) reported that the activity of vancomycin against both a beta hemolytic streptococcus and a strain of staphylococcus was characterized by an immediate bactericidal action without any preceding lag phase. This rapid bactericidal action of vancomycin (also observed by Geraci *et al.*, 1957) could account for the slow development of bacterial resistance toward it. However, recent work (Nieto *et al.*, 1972) has cast doubt on whether the action of vancomycin is, in itself, truly bactericidal. Vancomycin binds firmly to bacterial peptidoglycan and prevents the insertion of new sub-units, thus inhibiting cell wall extension and cellular growth. Removal of vancomycin from bacteria whose growth has been inhibited by it, by addition of an excess of synthetic peptide which also complexes with the antibiotic, allows cells to recover from the effects of vancomycin even after as much as three hours incubation with it. This suggests that any bactericidal action of vancomycin is dependent also on the continued activity of autolytic enzymes.

Reynolds (1961) and Jordan (1961) simultaneously discovered that vancomycin was a potent inhibitor of the synthesis of cell wall peptidoglycan in sensitive bacteria. This inhibition was accompanied (Reynolds, 1961; 1964a; 1964b) by the intracellular accumulation of wall precursors (N-acetylmuramyl-peptides) and there was no prior inhibition of protein, lipid, RNA or DNA synthesis (Jordan, 1961; Reynolds, 1961; Jordan and Innis, 1962).

Although the wall-inhibiting activity of vancomycin appears similar to that of certain other antibiotics such as penicillin, ristocetin, bacitracin and D-cycloserine (oxamycin), there are distinct differences among these antibiotics, which indicate that they act at different sites in the sequence of events leading to wall production. Reynolds (1962) found that exponentially-growing cells of *S. aureus* lost the ability to accumulate ^{14}C-labelled amino acids from the medium when treated with either penicillin or vancomycin, but, whereas the penicillin effect was prevented by molar NaCl or molar NH_4Cl, the vancomycin effect was not. With *Bacillus megaterium* KM penicillin caused a 50% conversion of the rods to spheroplasts in the presence of 10% sucrose with a simultaneous reduction in amino acid uptake, but osmotic stabilization did not protect these cells against vancomycin since the inhibition of amino acid uptake was not abolished, and the culture lysed in several hours. In addition, pretreatment of *S. aureus* and *B. megaterium* at 0° C with non-labelled penicillin or bacitracin prevented the subsequent uptake of radioactive penicillin, but similar pretreatment with vancomycin did not reduce the secondary penicillin uptake. Consequently it appears that vancomycin and penicillin have different cellular binding sites. This has been confirmed by HANCOCK and FITZ-JAMES (1964).

Several workers have suggested that vancomycin may only affect wall synthesis indirectly and that the plasma (cytoplasmic) membrane may represent the primary site of action. SHOCKMAN and LAMPEN (1962) examined the growth of *Strepto-coccus faecalis* protoplasts by a turbidimetric method and reported that although penicillin had no effect, certain other antibiotics, such as vancomycin, inhibited the growth at concentrations which also inhibited growth of intact cells. HANCOCK and FITZ-JAMES (1964) used both turbidity and ^{14}C-leucine incorporation as a measure of growth and found that vancomycin and bacitracin (in contrast to penicillin and D-cycloserine) inhibited the growth of *B. megaterium* KM proto-plasts. However, REYNOLDS (1971) pointed out that protoplasts are fragile entities and that a period of adaptation may be necessary before they are completely stabilized in a medium in which the mechanical support normally afforded by the peptidoglycan is replaced by an osmotic support. The addition of an antibiotic during this period of adaptation may affect the stabilization process and lead to lysis even though the same substance may have no effect on the fully-adapted protoplast. If the addition of vancomycin (or bacitracin) to a suspension of protoplasts of *B. megaterium* is delayed until "growth" of the protoplasts has become exponential (as measured by turbidity increase or by incorporation of radioactive amino acids into protein) then vancomycin has little, if any, effect on growth of the protoplasts (REYNOLDS, 1971). In this respect protoplasts of *B. megaterium* appear to differ from those of *S. aureus* in their susceptibility to vancomycin action (JORDAN, 1965; JORDAN and MALLORY, 1965).

In an effort to determine which of the two entities, the cell wall or the plasma membrane, was the primary locus of vancomycin action, JORDAN (1965) and JORDAN and MALLORY (1965) used protoplasts of *S. aureus* prepared by the use of *Chalaropsis* B enzyme. Vancomycin at a concentration of 83 µg/mg cell dry weight caused a rapid inhibition of the incorporation of ^{14}C-amino acids and inorganic ^{32}P into the plasma membranes. The incorporation of ^{14}C-leucine was inhibited strongly within 10 minutes of antibiotic addition, while the inhibition of ^{32}P incorporation, which could not be detected prior to 1.5 minutes after vancomycin addition, was also severe but was not complete for at least 30 minutes. Since this was in contrast with the prevention of glycine uptake into the wall mucopeptide of intact cells, which became absolute in about 0.9 minutes, it seemed that wall inhibition occurred earlier than the membrane inhibition. How-ever, such results cannot constitute conclusive evidence for the cell wall hypothesis since a minute undetected change in plasma membrane composition might result in a major change in wall manufacture. Using vancomycin iodinated with ^{125}I, and still biologically active, PERKINS and NIETO (1970) found that, when used at the M.I.C. level, the antibiotic was predominately located in the cell wall of *Micrococcus lysodeikticus* and *B. subtilis,* although about 8% of the adsorbed antibiotic was present in the plasma membrane after protoplasting the cells. The proportion of antibiotic in the membrane fraction increased with extended incubation but this observation is probably not important since preformed proto-plasts of *M. lysodeikticus* (PERKINS and NIETO, 1970) or *B. megaterium* (REYNOLDS, unpublished observations) absorbed very little iodinated vancomycin; further-more the inhibition of peptidoglycan synthesis by vancomycin is immediate and not dependent on continued incubation of the bacteria with the antibiotic. Thus most of the membrane-associated antibiotic would appear to be fixed there only after prior attachment on or within the cell wall. Unfortunately the significance of such binding relative to the inhibition of peptidoglycan synthesis is unknown.

Additional support for the idea that vancomycin has a primary effect on cell wall synthesis comes from the fact that, for a period of at least 10 minutes, the antibiotic did not prevent the synthesis in *S. aureus* of a trypsin-insoluble portion of the wall mucopeptide which is believed to be a portion of the plasma membrane (JORDAN, 1965; JORDAN and MALLORY, 1965). Of even greater interest was the observation that when intact cells were pre-exposed to vancomycin, their walls become resistant to attack by *Chalaropsis* B enzyme, and this effect was used to show that vancomycin binds firmly to the cell wall within 6–10 sec of its addition. It has also been found that pretreatment of isolated walls of *B. stearothermophilus* or *B. megaterium* with vancomycin prevents their solubili-zation by lysozyme (GALE *et al.,* 1972). The inability to remove the antibiotic from pre-exposed cells by washing in various buffers and the rapidity of the binding suggest that vancomycin is bound to sites in or on the wall but is not metabolically incorporated into the wall structure. The amount of antibiotic bound can be considerable since *S. aureus* can remove about 99% of the vancomy-cin from the growth medium when it is present at the minimum growth inhibitory concentration (REYNOLDS, 1966). It would seem, therefore, that vancomycin binds to the wall either at, or near, the potential sites of action of *Chalaropsis* B enzyme and lysozyme. Such sites, apparently, are the glycosidic linkages between N-acetylmuramic acid or N,O-diacetylmuramic acid N-acetylglucosamine in the wall peptidoglycan (TIPPER *et al.,* 1964). Consequently, results obtained with protoplasts, where direct contact between antibiotic and cytoplasmic membrane is possible, may not be comparable to results obtained with intact cells where a rigid wall could prevent the influx of antibiotic molecules.

Binding of vancomycin directly to isolated cell walls of *B. subtilis* was demon-strated by BEST and DURHAM (1965). These workers calculated that each mg dry weight of isolated *B. subtilis* walls absorbs more than 750 µg of vancomycin, an amount greatly in excess of that required for growth inhibition of intact cells. The corresponding value for *S. aureus* appears to be about 325 µg/mg dry weight of walls, based on the data of REYNOLDS (1966) that as much as 65 µg can be bound per mg cell dry weight and assuming that the walls of *S. aureus* represent about 20% of the cell dry weight. REYNOLDS (1966) suggested that *S. aureus* cells possess about 10^7 binding sites for vancomycin, in contrast to the 10^3 sites involved in penicillin binding. However, there is no need for all the sites to be occupied by vancomycin for wall synthesis to be inhibited, and in fact only 50% of the sites need to be occupied for nucleotide-peptide accumulation to commence (REYNOLDS, 1964b). More recently, BEST *et al.* (1968) showed that several Gram-positive, vancomycin-sensitive bacteria adsorbed con-siderable quantities of this antibiotic, whereas a resistant *Flavobacterium* species bound no detectable amounts. *Pseudomonas fluorescens*, although Gram-negative, was relatively sensitive and bound significant quantities of vancomycin. Therefore the degree of binding of vancomycin, within limits, is proportional to the degree of sensitivity of the cell, as might be expected.

The protective effects of Mg^{2+} against vancomycin action on Gram-negative bacteria (RUSSELL and THOMAS, 1966; RUSSELL, 1967) as well as the augmentation of vancomycin inhibition by the metal chelator EDTA (RUSSELL, 1967; BEST et al., 1968; HASLAM et al., 1970) are probably indicative of the stabilizing effects of Mg^{2+} on the outer lipoprotein-lipopolysaccharide layer of the cell wall. The addition of Mg^{2+}, for example, may result in decreased penetration of the drug to the sensitive sites in the inner peptidoglycan layer. Removal of Mg^{2+} would increase penetration.

BEST and DURHAM (1965) demonstrated that the binding of vancomycin to isolated cell walls of B. subtilis during a 90 minutes contact period could be reduced by magnesium, manganese, calcium and ferrous ions (and by the polycationic compounds spermine and polylysine), but not by sodium ions. The effective levels of magnesium were similar to those reducing the vancomycin-induced inhibition of wall peptidoglycan synthesis. The unexpected finding that manganese, calcium and ferrous ions appeared unable to prevent growth inhibition by vancomycin is a consequence of these ions being toxic to the test organism when used at the same concentration as magnesium (DURHAM, personal communication). However it is not clear whether these observations have any bearing on the mode of attachment of vancomycin to its specific binding sites, since the concentrations of antibiotic used were several fold in excess of the M.I.C. for this organism (2 µg/mg dry wt.). The amount of vancomycin remaining adsorbed to the walls after elution with metal ions was still greatly in excess of the maximum amount that could be bound to cells at the M.I.C., so it would appear that divalent cations were either preventing adsorption at, or eluting absorbed vancomycin from, non-specific binding sites.

Competition between Mn ions and vancomycin for binding sites on the walls of M. lysodeikticus has been reported by SINHA and NEUHAUS (1968). The amount of divalent cation used in this study approximately halved the number of adsorption sites for the antibiotic. In this connection PERKINS and NIETO (1970) found that 100 mM Mg^{2+} removed ^{125}I-labelled vancomycin from the wall mucopeptide of B. licheniformis but had little effect on vancomycin adsorption by M. lysodeikticus, either in vivo or on isolated walls. Thus the nature of antibiotic binding differs between these two microorganisms, perhaps because of differences in the degree of cross-linkage in the peptide chains of the wall peptidoglycan (HUGHES, 1968). Because of the ease of removal of pre-adsorbed vancomycin from the walls and peptidoglycan of B. licheniformis and B. subtilis by Mg^{2+} ions, PERKINS and NIETO (1970) suggest that much of the vancomycin binding in these two microorganisms is relatively non-specific.

The binding of vancomycin to the cell wall of intact sensitive bacteria, the immediate and marked depression in the incorporation of certain amino acids into the wall peptidoglycan and the concurrent accumulation of wall precursors strongly suggest that the primary site of vancomycin activity lies in the area of cell wall synthesis. If the wall represents the surface for the collection of enzymes concerned in its own manufacture (ROGERS, 1963) then vancomycin could inhibit these enzymes without any early change in plasma membrane structure. However, if such enzymes are located in or on the plasma membrane, as seems likely since protoplast membrane preparations containing less than 0.1% of the original wall material catalyse all the terminal reactions of peptidoglycan synthesis including transpeptidation (Reynolds, 1971 and unpublished observations), then vancomycin could still exert a direct effect on the wall via adsorption, thus preventing the introduction of new structural molecules into the existing wall during growth. The weakening of the cell wall, coupled with the stretching action brought about by the continued synthesis of protein and other protoplasmic constituents could eventually damage the plasma membrane.

With respect to a possible direct effect by vancomycin on enzymes concerned in wall manufacture, there have been several studies made on the effect of this antibiotic on the synthesis of wall components in cell-free systems. With respect to teichoic acid synthesis Burger and Glaser (1964), using a particulate enzyme (polyglycerophosphate synthetase) obtained from the plasma membranes of *B. subtilis* and *B. licheniformis*, noticed that vancomycin, at a level of 150 µg/mg dry weight of enzyme preparation (1.5 mg/ml reaction mixture), caused a 50% inhibition in the synthesis of polyglycerophosphate from ^{14}C-labelled cytidine diphosphate glycerol. In addition Glaser (1964) reported that 250 µg vancomycin/ml reaction mixture inhibited the enzymatic synthesis of polyribitol phosphate from cytidine diphosphate ribitol by a particulate enzyme closely related with the cell wall of *Lactobacillus plantarum*, an organism not inhibited by vancomycin. This rather unexpected finding, again coupled with high antibiotic concentrations, even when expressed as a ratio of enzyme dry weight to inhibitor weight, makes the interpretation of these results somewhat uncertain.

In view of the probable direct effect of vancomycin on peptidoglycan synthesis it is important that the sensitivity of the different enzymic steps to vancomycin should be determined. Anderson et al. (1965) studied a number of distinct enzymic activites concerned in peptidoglycan synthesis and put forward a scheme (Fig. 3) that was consistent with their results. It is believed that N-acetylmuramyl-pentapeptide and N-acetylglucosamine are added sequentially to a lipid acceptor to give rise to a lipid-pyrophospho-disaccharide-pentapeptide. It is the first stage in this reaction that releases UMP and is sensitive to high concentrations of vancomycin (Struve and Neuhaus, 1965). Anderson et al. (1965) confirmed the sensitivity of this reaction to vancomycin but were unable to detect any inhibition of the formation of the lipid intermediates at the limiting growth inhibitory concentration of the antibiotic; at this concentration incorporation of muramyl-pentapeptide into the lipid intermediates was stimulated twofold on account of the inhibition of a subsequent reaction. Any modifications to the pentapeptide chain such as the addition of ammonia or glycine to the α-carboxyl group of glutamic acid or the addition of amino acids found in the eventual cross-link are generally made at this stage. The addition of these substances is not inhibited by vancomycin at concentrations near the M.I.C. so the reactions have been omitted from Fig. 3. The N-acetylglucosamine-N-acetyl-muramyl-(pentapeptide)-pyrophospho-lipid, apparently reacts with an acceptor (presumably the growing point of a peptidoglycan backbone chain) with the release of pyrophospho-lipid and the formation of lysozyme-sensitive glycopeptide material. It is the final polymerisation reaction that is inhibited by vancomycin

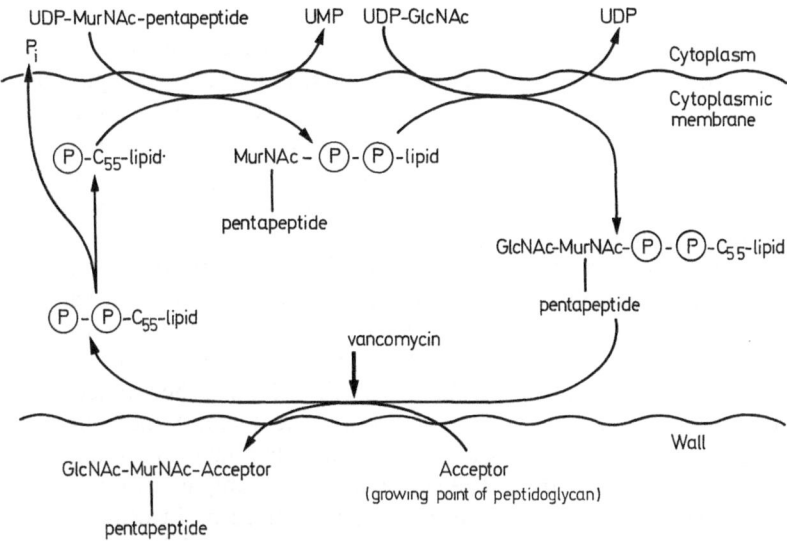

Fig. 3. Incorporation of N-acetylmuramyl-pentapeptide and N-acetylglucosamine into membrane-bound lipid intermediates and their subsequent transfer as a disaccharide-pentapeptide to the growing point of the peptidoglycan. (Reproduced with permission from GALE *et al.*, 1972). The Molecular Basis of Antibiotic Action, Fig. 3, 4, p. 56. J. Wiley and Sons Ltd. London)

(and ristocetin) at the limiting growth inhibitory concentration of the antibiotic. The inhibition by vancomycin is obtained when the antibiotic is present in the cell-free system or when the enzyme used in the *in vivo* incorporation system is prepared from cells that have been treated with vancomycin *in vivo*. However, REYNOLDS (1971) found that membranes prepared from protoplasts that had been reconditioned in the presence of vancomycin retained 95–100% of the glyco-peptide polymerase activity. Furthermore, when the membrane/wall fraction of intact cells that had either been grown or broken mechanically in the presence of vancomycin was further fractionated into membrane fractions either containing or lacking wall material it was shown that only the fraction with wall was inhibited in the subsequent peptidoglycan synthesis assay (NIETO *et al.*, 1972). This inhibition was shown to be due to vancomycin that had bound firmly to the wall during growth of the cells with vancomycin or during the breakage in the presence of vancomycin, and which was brought into solution again by the presence of UDP-N-acetylmuramylpentapeptide in the *in vitro* assay. From these results it seems possible that the *in vivo* mode of action of vancomycin may well be concerned with the final polymerisation reaction referred to above though it is not necessary to postulate an inhibition of the glycopeptide synthetase itself; blockage of the addition sites on the acceptor molecule would have the same effects as indicated by the results of NIETO *et al.* (1972) referred to above. In this connection SINHA and NEUHAUS (1968) used membrane particles from *M. lysodeikticus* which catalyzed peptidoglycan synthesis, and discovered that vanco-mycin, at a level of 50 µg/ml, inhibited synthesis by 93% in 3 hr. However, when isolated cell walls of the same organism were added to the incubation mixture, either at zero time or 60 min. later, the inhibition was decreased markedly. Walls previously satutated with vancomycin were inactive in this respect. In addition, the previously observed vancomycin stimulation of phospho-N-acetylmuramyl-pentapeptide translocase in *S. aureus* (Struve and Neuhaus, 1965; Struve *et al.*, 1966) was suppressed by cell walls of *S. aureus*. Consequently cell walls of *M. lysodeikticus* in high concentration have sufficient affinity for vancomycin to remove previously-adsorbed antibiotic from the sensitive sites to which it was bound previously.

Chatterjee and Perkins (1966) observed that vancomycin binds to peptidoglycan precursors as well as to isolated walls. They noted that in vancomycin-treated cells of *S. aureus, M. lysodeikticus* and certain plant-pathogenic corynebacteria, there accumulated not only UDP-N-acetylmuramyl-peptide but also a complex of this same compound with the antibiotic. The attached vancomycin was identified by the presence of glucose, aspartic acid and N-methylleucine and did not appear to be bound ionically. Ristocetin was bound to a similar complex in *Corynebacterium poinsettiae*. The suggestion that complex formation resulted from the activity of cellular enzymes was later reported by Perkins (1968) to be incorrect, since by the use of several techniques, including differential ultraviolet adsorption, it could be shown that complexing occurred upon mixing equimolar amounts of the two components. However, nucleotide-sugar-peptides lacking a terminal D-alanyl-D-alanine did not combine with the antibiotic. Later, Perkins (1969) found that the smallest peptide which reacted with vancomycin was acetyl-D-alanyl-D-alanine. Since vancomycin could be acetylated or formylated and still undergo complex formation it was considered that the free amino group was not directly involved. Best *et al.* (1970), using ^{14}C-labelled UDP-N-acetylmuramyl-pentapeptide (from *S. aureus*) complexed to vancomycin (or ristocetin), found that when mixed with cell walls of *M. lysodeikticus* the antibiotic could be sedimented with the walls, whereas the ^{14}C remained in the supernatant fluid. This was taken to imply that vancomycin had a greater affinity for wall sites than for wall precursor and that the complex itself does not bind to walls. Such wall sites could include the acyl-D-ala-D-ala termini of peptides at the growing point of the peptidoglycan. Complex formation between vancomycin (or ristocetin) and compounds containing acyl-D-alanyl-D-alanine has been studied in detail by Nieto and Perkins (1971 b, c). As a result of their observations they were able to define the types of molecular arrangements to which vancomycin would bind (see reviews by Gale *et al.*, 1972; Perkins and Nieto, 1972). It is apparent that in addition to binding to compounds in the cell which contain acyl-D-ala-D-ala (UDP-N-acetylmuramyl-pentapeptide, N-acetylmuramyl-(pentapeptide)-pyrophospholid, N-acetylglucosamine-N-acetylmuramyl-(pentapeptide)-pyrophospholipid, and the growing point of the peptidoglycan) vancomycin will bind also to other portions of the wall, although with a lower affinity (e.g. -L-ala-D-glu-α-gly in *M. lysodeikticus* and -(L-centre of meso DAP)-D-ala-(D-centre of meso DAP)- in the cross link which is present in the peptidoglycan of many Gram-negative organisms and the Gram-positive bacilli). When vancomycin first comes into contact with a bacterial cell it is much more likely to bind to these groups in the peptidoglycan, including the peptide chains at the growing points which end in acyl-D-ala-D-ala, than to lipid intermediates or to UDP-N-acetylmuramyl-pentapeptide to which it probably does not have access. Combination with the acyl-D-ala-D-ala termini at the growing point of the peptidoglycan would probably inhibit both the transpeptidase involved in crosslinking the newly formed peptidoglycan, and the D,D-alanine carboxypeptidase that removes the terminal D-alanine residue from the pentapeptide. Furthermore, since saturation of walls with vancomycin prevents the lytic action of lysozyme and of the *Chalaropsis* B enzyme, it seems likely that it would also inhibit the action of glycopeptide synthetase (polymerase) which catalyses a similar reaction to lysozyme but in reverse. These observations provide a possible explanation for the apparent interference by vancomycin with membrane function in inact bacteria. Most of the observed effects (e.g. inhibition of glutamic acid uptake) occurred much later than the inhibition of peptidoglycan synthesis. If the normal wall lytic enzyme is not active in the presence of vancomycin, the intergrity of the murein sacculus should be maintained during the incubation with antibiotic. The synthesis of cellular components will continue until all the elasticity of the peptidoglycan has been taken up; beyond this point the cell will burst if synthesis continues or will remain potentially viable (but non-growing) if the rigidity of the peptidoglycan framework can withstand the physical forces generated within the bacterium. Protein and nucleic acid synthesis and the entry of small molecules would be inhibited in such bacteria whereas the same processes in a growing protoplast suspension of the same organism would be unaffected.

A second hypothesis has been put forward by REYNOLDS (1969) who employed membrane preparations of protoplasts of *B. megaterium KM* which actively incorporated radioactivity from UDP-N-acetylmuramyl-ala-glu(^{3}H) diaminopimelic acid-(^{14}C) ala-(^{14}C) ala into mucopeptide in the presence of UDP-N-acetyl-glucosamine and Mg^{2+} ions. This reaction was inhibited by vancomycin, but since the membrane preparation contained less than 0.1% of the original peptidoglycan as a contaminant it was believed unlikely that vancomycin inhibited the reaction by binding to pre-formed mucopeptide "acceptor". In addition, the removal of equimolar amounts of UDP-muramyl-pentapeptide corresponding to any particular vancomycin concentration was insufficient to account for the observed inhibition. Since this antibiotic did not affect the incorporation of radioactivity into the lipid intermediates, it was argued that vancomycin directly affects the enzyme involved in the polymerization of wall components. This hypothesis is somewhat different from that outlined above but it is possible that peptidoglycan synthesis *in vitro* can be blocked by binding of the antibiotic to the lipid intermediate (N-acetylglucosamine-N-acetyl-muramyl-(pentapeptide)-pyrophospholipid), whereas the inhibitory effect *in vivo* is almost certainly a result of the binding of vancomycin to the acyl-D-ala-D-ala groups at the growing points of the peptidoglycan. Although it is not yet possible to define exactly how the reactive groups in vancomycin react with the acyl-D-ala-D-ala groups in the peptidoglycan it seems probable that at least three of the phenolic chromophores present in vancomycin are involved (NIETO and PERKINS, 1971a). If the structure of the antibiotically-active aglycone of vancomycin is similar to that of ristomycin A (see Fig. 1) it is likely that the wall peptide would sit in the cleft between the two groups of phenols (PERKINS and NIETO, 1972), though it must be remembered that two of the phenolic groups in vancomycin are chlorinated. If a phenolic cleft is present in vancomycin it would explain many of the observations obtained by optical rotatory dispersion, circular dichroism and UV difference spectroscopy that arise on formation of a complex between antibiotic and specific peptide. PERKINS and NIETO (1972) suggest that since the vancomycin molecule has little conformational flexibility, the binding of the antibiotic to the groups in the wall can be considered in terms of a lock and key model. "The peptide may be assumed to lie in a cleft in the vancomycin molecule, the cleft being of the appropriate geometry to provide the necessary steric restrictions."

References

ANDERSON, J.S., M. MATSUHASHI, M.A. HASKIN, and J.L. STROMINGER: Lipid-phosphoacetylmura-myl-pentapeptide and lipid-phosphodisaccharide-pentapeptide: presumed membrane transport intermediates in cell wall synthesis. Proc. Natl. Acad. Sci. U.S. **53**, 881 (1965).

ANDERSON, R.C., H.M. WORTH, P.N. HARRIS, and K.K. CHEN: Vancomycin, a new antibiotic. IV. Pharmacologic and toxicologic studies. Antibiotics Ann. **1956/57**, 75.

BEST, G.K., N.N. BEST, and N.N. DURHAM: Chromatographic separation of the vancomycin complex. Antimicrobial Agents Chemotherapy **1968**, 115.

BEST, G.K., N.H. BEST, D.V. FERGUSON, and N.N. DURHAM: Adsorption of vancomycin by sensitive and resistant organisms. Biochim. Biophys. Acta **165**, 558 (1968).

BEST, G.K., and N.N. DURHAM: Vancomycin adsorption to *Bacillus subtilis* cell walls. Arch. Biochem. Biophys. **111**, 685 (1965).

BEST, G.K., M.K. GRASTIE, and R.D. McCONNELL: Relative affinity of vancomycin and ristocetin for cell walls and uridine diphosphate-N-acetylmuramyl pentapeptide. J. Bacteriol. **102**, 476 (1970).

BEST, G.K., and R.D. McCONNELL: Chemical comparison of the vancomycin complex. Bacteriol. Proc. **52**, (1970).

BOYLE, A.M., and R.M. PRICE: Vancomycin prevents crown gall. Phytopathology **53**, 1272 (1963).

BURGER, M.M., and L. GLASER: The synthesis of teichoic acids. I. Polyglycerophosphate. J. Biol. Chem. **239**, 3168 (1964).

CHATTERJEE, A.N., and H.R. PERKINS: Compounds formed between nucleotides related to the biosynthesis of bacterial cell wall and vancomycin. Biochem. Biophys. Res. Commun. **24**, 489 (1966).

DAIKOS, G.K., M. ATHANASIADOU u. E. PAPADAKIS: Empfindlichkeit pathogener Staphylokokken und gramnegativer Bakterien gegenüber Vancomycin und anderen Antibiotica. Arch. Mikrobiol. **35**, 248 (1960).

DUTTON, A.A.C., and P.C. ELMES: Vancomycin: Report on treatment of patients with severe staphylococcal infections. Brit. Med. J. **1959**, 1144.

EHRENKRANZ, N.J.: The clinical evaluation of vancomycin in treatment of multiantibiotic refractory staphylococcal infections. Antibiotics Ann. **1958/59**, 587.

FAIRBROTHER, R.W., and B.L. WILLIAMS: Two new antibiotics. Antibacterial activity of novobiocin and vancomycin. Lancet **1956-II**, 1177.

GALE, E.F., E. CUNDLIFFE, P.E. REYNOLDS, M.H. RICHMOND, and M.J. WARING: The molecular basis of antibiotic action. London: J. Wiley, 1972.

GARROD, L.P., and P.M. WATERWORTH: Behaviour in vitro of some new antistaphylococcal antibiotics. Brit. Med. J. **1956**, 61.

GERACI, J.E., F.R. HEILMAN, D.R. NICHOLS, and W.E. WELLMAN: Antibiotic therapy of bacterial endocarditis. VII. Vancomycin for acute micrococcal endocarditis. Proc. Staff Maetings Mayo Clinic **33**, 172 (1958).

GERACI, J.E., F.R. HEILMAN, D.R. NICHOLS, W.E. WELLMAN, and G.T. ROSS: Some laboratory and clinical experiences with a new antibiotic, vancomycin. Antibiotics Ann. **1956/57**, 90.

GLASER, L.: The synthesis of teichoic acids. II. Polyribitol phosphate. J. Biol. Chem. **239**, 3178 (1964).

GRIFFITH, R.A., and F.B. PECK, JR: Vancomycin, a new antibiotic. III. Preliminary clinical and laboratory studies. Antibiotics Ann. **1955/56**, 619.

HANCOCK, R., and P.C. FITZ-JAMES: Some differences in the action of penicillin, bacitracin, and vancomycin on *Bacillus megaterium*. J. Bacteriol. **87**, 1044 (1964).

HASLAM, D. F., G. K. BEST, and N. N. DURHAM: Quantitation of the action of ethylenediaminetetraacetic acid and Tris (hydroxymethyl) aminomethane on a gram-negative bacterium by vancomycin. J. Bacteriol. **103**, 523 (1970).

HIGGINS, H. M., W. H. HARRISON, G. M. WILD, H. R. BUNGAY, and M. H. MCCORMICK: Vancomycin: A new antibiotic. VI. Purification and properties of vancomycin. Antibiotics Ann. **1957/58**, 906.

HUGHES, R. C.: The cell wall of *Bacillus licheniformis* N.C.T.C. 6346. Biochem. J. **106**, 41 (1968).

JORDAN, D. C.: Effect of vancomycin on the synthesis of the cell wall mucopeptide of *Staphylococcus aureus*. Biochem. Biophys. Res. Commun. **6**, 167 (1961).

JORDAN, D. C.: Effect of vancomycin on the synthesis of the cell wall and cytoplasmic membrane of *Staphylococcus aureus*. Can. J. Microbiol. **11**, 390 (1965).

JORDAN, D. C., and W. INNISS: Mode of action of vancomycin on *Staphylococcus aureus*. Antimicrobial Agents Chemotherapy **1961**, 218.

JORDAN, D. C., and H. D. C. MALLORY: Site of action of vancomycin on *Staphylococcus aureus*. Antimicrobial Agents Chemotherapy **1964**, 489.

KAVANAGH, F.: Vancomycin. In: F. Kavanagh (ed.), Analytical microbiology, p. 375. New York: Academic Press Inc. 1963.

KIRBY, W. M. M.: Vancomycin therapy of staphylococcal infections. Antibiot. & Chemotherapy **11**, 84 (1963).

KIRBY, W. M. M., and C. L. DIVELBISS: Vancomycin, clinical and laboratory studies. Antibiotics Ann. **1956/57**, 107.

KIRBY, W. M. M., D. M. PERRY, and J. L. LANE: Present status of vancomycin therapy of staphylococcal and streptococcal infections. Antibiotics Ann. **1958/59**, 580.

LEE, C.C., and R.C. ANDERSON: Toxicologic studies on vancomycin and polyethylene glycol 200. Toxicol. Appl. Pharmacol. **4**, 206 (1962).

LEE, C.C., R. C. ANDERSON, and K. K. CHEN: Vancomycin, a new antibiotic. V. Distribution, excretion, and renal clearance. Antibiotics Ann. **1956/57**, 82.

LOMAKINA, N. N., R. BOGNAR, M. G. BRAZHNIKOVA, F. SZTARICSKAI, and L. I. MURAVYEVA: On the structure of ristomycin A.
Abstr. 7th Intern. Symp. Chem. Natural Products, Riga, p. 625 (1970a).

LOMAKINA, N. N., I. MURAVIEVA, and M. S. YURINA: Molecular weight and the number of ionogenic groups of ristomycins and close antibiotics. Antibiotiki **15**, 21 (1970b).

LOMAKINA, N. N., I. A. SPIRIDONOVA, R. BOGNAR, M. PUSKAS, and F. SZTARICSKAI: Desoxyamino sugar from ristomycin. Its isolation and properties. Antibiotiki **13**, 975 (1968).

LOURIA, D. B., T. KAMINSKI, and J. BUCHANAN: Vancomycin in severe staphylococcal infections. Arch. Internal Med. **107**, 225 (1961).

MARSHALL, F. J.: Structure studies on vancomycin. J. Med. Pharm. Chem. **8**, 18 (1965).

MCCARTHY, C., and M. L. SNYDER: Selective medium for *Fusobacterium* and *Leptospira*. J. Bacteriol. **86**, 158 (1963).

MCCORMICK, M. H., W. M. STARK, G. E. PITTENGER, R. C. PITTENGER, and J. M. MCGUIRE: Vancomycin, a new antibiotic. I. Chemical and biologic properties. Antibiotics Ann. **1955/56**, 606.

MEHTA, P. P., D. GOTTLIEB, and D. POWELL: Vancomycin, a potential agent for plant disease prevention. Phytopathology **49**, 177 (1959).

NIETO, M., and H. R. PERKINS: Physicochemical properties of vancomycin and iodovancomycin and their complexes with diacetyl-L-lysyl-D-alanyl-D-alanine. Biochem. J. **123**, 773 (1971a).

NIETO, M., and H. R. PERKINS: Modifications of the acyl-D-alanyl-D-alanine terminus affecting complex formation with vancomycin. Biochem. J. **123**, 789 (1971b).

NIETO, M., and H. R. PERKINS: The specificity of combination between ristocetins and peptides related to bacterial cell wall mucopeptide precursors. Biochem. J. **124**, 845 (1971c).

NIETO, M., H.R. PERKINS, and P.E. REYNOLDS: Reversal by a specific peptide (Diacetyl-$\alpha\gamma$-L-diaminobutyryl-D-alanyl-D-alanine) of vancomycin inhibition in intact bacteria and cell-free preparations. Biochem. J. **126**, 139 (1972).

NISHIMURA, H., S. OKAMOTO, K. NAKAJIMA, M. SHIMOHIRA, and N. SIMAOKA: Studies on vancomycin. I. Physical, chemical properties and in vitro antibacterial studies. Shionogi's Ann. Rep. **7**, 465 (1957).

PERKINS, H. R.: Vancomycin and mucopeptide precursors. Biochem. J. **106**, 35 P (1968).

PERKINS, H. R.: Specificity of combination between mucopeptide precursors and vancomycin or ristocetin. Biochem. J. **111**, 195 (1969).

PERKINS, H. R., and M. NIETO: The preparation of iodinated vancomycin and its distribution in bacteria treated with the antibiotic. Biochem. J. **116**, 83 (1970).

PERKINS, H. R., and M. NIETO: The molecular basis for the antibiotic action of vancomycin, ristocetin and related drugs. In: Proceedings of the Symposium on Molecular Mechanisms of Antibiotic Action on Protein Biosynthesis and Membranes, Granada, 1971. Amsterdam: Elsevier, 1972.

PITTENGER, R. C., and R. B. BRIGHAM: *Streptomyces orientalis n. sp.*, the source of vancomycin. Antibiot. & Chemotherapy **6**, 642 (1956).

PRIDHAM, T. G., H. H. HALL, and R. W. JACKSON: Effects of antimicrobial agents on the milky disease bacteria *Bacillus popilliae* and *Bacillus lentimorbus*. Appl. Microbiol. **13**, 1000 (1965).

REYNOLDS, P. E.: Studies on the mode of action of vancomycin. Biochim. Biophys. Acta **52**, 403 (1961).

REYNOLDS, P. E.: A comparative study of the effects of penicillin and vancomycin. Biochem. J. **84**, 99 P (1962).

REYNOLDS, P. E.: Mucopeptide synthesis in *Staphylococcus aureus* after treatment with vancomycin. J. Gen. Microbiol. **35**, v (1964a).

REYNOLDS, P. E.: The mode of action of vancomycin. Ph. D. thesis, University of Cambridge (1964b).

REYNOLDS, P. E.: Antibiotics affecting cell wall synthesis. In: Sixteenth Symposium of the Society for General Microbiology. p. 47 (1966).

REYNOLDS, P. E.: Mucopeptide synthesis by protoplast membrane preparations of *Bacillus megaterium*. J. Gen. Microbiol. **58**, vi (1969).

REYNOLDS, P. E.: Peptidoglycan synthesis in bacilli. II. Characteristics of protoplast membrane preparations. Biochim. Biophys. Acta **237**, 255 (1971).

ROGERS, H. J.: The bacterial cell wall. The result of adsorption, structure or selective premeability. J. Gen. Microbiol. **32**, 19 (1963).

ROGOSA, M., R. J. FITZGERALD, M. E. MACINTOSH, and A. J. BUCHANAN: Improved medium for selective isolation of *Veillonella*. J. Bacteriol. **76**, 455 (1958).

RUSSELL, A. D.: Effect of magnesium ions and ethylene diamine tetra-acetic acid on the activity of vancomycin against *Escherichia coli* and *Staphylococcus aureus*. J. Appl. Bacteriol. **30**, 395 (1967).

RUSSELL, A. D., and I. L. THOMAS: Effect of Mg^{++} on the activity of vancomycin against *Escherichia coli*. Appl. Microbiol. **14**, 902 (1966).

SCHNEIERSON, S. S., and D. AMSTERDAM: In vitro sensitivity of erythromycin-resistant strains of staphylococci and enterococci to vancomycin and novobiocin. Antibiot. & Chemotherapy **7**, 251 (1957).

SHOCKMAN, G. D., and J. O. LAMPEN: Inhibition by antibiotics of the growth of bacterial and yeast protoplasts. J. Bacteriol. **84**, 508 (1962).

SINHA, R. K., and F. C. NEUHAUS: Reversal of the vancomycin inhibition of peptidoglycan synthesis by cell walls. J. Bacteriol. **96**, 374 (1968).

SMITH, R. M., A. W. JOHNSON, and R. D. GUTHRIE: Vancosamine. A novel branched chain amino-sugar from the antibiotic vancomycin. J. Chem. Soc. (Chem. Commun.) **1972**, 361.

STRUVE, W. G., and F. C. NEUHAUS: Evidence for an initial acceptor of UDP-NAc-muramyl-pentapeptide in the synthesis of bacterial mucopeptide. Biochem. Biophys. Res. Commun. **18**, 6 (1965).

STRUVE, W. G., R. K. SINHA, and F. C. NEUHAUS: On the initial stage in peptidoglycan synthesis. Phospho-N-acetylmuramyl-pentapeptide translocase (uridine monophosphate). Biochemistry **5**, 82 (1966).

TIPPER, D. J., J. L. STROMINGER, and J. M. GHUYSEN: Staphylolytic enzyme from *Chalaropsis*: mechanism of action. Science **146**, 781 (1964).

WAISBREN, B. A., and C. L. STRELITZER: A five year study of the antibiotic sensitivities and cross resistance of staphylococci in a general hospital. Antibiotics Ann. **1957/58**, 350.

WERINGA, W. D., D. H. WILLIAMS, J. FEENEY, J. P. BROWN, and R. W. KING: Structure of an amino-sugar from the antibiotic vancomycin. J. Chem. Soc. (Perkin I) **1972**, 443.

ZIEGLER, D. W., R. N. WOLFE, and J. M. McGUIRE: Vancomycin, a new antibiotic. II. In vitro antibacterial studies. Antibiotics Ann. **1955/56**, 612.

Subject Index

f = also following page; ff = following pages; *italics* = figure, scheme, table

Antibiotics

Editors: J. W. Corcoran; F. Hahn

Volume 1: **Mechanism of Action**
197 figures. XII, 785 pages. 1967
Cloth DM 172,—; US $66.30
ISBN 3-540-03724-1

Volume 2: **Biosynthesis**
115 figures. XII, 466 pages. 1967
Cloth DM 106,—; US $40.90
ISBN 3-540-03725-X

The goal of the editors of these volumes was to present
as complete coverage as possible of all information
currently available on the mechanism of action and the
biosynthesis of antibiotics. It includes not only antibiotics
on which there is relatively definitive knowledge, but also
subjects on which the data is fragmentary and less
certain. The range of subjects encompasses antibiotics
of great medical importance as well as those that are
tools for the laboratory investigator. Also included are
antibiotics for which no utility is known but which yet
contribute to the general state of knowledge in the field.
For almost all subjects, authors were enlisted who had
themselves worked with the specific antibiotics. The
authors were encouraged to discuss the antibiotic as part
of a general biosynthetic process or metabolic function,
yet to keep the information specific enough to convey in
detail the present state of knowledge for the antibiotic.
To overcome the difficulties of comprehending specialized information resulting from the fragmentation of
scientific knowledge, the authors sought the antibiotic
effects or syntheses to normal metabolic processes.
Exposition of overall processes such as cell wall synthesis, membrane permeability, and protein synthesis are
included. In the case of Penicillin the mechanism of its
action in the actual infection has been examined and the
relation of drug action to L forms of bacteria has been
discussed. The reveal information on different antibiotics
takes various roles: 1) as part of the general scientific
knowledge that is vital to all future advances in this area,
2) as tools for the investigation of normal metabolic
behavior, 3) as the rational for the use of antibiotics by
the clinical and general medical profession.

Springer-Verlag
Berlin
Heidelberg
New York
München Johannesburg
London Madrid New Delhi
Paris Rio de Janeiro
Sydney Tokio Utrecht
Wien

Progress in Molecular and Subcellular Biology

Managing Editor: F. E. Hahn
Editorial Board: T. T. Puck; G. F. Springer; W. Szybalski; K. Wallenfels

Vol. 1:
32 figures. VII, 237 pages. 1969
Cloth DM 58,–; US $22.40
ISBN 3-540-04674-7

Volume 1 contains contributions by Woese, Davies and Mandel who are concerned either theoretically or experimentally with the nature of the genetic code and its correct translation. Smillie and Scott write about the biosynthesis of chloroplasts and the photoregulation of the underlying basic processes. Agranoff reviews critically current ideas concerning the molecular basis of higher nervous activities. The managing editor, Hahn, makes a critical analysis of the origin, conceptual content and probable future developments in molecular biology.

Vol. 2:
Proceedings of the Research Symposium on Complexes of
Biologically Active Substances with Nucleic Acids and
Their Modes of Action
Held at the Walter Reed Army Institute of Research,
Washington, 16-19 March 1970
158 figures. IX, 400 pages. 1971
Cloth DM 72,–; US $27.80
ISBN 3-540-05321-2

These 28 contributions are a definite account of the most recent research results on selected aspects of the title topic. The major topical subdivisions are: Antibiotics and Nucleic Acids – Antimaterials and Nucleic Acids – Alkaloids, Natural Polyamines and DNA – Intercalation into Supercoiled DNA – Synthetic Drugs and Dyes Binding to Nucleic Acids – Antimutagens – Carcinogens and Nucleic Acids – The Natural State of DNA.

Among the contributors are
D. M. Crothers, G. F. Gause, W. and
H. Kersten, L. S. Lerman, H. R. Mahler,
P. O. P. Ts'o, J. Vinograd and M. Waring.

Vol. 3:
58 figures. VII, 251 pages. 1973
Cloth DM 66,–; US $25.50
ISBN 3-540-06227-0

The book signifies the advancement of molecular biology from the study of bacteria and their viruses to the consideration of events and entities in eukaryotic organisms. Reverse transcription is reviewed from the point of view of its theoretical importance for molecular biology and its practical importance in cancer research. Translation, i.e. protein biosynthesis, is analyzed in the light of knowledge of numerous inhibitors of individual reaction steps in the overall sequence. The transcription of transfer RNA is surveyed in detail as one aspect of transcription of genetic information. Thalassemia and Bence-Jones proteins are selected as examples of molecular pathology in eukaryotes. Finally the nature of mitochondrial DNA in neoplastic cells is discussed as one contribution to the molecular biology of cancer.

Prices are subject to change without notice

Springer-Verlag
Berlin
Heidelberg
New York